International Conference on Advanced Mechanical and Industrial Engineering (ICAMIE 2020)

IOP Conference Series: Materials Science and Engineering
Volume 909

Online
8 - 9 July 2020

ISBN: 978-1-7138-4755-7
ISSN: 1757-8981

Printed from e-media with permission by:

Curran Associates, Inc.
57 Morehouse Lane
Red Hook, NY 12571

Some format issues inherent in the e-media version may also appear in this print version.

Institute of Physics
Dirac House, Temple Back
Bristol BS1 6BE UK

Phone: 44 1 17 929 7481
Fax: 44 1 17 920 0979

techtracking@iop.org

Additional copies of this publication are available from:

Curran Associates, Inc.
57 Morehouse Lane
Red Hook, NY 12571 USA
Phone: 845-758-0400
Fax: 845-758-2633
Email: curran@proceedings.com
Web: www.proceedings.com

TABLE OF CONTENTS

PREFACE

On behalf of organizing committee of International Conference on Advanced Mechanical and Industrial Engineering (ICAMIE) 2020, I would like to express my sincere gratitude to all participants and speakers who contribute to make this event a great success. This is the first time for this biennial series of conference which is hosted by Mechanical Engineering Department and Industrial Engineering Department – Engineering Faculty – Sultan Ageng Tirtayasa University, Cilegon – Banten Indonesia. This event is also supported by IOP Conference Series: Materials Science and Engineering.

ICAMIE 2020 was originally scheduled on July 8-9th at Cilegon city, Banten province, Indonesia. However, due to the government's regulation on social distance related to the pandemic of COVID-19 and the limitation of overseas flights, the onsite presentation was replaced by one-day online presentation at July 8th 2020 via Zoom, organized at Faculty of Engineering, Sultan Ageng Tirtayasa University, Cilegon – Indonesia in collaboration with a professional provider of online video conference .

It is well-known that the concept of Industry 4.0 spread out in every field, including the modern technology, energy, and system which are considered as the key factors for smart industrial transformation. However, this transformation is not without challenges which are required to be solved. For this purpose, ICAMIE 2020 addresses innovations, trends, and recent developments on *Sustainable Technology, Energy and System toward Smart and Green Industrial Transformation* which covers mainly in the fields of Mechanical and Industrial Engineering. There are two sessions of keynote presentations, each consist of two keynote speakers. At first session, an efficient modeling and simulation method and fuel economy standards and labels as the tools for energy efficiency and optimization are presented. Whilst at second session, comprehensive aspects for the implementation of industry 4.0 are discussed. ICAMIE 2020 also have over 100 oral presentations in video format from participants coming from different countries.

Over 100 abstracts have been selected after peer review. Most of them are published in this proceeding (IOP Conference Series: Materials Science and Engineering) and the others are published in specific journals indexed by Scopus.

On behalf of the committee of ICAMIE 2020, we would like to express our thanks for all authors and reviewers who contribute to this proceeding.

Hadi Wahyudi, ST., MT., Ph.D — Chairman ICAMIE 2020
Prof. Dr-Ing Asep Ridwan, ST., MT., IPM — Co-Chairman ICAMIE 2020

IOP Conf. Series: Materials Science and Engineering **909** (2020) 011001 doi:10.1088/1757-899X/909/1/011001

PREFACE

I, on behalf of Universitas of Sultan Ageng Tirtayasa, warmly welcome all distinguished authors, guests and participants of International Conference on Advanced Mechanical and Industrial Engineering 2020 (ICAMIE 2020) held on July 8th, 2020 in Faculty of Engineering Universiatas Sultan Ageng Tirtayasa, Banten Province, Indonesia. It is a great honour for us to share knowledge to all of you.

ICAMIE 2020 is organized by both Mechanical and Industrial Engineering from Faculty of Engineering, Universitas Sultan Ageng Tirtayasa. The theme of this conference is "Development of Sustainable Technology, Energy, and Systems toward Lean and Green Industrial Transformation". This year, ICAMIE aims to raise awareness of industry revolution 4.0 and higher industrial efficiency in all major aspects, to address challenges of the implementation of industry 4.0 principles, and to provide solution for them. Besides, this international conference is expected to enhance the possibility of collaboration among the participants and gain benefits of industry 4.0 application in some advanced countries like Taiwan, Germany and many other countries.

In this book of abstract, there are 126 abstracts compiled from different interesting manuscripts of various disciplines of knowledge, not limited to mechanical and industrial engineering. All the 126 manuscripts had been reviewed by some experts from all around the world namely experts from Japan, Australia, Malaysia, Germany, and Indonesia. The published abstracts have passed all required improvement including originality, significance and clarity of the purpose. Our big appreciation is addressed to the participants that have participated to this international conference. We strongly believe that all the manuscripts have potential to be published in the SCOPUS Journal, Q2 and Q4 for journals and IOP publisher for proceeding.

Reviewing process of the submitted abstracts in ICAMIE 2020 was a challenging yet interesting process and largely depend on the goodwill of the reviewers to review the abstracts for conference presentation and publication. We would like to extend our gratitude for all reviewers for their time and effort in reviewing all the submitted abstracts. Moreover, we also thank to the presenters and attendees for their involvement in this ICAMIE 2020. We hope you gain some insights and partners to collaborate further in the future.

Finally, we would like to share our big appreciation to the committee of ICAMIE 2020 that has put considerable efforts for the success of this event. We hope all of us enjoy this event and continue to strengthen our collaboration, friendship and scientific exchange to bring benefits for industries and other related institutions in the future.

Serang, June 30th, 2020
Rector of Untirta
Prof. Dr. H. Fatah Sulaiman, S.T., M.T.

IOP Conf. Series: Materials Science and Engineering **909** (2020) 011001 doi:10.1088/1757-899X/909/1/011001

COMMITTEES

Advisory Board

Prof. Dr. Fatah Sulaiman — Universitas Sultan Ageng Tirtayasa
Prof. Dr. Eng A. Ali Alhamidi — Universitas Sultan Ageng Tirtayasa
Prof. Dr. Ing Herman Sasongko — Institut Teknologi Sepuluh Nopember
Prof. Dr. Adi Surjosatyo, M.Sc — Universitas Indonesia
Prof. Dr. Tech Suyitno — Universitas Sebelas Maret
Prof. Kaiwei Chu — Shandong University, China
Prof. T.M. Indra Mahlia — University of Technology Sydney, Australia
Prof. Shuo-Yan Chou, Ph.D — National Taiwan University of Science and Technology (NTUST), Taiwan
Prof. Dr. Ing. Hendro Wicaksono — Jacobs University Bremen, Germany
Prof. Dan Tomohisa — Kobe University, Japan
Prof. Talal Yusaf — University of Southern Queensland, Australia
Prof. Dr. Dato Rosli bin Abu Bakar — University Malaysia Pahang, Malaysia
Prof. Dr. Norzaidi bin Mohd Daud — Universiti Teknologi MAR, Malaysia

Scientific Committee

Prof. Dan Tomohisa — Kobe University, Japan
Prof. Talal Yusaf — University of Southern Queensland, Australia
Prof. Dr. Dato Rosli bin Abu Bakar — University Malaysia Pahang, Malaysia
Prof. Dr. Norzaidi bin Mohd Daud — Universiti Teknologi MARA, Malaysia
Assoc. Prof. Oki Muraza — King Fahd University of Petroleum & Mineral, Saudi Arabia
Assoc Prof Klodian Dhoska — Polytechnic University of Tirana, Albania
Assoc. Dandi Bachtiar — University Putra Malaysia
Prof. Dr.-Ing. Bernd Noche — Universitaet Duisburg-Essen, Germany
Prof. Dr.-Ing. Yilmaz Uygun — Yacob University, Bremen, Germany
Assoc. Prof. Dr. Eng Agung Sudrajad — Universitas Sultan Ageng Tirtayasa
Assoc. Prof. Dr. Ni Ketut Caturwati — Universitas Sultan Ageng Tirtayasa
Assoc. Prof. Oki Muraza — King Fahd University of Petroleum & Mineral, Saudi Arabia
Assoc. Prof. Dr. Lovely Lady — Universitas Sultan Ageng Tirtayasa
Assoc. Prof. Dr. Jamari — Universitas Diponegoro
Assoc. Prof. Dr. Wahyu Susihono — Universitas Sultan Ageng Tirtayasa
Assoc. Prof. Dr. Erwin — Universitas Sultan Ageng Tirtayasa
Assoc. Prof. Dr. Edi — Universitas Sultan Ageng Tirtayasa
Assoc. Prof. Dr. Eng Hendra — Universitas Sultan Ageng Tirtayasa
Assoc. Prof. Dr. Rina Lusiani — Universitas Sultan Ageng Tirtayasa
Assoc. Prof. Dr. Maria Ulfah — Universitas Sultan Ageng Tirtayasa
Assoc Prof. Klodian Dhoska — Polytechnic University of Tirana, Albania
Assoc. Prof. Dr. Hamdan Akbar N — Universitas Sultan Ageng Tirtayasa
Assoc. Prof. Dr. Budhi Sholeh Wibowo — Universitas Gajah Mada
Prof. Yassierli, Ph.D.,CPE — Institut Teknologi Bandung
Assoc. Prof. Dr. Parida Jewpanya — Rajamangala University of Technology Lanna, Thailand

Organizing Committee

Chairman:

Hadi Wahyudi, Ph.D, — Universitas Sultan Ageng Tirtayasa

Co-Chairman:

Prof. Dr. Ing Asep Ridwan, ST., MT., IPM. — Universitas Sultan Ageng Tirtayasa

IOP Conf. Series: Materials Science and Engineering **909** (2020) 011001 doi:10.1088/1757-899X/909/1/011001

KEYNOTE SPEAKERS

Prof. Kaiwei Chu

(Shandong University)

Topic:
Discrate Particle Simulation of Complex Particle-Fluid Flows from Fundamentals to Applications.

Prof.T M Indra Mahlia

(University of Technology Sydney)

Topic:
Fuel Economy Standards and Labels: The Way Towards an Energy Efficient Automotive Transportation.

Prof. Dr.-Ing Hendro Wicaksono

(Jacobs University Bremen)

Topic:
How can Industry 4.0 Support Global Sustainable Development Goals?

Prof. Shuo-Yan Chou

(National Taiwan University of Science and Technology)

Topic:
The Implementation of Smart Industry and Smart Factory in Taiwan.

IOP Conf. Series: Materials Science and Engineering **909** (2020) 011001 doi:10.1088/1757-899X/909/1/011001

ICAMIE 2020

"DEVELOPMENT OF SUSTAINABLE TECHNOLOGY, ENERGY, AND SYSTEMS TOWARD GREEN AND SMART INDUSTRIAL TRANSFORMATION"

International Conference on Advanced Mechanical and Industrial Engineering 2020 (ICAMIE 2020)

IOP Conf. Series: Materials Science and Engineering **909** (2020) 011001 doi:10.1088/1757-899X/909/1/011001

SPONSORS

Host:

RISTEK-BRIN

UNIVERSITAS SULTAN AGENG TIRTAYASA

Supported by:

ADHI - HK, JV

BAS BINA AGUNG SENTOSA

CIRIAJASA CM

JITRASIND(PT. JELITA CITRA SOLUSINI

DYNATECH

KMS PT. KARYA MITRA SERAYA

INNOLA PT. DEKA SARI PERKASA

SAGUNG SETO

Total-Lal High Quality Lab Equipment Provi

Vivente innovative office system

toc TAMARA OVERSEAS CORPORINC

IOP Conf. Series: Materials Science and Engineering **909** (2020) 011001 doi:10.1088/1757-899X/909/1/011001

SCHEDULE

No	Time (WIB)	Event	PIC
1.	08.30 – 09.00	Registration	OC
2.	09.00 – 09.15	Conference Chair Report	Hadi Wahyudi, Ph.D
3.	09.15 – 09.30	Opening Speech by UNTIRTA's Rector	Prof. Dr. Fatah Sulaiman
4.	09.35 – 09.40	Break for Preparing Session 1	OC
5.	09.40 – 10.15	Speaker 1: **Prof. Kaiwei Chu** Topic: **"Discrete Particle Simulation of Complex Particle-Fluid Flows from Fundamentals to Applications"**	Moderator
6.	10.15 – 10.45	Speaker 2: **Prof. T.M. Indra Mahlia** Topic: **"Fuel Economy Standards and Labels: The Way Towards an Energy Efficient Automotive Transportation"**	Moderator
7.	10.45 – 11.15	Q & A Session 1	Moderator
8.	11.15 – 11.20	Break for Preparing Session 2	OC
9.	11.20 – 11.50	Speaker 3: **Prof. Hendro Wicaksono** Topic: **"How Can Industry 4.0 Support Global Sustainable Development Goals?"**	
10.	11.50 – 12.20	Speaker 4: **Prof. Shou-Yan Chou** Topic: **"The Implementation of Smart Industry and Smart Factory in Taiwan"**	Moderator
11.	12.20 – 12.50	Q & A Session 2	Moderator
12.	12.50 – 13.00	Closing remarks	Hadi Wahyudi, Ph.D

IOP Conf. Series: Materials Science and Engineering **909** (2020) 011002 doi:10.1088/1757-899X/909/1/011002

Peer review declaration

All papers published in this volume of *IOP Conference Series: Materials Science and Engineering* have been peer reviewed through processes administered by the Editors. Reviews were conducted by expert referees to the professional and scientific standards expected of a proceedings journal published by IOP Publishing.

- **Type of peer review: Single-blind**
  A single-blind type is employed in this peer review in which the Editors choose the reviewers. The review focuses on evaluating the papers in the term of their suitability to the topic, depth of analysis, language and writing format, and turnitin result on plagiarism level. The review result is then given to authors as their guidance to revise their papers accordingly, without knowing the name of the reviewers.

- **Conference submission management system:**

  Easychair tool is used to manage the submission system.

- **Number of submissions received:** 140

- **Number of submissions sent for review:** 140

- **Number of submissions accepted:** 98

- **Acceptance Rate (Number of Submissions Accepted / Number of Submissions Received X 100):** 70%

- **Average number of reviews per paper:** 3

- **Total number of reviewers involved:** 25

- **Any additional info on review process:**

  The reviewers are the researchers who are experts in their fields and have the same fields as the reviewed papers.

- **Contact person for queries:** Dr. Lovely Lady (lady@untirta.ac.id)

IOP Conf. Series: Materials Science and Engineering **909** (2020) 012001 doi:10.1088/1757-899X/909/1/012001

Esterification glycerol (by product in biodiesel production) with oleic acid using mordenite natural zeolite as catalyst: study of reaction temperature and catalyst loading effect

C C S Nindya[1], D R Anggara[1], Nuryoto[1*], and K Teguh[1]
[1,1*]Chemical Engineering Department, Engineering Faculty, University of Sultan Ageng Tirtayasa Serang, Banten, Indonesia

E-mail : nuryoto15@gmail.com

Abstract. In the heterogeneous system, reaction temperature and catalyst loading are very affected to the fluid diffusivities and molecular interaction in the active sites of catalyst. Theoritically, increasing both of them ensued to increases the chemical reaction rate. The aim of this research was to know how the effect of reaction temperature and catalyst loading to reaction rate in esterification of oleic acid with glycerol, based on oleic acid conversion result. The experiments were Carried out in a batch reactor by the reaction temperature from 160 to 200°C, catalyst loading of 0.5 to 2 wt% of solution, the reactants glycerol ratio of 3 mole / mole of oleic acid, reaction time of 80 minutes, agitation speed of 600 rpm, and mordernite natural zeolite as a catalyst. Observation result Showed that increasing the reaction temperature and catalyst loading were Accompanying by increases of oleic acid conversion. The highest conversion to reach at reaction temperature of 200°C and a catalyst loading of 2 wt% that was 75.09%.

1. Introduction

The reaction temperature and catalyst concentration generally affect the reaction rate. The reaction rate can be seen from the result of the reactants conversion, or reaction rate constants. The greater the reactant conversion and reaction rate constants generated, then the reaction rate will increased. Basically the increasing reaction temperature will have an impact on the improvement of the translation and rotation of molecules reacting substances, and reduce the viscosity of the fluid, thereby increasing the diffusion rate of the reactants to the active site of the catalyst. This condition will increase the collisions between the reacting substances [1]. On the other hand, increasing the concentration of the catalyst will increase the collisions between the reactant - catalyst that will going to affect the process of protonation of reactant by the catalyst. The right reaction temperature and concentration of the catalyst will affect the effectiveness and efficiency of the reaction process. The results of the literature review showed that the reaction process is done above boiling point, it will resulting a decrease of the reaction product [2], In reaction ketalisation of acetone and glycerol, the reaction temperature of 70 °C will produce a higher product than the temperature of 80°C (which is from 70% to 60%). Therefore, increasing the reaction rate by increasing the reaction temperature needs to be done carefully so that the chemical reaction can take place properly. Giving excessive catalysts will not be a proportional impact on the reaction rate and the acquisition of a reaction product since each catalyst has certain characteristics and specific. So when the contact process has already reached the maximum, the addition of another catalyst would not affect the acceleration of chemical reactions that occur. The study of glycerol reaction sealed ketalisation with zirconia catalyst at a catalyst concentration of 1 to 5% by mass of glycerol [3] and SnO2 catalyst at a catalyst concentration of 1 to 7% by mass of glycerol [4], showed that with increasing concentrations of the catalyst, followed by an increase in the conversion of glycerol produced. The highest glycerol conversion catalyst is produced at a concentration of 5% by mass of glycerol (Reddy *et al.*, 2011), and 7% by mass of glycerol (Mallesham *et al.*, 2014), respectively by 98% and 91% [3, 4]. But the trial Kirbaslar (2001) on the esterification of acetic acid and ethanol were performed at a concentration of 1.08 catalyst; 3.30; 5.40; and 10.08 grams show different results. At a concentration of 1.08 to 5.40 grams of acetic acid conversion has increased from 55 to 60%, but the catalyst concentration of 5.40 to 10.08 grams of conversions resulting fixed at 60% [5]. This means that increasing the conversion of reactants by increasing the supply of catalyst to the reaction system actually has a limit.

This study tried to observe how phenomena that occur esterification reaction between glycerol and oleic acid with temperature and catalyst concentration as a variable parameter. The catalyst used in this study is a natural zeolite type mordernite of Bayah-Banten (Indonesia) [6]. The selection of the esterification reaction between glycerol and oleic acid as the observation process is based on anticivation or preparatory steps alternative processing technologies glycerol which is a byproduct of biodiesel. From the literature mentions that glycerol generated in the process of making biodiesel is equal to 10% of the mass of the biodiesel production capacity itself [7, 8], so that the development of biodiesel production will be followed by the increase of byproducts such as glycerol. If there are not enough alternative technologies, while glycerol continues to increase, the market will bestow glycerol and glycerol price impact will be very cheap. the policy of large countries (America, China, Canada, and Brazil) Against fossil energy transition to renewable energy continues encouraged [4]. America as a country with the largest energy consumer in the world, in the range of 2011-2012 itself has increased its production of biodiesel significantly by 30 million pounds in 2011 to 120 million pounds in 2012 [9]. The abundance of glycerol byproduct of biodiesel is a serious challenge for further resolved, while the use of glycerol in the industry is still limited [10]. So, to anticipate that, alternative technologies for utilization of glycerol as intermediate products is necessary, so that the glycerol uptake will remain high. This study aims to study the impact of temperature and catalyst concentration affect the conversion of the reactant (oleic acid), and how it impacts the rate of reaction.

2. Research Methods

2.1. Materials research

The raw materials used are technical glycerol were purchased from PT. WILMAR VEGETABLE INDONESIA grading analysis results of 87%, technical oleic acid was purchased from PT. TnT Chemical with levels of 95%. Catalysts used are zeolite mordenite of Bayah, Banten (ZAB) which has been activated with sulfuric acid 1 N, sized 40 mesh. The washing process that has been modified acid zeolite using distilled water 4 times to ensure the impurity that has been released and the rest of the acid solution is no longer attached to the natural zeolite and heating it in the oven for 2 hours at a temperature of 110°C [6].

2.2. Research equipment

Observations were made using a stirrer batch reactor, as presented in Figure 1.

IOP Conf. Series: Materials Science and Engineering **909** (2020) 012001 doi:10.1088/1757-899X/909/1/012001

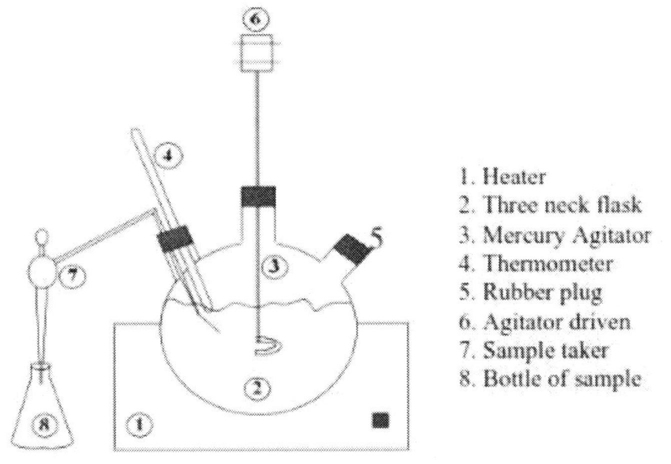

Figure 1. Scheme reactor at the esterification reaction of glycerol and oleic acid [6]

2.3. The course of the reaction

Glycerol with volumes that have been adjusted calculation (3 mol glycerol/mol oleic acid) are each heated in a separate place, and then inserted into the reactor. Heating the reactor to adjust the heating mantle button until close to the reaction temperature (160-200°C), while stirring run. When the temperature has been reached, take a sample to be analyzed early oleic acid (Ao) using an acid-base titration of 0.1 N NaOH. The catalyst is added, and every 20 minutes for analysis of samples taken for the rest of the oleic acid (As). The experiment was stopped until the reaction time of 80 minutes.

3. Result and discussion

3.1. Effect of temperature reaction

Increased reaction temperatures made a significant impact on the increase in the conversion of oleic acid (Figure 2). At the same reaction ranges of 0, 20, 40, 60 and 80 minutes, to the reaction temperature of 160 to 180°C an increase in the conversion of oleic acid from 0; 8.49; 17.76; 27.03; and 51.35% to 0; 10.44; 26.52; 47.39; and 70% (increase: 0; 1.95; 8.76; 20.36; and 18.65%), while at 200°C to 0; 43.43; 65.33; 70.07; and 72.26% (increase: 0; 32.99; 38.81; 22.68; and 2.26%).

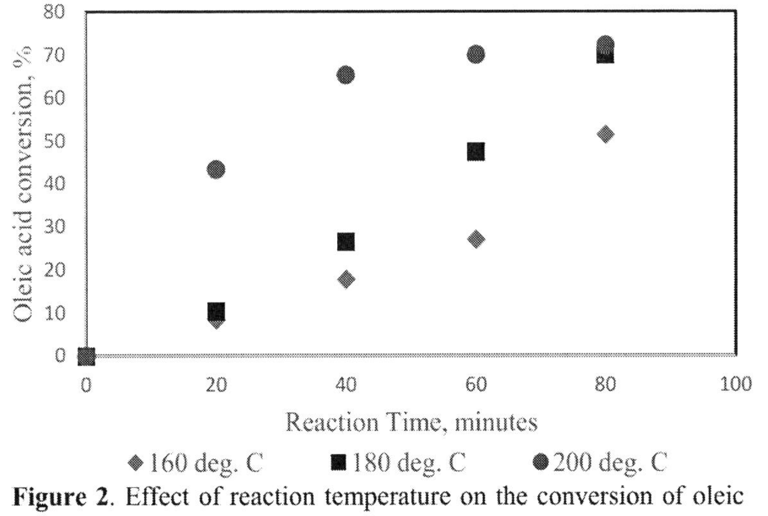

Figure 2. Effect of reaction temperature on the conversion of oleic acid at a concentration of 1% by mass glycerol catalyst.

In this case, when the conversion has reached around 70%, the increase is very small, only an increase of 2.26%, namely by increasing the reaction temperature from 180 to 200°C. This means at that condition it has approached the equilibrium point condition. Because of that, the increase in temperature made almost no change any more. or because of the large molecular size of the reactants. When the reaction temperature is raised, in theory, the diffusion rate of reactants to the active site of the catalyst contained on the inside of the catalyst increases. But due to the large size of reactant molecules [11,12] and the limited pore size of the catalyst, the rate of diffusion is held and the number of molecules that diffuse into the interior of the catalyst is limited. Under these conditions, the reaction rate will be determined by how many reactant molecules are able to enter the active site of the catalyst. The results in Figure 2 have the same tendency as the study conducted by Anggara *et al.*, 2019 in the same study using natural zeolite clinoptilolite from Lampung, where the increase in reaction temperature carried out no longer had a significant impact on the acquisition of the resulting conversion. At reaction temperatures from 160 to 170 °C the resulting conversion from 54.55 to 73.53%, but when the temperature is increased to 180°C the resulting oleic acid conversion results in only 73.65% [6]. This means that the natural zeolite catalyst is only able to convert oleic acid into a product in the esterification reaction between glycerol and oleic acid, only around 70%.

3.2. Effect of catalyst concentration

As shown in Figure 3, the increased concentration of catalyst has a positive effect on the acquisition of oleic acid conversion. The increase is quite significant when the catalyst concentration is increased from 0.5 to 1%, but when increased to 2%, the changes that occur are quite small. The result of oleic acid conversion produced in a row is 0; 37.45; 59.18; 66.29 and 67.42% (for concentrations of 0.5% mass of glycerol), 0; 43.43; 65.33; 70.07; and 72.26% (for concentrations of 1% mass of glycerol), and 0; 50.52; 66.44; 70.93; and 75.09% (for concentrations of 2% mass of glycerol). The data shows that the addition.

the concentration of catalyst to increase contact between reactants and their active site has a maximum limit. When the maximum conditions are exceeded, the addition of the catalyst to the reaction system will no longer have a significant impact.

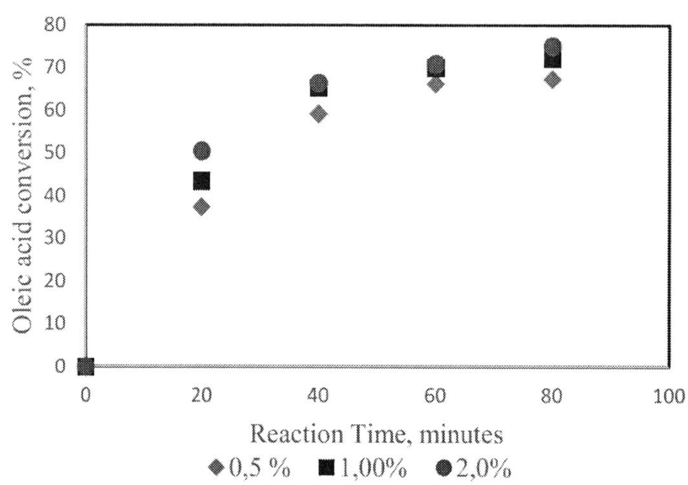

Figure 3. Effect Concentration catalyst to the conversion of oleic acid at the reaction temperature of 180°C.

The results of this research has some similarities with the trial of Ali and Merchant 2009 on research related to the hydrolysis of benzyl acetate. In these experiments, when the catalyst concentration increased from a concentration of 10 to 30 grams / liter of reactant conversion increased quite good, but when it increased again 30 to 50 grams / liter of benzyl acetate conversion is only increased by 4.2% (from 53.5 to 57.7 %). At the time of the catalyst concentration is increased again from 50 to 60 grams/liter of benzyl acetate conversion is generated tends to remain [13].

To illustrate the phenomena that occur during a reaction, a kinetics study of the reaction is carried out, namely by determining the reaction rate constants and the reaction order. The reaction rate and order constants are useful for the design of pilot plant and operational scales. The data used for kinetics studies are data on temperature variations. If referring to Figure 2, the possibility of a suitable model is to describe the phenomenon that occurs is homogeneous pseudo. In simple terms the esterification reaction between Glycerol and oleic acid is as follows:

$$A + G \rightleftarrows GMO + W \tag{1}$$

A : Oleic acid
G : Glycerol
GMO : Glycerol monooleate
W : Water

The assumptions used for the kinetics study are:

- The reaction rate is only affected by changes in oleic acid.
  This assumption is based on the number of moles of glycerol tend excess 3 time compared to oleic acid, so it is assumed to remain.
- The reaction is a reaction in the same direction and not back and forth, because the reactants (glycerol) made very excessive

Referring to these assumptions, the reaction rate of Equation (1) can be elaborated and simplified as in Equation (2).

$$-r_A = k\, C_A{}^n \tag{2}$$

With,
r_A = The rate of reaction of oleic acid, (molar / min) n
k = reaction rate constant, 1 minute
C_A = Concentration of oleic acid, the molar
n = order of the reaction

a. Trial : n = 1

$$-r_A = -\frac{dC_A}{dt} = kC_A \tag{3}$$

In integral be:

$$-ln\frac{C_A}{C_{Ao}} = k\, t \tag{4}$$

b. Trial : n = 2

$$-r_A = -\frac{dC_A}{dt} = kC_A{}^2 \tag{5}$$

In integral be:

$$(\frac{1}{C_A} - \frac{1}{C_{Ao}}) = k\, t \tag{6}$$

IOP Conf. Series: Materials Science and Engineering **909** (2020) 012001 doi:10.1088/1757-899X/909/1/012001

From the experimental data and the calculation results of the mathematical models presented in Figure 4, Figure 5, Table 1 and Table 2 it appears that the first-order mathematical model is better than the second-order to describe the phenomenon of the reaction between glycerol and oleic acid. This is evident from the price of R^2 generated. Price R^2 order one larger base than the second-order, even though the price of R^2 is still less than 1. In these conditions, there will be deviations between real data and use models when predicting the conversion of oleic acid produced. By making the deviation as a correction factor, then the usage of mathematical models that will be very helpful to predict the conversion of oleic acid at the reaction temperature range will be very helpful for future operations and design.

Table 1. Results of the calculation of the mathematical model approach order 1

	Temperature, deg.C		
	160	180	200
k. 1 / minutes	0.007	0.0124	0.0191
R2	0.8204	0.8839	0.8386

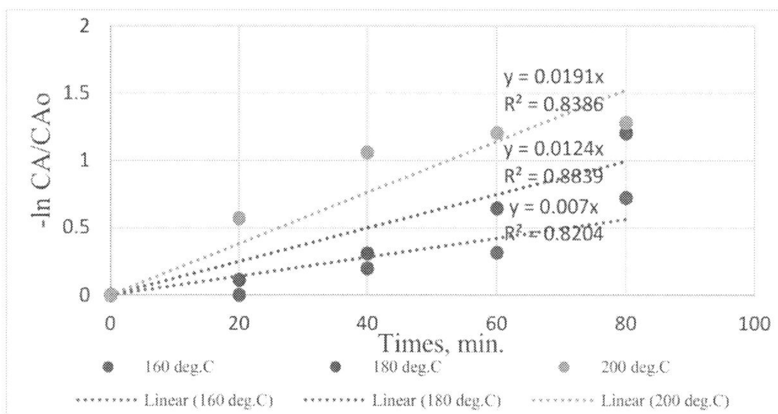

Figure 4. Relationship between versus time at various temperatures$-ln\frac{C_A}{C_{Ao}}$

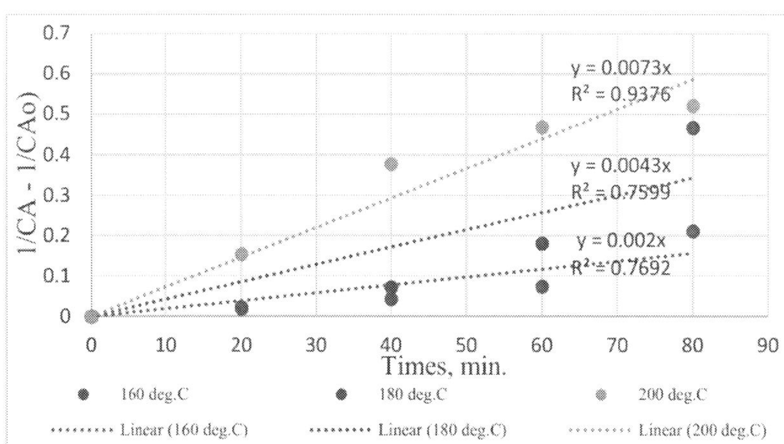

Figure 5. Relationship between versus time at various temperatures $(\frac{1}{C_A}-\frac{1}{C_{Ao}})$

Table 2. Results of the calculation of the mathematical model approach of order 2

	Temperature, deg.C		
	160	180	200
k. 1 / minutes	0.002	0.0043	0.0073
R2	0.7692	0.7599	0.9376

4. Conclusion

From the experiments, ZAB (Alam Bayah Natural Zeolite) that has modified with 1 N H_2SO_4, showing a good performance. Increasing concentrations of the catalyst and the reaction temperature, followed by increasing the conversion of oleic acid were obtained. The highest conversion was obtained at the reaction temperature200°C and catalyst concentration 2 wt% with a conversion of 75.09% oleic acid. The mathematical model approach better to approach a homogeneous pseudo-first-order at a price of R^2 average of 0.848.

Acknowledgements

The authors especially thank to the Faculty of Engineering, University of Sultan Ageng Tirtayasa who have funded this research through the Internal Grant skim through DIPA Associate Lecturer University of Sultan Ageng Tirtayasa 2019.

References

[1] Fogler S H 2006 Elements of chemical reaction engineering 4th edition *Prentice Hall International Series in the Physical and Chemical Engineering Sciences.*

[2] Maksimov A L *et al* 2011 Preparation of high-octane fuel oxygenate components from plant derived polyols *ISSN 0965_5441 Petroleum Chemistry 2011* **51** 61-69

[3] Reddy P S *et al* 2011 Acetalisation of glycerol with acetone over promoted zirconia and zirconia catalysts under mild reaction conditions *Journal of Industrial and Engineering Chemistry 17 Elsevier* 377-381

[4] Mallesham B P *et al* 2014 Eco-friendly additive synthesis of bio-fuels from renewable glycerol using nanocrystalline SnO_2-based solid acid *Catal Sci Journal 2014* 803-813

[5] Kirbaslar S I et al 2001 Esterification of acetic acid with ethanol catalysed by an acidic ion exchanger resins *J. Engin. Environ. Sci.Turk.* **25** 569-577

[6] Anggara D R *et al* 2019 The use of natural zeolite as a catalyst for esterification reaction between glycerol and oleic acid *Reactor Journal* **19 (4)** 172-179

[7] Ozorio P L *et al* 2012 Reactivity of glycerol/acetone ketal (solketal) and glycerol/ formaldehyde acetals toward acid-catalyzed hydrolysis *J. Braz. Chem. Soc* **23** 931-937

[8] Faria R P V *et al* 2013 Volarization gycerol as biofuel: thermodynamic and kinetic study of the acetalization pf glycerol with acetaldehyde *Journal of Industrial & Engineering Chemistry Research Publications Acs 52* 1538-1547

[9] US Energy Information Administration 2012 Biofuels issues and trends *Independent statistics & analysis Washington DC 20585*

[10] Tan H W *et al* 2013 Glycerol production anditsapplications asarawmaterial *Review Renewable and Sustainable Energy Reviews, Elsivier* 118-127

[11] Root L J and Stilinger F H 1988 Short range order in glycerol a molecular dynamics study *J Chem Phys* **90** 07 974

[12] Brohl K T *et al* 2015 Self assembly of magnetic nanoparticles at silicon surfaces *generous society of chemistry 201 5 Soft Matter*

[13] Ali H S and Merchant S Q 2009 Kinetic study of dowex 50 wx8-catalyzed esterification and hydrolysis of benzyl acetate *Ind. Eng. Res ACS Publications.*

Integration of opencv raspberry pi 3b+ and camera sensor in access control of vehicle ignition key system

Lukman Medriavin Silalahi[1*], Imelda Uli Vistalina Simanjuntak[2], Freddy Artadima Silaban[3], Setiyo Budiyanto[4], Heryanto[5], Muhammad Ikhsan[6]

[1,2,3,4,5] Department of Electrical Engineering, Universitas Mercu Buana, Jl. Raya Meruya Selatan, Kembangan, Jakarta 11650.

[6] Department of BaliFiber, PT Bali Towerindo Sentra, Tbk, Jl. Jenderal Sudirman Kav. 3,, Tanah Abang, Jakarta 10250.

*Email : lukman.medriavin@mercubuana.ac.id

Abstract. *The use of sensors for safety and comfort in vehicles has increased, as in several studies that are being discussed, including the use of pressure sensors for applications in the TPMS system or face recognition system security. The purpose of this research is to reduce vehicle theft that is happening right now. This system consists of a face recognition system. This reason is based on the ineffectiveness of the vehicle handlebar / steering lock. But in this study, a vehicle safety prototype was designed with a face recognition method in the hope that this prototype was designed to provide a sense of comfort, security and also the vehicle becomes a privacy vehicle. The method used is the Local Binary Path Histogram method to detect image objects and a small camera to capture images of the face, using the Raspberry Pi as a tool control center (in which the process occurs: Scanning, Extracting, Comparing, Matching, and all have been incorporated in OpenCV). The output of this study is the face that has been recorded in the database as access to open access to vehicle security. The conclusion of this research* is the process of facial recognition in real time activating the vehicle ignition lock system.

1. Introduction

At present many prototypes have been developed for a system in driving comfort and safety. In this research, we propose a prototype design of a face recognition system for safety in vehicles. Which in previous research have conducted research on tire pressure systems for driving comfort. [1]

Vehicle theft is the impact of the lack of vigilance of the vehicle owner and the security of the vehicle system which is influenced by several factors, namely security in the double lock of the vehicle or the intention to steal the vehicle. But in this research, we designed a prototype for a vehicle safety system by identifying faces using motorized vehicles. [1-3]

The development of digital technology today has influence in various fields. Among them are on computer technology, and application development from a microcontroller. Digital image processing is one type of technology to solve problems in image processing. Images will be processed in such a way that they are easier to process and convert into information. Face recognition is an example application of image processing. In this research, a face recognition and face detection application in the field of computer vision was created through image processing. The method used is the Local Binary Pattern Histogram method. This method was chosen to get fast, accurate, and efficient results in face detection on images. One of the benefits is the application of the Raspberry Pi microcontroller by using face recognition techniques as an automatic lock system that uses face detection. The sensor used in this module is a mini camera. [2-13]

The purpose of the Face Recognition System in this research is to identify and verify a person's face through a digital image. The process is to do the process of matching the texture of our facial curves to the face data stored in the database. This system is committed to developing a process of recognizing individuals who can be trusted. At present, there are several digital image processing techniques that are quite well known and can be used to identify the identity of the owner of face images such as the Eigen Classifier, Fisher Classifier, and Local Binary Pattern Histogram classifier. So in this study, the Local Binary Pattern Histogram method is used.

By using computer vision technology it is hoped that other advantages will be possible to enable communication between movements with a mini camera. So that when the vehicle wants to be used, the mini camera will take pictures. Then, if the captured image matches the dataset stored in the database, the ignition will automatically ON and the engine will start. By using a face recognition system it is expected to maintain the safety of motorbikes from the perpetrators of crime.

2. Research Model

2.1. Related research

Research conducted by Ajinkya Patil, Mrudang Shukla, 2014. "Implementation Of Classroom Attendance System Based On Face Recognition In Class". The research that was produced was a facial recognition system for the attendance of students in class to avoid wasting time in class so that the attendance process was automated based on image processing and face detection and facial recognition systems. The results of the research he has done is that the process of detecting faces to distinguish faces from non-faces has been successful by carrying out a number of scenarios that involve the process of facial recognition, namely marking the attendance of students based on existing databases in the collected student databases. The contents in the database are the student's name, student photo dataset and roll number. [4].

Research conducted by Ishita Gupta, Varsha Patil, Chaitali Kadam, Shreya Dumbre, 2016. "Face Detection and Recognition Using Raspberry Pi". The research produced is a face recognition system with the aim of security and surveillance by exploring the application of Raspberry Pi-based face recognition systems using conventional face detection and recognition techniques such as Haar Detection and PCA (Principal Component Analysis), so that this system can replace the use of passwords and RFID for high security access. [5]

Research conducted by Suchitra, Suja, dan Shikha Tripathi, 2016. "Real-Time Emotion Recognition From Facial Images Using Raspberry Pi II". The research produced is a method for direct recognition of emotions from facial images in order to identify human emotions. The method uses three steps of face detection including face detection using Haar-Cascade, feature extraction using ASM (Active Shape Model) of 26 extracted face points, and Adaboost classifier for the classification of five emotions including anger, disgust, happiness, neutral , to a surprise. So that the resulting interaction between humans - machines to improve work efficiency. Emotions that can be understood through the expression of text, voice and face. [6]

2.2. Method

In this research, facial images obtained came from the appearance of facial objects that have been captured until the face has been identified. The proposed method consists of several main process stages, namely:

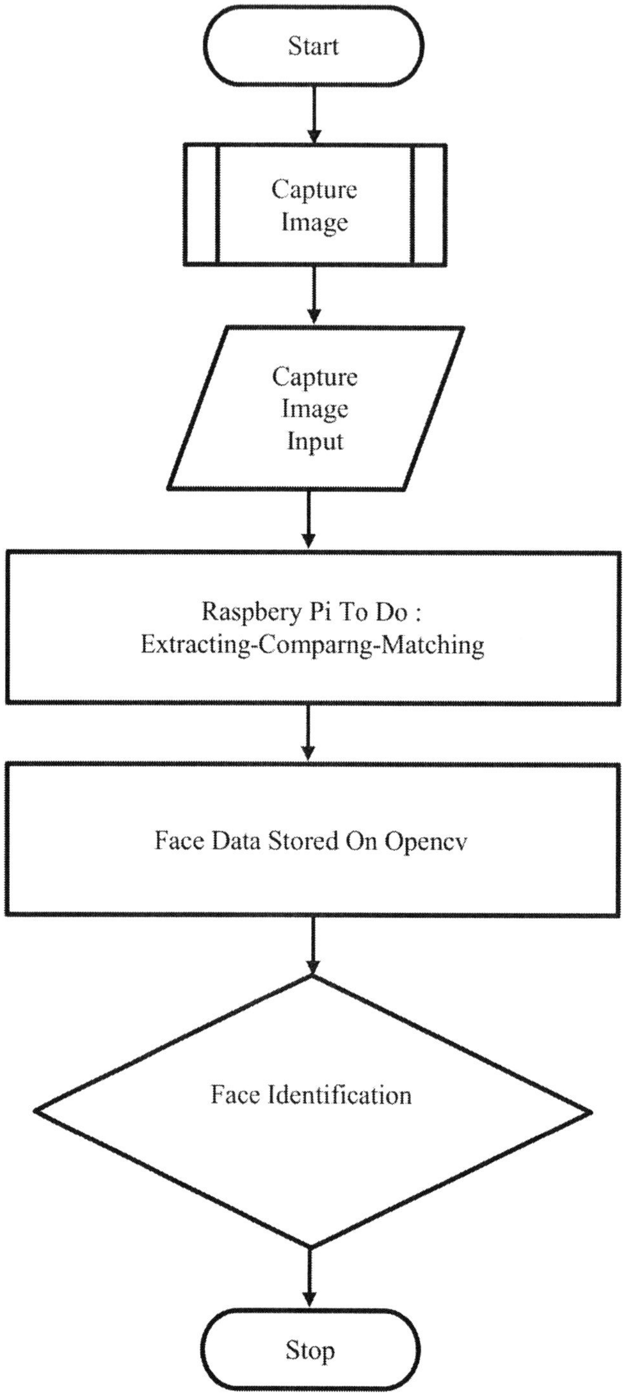

Figure 1. Flowchart proposed system

So in this research, a prototype was designed that integrates OpenCV, Raspberry Pi 3B + and a camera to control access to the vehicle ignition lock system. In Figure 1 above it can be explained that:

1. Raspberry Pi is a Single Board Computer (SBC) developed by the Raspberry Pi foundation in the United Kingdom (UK). Raspberry pi can work like a computer but mini-size that can create documents, process data with spreadsheets, watch movies, play games and of course coding. Raspberry Pi has two models, namely Model A and Model B, and in this study used

Model B which has a memory capacity of 512MB besides that Model B is also equipped with an Ethernet port. The Raspberry Pi design is based around the Broadcom BCM2835 SoC (System-on-Chip) which has embedded an ARM1176JZF-S processor with a frequency of 700MHz. Data storage is designed using an SD card for long-term boot and storage processes. (7–9)

2. The security system to be achieved, namely confidentiality (a key must be given security by using computer vision technology so that it is not easily opened by anyone), availability (computer-based information system which is a quality data processing center and the raspberry pi is used to a decision-making tool with the aim that the data and information is available to those who are authorized to use it), and integrity (all subsystems must provide an accurate picture of the physical system it represents). [2]
3. Face recognition is a computer application that runs automatically to identify or verify someone from digital images sourced from video. Face recognition technology is based on the principles of neural computing, which analyzes facial characteristics that have been inputted in the form of images through the camera and can be broadly classified into static and dynamic matching. [2]
4. The method used in this study consists of machine vision and knowledge-based methods. The machine vision method consists of a training phase and a testing phase in which images are input. Image pre-processing which includes histogram equalization and median filtering enhances images in feature extraction preparation. Like the features: Haar, and Local Binary Patterns (LBP), then extracted from the image used to set the machine vision classifier. [7][8]
5. OpenCV is an abbreviation of Open Computer Vision, which is open source libraries that are specialized for image processing. The aim is for computers to have abilities similar to those in visual processing in humans. This library is made for the C / C ++ language as an application realtime optimization, has API (Application Programming Interface) for high level and low level, there are functions that are ready to use for loading, saving, image and video acquisition. This OpenCV library has features including structural analysis, camera calibration, motion detection, object recognition, etc. [3]

So for the block diagram in this design consists of two block diagrams, namely a software application that will be used for Python and Linux-based control systems that are on Raspberry Pi and a hardware system that plays a role in the mechanism of the relay system. Software applications are made using the Python programming language that allows users to be able to access buttons and webcams that are used to recognize and identify faces that have been created in the database, so that if a face is recognized then the system will work to activate the ignition key automatically. The hardware system uses several important components, namely a PC / Laptop, Raspberry Pi 3, webcam, Monitor Screen, LED as indicator, and relay.

For input using a webcam camera connected to the Raspberry Pi 3B +, then processed by Raspberry Pi 3B + and the output is a relay that works connecting several cables in the ignition socket as a switch so that it can turn on electricity on the motorcycle. There are LED indicators, which means that the ignition lock process is accepted and welcome to start the vehicle engine. And to illustrate the diagram in Figure 2.

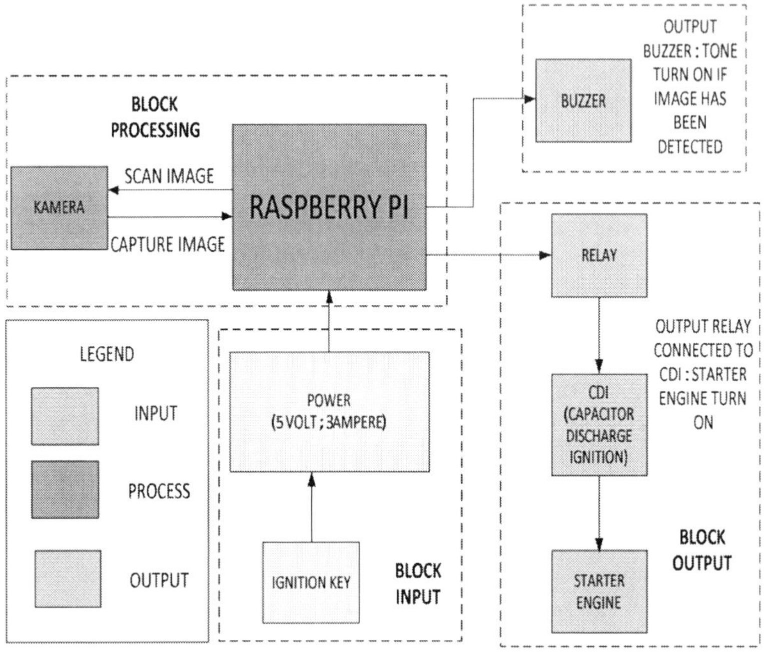

Figure 2. Blok diagram proposed system

3. Result And Analysis

3.1. Dataset program and face training program

In this section we will begin the first phase of the face recognition project using Raspberry Pi and OpenCV, which will start from the face detection step. In the first step, will create a dataset folder, where the folder will store for each id and some photos with a part that is used to detect faces. The results are as follows (Figure 3):

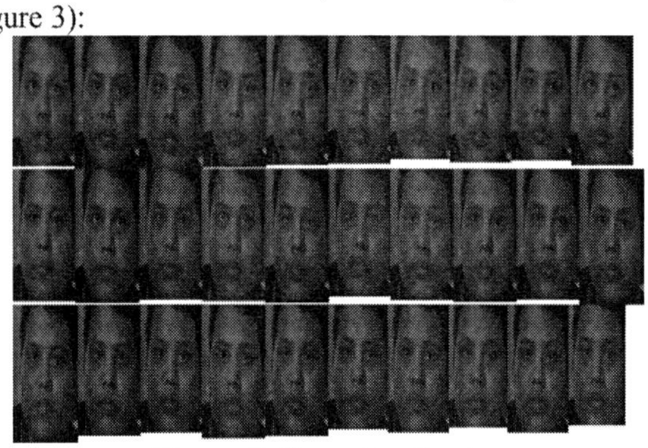

Figure 3. Dataset collection

Analysis and discussion for the face dataset program:

The face dataset command script has successfully run and outputs a total of 30 photos in 3 seconds. The results of the collection of photos in this dataset then we run the face_training command so that the dataset has been registered (recognize) into an algorithm where the file name is "trainer.yml". and both of these commands have a function as a predict recognizer. Predictive recognition is a comparison if the face does not match the dataset, the system will not recognize (unknown) and the ignition is inactive.

3.2. Face recognition program

In this section we will start the second phase of this research, which starts from running the command below:

```
names = ['None', 'Lukman']
********************************************
    # Check if confidence is less them 100 ==> "0" is perfect match
    if (confidence < 100):
      id = names[id]
      confidence = "  {0}%".format(round(100 - confidence))
    else:
      id = "unknown"
      confidence = "  {0}%".format(round(100 - confidence))
cv2.putText(img, str(id), (x+5,y-5), font, 1, (255,255,255), 2)
     cv2.putText(img, str(confidence), (x+5,y+h-5), font, 1, (255,255,0), 1)

  cv2.imshow('camera',img)

  k = cv2.waitKey(10) & 0xff # Press 'ESC' for exiting video
  if k == 27:
    break

# Do a bit of cleanup
print("\n [INFO] Exiting Program and cleanup stuff")
cam.release()
cv2.destroyAllWindows()
```

Figure 4. List program face recognition

Analysis and discussion for the face recognition program:

The face recognition command script has been successfully executed and every face detected by the camera will be identified (recognize predict) with the face dataset that has been recognized.

3.3. Running system program

In this section we will start the third phase of this research, where we get the following results :

1. The face dataset is recognized
   In this experiment it can be explained that the condition of the room is in a state of bright lighting when running the face recognition process. Then when the face recognition program is run, then the face recognition process is active, the camera sensor will turn on automatically and detect faces that are photographed on the camera and the results are displayed on the monitor as shown in Figure 5 below.

Figure 5. Face recognized

Analysis and discussion for the system program:
The system program has successfully been run against recognized face datasets. A percentage of> 51% is required for the process of predicting faces to be detected, and if a face is not recognized then the known percentage is <50%. And the distance of the camera captures the image as far as 1 meter. And in Figure 5 above we get 59% of the matching face photos stored in the face dataset directory stored in the system. then the system works with the ignition key turned on.

2. The face dataset is not recognized
In this experiment it can be explained that the condition of the room is in a state of bright lighting when running the face recognition process. Then when the face recognition program is run, then the face recognition process is active, the camera sensor will turn on automatically and detect faces that are photographed on the camera and the results are displayed on the monitor as shown in Figure 6 below.

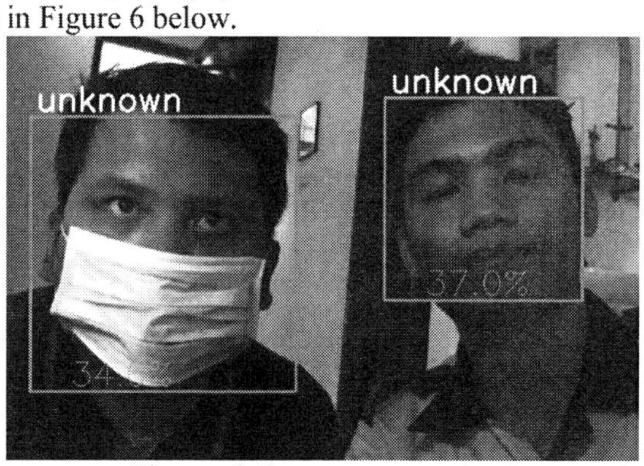

Figure 6. Face not recognized

Analysis and discussion for the system program:
The system program has successfully been run against recognized face datasets. A percentage of> 51% is required for the process of predicting faces to be detected, and if a face is not recognized then the known percentage is <50%. And the distance of the camera captures the image as far as 1 meter. In Figure 6 above the results obtained are 34% and 37% of the matching face photos stored in the face dataset directory stored in the system. So that the results are obtained that the system is not running well which results in ignition key still in the off state.

3. The face dataset that distinguishes between recognizable and not recognized
In this experiment it can be explained that the condition of the room is in a state of bright lighting when running the face recognition process. Then when the face recognition program

IOP Conf. Series: Materials Science and Engineering **909** (2020) 012002 doi:10.1088/1757-899X/909/1/012002

is run, then the face recognition process is active, the camera sensor will turn on automatically and detect faces that are photographed on the camera and the results are displayed on the monitor as shown in Figure 7 below.

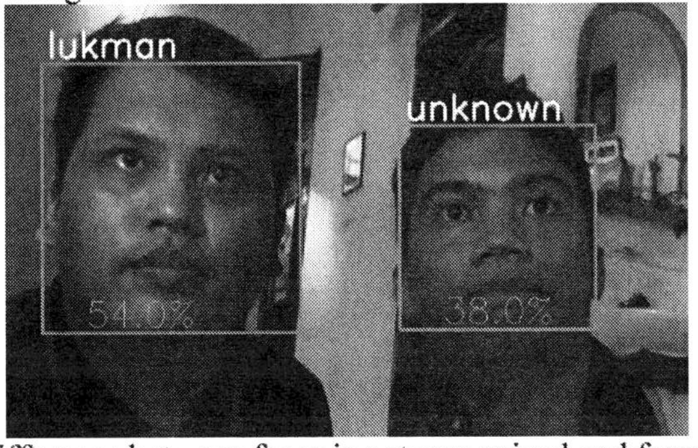

Figure 7. The difference between faces is not recognized and faces are recognized

Analysis and discussion for the system program :
The system program has successfully been run against recognized face datasets. A percentage of> 51% is required for the process of predicting faces to be detected, and if a face is not recognized then the known percentage is <50%. And the distance of the camera captures the image as far as 1 meter. In the picture 7 above we get the results of 54% and 38% of the matching face photos stored in the face dataset directory stored in the system. So that the results are obtained that the system is not running well even though one of the object images meets the standard face recognition criteria. This happens because the system is designed only for the process of recognizing one object, resulting in the ignition key still in the off state.

3.4. Design result

This section displays the results of the prototype display device before it is installed in the vehicle and after it is installed on the vehicle as follows:

1. Final prototype result
   In this experiment it can be explained that the prototype has not been installed to the vehicle as shown in Figure 8 below.

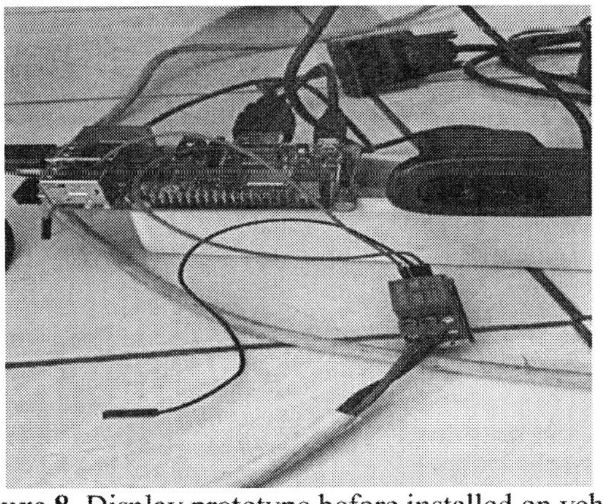

Figure 8. Display prototype before installed on vehicle

2. Prototype after being installed to the vehicle
   In this experiment it can be explained that the prototype has been successfully installed to the vehicle by jumping a CDI (Capacitor Discharge Ignition) jumper to the prototype as shown in Figure 9 below.

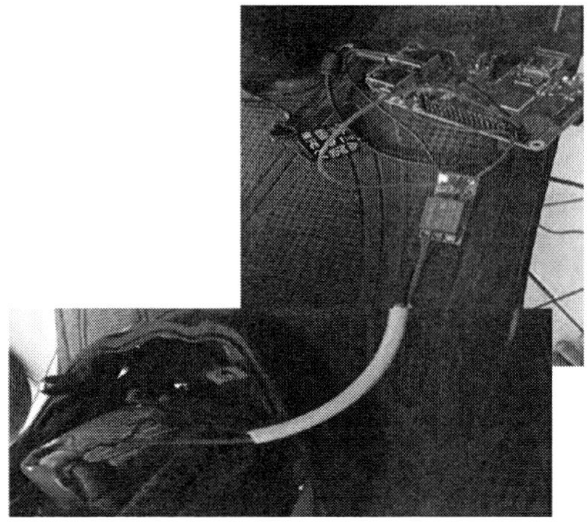

Figure 9. Display prototype after mounted to vehicle.

4. Conclusion

The purpose of this research is to design a prototype of a motorbike safety lock using raspberry pi with face recognition method. The existence of this prototype provides vehicle safety. For this reason, in-depth analysis has been carried out and its performance has been compared, so that the following conclusions can be drawn:

1. Optimal system capability of taking pictures up to 1 meter.
2. The ability of the camera to recognize the user's face is at a percentage> 51% and an unknown face is at a percentage <50%.
3. This prototype works with a face recognition system with the local binary pattern histogram method.
4. This prototype can set the contents of the dataset to increase vehicle ownership, and if there are unknown faces, the system will not work.

5. References

[1] Silalahi LM, Alaydrus M, Rochendi AD, Muhtar M 2019 Design of tire pressure monitoring system using a pressure sensor base *SINERGI* **23** 70-78

[2] Fauzi M 2018 Rancang bangun door lock face recognition dengan metoda eigenfaces menggunakan opencv2. 4.9 dan telegram messenger pada raspberry pi [Skripsi] Universitas Mercu Buana

[3] Asy'har AA 2015 Face recognition menggunakan metode viola & jones [Skripsi] Universitas Mercu Buana Jakarta

[4] Patil A, Shukla M 2014 Implementation of classroom attendance system based on face recognition in class *International Journal of Advances in Engineering & Technology* **7** 974-979

[5] Davies ER 2018 Face detection and recognition *Chapter in book: Computer Vision* 631–662

[6] Suchitra, Suja P, Tripathi S 2016 Real-time emotion recognition from facial images using raspberry pi ii 3rd *International Conference on Signal Processing and Integrated Networks (SPIN)* 666–670

[7] Adriansyah A, GM MR, Yuliza Y 2014 Rancang bangun dan analisa cctv online berbasis raspberry pi *SINERGI* **18** 105-110

[8] Adriansyah, A., Dani, A.W. and Nugraha, G.I 2017 Automation control and monitoring of public street lighting system based on internet of things *International Conference on Electrical Engineering and Computer Science (ICECOS)* 231-236

[9] Sirait F, Yoserizal Y 2016 Pemanfaatan raspberry pi sebagai processor pada pendeteksian dan pengenalan pola wajah *Jurnal Teknologi Elektro* **7** 146-150
[10] Sirait F 2015 Sistem monitoring keamanan gedung berbasis raspberry pi *Jurnal Teknologi Elektro* **6** 55-60
[11] Krishnan A, Hs A 2017 Design and implementation of robotic vision for face recognition 2nd *Internation Conference on Circuits, Controls, and Communications (CCUBE)* 135–138
[12] Umm-E-Laila, Khan MA, Shaikh MK, Bin Mazhar SA, Mehboob K 2017 Comparative analysis for a real time face recognition system using raspberry pi *IEEE 4th International Conference on Smart Instrumentation, Measurement and Application (ICSIMA)* 1–4
[13] Simanjuntak IU, Maulana G 2019 Rancang bangun kunci otomatis sepeda motor berbasis face recognition dengan metode eigenfaces opencv *Jurnal Teknik Elektro* **8** 122-128
[14] Budiyanto S, Asvial M, Gunawan D 2014 Implementation dedicated sensing receiver (DSR) in 3g-wifi offload *International Conference on Smart Green Technology in Electrical and Information Systems (ICSGTEIS)* 37-42
[15] Budiyanto S, Nugroho A. A new model of genetic zone routing protocol (GZRP): the process of load balancing and offloading on the umts-ieee 802.11g hybrid networks *Telkomnika* **15** 598-605
[16] Budiyanto S, Asvial M, Gunawan D 2016 Improved performance of hybrid algorithm for 3g–wifi offload networks *Jurnal Teknologi* **78** 5-9
[17] Budiyanto S, Sihombing HC, IM FR 2019 Depression and anxiety detection through the closed-loop method using dass-21 *Telkomnika* **17** 2087-2097

IOP Conf. Series: Materials Science and Engineering **909** (2020) 012003 doi:10.1088/1757-899X/909/1/012003

The effect of post-curing temperature on mechanical properties of coconut coir fiber reinforced polyester composite

Rizal S[1], Junaidi A[1], Gunawan I[1], Taufiqurrahman[1], Nasution J D[2], Mataram A[2*], Ilham A B[2], and Afriansyah[2]

[1] Department of Mechanical Engineering, Politeknik Negeri Sriwijaya, South Sumatra, Indonesia.
[2] Department of Mechanical Engineering, Faculty of Engineering, Universitas Sriwijaya, South Sumatra, Indonesia.

*E-mail: amataram@unsri.ac.id;

Abstract. This study used post-curing training as a method of strengthening. The smoking process was a process carried out in the form of heating the specimen at a certain temperature to improve the properties possessed by the composite. The purpose of this study was to study the effect of post-curing temperature on the impact strength of coconut coir fiber composites with temperature variations of 600°C, 700°C, 800°C, 900°C, 1000°C. Specimens were prepared using a mixing method for mixing ingredients and a hand lay-up where in the composite volume fraction is prepared according to each specimen, which is 70% polyester and 30% coconut coir fiber. Each specimen is heated for 60 minutes in a conventional oven. Average impact strength and impact energy per unit area increases with the increase in post-curing temperature due to the fact that the movement of composites as well as the reduction of holes in the composites and adhesion to one another (suturing) can improve mechanical testing. The highest strength values obtained in the specimen with an increase in T = 100°C, ie 18,509 J for the average impact energy and 0,1434 J/mm^2 for impact energy per unit area average.

1. Introduction

With the development of the industrial world this day, material requirements for a product are increasing but the use of metal materials in various product components is decreasing. This is caused by the weight of components made of metal, the formation process is relatively difficult, can experience corrosion and expensive production costs. Therefore, many other materials have been developed that have characteristic properties that are suitable with the desired metal material characteristics. One material that has been developed today is composite. Composite is a material combination between two or more components or materials that have a number of properties that cannot be owned by each of these components [1].

Natural fiber reinforced composite material is one of the right choices. Coconut with the Latin name Cocos nucifera is one of plant type that produces natural fibers. During this time coconut fibers are only used as a base for making doormats, brushes, and others. Things into consideration the use of this fiber is the use of coconut fibers are still minimal [2].

For this study, researchers used polyester resin as a matrix of composite materials. Polyester resins are the most widely used resin in a variety of applications that use thermoset resins, both separately and in the form of composite materials. Although mechanically, the mechanical properties possessed by polyester are not very good or just mediocre. This is because this resin is easy to obtain, the price is relatively affordable and most importantly it is easy in the fabrication process. The type of polyester resin used as a composite matrix is the unsaturated polyester type which is a thermoset that can undergo curing from the liquid phase to the solid phase when it gets the right treatment [3].

This study uses post-curing training as a method of strengthening. The smoking process is a process carried out in the form of heating the specimen at a certain temperature to improve the properties possessed by the composite. The smoking process is carried out by heating the specimen material at a certain temperature, but the temperature cannot exceed the glass transition temperature, because if it exceeds the temperature it will cause the material to soften and if the temperature of the material is increased, it will exceed the melting point of the material, and it will melt [1].

Several studies that have been conducted regarding the use of natural fibers by the method of post-curing reinforcement are investigating the effect of post-curing temperature on tensile testing of composite fibers reinforced epoxy resin woven banana fiber. The temperature variations used were 25°C, 70°C, 80°C, 90°C, and 100°C, from this study the highest tensile strength was obtained at a curing temperature of 100°C, with a value of 42.82 MPa. Increased tensile strength up to 40.26% of non-curing specimens [1].

Study on the effect of temperature in the mechanical properties of hybrid polyester fiber reinforced glass fiber and goal fiber composites. The temperature variations used are 100°C, 150°C, and 200°C with each variation of time for 1 hour, 2 hours and 3 hours. From the research results obtained, specimens with variations in temperature of 100°C with a test time of 1 hour have the highest tensile strength of 62.264 MPa [4].

The purpose of this research was to study the effect of post-curing temperature on mechanical properties of coconut coir fiber reinforced polyester composite.

2. Experimental

2.1. Materials

Polyester resins, Metyl Etyl Ketone Peroxide (MEKPO) catalysts, Sodium Hydroxide (NaOH) solutions obtained from Chemical Sources and coconut coir fibers obtained from coconut milk merchant waste at the Ogan Ilir Indralaya market, where existing coconut coir is usually disposed of it's useless The coconut husk is then separated from the coconut shell, then the fibers are washed thoroughly and dried in the sun. Then after drying, the coconut fiber is combed to separate the fiber from the coir meat using a wire comb. The fibers that have been combed are then cut to a size of ± 5 mm and soaked in an alkaline solution (NaOH) at 5% for 2 hours. The fiber which has been soaked for 2 hours is then dried.

2.2. Composite preparation and experimental method

In this study, dried coconut fibers were then mixed with polyester resin with a volume fraction of 30% coconut fiber and 70% polyester resin. Specimens that have been mixed are stirred manually until the mixture is solid. Then before going to the printing stage, we prepare a test mold that has been smeared with oil. Then the specimen is put into the mold to dry. The dried specimens were then removed from the mold and given heat treatment with temperature variations of 60°C, 70°C, 80°C, 90°C, and 100°C in a conventional oven for 60 minutes. Heated specimens can then be tested.

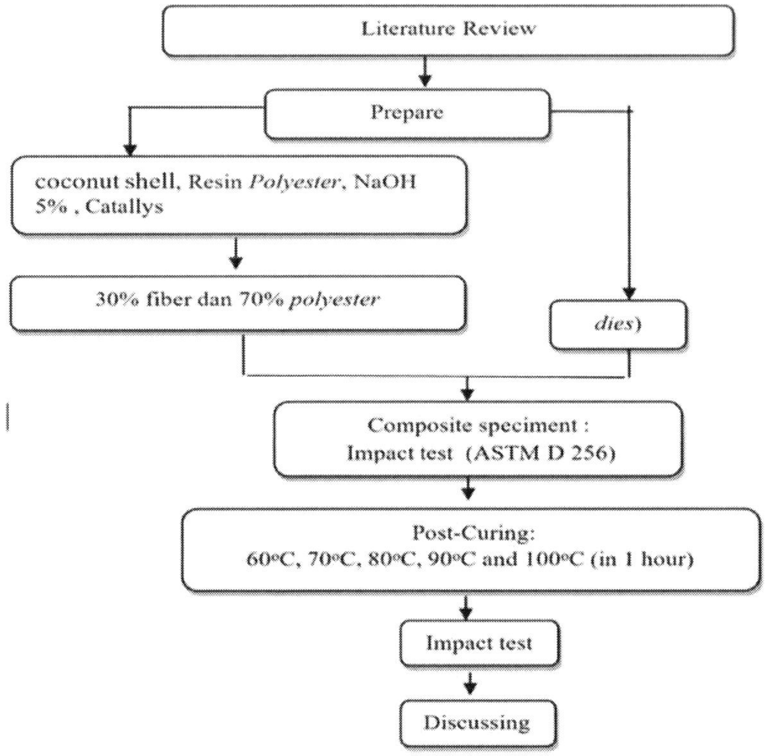

Figure 1. Experimental method

Figure 2. Composite specimen

2.3. *Impact testing*
The impact test is carried out by the Charpy impact test. The Charpy impact test measures the resistance of the material to the impact of the pendulum that vibrates. The impact strength is calculated from the impact energy divided by the cross-sectional area of the sample. In this study, tests using the standard Charpy and Charpy test methods for composites meet ASTM D 256 standart.

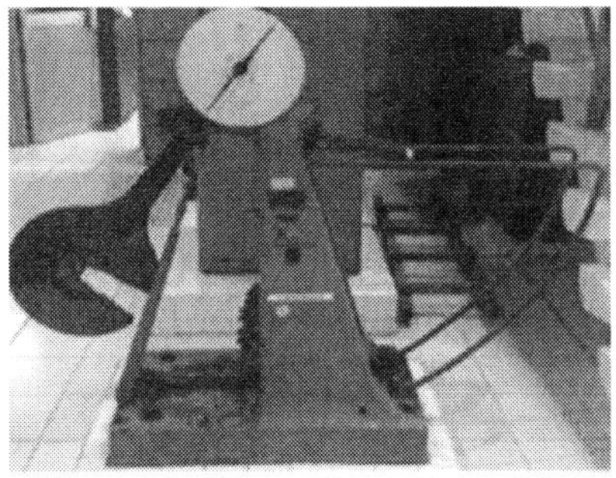

Figure 3. Charpy impact machine

3. Results and Disccusions

Here are the results of impact testing has been done is in Table 3 below:

Table 1. Impact testing results

Kode Spesimen *(T=Temperatur)*	$\alpha(^0)$	$\theta\ (^0)$
	90	85,9
Non post-curing	90	85,9
	90	85,7
	90	85,5
T=60⁰C	90	85,7
	90	85,7
	90	85,1
T=70⁰C	90	85,1
	90	85,2
	90	84,9
T=80⁰C	90	84,7
	90	84,9
	90	84,1
T=90⁰C	90	84,3
	90	84,1
	90	83,7
T=100⁰C	90	83,9
	90	83,7

The results of the test affect the data obtained and then perform the calculations manually by formula and are also statistically analyzed with Microsoft Excel. The average yield of impact energy (E) at the highest temperature is 100°C, ie 18,509 J and the minimum value for the non-curing sample is 12,375 J. The results for the average energy per unit area (W) with the highest grade are also when the heating temperature is 1000°C, ie 0.1434 J/mm^2 and the minimum value for samples non-curing 0.0959 J/mm^2.

Several studies that have been conducted regarding the use of natural fibers by the method of post-curing reinforcement were increase tensile strength up to 40.26% of non-curing specimens [1],[4].

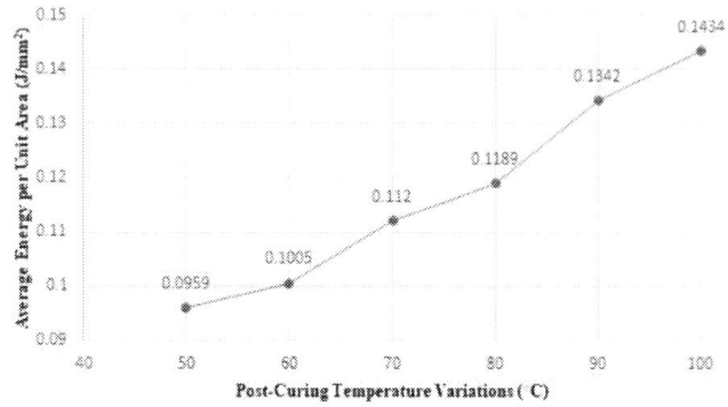

Figure 4 Graph of the relationship between the impact energy (E) on average by comparison post curing temperature variation and non post-curing

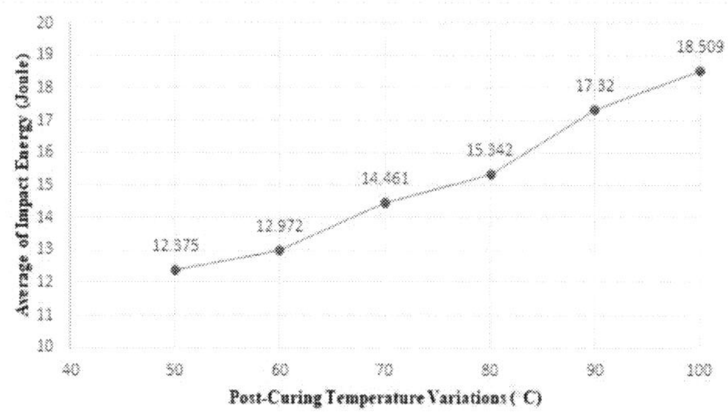

Figure 5. Graph of the relationship between the value of the impact energy per unit area (W) average temperature variation in the ratio post-curing and non-post-curing

4. Conclusion

The increase in impact strength in the average impact energy (E) and the impact energy per cross section (W) when you adjust the curing temperature of the heater is caused by the rise in temperature at Tg because the resin molecule reorganizes. The post-curing heating process can also increase the degree of crosslinking of the composite so that its mechanical properties are improved. The optimum strength value of the average impact energy sample (E) obtained at T=1000°C is 18.50 J, while the minimum strength of the non-post-curing samples is 12.375 J. With the impact energy per cross section (W), even at T=1000°C corresponds to 0.1434 J/mm^2 and the minimum non-curing sample value is 0.0959 J/mm^2.

5. References

[1] Bodja Suwanto 2010 Pengaruh temperatur post-curing terhadap kekuatan tarik komposit epoksi resin yang diperkuat woven serat pisang *e-jurnal wahana Politek. Negeri Semarang* 1–31

[2] Nst F S, Kimia D T 2013 Pengaruh penggunaan larutan alkali pada kekuatan berpengisi serbuk serabut kelapa **2** 14–20

[3] Bramantiyo A, D. Metalurgi, D. A. N. Material, F. Teknik, and U. Indonesia 2008 Sifat mekanik material komposit poliester – serat alam

[4] Sabuin A, Boimau K, Adoe D G H, Mesin J T, and Cendana U N 2015 Pengaruh temperatur pengovenan terhadap sifat mekanik komposit hibrid polyester berpenguat serat glass dan serat daun gewang **02** 69–78

IOP Conf. Series: Materials Science and Engineering **909** (2020) 012004 doi:10.1088/1757-899X/909/1/012004

A CFD simulation and experimental study: predicting heat transfer performance using SST k-ω turbulence model

C D Widiawaty[1,2a], A I Siswantara[2b], Budiarso[2], G G R Gunadi[1,2], H Pujowidodo[1,3], M H G Syafei[2]
[1]Department of Mechanical Engineering, Politeknik Negeri Jakarta, Depok 16424 Indonesia
[2]Department of Mechanical Engineering, Universitas Indonesia, Depok 16424, Indonesia
[3]Center for Thermodynamics, Engine, and Propulsion, BPPT, Serpong 15314, Indonesia

*Email : candra.damis.widiawati@mesin.pnj.ac.id

Abstract. The heat exchanger is the most common equipment in the industry. Nowadays, CFD simulation has been used widely to predict the performance of heat transfer. The accuracy of the CFD depends on several factors such as simulation parameters, models, and quality of the grid. The heat exchanger in industrial application has big heat transfer capacity, since it consists lot of tube and big shell size. In this study, the fluid heat transfer on the tube is represented by uniform wall tube temperature. The objective of this works is to evaluate the performance of the SST $k\omega$ turbulent and by applying heat transfer simulation of fluid on the tube side represented by uniform wall tube temperature. This present work compared CFD simulation and Experimental study. While the working fluids which are used in the experimental study are cooking oil in the shell and exhaust gas in the tubes. The constant of SST $k\omega$ turbulent are cmu: 0.09, c1: 1.44, c2: 1.92, and sigma epsilon 1.3. The CFD simulation result shows in 50 s simulation that the temperature of cooking oil is from 48 °C to 54 °C. This tendency has good agreement with the experimental study.

1. Introduction

Recently, in the industrial field shell and tube heat exchanger has been utilized extensively since it has a good ability in high temperature and pressure, good resistance in corroded fluid, easy in maintenance and manufacturing process [1]. The general procedure in designing heat exchanger based on A. Jones [2] (cited in[3]) started by collecting fluid parameters such as temperature and mass flow rate, determining the heat load at the tube and shell sides, evaluating effective corrected temperature difference, determining heat exchanger configuration such as shell and tube type, tube dimension, baffle distance and tube bundle. Also, the flow direction of cold and hot fluids influence on the performance of heat transfer. Both heat exchanger configuration and fluid flow direction impact the fluid dynamic and heat transfer which finally will have an impact on heat exchanger performance.

In its development, the design and reverse engineering of the heat exchanger can be done conventionally based on the basic calculation of heat transfer. For example, using Kern method, Delaware method, Effectiveness - Number of transfer unit method (ε-NTU Method), and Logarithmic mean temperature difference method (LMTD Method). Kern method has been used in reverse engineering of cooler at PLTA Jatiluhur [4], by changing the tube diameter and tube number. The existing equipment has 16 mm diameter and 88 tubes, the new design has 15 mm diameter and 124 tubes. The Kern Method used to predict the heat transfer and the pressure drop. LMTD is a simple method in designing a heat exchanger, it can be used if the only inlet and outlet fluid temperature is determined [5]. R. Rajesh et.all [5] used LMTD method for designing perforated heat exchanger at utilization of exhaust engine.

Advances in computer technology and modelling mathematics make it easier for engineers to design and reverse engineer using computational programs such as HTRI, Hysis, and Computational Fluid Dynamic. Recently, computational velocity and numerical techniques have been improved significantly. Furthermore, Computational Fluid Dynamics (CFD) was introduced by scientists gradually since it is meaningful to assess the operation of numerous heat exchangers in terms of heat and mass transfer [6]. Practically, CFD methods show uncertainties/errors which are similar to the experimental studies. Apart from the user errors, CFD includes technical errors which are parametric, numerical discretization (. e.g. spatial mesh size and temporal resolution) and modelling. In the mathematical model, overgeneralization of flow phenomena placed the most substantial modeling errors. Numerical discretization errors ranked second due to inadequate mesh resolution. While the parametric errors due to less comprehensive information of the geometry and properties of fluid[7]. The study which had been done by Candra et. al showed that grid quality influences the accuracy of fluid temperature on shell and tube heat exchanger[8]. Wangetal [9](cited in [6]) utilized commercial CFD FLUENT to inspect overall heat transfer rate and the pressure drop of the shell with helical baffles. The authors acquired that the CFD prediction tends to the experimental studies. Ozden and Tari [10] proposed the use of CFD simulation for predicting the heat and mass transfer behavior the shell side of a small heat exchanger.

Nowadays, lots of researchers evaluate the impact of turbulent model on the accuracy of CFD simulation results. There are various turbulent models on CFD program such as k-epsilon standard, k-ω epsilon, RNG and RSM [11]. Leutwyler and Dalton stated that compared to the Spalart-Allmaras model, the standard k-w model, the RNG k-e model and the standard k-w model, it is showed that the SST k-ω is the most appropriate model to evaluating the pressure distribution on the valve disk [7]. Additionally, an experiment on the turbulence model was conducted to prove that the standard k-e model performs excellent outcomes for the velocity profile together with heat transfer in shell and tube heat exchanger [6]. Another CFD model simulates the thermo-hydraulic performance by using the RNG k-e turbulence model employing the standard wall functions. The simulation studies showed that the pressure drop of the shell side is likely to the experimental results. The deviation is less than 15% between those simulation and experimental results [12]. Conversely, minor deviances were obtained with the predictions of the SST k-ω model. Moreover, in case of the cross-flow over tube bundles the standard k-ε and the SST k- ω model expects reasonably accurate within band +/- 5% [6].

Previous research showed that turbulent model impact on the accuracy of simulation results. Moreover, it is proven that the SST k-ω model gives a good result with +/- 5% error [6]. In this study, the prediction of heat transfer is represented by the temperature of cooking oil in the outlet and SST k-ω turbulent model is used in this simulation. The heat transfer of exhaust gas on tube is simplified by applying uniform tube wall temperature.

This simplification used to reduce the time and cost of the simulation process, especially in the large-dimensional STHE design. Hence, the novelty of this research is the accuracy test of the SST k-ω model and the use of wall temperature on the tube to the simulation results. The accuracy of the simulation result is known by comparing the cooking oil temperature on the outlet with the experimental result. Statistical approach is used to determine the uncertainty of measurements. Statistical method of repeated measurement for one measurement parameter is used in this study.

2. Experimental setup

The main component of the experiment setup consists of the heat exchanger 450 mm in length x 250 mm in width x 250 mm in thickness. A heat source is from the exhaust gas of the engine. The exhaust gas flows along the tube with 0.0011 kg/s. A pump flows the oil along with the shell with 0.015 kg/s in mass flow rate. K-type thermocouples made of Chromel/Alumel are used to measure temperature. There are four k-type thermocouples to measure the temperature, i.e. inlet and outlet of exhaust gas and inlet and outlet of cooking oil. Heat exchanger surface is isolated to reduce heat losses to the

IOP Conf. Series: Materials Science and Engineering **909** (2020) 012004 doi:10.1088/1757-899X/909/1/012004

surrounding.

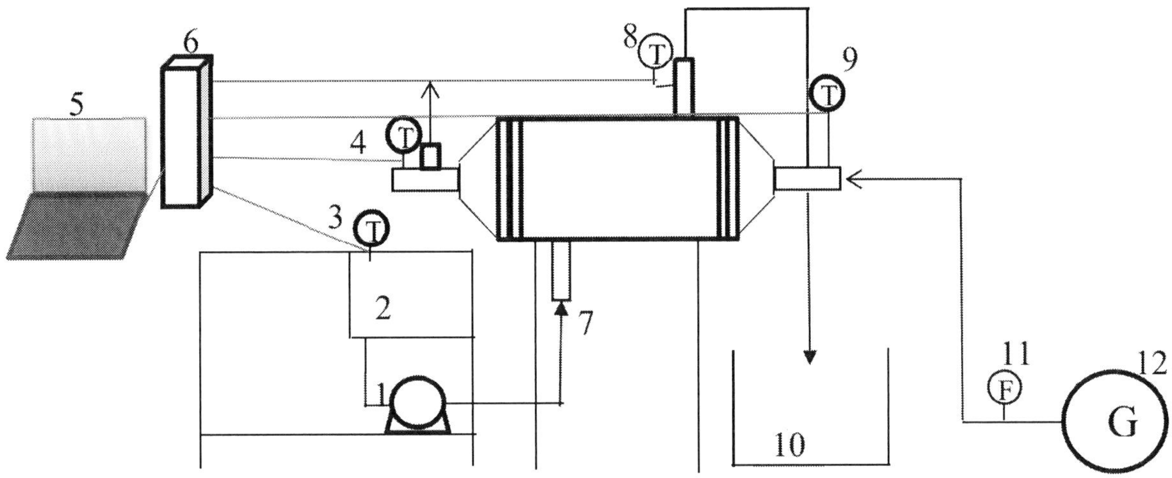

Figure 1. Schematic diagram of heat exchanger (1) oil pump (2) feed cooking oil storage (3) k-type thermocouple (4) k-type thermocouple (5) Laptop (6) DAQ (7) Oil inlet to He (8) k-type thermocouple (9) k-type thermocouple (10) Oil receiver (11) digital anemometer (12) engine

3. CFD Simulation

3.1. Governing Equation

This study analyzes heat transfer in 3-dimension flow, transient and turbulence. Governing equations as followed:

Governing transport equation[13]

Continuity:

$$\frac{\partial \rho}{\partial t}+\mathrm{div}(\rho u)=0 \quad (1)$$

Momentum:

x – momentum:

$$\frac{\partial(\rho u)}{\partial t}+\mathrm{div}(\rho u\mathbf{u})=-\frac{\partial p}{\partial z}+\mathrm{div}(\mu \text{ grad } u)+S_{M_x} \quad (2)$$

y – momentum:

$$\frac{\partial(\rho v)}{\partial t}+\mathrm{div}(\rho v\mathbf{u})=-\frac{\partial p}{\partial y}+\mathrm{div}(\mu \text{ grad } v)+S_{M_y} \quad (3)$$

z – momentum:

$$\frac{\partial(\rho w)}{\partial t}+\mathrm{div}(\rho w\mathbf{u})=-\frac{\partial p}{\partial z}+\mathrm{div}(\mu \text{ grad } w)+S_{M_y} \quad (4)$$

Energy:

$$\rho\frac{DE}{Dt}=-\mathrm{div}(\rho\mathbf{u})+\left[\frac{\partial(u\tau_{xx})}{\partial x}+\frac{\partial(u\tau_{yx})}{\partial y}+\frac{\partial(u\tau_{zx})}{\partial z}+\frac{\partial(v\tau_{xy})}{\partial x}+\frac{\partial(v\tau_{yy})}{\partial y}+\frac{\partial(v\tau_{zy})}{\partial z}+\frac{\partial(w\tau_{xz})}{\partial x}+\frac{\partial(w\tau_{yz})}{\partial y}+\frac{\partial(v\tau_{zz})}{\partial z}\right]+\mathrm{div}(k \text{ grad } T)+S_E \quad (5)$$

General equations of turbulence SST k - ω [6]:

$$\frac{\partial(k)}{\partial t}+\frac{\partial((uj)k)}{\partial x_j}=\frac{P_k}{\rho}-\beta^*\omega k+\frac{1}{\rho}\frac{\partial}{\partial x_j}\left[\left(\mu+\sigma_k\mu_t\right)\frac{\partial k}{\partial x_j}\right] \quad (6)$$

$$\frac{\partial(k)}{\partial t}+\frac{\partial((uj)k)}{\partial x_j}=\frac{P_k}{\rho}-\beta^*\omega k+\frac{1}{\rho}\frac{\partial}{\partial x_j}\left[\left(\mu+\sigma_k\mu_t\right)\frac{\partial k}{\partial x_j}\right] \quad (7)$$

$$F_1=\tanh(a_1{}^4) \quad (8)$$

$$a_1=\min\left[\max\left(\frac{\sqrt{k}}{0.09\omega y};\frac{500v}{y^2\omega}\right);\frac{4\rho\sigma_{\omega 2}k}{CD_{k\omega}y^2}\right] \quad (9)$$

IOP Conf. Series: Materials Science and Engineering **909** (2020) 012004 doi:10.1088/1757-899X/909/1/012004

$$vt=\frac{a1k}{\max(a1\omega;SF_2)},\ S=\sqrt{\left(\frac{\partial(u_i)}{\partial x_j}+\frac{\partial(u_j)}{\partial x_i}\right)\frac{\partial(u_i)}{\partial x_j}} \tag{10}$$

$$F_2=\tanh(a_1{}^4) \tag{11}$$

$$a_2=\max\left(2\frac{\sqrt{k}}{0.09\omega y};\frac{500v}{y^2\omega}\right) \tag{12}$$

$$CD_{k\omega}=\max\left(2\rho\frac{1}{\sigma w2}\frac{\partial k}{\partial x_j}\frac{\partial \omega}{\partial x_j},10^{-10}\right) \tag{13}$$

3.2. Boundary Condition

This study analyses the accuracy of the result of SST k - ω model and simplifying the process conditions of transferring tube flows into a constant temperature wall tube in the system (Fig 2). The outer wall is adiabatic. This simulation uses a 3D model. Cartesian grid is used with a total grid number of 2,000,000. Cooking oil with a temperature of 38 °C flows from the bottom of the nozzle towards the shell and exits on the top of the nozzle. Flue gas flows along the tube. The shell and tube flow is a cross-flow type. Grid and Boundary condition simulation can be seen in Figure 2 and simulation parameters can be seen in Table 1.

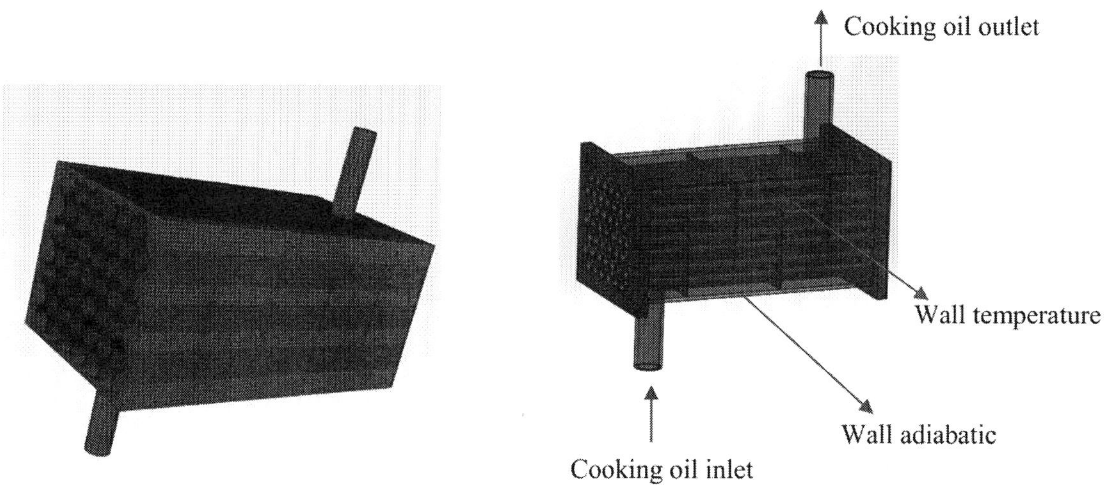

a. grid   b. boundary condition

Figure 2. (a) grid (b) boundary condition

Table 1. Simulation model and parameter

Description	Value	Note
Model		Transient, Incompressible, Subsonic
Time Set	0.0001 s	
Turbulence		SST-kω
Turbulence Model Coefficient		cμ=0.09 ; c1=1.44 ; c2 = 1.92 ; σ_ε =1.3
Fluid		Cooking oil
Density	880 kg/m^3	
Inlet Velocity	0.0456 m/s	
Outlet		Pressure
Temperature Inlet	38 °C	
Tube temperature	65 °C	

4. Result and Discussion

The main components of the experiment are the engine generator, heat exchanger, and a pump. The experiments are conducted until the fluctuation of gas and cooking oil temperature are quite stable. The temperature range of flue gas inlet temperature is 181°C - 189°C and cooking oil outlet temperature is 46°C - 47°C. The cooking oil temperature at the outlet are analysed by statistical approach to obtain the uncertainty of measurements [14]. Statistical calculation for cooking oil temperature at the outlet includes mean temperature 46.57 °C , standard deviation 0.306, and Standard error of the mean 0.0385. Based on level of confident 95%, then the uncertainty of temperature is 0.077, so T=(46.57 °C ± 0.0770).

Simulation results are verified and validated. Cooking oil flows on the shell side. Inlet cooking oil flows from the bottom of the nozzle, flows through the baffles, and then comes out on the upper side of the nozzle. The simulation flow phenomenon is under the real condition of the system. The streamline of the cooking oil can be seen in Figure 3.

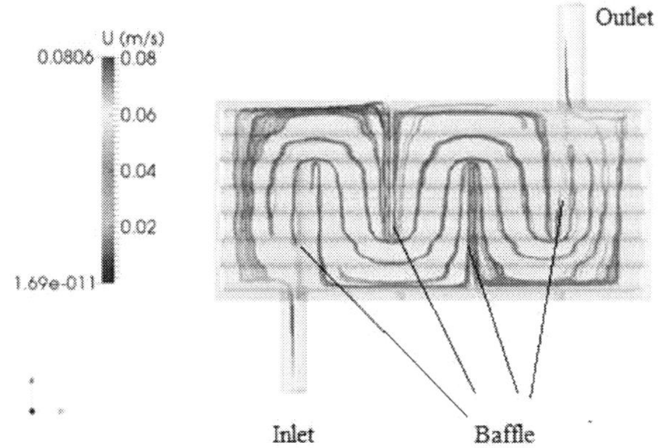

Figure 3. Stream line of cooking oil

Cooking oil comes from the inlet to the path between the tube bundles. It appears that the baffles lead the flow of cooking oil. At the inlet side, the cooking oil flows upwards then downward furthermore continues until the outlet. Figure 4 shows the tube bundle configuration and a cutting section.

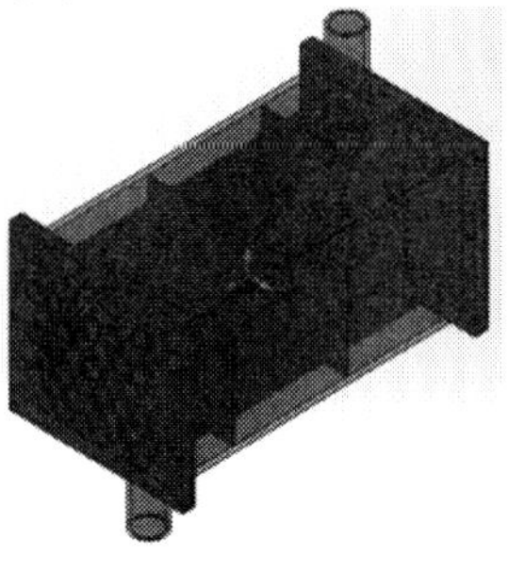

a. Isometric view

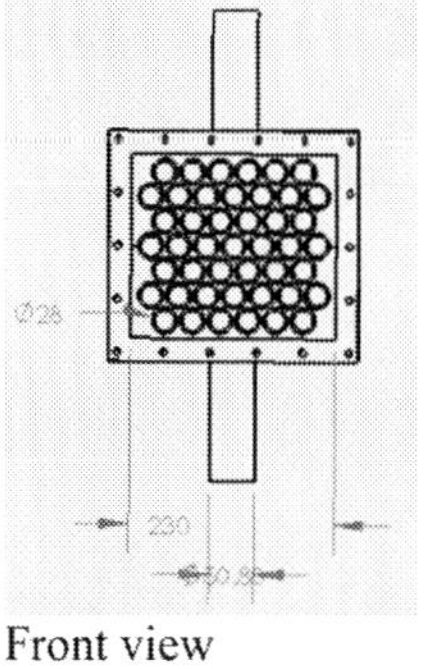

b. Front view

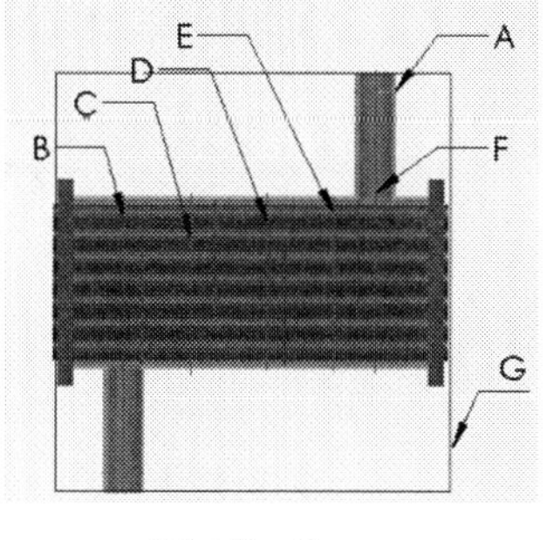

c. Plot Section

Figure 4. Tube bundle configuration

The velocity distribution in the heat exchanger affects the heat transfer process significantly. Therefore, the velocity distribution is verified. The simulation results of the velocity distribution can be seen in Figure 5 and Figure 6.

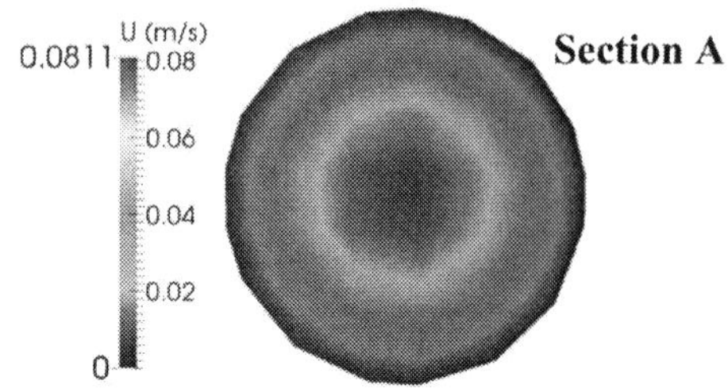

Figure 5. Graph of velocity distribution

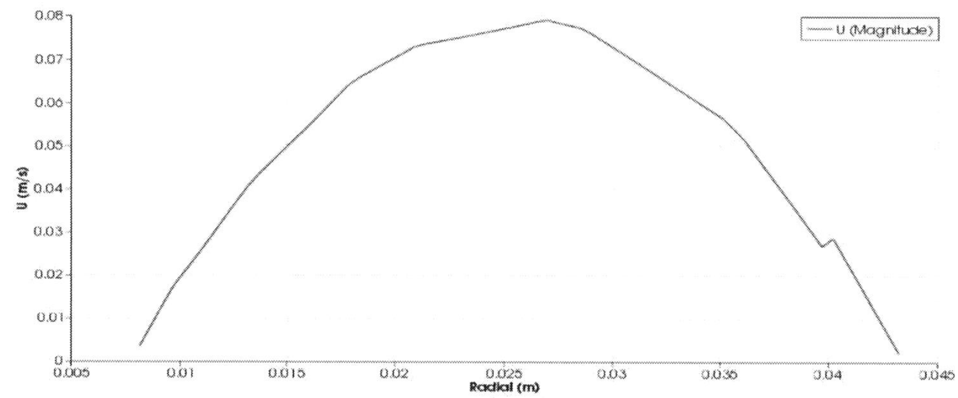

Figure 6. Contour of velocity distribution in the nozzle

Figure 6 shows the velocity distribution contour of the cross-section of the nozzle. It is shown that the largest velocity gradient is in the middle of the nozzle while the lowest is in the wall. The simulation shows that the velocity ranges from 0.005 to 0.08 m/s as shown in Figure 7. This phenomenon is following the concept of the boundary layer. It can have been seen that the viscous is dominant in nozzle. The fluid consists of layers, so there is a shift in each layer. Viscosity causes the fluid layer closest to the wall to rub against the wall, resulting in shear stress that causes a velocity gradient [14].

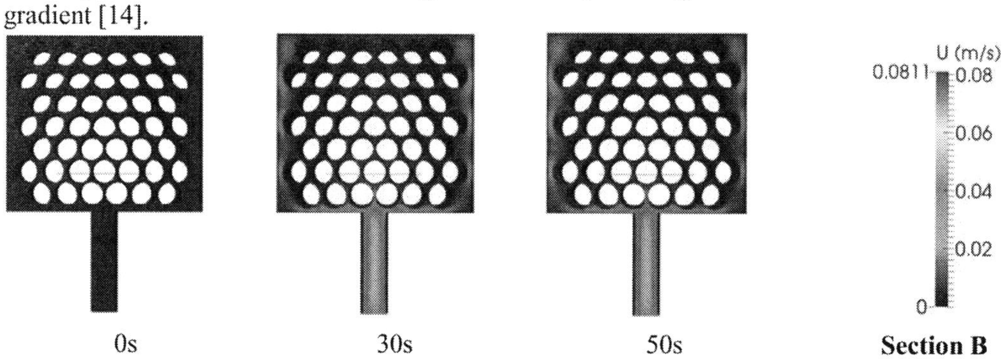

Figure 7. Velocity distribution in shell

The velocity of cooking oil on the shell is on average of 0.01 m/s which is smaller than the velocity on the inlet, this is caused by the area of the gap in the shell is greater than the area of the inlet of the nozzle. The difference in the cross-sectional area causes a change in velocity with a range of 0.001 to 0.0811 m/s. This phenomenon is in accordance with the theory of continuity. The velocity of cooking oil in the inlet and outlet nozzle is greater than in the tube bundles. To the heat exchanger geometry, the average cross-sectional area in the tube bundle is greater than the cross-sectional area at the inlet and outlet nozzle. The nozzle cross-section is 2025.08 mm2 in area and the bundle is 15241.9 mm2 in the average cross-sectional area of the tube.

IOP Conf. Series: Materials Science and Engineering **909** (2020) 012004 doi:10.1088/1757-899X/909/1/012004

Tube bundle configuration and mass rate parameters cause cooking oil flowing on the space between the tube bundle and the shell (Figure 7). This is indicated by the velocity that tends to be higher in this particular area. After 30 seconds of simulation, there is no significant velocity contour change. This shows that the flow began steadily. Therefore, it can be continued with temperature distribution analysis. Analysis of temperature distribution can be seen in Figure 8.

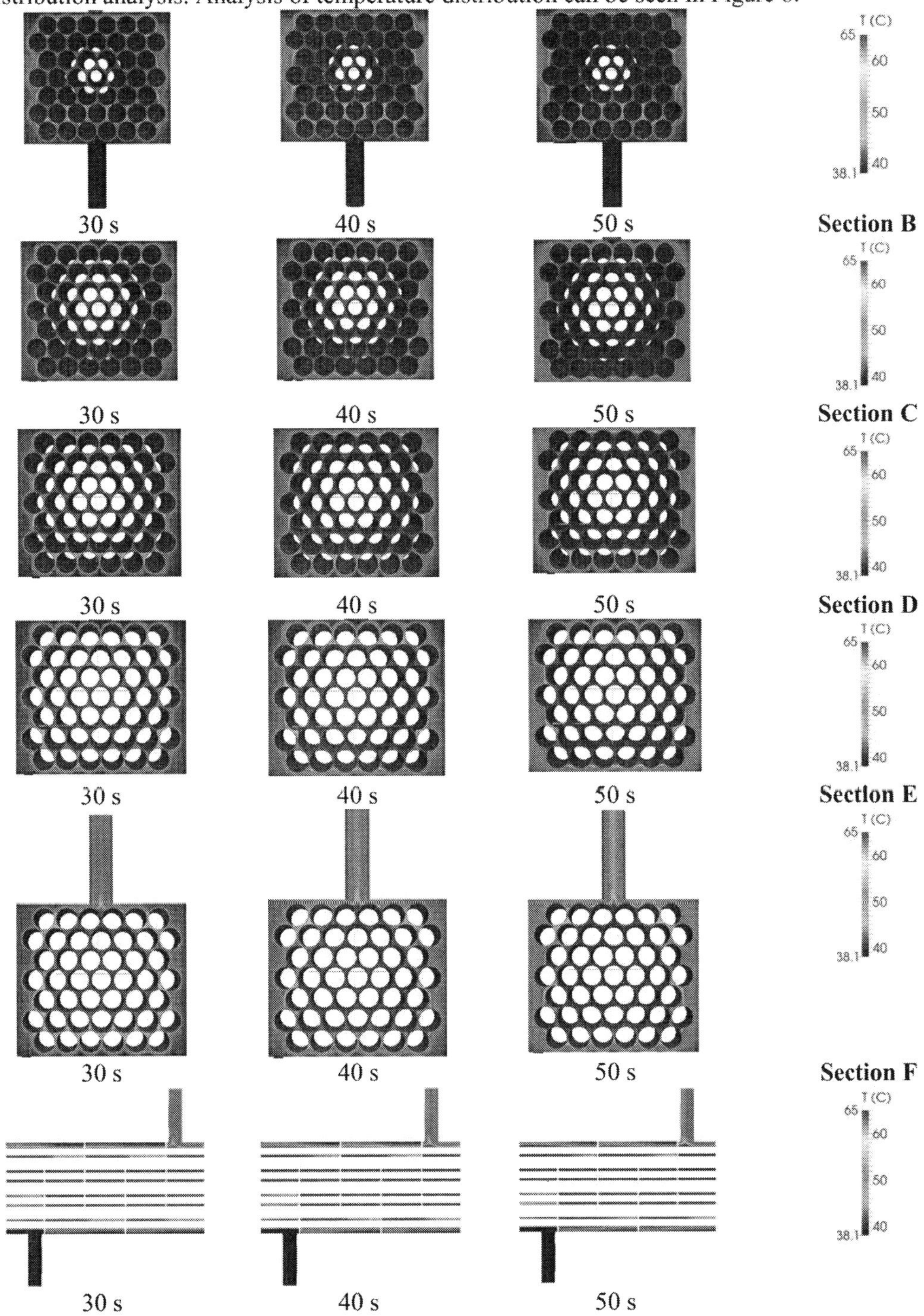

Figure 8. Cooking oil temperature distribution in shell

Figure 8. shows the temperature distribution in sections B, C, D, E, and F for 50 seconds. In general, changes in temperature distribution in each section after 40 seconds are not significant. In section B, there is a gradient temperature; further away from the inlet cooking oil temperature is increasing. However, high-temperature cooking oil fills in the tube gap compared to the area between the tube bundle and the shell. The temperature in section B reaches 60°C. This phenomenon occurs because the tube is a source of heat. Consequently, the fluid around the tube will transfer the heat between fluid molecules faster than the fluid that is far from the tube. Also based on the simulation of the velocity distribution [Fig 8], it is seen that the fluid velocity between the tube gaps is less than the fluid between the tube bundle gap and the shell. The slow-motion of fluid velocity causes the contact time of the fluid molecule with the heat source to be longer.

Cooking oil temperature distribution in sections C, D, E, and F tends to be uniformly distributed on the gap between the tubes. However, there is still a temperature gradient between the fluid in the tube gap and the fluid in the gap between the tube bundle and the shell. The temperature of the Cooking oil in the gap between the tube bundle and the shell is in the range of 45 – 48°C. While the Cooking oil temperature of in the tube gap is about 60 – 63°C. Distribution of temperature along the shell causes temperature gradient in the nozzle outlet [Section G]. Qualitatively the simulation results of temperature distribution are in accordance with the literature [6]. The temperature gradient in the nozzle is 48 – 54°C. The calculation method of control volume is discretization, then each cell has its temperature. Furthermore, the average cooking oil temperature at the outlet is 51 °C.

The average temperature based on simulation is validated with an experimental study which is T=(46.57 °C ± 0.0770) atau 46.49 - 46.65 °C). While the average fluid temperature distribution [Fig 8] is 51°C. Thus the use of SST k- ω turbulent models and wall temperatures of 65°C at 50 second gives quite high accuracy results in the analysis of the velocity distribution, but there is a different between simulation to experimental ± 9.7% for the temperature distribution.

5. Conclusion

This study analyzes the accuracy of the selection of SST k- ω turbulent models and wall temperatures in systems with heat source temperatures that fluctuate from 178 - 189°C. The 3D model is used in this simulation. The simulation of velocity distribution is in accordance with the boundary layer concept. Based on the statistical approach, then the temperature of the experimental result is T=(46.57 °C ± 0.0770) and the simulation average temperature is 51 °C. Thus the simulation parameters create a temperature difference of ± 9.7%.

Acknowledgement

The author would like to thank "KEMENTERIAN RISET, TEKNOLOGI, DAN PENDIDIKAN TINGGI INDONESIA" for funding this research through "PDUPT NKB-2834/UN2.RST/HKP.05.00/2020 and PT. CCIT for a licence of CFDSof®.

References

[1] Shah R K and Sekulic D P 2003 *Fundamentals of heat exchanger design* John Wiley & Sons.
[2] Jones A E, 2002 Thermal design of the shell-and-tube: exchanger sizing takes a trial and error approach. Use these guidelines to center your aim.(Cover Story) *Chem. Eng.* **109** 60–66.
[3] Nitsche M and Gbadamosi R O 2015 *Heat exchanger design guide: a practical guide for planning, selecting and designing of shell and tube exchangers* Butterworth-Heinemann.
[4] Gaos Y S and Widiawati C D, 2017 REVERSE ENGINEERING OIL COOLER DOUBLTUBE PLTA JATILUHUR *AME (Aplikasi Mek. dan Energi) J. Ilm. Tek. Mesin* **3** 1 41–44
[5] Rajesh R, Senthilkumar P and Mohanraj K 2018 Design of heat exchanger for exhaust heat recovery of a single cylinder compression ignition engine *J. Eng. Sci. Technol* **13** 2153–2165
[6] Pal E, Kumar I, Joshi J B and Maheshwari N K, 2016 CFD simulations of shell-side flow in shell-and-tube type heat exchanger with and without baffles *Chem. Eng. Sci.* **143** 314–340
[7] Duan Y, Jackson C, Eaton M D and Bluck M J 2019 An assessment of eddy viscosity models on

predicting performance parameters of valves *Nucl. Eng. Des.* **342** 60–77

[8] Widiawaty C D, Gunadi G G R and Syuriadi A, 2017 PEMODELAN DAN ANALISIS KINERJA SHELL AND TUBE HEAT EXCHANGER DENGAN METODE CFD *J. Poli-Teknologi* **16** 3

[9] Wang Q, Chen Q, Chen G and Zeng M 2009 Numerical investigation on combined multiple shell-pass shell-and-tube heat exchanger with continuous helical baffles *Int. J. Heat Mass Transf.* **525** 1214–1222

[10] Ozden E and Tari I 2010 Shell side CFD analysis of a small shell-and-tube heat exchanger *Energy Convers. Manag.* **51**1004–1014

[11] Davidson L 2015 Fluid mechanics, turbulent flow and turbulence modeling

[12] Lei Y, Li Y, Jing S, Song C Lyu Y and Wang F 2017 Design and performance analysis of the novel shell-and-tube heat exchangers with louver baffles *Appl. Therm. Eng.* **125** 870–879.

[13] Versteeg H K and Malalasekera W 2007 *An introduction to computational fluid dynamics: the finite volume method* Pearson education

[14] Harinaldi D I and Eng M 2005 Prinsip-prinsip statistik untuk teknik dan sains *Jakarta: Erlangga*

IOP Conf. Series: Materials Science and Engineering **909** (2020) 012005 doi:10.1088/1757-899X/909/1/012005

The cooling effect of polycrystalline type PV panels using perforated aluminum plates

I Bizzy,[1*], R Sipahutar[2], D Puspitasari[3], A Sofijan[4], M A Fajri[5]

[1,2,3]Department of Mechanical Engineering, Faculty of Engineering, Sriwijaya University, Indralaya 30662, South Sumatra, Indonesia

[4,5]Department of Electrical Engineering, Faculty of Engineering, Sriwijaya University, Indralaya 30662, South Sumatra, Indonesia

Corresponding Author's e-mail: irwin @unsri.ac.id

Abstract. A study has been conducted to improve the efficiency of Polycrystalline type PV panels. This research was conducted at the Energy Technology Research Laboratory, Engineering Faculty, Sriwijaya University, South Sumatra, Indonesia. The disadvantage of PV panel, when receiving solar radiation, it will increase the temperature of the PV panel but decrease the efficiency. This study uses a perforated aluminum plate with hole diameter of 2.5 mm. The plate is placed behind the PV panel. The dimensions of the plate are 960 mm x 600 mm x 20 mm and the number of holes of 1457. The specifications of PV panels are 1020 mm x 670 mm x 30 mm, Polycrystalline type PV panels, maximum power of 100 WP. The results showed that the PV panels given a perforated aluminum plate cooling had higher efficiency than those not given perforated plate cooling.

1. Introduction

PV panels have the potential to be developed as energy conversion equipment. These PV panels are able to convert solar energy into electrical energy for countries that have a hot climate all year round. Indonesia is one country that has this potential. South Sumatra Province is one of the regions in Indonesia that has this energy source. Although, this province is a storehouse of energy storage for the country of Indonesia, but the energy reserves taken today are still fossil-based (oil, coal, and gas). This type of energy decreases because it takes a long time to produce this energy again. Therefore, it is time to reduce the use of fossil energy and switch to using this abundant and environmentally friendly source of solar energy. Furthermore, the Indonesian government has made policies to increase the use of new and renewable energy, one of which is the use of solar energy to produce electricity for households, street lights, and traffic signs along the road. In fact, solar energy is one source of energy that can be used as electricity generation and to support transportation and industry.

Based on research results that increasing the temperature of the PV panel will reduce its efficiency [1]. Several ways have been investigated by researchers to improve the efficiency of this PV Panel, such as utilizing Phase Change Material (PCM) technology [2], [3], using water or air as a cooling medium [4], thermoelectric cooling [5], using program [6], etc. Cuce E. et.al., 2011 [7] have used fin cooling to produce better PV panel efficiency compared with without fin cooling. Amelia A. R. et.al., 2016 [8] have done research, the effect of PV panel temperature on its efficiency. The decline in the efficiency of PV panels can be caused by dusk factor on the surface [9]. Even Gholampour M. et.al.,

2014 [10] have research on perforated plates and PV panels. The results show that increasing the air mass flow rate using a fan can increase the PV panel cooling. It has been investigated cooling using a fan so that the temperature of the PV panel decreases and can increase the efficiency of the PV panel [11].

The use of PV panels is to produce electricity for household equipment, vehicles, to drive a pump for irrigation needs, etc. Some of these PV panel [12] consists of PV panel research to produce electricity for Air Conditioning equipment for preserve of agricultural products. Furthermore, Abdul Hadi N K et.al., 2018 [13] on thermal storage uses Phase Change Materials (PCM), in which the PV panel serves as a medium for receiving solar radiation and this system is used to produce a freezer. Adejuyigbe S. B. et.al, 2013 [14] utilize PV panels to conduct electricity for office equipment. Bhattacharjee B. et.al., 2019 [15] utilize PV panels to drive water a pump, and water is used for agricultural and household needs.

In this paper, the problem of PV panel efficiency has been reviewed again when PV panels reserve solar radiation and high temperatures as previous researchers have done. Cooling media used in the form of perforated plates made of aluminum without using a fan to suck hot air or cooling with free convection heat transfer. The perforated aluminum plate is placed at the below of the PV panel. The PV panel used is 100 Wp type Polycrystalline. This research is a continuation of the initial research that has been done but the power of PV panel used is still small, namely 8 Wp Polycrystalline type [16].

2. Experimental Procedure

Pollycritalline type PV panel mounted cooling media on the below to reduce the temperature of the panel. The cooling media used is perforated plates. This cooling media is made from aluminum. Figure 1 shows the structure of PV panels and perforated plates.

The PV panel tested consisted of 2 units, namely one unit without cooling media and one unit installed with cooling media in the form of perforated aluminum plates. In addition, some additional equipment was used in this test, such as solar charge control (SCC), batteries, and cables. Measuring instruments used consist of temperature measuring devices using K type thermocouple wire, measuring wind speed using anemometer, and measuring solar radiation intensity using pyranometer equipment, measuring electric current using current meters, and measuring voltage using voltage meters.

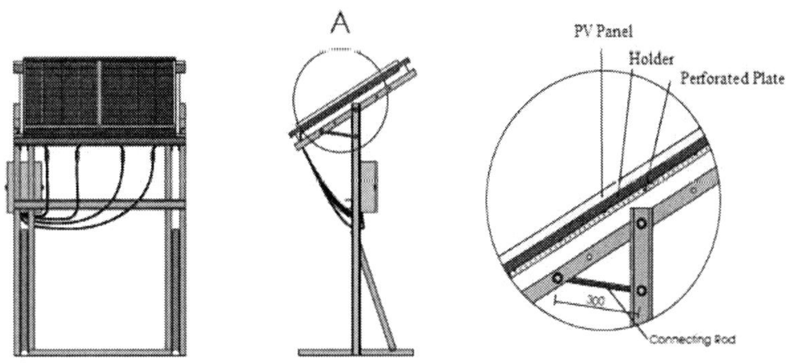

Figure 1. Structure of PV panels and perforated plates being tested

The specifications of the PV panels and the cooling media are explained in Table 1.

IOP Conf. Series: Materials Science and Engineering **909** (2020) 012005 doi:10.1088/1757-899X/909/1/012005

Table 1. PV panel specifications and perforated cooling media

	Spesification	Description
PV Panels:	2	units
Dimension (mm)	1020 x 670 x 30	mm
PV panel type	*Polycritalline*	-
Maximum power (P_{max})	100	W_p
Voltage at P_{max} (V_{max})	18	V
Current at P_{max} (I_{mp})	5.56	A
Maximum system voltage (DC)	1000	V
Perforated Aluminum Plate:	1	unit
Dimension	960 x 600 x 20	mm
Hole diameter	2.5	mm
Number of holes	1457	holes

3. Result and Discussion

Solar radiation that comes to a PV panel surface consists of two kinds, namely direct and diffuse solar radiation. When testing this PV panel, the climate in the South Sumatra Province of Indonesia is a hot climate. The following shows the data taken on February 4, 2020 (Figures 2 through 5).

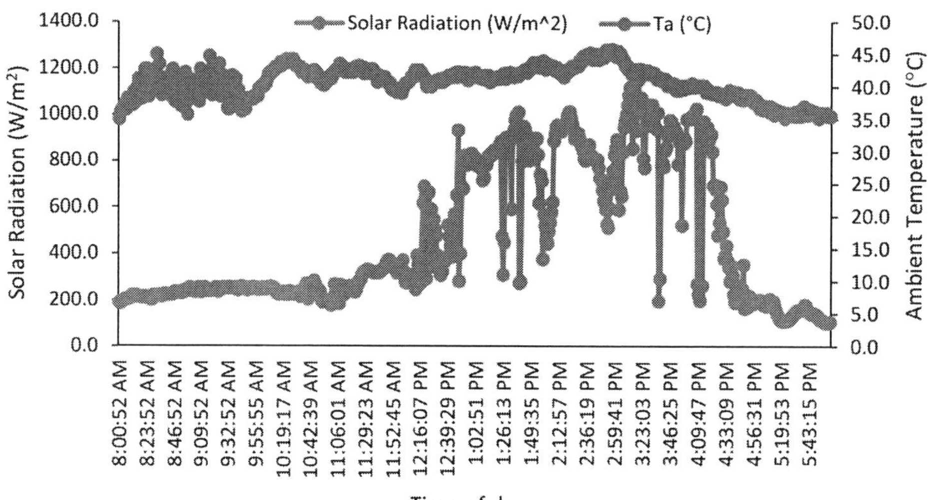

Figure 2. Comparison between ambient temperature and solar radiation on February 4, 2020

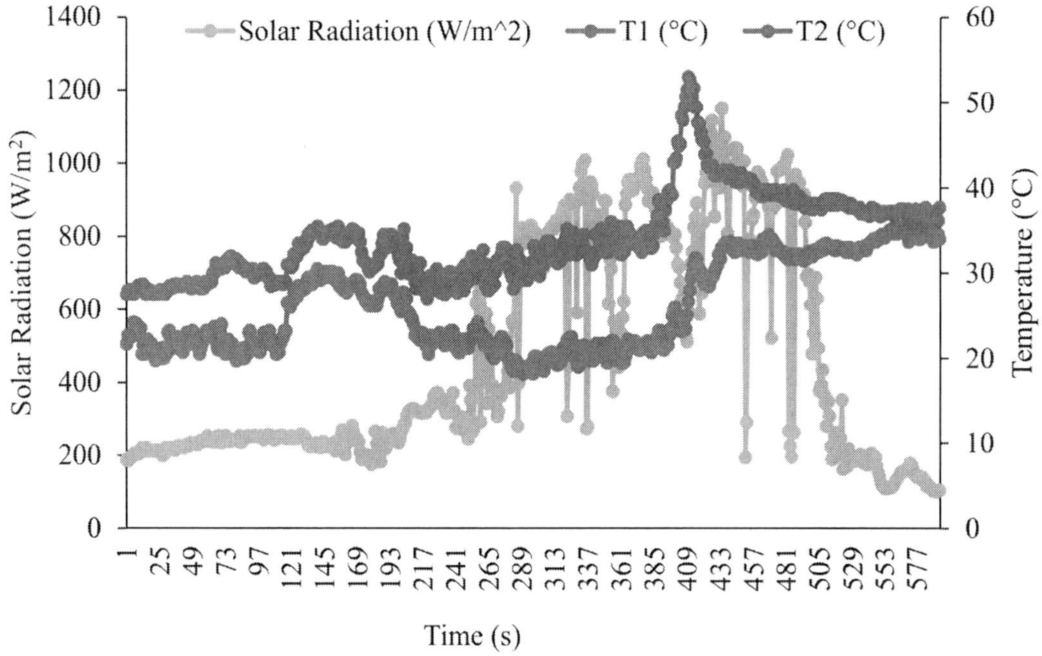

Figure 3. Comparison between PV panel surface temperature using perforated plate cooler (T1), PV panel surface temperature without cooler (T2) and solar radiation on February 4, 2020

As shown by the data taken in the field in Figure 3, the surface temperature of PV panels that do not use perforated aluminum plates (T2) is higher than the surface temperatures of PV panels that are cooled with perforated aluminum plates (T1). Likewise, in Figure 4 the opposite occurs, the tendency of the surface temperature under the PV panel not to use perforated aluminum plate (T2-2) is lower than the surface temperature below the panel which uses perforated aluminum plate cooler (T1-2). This event is interesting to study because the power produced by PV panels that use perforated aluminum plate cooler (P1) has higher compared to PV panels that do not use perforated aluminum plate cooler (P2) as shown in Figure 5.

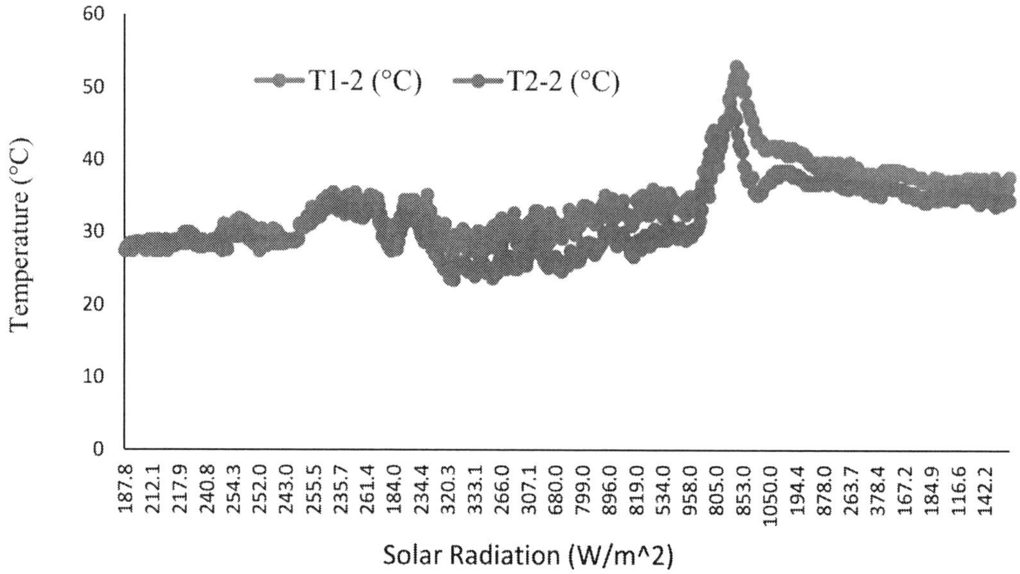

Figure 4. Comparison between the below surface temperature of the PV panel using perforated plate cooler (T1-2) and without perforated plate cooler (T2-2) on 4 February 2020

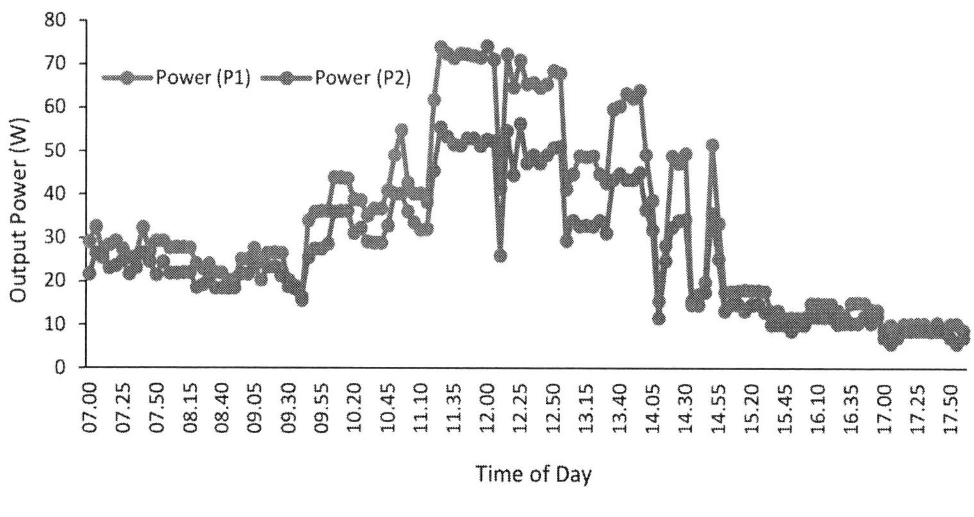

Figure 5. Comparison between PV panel output power using perforated plate cooler (P1) and PV panel output power without cooler (P2) on February 4, 2020

Not all of the solar energy coming to the surface of a PV panel can be used as the output power of a PV panel. Some of this energy will be lost to the atmosphere as lose of heat convection. The first, heat lose is the transfer of heat from the top surface of the PV panel to ambient air. The second, heat lose is the transfer of heat from the back surface of the PV panel to ambient air to holes in the perforated aluminum plate or to ambient air for without perforated aluminum plate.

The intensity of the sun or solar radiation (I_R, W/m^2) that comes to the surface of the PV panel (A_p, m^2) will effect the current (I, Ampere) and voltage (V, Volt) that produce by a PV panel. The efficiency of the PV panel (η) is reduced if the intensity of the sun is high as shown in the equation (1). The results of the efficiency calculation are shown in table 2 for Sunny weather conditions, namely January 21, 2020; February 4, 2020; and February 05, 2020.

$$\eta = \frac{V\, x\, I}{I_R\, x\, A_P} = \frac{P}{I_R\, x\, A_P} \qquad (1)$$

Table 2. PV panel efficiency

Date	I_R (W/m^2)	PV Panel with perforated aluminum plates			PV Panel without perforated aluminum plates		
		V_1 (V)	I_1 (A)	η_1 (%)	V_2 (V)	I_2 (A)	η_2 (%)
January 21, 2020	639.5	20.1	0.46	2.12	19.6	0.30	1.35
February 4, 2020	425.1	16.4	2.0	11.29	15.7	1.6	8.65
February 5, 2020	351.8	17.34	1.38	9.95	16.05	1.07	7.14

High average solar radiation has resulted in low PV panel efficiency, so cooling media is needed below the PV panel. Installation of perforated aluminum plate cooling media has increased the efficiency of 100 Wp Polycrystalline PV panels. This perforated aluminum plate media has been useful to increase the efficiency of PV panels. In addition, how to make it simple, inexpensive, and easy to install.

4. Conclusion

A study has been carried out for 100 Wp Polycrystalline PV panels. The PV panel used consists of 2 units, each with a perforated aluminum plate cooler and without a perforated aluminum plate cooler. The diameter of this plate under study is 2.5 mm. Some concluding observations from the investigation are given below.

PV panels that use perforated aluminum plate cooler are better due to an increase in current and voltage output of PV panels compared to without perforated aluminum plate cooler.

The higher solar radiation received by the PV panel will increase the surface temperature of the PV panel but decrease the efficiency of the PV panel. For this reason, a cooling media is needed below the PV panel.

Acknowledgment

The author would like to thank the Rector of Sriwijaya University and Dean of the Faculty of Engineering, Sriwijaya University for their motivation and assistance in conducting this research. To the staff of the Energy Technology Laboratory, Faculty of Engineering, Sriwijaya University, a thank you was also conveyed.

References

[1] Dincer F and Meral M E 2010 Critical Factors that Affecting Efficiency of Solar Cells *Smart Grid and Renewable Energy* **01** 47–50
[2] Machniewicz A, Knera D and Heim D 2015 Effect of Transition Temperature on Efficiency of PV / PCM Panels Effect of transition temperature on efficiency of PV / PCM panels *Energy Procedia* **78** 1684–9
[3] Rubab S M, Abbas M S, Balasaheb S M, Mateescu T D and Chereches N 2017 A Review Paper on Improving The Efficiency Of Solar Panel *International Research Journal of Engineering and Technology (IRJET)* **4** 1359–61
[4] Parida B, Iniyan S and Goic R 2011 A review of solar photovoltaic technologies *Renewable and Sustainable Energy Reviews* **15** 1625–36
[5] Borkar D S, Prayagi S V, Gotmare M J and Ambedkar B 2014 Performance Evaluation of Photovoltaic Solar Panel Using Thermoelectric Cooling *International Journal of Engineering Research* **5013** 536–9
[6] Zhe L W, Yusoff M I, Razak A A, Misrun M I, Ibrahim S, Fahmi M I and Rosmi A S 2019 Effect of Water Cooling Temperature on Photovoltaic Panel Performance by Using Computational Fluid Dynamics (CFD) *Journal of Advanced Research in Fluid Mechanics and Thermal Sciences* **1** 133–46
[7] Cuce E, Bali T and Sekucoglu S A 2011 Effects of passive cooling on performance of silicon photovoltaic cells **6** 299–308
[8] Amelia A R, Irwan Y M, Leow W Z, Irwanto M, Safwati I and Zhafarina M 2016 Investigation of the Effect Temperature on Photovoltaic (PV) Panel Output Performance *International Journal on Advanced Science, Engineering and Information Technology* **6** 682–8
[9] Reza M, Hizam H, Gomes C, Amran M, Ismael M and Hajighorbani S 2016 Power loss due to soiling on solar panel : A review *Renewable and Sustainable Energy Reviews* **59** 1307–16
[10] Gholampour M, Ameri M and Sheykh Samani M 2014 Experimental study of performance of Photovoltaic-Thermal Unglazed Transpired Solar Collectors (PV/UTCs): Energy, exergy, and electrical-to-thermal rational approaches *Solar Energy* **110** 636–47
[11] Yusoff M I, Irwanto M, Zhe L W and Nair G 2016 Cooling on Photovoltaic Panel Using Forced Air Convection Induced by DC Fan Cooling on Photovoltaic Panel Using Forced Air Convection Induced *International Journal of Electrical and Computer Engineering (IJECE)* **6** 526–34
[12] Sultan M and Miyazaki T 2018 Solar Assisted Evaporative Cooling Based Passive Air-Conditioning System for Agricultural and Livestock Applications *Journal of Engineering Science and Technology* **13** 693–703
[13] Abdul Hadi N K, Ahmed A M and Roni R T 2018 Experimental Study on Thermal Energy Storage Produced by Solar Energy for Driving Domestic Freezer *Journal of Engineering Science and Technology* **13** 2829–42
[14] Adejuyigbe S B, Ekiti O, Bolaji B O, Olanipekun M and Adu M 2013 Development of a Solar Photovoltaic Power System to Generate Electricity for Office Appliances *Engineering Journal* **17** 30–9
[15] Bhattacharjee B, Chakrabarti A and Sadhu P K 2019 Solar Photovoltaic Integrated Pump for Advanced Irrigation System *International Journal of Innovative Technology and Exploring Engineering (IJITEE)* **8** 3246–50
[16] Bizzy I and Mustafrizal L 2018 PV Panel Cooler to Enhance Output Performance Using Perforated Aluminium Plate *Symposium of Emerging Nuclear Technology and Engineering Novelty (SENTEN)* pp 1–4

IOP Conf. Series: Materials Science and Engineering **909** (2020) 012006 doi:10.1088/1757-899X/909/1/012006

Passive cooling using perforated aluminum plate to improve efficiency on monocrystalline of 100 Wp photovoltaic

A Sofijan[1], Z Nawawi[2], B Y Suprapto[3], I Bizzy[4], R Sipahutar[5]

[1,2,3] Department of Electrical Engineering, Engineering Faculty, Sriwijaya University, Indonesia
[4,5] Department of Mechanical Engineering, Engineering Faculty, Sriwijaya University, Indonesia

*Email: A_sofijan@ft.unsri.ac.id

Abstract. Many researchers had applied various cooling techniques to reduce the operating temperature of PV cells so it will not exceed the temperature limit. If the operating temperature is exceeded, it can worsen the performance of the PV panel system. Passive cooling is one of the cooling techniques to reduce the operating temperature and increase the efficiency of the PV panel. Cooling operation without the addition of electrical energy is a new original idea, namely passive cooling which uses perforated aluminum plates mounted behind the PV panels. Research that has been carried out using perforated aluminum plates, hole diameter of 5 mm, hole spacing of 200 mm, and total holes of 1551 can increase efficiency by up to 57.64 percent on 100 WP monocrystalline PV panels.

1. Introduction

Solar energy is one of the important renewable energy sources which has attracted the attention of many researchers in the world to be developed. There are two types of energy that can be produced from solar energy, namely electricity and heat energy. One way to convert solar energy into electrical energy is to use photovoltaic (PV) panels. The PV panel can directly convert the incident of solar radiation into electricity that is efficient, sustainable, and environmentally friendly [1] [2].

In general, PV panels are only able to convert a small portion of solar radiation into electricity. The remaining part of the solar radiation will be converted to heat, which can increase the temperature and reduce the performance of the PV panel. The maximum expected PV panel temperature in the presence of solar radiation is 1000 W/m^2, or about 70% of the absorption rate in the absence of wind conditions and the temperature is 60°C, whereas for wind speeds higher than 4 m/s, the PV panel temperature it will be lower than 40°C [3] [4]. Maximum power output, open circuit voltage, and short circuit current are some of the main parameters that are affected by variations in the temperature of the PV panel. In other words, the open circuit voltage and maximum output power decrease with temperature rise while short circuit current increases [5].

The extraordinary effect on photovoltaic performance is presented by cell temperature. They note that the temperature coefficient is negative for the charging factor, the open-circuit voltage, and the maximum output power while positive for the short-circuit current. Photovoltaic output power and ambient temperature are directly proportional. Therefore, the output power produced by photovoltaic is higher for low ambient temperature intervals than high temperature intervals.[6][7] investigate experimentally and numerically the impact of cell temperature on photovoltaic panel performance with a constant irradiance value.

The results show that the efficiency and output power decreases with increasing cell temperature, the decrease in efficiency is caused by temperature differences and hence the location of the PV panels is an important factor to consider. Cell temperature has a significant impact on the performance of photovoltaic cells.

Therefore, this research focuses on passive cooling techniques to reduce PV cell heat and improve PV cell efficiency using perforated aluminum plate.

2. Passive cooling techniques for PV panel

Passive cooling techniques can be divided into three main groups: air passive cooling, water passive cooling, and conductive cooling. Conductive cooling mostly ends with passive cooling of air, but an important difference is that the prevailing heat transfer mechanism of PV cells is conductive.[8] worked on experimental studies on polycrystalline PV cells under controlled conditions.

Two PV cells are used: one with aluminum fins as a heat sink, with thermal grease applied and one without a heat sink. Lighting varies from 200 to 800 W/m^2. A relative increase in electrical efficiency of 9% has been obtained through the use of passive cooling with coolant.[9][10] have shown that the depth of the flow channel under the PV cell has a significant effect on passive cooling, for a larger PV surface (1.95 m^2). It has been shown that, for a length-to-depth ratio of 0.085, the PV module heats up 5-6 °C when compared to the PV module in a normal installation. It was noted that the temperature difference increases with increasing insulation. In other words, passive flow channels can have the opposite effect on cooling the PV module.

Water passive cooling is somewhat more efficient, mainly because of the higher thermal capacity of water. Several studies have been made with front and back cooling.[11] used a submerged technique to cool down the mono-crystalline PV module in water. The effect had limited success: the temperature was maintained at 30 °C which in turn yielded a relative efficiency increase of 20 %, but insolation intensity dropped with depth. However, at a depth of 4 cm, relative efficiency is increased by 11 %.

3. Experimental

This research was carried out on two monocrystalline PV panels with a capacity of 100 wp in the energy technology laboratory. The research methodology was to experiment using a perforated aluminum plate mounted under a PV panel, testing carried out with and without cooling.

Data collection for short circuit current (Isc), open circuit voltage (Voc), panel temperature and light intensity, datalogger is done using temperature, current, voltage and intensity sensors. 4 temperature sensors are installed at the top and bottom of the panel, one sensor is installed for room temperature, while the current sensor is installed at the output panel when short circuit as Isc, the voltage sensor is mounted on the output panel in an open circuit as Voc.

The data logger in the program uses the Arduino mega-2560 by using SD-Ram as a data storage facility that is connected to a computer, so that the data obtained is real-time and better than manual data retrieval.

The lighting uses 12 halogen lamps, each with a capacity of 1000 watts installed at the top of the panel within 30 cm show on Figure 1.as .design experimental (indoor)

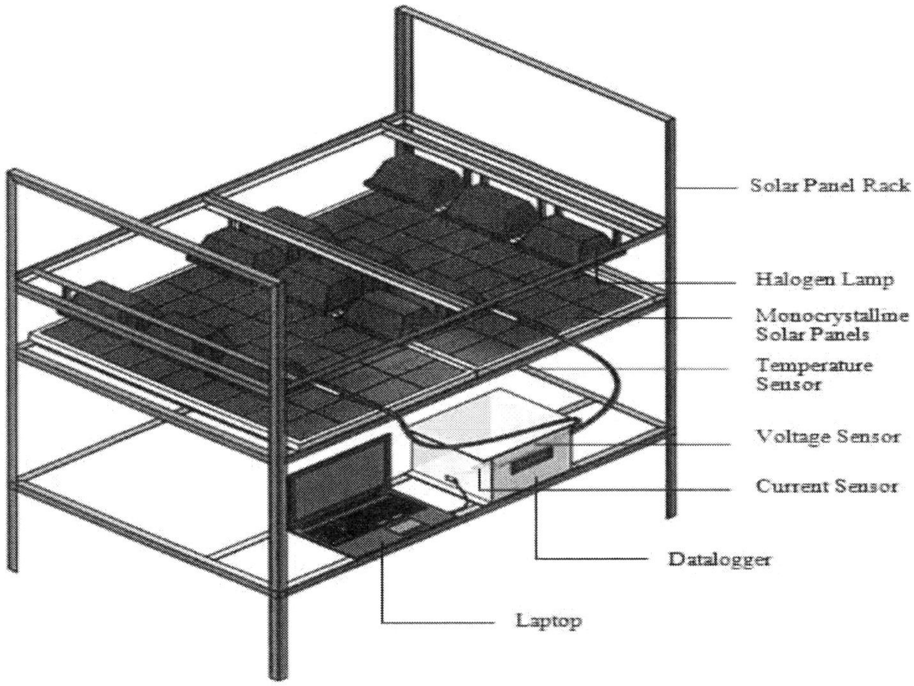

Figure1. Design experimental (indoor)

3.1. PV panel

Two units of monocrystalline photovoltaic panels with a capacity of 100 wp for 2 units, mounted parallel to the panel rack, one of which is installed perforated aluminum plate on the back

The specifications of the PV panels are explained in Table .1.

Table 1. The specifications of the PV panels

Electric Rating	Specification
Cell Type (2 units)	Monocrystalline (1020 x 670 x 30 mm)
Rated Maximum Power (P_{max})	100-Watt Peak
open-circuit voltage (Voc)	22.5 V
short-circuit current (Isc)	5.91 A
Voltage at Pmax	18.30 V
Current at Pmax	5.47 A
nominal operating cell temp (NOCT)	47+-2°C
Maximum System Voltage	1000 V_{DC}
maximum senes fuse rating	15A
operating temperature	-40°C-+85°C

3.2. Cooling media

Perforated aluminum plates as the cooling media used are respectively described.

Table 2. Dimensions of aluminum perforated plates.

Name of parts	Specification
Dimension	1020 x 670 x 30 mm
Hole diameter	5 mm
Distance between holes	20 mm
Number of holes	1551
Arrangement of holes	In line

4. Experimental results and Discussion

Experimental results show the comparison of PV panel temperatures in Figure 2. Clearly shows a decrease in PV panel temperature using perforated aluminum plate as a cooling medium, when 11.44.06 the upper surface of the PV panel temperature using perforated plate (T 1) shows 49.38 °C while the upper surface temperature PV panel without perforated plate (T 2) of 62°C as well as in T1.2=34°C and T2.2=41.19°C.

The increase in voltage and current is shown in Figure 3. where the highest V.1 is 18.48 V and the highest V.2 is 14.71 V. Current I.1 = 4.26 A, I.2 = 3.41 A.

In figure 4 the highest power P.1 = 78.38 W, P.2 = 49.48W and the highest efficiency η.1 = 7.63%, η 2 = 4.84%.

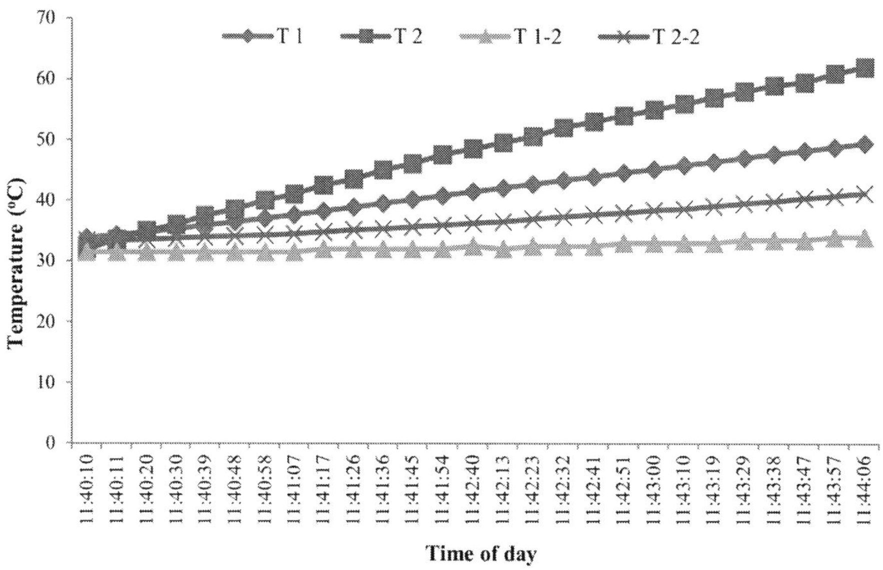

Figure 2. Comparison between upper surface temperature PV panel using perforated aluminum plate (T 1) PV panel without perforated plate (T 2) and below surface temperature PV panel using perforated plate (T 1-2) PV panel without perforated plate (T 2-2).

Figure 2 shows that the percentage decrease in temperature of the perforated plate T 1 compared to without the perforated plate T 2 to 20.35% on the top surface of the PV panel, while the percentage of temperature decrease brought to the surface of the PV panel using a perforated PV plate (T 1-2) without perforated plate (T 2-2), reaching up to 17.45%.

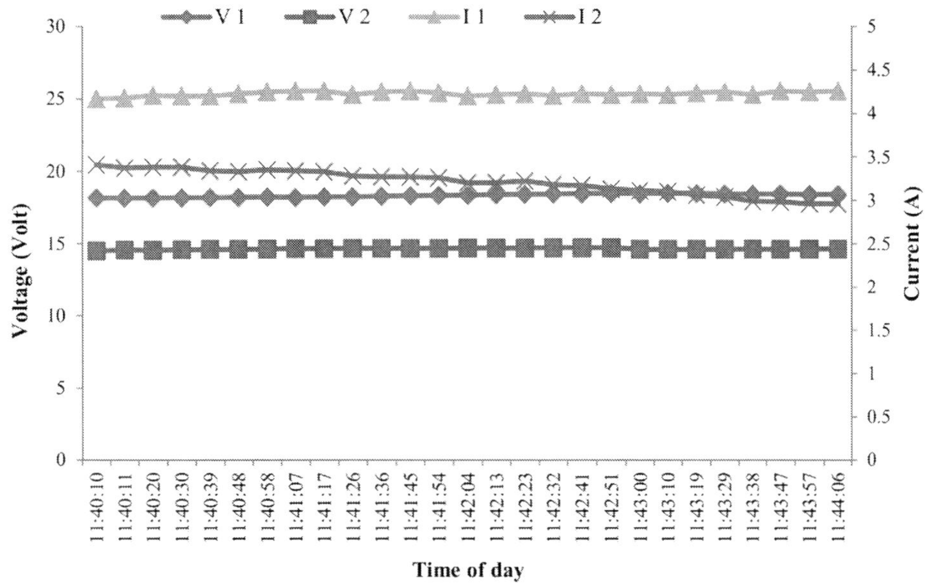

Figure 3. Comparison between voltage of PV panel using perforated plate (V 1) PV panel without perforated plate (V 2) and current of PV panel using perforated plate (I 1) PV panel without perforated plate (I 2)

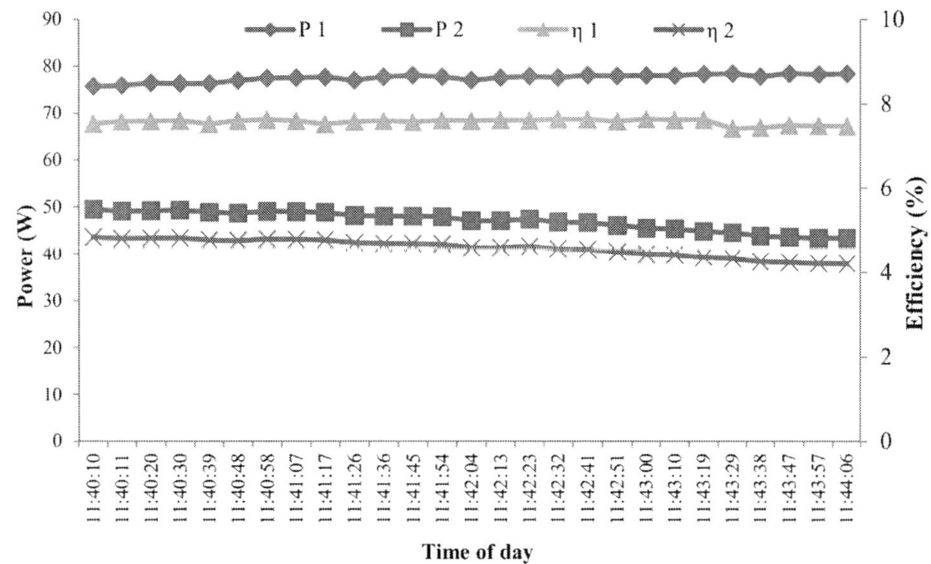

Figure 4. Comparison between power of PV panel using perforated plate (P 1) PV panel without perforated plate (P 2) and efficiency of PV panel using perforated plate (η 1) PV panel without perforated plate (η 2).

5. Conclusion

Passive cooling method using perforated aluminum plate with a hole diameter of 5 mm can increase efficiency by 57.64 % compared to PV panels without cooling.

IOP Conf. Series: Materials Science and Engineering **909** (2020) 012006 doi:10.1088/1757-899X/909/1/012006

The addition of perforated aluminum plate to PV panels mounted on the back surface of the PV panels can reduce the temperature of the upper panel by 12.62 °C or 20.35% compared to PV panels without cooling.

The conclusion of this experimental research shows a decrease in temperature on PV panels that use perforated aluminum plates and improve the efficiency of PV panels.

References

[1] Adeeb J, Farhan A and Al-Salaymeh A 2019 Temperature effect on performance of different solar cell technologies *J. Ecol. Eng.* **20** 249–54
[2] Al-Waeli A H, Sopian K, A Kazem H and T Chaichan M 2017 Photovoltaic thermal PV/T systems: A review *Int. J. Comput. Appl. Sci.* **2** 62–7
[3] Donne A Le, Scaccabarozzi A, Tombolato S, Binetti S, Acciarri M and Abbotto A 2013 Solar photovoltaics : a review solar photovoltaics : A Review *Am. Sci. Publ.* **2.** 1–9
[4] Makrides G, Zinsser B, Georghiou G E, Schubert M and Werner J H 2009 Temperature behaviour of different photovoltaic systems installed in Cyprus and Germany *Sol. Energy Mater. Sol. Cells* **93** 1095–9
[5] Chander S, Purohit A, Sharma A, Arvind, Nehra S P and Dhaka M S 2015 A study on photovoltaic parameters of mono-crystalline silicon solar cell with cell temperature e*nergy reports* **1** 104–9
[6] Zaoui F, Titaouine A, Becherif M, Emziane M and Aboubou A 2015 A combined experimental and simulation study on the effects of irradiance and temperature on photovoltaic modules *energy procedia* **75** 373–80
[7] Stutenbaeumer U and Mesfin B 1999 Equivalent model of monocrystalline, polycrystalline and amorphous silicon solar cells *Renew. Energy* **18** 501–12
[8] Cozzini M 2012 Solar cell cooling and heat recovery in a concentrated photovoltaic system *2012 COMSOL Conf. 18, 38123 T Milan*
[9] Sandip S. kale, V. W. Bhatkar M . D 2014 Performance evaluation of plate-fin-and tube heat exchanger with wavy fins- a review *J. Eng. Res. Appl.* **4** 154–8
[10] Ahmed I 2018 Enhancement the performance of PV panel by using fins as heat sink e*ng. technol. J.* **36**,798-805
[11] Wu S and Xiong C 2014 Passive cooling technology for photovoltaic panels for domestic houses *Int. J. Low-Carbon Technol.* **9** 118–26

IOP Conf. Series: Materials Science and Engineering **909** (2020) 012007 doi:10.1088/1757-899X/909/1/012007

Effect of cutting conditions on power demand and surface roughness through sustainable turning of mild carbon steel

Syah Risal[1], Rusdi Nur[2], Arthur Halik Razak[2], and Yan Kondo[2]
[1] Politeknik Ilmu Pelayaran Makassar, Jalan Tentara Pelajar No. 173 Makassar 90165, Indonesia
[2] Politeknik Negeri Ujung Pandang, Jalan Perintis Kemerdekaan Km. 10 Makassar 90245, Indonesia

Email: rusdinur@poliupg.ac.id

Abstract. Sustainable turning is a lathe machining process to carry out production in particular by paying attention to the sustainability of machinery without reducing production output. The machining process is done by optimizing the turning process parameters when carrying out the cutting process on medium carbon steel. Some parameters affect the lathe process, including speed, feed rate, and cooling. The purpose of this study was to determine the machining conditions in order to optimize the electric power consumption and surface roughness of the medium carbon steel turning process. The research method used was experimental and analysed using the ANOVA method. From the results of the research, it was concluded that the minimum electricity consumption was obtained in the process of turning steel St 42 without cooling with a speed of 237 rpm and a feeding of 0.157 mm/rev. The lower surface roughness was obtained in the turning process of St 60 steel with a rotation speed of 840 rpm and a feed rate of 0.157 mm/rev. The results of data analysis using ANOVA analysis method to get optimal conditions in order to minimum power demand and low surface roughness at 425 rpm of speed and 0.052 mm/rev of feed rate.

1. Introduction

Manufacturing or industry runs on the principle of sustainability as a major step in efforts to save energy. In the principle of a sustainable manufacturing industry there are efforts to reduce the amount of energy used in producing goods and services [1]. One of the big energy uses resulting from this production activity was to process waste and rubbish. In the principle of sustainable manufacturing this waste treatment must use minimal energy. In fact, according to him, waste management does not consume energy but produces new alternative energy. It efforts to recycle waste and garbage are also one of the principles in sustainable manufacturing [2].

The technology of the machining process or the manufacturing process should consider the concepts of research and development and cooling lubrication fluids, because both of these will produce quality products. The concept of research and development is a study that focuses in depth on objects, in this case the technology used in a sustainable manufacturing process. Likewise, the concept of cooling lubrication fluids is focused on new techniques and methods that can improve product quality and reduce production costs [3]. Whereas the sustainable concept in conventional machining focuses on environmentally clean work so that the use of the crystalline method is suitable for its development. Furthermore, efforts to implement sustainable development methods are intended so that quality systems, processes and final products can be achieved by implementing the 6R method.

The concept of sustainability offered should consider the following; production costs, energy consumption, waste management, influence on the environment, operational safety and operator health factors [4].

Savings in energy consumption are part of sustainability so that humans are able to balance their lives. Sustainability is an increasingly important need for human activities, this has led to making sustainable development a view that social, economic and environmental problems must be handled simultaneously and holistically in the development process. Sustainability has been applied in various fields, including engineering, manufacturing, and design. Producers are increasingly concerned about sustainability issues. For example, the recognition of the relationship between manufacturing operations and the natural environment has become an important factor in decision making among industrial communities [2].

Sustainability has become an important issue in the manufacturing sector. In the literature, it is generally agreed that sustainable development must cover three pillars, namely economic, social and environmental considerations [3]. Therefore, to achieve sustainable development, the industry must produce sustainable production. One of the ways to achieve environmentally friendly production is to reduce energy consumption through the use of products.

Sustainable production is a solution to overcome the problem of high electricity use which results in high costs. This applies to the field of engineering, including the machining process [5]. Machinery is an integrated part of the production. Thus, reducing energy consumption during machinery will contribute to reducing energy consumption to produce parts.

Optimizing energy demand in manufacturing is important to reduce the energy intensity of products and their vulnerability to rising energy prices, this is an important addition to reducing energy costs in manufacturing and to optimize energy from machine products. Production machines are one of the most widely used production processes and require an electricity supply. Several studies have been conducted to optimize cutting conditions based on machinery and economic considerations. For example, Hinduja and Sandiford present models and methodologies for selecting optimum cutting conditions based on minimum cost considerations in the milling process [6]. Lee and Tang developed a cutting model to maximize production levels and minimize production costs using polynomial networks [7].

This paper aims to determine the effect of machining parameters (rotation and feed rate) on various responses to electrical power consumption and surface roughness.

2. Experimental Design

The research will be carried out in the Mechanical Workshop and Mechanical Laboratory of the Mechanical Engineering Department. In this study, the equipment used was the lathe machine (PINDAD) with a maximum power of 5620 Watt and maximum rotation of 1500 rpm.

2.1. Materials and Tools

The material being machined was medium carbon steel (Strength of 42 N/mm^2 and 60 N/mm^2). The cutting tool used was an uncoated carbide tool with type tool holder of TCLNR2020K12. Surface roughness resulting from turning can be measured by surface roughness tester Surftest SJ-310 (Figure 1). Power consumption can be determined by measuring voltage and current using Clamp meter (Figure 2).

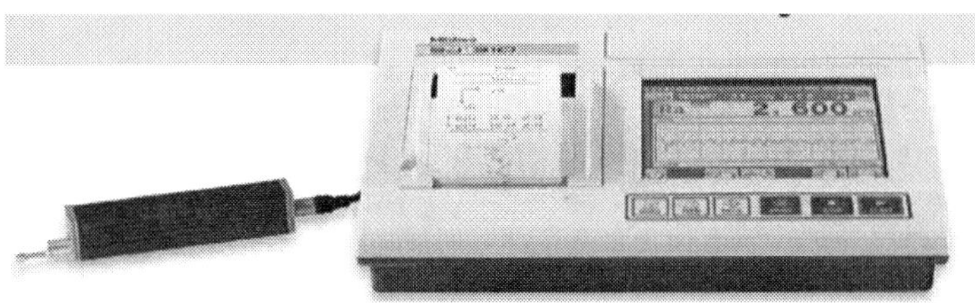

Figure 1. Surface roughness tester surftest SJ-310

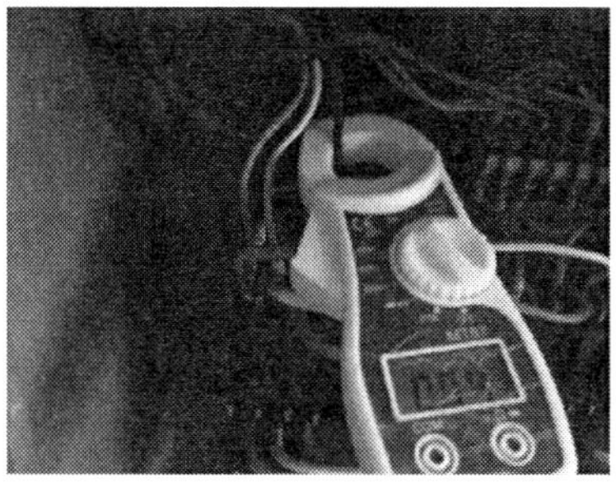

Figure 2. Clamp meter

2.2. Experimental Setup

Schematic design of the turning process on the workpiece is shown in Figures 3 and 4 below.

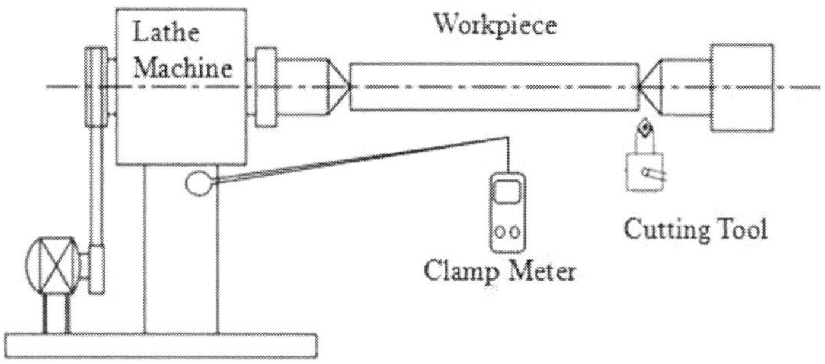

Figure 3. Schematic design of the turning process

Figure 4. Turning processes

The experimental design was based on a number of machining parameters that are varied as shown in Table 1.

Table 1. Machining parameters

Levels		Low (-1)	Centre (0)	High (+1)
Rotation Speed (Vc) [rpm]		237	425	840
Feed Rate (f)	[mm/rev]	0,052	0,105	0,157
Depth of cut (a_p)	[mm]		0,5 mm (constant)	
Coolant			No fluid	

3. Results and Discussion

The research will be carried out in the Mechanical Workshop and Mechanical Laboratory of the Mechanical Engineering Department.

3.1. Power Demand

The power demand was obtained from the measurement data of current and voltage. It was used by the clamp meter to measure the current and voltage. The result of the power demand can be described in Figure 5.

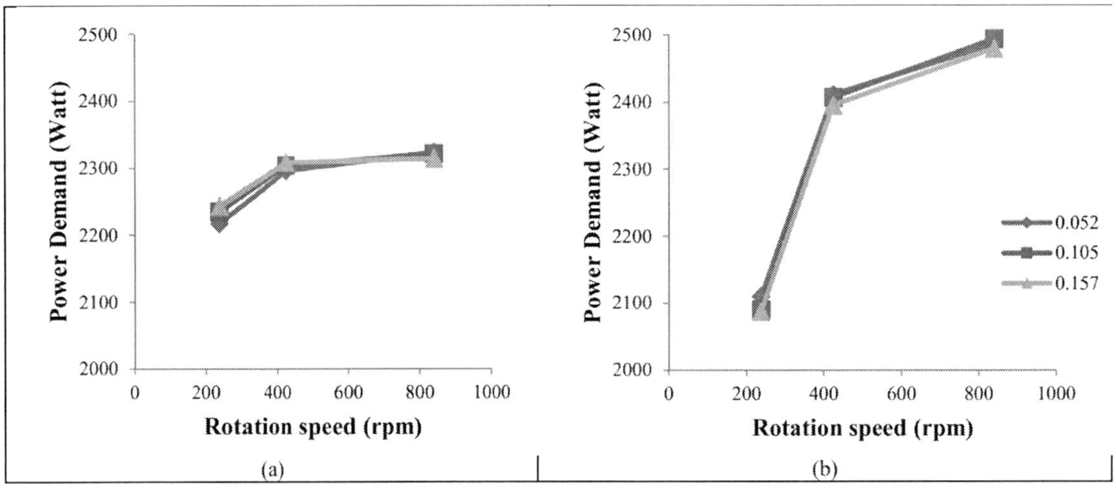

Figure 5. Graph of power demand for St 42 (a) and St 60 (b)

Based on the resulting the calculation of power demand with a variation of rotation speed and feed rate as shown in Figure 5(a), it shows that the highest power demand is 2324 Watt at 840 rpm and 0.052 rev/mm. The lowest power demand is 2217 Watt at a rotation speed of 237 rpm and the feed rate of 0.052 rev/mm. Likewise, for St 60, Figure 5 (b) shows that the highest power consumption is at 840 rpm and the feed rate is 0.105 rev/mm at 2494 Watt, and conversely, the lowest power consumption is 2087 Watt at 237 rpm of speed and 0.157 rev/mm of feed rate. Some researchers observed the effect of cutting speed on power consumption in the process of turning aluminum alloy [8,9].

3.2. Surface Roughness

The surface roughness was measured using Surftest SJ-310 (Mitutoyo). The data of surface roughness was repeated to 5 times for each measurement. The results of surface roughness were described in Figure 6.

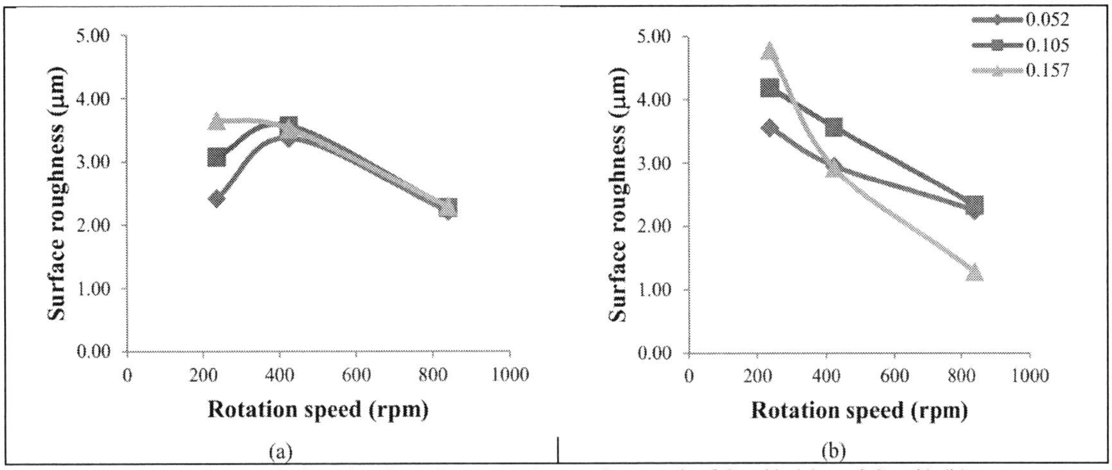

Figure 6. Graph of surface roughness for steel of St. 42 (a) and St. 60 (b)

Based on the results of the measurement of surface roughness as in Figure 6, it shows that the highest surface roughness is 3.655 µm at 237 rpm of speed and the 0.157 rev/mm of feed rate, and the lowest surface roughness is 2,215 µm at 840 rpm and 0.052 rev/mm. Likewise, on the material of St. 60, it shows that the highest surface roughness is 4.780 µm at 237 rpm and 0.157 rev/mm, and the lowest surface roughness is 1.288 µm at 840 rpm and the feed rate is 0.157 rev/mm. These responses were observed Nur et. when turning of 316L stainless steel [10].

4. Conclusion

Based on the results of research and data analysis, it can be concluded as follows:

1. Power demand is only affected by the variable rotation, where the greater the rotation used, the higher the value of power demand, whereas the smaller the rotation is used, the lower the power demand. While the feed rate does not affect power demand.
2. Surface roughness is influenced by the variable rotation speed and feed rate, where a large rotation and a small in feed rate will produce a low (smooth) surface roughness. Conversely, if a small round and large feed rate will get a high surface roughness.

References

[1] Amrina E, Vilsi AL 2015 Key performance indicators for sustainable manufacturing evaluation in cement industry In Procedia CIRP.

[2] Rosen MA, Kishawy HA 2012 Sustainable manufacturing and design: concepts, practices and needs *Sustainability* **4**(2)154–74

[3] Pusavec F, Krajnik P, Kopac J 2010 Transitioning to sustainable production–Part I: application on machining technologies *J Clean Prod* **18**(2)174–84

[4] Kopac J, Pusavec F, Ekinovic S 2006 High performance manufacturing – definition and aims *TMT*

[5] Hanafi I, Khamlichi A, Cabrera FM, Almansa E, Jabbouri A 2012 Optimization of cutting conditions for sustainable machining of PEEK-CF30 using TiN tools *J Clean Prod* **33** 1–9

[6] Hinduja S, Sandiford D 2004 An optimum two-tool solution for milling 2 1/2D features from technological and geometric viewpoints *CIRP Ann - Manuf Technol*

[7] Lee BY, Tarng YS 2000 Cutting-parameter selection for maximizing production rate or minimizing production cost in multistage turning operations *J Mater Process Technol* **105**(1-2)61–6

[8] Nur R, Noordin MY, Izman S, Kurniawan D 2014 The effect of cutting parameters on power consumption during turning nickel based alloy *Advanced Materials Research* **845** 799–802 p

[9] Nur R 2017 Sustainable machining by optimizing the power demand through dry turning of

316L stainless steel *ARPN J Eng Appl Sci* **12**(6)1885–9
[10] Nur R, Noordin MY, Izman S, Kurniawan D 2017 Machining parameters effect in dry turning of AISI 316L stainless steel using coated carbide tools *Proc Inst Mech Eng Part E J Process Mech Eng* **231**(4)676–83

IOP Conf. Series: Materials Science and Engineering **909** (2020) 012008 doi:10.1088/1757-899X/909/1/012008

Effect of gas pressure on conduit plasma atomization for fabricating spherical stainless steel powder

Dharmanto[1], Sugeng Supriadi[1], Ario Sunar Baskoro[1,*], and Bambang Suharno[2]

[1]Department of Mechanical Engineering, Faculty of Engineering, Universitas Indonesia, Depok, 16424, Indonesia.
[2]Department of Metallurgical and Materials Engineering, Faculty of Engineering, Universitas Indonesia, Depok, 16424, Indonesia.

Email: ario@eng.ui.ac

Abstract. In this study, spherical stainless steel powders are produced using the conduit plasma atomization. Parameters of the conduit plasma atomization process with gas variations of 1.5 bar pressure, 2 bar pressure, 2.5 bar pressure with 316L stainless steel raw material with a current of 25 Amperes, feed speed of 2 mm^3/s. The results of stainless steel powders were observed using a digital microscope (Dino-Lite AM4115), scanning electron microscopy (SEM-FEI-Inspect F50), and energy dispersive spectroscopy (EDS). To ensure the purity of the resulting 316L stainless steel spherical powder, EDS was used for qualitative and quantitative elemental analysis. The results showed that the 316L stainless steel spherical powder particles varied in size from 35 µm to 140 µm, making them ideal for powders metallurgy application. The effect of gas pressure on the powder weight percentages for particle sizes 50 µm – 100 µm for 1.5 bar pressure, 2.0 bar pressure, and 2.5 bar pressure were 67.14%, 78.71%, and 81.73%, respectively. It is possible that this could happen because to break down molten metal into smaller size droplets, it is needed the kinetic energy of larger gas pressure. So that large gas pressure can produce more small particle size compared to small gas pressure.

1. Introduction

Currently, the development of the manufacturing industry in the world is increasingly advanced, according to the industry, which is based on the metal powder. The preferred shape of the metal powder particles is the spherical metal powder. Spherical metal powder can be produced using gas atomization [1, 2] and plasma atomization[3]. Metal powder materials suitable for powders metallurgy application that have sizes between 5 µm to 200 µm [4]. Powder particles of this size are often used in additive manufacturing applications and metal injection molding applications.

There are two main types of classification using particle size distribution in additive manufacturing applications, including powders below 50 µm are used for most powder manufacturing systems. Powder particles with sizes below 10 or 20 µm must be avoided because it can reduce the powder's flowability. Powders with sizes between 50 and 100-150 µm are used for Electron Beam Fusion (EBM) and Laser Metal Deposition (LMD) technology. The favorable particle shape in the powder metallurgy among them is that it has a spherical shape because it has good powder flow properties and can help the process of forming a uniform powder layer. Especially applied to metal injection molding [5].

One of the most widely used powder metallurgy applications in the large scale manufacturing industry is Metal Injection Molding (MIM). MIM uses metal powder as the base material for making feedstock [6, 7]. MIM can produce small components with the right size with a smoother surface roughness quality compared to the investment casting process [5]. In addition, MIM has the advantage of saving production costs because there is almost no waste material that is wasted [8-10].

Metal powders widely used in the MIM process include stainless steel spherical powder with almost no satellites, homogeneous particle size distribution. It aims to improve the flowability of powder to enter the mold of MIM [11]. Stainless steel spherical powders with almost no satellites that have homogeneous particle size distributions can be made with plasma conduit atomization machines [12].

Plasma atomization previously used three plasma torch sources without a plasma conduit, so direct thermal energy spread to the entire tank, while this study was enough to use one plasma torch equipped with a plasma conduit. Plasma heat remains concentrated focused on the plasma conduit, and the molten metal can heat up longer so that there is sufficient heating time to produce perfectly spherical particles.

This study aims at investigating the effect of pressure on the conduit plasma atomization from 1.5 bar, 2 bar, and 2.5 bar pressure variations, in terms of the results of 316L stainless steel metal powder.

2. Methods

The scheme of the conduit plasma atomization is shown in Figure 1. The conduit plasma atomization uses a DC (Direct Current) plasma arc heat system. The plasma system consists of a high-frequency DC power supply, carbon graphite as an anode-cathode nozzle, reactor chamber, and 316L stainless steel wire as its initial raw material. The current used 25 amperes producing a number of high frequency periodic electric spark jumps between cathode and anode, which is driven by compressed air to the shrinking nozzle, which produces a thermal plasma jet to melt the wire material to form droplets of particles.

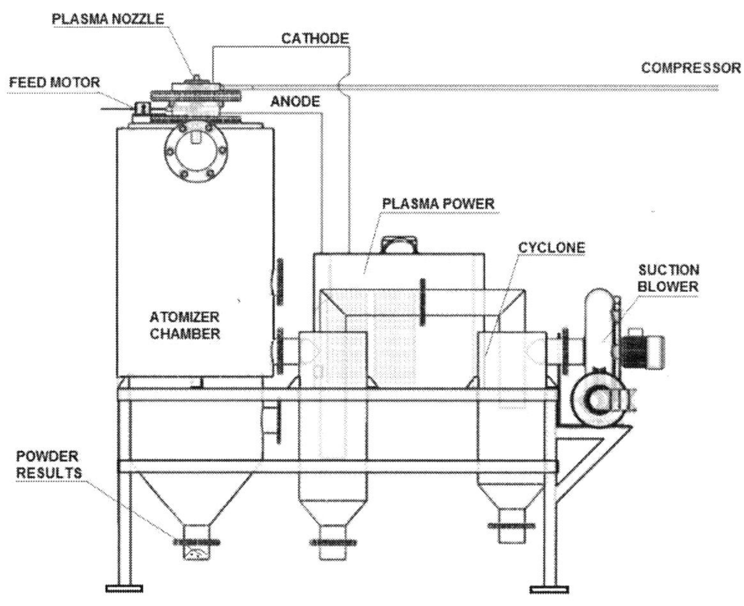

Figure 1. Schematic of a conduit plasma atomization

The feed material is 316L stainless steel wire with a diameter of 1.6 mm. The wire is inserted into the nozzle chamber at a constant speed of 2 mm^3/sec, which produces a particle droplet in the reactor chamber, then falls together in a collection container filled with water as a cooling. Testing of metal powder particles using the sieving method on the metal powders collectively using mesh sizes # 100, # 200, and # 325 with each opening 100-180 μm, 50-100 μm, and <50 μm and weighing method on according to ASTM standards E. 11-61. The sieving results were weighed using digital scales to determine the size characteristics and particle size distribution of metal powders.

3. Results and Discussion

The results of the plasma arc flame on the conduit plasma atomization at a pressure variations of 1.5 bar, 2 bar, and 2.5 bar can be seen in Figure 2.

(a) (b) (c)

Figure 2. Photograph of plasma arc flame at pressure: (a) 1.5 bar, (b) 2 bar, dan (c) 2.5 bar

Figure 2 shows that most plasma arc frames from a video at a pressure of 1.5 bar have a more stable plasma arc flame than a plasma arc at a pressure of 2 bar or 2.5 bar. The possibility of an unstable plasma arc flame can be caused by high pressure, which can increase cathode erosion or anode erosion. Higher the gas pressure, so greater the kinetic energy that will make erosion easier [13].

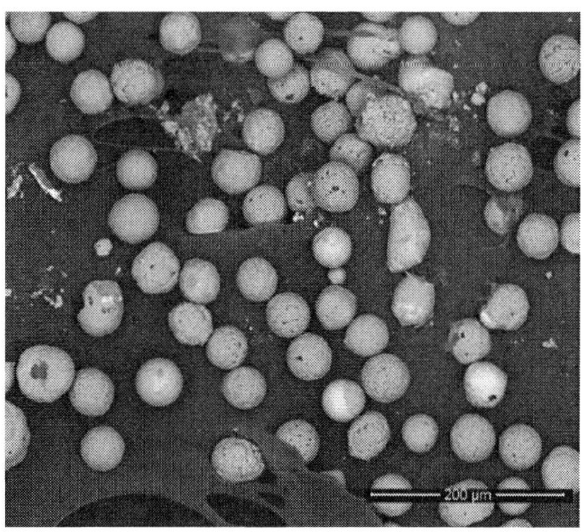

Figure 3. SEM of 316L stainless steel powder particle size <50 μm

Figure 3 shows SEM of 316L stainless steel powder from conduit plasma atomization at a particle size of <50 µm where particles show spherical shape with almost no satellites.

The results of conduit plasma atomization at a pressure variations of 1.5 bar, 2 bar, 2.5 bar with 316L stainless steel wire raw material at a current of 25 amperes, and feed speed of 2 mm3/s, as shown in Table 1.

Table 1. Digital microscope images of 316L stainless steel spherical powder of conduit plasma atomization

Particle size / **Gas Pressure**	**<50 µm**	**50-100 µm**	**100-140 µm**
1.5 bar	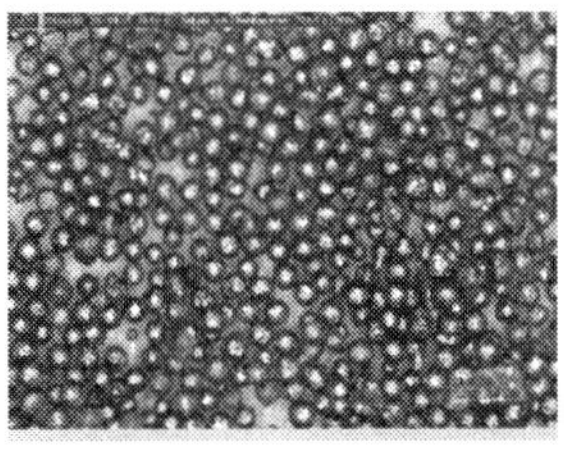	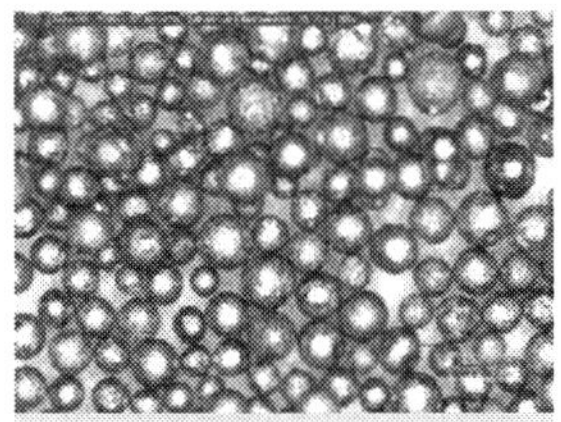	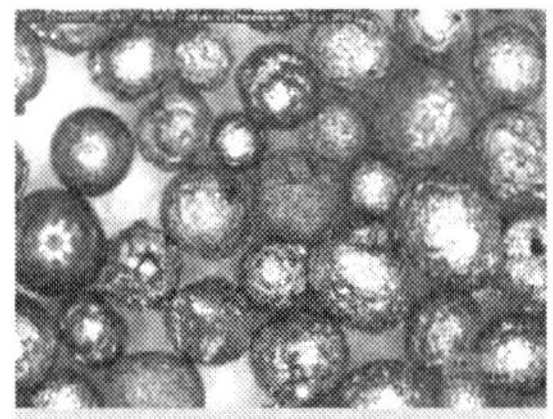
2.0 bar	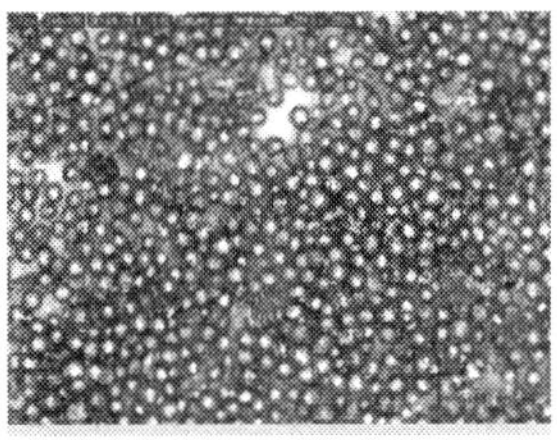	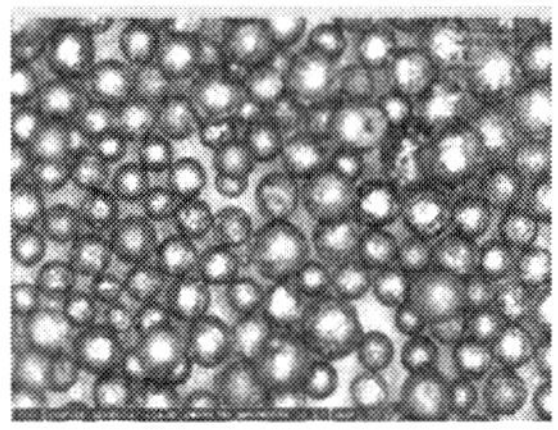	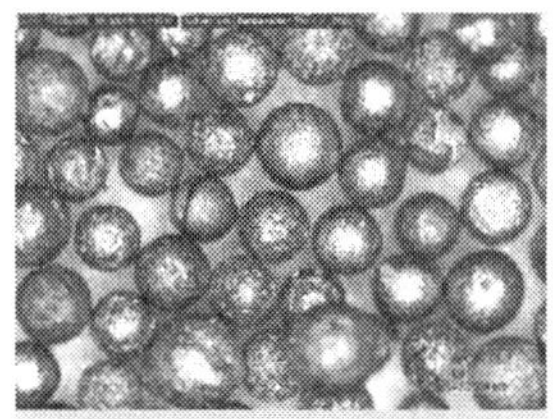
2.5 bar	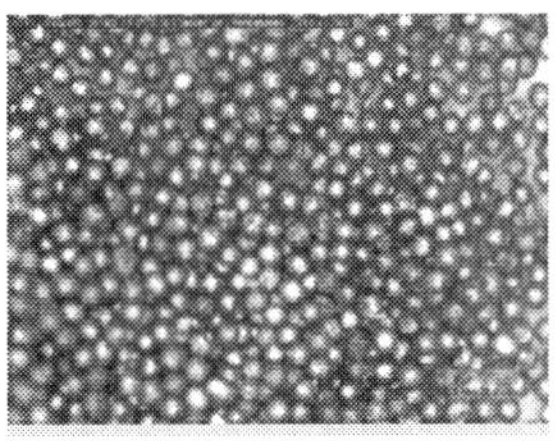	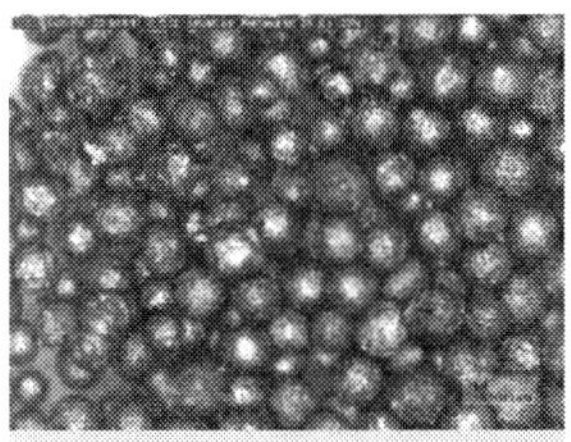	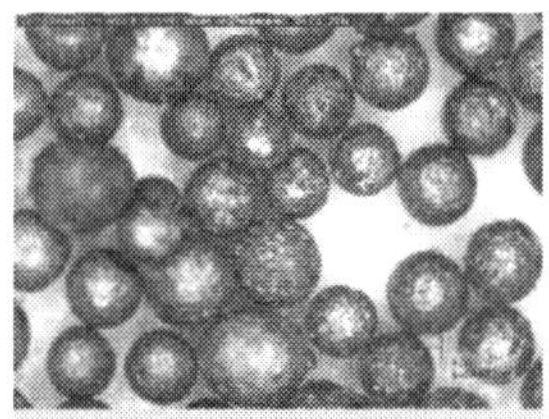

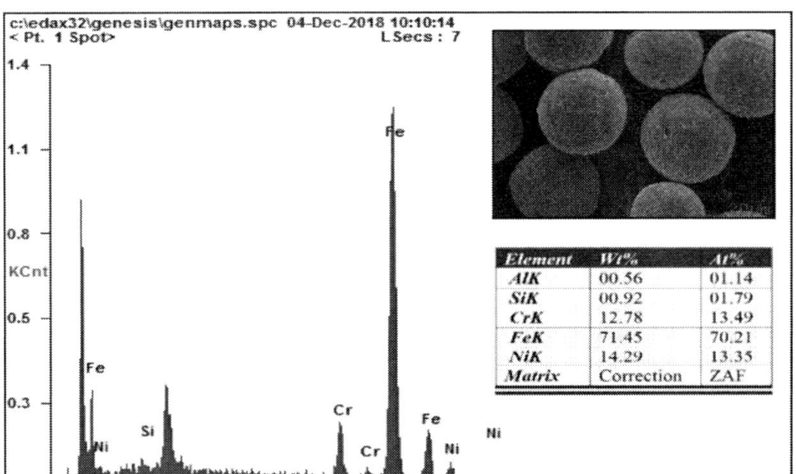

Figure 4. EDS of 316L stainless steel powder particle size <50 µm

Figure 4 shows the EDS results of 316L stainless steel powder, which has a chrome and nickel content of 12.78% and 14.29%. In comparison, the initial composition of 316L stainless steel wire raw material has a chrome and nickel content of 18.15% and 11.17 %, respectively.

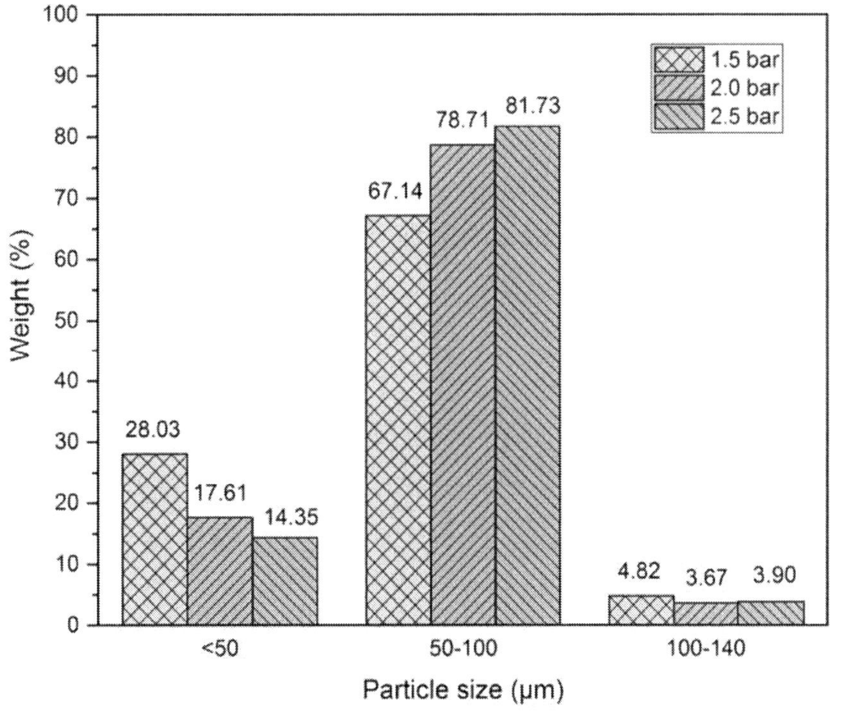

Figure 5. Particle size distribution of 316L stainless steel results of conduit plasma atomization variation of 1.5 bar pressure, 2 bar pressure, 2.5 bar pressure

Figure 5. shows the effect of gas pressure on the percentage of powder weight at a particle size of 50 µm - 100 µm at a pressure of 1.5 bar, 2 bar, and 2.5 bar are 67.14%, 78.71%, and 81.73% respectively. Higher gas pressure it can produce an increasing number of powders produced at a particle size of 50 µm - 100 µm. It is possible that this could happen to break down because the liquid metal into smaller size droplets need the kinetic energy of greater gas pressure so that large gas pressure can produce smaller particle sizes compared to small gas pressure. While the percentage of powder weight at particle size <50 µm at a pressure 1.5 bar, 2.0 bar, and 2.5 bar are 28.03%, 17.61%, and 14.35%, respectively, the higher gas pressure can produce a small amount of powder produced at particle sizes <50 µm. This is possible because the plasma arc flame 1.5 bar is more stable than the 2 bar or 2.5 bar plasma arc flame. Unstable plasma arcs can be caused by cathode erosion, or anode erosion, so high erosion is possible to reduce the plasma arc temperature [13]. The unstable plasma arc flame causes a decrease in heat, which can be indicated by a decrease in the amount of weight of the powder produced at a particle size <50 µm. Interesting future research is to make the plasma arc stable by adding an intelligent control system [14, 15].

4. Conclusion

The effect of gas pressure variation on the percentage of powder weight at a particle size of 50 µm - 100 µm at a pressure 1.5 bar, 2.0 bar, and 2.5 bar are 67.14%, 78.71%, and 81.73% respectively. Higher gas pressure it can produce an increasing number of powders produced. It is possible that this could happen to break down the liquid metal into smaller-sized droplets. The kinetic energy of greater gas pressure is needed so that large gas pressure can produce smaller particle sizes compared to small gas pressure.

Acknowledgment
This research was supported by the Ministry of Research and Technology of the Republic of Indonesia through the Excellent Applied Research in Higher Education scheme (contract number: NKB-1744/ UN2.R3.1/ HKP.05.00/ 2019).

References

[1] S. Supriadi, T. I. Susimah, M. H. S. Akbar, B. Suharno, and A. S. Baskoro 2020 Designs and evaluations of a gas atomizer to fabricate stainless steel metal powder to be applied at a metal injection molding in *Key Engineering Materials* 40-47
[2] S. Supriadi, T. I. Susimah, M. H. S. Akbar, B. Suharno, and A. S. Baskoro 2020 Effect of pressure on the gas atomizer to fabricate stainless steel metal powder in *Key Engineering Materials* 54-58
[3] G. Chen, S. Zhao, P. Tan, J. Wang, C. Xiang, and H. Tang 2018, A comparative study of Ti-6Al-4V powders for additive manufacturing by gas atomization, plasma rotating electrode process and plasma atomization *Powder technology* **333** 38-46
[4] P. K. Samal and J. W. Newkirk 2015 ASM Handbook Volume 7: Powder Metallurgy *ASM International* 1-907
[5] J. Džugan and Z. Nový 2017 Powder application in additive manufacturing of metallic parts *powder metallurgy: fundamentals and case studies* 183
[6] B. Suharno, F. Mawardi, S. Dewantoro, B. Irawan, M. Doloksaribu, and S. Supriadi 2019 Effect of powder loading on local feedstock injection behavior for fabrication process of orthodontic bracket SS 17-4 PH using metal injection molding in *AIP Conference Proceedings* 020030
[7] S. Supriadi, S. Dewantoro, F. A. Mawardi, B. Irawan, M. Doloksaribu, and B. Suharno 2019 Preparation of feedstock using beeswax binder and SS 17-4PH powder for fabrication process of orthodontic bracket by metal injection molding in *AIP Conference Proceedings* 020029
[8] G. Schieleper 2006 A manufacturing process for precision engineering components *Powder metallurgy association, United Kingdom*
[9] S. Supriadi, T. W. Sitanggang, B. Irawan, B. Suharno, G. Kiswanto, and T. Prasetyadi 2017 Orthodontic bracket fabrication using the investment casting process *Int. J. Technol,* 42015 613-621
[10] K. Higashitani, H. Makino, and S. Matsusaka 2019 *Powder technology handbook*: CRC Press
[11] A. Dehghan-Manshadi, M. Bermingham, M. Dargusch, D. StJohn, and M. Qian 2017 Metal injection moulding of titanium and titanium alloys: Challenges and recent development *Powder Technology* **319** 289-301
[12] Dharmanto, S. Supriadi, Baskoro A.S E 2019 Effect of feed metal mow rate on mow-cost Plasma Atomizer for Fabricating 316L Stainless Steel Powder," *International Journal of Technology,* **10** 1593-1601
[13] R. Knight, R. Smith, and D. Apelian 2020 Application of plasma arc melting technology to processing of reactive metals *International materials reviews* **36** 221-252
[14] A. S. Baskoro and S. Supriadi 2019 Review on plasma atomizer technology for metal powder," in *MATEC Web of Conferences* 05004
[15] A. S. Baskoro, R. Tandian, A. Edyanto, and A. S. Saragih 2016 Automatic tungsten inert gas (TIG) welding using machine vision and neural network on material SS304 in *International Conference on Advanced Computer Science and Information Systems (ICACSIS)* 427-432

IOP Conf. Series: Materials Science and Engineering **909** (2020) 012009 doi:10.1088/1757-899X/909/1/012009

The electrical energy usage of monitoring system at real-time using IoT as the primary policy of energy efficiency

Suhanto[1*], Achmad Setiyo Prabowo[1], Rifdian Indrianto Sudjoko[1], Wiwid Suryono[1]

[1]Department of Airport Electrical Engineering, Politeknik Penerbangan Surabaya, Surabaya 60236, Indonesia

*Email: nadiafather@gmail.com;

Abstract. The policy of electrical energy conversion requires an energy audit to be right on target. Energy audits require measuring devices and energy usage data recorders in every part of the system. Measurement and recorder integration is using the technology of the Internet of Things (IoT) can facilitate the collection, monitoring, and data analysis. This research is designed and implemented for monitoring the use of electrical energy in real-time by utilizing IoT at a low cost. This system includes a microcontroller that is can communicate via WiFi network type ESP32 Devkit V1, instrument system of measuring electrical energy type PZEM-004T, the connector and wrecker device use Solid State Relay (SSR) and open-source IoT platform. The finding showed that the developed energy monitoring system can successfully record voltage, current, active power, and accumulative power consumption. The monitoring system can be acquiring the power used by the load relatively fast is around 2 to 4 seconds, and the monitoring system uses relatively little power around 5 W.

1. Introduction

The growth of electrical energy usage has increased significantly, thus requiring an enhancement in energy supply in the coming future. One of the biggest consumers is the campus. The use of Air Conditioning (AC) and LCD projectors in lecturing room significantly increase the consumption of electrical energy consumption. Besides, students also have bad behaviors by ignoring the energy-saving program. Several wastes of electrical energy in a campus environment that most often happen such as turning the AC and lights on when the lecture room is not used, besides that some workroom such as meeting rooms and laboratories often sustain this condition.

This condition can be reduced by giving a firmer policy like strict penalties to institutes that make significant energy waste. Therefore, to be right on target, the policy must be supported by accurate data. One of the important data is energy usage data in each field based on time usage. The data can be collected from Automatic Meter Reading (AMR). Real-time measurement in several fields requires a considerable amount of AMR. The price of AMR which has the ability of a large capacity data logger is quite expensive, so the energy audit process needs a substantial cost. It can be solved by using AMR that has no data logger but it has been integrated with a database via internet network due to the development of technology Internet of Things (IoT) that mostly used today to connecting or transferring data through low bandwidth networks. IoT is used in every field such as the automotive industry, logistics, health care, smart grids, and smart cities.

Integrated AMR with IoT on this research is integration between module IoT type ESP32 with module measuring electrical energy PZEM-004T. The load connected to the power source is controlled by a Solid-state type switch. Data measurement is connected with the IoT platform through an internet Wifi connection. IoT platform has a datalogger and dashboard section as a Human Machine Interface (HMI) display.

2. Basic Theory

Here the explanation of the device used in this research. There are software and hardware

2.1. Energy audit

An energy efficiency program that succeeded energy audit has been conducted. The audit process is the first step to identify the potentials of energy saving. This audit will produce energy data usage that can be used for guidance on energy efficiency program. Automatically, the audit results will also provide information about the right steps to operate an energy efficiency program. This process becomes a reference in constructing the action plan that contains energy-saving recommendations. Historically study, it becomes fundamental to identify high sectors of its energy usage and influence toward energy use maps. This information is useful for determining energy-saving priorities as well as providing an overview of energy use patterns in a building.

2.2. Automatic meter reading (AMR)

Automatic Meter Reading (AMR) is a technology for collecting consumption, diagnostic and status data automatically from measuring energy devices (gas, electricity) and transferring the data to the data center for billing, problem-solving and analysis. This technology can economize the cost of regular flow to each physical location to read meters.

AMR system can be utilized for some needs like monitoring the energy supply to the customer, monitoring the network loss planning needs, billing or billing, and so on. Reading results data is saved on a database and can be used for analyzing, transaction and troubleshooting. Certainly, this technology can assist the enterprise of the electricity service provider to degrade operational costs and become the additional worth to customers, exactness and accuracy of data read, and give benefits to its user's service. In the beginning, the meter reading is carried out using a cable (wired) or direct dialing/ reading. A computer is connected to the meter using a communication cable (RS 232, RS 485 or RJ 45) or optical probe if it is conducted in the field.

Recently, many communication technologies that can be used by AMR systems such as PSTN (landline phone), GSM, radio wave, PLC (Power Line Carrier) and last meter reading is using LAN/WIFI for meters that supported by TCP / IP. AMR system consists of three components: (1) control Center it is a reading center, database center, and application program center, user, (2) electronic Meters it is an energy metering device with communication capabilities, (3) communication media, it is a means of communication between the control center and the electronic meter. The advantages of using the AMR System are below: (1) meter reading in real-time and accurately, (2) detect the early use of electrical energy illegally, (3) speed up the issuance process of account, (4) increase in electricity sales revenue, (5) State Electricity Enterprise (PLN) and customers can control electricity consumption through accurate data records, (6) improve the quality of service to customers.

2.3. ESP 32

ESP32 is a wifi module and as one of the devices created by DOIT to evaluate the ESP-WROOM-32 module. The basic component of this system is microcontroller ESP32 that has WIFI support, Bluetooth, Ethernet in one chip at low power. The module and address of pinout I/O can be seen in Figure 1.

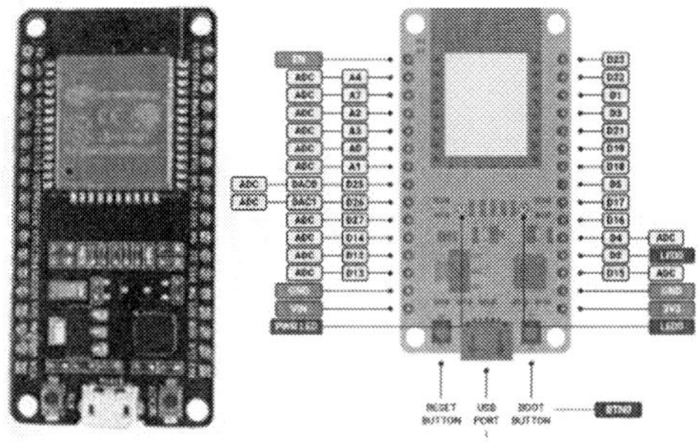

Figure 1. ESP32 and diagram of pinout module

ESP32 module has specifications below:

- Microcontroller: Tensilica 32-bit Single- / Dual-core CPU Xtensa LX6
- Operation Voltage: 3.3V, Input Voltage: 7-12V
- Digital I / O Pins (DIO): 25, Analog Input Pins (ADC): 6, Analog Output Pin (DAC): 2
- UARTs: 3, SPI: 2, I2Cs: 3
- Flash Memory: 4MB, SRAM: 520 KB
- Beat rate: 240 Mhz, Wi-Fi: IEEE 802.11 b / g / n / e / i

2.4. IoT platform

IoT is a concept that has a lot of advantages to internet connectivity. The purpose of each IoT device is connecting to the application device for delivering information using internet protocol transfer. The gap between censor device and data network is filled by IoT Platform that connects data networks to censor settings and provides insight using backend application to discover the amount of data produced by hundreds of sensors.

IoT platform is a multilayer technology that enables supply, management and automation devices that directly connected to IoT. That platform connects the hardware to the cloud using flexible connectivity options, enterprise-level security mechanisms, and extensive data processing power. IoT platform position towards hardware and software application in Figure 2.

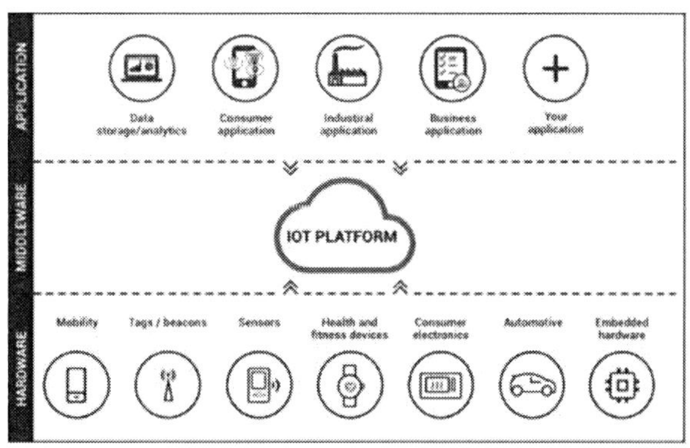

Figure 2. IoT platform position towards sensor hardware and software application.

IoT platform provides a set of features that extremely accelerates the application development for connected devices and maintains cross-device scalability and compatibility.

3. System Planning

The IoT-based electrical energy monitoring system has several main parts like the electrical energy usage instrument system, the load control switch system, the viewer system, the communication system, and the IoT platform.

3.1. Functional diagram system

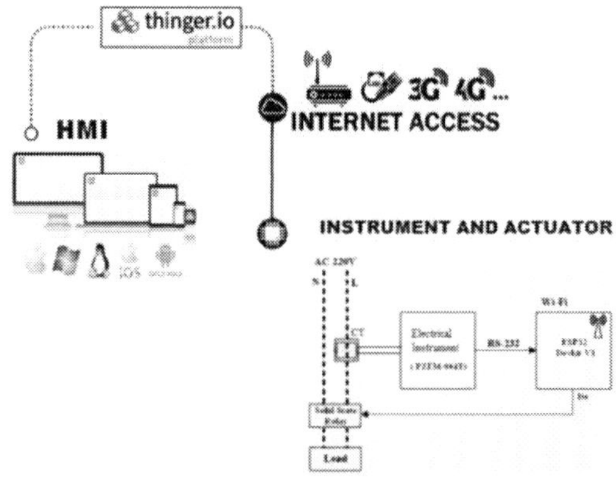

Figure 3. Whole concept system in IoT based electrical energy monitoring system

3.2. Instrument device and actuator

This instrument device system is integration between ESP32 module and Devkit V1 as its microcontroller, PZEM-004T Module as measuring instrument of electrical energy parameter and SSR as the actuator of connector and wrecker load. This functional diagram is like a figure in 4. PZEM-004T Module is a module of electrical energy used by the load. This module is relatively inexpensive. To measure current, this module uses CT (current transformer) non-invasive sensor, chip measuring the energy of SD3004 and microcontroller to measure voltage, current, active power, and power consumption accumulative.

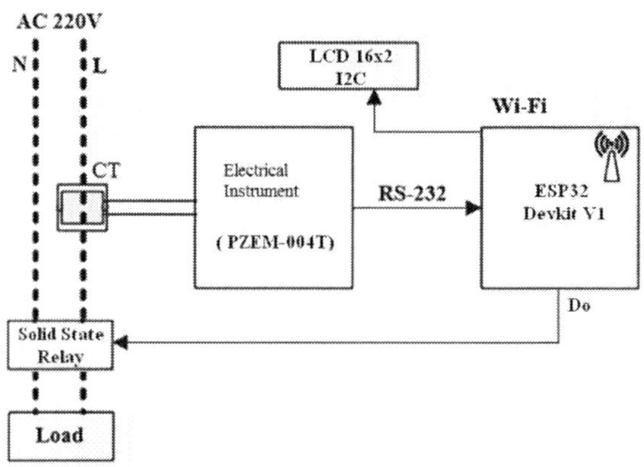

Figure 4. Diagram of block Building Automation System.

Some primary command scripts software on instrument devices include defining Usernames, ID and passwords device. These definitions should be appropriate with the IoT platform in Thinger.io. Other definitions are SSID name and SSID Password as an internet gateway. Those definitions are here below:

```
#define USERNAME "poltekbangsby"
#define DEVICE_ID "beban01"
#define DEVICE_CREDENTIAL "beban01x"
#define SSID "tlb"
#define SSID_PASSWORD "tlbok"
```

Another definition is to integrate a microcontroller with the PZEM-004T module. Communication between electrical energy module censor with ESP32 utilizes using asynchronous serial communication lines. Those script definitions are here below:

```
#define HardwareSerial Serial2(2);
PZEM004T pzem(&Serial2);
IPAddress ip(192,168,1,1);
```

Serial communication uses a second serial line with the PZEM004 module 192.168.1.1. Where this module can be connected multipoint with several devices that have different addresses. Another important command is a connection between instrument device with internet gateway and packet data command with protoson format as below:

```
thing.add_wifi(SSID, SSID_PASSWORD);
thing["Load_Control"] << digitalPin(ssr);
thing["Power_Measure"] >> [](pson& out)
 {
  out["voltage"] = tegangan;
  out["current"] = arus;
  out["power"] = daya;
  out["energy"] = kwh;
 };
```

That packet data has two resource name namely "Load Control" and Power Measure". Load Control is to control SSR through a pin output to wreck or connect load. In this sytem, pin SSR is connected with pin GPIO 23. Resource Power Measure has 4 variables for instance voltage, current, power and energy. These variables value are the result of data acquisition from PZEM-400T measurement module.

3.3. Communication data

Communication between the measuring device and the IoT platform is using an internet connection. The measuring device uses a wifi connection to connect microcontroller with an internet gateway. The communication protocol between measuring device or sensor device with IoT platform utilizes the protoson protocol. This protocol is different from the IoT protocol in general, namely Message Queuing Telemetry Transport (MQTT). The protoson protocol is an IoT protocol developed by the Thinger.io team. For users of microcontroller devices with Arduino IDE that has been provided a library of protoson protocol. There are other general categories related to Linux-supported devices such as Raspberry Pi, Intel Edison, or other Linux computers running on Ubuntu or macOS. The Wifi device in this system uses the ESP32 Devkit V1 module. This device has great power saving with a relatively high beat speed which is compatible with processing and delivering data via internet connection.

3.4. IoT platform

This system uses Thinger.io as Its IoT platform. Thinger.io has some features below: (1) cloud console, (2) mobile app, (3) server API, (4) server deployment. This monitoring system, measurement data display is designed on Cloud Console. The first design in this platform is an additional device. Each device that sends data to a platform is one device. The parameter device entered are Device Id, Device credential. Both parameters should be compatible with those described in the instrument device. The menu display for additional devices is shown in Figure 5.a, meanwhile, the description of the Arduino IDE is shown in Figure 5.b.

(a) (b)

Figure 5. The additional device at IoT platform, (a) Menu display of additional device on Thinger.io, (b) Parameter description device on Arduino IDE

The connectivity status between the instrument device and IoT platform can be seen in the Device List as shown in Figure 3.3.c. if the device is connected, it will be in a "Connected" status and when it is not connected it has a "Disconnected" status.

The measuring data display is designed at the menu dashboard. That menu provides some widgets that can visualize data. Display data design on this monitoring system in Figure 6. variables displayed are voltage, current, power, and energy consumption at the load. Besides, measurement variables also included a button that can be used to turn on or turn off the load.

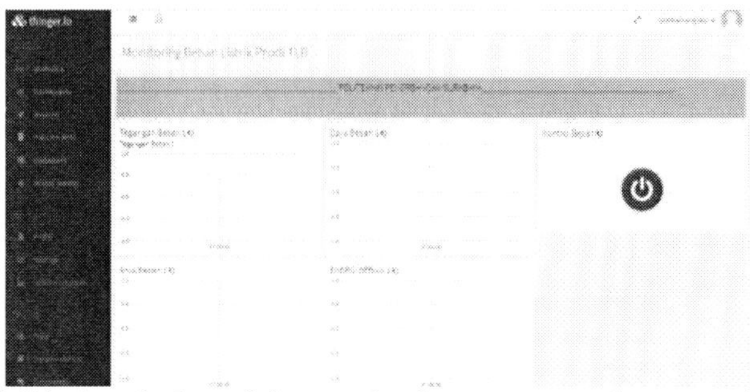

Figure 6. Display design of the monitoring system at the menu dashboard

4. Testing and Analysis

Testing the system is aimed to determine the characteristics of performance system. The testing types carried out includes measuring instruments testing, connectivity testing, and testing.

4.1. Measuring instrument testing

The measuring instrument test is intended to perceive the characteristics of measuring electrical energy. The measuring instrument system utilizes the PZEM-004T sensor module. The measurement results of the electrical energy sensor module with a lamp load as in table 1. The results of the testing system with the AC load have an outcome shown in table 2 based on the data testing result that the monitoring system can conduct measurement appropriately and in real-time. The time lag between measurement and display on the dashboard is greatly influenced by communication network quality. The electrical energy consumption in this monitoring device is very small at around 5 W.

Table 1. The results of the electrical energy sensor module with a lamp load

No.	Load	Voltage (V)			Ampere (A)			Power (Watt)		
		Clamp meter	PZEM-004T	% Error	Clamp meter	PZEM -004T	% Error	Clamp meter	PZEM -004T	% Error
1	Lamp 25 Watt	220,23	220,02	0,10	0,13	0,12	7,69	26,43	26,40	0,10
2		220,21	220,48	0,12	0,12	0,13	8,33	26,5	28,66	8,16
3		220,20	220,11	0,04	0,13	0,13	0	28,73	28,61	0,40
4		221,38	221,12	0,12	0,12	0,12	0	26,52	26,53	0,05
5		220,12	220,23	0,05	0,12	0,13	8,33	28,64	28,62	0,03

Table 2. The results of the testing system with the AC load have an outcome

No.	Load	Voltage(V)			Ampere (A)			Power (Watt)		
		Clamp meter	PZEM -004T	% Error	Clamp meter	PZEM -004T	% Error	Clamp meter	PZEM-004T	% Error
1	AC 1 PK	220,11	220,14	0,01	4,22	4,12	2,36	928,86	906,97	2,35
2		220,32	220,21	0,05	4,01	3,95	1,49	883,48	869,82	1,54
3		220,17	220,20	0,01	4,1	4,03	1,70	902,69	887,40	1,69
4		221,24	221,11	0,05	3,99	3,91	2,00	882,74	864,54	2,06
5		220,33	220,12	0,09	4,02	3,96	1,49	885,72	871,67	1,58

4.2. Connectivity test

The data connection test from IoT device to IoT platform is conducted with the time lag measurement from sending to displaying data. The average time needed to acquire data from the sensor to be displayed on the dashboard is around 2 to 4 seconds due to the acquisition data process in censor is around 1 second

5. Conclusion

The conclusion that may be drawn from our research are as follows:

1. The monitoring system can be acquiring the power used by the load relatively fast is around 2 to 4 seconds
2. IoT system cab displays the amount of voltage, current, and power at load
3. The monitoring system can be supervised everywhere as long as it has internet network access
4. Historical data on IoT servers can display the characteristics of electrical energy use for a certain period of time
5. The monitoring system uses relatively little power around 5 W

References

[1] Amin, M., and Wollenberg, B. F. 2005 *Toward a Smart Grid, october* 34–41
[2] Chooruang, K., and Mangkalakeeree, P. 2016 Wireless heart rate monitoring system using MQTT *Procedia Computer Science* **86** 160–163
[3] F. Benzi, N. Anglani, E. B. and L. F. 2011 Electricity smart meters interfacing the households *IEEE Transactions on Industrial Electronics* **58** 4487–4494
[4] Rashed Mohassel, R., Fung, A., Mohammadi, F., and Raahemifar, K. 2014 A survey on Advanced Metering Infrastructure. *International Journal of Electrical Power and Energy Systems* **63** 473–484
[5] Suhanto, S., Faizah, F., and Kustori, K. 2019 Designing a building automation system with open protocol communication and intelligent electronic devices *Journal of Physics: Conference*

Series **1381**
[6] Vega, A. M., Santamaria, F., and Rivas, E. 2015 Modeling for home electric energy management: A review *Renewable and Sustainable Energy Reviews* **52** 948–959

IOP Conf. Series: Materials Science and Engineering **909** (2020) 012010 doi:10.1088/1757-899X/909/1/012010

Metallurgical performance of column and mechanical flotation as a rougher circuit in sphalerite ore flotation with %solid and frother dose variation

S Oediyani[1,3*], A Muttaqin[1,2,3], D Haryono[1,3], R F Suwandana[1,3]

[1]Laboratorium Material Maju dan Tomografi, Universitas Sultan Ageng Tirtayasa, Gedung Center of Excellence (CoE) 2 Lt. 4, Jl. Jenderal Sudirman Km. 3 Cilegon, Banten, Indonesia 42435

[2]Pusat Penelitian dan Pengembangan Teknologi Mineral dan Batubara, Jl. Jend. Sudirman No.623, Wr. Muncang, Kec. Bandung Kulon, Kota Bandung, Jawa Barat 40211

[3]Jurusan Teknik Metalurgi, Fakultas Teknik, Universitas Sultan Ageng Tirtayasa, Jl. Jenderal Sudirman Km. 3 Cilegon, Banten, Indonesia 42435

*Email: s_oediyani@untirta.ac.id

Abstract. Column flotation is commonly used in cleaner circuits at the ore beneficiacy industry. However, this research wants to apply the column to the rougher circuit as an alternative to the mechanical flotation due to its energy efficiency which is 80% more efficient. This research was conducted using column and mechanical flotation cells to observe their metallurgical performance as a rougher circuit. The feed is sphalerite ore at size -140+230# with variables of % solid 7.5;10; 12.5; 15 and doses of frother 20;30;40; and 50ppm. Reagents used include OrePrep, Potassium Amyl Xanthate 500g/ton, copper sulfate 500g/ton, and soda ash 500g/ton. Flotation process is carried out 5 minutes for conditioning and 10 minutes for aeration. The results obtained the recovery value and grade. The results from the experiment, it can be concluded that the recovery obtained by mechanical is superior, and for column flotation grade is superior. In the variation of solid percent and frother dose, recovery increases to the maximum point and then decreases because no signifant effect after maximum point. The recommended variables for mechanical flotation are 12.5 solid% and 50ppm frother dose, because the recovery worth 75.462%. For column flotation, 15% and 40ppm, obtain the highest recovery 53.439%.

1. Introduction

Sphalerite (ZnS) is one of the minerals belonging to sulfide minerals with one of its properties, which is naturally having relatively high hydrophobicity. Sphalerite is also usually naturally bound to impurities that are classified as strong hydrophilic, so that with this difference in hydrophobicity the process of separation is suitable for sphalerite ore by using flotation (1). Flotation is a process of mineral separation based on differences in the ability of adhesion of particles to air bubbles. This process is complex because it includes interactions between mineral particles, air bubbles, reagents, water and the flotation machine so that the process of floating mineral particles occurs. Based on the regulation of Energy and Mineral Resources Ministry of Indonesia no. 5 of 2017 states that the levels for zinc concentrate to be exported have a value of ≥ 51%. For this reason, it is necessary to benefit the sphalerite ore in order to meet these requirements.

Flotation in the mineral processing industry is a process of concentration, and usually flotation within the scope of the industry is not only carried out in one stage of the process but rather several stages of the process which have their respective functions to obtain the targeted concentrate levels. Usually called a flotation circuit which includes 3 stages of the process namely rougher, cleaner, and scavenger. The rougher circuit with its input in the form of fresh feed or ore has a function to maximize recovery, the cleaner circuit with concentrated input from the rougher circuit has a function to get the target grades, and the scavenger circuit with the tailings input from the two previous circuits has a function to maximize gain valuable minerals that are still left in the tailings. Initially each of the circuits used a mechanical flotation cell system, but in a yahyaei research in 2006 it tried to replace mechanical flotation with column flotation on a cleaner circuit, because mechanical flotation has several disadvantages that affect process efficiency, namely high energy consumption, high operating costs, and difficult footprints and controls. After the application of column flotation on the cleaner circuit in yahyaei research, it was found that energy consumption decreased by 83%, operating costs by 76%, and separation efficiency increased by 7% (2).

Industrial-scale flotation columns are generally cylindrical with a height of 9-14 m and a diameter not exceed than 2 m. Ore in the form of slurry is feed from the top of column (about 2/3 the height of the column), flowing countercurrently with air bubbles produced by the sparger. Hydrophobic particles will attach to the bubbles, and move up to the top of column and form a froth zone. Hydrophilic particles will be left in the water and removed as tailings. The interaction between particles and air bubbles takes place in a zone called the collection zone. In addition, column cells have wash water to eliminate entrainment particles that carry over into the froth zone. Over the last few decades the concept of column cells has slowly been recognized (3). In the flotation circuit, the rougher circuit is used to obtain as many valuable minerals as possible and a cleaner circuit to increase grades of valuable minerals in the concentrate. Column cells are the most suitable cells for cleaner circuits, because their high selectivity results in higher levels of acquisition (4). Another advantage of column flotation is that it can produce thicker and more stable bubbles formation, so that a relatively higher recovery is obtained because it uses a taller cell and a smaller surface area than the column compared to conventional flotation cells (5). Fuerstenau et al., Considered column cells as an advantage for some reasons, namely column cells can provide better separation because column cells can obtain better recovery without reducing grades, require less capital and also lower operational costs than mechanical cells. In addition, column cells are also more adaptable to automatic control (6). Jena et al., stated that column cells can reduce amount of stages in operations and able to handle finer feeds (7). The number of collectors needed in a column cell is lower than a mechanical cell, and its construction and design is also simpler. The basic difference between column flotation and mechanical flotation is that column flotation does not have an agitator. Air bubbles in the column flotation are generated from the air flow through the sparger (1). Meanwhile, the mechanical flotation cell mechanism uses an agitator to produce air bubbles. In flotation circuits, mechanical flotation is generally used in rougher circuits because the mineral floatation mechanism is valuable in finding a high recovery, correspond with the function of the rougher circuit. However, mechanical flotation has several deficiencies, one of which is in terms of energy consumption. Mechanical cells consume 100-200 watts/volume. Meanwhile, column cells consume 20-30 watts/volume, or about 80% more efficient (8). These deficiencies led to the idea of applying column flotation cells to the rougher circuit as an alternative to mechanical flotation cells. So, by applying column cells to replace mechanical cells commonly used in rougher circuits, as well as the success of column cells applied to cleaner circuits in yahyaei research. Therefore, the use of column cells in the rougher circuit has the potential to increase the efficiency of the metallurgical performance and reduce production costs.

In previous research, column flotation cells were applied to a cleaner circuit and monitoring was carried out using an electrical capacitance volume tomography (ECVT) in the collection zone area to provide better results in imaging objects. However, this research is based on the application of column flotation in the rougher circuit with the aim of getting better %recovery than conventional flotation cells in the rougher circuit, so, in the future column flotation can be applied to the rougher circuit then can be monitored using ECVT. To do this research, mechanical or conventional flotation is needed as a comparison of column flotation, because basically mechanical flotation is used in the rougher circuit.

2. Experimental Methods

2.1. Sample Preparation

In this preparation stage sphalerite ore is prepared so the parameters are suitable to be carried out the flotation process, the first to do is comminution with the grinding method using a rod mill, in order to smooth the particle size in order to increase the degree of liberation as in Figure 1. after that do the sieving using a multilevel screen with a size of 100; 140; 230 # until you get the size -140+230#. Oversize +100# will be crushed again. With the best grinding parameters, 25 steel balls, 40 minutes grinding time, and 2 kg feeds succeeded in getting the desired fraction with the most mass of 550 grams from 2 kg feeds. After that, do washing and drying to reduce the slime contained in the ore. Finally, in the preparation phase, a coning-quartering method was used for all flotation feeds and samples for the initial characteristics of XRD and XRF.

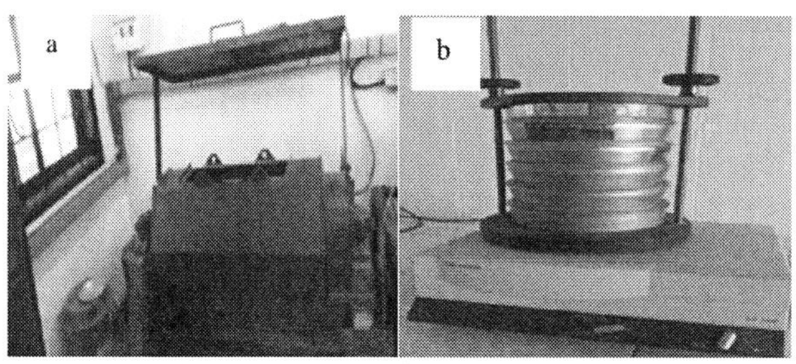

Figure 1. (a) Rodmill (b) Sieving.

2.2. Flotation Process Rougher Circuit

The main process stage wherein applying column flotation and mechanical flotation on the rougher circuit such as Figure 2. to compare metallurgical performance with the same scope and the same conditions, namely 5 minutes conditioning; variation of oreprep dose 20; 30; 40; and 50 ppm; variation of %solid 7.5; 10; 12.5; and 15%; PAX, Na2CO3 and CuSO4 500 ppm; aeration time of 10 minutes. Different parameters are the air flow rate at the 2.5 L/minute on column flotation and 1200 rpm agitator speed at the mechanical flotation. The products are in the form of concentrates and tailings and then carried out drying and sampling again for the final characteristics with XRF.

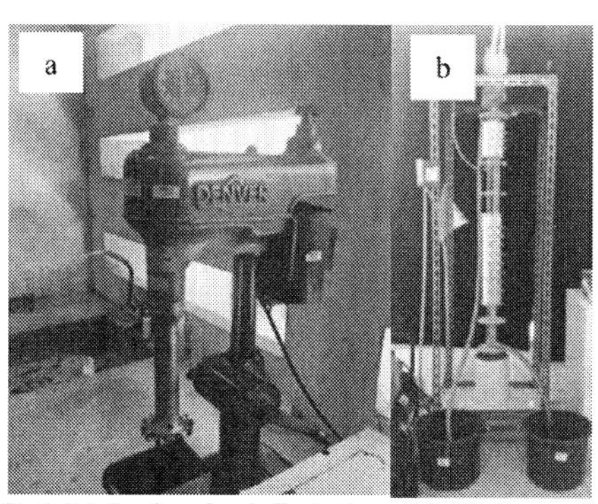

Figure 2. (a) Mechanical flotation (b) Column flotation.

3. Result and Discussion

Before the flotation process was carried out, the sphalerite and quartz samples were characterized to determine the compounds and elemental content respectively using the X-ray diffractometer (XRD) and X-ray fluorescence (XRF) methods. Data from XRD testing are compounds contained in the ore. The data is in the form of graphs with certain peaks according to the compounds characterized. The XRD reading method uses PANalytical X'Pert High Score Plus software. The top pattern of the sphalerite sample is shown in Figure 3. From the picture, it appears that the dominant mineral is sphalerite (ZnS) with galena (PbS), pyrite (FeS2), and quartz (SiO2) associations. XRD data obtained prove that the sample used is true sphalerite ore, because sphalerite generally in nature is often found to bind to FeS2, PbS, CuS, and SiO2 in the deposit. (Wills, 2016).

The elements contained in the ore are analyzed using Portable X-Ray Fluorescence Oxford X-MET 5100. XRF analysis is used as quantitative data to calculate %recovery. In the sample raw material of sphalerite ore, the Zn content was 25.78% and the S content was 10.95%. In the sphalerite sample there are also other dominant elements such as Fe of 15.06%, Pb of 7.67%, and Si of 27.83%, which can be seen in table 1.

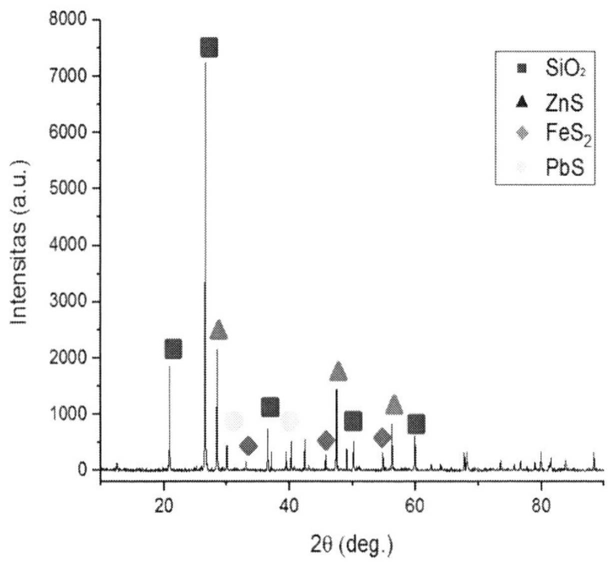

Figure 3. Results of XRD analysis of sphalerite.

Table 1. XRF feed analysis results

Elements	Zn	S	Si	Fe	Al	Pb	Cu	Mg	Lain
(%)	25,78	10,9	27,8	15,06	2,3	7,67	1,35	5,77	3,32

3.1. Pleminary Experiments

Preliminary experiments on column flotation for three phases namely, bubbles, water, and sphalerite were carried out within 5 minutes for the conditioning phase and 10 minutes for the aeration stage using particle size -230+325#, the flow rate is 2.5 l/min, variations of frother dose 10; 20; 30; 40 & 50 ppm, and 7.5 %solid. This is done to find out whether the column flotation process is going well, and to find out whether column flotation can separate valuable minerals from their impurities. The results obtained from the three phase experiment are shown in Figure 4.

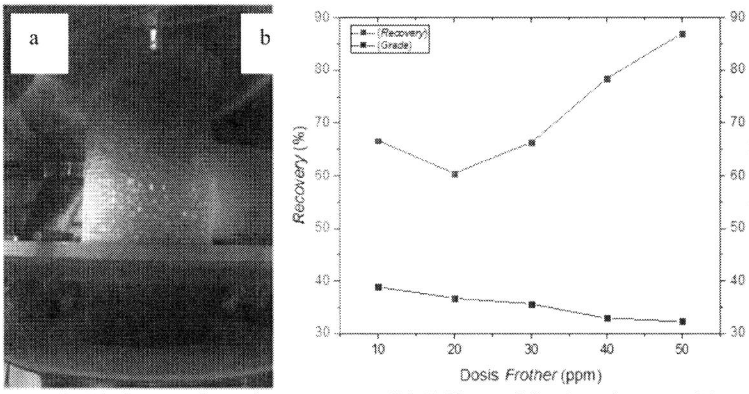

Figure 4. (a) Separation phenomenon (b) Effect of frother dose on %recovery and %grade graph.

It can be concluded from Figure 4b that the graph of the three-phase trial results proves that column flotation can separate valuable minerals and mineral gangue, as evidenced by the relatively high %recovery. Not only from the graph, from Figure 5a in the process of techinacal operation also shows that the phenomenon of bubble formation by frother can occur. The phenomenon that occurs in the froth zone, namely when valuable minerals are carried by bubbles and are held in the froth zone before falling into the concentrate outlet can be seen in Figure 4a.

3.2. Effect of Frother Dose and Solid Percent on Recovery and Grade of Mechanical Flotation

Recovery, which is one of the metallurgical performance parameters used as main data for comparison of mechanical flotation and column flotation processes. Frother doses are varied at 20, 30, 40, and 50 ppm. The flotation process (aeration) using a D-12 denver mechanical flotation cell lasts 10 minutes using an agitator rotation speed at 1200 rpm. Flotation feed in the form of 1.2-liter slurry with a variation of %solids 7.5; 10; 12.5; and 15%. Reagents other than frother, i.e. pH regulators (Na_2CO_3), activators ($CuSO_4$), and collectors potassium amyl xanthate (PAX) are added at a dose of 500 g/ton of feed at the conditioning stage. Conditioning duration is 5 minutes. Na_2CO_3 (soda ash) serves to maintain the slurry alkalinity in the range of 8-13 so that the xanthate collector is relatively stable (9). Copper ions from Cu^{2+} replace zinc ions on the surface of the sphalerite to optimize the performance of xanthate, producing metal-xanthate that is stable and difficult to dissolve. When compared with other types of xanthate, potassium amyl xanthate (PAX) provides a greater hydrophobicity effect because it has a longer carbon chain (C5) (1). Recovery and grade as a function of frother dose are shown in Figure 5 and the function of %solid is shown in Figure 6.

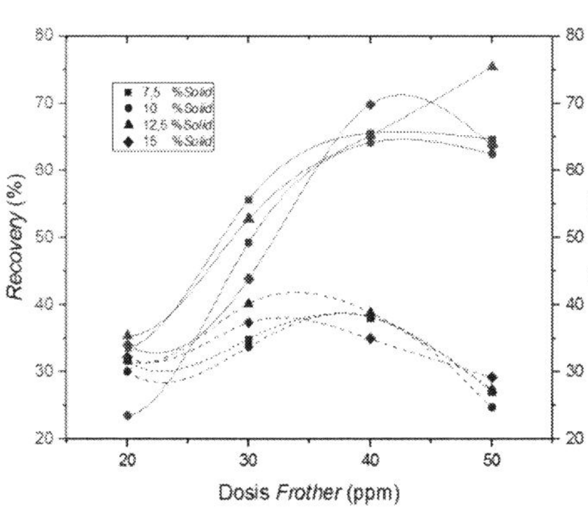

Figure 5. Graph of relationship between frother dose with recovery and grade in each solid percent.

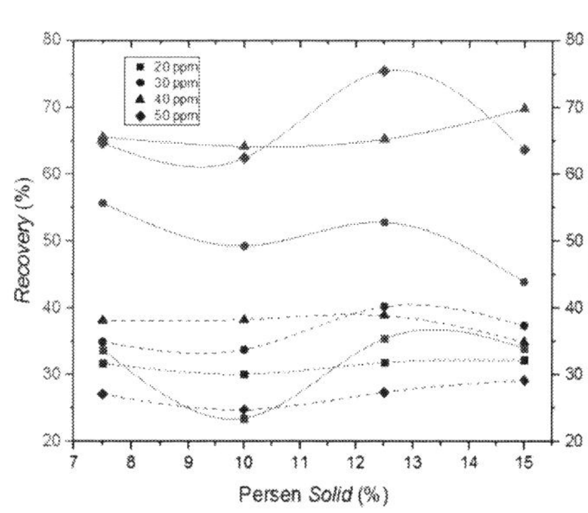

Figure 6. Graph of relationship between solid percentage with recovery and grade at each frother dose.

Figure 5 shows the maximum conditions achieved at a frother dose of 40 ppm, and there are some that have not yet found the maximum conditions, then the tendency decreases with increasing frother doses according to the theory (10), This phenomenon is caused by critical coalescence concentration (ccc) in the frother. CCC is the maximum point when the frother can no longer reduce the bubble size due to coalescene. When it reaches the ccc point, the frother no longer has a significant effect which causes coalescence or the phenomenon of small-sized bubbles unite into large-sized bubbles that are more easily collapse so that recovery decreases. The same thing happens with the addition of %solid when the addition no longer significantly influences after reaching its maximum condition, there is a maximum point that is at 12.5% solid percent. In Figure 6 that happens after the maximum conditions are achieved is the bubble overloading phenomenon shown in Figure 7 which causes recovery to decrease or no longer have a significant effect, with an increase in solid percentage, the number of particles carried and attached to the bubble also increases as much. Therefore, it causes the bubble overloading phenomenon so that the particles fall back and recovery decreases (11).

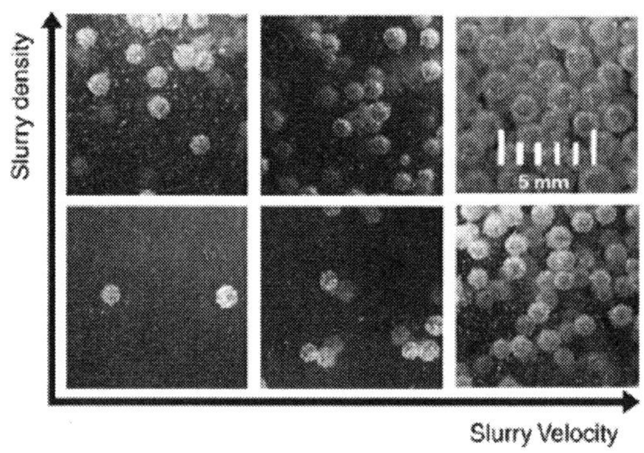

Figure 7. Bubble overloading phenomenon (12).

All of the variable combinations, the highest recovery was achieved at a %solid condition of 12.5% and the highest dose (50 ppm) resulted in a recovery of 75.462%. For the highest Zn grade obtained from the condition of 12,5 %solid and 30 ppm frother dose that is, 40.2%. In this case the Zn content still cannot meet the ESDM PERMEN requirement of ≥ 51%. Therefore, it is necessary to do further stage, that is circuit cleaner to reach the desired grade.

3.3. Effect of Frother Dose and Solid Percent on Recovery and Grade of Column Flotation

One of the objectives of this study is to determine the maximum effect of the flotation process variables in increasing recovery and grades of sphalerite ore. The same is done in mechanical flotation also done in the column flotation process with a solid percent variation of 7.5; 10; 12.5; and 15% and the variation of the frother dose are 20, 30, 40, and 50 ppm, with a slurry volume of 1 L and volume in the column cell is 1.7 L, and a constant air flow rate for each variable is 2.5 L/min. The products in the form of concentrates and tailings are produced from a combination of these 2 variables and are used as the main data for metallurgical performance calculations and compared with mechanical flotation. The function of frother dosage and percent solid against metallurgical performance is shown at the graphs in Figures 8 and 9. The increase in the pulp density will result in the decrease of recovery. It is caused by the fact that the increase in the pulp density results in the growing number of particles in the pulp and consequently in the growing resistance when the bubble-particle aggregates travel upwards to the froth. Under such conditions detachment of particles from the bubbles is easier (13). This may be also due to the reduction in selectivity due to increasing pulp density which increased non selective solids entrainment and bubbles-particles detachments(13). The most selective results and the lowest entrainment were obtained at the lowest solid concentration due to lower amount of water that was transferred to the froth(14).

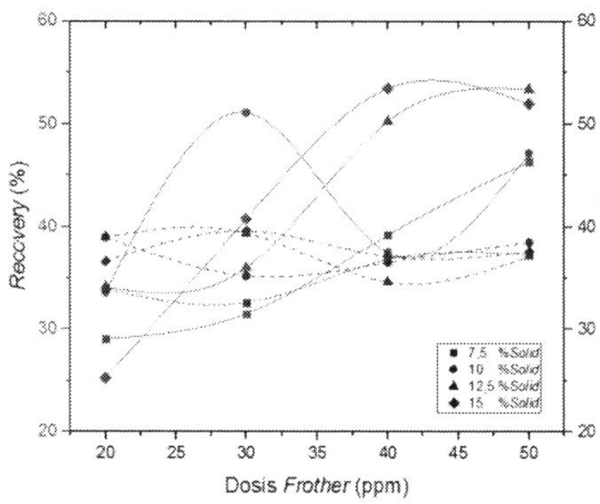

Figure 8. Graph of relationship between frother dose with recovery and grade in each solid percent.

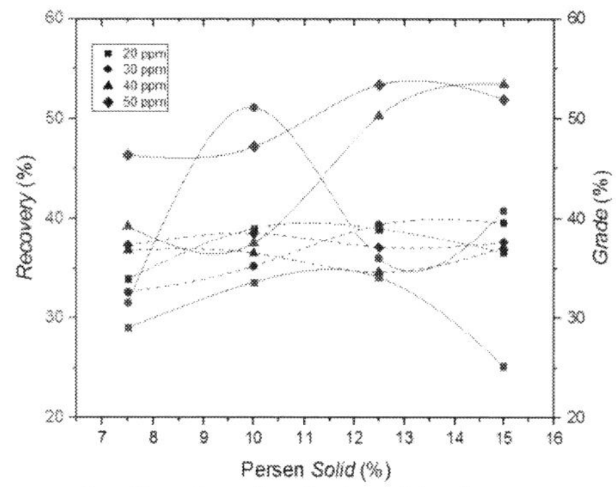

Figure 9. Graph of relationship between solid percentage with recovery and grade at each frother dose.

Based on Figure 8, the higher addition of the frother dose, the recovery value tends to increase or the acquisition obtained is higher, and there is also a maximum value of the addition of the frother dose. At 7.5 and 12.5 %solids from the graph do not yet have a maximum dose, for 15% solids have a maximum dose is 40 ppm, for 10% solids have a maximum dose is 30 ppm and more than that have a significant tendency to reduce the recovery value. After getting the maximum value of addition the dose no longer have a significant increase. This is due to the saturated concentration of frother which causes the addition of the dose does not greatly affect the formation and reduction of bubble size and another frother influences.

From all of combinations of variables, the highest recovery was achieved in the condition of 15 %solid and 40 ppm frother dose resulted in recovery of 53.439% with 37.1% Zn content. For the highest Zn levels obtained from the condition of 15% solid and 30 ppm frother dose that is, 39.6%. The grade obtained has a tendency to be inversely proportional to the recovery value, when the recovery value is high, it can be seen from the graph that the grade value has decreased. Factors that cause high or low Zn levels, one of which is the thickness of the froth zone, can be seen in Figure 10.

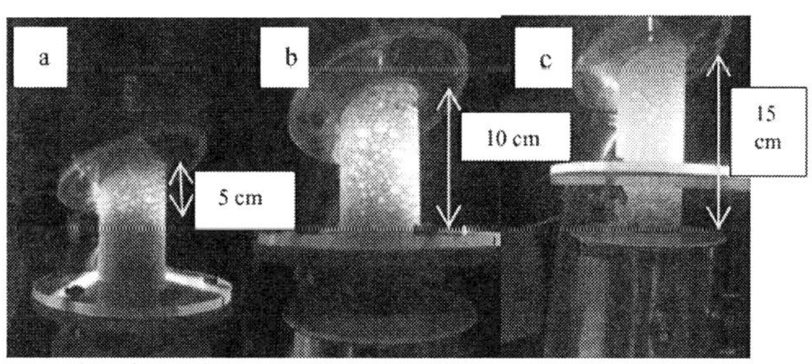

Figure 10. The thickness of the froth zone (a) 5 cm (b) 10 cm (c) 15 cm.

The recovery by entrainment decreases with the increased dilution of the pulp and the froth height, and increases with inc/reased agitation, rate of aeration, and froth stability. Also, it decreases with increasing particle size and its density, and depends on the particle shape, with lamellar shape particles being more easily entrained(15). The thick froth zone causes the residence time of entrainment particles to be longer in the froth zone and allows wash water to be removed before entrainment particle is carried to the outlet concentrate because entrainment particles that like water will be attracted to the flow of wash water that falls from the top side of column.

The effect of solid percent on column flotation, which is along with the addition of values will increase recovery to the maximum point and after that no longer has a significant effect. That is because when the percentage of solids gets higher the amount of minerals in slurry will increase, and small-sized air bubbles cannot float valuable minerals onto the froth surface because they are too heavy or have exceeded bubble carrying capacity which causes bubble overloading (12). In addition, the cause of the decrease after passing the maximum point at 15 %solid is more fast the slurry viscosity, the bubble rise velocity decreases. This is caused by, the more particles are carried and the longer the aggregate is raised up to reach the froth zone, causing the flotation kinetics to be longer or residence time in the cell much longer. This phenomenon causes the process to be stopped before all valuable minerals are lifted to the concentrate because the aeration time which is only 10 minutes is not enough to lift all the valuable particles, it may take more than 10 minutes to be able to lift all the valuable minerals thus reducing what's left in the tailings. This phenomenon is characterized by the darkness of the column flotation collection zone in the 10th minute. A high percent solid value can also make the flotation process difficult or even impossible.

3.4. Comparison of Metallurgy Performance of Column Flotation and Mechanical Flotation

The purpose of this research is to determine whether effective that column flotation is applied to the rougher circuit by looking at metallurgical performance when compared to mechanical flotation that has been approve commonly used on rougher circuits. The best tendency of each flotation cell for recovery and grade compared to the results is shown in Figures 11 and 12.

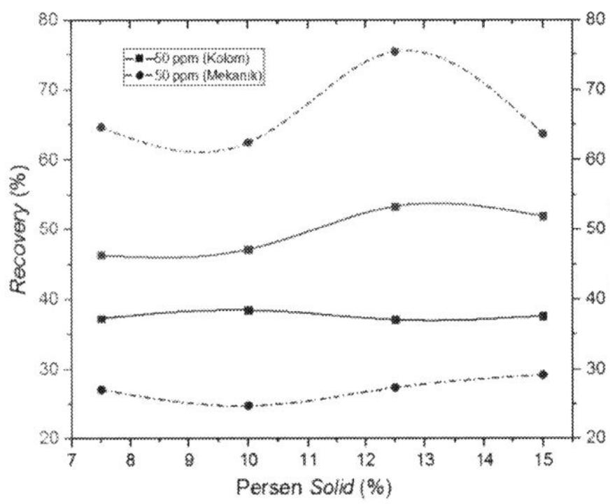

Figure 11. Graph of comparison column flotation with mechanical flotation with% solid effect on recovery and grade.

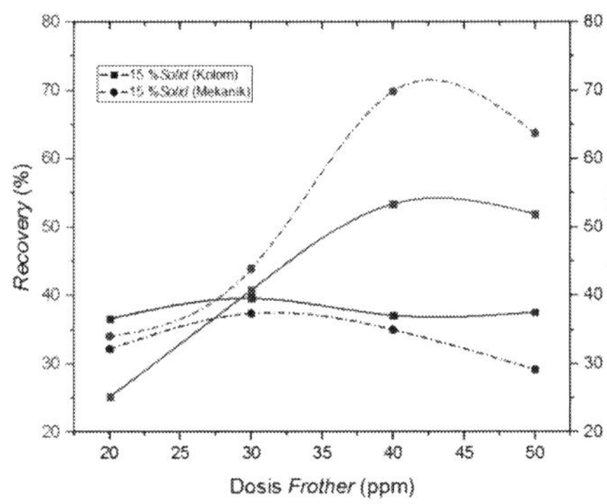

Figure 12. Graph of comparison column flotation with mechanical flotation with the effect of frother dose on recovery and grade.

Based on Figures 11 and 12 it can be concluded that the acquisition obtained is higher mechanical flotation than column flotation. Meanwhile, the resulting grade is superior to column flotation. It can be seen in the graph that the tendency obtained is in accordance with the previous research literature namely research (16) that the% grade has a tendency that is inversely proportional to the% recovery in the effect of the percent solid added. It can be concluded that column flotation is still not in accordance with the function of the rougher circuit which seeks the highest grade because it is still not better to mechanical flotation. The cause of column flotation cells is still not getting high recovery, the main cause is indeed the flotation cell mechanism, namely the column flotation system is more selective for lifting valuable minerals which makes the mass of concentrate is not too high and the mechanism depends on the thickness of the froth zone and the performance of wash water which cause %grade on column flotation is higher (17). Mechanical flotation can get high recovery, because mechanical flotation cells are more suitable at a high ratio of %solid due to good particle dispersion and perfect mixing due to conditions created by agitators (18). In column flotation it does not have agitator which makes it less superior in high %solid conditions less superior than mechanical flotation, but in low %solid with 7.5 and 10 %solid column flotation some recovery values are superior compared to mechanical flotation.

The cause of mechanical flotation has a high %recovery due to the influence of the zones formed and the cell mechanism that supports it to obtain high grades and is still effective at high viscosity of slurry. The turbulent zone in mechanical flotation with its agitator system causes a large amount of valuable particle mass to be lifted thus increasing %recovery because it can circulate high %solid under suspension but not too selective which causes many impurities to be carried to the concentrate (18). In contrast to column flotation where fluid flow is very influential on the increase in recovery, along with increasing flow rate to certain conditions a flow transition phenomenon from laminar (bubbly flow regime) to turbulent (churm-turbulent regime) (3), so that at certain conditions the flow rate value needs to be maintained so that the separation process occurs optimally. The increase in recovery along with the laminar flow can be explained according to the statement C.M. Goodall that in laminar conditions the bubble size is relatively smaller. The small bubble size causes the possibility of particles to fall (detachment) to be lower. This is because bubbles become more stable to be able to rise to the surface carrying particles so that recovery increases. Unlike the case when turbulent flow, it will cause bubbles to become unstable. This bubble instability results in the possibility of a bubble coalescence phenomenon and increasing bubble size, so that many particles undergo detachment and recovery decrease (19). Thus, the churn-turbulent flow regime is undesirable in processes that target high recovery. But in the process of column flotation with a flow rate of 2.5 L/min, the gas holdup formed still shows the churn-turbulent flow regime even though it is not too large. Maybe this is also one of the causes of column flotation recovery not being able to get the highest results(20).

4. Conclusion

Based on the comparative research of the metallurgical flotation performance that has been done, it can be concluded that for column flotation, solid percent affects the recovery and grade. As the solid percent value is added, recovery increases to the maximum point. After reaching the maximum point, adding value no longer has a significant effect. The maximum condition occurs in different percent solids, namely 10 and 12.5%. However, when the percentage of solid is too large at 15%, the addition of value no longer has a significant effect on condition 20; 30; and 50 ppm. But in the condition of a 40 ppm frother dose, the maximum point cannot be found. Meanwhile, the grade tendency will decrease after passing the 10% point of percent solid. For variable dose of the frother also affects the recovery value and grade, with the same tendency as the percent solid. The maximum condition occurs at frother doses of 30 and 40 ppm. When the frother dose is 50 ppm it will reduce the recovery value because it no longer has a significant effect. At the solid percent conditions of 7.5 and 12.5% the maximum point has not been found. The recommended variable to use based on this experiment is 15% solid under 40 ppm frother dose conditions because it gets the highest recovery value of 53.439%.

For mechanical flotation, solid percent affects the recovery and grade, which is same as the percent solid on column flotation. The maximum condition occurs at 12.5% solid, both at condition 20; 30; and 50 ppm. But in the condition of a 40 ppm frother dose, the maximum point cannot be found. Meanwhile, the grade tendency decreases if the solid persent is too high, such as 12.5 and 15% solid. The recommended variable to use based on this experiment with the condition of a 40 ppm frother dose and 12.5% solid has the highest recovery value is 75.462%. Then, for the dose of the frother also affects the recovery value and grade. The effect is the same as solid percent. The maximum condition occurs at 40 ppm frother dose after that will reduce the recovery value in several percent solid conditions, but in the 12.5% solid condition the maximum point cannot be found. Meanwhile, the grade tendency decreases at high frother doses such as 50 ppm.

The metallurgical performance comparison of the two flotation cells is divided into recovery and grade, for the recovery values obtained from the graph are superior to mechanical flotation. Whereas, the grade is superior to column flotation, which shows that the rougher circuit which functions to find recovery as high as possible is still mechanical flotation, which indicates less effective column flotation if applied to the rougher circuit. However, if in the search for high grades of the function of the cleaner circuit, column flotation is more effective if used on that circuit.

Acknowledgement
This research is granted by Ministry of Research and Technology of Indonesia. Special gratitude also aprreciated to Advanced Materials and Tomography Laboratory of Sultan Ageng Tirtayasa University and Puslitbang TekMIRA which have provided the necessities during the research process.

References

1. Wills BA, Finch JA Chapter 12 - Froth Flotation In: Wills BA, Finch JA, editors Wills' Mineral Processing Technology (Eighth Edition) Eighth Edi Boston: Butterworth-Heinemann; 2016 p. 265–380.
2. Yahyaei M, Banisi S, and Javani H Replacing mechanical flotation cells by a flotation column at the pilot plant of the Sarcheshmeh copper mine. *Sep Sci Technol* **41** 3609–3617.
3. Finch JA, Dobby GS Column flotation 1990
4. Hurdeyal R A Comparison between Column and Mechanical Cell Performance in Platinum Flotation In fulfilment of the degree Masters in Engineering 2012
5. Smith RJB. Experimental Studies on Column Flotation Cell. Int J ChemTech Res. 2012;4(3):1198–202.
6. Fuerstenau MC, Han KN. Principies of Mineral Processing. Society for Mining, Metallurgy, and Exploration, Inc. 2003. 1-586 p.
7. Jena MS, Biswal SK, Das SP, Reddy PSR. Comparative study of the performance of conventional and column flotation when treating coking coal fines. Fuel Process Technol [Internet]. 2008;89(12):1409–15. Available from: http://dx.doi.org/10.1016/j.fuproc.2008.06.012
8. Wang LK, Shammas NK, Selke WA, Aulenbach DB Flotation Technology Humana Press 2011 (Handbook of Environmental Engineering)
9. Roy K-M Xanthates In: Ullmann's Encyclopedia of Industrial Chemistry Weinheim Germany: Wiley-VCH Verlag GmbH & Co. KGaA; 2000.
10. Finch JA, Nesset JE, Acuña C. Role of frother on bubble production and behaviour in flotation. 2008;21:949–57.
11. Luo X, Feng B, Wong C, Miao J, Ma B, Zhou H. The critical importance of pulp concentration on the flotation of galena from a low grade lead – zinc ore. Integr Med Res [Internet]. 2015;5(2):131–5. Available from: http://dx.doi.org/10.1016/j.jmrt.2015.10.002
12. Uribe-Salas A, de Lira-Gómez P, Pérez-Garibay R, Nava-Alonso F, Magallanes-Hernández L, Lara-Valenzuela C. Overloading of gas bubbles in column flotation of coarse particles and effect upon recovery. Int J Miner Process. 2003 Sep;71(1–4):167–78.
13. Amin R, Abdelkhalek MA, Ibrahim SS, Yehia A. Statistical Design of Column Flotation Parameters for Beneficiation of Egyptian Oil Shale Statistical Design of Column Flotation Parameters for Beneficiation of Egyptian Oil Shale. 2018;
14. Wei T, Peng Y, Vink S. SC reagents in stabilizing froth in coal flotation. Int J Miner Process [Internet]. 2016; Available from: http://dx.doi.org/10.1016/j.minpro.2016.01.005
15. Ore K. Kaolin Ore. 2017;1–13.
16. Stephani Y. Monitoring proses flotasi kolom pada bijih sphalerite menggunakan kamera dengan variasi persen solid dan laju alir udara terhadap recovery. 2018;
17. Massinaei M, Kolahdoozan M, Noaparast M, Oliazadeh M, Yianatos J, Shamsadini R, et al. Froth zone characterization of an industrial flotation column in rougher circuit. Miner Eng [Internet]. 2009;22(3):272–8. Available from: http://dx.doi.org/10.1016/j.mineng.2008.08.003
18. Al-Fariss TF, El-Nagdy KA, Abd El-Aleem FA, El-Midany AA. Column versus mechanical flotation for calcareous phosphate fines upgrading. Part Sci Technol. 2013 Sep;31(5):488–93.
19. Goodall CM, O'Connor CT Residence time distribution studies in a flotation column. Part 2: the relationship between solids residence time distribution and metallurgical performance. Int J Miner Process. 1992 Oct;36(3–4):219–28.
20. Yan X, Shi R, Xu Y, Wang A, Liu Y, Wang L, et al Bubble behaviors in a lab-scale cyclonic-static micro-bubble fl otation column 2016

IOP Conf. Series: Materials Science and Engineering 909 (2020) 012011 doi:10.1088/1757-899X/909/1/012011

Low cost prototype simulation of spectrum analyzer base on GNU radio and RTL-SDR

B.B Harianto*, Ade Irfansyah, Yuyun Suprapto

Aviation Polytechnic of Surabaya, Indonesia

*Email: bambangfarzardy@gmail.com

Abstract— this paper describes the prototype spectrum analyzer simulation using Software Defined Radio technology. The use of this technology continues to grow very rapidly, this is what drives researchers to develop this system. SDR is a radio communication system where the modulation system contained in the hardware is replaced by implementing software on a computer or embedded system. The SDR system commonly used is GNU Radio. The use of SDR is very cheap (around $ 20), and is small in size so that it offers cost effectiveness and portability requirements added by GNU radio software that is open source. In this experiment, there were several parameters measured including power and signal to noise ratio (SNR) measured using GNU Radio and compared with measurements using the Spectrum Analyzer GSP-730 which was applied to prototype simulations using 60 MHz to 80 MHz frequency inputs from generator frequency. From the experimental results show the average GNU radio power is -15.28 dBm and the average spectrum analyzer power is -15.2 dBm, so the accuracy level of the power measurement on GNU Radio reaches 99.47%, while the average SNR spectrum analyzer is 54 , 2 and GNU radio SNR 54.32 so the accuracy of SNR measurement in GNU Radio reaches 99.78%.

1. Introduction

Spectrum analyzer is a multipurpose measuring instrument in measuring signals in the frequency domain. An analyzer that can be used to measure signals on a transmitter that requires measurement parameters such as frequency, power, gain, and noise[1]. Spectrum analyzer are also used to investigate the distribution of energy along the frequency spectrum of known electrical signals, from this investigation obtained very valuable information about the width of the frequency field (bandwidth), the effect of various types of modulation, the generation of false signals and so on all its benefits in RF frequency (radio frequency) planning and testing[2]. Spectrum analyzer have a special function to measure a number of signals in a limited frequency spectrum [3]. The many uses of spectrum analyzer do not be surprised if the price of spectrum analyzers is not cheap. Whereas GNU Radio is software that can be used free of charge, and is used to create a Software Defined Radio (SDR), without the need for hardware[4]. GNU Radio uses the C ++ or Python programming language in its implementation. The advantage of GNU Radio is open source software, and it doesn't cost much in terms of its operations. Researchers aim to measure the power of a frequency using GNU Radio then compare it to a spectrum analyzer[5].

Every individual or institution that uses on a frequency wave must use a spectrum analyzer and it will cost more to buy a spectrum analyzer. This prototype wants to simulate a spectrum analyzer using SDR and GNU Radio so as to reduce prices in the use of spectrum analyzers. The simulation carried out is measuring the power & Signal to noise ratio at frequencies 80, 75, 70, 65, 60 MHz, the difference in power units between the spectrum analyzer (dBm) and GNURadio (dB) requires that this simulation convert units into the same units in this case the simulation is done on dBm units, to change the dB units on GNU Radio using the principle of 30 dBm = 1000 mW = 1W[6] so it can be concluded to convert dB to dBm by subtracting 30 dB. GNU Radio is not only useful in research but also useful in the world of education, according to the research [7] it has been suggested that alternative methods for developing practical sessions on the Theory of Communication Theory have been widely presented. Many universities apply learning using an equation-based environment such as Matlab, or only focus on the signal processing capabilities offered by DSP but the incorporation of hardware-software systems in GNU Radio and HackRF allows learning to be more effective. SDR is the best solution for next generation technology, SDR is one of the solutions for researchers to test techniques in real time, SDR proves technological advancements especially in telecommunications and electronics[8].

2. Technical Description

This section will explain technically the concept material that will be used such as: Signal to Noise Ratio and Fourier analysis.

2.1. Signal to noise ratio

In several studies to find the direction of multiposition of radio sources with unknown parameters, continuous wave measurements can be determined by knowing the signal-to-noise ratio (SNR) at each receiver channel output. The signal to noise ratio is defined as the ratio between the desired signal power and noise power, and is widely used as a standard measure of signal quality for communication systems[9]. An information signal as a communication medium will experience a lot of interference by noise, so that it can damage the information signal. Signals that experience this interference experience a decrease in quality. The quality of this signal can be determined from the value of the Signal to Noise Ratio (SNR) measured in decibels (dB)[10]. Calculation of signal to noise ratio can be through a reduction between the noise value and the power of a frequency[11]. The steps to calculate SNR are as follows:

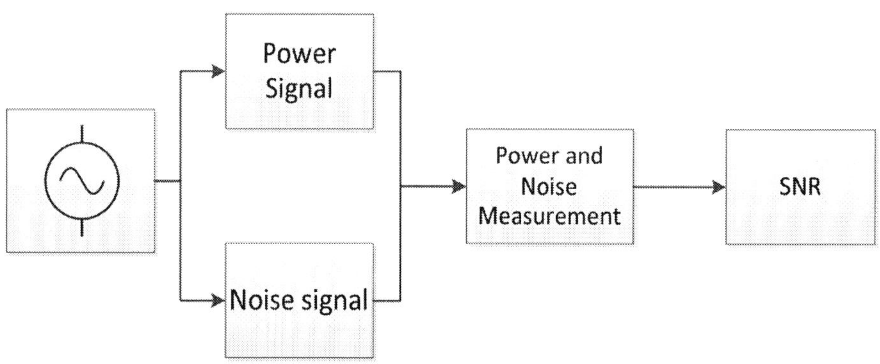

Figure 1. Calculate SNR in GNU Radio and spectrum analyzer

In figure 1 Block Counter SNR value of the signal obtained by calculating the Root Mean Square (RMS) value of each original signal and the noise signal. This RMS value is the signal power value. Thus, the original signal RMS value (RMS1) and the noise signal RMS value (RMS2) are obtained. To get the SNR value of the signal is by dividing RMS1 with RMS2 then made in decibels (dB).

2.2. Fourier analysis
Fourier analysis is a mathematical way to determine the frequency and amplitude of harmonics. This process is used to solve the problem of complex waveforms which decompose the waves into sinusoidal components [12]. Each complex wave consists of a number of frequencies, at various amplitudes and phase angles. Analyzing the effects (eg gain / attenuation and phase angle) of electronic circuits on such functions can be simplified if each component can be considered separately. This periodic function analysis method is called Fourier analysis which was developed by French mathematicians. In figure 2 below shows the polar representation (e.g., Phase) of a single frequency signal, amplitude R and angular velocity[13].

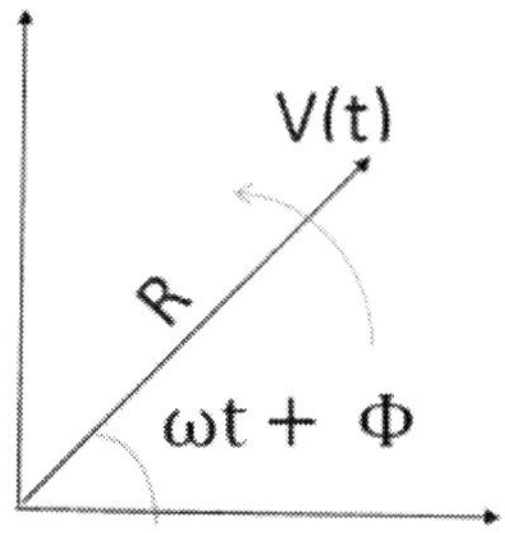

Figure 2. Polar representation of a single frequency signal[13]

Fourier analysis can also be implemented in equation 1 [13]:

$$v(t) = RE^{(\omega t+\Phi)} = R\sin(\omega t + \Phi) \quad (1)$$

If there are other signals added to other frequencies and phases so the sum of the two signals is implemented by equation 2 [13]:

$$f(t) = v_1 + v_2 = R\sin(\omega t + \Phi_1) + R\sin(\omega t + \Phi_2) \quad (2)$$

The addition of more frequencies will produce more complex wave forms. The expansion shown in equation 3 below is called the Fourier Series, where 'a and' b are called the Fourier Coefficients and are usually written in the format[13]:

$$f(t) = a_0 + \sum_{n=1}^{\infty} a_n\cos(n\omega t) + \sum_{n=1}^{\infty} b_n\sin(n\omega t) \quad (3)$$

Fast Fourier Transform (FFT) uses the repeated properties of Discrete Fourier Transform (DFT) and reduces the number of additions to produce the same results by evaluating DFT directly; the most significant difference is that FFT is a more efficient method for achieving the same results. Whereas IFFT (Inverse Fast Fourier Transform) is a method of solving discrete signals that works the opposite of FFT. IFFT is a fast computational algorithm for calculating IDFT (Inverse Discrete Fourier Transform). The subcarrier in IFFT has a frequency integer multiple of its base frequency as well as the Fourier series component in the composite signal. One of the uses of IFFT is the OFDM (Orthogonal Frequency Division Multiplexing) system[14].

3. Hardware and Software for Simulation
Hardware & Software for simulation requires some hardware such as a waveform generator, SDR, Spectrum Analyzer and laptops with the specifications which will be explained below. While the software used is GNU Radio to process the incoming signal. The simulation will be performed as in Figure 3 below.

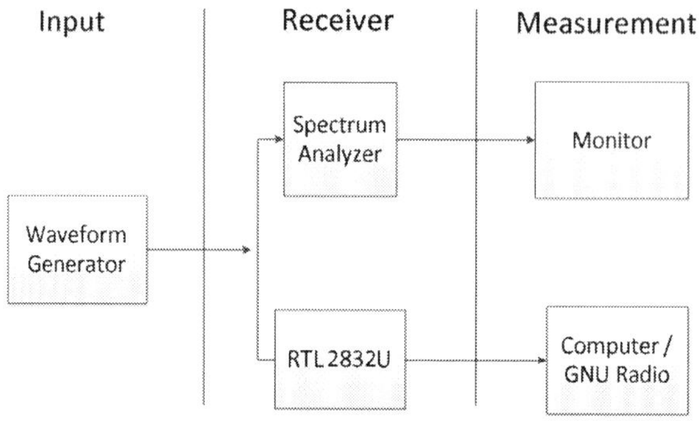

Figure 3. Block diagram of simulation

3.1. Spectrum analyzer (hardware)
In this study the spectrum analyzer used is the spectrum of the GW-Instek gsp-730, measurements made in the frequency range of 60-80 MHz. Specifications on the spectrum analyzer used are, frequency range 150 kHz-3GHz, Noise floor 100 -100dBm, RBW (30kHz, 100kHz, 300kHz, 1 MHz) and support interfaces can support USB Device / Host, RS-232C[15]. Spectrum analyzer is used to compare signal results with GNU Radio from audio waveform. The spectrum analyzer used is shown in Figure 4.

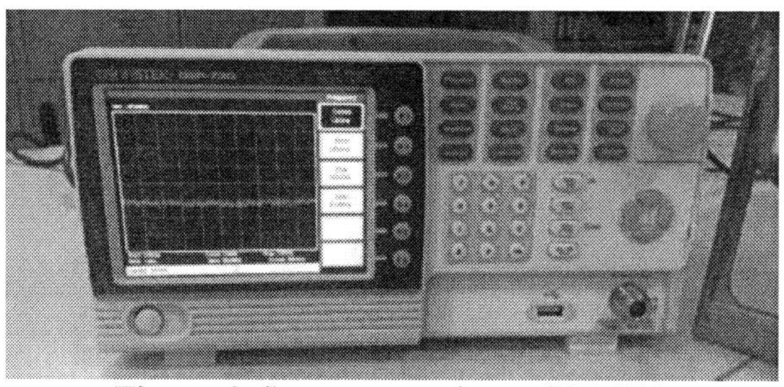

Figure 4. Spectrum analyzer GSP-730

3.2. Waveform generator (hardware)
Waveform generator functions to make waves with a certain frequency, the waveform generator used is owned by Agilent 33250A with specifications, 80 MHz sine, square, ramp, triangle, noise and DC, 12bit, 200 Msa / s, 64 K point arbitrary waveform, GPIB & RS -232 Interfaces, BenchVue Enabled[16]. The waveform generator can be seen in Figure 5.

Figure 5. Waveform generator

3.3. RTL-SDR

In this measurement the link between the waveform generator and the computer / GNU Radio uses the RTL2832U device with frequency specifications 24 - 1766 MHz bandwidth 3.2 MHz ADC Resolution 8 bits [17]. RTL devices have block diagrams like Figure 6.

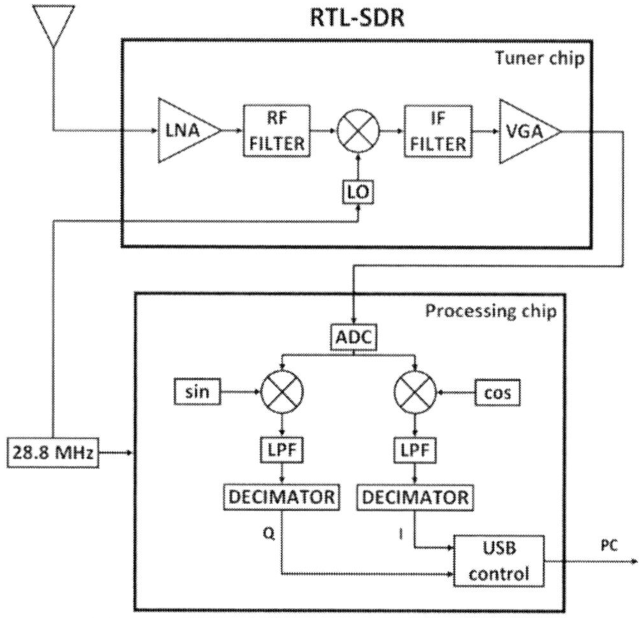

Figure 6. Block diagram of RTL-SDR[18]

3.4. GNU radio (software)

Simulation using GNU Radio tiper software 3.7.11.1 in this type already has various blocks such as source, sinks, and graphical sinks to misc. This simulation uses several blocks, namely RTL-SDR Source, Multiply Cons, and WX GUI FFT Sink. With the functions as follows in table 1.

Table 1. Function GNU Radio block

Block	Function
RTL-SDR Source	Link RTL-SDR Hardware with GNU Radio Software
Multiply Cons	Multiplying the input data with the desired number variable
WX GUI FFT Sink	Displays a signal graph with Power and Frequency parameters

4. Simulation Spectrum Analyzer

Spectrum analyzer simulation measures 2 parameters namely power and noise. Noise measurement is used to determine the value of the Signal to Noise Ratio (SNR).

4.1 Power simulation with spectrum analyzer

The first simulation uses the Spectrum Analyzer as a comparison of GNU Radio with the frequency described. Figure 7a-7e is a simulation result for measuring power in the spectrum analyzer at various frequencies.

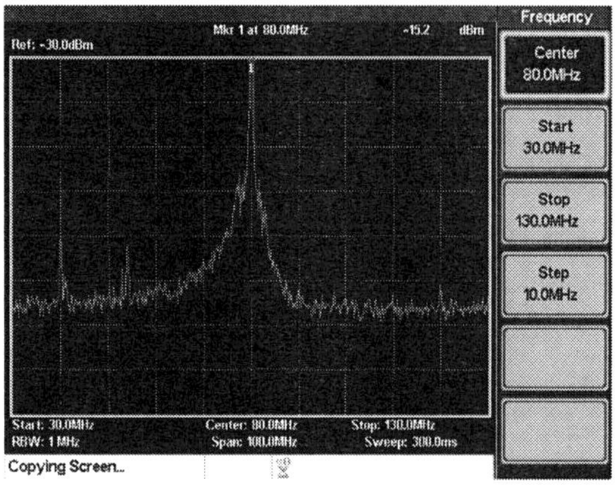

7a.

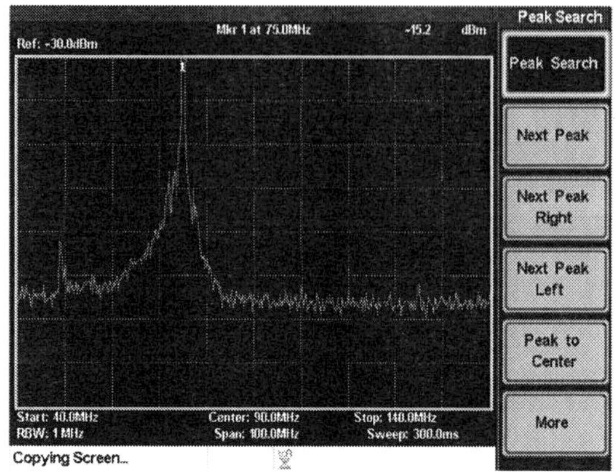

7b.

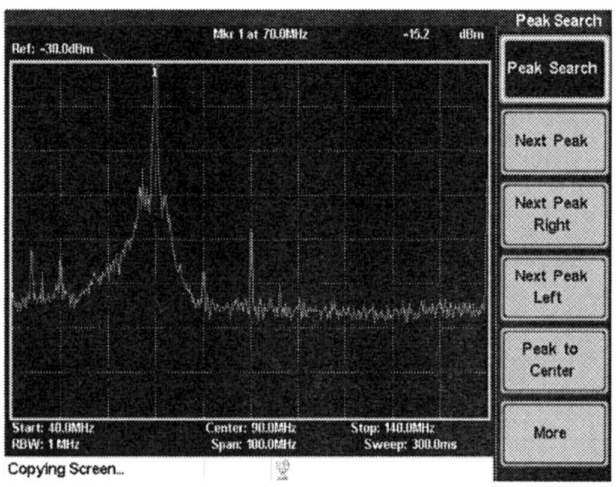

7c.

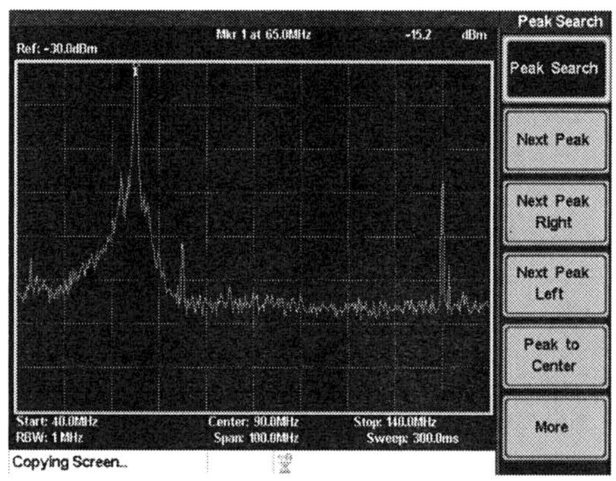

7d.

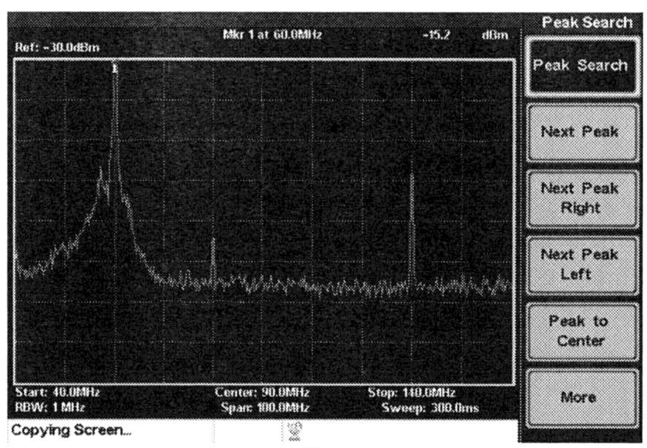

7e.

Figure 7. Power measurement on spectrum analyzer monitor

In figure 7a measurement of power value with a frequency of 80MHz, 7b measurement of power value with frequency of 75 MHz, 7c measurement of power value with frequency of 70 MHz, 7d measurement of power value with frequency of 65 MHz and 7e measurement of power value with frequency of 60 MHz.

IOP Conf. Series: Materials Science and Engineering **909** (2020) 012011 doi:10.1088/1757-899X/909/1/012011

4.2 Noise simulation with spectrum analyzer

The second simulation is measuring noise in a spectrum analyzer at the same frequency as power measurements. In figure 8a-8e is the result of simulation for noise measurement to determine the signal to noise ratio.

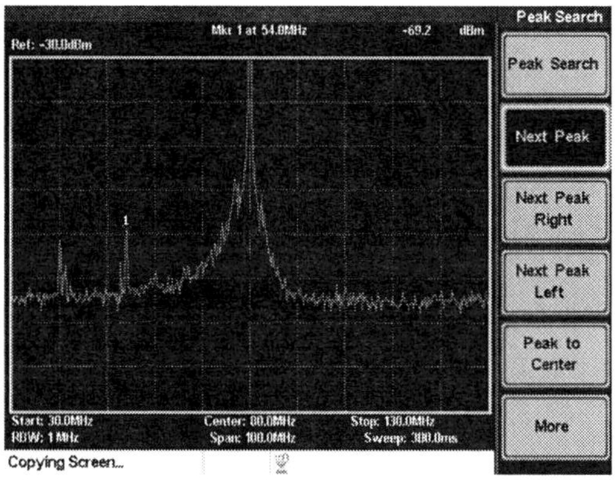

8a.

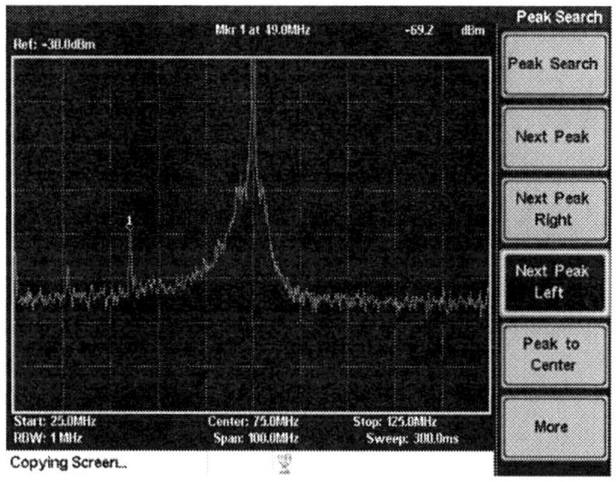

8b.

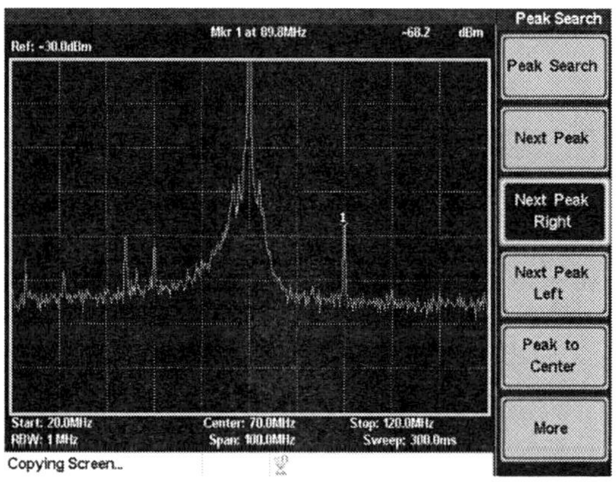

8c.

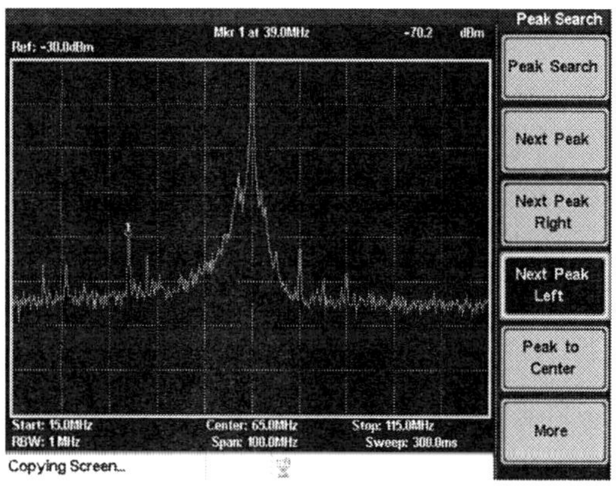

8d.

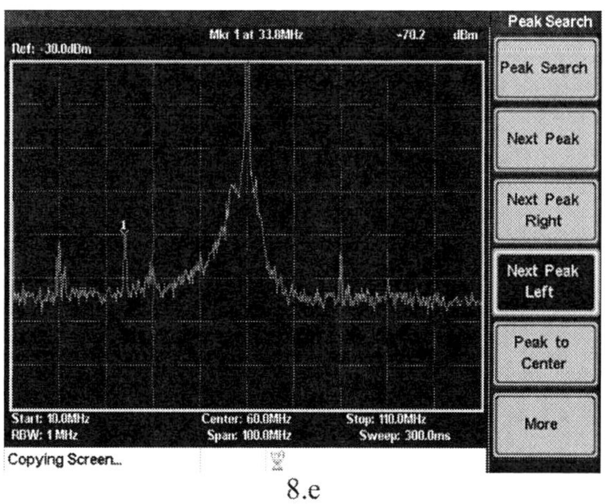

8.e

Figure 8. Noise measurement on spectrum analyzer monitor

In figure 8a the measurement of noise values with a frequency of 80 MHz, 8b is the measurement of noise values with a frequency of 75 MHz, the measurement of noise 8c with a frequency of 70 MHz, the measurement of noise 8d with a frequency of 65 MHz and the measurement of noise 8d with a frequency of 60 MHz.

5. Simulation GNU Radio

The third simulation uses the GNU Radio as a comparison of the Spectrum Analyzer with the frequency that has been explained, before carrying out the simulation, it needs several blocks that must be in GNU Radio, for more details, shown in Figure 9.

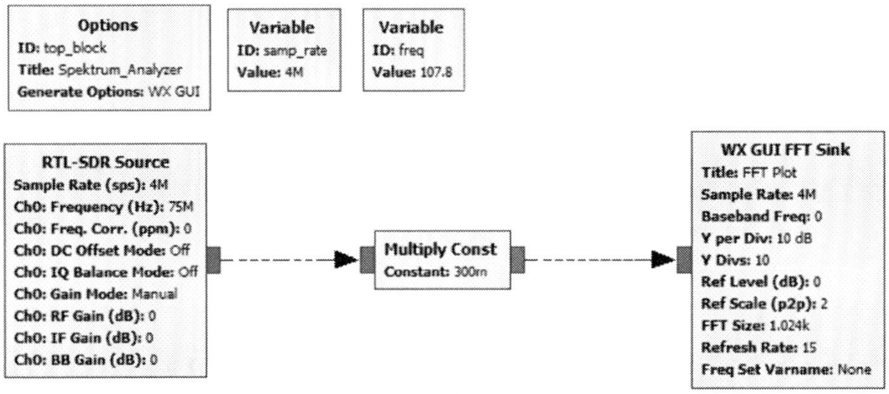

Figure 9. RTL-SDR blok for spectrum analyzer

5.1 Power simulation with GNU radio

The simulation is e using RTL-SDR which is connected to the waveform generator using a BNC connector to see the power value in GNU Radio. Figure 10a-10e is the result of experiments on GNU Radio on various frequencies.

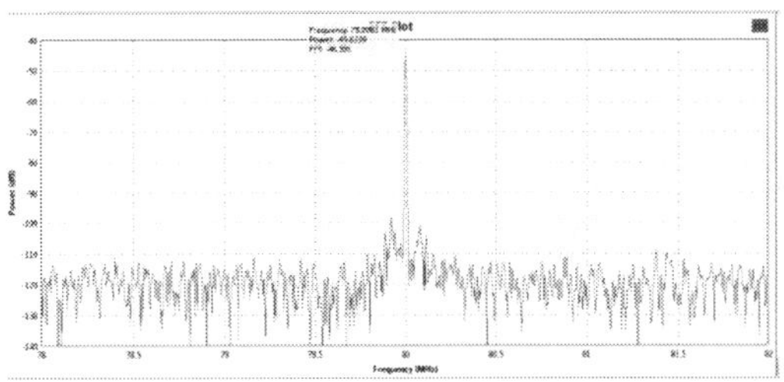

10a.

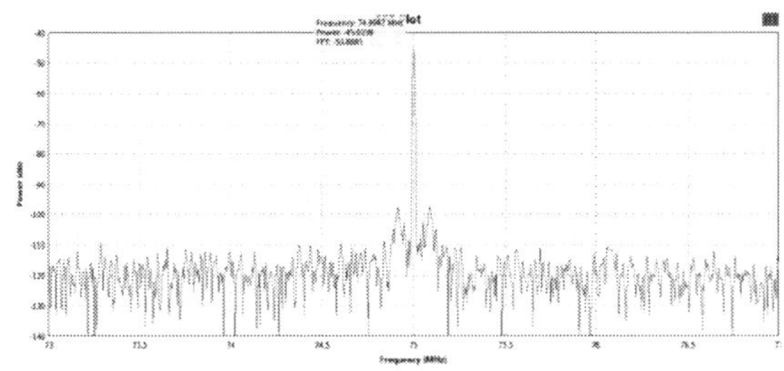

10b.

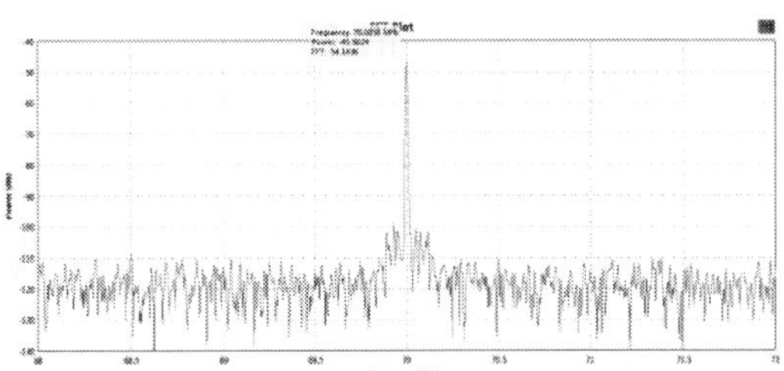

10c.

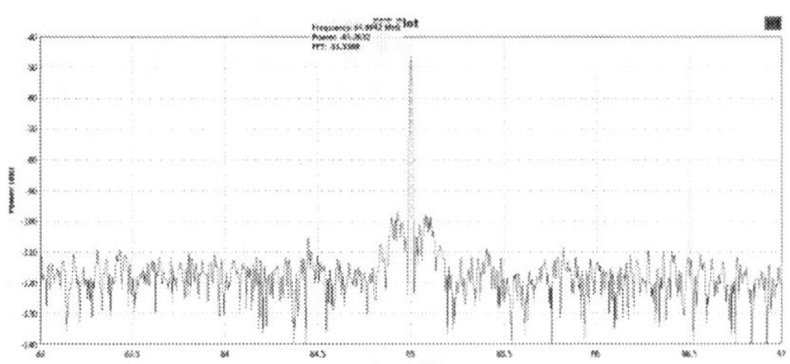

10d.

IOP Conf. Series: Materials Science and Engineering **909** (2020) 012011 doi:10.1088/1757-899X/909/1/012011

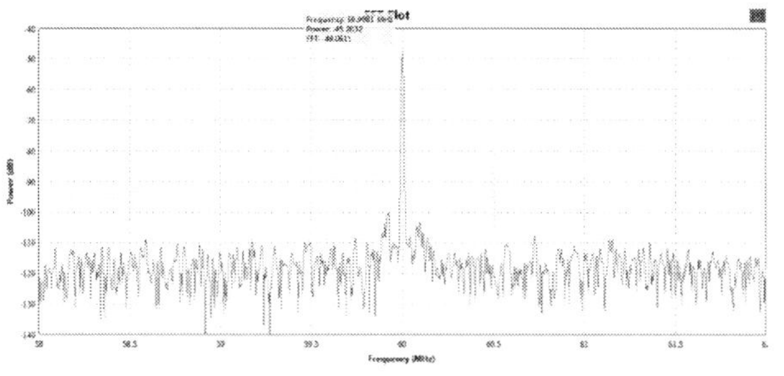

10e

Figure 10. Power measurement on GNU Radio

In figure 10a shows a measurement with a frequency of 80 MHz, 10b shows a measurement of 75 MHz, 10c shows a measurement with a frequency of 70 MHz, 10d shows a measurement with a frequency of 65 MHz, 10e shows a measurement with a frequency of 60 MHz

5.2 Noise simulation with GNU radio

The fourth simulation is measuring noise in the spectrum analyzer at the same frequency as the power measurement. Figure 11a-11e is a simulation result for noise measurement to determine the signal to noise ratio.

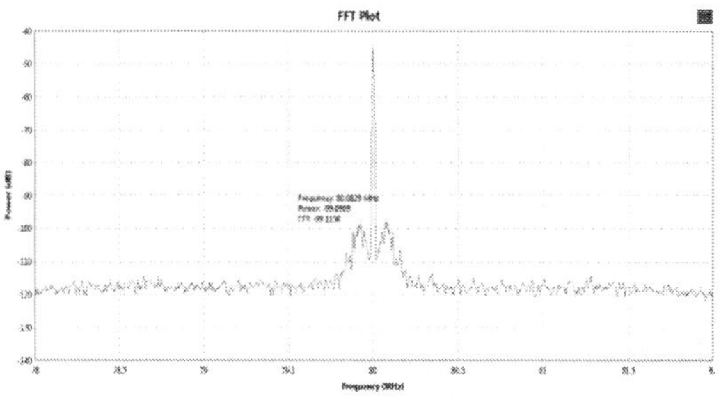

11a.

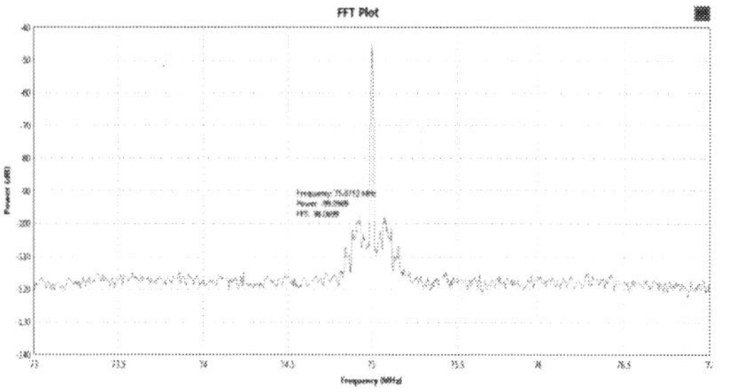

11b.

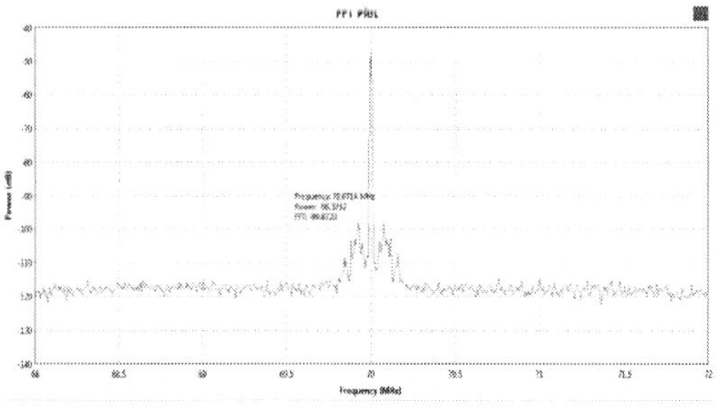

11c.

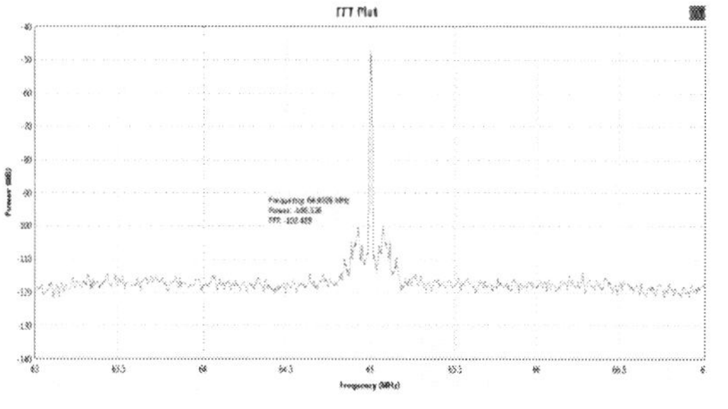

11d.

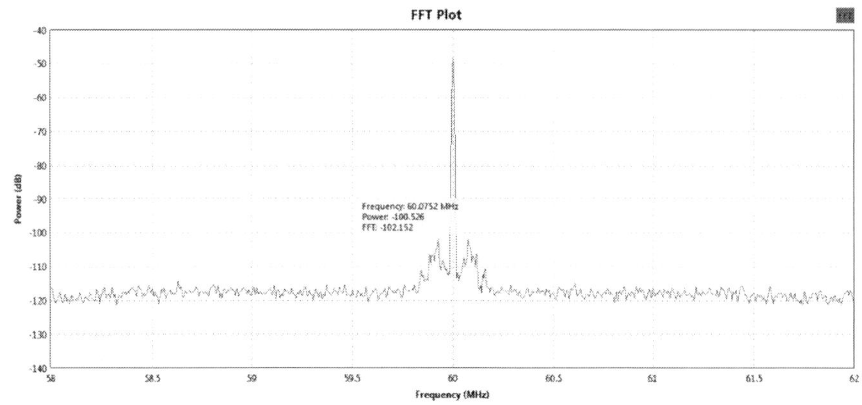

11.e

Figure 11. Noise measurement on GNU Radio

Figure 11 shows the measurement of noise on GNU Radio with, 11a using 80 MHz frequency, 11b using 75 MHz frequency, 11c using 70 MHz frequency, 11d using 65 MHz frequency, 11e using 60 MHz frequency.

6. Analysis of Simulation for GNU Radio & Spectrum Analyzer

The value that has been collected from the above simulation will be displayed will be analyzed namely the difference in power, signal to noise ratio in the spectrum analyzer and GNU Radio.

6.1 Power measurement analysis

The power measurements are shown in table 3 below, because the power units in the spectrum analyzer and GNU Radio are different so they must be converted to the same unit, in this case the unit will be equated to dBm.

Table 2. Power measurement GNU Radio and spectrum analyzer

Frequency (MHz)	GNU Radio (dB)	GNU Radio (dBm)	Spectrum Analyzer (dBm)
80	-45.0	-15.0	-15.2
75	-45.0	-15.0	-15.2
70	-45.5	-15.5	-15.2
65	-45.2	-15.2	-15.2
60	-45.2	-15.2	-15.2

The validity of the spectrum analyzer and GNU Radio will be taken with an average of the five values in table 2 above. The first is on GNU Radio media $\overline{X}=\frac{(-15.0)+(-15.0)+(-15.5)+(-15.2)+(-15.2)}{5}$ so that the GNU Radio has an average value of -15.28 for the second one using the spectrum analyzer media $\overline{Y}=\frac{(-15.2)+(-15.2)+(-15.2)+(-15.2)+(-15.2)}{5}$ so the spectrum analyzer has an average of -15.2. The average deviation *(D)* is taken from $\frac{|(\overline{X}-\overline{Y})|}{\overline{Y}} \times 100\%$ the equation is:

$$D = \frac{|-15.28 - (-15.2)|}{15.2}\ x\ 100\%$$

$$D = 0.5\%$$

6.2 Signal to noise ratio measurement analysis

Noise measurements in the GNU Radio are shown in table 3, after the noise exchange is carried out, the signal to noise ratio (SNR) value can be obtained.

Table 3. SNR measurement with GNU Radio

Frequency (MHz)	*Power* (dB)	*Noise* (dB)	*Signal to Noise Ratio*
80	-45.1	-99.2	54.1
75	-45.0	-99.0	54.0
70	-45.4	-98.4	53.0
65	-45.5	-100.4	55.4
60	-45.5	-100.6	55.1

Noise measurements in the spectrum analyzer are shown in table 4 below. Reduction between power and noise will get the value of the signal to noise ratio (SNR).

Table 4. SNR Measurement with spectrum analyzer

Frequency (MHz)	*Power* (dBm)	*Noise* (dBm)	*Signal to Noise Ratio*
80	-15.2	-69.2	54
75	-15.2	-69.2	54
70	-15.2	-68.2	53
65	-15.2	-70.2	55
60	-15.2	-70.2	55

The validity of the spectrum analyzer and GNU Radio will be taken with an average of the five values in table 2 above. The first is on GNU Radio media $\overline{X}=\frac{(54.1)+(54.0)+(53.0)+(55.4)+(55.1)}{5}$ so that the GNU Radio has an average value of 54.32, for the second one uses the spectrum analyzer media $\overline{Y}=\frac{(54)+(54)+(53)+(55)+(55)}{5}$ so the spectrum analyzer has an average of 54.2. The average deviation *(D)* is taken from $\frac{|(\overline{X}-\overline{Y})|}{\overline{Y}} \times 100\%$ the equation is:

$$D = \frac{|54.32 - 54.2|}{54.2} 100\%$$

$$D = 0.22\%$$

7. Conclusion

This paper proposes measurement of power and signal to noise ratio using RTL-SDR and GNU Radio as a substitute for the spectrum analyzer function because costs are more affordable and RTL-SDR is easily found on the market while GNU Radio can be downloaded on the GNU Radio website for free..

References

[1] Adnani A Al, Duplicy J and Philips L 2013 Spectrum analyzers today and tomorrow: Part 2 0–4

[2] Lu T and Chen F 2012 Multiwfn: A multifunctional wavefunction analyzer *J. Comput. Chem.*

[3] Jose J, Edison R and S S S 2017 Low cost android radio spectrum analyzer using RTL-SDR dongle 98–101

[4] Vachhani K 2015 Experimental study on wideband FM receiver using GNURadio and RTL-SDR 1810–4

[5] Fuxjäger P, Costantini a, Valerio D, Castiglione P, Zacheo G, Zemen T and Ricciato F 2007 11p transmission using GNU radio Spectrum *IEEE* 802

[6] Peluso V, Vancorenland P, Marques A M, Steyaert M S J and Sansen W 1998 A 900-mV low-power ΔΣ A/D converter with 77-dB dynamic range *IEEE J. Solid-State Circuits* **33** 1887–96

[7] Del Barrio A A, Manzano J P, Maroto V M, Villarín Á, Pagán J, Zapater M, Ayala J and Hermida R 2019 HackRF + GNU Radio: A software-defined radio to teach communication

theory *Int. J. Electr. Eng. Educ.* 1–18

[8] Sowjanya P-_uts. pd. and Satyanarayana P 2019 Real-time data transfer based on software defined radio technique using GNU radio/usrp *Int. J. Eng. Adv. Technol.* **9** 279–88

[9] Romano G 2019 Blind signal-to-noise ratio estimation of real sinusoid in additive noise *2019 42nd Int. Conf. Telecommun. Signal Process.* **1** 224–7

[10] Welvaert M and Rosseel Y 2014 On the definition of signal-to-noise ratio and contrast-to-noise ratio for fmri data on the definition of signal-to-noise ratio and contrast-to-noise ratio for fmri data

[11] Finder M D, Kalashnikov K S, Bugaev Y N and Ivanov V A 2018 The measurement of the signal-to-noise ratio in the reception channels of a the measurement of the signal-to-noise ratio in the reception channels

[12] Soares L R, Member S, Oliveira H M De and Cintra R J S 2006 Analysis using fourier-like wavelets **00**

[13] Dongle S 2015 Low cost sdr spectrum analyzer and analog radio receiver using GNU Radio , Raspberry Pi2

[14] Jang I and Jo G 2017 Low latency IFFT design for OFDM systems supporting full-duplex FDD 642–6

[15] GW-Instek Company Spectrum Analyzer GSP-730 User Manual

[16] Aglient Company Waveform Generator 33250A User Manual

[17] Santiago Rivera I, Vidal Beltran S and Martinez-Pinon F 2018 Spectrum analyzer by software defined radio *Proc. - 2018 Int. Conf. Mechatronics, Electron. Automot. Eng. ICMEAE 2018* 93–7

[18] Bnilam N, Joosens D, Steckel J and Weyn M 2019 Low cost AoA unit for IoT applications *13th Eur. Conf. Antennas Propagation, EuCAP 2019* 3–7

IOP Conf. Series: Materials Science and Engineering **909** (2020) 012012 doi:10.1088/1757-899X/909/1/012012

Battery energy storage system as the optimal solution for underdeveloped, remote, outermost region: Pasi island study case

Donny Yoesgiantoro[1,2*]

[1] Defence Management Faculty, Indonesia Defence University, Bogor 16810, Indonesia
[2] Energy Security Center for Excellence, Menara Sentraya A3, Jakarta 12160, Indonesia

*Email: energyprogram@gmail.com

Abstract. Underdeveloped, remote and outermost (3T) regions are areas that are becoming the priority of development by the Government of the Indonesian Republic in strengthening national security. However, the constraints to development in the 3T region are immense, ranging from great distances, isolated areas to its geographical location in the mountainous area of even in small islands. Pasi Island is home to three villages that fall into the 3T area category. Battery ESS is an energy storage system in the form of giant batteries with many advantages, such as inexpensive, mobile, has a large capacity, long life cycle, easy to operate and zero waste. Seeing the advantages offered by the battery ESS so that it has the potential to be an optimal solution to meet the electrification ratio in the 3T region. Cost-benefit analysis is chosen because the results can provide a clearer picture of the costs and benefits going forward. So that the calculation of the resulting cost-benefit is more optimal then two calculation scenarios are used using the basic difference in electricity sales rates. In the first scenario, it uses BPP tariff for the Selayar Regency area of Rp 2,445, and in the second scenario, it uses the BPP tariff of Rp 3,041 for small subsystem areas. The calculation results indicated that the second scenario provided immense benefits with a short payback time. The NPV value generated by scenario II is 116.5 billion rupiah with an IRR percentage of 100.33% and a cost-benefit ratio of 9.332.

1. Introduction

Underdeveloped, Remote and Outermost Regions (3T) are regions that have a low level in the human development index (HDI); availability of facilities and infrastructure; economic growth rate; and other socio-economic indicators [1] compared to other regional averages on a national basis. From a macroeconomic perspective, to be able to develop 3T regions so that they can be equal to other regions in Indonesia, it needs to be strengthened from the level of the local economy, HDI development, and improvement of adequate facilities and infrastructure. But to be able to do it all, it must be supported by an adequate supply of energy, funds and Human Resources (HR) support from the central government.

According to the National Energy General Plan [2] (RUEN), in 2025 it is mandated that Indonesia's energy mix consists of 23% renewable energy (EBT), 25% petroleum, 30% coal and 22% natural gas.

Also, the RUEN mandates that by 2020 Indonesia's electrification ratio must reach 100% with the installed plant capacity in 2025 reaching 115 GW. But until mid-2019, the energy mix ratio from the EBT sector is still very low, far from the target mandated by the RUEN itself so this can result in an electrification ratio target of 100% by 2020 will be difficult to fulfill if it does not reach 3T areas that are geographically far from an existing utility network. Therefore, an optimal solution is needed, to be able to quickly electrify the 3T region to achieve the target of 100% electrification ratio by 2020.

Energy Storage System (ESS) is an important part of the new renewable energy generation system due to the nature of the EBT itself which is still largely intermittent so that for reliable electricity supply, reliable ESS is needed [3]. The size, cost, and reliability of ESS depend on the form of energy stored in it, be it potential energy, kinetic, chemical, electromagnetic, thermal, and so on [4]. There are many types of ESS itself, such as Electric Double Layer Capacitor (EDLC); Battery ESS; Superconducting Magnetic Energy Storage (SMES); Flywheel; Plug-in Electric Vehicle (PEV); etc. Battery ESS is one type of ESS that is widely used in EBT generation. This type of ESS is based on a secondary battery that can be charged and discharged repeatedly with a certain life span [5].

Pasi Island is an island located in Selayar Regency, South Sulawesi Province. There are three (3) villages on the island located in the Java Sea region, which had a total population of 4454 people in 2010 and reached 4747 inhabitants per 2019 [6]. Selayar Islands Regency, in general, has an electrification ratio of 68.46% which means 31.36% of the population has not enjoyed electricity, especially in villages located outside the main island (Selayar). The existing electricity grid on Pasi Island only has a capable power capacity of 235 KW with a peak load of 116 KW and a reserve load of 119 KW. But unfortunately, with the existing network, PLN is only able to illuminate the island for only six (6) hours per day, this has reduced the interest of the public to subscribe to PLN electricity.

Therefore, we need an optimal solution so that it can quickly, cheaply and efficiently meet the national electrification target of 100% as announced by President Jokowi and also PT. PLN (Persero). Battery ESS is the optimal solution in responding to the needs of electrification in the 3T region, because of low procurement costs, fast and efficient commercial operation date-time both in terms of charging, operating and also without waste.

2. Cost-Benefit Analysis Variables

To simplify the calculation, a basic parameter must be set to get reliable and accountable results according to scientific rules. The calculation compares between the two (2) scenarios using Battery ESS as a baseload with a project period of thirty years (30) according to the Build, Own, Operate and Transfer (BOOT) scheme in the Power Purchase Agreement (PPA) [7]. In first the scenario, the selling price of electricity used is the BPP tariff in the Selayar Islands Regency and the second scenario, the BPP tariff used is a small subsystem BPP tariff.

The 3T areas used as the basis for the calculation are three (3) villages in Pasi Island in the Regency of Selayar Islands with a total of 1356 [FN-1] heads of households (HH) with the assumption that the installed capacity in each HH is 1,300 watts. The benefit of electricity sales is 85% of the BPP tariff used in each scenario because the magnitude of the two BPP tariffs in each scenario is above the national BPP tariff [8].

2.1. General variables

General variables in the PPA project have a period of thirty (30) years with Commercial Operational Date (COD) in year one (see table 1). Each scenario uses the assumption of using ten (10) ESS battery units, where each village has three (3) ESS battery units with a total of nine (9) used and one (1) ESS battery becomes a backup unit if there is a problem in one unit battery in one of the villages. There are twenty (20) workers in the two scenarios, with a division of two (2) shifts consisting of two (2) ESS battery operators and one (1) truck driver in each village and one (1) daily operating supervisor.

Table 1. General variables

Variable	Value	Unit
Number of Customers	1356	Household
Installed Capacity	1300	VA/household
Project period	30	Year
Number of ESS Battery	10	Unit
The number of workers	20	People
Number of Trucks	3	Unit
Annual Power Generation	8.760.000	kWh/year
Commercial Operation Date	One (1)	Year
Discount rate	10	%

Furthermore, to expedite the charging process on the ESS battery, each village has prepared one (1) unit of battery transport truck from each village to the South Sulawesi PLN charging depot using a ferry from Pasi Island to Sulawesi Island through Bira Port, Bulukumba Regency. Each battery has a capacity of 100 kV so overall for one year all battery units can produce 8,760,000 kWh. Finally, the discount rate assumption used is 10%.

2.2. Benefit variables

In the benefits variable, there are four (4) variables, namely the BPP Selayar Islands Regency tariff; small subsystem BPP rates; electrification benefits; and the residual value of the project (see table 2). Tariff BPP Kepulauan Selayar Regency is used in scenario I with a magnitude of Rp 2,445 per kWh. In scenario II, the electricity sales tariff used is a small subsystem BPP tariff with a value of Rp 3,041 per kWh. Because both of these rates are above the national BPP tariff, and also considering that ESS batteries are included in renewable energy, therefore the maximum sales tariff percentage used is 85% of the BPP tariff used in each scenario.

Table 2. Benefit variables

Variable	Value	Unit
BPP Tariff of Selayar Regency	2445	Rupiah/kWh
BPP Tariff of Small Sub-system	3041	Rupiah/kWh
Electrification Benefit	2.012.931.737,76	Rupiah/year
Residual Value	10	%

In addition to the benefits received from the sale of electricity, there are benefits from fulfilling electrification on Pasi Island, because these benefits are not value that has real money, so it needs to be quantified into the applicable monetary value. In economics, this quantification is called a shadow price. Shadow prices are prices of components that do not have real value but have value in terms of economic analysis so that shadow prices can also be called social prices.

Because the price is valued by the community, therefore the amount of the shadow price is based on the assumption of people's willingness to pay both benefits and costs [9] [10]. Therefore, we can calculate the amount of the shadow price of the benefits obtained by the community if a reliable electricity infrastructure can enter the 3T region. This benefit can also be borne by a fee that the community is willing to pay so electricity can enter the 3T region.

We assume that all households on Pasi Island subscribe to electricity using PLN's diesel, then combine it with preliminary data which states that PLN's services on Pasi Island only last a maximum of six (6) hours. Then both scenarios use the depreciation value of the residual value at the end of the project with a large percentage of 10%. The value of electrification benefits can be obtained by combining the two variables above into mathematic formula as in equation (1).

$$B = ((k \times P \times t)^1 \times p) \times 365 \text{ hari} \tag{1}$$

B = diesel savings benefit (rupiah)

k = fuel specific consumption of diesel generators (liters / kwh)
P = diesel generator power (kVA)
t = generator daily operation (hours)
p = price of diesel fuel (rupiah)

Small-scale PLN diesel generators under 250 KW with 75% to 100% load have an SFC value of 0.273 liters / KWh. The generator in Pasi Island has a capacity of 235 KW with a long time the community enjoys electricity generated from the PLTD is only six (6) hours. The P-value of the generator is 261.11 KVA, this value is obtained by multiplying the magnitude of the generator power that is 235 KW to the value of the power factor (power factor) with a general assumption of 0.8. Pertamina's high speed diesel (HSD) for the industry as of August 2019 has a price of Rp 12,894.33 after VAT, PPh and PBBKB for Area III or Sulawesi and NTB Areas. By distributing known variables into the equation (1), the resulting B-value is
B = ((0.273 x 261.11 x 6) x 12.894.33) 365 days
B = (427,698 x 12,894.33) x 365 days
B = 5,514,879.15 x 365 days
B = Rp 2,012,931,737.76 per year

2.3. Cost variables

2.3.1. Fixed costs

Fixed costs are costs incurred either in one time or in a certain period with the same amount of value or static. In the fixed cost parameters, there are six (6) parameters (see table 3). The cost of procurement and construction of ESS batteries is the US $ 900 per kWh, these costs include land acquisition, equipment and material procurement, construction and network construction. In addition to CapEx costs, another procurement cost is the cost of procuring trucks for mobility from the village to the charging depot assuming a procurement cost of Rp. 400,000,000.00 per truck. The type of truck needed is the type that is capable of carrying trailers or battery containers in off-road conditions. The types of trucks that fit the needs above such as the Mitsubishi Colt Diesel 74 Long with a price range of 328-360 million rupiahs, Hino Dutro Cargo 130 MD with a price range of 312-330 million rupiahs and Tata Ultra in the price range of 400-430 million rupiahs.

Table 3. Fixed cost variables

Variable	Value	Unit
Capital Expenditure	900	US Dollar/kwh
Truck Procurement Cost	400.000.000	Rupiah/truk
ESS Charging Cost	1.115	Rupiah/kwh
Ferry Cost	1.280.000	Rupiah/trip
Truck Fuel Cost	12.240.000	Rupiah/tahun/truk
Corporate Overhead	0,5	% dari *capex*

The next fixed cost parameter is the charge for charging the ESS battery, charging costs using class I-3 tariffs as of August 2019, which is Rp 1,115, because the power needed to charge three (3) batteries every day is 375 kVA. This figure is obtained through multiplication of factors of the number of batteries 3 units x 100 KW with also the power factor (PF) of batteries 0.8. Other fixed costs incurred in the process of the charging location are the cost of the ferry crossing from the Selayar Regency to Bulukumba Regency amounting to IDR 1,280,000 per trip.The ferry that serves the Selayar Islands Regency route and the Bulukumba Regency is KMP Bontoharu with the intensity of traveling two (2) times a day. The annual fuel cost of the transport truck is Rp 12,240,000, assuming a full tank of Mitsubishi Colt FE 74 L fuel tank with a capacity of 100 liters. Fully refueling is done monthly using Dexlite diesel at a fixed price of Rp10,200 as of August 2019. Finally, on administrative or corporate overhead variables, assuming the annual costs incurred are 5% of the total CapEx costs.

2.3.2. Variable costs
Maintenance costs use an assumption of 0.5% of capital expenditure and increase every ten years by 0.5%. So in scenarios, I and II, the maintenance costs needed in the first ten years amounted to Rp 63,900,000, in the second ten years amounted to Rp 127,800,000 and in the last ten years amounted to Rp 191,700,000. As for salaries using the South Sulawesi provincial minimum wage (UMP) of Rp. 2,860,382 [11] with an increase of 8% every five (5) years.

Table 4. Var. cost variables

Variable	Value	Unit
Maintenance		
Year 1-9	0,5	
Year 10-19	1	% of CapEx
Year 20-29	1,5	
Wages		
Year 1-4	2.860.382	
Year 5-9	3.089.213	
Year 10-14	3.336.350	Rupiah/worker/month
Year 15-19	3.603.258	
Year 20-24	3.891.518	
Year 25-29	4.202.840	

In the first five years, the worker's wages were IDR 2,860,382 per worker/year, increasing in the second five years to IDR 3,089,213. In the next five years, the wages of each employee increased again to Rp 3,336,350. In the fourth five years, namely the years 15-19, employee wages to Rp 3,603,258. In the fifth and sixth years, employee salaries increased respectively by 8% to Rp 3,891,518 and Rp 4,202,840. Both of these parameters are included in the variable cost parameter because their properties change, whether they go up or down, in this case, the value rises dynamically (see table 4).

3. Cost-Benefit Analysis
Economic cost-benefits analysis is an analysis from a macro point of view so that the benefits and costs arising from a policy or project have an impact on society, both positively and negatively. In this study, a cost-benefit analysis is carried out by utilizing Microsoft Excel to make it easier to find the value of NPV, IRR and its cost-benefit ratio. Similar to financial appraisal, some of the components or variables calculated are the same, but in economic calculations, variables that are genuine money transfers between two sides without goods or services, such as taxes, are not counted in the economic cost-benefit calculation. This is because the movement of money does not reduce the resources or wealth that exists in the community.

In the cost-benefit analysis, there are three (3) things that are used as a benchmark in the comparison between alternatives, namely the NPV value, the percentage of IRR and also the cost-benefit ratio. The alternative with the highest NPV, IRR and cost-benefit ratio is the most viable option to be carried out, continued or further calculated. In general, if the NPV value of a project or alternative shows a positive number, then the value of the IRR and B / C ratio will also follow, so it can be concluded if in the two alternatives there is one that has the highest NPV, it is certain that the cost-benefit ratio and the IRR percentage is also high.

3.1. Scenario I
The cost-benefit calculation results in the scenario I show good NPV, IRR, and B / R ratio results. Scenario I using the Selayar Regency BPP tariff as the basis for electricity sales generates an NPV value of Rp. 74,900,608,862.00. There are two (2) main benefit components in the calculation of these cost-benefits, namely the benefits of electricity sales and the benefits of electrification (see table 5).

Table 5. Scenario I cost-benefit analysis

Variable	Value	Unit
Capital Expenditure	12.780.000.000	Rupiah
Power Sales Benefits	18.205.470.000	Rupiah/year
Electrification Benefits	2.012.931.738	Rupiah/year
Discount Rate	10	%
Present Value of Bt	88.880.608.862	Rupiah
Present Value of Ct	13.980.000.000	Rupiah
Net Present Value	74.900.608.862	Rupiah
Payback	Year 2	
IRR	68,56	%
B/C Ratio	6,358	

In the electricity sales benefit component, each year after entering the Commercial Operation Date (COD), the benefits received from electricity sales amount to Rp 18,205,470,000. In the electrification benefit component, the benefit received by the community from the inclusion of electricity using an ESS battery as a baseload every year is Rp. 2,012,931,738.00. The cost-benefit calculation in the scenario I use a discount rate of 10%. Overall, the present value of all benefits after deducting the discount rate is IDR 88,880,608,862 and the present value of all costs after deducting the discount rate is IDR 13,980,000,000.00.

NPV value is obtained by reducing the value of the present value of Bt with the value of the present value of Ct so that the results obtained are Rp. 74,900,608,862.00. Payback from the scenario I occur at the point of the second year after the COD (see figure 1). The value of the cost-benefit ratio can be determined by using equation (2)

$$B/C = PVbt - PVct \tag{2}$$

B/C = Benefit Cost Ratio
PVbt = Present Value of All Benefits
PVct = Present Value of All Costs

So that the known PVbt and PVct values can be distributed into the formula so that the scenario I cost-benefit ratio value of 6.358 is obtained. The value of IRR can be determined by utilizing the net benefits from each of the available years, then utilizing the formula features in Microsoft Office Excel in the Finance sub-menu section. The results of calculations from Excel show the IRR value of the scenario I is 68.56%.

Based on the scenario I graph (see figure 1), the NPV value in scenario I in the first ten years the value rises quite rapidly, this is seen in year 0 where the NPV value is at a negative rate of 13.9 billion rupiahs and rises significantly to the point of 40 billion rupiahs in year 9. In the next ten years, the value of the NPV began to experience a slight slowdown, the lines on the graph began to slope. At year 10, the NPV is in the range of 45 billion rupiahs and at the end of the second ten years, it is at 66 billion rupiahs. In the final ten years of the project, the increase in NPV was very slow with the accumulation of the last ten years of only around 10 billion, with the 20th year at the point of 68 billion rupiahs and closing at 75 billion rupiahs at the point of 29.

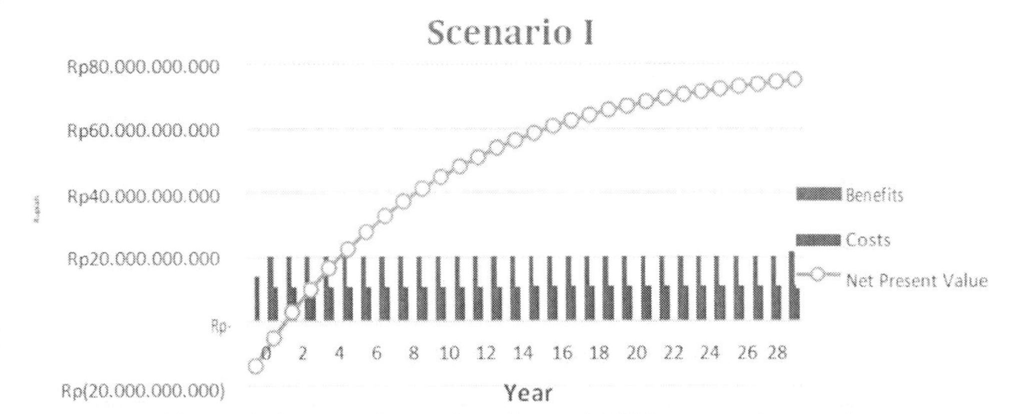

Figure 1. Scenario I costs, benefits, and NPV comparation graph

3.2. Scenario II

Scenario II using the Small Sub-system BPP tariff of Rp 3,041.00 as the electricity sales price generates an NPV value of Rp 116,481,195,900.00 (see table 6). Annual power sales in scenario II are Rp. 22,643,286,000.00. The benefits of electrification in scenario II also have the same amount as in scenario I, which is Rp. 2,012,931,738.00. The cost of CAPEX and discount rate used in scenario II also have the same amount as scenario I. This is because the parameters or variables that affect are generally the same, only differing in terms of power sales tariffs. So that there are differences in the present value of costs and benefits, net present value, IRR and also the cost-benefit ratio.

In scenario II, the project reaches a break-even point in the second year after COD. The present value of Bt is IDR 130,461,195,900.00 and Ct is IDR 13,980,000,000.00. The present value of Ct from the scenario I and II has the same value because the largest expenditure is in year 0 at the time of development (before the COD point), and after COD, the benefits received are far above the costs incurred. The net present value of Bt is reduced by the present value of Ct, so the NPV value from scenario II is Rp. 116,481,195,900.00.

With the amount of benefits received in scenario II, the project reaches a break-even point in the second year (2). The present value of Bt and the present value of Ct known above can be used again to calculate the cost-benefit ratio value from scenario II. The present value of Bt is divided by the present value of Ct, resulting in a ratio value of 9.332. The percentage of IRR can be found easily using the "formula" feature in Microsoft Office Excel. First, select the "formula" tab, then select the "financial" sub-menu, finally select "IRR". The percentage value of IRR generated using the excel calculation in this scenario shows an amount of 100.33%.

Table 6. Scenario II cost-benefit analysis

Variable	Value	Unit
Capital Expenditure	12.780.000.000	Rupiah
Power Sales Benefits	22.643.286.000	Rupiah/year
Electrification Benefits	2.012.931.738	Rupiah/year
Discount Rate	10	%
Present Value of Bt	130.461.195.900	Rupiah
Present Value of Ct	13.980.000.000	Rupiah
Net Present Value	116.481.195.900	Rupiah
Payback	year 2	
IRR	100,33	%
B/C Ratio	9,332	

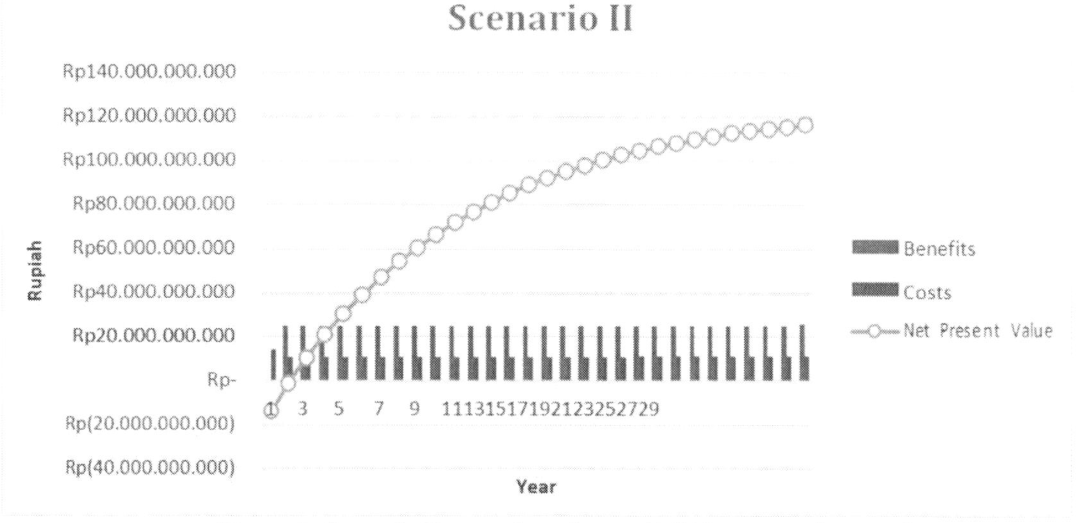

Figure 2. Scenario II costs, benefits, and NPV comparation graph

Based on the coordinates of the graph in scenario II (see figure 2), the NPV value in scenario II rises significantly from the negative point to the range of years 15-17, and then the increase starts to slow down. In general, in the first ten years, the NPV value moved up very significantly, with the initial value at a negative point until in the tenth year it reached a positive point at 66 billion rupiahs. In the second ten years, the graph still shows an increase but starts to slow down, precisely at the point between the 15-17 curve lines start to show a greater slowdown. At the end of ten, the movement of the NPV value was slow with a margin of difference in year 21 and year 30 of only 14 billion rupiahs. This is very different from the first ten years which had a margin of difference of nearly 80 billion rupiahs.

3.3. Analysis

Decisions about the feasibility of a project based on several alternatives can only be based on the NPV value, although the percentage of IRR shows a positive value, sometimes it can result in inconsistencies in the alternative ranking. However, several large corporations and well-known financial institutions prefer IRR because it is easier to compare in terms of capital expenditure alone [12]. In scenario I, with a CAPEX of Rp. 12,780,000,000.00 an NPV result of Rp. 74,900,608,862.00, while in scenario II, with the same amount of CAPEX can produce a much larger NPV value of Rp. 116,481,195,900.00. This shows that with a small subsystem BPP tariff as a benchmark price of electricity sales to PLN will greatly benefit the ESS battery project in three (3) villages located on Pasi Island.

Although the best alternative selection can only use NPV as a benchmark, the authors will also use the IRR percentage and cost-benefit ratio values to strengthen the basis of the selection. In scenario I, the percentage value of IRR is 68.56% and in scenario II the value of the percentage of IRR is at 100.33%, this shows that scenario II is again superior to scenario I. Similarly, the value of the cost-benefit ratio, scenario II has a superior result at 9,332 level than scenario I at point 6,358.

The conclusion of the three comparisons between NPV value, IRR percentage and also the cost-benefit ratio value shows that scenario II is the optimal choice that can be implemented as a project to fulfill electrification ratios in three (3) in the Pasi Island region, South Sulawesi. After the best choice has been determined, each component or variable that influences the NPV value is tested for sensitivity to mitigate the volatility of NPV values that are too extreme or that risks making the project not viable.

4. Sensitivity Testing with Switching Value Table and Break-Even Analysis

Break-Even Analysis (BEA) is a method of analysis used to see the relationship between fixed costs, variable costs and also revenue so that an investment or project can reach break-even point [13]. This method can also be used to compare several alternatives to find the most optimal one to be implemented [14].

Furthermore, this analysis method can be used to see how the changes in fixed costs, variable costs, commodity prices or income can affect the level of profit and the break-even point of the project or policy. Switching Value (SV) is a change in the value of each variable that affects the break-even point or NPV = 0 with a predetermined discount rate and SV value expressed in percentage [15]. Mathematically, the equation (3) can be used to find the SV values

$$SV = 100\% \times (vt - v0) / v0 \quad (3)$$

SV = switching value
vt = final value
v0 = initial value

Table 7. Switching value table

NPV 116.481.195.900	Initial Value	Switching Value	Persentase Sensitifitas
Small Sub-system BPP	Rp 3.041,00	Rp 1.371,40	-55%
Electrification Benefits	Rp 2.012.931.737,76	Rp -10.418.882.041,00	-618%
Residual Value	10%	-14.448%	-144.580%
CapEx			
Battery ESS unit price	$ 900	$ 8.286,88	821%
The Amount of ESS unit	10	92	821%
Charging Cost			
Charging Cost	Rp 1.115,00	Rp 2.534,00	127%
Annual Power Production	8.760.000 kwh	302.120,78 kwh	-97%
Transportation			
Dexlite Price	Rp 10.200,00	Rp 3.463.481,61	33.856%
Ferry Cost	Rp 1.280.000,00	Rp 2.073.248.963,00	161.873%
Corporate overhead	0,50%	97,78%	19.456%
Truck Price	Rp 400.000.000,00	Rp 39.227.065.300,00	9.707%
Annual Maintenance	0,50%	65,67%	13.034%
Annual Wages	8%	193,71%	2.321%

But to ease the calculation, we can use the function of excel, 'what-if analysis' by targeting NPV cells with a value of 0 and correlating with the initial value of the variable you want to test. The smaller the percentage value that comes out, the higher the sensitivity value. The results of the SV which show the highest sensitivity value indicate that these parameters greatly affect changes in NPV and break-even points, making it easier for stakeholders to take preventative actions.

The sensitivity test shows the two components that have the most sensitive value for changes in NPV values, with the SV percentage below 100%, namely the BPP tariff component and also the annual power generation (See table 7). At the BPP tariff (small sub-system BPP), if the tariff reduction is greater than 55%, the battery ESS project is not feasible to proceed. The second variable that is sensitive to changing the NPV value is the amount of annual electricity production by battery ESS. If the annual power production drops to > 97%, the electrification of the Pasi Island project with an ESS battery will not be feasible.

As one of the most sensitive variables, the BPP rate has a very small probability of decreasing, because the large BPP rate has been determined and agreed before the signing of the PPA between the IPP and PLN. However, if there is a change in policy in the future or there are discussions such as BPP tariff adjustments, a reduction may occur but still has a very small percentage. One of the reasons for the BPP tariff adjustment is the increase in the level of the economy in the Selayar Regency so that it is no longer classified as a 3T region. Another factor that might occur is the decrease in the variable costs of the operating tariff compiler so that the magnitude of the sub-system BPP also decreases to adjust.

In the annual electricity production variable, a significant reduction in electricity production has a very low likelihood ratio. This is because the shape of the ESS battery itself is in the form of a modular unit so that if there is damage to one unit, it will not affect the overall electricity production. Also, in this scenario, the procurement of extra ESS battery units has been carried out to mitigate the possibility of damage by preparing one (1) reserve in each village and one (1) main reserve if damage occurs to more than one (1) unit in the village certain. But seeing the possibility of force majeure or natural disasters and riots can occur, the operator can prepare a backup plan to mitigate these possibilities. The mitigation effort can be in the form of a contract with the procurement of ESS batteries so that if many units are damaged, it can immediately (under a span of one (1) week) be replaced by a new unit.

In general, other components besides the two components above do not greatly affect changes in NPV value due to the very high SV percentage, above or below 100%. The ferry tariff component and the residual value are the two components that have the least influence, with the SV percentage at the residual value of -144,580% and the ferry rate of 161,873%. Seeing the number of components that have a very large percentage change in variables, then these components can be removed from the SV table. The use of sections to divide the text of the paper is optional and left as a decision for the author. Where the author wishes to divide the paper into sections the formatting shown in table 2 should be used.

5. Conclusion and Recommendation

5.1. Conclusion

Three (3) villages in Pasi Island need 4,632,638.4 KWh / year in the first five (5) years assuming a load of 30% which has an installed capacity of 1300 Watt per customer with a total of 1356 households. Customer load increases by 5% every five years until assuming a maximum load of 50% in the last ten

(10) years at the PPA with the amount of electricity demand each year of 7,721,064 KWh. Therefore ten (10) ESS battery units are the optimal amount to meet the needs of each year with a maximum potential annual electricity production of 8,760,000 KWh / year.

Existing power on Pasi Island has a capacity of 235 KW with a peak load of 116 KW and a reserve load of 119 KW which serves a total of 638 households. However, this power plant is only able to supply a maximum of six (6) hours, so it can be assumed that the average per house uses 200 watts/day/household. Battery ESS with scenario II or a small sub-system BPP rate of Rp 3,041.00 is the optimal choice for the electrification of Pasi Island. This is indicated by the largest NPV value, along with the IRR percentage and the cost-benefit ratio. Scenario II shows the results of the calculation of the NPV value of Rp 116,481,195,900.00, the percentage of IRR 100.33% and the cost-benefit ratio at the level of 9,332. Besides, with scenario II, the battery ESS project reaches a break-even point in the second year after COD.

There are two (2) variables that are most sensitive to affect the NPV value, this is indicated by the value of switching values with a percentage in the range of -100% to 100%, namely the BPP rate with - 55% and annual electricity production with -97%. However, both components can be mitigated well with the certainty of the basic tariff on the PPA and the advantages of the modular ESS battery itself.

5.2. Recommendation

Author recommend more studies to see the probability of risk occurring for each component that is sensitive to the amount of NPV. Probability testing of each risk can be carried out using a risk simulation program such as Monte Carlo. Seeing the possibility of developing discourse of distribution through submarine cables by PLN, further research can see a comparison between electrification using submarine cable lines or with ESS batteries. This advanced study can use a financial analysis or analytic hierarchy process or decision matrix.

The development of renewable energy (EBT) in Indonesia is more concrete with the issuance of regulation [2] on RUEN, which is part of the National Energy Policy (KEN), in addition to the National Electricity General Plan (RUKN). RUKN downgraded the Electricity Supply Business Plan (RUPTL). Utilization of EBT has the aim of creating energy independence and energy security. Energy security is the ability to answer the dynamics of global (external) energy changes, as well as the ability to guarantee the availability of energy at a fair price (internal), while energy independence consists of the elements of Availability, Accessibility, Affordability, Acceptability, and Sustainability (4A + S). Indonesia has a great opportunity to develop EBT. This is not only due to the large potential of resources, but also its varied types (Indonesia has almost all types of EBT).

EBT development has also been included in the EBT Roadmap in 2025 and 2050. The electricity utility sector has a major role in the development of EBT. But now, the development of EBTs still seems expensive because they have not included environmental costs in the calculation of energy production costs in Indonesia. Another challenge is the economics of EBT which still cannot compete with fossil energy. EBT has not fully met the 4 A + S criteria, especially Accessibility and Affordability. While fossil energy subsidies are a strong barrier to the development of EBT, because it creates a huge price gap, it is politically difficult to eliminate because it is related to industry and consumers. President Joko Widodo has only succeeded in eliminating subsidies for fuel prices.

EBT development in Indonesia is still Adhoc, so it requires "Political Will". The problem of EBT will be more complicated in the future because globally it will still depend on fossil energy, especially petroleum, while the future trend of gas and EBT consumption will increase. National energy security needs to be improved in the form of operational policies and strategies and requires a New and Renewable Energy Agency (EBT Agency) that has the main tasks and functions (tupoksi) at the micro-management level (operational) of new and renewable energy. His job is to supervise and implement micro regulations on the EBT power plant. On the other hand the Ministry of Energy and Mineral Resources (ESDM), in this case the Directorate General of New and Renewable Energy (EBT) has its duties and functions at the macro-management level (policy) of EBT, carries out policies and the implementation of macro regulations on the EBT sector.

EBT Agency is a State-Owned Legal Entity (BHMN) as is the case with Non-Departmental Government Institutions (LPND) reporting directly to the President of the Republic of Indonesia. Coordination with various sectors is carried out according to the main tasks and functions of each sector, including the Ministry of Energy and Mineral Resources in energy policy, the Ministry of Finance in financial policy, the Ministry of Environment and Forestry in terms of the EBT operations in forest areas.

Acknowledgments
The Author would like to thank all colleagues from Indonesia Defence University and Energy Security Centre of Excellence who provided immense insight, data and expertise that greatly assisted the research.

References

[1] Bratakusumah D S 2006 *Daerah tertinggal* Jakarta: Kementerian Pendayagunaan Aparatur Negara dan Reformasi Birokrasi Republik Indonesia
[2] Presiden Republik Indonesia 2017 *Peraturan Presiden Republik Indonesia No 22 Tahun 2017 tentang Rencana Umum Energi Nasional* Jakarta: Sekretariat Negara
[3] Funabashi T 2016 *Integration of distributed energy resources in power systems: implementation, operation and control* Cambridge: Academic Press
[4] Novakovic B & Nasiri A 2015 Introduction to electrical energy systems *Electric Renewable Energy Systems* **1**
[5] Breeze P 2018 *Power system energy storage technologies* Cambridge: Academic Press
[6] Badan Pusat Statistik Kabupaten Kepulauan Selayar 2019 *Kabupaten Kepulauan Selayar dalam angka 2019* Selayar: BPS

[7] Menteri ESDM 2017 *Peraturan Menteri Energi dan Sumber Daya Mineral Republik Indonesia No 10 Tahun 2017 Tentang Pokok-pokok Dalam Perjanjian Jual Beli Tenaga Listrik* Jakarta: Kementrian Energi dan Sumber Daya Mineral Republik Indonesia

[8] Menteri ESDM 2019 *Keputusan Menteri Energi dan Sumber Daya Energi Republik Indonesia No 55 K/20/MEM/2019 Tentang Besaran Biaya Pokok Penyediaan Pembangkitan PT. Perusahaan Listrik Negara (Persero) Tahun 2018* Jakarta: Kementrian Energi dan Sumber Daya Mineral Republik Indonesia

[9] Hirth R A, Chernew M E, Miller E, Fendrick A M & Weissert W G 2000 Willingness to pay for a quality-adjusted life year: in search of a standard *Medical decision making* **20** 332-342

[10] Van Soest D P, List J A & Jeppesen T 2006 Shadow prices, environmental stringency, and international competitiveness *European Economic Review* **50** 1151-1167

[11] Gubernur Sulawesi Selatan 2018 *Surat Keputusan Gubernur Sulawesi Selatan Nomor 2877/X/2018* Makassar: Pemerintah Daerah Provinsi Sulawesi Selatan

[12] Osborne M J 2010 A resolution to the NPV–IRR debate? *The Quarterly Review of Economics and Finance* **50** 234-239

[13] Gutierrez P H & Dalsted N L 1990 Break-even method of investment analysis *Farm and ranch series Economics* **3** 759

[14] Ekren O, Ekren B Y & Ozerdem B 2009 Break-even analysis and size optimization of a PV/wind hybrid energy conversion system with battery storage–A case study *Applied Energy* **86** 1043-1054

[15] Asian Development Bank 2010 *Introductory course on economic analysis of investment projects session 3.3 sensitivity and risk analysis* Manila: Asian Development Bank

Mathematical model for calculating the equilibrium point of the refrigerant circuit

A Daci[1*], **J Bundo**[2]

[1]Department of Mathematical Engineering, Faculty of Mathematical and Physical Engineering, Polytechnic University of Tirana, Albania

[2]Department of Automation, Faculty of Electrical Engineering, Polytechnic University of Tirana, Albania

*alfreddaci@gmail.com

Abstract. Calculating the equilibrium point of the refrigerant circuit is one of the most important processes in the air treatment industry. Knowing the evaporation and condensation temperatures serves not only to obtain information on the performance and operating point of the compressor, but also to evaluate the performance of the machine under different climatic conditions. Often, the solution of the refrigerant circuit is accomplished by empirical methods or numerical methods such as the Newton – Raphson method or the medium method. Since the iteration time with these methods is unknown at the time of the iteration start, it is very important to implement algorithms that provide convergence information or not, and which can bring the system to a solution in the shortest possible time. The implementation of such a method would pave the way for simulating the behaviour of machines in the air treatment industry as an advanced verification process for real evaluation of machine performance. In this article, experimental results and mathematical model calculations will be presented and discussed.

1. Introduction

Calculating the equilibrium point of the refrigeration circuit is a critical operation in machines containing such a circuit, especially in the air treatment industry. We can say this because we often need to control parameters such as temperature, humidity or both simultaneously.

For different application purposes the refrigeration circuit may be in different configurations. Here we mention configurations with variable gas flow regulated by inverters, circuit with more than one compressor located in parallel or cascade or refrigeration circuit where the condensing or evaporating element can be composed of more than one part. In this paper we will refer to the simplest model of the refrigeration circuit, although the following reasons apply to each model of the refrigeration circuit. [1] Below we will schematically present the refrigeration circuit that will be studied.

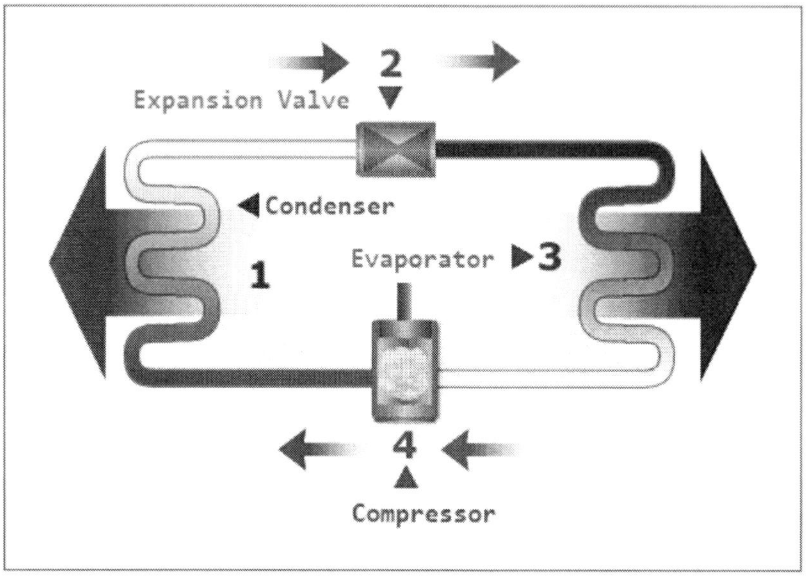

Figure 1. Schematic refrigeration circuit

In the figure 1 above we distinguish that the circuit is composed of 5 main elements that are:

- Compressor
- Condensing element
- Gas expansion valve
- Evaporative element
- Gas and liquid passage pipes

The compressor set in motion would compress the gas. The latter according to the law of ideal gas in an isohoric process would increase its temperature in a linear relationship with the ratio of temperature change as follows [2]:

$$T_2 = \frac{P_2}{P_1} * T_1 \quad (1)$$

Gas at high temperatures (about 70 C), condenses having contact with air at much lower temperatures and according to the second law of thermodynamics the air temperature rises. This temperature exchange is determined by various factors such as the geometry of the exchanger, air and gas temperature, humidity and air flow.

Condensed gas passes through the gas expansion valve where its pressure decreases greatly. Once again referring to equation (1) we determine that the gas temperature decreases too much as the fluid passes into the gas expansion valve. Here we again have the thermal exchange of gas with the air where the latter has a much higher temperature value than the temperature of the fluid. Even in this case the amount of thermal exchange depends on the factors mentioned above.

2. Equilibrium of the refrigeration circuit

The process of condensation and evaporation both occur at certain temperatures. These temperatures are the product of several factors where we can mention the geometry of the exchangers, the climatic conditions of the air and the air flow. During the heat exchange in the condensing and evaporating phase each of the elements presents a certain power.

Thermal power will be called the full power developed by the condensing element, while the refrigeration power will be called the total power developed by the evaporating element.

If we accept all other invariant factors, the power of the condensing element can be modelled as a function of the condensing temperature, and the power of the evaporating element as a function of the

evaporation temperature [3]. At the same time, one of the most classic forms of calculating the working parameters of the compressor is the polynomial model as a function of two variables, which are the evaporation temperature and the condensation temperature. This model applies to the compressor refrigeration power, electric power, electric current and gas mass per hour.
The model is presented as follows [1]:

$$F(tev,tco) = A + B * tev + C * tco + D * tev^2 + E * tev * tco + F * tco^2 + G * tev^3 + \\ + H * tco * tev^2 + I * tev * tco^2 + L * tco^3 \qquad (2)$$

We say that the circuit is in equilibrium at the moment that the cooling power of the compressor is equal to the cooling power of the evaporating element and the thermal power of the compressor is equal to the thermal power of the condensing element. The thermal power of the compressor will be calculated as the sum of the cooling power and the electric power. In these conditions, finding the equilibrium conditions of the refrigeration circuit consists in detecting the evaporation and condensation temperature for which the circuit is in equilibrium. From the recognition of these temperatures it can be calculated the power at the output of each element and as a result also the climatic conditions at the output in the case of air treatment machines. There are several studies that treat this concept, but it is difficult to find algorithms that can be programmed or procedures that can help on finding the equilibrium point of refrigerant circuit in a non empirical way. In this paper we are presenting two different approaches (graphical and analytical) that can be implemented in a program or can be used to solve these kind of problems manually. We think that this work can also help on improving the intuition of finding equilibrium point of a refrigerant circuit.

3. Graphic solution

As a problem that cannot be specifically solved, the solution to this problem is iterative. As a two-dimensional function, the graphical representation of the compressor's refrigeration and thermal power would be presented as a network of possible values within which the compressor could operate. Below we present the graphical solution of this problem based on an air treatment machine which has a configuration of the refrigeration circuit as the one considered in this study.

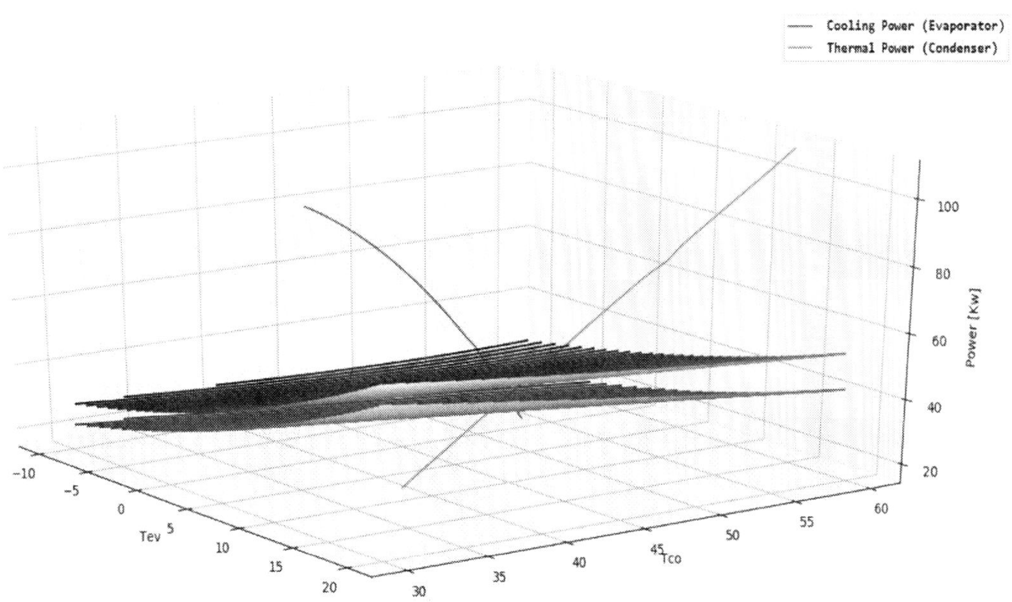

Figure 2. Graphical presentation of the equilibrium of the refrigeration circuit

In the graph above we have graphically presented the two work surfaces of the compressor and the respective curves of the condensing and evaporating element. Reiterating that the thermal power of the compressor is the sum of the refrigeration power and the electric power, we can easily distinguish from the graph that the surface positioned below represents the refrigeration power, while the one above represents the thermal power. Graphically we say that we are in equilibrium when the curve of the condensing element pierces the surface of the thermal power in the pair of condensing and evaporating temperatures for which the curve of the evaporating element pierces the surface of the refrigeration power.

Starting from the refrigeration power, we first build the surface and the respective curve presented as follows.

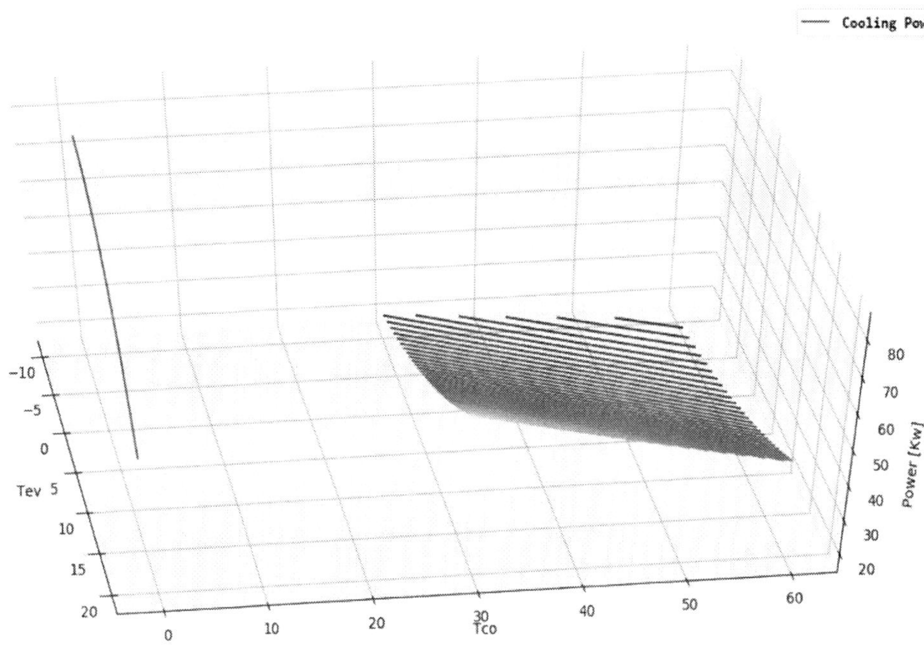

Figure 3. Graphical representation of the surface and curve of refrigeration power

By moving the evaporator curve towards the surface of the compressor's refrigeration power (according to the condensing temperature axis) until the curve touches the surface. At the cutting point we read the value of the condensation temperature from the surface. Further moving the condenser element curve to the evaporation temperature point where we have the point of intersection with the surface, we see if this curve intersects the surface of the thermal power at the value of the condensation temperature detected on the refrigeration surface. If not, this process is repeated with another point by moving the refrigeration power curve until the solution is reached as shown in figure 2.

This process, if performed graphically, would take a long time and would not be efficient. The first problem would be to build surfaces and curves for system components and then to perform iterative steps until a solution is reached. For this reason, we will look at the analytical solution and the problem-solving algorithm analytically.

4. Analytical solution

As stated above, if we accept that the climatic and geometric data of heat exchangers are constant, then the resultant power of each of these elements remains a function of the evaporation and condensation temperature. So we would have:
with a dot following the number and then separated by a single space:

- Refrigerant power of the evaporating element: $Eva_{refr}(te)$
- Thermal power of the condensing element: $Con_{therm}(tc)$
- Compressor refrigeration power: $Com_{refr}(te,tc)$
- Compressor thermal power: $Com_{therm}(te,tc) = Com_{refr}(te,tc) + Com_{el}(te,tc)$

where with "el" we indexed the electrical power obtained from the network. Since the following solution will be a mathematical solution, we will physically accept the temperature values in a certain range of values.

If we accept that we are in equilibrium we would have:

$$Eva_{refr}(te) = Com_{refr}(te,tc) \tag{3}$$

If we accept that the mathematical function for calculating the vaporizing and condensing element are known functions, the equation (3) would turn us into a third order equation with one variable. Thus, according to function (2) we would have:

$$\begin{aligned} Eva_{refr}(te) = A + B*te + D*te^2 + G*te^3 + tc*\left(C + E*te + H*te^2\right) + \\ + tc^2*\left(F + I*te\right) + tc^3*L \end{aligned} \tag{4}$$

Recognizing that we are iterating with evaporation temperature it follows that:

$$\begin{aligned} & A + B*te + D*te^2 + G*te^3 \rightarrow const.(C_{te1}) \\ & C + E*te + H*te^2 \rightarrow const.(C_{te2}) \\ & F + I*te \rightarrow const.(C_{te3}) \\ & 0 = \left(C_{te1} - Eva_{refr}(te)\right) + tc*C_{te1} + tc^2*C_{te2} + tc^3*L \end{aligned} \tag{5}$$

We see that the above equation has the general form of the third order equation with one variable . The solution of this equation can be achieved with different numerical methods such as Newton-Raphson method, Bisection method or referring to the general solution of the equation presented by N. Tartaglia, L. Ferrari and S. del Ferro [4].

From the equation solutions we will filter the resulting values in the range of acceptable values of condensation temperature. By substituting this value for the function that calculates the thermal power of the condensing element and for the function that calculates the thermal power of the compressor we would have 2 respective thermal powers. At this point the assessment of the relative difference of these powers would give us the information needed to judge whether we are in equilibrium or not. Hence, we have:

$$d = \frac{\left|Com_{therm}(te,tc) - Con_{therm}(tc)\right|}{Com_{therm}(te,tc)} * 100$$

If the value of the percentage deviation were less than the tolerance (usually 5%) we say that we are in equilibrium conditions, otherwise we would increment the value of the evaporation temperature.
In this analytical solution we will need to define some important parameters that are:

- Initial value of evaporation temperature
- The step of iteration
- Tolerance
- Maximum number of iterations

Based on the experience and experiments below we will give some recommendations for each of these parameters. To determine the evaporation temperature, experience dictates that it is recommended to start from an expected value of equilibrium evaporation temperature or from a value calculated as the difference in air temperature at the inlet of the minus K, where K is located in the segment [15, 20]. This is a parameter for which it is difficult to suggest specific values, but it is recommended not to increment with whole numbers, but with steps that can bring as much convergence as possible such as 0.07, 0.12, 0.17, etc. The tolerance required in air treatment machines usually varies in the interval 5 % - 10 %. This can be increased in situations where we know that the compressor polynomial model carries errors in polynomial coefficients or when the power function for each of the exchanges does not represent high accuracy. The maximum number of iterations should usually be defined as a value in the range of 50-100 iterations.

5. Conclusions

- It was noticed that when we used the analytical approach, high accuracy levels were achieved, considering a maximum tolerance of 5 %, this in confront with other commercial softwares which usually use algorithms that iterate both temperatures. Using these softwares we had to define a tolerance of 15 – 20 % for iteration to converge.
- Using the graphical solution of the problem is simple. Nevertheless, it is more complex due to the fact that we need to calculate a wide range of values. This makes the solution less usable.
- Using the compressor model to find the condensation temperature when iterating with evaporating temperature with cooling power treated as a known variable, helps in increasing the accuracy levels and accelerates the iterative process.
- From the experiment, we came to thie conclusion that this method grants a better performance in simulating the hvac/r machines, in comparison to the iterative bidirectional methods.

6. Appendices

Pseudocode:

```
Set numIter to 0
Set Success to 0
Set actualEvapTemp to (EvapCoilAirIn – Delta)
While numIter < 100 and not Success
        Set EvapCoil evaporation temperature to actualEvapTemp
        If EvapCoil calculation is valid
                //finding condensing temperature
                Set condensTemp to
                CondensCoilTemp = condensTemp
                If CondensCoil calculation is valid
                        Calculate compressor thermal power for actualEvapTemp and condensTemp
                        Calculate relative error
                        If Error < tolerance
                                Set Success to 1
```

```
                Else
                        Increment actualEvapTemp
                        Increment numIter
Else

        Increment actualEvapTemp
        Increment numIter
```

7. References

[1] Santa R and Garbai L 2013 *Annals of Faculty Engineering Hunedoara* 271 – 80
[2] Li S, Zhang Y D, Li Y and Liao Q R 2015 *International Journal of Heat and Technology* **33** 123 – 28
[3] Trutassanawain S and Groll E 2004 *International Refrigeration and Air Conditioning Conference at Pardue* 679
[4] https://mathworld.wolfram.com/CubicFormula.html

IOP Conf. Series: Materials Science and Engineering **909** (2020) 012014 doi:10.1088/1757-899X/909/1/012014

Monitoring and controlling electricity consumption using Wemos D1 Mini and smartphone

Wiwid Suryono, Achmad Setiyo Prabowo, Suhanto and Abdul Mu'ti Sazali

Poltekbang Surabaya, Jalan Jemur Andayani I No 73 Surabaya, 60236, East Java, Indonesia

*Email: widsuryono@gmail.com

Abstract. The purpose of this study is utilizing the system that can monitor and control electricity consumption using a smartphone remotely. Finding in this study related monitoring and controlling electricity consumption using wemos D1 Mini and blynk app on a smartphone is quite effective. The electricity consumption information is displayed on LCD of wemos D1 mini and blynk app on a smartphone in the form of voltage, ampere, watt, and frequency. This device has two control function namely local mode and remote mode. Using local mode will generate output load turned on by switch and using remote mode will generate output load activated via blynk app on a smartphone.

1. Introduction

The electricity provider in Indonesia that is held by an Indonesian state-owned electricity company (PLN) applies the pre-paid payment system for using electrical energy. People are required to top up the credit with a certain amount so the electrical energy can be used [1]. Excessive electricity consumption is caused by unmonitored electricity consumption leading to electrical energy waste. Besides, the uncontrolled use of electrical equipment has a big risk of fire due to excessive current load [2]. One of the ways to solve these problems is by utilizing a system that can monitor and control electricity consumption [3].

In the previous researches, Yun J. et.al. had presented a way of monitoring and controlling energy consumption of electronics and devices using smart sockets and smartphones with a Java-based program. They developed wireless smart power sockets which can measure power consumption of electrical devices, and transmit the collected data set to a gateway connected to a host server.

This study aims to develop a system that can monitor and control electricity consumption using a smartphone remotely. The design of this device uses a common power socket in Indonesia with the purpose that the installation of this device is effortless. The information provided on displays are voltage, ampere, watt, and frequency. Besides, users can also control electricity consumption using a smartphone. The system of monitoring and controlling electricity has a big impact on society for economizing electricity consumption by turning off the unnecessary electrical device using a smartphone.

2. Literature review

2.1. Wemos D1 Mini

Wemos D1 Mini is a microcontroller based on ESP8266 module development which can connect to other microcontroller devices like arduino using the internet via Wi-Fi. The advantage of Wemos D1

Mini is this device can stand alone processing each code that comes in without using arduino as its microcontroller since wemos D1 mini has a Wi-Fi bundling module. There are digital pin and analog pin inside wemos D1 mini. Digital pin can be configured as input or output. Analog pin has a 10-bit resolution at 3.2-volt maximum [4].

Figure 1. Wemos D1 Mini

2.2. PZEM 004-T

PZEM 004-T module is a multifunction sensor module that is used to measure watt, voltage, ampere, energy, and frequency of the electric current. This module has been equipped with an integrated voltage sensor and a current sensor (CT). This device is designed for indoor usage with the load installed not more than 80-260 VAC and 100A [5].

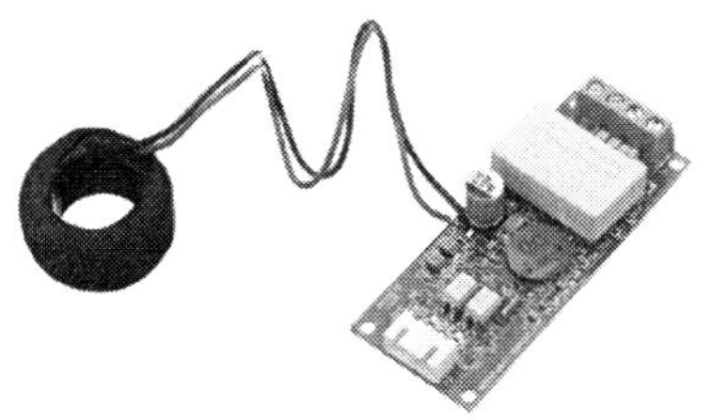

Figure 2. PZEM 004-T

2.3. Relay

Relay is a switch that operated by electricity and electrochemical component that consists of 2 main parts such as electromagnet (Coil) and mechanical (Switch). Relay uses the electromagnetic principle to move the switch so it can deliver high voltage at low power. For example, 5V and 50 mA of an electromagnetic relay can move armature relay to deliver electricity at 220V 2A [6].

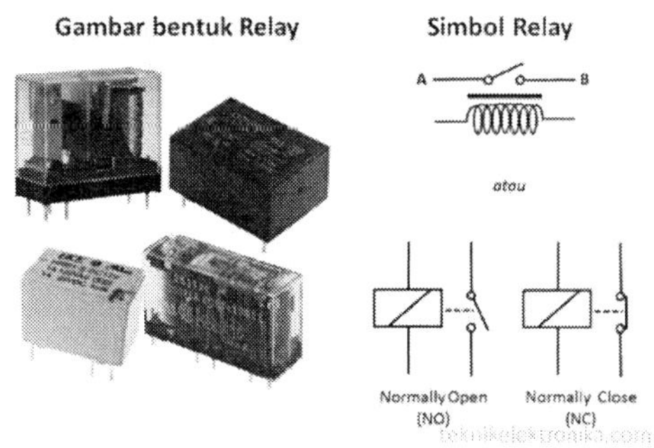

Figure 3. Relay

2.4. Alphanumeric LCD

LCD (Liquid Crystal Display) is one of the displays made from liquid crystal which its operation uses a dot matrix system. LCD mostly used for electronic devices display such as calculator, digital multitester, digital clock, etc. LCD is easily connected to Arduino. The LCD is 2x16 with 2 rows and 16 columns wide display. This LCD has been equipped with its controller so that it is easy to use and adjust the data pin of its LCD with a microcontroller [7].

Figure 4. Alphanumeric LCD

2.5. Blynk

Blynk is a server service that is utilized to provide internet of things project (IoT). This server service has mobile users both android or iOS. Blynk is a digital dashboard with graphic user interface facilities in the making of its projects. Adding components at Blynk app with drag and drop simplifies the adding of input/output components without programming ability.

Blynk was created for controlling and monitoring remotely using the internet or intranet (LAN network). The ability to store data and display data visually using the number, color or graphics facilitates creating projects in the field of the Internet of Things (IoT) [8].

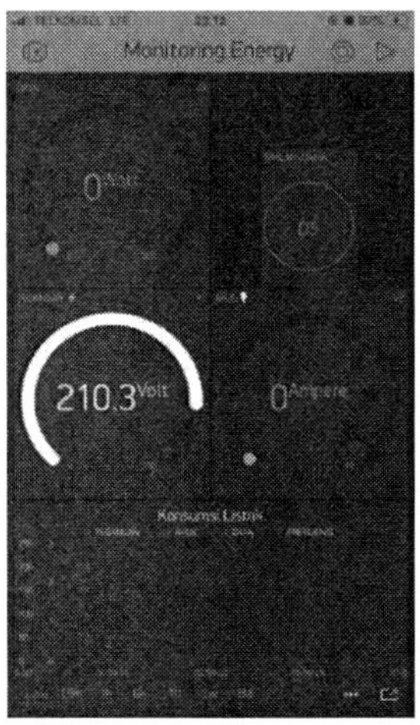

Figure 5. Blynk app.

3. Methods

The designing, manufacturing, and testing of this device need to be organized into a complete and clear design scheme. The design process involves the selecting components process, making frame design, designing the physical device form, and creating an interface of blynk application. Several steps are made to make this system well functioned and facilitated in making this device, it is shown as the flow chart in figure 6.

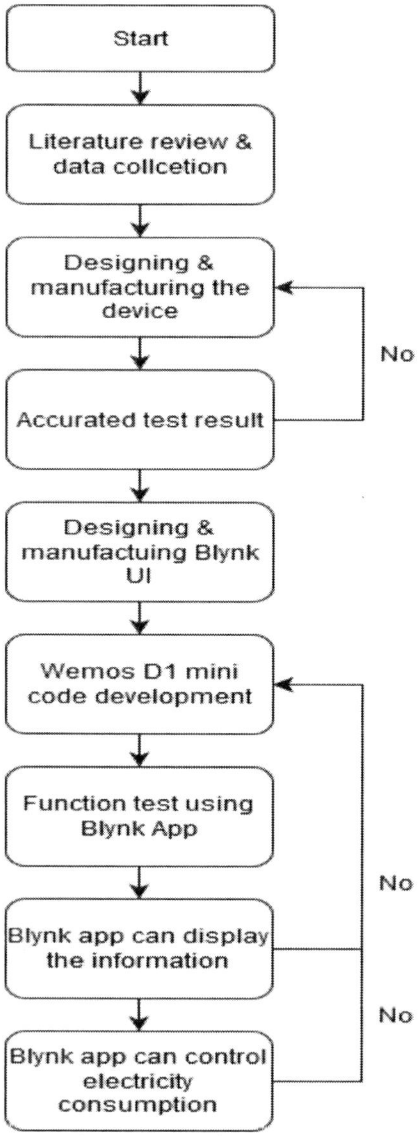

Figure 6. Method flowchart.

This research was started by collecting information associated with this study. The next step is designing and making a device that can measure the electric load. If the test shows good results so the working process will be continued to design and manufacturing graphic user interface in Blynk app. When those processes have been over so the development of wemos D1 mini code program will be carried out based on monitoring and controlling devices. If the test does not show good result so it needs repair and evaluation to code the program that has been made.

Wemos D1 mini is the main component in this study as a processor for the whole system and connector to the internet, PZEM-004T as voltage, ampere, watt, frequency (Hz) sensor and relay module as an electric switch that controls the electric load. Data processing in the cloud server is using Blynk platform as one of the IoT services that have complete tools in a smartphone. The whole system design of the monitoring device is shown in figure 7.

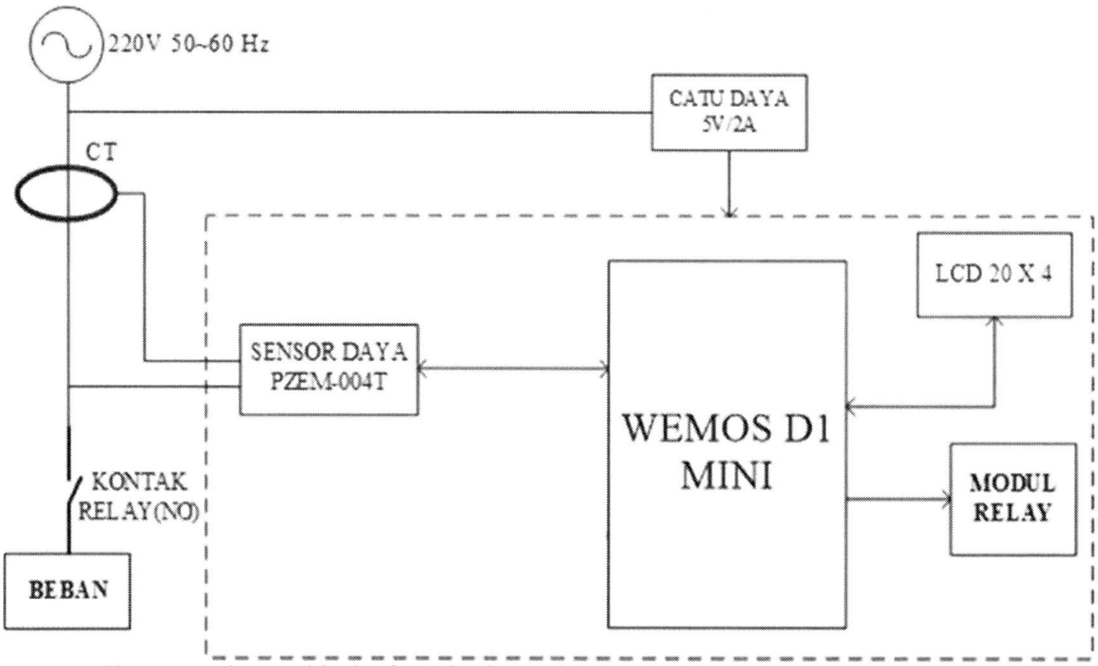

Figure 7. Diagram block of monitoring and controlling electricity consumption system.

4. Result and Discussion

The results of assembling monitoring devices and electric power control are shown in Figure 8.

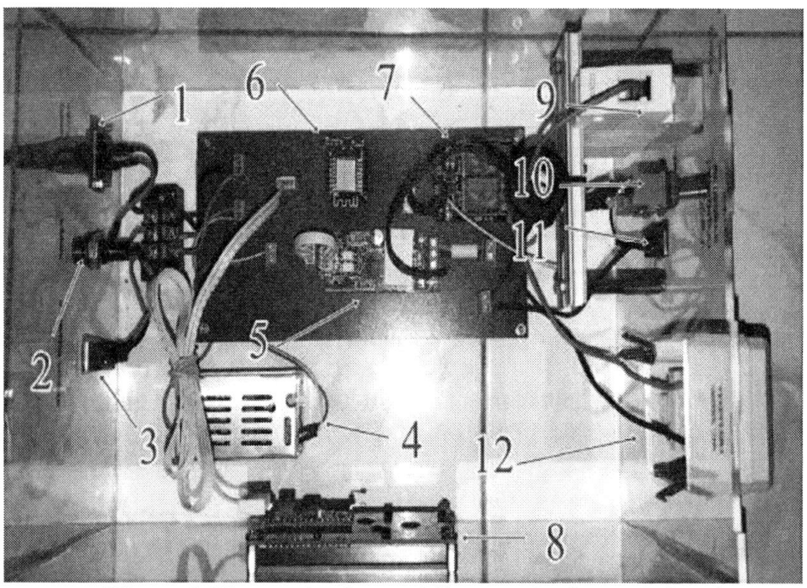

Figure 8. The result of assembling monitoring and electric power control with Wemos D1 Mini using a smartphone.

The detail explanation of each part of the assembling result as shown in Figure 8 is explained in the following table:

IOP Conf. Series: Materials Science and Engineering **909** (2020) 012014 doi:10.1088/1757-899X/909/1/012014

Table 1. Parts of hardware systems

No.	Parts of hardware system
1	AC Input 220 Volt AC
2	Fuse 5 Ampere
3	Switch power On/Off
4	Power supply 5VDC/2Ampere
5	Sensor PZEM 004-T
6	Wemos D1 Mini
7	Relay module 5Volt Dc
8	Alphanumeric LCD 20x4 Character
9	Miniature circuit breaker 4 Ampere
10	Switch local/remote
11	Switch output load On/Off
12	Socket 220 Volt AC

Figure 8 and Table 1 explain that the AC 220VAC voltage input transmits a fuse of 5Ampere, then the power supply voltage stand by the power switch. When the switching power is turned on, the voltage to the power supply is reduced until 5VDC / 2Ampere. The 5VDC voltage is used to switch on the system in this series of devices such as Wemos D1 Mini, 5Vdc Relay Module, and 20x4 Character LCD. Meanwhile, when the Local Switch and Output load are turned on then the voltage is supplied to the 220VAC socket for distribution to the load. However, when the switch is in the remote position then the voltage output at the 220VAC socket is controlled via the *blynk* application on the smartphone.

Power Supply test uses a 5Vdc power supply because wemos D1 mini requires 5 Vdc power supply. Conducting a power supply test in advance ensures the amount of voltage issued following the required voltage.

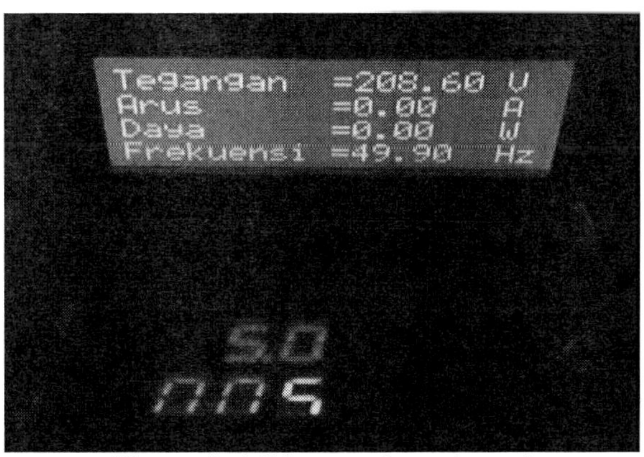

Figure 9. AC input voltage (upper) and DC output (lower) Power Supply

Figure 9 is the testing result of the 5 Volt DC power supply that when the input voltage is 208.60 Volt AC, then the output voltage of the Power Supply remains constant at 5 Volt DC. The details are shown in table 2.

Table 2. PZEM Sensor Measurement Accuracy Testing

AC input voltage	DC output voltage
205.20 VAC	5 VDC
206.70 VAC	5 VDC
207.20 VAC	5 VDC
208.60 VAC	5 VDC

After conducting a test with the result shown in table 2, it can be concluded that the power supply is functioning normally even though the input voltage rises or falls but the result at the output voltage of the power supply remains constant at 5Vdc.

The result of the PZEM sensor reading test is measuring voltage value, ampere, effective watt, and frequency. The measured data by the PZEM sensor will be displayed into LCD then it is compared to the measured score in Avometer. Data result of PZEM Sensor Measurement Accuracy Testing is displayed in table 3.

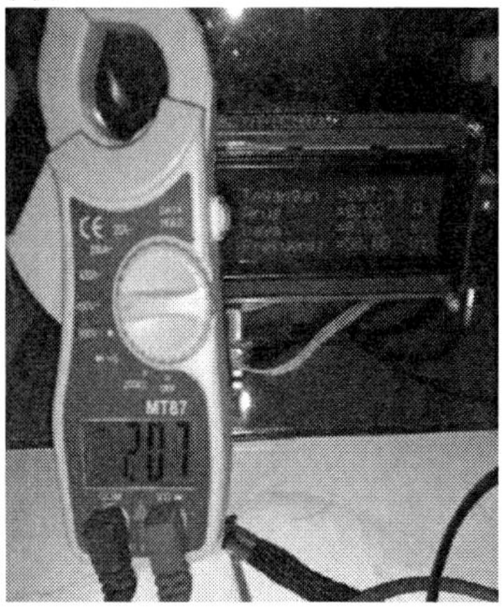

Figure 10. Voltage measurement using PZEM and Avometer.

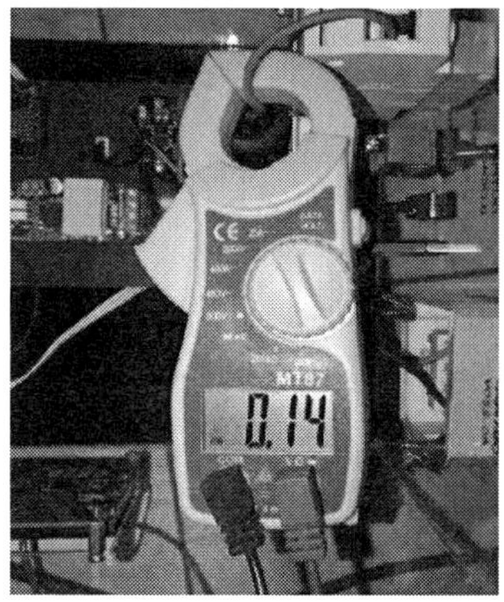

Figure 11. Ampere measurement using a Clamp meter

Table 3. PZEM Sensor Measurement Accuracy Testing

Type of Load Measured	Avometer Voltage (V)	Avometer Ampere (A)	Avometer Watt (W)	LCD Voltage (V)	LCD Ampere (A)	LCD Watt (W)
Solder	209	0,13	27,17	208,95	0,13	27,16
HP Charger	209	0,06	12,54	207,50	0,06	12,45
HP Charger & Solder	209	0,18	37,62	207,60	0,17	35,29
Solder & Charger HP	209	0,14	29,26	209,00	0,14	29,26
Laptop v	209	0,19	39,71	206,55	0,19	39,24
Fan	209	0,05	10,45	208,45	0,05	10,42
Printer	209	0,32	66,88	208,12	0,31	64,52
Laptop Charger, Fan & Printer	209	0,46	96,14	208,87	0,45	93,99

Relative error percentage is counted in each sensor measurement results data. The calculation data of relative error percentage of PZEM sensor measurement results are presented in Table 4, using the following formula.

$$\%error = \frac{(avometer\ measurement - PZEM\ measurement)}{avometer\ measurement} \times 100\%$$

Table 4. PZEM Sensor Measurement Accuracy Testing.

	Relative error percentage		
Type of Load Measured	Voltage (%)	Ampere (%)	Watt (%)
Solder	0.000	0.000	0.000
HP Charger	0.007	0.000	0.007
Solder & Charger HP	0.007	0.056	0.062
Laptop v	0.000	0.000	0.000
Fan	0.012	0.000	0.012
Printer	0.003	0.000	0.003
Laptop Charger, Fan & Printer	0.004	0.0031	0.035

The calculation results of the average relative error percentage value for each measurement are shown in Table 5.

Table 5. Calculation of average error percentage

Average error percentage		
Voltage (%)	Ampere (%)	Watt (%)
0.004	0.014	0.018

Blynk application is provided monitoring load data in the form of voltage, ampere, watt, and frequency. The testing result of the load monitoring function as shown in table 6.

Table 6. Data Testing Results of the Load monitoring function

LCD				Blynk app				
Voltage (V)	Ampere (A)	Watt (W)	Frequency (Hz)	Voltage (V)	Ampere (A)	Watt (W)	Frequency (Hz)	Delay Time Data Updated (mS)
208,95	0,13	27,16	50	208,95	0,13	27,16	50	305
207,50	0,06	12,45	50	207,50	0,06	12,45	50	365
207,60	0,17	35,29	50	207,60	0,17	35,29	50	315
209,00	0,14	29,26	50	209,00	0,14	29,26	50	300
206,55	0,19	39,24	50	206,55	0,19	39,24	50	332
208,45	0,05	10,42	50	208,45	0,05	10,42	50	310
208,12	0,31	64,52	50	208,12	0,31	64,52	50	335
208,87	0,45	93,99	50	208,87	0,45	93,99	50	354
Delay time average							325,3	

Table 6 above shows the testing result of the load monitoring function with a Smartphone that the average delay time is 325.3 mS. With the longest delay time of 365 mS and the fastest delay of 300 mS then the response of the monitoring function is quite good.

The load control function in Blynk application is on the menu tab containing widgets that can be utilized to control the output load. Load output control is a tool to stream (On) or cut (Off) the ampere from source to load. Data Testing Results of the Load control function are displayed in table 7.

Table 7. Data Testing Result Load Control Function

Button status of Blynk application	Relay Indicator in the monitoring device	Voltage measured in Voltmeter	Delay time of load control from Blynk application (mS)
ON	ON	209	803
OFF	OFF	0	856
ON	ON	209	914
OFF	OFF	0	855
OFF	OFF	0	845
ON	ON	209	945
ON	ON	209	852
ON	ON	209	795
OFF	OFF	209	745
ON	ON	209	756
	Delay time average		845

Table 7 explains the testing results of the load control function. The obtained result is the average delay of 845 mS. With the longest delay of 945 mS and the fastest delay of 745 mS, then the load control function response is quite good.

5. Conclusion

Based on design, manufacturing, and testing process that can be resumed as below:

1. Monitoring and electric power control can be made from Wemos D1 Mini and blynk platform. LCD as a monitoring function and Blynk application provides voltage, ampere, watt, and frequency. 325,3 mS is the average time that is required by the system to update data in blynk application and the fastest time is 300 mS and the longest delay time is 365 mS.
2. Load control function has been created with 2 types of function control, local control and remote-control function that can be managed by Blynk application. The average time required by the system is 845 mS, the fastest process is 745 mS and the longest delay time is 945 mS when it is using to control the load from Blynk application.
3. Measurement accuracy testing by the sensor has a relative error percentage in 0,004 % to measure voltage, 0,014 % to measure ampere and calculate the power is 0,018%.

Finally, to accomplish monitoring and electric power control better there are several things to be concerned in designing and operation including (1) a good internet connection network is required to communication from wemos D1 Mini to Blynk platform is not being hampered and (2) using Avometer True RMS for better measurement accuracy.

References

[1] Budianto A and Saragih H 2011 Penerapan system listrik pln prabayar dengan penggunaan dan pengoperasian kwh meter prabayar secara it dalam e-payment sistem pulsa listrik *Journal Sistem Informasi* 7 77

[2] Lee B W Sim J Park K B and Oh I S 2008 Practical application issues of superconducting fault current limiters for electric power systems *IEEE Transactions on Applied Superconductivity* 18 620–3
[3] Amir A Marwanto A and Nugroho D 2018 Rancang bangun purwarupa alat monitoring dan kontrol beban satu fasa berbasis IoT (internet of things) *Transmisi* 20 29
[4] Chooruang K and Meekul K 2019 *Design of an IoT energy monitoring system International* Conference on ICT and Knowledge Engineering November 2018 (IEEE Computer Society) 48–51
[5] Alqinsi P Matheus Edward I J Ismail N and Darmalaksana W 2018 IoT-based UPS monitoring system using MQTT protocols *Proceeding of 2018 4th International Conference on Wireless and Telematics ICWT 2018* (Institute of Electrical and Electronics Engineers Inc.)
[6] Gurevich V 2018 Electric relays: principles and applications (CRC Press)
[7] Wilcher D and Wilcher D 2012 LCDs and the Arduino *Learn Electronics with Arduino (Apress)* 179–204
[8] Jesani P P Raval P T J and Chaudhary P K A 2017 A review on iot based smart home using blynk framework *Ijariie* 3 624–32
[9] Zienkiewicz O C and Zhu J Z 1987 A simple error estimator and adaptive procedure for practical engineerng analysis *Int. J. Numer. Methods Eng.* 24 337–57
[10] Yun J, Lee SS, Ahn I Y, Song M H, Ryu M W 2012 Monitoring and control of energy consumption using smart sockets and smartphones in Kim T et al (eds) Computer applications for security, control and system engineering. *communications in computer and information science* Springer 339 Berlin Heidelberg

Effect of the surface of the rough pipe on the fluid flow rate

Oktarina Heriyani[1*], Dan Mugisidi[2], Irfan Hilmi[2]
[1]Electrical Engineering, Faculty of Engineering, University of Muhammadiyah Prof. Dr. HAMKA
[2]Mechanical Engineering, Faculty of Engineering, University of Muhammadiyah Prof. Dr. HAMKA

*Email: oktarina@uhamka.ac.id

Abstract. The concept of controlling the flow over the circular cylinder to reduce drag force and vibration caused by vortex flow has been explored by many researchers. One of them is by adding plates and helical strakes. However, the splitter plate is only useful for determining the direction of the current, while helical strakes increase drag force compared to cylinder forces of the same diameter. The problem that arises is that the surface roughness of the pipe will affect the characteristics of the fluid flow through it, so this study aims to analyze the characteristics and effects of fluid flow through the rough surface of the outer wall of the pipe by visualizing using Particle Image Velocity (PIV). The variation of pipe roughness used in this study is the roughness of sandpaper 100 and sandpaper 400 that encloses the pipe. The results showed that the addition of surface roughness resulted in the addition of the distance of the formation of vortex where the distance of 0.07 m was the longest distance from the bottom vortex formation on the pipe wrapped in sandpaper 100.

1. Introduction

In the field of marine engineering, especially applications for offshore areas, such as the installation of platform pillars and pipelines still use a cylindrical structure. Uncontrolled flow over the fluid pipe causes the researchers to explore. Many researchers have explored the concept of controlling flow over a circular cylinder to reduce shear forces and vibrations caused by vortex flow, using several additional forms such as spiral grooves [1]. The researchers reduced drag force and vibration caused by the vortex structure by adding plate splitters [2] and helical strakes [3]. Some research results show that the addition of these tools can reduce the interaction between the two sliding layers separated from both sides of the cylinder. This results in partial or total removal of the vortex wave. However, the splitter plate is only useful for determining the direction of the current, while helical strakes increase the drag force compared to the cylinder force at the same diameter [4].

In fact the fluid flow experiences a decrease in pressure as the length of the pipe through which the fluid passes. Pressure drop is closely related to surface roughness of one of the pipes. Commercially available pipes, roughness is not uniform and not passed with certainty. The use of smooth pipe surfaces is very rarely found, whereas rough pipe surfaces are used. Likewise, the pipes are very old, usually quite a lot of crust or corrosion and dirty arise so that it has a very large roughness value. Surface roughness not only reduces the average resistance in the cylinder, but also leads to a lower critical Reynolds number [5] [6]. On the bridge pillar, vortex flow that occurs has eroded properties [7]. The problem that arises is the surface roughness of the pipe will affect the characteristics of the fluid flow through it. This study aims to analyze the characteristics and effects of fluid flow through the rough surface of the outer wall of the pipe by visualizing using Particle Image Velocity (PIV).

2. Methodology

This research was conducted experimentally by simulating the results with Particle Image Velocity (PIV). The test is carried out using a tunnel system, like Figure 1 below:

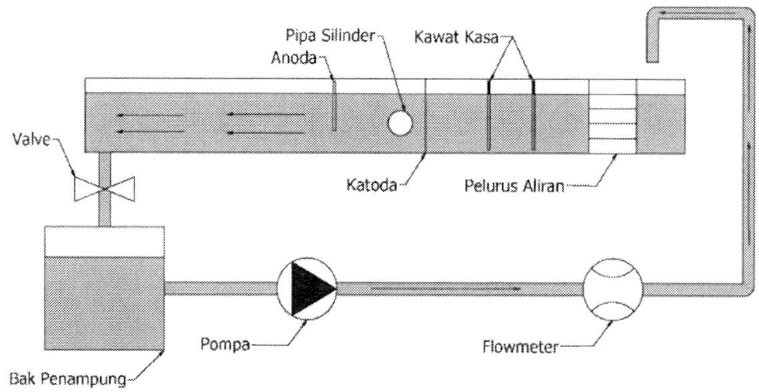

Figure 1. Testing system

In this study, the pump takes water from the reservoir and then flows into the tunnel. Water that falls into the tunnel will experience turbulence. In order to reduce the effects of turbulence, three sections of flow straighteners are provided to reduce the effect. In the part closest to the position of the fall of water, given straighteners made of straw and stream straighteners 2 and 3 made of wire mesh so that the water flowing toward the object is laminar flow. To maintain the height of the water in the tunnel, a valve is given downstream so that the outflow of water can be likened to the incoming water flow. This research was carried out at the speed of one flow by comparing two cylindrical pipes which have different surface roughness. Water flows through a flow breaker made of straws and wire mesh. Then, water flows through the cathode in the form of a 0.15 mm diameter stainless steel wire, and the anode in the form of alumunium rod, object, and returns to the reservoir. the object through which the flow passes is a pipe that has been wrapped in outer surface with sandpaper 100 and sandpaper 400. The tunnel building used has the same geometry as used in the study as figure 2 follows:

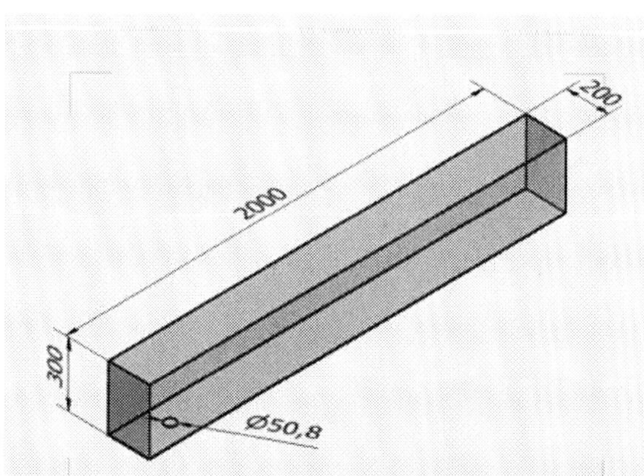

Figure 2. Tunnel geometry

Data is collected by recording phenomena that occur when bubbles are carried by flow and pass through pipes in tunnels using a 60 fps camera. Then, the video format is converted to an image using the Video to JPG Converter software. To be simulated and analyzed, the image is reprocessed using PIVlab from MATLAB 2017.

3. Results and Discussion

Based on the results obtained that the test on the pipe wrapped in sandpaper 100 shows that a small vortex is formed on the surface of the bottom pipe, the results can be seen directly before data processing, for the process of solving the vortex can be seen in Figure 3.

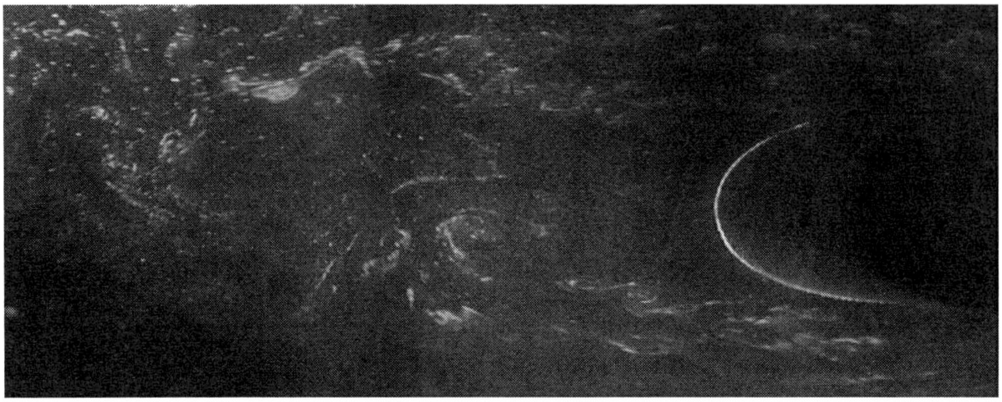

Figure 3. Vortex breaking process in pipes wrapped in 100 sandpaper

The direction of fluid movement and the recirculation zone or vortex vortex formation area can be seen in Figures 3 and 4 below.

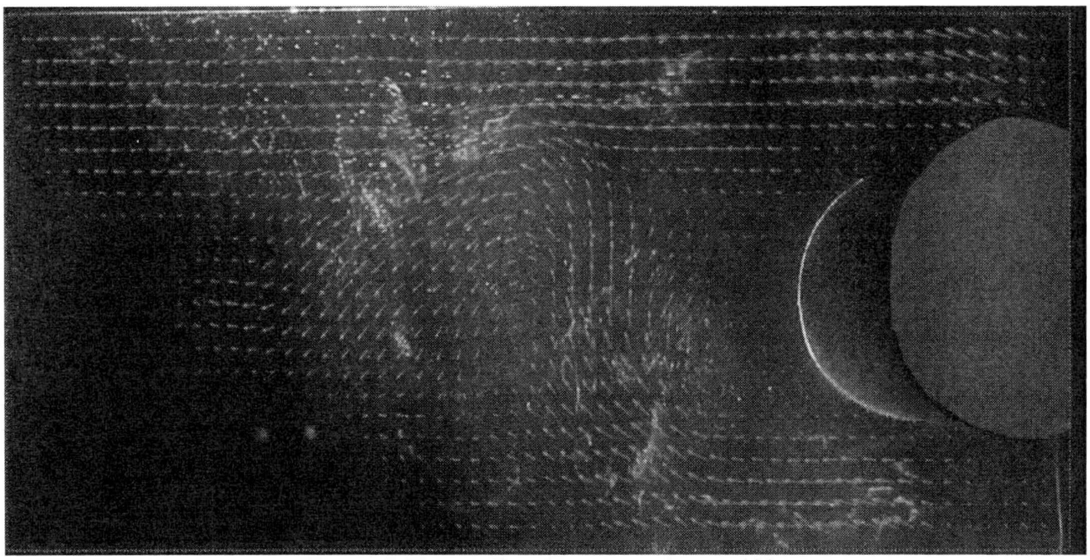

Figure 4. Vector direction on sandpaper 400

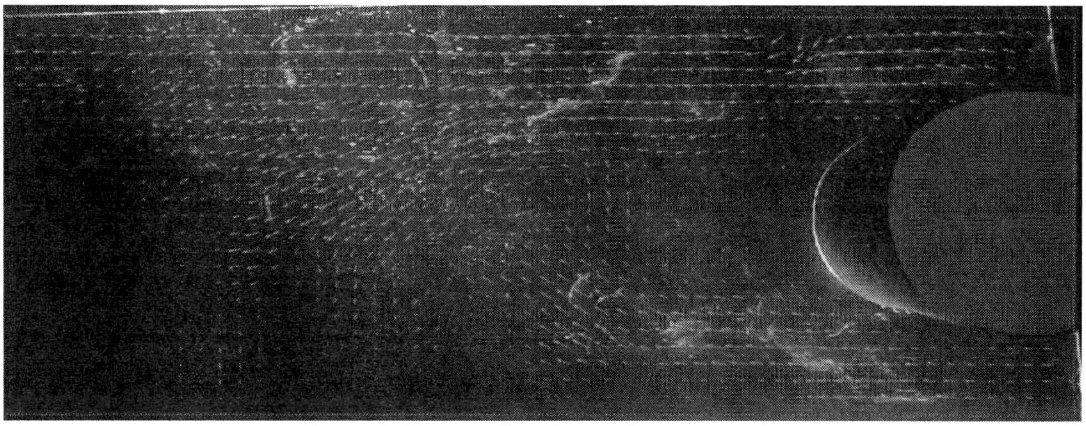

Figure 5. Vector Direction on Sandpaper 100

Figures 4 and 5 show that the direction of the vector formed has similarities in the process of vortex formation. The lower vortex is formed at a shorter distance than the upper vortex. Lower vortexes are shorter because they are affected by buoyancy [8].

At the measurement of recurlation zone, sandpaper 400 produces vortex at the bottom with a distance of 0.07 m and the vortex at the top has a distance of 0.11 m from the midpoint of the cylindrical pipe, as shown in Figure 6 below.

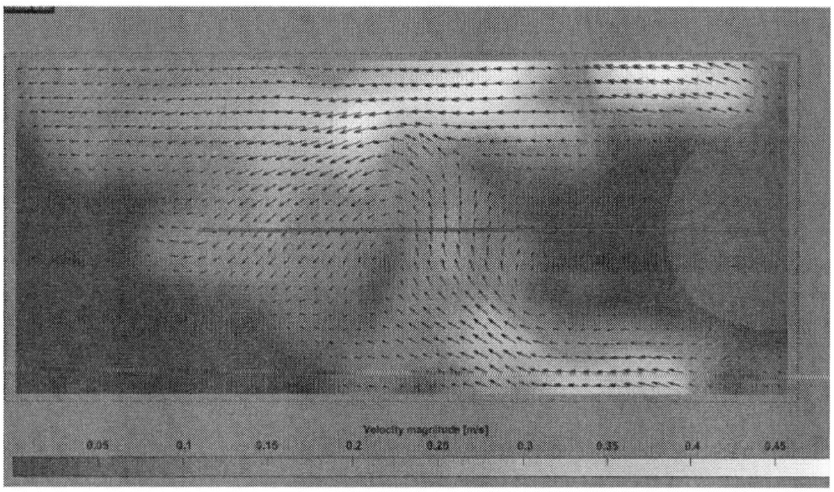

Figure 6. Velocity magnitude sandpaper 400

Measurements from the middle of the cylinder pipe to a distance of 0.16 m indicate that the highest speed profile is shown at 0.27 m / s and the lowest speed is almost 0 m / s shown in figure 7 below.

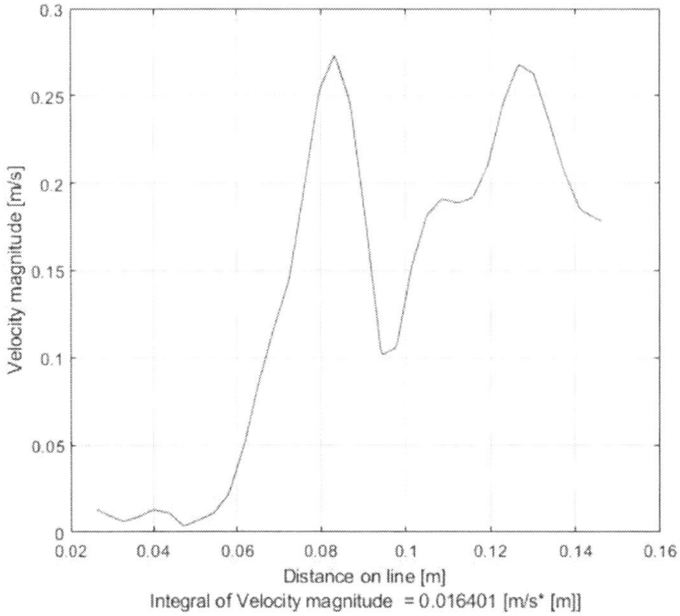

Figure 7. Sandpaper graph 400

On sandpaper 100 produces vortex on the bottom with a distance of 0.08 m and vortex on the top has a distance of 0.12 m from the midpoint of the cylinder pipe, as shown in the following figure 8.

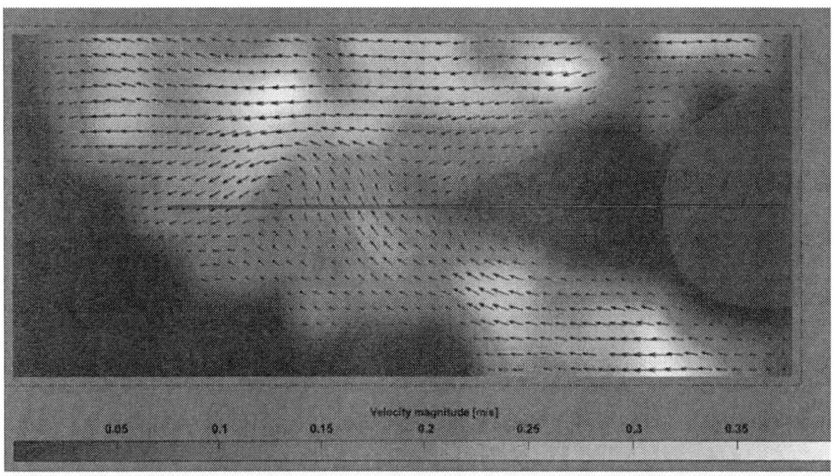

Figure 8. Velocity magnitude sandpaper 100

Measurements from the middle of the cylinder pipe to a distance of 0.16 m indicate that the highest speed profile is shown at 0.25 m / s and the lowest speed 0 m / s is shown in Figure 9.

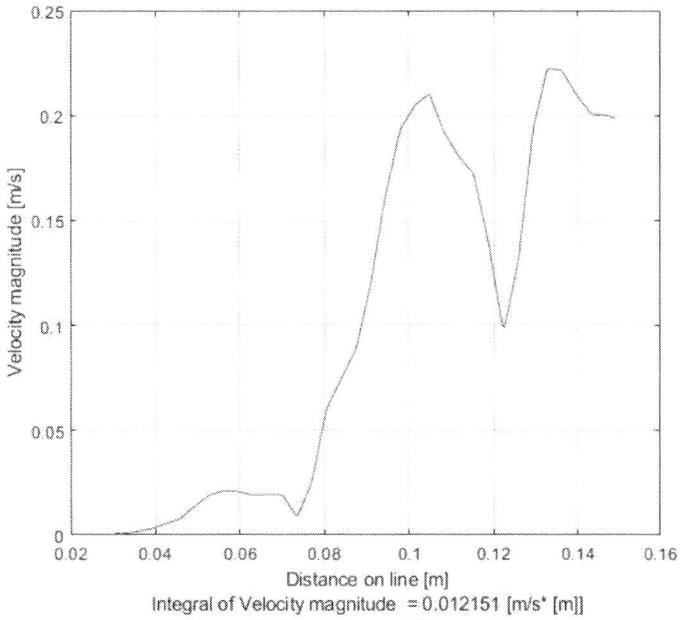

Figure 9. Sandpaper 100 graph

From the results of the analysis above shows that the vortex occurs in all the pipes at the top and bottom, the bottom vortex always forms the closest compared to the top vortex. Measurements made from the midpoint of the pipe to a distance of 0.16 m show the results of the formation of the bottom vortex closest to the pipe is a pipe wrapped in sandpaper 400 with a distance of 0.07 m, shown in figure 6. While the formation of the farthest bottom vortex, occurs in pipes wrapped in sandpaper 100 with a distance of 0.08 m, shown in Figure 8.

4. Conclusion

Fluid flow passing through the rough surface of the cylinder pipe will break the vortex flow so that at the bottom of the cylinder pipe a small vortex vortex is formed. The addition of surface roughness will result in an additional distance for the formation of vortex where the distance of 0.07 m was the longest distance from the bottom vortex formation on the pipe wrapped in sandpaper 100.

Acknowledgments

Thank you to LEMLITBANG UHAMKA for funding this internal research with contract number 667 / F.03.07 / 2019 and the UHAMKA Faculty of Engineering for facilitating its research infrastructure.

Reference

[1] T. Zhou, S. F. M. Razali, Z. Hao, and L. Cheng 2011 On the study of vortex-induced vibration of a cylinder with helical strakes *J. Fluids Struct* **27** 903–917

[2] H. Akilli, B. Sahin, and N. F. Tumen 2005 Suppression of vortex shedding of circular cylinder in shallow water by a splitter plate *Flow Mean Instrum* **16** 211–219

[3] I. Korkischko and J. R. Meneghini 2010 Experimental investigation of flow-induced vibration on isolated and tandem circular cylinders fitted with strakes *J. Fluids Struct* **26** 611–625

[4] B. Zhou, X. Wang, W. M. Gho, and S. K. Tan 2015 Force and flow characteristics of a circular cylinder with uniform surface roughness at subcritical Reynolds numbers *Appl. Ocean Res* **49** 20–26

[5] F. K. Fuss 2011 The effect of surface skewness on the super/postcritical coefficient of drag

of roughened cylinders *Procedia Eng* **13** 284–289
[6] K. Y. Kiu, B. Stappenbelt, and K. P. Thiagarajan 2011 Effects of uniform surface roughness on vortex-induced vibration of towed vertical cylinders *J. Sound Vib* **330** 4753–4763
[7] J. Ikhsan, W. Hidayat J. T. Sipil, and U. M. Yogyakarta 2006 Pengaruh bentuk pilar jembatan terhadap potensi gerusan lokal **9** 124–132
[8] M. Bazargan D. Fraser and V. Chatoorgan 2005 Effect of buoyancy on heat transfer in supercritical water flow in a horizontal round tube *J. Heat Transfer* **127** 897–902

IOP Conf. Series: Materials Science and Engineering **909** (2020) 012016 doi:10.1088/1757-899X/909/1/012016

The effect of vortex generator on the approach value on forced draft type cooling tower

Oktarina Heriyani[1*], Dan Mugisidi[2], P H Gunawan[2], Dwi Apriliana Putri[2]
[1]Electrical Engineering, Faculty of Engineering, University of Muhammadiyah Prof. Dr. HAMKA
[2]Mechanical Engineering, Faculty of Engineering, University of Muhammadiyah Prof. Dr. HAMKA

*Email: oktarina@uhamka.ac.id

Abstract. Cooling tower is needed right now in the context of efficiency and energy conventions for cooling circulation. The type of cooling tower used depends on the level of performance. One type of cooling tower used in this study is forced draft. Cooling tower in this study was added by a rectangular winglet generator vortex with the aim to increase the effectiveness of performance by reducing the value of the approach. This is because forced tower cooling draft on its utilization results in performance that is not commensurate with the costs incurred. The results obtained in this study are a comparison of the value of the forced draft cooling tower approach with the addition of a vortex generator and without the addition of a vortex generator. This research was conducted in the workshop and laboratory of Mechanical Engineering, Faculty of Engineering UHAMKA with stages: designing tools, making prototypes, assembling, and testing. Based on the results of the study it was found that the approach value obtained on cooling towers that use 0.86% lower than without a vortex generator.

1. Introduction

Cooling tower is defined as a heat exchanger used as an air conditioner with air as the cooling medium. To produce maximum work, fillers are needed that function to inhibit the rate of air flow so that fluid and air contact will be longer. The increased effectiveness of cooling tower cooling will affect the removal of some of the water into the air due to the evaporation process [1]. However, water loss during evaporation is a better assessment of cooling tower characteristics.

The type of cooling tower used depends on the level of performance produced by the cooling tower. D. Pearlmutter, E. Erell, Y. Etzion, and I.A. Meir, H. Di revealed about the use of preparation in the down-draft cooling tower [2]. The performance of a cooling tower depends on the characteristics of its material, such as fluidized [3], corrugated plastic [4], stone sponges and rubber foam [5], and bamboo wulung [6]. In addition to its content characteristics, cooling tower performance is affected by wind speed [7].

The coolingg tower performance is evaluated to discuss the design value, identify energy waste, and to improve the cooling tower engine so that it has a better impact on the cooling tower [8]. The parameters to measure cooling tower performance are range and approach. The level of approach and cooling range are assessed to identify areas where energy waste is occurring by providing improvements. Although, the range and approach must be controlled, but the approach is a better indicator for cooling tower performance. The lower the temperature of the wet air ball entering, the more effective the cooling tower [9], the cooling tower performance is assessed at the approach and cooling range level to identify areas where energy waste occurs by providing improvements. Although, the range and approach must be controlled, but the approach is a better indicator for cooling tower performance. The lower the temperature of the wet air ball entering, the more effective the cooling tower is. There are several factors that influence the effectiveness of cooling tower performance, namely the amount of water surface that comes in contact with air, and the length of time when contacting water with air, the speed of air through the tower, and the direction of air flow associated with the surface of water contact (parallel, perpendicular or opposite).

Cooling tower performance analysis obtained value range 16,4830C - 18,4240C, approach value 5,0170C - 5,7280C and effectiveness value 76,688% [8]. Researchers do a lot of research to get cooling tower performance. Previous research was only conducted in 2019 to obtain novelty from the cooling tower performance [10], the experimental results show that the effectiveness increases by 22% and 30% for NDWCT and FDWCT fill thickness of 20 cm, respectively. Cooling tower optimization is solved by using a counterflow particle optimization algorithm involving the effectiveness and rate of evaporation of water, while the air flow rate and water as a decision variable [11]. A field test study was carried out on the design of a natural wet cooling tower to investigate quantitatively the distribution of diameter of water droplets in the rain zone for the first time. The results show that it is spread between 0.312 mm and 15 mm, regardless of the cross wind speed [12].

One type of cooling tower that is rarely used in industry is forced draft [13]. This is due to the high costs incurred, while the effectiveness produced is not commensurate. This is a problem that will be examined in this study. To overcome this, this study uses a vortex generator to increase its effectiveness by reducing the value of the approach to the cooling tower.

2. Methodology

This research begins with designing tools, making prototypes, assembling, and testing. The prototype is designed as shown figure 1. The design of the study or experiment is to use a cooling tower with the addition of vortex generators as flow inhibitors and without vortex generators as in Figure 1. The amount of water and air entering the test equipment is of equal value [14], at 1 liter per minute. Water from the basin is pumped through a half inch diameter pvc pipe to the storage area. Then, the water is heated in storage using a heater to a stable temperature, which is 60.1 °C. Furthermore, it is streamed through a flowmeter. Measurement and regulation of release using valves. Water through the flowmeter is released at the top of the tower through a sprinkler to divide the water into small beads. Excess water in storage is thrown back into the water bath. At the same time air from the compressor is channeled through pipes to the bottom of the tower.

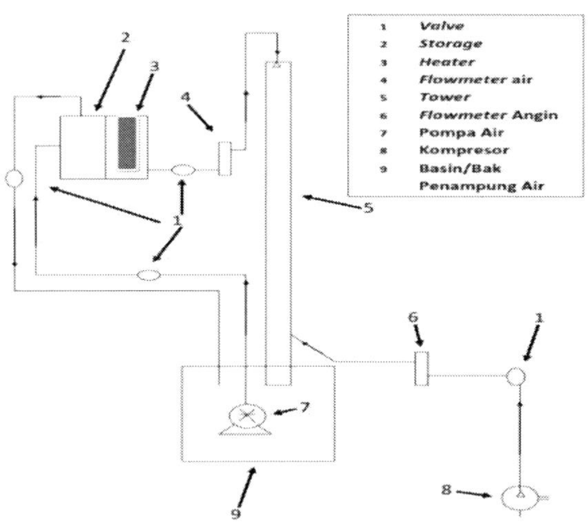

Figure 1. Cooling tower schematic diagram

The test equipment is based on the Ramkumar and Raguphaty test model [15]. The thickness of the acrylic used is 5 mm with a height of 195 cm. The size of the tower from 30 cm × 30 cm to 3 cm × 3 cm. This is because the cross-sectional area of the Ramkumar and Raguphaty 900 cm ^{2} simulation devices is divided by the number of filler holes of 56 which are assumed to have other holes of equal value. The type of vortex generator used in this study is a rectangular winglet made of aluminum with a size of 1 cm × 5 cm. The number of vortex generators used was 3 pairs with a slope of 15 ° and a distance of 35 cm in each pair. Data collection was carried out for 2 hours for 6 days. Before data collection, a leak test is performed. Leakage tests are carried out to observe the connection of the pipe with the compressor and the connection of the inlet water with the pump. Furthermore, observing data includes the cooling capacity (Q), the temperature of the inlet air ($T_{in\ air}$) and outlet ($T_{out\ air}$), the temperature of the inlet water ($T_{in\ water}$) and outlet ($T_{out\ air}$), and the temperature of the wet ball (T_{wb}). Data collection was carried out for 2 hours with a distance of observation every 15 minutes. The results of the observation data will be processed to get the approach value. Then, the approach values using the vortex generator on the cooling tower and without the vortex generator will be compared

3. Results and Discussion

Based on the test results obtained that the reynold number on cooling tower with vortex is $1,1887 \times 10^{-11}$ for air and 143.505.9 for water. While the reynold number in cooling towers without vortex generators is $3,566 \times 10^{-11}$ for air and $2,090 \times 10^{-9}$ in water. That is, cooling tower with vortex generator can bring up vortex. This is due to vortex occurring at reynold numbers more than 50 (Ciptoadi, 2011).

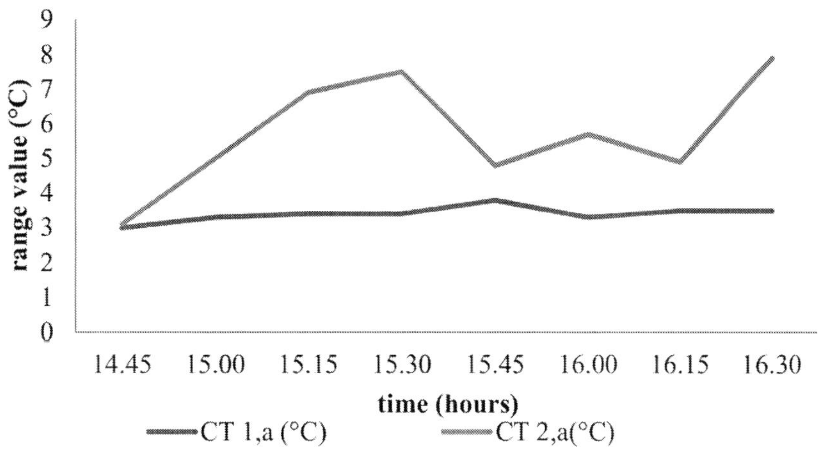

(a)

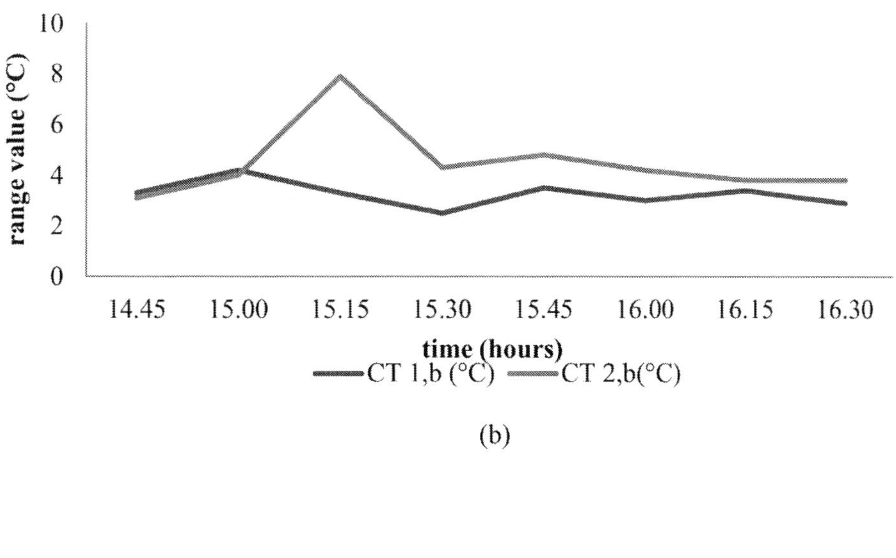

(b)

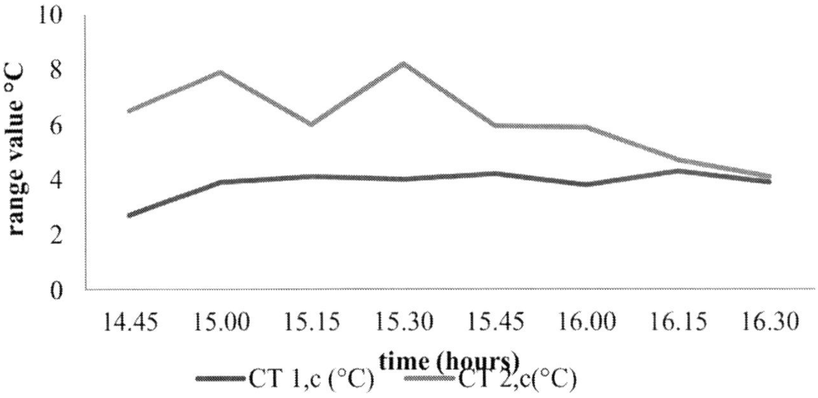

(c)

Figure 2. Comparison of range value (a) range value the first day (b) range value the second day (c) range value the third day

The figure 2 (a), (b), (c) shows that the vortex added to the cooling tower (blue line) has an average range greater than the range of the cooling tower without the vortex generator (red line). The largest range of cooling towers with vortex generators is 8.2 °C while the cooling towers without vortex generators have a range of 4.3 °C. This is because cooling towers with vortex generators lower the temperature more than cooling towers without vortex generators. High Range values indicate cooling tower can reduce temperature efficiently with good performance [14]. Therefore, the vortex generator added to the forced draft type cooling tower can increase the temperature decrease in the cooling tower.

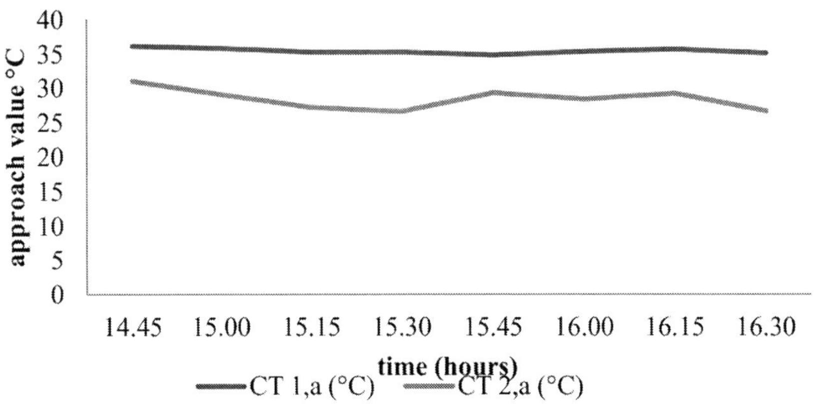

(a)

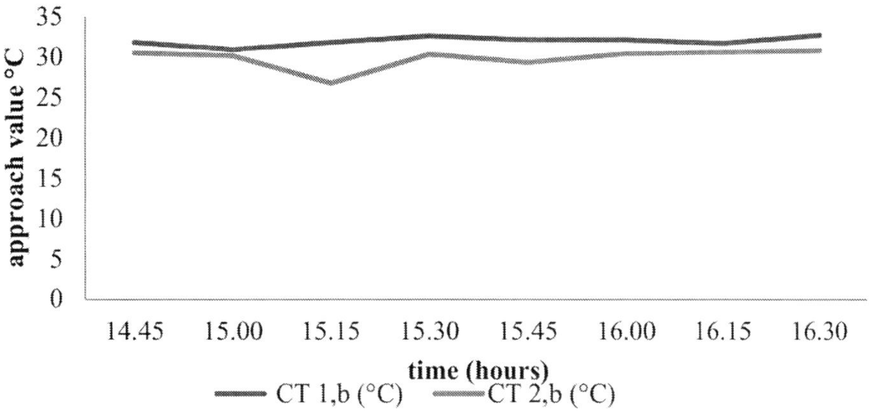

(b)

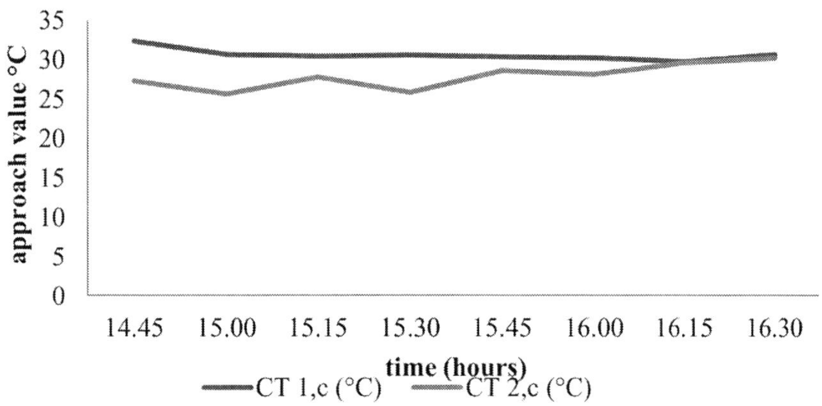

(c)

Figure 3. The Comparison Approach Value (a) approach value the first day (b) approach value the second day (c) approach value the third day

Figure 3 (a), (b), (c) shows the average cooling tower approach with vortex generator has a smaller approach than cooling tower without vortex generator. The smallest approach to cooling tower with vortex generator is 25.7 °C, while the smallest approach to cooling tower without vortex generator is 29.8 °C. That is, the approach with the vortex generator is lower than the approach without the vortex generator. Therefore, the performance of cooling towers with vortex generators is closer to the ambient temperature according to the main functions of cooling towers. Cooling tower with vortex generator is better than cooling tower without vortex generator because the lower the approach value, the better the cooling tower performance [15]. So it can be said that the application of the vortex generator is able to reduce the value of the approach to the forced draft type of cooling tower.

Cooling tower performance is assessed at the approach and cooling range level to identify areas where energy waste is occurring by providing suggestions for improvement. Although, the range and approach must be controlled, but the approach is a better indicator for cooling tower performance. The lower the temperature of the wet air ball entering, the more effective the cooling tower is. [8], the cooling tower performance is assessed at the approach and cooling range levels to identify areas where energy waste occurs by providing improvements. Although, the range and approach must be controlled, but the approach is a better indicator for cooling tower performance. The lower the temperature of the wet air ball entering, the more effective the cooling tower is.

4. Conclusion

The smallest approach to cooling tower with vortex generator is 25.7 °C, while the smallest approach to cooling tower without vortex generator is 29.8 °C. The application of vortex generators is able to reduce the approach value in forced draft type cooling towers because cooling towers with vortex generators are better than cooling towers without vortex generators because the lower the approach value, the better the cooling tower performance

Acknowledgments

Thank you to LEMLITBANG UHAMKA for funding this internal research with contract number 734 / F.03.07 / 2019 and the UHAMKA Faculty of Engineering for facilitating its research infrastructure.

Reference

[1] E. EL-Wakil, M.M., dan Jasjfi 1992 *Power Plant Technology* Erlangga
[2] D. Pearlmutter, E. Erell, Y. Etzion, I. A. Meir, and H. Di, 1996 Refining the use of evaporation in an experimental down-draft cool tower *Energy Build* **23** 3 191–197 doi: 10.1016/0378-7788(95)00944-2.
[3] P. Mulyono and sugeng wahyudi 1998 Karakteristik menara pendingin dengan bahan isian terfluidisasi
[4] P. Mulyono and K. Baskoro 2000 Karakteristik menara pendingin dengan bahan isian plastik bergelombang
[5] Wibisono 2005 Perbandingan unjuk kerja antara bahan pengisi pada menara pendingin tipe induced counterflow 152–162
[6] A. T. H, D. L. S, and H. Sutjahjono 2014 Analisis beban kalor cooling tower induced draft counterflow dengan bahan pengisi bambu wulung (heat load analysis of induced draft counterflow cooling tower with bamboo filler wulung) abstrak pendahuluan metode penelitian
[7] M. Gao, F. Sun, K. Wang, Y. Shi, and Y. Zhao 2008 Experimental research of heat transfer performance on natural draft counter flow wet cooling tower under cross-wind conditions **47** 935–941 doi: 10.1016/j.ijthermalsci.2007.07.010.
[8] Danial Ahmad Fauzi1 and B. Rudiyanto 2016 Analisa performa menara pendingin pada pt. Geo dipa energi unit dieng *J. Ilm. Rotari* **1** 125–32
[9] Ilvana M 2000 Analisa perhitungan karakteristik dan efektifitas homon cooling towers *Univ. Diponogero, Semarang*
[10] F. Hamad 2019 Evaluation of thermal performance for natural and forced draft wet cooling tower *J. Mech. Eng. Sci* 17–29
[11] Y. Zhang and et all Zhang, Huan 2019 Optimal configuration and operating condition of counterflow coolingtowers using particle swarm optimization algorithm *Appl. Therm. Eng* 318–327
[12] X. Chen and F. Sun 2019 Field test study on water droplet diameter distribution in the rain zone of anatural draft wet cooling tower *Appl. Therm. Eng.* 1–8
[13] H. P. Siallagan 2017 Analisis kinerja cooling tower 8330 ct01 pada water treatment plant-2 pt krakatau steel (persero). tbk *J. Tek. Mesin* **6** 215–219
[14] R. Ramakrishnan and R. Arumugam 2012 Optimization of operating parameters and performance evaluation of forced draft cooling tower using response surface methodology (RSM) and artificial neural network (ANN) *J. Mech. Sci. Technol* **26** 1643–1650 doi: 10.1007/s12206-012-0323-9.
[15] T. Indrawati and R. R. Ginanjar 2018 Perancangan mini cooling tower sederhana sebagai pendingin air kondensor pada proses refluks uji chemical oxygen demand (cod) **1**. 16–22

IOP Conf. Series: Materials Science and Engineering **909** (2020) 012017 doi:10.1088/1757-899X/909/1/012017

Preliminary study to determine the maximum safe speed of vehicles based on rolling moments and vehicle skid to reduce of driving accidents

D A Sumarsono[1], R Siregar[1,2,*], M Adhitya[1], N Nazaruddin[1,3], S Prasetya[1,4], F Zainuri[1,4], and G Heryana[1,5]

[1]Department of Mechanical engineering, Universitas Indonesia, 16424, Indonesia
[2]Department of Mechanical Engineering, Darma Persada University, 13450, Indonesia
[3]Department of Mechanical Engineering, Universitas Riau, 28293, Indonesia
[4]Department of Mechanical Engineering, Politeknik Negeri Jakarta, 16425, Indonesia
[5]Department of Mechanical Engineering, STT Wastukancana, 41151, Indonesia

*Email: rolansiregar@ft.unsada.ac.id

Abstract. Indonesia is one of the countries with the highest death rate due to road traffic accidents. Research on vehicle safety needs to be done immediately. Some travel companies, rental vehicles, and tourism vehicles have made permit speed regulations, but the basis for selecting these limits is not scientific. In this study will be presented how to determine the maximum speed allowed in a vehicle scientifically, and how to calculate the speed of the permit can be a reference to other vehicles. This research method is analytical and experimental methods. The normal force of the wheel when the vehicle turns will be different, where the normal force of the wheel inside will first be lifted due to rolling force, and will have an impact on the skid that might occur. So, this is becoming a reference in calculating the vehicle's maximum clearance speed. The maximum speed allowed of the vehicle under study depends on the turning radius. Therefore, knowing the wheel's normal force from various conditions and the maximum speed of the vehicle, a safety warning system can be developed in the future to avoid vehicle accidents.

1. Introduction

Data from the Traffic Division of the Indonesian National Police states that every year around 30,000 people die from traffic accidents in Indonesia[1]. Various ways have been carried out by the Indonesian government in reducing accidents such as vehicle speed restrictions that have begun to apply to the travel company vehicles and expedition trucks. This research presents about determining the maximum speed of vehicles that are still in the safe range. Because of the accident problem, this research takes an important role and must be followed up immediately in the driving accident reduction program. Vehicle speed will be related to the normal force of the wheel and automatically loaded brake in-vehicle unit[2]. Besides, the vehicle speed affects the stability of the vehicle motion, such as the risk of occurring oversteer and understeer. The instability of vehicle motion often occurs in turn conditions when the vehicle speed is outside the safe limit. Centrifugal force will pull the vehicle when circular motion and cause a decrease in normal force on the wheels inside. So, this maximum speed limit needs to be applied to public vehicles such as rental cars, tourism cars, and travel vehicles, so the risk of accidents is reduced. The aim of the study is basic for the development of a safety warning system so that drivers use vehicles with safe speeds. This speed limitation is being widely applied in public passenger companies in Indonesia to avoid traffic accidents such as understeer/oversteer, brake failure, and crash. To achieve this aim, the objectives of this research are to determine the vehicle speed maximum in turning radius and its relation to skid and rolling.

2. Literature review and problem statement

A vehicle's safety level should be known so that the driver can use the vehicle at a safe speed. Driving comfort felt by the driver is not only related to individual satisfaction but also driving safety and the long-term health of the driver and passengers [3]. This research has been widely carried out by previous researchers regarding driving comfort such as driver speed [4], aerodynamic, road surface[5], driver character, suspension system [6][7], power train [8], brake system[2], and etc. Driver speed is related to the cause of traffic accidents [9]. Many traffic signs have been made to show safe speed to avoid accidents. Road roughness has also been studied because it is related to driving comfort such as noise from wheels, slips, road slope in turn conditions, and vehicle vibrations [10].

On the other, the most significant cause of accidents is human error, so many researchers are interested in creating a safety warning system [11]. The comfort of controlling vehicles at high speeds is also related to vehicle aerodynamics, and therefore vehicle manufacturers are competing to create more aerodynamic [12]. The braking system on a vehicle is one of the essential things, so drivers and passengers can drive safely [13]. Some articles that have been reviewed are related to the safe speed of the vehicle. This study presents preliminary studies on how to connect skid, rolling, and safe vehicle speed, so there is no vehicle motion instability.

3. Methodology

Schematic determination of vehicle speed limits can be done by calculating the rolling moment, skid, and speed of a vehicle. This research, the calculation approach for Toyota Avanza 1.5 is displayed. A more detailed schematic calculation of the vehicle's maximum speed is shown in Figure 1.

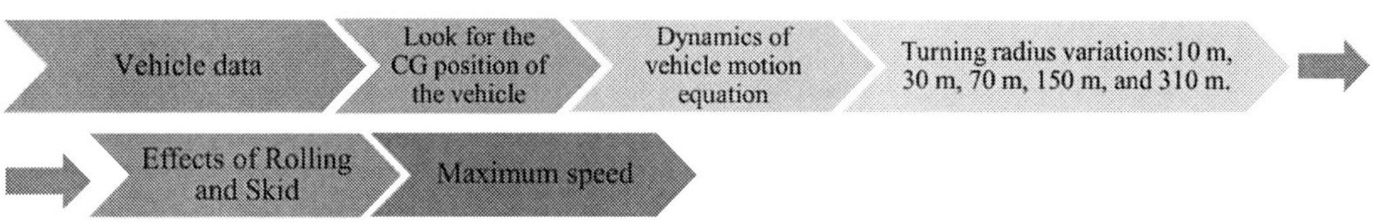

Figure 1. Schematic of determining the maximum speed

From the schematic, Figure 1 has shown the main sequence in determining the maximum speed limit of vehicles that can be applied to public passenger vehicles so that the risk of loss due to accidents can be reduced. The vehicle under review was a passenger vehicle Toyota Avanza 1.5. This type is chosen, assuming it has referenced other vehicle braking systems. These specifications can be seen in Table 1.

Table 1. Specifications of the vehicle being review

No	Item	Symbol	Value	Unit
1	Wheelbase	L	2655	mm
2	Wheel track front	tf	1425	mm
3	Wheel track rear	tr	1435	mm
4	Curb weight	W	1175	kg

Curb weight is the empty weight of the vehicle, including standard accessories and a full fuel tank, but does not include drivers or passengers as well as luggage items. This research will focus on the assessment of vehicle weight, i.e. driver only weight (DoW).

3.1. Calculation of the position of the Center of Gravity (CG)

In calculating the CG position obtained is the CG distance from the front wheel axle (p), the CG distance from the rear wheel axle (r), and the CG height from the ground (h). CG calculation can be made in the form of programming Simulink as in Figure 2.

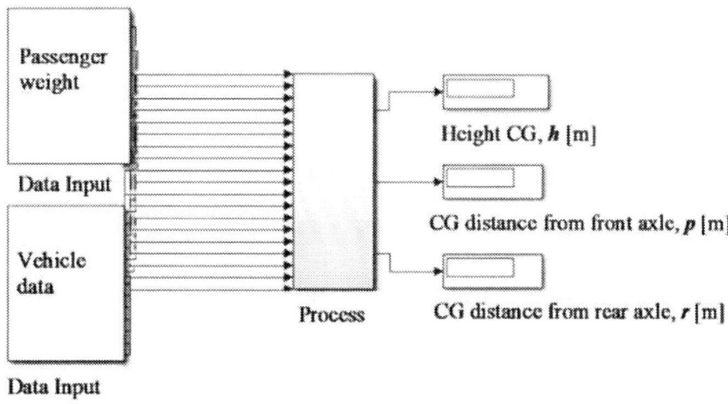

Figure 2. Simulink diagram of CG calculations

Data input is the beginning of the process to enter the number of passengers, passenger weight, passenger position following the seat, as well as vehicle geometry such as curb weight, wheelbase distance, wheel track, and etc. In the process section, there is a mathematical formula to process the input data so that the results obtained in the form of CG. Weighing is done with a slope of 13.14 degrees with the specifications of a medium-sized passenger vehicle, as shown in Figure 3.

a *b* *c*

Figure 3. Weighing the vehicle weight: *a* –Vehicle scale; *b* – Display; *c* – Vehicle type

Figure 3 can be explained that the weighing is done by thin and precision tools to facilitate the data collection process. The vehicle is weighed in a tilted condition with a standard vehicle.

3.2. The effect of rolling to normal force and allowable speed

The braking load on each wheel will be different when the vehicle turns, and the body roll causes this. Vehicles with critical speeds on turning roads can be overturned due to the body roll. The load on the outside wheel will be much higher than the load on the inside wheel. To avoid the car overturning, the driver can brake so that the speed decreases. The load distribution on each brake will be different when braking. Figure 4 shows the modeling of forces acting on the vehicle turning in the conditions of Ackerman [14], where the road is assumed to be flat.

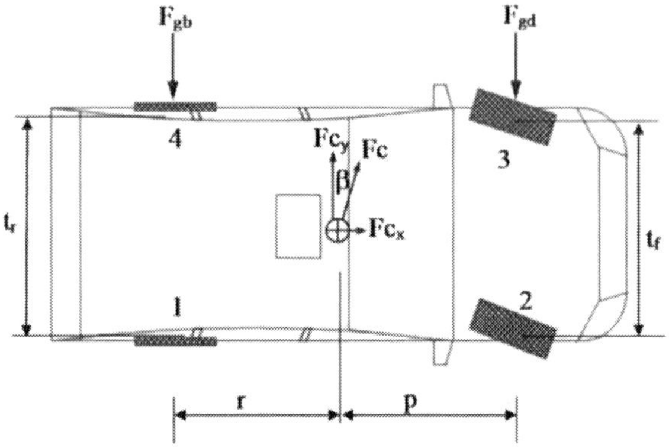

Figure 4. The force acting on a vehicle turn with ideal conditions

Based on Figure 4, the sign notation can be explained Fgd and Fgb shows the frictional force of the road surface roughness for front and rear wheels, β show the angle side slip vehicle when turning, tr shows the width of the rear axle wheel track, and on the front wheels is tf. The p and r notations show the distance between the axle to the center where p plus r is called the wheelbase (l), wheels 1 and 2 are grouped into wheels inside, and wheels 3 and 4 are grouped into wheels outside. Bolt / rolling analysis is carried out to obtain a reasonable force on each wheel. The wheel will lift if the normal force on one of the wheels inside is zero. The normal force equation on each wheel can be displayed in equations 1, 2, 3, and 4 assuming the effect due to the wind is very small, so it does not count as well as the effect of suspension [15][16].

The normal force on wheel 1 can be seen in (1):

$$F_{Z1} = \frac{W \cdot p}{2(p+r)} - \frac{\left(\frac{p}{p+r}\right) \cdot F_c \cos\beta \cdot h}{tr} - \frac{F_c \sin\beta \cdot h}{2(p+r)}, \tag{1}$$

Then the normal force on the wheels 2 can be seen in (2):

$$F_{Z2} = \frac{W \cdot r}{2(p+r)} - \frac{\left(\frac{r}{p+r}\right) \cdot F_c \cos\beta \cdot h}{t_f} + \frac{F_c \sin\beta \cdot h}{2(p+r)}, \tag{2}$$

The normal force on the wheels 3 can be seen in (3):

$$F_{Z3} = \frac{W \cdot r}{2(p+r)} + \frac{\left(\frac{r}{p+r}\right) \cdot F_c \cos\beta \cdot h}{t_f} + \frac{F_c \sin\beta \cdot h}{2(p+r)}, \tag{3}$$

The normal force on the wheels 4 can be seen in (4):

$$F_{Z4} = \frac{W \cdot p}{2(p+r)} + \frac{\left(\frac{p}{p+r}\right) \cdot F_c \cos\beta \cdot h}{tr} - \frac{F_c \sin\beta \cdot h}{2(p+r)}. \tag{4}$$

The critical condition of the vehicle to be rolled over is when one of the forces on the inside wheel is zero. If the review is $F_{Z1} = 0$ we will get the maximum allowable turning speed as in (5) assuming the angled side slip β close to zero, and the wind load is so small that it doesn't count [15][16].

$$v_{rg} = \sqrt{\frac{R \cdot g \cdot t_r}{2h}} \tag{5}$$

To get the value of the centrifugal force can be done by calculation using (6).

$$F_c = \frac{m \cdot v^2}{R} \tag{6}$$

And the value of R is the turning radius, which is approached under ideal conditions as in (7), and *v* is the vehicle speed in units (m/s).

$$R = \frac{p+r}{\delta_f} \cdot 57,29 \tag{7}$$

From the equations above can be explained notation δ_f is the steering angle in degrees, *m* is the mass of the vehicle (kg), *g* is the acceleration of gravity (m/s^2), *h* is the height of the center of mass to *ground.*

3.3. The effect of the skid on the normal force of wheel and allowable speed

The skid phenomenon in question is a shift in the car towards the outside body of the vehicle, which is not appropriate. Skid will occur if the lateral force, such as centrifugal force is greater than the lateral force of the wheel that holds it (Figure 5) [15].

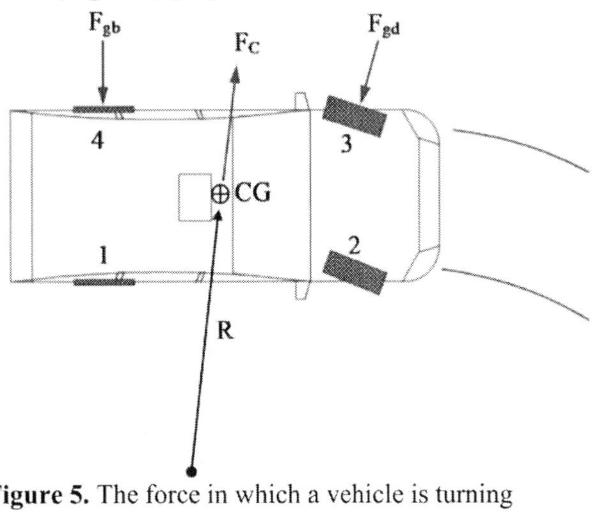

Figure 5. The force in which a vehicle is turning

The critical condition of a vehicle is when the side forces (Fc) of the vehicle are the same as the lateral forces (Fgd+Fgb) acting on the wheel that holds it. The side forces acting on the front and rear wheels are the centrifugal force that has been distributed that is the centrifugal force on the front wheels and the centrifugal force on the rear wheels. The wind force that works is considered so small that it is ignored. Centrifugal force equation on the front wheels (Fcd) and rear (Fcb) can be displayed in (8) and (9) below [15] [16].

$$F_{cd} = \frac{r}{p+r}(F_c \cos\beta) \tag{8}$$

$$F_{cb} = \frac{p}{p+r}(F_c \cos\beta) \tag{9}$$

A critical condition where the front wheel will be skid for a moment if the centrifugal force on the front wheel is equal to the coefficient of road friction times the normal front wheel force, as shown in (10).

$$F_{cd} = \mu \cdot Fzd \quad (10)$$

Assuming angles side wall $\beta = 0$, the maximum speed of the vehicle turns so that the front wheels don't skid can be formulated in (11).

$$v_d = \sqrt{\mu R g} \quad (11)$$

Where R is the turning radius of the road (m) and μ is the coefficient of road friction. Furthermore, the critical condition where the rear wheel will skid if the centrifugal force of the rear wheel is equal to the multiplication coefficient of road friction with the normal force of the wheel (12).

$$F_{cb} = \mu \cdot Fzb \quad (12)$$

If the force and moment of the wind are ignored and the angle $\beta = 0$ then the maximum speed of the vehicle turns so that the rear wheels do not skid can be shown in (13).

$$v_b = \sqrt{\mu R g} \quad (13)$$

From (11) and (13) it can be seen that $vd = vb$, it can be concluded that the vehicle will experience skid on the front or rear wheels together, but this rarely happens because in actual conditions the angle β influence, so skid will occur on one of the wheels inside. If the vehicle goes beyond the maximum allowable speed, the vehicle will skid. Skid on the front wheels occurs earlier than the rear wheels, the vehicle will tend to understeer, and if the rear wheels are skid, the vehicle will tend to be oversteer. Vehicles that have skid will make the vehicle difficult to control so that it can cause accidents. The turning radius scenario starts from 10 m, 30 m, 70 m, 150 m, and 310 m.

4. Results and discussion

Figure 6 shows the normal force chart of the wheel when the vehicle is in an unladen condition or often called a driver only weight (DoW) where the vehicle is moving at the maximum allowable speed. This calculation is carried out in turn motion, where the centrifugal force will cause a rolling moment in the vehicle.

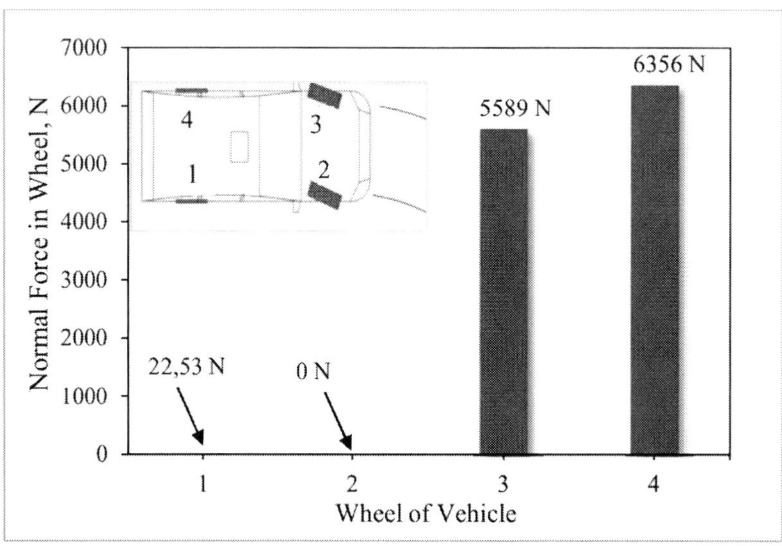

Figure 6. The normal force of wheels when the vehicle turns

From Figure 6 it can be explained that the first wheel lifted from the ground is the second wheel (inside) which is marked with zero normal force. Rolling occurs when a normal force on a wheel has a smaller value equal to zero on one or more wheels of the vehicle (it can be seen in section 3.2.). Based on calculations that have been made that the first wheel experiences a normal force equal to zero due to rolling is the inner wheel which is the second wheel, and the next smallest normal force value is wheel 1, and continued with one of the outer wheels depending on the passenger vehicle. Then the permitted speed due to skid is the maximum speed of the vehicle that is permitted when moving on a turn where one of the vehicle's wheels will momentarily experience a skid to the side of the vehicle due to the centrifugal force that is pulling it. The graph in Figure 7 shows the maximum allowable speed of the vehicle at a certain turning radius.

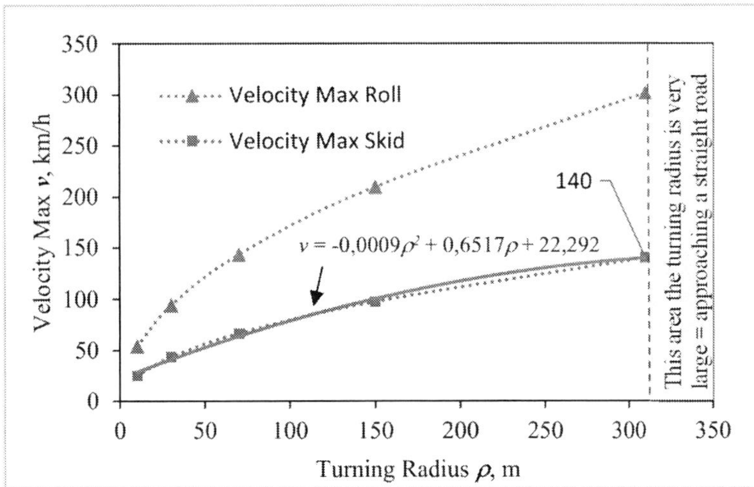

Figure 7. Graph of the maximum allowable speed

Based on this graph, before one of the wheels of a vehicle is lifted, the vehicle has skid early, which means that the maximum speed allowed at a certain radius is the maximum speed of the skid effect. Then the reference to determine the maximum speed is the effect of skid not rolling.

The relationship of the maximum speed of the vehicle to the turning radius can be displayed in the form of an equation 2nd order polynomial (14).

$$v = -0{,}0009\rho^2 + 0{,}6517\rho + 22{,}292 \qquad (14).$$

The greater the turning radius, the maximum vehicle speed will be higher. On highways or toll roads, the road is not always straight, and sometimes it turns sharply, sometimes turns slightly, therefore the maximum speed on a straight road needs to be made regulation. Vehicles that move straight when you find the road turn even though the turn is not significant, the vehicle can skid, so the vehicle is unstable. That large or insignificant turn can be assumed to be around 310 m with the assumption that it is approaching straight, then determining the maximum speed of the vehicle on the straight road is 140 km/h. That is the reason 140 km/h was chosen to be the maximum allowable speed so that when a vehicle moves straight at high speed and suddenly meets a turning road, it does not have an accident due to the instability of the vehicle's motion. So, the vehicle under review can be set so that the speed cannot exceed 140 km/h on any road.

5. Conclusion

The normal force of the wheel when the vehicle turns will vary where the normal force of the wheel (2) the inside will first be lifted due to rolling force. This becomes a reference in calculating the vehicle's maximum clearance speed. The speed of a vehicle moving on a turning road depends on the turning radius as in the following equation $v = -0{,}0009\rho^2 + 0{,}6517\rho + 22{,}292$, where ρ (*m*) is the turning radius, and v is speed (*km/h*). With various considerations, the vehicle's maximum speed on a straight road was chosen 140 km/h. This was chosen to avoid vehicle instability when it met at an insignificant turning radius. The basics of calculating the safe limit of vehicle speed can be a reference in determining the maximum speed of certain vehicles such as travel vehicles, rental cars, cargo cars, etc.

6. References

[1] Suhariyanto 2019 Land Transportation Statistics 2018 *BPS-Statistics Indonesia* Jakarta 31
[2] R. Siregar, M. Adhitya, D. A. Sumarsono, G. Heryana, and F. Zainuri 2020 Study the Brake Performance of a Passenger Car Based on the Temperature that Occurs in Each Brake Unit *AIP Conf. Proc* **020013** 1–7
[3] Y. Zhou and S. Chen 2016, Vehicle ride comfort analysis with whole-body vibration on long-span bridges subjected to crosswind, *J. Wind Eng. Ind. Aerodyn*
[4] Misbakhul anwari 2018, *Speed and Speed Management 2018*
[5] G. Heryana *et al* 2020 Plug in hybrid electric vehicle power consumption analysis in tropical area *AIP Conf. Proc* **2230** 50001
[6] Y. Wang, W. Zhao, G. Zhou, Q. Gao, and C. Wang 2018 Suspension mechanical performance and vehicle ride comfort applying a novel jounce bumper based on negative Poisson's ratio structure *Adv. Eng. Softw* **122** 1–12
[7] N. Nazaruddin, F. Zainuri, R. Siregar, G. Heryana, M. Adhitya, and D. A. Sumarsono 2019 Electric power steering: an overview of dynamics equation and how it's developed for large vehicle *IOP Conference Series: Materials Science and Engineering* **673** 12112.
[8] F. Zainuri, D. A. Sumarsono, M. Adhitya, and R. Siregar 2017 Design of synchromesh mechanism to optimization manual transmission's electric vehicle in *AIP Conference Proceedings*
[9] S. Doecke 2018 Investigation of pre-crash vehicle speed
[10] G. Cantisani and G. Loprencipe 2010 Road roughness and whole body vibration: evaluation tools and comfort limits *J. Transp. Eng.*
[11] Y. Liang, J. D. Lee, and L. Yekhshatyan 2012 How dangerous is looking away from the road? Algorithms predict crash risk from glance patterns in naturalistic driving in *Human Factors*
[12] V. Sirenko, R. Pavlovs'ky, and U. S. Rohatgi 2012 Methods of reducing vehicle aerodynamic drag in *American Society of Mechanical Engineers, Fluids Engineering Division (Publication) FEDSM*
[13] K. D. Kusano and H. C. Gabler 2012 Safety benefits of forward collision warning, brake assist, and autonomous braking systems in rear-end collisions *IEEE Trans. Intell. Transp. Syst.*
[14] R. N. Jazar 20118 *Vehicle dynamics: Theory and applications.*
[15] I. N. Sutantra and B. Sampurno 2010 *Automotive of Technology*. Surabaya: Penerbit Guna Widya
[16] A. Day 2014, *Braking of Road Vehicles* 1st ed. Oxford: Elsevier

IOP Conf. Series: Materials Science and Engineering **909** (2020) 012018 doi:10.1088/1757-899X/909/1/012018

Extremely high surface area of activated carbon originated from sugarcane bagasse

Murie Dwiyaniti[1], Elang Barruna A.G[1], Muhamad Naufal R[1], Iyan Subiyanto[1], Rudy Setiabudy[1], Chairul Hudaya[1]

[1]Department of Electrical Engineering Faculty of Engineering, Universitas Indonesia, Depok 16424, Indonesia

Email: murie.dwiyaniti82@ui.ac.id

Abstract. In this study, the extremely high specific surface area of activated carbon originated from sugarcane bagasse was produced by dry chemical activation. In this process, KOH was used as a chemical agent and directly mixed sugarcane bagasse carbon with KOH powder, which has been treated in a solid form. In particular, the influences of the impregnation ratio and activation temperature were investigated. The as-prepared activated carbons were characterized by Brunauer-Emmet-Teller (BET), which aims to determine the surface area of activated carbon and Raman Spectra analysis to examine the vibration modes of material characteristics. The maximum specific surface area of the activated carbon reached 3554 m^2/g at 800 ^0C and an impregnation ratio of 1:4. Furthermore, the Raman spectra of activated carbon exhibited graphite structure for sample in impregnation ratio 1:2 and 1:3 due to having of G and G^{-1} band. This material is essential for make battery.

1. Introduction

Sugarcane bagasse (SB) is waste generated from the process of extracting sugar water from sugarcane stems. Sugarcane is the primary raw material of the sugar industry, which is one of the plantation commodities that has a strategic role in the economy in Indonesia. With an area of around 420.15 thousand hectares in 2017, thousands of sugar cane farmers and workers are very dependent on the sugar industry for their survival [1]. However, to produce sugar at 8% will produce as much as 30% bagasse waste [2]. Therefore, this is an opportunity for researchers to be able to change their inconvenience in nature into a very useful product.

In general, sugarcane bagasse has been used as raw material for a pulp and paper [3, 4], particleboard [5], animal feed, and fertilizer. However, this utilization is very limited and has low economic value. New technology is needed to diversify the utilization of sugarcane bagasse into high economic value products. One of the alternatives is to be processed into activated carbon. The carbon content in sugarcane bagasse is quite high, which is 24.7% [6]. Also, sugarcane bagasse has a chemical composition of 25% hemicellulose, 50% cellulose, and 25% lignin [7]. Cellulose has a high degree of crystallinity, while hemicellulose and lignin are amorphous. High cellulose and low lignin content are indicated to produce micropore structure on activated carbon [8, 9]. So that sugarcane bagasse is very potential to be an excellent active carbon material.

Activated carbon, which has a porous structure and a large surface area > 3000 m^2/g, is needed in the manufacture of lithium-ion capacitor batteries [10] to accommodate more lithium-ion during charge and discharge. Until now, the activated carbon material that produces the largest surface area is graphene, about 3355 m^2/g [11]. However, graphene is very expensive, and the manufacturing process is complicated and dangerous for health. Active carbon originated from SB is an alternative, environmentally friendly material that is expected to reach a large surface area.

The process of making SB into activated carbon consists of three stages, namely dehydration, carbonation, and activation. The activation process can be carried out through two methods, namely physical activation with different oxidizing gases, such as air, O2, CO2, or steam, and ii) chemical activation with chemical compounds [12]. Chemical activation results in better pore structure and surface area. Many researchers have been activated carbon from SB using an activator agent H3PO4 [13, 14], ZnCl2 [6, 15, 16], NaOH [17], and KOH [18-20]. The high surface area of activated carbon is produced by NaOH in temperature 700^{0}C, result around 2400 m^2/g [17]. However, NaOH is only effective for the activation of irregular materials and irregular intercalation processes [21]. While the activated carbon desired in this study must have high porosity. One aspect that influences the process of forming a porosity is intercalation. KOH activator has an advantage over NaOH; namely, potassium is more easily intercalated between graphene layers [21], because potassium has a lower boiling point [22]. KOH can react effectively to a variety of polymer and carbon materials, including the high arrangement of hexagonal carbon structure and the formation of nano-sized pores.

So in this study proposes KOH in a solid form as an activator agent to directly mixed with carbon originate from SB. This method called dry activation. The dry activation method on carbon originated from sugarcane bagasse is estimated to be able to produce a large surface area. The carbon activation process becomes simple and easy.

2. Research Method

The experimental works were conducted to synthesize the activated carbon starting from preparation, carbonation, and activation process. As the samples, SB was obtained from the sugar factory of PT Gendhis Multi Manis (GMM), which is located at Tinapapan, Tadonan, Blora, Central Java, Indonesia - 58256.

2.1. Activated carbon synthesis

The synthesis of activated carbon material from SB will be carried out using the thermal carbonization method. First, SB is drained for 24 hours in the oven at 150 ^{0}C to ensure that there is no H_2O compound left. After drying, the SB is mashed so that the material particles become very small by blending it. Then, the material is carbonized using a furnace at 500 °C for one hour with an increase in temperature of 10 °C/minute under the Ar atmosphere at 200 cc/min. After carbonized, the carbon is rinsed with distilled water. Then the final product is put into an oven at 80 °C for 24 hours. The carbonation yield of SB after this referred to as sugarcane bagasse carbon (SBC). The stages of carbonization can be seen in Figure 1.

The carbon activating process is conducted through the dry chemical activation method using KOH as an activator agent. Carbon and KOH are mixed with a ratio of 1:2, 1:3, 1:4, and 1:5. After the stirring process, the carbon is heated using a furnace to a temperature of 800 °C for one hour, with a 10 ^{0}C temperature increase per minute. The carbon is neutralized using HCL solution and rinsed with distilled water. Then the final product is put into an oven at 80 °C for 24 hours. Activated carbon resulting from KOH activation of 1: 2, 1:3, 1:4, and 1:5 hereinafter referred to as SBAC12, SBAC13, SBAC14, and SBAC15, respectively. The stages of activating carbon can be seen in Figure 2.

Figure 1. The stages of the carbonization process of sugarcane bagasse

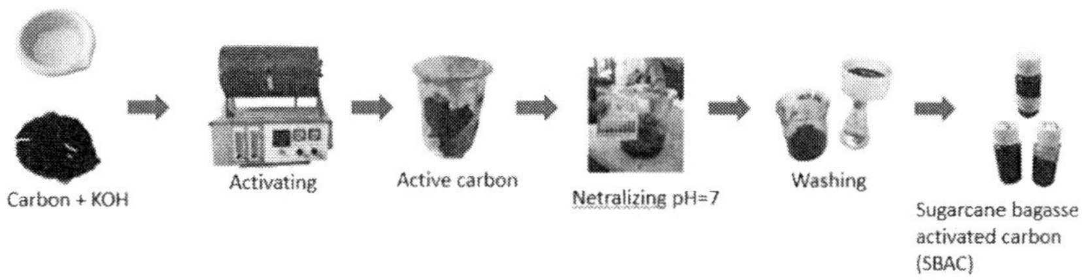

Figure 2. The stages of activating carbon process

2.2. Characterization

Two methods characterized the samples. First, Brunauer-Emmet-Teller (BET) using a Quantachrome NovaWin instrument, with degassing nitrogen for four hours, which aims to determine the surface area of activated carbon. Second, Raman Spectroscopy using a micro-RAMAN spectrometer with 532 nm of laser source wavelength to identifying the structure crystal of material.

3. Results and Discussion

3.1. The process of making carbon from sugarcane bagasse

The first stage to make SB activated carbon is to dry and grind sugarcane bagasse to form a powder. Then the powder is filtered using 100 mesh. In Figure 3, we can see changes in SB.

A B C

Figure 3. Sugarcane bagasse dry (A), after grind (B), and after filter 100 mesh (C)

In this study, the carbonization temperature used was 500^{0}C. This temperature was chosen based on the process of degradation of the chemical composition of sugarcane bagasse consisting of hemicellulose, cellulose, and lignin. Hemicellulose is degraded by heat at temperatures of 220–315^{0}C with the fastest decomposition at temperatures around 270^{0}C. Cellulose is degraded at temperatures of 315-400^{0}C with the fastest decomposition at temperatures around 350^{0}C. Lignin is slowly degraded with a temperature range of 160 - 900^{0}C [23]. The result of SBC is shown in Table 1.

Table 1. The yield of SBC

Mass SB (gram)	Mass SBC (gram)	% yield	% lost
2.3012	0.6889	29.94	70.06
2.5483	0.6295	24.70	75.30
2.5099	0.5982	23.83	76.17
2.4186	0.5731	23.70	76.30
2.5400	0.6212	24.46	75.54
2.5089	0.5972	23.80	76.20
2.4185	0.5730	23.69	76.31
2.2763	0.5641	24.78	75.22

Average % yield = 24.86
Average % lost = 75.14

In the carbonization process, mass loss occurs when the temperature is increasing. Most lost mass occurs in the cellulose content and the least lost in lignin [23]. It has happened in our experiment, as shown in Table 1. Lost SBC mass is an average of 75.14 %. While the yield of SBC is an average of 24.86%. It is similar to the result of the carbonization process of SB from several studies where yield is around 20-30% and lost mass is around 70-80% [6], [7], [24], [25], [26].

3.2. The process of activating the carbon

The next process is activating SBC. The activation method used is chemical activation (pyrolysis), with KOH as an activation agent. The method is by directly mixing SBC with KOH powder, which has been processed from a solid form. The purpose of this activation is to produce higher activated carbon yields and form more pores so that a higher surface area of carbon is produced.
In this research used four samples, SBAC12, SBAC13, SBAC14, and SBAC15. SBC activation results are in Table 2.

Table 2. SBC activation result

	Activation Variable		SBC + KOH (gram)	after activated (gram)	% lost	% Yield
	Temp ^0C	SBC:KOH				
SBAC12	800	1:2	6.379	4.974	21.96	78.04
SBAC13	800	1:3	6.353	4.912	22.69	77.31
SBAC14	800	1:4	6.323	4.808	23.97	76.03
SBAC15	800	1:5	6.046	4.489	25.76	74.24

From table 2, it can be seen that the activating process decreases the mass. The percentage of mass loss tends to increase with increasing ratio KOH. The most significant percentage of mass loss occurs in SBAC15, with a percent loss of 25.76%. This is caused by the process formation of new pores of carbon [27].

3.3. BET and Raman Spectra

3.3.1 BET analysis

Having a high surface area is one indication of a good quality of activated carbon. The amount of adsorption power depends on the surface area of activated carbon. The surface area can be determined by the BET method. The surface area test results are in Table 3.

Table 3. The surface area test results of sugarcane bagasse active carbon (SBAC)

	Temperature (^{0}C)	ratio SBC: KOH	Specific surface area
SBAC12	800	1:2	2136.828 m^2/g
SBAC13	800	1:3	2236.926 m^2/g
SBAC14	800	1:4	3554.820 m^2/g
SBAC15	800	1:5	2747.290 m^2/g

A comparison of the largest specific surface area (SSA) due to the influence of KOH activation can be seen in Table 3. Activation using KOH on carbon material is used to increase surface area and improve its electrochemical performance [28]. The results show that SSA increases with the addition of KOH because the amount of KOH can affect pore development. The smallest SSA is in SBC12, about 2136 m^2/g, and the largest is given by SBC14, which is 3554 m^2/g. Kalderis and Shofa also reported the same study of using KOH as an activator on carbon from sugarcane bagasse where the presence of KOH activator will reduce pore diameter and expand pore volume and surface area with an SSA value of 864 m^2/g by Kalderis [29] and Shofa with a value of 1135 m^2/g [18]. However, the SSA value obtained in our study is extremely high, even surpassing activated carbon from graphene material, which has an SSA of 3523 m^2/g [30]. Graphene is a superior material for numerous applications.

3.3.2 Raman analysis

All samples are characterizing by Raman spectroscopy to know vibration modes of material characteristics, which is plotted as a Raman shift. The result of Raman spectroscopy can be seen in fig 4. From figure 4, we can see that all samples have the sharpness band in value of 1575-1583 cm^{-1}, it is known as G band that is composed of sp^2 bonded carbon and graphene material. The sharpness of the band also tells us that the bonds are, for the most part, very uniform [31]. Furthermore, in fig 4, we also see a prominent band around 1346 – 1362 cm^{-1}. This band is called the D band. The D band shows some disorder in the graphene structure originating from the hybridization vibrational mode associated with the graphene edge [31]. The results of the Raman show that the addition of KOH did not cause a significant change in the shifts in the G and D bands. However, the small KOH ratio (1:2 and 1:3) caused an additional band, in around 2600-2700 cm^{-1}. It is known as the G^{-1}, referred to as a 2D band, which indicates the number of graphene layers [32]. It is also indicated that SBAC12 and SBAC13 have graphite structure [33].

The quality of the sample is determined by the intensity ratio of the ID / IG band. For high-quality samples, the ID / IG ratio is generally less than 2%. In this study the ID / IG results for SBAC12, SBAC13, SBAC14, and SBAC15 were 0.70, 0.83, 1.06, and 0.92 respectively. It shows that the ID / IG ratio for all samples is less than 2%, which indicated the sample has a good quality.

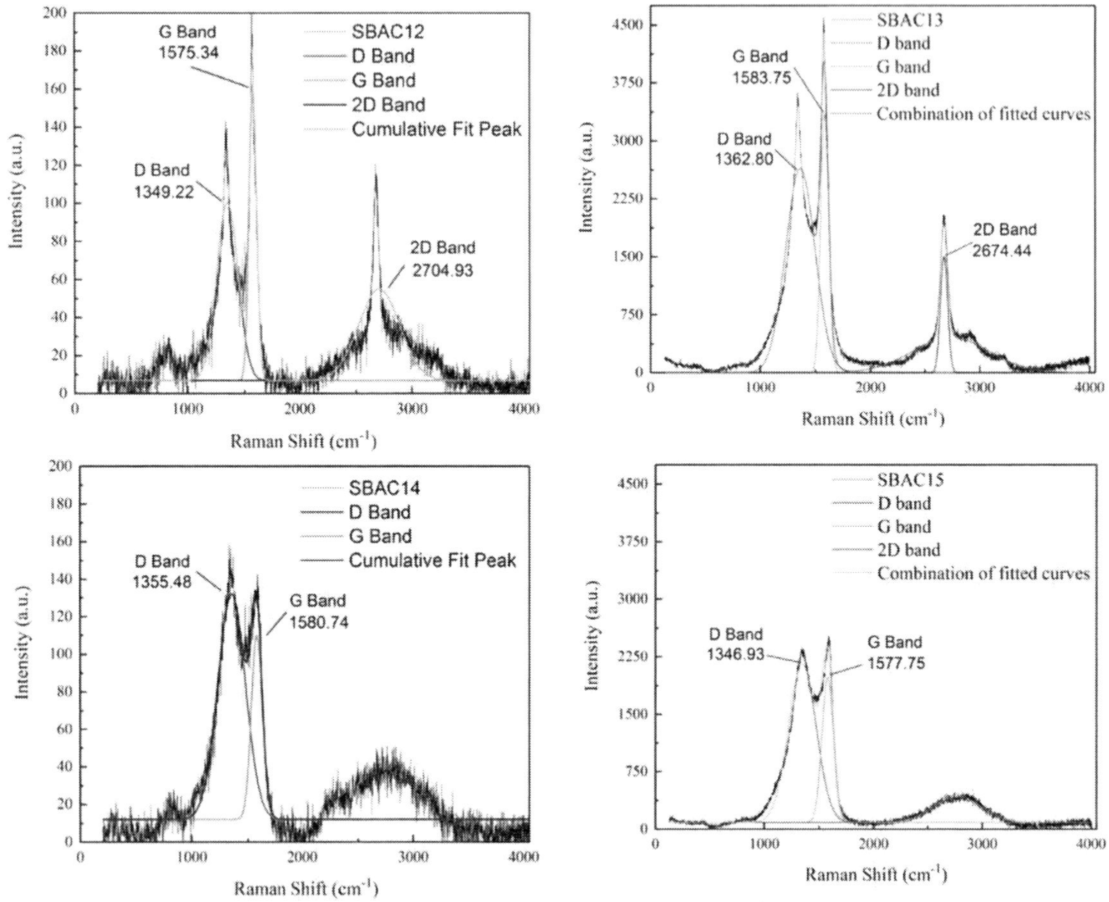

Figure 4. Raman spectra of four samples at temperature 800^{0}C, SBAC12, SBAC13, SBAC14 and SBAC15

4. Conclusion

Activated carbon originates from SB has been produced using a dry chemical activation method with KOH as an activator agent. This method provides an extremely surface area of about 3554 m^2/g and graphene carbon structure for sample SBAC14. While another sample SBAC12 and SBAC13 have graphite structure even though the surface area is less then SBAC14. Thus, sugarcane bagasse can be an alternative material for activated carbon because it has a large surface area and contains graphite material.

Acknowledgments

The authors would like to acknowledge the financial support provided by KIST (Korea Institute of Science and Technology) School partnership program.

References

[1] Ditjenbun 2017 Statistik perkebunan Indonesia 2016-2018 *http://ditjenbun.pertanian.go.id*

[2] T. Arief 2015 Panduan penilaian potensi biomassa sebagai sumber energi alternatif di indonesia *Penabulu Alliance Jakarta*

[3] A. Vitaloka, A. Rohanah, and A. Rindang 2017 Characteristics of paper from sugar cane biomass and waste paper *Jurnal Rekayasa Pangan dan Pertanian* **5** 140-144

[4] C. Purnawan, D. Hilmiyana, W. Wantini, and E. Fatmawati 2012 Pemanfaatan limbah ampas tebu untuk pembuatan kertas dekorasi dengan metode Organosolv *Ekosains* **4** 2-7
[5] M. Iskandar and A. Supriadi 2013 Pengaruh kadar perekat terhadap sifat papan partikel ampas tebu *Jurnal Penelitian Hasil Hutan* **31** 19-26
[6] D. Kalderis, S. Bethanis, P. Paraskeva, and E. Diamadopoulos 2008 Production of activated carbon from bagasse and rice husk by a single-stage chemical activation method at low retention times *Bioresource technology* **99** 6809-6816
[7] A. Pandey, C. R. Soccol, P. Nigam, and V. T. Soccol 2000 Biotechnological potential of agro-industrial residues. I: sugarcane bagasse *Bioresource technology* **74** 69-80
[8] P. Simon and A. Burke 2008 Nanostructured carbons: double-layer capacitance and more *The electrochemical society interface* **17** 38-44
[9] D. J. Díaz 2010 Electrochemistry of Porous Materials *ACS Publications*
[10] B. Li *et al* 2016 Activated carbon from biomass transfer for high -energy density lithium- ion supercapacitors *Advanced Energy Materials* **6** 1600802
[11] F. Zhang *et al* 2013 A high-performance supercapacitor-battery hybrid energy storage device based on graphene-enhanced electrode materials with ultrahigh energy density *Energy & Environmental Science* **6** 1623-1632
[12] R. C. Bansal and M. Goyal 2005 Activated carbon adsorption *CRC press*
[13] T. Adinaveen, L. J. Kennedy, J. J. Vijaya, and G. Sekaran 2013 Studies on structural, morphological, electrical and electrochemical properties of activated carbon prepared from sugarcane bagasse *Journal of industrial and Engineering Chemistry* **19** 1470-1476
[14] A. D. S. N. Hidayati, S. Kurniawan, N. W. Restu, and B. Ismuyanto 2016 Potential of sugar cane as an alternative raw material for making activated carbons *Natural B* **3** 318-322
[15] H. Demiral and G. Gündüzoğlu 2010 Removal of nitrate from aqueous solutions by activated carbon prepared from sugar beet bagasse *Bioresource technology* **101** 1675-1680
[16] T. E. Rufford, D. Hulicova-Jurcakova, K. Khosla, Z. Zhu, and G. Q. Lu 2010 Microstructure and electrochemical double-layer capacitance of carbon electrodes prepared by zinc chloride activation of sugar cane bagasse *Journal of Power Sources* **195** 912-918
[17] K. Konno, Y. Ohba, K. Onoe, and T. Yamaguchi 2008 Preparation of activated carbon having the structure derived from biomass by alkali activation with NaOH, and its application for electric double-layer capacitor *Tanso* **2008** 2-7
[18] Shofa 2012 Pembuatan karbon aktif berbahan baku ampas tebu dengan aktivasi kalium hidroksida *Teknik Kimia* Universitas Indonesia
[19] B. S. Pratama 2018 Konversi ampas tebu menjadi biochar dan karbon aktif untuk penyisihan Cr (VI) *Jurnal Rekayasa Bahan Alam dan Energi Berkelanjutan* **2** 7-12
[20] K. Foo, L. Lee, and B. Hameed 2013 Preparation of activated carbon from sugarcane bagasse by microwave assisted activation for the remediation of semi-aerobic landfill leachate *Bioresource technology* **134** 166-172
[21] E. Raymundo-Pinero, P. Azaïs, T. Cacciaguerra, D. Cazorla-Amorós, A. Linares-Solano, and F. Béguin 2005 KOH and NaOH activation mechanisms of multiwalled carbon nanotubes with different structural organisation *Carbon* **43** 786-795
[22] Y. Guo, S. Yang, K. Yu, J. Zhao, Z. Wang, and H. Xu 2002 The preparation and mechanism studies of rice husk based porous carbon *Materials chemistry and physics* **74** 320-323
[23] H. Yang, R. Yan, H. Chen, D. H. Lee, and C. Zheng 2007 Characteristics of hemicellulose, cellulose and lignin pyrolysis *Fuel* **86** 1781-1788
[24] T. A. Mulyati 2008 Preparasi dan karakterisasi karbon aktif dari limbah ampas tebu menggunakan aktivator KOH *Indonesian Chemistry and Application Journal* **1** 61-67
[25] M. Samsuri *et al* 2007 Pemanfaatan sellulosa bagas untuk produksi ethanol melalui sakarifikasi dan fermentasi serentak dengan enzim xylanase *MAKARA* **11** 17-24
[26] M. Christiyanto and A. Subrata 2005 Perlakuan fisik dan biologis pada limbah industri pertanian terhadap komposisi serat *e-print.undip.ac.id*

[27] J. S. Im, S.-J. Park, T. J. Kim, Y. H. Kim, and Y.-S. Lee 2008 The study of controlling pore size on electrospun carbon nanofibers for hydrogen adsorption *Journal of colloid and interface science* **318** 42-49
[28] C. Peng, X.-b. Yan, R.-t. Wang, J.-w. Lang, Y.-j. Ou, and Q.-j. J. E. A. Xue 2013 Promising activated carbons derived from waste tea-leaves and their application in high performance supercapacitors electrodes *Electrochimica Acta* **87** 401- 408
[29] D. Kalderis *et al* 2008 Adsorption of polluting substances on activated carbons prepared from rice husk and sugarcane bagasse *Chemical Engineering Journal* **144** 42-50
[30] L. Zhang *et al* 2013 Porous 3D graphene-based bulk materials with exceptional high surface area and excellent conductivity for supercapacitors *Scientific reports* **3** 1-9
[31] J. Hodkiewicz 2010 Characterizing carbon materials with Raman spectroscopy Scientific **51946**
[32] L. Malard, M. Pimenta, G. Dresselhaus, and M. Dresselhaus 2009 Raman spectroscopy in graphene *Physics reports* **473** 51-87
[33] A. C. Ferrari, Robertson 2004 Raman spectroscopy of amorphous, nanostructured, diamond–like carbon, and nanodiamond *Physical and E. Sciences* **362** 2477-2512

IOP Conf. Series: Materials Science and Engineering **909** (2020) 012019 doi:10.1088/1757-899X/909/1/012019

Performance analysis and evaluation of a 10.6 kWp grid-connected photovoltaic system in Serpong

E Nurdiana[1,2], I Subiyanto[1], A Indarto[1], Riza[2], G Wibisono[1], C Hudaya[1*]
[1]Department of Electrical Engineering, University of Indonesia
[2]National Laboratory of Energy Conversion Technology, Agency for Assessment and Application of Technology

*Email: c.hudaya.eng.ui.ac.id

Abstract. The grid-connected Photovoltaic (PV) System was installed in the rooftop of Energi Building, PUSPIPTEK in 2017 located in Serpong, South Tangerang. The System consists of 40 modules @265 Wp, a 10 kWh battery system, and a 10250 kW grid-tie inverter. This study presents the performance analysis of the PV system based on the eight-months monitored data from July 2019 to February 2020. The evaluated parameters of the PV system include yields, performance ratio, capacity factor, energy efficiency, and average daily energy output of the PV system. The results show that the average value of the performance ratio and capacity factor during the eight-months monitored period were 82.42% and 14.07% with average values of the array (Y_a), final (Y_f) and reference (Y_r) yields were 3.49 kWh/kWp/d, 3.38 kWh/kWp/d, and 4.12 kWh/kWp/d, respectively. Meanwhile, the average daily energy output of the PV system during the eight-months monitored period was 36.92 kWh per day with the average values of the module (η_{PV}), system (η_{sys}), and inverter (η_{inv}) efficiency was found to be 15.29%, 14.77%, and 96.63%, respectively. Compared with previous studies in the literature, these results indicated that the grid-connected PV system installed in Energi Building PUSPIPTEK Serpong exhibited good performance.

Nomenclature	
E_{DC}	DC energy output, kWh
E_{AC}	AC energy output, kWh
$E_{AC,m}$	Monthly total AC energy output, kWh
Y_a	Array yield, kWh/kWp/day
Y_f	Final yield, kWh/kWp/day
Y_r	Reference yield, kWh/kWp/day
$P_{PV,rated}$	Nominal power of PV system, kWp
H_t	Global solar radiation, kWh/m^2
H_R	Reference irradiance of the PV, kW/m^2
PR	Performance ratio, %
CF	Capacity factor, %
CF_m	Monthly capacity factor, %
η_{PV}	PV module/array efficiency, %
η_{sys}	PV system efficiency, %
η_{inv}	Inverter efficiency, %
S	PV module/array area, m^2
L_C	Array capture losses, kWh/kWp/day
L_S	System losses, kWh/kWp/day

1. Introduction

The implementation of solar photovoltaic (PV) energy in the world has developed rapidly. REN21 reported that the global capacity of the PV system reached 405 GW in 2017, and increased to 505 GW in 2018. The top five countries with the highest capacity of the PV system are China, the United States, Japan, Germany, and India with a capacity of 176.1 GW, 62.4 GW, 56 GW, 45.3 GW, and 32.9 GW, respectively [1]. The total capacity of the PV system in Indonesia reached 135 MW until 2019 with solar energy potential reached 207,8 MW. This amount is around 0.02 % of the total solar energy potential in Indonesia [2]. This capacity will be increased to pursue the target of RUEN scenario with targeted PV system installation was 6.379 MW in 2025 and 45.000 MW in 2050 [3].

Evaluation of the PV system's performance is essential to study. This study was needed to know how a good PV system works. In recent years, the performance of the PV system was evaluated and investigated in many countries in the world. The performance of a grid-connected PV system depends on various things such as cell technology, inverters, and installation configuration. It also depends on the weather parameters as global irradiance, ambient temperature, and soiling losses as analyzed by the investigator of the performance of PV grid-connected in Morocco and Brazil [4-5]. Also, the performance of various solar PV module technology in Malaysia and Peru was evaluated [6-7]. On the other hand, research on the performance of PV systems has never been found in Indonesia, especially research that assesses the parameters of PV system performance standards.

This paper assesses and analyzes the performance of a 10.6 kWp grid-connected PV rooftop system installed in the Energi Building, PUSPIPTEK located in Serpong, South Tangerang. The objective of this study is to analyze the performance parameters of the PV rooftop system located in Serpong in terms of performance parameters standard such as yields, performance ratio, capacity factor, energy efficiency, and average daily energy output. In the end, the results of this study are compared with the results of the previous study located in other countries.

2. PV System Description

2.1. Configuration of PV System

In 2017, a grid-connected PV rooftop system was installed on the top of the Energi Building, PUSPIPTEK, Serpong, South Tangerang. This rooftop PV system started to operate in December 2017. The PV system consists of 40 units of 265 Wp polycrystalline PV modules with specifications as shown in Table 1. The 20 modules connected in series to form one PV string, then 2 PV strings are connected in parallel to form a PV system with a capacity of 10.6 kWp. The configuration of the PV system is shown in Fig. 1.

Table 1. PV module specification under standard test condition (STC)

Parameter	Spesification
Nominal max. power (Pmax)	265 W
Opt. operating voltage (Vmp)	30,6 V
Opt. operating current (Imp)	8,66 A
Open circuit voltage (Voc)	37,7 V
Short circuit current (Isc)	9,23 A
Module efficiency	16,47 %
Operating temperature	-40 °C~+80 °C
Maximum system voltage	1000 V (IEC)1000 V (UL)
Power tolerance	0 ~ +5 W

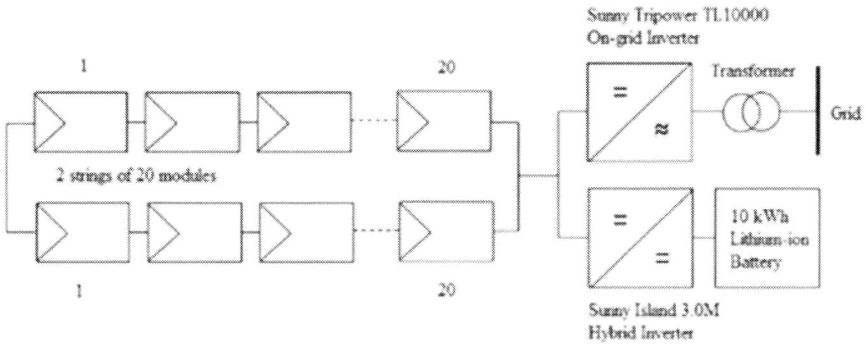

Figure 1. The configuration of 10.6 kWp grid-connected PV rooftop system in Energi Building

The PV system works at 600 VDC voltage. This system equipped with a 10 kWh battery system. The PV system connected to the Sunny Tripower TL10000 grid-tie inverter with a maximum AC power of 10250 kW to convert DC voltage, the power output from the PV module to AC voltage. The power output from the inverter connected directly to the PLN network. This system is also equipped with a weather station to record the data of solar intensity, module temperature, ambient temperature, and wind speed [8].

2.2. Weather Conditions at Location of PV System
The PV system installed on the rooftop of Energi Building, PUSPIPTEK located in Serpong with a longitude 106'38'-106'47' °E and latitude 06'13'30'-06'22'30' °S. Based on the data of Photovoltaic Geographical Information System (PVGIS), the global horizontal irradiation and ambient temperature at Serpong during 2016 are shown in Figure 2.

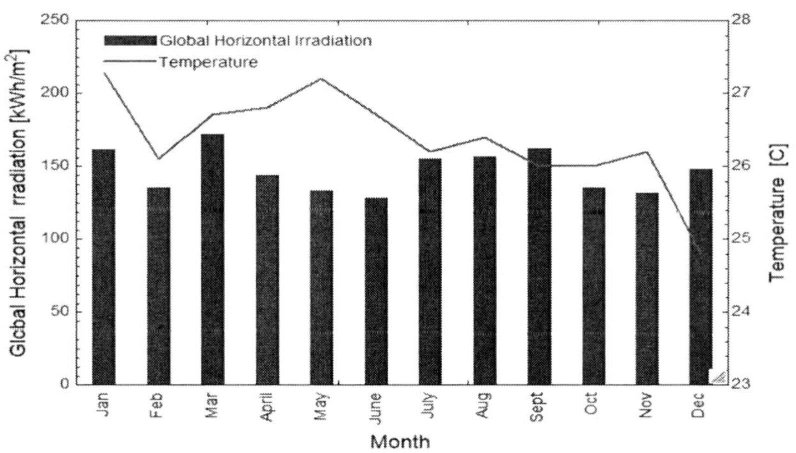

Figure 2. Monthly irradiance and ambient temperature in Serpong during 2016

3. Performance Parameters of PV System

The performance of the PV system is evaluated according to several parameters. Based on IEC61724, standard parameters for evaluating the performance of grid-connected PV systems are energy production of the PV system (E_{DC}), array yield (Y_a), final yield (Y_f), reference yield (Y_r), performance ratio (*PR*), capacity factor (*CF*), energy efficiency and system losses [4-5].

3.1. Energy Productions
The energy production of the PV system is the energy generated by the PV array during a specific time. The amount of this energy affected by solar radiation and module temperature. In this study, the energy production calculated in average daily energy during the monitored period.

3.2. Yields

Yields of the PV system divided into array yield, final yield, and reference yield. The yields indicate the actual operation of the array relative to the nominal capacity of the PV system [5].

3.2.1. Array yield (Y_a)

can be interpreted as the amount of DC energy delivered by the PV system divided by the nominal power of the PV system. The Y_a represents the time that the PV system must be operating with its nominal power to produce the energy generated, measured in kWh/kWp [5]. The array yield equation is given as [4-5]:

$$Y_a = \frac{E_{DC}}{P_{pv,rated}} \quad (1)$$

Where E_{DC} is a total of DC energy delivered by the PV system (kWh) and $P_{PV,rated}$ is the nominal power of the PV system (kWp).

3.2.2. Final yield (Y_f)

Final Yield is a ratio of total AC energy delivered by the PV system during a specific period and nominal power of the PV system. It exhibits how many hours a day the PV system must operate at its nominal power to generate the same amount of energy as was recorded. It measured in kWh/kWp [4]. The final yield is given as [4-5]:

$$Y_f = \frac{E_{AC}}{P_{pv,rated}} \quad (2)$$

Where E_{AC} is a total of AC energy generate by PV system (kWh) and $P_{PV,rated}$ is the nominal power of the PV system (kWp).

3.2.3. The reference yield

Reference yield is the comparison of the global solar radiation (H_t) and the PV system's reference irradiance (H_R). The value of H_R is 1 kW/m^2 [4-5]. Reference yield is a quantity of the theoretical energy available at a certain location during a specified period time, measured in kWh/kWp [5]. The reference yield equation is below:

$$Y_r = \frac{H_t}{H_R} \quad (3)$$

3.3. Performance Ratio (PR)

The performance ratio (*PR*) represents the whole losses effect on the normal power output of a PV system. The *PR* values depending on module temperature, imperfect utilization of incident solar radiation also the failures of the system components. The *PR* value represents how close the PV system to its ideal performance during real operation. *PR* of the PV systems independent of location, tilt angle, orientation, and their nominal power capacity [4-5]. *PR* value varies from 0.6 to 0.8. It does not indicate the number of energy produced by the PV system. The PV system with high *PR* in low solar irradiation place may produce less energy than a PV system with low *PR* in a high solar irradiation location vice versa [9]. *PR* is defined as a ratio of final yield and reference yield [4-5]. *PR* also defined as the energy output E_{AC} divided by the nameplate DC power E_{DC} obtained in the Standard Test Condition [10]. *PR* obtained by the equation below

$$PR = \frac{Y_f}{Y_r} x100\% \quad (4)$$

3.4. Capacity Factor (CF)

Capacity Factor (*CF*) is the comparison between real energy generated by the PV system (E_{AC}) and theoretically rated energy or the number of energy produced by the PV system if it is operated at full nominal power for 24 h per day. The monthly capacity factor (CF_m) can be calculated using equations below

$$CF_m = \frac{E_{AC,m}}{P_{PV,rated} * 24 * D} \quad (5)$$

Where $E_{AC,m}$ is the amount of the PV system energy delivered to the grid in a certain month, and D is the number of days in a certain month.

3.5. Energy Efficiency

PV system efficiencies include a PV module, system, and inverter efficiency. The equation of PV module, system, and inverter efficiency, respectively are shown below [4-5].

$$\eta_{PV} = \frac{E_{DC}}{H_t * S} x100\% \quad (6)$$

$$\eta_{sys} = \frac{E_{AC}}{H_t * S} x100\% \quad (7)$$

$$\eta_{inv} = \frac{E_{AC}}{E_{DC}} x100\% \quad (8)$$

3.6. Array Capture Losses (L_C) and System Losses (L_S)

Array capture losses are due to the losses on PV array [4]. While system losses are due to DC into AC conversion by inverter [11]. These losses are given as

$$L_c = Y_r - Y_a \quad (9)$$

$$L_s = Y_a - Y_f \quad (10)$$

4. Results and Discussion

Performance of a 10.6 kWp grid-connected PV system analyzed based on monitored data during eight months from July 2019 to February 2020. Data of the PV and weather station system recorded on 5 minutes intervals. The data are available on the monitoring system and able to download in CSV file. The data include PV power (DC power), inverter output power (AC power), and radiation. These data used to calculate average daily energy, yields, performance ratio, capacity factor, efficiencies, and losses. The results of DC and AC average daily energy, performance ratio, capacity factor, also a module, system, and inverter efficiency are shown in Table 2.

Table 2. The calculation results of daily energy, performance ratio, capacity factor and efficiencies

Months	Average daily DC energy ($E_{DC,d}$) kWh/day	Average daily AC energy ($E_{AC,d}$) kWh/day	Performance Ratio (PR) %	Capacity Factor (CF) %	Module Efficiency (η_{PV}) %	System Efficiency (η_{sys}) %	Inverter Efficiency (η_{inv}) %
Jul-19	38.62	38.15	87.58	15.00	15.89	15.69	98.77
Aug-19	39.09	38.33	82.69	15.07	15.12	14.82	97.99
Sep-19	42.32	41.59	83.59	16.35	15.25	14.98	98.23
Oct-19	41.21	40.52	81.66	15.93	14.88	14.63	98.33
Nov-19	41.75	41.00	79.45	16.12	14.52	14.24	98.07
Dec-19	34.52	33.47	80.03	13.16	14.76	14.34	96.89
Jan-20	29.77	27.57	82.50	10.84	15.95	14.78	92.74
Feb-20	30.28	28.15	83.88	11.06	16.23	15.03	92.64

Average	37.20	36.10	82.67	14.19	15.32	14.81	96.71

4.1. Energy Production

The average daily DC energy of the PV system is the amount of energy generated by the PV system on average in one day. While average daily AC energy is the amount of output energy from inverter on average in one day. The value of monthly average daily DC energy during monitored period varied from 29.77 kWh/day in January 2020 to 42.32 kWh/day in September 2019 and the value of monthly average daily AC energy of the PV system is varied from 27.57 kWh/day in January 2020 to 41.59 kWh/day in September 2019. The average daily energy during the monitored period was found 37.20 kWh/day for DC energy and 36.10 kWh/day for AC energy. The amount of energy output influenced by the decreasing number of sun hours because of the number of cloudy days during the monitored period [4]. In September 2019, the PV system produced the highest energy because it is on the dry season with high solar intensity. While in January 2020, the amount of PV system energy production is lowest among other months because the season was changed to rainy season with high-intensity rainfall. The amount of output energy also affected by global solar radiation. Figure 3 shows the trend of average daily energy following the trend of global solar radiation.

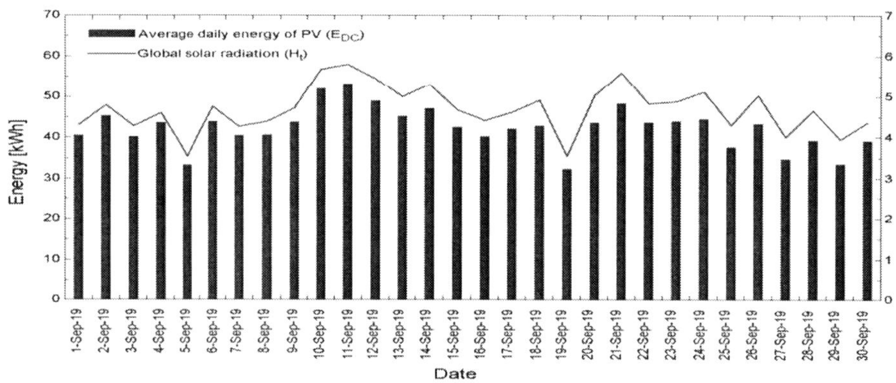

Figure 3. Average daily energy production of a 10.6 kWp grid-connected PV rooftop system in the Energi Building and global solar radiation during September 2019

The other parameter influencing PV energy production is module temperature. Figure 3 shows that the higher PV energy production on September, 11th with the close irradiance to September, 21st. This is due to the high irradiance combined with lower temperature on September, 11th compared to the high irradiance with higher temperature in September, 21st. On September, 11th, global solar radiation reached 5.80 kWh/m^2 and module temperature reached 43.04 ^{0}C, while in September, 21st, global solar radiation reached 5.60 kWh/m^2 and module temperature reached 45.16 ^{0}C. Energy production on September, 11th reached 53.20 kWh, while on September, 21st reaches 48.38 kWh. The trend of module temperature during September 2019 showed in Figure 4.

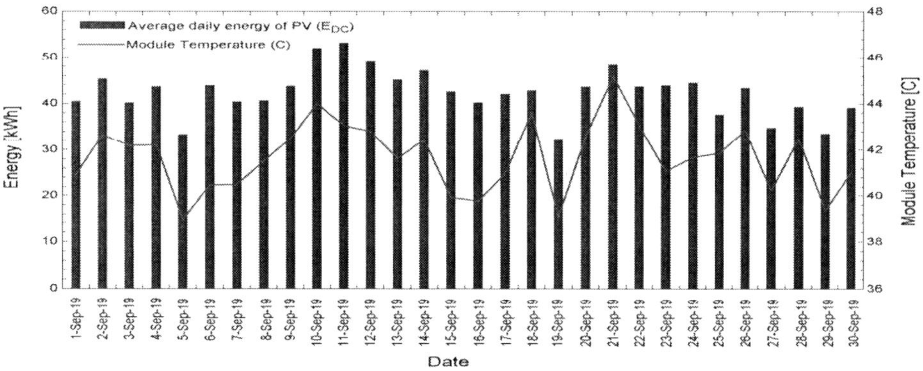

Figure 4. Average daily energy production and module temperature of a 10.6 kWp grid-connected PV rooftop system in the Energi Building during September 2019

4.2. Yields

The monthly average daily array yield, final yield, and reference yield of the 10.6 kWp grid-connected PV system in the Energi Building from July 2019 to February 2020 are shown in Figure 5. The figure shows that the highest array yield and final yield happened in September 2019, while highest reference yield happened in November. The monthly average daily array yield varied from 2.80 kWh/kWp/d in January 2020 to 3.99 kWh/kWp/d in September 2019. The monthly average daily final yield varied from 2.60 kWh/kWp/d in January 2020 to 3.92 kWh/kWp/d in September 2019. The monthly average daily reference yield varied from 3.17 kWh/kWp/d in February 2020 to 4.87 kWh/kWp/d in November 2019. The array yield and final yield values in January and February are lower than other months, because of the low irradiance and the reduced sun hours in this month. The highest array and final yield recorded in September 2019, but the highest reference yield recorded in November 2019. This condition shows that in November, the PV system has higher losses so that causes the lower array and final yield in November than September. The average value of array yield, final yield, and reference yield during the eight-month monitored period are 3.51 kWh/kWp/d, 3.41 kWh/kWp/d, and 4.14 kWh/kWp/d, respectively. In previous research in Brazil, the monthly average value of array yield, final yield and reference yield during the one-year monitored period are 4.9 kWh/kWp/d, 4.6 kWh/kWp/d, and 5.6 kWh/kWp/d, respectively [5].

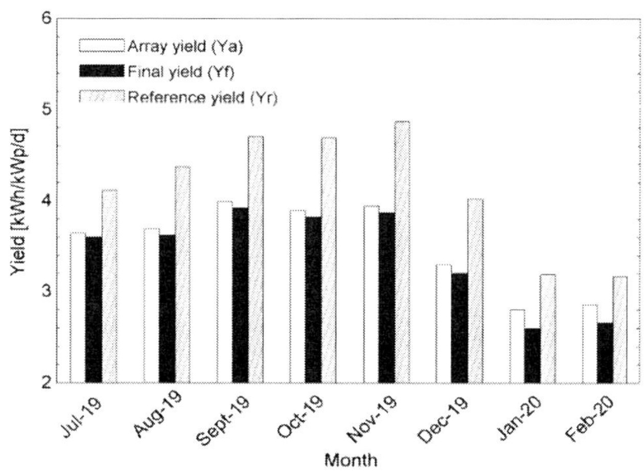

Figure 5. The value of array yield (Y_a), final yield (Y_f) and reference yield (Y_r) of a 10.6 kWp grid-connected PV rooftop system in Energi Building from July 2019 to February 2020

4.3. Performance Ratio (PR)

PR shows the ratio of the energy transferred to the grid after the reduction of energy losses such as soiling, shadowing, and thermal losses, etc. It does not indicate the number of energy generated by the PV system. Figure 6 shows that *PR* does not represent the number of energy production of the PV system. The PV system with high *PR* in low solar irradiation place may produce less energy than a PV system with low *PR* in a high solar irradiation location vice versa [9]. The value of *PR* affected by losses in conversion of solar energy into electricity (DC energy) and the losses in conversion of DC into AC energy. In this study, monthly average *PR* of the PV system varied from 79.45% in November 2019 to 87.58% in July 2019 with an average of 82.67%. It means that the PV system able to convert 82.67% of incident solar energy into useful electricity. While 17.33% of the solar energy concerning the PV array is not converted to electricity due to losses or component's defect. In the other study, the *PR* of a grid-connected PV system in China was varied from 80% to 85% [12]. While the annual average monthly *PR* of the PV system in Eastern India was 78% [9].

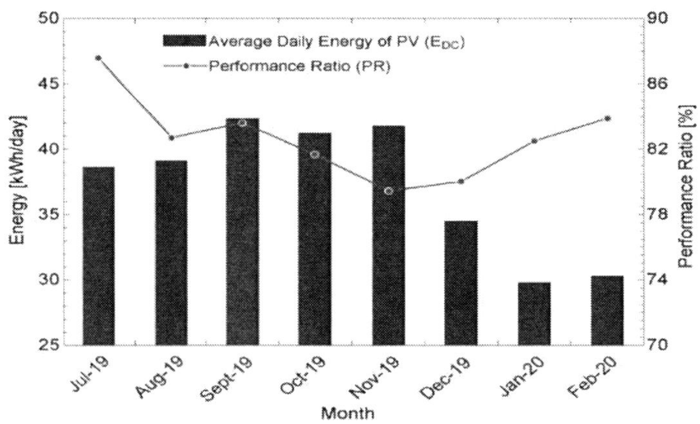

Figure 6. The value of Perfor-mance Ratio (*PR*) of a 10.6 kWp grid-connected PV rooftop system in Energi Building compared to energy production from July 2019 to February 2020

4.4. Capacity Factor (CF)

In this study, CF calculated monthly and its average during the eight-month monitored period. The monthly CF of the PV system in Serpong was varied from 10.84% in January 2020 to 16.35% in September 2019 with an average of 14.19%. This value illustrated the amount of time during a monitored period if the PV system is producing energy at its full power output. It means that the PV system capable of generating full power energy in about 34.62 days of the eight-month monitored period (July 2019-February 2020). The value of the capacity factor compared to energy production shows in Figure 7. The capacity factor of the PV system varies depending on the location and affected by solar radiation. The typical capacity factor is in the range of 0.15 to 0.4 [5]. In Morocco, the capacity factor varied from 6.55% to 21.42% [4]. In northeastern Brazil, capacity factor varied from 15.5% to 23.1% [5].

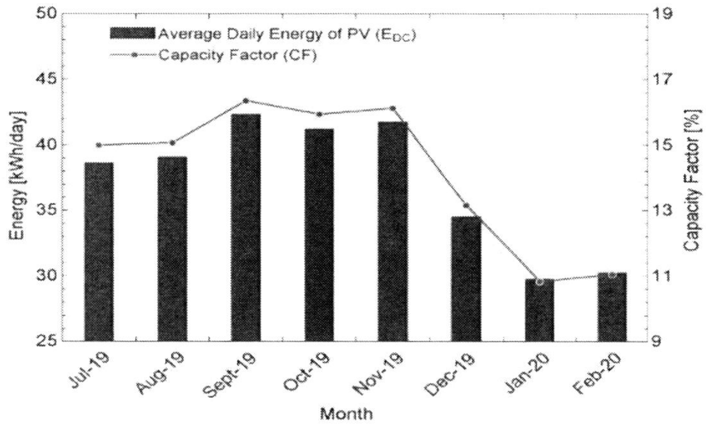

Figure 7. The value of Capacity Factor (CF) of a 10.6 kWp grid-connected PV system in Energi Building compared to energy production from July 2019 to February 2020

4.5. Energy Efficiencies

The calculated efficiencies of this PV system included module (η_{PV}), system (η_{sys}), and inverter (η_{inv}) efficiency, the results have shown in Figure 8. Module efficiency ranged from 14.52% in November 2019 to 16.23% in February 2020, system efficiency ranged from 14.24% in November 2019 to 15.69% in July 2020 and inverter efficiency ranged from 92.64% in February 2020 to 98.77% in July 2019. The average of the module, system, and inverter efficiency are 15.32%, 14.81%, and 96.71%, respectively. The module efficiency value decreased from July 2019 to November 2019, because of the dry season so that the PV system covered by dust. Whereas, the value of module efficiency increased from December 2019 to February 2020 because the dust on the PV module was swept away by rainwater that causes the PV module more efficient. The inverter efficiency tends to decrease from July 2019 to December 2019 and drastically decrease in January and February 2020. This caused by the failure of the inverter component that causes higher system losses on the inverter. The value of system efficiency influenced by module and inverter efficiency. The results show that system efficiency smoothly decreases in July until November 2019 and smoothly decreases in December 2019 until February 2020. These values are higher than the efficiency value of the grid-connected PV system in Morocco, where the lowest value of modules efficiency is 11.80% in September and the highest value is 13.22% in February, the lowest value of system efficiency is 11.41% in September and the highest value is 12.93% in February, whereas the lowest value of inverter efficiency is 96.7% in February and the highest value is 96.8% in December [4].

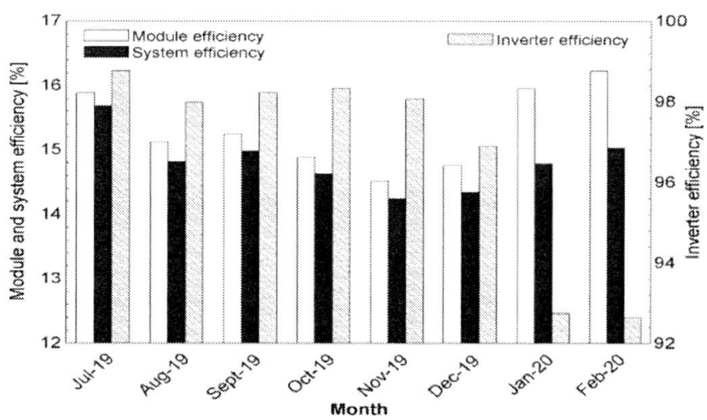

Figure 8. Module, system and inverter efficiency of the 10.6 kWp grid-connected PV system in Energi Building from July 2019 to February 2020

4.6. Losses

The losses evaluated in this study are array capture (L_C) and system (L_S) losses. Monthly Y_f with corresponding L_C and L_S have shown in Figure 9. The figure shows that the L_C value increased in July 2019 and reached the highest in November 2019. This increasing due to the dry season which started in July 2019 and ended in November 2019, so that dust and soiling covered the array PV during this period. In January and February 2020, high-intensity rain swept the dust and soil in the array PV, so that the array PV became clean and caused lower L_C in these months. The L_S value influences the inverter efficiency, where the higher L_S causes lower efficiency. It can be seen that in January and February 2020, the L_S reached the highest value so that caused the lower inverter efficiency in these months (see Figure 8). The L_C values of the PV system in this study varied from 0.32 kWh/kWp/d in February 2020 to 0.93 kWh/kWp/d in November 2019. The L_S values of the PV system varied from 0.04 kWh/kWp/d in July 2019 to 0.20 kWh/kWp/d in February 2020. The average L_C and L_S are 0.63 kWh/kWp and 0.10 kWh/kWp. Losses in other PV systems in Morocco have observed: the L_C varied from 0.025 kWh/kWp in February to 0.76 kWh/kWp in September. The L_S ranged from 0.14 kWh/kWp in January and 1.89 kWh/kWp in May [4].

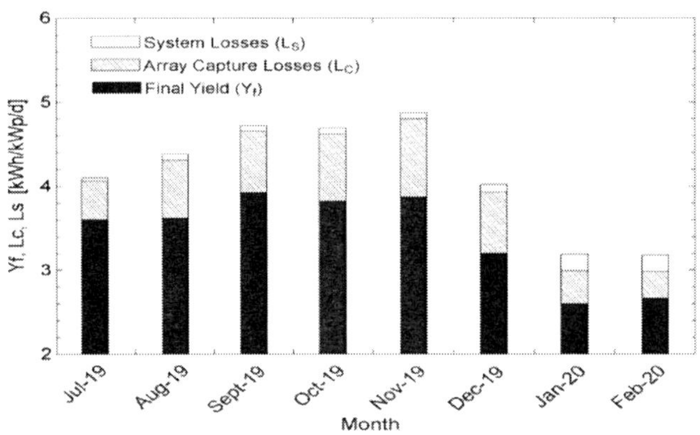

Figure 9. Monthly final yield (Y_f) with corresponding array capture losses (L_C) and system losses (L_S) of the 10.6 kWp grid-connected PV system in Energi Building from July 2019 to February 2020

5. Conclusion

In this paper, the performance of a 10.6 kWp grid-connected PV rooftop system installed in the Energi Building, PUSPIPTEK Serpong, South Tangerang has been observed and evaluated. The average daily energy during the eight-months monitored period was 37.20 kWh/day for DC energy (PV output) and 36.10 kWh/day for AC energy (inverter output) with the average values of the array (Y_a), final (Y_f), and reference (Y_r) yields were 3.51 kWh/kWp/d, 3.41 kWh/kWp/d, and 4.14 kWh/kWp/d, respectively. The average performance ratio (PR) during the eight-months monitored period was found 82.67% with the average of the module (η_{PV}), system (η_{sys}), and inverter (η_{inv}) efficiency were found to be 15.32%, 14.81%, and 96.71%, respectively. Meanwhile, the average capacity factor (CF) during the eight-months monitored period was about 14.19%. Compared with existing studies in the literature, these results indicated that the grid-connected PV system installed in the Energi Building, PUSPIPTEK Serpong exhibited good performance.

Acknowledgment
The authors would like to thank the National Laboratory for Energy Conversion Technology (B2TKE) BPPT who supported this study. We acknowledge the financial support provided by Universitas Indonesia through the Hibah Publikasi International Terindeks 9 (PIT9) funding scheme managed by the Directorat of Research and Community Engagement (DRPM) Universitas Indonesia.

References
[1] REN21 2019 *Renewable 2019 Global Status Report* (Paris: REN21) ISBN 978-3-9818911-7-1
[2] Dirjen EBT 2019 *Kebijakan, Regulasi dan Inisiatif Pengembangan Energi Surya di Indonesia* (Jakarta: ESDM)
[3] IESR 2017 *ENERGI TERBARUKAN: Energi untuk Kini dan Nanti* (Jakarta: IESR)
[4] Kamal A, Ali E and Adel E 2016 Performance analysis and investigation of a grid-connected photovoltaic installation in Morocco *Energy Reports* **2** 261–66
[5] Lutero C de L, Leonardo de A F and Francisco H B de L M 2017 Performance analysis of a grid connected photovoltaic system in northeastern Brazil *Energy for Sustainable Development* **37** 79–85
[6] Hayder S M, Chin K G and Kyairul A B 2019 Performance evaluation of various solar photovoltaic module technologies under tropical climate condisions at Melaka, Malaysia *Journal of Engineering and Applied Sciences* **14** 336-41
[7] Irene R F, Emilio M C, Rafael E P, Gustavo N and Juan de la C 2019 Analysis of the Performance of Various PV Module Technologies in Peru *Energies* **12** 186
[8] Eka N, Riza, Ifanda and Abduh A B 2019 Performance of 10 kWp PV Rooftop System Based on Smart Grid in Energy Building PUSPIPTEK *IEEE* 978-1-7281-3931-9/19/$31.00
[9] Renu S and Sonali G 2017 Performance analysis of a 11.2 kWp roof top grid-connected PV system in Eastern India *Energy Reports* **3** 76–84
[10] Akash K S, K. Sudhakar and Prashant B 2016 Simulation and performance analysis of 110 kWp grid-connected photovoltaic system for residential building in India: A comparative analysis of various PV technology *Energy Reports* **2** 82–8
[11] Chandrakant D, Deepak P, Anshul A, Akash K S, K. Sudhakar, Murali M S R and Amit B 2018 Performance simulation of grid-connected rooftop solar PV system for small households: A case study of Ujjain, India *Energy Reports* **4** 546–53
[12] Chong L 2018 Comparative Performance Analysis of Grid-Connected PV Power Systems with Different PV Technologies in the Hot Summer and Cold Winter Zone *International Journal of Photoenergy* **2018** ID 8307563 1-9
[13] Canadian Solar *Data Sheet PV Module CS6P-260| 265P*
[14] SMA *Installation Manual PV Inverter : Sunny Tripower 10000TL /12000TL / 15000TL / 17000TL*

Design of control and monitoring tools for electricity use loads, and home security systems with internet of things system based on Arduino Mega 2560

Setiyo Budiyanto[1*], Lukman Medriavin Silalahi[2], Freddy Artadima Silaban[3], Ketty Siti Salamah[4], Fajar Rahayu[5], M. Iwan Wahyuddin[6], Septi Andryana[7]
[1,2,3,4]Department of Electrical Engineering, Universitas Mercu Buana, Jl. Raya Meruya Selatan, Kembangan, Jakarta, 11650
[5]Department of Electrical Engineering, Universitas Pembangunan Nasional "Veteran" Jakarta, Jl. RS. Fatmawati Raya, Pondok Labu, Jakarta 12450
[6,7]Department of Informatics Engineering, Universitas Nasional, Jl. Sawo Manila, Pasar Minggu, Jakarta 12520

*Email : sbudiyanto@mercubuana.ac.id

Abstract. High economic growth makes the demand for comfortable and safe houses increasingly increase and the most obvious application of technology is technology with automated systems. With this technology, the use of electricity in homes can be monitored remotely. With the aim of optimizing and time efficiency, this research created a control system based on IoT system. The control system with the function of providing security and comfort for the homeowner and the people who live in it, because it can facilitate the work to be faster, effective and efficient. In this regard, in this research the title "Design of Controls and Monitoring of Electricity Loads, and Home Safety with IoT-based Arduino Mega 2560 Systems" is adopted. This application is used as consideration for someone applying the concept of smart homes and the Internet of Things in real life. The results of this research are testing the camera to process the results of taking pictures until the upload process is completed in storage on a computer that is made as a client server is 30 seconds and the entire system works properly and normally.

1. Introduction

Along with the times, technology is growing very fast, even things that were considered impossible or impossible can now be realized with the help of technology. High economic growth makes the demand for comfortable and safe houses increasingly increase and the most obvious application of technology is technology with automated systems. With this technology, the use of electricity in the home can be monitored and offers convenience in controlling the house remotely. Sometimes we forget to turn off the lights when we are outside so we have to go back and do a very inefficient check both in terms of time and financially such as the cost of gasoline to return home. With the goal of efficiency, an idea emerged to create an iot-based control system. Control system with various functions, will provide security and comfort for homeowners and the people who live in it, because it can facilitate the work to be faster, effective and efficient [1-8].

In this regard, the author will try to compile the final project with the title "Design of Controls and Monitoring of Electricity Loads, and Home Safety with IoT System based on Arduino Mega 2560". This application is used as consideration for someone applying the concept of smart home and the Internet of Things in real life.

The purpose of this research is to implement control of home appliances using Arduino with a web server, introducing smart homes more interactively, minimizing the use of home electricity with control applications via the web and becoming recommendations for implementing smart homes with control via the internet in the real world.

2. Research Model

2.1. Literature Review

Literature review conducted by comparing research of the same type with the same literature is used to be made into consideration and is expected to help in making a new system [9-13].

Table 1. Literature review

No	Title	Author	Method	Controller
1	Remote Control System for cloud computing based home electronic devices	Dede Kurniadi	Rapid Application Development	Arduino Uno
2	Web-based Smart Home Controller for Home Electronic Equipment	Fauzan Masykur	Literature Review and Study Experiments	Raspberry Pi
3	Implementation of wireless communication on Arduino-based smart home	Fathur	Study Experiments	Arduino Uno
4	Designing Prototype Security System for Smart Home Monitoring	Muhammad Ilham	Implementasi Universal	Incomsat SIM900
5	Smart home automation with Raspberry Pi and Android smartphones	Erick Fernando	Implementasi Universal	Raspberry Pi

2.2. Method

In this chapter describes the system requirements needed in the design of the connecting system design, circuit breaker, and monitoring of the flow of electrical equipment, discussing the design of the system to be made that has been adapted to user needs.

Basically this system will work to adjust the desires of consumers, which equipment will be arranged and connected with Arduino so that it can be controlled and monitored remotely. However, this research will take some basic electrical equipment commonly used in the home. The working principle of this system is to connect the server network and the electrical terminal to Arduino so that any electronic device connected to the terminal can be controlled and monitored using applications that have been made and connected to the server network.

Here is a block diagram of "Design of Electric Load Control and Monitoring Devices, and Home Safety with an IoT system based on Arduino Mega 2560".

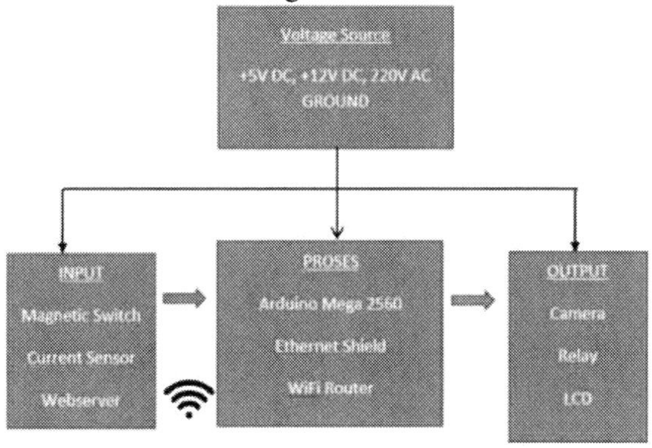

Figure 1. Diagram block system

Based on Figure 1, it can be seen that the circuit design is block diagram consisting of a voltage source block, input block, process block and output block. the input block describes the input for the microcontroller and the input media, the process block describes the process after input comes in and the components that act as input processors, while the output block explains the output produced and the output media.

Based on Figure 1, the working principle of each block is explained in table 2:

Table 2. Work principle

No	Components	Details
1	Current Sensor PZEM - 004T	This sensor is used to read the power used by the load.
2	Arduino Mega 2560	This module is used as a controller of the whole system.
3	Module Relay 4 Channel	This module is used to control the load that you want to use.
4	Ethernet Shield	This module is used so that Arduino can communicate with wireless servers
5	Camera Module VC0706	This module is used for security. This module will work when the door is opened and can also be controlled manually.
6	Magnetic Switch	This switch is installed to activate the camera module when this magnetic switch detects an open door.
7	WiFi Router	This tool works to communicate with the server and transmit via wireless

2.3. Design

The design of the input block can be seen in Figure 2.

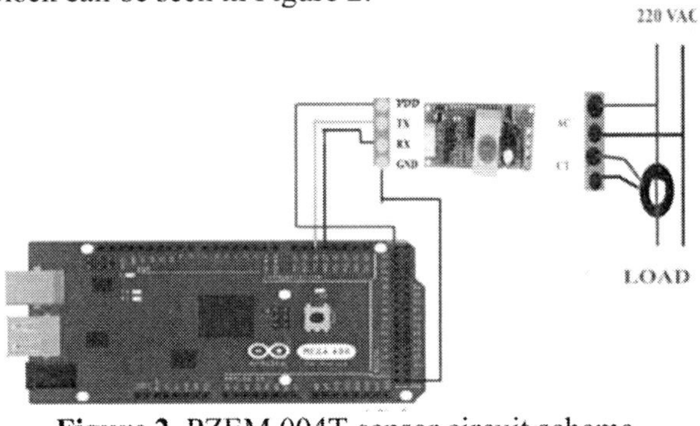

Figure 2. PZEM 004T sensor circuit scheme

The output block design can be seen in Figure 3.

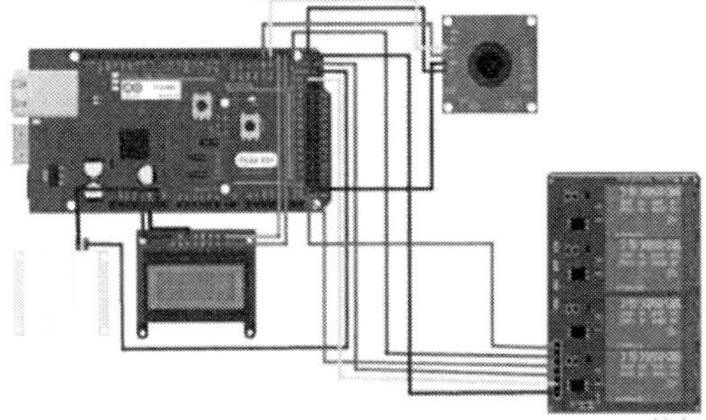

Figure 3. Output circuit

The block design process can be seen in Figure 4.

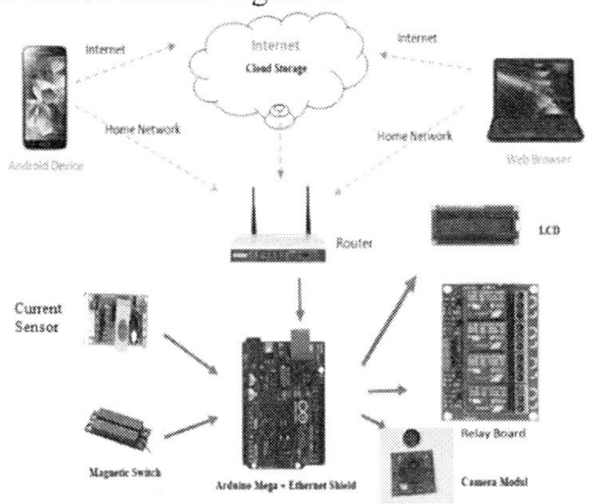

Figure 4. Arduino connection to the Internet

3. Result And Analysis

3.1. Prototype Result

The manufacture of mechanical parts includes the process of working boxes (boxes) where this process must support electronic parts so that the desired tool is realized. This process includes drilling, wearing, and mounting the box and the results of the design can be seen in Figure 5.

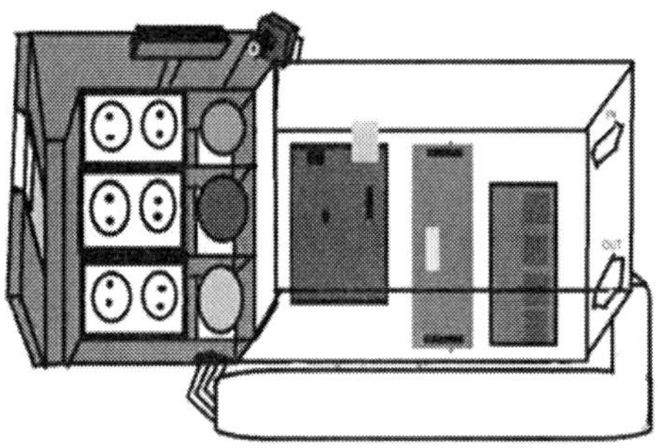

Figure 5. Design results

3.2. Network Connection Testing

In testing the network connection it is necessary to find out the ip address of the client computer connected to the access point on the server using the command prompt. Type "ipconfig" at the command prompt as can be seen in Figure 6.

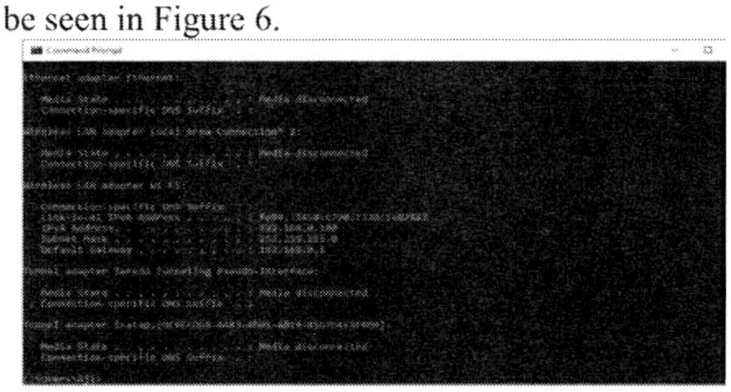

Figure 6. Command prompt

Figure 7 shows that the ip address connected to the access point server is 192.168,0.100, then the address can be entered in the Arduino program for Arduino communication with the client server computer, as shown in Figure 7.

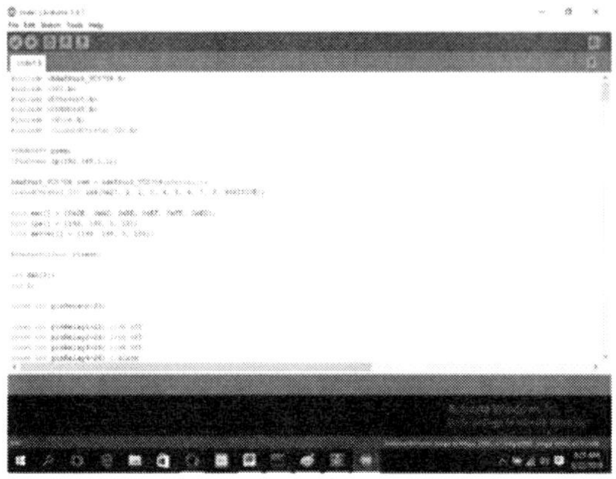

Figure 7. Arduino program

After completing, and successfully uploading the program, open the Arduino serial monitor and look at the LCD on the panel to see the client computer successfully connected to the server, as shown in Figure 8.

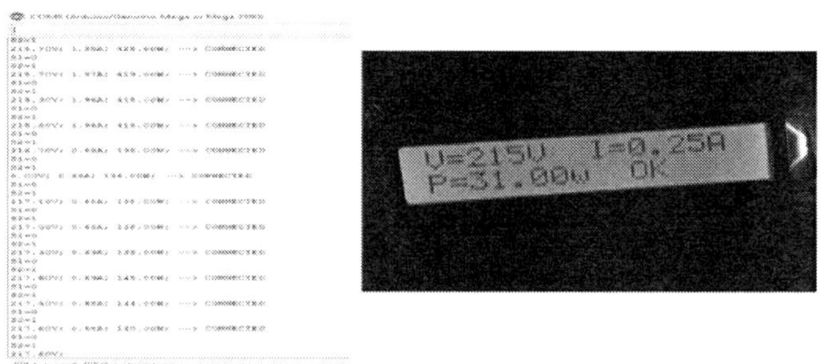

Figure 8. Server connections on the Arduino serial monitor and LCD panel

3.3. IoT System Testing

When finished, and successfully uploaded the program, open the Arduino serial monitor and look at the LCD on the panel to find out the client computer successfully connected to the server, as shown in Figure 9.

Figure 9. Switch testing no. 1

The test results in Figure 9 get that when the switch button 1 is "OFF" and orange on the web server when pressed it will change to "ON" and the symbol changes to green, indicating that the webserver ordered coil relay 1 to be connected to load no. 1.

Figure 10. Switch testing no. 2

The test results in Figure 10 get that when the switch button 2 is "OFF" and orange on the web server when pressed it will change to "ON" and the symbol changes to green, indicating that the webserver ordered coil relay 2 to be connected to load no. 2.

Figure 11. Switch testing no. 3

The test results in Figure 11 get that when the switch button 3 which is "OFF" and orange on the web server when pressed will change to "ON" and the symbol changes to green, indicating that the webserver ordered coil relay 3 to be connected to load no. 3.

Table 3. IOT System Testing Result

No	Button	Result	Decision
1	Switch Button 1	Lamp 1 lights up and stop contact no. 1 can be used	Matching
2	Switch Button 2	Lamp 2 lights up and stop contact no. 2 can be used	Matching
3	Switch Button 3	Lamp 3 lights up and stop contact no. 3 can be used	Matching
4	Switch Button 4	Relay is on, for spare load	Matching

The results of the data in table 3 of the load usage on the socket in realtime can be seen in the Arduino serial monitor and also in the server application in Figure 12.

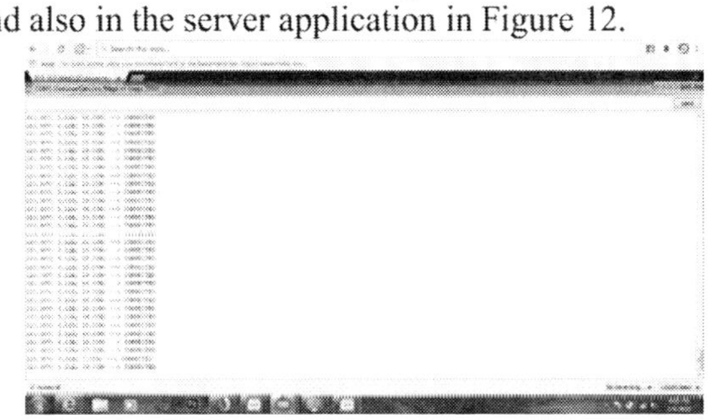

Figure 12. Arduino serial monitor monitoring the load

The test results in Figure 12 shows a serial monitor image that shows the value of the voltage that is read, the value of the current connected, the value of the power used, and "connected" which means that the webserver is connected to the client server. For testing the use of the load can be seen in Figure 13.

Figure 13. Load testing

In Figure 13 shows the power outlet is connected to the installed load, in the form of a rice cooker and an electric iron, as well as 3 LED lights, so that the sensor reads the load power that will be displayed on the webserver which can be seen in Figure 14.

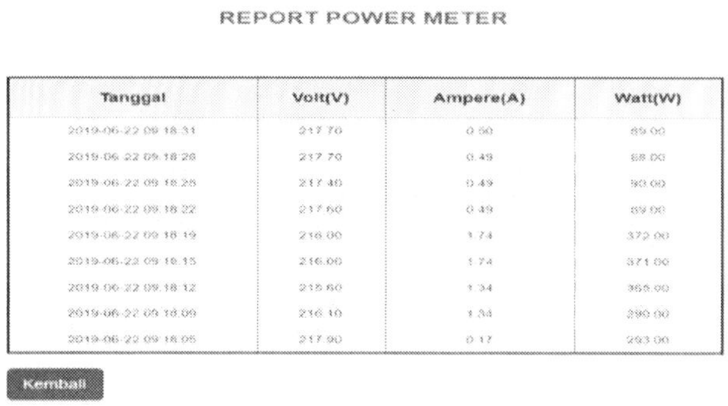

REPORT POWER METER

Tanggal	Volt(V)	Ampere(A)	Watt(W)
2019-06-22 09:18:31	217.70	0.50	89.00
2019-06-22 09:18:28	217.70	0.49	68.00
2019-06-22 09:18:25	217.40	0.49	90.00
2019-06-22 09:18:22	217.60	0.49	89.00
2019-06-22 09:18:19	216.00	1.74	372.00
2019-06-22 09:18:15	216.00	1.74	371.00
2019-06-22 09:18:12	215.60	1.34	365.00
2019-06-22 09:18:09	216.10	1.34	290.00
2019-06-22 09:18:05	217.90	0.17	293.00

Kembali

Figure 14. Report power meter on the web server

On the results of the power meter report in Figure 14 shows the load data that is read in the form of the connected voltage, the current connected, and the power used in realtime. And the data will be stored in the database server.

3.4. Manual System Testing

Testing is done to find out whether the system manual is running well to anticipate if the IoT system has a problem. In the manual system there are switches "OFF" and "ON" on the switch, which indicates that the "OFF" switch will disconnect the relay from being connected to the load, and the "ON" switch will connect the relay to the load so that the entire load can be used.

Figure 15. Switch testing OFF

The test results in Figure 15 show that when the switch in the panel box is in the middle position, the switch will instruct the relay to disconnect from the load, so that all loads cannot be used.

Figure 16. Switch testing ON

The test results in Figure 16 show that when the switch in the panel box is in the down position, the switch will instruct all relays to connect to the load, so that all loads can be used.

Table 4. Manual System Testing Results

No	Switch	Result	Decision
1	OFF	All switches and sockets are off	Matching
2	ON	All switches and sockets are on	Matching

3.5. Security System Testing

Testing is done to find out the security system that is designed automatically or manually control on the camera running well.

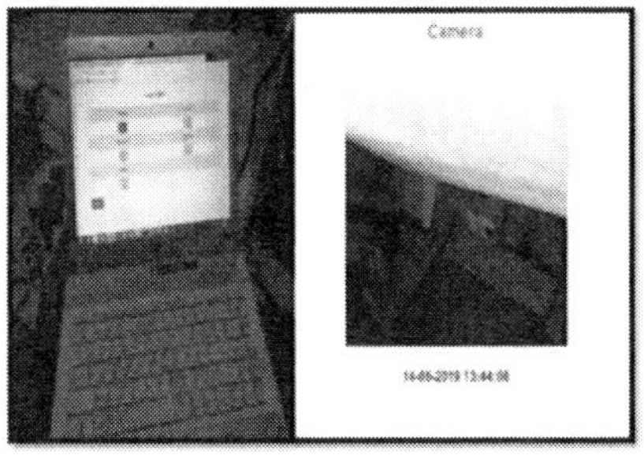

Figure 17. Camera button manual testing

The test results in Figure 17 show that when the "OFF" and orange capture button on the webserver when pressed will change to "ON" and the symbol turns green then turns orange again, indicating that the webserver commands the camera to take pictures.

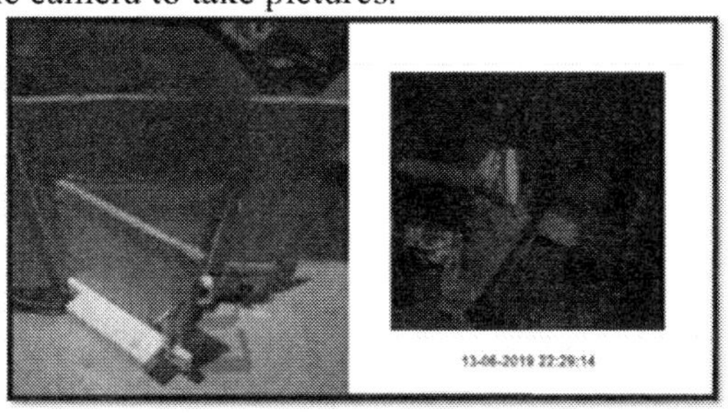

Figure 18. Door sensor testing

The test results in Figure 18 show that when the door is open, the magnetic sensor will instruct the camera to take pictures in real time so that someone who has entered the room can be identified.

Table 5. Security System Testing Results

No	System Mode	Result	Decision
1	Manual Button	The camera takes pictures in realtime when after pressing a button	Matching
2	Door Sensor	The camera takes pictures in realtime when the door is open.	Matching

To find out the camera taking pictures can be done by looking at the serial monitor on Arduino, and the results of the image will be stored on the client computer server storage.

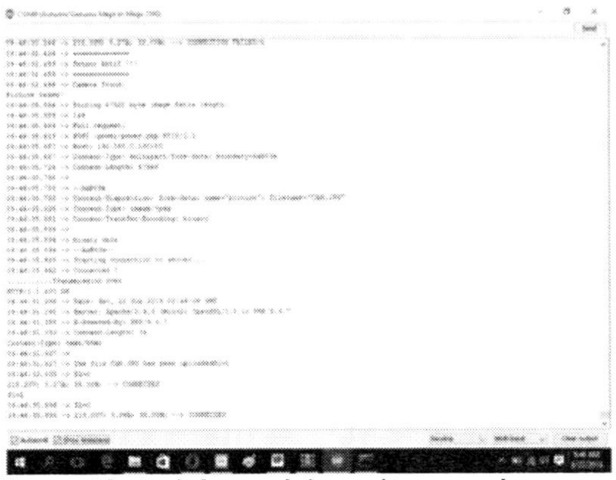

Figure 19. Arduino serial monitor security system

Figure 19 shows that the camera has succeeded in taking pictures and storing images on the server storage with a processing time of 30 seconds until the upload process is complete which is shown on the Arduino serial monitor.

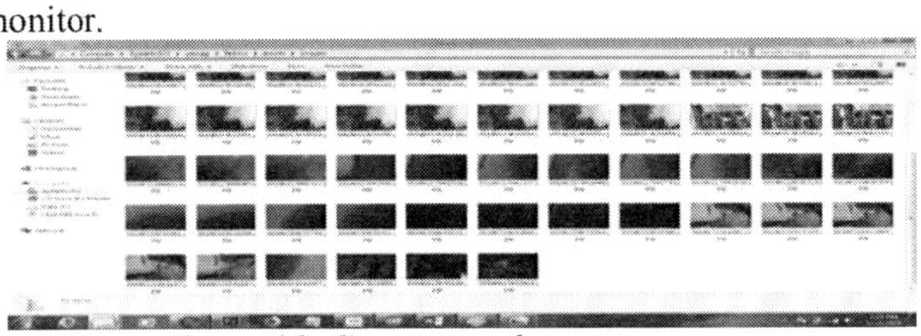

Figure 20. Camera results storage server

In picture 20 shows the results of taking a camera that has been stored on server storage, so users can see someone who has entered the room.

3.6. Application Testing

Testing is done to find out whether the application can connect to the web server when running. Test results can be seen in table 6

Table 6. Application Testing Results

No	First Display	Result	Test Result	Decision
1		Enter the web server application when clicking enter		Matching

In Table 6 shows the application can be connected directly to the server display so that users can quickly access the webserver through the application.

4. Conclusion

From the results of the design and testing of electrical load control and monitoring devices, as well as home security with the IoT based Arduino Mega 2560 system that has been done, it can be concluded that:

1. The MAN switch on the panel box instructs all relays to connect to the load, so that all loads can be used, and the camera is not functioning.
2. The OFF switch on the panel box instructs all relays to be disconnected with the load, so that all loads cannot be used, and the camera is not functioning.
3. The IoT switch on the panel box instructs the relay to be controlled through the server display, so that all loads can be controlled and the load power usage can be monitored, and the security system is activated manually through the webserver or sensor on the door.
4. In the camera test, the time to process the results of taking pictures until the upload process is completed is stored on a computer storage that is made as a client server is 30 seconds.

5. References

[1] Istiyanto JE Purwarupa sistem rumah cerdas berbasis arduino uno yang dikendalikan dengan smartphone android Doctoral dissertation Yogyakarta: Universitas Gadjah Mada

[2] Arham AM, Jati AN, Mulyana A Desain dan implementasi panel surya pada sistem pendeteksi banjir yang menggunakan wireless sensor network *eProceedings of Engineering* **2**

[3] Solichin A 2016 Pemrograman web dengan PHP dan MySQL Penerbit Budi Luhur

[4] Silvia AF, Haritman E, Mulyadi Y 2014 Rancang bangun akses kontrol pintu gerbang berbasis arduino dan android *Electrans* **13** 1-10

[5] Riza A Modul pelatihan pembuatan aplikasi Android menggunakan MIT App Inventor.

[6] PrasetiyowatI F. Aplikasi rumah pintar (smart home) pengendali peralatan elektronik rumah tangga berbasis web (Doctoral dissertation, Universitas Muhammdiyah ponorogo

[7] Sianipar RH 2015 Membangun Web dengan PHP dan MySQL Untuk Pemula dan Programmer Penerbit INFORMATIKA

[8] Safwan S, Zulfikar Z, Zulhelmi Z 2016 Desain sistem keamanan rumah melalui penyalaan peralatan listrik berdasarkan kebiasaan pemilik rumah *Karya Ilmiah Teknik Elektro* **1**

[9] Kurniadi D 2017 Sistem kendali jarak jauh perangkat elektronik rumah berbasis cloud computing. *Jurnal Algoritma* **14** 333-42.

[10] Fernando E 2014. Automatisasi smart home dengan raspberry pi dan smartphone android. *Konferensi Nasional Ilmu Komputer* (KONIK) **5**

[11] Fattah MI, Siregar S, Soegiarto D 2015 Rancang bangun prototype sistem keamanan untuk smart home monitoring *eProceedings of Applied Science* **1**

[12] Masykur F dan Prasetiyowati F 2016 Aplikasi rumah pintar (smart home) pengendali peralatan elektronik rumah tangga berbasis Web. Jurnal Teknologi Informasi dan Ilmu Komputer. Mar **3** 51-58.

[13] Rachman FZ 2016 Implementasi komunikasi nirkabel pada smart home bebasis arduino. *SNTI* 233-238.

[14] Budiyanto S, Asvial M and Gunawan D 2014 Implementation dedicated sensing receiver (DSR) in 3G-WiFi Offload. In2014 International Conference on Smart Green Technology in Electrical and Information Systems (ICSGTEIS) *IEEE* 37-42

[15] Budiyanto S and Nugroho A 2017 A new model of Genetic Zone Routing Protocol (GZRP): The process of load balancing and offloading on the UMTS-IEEE 802.11 g Hybrid Networks. *Telkomnika* **15**

[16] Budiyanto S, Asvial M and Gunawan D 2016 Improved performance of hybrid algorithm for 3G–WiFi offload networks *Jurnal Teknologi* **78** 5-9

[17] Budiyanto S, Sihombing HC and IM FR2019 Depression and anxiety detection through the closed-loop method using DASS-21 *Telkomnika* **17**

[18] Silalahi LM, Alaydrus M, Rochendi AD dan Muhtar M 2019 Design of Tire Pressure Monitoring System Using a Pressure Sensor Base *Sinergi* **23** 70-78.

IOP Conf. Series: Materials Science and Engineering **909** (2020) 012021 doi:10.1088/1757-899X/909/1/012021

Improvement of small scale turbine performance: in West Aceh region using a clark-y type of turbine blades

Maidi Saputra[1*] **and Rika Sartika**[1]

[1]Mechanical Eng. Dept., Teuku Umar University, Kampus Alue Peunyareung Aceh, Indonesia

*Email: maidisaputra@utu.ac.id

Abstract. This study aims to improve the performance of small scale wind turbines in the West Aceh region using a Clark-Y type turbine blade. Aceh Barat is one of the districts in Aceh province which is located at coordinates 04°06′ - 04°47′ NL and 95°52′ - 96°30′ EL and an area is located on the west coast of Aceh province bordering directly on the Indian Ocean, and one of the areas are worst affected by the tsunami disaster in 2004. The reference data shows that the potential of wind energy in the West Aceh region on average is quite varied, which is the lowest of 3.4 m/s to the highest speed of 5.8 m/s. The results of tests on the turbine blades before and after optimization provide increasing the ability of the turbine to produce energy and mechanical power. Based on the characteristics of the rotor rotation, the highest rotation data that occurs on the blades before the optimization is 316.6 rpm, and the highest rotor rotation on the blades after optimization is 361.3 rpm. Accordingly, there is an increase in turbine capability by 12.4%.

1. Introduction

1.1 Potential and wind energy conversion in the West Aceh region

The potential of wind energy in the West Aceh region has the potential to be able to rotate small-scale wind turbines. The average wind speed produced in the Meulaboh region is quite varied; it is as low as 3.4 m/s to the highest speed of 5.8 m/s [1], [2]. Wind turbines convert wind power captured by the turbine blades in the form of mechanical force. Wind turbines convert energy into electrical energy through generators that take the rotational power through the turbine shaft, and then the power generated by the turbine shaft must go through a gear system that regulates the spin power to the generator [3]. The power generated by wind turbines is stated as [3]:

$$P = \frac{1}{2}\rho\pi R^2 V^3 C_p \tag{1}$$

ρ is the density of air in kilograms per cubic meter, R is the radius of the wind turbine rotor, V is the air velocity, and Cp is the power coefficient of the wind turbine, which is a function of the tip speed ratio (TSR). The theoretical maximum Cp limit, according to Betz theory, is 16/27, but in practice, the maximum Cp efficiency that can be achieved by wind turbines is 40% - 45%. The efficiency curve of the rotor Cp (λ) is a nonlinear function of the TSR, which is determined by the design of the turbine blades and pitch angle, as shown in Figures 11 [3], [4].

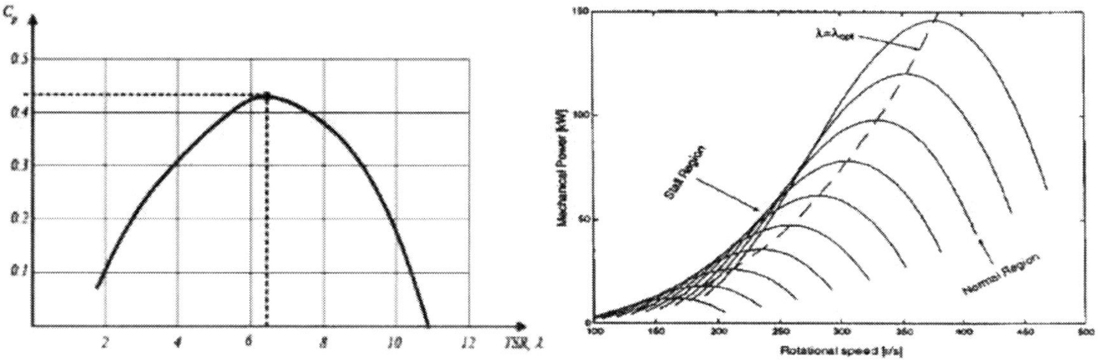

Figure 1. Relationship between TSR and power coefficient (Cp), and between maximum Cp power and rotor speed.

1.2 The power efficiency of available wind turbines

According to Betz, the maximum amount of energy that can be converted by a wind turbine is 0.59259 from the total energy available [5]. Betz's momentum theory explains the modelling of the flow of wind energy through the turbine blade. The kinetic energy contained in the wind will decrease when pounding the turbine blade, and the airflow will turn and scattered which a much reduced kinetic energy value, and there is a relationship between the ideal power coefficient, C_p with the value of the ratio between the wind speed before and after pounding the turbine blade, such as shown in figure 2 [5], [6].

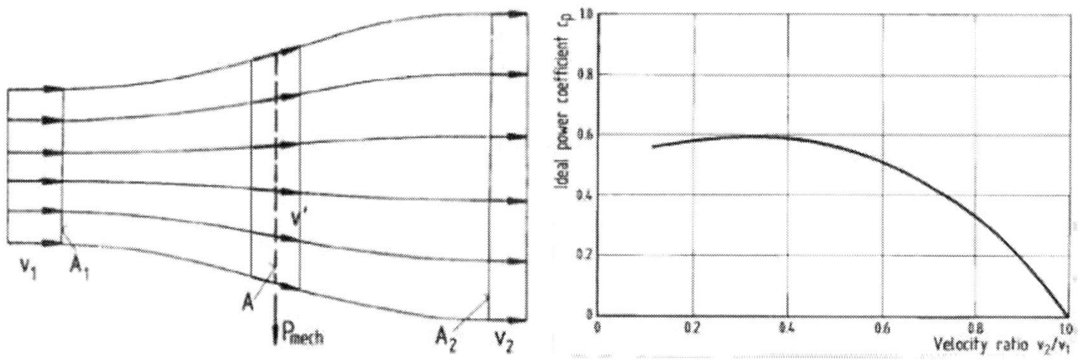

Figure 2. Flow model and power coefficient are proportional to wind speed before and after pounding the turbine blade.

1.3 Optimizing the performance of wind turbines

Enhancement of wind turbine performance can be done by improving and testing turbine blade design and developing types of airfoils. Blades and airfoil designs can be done one of them using the Genetic Algorithm method, which is useful for improving the aerodynamic performance of the blade [7], [8]. The process of developing a wind turbine blade as a basis for improving the performance of a wind turbine is divided into two stages, namely aerodynamic and structural design [9], [10]. From these two stages, at the aerodynamic stage, it is necessary to pay attention to the application of new airfoil models that support the performance of the blade. From the formative stage, it is necessary to consider the durability and flexibility of the blade designed for the blade's ability to withstand operating fatigue [9], [11].

Research conducted by Liu et al. [12] provides results of the development of wind turbine blades design that pays attention to the optimization and improvement of the aerodynamic functions of the blades, which provides an increase in wind turbine performance in annual power production by 7.5%. Other relative studies give results in an increase in annual power production by 8.51% [13]. Other studies have shown that the performance ratio of the maximum wind turbine blade is provided by the blade with a maximum thickness of 40% of the length of the chord and the attack angle of the turbine by 20% [8],

[14]. Based on the conclusions of a study conducted by Wiratama, I.K, et al. showed that the difference in profile and number of blades greatly influenced the performance of wind turbines. The results of this study indicate that the profile of a sizeable tapered blade gives maximum performance results [8], [15].

One form of the airfoil that is simple in manufacturing and providing maximum results is a Clark-Y type wind turbine blade. This type of wind turbine blade can be made with a simple method because all manufacturing processes are carried out manually. The test results give a TSR value of 7, rotor solidity of 0.57, and the resulting output power of 165.924 Watt [16].

2. Research methodology

2.1 Application area

Aceh Barat is one of the districts in Aceh province which is located at coordinates 04° 06′ - 04° 47′ NL and 95° 52′ - 96° 30′ EL. Aceh Barat is an area which is located on the west coast of Aceh province, and it borders directly with the Indian Ocean and is one of the areas worst affected by the 2004 tsunami [17]. The West Aceh region has an enormous wind power potential which is resulted from the advantage of its geographical location, and it is directly adjacent to the Indian Ocean. The location and map image of the application area of this study are shown in Figure 3.

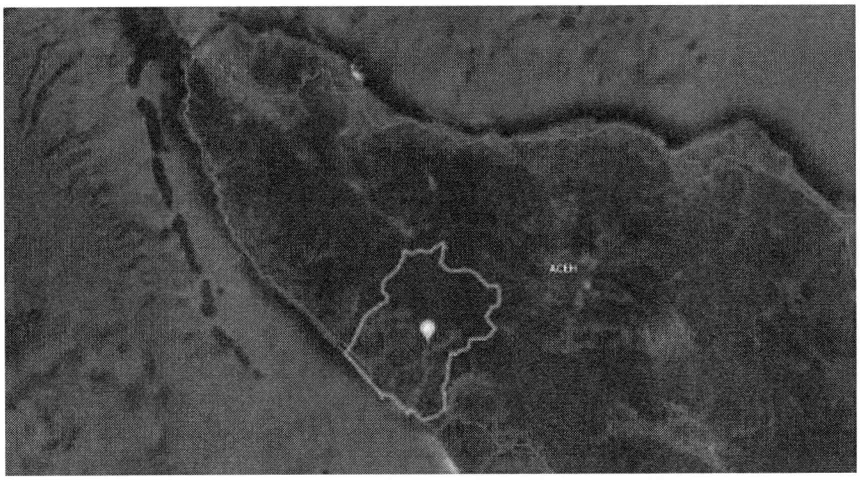

Figure 3. Areas of research application [18].

2.2 Estimated energy generation

West Aceh Regency has an average temperature of 26.5 °C, air pressure of 1,010 atm, air humidity of 86%, an average monthly solar radiation of 41.17%, an average rainfall of 255.1 mm and wind direction and speeds an average of 3 knots westward. From these data, West Aceh district is very suitable to be developed as a central area for wind power generation [19].

Wind energy sources as a renewable energy source provide wind data that will be used by power plants. Measurement of wind energy potential is done by measuring wind speed using anemometer and for annual data obtained from the Metrology and Geophysics Agency of Meulaboh Station. The wind data is converted at the height of 15 meters, so it follows the height of the wind turbine hub at the time of testing using equation (2) [20], [21].

$$V_{hub} = V_{ane} x \left(\frac{H_{hub}}{H_{ane}}\right)^{\alpha} \tag{2}$$

Where V_{hub}, V_{ane}, H_{hub}, H_{ane}, and α are the speed and height of the hub at the time of testing.

The condition of the wind turbine installation is in the "High and Protected Plant" area with a friction coefficient (α) of 0.20. An annual average wind speed of 4.8 m/s is obtained with maximum values occurring in October and December of 5.8 m/s, and the minimum value occurred in September of 3.4 m/s, as shown in tables 1 [1], [21], [22].

Table 1. Average wind speed per month for a year after correction.

No.	Month	Average wind speed per month (m/s)
1	January	4.1
2	February	5.6
3	March	4.5
4	April	4.0
5	May	4.8
6	June	5.6
7	July	4.9
8	August	4.7
9	September	3.4
10	October	5.8
11	November	4.1
12	Desember	5.8
The average wind speed overall m/s		4.8

2.3 Design and optimization of turbine blades

The theory about the working principle of the turbine blades has been developed by Glauret [23]. The theory explains the theory of blade elements and momentum [24]. The BEM theory is divided into two parts in designing of the aerodynamics blade. For the first part, a blade model is divided into the smallest sections, as shown in Figure 4. In the second part, the aerodynamic force calculation of the blade is calculated as a two-dimensional airfoil, which is linked to the flow conditions, as shown in Figure 5 [13].

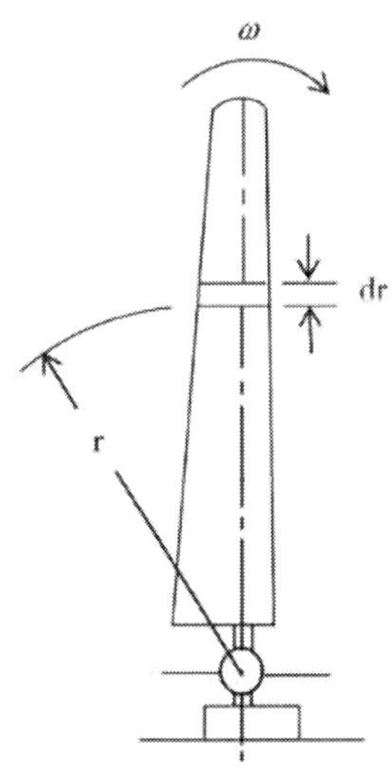

Figure 4. The smallest parts of the blade [13].

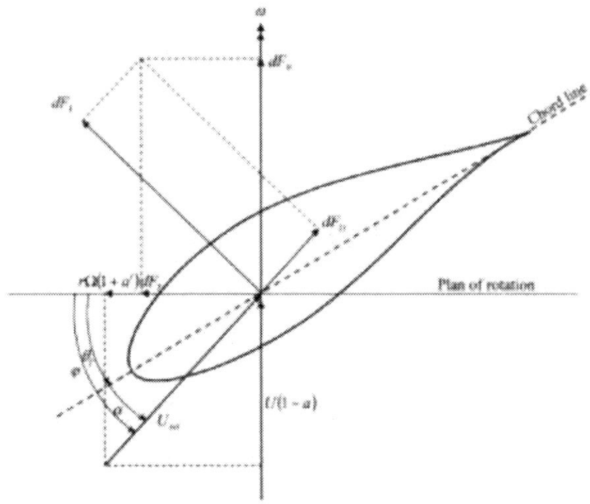

Figure 5. The aerodynamic force of the blades [13].

The design of a wind turbine is always aimed at achieving maximum turbine performance. The performance of the wind turbine is influenced by the blade design and operating conditions that allow the wind turbine to work with excellent efficiency, as shown in Figure 6 [25]. A genetic algorithm theory has been developed in perfecting a product. A genetic algorithm is a non-gradient-based algorithm to solve the optimization problems which are found in a product. In general, optimization is given by the following equation [13]:

$$x = (x_i) \qquad \forall i = 1,2, \dots N. \tag{3}$$

Minimize $f(x)$

$$h_j(x) \leq 0 \qquad \forall j = 1,2, \dots n. \tag{4}$$

Where x, N, f(x) and hj(x) are vector design parameters, the number of design parameters, objective functions, and design constraints.

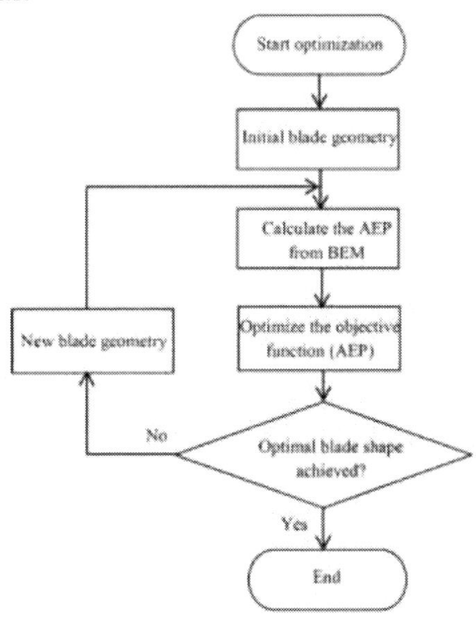

Figure 6. Flowchart of the optimization process [13].

3. Results and Discussion

The research results are the optimization values of the blade, which is designed to increase the value of Annual Energy Production (AEP). Before the optimization is done, a test is performed to get the characteristic value of the bar, which is then plotted in a graph. Afterward, the blade design optimization process is carried out concerning Annual Energy Production (AEP), which is done by looking at the characteristics of the blade performance that are plotted in a graph as well.

3.1 Design of Blade

The design of Blade starts with determining the type of blades that are designed. For this study, the airfoil model, which is used, is the Clark-Y type. Then, from the literature analysis and reference has taken a rotor diameter of 2 m, a TSR value of 7, and a rotor solidity value of 0.57. CAD design and animation are given in Figure 7 [16].

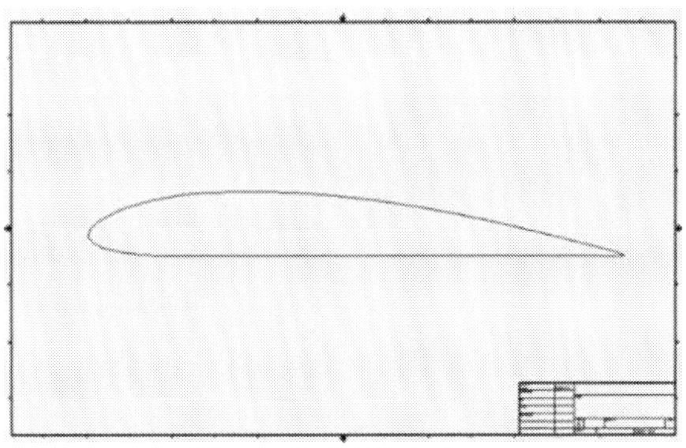

Figure 7. CAD design.

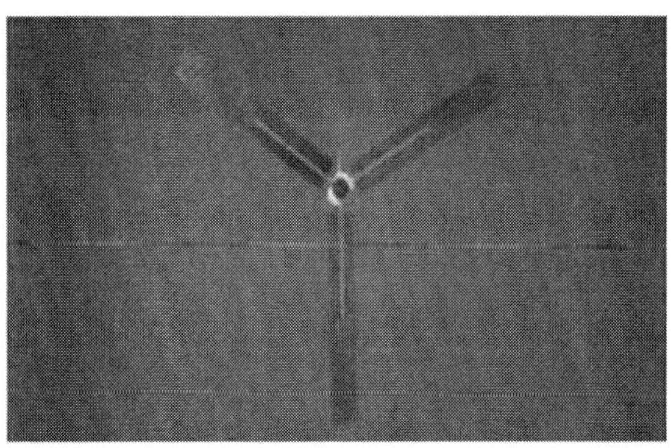

Figure 8. Blade type Clark-Y animation.

3.2 Test results of the blade before and after optimization

Wind turbine testing using blades before the optimization is done within a few days of testing. The testing is carried out to get the value of the characteristics of the wind turbine that illustrates its ability to produce Annual Energy Production (AEP). The test results are given in Figure 8 below.

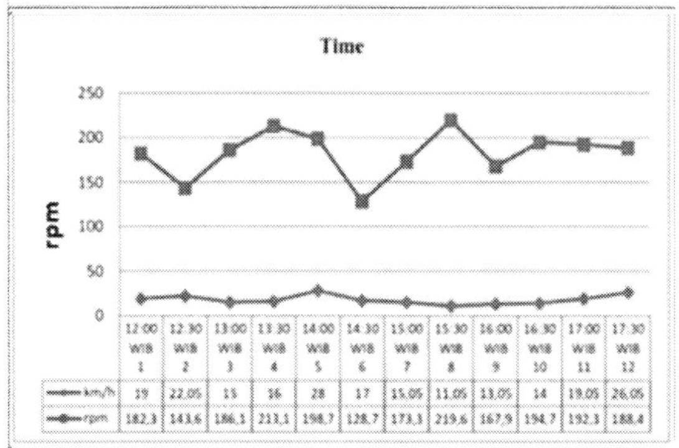

Figure 9. Results of the blade test before optimization for data 1.

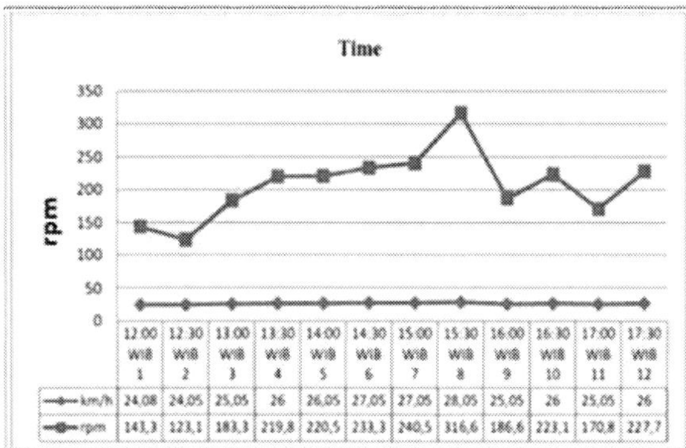

Figure 10. Results of the blade test before optimization for data 2.

Testing on the blades that have been carried out optimization gives different values. The process of optimizing the turbine blades has given the addition of mechanical power from the wind turbine in energy production, as shown in Figure 9.

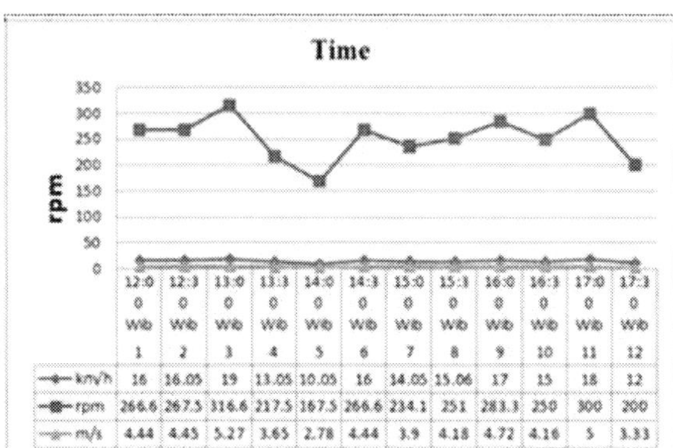

Figure 11. Results of the blade test after optimization for data 1.

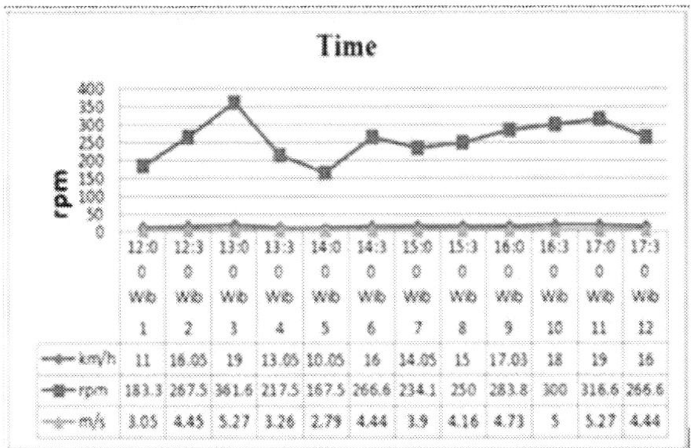

Figure 12. Results of the blade test after optimization for data 2.

From the results of tests on the turbine blades before and after optimization provides data on the increase in the ability of the turbine to produce energy and an increase in mechanical power. Based on the characteristics of the rotor rotation, the highest rotation data that occurs on the blades before the optimization is 316.6 rpm, and the highest rotor rotation on the blades after optimization is 361.6 rpm. As a result, there is an increase in turbine capability by 12.4%.

4. Conclusion

The theory about the working principle of the turbine blades has been developed by Glauret. The theory explains the theory of blade elements and momentum. The design of a wind turbine is always aimed at achieving maximum turbine performance. The performance of the wind turbine is influenced by the blade design and operating conditions that allow the wind turbine to work with good efficiency. Therefore, a genetic algorithm theory has been developed in product optimization. From the results of tests on the turbine blades before and after optimization provides data on the increase in the ability of the turbine to produce energy and an increase in mechanical power. Based on the characteristics of the rotor rotation, the highest rotation data that occurs on the blades before the optimization is 316.6 rpm, and the highest rotor rotation on the blades after optimization is 361.6 rpm. As a result, there is an increase in turbine capability by 12.4%.

5. References

[1] M. Saputra, A. Syuhada, and R. Sary 2018 Study of solar and wind energy using as water pump drive-land for agricultural irrigation *in proceedings-2018 4th International Conference on Science and Technology ICST 2018* doi: 10.1109/ICSTC.2018.8528643

[2] M. Saputra 2016 Homer sebagai analisator energi angin di kota meulaboh **2** 48–57

[3] Z. Chen, J. M. Guerrero, and F. Blaabjerg 2009 A review of the state of the art of power electronics for wind turbines *IEEE Trans. Power Electron* **24** 1859–1875

[4] J. Twidell J Wind turbines: technology fundamentals *Renew. En- ergy World* **6** 102–111

[5] F. Aryanto, M. Mara, and M. Nuarsa 2013 Pengaruh kecepatan angin dan variasi jumlah sudu terhadap unjuk kerja turbin angin poros horizontal *Din. Tek. Mesin* **3** 50–59

[6] E. Hau 2006 Wind turbines: fundamentals, technologies, application, economics

[7] H. Snel 2003 Review of aerodynamics for wind turbines wind energy **6** 203–211

[8] I. K. Wiratama, I. D. K. Okariawan, I. G. N. Yudhyadi, I. M. Mara, and A. Juliansah Analisis 2019 Pengaruh profil sudu dan jumlah sudu terhadap unjuk kerja turbin angin poros horizontal pada daerah kecepatan angin rendah *Din. Tek. Mesin* **9** 71

[9] J. Zhu, X. Cai, and R. Gu 2016 Aerodynamic and structural integrated optimization design of horizontal-axiswind turbine blades *Energies* **9** 1–18

[10] Tony Burton, Nick Jenkins, David Sharpe, Ervin Bossanyi Wind energy handbook-Google Books [Online]. Available: https://books.google.co.id/books?hl=en&lr=&id=dip2LwCRCscC&oi=fnd&pg=PT16&dq=Burton,+T.%3B+Jenkins,+AND.%3B+Sharpe,+D.%3B+Bossanyi,+E.+Wind+Energy

+Handbook%3B+John+Wiley+%26+Sons:+Chichester,+West+Sussex,+UK,+2011.&ots=IdFJXqMpEb&sig=Ns7wPpo7PJrLzuioc-uDk06SakM&redir_esc=y#v=onepage&q&f=false. [Accessed: 09-Apr-2020].

[11] B. Kim, W. Kim, S. Lee, S. Bae, and Y. Lee 2013 Developement and verification of a performance based optimal design software for wind turbine blades *Renew Energy* **54** 166–172

[12] X. Liu, Y. Chen, and Z. Ye 2007 Optimization model for rotor blades of horizontal axis wind turbines *Front. Mech. Eng. China* **2** 483–488

[13] A. Hassanzadeh, A. Hassanzadeh Hassanabad, and A. Dadvand 2016 Aerodynamic shape optimization and analysis of small wind turbine blades employing the Viterna approach for post-stall region *Alexandria Eng. J* **55** 2035–2043

[14] Pengaruh lokasi ketebalan maksimum airfoil simetris terhadap koefisien angkat aerodinamisnya|Nurcahyadi| *Semesta Teknika [Online]* Available: https://journal.umy.ac.id/index.php/st/article/view/799. [Accessed: 07-Apr-2020]

[15] N. H. Sari and W. G. Laksamana 2019 Perancangan bilah tipe taperless pada kincir angin: studi kasus di PT. Lentera Bumi Nusantara Tasikmalaya *Din. Tek. Mesin* **9** 104

[16] A. Sayogo, N. Caroko, and Wahyudi 2016 Perancangan dan pembuatan kincir angin tipe horizontal axis wind turbine (HAWT) untuk daerah pantai selatan jawa *UMY Repos* 1–5

[17] A. Ibrahim 2010 Profil kabupaten Aceh Barat

[18] West Aceh Regency-Google Maps. [Online] Available: https://www.google.com/maps/place/West+Aceh+Regency,+Aceh/@4.557125,96.5801166,317508m/data=!3m1!1e3!4m5!3m4!1s0x303f265aa5ffff8d:0x3039d80b220ccc0!8m2!3d4.4542745!4d96.1526985. [Accessed: 12-May-2020]

[19] Kabupaten Aceh Barat dalam angka 2018

[20] K. S. R. Murthy and O. P. Rahi 2016 Preliminary assessment of wind power potential over the coastal region of Bheemunipatnam in northern Andhra Pradesh, India *Renew. Energy* doi: 10.1016/j.renene.2016.08.017

[21] A. M. Patel and S. K. Singal 2018 Economic analysis of integrated renewable energy system for electrification of remote rural area having scattered population *Int. J. Renew. Energy Res* **8(1)** 523–539

[22] Wind and solar power systems: design, analysis, and operation Second Edition-Mukund R. Patel-Google Books [Online] Available: https://books.google.co.id/books?hl=en&lr=&id=CfbLBQAAQBAJ&oi=fnd&pg=PP1&dq=Patel+MR.+Wind+and+solar+power+systems:+Design,+Analysis,+and+Operation.+2nd+ed.+Boca+Raton:+CRC+Press%3B+2006.&ots=V1BXmhh3Zv&sig=voU2bh-jQOlgBx8Ta0YTWarGgb0&redir_esc=y#v=onepage&q&f=false. [Accessed: 13-May-2020]

[23] H. Glauert 1963 Airplane propellers in Aerodynamic Theory *W. F. Durand, Ed., Dover, New York, NY, USA*-Open Access Library [Online] Available: http://www.oalib.com/references/14107977. [Accessed: 13-May-2020]

[24] Q. Wang, J. Wang, J. Chen, S. Luo, and J. Sun 2015 Aerodynamic shape optimized design for wind turbine blade using new airfoil series *J. Mech. Sci. Technol* **29** 2871–2882

[25] X. Shen, H. Yang, J. Chen, X. Zhu, and Z. Du 2016 Aerodynamic shape optimization of non-straight small wind turbine blades *Energy Convers. Manag* **119** 266–278

IOP Conf. Series: Materials Science and Engineering **909** (2020) 012022 doi:10.1088/1757-899X/909/1/012022

Development of leaf spring design in large vehicles made from material type 65 si7 using static analysis with reverse engineering

N Nazaruddin [1,2a], F Zainuri[1,3], S Prasetya[1,3], R Siregar[1,4], G Heryana[1,5], M Adhitya[1] and DA Sumarsono[1]

[1]Research Center for Advanced Vehicle (RCAVe), Universitas Indonesia, 16424, Indonesia

[2]Department of Mechanical Engineering, Universitas Riau, 28293, Indonesia

[3]Department of Mechanical Engineering, Politeknik Negeri Jakarta, 16425, Indonesia

[4]Department of Mechanical Engineering, Darma Persada University, 13450, Indonesia

[5]Department of Mechanical Engineering, STT Wastukancana, 41151, Indonesia

[a)]Corresponding author: nazaruddin@eng.unri.ac.id

Abstract. University of Indonesia Electric Bus was first developed using Hino R260 chassis. For further development, especially to fill out local content, it must be carried out engineering of its components. One of which will be discussed in this paper is the spring component. Reverse Engineering method is used to redesign in such a way. Measurement of the existing leaf springs in the chassis is done. The dimensions and number of leaf springs for the front and rear wheels of the chassis are produced. With CAD software and finite element analysis, a static analysis is performed on each of these springs. Front axle leaf spring arrangement consists of 8 spring sheets with dimensions of width 80 mm and thickness of 10 mm with stratified length. Rear axle leaf spring arrangement consists of 5 spring springs with dimensions of 90 mm width and 12 mm thickness with evenly distributed length. The theoretical stress on the front leaf springs is 512.74 MPa and finite element analysis is 531.5 MPa. The theoretical stress behind the leaf spring which occurs is 721.67 MPa and elemental analysis up to 728.4 MPa. The material for both of these springs is chosen is 65 si7 or SUP9.

1. Background

Before the issuance of presidential regulation No. 55 of 2019 which contains the acceleration of the battery-based motor vehicle program, Universitas Indonesia has conducted research on electric cars. Some types that have been launched in 2017 are City Car 01 to 03 and Electric Bus. These electric cars are innovations by FTUI lecturers and students who are members of the UI Molina Team. This research is about primary car electric design[1] at the University of Indonesia, energy optimization[2], chassis design[3], brakes[4][5], transmission[6] and steering systems[7][8]. The EV bus is a vehicle with a capacity of 60 passengers with motor power of 120 kW and 300 Ah. In the future this bus is expected to gradually be able to replace the overall function of the UI yellow bus as an internal transportation tool for the UI campus[9].

One of important structure in vehicle is chassis. At chassis is attached to components such as the transmission system, brake system and suspension system. The system on the chassis is very directly influential on comfort when driving, stability and others[10], [11]. Static analysis on the ladder frame chassis[12][13][14] with any type of material[15] and various thickness [16] with FEA method[17] been done. While research on leaf springs is a static analysis of leaf springs just for light commercial vehicles[18], but for large vehicles still need challenges for research on this leaf spring.

In this paper, the author wants to explain the static analysis for leaf spring of the University Indonesia's electric bus (large vehicle) by reverse-engineering method. The suspension system that was installed on this bus with type R260 is leaf spring. The goal is to analyze this type of suspension, the leaf spring, with the structure attached to this chassis. The material to be used is the standard material used for leaf spring. The material chosen is to see availability in the country, to fulfill local content as mandated by the president's degree.

2. Basic Theory: Leaf Spring

Leaf springs are springs that focus on strength so they can receive large loads. Springs are used using spring-eptic springs; there are 5 springs in each wheel. The thickness of each spring sheet is around 28 mm. of different lengths. The spring sheet is joined using a bolt in the middle of the spring and clamp. At both ends of spring no. 1 or the longest spring made a hole as a mounting place on the frame hanger. Leaf springs are mounted on the side frame using 2 bolts and U plates. At both ends of the spring are given a special rubber coating to eliminate noise due to friction between the plates when the leaf springs are working to take the load, so that the rubber does not come loose then tighten.

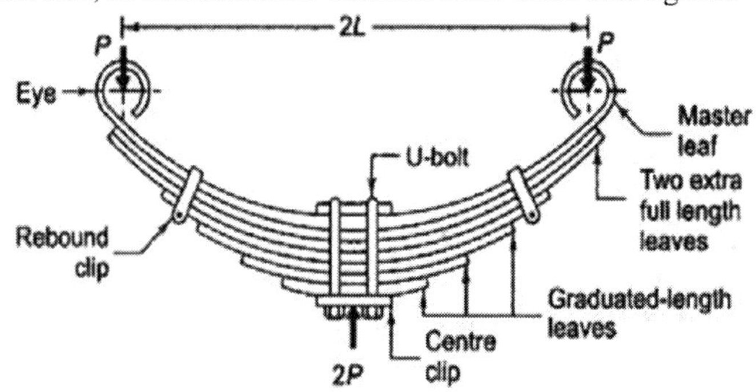

Figure 1. Component of leaf spring [18]

The stress and deflection equation for leaf springs can be used the following formula:

$$\sigma_{max} = \frac{pFL}{Nb_N h^2} \quad (1)$$

$$\delta_{max} = \frac{qFL^3}{ENb_N h^3} \quad (2)$$

Where the constants p and q are 3 for simple beam and 6 for fixed beam and N is the number of leaf spring sheets. Idealization of this equation is with assumptions like figure 2.

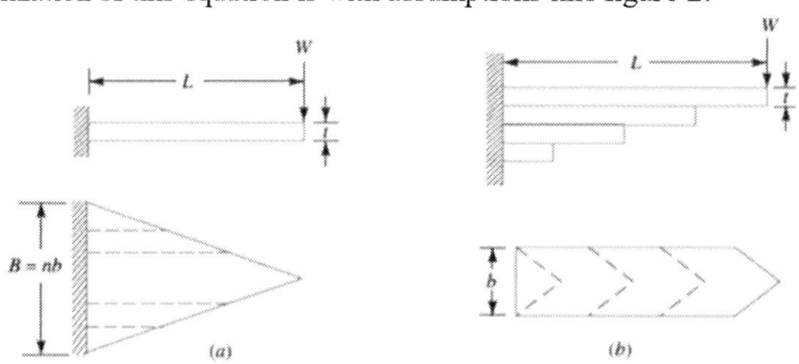

Figure 2. Ideal for leaf springs[19]

3. Methodology

3.1. Reverse Engineering

The University of Indonesia succeeded in converting conventional buses to electric buses (figure 3) through the National Electric Car (Molina) project in 2013. The chassis used on this electric bus still uses the R260 frame (figure 4) from one of the well-known bus brands today. This chassis frame consists of two main beams with four cross beams and two pipe beams. There are several systems on this bus vehicle that depend on this chassis, one of which is the suspension system.

Figure 3. Universitas Indonesia's electric bus

Figure 4. Rolling chassis model of electric bus

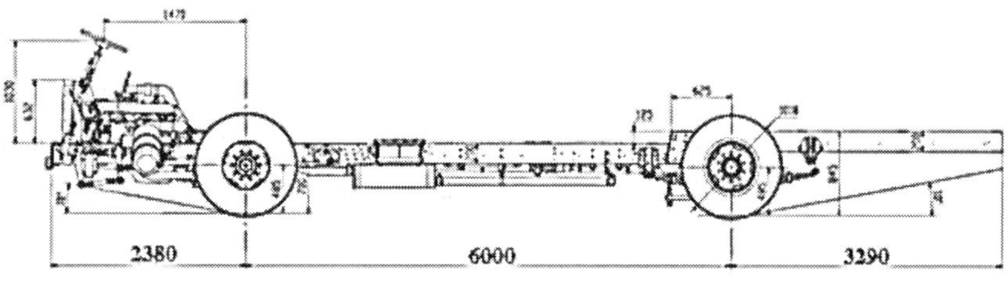

Figure 5. Main dimension of EV-bus rolling chassis.

The chassis has a main size is 2380 for front overhangs, 6000 mm for wheel base and 3290 mm for rear overhangs (figure 5)

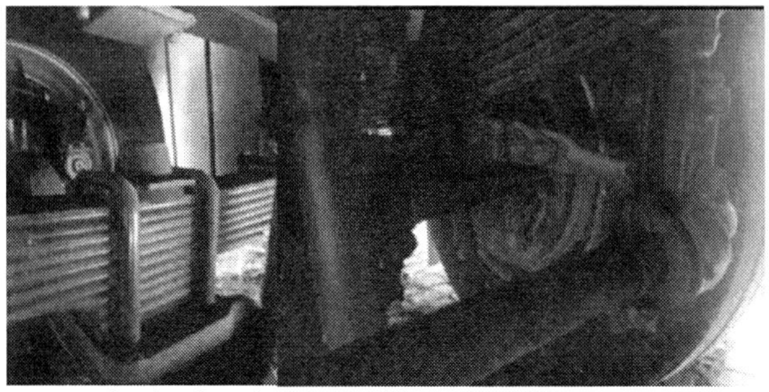

Figure 6. The leaf spring on rolling chassis parts are measured

The type of suspension system used in this chassis is a leaf spring and a shock absorber (figure 6). Measurement of the main dimensions, number and arrangement of leaf springs for front leaf spring and rear overhang are measured directly is one of the reverse engineering efforts on this suspension system.

3.2. Modeling with CAD software
The process of re-drawing with CAD software is done after all the dimensions of the leaf spring have been obtained. The results of this re-drawing can be seen in Figure 7 below.

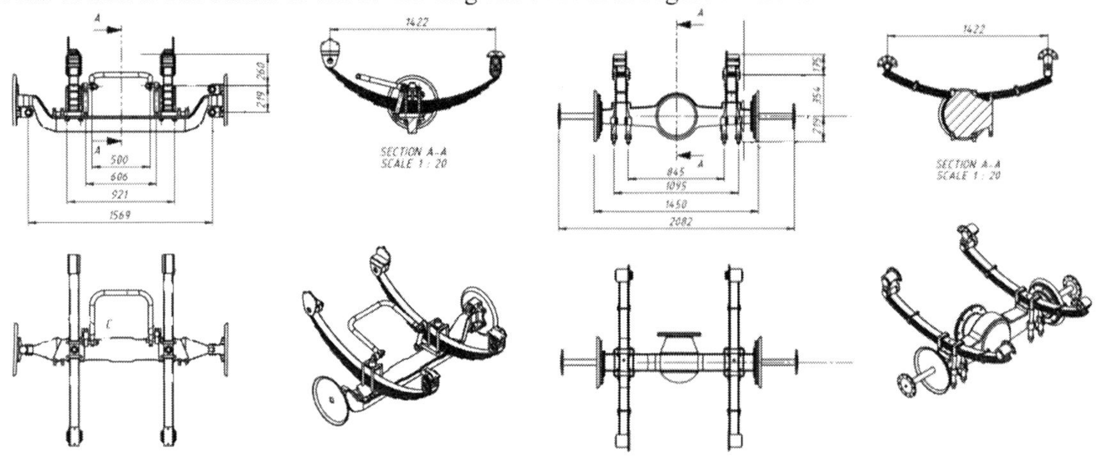

(a) (b)

Figure 7. Model 3D sub assembly leaf spring (a) front axle (b) rear axle

3.3. Material Selection
One of the mandates of the Presidential Decree No. 55 of 2019 is to increase local supply of electricity. Therefore, materials that will be selected according to code 65 si7 1095 with one of the products from PT. Krakatau Steel is a well-known producer of sheet steel or steel bars in Indonesia. Following are the mechanical and chemical properties of leaf material 65 si7 1095 obtained from various sources[20] (see Table 1).

Table 1. Mechanical and thermal properties of SUP9 material.

Mechanical Properties			Thermal Properties		
• Elastic modulus	190	GPa	• Melting point	1450	°C
• Poisson ratio	0,29		• Specific heat capacity	470	J/kg-K
• Tensile strength	1225	MPa	• Thermal conductivity	48	W/m-K
• Yield strength	1080	MPa	• Thermal expansion	13	μm/m-K
• Shear modulus	73	GPa			

IOP Conf. Series: Materials Science and Engineering **909** (2020) 012022 doi:10.1088/1757-899X/909/1/012022

The values of these 65 si7 material properties will be made as custom materials in finite element analysis with software computers.

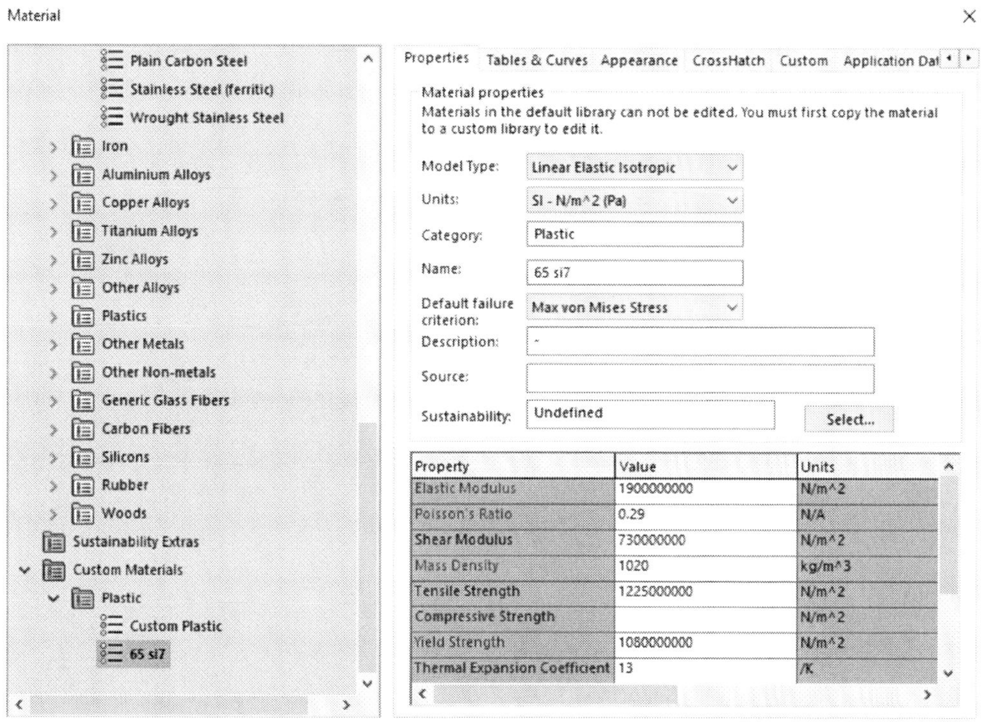

Figure 8. Input new material in CAD software

4. Result and Discussion

4.1. Determination of Load on Leaf Spring

For chassis analysis, it is modeled with a beam with two overhangs. The support of the chassis is on front and rear wheels. So the Chassis is a beam that is only supported with evenly distributed loads. The load acting on the entire beam range is 69651 N (from R260 data specifications). The length of the beam is 11670 mm, the uniformly distributed load is 69,651 / 11,670 – 5.9684 N / mm. Now taking the reaction around the front wheels is 29,543.89 N and the rear wheels around 40,107.71 N[3]. Modeling of these forces on the front leaf spring and rear leaf spring can be seen in Figure 9 and Figure 10. The support model for leaf springs is modeled as a fixed support[19].

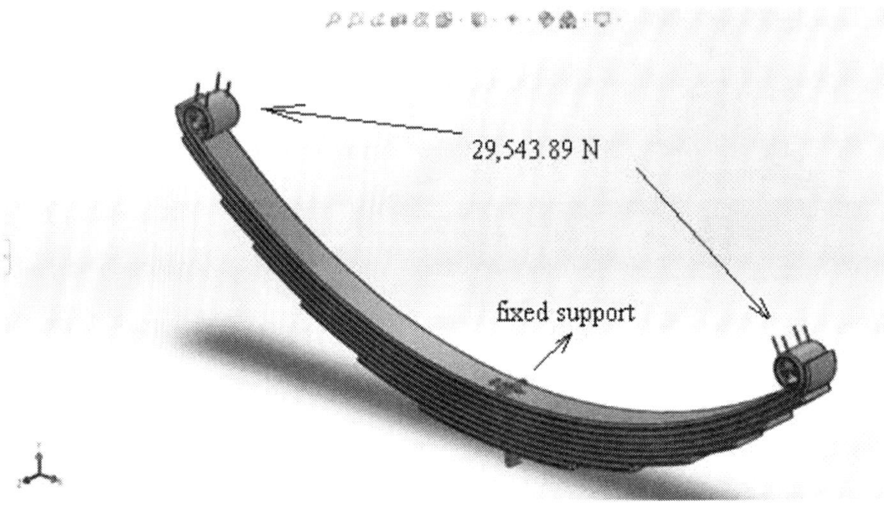

Figure 9. Front leaf spring with support and force position

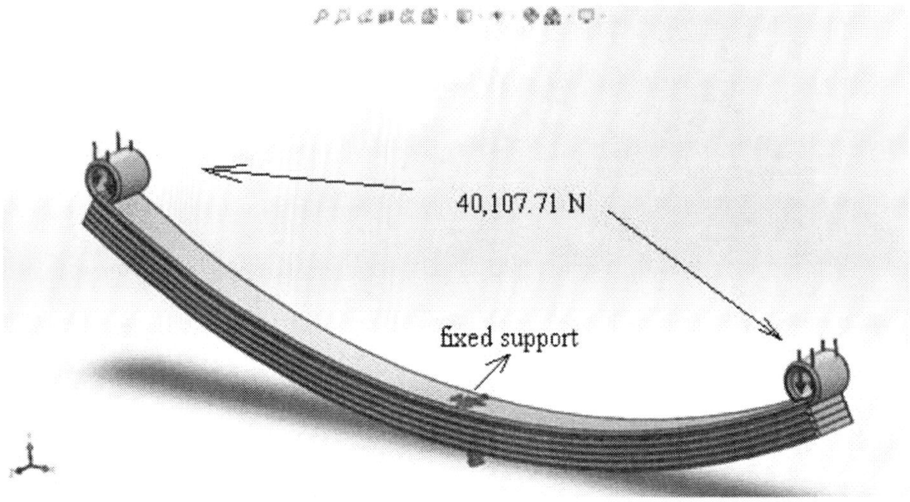

Figure 10. Rear leaf spring with support and force position

4.2. Meshing on Leaf Spring

Conducting the process of determining the type and size of loading and type of support on leaf springs has been completed at an early stage. Before conducting static analysis, the meshing process is an absolute thing to do to be able to do the finite element process properly and correctly. The form of meshing results for the rear leaf spring and front leaf spring can be seen in the following two figures (figure 11 and figure 12).

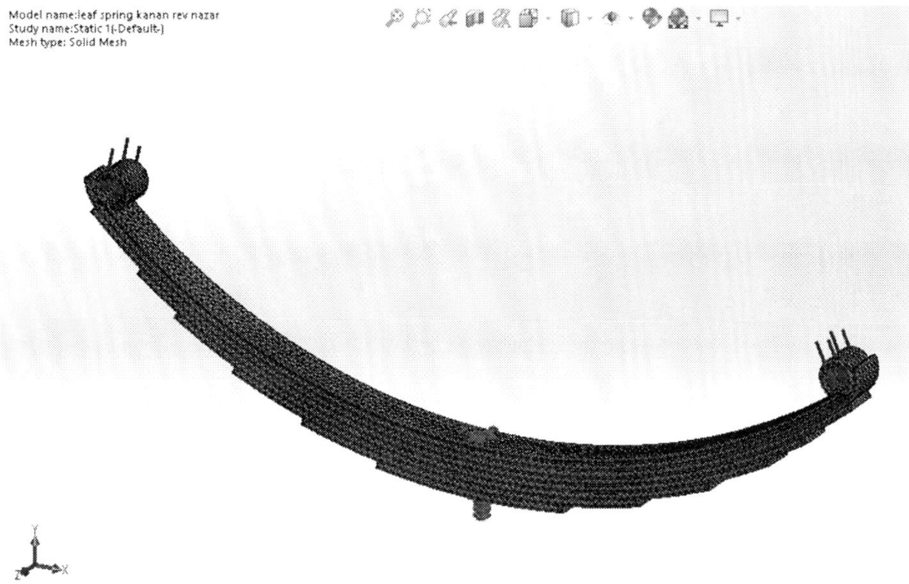

Figure 11. Meshing for front leaf spring

Figure 12. Meshing for rear leaf spring

4.3. Static Analysis of Leaf Spring with CAD Software

The next step can be performed static analysis on CAD software. Contact modeling between leaf springs is assumed to be global in contact only. Looping is needed in this analysis, where if the comparison with the results of theorically differs greatly, then the model of each leaf spring sheet will be improved.. The following result for front and rear axle are displayed (see figure 13 and figure 14).

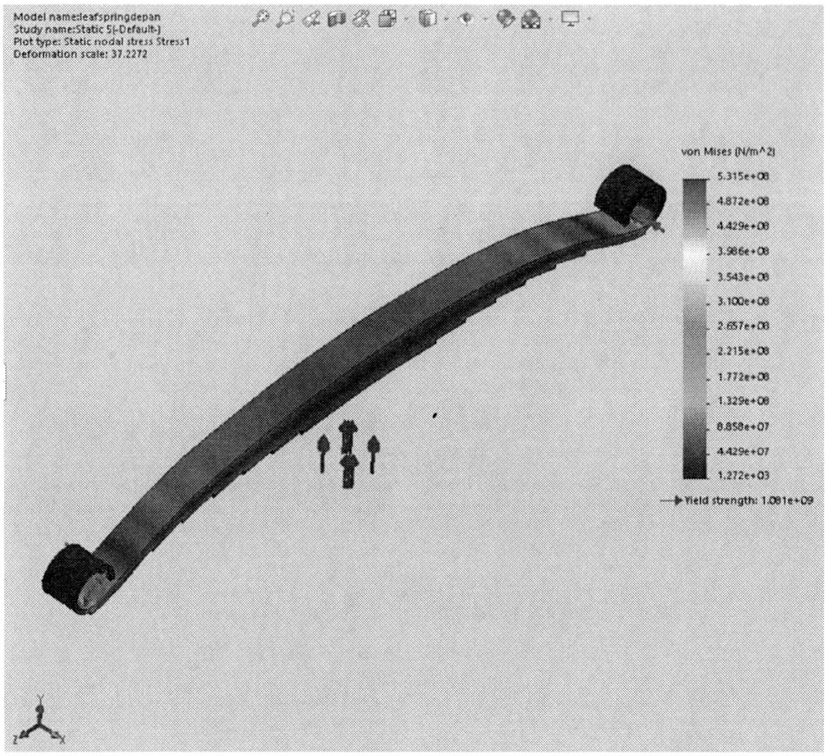

Figure 13. The results of finite element analysis of the front axle leaf spring

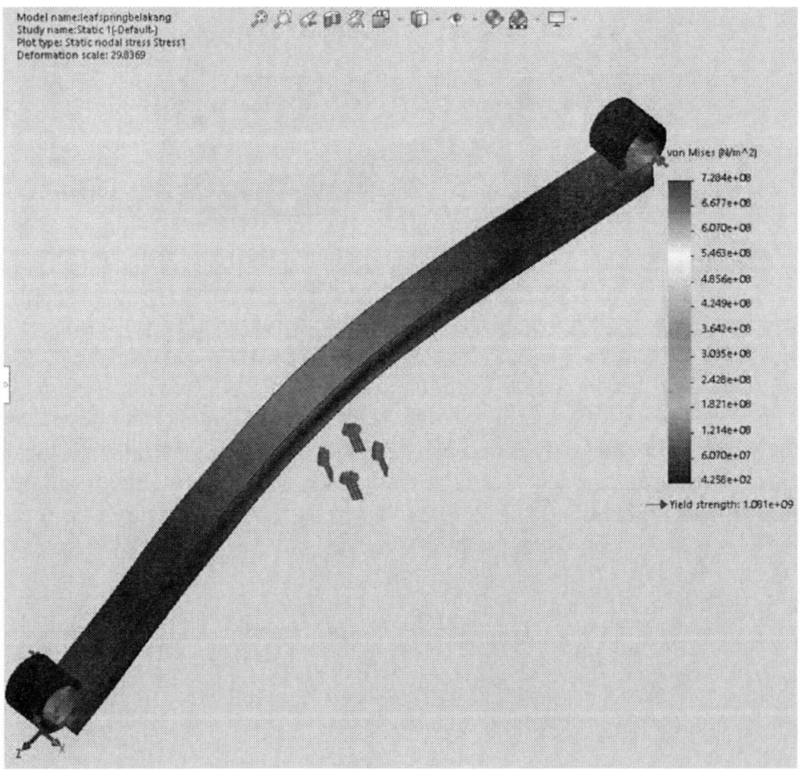

Figure 14. The results of finite element analysis of the rear axle leaf spring

The stress that occurs theoretically in the front leaf spring and the rear leaf spring can be calculated with the equation number (1) as follows:

$$\sigma = \frac{pFL}{Nb_N h^2} = \frac{3Pl}{nbt^2} = \frac{3 \cdot 14.772 \cdot 759}{8 \cdot 80 \cdot 10^2} = 512{,}74 \text{ MPa (Front Axle Leaf Spring)} \quad (3)$$

$$\sigma = \frac{3Pl}{nbt^2} = \frac{3 \cdot 20.054 \cdot 777.3}{5 \cdot 90 \cdot 12^2} = 721{,}67 \text{ MPa (Rear Axle Leaf Spring)} \quad (4)$$

and the deflection that occurs theoretically in the front leaf spring and the rear leaf spring can be calculated with the equation number (2) as follows :

$$\delta_{max} = \frac{qFL^3}{ENb_N h^3} = \frac{3 \cdot 14.772 \cdot 759^3}{2 \cdot 10^{11} \cdot 8 \cdot 80 \cdot 10^3} = 0{,}15 \text{ mm (Front Axle Leaf Spring)} \quad (5)$$

$$\delta_{max} = \frac{qFL^3}{ENb_N h^3} = \frac{3 \cdot 20.054 \cdot 777.3^3}{2 \cdot 10^{11} \cdot 5 \cdot 90 \cdot 10^3} = 0{,}31 \text{ mm (Rear Axle Leaf Spring)} \quad (6)$$

So, the safety factor that occurs theoretically in the front leaf spring and the rear leaf spring can be calculated with the equation as follows :

$$F_s = \frac{\sigma_u}{\sigma} = \frac{1470}{512{,}74} = 2{,}87 \quad \text{(Front Axle Leaf Spring)} \quad (7)$$

$$F_s = \frac{\sigma_u}{\sigma} = \frac{1470}{721{,}67} = 2{,}04 \quad \text{(Rear Axle Leaf Spring)} \quad (8)$$

From the calculation results above, the theoretical and analytical results using finite element software are not much different (see Figure 12 and Figure 13, as well as the two calculation results). The maximum deflection that occurs is relatively small at 0.15 for front axle leaf spring and 0.31 mm for rear axle leaf spring. The smallest safety factor that occurs is quite safe above number 2 for the both leaf springs. Dimension results from both leaf springs with reverse engineering method, such as number of leaves, width and length from the results of direct measurement by reverse engineering can be used. By taking standard leaf spring material, the appropriate material is 65 si7 or SUP9. Hopefully this analysis can be used as a reference for domestic production of leaf springs for electric buses such as those owned by the University of Indonesia.

The development of further research is the possibility of developing the use of other types of springs for electric cars, especially buses. Air suspension is a more challenging type for this type of large bus. Air spring flexibility is better because the stiffness can be regulated in such a way as the air pressure contained therein. Researchers are very interested in continuing this research in the direction of air springs and may also be produced domestically.

5. Conclusion

After a static analysis of the two leaf springs (front and rear leaf spring), as one of the results of the reverse engineering process, it can be concluded several things, namely:

1. Front axle leaf spring arrangement consists of 8 spring sheets with dimensions of width 80 mm and thickness of 10 mm with stratified length.
2. Rear axle leaf spring arrangement consists of 5 spring springs with dimensions of 90 mm width and 12 mm thickness with evenly distributed length.
3. The theoretical stress on the front leaf springs is 512.74 MPa and finite element analysis is 531.5 MPa.
4. The theoretical stress behind the leaf spring which occurs is 721.67 MPa and elemental analysis up to 728.4 MPa.
5. The material for both of these springs is chosen is 65 si7 or SUP9
6. The smallest safety factor that occurs above number 2.

IOP Conf. Series: Materials Science and Engineering **909** (2020) 012022 doi:10.1088/1757-899X/909/1/012022

Acknowledgments
We express our thanks to the University of Indonesia through the "PUTI Doktor UI 2020" scheme with contract number NKB-678/UN2.RST/HKP.05.00/2020 and the Indonesian government through the Indonesian Endowment Fund for Education (LPDP) for funding this research.

References
[1] S. Prasetya, G. Heryana, Y. Whulanza, M. Adhitya, and R. D. A. Sumarsono 2018 A preliminary case study of a electric commuter bus in Universitas Indonesia *The 15th International Conference on Quality in Research (QiR 2017)*
[2] G. Heryana *et al.* 2020 Plug in hybrid electric vehicle power consumption analysis in tropical area *AIP Conference Proceedings* **2230** 1 50001
[3] N. Nazaruddin, A. Syehan, G. Heryana, M. Adhitya, and D. A. Sumarsono 2020 Mode shape analysis of EV-bus chassis with reverse engineering method *IOP Conference Series: Materials Science and Engineering* 2019 **694** 1 12002
[4] R. Siregar, M. Adhitya, D. A. Sumarsono, Nazaruddin, G. Heryana, and F. Zainuri 2020 Study the brake performance of a passenger car based on the temperature that occurs in each brake unit *AIP Conference Proceedings* **2227** 1 20013
[5] S. Prasetya, M. Adhitya, H. D. S. Budiono, and D. A. Sumarsono 2018 A investigation of braking system actuators for electric shuttle bus *E3S Web of Conferences* **67** 1023
[6] R. Siregar, F. Zainuri, and M. Adhitya 2017 Design a new generation of synchromesh mechanism to optimization manual transmission's electric vehicle *Proceeding of the 15th International Conference on QIR (Quality in Research) ISSN* **1411** 1284
[7] N. Nazaruddin, F. Zainuri, R. Siregar, G. Heryana, M. Adhitya, and D. A. Sumarsono 2019 Electric power steering: an overview of dynamics equation and how it's developed for large vehicle *IOP Conference Series: Materials Science and Engineering* **673** 1 12112
[8] Nazaruddin, M. Adhitya, D. A. Sumarsono, R. Siregar, and G. Heryana 2020 Review of electric power steering type column steering with booster motor and future research for EV-Bus *AIP Conference Proceedings* **2227** 1 20016
[9] W. Ayu 2016 UI Perkenalkan mobil listrik di dies natalis ke-52 FTUI
[10] W. F. Milliken, D. L. Milliken, and M. Olley 2002 Chassis design: principles and analysis *Society of Automotive Engineers Warrendale* **400**
[11] W. Ribbens *et al.* 2009 Automotive engineering: powertrain, chassis system and vehicle body *Butterworth-Heinemann*
[12] V. V Patel and R. I. Patel 2012 Structural analysis of a ladder chassis frame *World J. Sci. Technol.* **2** 4 5–8
[13] A. Singh, V. Soni, and A. Singh 2014 Structural analysis of ladder chassis for higher strength *Int. J. Emerg. Technol. Adv. Eng* **4** 2 253–259
[14] V. Francis, R. K. Rai, A. K. Singh, P. K. Singh, and H. Yadav 2014 Structural analysis of ladder chassis frame for jeep using Ansys *Int. J. Mod. Eng. Res. ISSN* 2249–6645
[15] M. R. Chandra, S. Sreenivasulu, and S. A. Hussain 2012 Modeling and structural analysis of heavy vehicle chassis made of polymeric composite material by three different cross sections *Int. J. Mod. Eng. Res* **2** 4 2594–2600
[16] H. B. Patil, S. D. Kachave, and E. R. Deore 2013 Stress analysis of automotive chassis with various thicknesses *IOSR J. Mech. Civ. Eng* **6** 1 44–49
[17] R. Rajappan and M. Vivekanandhan 2013 Static and modal analysis of chassis by using FEA *Int. J. Eng. Sci* **2** 2 63–73
[18] R. T. Patil, V. B. Patil, A. A. Patil, and A. R. Patil Static analysis of leaf spring for light commercial vehicle
[19] J. E. Shigley 2011 Shigley's mechanical engineering design *Tata McGraw-Hill Education*
[20] H. Chandra, D. K. Pratiwi, and M. Zahir 2018 High-temperature quality of accelerated spheroidization on SUP9 leaf spring to enhance machinability *Heliyon* **4** 12 e01076

Experimental study of cascaded thermoelectric generators with differences in focal length using LED lights energy radiation

Tutang Muhtar Kamaludin[1], Awal Syahrani S[2], Danny Syamsu W[3], Basri[4] and Mustofa[5*]

[1] Civil Engineering Department, Universitas Tadulako, Palu, Indonesia, 94118
[2,3,4,5]Mechanical Engineering Department, Universitas Tadulako, Palu, Indonesia, 94118

[*] Correspondence author: mustofauntad@gmail.com

Abstract. The length of concentration focus to the cascaded thermoelectric generator (TEG) module determines the power of the output along the entire surface of the heated side receiving the thermal radiation. LED-type (*Lighting Emitting Diode*) bulb with low power 10, 15, and 20 Watt shown that the emitted lights also contain the thermal that can generate electricity on the TEG. The experiment results show that the 20 Watt lamp with a concentration focus of 110 mm has the highest power and efficiency (2.5%), while the efficiency of 15 and 10 Watt bulbs is 1.2 and 1.6%, respectively. The significant amount of bulb power used by consumers today has the potential to be the source of new and renewable energy to produce electricity. Although the amount of power still small, it can be used to charge mobile phones and power banks.

1. Introduction

Discourse over clean and renewable energy becomes the topic that many people hold as one of the solutions to anticipate the fossil energy crises, which diminishes over time. One of the less-researched sources of thermal energy is indoor bulb radiation. The bulb known to people as energy-efficient lighting is LED (*Lighting Emitting Diode*). LED's primary function is as the source of lighting, but the bulb also produces thermal radiation. The heat can be utilized by a thermoelectric generator (TEG). In general, research on TEG focused on the calefaction was sourced from the heat dissipated from the engine's combustion chambers, radiator chamber, and other combustion mechanisms, for example, the heat from the organic and inorganic waste incineration. TEG module as the technological tool working under the *Seebeck effect* can utilize exhaust heat as the thermal input to generate electric current due to the voltage difference in the type p and n semiconductor materials TEG pin. TEG has many advantages, among others, direct energy conversion, absence of moving parts, and operating without noise and environmentally friendly. Despite its low-efficiency value, it does not prevent the researchers from losing their motivation to continue the investigation and modification in different applications. Some experimental investigations related to the use of exhaust heat, such as Hsiao et al.,[1] have put TEG at the exhaust pipe and radiators chambers of the vehicle engine. The result showed a better TEG performance in the exhaust pipe. A group of researchers from Komatsu company has made a large-scale development on this exhaust heat, like Snyder [2], who utilized the exhaust heat in the vehicles. He installed the BiTe/Silicide cascade-type module, which produced DC electricity power as the power charger to the automotive battery that can reduce and even replace the alternator functions in the cars.

Kaibe et al., [3-4] have developed 16 parallel and serial Bi-Te cascade-type TEG modules. Four batteries with 12V-65Ah capacity stores the generated power. Mahmud et al. [5] developed the model and prototype of thermal electric modular utilization for TEG from various types of materials; Bi_2Te_3 (Bismuth Telluride), PbTe-Bite, CMO and CMO Cascade-32-62S-32-62S [Calcium Mangan oxide]. The output powers of the TEG produced 0,091 Watt, enough to charge a mobile phone.

Concerning the utilization of bulb heat radiation, Mustofa et al. [6] performed an early investigation using halogen, incandescent, and xenon light bulbs with power on a single hybrid of mini PV and TEG. Their research merely described the light radiation and wavelength without explaining the characteristics of PV and TEG. Meanwhile, Piarah et al. [7], in his research, used a 50-Watt Halogen light bulb with hot mirror spectrum separator. The source of the radiation spectrum of the Halogen light bulb did not reflect the reality on the ground, and that type of light bulb was used merely for experiments, not reflect the daily usage needs. Therefore, the utilization of LED light bulb type, dubbed in electronic media as an energy-efficient light bulb, as the source of energy provides an exciting option to test the utilization of TEG in the household, particularly at night. Therefore, the purpose of this research is to optimize the LED light leading to reduce the wasting energy. Moreover, the length of the Fresnel lens focus will be varied to obtain optimum TEG output power.

2. Materials and Methods

2.1 Experimental set-up

This experiment uses the Peltier module made from ceramic material/Bismuth Telluride with TEC1 12706 model with dimensions of 40x40x3.4 mm and an operating temperature of 29-50°C. Meanwhile, to focus the LED light radiation used Fresnel lens from Polymethyl methacrylate (PMMA) material with the dimension of 112x73 mm and 110 mm focal length, 2 mm thick lens with 92% light transmittance. As displayed in Fig 1, a thermoelectric module is placed under the Fresnel (Fl) lens, while the LED light bulb is positioned at the top and covered. Its light beam is focused on the FL and further transmitted to the TEG. The TEG module is stacked and connected in series. The focus length of the FL and TEG module is placed in three positions (110, 120, and 130) mm. The objective of the placement is to find out the parameters affecting the performance of the Peltier module with the radiation of the low intensity LED light bulbs.

During the observation, a rotary potentiometer is used to regulate the R load of the electricity produced by the TEG. As for the commercial power of the LED light bulbs, variations of 20, 15, and 10 Watt with different bulb diameters were used (Fig 2). The selection of the power is made based on the intensity that people generally use so that the bulb will have a double function mainly in the *peak-hours* if further development is aimed to conserve household electricity consumption supplied by the state electricity company (PLN).

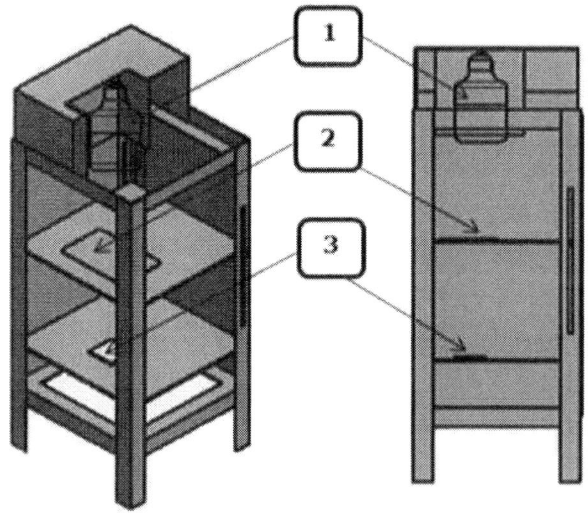

Figure 1. Sketch on the experimental set-up

Description
1 LED bulb light
2 Fresnel lenses
3 TEGs module

Figure 2. Muxindo LED light bulbs with 20, 15, and 10 Watt of power

2.2 Model equation

The calculation of TEG (P_{TEG}) power refers to the output power resulted from the calculation of the current flows through the resistor in the circuit. The following equation defines the output power:

$$P_{TEG} = I_{TEG}^2 R_{Load} \tag{1.1}$$

with,

$$I_{TEG} = \frac{\alpha(T_{hot} - T_{cold})}{R_{TEG} + R_{Load}} \tag{1.2}$$

because 2 TEG modules are connected in series, therefore $R_{TEG} = R_1 + R_2$

while ∝ is the *Seebeck* coefficient obtained from the following equation:

$$\alpha = \frac{\Delta V}{T_{hot} - T_{cold}} \quad (\text{V/°C}) \qquad (1.3)$$

and,

R_{TEG} = integral load of TEG (3.96 Ω)
R_{Load} = circuit load (Ω)

The electricity current is calculated with the following equation

$$I = \frac{\propto(T_{hot} - T_{cold})}{R_{TEG} + R_{LOAD}} \qquad (1.4)$$

While for the absorbed heat (Q_h) is determined from the below equation

$$Q_h = \alpha I T_{hot} + K(T_{hot} - T_{cold}) - 0.5 I^2 R_{TEG} \qquad (1.5)$$

$$K = \frac{k A_{TEG}}{t} \qquad (1.6)$$

where k is the module's thermal conductivity, T_{hot} means the temperature of the heated side, T_{cold} is the temperature of cold side, I means the TEG current, α is the Seebeck coefficient, R_{TEG} refers to the internal resistance, A_{TEG} size of the TEG surface and t is TEG. For efficiency calculation (η), the TEG modules can be calculated with the equation below::

$$\eta_{TEG} = \frac{P_{TEG}}{Q_h} x100 \qquad (1.7)$$

3. Results and Discussion

3.1. The size of LED light area and characteristics of TEG power

3.1.1 Focus length with TEG module (110 mm)

Fig. 3 showed the diameter of the LED light size transmitted by Fl to the heated side of the TEG stacked and connected in series. The diameter of the light is located at the optimum focal point based on the dimension of the heated side of the module's surface. It shows that the thermal radiation is within the optimum intensity that the bulb can emit.

Figure 3. The diameter of LED lights to TEG, 110mm

Subsequently, the thermal radiation of LED light bulb with 10, 15, and 20 Watt of power generates the characteristics of power-voltage P-V and current-voltage I-V of the TEG, as shown in Fig 4. It is

IOP Conf. Series: Materials Science and Engineering **909** (2020) 012023 doi:10.1088/1757-899X/909/1/012023

noticeable that the maximum output power of TEG is generated by using a 20 Watt light bulb by 10x10^{-3} µWatt at 2.5 Volt (Fig 4c). The result also shows that the higher the radiation power of the LED, the bigger the output power of the TEG at a focal length of 110 mm from the Fresnel lens.

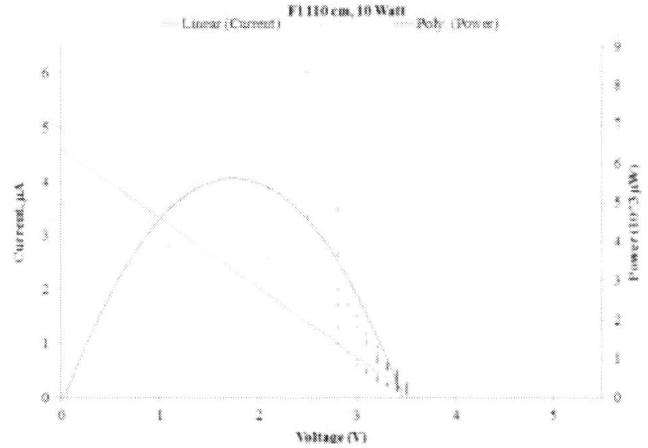

(a)

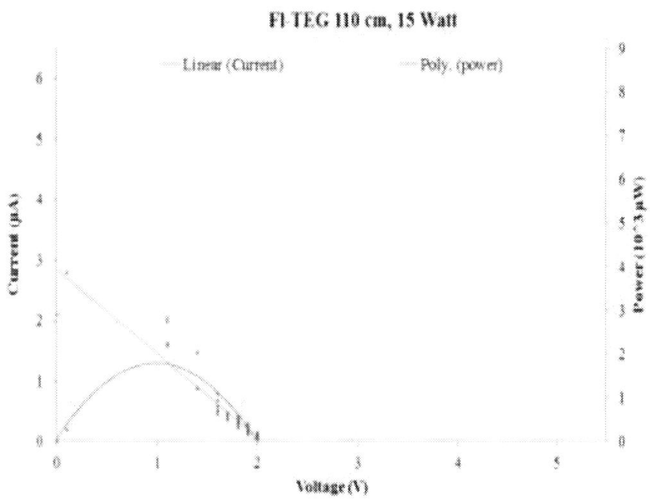

(b)

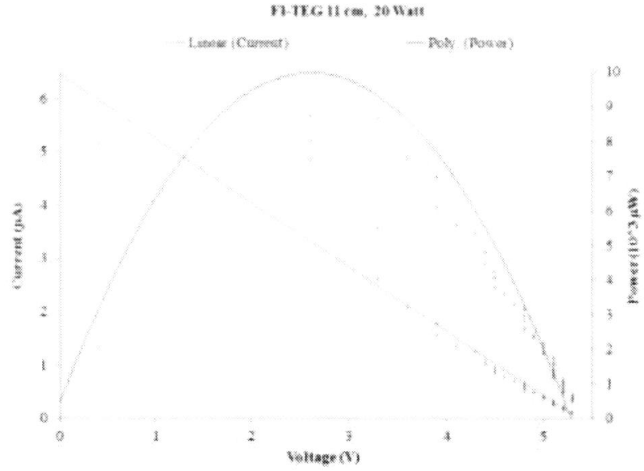

(c)

Figure 4. Characteristics of P-V and I-V in cascaded TEG, 110 mm(a) 10, (b) 15, and (c) 20 Watt

3.1.2 Focal length with TEG module (120 mm)

Unlike the Fig 3, the LED light bulb's diameter transmitted by Fl to the heated side of the TEG is somewhat higher with the lens distance and the heated side of the module's surface of 120 mm. The explanation is because since it moves away from the focus area of the lens, diminishing the intensity and thermals.

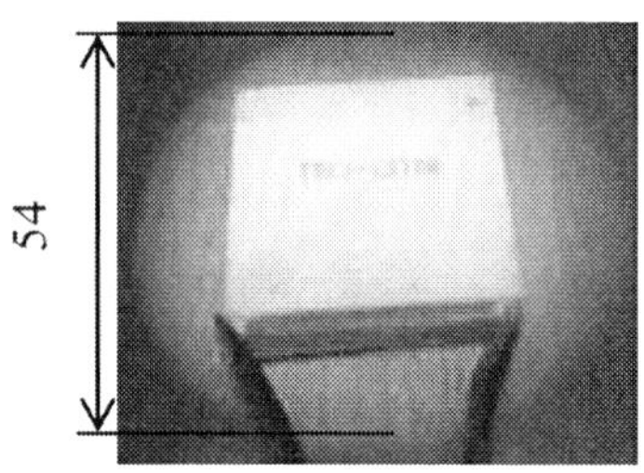

Figure 5. The Diameter of LED light to TEG, 120 mm

The characteristics of the current, the voltage of TEG module electricity show a similar trend with the focal length of the Fresnel lens of 110 mm and 120 mm, respectively. Although the 20-Watt bulb has higher maximum power compared to the 15-Watt bulb, the value is almost similar by 3.5 with $3x10^{-3}$ μWatt or 0.5 difference.

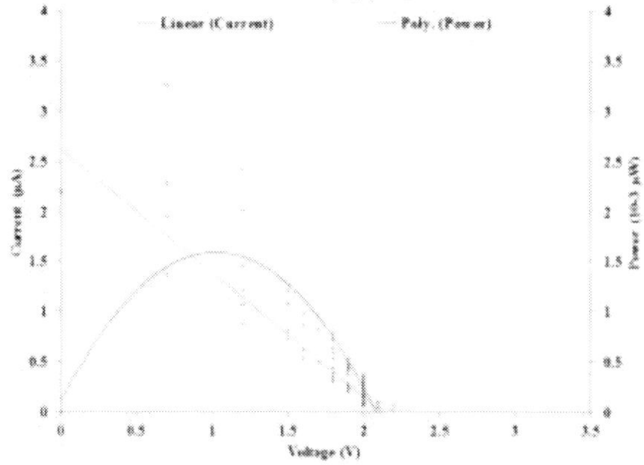

(a)

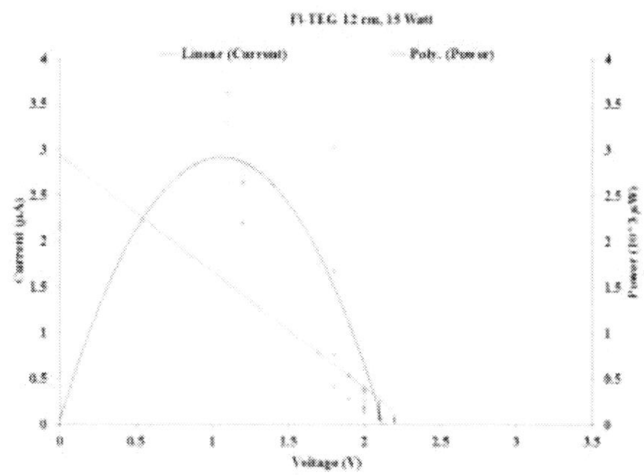

(b)

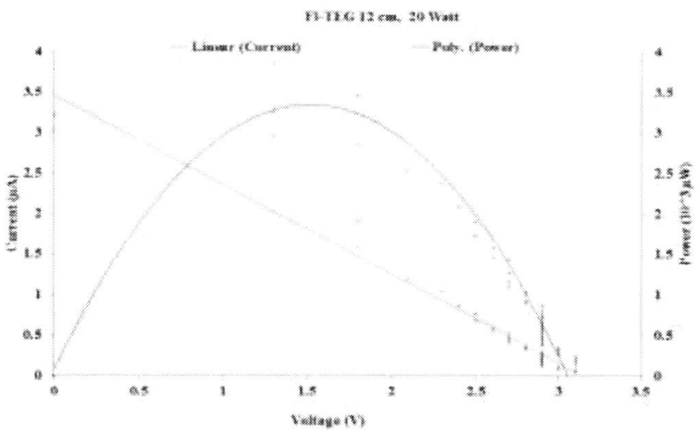

(C)

Figure 6. Characteristics of P-V and I-V in cascadedTEG, 120 mm; (a) 10, (b) 15, and (c) 20 Watt

3.1.3 Focal length with TEG module (130 mm)

It is evident in Fig 7 that the light emitted by the LED as transmitted by Fl to the heated side of the TEG expands to the edges of the TEG and shows a diminishing thermal intensity and radiation. It is consistent with the amount of electric power released because of the Seebeck effect of the TEG modules. Interestingly, the maximum power at the farthest focal length from the Fresnel lens is produced by thermal radiation of 15-Watt light bulb and not 20-Watt (Fig 8). It is different in a focal length of 110 and 120 mm, where both produce the optimum power in 20-Watt light bulbs. One of the causes is the variations in the bulb diameter of 10, 15, and 20-Watt in 50, 54 and 57 mm, respectively affecting the thermal radiation transmitted by the Fresnel lens.

Figure 7. The diameter of LED lights to TEG, 130 mm

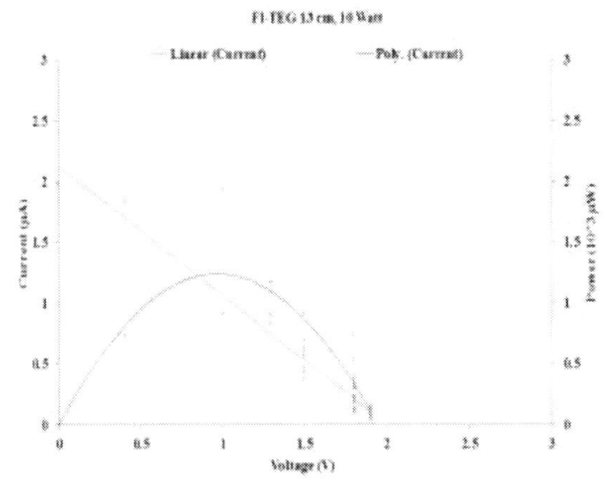

(a)

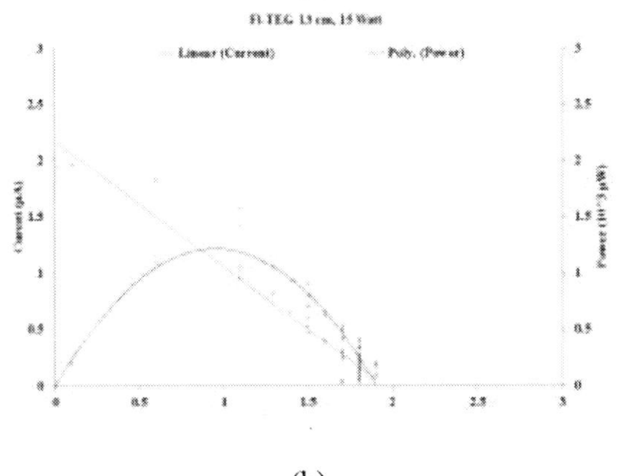

(b)

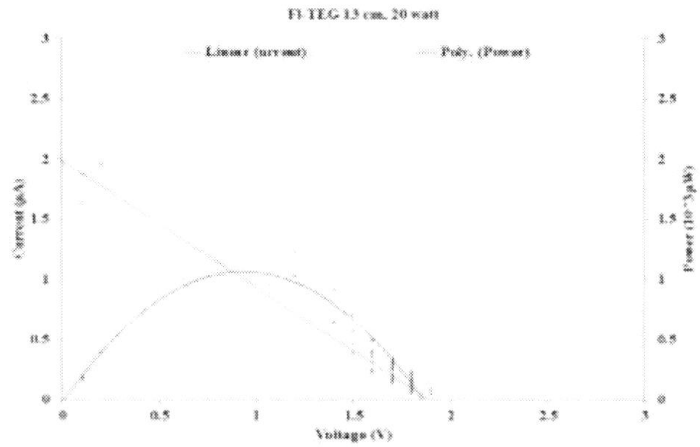

(c)

Figure 8. Characteristics of P-V and I-V in cascadedTEG, 130 mm (a) 10, (b) 15, and (c) 20 Watt

3.2 TEG Efficiency

The experiment results and calculation analysis show that the value most efficient (2.5%) is obtained at the focal length, the distance between the Fresnel lens with TEG modules, of 110 mm, using 20-Watt light bulbs. The next most efficient is a 120 mm focal length with the same light bulb power. At a focal length of 130 mm, the efficiency value is almost the same for the three light bulb powers (10, 15, and 20) Watt (Fig 9).

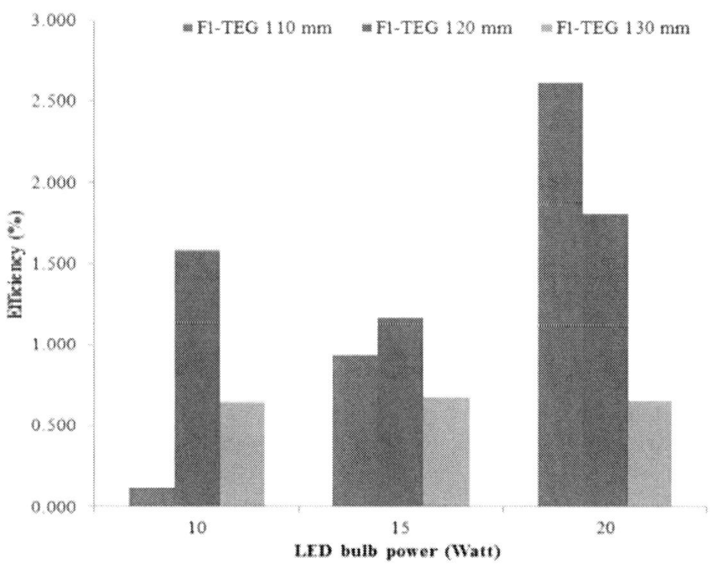

Figure 9. The efficiency of cascaded TEG series

4. Conclusions

The heat energy emitted by the LED light bulb is not significant enough to produce a temperature difference between the hot and cold sides of the thermoelectric generator module, consequently the resulting voltage is less than 1 Watt. Furthermore,the amount of power and efficiency produced by the TEG is determined by the focal length of thermal radiation size received by the heated surface of the TEG and the thermal radiation power of the light bubs. The bigger the light bulb power, the bigger the thermal radiation emitted, as reported by Piarah [8] research. It is interesting to conduct a further investigation to find out the spectral power of the light spectrum transmitted to the TEG [7] and its implication against the additional heat absorbers on the TEG. More LED light bulb radiation is also needed for a further research.

Acknowledgments

The authors acknowledge the Faculty of Engineering of the University of Tadulako for supporting this research. We also feel indebted and want to express our appreciation to Haslan and Rahim Ma'ruf who helped collect this research data.

References

[1] Hsiao Y Y, Chang W C, and Chen S L 2010 A mathematic model of the thermoelectric module with applications on waste heat recovery from an automobile engine *Energy* Elsevier **35** 1447–1454

[2] Snyder G J 2008 Small Thermoelectric Generators *Interface* The Electrochemical Society, 54-56

[3] Kaibe H, Kajihara T, Fujimotot S, Makino K, and Hachiuma H 2011 Recovery of Plant Waste Heat by a Thermoelectric Generating System Komatsu Technical Report 26-30

[4] Kaibe H T, Rauscher I, Fujimoto S, Kurosawa T, Kanda T, Mukoujima M, Aoyama I, Ishimabushi H, Ishida K, and Sano S, Development of thermoelectric generating cascade modules using silicide and Bi-Te *Technology Research Center Research Division* Komatsu Ldt

[5] Mahmud K H, Yudistirani S A, Ramadhan A I 2017 Analysis of Power Characteristics Of Model Thermoelectric Generator (TEG) Small Modular International Journal of Scientific & Technology Research **6** 4 161-167

[6] Mustofa, Djafar Z, Syafaruddin, and Piarah W H 2018 A new hybrid of the photovoltaic-thermoelectric generator with hot mirror as spectrum splitter, *Journal Physical Science* **29** 2 63–75

[7] Piarah W H, Djafar Z, Syafaruddin, and Mustofa 2019 The Characterization of a Spectrum Splitter of TechSpec AOI 50.0 mm Square Hot and Cold Mirrors Using a Halogen Light for a Photovoltaic-Thermoelectric Generator Hybrid *Energies* **12** 353 1-13

[8] Piarah W H, Djafar Z, Hariyanto, and Mustofa 2019 A New Simulation of Photovoltaic and Thermoelectric Generator Hybrid System with a Beam Splitter Cold and Hot Mirror for Low Intensity *International Review of Mechanical Engineering (I.RE.M.E.)* **13** 9 559-567

Development of control systems on vacuum evaporators with 50 l capacity to reduce honey water content accordance to sni 01-3545-2004

N. Nazaruddin [1a], Hapsoh[2], A I Darmawan[1]

[1]Department of Mechanical Engineering, Universitas Riau, 28293, Indonesia
[2]Department of Agricultural Cultivation, Universitas Riau, 28293, Indonesia

Email: nazaruddin@eng.unri.ac.id

Abstract. Riau still has many protected forests and nature reserves. One of the wealth of protected debt is the forest's Honey. One of the weaknesses of forest honey is the high water content in honey. The permissible water content in honey is 22% based on SNI 01-3545-2004. This paper presents a vacuum evaporator of control system design as a tool for reducing the water content in Honey. The control system design is made such that the honey temperature in the evaporation process is no more than 40°C. There are three quantities that are controlled in this process, namely vacuum chamber pressure, temperature and screw rotation. Temperature control system and pressure control system use close loop system, while the control system for screw rotation uses open loop system. The three main components of this control system are the VE 115N vacuum pump, the stirring screw motor with 1 HP power, 2 heaters with 250 watts of power each, Temperature Control (Eliwell IC901 thermostat and thermistor sensor), 35x25 panel box. To meet the quality of honey, the control system is set with a maximum temperature of heating water 40°C and vacuum pressure at 7.4 kPa and screw rotation at 60 rpm

1. Background

One of the forest products in Riau Province is honey bee. One of the favorite trees for nesting is the Sialang Tree [1]. The lack of honey quality in Indonesia is partly due to the high water content in the honey [2]. The water content makes it easy to ferment honey. Therefore we need a tool to reduce the water content. The quality of honey as a food ingredient has been regulated in SNI 01-3545-2004. The maximum water content is permitted at 22% and honey heating should not exceed 40°C [3].

Research on the tools used in order to reduce the water content in honey has been widely studied. The condenser evaporator [4], the effect of thickening honey by using a vacuum evaporator on the quality of honey produced, and to increase the variant of honey products [5]. The energy performance of vacuum evaporators for liquid digestate treatment[6], scale up production indonesian liquid propolis from raw propolis and wild beehive using bubbling vacuum evaporator[7], experimental study on water separation process in a novel spray flash vacuum evaporator with heat-pipe[8], analysis of the residual gases in several types of high-vacuum evaporators[9] and changes in properties of palm sugar syrup produced by an open pan and a vacuum evaporator during storage[10] are some other recent examples of research on vacuum evaporators that can be used as a reference for this research.

Designing a vacuum evaporator that we do with a capacity of 50 liters [11]. The results of the design and manufacturing review [12] were taken to suit the conditions of people in areas with low electricity consumption. This design is made in such a way, where the water used as a heating medium is placed around the walls of the vacuum evaporator. The hot water that comes from two heaters is conduction through the vacuum evaporator wall. Honey in the evaporator room is conditioned under 1 atm. Stirring in a vacuum evaporator chamber using a screw.

To get better results, the evaporator must be equipped with a control system [13]–[15]. In this paper we will present the design of the control system used in the vacuum evaporator we have made. There are three components that are controlled on this evaporator, namely temperature, pressure and screw rotation. The temperature is maintained below 40°C, the pressure is about 7.4 kPa atm and the screw rotation of the electric motor is at a certain speed. The control system for temperature and pressure will use a closed loop system, while the screw rotation with an open loop system

2. Methods

In the previous two papers we have presented about this vacuum evaporator which is about the design and manufacturing process. The design results can be seen in Figure 1(a) and the manufacturing results can be seen in Figure 1(b). The design of this vacuum evaporator based on pressure vessel principle refers to ASME section VIII Boiler and Pressure Vessel code standards.

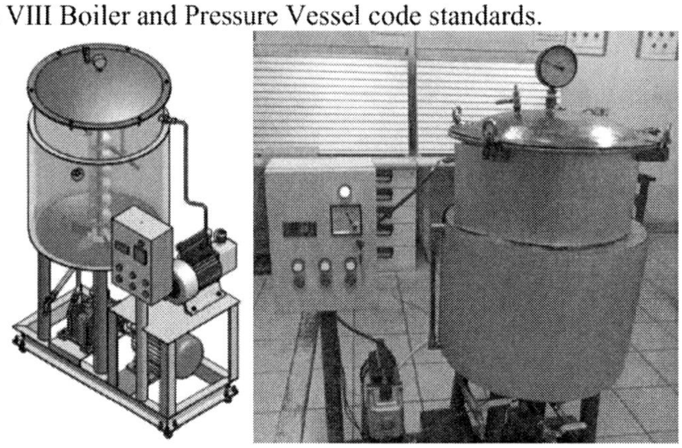

(a) (b)

Figure 1. (a) Design result[11], (b) Results of making vacuum evaporators[12]

In this paper is to make a control system of the engine vacuum evaporator which consists of three main process control, temperature control, pressure control and rotation control screw. The control system of each process has a different purpose is certainly the type of control system used and application methods are also different. The following are the planning and control system making process of each instrument (figure 2).

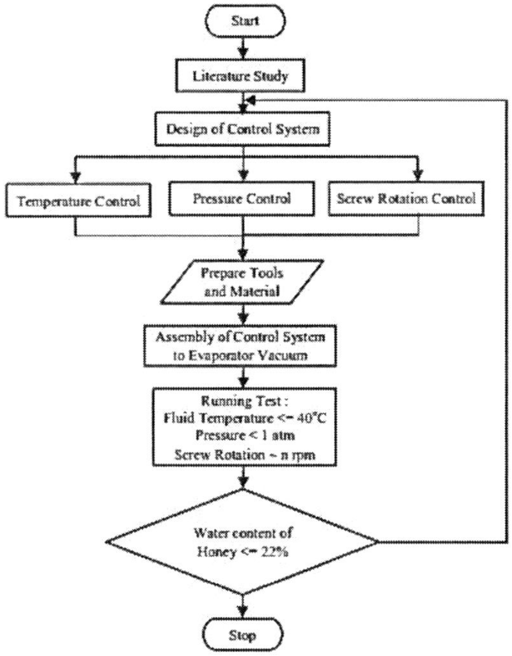

Figure 2. Flow chart of control system

The purpose of the planning and manufacturing of a temperature control system that is in accordance to the principle of vacuum evaporator. Evaporating the water at low temperatures with pressures below 1 atm. Who wants to be controlled is the temperature at 40°C figures, which needed a system that can regulate a constant temperature in the Honey. The system can be ordered working electric heaters and provide temperature information processing.

The purpose of the planning and manufacturing of pressure control system is to determine and set the pressure to be achieved and can be controlled during the process is underway. Machine working vacuum evaporator under vacuum, which is under a pressure of 1 atm. The pressure to be achieved is 7.4 kPa using a vacuum compressor.

Engine vacuum evaporator equipped with a stirrer and a screw that serves as a tool to condition the Honey that is in the vacuum chamber in order to keep it moving. So that the screw rotation honey continues to flow from the bottom to the top to make the Honey will be faster uneven heating and evaporation process. The control system in the round screw using a type of open control system.

Tools and materials used include (figure 3):

a. Vacuum Compressor VE 115N
b. Electric Motor 1 HP
c. Pressure Gauge Vacuum
d. Panel boxes 35x25 and electrical cables
e. Heater (2x250 watt) and Temperature Control (Eliwell thermostat IC901 and sensor termistor)
f. Measuring instrument

(a) (b) (c)

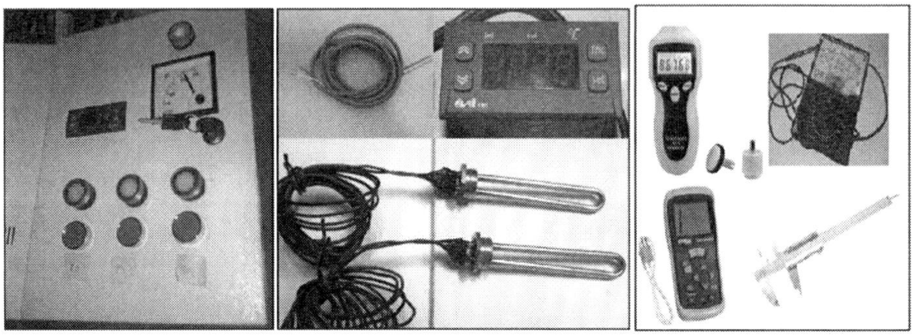

(d) (e) (f)

Figure 3. Tools and material used for evaporator vacuum

3. Result and Discussion

3.1 Temperature control system block diagram

The block diagram of the evaporator vacuum engine temperature control can be seen in Figure 4 where feedback occurs in a system that has automatic control. To measure the temperature of Honey using an appropriate sensor. The sensor will send the temperature measurement results to be read by the controller, then the controller will calculate through a certain algorithm to produce a signal which is then used as a relay contact action on the controller.

Changes in temperature read will affect the controller command of the relay. Temperature measurement by a sensor below 40°C, the controller will give the relay command to be an NC circuit so that the flow of electricity is connected to the electric heater. Conversely, if the temperature is above 40°C, the relay will become a NO circuit and the electricity will be cut off. Figures 40°C according to the setpoint value that has been adjusted on the controller

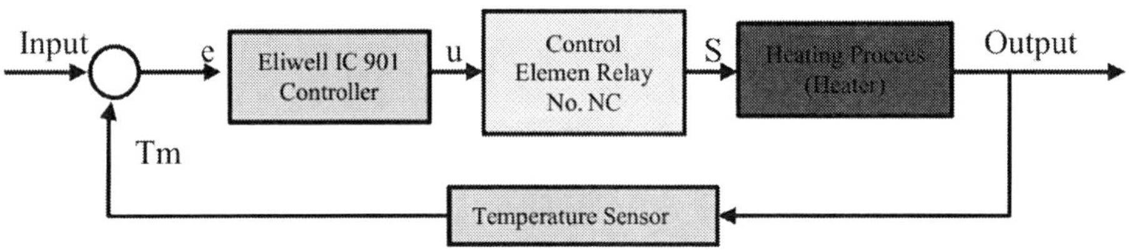

Figure 4. Close loop system for temperature control

The work of the temperature control system with the feedback method has functions such as measuring, comparing, calculating and acting. The temperature control function can be explained in the following table 1:

Table 1. Control temperature function

Measure	Compare with	Increase Temperature	Action
Honey temperatur (Output)	Honey temperatur setpoint (input 40°C)	If output > 40°C heater OFF	NC
		If output < 40°C heater ON	NO

The following will explain the circuit in the temperature control system components can be seen in Figure 5.

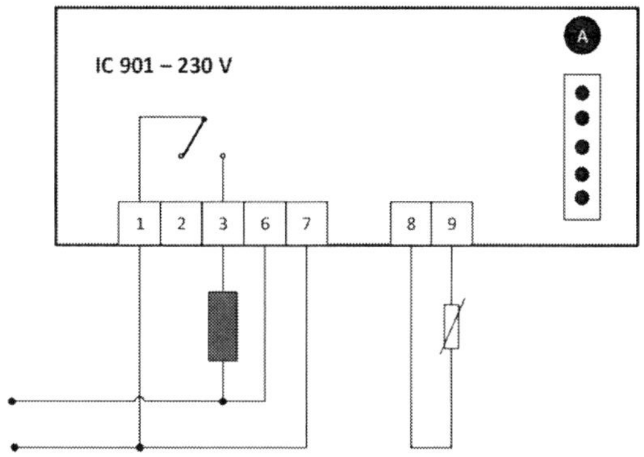

Figure 5. Temperature control system circuit

Circuit description Eliwell IC 901 230V
1 - 2 N.C regulator relay output
1 - 3 N.O regulator relay output
6 - 7 Power supply
8 - 9 Sensor input (termistor)

The components of the temperature control system are arranged according to the circuit above after which they are joined in the control system panel box. The entire assembly process of the temperature control system instrument and supporting electrical components in this study can be seen in Figure 6. Some electrical components such as PTM and PTB are used as main contacts and relays as a safety in the electrical system.

Figure 6. Temperature control system assembly

The picture above explains the process of assembling the control system instruments and electrical components in the vacuum evaporator engine control system panel box.

3.2 Block Diagram for Pressure Control System

The block diagram of the evaporator vacuum machine pressure control can be seen in Figure 7. where there is feedback in a system that has manual control by humans. To measure the pressure of a vacuum chamber seen or read by humans through a vacuum pressure gauge. The human eye as a sensor to read the pressure gauge and the brain as a controller will command the hand to press the on or off button so that the actuator will work.

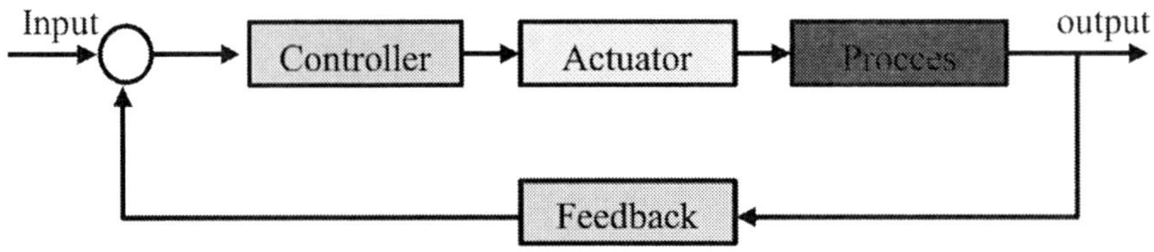

Figure 7. Pressure control system block diagram

The pressure to be achieved in the processing process using a vacuum evaporator is 7.4 kPa. Vacuum compressors that function as air suction in the vacuum room to be removed are selected as actuators in the pressure control system. And the electrical components used are PTM and PTB as contact on and off the pressure control system. Humans as an act of feedback or also called manual control will take action by reading a pressure gauge.

3.3 Block Diagram for Screw Rotation Control

The block diagram of the evaporator vacum machine screw control can be seen in Figure 8 where the control system used is an open control system.

Figure 8. Block diagram of a screw rotation control system

The screw rotation which is controlled on the vacuum evaporator machine uses an electric motor as an actuator, the same as the previous control using the electrical components and the actuator, the process will take place. However, the screw rotation control output is not measured, because it uses an open type control system. So with the PTM contact and PTB control system will be active during the process, the control will be deactivated after the process is complete. The screw rotation speed has been determined to be 60 rpm.

After the three control systems have been prepared, the electrical system is designed to assemble the control system and the vacuum evaporator can function. The electrical system can be seen in Figure 9.

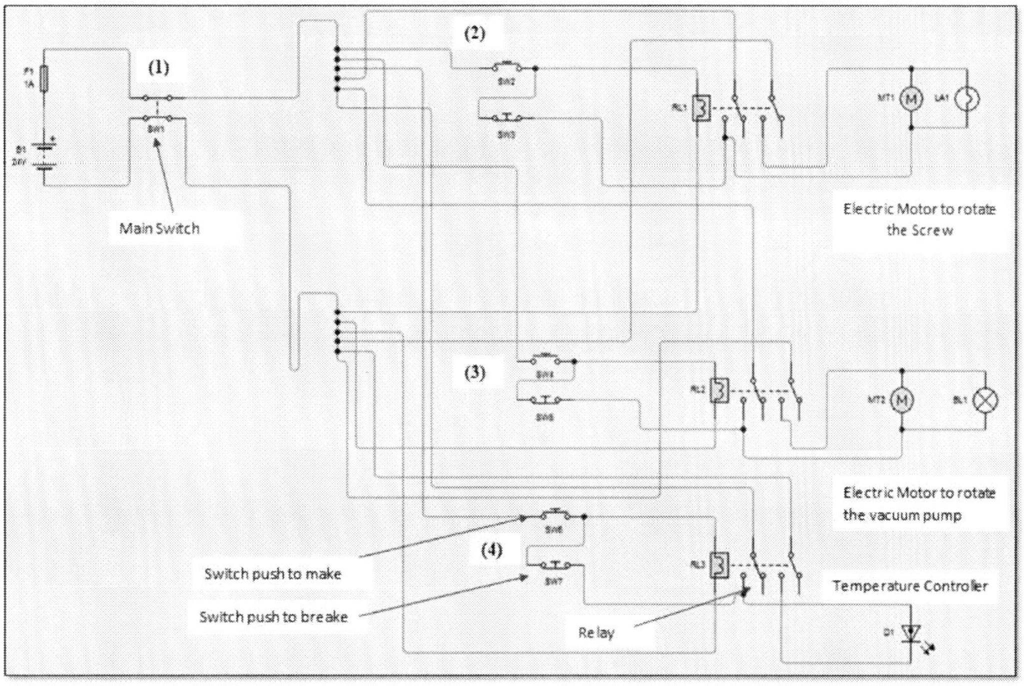

Figure 9. Electrical circuit for support the control system of vaccum evaporator

The working principle of the control system and the electricity system is as follows: (1) In the first position the main sweet is activated to activate the control system of the water-lowering machine in honey. (2) Next the green button on the screw mover motor is activated. then the motor will rotate. And there will be stirring in a vessel that has been filled with honey. During this process the motor will always move to do stirring (3). The green button on the vacuum pump motor is activated so that the vacuum pump operates to suck the pressure inside the vessel. This process lasts until the pressure in the vessel reaches 7.4 vacuum pressure. Pressure figures can be seen on the gauge pressure. After the pressure in the honey reaches 7.4 cmHg. Stirring continues but the red button on the vacuum pump is turned off and on. (4) The process of heating in honey begins by pressing the green button on the temperature of the controller in this system has been set the standard temperature of honey heating according to SNI maximum 40°C. The heater will start working on heating the water on the wall of the vessel. In this system there is a temperature sensor placed in the vessel to directly measure the temperature of honey when the temperature of the honey reaches 40 ° Celsius. to heat the water so that the honey is maintained.

Further research that can be done is to study an experimental vacuum evaporator. These experimental results can be validated with a number of similar studies with this research, especially in terms of use to reduce water content in certain fluid content. It is hoped that the research can be useful especially for communities around the forest with honey bee farmer income. The final goal is that the quality of honey in accordance with SNI standards can be of export quality and also in the process of storing honey with low water content can last a long time.

4. Conclusion

After carrying out the making and testing of research with the title review of making a temperature control system, pressure and screw rotation of the vacuum evaporator machine to reduce the honey content of 50 liters of honey. The author can draw the following conclusions:

1. The control system used is of three types, namely a temperature control system, a control system for vacuum pressure and a control system for the honey stirrer screw rotation in the vacuum evaporator chamber.
2. The temperature control system and pressure control system use the close loop system, while the control system for screw rotation uses the open loop system.
3. The three main components of this control system are the VE 115N vacuum pump, the stirring screw motor with 1 HP power, 2 heaters with 250 watts of power each, Temperature Control (Eliwell IC901 thermostat and thermistor sensor), 35x25 panel box.
4. To meet the quality of honey in accordance with SNI 01-3545-2004 the control system is set with a maximum temperature of honey 38.5°C and in heating water 40°C and vacuum pressure at 7.4 kPa and screw rotation at 60 rpm

Acknowledgments
We express our thanks to University Riau through the "IbPE" scheme from 2013 to 2015 for funding this research and people as bee farmer in Air Molek Indragiri Hulu region for survey location.

References

[1] E. Suhesti and H. Hadinoto 2015 Hasil Hutan Bukan Kayu Madu Sialang Di Kabupaten Kampar (Studi Kasus: Kecamatan Kampar Kiri Tengah) *Wahana For. J. Kehutan.* **10** 2, 16–26.
[2] S. Bogdanov 2012 Honey as nutrient and functional food *Proteins* **1100** 1400–2700.
[3] B. S. Nasional 2004 SNI 01-3545-2004 Madu *Badan Stand. Nas. Jakarta, Indonesia.*
[4] B. S. Amanto, N. H. R. Parnanto, and B. Basito 2012 Kajian Karakteristik Alat Pengurangan Kadar Air Madu Dengan Sistem Vakum Yang Berkondensor *J. Teknol. Has. Pertan.* **5** 1.
[5] S. Kalimah, M. Akbar, and A. Jauhari 2018 Penggunaan Alat Evaporator Bagi Peningkatan Ragam Varian Kemasan Produk Madu *Seminar Nasional Hasil Pengabdian Kepada Masyarakat* **2** 1 134–137.
[6] M. Vondra, V. Máša, and P. Bobák 2018 The energy performance of vacuum evaporators for liquid digestate treatment in biogas plants *Energy* **146** 141–155.
[7] M. Sahlan, A. Akhmariadi, D. K. Pratami, H. Hermansyah, and A. Wijanarko 2018 Scale Up Production Indonesian Liquid Propolis From Raw Propolis and Wild Beehive Using Bubbling Vacuum Evaporator *MATEC Web of Conferences* **248** 4001.
[8] W. Gao, C. Li, C. Xu, and D. Wang 2016 Experimental study on water separation process in a novel spray flash vacuum evaporator with heat-pipe *Desalination* **386** 39–47.
[9] H. L. Caswell 1960 Analysis of the residual gases in several types of high-vacuum evaporators *IBM J. Res. Dev.* **4** 2 130–142.
[10] P. Naknean, M. Meenune, and G. Roudaut 2013 Changes in properties of palm sugar syrup produced by an open pan and a vacuum evaporator during storage *Int. Food Res. J.* **20** 5 2323.
[11] N. Nazaruddin, H. Hapsoh, and A. Afrian 2015 Perancangan Vacuum Evaporator Penurun Kadar Air Dalam Madu Kapasitas 50 Liter *Proceeding Seminar Nasional Tahunan Teknik Mesin XIV (SNTTM XIV)* Universitas Lambung Mangkurat Banjarmasin
[12] A. I. Darmawan and N. Nazaruddin 2015 Kaji Pembuatan Sistem Kontrol Mesin Vacuum Evaporator Penurun Kadar Air Madu Kapasitas 50 Liter *J. Online Mhs. Fak. Tek. Univ. Riau*, **2** 2 1–6.
[13] I. Eisele, B. Bullemer, and A. Beck 1985 Effusion type evaporator cell for vacuum evaporators *Google Patents.*
[14] N. Yoshida, T. Saito, and T. Kawakami 1988 Rotary evaporator with control system *Google Patents.*
[15] R. F. Potter and G. G. Kretschmar 1965 Vacuum evaporator apparatus *Google Patents.*

IOP Conf. Series: Materials Science and Engineering **909** (2020) 012025 doi:10.1088/1757-899X/909/1/012025

Hybrid electric vehicle power consumption analysis in tropical area

Ghany Heryana[1,2,*] , DA Sumarsono[1], Mohammad Adhitya[1,*], N Nazaruddin[1,3], Rolan Siregar[1,4], Fuad Zaenuri[1,5], Sonki Prasetya[1,5]

[1]Research Centre for Advanced Vehicle (RCAVe), Department of Mechanical Engineering, Universitas Indonesia, 16424, Indonesia
[2]Department of Mechanical Engineering, STT Wastukancana, 41151, Indonesia
[3]Department of Mechanical Engineering, Universitas Riau, 28293, Indonesia
[4]Department of Mechanical Engineering, Darma Persada University, 13450, Indonesia
[5]Department of Mechanical Engineering, Politeknik Negeri Jakarta, 16425, Indonesia

Email: madhitya@eng.ui.ac.id

Abstract. Internal Combustion Engine (ICE) is one of the biggest fossil fuel consumers. Many attempts have been made by researchers to replace ICE with electric motors and batteries as a source of energy. On the other hand, the dependence on ICE cannot be eliminated. One of the better ideas is to combine an ICE with an electric motor or called a hybrid. This study aims to determine the amount of energy consumption used by Hybrid Electric Vehicles (HEV) in urban areas in a tropical megapolitan city. The method used is to monitor the use of energy in hybrid cars directly. Sensors and data acquisition are installed on the vehicle to find out the route, speed, fuel consumption, electricity consumption, etc. Data acquisition is also placed on pure ICE vehicles with comparable engine capacity. Previous research, namely the analysis of energy consumption in the Plug-In Hybrid Electric Vehicle (PHEV), will also be used as a comparison. The results obtained indicate that HEV could travel 22 kilometers with one liter of gasoline. ICE car could cover 13 kilometers by 1 liter of petrol. And PHEV could travel 34 kilometers with 1 liter of gasoline. Thus it can be concluded that HEV has the advantage of saving gasoline usage, but PHEV is the best of them.

1. Introduction

Indonesia is a developing country with quite a large population of motor vehicles. In certain areas, such as Jakarta and its surrounding supporting cities, the number of motorized vehicles is the most congested, because Jakarta is still a center of concentration in economic activity, not only on Java but in Indonesia. According to statistical data in 2018, the number of gasoline four-wheeled vehicles in Jakarta is more than four million units[1]. As a consequence, of course, congestion and air pollution cannot be avoided[2].

In 2019, Jakarta was the city with the worst level of air pollution in the world. The most significant contributor to pollution is two-wheeled motor vehicles, buses, and four-wheeled vehicles[3]. Four-wheeled motor vehicles accounted for 16.11% of Jakarta's air pollution. In this percentage, there are substances such as PM, HC, CO, NOx, and Sox[4]. Then the 2020 data, quoted from IQAir [5], Jakarta ranks ninth in the city with the highest air pollution in the world. It means that Indonesia is not yet out of the top 10 cities with the worst air quality.

The conversion of ICE vehicles into electric vehicles is one of the ideal solutions[6]. Of course, this requires time for preparing facilities, infrastructure, and mature regulation. This step needs to be bridged subtly, one of which is to use hybrid vehicles. The vehicle engine is a combination of ICE with an electric motor.

In Indonesia, hybrid vehicles are still classified as luxury and expensive vehicles. Two primary movers in a car also make the purchase taxes are high. Therefore, at present, hybrid cars can only be purchased by the upper-middle class. The purpose of the purchase was more prestigious. Apart from that, the movement to reduce air pollution must be started and encouraged. If demand and population increase, it is believed that the price of hybrid vehicles will decrease, especially if accompanied by supporting taxation policies[7].

2. Literature

Hybrid vehicles are vehicles that have more than one power source to drive. The most common examples are hybrid cars with ICE power and electricity, called the Hybrid Electric Vehicle (HEV). HEV energy is obtained from the battery. In HEV, the battery can only be charged while the car is driving. The charge process occurs when the function of an electric motor turns into an electric generator, i.e., during ICE work (Figure 1). Also, the charge occurs when there is a braking[8] or deceleration process. This charging process is called regenerative braking. Another type of HEV is PHEV[9]. PHEV has an additional feature where the battery can be charged from an outside power source[10].

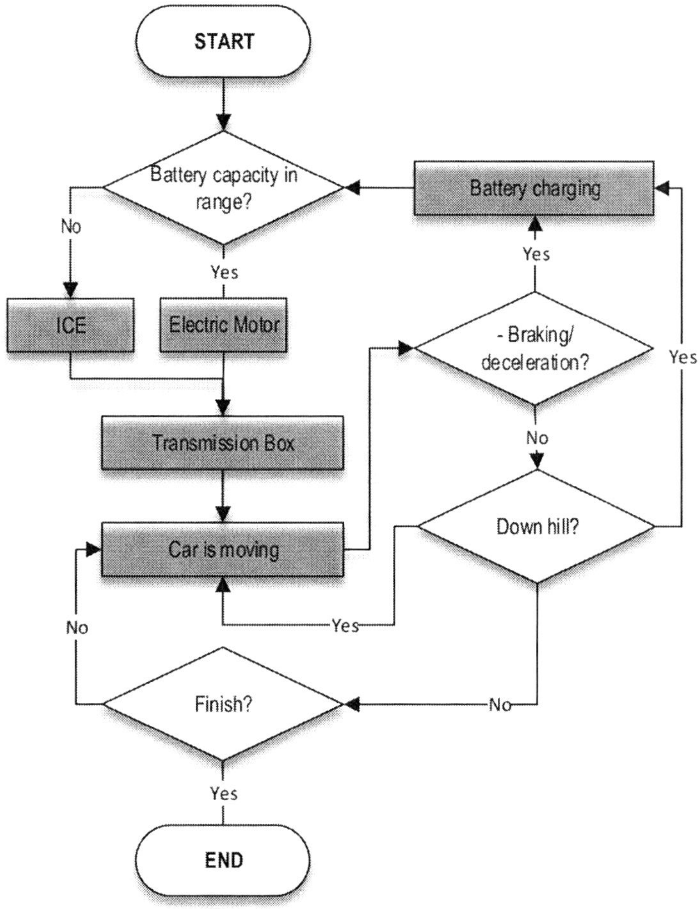

Figure 1. Operation of HEV

In general, HEV is divided into three types. The first is the HEV series. In this type, ICE is only used to rotate the alternator. The alternator charges the battery. The battery is then used to drive the electric motor. When the battery capacity is insufficient, ICE will charge the battery. There is no direct mechanical connection between ICE and the vehicle transmission box.

The second type is the parallel HEV. In this type, both ICE and electric motors are connected to the transmission box. When the battery is sufficient, the electric motor will work. When the battery is inadequate, ICE will drive the vehicle while turning the electric motor. This rotation makes the electric motor function as a generator to charge the battery. Under certain conditions, where higher power or torque is needed[11][12], the algorithm will instruct the ICE and electric motor to work at the same time (parallel). In these conditions, the battery must be at a level that meets the requirements. Thus the parallel type is more potent than the serial HEV with the same capacity.

The third type is a combination of serial and parallel HEV. The characteristic of this type, ICE, is connected to the transmission box and a generator as well. This generator is connected to a power converter to drive the motor. The motor is connected to the transmission box.

In this study, the 2017 HEV Toyota Prius was used. This vehicle is a series-parallel HEV type, with a 134 bhp/100 kW output system of ICE. The HEV electric motor output is 80 bhp.

The Prius uses 168 NiMH (Nickel-metal hybride) batteries, where the voltage per cell is 1.2 V. The maximum battery voltage is 201.6 volts and 6.5 Ah (Table 1). Thus the energy capacity is 1310.4 Wh (1.3 kWh)[13].

Table 1. ICE and EV Toyota Prius 2017 specification

HYBRID SYNERGY DRIVE	
Type	Series/parallel, full hybrid
System output (bhp/kW)	134/100
ENGINE	
Engine type	2ZR-FXE (Atkinson cycle)
No. of cylinders	Four in-line
Valve mechanism	16-valve DOHC with VVT-i
Bore x stroke (mm)	80.5 x 88.3
Displacement (cc)	1,798
Compression ratio	13.0:1
Fuel system	EFI
Octane No.	95 or greater
Max. power (bhp/kW @ rpm)	98/73 @ 5,200
Max. torque (Nm @ rpm)	142 @ 4,000
Emissions level	Euro 5
ELECTRIC MOTOR	
Motor type	Permanent magnet, synchronous
Max. voltage (CD V)	650
Max. power (bhp)	80
Max. torque (Nm)	207
HIGH-VOLTAGE BATTERY	
Battery type	Nickel-metal hydride
Nominal voltage (SC V)	201.6 (168 x 1.2V cells)
No. of battery modules	28
Battery capacity (Ah)	6.5
System voltage (V)	650

Source: https://media.toyota.co.uk/wp-content/files_mf/1329489972120216MTOYOTAPRIUS TECHNICALSPECIFICATIONS.pdf

The electrical energy capacity of 1.3 kWh can also be converted as 78 kW minutes. It means that if within 1 minute required 78 kW of power, the battery would run out within 1 minute, then ICE will work fully replace the electric motor. If the average energy needed to run a vehicle in EV mode is 15.6 kW, the battery is only able to drive an electric motor for 5 minutes. This is assuming that the battery can be discharged up to 0%. In fact, the battery has never been or is not allowed to reach 0%. At the point of 20-30% remaining capacity, the battery will be recharged[14].

3. Methodology/Experimental

In this study, conducted experimentally. HEV is installed with a tool for data acquisition that can record data on the use of fuel and electricity within a specific time. Besides that, the vehicle position data is also recorded while driving. Vehicles are used every day under normal conditions[15]. The study was conducted in areas of Jakarta, Bogor, Depok, Tangerang, Bekasi, with congested road conditions combined with toll roads. Because the average temperature in the tropics is quite hot, air conditioning is always used. Unlike the PHEV, in HEV, there is no selection of fully BEV or fully ICE modes. Thus there is only one mode, hybrid. AC on HEV is driven by a separate electric motor, in contrast to AC on ICE vehicles driven by crankshaft rotation.

The vehicle is driven according to driver behavior. The route taken is from home to work and back home. Thus it is expected to collect data on what is the average distance per day. From here, it can be developed to determine fuel consumption. The results of the analysis are used to conclude whether hybrid vehicles are significant enough to save fuel use, which indirectly shows a contribution to the reduction in air pollution levels (Figure 2).

This research will focus on HEV fuel consumption. Data obtained from the data logger will also be used for further study.

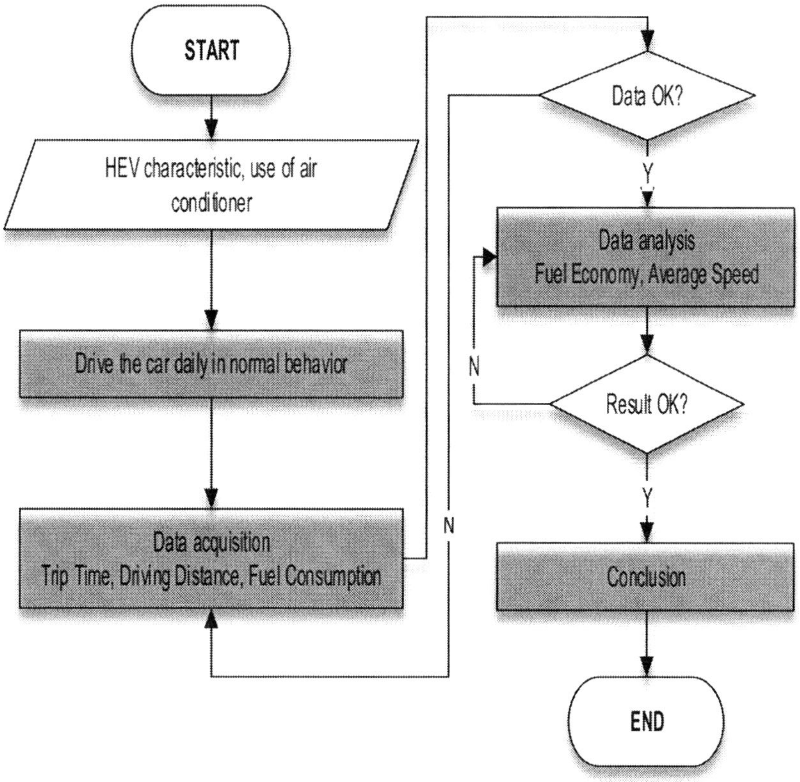

Figure 2. Experiment flow chart

4. Result and Discussion

Data was taken from June 26 to September 28, or about three months. From Table 2, it can be seen that in the driving range between 0-10 km, fuel consumption was the most wasteful at 8.6 km/liter. The total distance in the range of 0-30 km was more than 200 km. The best fuel consumption in the range was 17.1 km/liter. Also, the data shows that more than 1200 minutes of the time needed for a range of 0-30 km. In the range of 30-60 km, the total distance of about 600 km and the whole travel time of about 2000 minutes. The best fuel consumption was in the range of 60-90 km, which was 23.1 km/liter.

For a range of 90-110 km, the total mileage was less than 320 km. The best fuel consumption was in the range of 100-110 km, which was 24.9 km/liter. The highest average speed was in the range of more than 110 km, which was 27.8 km/hr.

Mileage of more than 110 km gave average fuel consumption of 22.4 km/liter. The total cruising distance in this range was more than 2300 km.

From several groups the range can be seen, within three months, with everyday use, if the conditions of use were at a speed range of 0-10 km, fuel consumption was 8.6 km/liter. The best conditions were in the range of 100-110 km, which was 24.9 km/liter. The average fuel use during the experiment period (three months) was 21.95 km/liter, with a total mileage of more than 5600 km.

Behavior driving on this vehicle, the longest distance was in the range 70-80 km, 80-90 km, and more than 110 km. The miles of more than 110 km are classified as out-of-town trips because the Jabodetabek city range itself is between 30-40 km. Thus the round trip around 60-80 km. If the range of more than 110 km is ignored, then the longest distance was in the range of 70-80 km. It was 1266.6 km.

Table 2. HEV energy consumption – experiment data of Prius

Range	Trip Time [min]	Driving Distance [km]	Fuel Consumption [L]	Fuel Economy [km/L]	Average Speed [km/h]
0-10 km	266.8	32.1	3.7	8.6	7.23
10-20 km	634.4	132.8	7.8	17.1	12.56
20-30 km	519.4	102.7	6.0	17.0	11.87
30-40 km	1119.5	281.0	14.3	19.7	15.06
40-50 km	302.7	92.2	4.4	20.8	18.28
50-60 km	624.0	224.4	10.0	22.3	21.58
60-70 km	445.5	138.3	6.1	22.6	18.63
70-80 km	3073.5	1266.6	54.8	23.1	24.73
80-90 km	1897.2	689.7	30.7	22.4	21.81
90-100 km	287.6	97.2	4.4	22.3	20.28
100-110 km	571.6	214.4	8.6	24.9	22.51
>110 km	5087.2	2356.9	105.4	22.4	27.80
Overall	14829.2	5628.3	256.4	21.95	22.8

In Figure 3 can be seen more simply. The tendency of fuel consumption will save between 50 to 110 km. After that, fuel efficiency cannot be increased anymore. Besides, Figure 3 also shows that in a range below 50 km, fuel efficiency is not very good.

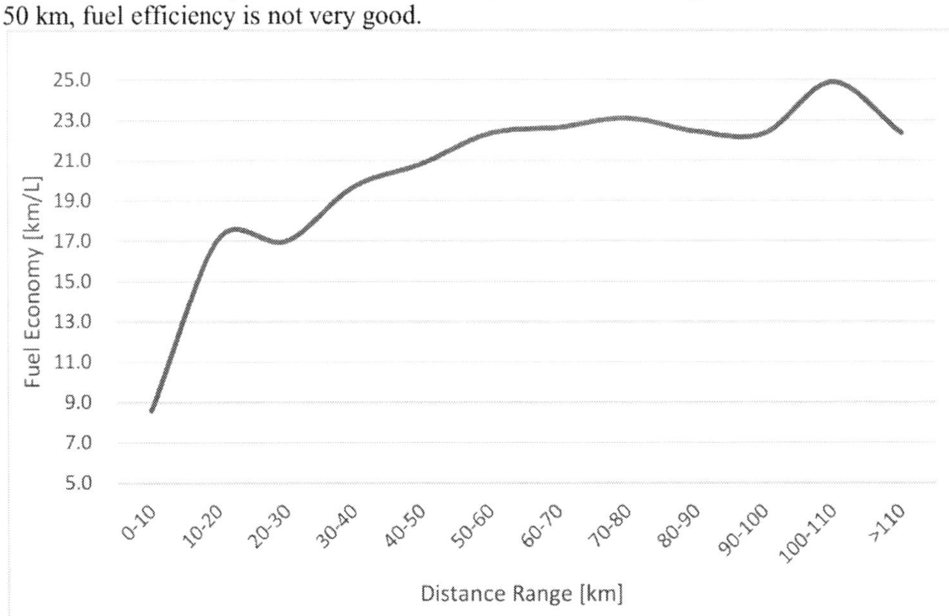

Figure 3. Fuel efficiency including weekend - HEV

Referring to Table 3, which claims that the fuel consumption used ranges from 72.4-76.4 mpg, or about 25.6-27 km/liter, then these conditions approach the fuel efficiency of the experiment in the ranges of 70-80 km and 100-110 km, but it still below the claim value from Toyota.

Table 3. Prius HEV claim

FUEL CONSUMPTION	15in wheel
Combined	72.4 70.6
Extra urban (mpg)	76.4
Urban (mpg)	72.4

Source: https://media.toyota.co.uk/wp-content/files_mf/1329489972120216MTOYOTAPRIUS TECHNICALSPECIFICATIONS.pdf

5. Conclusion and Suggestion

The results of experiments in dense tropical regions, illustrate that actual fuel consumption is higher than the claims of the brochure.

The air conditioning is mandatory in the tropics. It should be suspected that energy is also channeled large enough to cool the cabin. It is consistent with previous research, which states that AC is the second-largest power consumption after the main motor in tropical electric vehicles [2].

The use of HEV in big city areas with almost always traffic jams like Jakarta only gives a small added value to fuel savings.

The dominant driving distance of the driver in this experiment is between 70-80 km, and in this range also shows the most efficient fuel consumption. The range can be a reflection of the distance from home - office and offices every day and represents the distance from Bekasi - Jakarta, Bogor-Jakarta, and Tangerang-Jakarta.

For tropical areas and big cities like Jakarta, traffic conditions are not favorable. But if HEV is used to cover medium distances (70-80 km), the benefits will be optimal.

In subsequent studies, it is very strategic to discuss the relationship between HEV, PHEV, BEV prices, local fuel prices (Indonesia), taxation, and battery prices. The research is essential so that it can provide suggestions to policyholders, consumers, and producers related to the type and price of the best electric vehicles for developing countries. Greenhouse issues, scarcity and rising fuel prices, electricity unit prices, and the amount of tax will determine the acceptance of electric cars in Indonesia.

Acknowledgment

The authors acknowledge the financial support from the PUTI UI (Publikasi Terindex Internasional Universitas Indonesia) Grant through Contract No. NKB-643/UN2.RST/HKP.05.00/2020.

References

[1] H. Widowati 2019 Berapa jumlah kendaraan di DKI Jakarta? *databoks* [Online]. Available: https://databoks.katadata.co.id/datapublish/2019/08/02/berapa-jumlah-kendaraan-di-dki-jakarta.

[2] H. C. Frey, X. Zheng, and J. Hu 2020 Variability in measured real-world operational energy use and emission rates of a plug-in hybrid electric vehicle *Energies*, **13**, no. 5, 1140.

[3] D. H. Jayani 2019 Kualitas udara jakarta terburuk di dunia *databoks* [Online]. Available: https://databoks.katadata.co.id/datapublish/2019/07/29/kualitas-udara-jakarta-terburuk-di-dunia.

[4] R. Arifin 2019 Motor penyumbang polusi tertinggi DKI Jakarta *oto detik* [Online]. Available: https://oto.detik.com/berita/d-4668990/motor-penyumbang-polusi-tertinggi-dki-jakarta.

[5] Admin 2020 Air quality and pollution city ranking *IQAir* [Online]. Available: https://www.iqair.com/world-air-quality-ranking.

[6] G. Heryana, S. Prasetya, M. Adhitya, and D. A. Sumarsono 2018 Power consumption analysis on large-sized electric bus *IOP Conf. Ser. Earth Environ. Sci.* **105** 12041.

[7] W. YS, C. KT, and C. CC 2006 Battery sizing for plug-in hybrid electric vehicles *J. Asian Electr. Veh.* **4, no. 2** 899–904.

[8] R. Siregar, M. Adhitya, D. A. Sumarsono, G. Heryana, and F. Zainuri 2019 Study the brake

performance of a passenger car based on the temperature that occurs in each brake Unit **2–9**.
[9] K. J. Holmes, E. Mantus, J. Kassakian, and E. Zeitler 2016 The plug-in electric vehicle system from technologies to consumers *World Electr. Veh. J.* **8, no. 4** 721–732.
[10] G. Heryana *et al.* 2020 Plug in hybrid electric vehicle power consumption analysis in tropical area *AIP Conf. Proc.* **2230, no. 1** 50001.
[11] Nazaruddin, M. Adhitya, D. A. Sumarsono, R. Siregar, and G. Heryana 2016 Review of electric power steering type column steering with booster motor and future research for EV-Bus *AIP Conf. Proc.* **2227, no. 1** 20016.
[12] F. Zainuri, D. A. Sumarsono, M. Adhitya, and R. Siregar 2017 Design of synchromesh mechanism to optimization manual transmission's electric vehicle *AIP Conf. Proc.* 1823.
[13] J. Thomas, S. Huff, B. West, and P. Chambon 2017 Fuel consumption sensitivity of conventional and hybrid electric light-duty gasoline vehicles to driving style *SAE Int. J. Fuels Lubr.* **10 no. 3** 672–689.
[14] I. Cho and J. Lee 2020 Characteristics of battery soc according to drive output and battery capacity of parallel hybrid electric vehicle *Appl. Sci.* **10 no. 8** 2833.
[15] Y. Zhou *et al.* 2016 An assessment of causes of PEV success across US Metro areas *World Electr. Veh. J.* **8 no. 4** 818–830.

IOP Conf. Series: Materials Science and Engineering **909** (2020) 012026 doi:10.1088/1757-899X/909/1/012026

A Performance study of a single basin double slope solar still with 45° of glass-cover slope angle in Indonesia

Caturwati Ni Ketut, Yusvardi Yusuf, Zaenal Arifin, Muhamad Fajar Komara
Mechanical Engineering Department- Universitas Sultan Ageng Tirtayasa,
Jl. Jend.Sudirman km.3, Cilegon 42435, Indonesia, Orcid id : 0000-0003-1107-1499
e-mail : n4wati@untirta.ac.id

Abstract. Solar energy is a clean energy, pollution free and available in abundance that can be used directly for heating. Utilization of solar heat in the supply of clean water through distillation is very suitable for use in remote coastal and swampy areas. In this study the performance test of a single basin type solar distillation system with a double slope cover that has 45° slope angle is operated in Cilegon-Indonesia. With rare rainy conditions the average solar distillation efficiency reaches 34.3%. The formation of condensate at night is almost five times greater than during the daytime, because temperature of water in the basin is higher than the inner surface temperature of the glass-cover. Thus the use of material as a thermal energy storage when the intensity of radiation is high and releasing it into the water at night will help a lot to increase condensate production.

1. Introduction

The lack of fresh water becoming a big problem in the world. Because water is a primary need not only for humans, but also a main need for all living things in this world. The surface of the earth is more covered by sea water than covered by land, about 71% of the earth's surface is covered by sea water [1]. However, not all water available on earth can be used for humans and other living things. A number of requirements are needed to maintain health and avoid the various disturbances stated in the water quality standards, among others; non-toxic, does not contain substances that cause health problems, does not contain excess chemicals.

Increasingly scarcity of clean water is getting worse given the limitations of water sources caused by the diminishing forest area as a supplier of natural water and the growing human population. To solve this problem water treatment efforts need to be done that can improve the quality of existing water to match the water quality needed by humans and other living things.

Desalination is one of the earliest forms of water treatment by humans to convert sea water into fresh water. Desalination methods are still widely used throughout the world today. Solar desalination utilizes heat energy from sunlight to evaporate pure water from sea water, then water vapor is condensed to obtain pure water and waste in the form of water with high salt content.

Utilization of solar energy in the distillation process can be direct or indirect. Direct solar distillation uses solar heat to produce distillates in the solar colector. Whereas the indirect method combines conventional distillation techniques with other systems such as vapor compression desalination (VC), multilevel flash desalination (MSF), membrane distillation (MD) and electrodialysis, Reverse Osmosis (RO) and the addition of solar collectors for heat generation. Direct solar distillation requires large land with relatively low productivity compared to indirect solar distillation [2].

Utilization of renewable energy sources is highly recommended for distillation systems, especially in remote areas. Geothermal energy is also effectively used in seawater desalination [3]. Direct solar distillation is considered the best solution for water production in small communities, remote area, where fresh water is not available. But the amount of fresh water produced per unit area is very low. Several attempts were made to increase the productivity of distilled water produced, including by adding phase change material as an thermal energy storage [4 -8].

Solar distillation is very effective in producing pure water, removing pathogenic elements, salts, heavy metals found in raw water. This is very beneficial for the user and avoids health problems contained in water.[9]

The position of the sun changes continuously throughout the day, so the intensity of solar radiation reaching the earth depends on the time and geographical position. Indonesia is located around the equator, so solar radiation can be obtained throughout the day. In this research, a double slope type solar distillation performance test was performed with a 45 degree slope angle.

2. Experimental Set Up

The study was conducted experimentally by using an equipment called solar distillation, which consists of a basin made of aluminum with a coat of black paint having a length of 106 cm and a width of 60 cm with a height of 30 cm. The outside wall and the bottom of the basin are placed insulating material (Styrofoam) to prevent heat leakage, and placed above a wood basin. Basin filled with raw water as deep as 10 cm. A double slope cover glass is mounted above the basin with a 45° tilt angle.

Transparent glass cover will allow solar radiation to enter towards the water in the basin. Solar radiation heats water until it starts to evaporate and collect at the bottom of the glass cover as a layer of dew. The dew layer moves along the surface of the glass toward the condensate channel.

Certain important parameters must be measured simultaneously among others is the temperature in some parts, global solar radiation and the condensate produced. Figure 1. Shows the test equipment accompanied by the position of the measuring instrument.

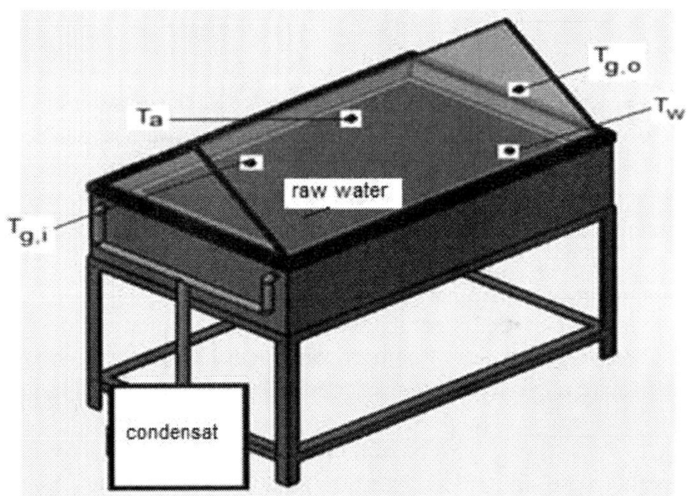

Figure 1. Arrangement of solar distillation testing instruments.

Where :

T_W = Water temperature in the basin (°C)
$T_{g,i}$ = Lower surface temperature of cover glass (°C)
$T_{g,o}$ = Upper surface temperature of cover glass (°C)
T_a = Air temperature in the test equipment (°C)
I_R = Intensity of global radiation (klx - kilolux)
Q_c = Volume of condensate produced (ml)

3. Methodology

The experiment was carried out 5 times with the observation time between 8:00 and 17:00. Whereas the condensate produced is observed until 8:00 the next day considering that condensate production also occurs at night. Tests carried out in May 2019 during the dry season. This site is 6° 0' 56" South, 106° 3' 11" East. The climate is non-semi-arid and is characterized by bright sunlight and high annual rainfall during the rainy season.

The performance of solar distillation is expressed in terms of efficiency which is the ratio of heat to vaporizing water to the total heat of radiation received by the surface of the solar distillation device which can be expressed in (1).

$$\eta = \frac{m_c . h_{fg}}{A_c x I_R x t} \; x \; 100\ \% \qquad (1)$$

Where :

η = solar distillation efficiency (%)
m_c = condensate mass (kg)
h_{fg} = latent heat of evaporation (kJ/kg)
A_c = heat absorbent area (m^2)
I_R = radiation Intensity (kW)
t = radiation time (s)

Radiation time in one day is expressed as a time of high enough sunlight from 8.00 to 17.00. While the condensate mass is determined from the volume of condensate produced in one day from 08.00 to 08.00 the next day.

$$m_c = Q_c x \rho_w \qquad (2)$$

Where :

ρ_w = water density (1000 $^{kg}/_{m^3}$)

4. Results and Discussion

4.1 Temperature Distribution and Intensity of Solar Radiation.

Temperature distribution in the solar water distillation system and global intensity of solar radiation for 5 repetition of data collection shown in Figure 2.

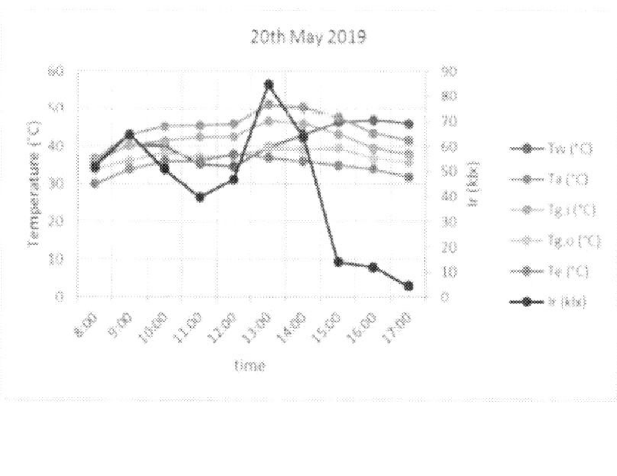

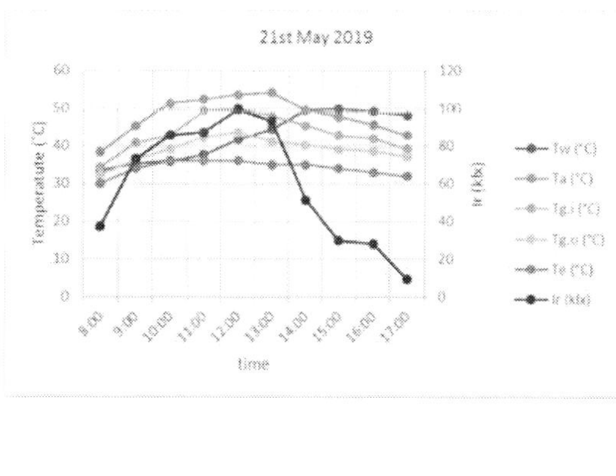

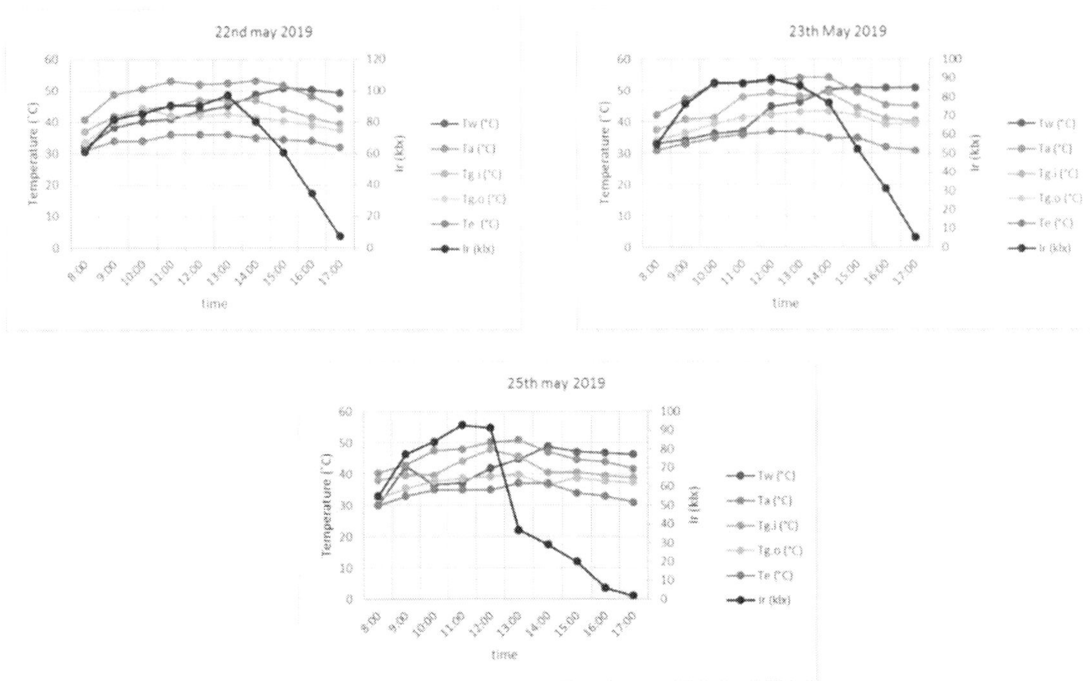

Figure 2. Temperature distributions and global intensity of solar radiation

From observations when the intensity of solar radiation is high, the air temperature in the solar distillation chamber (Ta) has the highest value, while the environmental temperature (Te) has the lowest value. This is due to heat leakage from the system to the environment is very low. All radiant heat received is used to optimally heat the distillation system parts. Furthermore, when there is a decrease in the intensity of solar radiation, the temperature of the water in the basin seems to continue to increase some time before it goes down, this is due to the heat obtained from the constituent material of the basin which is coated with Styrofoam as an insulator that is quite effective in resisting heat leakage.

4.2 Distillate Production

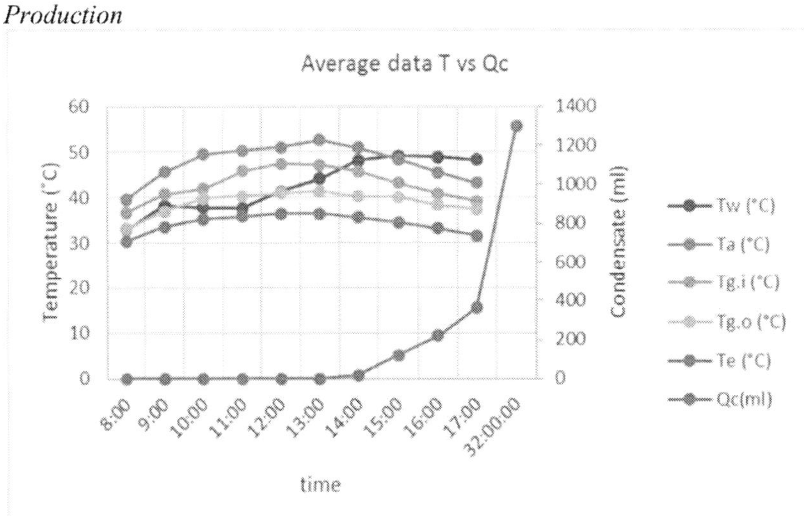

Figure 3. The average temperature distribution and condensate production

From 5 times the test data obtained the average value of the temperature distribution on the solar distillation test equipment. Figure 3 shows the average temperature: water in the basin (Tw), air temperature above the water basin (Ta), surface temperature in the glass cover (Tg, i), average environmental temperature (Te) and the production of distillate water produced up to 8:00 the next day (in table 32:00:00).

Distillate production on average began to appear at 14:00 with an amount of 18 ml and until 08.00 the next day the production of distillates reached 1296 ml. Distillate production begins to be detected when the temperature of the bottom cover glass begins to decrease, this makes it easier for water vapor contained in the air above the basin to condense when touching the surface of the cover glass. Distillate production, Qc, at 32:00:00 states that distillate production until 08.00 the next day.

The distillate production in double slope solar distillation with 45 degree slope of the cover glass is as follows.

- Average of distillate production : 1296 ml/day
- Distillate production during the day (08.00 - 16.00) : 220 ml
- Average nighttime distillate production (16.00 - 08.00 the next day) : 1076 ml.

Distillate production to the global solar radiation intensity is shown in Figure 4. From 5 times observations, the highest average solar radiation intensity was 83,560 lux at 12.00. While the average value of the intensity of solar radiation from 8:00 to 16:00 is 62,898 lux.

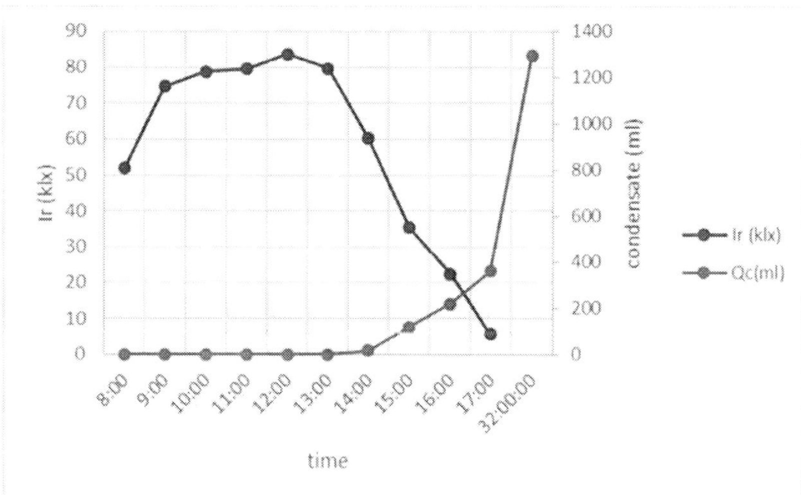

Figure 4. Average of distillate production and global solar radiation intensity.

4.3 Solar Distillation Efficiency

For an average water temperature of 40 °C, then :

- value of latent heat for water evaporated, h_{fg}, = 2406.8 kJ / kg
- water density , ρ , = 1000 kg / m3

distillate production :

Q_c = 1296 ml

m_c = 1.296 kg

Average of global solar radiation intensity for eight hours (08.00-16.00) :

I_R = 62,898 lx = 496.8942 W/m^2.

Basin Surface area :

Ac = 106 cm x 60 cm = 0.636 m2

So the average solar distillation efficiency for a double slope with a slope angle of 45° is :

$$\varepsilon = \frac{1.296\ kg\ x\ 2406.8\ kJ/kg}{0.636\ m^2 x 496.8942\ {}^{W}/_{m^2}\ x\ 8\ h} = \frac{3,119,212.8\ J}{9,101,511.68 J} = 34,3\ \%$$

5. Conclussions

The purpose of this experiment is to evaluate of effectivity double slope solar distillation with 45 °slope angles, operated in Indonesia with semi-arid weather.

Based on experiments and discussions conducted, the following conclusions can be drawn :

1. The temperature of the water in the basin and the air temperature above it does not necessarily decrease following the decrease in the intensity of global solar radiation, due to the lack of heat leakage out of the system.
2. Condensate starts to be produced when the water temperature in the basin is higher than the inner surface temperature of the glass cover.
3. The formation of condensate at night is almost five times greater than during the daytime. So the difference between the temperature of the basin water and the temperature of the glass cover plays an important role. For this reason, needs to be carried out to maintain the temperature of the basin water higher than the inner surface of the glass cover, for example by adding heat energy storage material in the basin.
4. The results of performance tests on double slope solar distillation with a 45 degree slope of the cover glass that is operated in rare rain conditions in Indonesia shows an efficiency value of 34.3 %.

References

[1] Abduldaem Al-Kaheel http://kaheel7.com/id/
[2] Hazim MQ and Fawzi B 2008 Solar thermal desalination technologies Desalination **220** 633–644
[3] Aiman Eid AR et al. 2017 Desalination by renewable energy: a mini review of the recent patents Hemijska industrija DOI: 10.2298/HEMIND160308008A.
[4] Ismaila H. Zarma et al. 2017 Thermal energy storage in phase change materials ICCEEE 1st International Cnferecne of Chemical, Energy and Environmental Engineering, Hilton Alexandira Green Plaza Smouha Alexandria Egypt.
[5] Kantesh.D.C. 2012 Design of solar still using phase changing material as a storage medium, International Journal of Scientific & Engineering Research **3** issue 12
[6] Anand Patel and Sadanand Namjoshi 2016 Phase change material based solar water heater International Journal of Engineering and Science Invention **5** 31-34
[7] Atul Sharma, C. R. Chen 2009 Solar water heating system with phase change materials International Review of Chemical Engineering (I.RE.CH.E.) **1**
[8] S.Naga Sarada et al. 2014 Solar water distillation using two different phase change materials Applied Mechanics and Materials **592-594** 2409-2415
[9] Chalapathi K, Dr.M.Inayathulla 2020 Solar desalination by using phase changing material, IJARIIE-**6** Issue-2

IOP Conf. Series: Materials Science and Engineering **909** (2020) 012027 doi:10.1088/1757-899X/909/1/012027

Camshaft modification on gasoline single cylinder engine to increase engine performance

Agung Sudrajad, Yusvardi Yusuf, Irwan Prasetyo
Department of Mechanical Engineering, Faculty of Engineering
Universitas Sultan Ageng Tirtayasa

Email: agung@untirta.ac.id

Abstract. Motorcycle is the main transportation in urban and rural areas nowadays. Many users have not satisfied with performance of standard manufactured. The objective of this research is to improve motorcycle performance by modifying the camshaft duration. The research was conducted at Combustion Laboratory of Department of Mechanical Engineering Faculty of Engineering. The standard of camshaft duration is intake valve 221° LSA, 96.75^{O} and exhaust valve 202° LSA 84°. The first modification camshaft is intake valve 235° LSA 100.5^{O} and exhaust valve 235° LSA 102.5^{O}. Second modification camshaft is intake valve 260° LSA 104° and exhaust valve 260° LSA 107°. The results of experiment show the good engine performance was in the second modification. Increased of torque are 19.182% and 22.902% for the first and second modification respectively. Moreover, increased of power are 7.378% and 13.45% for the first and second modification respectively.

1. Introduction

Motorcycle as transportation is very important for people. Recently, the statistic data of sales of motorcycle show increased significantly [1]. According to technological developments, motor vehicle companies apply high technological standards for their products. Some motorcycle users consider that the standard engine performance is still not suitable, that makes motorcycle users decide to make engine modifications to improve performance. One of the ways to improve motorcycle engine performance is to maximize camshaft performance. Camshaft modification can increase engine torque and power. According to [3], by changing the duration of the camshaft and increasing the size of the valve opening, the performance of the engine will increase. Changing the duration of the camshaft aims to change the opening and closing times of the valve. Improve volumetric efficiency, while the change in Lobe Separation Angle (LSA) will increase power in the upper rotation [4].

Generally, modification of camshaft is changing the duration and LSA to improve engine performance. However, according to [6] the change in duration and LSA on each camshaft is not as long as it is directly proportional between torque and speed of engine. Recently, the duration and LSA modifications are carried out using Computer Numerical Control (CNC) software and machines. The objective of this study is to improve the performance of a single cylinder engine capacity of 155 cc by modifying the duration and LSA camshaft.

2. Methodology

The modification of camshaft was conducted by changing the camshaft duration and Lobe Separation Angle (LSA). The standard of camshaft duration is intake valve 221° LSA, 96.75^{O} and exhaust valve 202° LSA 84°. The first modification camshaft is intake valve 235° LSA 100.5^{O} and exhaust valve 235° LSA 102.5^{O}. Second modification camshaft is intake valve 260° LSA 104° and exhaust valve 260° LSA 107°. The study was conducted at the Motor Fuel Laboratory, Department of Mechanical Engineering, Faculty of Engineering, University of Sultan Ageng Tirtayasa. The comparison of the performance of a standard camshaft with a modify camshaft 1 and a modify camshaft 2 was examined. The performance to be known are power, torque and fuel consumption. Variable engine speed at 2000, 4000, 6000 and 8000 with a maximum load were setup. The loading test was carried out by brake mechanism. The engine specification is shown in table below.

Table 1. Specification of Engine Test

Engine Specification	
Engine type	4 stroke, SOHC 2 Van Cooler
Bore / Stroke	50.0 x 57.9 mm
Volume Cylinder	115 cc
Compression Ratio	9.3 : 1
Engine Lubrication	0.9 Liter
Fuel	RONn 92

Table 2. Mathematical Formula for Calculation

	Formula
Force moment at *handle*	$F_1 = m \times g = N$
Force at piston brake	$F_2 = \frac{T_1}{X_2}$
Force at piston *Kaliver*	$F_3 = \frac{F_2 \; x \; A_2}{A_1}$
Constanta of Spring	$k = \frac{G \; x \; d^3}{8.N.D\text{^}2}$
Center of Friction	$rm = \frac{(rb+rk)}{2}$
Force friction at disk brake	$F_{gesek} = F3 \times \mu$
Torque	$T = F_3 \times \mu \times kt \times rm$
Power	$P = 2\pi \times \frac{N_{rev}}{s} xTx10^{-3}$
Fuel Consumption	$Sfc = \frac{Fc}{P}$

n	= Engine speed (rpm)	X_1	= Distance of brake handle (m)
P	= Engine Power (hp)	X_2	= Distance to the piston (m)
T	= Torque (Nm)	m	= load (kg)
Sfc	= Specific Fuel	g	= gravitation (9,8 m/s^2)
p	= Hydraulic Pressure	G	= Modulus (Gpa)
F_1	= Brake Force in Handle	K	= Constanta
F_2	= Piston trust (N)	d	= thin of braking (m)
F_3	= Breaking Force at Piston	D	= diameter of braking (m)
N	= Number of coil spring	F_{pegas}	= Spring Force (N)
A_1	= Area of Piston Brake (m^2)	F_1’	= Force at Brake (N)

3. Results and Discussions

The test conducted with variable of engine speed on the modification of camshaft. Figure 1 shows the force of brake by variable of engine speed at 2000, 4000, 6000 and 8000 rpm.

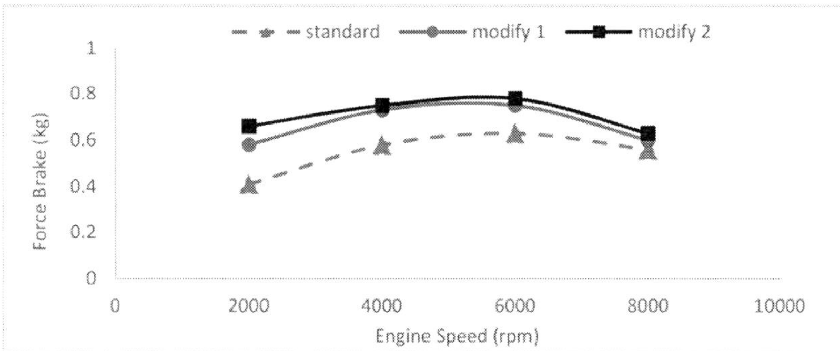

Figure 1. Force Brake by variable of engine speed

The experimental result shows the force of brake on second modification of camshaft higher than standard and modification 1. The peak of data is on the 6000 rpm, its shown that the best performance of engine is on the 6000-rpm engine speed with second modification of camshaft. Increased of force brake is shown the increased of engine performance. Increased of force brake comparing between first modification and standard camshaft are 50%, 22%, 22.6% and 3.5% at 2000, 4000, 6000 and 8000, respectively. Meanwhile, comparing standard and second modification are 65%, 25.4%, 23.8% and 8.6% at 2000, 4000, 6000 and 8000 rpm, respectively. The increase of brake force on the second modified camshaft is caused by the high angle of the camshaft duration. Its result in a longer valve opening time and causes a greater amount of fuel and air carburized in the combustion chamber

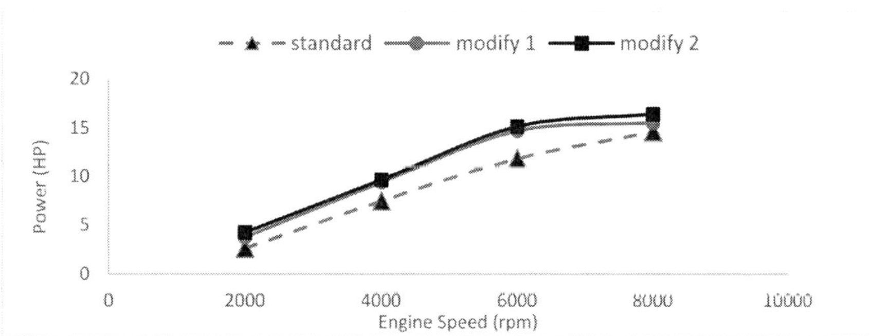

Figure 2. Power by variable of engine speed

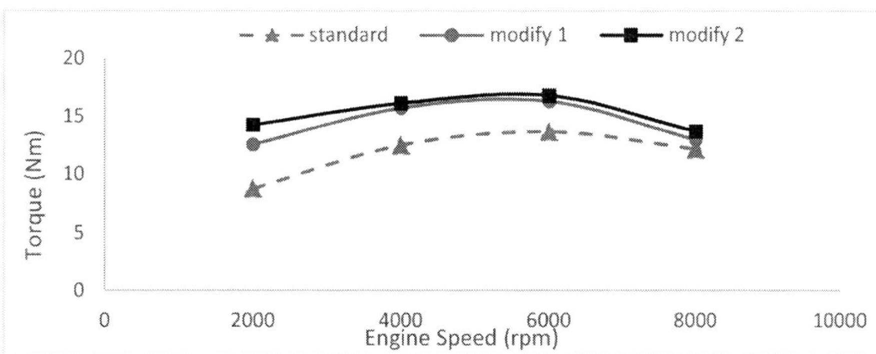

Figure 3. Torque by variable of engine speed

Figure 2 and 3 shows the power and torque comparison between standard, first modification and second modification of camshaft by variable of engine speed. The complete results of comparison between standard, first and second modification is shown in table 3 and 4.

Table 3. Results of Power Comparison

RPM	Power (*Horse Power*)			Percentage of Power Increased (%)	
	Camshaft Standard	*Camshaft* 1	*Camshaft* 2	Cam std vs Cam 1	Cam std vs Cam 2
2000	2.63	3.76	4.27	30.1	38.4
4000	7.49	9.4	9.66	20.3	22.5
6000	11.86	14.69	15.15	19.3	21.7
8000	14.62	15.53	16.41	5.9	10.9

Table 4. Results of Torque Comparison

RPM	Torque (*Nm*)			Percentage of Torque increased (%)	
	Camshaft Standar	*Camshaft* 1	*Camshaft* 2	Cam std vs Cam 1	Cam std vs Cam 2
2000	8.8	12.58	14.28	30.0	38.4
4000	12.53	15.72	16.14	20.3	22.4
6000	13.68	16.3	16.81	16.1	18.6
8000	12.2	12.97	13.7	5.9	10.9

Based on the experimental results on table 3 and 4, we can conclude that the optimum performance is in second modification of camshaft. Percentage of increased in power and torque is greater than first modification.

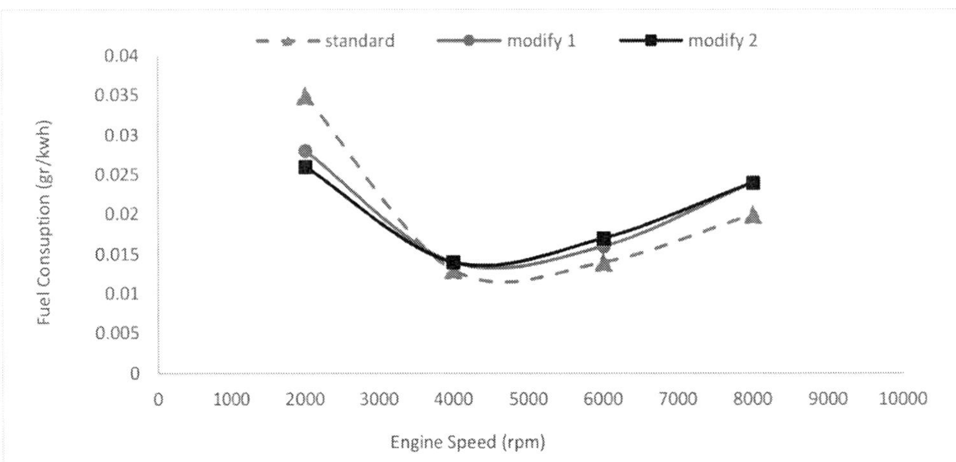

Figure 4. Fuel Consumption by variable of engine speed

Figure 4 shows the results of consumption of fuel by engine speed variable and modification of camshaft. Based on the calculation of fuel air ratio, the second camshaft modification is slightly efficient compare to the others. the optimum ratio of air and fuel will produce good fuel consumption

4. Conclusions

In this study, the engine performance output of a single cylinder gasoline engine running with two different camshaft is analysed. The samples include three kinds of camshaft, standard, first modification and second modification. The engine is running at four different engine speeds which are 2000 rpm, 4000 rpm, 6000

IOP Conf. Series: Materials Science and Engineering **909** (2020) 012027 doi:10.1088/1757-899X/909/1/012027

rpm, and 8000 rpm with the constant load. In the paper, brake force, the engine power and torque, fuel consumption was taken and analysed. From the results obtained, several conclusions are made:

1. Second camshaft modification, intake valve 260° LSA 104° and exhaust valve 260° LSA 107°, gives highest performance output compare to other camshaft
2. The percentage of torque and power decrease when the engine speed is increases

References

[1] Badan pusat statistik. (2020). Perkembangan Jumlah Kendaraan Bermotor Menurut Jenis. [Online]. Available: https://www.bps.go.id/linkTableDinamis/view/id/1133.

[2] Irashi, Zidni. (2018).Uji Performa Generator Listrik Berpenggerak Motor Diesel Satu Silinder Di Laboraturium Fakultas Teknik Untirta.p39.

[3] Kristanto, Dwi. (2017). Pengaruh perubahan durasi camshaft terhadap performance mesin FD 110 cc.p57

[4] Anam Choirul, M, dan Muhaji. (2018). Pengaruh Variasi Perubahan Kontur CamshaftTerhadap Unjuk Kerja Pada Honda Megapro 160D Advanced.p86.

[5] Stevansa, Adrianto Priyo. (2014).Pengaruh Penggunaan Camshaft Standar dan Camshaft Racing TerhadapUnjuk Kerja Motor Bensin Empat Langkah".p2.

[6] Alam, Perkasa Teguh. (2014). Analisa Unjuk Kerja Mesin Diesel Kapasitas 132cc pada Prototipe CulaSatuUntitra.p30.

[7] Graham bell, Alexander. (1981). *Performance tuning in theory and practice four stroke. England* :Haynes Publising Group

[8] Pudjanarsa, Astu., and Nursuhud, Djati. (2006) *Mesin Konversi Energi.*Yogyakarta: C.V ANDI OFFSET

IOP Conf. Series: Materials Science and Engineering **909** (2020) 012028 doi:10.1088/1757-899X/909/1/012028

Particleboard characterization using sawdust from sengon wood, mahogany wood, bayur wood, and rice husk ash as composite fillers

Sunardi[1], Rina Lusiani[1], Iman Saefuloh[1], Erny Listijorini[1], Anggit Eka Sumarna[1], Moh. Fawaid[2], Yenny Meliana[3]
[1] Department of Mechanical Engineering, Universitas Sultan Ageng Tirtayasa, Cilegon 42435 Indonesia
[2] Department of Mechanical Engineering Education, Universitas Sultan Ageng Tirtayasa, Serang 42117 Indonesia
[3] Research Center for Chemistry, Indonesian Institute of Science, Tangerang Selatan 15314 Indonesia

Email: sunardi@untirta.ac.id

Abstract. Banten Province, Indonesia, has abundant natural resources. Waste from processing natural resources has not been utilized maximally. Several types of wood and plants used in this study were sengon (*albizia Chinensis*), bayur (*pterspermum javanicum*), mahogany (Swietenia mahogany) and rice husk ash as filler. The reference particleboard characteristics used are Indonesian Standard, SNI 03-2105-2006. The characteristics observed were pore diameter, hardness, impact toughness, flexural strength, and thickness development. The particleboard composition is 50% filler, 15% oil palm empty fruit bunches, 20% polyvinyl acetate adhesive, and 15% epoxy resin. The length of fiber used is 5 mm. This research shows that particleboard with mahogany filler has optimum characteristics, namely pore diameter 2.0288 nm, hardness 43 N/mm^2, flexural strength 641.65 N/mm^2, toughness 3.54 kJ/m^2, and thickness swelling 1.20%.

1. Introduction

The area of forest and water conservation in Banten Province is 24.03 million hectares. Production of teak wood is 17,033.71 m^3 and the other wood is 13,711.34 m^3-processed wood products in the form of sawn wood 39,311.37 m^3 and plywood 196,894 m^3 [1]. The effect of a high production rate is environmental problems. Sawdust from these wood products must be utilized to reduce pollution. For this reason, wood waste can be used as a filler in particleboard composites, such as sengon [2], iroko [3], oak [4], acacia [5], pine [6] and the others wood [7]–[9]. On the other hand, wood can also be used as a material for biogas products [10]. The other material that can be used as filler includes the orange peel [11], sycamore leaf [9], rice husk [12], water hyacinth petiole [13], palm kernel cake [14], sago [15] and mango shell [16]. Fibers that are often used for reinforcement in composite include pineapple fiber, sisal [17], areca nut [18], coconut [19], bamboo [20], jute [21], oil palm empty fruit bunches [22], and cellulose material [23].

The quality of particleboard is influenced by the characteristics of the matrix, filler, and fiber used. The particle size used in composites has a significant effect on the particleboard produced. Terciu et al. used oak wood particles as fillers in the manufacture of composites of various sizes, namely 1-2mm,

0.4-1mm, and 0.2-0.4mm. Particle size causes changes in the value of the composite density, tensile strength, elongation when it breaks, and energy absorbed. Research conducted shows that oak wood particles with a size of 0.4-1 mm can absorb maximum energy. The long shape of oak wood particles also increases fracture resistance when compared to spherical or rectangular particles [4]. Coconut shell particle size of 0.25 mm can provide the optimum properties, like tensile strength, flexural strength, impact strength, and hardness compared to larger particle sizes [24]. The particle size of pine bark with 50 mesh was obtained density, moisture content, modulus of rupture, and bonding strength corresponded to SNI 03-2105-2006 standards for medium density particleboard type 8 [25]. The rice straw with mesh 50 provides a density of 0.76 gr/cm^3. Smaller rice straw particles will increase density, but the thickness swelling is decreased [26]. The same phenomenon also occurs in acacia wood [5]. The smaller the particle size tends to hold the water and the bond between the particles will be more durable [27]. The sengon wood powder, which is used as particleboard, also has the same phenomenon. Smaller particle size increases the density, modulus of elasticity, and hardness of the particleboard surface, but the impact strength and thickness swelling are decreases [2]. Combining particle size of sengon wood will affect the quality of particleboard. From this research, a 40-60-80 mesh filler combination provides the particleboard characteristics met the SNI standard 03-2105-2006 [28]. Areca nut fiber used as reinforcement in particleboard shows that the smaller the fiber size, the modulus of rupture and screw bonding increases [29].

The pre-treatment of the fiber significantly affects the quality of particleboard. From studies show that the longer pre-treatment of fiber or filler in a solution will reduce its strength. The alkali treatment of sisal and bagasse fibers decreases tensile strength, tensile modulus, and water absorption, but the flexural strength, impact, and hardness increases [30]. Fibers treated with 10% KOH for 36 hours at 30^{0}C can provide a bond between the matrix and the fiber better than those without treatment [18]. The more extended immersion of fiber in the alkali solution will increase water absorption [31]. Oil palm empty fruit bunched fiber immersed in 5% NaOH solution for 2 hours provides the optimum particleboard quality compared to the untreated fiber or immersed for more than 2 hours. Immersing fiber for more than 2 hours can damage the cellulose content [32]. The higher fiber immersion temperature can cause water absorption, and thickness swelling of the particleboard will increase [33].

The composition and ratio of the constituent elements in the composite have a significant effect. The ratio of sawdust and polyethylene terephthalate (PET) is very influential on the quality of wood-plastic composite [33]. Rice straw with a composition of 80, 85, and 90%vol produces particleboard that meets SNI 03-2105-2006, such as density, modulus of rupture, and modulus of elasticity [34]. The composition of wood particles and sycamore leaves on a composite can reduce mechanical properties if the number of sycamore leaves is higher [9]. The addition of black rice husk ash can reduce its tensile strength and elongation, but the modulus of elasticity will increase. From this research, it is known that the ratio between filler and matrix 95/5 gives the best performance particleboard[35]. The composition of rice straw and polypropylene (60/40) produce the particleboard with a density of 0.76 gram/cm^3[26]. The maximum content of the polystyrene-based resin matrix is 30% can produce particleboards with the best performance when mixed with bamboo fiber [20]. The addition of rice husk ash will decrease the density and flexural strength, compression stress, and hardness, but the water absorption and modulus of elasticity will increase [31]. The particleboard characteristics are affected by density and particle size, and distribution of filler and binder. From this research, it is known that the content of mango shell particle 40-50% has optimum characteristics[16]. The composition of jute fiber and glass fiber (50/50) produces optimum tensile strength, flexibility, and toughness [21]. Composites with a composition ratio of resin, sisal, and bagasse (60/20/20) provide optimum performance [30]. Particleboard has maximum values and complies with SNI 03-2105-2006 if the composition between matrix and oil palm empty fruit bunch fiber 50:50 [36]. The oil palm empty fruit bunch fiber can be used as a particle reinforcement with sengon wood powder as filler. This study shows that the composition of 15% oil palm fruit bunches provides excellent mechanical properties [37]. Tensile strength, flexural, and impact decreases with the increasing volume of coconut filler fraction. This study states that the 40% volume fraction, coconut shell particle reinforced epoxy produced, has the most optimum value [24].

The type of adhesive is a crucial factor in composite characteristics. The composition and characteristics of the adhesive determine the bonding strength between filler, fiber, and matrix. Cellulose in the form of cellulose nanofibers by 1% to urea-formaldehyde can improve the fracture toughness of composites, reduce thickness swelling, and increase internal bonding and flexural strength [7]. The adhesive content will produce different types of particleboard. In general, the 11% binder content met is the minimum requirement for modulus of rupture and modulus of elasticity in particleboard recommended by US Standard ANSI/A208.1[38]. Adding nano-TiO_2 and nano-SiO_2 to polyvinyl acetate can improve the quality of the joints on the particleboard [39].

The gap between the previous research and the topic discussed in this paper is degradable property. This property is due to fillers and fibers used are a natural material. The matrix also used still allows degraded because the resin used has a slight percentage. Particleboard refers to SNI 03-2105-2006. Another purpose of this study is to use wood waste, thereby reducing environmental pollution.

2. Method

Particleboard samples in this research were made consist of palm oil empty bunches fibers with 5 mm length (15%), polyvinyl acetate (20%), epoxy resin (15%), and filler (50%). Fillers used in this sample had mesh 40 from any wood in Indonesia, especially Banten Province. The wood particles used were sengon wood, mahogany wood, and bayur wood obtained from a sawmill in Pandeglang. The sample's initial shape is a beam with a size of 15 x 70 x 50 mm. A single punch made the samples of the cold press using a hydraulic press machine with 30 bar.

2.1. Material

The material used as a constituent of particleboard is wood and rice husk ash as filler, oil palm empty fruit bunches as reinforcement, epoxy resin, and polyvinyl acetate adhesive as a matrix. Particleboard composition using volume fraction, which is 15% fiber, 50% filler, 20% polyvinyl acetate, and 15% epoxy resin. The parameters for determining the volume fraction are:

$$V_f = \frac{\rho_m W_f}{\rho_f W_m + \rho_m W_f} \dots \dots (1)$$

Where V is the volume fraction, ρ is the density of material, and W is the weight of fibers. Alphabet m and f indicate matrix and fiber.

(a) (b) (c) (d)

Figure 1. Filler from wood (a) sengon, (b) mahogany, (c) bayur, and (d) rice husk ash

2.2. Preparation of Filler

Sawdust is obtained from sawmills in Pandeglang, Banten. The sawdust is dried in the sun and then sieved with mesh 40. Figure 1 shows that the sawdust used in this research from sengon wood, bayur wood, mahogany wood, and rice husk ash. The density of sengon wood is 250 kg/m^3[40], mahogany 800 kg/m^3[41], bayur 530 kg/m3[42] and density of rice husk ash is 90-150 kg/m^3[43].

2.3. Preparation of Fiber

The fiber used is the base of oil palm empty fruit bunches. The fiber is cleaned and dried to decompose the fiber. The length of the fiber used is 5 mm. Initial treatment on the fiber is immersing in a 5% NaOH

IOP Conf. Series: Materials Science and Engineering **909** (2020) 012028 doi:10.1088/1757-899X/909/1/012028

solution for 2 hours. After 2 hours, the fibers are cleaned with pure water to eliminate the NaOH solution's effects and dry in the sun for 24 hours.

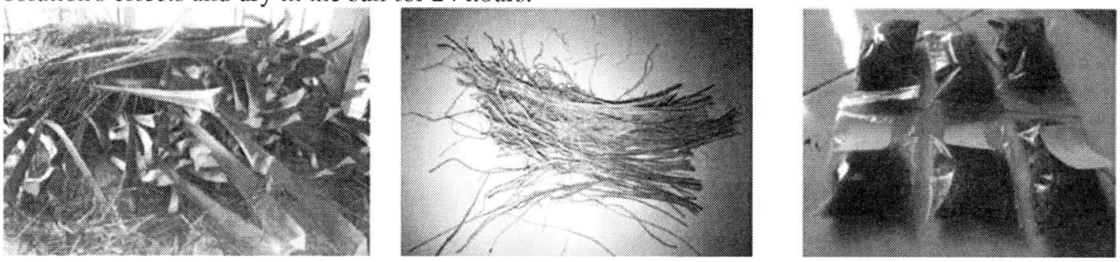

Figure 2. Fiber separation of oil palm empty fruit bunches

2.4. Preparation of Specimens
The mixing process between fillers, fiber, and the matrix is carried out with an electric mixer for 20 minutes at speed 700 rpm for getting a homogeneous mixture. After that, the mixture is allowed to stand for 10 minutes before the compacting process. Compacting was carried out by a single punch of cold press with 30 bar pressure and a holding time of 120 minutes. From this beam-shaped sample, specimens are then made according to refer to testing standards. Hardness test (ISO 2039-1 Standard), bending strength (ASTM D 790), impact toughness (ISO 179-1), pore diameters with surface area analyzer, a microstructure using a scanning electron microscope, and thickness swelling with SNI 03-2105-2006.

3. Result and Discussion
The quality of particleboard is determined when the particleboard is in contact with external conditions. The properties that will be discussed in this paper are pore diameter, hardness, bending strength, impact toughness, and thickness swelling.

3.1. Pore Diameter
Surface area analyzer (SAA) can be used to determine the surface area, pore volume in the nano-scale. By using SAA, the required sample is only 0.01-0.1 grams. From the SAA test obtained the data, as shown in Figure 3. From Fig. 3, it is known that the particleboard with mahogany filler has the smallest pore diameter. This condition shows that the particleboard has the lowest porosity or the highest density. Porosity will affect the properties of particleboard, such as hardness, toughness, flexural, and thickness swelling.

The mahogany wood has the highest density compared to the other fillers. This density causes the mahogany filler to be distributed and mixed evenly on the sample so that the composite constituents can be compressed easily. If the wood density is low, it isn't easy to mix the constituent materials homogeneously. This condition can cause the composite to have higher porosity. From this condition, it can be said that the density of wood will significantly affect the resulting particleboard's porosity. The higher the density of wood, the smaller the porosity of particle boards. The smaller the porosity, the higher the particleboard density. This result is in line with research conducted by Maharani et al [27].

Pore diameter needs to be known because it will be closely related to density, hardness, tensile strength, impact toughness, and thickness swelling. That is why the characterization of pore diameter is placed at the beginning of this discussion.

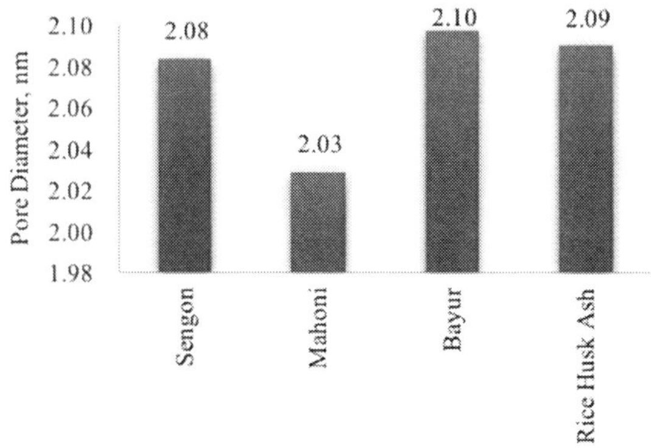

Figure 3. Pore diameter

3.2. Hardness

Hardness testing is carried out to determine how strong the particle board is when other materials scratch it. Because in its application, particleboard will come into contact with other objects. The hardness testing is done by the ball indentation method or equal to Rockwell hardness type H. The value acquired is called ball indentation hardness (N/mm²). Composite with mahogany filler has the highest hardness, which is 43 N/mm². This surface hardness correlates directly with porosity formed during the manufacturing process. The filler distribution pattern in the composite will affect the mechanical properties of the particleboard. The homogeneous distribution of fillers can provide higher hardness and strength than the distribution of fillers that are not uniformly and lumpy. This filler distribution can be studied from hardness at any point in the material surface. If the hardness at any point has the same value, it is said that filler distribution is uniform.

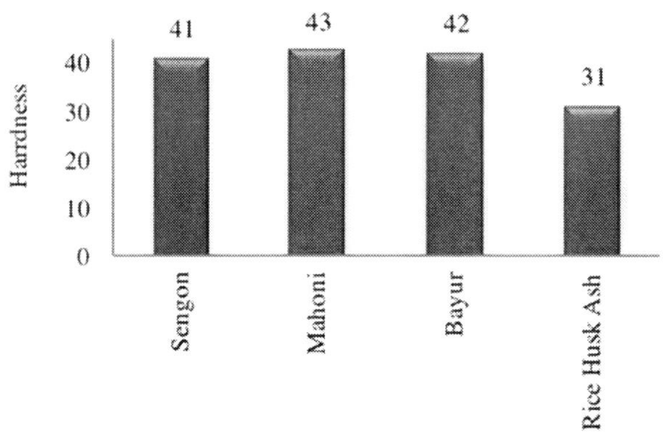

Figure 4. Correlation between filler type and hardness

A homogeneous bonding between filler and matrix can improve the mechanical properties of the composite. Force subjected to the surface area will be continued and retained by the constituent elements of particleboard simultaneously. Particleboard with rice husk ash filler has the lowest hardness. The filler distribution causes this condition [31]. The formation of filler lumps in certain parts causes different hardness. When indenter presses on the surface of filler lump, the more profound deformation will occur. Whereas if the indenter presses the matrix area, the surface will be less deformation.

Unequal distribution of filler is caused during the mixing processes between constituent material. The density of rice husk ash is the main factor that causes the rice husk ash to clot so that there is no strong bond between the fiber, filler, and matrix. The effect is the clumping of rice husk ash cannot bind fiber properly. This binding can be seen from the phenomenon of pull out on particleboard using rice husk ash as filler.

3.3. *Bending Strength*

Bending strength is one of the fundamental properties that particle board must possess. When used, the particleboard will have a long span. The particleboard must be able to withstand the applied load. Large deflection due to force acting on the particleboard can interfere with the stability of other objects that are placed. Bending testing is carried out using the 3 points bending method (ASTM D790). Specimen with size 100x14x5mm is subjected to loading to the center. The loading speed of 2.5 mm/min towards the bottom of the specimen. The filler type influences the bending strength of the composite. The highest bending strength occurs in composites with rice husk ash filler, which is 1,160.61 N/mm^2. This strength is caused by composites that are dominated by the matrix. The bending strength of composites is inversely proportional to its hardness. This composite tends to be brittle, characterized by high pore diameter and low deformation in the type of rice husk ash as filler [31].

Particleboard with rice husk ash filler has a unique behavior, that is high bending strength, but hardness, toughness, porosity, and thickness swelling is low. of mechanical and physical properties. The temporary hypothesis is the force subjected to composite held by matrix and its fiber. The bonding between matrix and fiber is appropriately, so this composite can withstand the external force that subjected.

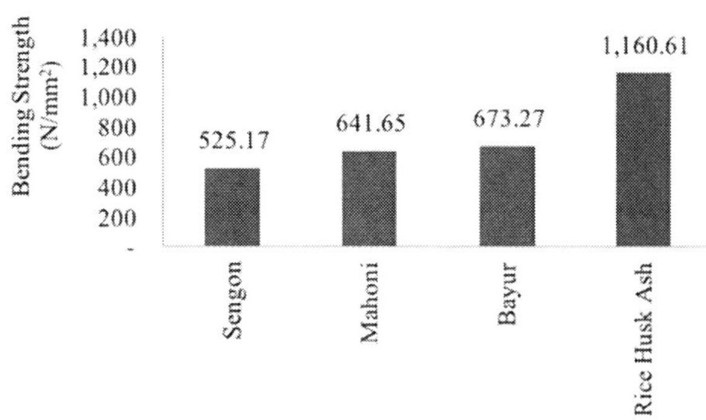

Figure 5. Correlation between filler type and bending strength

3.4. *Impact Toughness*

Impact tests are carried out to determine the particleboard's toughness when it falls, or there is a collision with another object. That's the reason why this test needs to be done. Impact testing was conducted in the Polymer Technology Centre, LIPI. The standard used is ISO 179-1, which sample has a size 80 x 10 x 5 mm. The sample is placed horizontally on the instrument and then given the pendulum's impact with an energy of 2 joules and a speed of 2.9 m/sec^2.

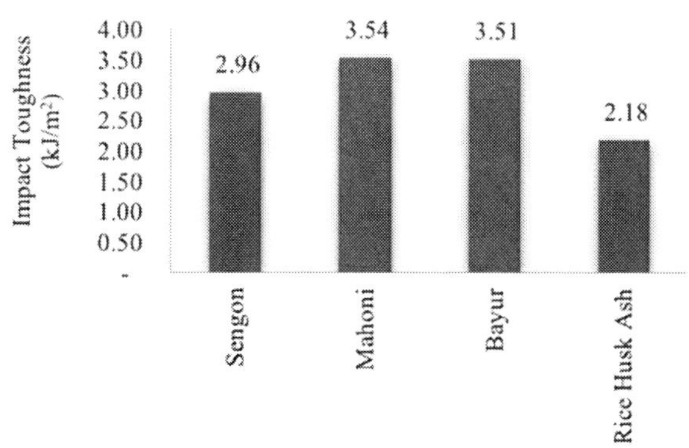

Figure 6. Correlation between filler type and impact toughness

From this experiment, it is known that the highest impact strength occurs in particleboard with mahogany wood filler, which is 3,54 kJ/m², followed by bayur, sengon, and rice husk ash filler 3.51 kJ/m², 2,96 kJ/m² and 2,18 kJ/m², respectively. Fig. 6 shows that the type of filler influences the impact strength of the particleboard significantly. Composite with mahogany filler has high impact strength caused by mahogany wood particles mixed with matrix and fiber homogeneously. When a fast loading is given to the specimen, the force will be distributed evenly to the particleboard directly so that each constituent material will receive loading. This condition can be caused by the bonding between the composite materials that formed adequately.

The toughness of particleboard with rice husk ash filler has the lowest value. From the previous discussion that porosity is high, and elongation is low. This low elongation shows that this material tends to brittle so that particleboard characteristics are natural to break when the load exceeds its strength. The toughness of particleboard with rice husk ash and sengon wood filler has toughness under the existing particleboard.

3.5. Thickness Swelling

Thickness swelling testing is carried out to determine the performance of the particleboard when in contact with water. Usually, the particle board exposed to water will expand due to the release of bonds between the composite components. It is known that the particle board on the market is very quickly damaged/destroyed when exposed to water. While the particleboard developed in this study aims to improve the thickness swelling characteristics for the better, this composite still has biodegradable properties.

Thickness swelling is a physical property that can be seen directly. The particleboard, when in contact with water, will cause dimensional changes. Thickness swelling is strongly influenced by filler type characteristics—the higher percentage of thickness swelling, the worse particleboard performance. Composite with low density tend to have high porosity. This low density is why water quickly enters the pores of particleboard so that that water will break the bond between filler, fiber, and matrix.

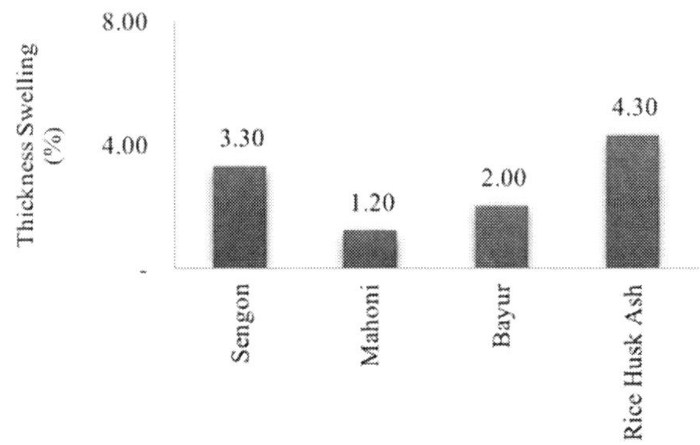

Figure 7. Correlation between filler type and thickness swelling

Figure 7 shows that particleboard with mahogany filler has the lowest thickness swelling, which is 1.20%. This thickness swelling is lower than the other filler and the existing particleboard in the market. The lowest thickness swelling is due to the smallest pore diameter. Particleboard with minimal porosity will make the water trying to enter the pores. This minimal porosity causes the percentage of thickness swelling to be smaller. On the other hand, composite with rice husk ash filler will be very different, and thickness swelling is very high, which is 4.30%.

Thick development is closely related to porosity and filler characteristics when in contact with water. The higher the pore diameter, the more water comes in contact with the surface of the material. This phenomenon will cause the development of thicker particle board with husk ash filler to be higher. This condition is in line with research conducted by Baharuddin et.al [5].

3.6. Microstructure
Microstructure observation using a scanning electron microscope can answer the phenomenon of particleboard characteristics. From Figure 8, it can be seen that the sengon and rice husk ash filler have high porosity. The figure also shows that the bond between matrix, filler, and fiber does not occur ideally. Perfect binding occurs on particleboard with mahogany and bayur filler. This perfect binding is related to why the hardness, toughness, porosity, and thickness swelling are better than other fillers directly.

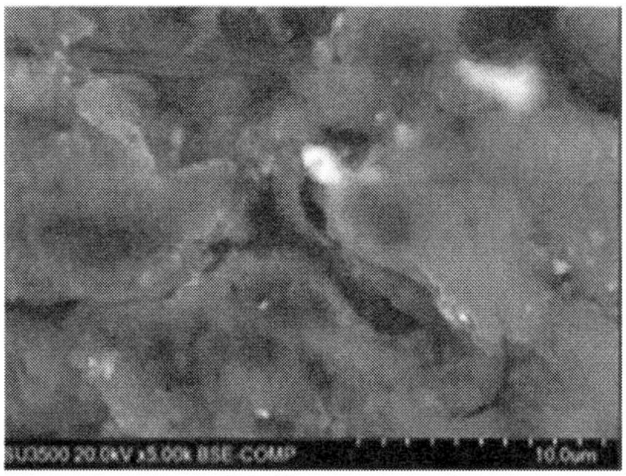
(a) Sengon wood

(b) Mahogany Wood

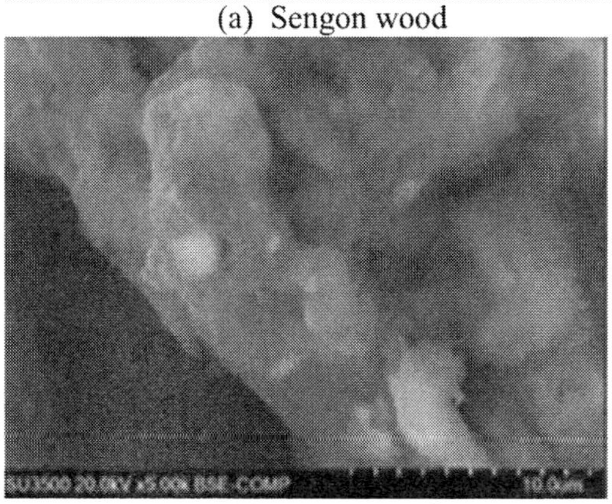
(c) Bayur wood

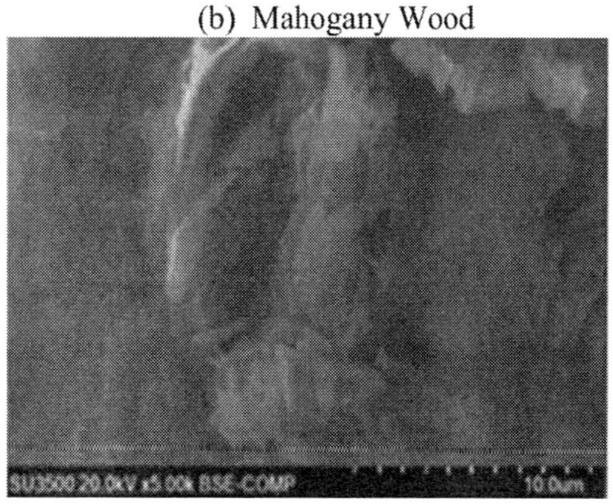
(d) Rice husk ash

Figure 8. Correlation between filler type and thickness swelling

4. Conclusion
The utilization of wood waste is one way to reduce environmental problems and look for alternative materials to replace logs as furniture materials and other household needs. From this research, it is known that this type of filler significantly affects particle properties. The density of wood sawdust is in line with hardness and toughness, but inversely proportional to pore diameter and swelling thickness. Particleboard with Mahogany filler has excellent properties and meets Indonesian standards, SNI 03-2105-2006. The characteristics obtained are pore diameter 2.0288 nm, hardness 43 N/mm^2, bending strength 641.65 N/mm^2, toughness 3.54 kJ/m^2, and swelling thickness of 1.20%.

Acknowledgments
This research was supported by the Faculty of Engineering, Universitas Sultan Ageng Tirtayasa, and Indonesian Institute of Science, Serpong. For this reason, the author would like to thank the partners for conducting this research.

References

[1] BPS 2020 Provinsi banten dalam angka

[2] Sunardi S, Moh. F, Rina L and Rumodang P 2017 Pengaruh butiran filler kayu sengon terhadap karakteristik papan partikel yang berpenguat serat tandan kosong kelapa sawit *J. Mesin Teknol.* **11** 28-32

[3] Engr O P and Hilary U 2018 Effect of fillers loading on the mechanical properties of hardwood sawdust oil bean shell reinforced epoxy hybrid composites *Int. J. Sci. Sci. Eng. Technol.* **4** 620-626

[4] Terciu O M, Curtu I and Teodorescu-Draghicescu H 2012 Effect of wood particle size on tensile strength in case of polymeric composites *8th International DAAAM Baltic Conference Industrial Engineering* 1–6

[5] Baharuddin M N M, Norazwani M Z, Eida N R, Harun W S W and Zaeime S Z 2019 Physical properties of homogeneous particleboard based on acacia tree and polyurethane adhesive as a resins *Int. J. Eng. Adv. Technol.* **8** 3382–3387

[6] Dorota D, Radoslaw M, Dorota D and Adam D 2015 Possibility of using the expanded polystyrene and rape straw to the manufacture of lightweight *Maderas, Cienc. y Tecnol.* **17** 647–656

[7] Stefan V, Jorn R, Martin W and Wolfgang G 2012 Particle board and oriented strand board prepared with nanocellulose-reinforced adhesive *J. Nanomater.* **2012** 1–8

[8] Ejiogu I K, Odiji M O, Ayejagbara M O, Shekarri T N B and Ibeneme U 2018 Mechanical properties of urea formaldehyde particle board composite *Am. J. Chem. Biochem. Eng.* **2** 10–15

[9] Hamidreza P, Payam M and Sima S 2014 Particleboard from wood particles and sycamore leaves: physico-mechanical properties *Eng. Agric. Environ. Food* **xxx** 1–6

[10] Iyiola O O and Omojola A 2013 Comparative study of the effects of sawdust from two species of hard wood and soft wood as seeding materials on biogas production *Am. J. Eng. Research* **02** 16–21

[11] Mehmet T, Alpay S, Ahmet F K and Metin G 2019 Production of useful composite particleboard waste orange peel *Cellul. Chem. Technol.* **53** 517–526

[12] Alao K T, Alao T O, Adeleke A O and Olalere O A2015 Recycling of rice husk into a locally-made water-resistant particleboard *Ind. Eng. Manag.* **4** 1–6

[13] Adela S R, Maria M B B and Trinidad R T 2019 Microstructural and thermo-physical characterization of a water hyancinth petiole for thermal insulation particle board manufacture *Materials* **12** 1–16

[14] Bono A, Anisuzzaman S M, Ismail N M and Haziami R 2013 Performance of particleboard with palm kernel cake as filler 337-345

[15] Tay C C, C. Mohd S O and Sinin H 2014 Water absorption and thickness swelling behavior of sago particles urea formaldehyde particleboard **3** 1375–1379

[16] Adewuyi M A, Oyinlola K A and Akindapo J O 2017 Suitability of mango seed shell particles and recycled high density polyethylene (rhdpe) composites for production of particleboard *Am. J. Eng. Research* **6** 314–325

[17] Supriya M, Amar K M, Lawrence T D, Manjusri M and Georg H 2004 A review on pineapple leaf fibers, sisal fibers and their biocomposites *Macromol. Mater. Eng.* **289** 955–974

[18] Srinivasa C V, Basavaraju B, Mownesh G K and Raghu P G R 2010 Flexural behaviour of areca fibers composites *Bioresources.com* **5** 1846–1858

[19] Norhayati N, Masiri K, Aslila A K, Suhaila S, Aziza Z, Siti R N Z, Nur A A and Wan H A H A 2018 Panel board from coconut fibre and pet bottle *Web Conf.* **34** 1–9

[20] Abdulkareem S A and Adeniyi A G 2017 Production of particleboards using polystyrene and bamboo wastes *Niger. J. Technol.* **36** 788–793

[21] Sanjay M R, Arpitha G R and Yogesha B 2014 Investigation on mechanical property evaluation of jute - glass fiber reinforced polyester *J. Mech. Civ. Eng.* **11** 50–57

[22] Rina L, Sunardi S and Yogie A 2014 Pemanfaatan limbah tandan kosong kelapa sawit sebagai papan komposit dengan variasi panjang serat *Sem. Nas. Int. Pros.* 240–248

[23] Han-Seung Y and Douglas J G 2011 Mechanical properties of cellulose nanofibril-filled

polypropylene composites *Wood Fiber Sci.* **43** 143–152
[24] Manjunatha C G H and K Sabeel Ahmed 2017 Experimental characterization of coconut shell particle reinforced epoxy composites *J. Mater. Environ. Sci.* **8** 1661–1667
[25] Asfarizal S, Anwar K, Gunawarman and Santosa 2019 Effect of pressing time and particle size of pine bark on the quality of oil palm empty fruit bunches particleboard *J. Penelit. Has. Hutan* **37** 171–184
[26] Fitri N, Mursal, and Ismail I 2020 Physical and chemical properties of particleboard made of rice straw and plastic waste *IOP Conf. Series: Journal of Physics* ***1460*** 1–8
[27] Rizki M, Tamai Y, Takashi Y and Terazawa M 2010 Scrutiny on physical properties of sawdust from tropical commercial wood species: effects of different mills and sawdust's particle size *J. For. Res.* **7** 20–32
[28] Sunardi S, Moh. F, Rina L, Setyo B A K and Teguh D W 2019 The effect of wood sawdust mesh combination on mechanical behaviour of particle board *IOP Conference Series: Materials Science and Engineering* **494** 1-7
[29] Maulida, Eka R J and Hendry S 2013 Characteristic mechanic of polyester composite particle board with filller particle of areca nut fiber *The 11 th International Conference of Mining, Materials and Petroleum Engineering* 23–27
[30] Prashant D, Saurabh K S and Anurag S 2019 Study of mechanical behaviour & water-absorption characteristics of sisal-bagasse fibre reinforced hybrid epoxy composites *Int. J. Eng. Tech. Research* **9** 23–32
[31] Md. Mahfujul I, Humayun K, Md. Abdul G, Md. Mahbubur R B, Md. Alamgir K, Md. Rakibul Q and Farid A 2015 Study on physio-mechanical properties of rice husk ash polyester resin composite *Int. Lett. Chem. Phys. Astron.* **53** 95–105
[32] Sunardi S, Rina L, Aditya Y N and Moh. F 2018 Perlakuan alkali pada serat tandan kosong kelapa sawit terhadap mutu papan partikel *Semnas Sains & Terapan IV* 84–89
[33] Khandkar S R, Md Nazrul I, Md Mushfiqur R, Md Obaidullah H, Rudi D and Khalil HPS 2013 Flat-pressed wood plastic composites from sawdust and recycled polyethylene terephthalate (pet): physical and mechanical properties *a Springer Open J.* **2** 1–7
[34] Ismail I, Quratul A, Zulfalina, Zulkarnain J and Siti H S M F 2018 Mechanical and physical properties of the rice straw particleboard with various compositions of the epoxy resin matrix *IOP Conf. Series: Journal of Physics* **1120** 1–7
[35] Rotua A and Maulida 2014 Pengaruh ukuran partikel dan komposisi abu sekam padi hitam terhadap sifat kekuatan tarik komposit poliester tidak jenuh *J. Tek. Kim. USU* **3** 31–36
[36] Mima R L, Teuku M, Jon K and Rifdullah K 2018 Characterizing particle board made of oil palm empty fruit bunch using central composite design *Makara J. Sci.* **22** 17–28
[37] Sunardi S, Moh. F and Muhamad C 2016 Pemanfaatan serat tandan kosong kelapa sawit sebagai penguat papan partikel dengan variasi fraksi volume serat *Mach. J. Tek. Mesin* **2** 36–39
[38] Ciannamea E M, Marin D C, Ruseckaite R A and Stefani P M 2017 Particleboard based on rice husk: effect of binder content and processing conditions *J. Renew. Mater.* **5** 19–22
[39] Timucin B, Ali N T, Nurgul T, Deniz A and Eser S 2017 The bending and tension strength of furniture joints bonded with polyvinyl acetate nanocomposites *Maderas, Cienc. y Tecnol.* **19** 51–61
[40] Wayan D, Dodi N, Istie R, Meriem F and Remy M 2013 Determination of juvenile and mature transition ring for fast growing sengon and jabon wood *J Indian Acad Wood Sci.* **10** 39–47
[41] Agus N, Ragil W and Ganis L 2017 Karakteristik papan partikel limbah kayu mahoni dengan perlakuan pengawetan asap cair *J. Nas. Teknol. Terap.***1** 1–7
[42] Syamsul H 2014 Distribution patterns of bayur (pterospermum javanicum jungh) and its associated plantsin the gunung rinjani national park *J. Penelit. Hutan dan Konserv. Alam* **11** 225–237
[43] Bhupinder S 2018 Waste and supplementary cementitious materials in concrete," in *waste and supplementary cementitious materials in concrete* 417-460

IOP Conf. Series: Materials Science and Engineering **909** (2020) 012029 doi:10.1088/1757-899X/909/1/012029

Biodiesel characteristics of tuna fish bio-oil waste in the transesterification process with variation of reaction time and stirring speed

I Rosyadi[*], Ni Ketut Caturwati , Dhimas Satria, Haryadi and Muzaky
Department of Mechanical Engineering, Universitas Sultan Ageng Tirtayasa, Banten, Indonesia

[*]E-mail: imron_hrs@yahoo.co,id

Abstract. Waste from fishery production from canned fish production in Indonesia is very abundant, this waste can be used as raw material for making biodiesel. In this research, the production of biodiesel from tuna oil waste is carried out through the transesterification process. The purpose of this study was to determine levels of Free Fatty Acid, the effect of reaction time and stirring speed on the transesterification process on the characteristics of biodiesel. Fixed variables used consisted of methanol plus bio-oil in a ratio (6: 1), methanol at a concentration of 98%, reaction temperature of 60 oC, KOH catalyst weighing 3.5 grams, while the independent variable consisted of the time of transesterification reaction and stirring speed. The variation of transesterification reaction time used is 1, 2, and 3 hours while the variation of the speed of stirring the transesterification reaction is 160, 245, and 360 rpm. The optimal conditions obtained were 2 hours reaction time and 360 rpm stirring speed with biodiesel characteristics, namely a density of 849 kg / m ^ 3, viscosity of 3.53 cSt, water content of 0.6%, calorific value of 9,390 cal / gr, flash point 84 ° C, and acid number 3.6343 mg KOH / gr.

1. Introduction

Indonesia is the second largest capture fisheries producer in the world after China (12.24 million tons) with a production volume reaching 6.10 million tons. Followed by the United States (4.89 million tons), Russia (4.46 million tons), and Peru (3.7 million tons) [1]. One type of fish that is widely produced is tuna, in 2016 tuna production in Indonesia amounted to 1,129,375 tons [2].

A large amount of production surely has an impact on the residual waste of the production process which is also large and often causes environmental problems. Waste generated from fisheries activities is around 20-30% of annual fish production, fishery waste products can be in the form of solids, liquids or gases. Fish oil as one of the fisheries processing waste is a material that has the potential for making biodiesel (environmentally friendly fuel) because it has a high fat content [3].

Research on making biodiesel from baung fish waste with solid clay catalyst with variable levels of catalyst and methanol from fish oil. The results of this study were obtained Biodiesel can be directly used on diesel engines because physical and chemical properties have met the characteristics of biodiesel or Standar Nasional Indonesia (SNI) [4]

Research on the amount of reaction and reaction time on biodiesel production from pangasius waste, the hypothesis with variables is the reaction time and the amount of catalyst. The results of this study are, The longer the reaction time, the less the yield and conversion of biodiesel produced due to the reversible nature of the transesterification reaction [5].

This research has a high urgency, the need for this Biofuel is very large and demands fast action steps in accordance with the Minister of Energy and Mineral Resources Regulation No.12 / 2015, which is about the Government's efforts to encourage the use of Biofuel to reduce the consumption of fossil fuels derived from fossil fuels. . The government is targeting biofuel to be 30% biodiesel and 20% bioethanol from the total needs of diesel oil and gasoline by 2025. The aim of this research is to determine the effect of reaction time and speed of stirring in the characteristics of biodiesel fuels resulting from the transesterification process.

2. Methodology

2.1 Fuel Preparation and Properties

Based on research that has been done, that biodiesel fuel has physical and chemical properties that are almost the same as conventional diesel fuel and also has an energy value that is almost equivalent without making modifications to diesel engines. The use of biodiesel in Europe is done by mixing biodiesel fuel with conventional diesel with a certain ratio which is more due to maintaining the technical factors in the engine for new products and maintaining the quality of the biodiesel cetane number which must be the same or greater [6].

Biodiesel can be used as a diesel mixture because it can be mixed with High Speed Diesel in any composition and can be applied. without engine modification. The following is a comparison of the characteristics of biodiesel with diesel oil. There are several standards in the world that regulate the quality of biodiesel for example ASTM (American Society for Testing Material), EN (European Commite for Standadization), and SNI (Standar Nasional Indonesia). The applicable , standard for the quality standard of biodiesel in Indonesia is SNI presented in table 1.

Table 1. Requirements biodiesel SNI 7182:2015 .

Properties	Standard	Test Methods
Density [kg/m3] at 40 oC	850 -890	ASTM D 1298 or D 4052
Kinematic Viscosity [cSt] at 40oC and 1 Atm	2,3 - 6,0	ASTM D 445
Cetane number , min	51	ASTM D 613 atau D 6890
Flash Point oC, min	100	ASTM D 93
Acid Number mg-KOH/g, max	0,5	ASTM D 664
Water Content % Vol, Max	0,05	ASTM D 2709
Higher heating value [kj/kg]	46,10	ASTM D 5865

Source : ASTM International

2.2 Biodiesel Process

Basically the process of making biodiesel is very simple. Biodiesel is produced through a process called free fatty acid esterification reaction or triglyceride transesterification reaction with alcohol with the help of a catalyst and from this reaction methyl esters / ethyl esters of fatty acids and glycerol are produced. The process of esterification with an acid catalyst is required if the oil contains an FFA above 5%, if the oil that has a FFA content of more than 5% is transfetrified directly with a basic catalyst [7].

Esterification reaction is a reaction between fatty acids with alcohol to produce esters and water with the help of an acid catalyst. The esterification reaction is also a reaction that is reversible (reversible). The purpose of the esterification reaction is to reduce the content of free fatty acids in the ingredients [5].

$$RCOOH + CH_3OH \rightleftharpoons RCOOCH_3 + H_2O$$

Fee Fatty Acid Methanol Acid Catalyst Metil ester Water

Figure 1. Esterification reaction

Transesterification reaction is a reversible reaction, so if the reaction has reached equilibrium, the reaction can shift back toward the reactants. Transesterification is the conversion stage of triglycerides (vegetable / animal oils) to alkyl esters, through reaction with alcohol, and produces a by-product, glycerol. Triglycerides through the transesterification reaction with glycerol are converted into monoglycerides and diglycerides with the help of catalysts such as sodium methoxide and other Lewis bases [5].

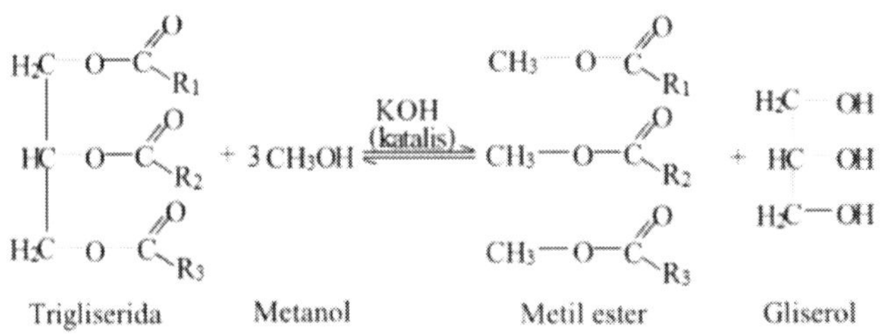

Figure 2. Transesterification reaction

Factors that influence biodiesel production include reaction time, ratio of alcohol to oil, catalyst type, temperature, stirring, settling time, water content, methanol, biodiesel quality standard.

2.3 Properties Biodiesel Testing
2.3.1 Free Fatty Acid Testing

The raw material is first analyzed its free fatty acid levels, if the raw material used contains high free fatty acid levels (FFA) which is more than 5%, then the esterification process is necessary to reduce the levels of free fatty acids to less than 5%. The test is carried out by titration using KOH 0.1 N. The first stage is to put as much as 10 ml of sample into the erlenmeyer. Next we add 10 ml of 95% alcohol and heat it to a temperature of 50°C while stirring. After heating, drops with a solution of phenophthalein indicator as much as 2-3 drops, shake the mixture. Final titration with 0.1N KOH until the color of the sample turns pink which does not disappear for 10 seconds indicates the end point of the titration. Calculate the levels of FFA contained in raw materials using the following formula :

$$FFA\ (\%) = \frac{V \times N \times M}{m} \times 100\% \quad (1)$$

2.3.2 Density Testing

Density testing was carried out at the Basic Chemistry Laboratory, Department of Chemical Engineering, Faculty of Engineering, Sultan Ageng Tirtayasa University. The first stage of the 100 ml erlenmeyer is weighed empty. Then the sample is put into 25 ml erlenmeyer, then the sample is weighed. The result of density testing is the scale of the sample divided by 25 ml.

2.3.3 Viscosity Testing

Viscosity testing is carried out using a viscometer. The obtained viscosity value is the value of dynamic viscosity, it is necessary to convert to kinematic viscosity with the following equation:

$$\upsilon = \mu / \rho \qquad (2)$$

where υ is kinematic viscosity (cSt), μ is dynamic viscosity (poise) and ρ is density (g/cm3)

2.3.4 Water Content Testing

Water Content testing is carried out using the toluene distillation method. The first stage is by weighing a sample of 25 ml, then the sample is inserted into the distillation flask. Then 75 ml of toluene solution is added, then put the distillation flask in a special distillation device with a reservoir of evaporating water. The next stage is to set the distillation heating temperature to 80 ° C. The distillation process is carried out for 1 hour, the water will drip and be collected by a measuring cup.

2.3.5 Caloric Value Testing

Analysis of the heating value of a fuel is intended to obtain data about the heat energy that can be released by a fuel by the reaction or combustion process. The amount of heat is measured in calories and is produced when a briquette is oxidized completely in a bomb calorimeter. The size of the sample does not exceed 1.5 mg, testing is done by measuring the temperature of the initial and final conditions of the sample on the test equipment.

2.3.6 Flash Point Testing

Flash point testing is performed using the SYD-261 PMCC Flash Point Tester. Where the principle of analysis is a number of samples heated with a certain heating speed while stirring in a closed cup. The ignition test starts at every 3 °C temperature rise by bringing the igniter closer to the sample surface, until the flash point is detected.

3. Results and discussion

3.1. FFA Content

Table 2. Free Fatty Acids (FFA) after transterification

No	V (ml)	N	M	M (gr)
1	29	0,1	28,2	2,5
2	31	0,1	28,2	2,5
3	30	0,1	28,2	2,5

From the table 2, using equation 2.3.1, the FFA value is 33.84%. High levels of FFA are caused by spoilage that occurs in the raw material, so the esterification process is needed to reduce the free fatty acid levels.

The esterification process was carried out by adding a 98% mixture of methanol as much as 30% by volume of bio-oil and HCL 32% by 1% by weight of bio-oil into 1400 ml bio-oil. The esterification

IOP Conf. Series: Materials Science and Engineering **909** (2020) 012029 doi:10.1088/1757-899X/909/1/012029

reaction time is 60 minutes with a stirring speed of 245 rpm at 60 ° C. The results of FFA levels of raw materials after esterification process are as follows:

Table 3. Free fatty acid (FFA) after esterification

No	V (ml)	N	M	M (gr)
1	4,60	0,1	28,2	2,5
2	4,60	0,1	28,2	2,5
3	4,59	0,1	28,2	2,5

Using equation sub bab 2.3.1, the FFA value is 5.18%. The esterified bio-oil is then transesterified.

3.2 Biodiesel Production Results (Transesterification)

The biodiesel process is carried out with a composition of 250 ml bio-fish oil and 98% methanol mixture of 1500 ml and KOH weighing 3.5 grams. Transesterification reaction carried out with fixed variables is the reaction temperature of 60 ° C with the independent variables being the reaction time and stirring speed. The reaction time used is 1, 2 and 3 hours, the stirring speed used is 160, 245 and 360 RPM. There are 9 samples where each sample has a different treatment, the results of biodiesel and byproducts can be seen on the table as follows:

Table 4. Experimental data transesterification production.

Sample	Time (hours)	RPM	Biodiesel (ml)	Gliserol (ml)	Visual Appearance
1	1	160	800	50	Turbid-light
2	1	245	800	75	Turbid-light
3	1	360	650	100	Dark-Turbid
4	2	160	430	120	Turbid-light
5	2	245	450	130	Light
6	2	360	380	150	Light
7	3	160	330	130	Light
8	3	245	350	150	Turbid-light
9	3	360	320	170	Dark-Turbid

From the visual appearance in table 4, with a stirring time of 2 hours showed a lighter appearance close to SNI standard biodiesel and considered the most ideal time for stirring. The longer the stirring time of biodiesel produced is reduced and the glycerol production increases.

3.3 Biodiesel Characteristics

3.3.1 Density Analysis

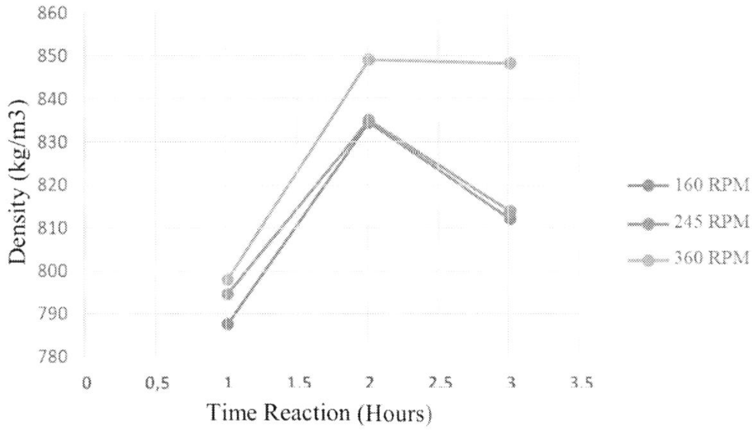

Figure 3. Time reaction and RPM effect to density

Based on these figure 3, it can be seen that the reaction time fluctuates with the density of biodiesel produced because it is caused by the nature of the transesterification reaction which when it reaches an equilibrium point then the reaction can reverse. While the stirring speed affects the resulting density increases. Increasing the stirring speed will accelerate the movement of molecules and increase the chance of collisions between molecules so that the reaction speed increases. The biodiesel density that is closest to the requirements of biodiesel SNI 7182: 2015 (850-890 kg / m3) is sample 6 with a variation of 2 hours, 360 RPM.

3.3.2 Viscosity Analysis

Table 5. Biodiesel Kinematic Viscosity Produced

Sample	Dynamic Viscosity (mPa.s)	Kinematic Viscosity (cSt)
1	2	2,54
2	2	2,52
3	2	2,51
4	3,5	4,19
5	3	3,59
6	3	3,53
7	3	3,69
8	2,8	3,44
9	3	3,54

Based on the table 5 The effect of reaction time on the viscosity of biodiesel produced fluctuates to form an up and down pattern caused by the nature of the reversible transesterification reaction. While the effect of stirring speed on the viscosity of biodiesel is likely not to affect the viscosity produced because the density of biodiesel produced is not too much different. The viscosity value of the biodiesel produced meets the SNI 7182: 2015 biodiesel standard (2.3 - 6.0 cSt).

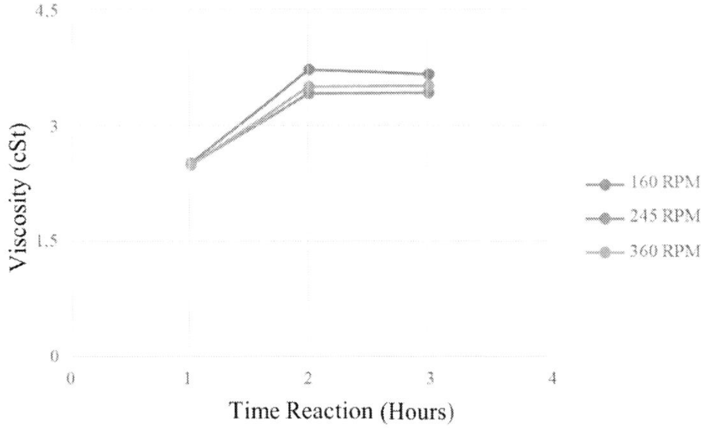

Figure 4. Time reaction and RPM effect to viscosity value

3.3.3 Water Content Analysis

Reaction time affects the water content in biodiesel, the longer the reaction time the smaller the water content contained in biodiesel. Presented in figure 5, while the stirring speed does not significantly affect the biodiesel produced. The biodiesel water content is more influenced by the water content in the raw material, the accuracy of the washing process in the separating funnel, and the process of removing the water content in the biodiesel purification process. Based on measurements of water content of samples 7.8, and 9 have met the SNI 7182: 2015 biodiesel standard (max 0.05). However, samples 1,2,3,4,5, and 6 have not met.

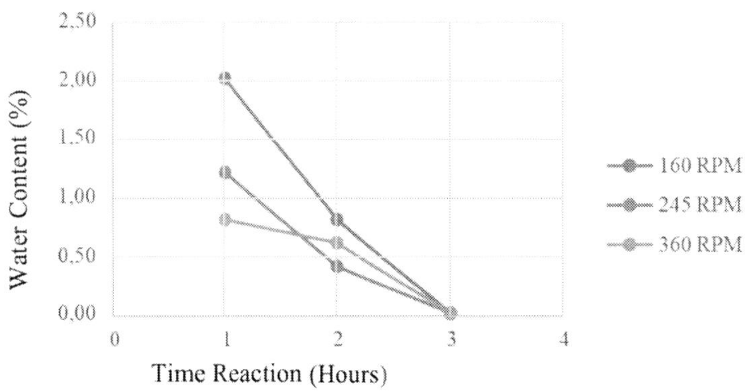

Figure 5. Time reaction and rpm effect to water content

3.3.4 Caloric Value Analysis

The difference in the heating value of each biodiesel treatment but not too significant. The heating value is influenced by the density and composition of the fatty acid constituents. The heating value for Samples 4,5,6, and 9 has fulfilled the SNI 7182: 2015 biodiesel standard (8,880 cal / gr),

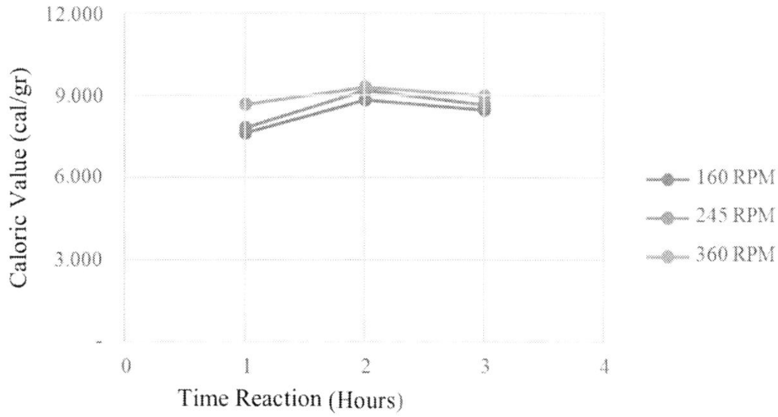

Figure 6. Time reaction and RPM effect to caloric value

3.3.5 Flash Point Analysis

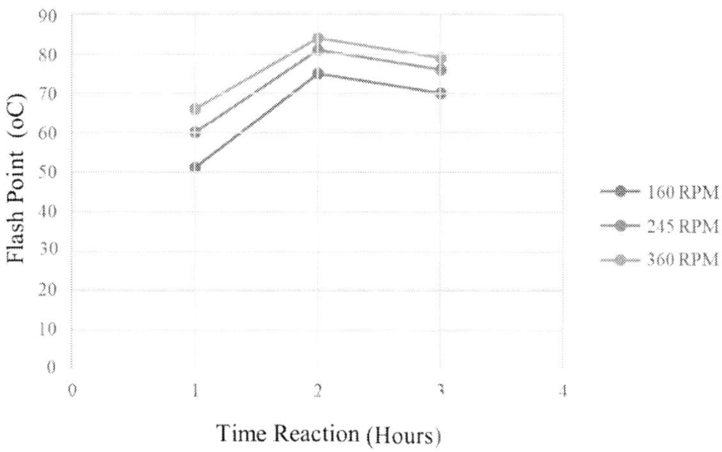

Figure 7. Time reaction and RPM effect to flash point

Based on figure 7, it can be seen that the reaction time fluctuates to the flash point of biodiesel produced due to transesterification reactions that can be reversed when the reaction experiences equilibrium. While the stirring speed influences the flash point of the biodiesel produced. The flash point value produced by the biodiesel does not meet the standards of biodiesel SNI 7182: 2015 (min 100 °C).

3.3.6 Acid Number Analysis

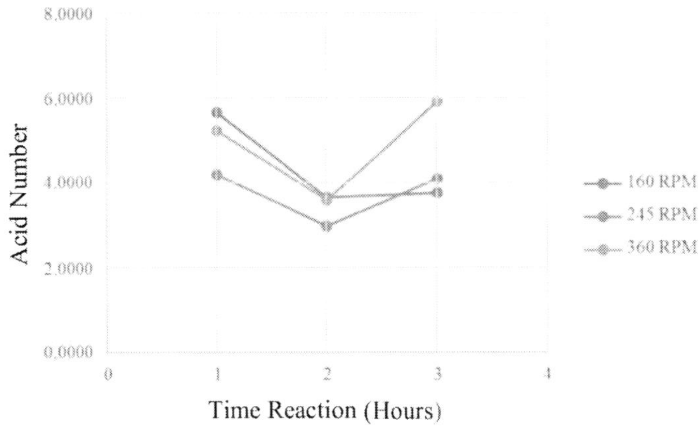

Figure 8. Time reaction and RPM effect to flash point

Based on the figure 8, it can be seen that the reaction time fluctuates with the number of biodiesel acids produced because the transesterification reaction can be reversed when the reaction is in equilibrium. While the stirring speed has no effect on the number of biodiesel acids produced. Acid numbers are influenced by the composition of fatty acid constituents of raw materials. The value of acid numbers produced by biodiesel does not meet biodiesel standards SNI 7182:2015 (max 0.5).

4. Conclusion

Based on the research that has been done, the following conclusions can be drawn :

1. The level of Free Fatty Acid (FFA) contained in the bio-oil raw material from Tuna fish waste is quite high at 33.84%.
2. Biodiesel from Tuna fish bio-oil cannot be used in diesel engines because the acid level of the raw material does not meet the SNI 7182: 2015 biodiesel standard and Pertamina's diesel standard which can damage the engine. Utilization of Bio-oil as a mixture of making biodiesel fuels can improve the quality characteristics of methanol as fuel.
3. The optium treatment in this study was 2 hours, 360 RPM.

Acknowledgments

This research was supported by the Research of Lecturer of Beginners (PDM), organized by the ministry of research and technology (Kemenristekdikti), through a research faculty of engineering in 2020.

References

[1] Food And Agriculture Organization Of The United Nations 2018 The state of world fisheries and aquaculture ISBN 978-92-5-109185-2.

[2] Batungbacal, Ephraim Dan Mitchell, James 2018 From sea to can 2018 southeast asia canned tuna ranking *GREENPEACE*

[3] Tri Nugroho 2010 Pemanfaatan minyak ikan untuk produksi biodiesel squalen

[4] Hernanda, Sukiman, Dkk 2014 Pembuatan biodiesel dari limbah ikan baung dengan katalis padat lempu *Jurnal Online Mahasiswa (JOM) Bidang Teknik Dan Sains* **1**

[5] Moeksin, Rosdiana, Dkk. 2016. "Pengaruh Jumlah Katalis Dan Waktu Reaksi Terhadap Produksi Biodiesel Dari Limbah Pangasius Hypothalamus".

[6] Arpiwi, Ni Luh 2015 Diktat kuliah bioenergi: biodiesel dan Bioetanol *Prodi Biologi Fakultas Matematika Dan Ilmu Pengetahuan Alam Universitas Udayana*

[7] Hikmah, M.N., Zuliyana 2010 Pembuatan metal ester (biodiesel) dari minyak dedak dan metanol dengan proses esterifikasi dan transesterifikasi *Skripsi* Semarang Universitas Diponegoro.

IOP Conf. Series: Materials Science and Engineering **909** (2020) 012030 doi:10.1088/1757-899X/909/1/012030

The study of corrosion behaviour api 5l steel in sea water using immersion test method

I Saefuloh[*1], N Kanani[2], Y Rukmayadi[1], M I Noor[1]. Y Yusuf[1], I Setiawan[1], S ula[1], S Abdullah[1], H A Notonegoro[1].

[1]Mechanical Engineering Department, Universitas Sultan Ageng Tirtayasa, Banten, Indonesia

[2]Chemical Engineering Department, Universitas Sultan Ageng Tirtayasa, Banten, Indonesia

E-mail: iman.saefuloh@untirta.ac.id

Abstract. API 5L steel, a low carbon steel with a high alloy content, is often used as an underwater construction material or oil and gas pipeline. Banten has a wide sea area and has a long coastline where part of the coastline is used for petrochemical industries. There are three coast areas of Banten, namely Bayah Rangkasbitung, Anyer and Merak, that are used for petrochemical industries. Most of those industries use API 5L steel for underwater construction materials and pipes to channel seawater, chemical compounds, gas and oil. Whereas, seawater will greatly affect the lifetime of API 5L construction steel because it can cause corrosion. The purpose of this study is to analyze the corrosion characteristics of API 5L steel affected by the environment of seawater media and the temperature of seawater in accordance with its natural conditions. The samples of API 5L steel were immersed in seawater media with variation of immersion time of 5, 10, 15 Days and variation of immersion temperature of 20°C, 40°C, and 60°C. In this study, a series of tests were carried out, namely immersion test, weight loss test, and SEM test (Scanning Electron Microscopy). The highest corrosion rate on API 5L steel is 0.39 mmpy which belongs to Merak's seawater media in the immersion time of 15 Days and the temperature of 60°C, while the lowest corrosion rate on the API 5L steel is 0.013 mmpy which belongs to Bayah's seawater media with a time of immersion of 10 Days and a temperature of 20°C.

1. Introduction

Corrosion is a form of metal damage caused by metal contact with the surrounding environment [1]. Corrosion, in other definition, is a damage or degradation of metal due to redox reactions between a metal and various substances in its environment that produce compounds that are not desired [2]. Almost all industrial sectors have problems with corrosion, for example; the metal industry sector, the transportation industry, the mining and energy industry, public works, and the agricultural industry [3]. Problems that arise can be in the form of damage, the life of goods that do not meet expectations to the inadequate safety factor, and a damage to equipment or a failure in operations that cause no small losses [1-3].

Banten has a large sea area with a long coastline where part of the coastline is used for petrochemical industries. There are 3 areas of Banten that are used for petrochemical industries, namely; Bayah (Indian Ocean), Anyer (Indian Ocean), and Merak (Sunda Strait). Most of these industries use API 5L steel as underwater construction material and pipes to channel seawater, chemical compounds, gas and oil. Sea water will greatly affect the service life of API 5L steel because it can cause corrosion [4-5].

Many factors can cause material corrosion, one of them is the influence of the concentration of corrosion media [1,3]. The concentration of solute in a solution, which causes more and more free ions in the solution. In seawater media, chloride ions become a source of problems to corroded materials [4-6]. The more concentration of chloride ions the faster corrosion occurs to the materials. In addition, the existence of bacterial chloride ions in seawater can also degrade heavy metals due to bacterial corrosion or bio-corrosion. Besides, the different temperatures between the surface of the middle and in the sea are catalysts that can accelerate the corrosion of AISI API 5L steel material [7-9].

Nowadays, many studies have been conducted in relation with the corrosion rate of API 5L steel affected by chemical compounds, the media of gas corrosion, oil and sea water, for example; the corrosion which is affected by chemical compounds such as; sulfuric acid and sodium chloride [10-12], both of these compounds are strong electrolytes. The study conducted by showed that the constant interaction between the material and its environment can cause a decrease in the value of the material both slowly and significantly [4-5].

According to the reason above, it is interested to examine the corrosion rate of API 5L steel which is widely used in construction and oil and gas pipelines using seawater media taken from various industrial areas in Banten. In this research, temperature and immersion time in the sea water will be added to calculate the corrosion rate in API 5L steel, bearing in mind that the implementation of this research is in offshore building construction. It is hoped that this research can provide information about the effect of seawater in various areas on the corrosion rate of the material most widely used for the petrochemical industry

2. Method

This research was conducted in 2 stages. The first stage involved the process of preparation of API 5L Steel samples. The process included:

1. Sample Cutting
   In this study, API 5L steel was cut with dimensions of 30 mm x 20 mm x 2 mm
2. Sanding Samples
   In this process, the samples were sanded to lose the metal oxides on the surface of the sample. The sanding process was carried out by using sandpaper with grid roughness values of 180. 300 and 800.
3. Sample Cleanup
   The samples were dipped in liquid acetone to clean the impurities that stuck on the steel, then the samples were washed using liquid distilled water.

The second stage involved the weight loss test method which aimed to determine the corrosion rate (mmpy) of the samples by weighing the initial weight and final weight after the test [13]. The steps of weight loss testing were as follows:

1. Measuring the dimensions of the samples.
2. Weighing the samples to find out the initial weight of the samples.
3. Immersing the samples into corrosion media in the form of sea water with the provisions of immersion time 5, 10, and 15 days at a temperature of 20oC, 40oC and 60oC.
4. The samples, that had been immersed, were then removed from the liquid. Afterwards, the samples were dried out and weighed to get their final weight which aimed to determine the corrosion rate

3. Results and Discussion

3.1. The Characteristic of Corrosion (chemical and biological content) of each water area media.

The results of this corrosion rate test can be analyzed in the seawater as the corrosion media which is divided into 3 different types of seawater, namely Bayah Rangkasbitung seawater (Indian Ocean), Anyer seawater (Indian Ocean) and Merak seawater (Sunda Strait), by using the variations of the temperature of each seawater. Based on the results of the calculation of the corrosion rate, the results of the corrosion rate for time variations of 5, 10, 15 days and temperature variations of 20°C, 40°C, 60°C can be seen in the following graphic image:

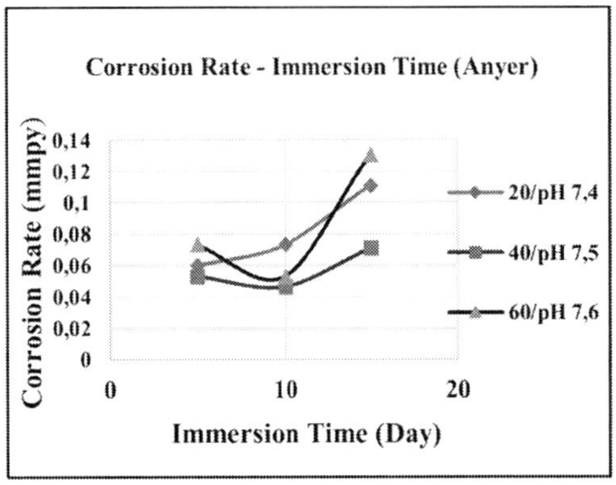

(a)

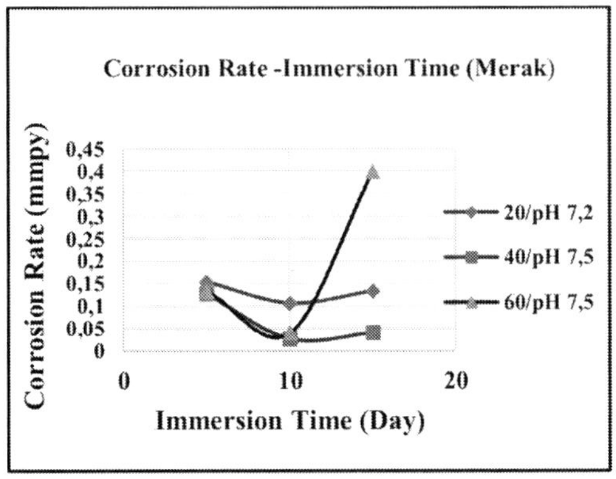

(b)

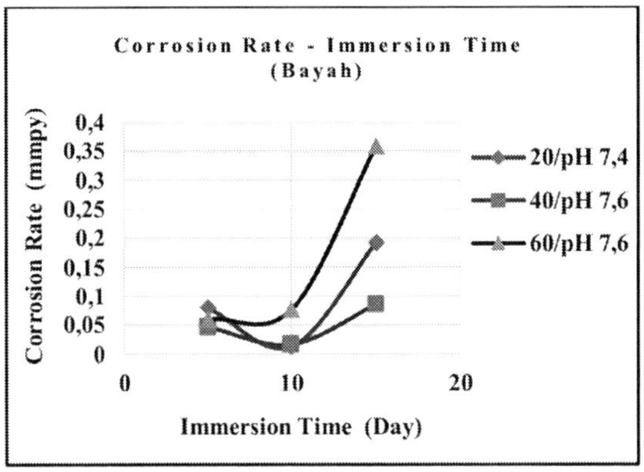

(c)

Figure 1. Relationship Among Corrosion rate and immersion time and temperature variation to API 5L steel corrosion rate (a) Anyer seawater (Indian Ocean), (b) Merak seawater (Sunda Strait), and (c) Bayah seawater (Indian Ocean).

The picture above shows the value of the corrosion rate (mmpy) of each test sample material as well as for each variation of the test. The corrosion rate values in the seawater immersion media has obtained the data that the value of the corrosion rate has a tendency to rise in line with the increase of each temperature and each additional time of immersion.

According to the seawater temperature variation, The highest rate of corrosion, that has a value of 0.392 mmpy, occurs to Merak seawater (Sunda Trait) at the temperature of 60 ° C and the lowest rate of corrosion, which is 0.136 mmpy, occurs to Merak seawater (Sunda Strait) at the temperature of 20° C. This is because the temperature is an important factor in the chemical reaction process and in the corrosion rate of the material. Moreover, the increase in temperature will also increase the rate of corrosion due to the higher temperature will accelerate the diffusion of oxygen through the cathodic layer of oxide formed. If the temperature rise higher, it will cause the particles to move more actively so that collisions between particles also occur frequently and cause the rate chemical reactions are getting faster, the percentage increase in the rate of corrosion that is affected by the temperature reaches 60%. For this reason, it can be concluded that the higher the temperature, the higher the corrosion rate.

Based on the immersion time variation, the highest corrosion rate, which has a value of 0.392 mmpy, occurs to Merak seawater (Sunda Trait) with the immersion time of 15 days and the lowest corrosion rate occurs to Anyer seawater (Indian Ocean) with the immersion time of 5 days (0.136 mmpy). This is due to the increased surface activation level of API 5L steel where activation is the process of forming metal oxide compounds on the metal surface. The rate of corrosion increases because metal oxides react more actively with the metal surface. In addition, the rate of chemical reaction between the metal surface and seawater will increase slowly so that the corrosion rate on the material also rises, the percentage increase in the corrosion rate which is affected by the immersion time also reaches 60%. For the influence of the immersion time, it can be concluded that the longer the immersion time, the more corrosion rate will increase.

3.2. The Comparison of the Corrosion Rate of Three Areas (Bayah, Anyer, and Merak Seawater)

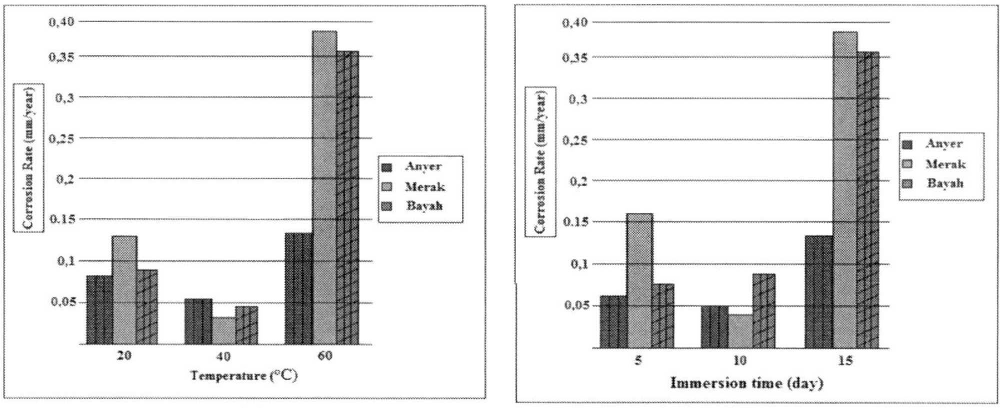

(a) (b)

Figure 2. Comparison graph of the corrosion rates of the three areas (a) temperature (b) immersion time

Based on the figure above, it can be analyzed that Merak Seawater is more corrosive compared to the two regions (Anyer and Bayah). This is because Merak sweater has higher concentration of sea chloride ion than two others. The chloride ion content can accelerate corrosion reactions on metal surfaces. Moreover, Merak seawater contains a lot of bacteria or microbiology which can cause API 5L steel surface to degrade due to microbiological corrosion or bacterial corrosion.

3.3. The Observation of API 5L Steel Surface using SEM (Scanning Electron Microscopy)

Based on observations of microstructure using scanning electron microscopy (SEM) shown in Figure 3, the corrosion forms that occur on the surface of API 5L steel from the three areas are the same, namely uniform corrosion

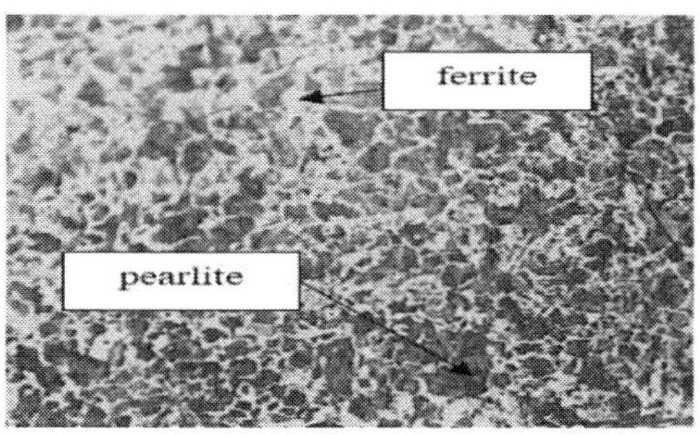

(a)

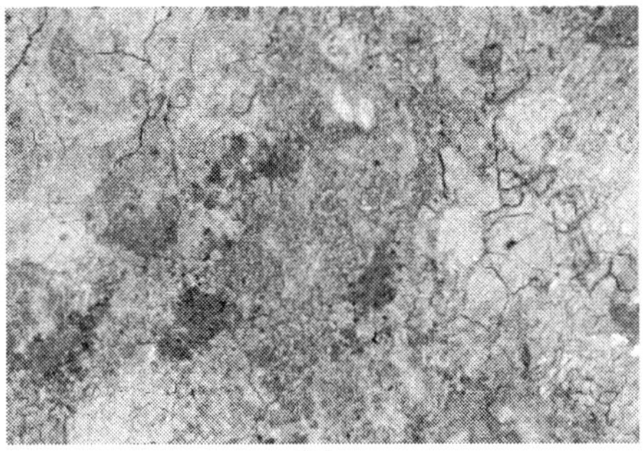

(b) (c)

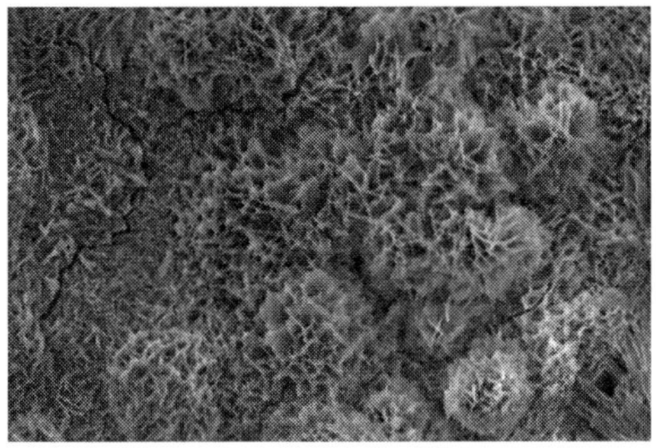

(d) (e)

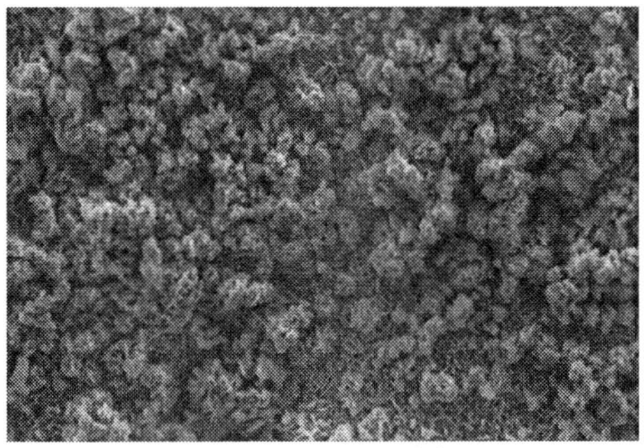

(f) (g)

Figure 3. Scanning Electron Microscopy (SEM) Testing (a). 1000x magnification before immersion, 100x magnification at temperature of 60°C (b) Anyer Seawater, (d) Merak Seawater and (f) Bayah Seawater and 1000x magnification at temperature of 60°C (c) Anyer Seawater (e) Merak Seawater (f) Bayah Seawater.

Figure 3 (c) shows the form or pattern of corrosion that occurs for types of seawater originating from Merak. The figure resembles the shape of sea corals due to the corrosion reaction that occurs between the surface of API 5L steel is dominantly in the form of bacterial corrosion or microbiology corrosion, but for the two other areas (Anyer and bayah) which are indicated by figures 3 (c) and (g), the corrosion occurs because of the interactions between the acidic chloride ion compounds contained in seawater and the metal surface.

4. Conclusion

Based on the results of testing and analysis it can be concluded that the temperature and immersion time affect the rate of corrosion. The higher the temperature, the higher the corrosion rate will occurs, and the longer the immersion time, the higher the corrosion rate will have. Based on the comparison of corrosion rates in each area, seawater originating from Merak (Sunda Strait) is more corrosive compared to seawater originating from Anyer and Bayah (Indian Ocean). The types of corrosion that occur in both processes and the three regions are the same, namely uniform corrosion.

5. Acknowledgements
The authors wish to thank for the financial support provided through the research grant scheme by from Directorate of Higher Education, Ministry Education and Culture of the Republic of Indonesia and Universitas Sultan Ageng Tirtayasa.

Reference

[1] Roberge, P. R. 2000 *Handbook of Corrosion Engineering* (McGraw-Hill, New York) 333-351, 1093.
[2] ASM International, 2003. ASM Metals Handbook *Corrosion: Fundamentals Testing and Protection* United State of America 13A.
[3] Jones, D. A. 1996 Principles and Prevention of Corrosion Second Edition (Pretice Hall Inc. Unites States of America) 34 – 390.
[4] ShaojunYuan, et al. 2013 Surface chemistry and corrosion behaviour of 304 stainless steel in simulated seawater containing inorganic sulphide and sulphate-reducing bacteria *Corrosion Science* **74** 353-366
[5] Abd Alrahim Al Shikshak, et al. 2015 Effect of flow velocity of sea water on corrosion rate of low carbon steel *Applied Mechanics and Materials* **799-800** 232-236.
[6] MS Paula, et al. 2015 Carbon steel corrosion induced by sulphate-reducing bacteria in artificial seawater: electrochemical and morphological characterizations *Revista Matéria* **21 (4)** 987-995
[7] S Paul, A Pattanayak, SK Guchhait 2014 Corrosion behavior of carbon steel in synthetically produced oil field seawater *International Journal of Metals* 1-11.
[8] A Abdullah, N Yahaya, et al. 2014 Microbial corrosion of API 5L X-70 carbon steel by ATCC 7757 and consortium of sulfate-reducing bacteria *Journal of Chemistry* 1-7.
[9] L.M. Quej-Aké, A. Contreras and J. Aburto 2018 Electrochemical Study on Corrosion Inhibition of X52 Steel by Non-ionic Surfactant in Substitute Ocean Water, *Int. J. Electrochem*. Sci. **13** 7416–7431
[10] Yunan Prawoto, WB wan nik et al. 2009 Effect of pH and Chloride Concentration on The Corrosion of Duplex Stainless Steel *The Arabian Journal for Science and Engineering* **34 (2C)** 116-127.
[11] Iman Saefuloh, Nufus Kanani, et al. 2020 The Corrosion Inhibition of API 5L Steel Using Natrium Acetate and Natrium Nitrite in Natrium Chloride Acid Solution *Materials Science Forum* **988** 3-10.
[12] Iman Saefuloh, Nufus Kanani, et al. 2020 The Study of Corrosion Behavior and Hardness of AISI Stainless Steel 304 in Concentration of Chloride Acid Solution and Temperature Variations *Journal of Physics: Conference Series* 1477, 052058.
[13] ASTM Internasional. 2004 ASTM G31 – 72: *Standard Practice for Laboratory Immersion Corrosion Testing of Metals,* Philadelphia 5 .

IOP Conf. Series: Materials Science and Engineering **909** (2020) 012031 doi:10.1088/1757-899X/909/1/012031

Numerical and experimental analysis of drag force in medium speed train design

Jean Mario*[1], Beny Halfina[2], Dimas Bahtera[3], Lukman Shalahuddin[4], Agus Windharto[5]

[1,2,3,4] Agency for Assessment and Application of Technology - BPPT, Indonesia
[5] Center for Science and Technology – PUI ITS, Surabaya, Indonesia

*Email: jean.mario@bppt.go.id

Abstract. Aerodynamic drag is the highest portion of the overall resistance force in a high speed train. More streamlined design will have lower coefficient of drag. Therefore, improving the mask of car and carbody is essential to minimise the drag force. This study aims to obtain the value of the drag coefficient on the design of Indonesia's Medium Speed Train which has an operational speed of 160 km/h by the numerical method and then to validate it with the wind tunnel experimental test. In the initial stage, two designs was developed using a 3D CAD software. Subsequently, a CFD simulation was performed to determine the value of the drag coefficient. Two test models were built with a scale of 1:25, then both were tested in a wind tunnel. Based on the test results, the coefficient of drag of both models have relatively similar value. The graph data from the CFD simulation results then compared to the wind tunnel results. In addition, the wind speed variations was conducted to observe the effect of the Reynolds number on the drag coefficient.

1. Introduction

This research was conducted to support the Jakarta-Surabaya Corridor Rail Speed Improvement Project. The results of the feasibility study are the implementation of the Jakarta-Surabaya semi- high-speed train (Medium Speed Train). It would cut the currently 9 hours travel time to only 5.5 hours with an operational speed of 160 km/hour. In fact, train speed in Indonesia is limited by the infrastructure issue, which is narrow gauge of 1067 mm. This condition could lead the possibility of derailment due to instability of train motion. Despite the shorter travel time, speed increase affects the aerodynamics of the train. It is imperative to observe the induced drag, noise generated by the aerodynamic force, and vibrations generated from the wind (Cui, Wang, Hu, Gao, & Yang, 2012). Due to the movement of the train, there is pressure difference between the front area or Mask of Car and the tail end of the train which will cause the drag force (Banga, Zunaid, Ansari, Sharma, & Dungriyal, 2015). On previous research, the air flow velocity in the wind tunnel was determined after observing the Reynolds number consequences on the coefficient of drag (Zhang, Yang, & Liang, 2018). Others, the result between physical and computational modelling techniques, whilst predicting the average value of the force coefficient can be accounted as the approached method and techniques of modelling is close to actual conditions (Gallagher et al., 2018).

To achieve minimum drag resistance, the Mask of car and carbody of the train are the main component that must have good aerodynamic performance. The basic tools in the study of train aerodynamics are CFD calculation and wind tunnel test. The numerical CFD simulation is beneficial to cut down the modelling cost and reduce the time of the model development. It enables the design team to do rapid testing of the drag coefficient of the model and provides guidance for the iterative method of refining the 3D model to get the optimum coefficient of drag. Moreover, the CFD simulation need to be validated using wind tunnel test as the proven method who have been adopted by the aerodynamic experts to assess the train aerodynamic characteristics. In this study, wind tunnel test was conducted to obtain the aerodynamic coefficient, particularly the coefficient of drag. In the case of the study of train aerodynamics, all of these approaches are challenged with difficulties (Baker, 2010). This study aims to determine the value of the drag coefficient on the design of Indonesia's Medium Speed Train (MST) which has an operational speed of 160 km/h both numerically and experimentally.

2. Method

2.1. 3D CAD Model Preparation

The initial stage of this research is to prepare the 3D CAD Model of MST based on the required specification. The fundamental idea in designing MST that has good aerodynamics is the form of a streamlined mask of car and carbody that will result in easier freestream flow from the upstream to the downstream. The slender shape of the train nose has been used in the modern high-speed train to decrease in the drag that occurs on the body which will increase the aerodynamic efficiency of the train body. However, with the main propulsion being used is a diesel electric which requires a larger space, the mask of car train has limited dimensions to be built with longer nose shape. In this research, there are two different models of MST, Model A and Model B which have different profile of mask of car as shown in Figure 1 and Figure 2.

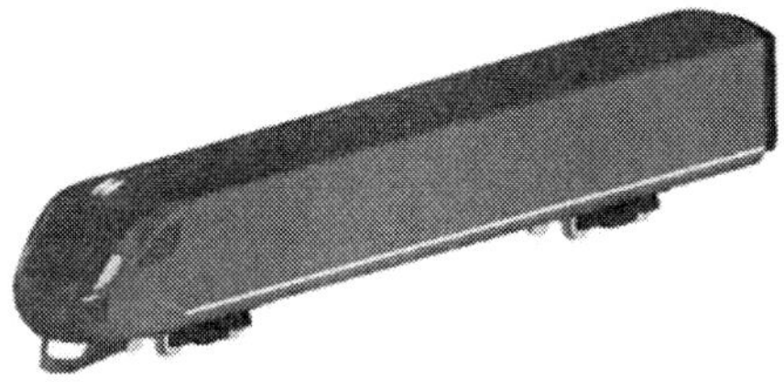

Figure 1. 3D CAD MST model A

Figure 2. 3D CAD MST model B

Figure 3 shows the configuration of Medium Speed Train model for the wind tunnel test which has 3 cars. This configuration allows one to observe the effect of air flow in the head, the carbody and after leaving the tail end of the MST.

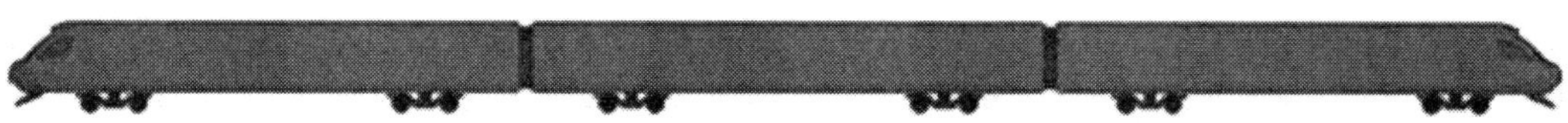

Figure 3. Configuration of medium speed train model for wind tunnel test

2.2. CFD Preparation

Following the design stage in a 3D CAD model, numerical simulations are performed using CFD. From this step, the governing equations which is consist of mass conservation, momentum equation and *k*-e standard turbulent model is solved using computational method (Utomo, Fajar, & Arpriyanto, 2018). However, since the simulation based on a theoretical calculation, the accurate result will always depend on the correct configuration and simulation set up. Based on the actual operating condition, the simulation setup was built. For each model, the setup is similar as illustrated in Figure 4 below.

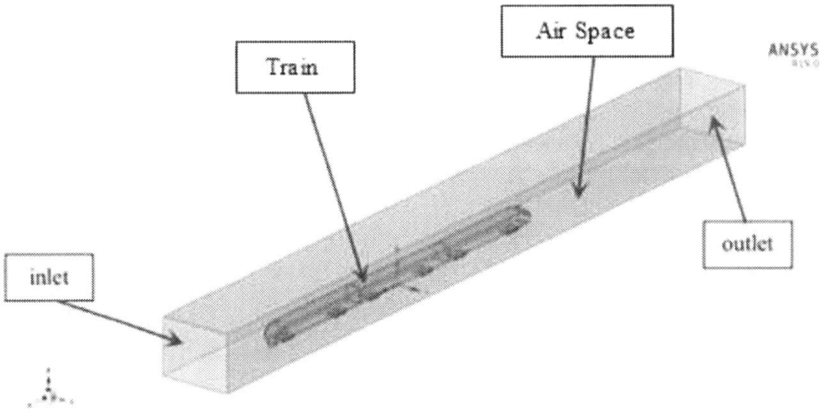

Figure 4. Computational domain

Figure 4 illustrates the MST model within the computational domain. Boundary condition of the front surface of the train is set as velocity inlet and the downward surface behind the train is set as pressure outlet. The boundary type of boundary condition of frontal surface is set as inlet and has distance of 20,000 mm from the frontal train nose, while the pressure outlet downward the tail end of the MST model is set as 50,000 mm. The width of the enclosure is set to be 5,000 mm from the outer edge of the MST, while the height is set to be 10,000 mm. The space between ground and bogie of the MST is set to be 500 mm. The size of the domain was determined to meet the actual size of the wind tunnel test chamber. Thus, the computational domain is assumed to have similar size and condition with wind tunnel.

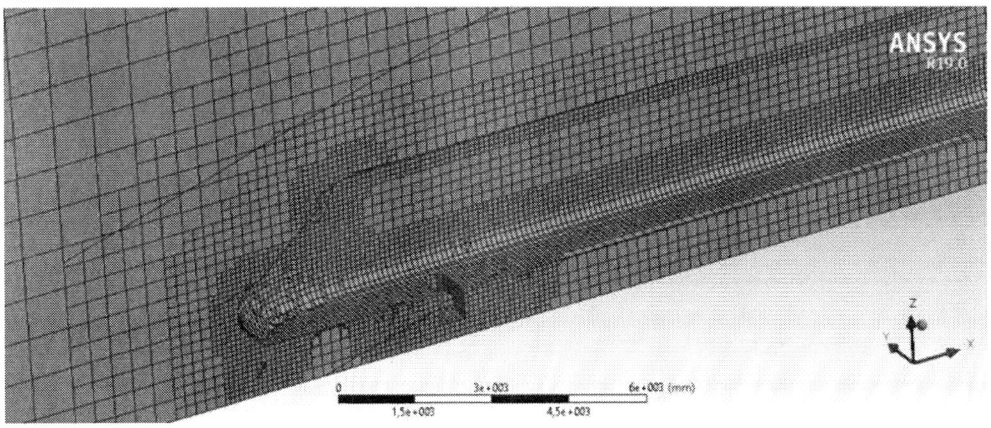

Figure 5. Meshing of the air space surrounding the model

Figure 5 shows that default set of meshing was used for wind tunnel area due to the simplification of simulation which is for Model A has maximum aspect ratio 3.303057e.+01 with number of nodes 577,294 and elements 502,455 while Model B has maximum aspect ratio 4.50723e.+01 with number of nodes 660,834 and elements 566,649. For detailed analysis needs, the meshing was fined specifically in the mask of car area which was set to 100 mm. Then the boundary conditions parameters determined as shown in Table 1:

Table 1. Boundary conditions properties

Parameters	Value
Wind Speed	44 m/s
Method	3 Dimension
Solver type	Pressure based
Model viscous	k-epsilon SST

In the boundary parameters, the wind speed was set at 160 km/hour or 44 m/s and has no increment considering the Reynolds Number check in wind tunnel which will be discussed on the result and discussion, density of air is 1.225 kg/m^3 and reference area is 9 m^2 based on the geometry of mask of car of MST model. Thus, the CFD simulation determined drag coefficient as the value of air resistance of MST which is calculated using the following equation:

$$C_D = \frac{F_D}{\frac{1}{2}\rho U^2 A} \quad (1)$$

This coefficient of drag is influenced by the Reynolds numbers (Re). Reynolds numbers are a function of wind speed which is calculated with the formula below:

$$N_{Re} = \frac{\rho v d}{\mu} \quad (2)$$

2.3. Wind Tunnel Model Preparation
It is important to always validate the result of numerical simulation with the actual condition. Thus, in this study, the wind tunnel test was used to validate the results. The Experiments in wind tunnels are a reliable method and have long been used by aerodynamics experts to study the aerodynamic characteristics of trains. The trainset was modelled with a certain scale adjusting the wind tunnel test section. Several wind flow velocities were blown into the chamber. A railroad study in a wind tunnel is conducted to find the aerodynamic coefficient values of the model, especially the coefficient of drag. In addition to finding the coefficient of drag, the pattern of air flow which forms around the body of the model can be determined by visualizing the flow. The test facility is the Indonesian Low Speed Tunnel (ILST) as shown in Figure 6 equipped with 6 component internal balance, data acquisition and processing system, and a set of train models.

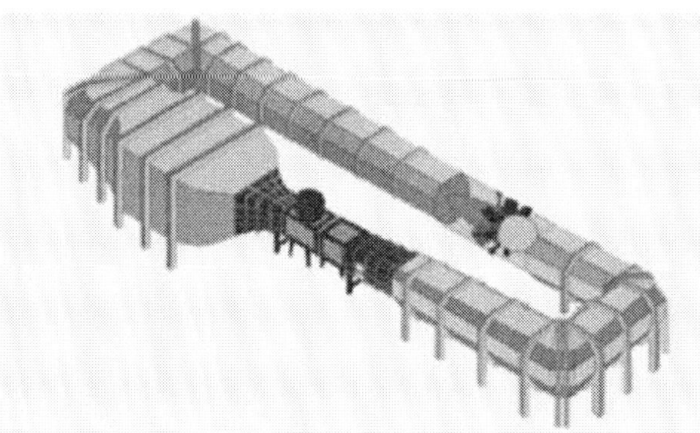
Figure 6. Indonesian low speed tunnel (ILST)

Figure 7. ILST test section

ILST is a wind tunnel with a closed test section and has an atmospheric type. The test section has a rectangular shape and measures 4 m x 3 m x 8 m as shown in Figure 7. On the left and right sides of the end of the test section there are air holes to neutralize the air pressure in the test section to the conditions of standard air pressure (standard atmosphere).

The model tested in the ILST wind tunnel is a set of train models that are scaled 1:25 with dimensions of length 2800 mm, width 120 mm and height 120 mm. A set of models consists of three train cars namely head car, middle car and tail car which are equipped with central strut system. The central strut connects the train model to the internal balance mounted under the ILST turntable. In addition, there is an additional obstacle barrier feature (cowcatcher). At full scale conditions, the train is planned to operate at speeds of 160 km/h or around 44 m/s. The configuration of wind tunnel test described in Figure 8.

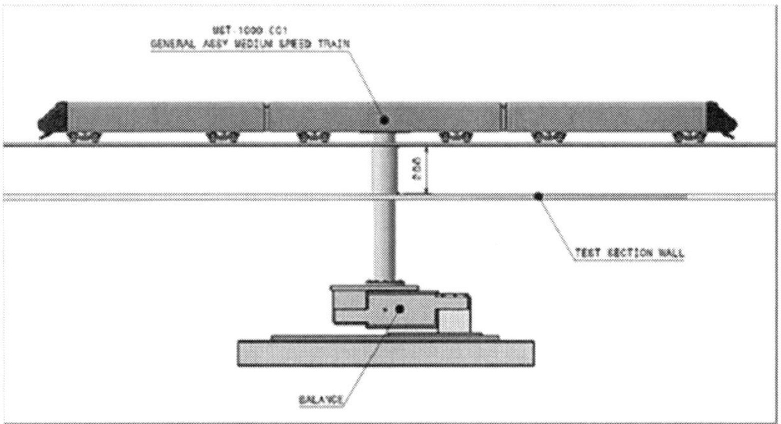

Figure 8. Design configuration for wind tunnel test

Sequences of testing the model in a wind tunnel include internal balance check load, ground board and model installation, flow checking on the ground board around the model, checking Reynolds numbers, measurement of drag coefficient on the model, flow visualization on the model. Detailed test configurations are listed in the test run log. The actual model setup inside the wind tunnel is shown in Figure 9, Figure 10, and Figure 11.

Figure 9. Model A

Figure 10. Model B

Figure 11. MST model on test section

3. Result & Discussion

3.1. CFD Result

CFD simulation of the two MST configurations was run until achieving convergence, the drag force and coefficient of drag result is shown in Table 2 below. Parameters in the numerical simulation using CFD are wind speed 160 km/hour or 44 m/s, density of air 1.225 kg/m^3 and reference area is 9 m^2. After the simulation reached convergence, the drag force result is 9797. 41 N with the coefficient of drag (CD) is 0.902.

Table 2. Result of model A simulation

Train Configuration	Velocity (m/s)	Density of Air (kg/m^3)	Reference Area (m^2)	Drag Force (N)	Coefficient of Drag
Single Car	44.4	1.225	9	7029.63	0.647
Triple Cars	44.4	1.225	9	9797.41	0.902

Table 3. Result of model B simulation

Train Configuration	Velocity (m/s)	Density of Air (kg/m^3)	Reference Area (m^2)	Drag Force (N)	Coefficient of Drag
Single Car	44.4	1.225	9	5981.81	0.550
Triple Cars	44.4	1.225	9	8520.30	0.784

Table 2. show the two model configurations of Model A which resulted different drag force and drag coefficient depending on the number of cars. Drag force of single car has a value 7029.63 N, while the triple cars have higher value in the similar condition. It is proportional to the drag coefficient value from both of configurations. Single car attained a value 0.647 and triple cars reached 0.902.

Table 3 illustrates two model configurations of Model B which resulted similar trend of drag force and drag coefficient value. The single car of Model B has drag force and drag coefficient sequentially 5981.81 N and 0.550. While the triple cars have drag force and drag coefficient sequentially 8520.3 N and 0.784. This phenomenon reveals that the number of cars has influence to drag force and coefficient of drag, which mean length of train affect the downstream and upstream air flow.

Air velocity distribution and pressure distribution along the trainset are shown in the series of figures below. Figure 12 and Figure 13 show the contour plot for velocity in model A and B while pressure distribution are shown in Figure 14 and Figure 15. The nose region of the MST has the high-pressure area. While relatively lower pressure shown in the window of the train. As the train reached maximum speed the pressure drag escalated and so does the drag force. High-pressure region is also observed at the gaps between cars contributing the total drag resistance of the trainset. Previous research (Utomo et al., 2018) mentioned that the total drag coefficient in the high-speed train consists of pressure drag coefficient and viscous drag coefficient. However, in the case of a train the viscous drag effect is larger than the pressure due to the geometry of the train which has very long body compared to width or height. From this simulation, the designer can get the insight to improve the design of the Mask of Car. Optimized shape design will improve the aerodynamic performance thus reducing the pressure drag and total drag force which eventually reducing fuel consumption.

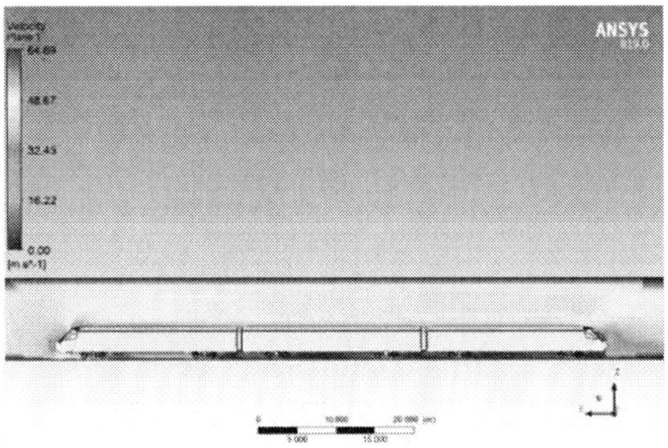

Figure 12. Air velocity distribution model A

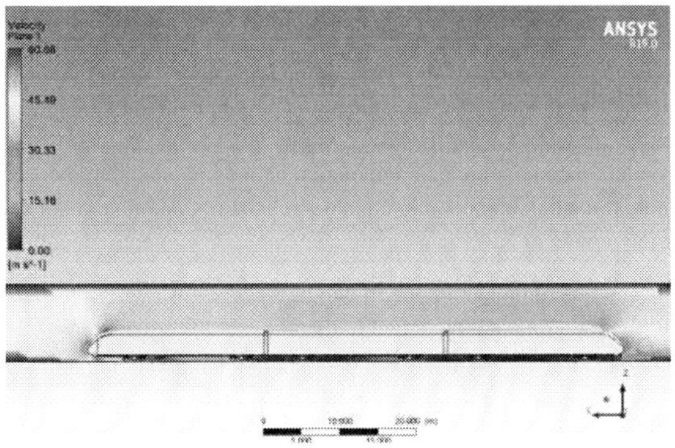

Figure 13. Air velocity distribution model B

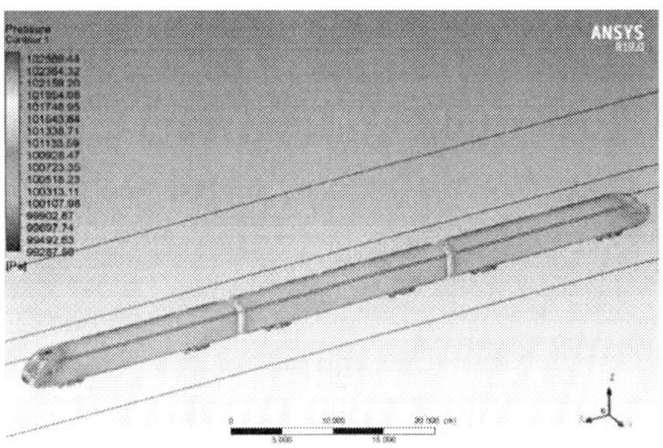

Figure 14. Pressure distribution model A

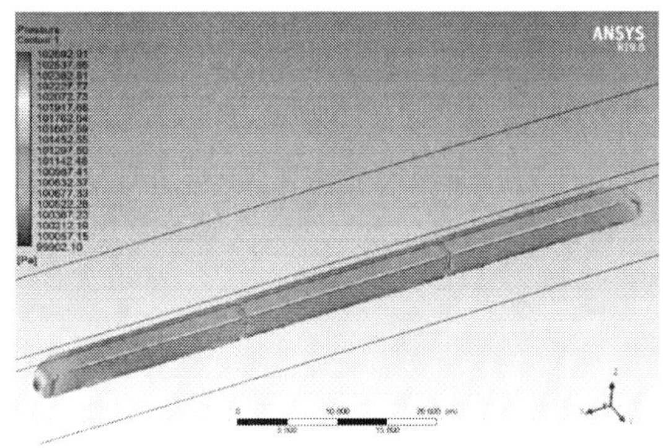

Figure 15. Pressure distribution model B

More detail image on the difference in the vortices around the train model A and B are shown in Figure 16 and Figure 17 which explain why there is a difference between the aerodynamic forces. Due to more complex profile at the area of windshield in model A, there are more vortices around the windshield and under the nose. Since vortex drag is the main component of aerodynamic drag, the aerodynamic drag of the model A is bigger than those in model B.

The velocity contour shows a blue region at the nose of the train which indicates that when the air strikes the train its velocity rapidly decrease or became stagnant. The air particle gets denser thus result in the higher pressure. According to (Banga et al., 2015), air has a tendency travelling upwards from the high pressure region the low-pressure region on top. Contrast to the nose area, at the top of the train head or above the windshield the velocity contour shown red region which means that the air velocity increases more than 60 m/s.

Figure 16. Model A velocity streamline

Figure 17. Model B velocity streamline

In the CFD simulation, a pressure-based solver in Fluent was used for the numerical simulation. In the control volumes of the cells, the pressure gradients were calculated. Surface pressure on the train body was mainly concentrated at the front and rear of the train. The plot contour compares the pressure distribution of the train. There is different pressure contour due to the frontal shape of the train as shown in Figure 18 and Figure 19. Model A with relatively complex shape and flat surface of the windshield results in the negative pressure. While in model B, the pressure is only at the tip of the nose.

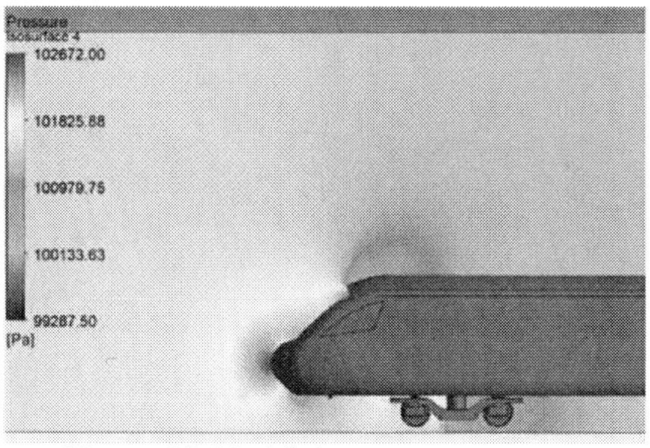

Figure 18. Model A pressure contour

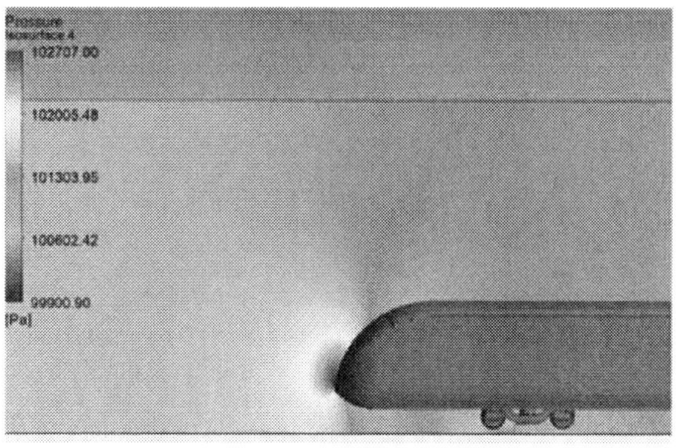

Figure 19. Model B pressure contour

3.2. *Wind Tunnel Result*

In the wind tunnel test of MST models, it was conducted the Reynolds Number check which is needed to review the dependence of air velocity to coefficient of drag. In the wind tunnel, the static ground is not like the actual operating condition. There is a boundary layer formed at the static ground affecting the Reynolds Number (Chen, Yao, Lin, & Han, 2017). Theoretically, aerodynamic coefficient value is similar between model and prototype. This coefficient is influenced by Reynolds Number which is the function of air velocity (the formula is shown as Eq. 2). In this wind tunnel test, the aerodynamic coefficient value can be used following the air velocity has slightly affect the coefficient of drag. It means that varying air velocity which are blown, the aerodynamic coefficient will be steady. Reynolds Number check for models used 11, 22, 33, and 44 m/s as shown in the Table 4.

Table 4. Reynolds number

Air velocity (m/s)	Reynolds Number
11	8.74×10^4
22	1.75×10^5
33	2.62×10^5
44	5.5×10^5

While the Reynolds Number of full scaled MST on operation speed 44 m/s is 8.74 x 10^6. Figure 20 illustrates that there is no effect between Reynolds Number to coefficient of drag of both of models. Thus, the wind tunnel test is be able to measure the coefficient of drag with multiple variation of air velocity with constant result.

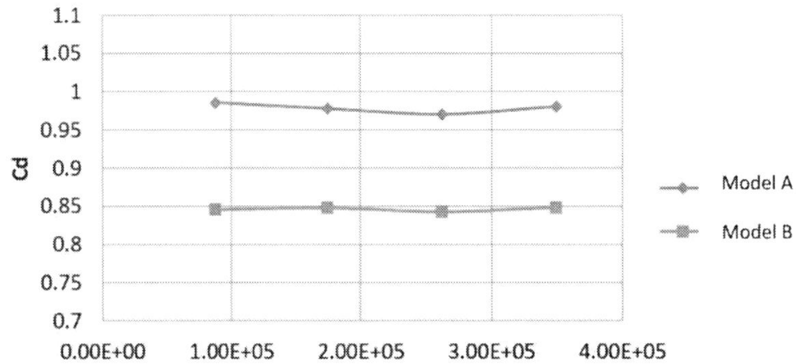

Figure 20. Reynolds number check

The wind tunnel test discovered that the drag coefficient of model A as shown in Figure 21 is 0.99 and drag coefficient of model B as shown in Figure 22 is 0.84.

Figure 21. Wind tunnel model A

Figure 22. Wind tunnel model B

Comparison of graph data from CFD simulation results with wind tunnel test results shows relatively small difference of 7%. The errors in the CFD compared to the actual measurement caused by some components, biggest proportion is the front stagnation region (Perzon, Janson, & Höglin, 1999). The wind tunnel test validates the result of numerical simulation.

4. Conclusion

The CFD simulation was run for two model type, single car and three cars configuration. In the single car model, the coefficient of drag of model A and B is found to be 0.647 and 0.550 respectively. The three cars configuration has bigger Cd, 0.902 for model A and 0.784 for model B. The value of CD shown consistently to the model shape regardless of the length of configuration. However, the three cars configuration is closer to the actual trainset. Thus, this configuration is used in the experimental wind tunnel test. The results from the numerical simulation then validated using the wind tunnel test. It is found that there is 7% difference between the numerical simulation and the experimental test. One factor affecting the difference was possibly caused by the ground board where the train model is mounted. It has its own drag coefficient; therefore, the drag coefficient of the wind tunnel test results is bigger. In addition, the support system that connected the model to the balance also contributed to the drag coefficient. There needs to be a change or improvement in the method of testing in the wind tunnel so that the results are more accurate. Nevertheless, the difference is relatively small, which proves that the

IOP Conf. Series: Materials Science and Engineering **909** (2020) 012031 doi:10.1088/1757-899X/909/1/012031

numerical simulation has a reliable result and the setting parameters can be used for further development of the models.

Acknowledgment
This research was funded by Badan Pengkajian dan Penerapan Teknologi (BPPT). We thank our colleagues from Balai Besar Teknologi Aerodinamika, Aeroelastika dan Aeroakustika (B2TA3) who provided insight and expertise that greatly assisted the research and thank to our colleagues from Institut Teknologi Sepuluh November (ITS) Surabaya for providing the 3D model of the MST. We also thank to PT. INKA (Persero) who provided the ANSYS 19.0 Software as tool for this research.

Reference
[1] Baker C 2010 The flow around high speed trains *Journal of Wind Engineering and Industrial Aerodynamics* **98** 277–298
[2] Banga S, Zunaid M, Ansari N A, Sharma S and Dungriyal R S 2015 CFD simulation of flow around external vehicle *IOSR Journal of Mechanical and Civil Engineering* **12** 87–94
[3] Chen D, Yao S, Lin P and Han Y 2017 Wind tunnel test study for the influence of reynolds number on the aerodynamic performance of high-speed trains *DEStech Transactions on Engineering and Technology Research* 131–139
[4] Cui K, Wang X P, Hu S C, Gao T Y and Yang G 2012 Shape optimization of high-speed train with the speed of 500kph *Lecture Notes in Electrical Engineering* **148** 187–197
[5] Gallagher M, Morden J, Baker C, Soper D, Quinn A, Hemida H and Sterling M 2018 Trains in crosswinds – comparison of full-scale on-train measurements, physical model tests and CFD calculations *Journal of Wind Engineering and Industrial Aerodynamics* **175** 428–444
[6] Perzon S, Janson J and Höglin L 1999 On comparisons between CFD methods and wind tunnel tests on a bluff body *SAE Technical Papers* **724**
[7] Utomo T, Fajar B and Arpriyanto H 2018 CFD analysis on the aerodynamics characteristics of Jakarta-Bandung high speed train *MATEC Web of Conferences* **159**
[8] Zhang L, Yang M zhi and Liang X feng 2018 Experimental study on the effect of wind angles on pressure distribution of train streamlined zone and train aerodynamic forces *Journal of Wind Engineering and Industrial Aerodynamics* **174** 330–343

IOP Conf. Series: Materials Science and Engineering **909** (2020) 012032 doi:10.1088/1757-899X/909/1/012032

Development of liquid smoke production process as a latex coagulant by utilizing a refrigeration machine

Riman Sipahutar[1,*], Diah Kusuma Pratiwi[2], Irwin Bizzy[3], Armin Sofijan[4], Baiti Hidayati[5]
[1,2,3,5]Mechanical Engineering Department, Engineering Faculty, Sriwijaya University
[4]Electrical Engineering Department, Engineering Faculty, Sriwijaya University

Email: rimansipahutar@ft.unsri.ac.id

Abstract. This study aims to obtain a method of producing liquid smoke as a latex freezing material using a pyrolysis reactor equipped with a counter flow condenser with a cold water cooling medium. The raw materials as local wisdom studied were coconut shells and oil palm shells, abundant in South Sumatra. The condensation process uses a shell and tube type heat exchanger with counter flow. The raw materials were pyrolyzed with time variations of 180, 240 and 300 minutes. The production of liquid smoke with a pyrolysis duration of 300 minutes has the highest yield values of 36,66% and 29,74%, respectively, for the the raw materials of oil palm shell and coconut shell. The average amounts of tar separated from liquid smoke in a settling tank are 8.13% and 8.08%, respectively, for raw materials of oil palm shell and coconut shell.

1. Introduction

Indonesia is known as the Equatorial Emerald, an island nation and a country with abundant natural resource potential. Potential natural resources that are quite large include coal, oil and gas, Coal Bed Methane (CBM), geothermal, oil palm and rubber plants. Indonesia is estimated to have the opportunity to increase production while also increasing export volumes. estimated to reach 2.2% per year. In an effort to take advantage of these opportunities, Indonesia still faces various problems related to poor quality. To overcome this problem, several policies and developments need to be researched comprehensively [1].

Rubber plant (Havea brasiliensis) is one of the important plantation commodities and also a source of non-oil and gas foreign exchange for the Indonesian country. Most of Indonesia's rubber fields are smallholder plantations that contribute significantly to national natural rubber production. Indonesian rubber plantations spread across Sumatra island (70%), Kalimantan island (24%) and Java island (4%). In fact, Indonesia has the potential to become a major producer of global rubber if various major problems faced in managing rubber plant products can be overcome and agribusiness developed professionally. Expansion of the area of rubber plantations in Indonesia is expected to continue due to the relatively stable and high condition of world rubber prices. This needs the support of the government to achieve higher productivity and quality [1].

Rubber tree sap or commonly called latex is a rubber raw material used for manufacturing various kinds of tools for domestic or outdoor use such as shoe soles, car tires and various other products. One way to convert liquid rubber into crumb rubber is to use liquid smoke. Technology to make liquid smoke is needed especially by small farmers.

In general, rubber farmers in Indonesia still use frozen materials such as vinegar and potassium alum which can damage the quality of rubber. Besides damaging the quality of the latex produced, its use also produces a very bad odor that can interfere with the comfort of life of the people around the processing location. Currently, for the freezing of rubber tree sap is used liquid smoke produced from the raw materials of coconut shell or oil palm shell that can prevent and kill bacterial growth and also function as an antioxidant. The use of 1 liter of liquid smoke concentrate can freeze 200-250 liters of latex or the equivalent of 75 kg of dry rubber.

One method for producing liquid smoke is by pyrolysis of raw materials in the form of coconut shells, oil palm shells, rubber-wood or rice husks in a pyrolysis reactor. The gas produced from the reactor is then channeled through a long pipe to a condenser. In the condenser, the gas resulting from the pyrolysis process changes phase from gas to liquid or condensate [2]. The condensate liquid is then stored in a settling tank to separate tar from liquid smoke.

Tar which has a greater density will settle to the bottom of the tank while the condensate above it which has a lower density is liquid smoke (liquid smoke). There are three types of liquid smoke along with its benefits, namely: Grade-3 liquid smoke that is used in rubber processing to remove odors and wood preservatives to be resistant to termites, Grade-2 liquid smoke that is used for preserving smoked food (for example smoked meat or fish smoke) so it's safer than the use of formalin, and Grade-1 liquid smoke that is used as a preservative for fast food like meatballs, noodles, and tofu.

2. Literature Review

2.1 Refrigeration System

The refrigeration system can be applied to the field of refrigeration and air conditioning. At present the application covers a very broad range of fields, ranging from household, agricultural, to the gas, petrochemical, petroleum and even industrial uses such as manufacturing and construction. The fields of refrigeration and air conditioning are interrelated, but each has a different scope. In air conditioning not only covers the field of cooling but also includes heating, but both of them equally discuss the problem of air humidity, air quality and its distribution [3].

Refrigeration is the process of releasing heat from unwanted places. The heat taken from food aims to maintain the quality and taste of the meal. While the heat taken from a room aims to maintain human comfort in it. There are so many applications in the industrial world where heat has been released from several places or materials for the desired purpose [4].

In general, the field of refrigeration covers a temperature range up to 125 K, whereas processes that take place below 125 K are often called cryogenics (Cryogenics). This difference is due to the typical phenomena that occur at temperatures below 100 K, where in this temperature range gases such as nitrogen, oxygen, hydrogen and helium can melt [5].

The vapor compression cycle is the cycle most widely used in refrigeration systems. In this cycle the vapor is compressed, and then condensed into a liquid, then the pressure is lowered so that the liquid can evaporate again.

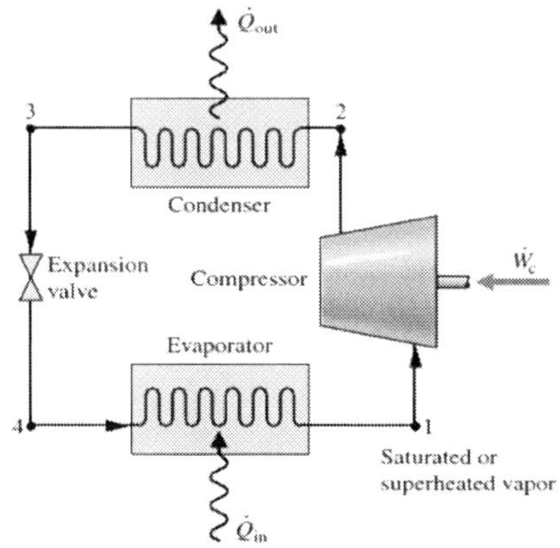

Figure 1. Vapor compression system

The standard vapor compression system can be seen in Figure 1 above. The cycle of a standard vapor compression system is as follows:

1-2 : Adiabatic and reversible compression, from saturated vapor to condenser pressure. This process takes place in a compressor where refrigerant vapor with low pressure and temperature enters the compressor through the suction line. The refrigerant vapor is then compressed in the compressor cylinder so that the temperature and pressure of the refrigerant vapor coming out of the compressor through the discharge line rises. The process that occurs in a compressor is assumed to be an isentropic process and the amount of compression work can be expressed by the following equation [6]

$$q_w = (h_2 - h_1) \qquad (1)$$

with:
q_w = the amount of compression work (kJ/kg)
h_1 = enthalpy of refrigerant when entering the compressor (kJ/kg)
h_2 = enthalpy of refrigerant when exiting the compressor (kJ/kg)

2-3 : Release of reversible heat at a constant pressure, causing a decrease in heat further (desuperheating) and condensation of refrigerant. This process takes place in the condenser where the high temperature and pressure refrigerant vapor entering the condenser is condensed so that the refrigerant vapor undergoes phase change from the gas to liquid. The amount of heat released in the condenser can be expressed by the following equation [6].

$$q_c = (h_2 - h_3) \qquad (2)$$

with:
q_c = The amount of heat released by the condenser (kJ/kg)
h_2 = Enthalpy refrigerant when entering the condenser (kJ/kg)
h_3 = Enthalpy refrigerant when exiting the condenser (kJ/kg)

3-4 : Non-reversible expansion at constant enthalpy, from saturated liquid to evaporator pressure.

4-1 : Addition of reversible heat at constant pressure, which causes evaporation of mixture refrigerant to saturated vapor refrigerant. This process occurs in the evaporator where the mixture refrigerant entering the evaporator absorbs heat from the cooled room or media. The amount of heat absorbed by the refrigerant in the evaporator is expressed by the following equation [7].

$$q_e = (h_1 - h_4) \qquad (3)$$

with:
q_e = the amount of heat absorbed by the evaporator (kJ/kg)
h_1 = enthalpy of refrigerant when exiting the evaporator (kJ/kg)
h_4 = enthalpy of refrigerant when entering the evaporator (kJ/kg)

2.2 Liquid Smoke

Liquid smoke is the result of condensation of pyrolysis vapor of raw materials containing lignin, cellulose, hemicellulose and other carbon compounds. Liquid smoke is produced by the pyrolysis process, ie compounds which evaporate simultaneously will be condensed in the cooling system. During the condensation process a coarse smoke condensate will be formed which will separate into three phases, that is, the soluble phase in water can be directly used, while the purified high-level tar phase extract can be used again for the production of liquid smoke and is usually called the primary tar fraction [8].

Liquid smoke can be applied in various ways such as mixing, spraying, dipping or mixed directly into food, soaking and injection. Besides heating liquid smoke to produce smoke-containing vapors is one of the methods used for food smoke.

Lots of materials that can be used as raw materials for making liquid smoke include wood, coconut shells, oil palm shells, coconut fiber, and cassava stems.

2.2.1 Coconut shell

Coconut shell is the hardest part of coconut fruit. Located next to the coir with a thickness of 3-5 mm and serves as a rotector of coconuts from damage from external influences. Coconut fruit needs to consist of fruit flesh of 30%, fiber of 33%, shell of 15%, and coconut water of 22%.

Coconut shells have as much lignin and a small amount of cellulose. The content of hemicellulose of coconut shell is almost the same as wood, and the water content varies according to environment, and fruit maturity. Coconut shell derived from ripe fruit in a dry state of air with a water content of about 6-9%.

The contents of lignin, cellulose, hemicellulose, and water of Indonesian coconut shell are 23.84%, 30.44%, 25.64% and 14.00%, respectively. The chemical composition of Indonesian coconut shell can be seen in Table 1 below.

Table 1. Chemical composition of Indonesian coconut shell [9]

Chemical composition	(%)
Lignin	23.84
Cellulose	30.44
Hemicellulose	25.64
Water content	14.00

2.2.2 Oil Palm shell

Oil Palm is a material for the processing of palm oil that produces a lot of waste, especially in the form of shells. As industrial waste, palm kernel shells are produced by oil processing plants up to tens of tons every day. In fact, this shell can be used as raw material for processing liquid smoke.

The most cultivated oil palm plantation in Indonesia is Elaeis Guineensis because this species has higher productivity. Based on the thickness of the shell, oil palm is divided into three types, namely: Dura, Pisifera, and Tenera. Dura is an oil palm whose fruit has a thick shell. Pisifera is oil palm that does not have a shell so it does not have a core that produces economical oil and sterile female flowers. Therefore Pisifera very rarely produce fruit while Tenera is a cross between female Dura and male Pisifera. This species is considered a superior seed that produces thin fruit skin and fertile female flowers [10].

3. Research Methodology

The research method chosen was experimentation by designing pyrolysis equipment for raw materials of coconut and oil palm shells and liquid smoke resulting from the pyrolysis process will be applied to the freezing process of liquid rubber sap (Latex).

3.1 Research Flowchart

The research began with the preparation of research equipment, which is a pyrolysis reactor equipped with a condenser which is cooled by cooling water from a refrigerated water tank. The next step is testing the equipment and continuing with the collection of test results on the raw material for coconut shells and oil palm shells. Then, data analysis and processing are carried out before drawing conclusions and suggestions as the final stage. The stages of research conducted can be seen in the research flowchart shown in Figure 2 below.

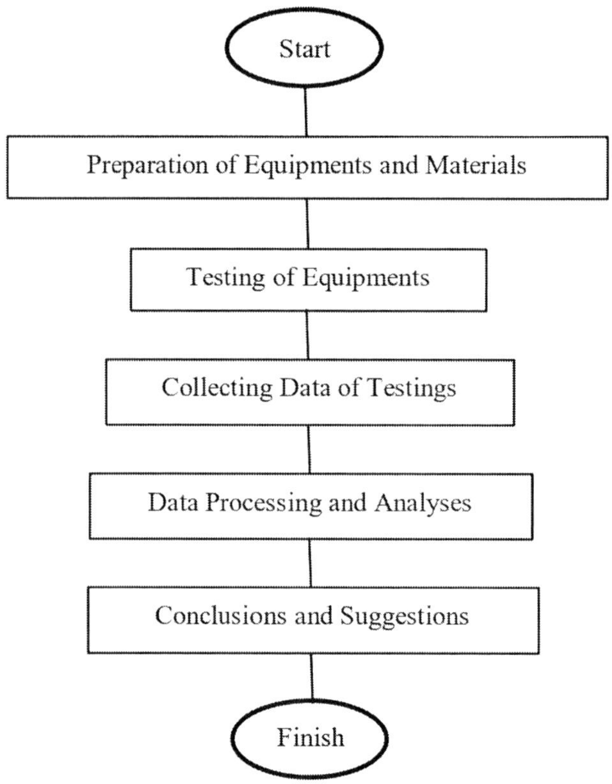

Figure 2. Research flowchart

3.2 Research Equipment

The equipment used in this study is a pyrolysis reactor equipped with a condenser cooled by cold water coming from a cooling water tank cooled by the evaporator pipes of a refrigeration machine. The temperature of the cooling water can be controlled through the evaporator pipes from the refrigeration machine which are passed into the cooling water tank. The vapor from the pyrolysis reactor will be condensed in the condenser and then the condensate from the condenser will be flowed to the settling tank to separate the tar from liquid smoke. The arrangement of equipment used is shown in Figure 3 below.

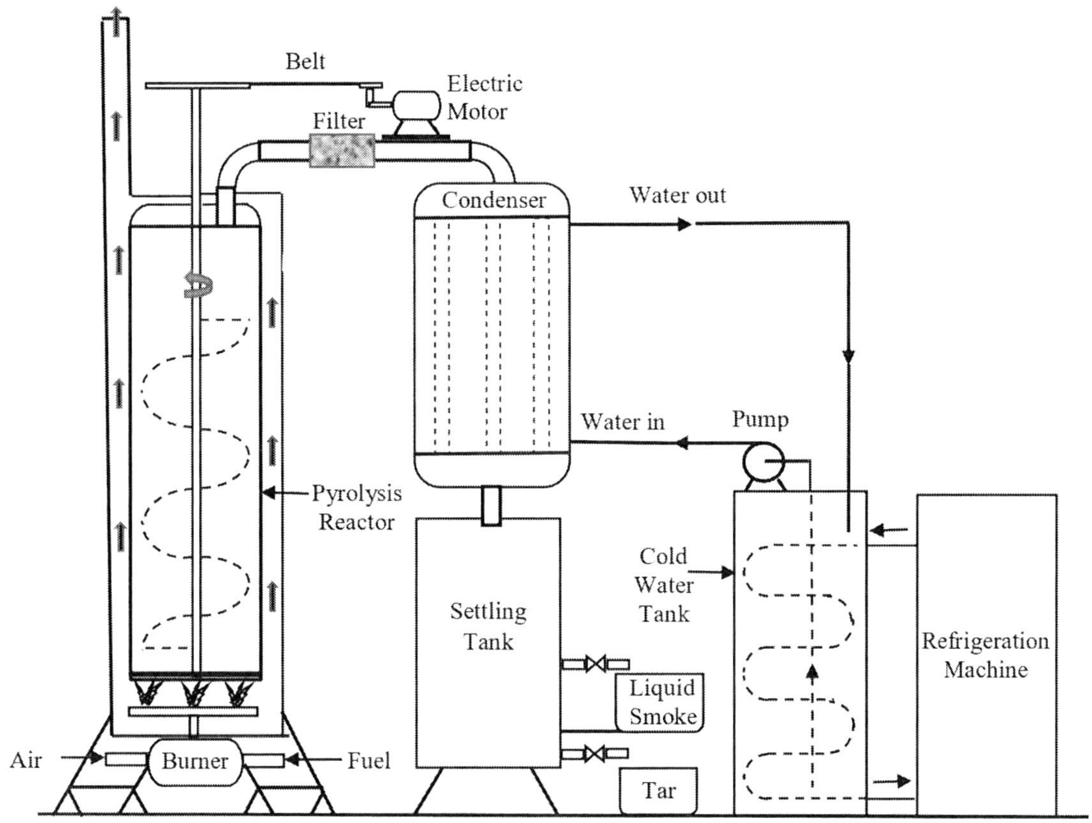

Figure 3. A pyrolysis reactor equipped with a condenser cooled with evaporator pipes of refrigeration machine

4. Results and Discussion

In this study, the raw materials used are coconut shells and oil palm shells. The two raw materials were pyrolyzed with time variations of 180, 240 and 300 (min). The fuel used is LPG gas and the amount of raw material for each cycle is 25000 (g) and the resulting liquid smoke is presented in Table 2 below.

Table 2. Data from the pyrolysis of raw materials of coconut shells and oil palm shells on the pyrolysis duration of 180, 240 and 300 (min).

No.	Raw Materials	Raw material weight (g)	Duration of Pyrolysis (min)	Total Distillation (cc)	Total Liquid Smoke (cc)	Tar Percentage (%)
1	Coconut Shell	25000	180	6987	6422	8.09
		25000	240	8175	7516	8.06
		25000	300	8439	7758	8.07
2	Oil Palm Shell	25000	180	8598	7898	8.14
		25000	240	10046	9229	8.13
		25000	300	10412	9565	8.13

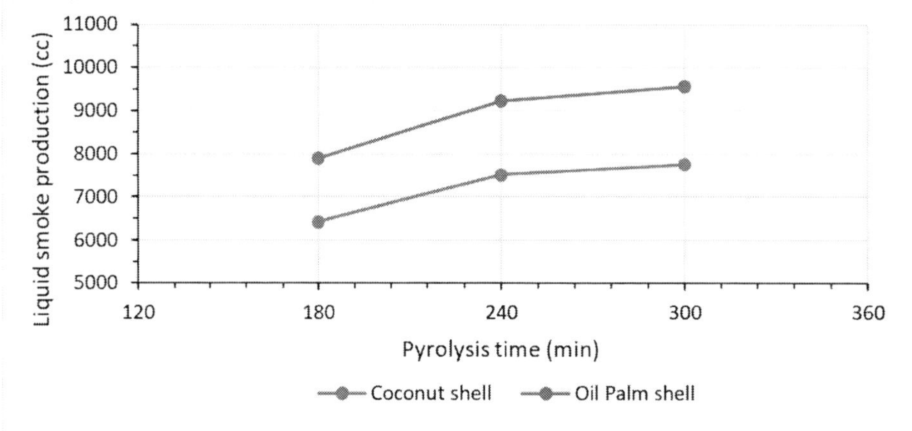

Figure 4. The effect of varied pyrolysis time on liquid smoke production for raw materials of coconut shell and oil palm shell at average pyrolysis temperature of 305°C and raw material weight of 25000 g.

Figure 4 shows that an increase in pyrolysis time from 180 (min) to 300 (min) resulted in an increase in the production of liquid smoke for the raw material of coconut shells and oil palm shells by 20.80% and 21.11%, respectively. It can also be seen that the percentage increase in the production of liquid smoke decreases between 180 to 240 minutes compared to between 240 to 300 minutes ie. from 17.04 down to 2.22 and from 16.85 down to 3.64, respectively, for raw materials coconut shells and oil palm shells. This is due at the 300th minute, the element of liquid smoke is almost completely out of the raw material.

The production of liquid smoke in the pyrolysis process with a pyrolysis duration of 300 minutes has the highest yield values of 29.74% and 36.66%, respectively, for the raw material of coconut shell and oil palm shell. In addition, the production of liquid smoke in the process of pyrolysis with a duration of 180 minutes has the lowest yield values of 24.62% and 30.27%, respectively, for the raw material of coconut shell and oil palm shell, as shown in Figure 5 below.

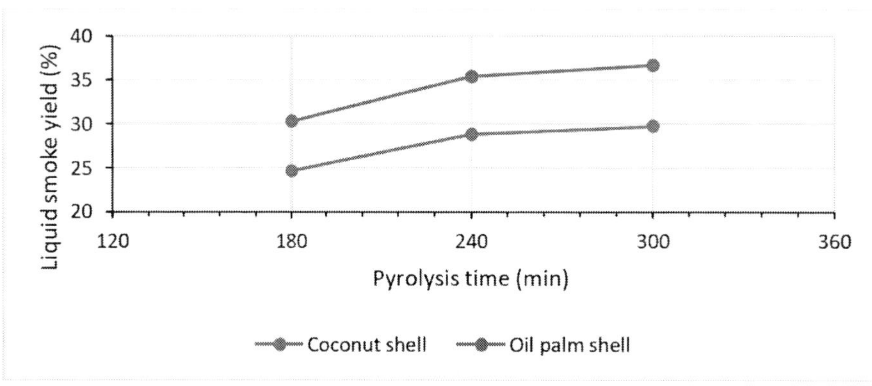

Figure 5. The effect of varied pyrolysis time on liquid smoke yield for raw materials of coconut shell and oil palm shell at average pyrolysis temperature of 305°C and raw material weight of 25000 g.

The total production of liquid smoke is highly dependent on the type of raw material used and the length of the pyrolysis process. This is consistent with the results of research conducted by Budaraga, et al. (1996) on the raw materials of coconut fibre, coconut shell and cinnamon [11]. The average amount of tar separated from the distillation liquid from the pyrolysis process of raw materials for coconut shells and oil palm shells is 8.07% and 8.13 as shown in Figure 6 below.

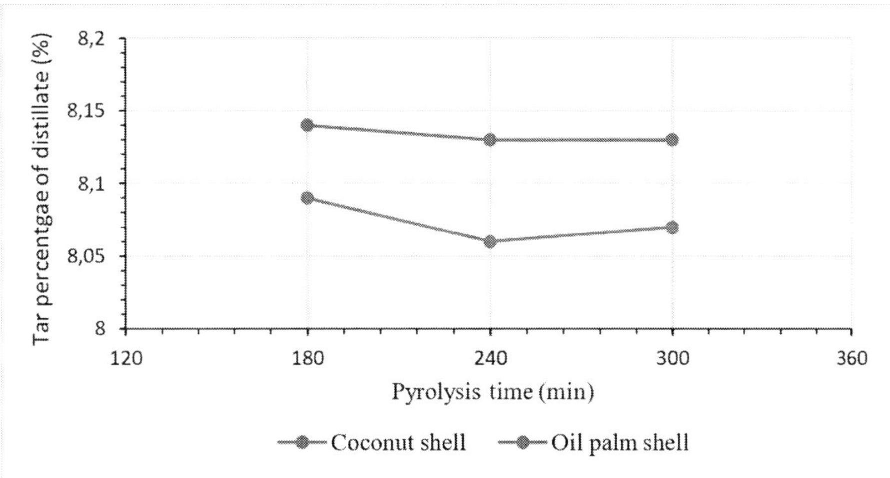

Figure 6. The effect of varied pyrolysis time on tar percentage of distillate (%) for raw materials of coconut shell and oil palm shell at average pyrolysis temperature of 305°C and raw material weight of 25000 g.

5. Conclusions

From the results of research on the development of the liquid smoke production process utilizing a refrigeration machine can be concluded as follows: 1). The raw material of oil palm shell has more potential to produce liquid smoke compared to the raw material of coconut shell, in the pyrolysis process. 2). The highest yield of liquid smoke occurred at a pyrolysis duration of 300 minutes, that is 29.74% and 36.66% for the raw materials of coconut shells and oil palm shells, respectively. 3). Condensate derived from the condensation process on vapor originating from the pyrolysis process of raw material of oil palm shell consists of liquid smoke of 91.87% and tar of 8.13%. 4). Condensate derived from the condensation process on vapor originating from the pyrolysis process of raw material of coconut shell consists of liquid smoke of 91.93% and tar of 8.07%.

References

[1] Damanik S 2012 *Pengembangan Karet (Havea brasiliensis) berkelanjutan di Indonesia* **11**
[2] Rezaiyan J and Cheremisinoff N P 2005 *Gasification Technologies A Primer for Engineers and Scientists* (Boca Raton: Taylor & Francis Group)
[3] Stoecker W F and Jones J W 1982 *Refrigeration and Air Conditioning* (New York: McGraw-Hill, Inc.)
[4] Dossat R J 1997 *Principles of Refrigeration* (New York: Pearson Education)
[5] Arora C P 2000 *Refrigeration and Air Conditioning* (New Delhi: Tata McGraw-Hill Publishing Company Limited)
[6] Chadderton D V 1993 *Air Conditioning A Practical Introduction* (London: E & FN SPON)
[7] Jordan R C and Priester G B 1981 *Refrigeration and Air Conditioning* (New Delhi: Prentice-Hall, Inc.)
[8] Basu P 2013 *Biomass Gasification, Pyrolysis and Torrefaction* (London: Academic Press)
[9] Darmadji P 2002 Optimasi Proses Pembuatan Tepung Asap *Agritech* **22** 172–7
[10] Utomo B S D, Wibowo S and Widianto T N 2012 *Asap Cair* (Jakarta: Penebar Swadaya)
[11] Budaraga I K, Arnim., Marlinda Y and Bulanim U 2016 Liquid Smoke Production Quality from Raw Material Variation and Different Pyrolysis Temperature *Int. J. Adv. Sci. Eng. Inf. Technol.*

IOP Conf. Series: Materials Science and Engineering **909** (2020) 012033 doi:10.1088/1757-899X/909/1/012033

Stress analysis on the frame holder of generator translation motion on horizontal direction for sea wave power plant using finite element method (fem)

A. Indriani [1,*], Hendra [2], Yenni S.[1], Aswata, W.[2], Rispandi[3], Hernadewita[4] and A. Tanjung[5]

[1]Electrical Engineering Dept. University of Bengkulu, Indonesia

[2]Mechanical Engineering Dept. University of Sultan Ageng Tirtayasa Banten, Indonesia

[3]Mechanical Engineering Dept. University of Bengkulu, Indonesia

[4]Industrial Engineering Dept. University of Mercubuana, Indonesia

[5]Polytechnic State of Padang, Indonesia

Jl. W. R. Supratman Kandang Limun, Bengkulu, Indonesia

Email: h7f1973@yahoo.com

Abstract. The generator is a machine component that converts mechanical energy into electrical energy. Generators can be found in power generation systems. In this paper, generators are used on the sea wave power plants [1-4]. This generator works by utilizing the effects of ocean waves that rise and fall so as to push the piston to move in the same direction with ocean waves. The performance of the generator is influenced by the magnitude of the force of the piston, rotors and stator. Stable, rigid and smooth piston movements also can increase generator output performance. The design and material of the generator holder is very influential on the generator stiffness. Generator stiffness includes deformation and stress that occur and can be obtained using the finite element method (FEM) [5-13]. In this paper focus on simulating the generator holder shape horizontal direction and the amount of deformation and stresses by varying frame holder generator shape and inclination (75^0, 80^0, 90^0). In previous studies, the translational motion generator frame holder design is less rigid (stable). From the simulation results, the maximum voltage value (σ_{max}) of von mises is 6,427 MPa, the deflection value is 61.29 mm with 90^0 generator stand design positions.

1. Introduction

The generator is a machine component that converts mechanical energy into electrical energy. Generators can be found in power generation systems such as hydropower, wind, diesel, ocean waves and others. In this paper, generators are used on the sea wave power plants [1-4]. The generator works by utilizing the effects of ocean waves that rise and fall so as to push the piston to move in the same direction with ocean waves (pneumatic). Generator for ocean waves can be found in the pelamis mechanism, pneumatic system [1-4], and etc. [1]. Pneumatic system has been use in generator by translation motion system on vertical direction [2-4]. The working principle of translational and rotation motion generator in horizontal direction is to utilize the mechanism of piston work down and up due to thrust force or load. Piston in the generator translation and rotation consist on shafts that have been fitted with pole winding magnet (coil) that move on down and up (rotor). Up and down movements of the coil meet with neodymium magnets which are arranged along the generator tube (stator) will be produce magnetic electric energy. In the rotation generator motion system uses a shaft rotation which is connected to the gears of the translator generator piston motion.

The performance of the generator is influenced by the thrust force of the piston up and down, the rotor and stator material, the number of rotors and stators, magnets, coils and others. Stable, rigid and smooth piston movements also can increase generator output performance. The mechanism of generator of translation and rotation motion requires a strong, stiffness, toughness, constant and stable generator holder to get the good output of generator. The movement and strength of piston up and down is influenced by the thrust force of piston, piston material, generator holder and generator holder material.

The design of the Frame of generator holder and the type of material is very influential on the generator stiffness. Generator stiffness includes deformation and stress that occur where the rigid generator will produce small stress and deformation. The amount of deformation and the voltage that occurs at the frame holder of generator translation and rotation can be obtained using the finite element method (FEM) [5-13].

In this study focus on stress and deformation happen on the frame holder of generator translation and rotation motion on horizontal direction and amount of deformation and stresses that occur using variations a generator holder material, generator holder shape, inclination angles of frame holder of generator. Inclinations angles of frame holder of generator 75^0, 80^0, 90^0 and random are used to evaluate of the performance of generator translation and rotation on horizontal direction and also to solving problem for low skill operator in manufacture the frame of generator holder. In previous studies the inclination angle of frame of generator is random and the result of generator translation and rotation motion generator design is less rigid (unstable) and low precision so that it affects the performance of the generator [2]. The research focus on the stress analysis using finite element method (FEM) analysis is applied to calculate stress and deformation on the design of frame holder of generator before manufacturing process [5-13]. By FEM analysis its obtain good design of frame holder of generator depend on the load and maximum stress that occurs on the frame holder of generator and also reduce of cost production for manufacturing.

Table 1. Properties of material frame holder of generator [5-13]

Material	Young's Modulus (GPa)	Poisson's Ratio	Tensile Strength (MPa)
Steel	210	0.3	330
Cast Iron	120	0.3	758

2. Method

FEM analysis is applied to calculate stress in the frame holder of generator translation and rotation motion on horizontal direction. 3D model has been used for this analysis [5, 10] and variation of inclination angle are 75^0, 80^0, 90^0 and random is applied in this frame holder of generator. Material of frame holder of generator is steel and cast iron. Figure 1 show the generator translation and rotation motion on horizontal direction and Fig. 2 show the design of frame holder of generator by 3 type depend on variation of inclination angles (random, 75^0 and 90^0). Variation load is applied at the frame holder of generator using 1000 N to 10 kN. Table 1 shows the material properties of steel, and cast iron. Figure 2 shows the 3D model of design of generator translation and rotation on horizontal direction with total number of elements is 245832 and nodes is 443465.

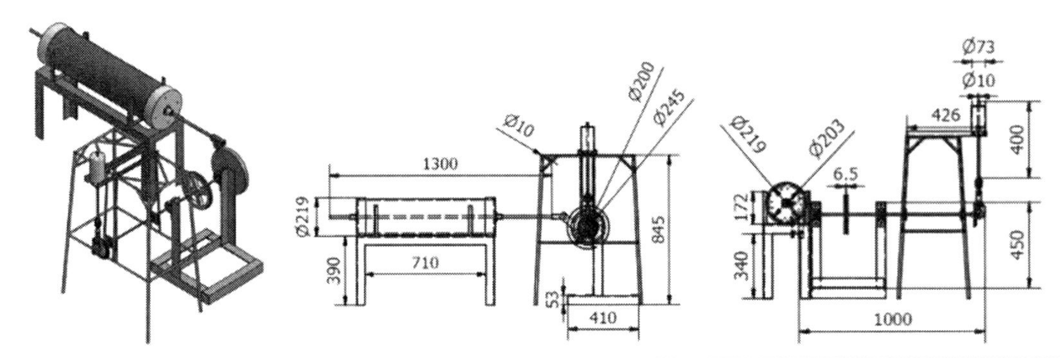

Figure 1. Design of generator translation and rotation motion on horizontal direction

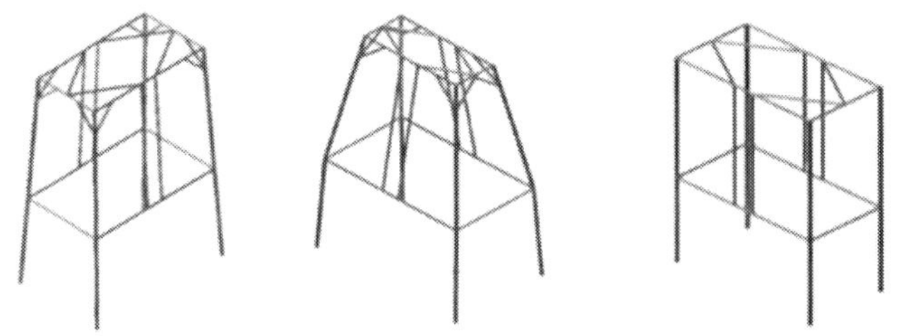

a. Random **b. Inclination angle 80⁰** **c. Inclination angle 90⁰**

Figure 2. Design of frame holder of generator by varying inclination angles.

	Design of Frame of Generator Horizontal Motion by Varying of Inclination Angles		
	Random	**80^0**	**90^0**
	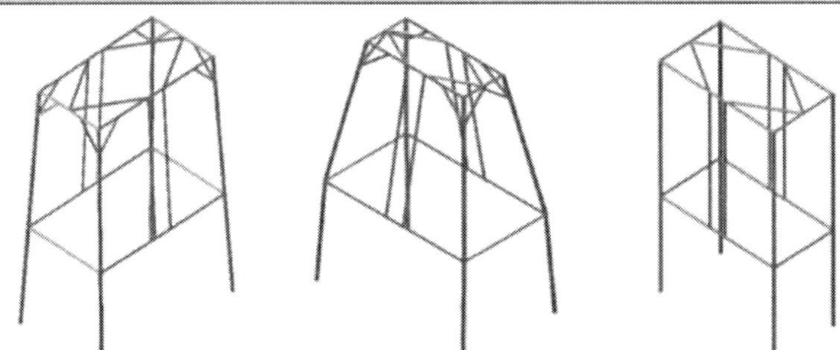		
Stress (σ_{max})	6.44 MPa	6.40 MPa	6.37 MPa
Displacement (δ_{max})	(0,062 mm)	(0,057 mm)	(0,056 mm)

Figure 3. Maximum stress and deformation value on design of frame of generator translation and rotation motion on horizontal direction

3. Result and Discussions

Results of simulation of stress and deformation for design of frame holder of generator translation and rotation motion on horizontal direction by FEM analyze can be seen at Figs. 3-8. Stress and deformation value of frame holder of generator by varying of inclination angle of frame holder can be seen in Table

2. As shown in Table 2, the value of maximum stress and deformation of frame holder by inclination angle 90^0 is smaller than value stress on random inclination angle. The maximum stress value is 6.34 MPa and deformation value is 0.056 mm for inclination angle 90^0. And for random inclination angle is 6.44 MPa and 0.062 mm. This results show that the stress and deformation for inclination angle 90^0 is better than other varying of inclination angle of frame holder of generator.

Table 2. Maximum stress and deformation value on design of frame of generator translation and rotation motion on horizontal direction

NO	Force (N)	σ_{max} (MPa)						
		σ_{xx}	σ_{xy}	σ_{xz}	σ_{yy}	σ_{yz}	σ_{zz}	σ_{vm}
1	1000	54,49	16,98	21,27	27,82	9,13	23,51	64,41
2	2000	108,8	33,95	42,54	55,53	18,26	47,01	128,7
3	4000	217,2	67,91	85,07	110,5	36,54	94	257
4	5000	271,8	84,89	106,3	138,5	45,66	117,6	321,5
5	6000	327,7	101,9	127,5	167,6	54,76	141	386,9
6	8000	435,1	135,8	170,2	222	73,05	188	514,7
7	10.000	544	169,8	212,7	277,6	91,3	235,1	643,4

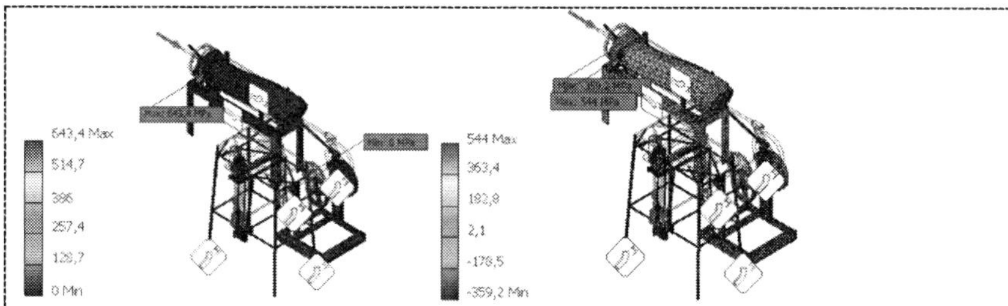

a. Maximum Stress σ_{vm} b. Maximum Stress σ_x

Figure 4. Stress distribution of drum rotary.

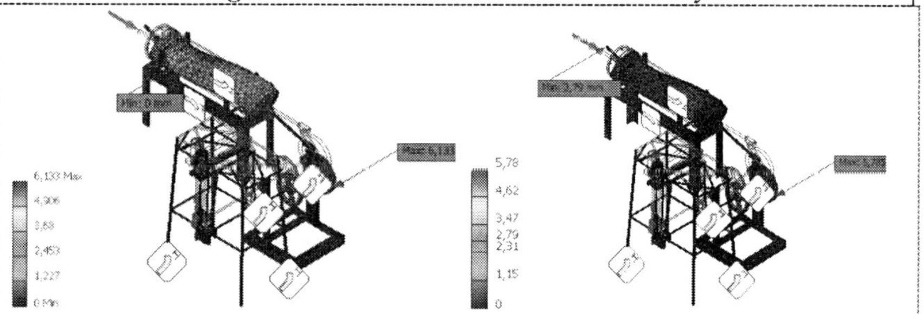

a. Maximum displacement b. Maximum displacement δz

Figure 5. Displacement distribution of drum rotary for fem

Table 2 show the value of maximum stresses for frame holder of generator with random inclination angle. The maximum stress Von Misses value is 64.41 MPa with load 1000 N and increasing to 643.4 MPa with load 10 kN. Maximum stress appear at σ_{xx} with value 54.49 MPa with load 1000 N on Z direction. When the load is increasing maximum stress value become increase 544 MPa for load 10 kN and also deformation is increasing with higher of load. This happen due to the load give significant effect on the stress and deformation, when the load is high then stress and deformation become increase. To reduce the stress and deformation can be done by using ribbing, selection of material for frame holder of generator and using inclination angle 90^0.

The stress and deformation distribution of frame holder of generator translation and rotation motion on horizontal direction with random inclination angle can be seen in Fig. 3. Figure 3 show that the position of maximum stress and deformation appear at the insert of tube generator (σ_{xx}). The blue colour show the condition of stress have minimum value of stress (compression) and red colour is maximum value of stress (tensile). The maximum stress appear in the insert tube of generator due to the thrust force as a input for driven a generator that start from this part. The maximum deformation appear at the Z direction with value 5.78 mm by using load 10 kN.

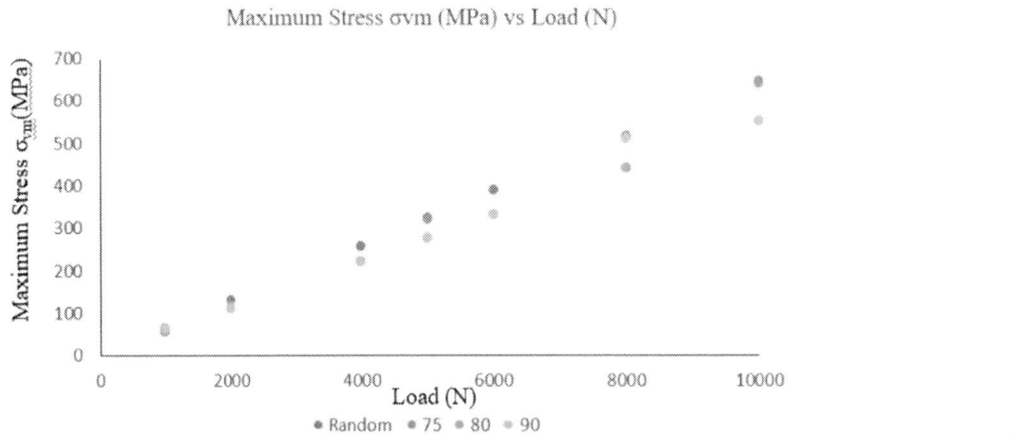

Figure 6. Maximum stress for varying load and inclination angle of frame holder of generator

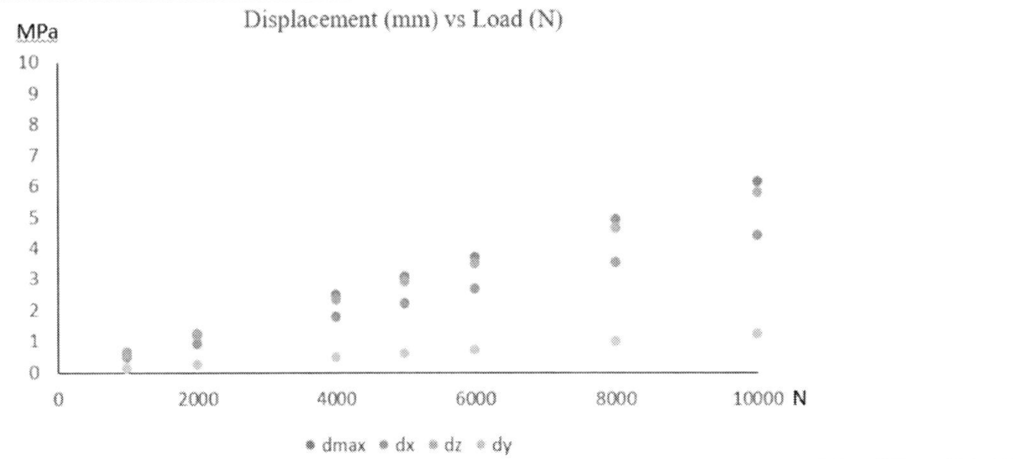

Figure 7. Maximum deformation for varying load and inclination angle of frame holder of generator.

Figure 5 show that maximum stress for frame holder of generator depend on variation load and frame holder inclination angles. Maximum stress have bigger value at frame holder inclination angle is random and become smaller using 90^0. For addition of load, the maximum stress become higher when the value load is big. Maximum stress for random inclination angle is similar with 75^0 compare with 80^0 and 90^0. The maximum deformation appear at the random inclination angle of frame holder on z direction as shown in Figs. 6 and 7. This figure show that when the value of inclination angle smaller than 90^0 and the load increase then maximum stress and deformation become increase. This result show that load and frame holder inclination angle for structure of generator is significant give effect on the stress and deformation.

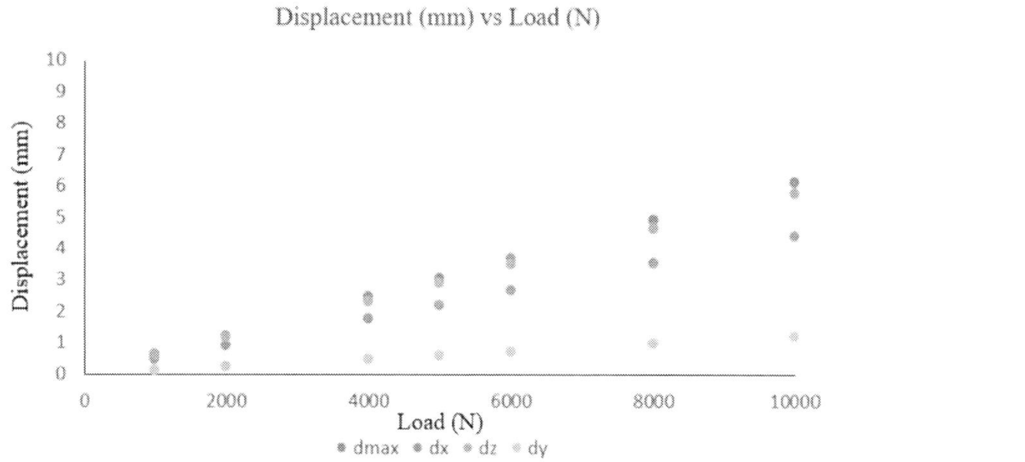

Figure 8. Position maximum deformation for varying load and inclination angle of frame holder of generator.

4. Conclusions

Steel and cast iron material been used in frame holder of generator translation and rotation motion on horizontal direction. Frame holder of generator have some requirement such as stiffness, toughness, stable, stress, deformation and etc. The stress and deformation of frame holder of generator depend on the load, shape of frame, inclination angle and etc. To get this requirement therefore finite element method are used to calculate design of frame holder of generator by variation load and inclination angle of frame holder. The conclusions can be made in the following:

1. The maximum stress of frame holder of generator translation and rotation motion on horizontal direction for random inclination angle is higher than 90^0. The maximum stress value is 644 MPa with load 10 kN. For 90^0 inclination angle of frame holder maximum stress is 637 MPa. The maximum deformation for 90^0 inclination angle is smaller compare with other inclination angle (random, 75^0 and 80^0).
2. Maximum stress and deformation become higher when the load is increasing and value of inclination angle smaller than 90^0.
3. Maximum stress in the frame holder of generator appear at insert of tube piston (σ_x) and maximum deformation in Z direction on the disk of piston.

Acknowledgments

This research support by Ministry Research, Technology/ Head of the National Research and Innovation Agency Republic of Indonesia through *Hibah Penelitian Dasar Unggulan Perguruan Tinggi* (PDUPT) 2020 Number 165/SP2H/LT/DRPM/2019 and Number 791/UN30.15/AMD/LT/2020, and Institution of Research and Community Service University of Bengkulu.

5. References

[1] Hendra, Indriani, A., Hernadewita Applying of Piston Mechanism Design In The Wavelength Electrical Generating of Ocean For Fishing Comunity *Advanced Materials Research* **918** 2014-01-01.
[2] Indriani, A., Darmana, I., Hernadewita, Setiawan, A., Hendra 2020 Control Output of Generator Translation and Rotation Using Buck Boost Converter for Sea Wave Power Plant *International Journal of Mechanical Engineering and Technology* **11** Issue 5 61-74.
[3] Indriani, A., Dimas, S., Hendra 2018 Effect of Dimension and Shape of Magnet on the Performance AC Generator with Translation Motion *IOP Conference Series: Materials Science and Engineering* **307** Issue 1.

[4] Indriani, A., Hendra, Suhartini, Y., Tanjung, A. 2019 Performance of Generator Pneumatic for Power Plant of Ocean Wave *IOP Conference Series: Materials Science and Engineering* **505** Issue: 1.
[5] Hendra, Syukriah, Silalahi, M., Hernadewita, Hermiyetti 2019 Finite Element Method Analysis for Manufacturing Design Drum Dryer of Rotary Dryer Machine *IOP Conference Series: Materials Science and Engineering* **505** Issue: 1.
[6] Tsuyunaru, M., Noda, N. A., Hendra and Takase, Y. 2008 Maximum Stress for Shrink Fitting System used for Ceramics Conveying Rollers *Transactions of the Japan Society of Mechanical Engineering* **74** 919-925.
[7] Tsuyunaru, M., Noda, N. A., Hendra and Takase, Y. Maximum stress for shrink fitting system used for ceramics conveying rollers, Nihon Kikai Gakkai Ronbunshu *A Hen Transactions of the Japan Society of Mechanical Engineers* **74** Issue : 7.
[8] Noda, N. A., Hendra, Oosato, M., Suzumoto, K., Takase, Y., and LI, W. 2011 Strength Analysis For Shrink Fitting System Used For Ceramics Rolls In The Continuous Pickling Line *Key Engineering Materials* **462-463** 1140-1145.
[9] Harada, S., Noda, N. A., Uehara, O. and Nagano, M. Tensile Strength of Hot Isostatic Pressed Silicon Nitride and Effect of Specimen Dimension *Transactions of the Japan Society of Mechanical Engineering* **57** No.539 1991.
[10] Takase, Y., Li, W., Hendra, Ogura, H., Higashi, Y. And Noda, N.A. 2011 Three-Dimensional Surface Heat Transfer Coefficient and Thermal Stress Analysis for Ceramics Tube Dipping into Molten Metal *Advances in Fracture and Damage Mechanics IX, Key Engineering Materials* **452-453** 233-236 (Trans Tech Publications, Switzerland).
[11] Noda, N. A., Hendra, Takase, Y., and Li, W. 2009 Thermal Stress Analysis for Ceramics Stalk in the Low Pressure Die Casting Machine *Journal of Solid Mechanics and Material Engineering* **3** No.10 1090-1100.
[12] Noda, N. A., Yamada, M., Sano, Y., Sugiyama, S., and Kobayashi, S. 2008 Thermal Stress for All-ceramics Rolls used in Molten Metal to Produce Stable High Quality Galvanized Steel Sheet *Engineering Failure Analysis* **15** 261-274.
[13] Noda, N.-A., Hendra, Li, W., Takase, Y., Ogura, H., Higashi, Y. 2013 Thermal stress and heat transfer coefficient for ceramics stalk having protuberance dipping into molten metal, Nihon Kikai Gakkai Ronbunshu *A Hen Transactions of the Japan Society of Mechanical Engineers* **79** Issue 802.

Design of machine element in alpha type stirling machine innovation design based on plastic waste

Dhimas Satria[1], **Imron Rosyadi**[1], **Rina Lusiani**[1], **Erny Listijorini**[1]

[1]Mechanical Engineering Department, Universitas Sultan Ageng Tirtayasa, Indonesia

Email: dhimas@untirta.ac.id

Abstract. One effort to overcome the use of fossil fuels is to use alternative energy sources, including plastic waste. Stirling engine is an external combustion engine that can utilize heat from various heat sources. In this research, an analysis of the design of the engine elements in the alpha-type Stirling engine design innovation is based on plastic waste, so it is expected to get the best and safest engine element design specifications. Based on the analysis results of calculations based on the Kwurmi method, it has been obtained the specifications of the design results of the engine elements are as follows: shaft design (length = 250 mm, diameter = 15 mm), post design (length = 20 mm, width = 6 mm, thickness = 6 mm), pulley design (driver diameter = 100 mm, diameter driven = 60 mm), belt design (length = 987 mm, width = 30 mm, thickness = 5 mm) , Stirling engine (power = 94.3 to 205.4 watts, engine speed = 300 to 1305 rpm, heat temperature = 623 ° k, cold temperature = 305 ° k, phase angle = 180 °, dimension = 580 mm × 250 mm × 700 mm, thermal efficiency = 0.51).

1. Introduction

Energy demand that continues to increase is closely related to energy sources that continue to decrease. Because when viewed from the current source of energy procurement, world energy sources are still much dependent on fossil energy. Where fossil energy has a minimal amount. According to Christ Lewis, in his book Biological Fuels estimates that natural gas is not available in 2047, petroleum in 2080, and coal in 2180. The depletion of fossil energy is caused by the exploration of fossil energy on a large scale and is not proportional to the time the formation of fossil energy [1]. The solution to the significant energy needs and limitations of fossil fuels is to utilize alternative energy that can be renewed and easy to obtain [2]. The appropriate technology for utilizing alternative energy sources is the use of an external combustion engine because it has the advantage of being able to use all types of fuel. Stirling engine is an external combustion engine that can utilize heat from a variety of heat sources. Research on Stirling engine design has been carried out by several previous researchers, research on the design and manufacture of alpha type Stirling engines in general [3], modified alpha type Stirling engine using diesel concentration conjunctions [4], and alpha type Stirling engine by utilizing solar energy [5]. Then as a comparison of the alpha type Stirling engine, is research on other types of Stirling engines, namely the effect of phase angle on the efficiency of the beta type Stirling engine [6] and manufacturing and testing of gamma type Stirling engines [7].

This research is a development from previous research, namely alpha type Stirling engine biomass-based design innovation [8][9].

IOP Conf. Series: Materials Science and Engineering **909** (2020) 012034 doi:10.1088/1757-899X/909/1/012034

In this research, an analysis of the design of the engine elements in the alpha-type Stirling engine design innovation is based on plastic waste, so it is expected to get the best and safest engine element design specifications.

2. Methodology

This study's research method is to calculate and analyze the safe dimensions of the engine elements used in the Stirling engine, namely the shaft, tide, and belt, using the Kwurmi method [10].

3. Result and Discussion

Alpha type Stirling engine is capable of producing power up to 100 Watts. By using a source of plastic waste materials. The Stirling engine heat cylinder will utilize the energy from burning plastic waste, which will then provide the Stirling engine's work so that the Stirling engine can spin. The rotation of the Stirling engine shaft is then transmitted using a belt and pulley to the generator. The generator will produce electrical energy that can be useful for various light needs such as turning on lights, charging batteries, and so forth.

3.1. Shaft Calculation

Known:

P = 100 watt; n = 300 rpm; $\sigma = \tau . \sqrt{3} = 10{,}4\ MPa$; $\tau = 6\ MPa$; W_1 = 20 N (Flywheel); W_2 = W3 = 5 N (piston); W_4 = 10 N (pulley)

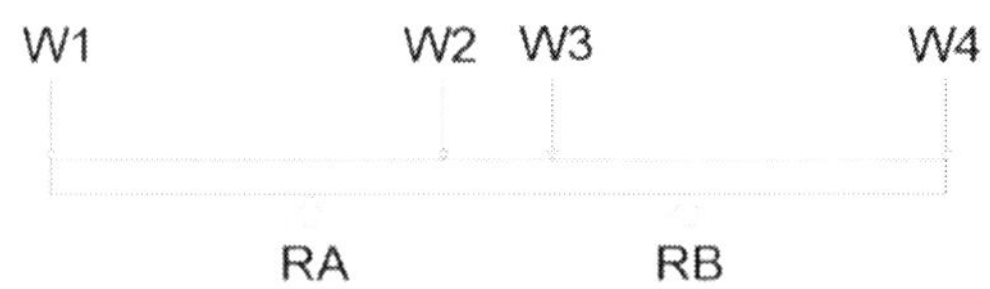

Figure 1. Illustration of shaft loading

Torque:

$$T = \frac{P \times 60}{2 \times \pi \times n} \quad (1)$$

$$T = \frac{100\ W .\ 60}{2 \times 3.14 \times 300\ rpm} = 3.19\ Nm$$

The biggest pedestal:

$$\sum M_A = 0$$

$$(5N \times 37.5mm) + (5N \times 67.5\ mm) + (10N \times 177.5\ mm) - (R_B \times 105mm) - (20N \times 72.5mm) = 0$$

$$R_B = 8.1\ N$$

$$\sum M_B = 0$$

$$(R_A \times 105mm) + (10N \times 72.5mm) - (5N \times 37.5mm) - (5\ N \times 67.5mm) - (20N \times 177.5\ mm) = 0$$

$$R_A = 31{,}9\ N$$

Bending moments:

$$M = \frac{R_A \text{ x a x b}}{L} \tag{2}$$

$$M = \frac{31.9 \text{ N x } 0.0725 \text{ m x } 0.1775 \text{ m}}{0.25 \text{ m}} = 1.64 \text{ Nm}$$

Equivalent torque:

$$Te = \sqrt{T^2 + M^2} \tag{3}$$

$$Te = \sqrt{(3.19 \text{ Nm})^2 + (1.64 \text{ Nm})^2} = 3.587 \text{ Nm} = 3857 \text{ Nmm}$$

Shaft diameter based on equivalent torque:

$$d = \sqrt[3]{\frac{16 \text{ x Te}}{\pi \text{ x } \tau}} \tag{4}$$

$$d = \sqrt[3]{\frac{16 \text{ x } 3857 \text{ Nmm}}{3.14 \text{ x } 6 \frac{N}{mm^2}}} = 14.85 \text{ mm}$$

The bending moment is equivalent:

$$Me = \frac{1}{2}\left(M + \sqrt{T^2 + M^2}\right) \tag{5}$$

$$Me = \frac{1}{2}\left(1.64 \text{ Nm} + \sqrt{(3.19 \text{ Nm})^2 + (1.64 \text{ Nm})^2}\right) = 2.114\, Nm = 2114\, Nmm$$

Shaft diameter based on equivalent bending moment:

$$d = \sqrt[3]{\frac{32 \text{ x Me}}{\pi \text{ x } \sigma}} \tag{6}$$

$$d = \sqrt[3]{\frac{32 \text{ x } 2114 \text{ Nmm}}{3.14 \text{ x } 10.4 \frac{N}{mm^2}}} = 12.75 \text{ mm}$$

Thus, the diameter is chosen based on equivalent torque because it is higher than found on the same bending moment. With the standard shaft diameter, the diameter of the shaft used is 15 mm.

IOP Conf. Series: Materials Science and Engineering **909** (2020) 012034 doi:10.1088/1757-899X/909/1/012034

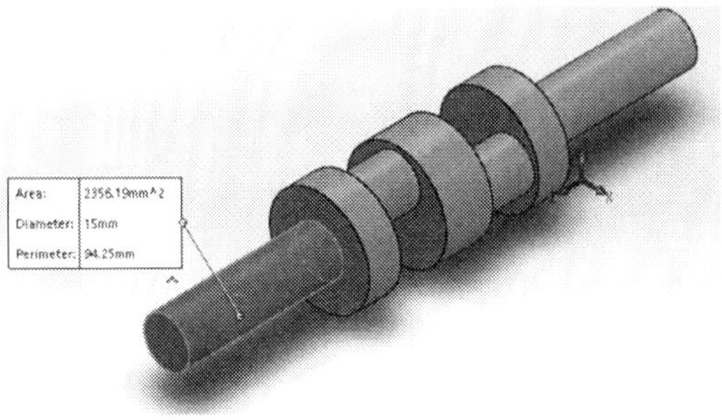

Figure 2. Crankshaft planned

3.2. Key Calculation

Known: $\tau_S = 6$ MPa; $\sigma_C = 9$ MPa

Finding the width (b) and thickness (t) key:

Based on the standard peg table, a shaft diameter of 17 mm was chosen as the standard because a diameter of 15 mm is not in the table, the width of the post (b) = 6 mm, and the thickness of the post (t) = 6 mm.

Table 1. Standard key

Shaft diameter (mm)	Key cross-section		Shaft diameter (mm)	Key cross-section	
	Width (mm)	Thickness (mm)		Width (mm)	Thickness (mm)
6	2	2	85	25	14
8	3	3	95	28	16
10	4	4	110	32	18
12	5	5	130	36	20
17	6	6	150	40	22
22	8	7	170	45	25
30	10	8	200	50	28
38	12	8	230	56	32
44	14	9	260	63	32
50	16	10	290	70	36
58	18	11	330	80	40
65	20	12	380	90	45
75	22	14	440	100	50

Length of key due to shear stress (assuming the post material is the same as the shaft material):

$$T = L \text{ x } b \text{ x} \frac{d}{2} = \frac{\pi}{16} \text{x } \tau_s \text{ x } d^3 \tag{7}$$

$$L = \frac{\pi \text{ x } d^2}{b \text{ x } 8} = \frac{3.14 \text{ x } (15 \text{ mm})^2}{6 \text{ mm x } 8} = 14.72 \text{ mm}$$

Length of key due to stress crushing:

$$T = L\,x\,\frac{t}{2}\,x\,\sigma_c\,.x\frac{d}{2} = \frac{\pi}{16}x\,\tau_s\,x\,d^3 \quad (8)$$

$$L = \frac{\pi\,.\,d^2\,.\,\tau_s}{4\,.\,t\,.\,\sigma_c} = \frac{3{,}14\,.\,(15mm)^2\,.\,6\,{}^{N}/_{mm^2}}{4\,.\,6\,mm\,.\,9\,{}^{N}/_{mm^2}} = 19{,}63\text{ mm}$$

Dimensions of key:

Key length is selected based on crushing stress = 19.63 mm = 20 mm (recommended key length). So, the dimensions of the post L = 20 mm, b = 6 mm, and t = 6 mm.

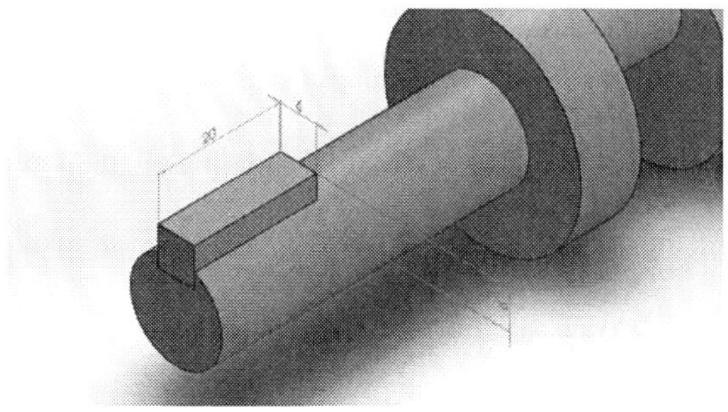

Figure 3. Key planned

3.3. Belt Calculation

Known: P = 100 Watt; n = 300 rpm; D_1 pulley = 100 mm; D_2 pulley = 60 mm; x = 0,367 m; slip = 2; Friction is lost on each shaft = 5%

Figure 4. Schematic pulley and belt

Speed of belt through drive pulleys:

$$v = \frac{\pi.D1.n1}{60}\left(1 - \frac{slip}{100}\right) \quad (9)$$

$$v = \frac{3.14\text{ x }0.1\text{ m x }300\text{ rpm}}{60}\left(1 - \frac{2}{100}\right) = 1.54\ {}^{m}/_{s} = 92.4\ {}^{m}/_{menit}$$

The coefficient of friction:

$$\mu = 0.54 - \frac{42.6}{152.6 + v} \tag{10}$$

$$\mu = 0.54 - \frac{42.6}{152.6 + 92.4} = 0.37$$

Belt rotation on pulleys driven:

$$n2 = v\left(1 - \frac{slip}{100}\right).\frac{60}{\pi.D2} \tag{11}$$

$$n2 = 1.54\ {}^{m}/_{s}\left(1 - \frac{2}{100}\right).\frac{60}{3.14 \times 0.06\ m} = 480.7\ rpm$$

Torque on the pulleys driven:

$$Torque = \frac{P.60}{2.\pi.n2} \tag{12}$$

$$Torque = \frac{100\ {}^{J}/_{s}.60}{2 \times 3.14 \times 480.7\ rpm} = 1987.6\ Nmm$$

Because there is a friction that is lost on each shaft

$$Torque = 1.05 \times 1987.6\ Nmm = 2086.98\ Nmm$$

Look for T1 and T2 values:

$$(T1 - T2)\ r2 = (T1 - T2)\ r1\ mm = Torque \tag{13}$$

$$(T1 - T2) = \frac{2086.98\ Nmm}{30\ mm} = 69.6\ N$$

$$\sin\alpha = \frac{r1 - r2}{x} = \frac{50 - 30}{367} = 0.055, \alpha = 3.2°$$

$$\theta = (180 - 2\alpha) = 173.6 \times \frac{\pi}{180} = 3.03\ rad$$

$$2.3 \log\left[\frac{T1}{T2}\right] = \mu.\theta \tag{14}$$

$$\log\left[\frac{T1}{T2}\right] = \frac{0.37 \times 3.03}{2.3} = 0.5$$

$$\left[\frac{T1}{T2}\right] = 3.2$$

Substituted to $(T1 - T2) = 69.6\ N$

$$(3.2T2 - T2) = 69.6\ N$$

$$T2 = 31.63 \text{ N}$$

$$T1 = 3.2 \times 31.63 \text{ Nmm} = 101.22 \text{ N}$$

Maximum belt tension:

Belt width (b) = 30 mm dan thick belt (t) = 5 mm (assumption)

$$\sigma = \frac{T1}{b \,.\, t} \tag{15}$$

$$\sigma = \frac{101.22 \text{ N}}{30 \text{ mm} \,.\, 5 \text{ mm}} = 0.675 \text{ MPa}$$

Belt length:

$$L = \pi(r1 + r2) + 2x + \frac{(r1 - r2)^2}{x} \tag{16}$$

$$L = 3.14(0.05 + 0.03)\text{m} + 2 \times 0.367 \text{ m} + \frac{(0.05-0.03)^2 \text{ m}^2}{0.367 \text{ m}} = 0.987\, m$$

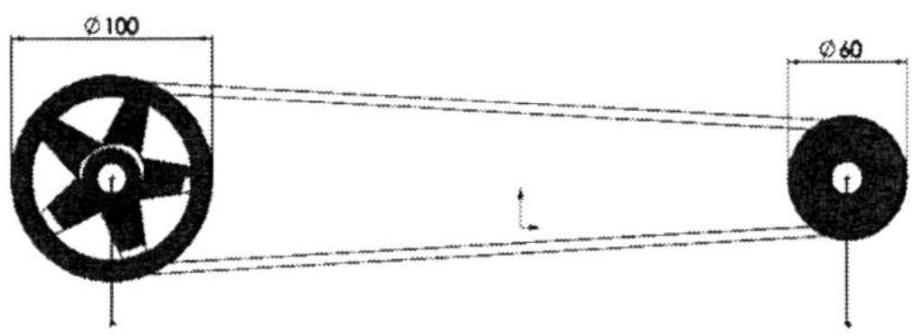

Figure 5. Pulleys and belts design

4. Conclusions

Based on the analysis results of calculations based on the Kwurmi method, it has been obtained specifications of the results of the design of engine elements on the alpha type stirling engine design innovation based on biomass that is suitable and safe is as follows: Shaft Design (Length = 250 mm, Diameter = 15 mm), Stake Design (Length = 20 mm, Width = 6 mm, Thickness = 6 mm), Pulley Design (Driver Diameter = 100 mm, Driven Diameter = 60 mm), Belt Design (Length = 987 mm, Width = 30 mm, Thickness = 5 mm), Stirling Engine (Power = 94.3 to 205.4 Watt, Engine Speed = 300 to 1305 Rpm, Heat Temperature = 623 ° K, Cold Temperature = 305 ° K, Angle Phase = 180 °, Dimension = 580 mm × 250 mm × 700 mm, Thermal Efficiency = 0.51

Acknowledgments

The authors would like to send the gratitude to Faculty of Engineering Universitas Sultan Ageng Tirtayasa who has funded this research in the 2020.

References

[1] C. Lewis 1983 Biological Fuel London: Arnold.

[2] D. Satria, E. Listijorini, and M. R. Nurghodan 2015 Perancangan Sistem Kendali Suhu Pada Mesin Pengering Hybrid Menggunakan Metode Fuzzy Logic *J. Ilm. Rekayasa Pertan. dan*

Biosist **3** (2): 179–185.

[3] V. Gehlot, A. Nigam, and K. Marmat 2014 Development and fabrication of Alpha Stirling *Engine OSR J. Mech. Civ. Eng.* **11** (6): 69–71.

[4] A. Bhagat, A. Modi, P. Hinganikar, P. Tambekar, N. . Kakade, and B. . Kale 2016 Design of Alpha Stirling Engine in Conjunction with Solar Concentrator *Int. Res. J. Eng. Technol* **3** (4): 1109–1114.

[5] C. Roldan, P. Pieretti, and L. Rojas-Solorzano 2010 Conceptual And Basic Design Of A Stirling Engine Prototype For Electrical Power Generation Using Solar Energy *ASME*: 1–11.

[6] F. . Siddiqui, N. Hayat, M. Farhan, M. Farooq, and H. Bilal 2015 Effect Of Phase Angle On The Efficiency Of Beta Type Stirling Engine *J. Fac. Eng. Technol* **22** (2): 99–109.

[7] I. Yuliyani and M. Irwan 2013 Pembuatan dan Pengujian Prototipe Mesin Stirling Tipe Gamma *IRWNS*: 215–219.

[8] D. Satria, S. Susilo, R. Lusiani, and Y. Hermawan 2019 Design of alpha type stirling machine biomass-based innovation design with the capacity of 100 watt.

[9] D. Satria, R. Lusiani, E. Listijorini, A. Aswata, and Y. Hermawan 2020 Analisa Performa Mesin Stirling Tipe Alpha Inovasi Desain Berbasis Biomassa *J. Rekayasa Mesin* **15** (1): 33–41.

[10] R. . Khurmi and J. . Gupta, *Machine Design*. New Delhi: Eurasia Publishing House (PVT.) LTD., 2005.

[11] L. Cohen 1995 *Quality Function Deployment, How to make QFD Work for You*. New York: Addison-Wesley Publishing Company.

[12] D. Satria, D. P. A. L. Esiswitoyo, N. K. Caturwati, E. Listijorini, and R. Lusiani 2018 Body Design Concept of Remotely Operated Vehicle (ROV) of Observation Class with the Method of Concept Screening and Concept Scoring in *MATEC Web Conf. Volume 218, 02009, The 1st International Conference on Industrial, Electrical and Electronics (ICIEE 2018)*.

Design of pneumatic press machine for 250 page bundle book based on microcontroller arduino uno

Erny Listijorini[1], Haryadi [1], Dhimas Satria[1], Sunardi[1], Yusuf Azizi Rahman[1]
[1]Mechanical Engineering Department, Engineering Faculty, Sultan Ageng Tirtayasa University
Jl. Jenderal Sudirman KM 03 Cilegon 42435

Email: ernylisty@untirta.ac.id

Abstrak. Book is a collection of recorded papers that were put together by binding technique. Several types of binding techniques are carried out in the book bundle suppression stage. In this research, designing and building a book bundle press machine. The design starts with the demand for printing, house of quality, determination of specifications, functions, morphology, concepts to produce the best variant selection. The book bundle suppression system used follows the needs of printing with a semi-automatic system. This Research, a pneumatic press system with 250 pages capacity was controlled by Arduino UNO. This study discusses the selection and counting of pneumatic components and their framework, the selection of control components, simulating the results of component selection and control. From the design results obtained component costs of IDR 2,565,500.

1. Introduction

The graphic industry is a company in the printing of books, tabloids, banners, notes, letters and others. According to the Ministry of Industry the growth of the graphics industry this year increased by 5.3%. Indonesia has become one of the countries of Asia whose graphic industry growth is very good.

There are now 1,317 publishers registered as members of the Indonesian Publisher Association . Among those publishers, 94% are listed as active publishers. Active publishers are centered in 24 provinces out of 33 provinces in Indonesia. Around 1,182 publishers are in Java.

This means that around 90% of publishers are concentrated in Java, especially DKI Jakarta, West Java, Central Java and East Java. 1 , The printing industry in Indonesia continues to experience growth until the end of 2012. The graphic printing industry, which rose 40% in 2011 to US $ 392 million compared to imports in 2010 which were only US $ 280 million. Thus industry competition not only attracts consumers to believe in low priced products..

2. Research Methodology

2.1 Research Flow Chart

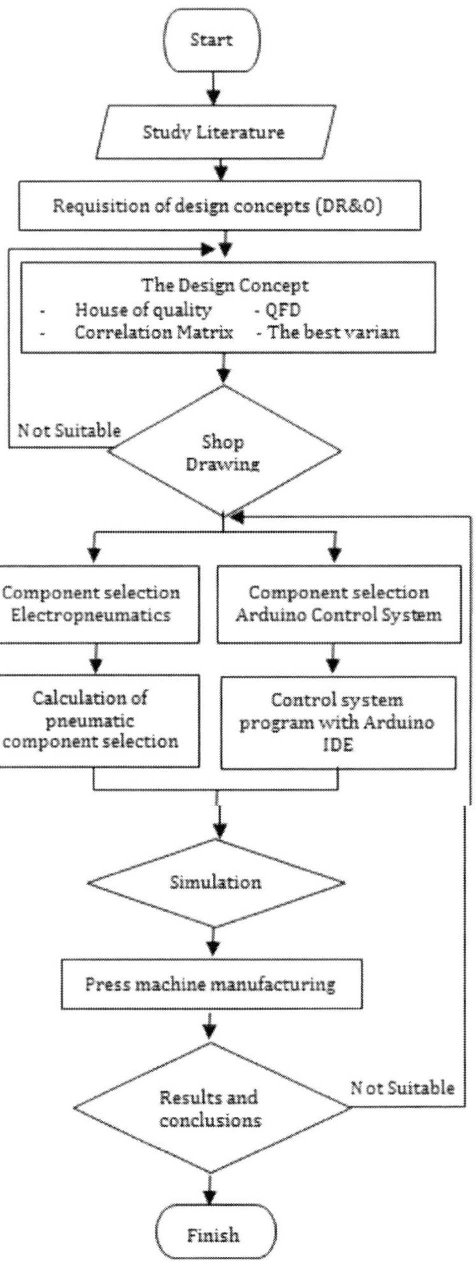

Figure 1. Research flow chart

2.2 Requirement list

Request for machine specifications was obtained from one of the printing companies in the city of Serang. Requests for specifications with interviews produced the following list of requests:

1. The price of production is smaller than the price of the machine on the market.
2. Semi-automatic tool system
3. The operator is as minimal as possible work
4. Does not damage paper

3. Results

IOP Conf. Series: Materials Science and Engineering **909** (2020) 012035 doi:10.1088/1757-899X/909/1/012035

3.1 Determination of the Best Variant

The best variant determination stage is the translation phase of variants that might be applied to this paper bundle press from function variants, shape variants, to physical variants taking into account the advantages and disadvantages of several variants.

In designing the hydraulic system of this paper bundle press machine, the following variants will be used in the table below:

Tabel 1. Variant of book press machine

No	Variant	A	B
1	Drive	fuel motor	Electric motor
2	Press Mechanism	Hydraulic	Pneumatic
3	Power transfer	Pulley & belt	Coupling
4	Control	PLC	Arduino

Of the several variants of this press machine design, to get the best variant then compare the functions of each variant.

From the results of the selection of the best variants, there are 2 best specifications, the choice of thinking about the working system tool. Then the variant was chosen with a pneumatic system, electric motor, clutch and Arduino. The use of hydraulics in addition to expensive component prices and the use of fluid or oil can damage the product.

3.2 Data Operations Press Book Bundle

3.2.1 Open side height and back of book

The first operation data is knowing the height of each bundle to determine the height of the frame and the length of the cylinder stroke with the desired bundle capacity.

Table 2. Tinggi sisi buka dan punggung bundel buku

NO	Number of page	Thick side open	Thick side back	Weight
1	50	5 mm	15 mm	25 Kg
2	100	10 mm	30 mm	50 kg
3	150	15 mm	45 mm	75 kg
4	200	20 mm	60 mm	100 kg
5	250	25 mm	75 mm	125 kg

3.2.2 Pressure Force

In the experiment to find the required compressive force using a sample of 50 sheets of paper under pressure from above. Below there are scales to indicate the amount of weight that is converted to a magnitude of force. The following is an illustration of paper press force taking data.

Figure 2. 250 page emphasis paper sample

In the suppression experiment with 50 sheets of paper 3 times the data was taken and calculated the average weight of the emphasis, on 50 pages of paper required weight of 25 kg of pressure. In the number of each bundle desired the straight comparison calculation is performed as follows:

$$x = 125\ kg$$

The calculation continues on the number of papers 100, 150 and 200 pages, the following results are obtained:

Press force 50 page = 25 kg x 9.86 m/s^2 = 246,5 N
Press force 100 page = 50 kg x 9.86 m/s^2 = 493 N
Press force 150 page = 75 kg x 9.86 m/s^2 = 739,5 N
Press force 200 page = 100 kg x 9.86 m/s^2 = 986 N
Press force 250 page = 125 kg x 9.86 m/s^2 = 1232,5 N

3.2.3 Pressure needed for each book bundle

The area of design pressure on the smallest size paper is A5. Emphasis on paper is desired with an area of paper length A5 = 210 mm and width from the back to the back side = 75 mm.

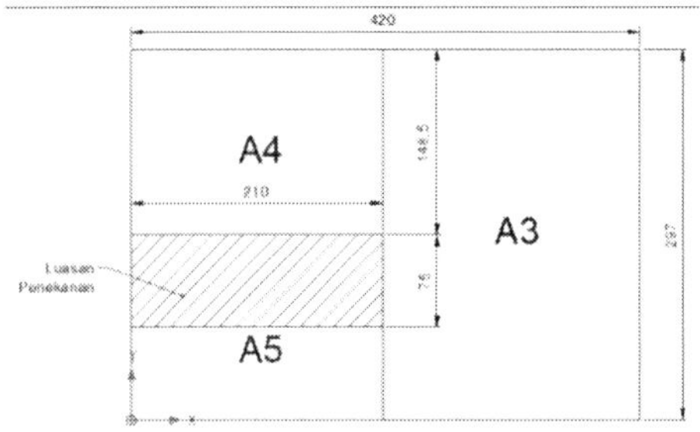

Figure 3. Extent of bundle suppression

With the equation P = F / A, where A = 0.075 m x 0.21 m = 0.01575 m^2. The resulting paper pressure is needed for each amount of paper as follows:

1. Pressure 50 page = $\frac{246{,}5\ \text{N}}{0.01575\ m^2} = 15.650\ N/m^2$
2. Pressure 100 page = $\frac{493\ \text{N}}{0.01575\ m^2} = 31.301\ N/m^2$
3. Pressure 150 page= $\frac{739{,}5\ \text{N}}{0.01575\ m^2} = 46.952\ N/m^2$
4. Pressure 200 page= $\frac{986\ \text{N}}{0.01575\ m^2} = 62.605\ N/m^2$
5. Pressure 250 page= $\frac{1232{,}5\ \text{N}}{0.01575\ m^2} = 78.253\ N/m^2$

3.3 Component Selection Pneumatic

3.3.1 Sylinder pneumatic

The need for compressive force with a number of 250 pages is 1232.5 N and a maximum pressure of 8 bar design with a double working cylinder type. Then the selection of a pneumatic cylinder can be determined by the following calculation:

$$A = 1{,}69x10^{-3}\ m$$

$$D = 0{,}0463\ m = 46\ mm$$

After the cylinder is selected, it calculates whether the cylinder can be used and the air capacity of the cylinder. The calculation of the force and capacity produced on the pneumatic system as follows:

1. The area of the cylinder steps down
$$A = 1{,}96x10^{-3}\ m^2$$
2. Cylinder working force steps down
$$F_{th} = 1570\ N$$
3. The force is able to work effectively the cylinder steps down
$$F_{eff} = 1413\ N$$
4. Extensive cylinder step up
$$A' = 1{,}64x10^{-3}\ m^2$$
5. Cylinder working force step up
$$F_{th} = 1.318{,}8\ N$$
6. The force is able to work effectively the cylinder steps down
$$F_{eff} = 1.187\ N$$
7. Position cylinder volume down
$$V_{ex} = 147\ cm^3$$
8. Position cylinder volume up
$$V_{re} = 1.230\ cm^3$$
9. Total volume
$$V_{ex} = 199{,}9\ cm^3$$

The resulting force of the selected cylinder is 1570 N and the demand force is 1232.5 N. then the selected cylinder can be used. In a working press system, there are a number of different pages for each bundle. In this case, the working pressure on each bundle is also different. Therefore, it is necessary to calculate the working pressure in each bundle with the following calculation:
Pressure for 30 pages (150 N)

$$P = 76.433{,}1\ Pa$$

With the same calculation, each bundle is known by the following table:

Table 3. Press on the number of pages in cm

No	Number of Page	Range sensor(cm)	Compressive weight (Kg)	Press force (N)	Pressure (Pa)
1	30	13	15	150	76.433,1
2	60	12	30	300	152.866,2
3	90	11	45	450	229.299,3
4	120	10	60	600	305.732,4
5	150	9	75	750	382.165,6
6	180	8	90	900	458.598,7
7	210	7	105	1050	535.031,8
8	240	6	120	1200	611.464,9
9	270	5	135	1350	687.898,0

3.3.2 Pneumatic Hose

To drain fluid from the compressor to the cylinder using a pressurized hose. The minimum diameter used in this design can be calculated as follows:

$$d = 2{,}5x10^{-3}\ m = 2{,}5\ mm$$

From the calculation results obtained a minimum internal diameter of 1.35 mm. from the hose selection results obtained hose with the following specifications:

- Internal diameter : 5 mm
- Outer diameter : 8 mm
- Max. pressure : 10 bar

Calculation of corrections to pressure drop

$$\Delta P = 158{,}8\ N/m^2\ < 5000\ N/m^2$$

then the diameter of the hose can be used.

3.3.3 Compressor Selection

In the selection of the compressor must weigh the flow rate, pressure and power needed in the design. The calculation of compressor power needed is as follows:

- Compressor discharge

$$Q_s = 3{,}14x10^{-5}\ m^3/2\ = 1{,}88\ l/min$$

- Power Compressor

$$N_s = 1{,}5\ KW$$

- Drive motor power

$$N_m = \ 1{,}435\ KW$$

3.4 Frame Material Selection

This material selection is intended for the type of material and profile iron that is right for the frame on the press.

Function : Frame mesin press
Constrain : Yieldstrength min=78.253 N/m^2
Modulus of elasticity : 100 – 300 Gpa
Objective : Minimum cost
Free Variabel : Pemilihan material

From the screening results that match the specifications and the lowest price, low carbon steel material is obtained. The profile steel used as the frame is UNP 50 channel iron (50 mm x 38 mm x 5 mm). 2016 SP1 solidwork application simulation was carried out in order to produce a maximum working stress of 3.125 x $〚10〛^7$ N / m ^{2} and yield strength of cast carbon steel material = 2,482 x $〚10〛^8$ N / m ^{2}. Obtained safety factor (SF) = permit voltage divided by work voltage resulting in 7.9 Maximum displacement occurred = 0.12 mm. The following is a picture of the static frame simulation results.

Figure 4. Static simulation of a pneumatic press frame with a force of 1232.5 N

3.5 Calculation of Welding Connection Strength

In the process of making the framework of this press machine using UNP 8 iron with the welding connection method. Calculation only on stems subject to large forces. As for the bars that counted bars 1,2 and 3 from the following image:

Welding uses type SMAW (shield metal arc welding) with AWS E6013 standard welding wires. Then the welding wire specifications are based on AWS standards namely;

$$F_{t1} = 330.947{,}2\ N$$

$$F_{t2} = 330.947{,}2\ N$$

$$F_{maks\ las} = 661.894{,}3\ N$$

$$F_{press} < F_{weld\ maks}$$
$$1232{,}5\ N < 661.894{,}3\ N\ \text{(Safe)}$$

3.6 Calculation of Cylinder Thread Connection with Frame

Cylinder specifications use 10 bolt heads with M6 threads. The bolt eyes are 4 and are made of black steel with a tensile stress of 1600 kg/cm^2. Then the maximum tensile load capable of a cylindrical screw is:

$$P = 11.445\ N$$

Tensile load operation < Maximum screw tensile load

1232,5 N < 11.445 N (Safe)

3.7 Simulation Elektropenumatik

In this electrop pneumatic system there are 2 conditions as control inputs so that the working pressure adjusts the number of pages in a book bundle. Input 1 is to find out the number of pages to be pressed and pressure 2 input that has worked on the pneumatic system When the on / off button is pressed, the electrical circuit will flow current to the active inductive proximity due to the extended cylinder position, then flow to relay Y1 which activates the NO Y1 contactor. The pressure sensor will be active if the pressure is as desired, thus activating the Y3 relay to activate the NO S3 contactor.

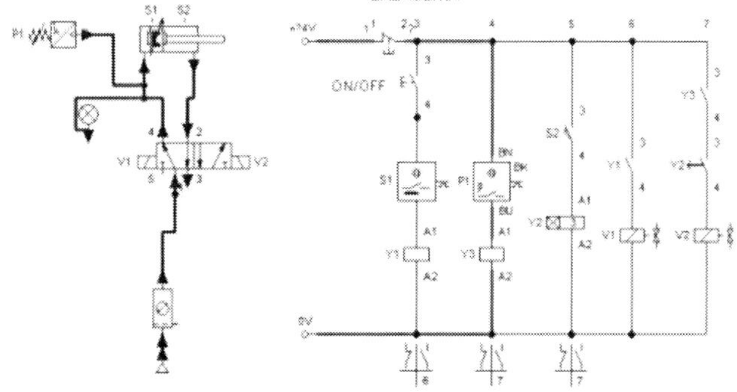

Figure 5. Pneumatic system and circuit extend condition

When the cylinder reaches the max stroke, the contactor Y2 will be active and the pressure sensor will detect whether the pressure has been reached. If it has been reached, the Y3 relay will be active and wait for a countdown of 3 seconds so that the Y2 solenoid valve will flow through. Then the direction valve will move to the left and the cylinder will extend.

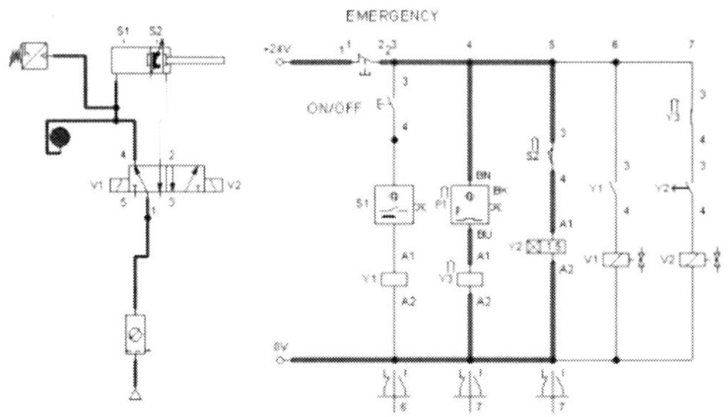

Figure 6. Pneumatic system and circuit condition reaction

3.8 Arduino Simulation of Pneumatic Control with Proteus

The position of the ultrasonic sensor is above the bundle of books you want to press. To determine the thickness of the book, it can be calculated the thickness of the book = 14 cm - the distance readable sensor. After obtaining the thickness of the book which is used as a reference for a large pressure that must be pressed. As for each sensor distance that is read on each book thickness from 0 to 250 pages for the control program follows table 4.2. This controller uses arduino with C. The logical function used is IF-ELSE. The logic statement used is "if the sensor read distance is smaller than 14 cm, it will flow through the current to relay Y1. Otherwise relay Y1 does not flow. In the presentation of the simulation using lights as an indicator of current flow or not.

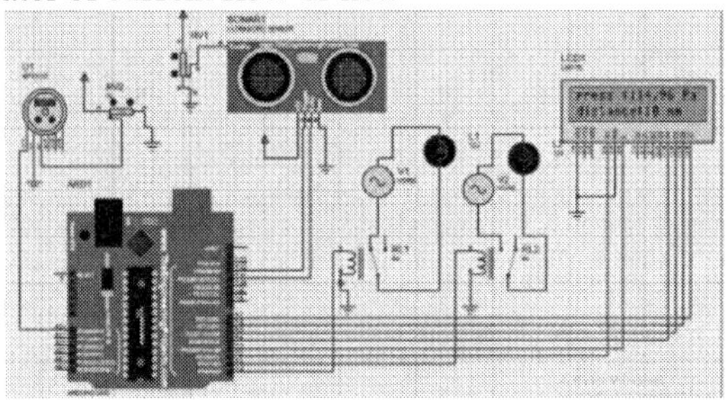

Figure 7. Control and coding diagram with arduino UNO Y1 relay active condition

If the conditions do not match the program it will activate relay Y2 (cylinder extend). The logic statement used is ". If the sensor distance is greater than 14 cm, the relay Y2 is active" and "if the distance is less than 14 cm (for example 13 cm) and the pressure is greater than 76,443.1 pa, then the Y2 relay is active"

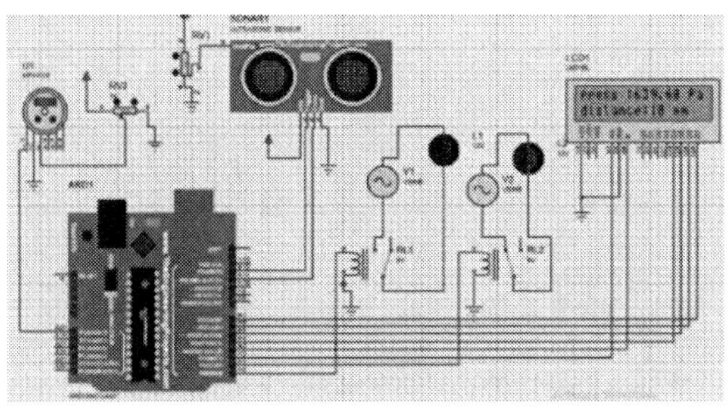

Figure 8. Control and coding diagram with arduino UNO Y2 relay condition active

4. Conclusion

From the results of the design selection and calculation of the design of a 250 page book pneumatic press machine capacity based on the Arduino UNO microcontroller, the following conclusions are obtained:

1. The design uses a 250-page Pneumatic system with an arduino control system. Hydraulic components used are:
   a. 8.75 MPa double acting pneumatic cylinder with 40 mm cylinder
   b. Compressors with 1.5 KW of power and minimum discharge of 1.88 L / min
   c. The framework uses UNP 8 with the dimensions attached
   d. Direction valve dengan seleniod 5/2 double seleniod AC 220 volt
2. The framework uses UNP 8 with the dimensions attached
3. The cost of components for producing pneumatic book bundle presses costs Rp. 2,565,500.

References

[1] Croser, Peter dan Frank Abel 2002 Pneumatik basic level festo deditac Denkendorf: Festo Deditac GmbH & Co

[2] Fadillah, Faqih 2017 Rancangan mesin cetak dan press tahu sistem pneumatic dengan kendali plc untuk meningkatkan mutu produksi home industry tahu Malang: Universitas Negeri Malang

[3] Hidayat,Taufan 2012 Analisis kinerja ketahanan tekan lingkar sebagai parameter mutu kertas lainer dan medium Bandung; Balai Besar Pulp dan Kertas

[4] J, K, Gupta dan R, S, Khurmi 1982 Machine Design New Delhi: New Azad Book Binding House

[5] Phal,G. and Beitz, W., Feldhunsen, J., and Grote K.H 2004 Engineering Design Thrid edition

[6] R Majumdar 1995 System pneumatic principle and maintenance New Delhi: Tata-Mc Graw – hill

[7] Santoso, Hari 2015 Panduan praktis arduino untuk pemula

[8] Satria, Dhimas 2014 Elemen mesin 1 Serang Untirta Press

[9] Standar Cetak Indonesia 2006 *Kertas Cetak A* https://www.academia.edu/21241309/Standar_Nasional_Indonesia_Kertas_cetak_A

IOP Conf. Series: Materials Science and Engineering **909** (2020) 012036 doi:10.1088/1757-899X/909/1/012036

Development of drying machine to improve palm sugar quality

Erwin Erwin[1,3*], Meutia[2,4]

[1] Mechanical Engineering Department, Universitas Sultan Ageng Tirtayasa
[2] Agribusiness departemen, Universitas Sultan Ageng Tirtayasa
[3] Renewable energy research laboratory, Universitas Sultan Ageng Tirtayasa
[4] Indonesia center of excellence for food security (local food innovation), Universitas Sultan Ageng Tirtayasa

*Email: erwin@untirta.ac.id

Abstract. One variant of palm sugar products is sugar in the form of powder or often called brown sugar, which is of good quality is bright yellow sugar, with a strong special aroma and a uniform and dry crystal size according to (Indonesian nation standardized) SNI 01-3743-1995 maximum allowable water content is 3%. An artificial drying machine is needed to speed up the process of separating the water content in sugar. This machine is designed with a tiered rack with a heating source. The temperature regulation uses thermostat, the way it works is by electrically charged elements to heating air, and directing hot air into the drying chamber by using fan (fan blower) and (exhaust fan) for air circulation. The quality to be achieved in the experiment is sugar with a maximum water content of 3%. Experiments were carried out several times with brown sugar that has been matched in grain size. Drying time is very important, because this time is related to production capacity and energy consumption. The remaining water content is less than 3% according to the needs of the planning, the tool is functioning properly.

1. Introduction

Palm sugar is one of the commodity products in Lebak-Banten district whose production has penetrated national and international markets, with export volumes reaching 30 tons per month. And according to data from the district's industry and trade service, the number of palm sugar makers is 5,815 units.

Good quality palm sugar, which has a bright yellow color, has a strong characteristic aroma and a uniform and dry crystal size according to the SNI 01-3743-1995 quality standard. The maximum permissible water content is 3% [1]. Based laboratory test on palm sugar contains water ranges from 4.29% by weight of sugar.

There are two ways that can be used, the first by drying in the sun the advantages of this method is cheap from the cost of unnecessary expensive support, and the drawback is the amount of heat obtained is not fixed so long drying time is difficult to predict, ant sugar can be contaminated with wind-borne dust, and also requires a large area.

The second method with an artificial dryer (oven) excess temperature can be controlled, cleanliness is maintained and the heat obtained remains unaffected by weather conditions so that the drying time can be predicted, the shortage of this method is the need for additional costs for the manufacture and operation of the equipment.

This research aims to produce a product that can reduce the water content of palm sugar so that it can meet the requirements of SNI standards, and export standards.

2. Research Methodology

When the drying process takes place evaporation of water from the sugar granules, the water moves to the surface. The mechanism of movement of water affects drying during periods of constant drying rate and when the drying rate decreases [2]. There are three forms that explain the rate of drying figure 1:

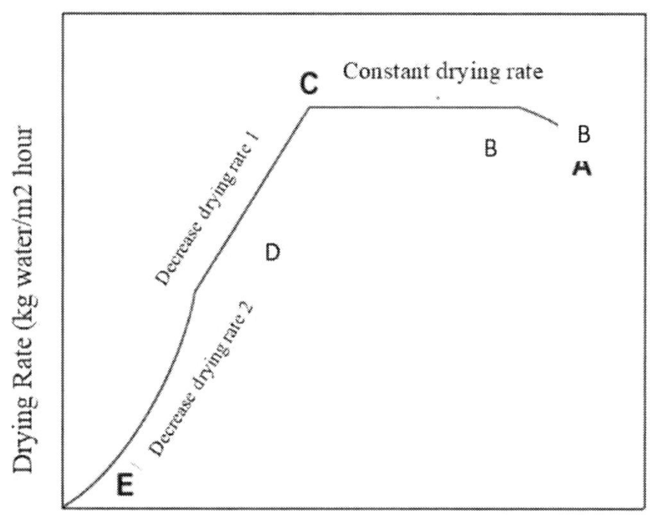

Figure 1. Drying speed curve

To determine the moisture content of the material determined based on dry weight (dry basis) and wet weight (wet basis). Calculation of water content (wet basis) can be formulated as follows:

$$M\_wb = (W1 - W2)/W1 \times 100\% \quad (1)$$

Where:
M_wb = water content (%)
W1 = Wet Sugar Weight
W2 = Dry sugar weight

To find out the dry basis water content, formula is used:

$$M\_db = (100 \cdot M\_wb)/(100 - W\_wb) \quad (2)$$

2.1. Heat for drying

When the heat energy is received or released by a substance, then the object will experience two things, namely changes in temperature and form. An increase in temperature on an object can be determined using an equation that relates to the specific heat or heat capacity.

Drying load of sugar is calculated by the equation below:

$$Q_{tot} = Q_1 + Q_2 + Q_3 + \ldots + Qn \quad (3)$$

$$Q_1 = m_k.Cp.\Delta T \quad (4)$$

$$Q_2=m_a.Ca.\Delta T \tag{5}$$

To measure the amount of heat needed for evaporation of water in ant sugar, the following equation is used:

$$Q_3= m_{water\rightarrow steam}\cdot hfg \tag{6}$$

determine the amount of heat given by the heating air to the dried material used in the equation:

$$q = \rho \cdot v \cdot c_u \cdot (T_1 - T_2) \tag{7}$$

To determine the drying efficiency used equation:

$$\eta = \frac{Q}{q} \times 100\% \tag{8}$$

while the equation to find out the amount of heat needed during drying is:

$$q = Q_{tot} \times t \tag{9}$$

To find out the drying mass rate, a formula is used:

$$m_{udara} = \rho \times Q_{debit} \tag{10}$$

while the formula for calculating total energy is used

$$E_{total\ terpakai} = E_{elemen} + E_{kipas} \tag{11}$$

Heat for drying can come from heating elements that use electricity or use heat from other sources such as renewable energy [3].

2.2. Drier design and testing

The dryer must be designed to fulfill the function so that the work process runs smoothly and in line with expectations. There are 4 systems that must be designed, namely the frame, heater, drying chamber and air circulation unit. The frame design is made as shown below:

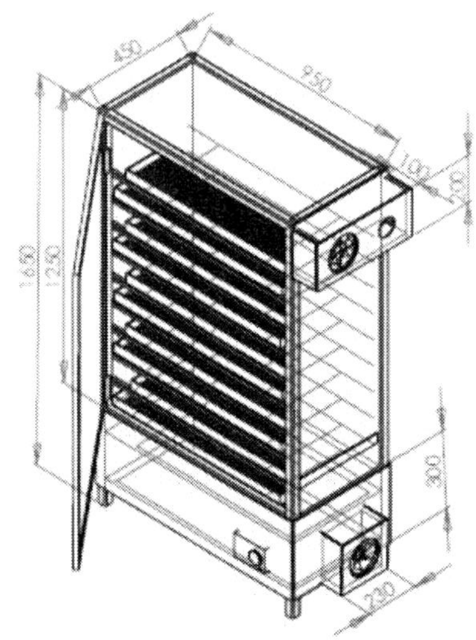

Figure 2. Drying machine rack design

In designing the product specification target is the capacity of drying is ± 100 kg / hour, the available electric power is limited to 900 watts, and is able to reduce water content up to 3% for 60 minutes per batch.

From the results of laboratory tests, it is known that the percentage of water content is wet sugar base (m_b)= 4.29% and dry(m_k)=2.84%. To find out the mass of palm sugar can be found by the equation (for 100gr sugar):

$p = \frac{m}{v}$ (12)

so: $p = \frac{100}{109,65} = 0,911$ gr/cm^3

and to find out the mass of sugar water content that must be evaporated (m_{water}) the 9 kg trial material is as follows:

amount initial water mass$= \frac{4,29}{100} \times 9$ kg $= 0,386$ kg

m_k = 9 kg - 0,386 = 8,614 kg

sugar weight for 2,84% $= \frac{8,614 \times 100}{100 - 2,84} = \frac{861,4}{97,16} = 8,865$ kg

M_{water} steam = 9 – 8,865 kg

M_{water} steam = 0,135 kg

For a drying load of 30 kg:

Initial water mass $= \frac{4,29}{100} \times 30$ kg = 1,287 kg

Dry weight = 30 kg - 1,287 = 28,713 kg

Final water weight 2,84% $= \frac{28,713 \times 100}{100 - 2,84} = \frac{2871.3}{97,16} = 29,55$ kg

M_{water} = 30 – 29,55 kg

M_{water} steam = 0,45 kg

Calculate the drying heat energy requirements.

Design specification:

Weight of dried sugar	= 9 kg
Drying air temperature	= 60°C
Sugar end temperature	= 54 °C
Outlet air temperature	= 53 °C
Air speed	= 150 Cfm
sugar specific heat	= 0,9072 kj/kg. °C
initial water percentage	= 4,29%
expected water percentage	= 2,84%
time	= 60 minutes

- Sensible heat calculation of sugar (Q_1)

$$Q_1 = m_k \,.Cp\,.\,\Delta T \quad (12)$$

- Calculation of the sensible heat of water in sugar (Q_2)

$$Q_2 = m_a . C_a . \Delta T \quad (13)$$

- Calculation of heat for evaporation (Q_3)

$$Q_3 = m_{water\rightarrow steam} \cdot hfg \quad (14)$$

- Heat absorbed by the wall heater tube (Q_4)

$$Q_4 = m_{tube}.Cp_{tube}.\Delta t \quad (15)$$

- Heat absorbed by the glass wall (Q_5)

$$Q_5 = m_{glass}.Cp_{glass}.\Delta t \quad (16)$$

- Heat absorbed by the tray plate (Q_6)

$$Q_6 = m_{tray} . Cp_{tray} . \Delta t \quad (17)$$

- Heat absorbed by the drying frame (Q_7)

$$Q_7 = m_{kp} . Cp_{kp} . \Delta t \quad (18)$$

- The total need for drying heat is:

$$Q_{tot} = Q_1 + Q_2 + Q_3 + Q_4 + Q_5 + Q_6 + Q_7 \quad (19)$$

Q_{tot} = 187,55 kj + 38,9 kj + 326,565 kj + 19,8 kj + 10,75 kj + 24,19 kj + 86,4 kj

Q_{tot} = 694,155 kj = 92.77 watt/hours

- Mass drying rate:

$$M_{air} = \rho \times Q_{rate} \quad (20)$$

- The heat supplied by air for the evaporation of material water:

$$q = \rho \cdot V \cdot c_u (T_1 - T_0) \quad (21)$$

- So to find out the amount of heat during drying are :

$$q = Q_{tot\ drying} \times t \quad (22)$$

- Drying efficient

$$\eta = \frac{\text{the amount of heat used}}{\text{heat given by air}} \quad (23)$$

- Total energy used

$$E_{total\ used} = E_{heater} + E_{fan} \quad (24)$$

If the rack is full of 30kg wet sugar, the drying load is:

$$Q_1 = m_k . Cp . \Delta T \quad (25)$$

30kg load divided by 10 shelves

Q_1 = 28,713 kg · 0,9072kj/kg °C · (52-30)°C

Q_1= 573,06 kj

Q_2 = 1,287 kg · 4,2 kj/kg °C · (54-30)°C

Q_2 = 128 kj

Q_3 = 0,45 kg · 2419 kj/kg

$Q_3 = 1088{,}55$ kj

- The total need for drying heat is:
  $Q_{tot} = Q_1 + Q_2 + Q_3 + Q_4 + Q_5 + Q_6 + Q_7$

  $Q_{tot} = 573{,}06$ kj + 128 kj + 1088,55 kj + 19,8 kj + 10,75 kj + 24,19 kj + 86,4 kj

  $Q_{tot} = 1930{,}75$ kj

  = 536.11 watt/hours

- Total drying time for 30 kg sample:
  Known load 9 kg and range time is 60 minutes

  Average time per shelf $= \frac{60}{3} = 20$ minutes

  30kg load divided by 10 shelves $= 20 \times 10$

  $= 200 \text{ minutes} = \frac{200\text{minutes}}{60\text{minutes/hours}}$

  = 3,3 hours

- So to find out the amount of heat during drying is:

$$q = Q_{tot\ drier} \times t \tag{26}$$

  q = 536.11 wh · 3,3hours

  q = 1769,16 watt/hours

- Efisiensi pengeringan

  $\eta = \frac{1769{,}16}{6{,}36\ m^3/s \cdot 200 \cdot 60} \times 100\%$

  $= 2{,}31\ \%$

And to reduce heat losses due to leakage, every corner of the connection, air gap as much as possible to be closed with silicon glue.

Table 1. Heat absorbed when drying 9kg

Heat absorbed	Amount
Heat to heat sugar (Q_1)	187,55 kJ
Heat to heat water in sugar (Q_2)	38,9 kJ
Heat to evaporate water (Q_3)	326,565 kJ
The heat absorbed by the heater tube (Q_4)	19,8 kJ
Heat absorbed by the glass wall (Q_5)	10,75 kJ
The heat is absorbed by the tray plate (Q_6)	24,19 kJ
The heat absorbed by the drying frame (Q_7)	86,4 kj
Total heat absorbed (Qtot)	694,155 kJ

Table 2. Heat absorbed when drying 30 kg palm sugar

Heat absorbed	Amount

Heat to heat sugar (Q_1)	1079,665 kJ
Heat to heat water in sugar (Q_2)	128 kJ
Heat to evaporate water (Q_3)	1088,55kJ
The heat absorbed by the heater tube (Q_4)	19,8 kJ
Heat absorbed by the glass wall (Q_5)	10,75 kJ
The heat is absorbed by the tray plate (Q_6)	24,19 kJ
The heat absorbed by the drying frame (Q_7)	86,4 kj
Total heat absorbed (Qtot)	2437,355 kJ

3. Experiment Result

Each rack contains 3 kg of palm sugar with a water content of 4.3%, drying temperature is set at 60°C, and the result can be seen at table 3.

Table 3. Experiment result

No	Time (minutes)	Temperature reached (°C)		
		Rack 1 (lower)	Rack 2 (middle)	Rack 3 (upper)
1	20	38	37	38
2	30	43	43	42
3	40	44	44	43
4	50	51	50	50
5	60	54	53	53
Final weight (kg)		2.954	2.955	2.955

Figure 3. Dried palm sugar

4. Result

The design of a palm sugar dryer has succeeded in reducing the water content from 4.3% to 2.95% in 60 minutes. There are some things that still need further development. Durability testing of the device also needs to be done to determine it's resilient. Utilization of renewable energy can be done in the heating system as it is done in the liquid sugar evaporation process [4].

5. Reference

[1] BSN 1995 *Gula Palma*

[2] Geankoplis, C.J. 1983 *Transport processes: Momentum, heat and mass*United States: Allyn and Bacon Inc.

[3] Lusiani, R., A. Sudrajat, and Erwin 2019 Effect of air fuel ratio (AFR) on performance of gasification machine TG30-1 using water boiling test method on wooden powder biomass *IOP Conference Series: Materials Science and Engineering* **673** 012123

[4] Listijorini, E. and S. Wiyono, 2019 Rice husk gasification performance to decrease water content in liquid palm sugar *IOP Conference Series: Materials Science and Engineering* **673** 012121

IOP Conf. Series: Materials Science and Engineering **909** (2020) 012037 doi:10.1088/1757-899X/909/1/012037

Characteristics of aluminium AC4B reinforced with nano -SiC composites through stir casting methods

T N Fitri[1], W Narottama[1], and A Zulfia[1*]

[1] Department of Metallurgical and Materials Engineering, Faculty of Engineering, University of Indonesia, UI Campus, Depok, 16424, Indonesia

*Email: anne@metal.ui.ac.id;

Abstract. Aluminium with a composition of 7-10% Silicon (Si), 2-4% Copper (Cu) is a matrix used by adding Nano SiC reinforcement particles with variations of 0.25; 0.3; 0.4*%Vf* with the purpose of making bullet-proof materials that have high toughness, impact resistance, and good strength. In this study, fabricated composites used a stir casting process to produce superior mechanical properties which was caused by the bond between an effective matrix and reinforcement particles. Besides, the addition of Nano SiC reinforcement particles, it was also followed by adding 5*%wt* Magnesium (Mg) as a wetting agent, 0.04*%wt* Strontium (Sr) as modifier agent, and 0.15*%wt* Al-5Ti-1B as a grain refining agent. Fabricated composite specimens will be carried out several tests to determine mechanical properties such as density, porosity, hardness, tensile strength, impact resistance, and ballistic test. The result of the test will be compared to the results of that without the addition of reinforcement particles. Then, using the *Optical Microscope* (OM), *Scanning Electron Microscope* (SEM), *Dispersive X-Ray Spectroscopy* (EDX), and *X-Ray Diffraction* (XRD) to analyse the dispersion of reinforcement particles in the selected alloy matrix.

1. Introduction

As one of the developing countries, Indonesia has now entered a new phase in the development of domestic infrastructure. Improving the quality of domestic infrastructure is considered as one of the foundations of a country's progress. Almost all infrastructure in this country uses materials that have high strength, endurance and toughness to support all activities. Of the many materials used, there is currently an increasing use of composite materials to obtain superior properties, such as one of the studies that will be conducted is the use of composite materials. This composite material can be said as a substitute for iron or steel as the main source of infrastructure in this country. Steel has long been used for application because of the ease of the process and lower production costs and has a variety of mechanical properties that vary. However, steel has a high specific gravity so it is not suitable for application with light specific gravity requirements.

The material used as an alternative matrix is Aluminium which has a low density of 2.7 g/cm^3 (1/3 mass of steel), corrosion resistant properties, cast ability properties, machinability properties, and a low processing temperature that facilitates the fabrication process in the Metal Matrix Composite. Because it has low strength, Aluminium is added with the elements of alloys to increase its strength in an efficient manner, for example is copper, manganese, silicone, zinc, magnesium, and etc. In addition, Aluminium has characteristics that are low density, melting point, resilience, and good ability to be moulded.

In this study, it was used a composite with an Aluminium matrix AC4B with micro Nano SiC reinforcement that have a superior properties compared to micro-sized particles because of their small size so that they have a high surface area to volume ratio. In addition of Nano SiC particles also have advantages when compared with the SiC particle, although the tendency for agglomeration is quite high. Beside that, there is addition elements of alloys such as Magnesium (Mg) to increase the wetting behaviour or the elasticity of SiC particles and matrices and to increase bonding strength, Strontium (Sr) as a modifier to change the microstructure of Aluminium to be more spherical so as to reduce stress concentration from material, TiB as a grain refiner that serves to nucleate grains so that a heterogeneous process will occur nucleation and make the solidification process run quickly. The addition of Nano SiC reinforcement will increase the mechanical strength of the composite material with the Orowan Strengthening mechanism[1]. Orowan mechanism is dislocation that also bow around a precipitate particle through. The effect of a precipitate on the movement of dislocations essentially depends on it's strength and size. The two boundary cases in the interaction with dislocation are called by the Orowan mechanism. The formation of precipitation is dependent on temperature and time and usually occurs in discontinuous manner. While the alloy components first exhibit an irregular distribution, they accumulate at preferred positions in the lattice at elevated temperatures, leading to the formation of short-range separation and clustering.

Fabrication for the manufacture of composites is done by stir casting techniques, wherein the method is easy to do because it is economical and can produce defects in the porosity of the material and the uncomplicated process. With the stir casting method, the homogenous distribution of reinforcing particles is the final aim to be obtained in the composite final product to achieve maximum strength. Therefore, the rotation speed and length of the stirring process must be adjusted to achieve optimal conditions. This process is still widely used in aluminium casting because the process is simple, flexible, and applicative to produce in large quantities, so this method fit in with the current environment.

2. Methods

The study was carried out by a 2-step process. Those were fabrication of composite samples with stir casting methods and testing of samples from chemical composition testing to mechanical properties. In the fabrication process, it began by melting Aluminium AC4B on the melting furnace tilting with a temperature of 850°C and pre-heating Nano SiC reinforcement particles for 1 hour at a muffle furnace at 950°C, after which the reinforcing particles are inserted into melting Aluminium. Then, the melting Aluminium was stirred using a stirrer for 40 seconds and followed by a degassing process using argon gas for 2 minutes. After that, elements of alloys were added, namely, 5wt% of Mg, 0.15wt% of TiB, and 0.04wt% of Sr on melting Aluminium which was then carried out by stirring with a stirrer followed by degassing again and followed by slagging which floated on melting Aluminium composite with a mixture of Nano SiC reinforcement particles. Then, the melting Aluminium composite was poured into a mould that has been preheated at 400°C, followed by ice-water cooling process. AC4B/Nano SiC composite made with variations of 0.25; 0.3; 0.35; 0.4; and 0.5%Vf. The composition of pure Aluminium AC4B can be seen in table 1.

Table 1. Al AC4B composition determined by Optical Electron Spectroscopy

Element	Al	Si	Cu	Mg	Fe	M N	Sr.
%	86.5	7.65	2.51	1.05	1.01	0.233	<0.0001

Destructive and identification of elements and compounds test are carried out to determine the characteristics of the composite sample made. Tensile test is carried out to determine the tensile strength and elongation with the GoTech 27-7000 LA 10 according to ASTM E8 standard. Hardness test is carried out using the Rockwell B method according to ASTM E18-11 standard using a 1/16-inch diameter steel ball indenter. The impact test is carried out using the Charpy method with the GoTech Testing Machine according to the E23-01 standard. Density testing was carried out to determine the effect of adding Nano SiC to composite density using the Archimedes law and then calculate the amount of porosity in the sample. Microstructural observations are carried out using an optical microscope according to ASTM E3-11 standard using ETSA HF.

Identification of phase, interface, and distribution of Nano SiC particles were carried out using SEM and EDS. OES testing is done to observe the elements contained in the composite. XRD is carried out to determine the compounds that can be formed in the AC4B/Nano SiC composite.

3. Results and Discussion

3.1 Chemical composition on Al AC4B/Nano SiC composites

Table 2. shows the comparison of the chemical composition of pure AC4B alloys with that with the addition of Nano SiC reinforcement particles in each variation, which is 0.25; 0.3; 0.35; 0.4; and 0.5%Vf. AC4B is an Aluminium alloy with the main alloy 7-10% Si and 2-4% Cu. If it is seen in table 2., the results of the test, there are less levels of Si and Cu in the composite material caused by the use of scrap AC4B composites as a matrix. The Mg element increased significantly in the composite because it was added to increase the wetness for 5%, that is increase the wettability between the AC4B matrix and reinforced Nano SiC for make the better interface [6]. While for the high Fe element it was caused by the stirrer rod of SKD 61 which melted during the stirring process. This can also be explained by the fading mechanism that occurs due to the too long holding time that is set when casting and the lack of stirring so that in the presence of deposition of grain refiner particles on crucible.

Table 2. Composition of AC4B/Nano SiC Composite determined by Optical Electron Spectroscopy

Element	Composition (%wt.)					
	AC4B ingots (Matrix)	AC4B / 0.25% Nano SiC	AC4B / 0.3% Nano SiC	AC4B / 0.35% Nano SiC	AC4B / 0.4% Nano SiC	AC4B / 0.5% Nano SiC
Al	86.5	82,300	57,473	82,067	82,567	79,233
Si	7.65	7.007	5.617	6.617	6,900	7,013
Fe	1.01	4,403	3,763	4,013	4,677	3,600
Cu	2.51	2,040	1,471	2.107	2,037	1973
M N	0.233	0.218	0.162	0.310	0.225	0.202
Mg	1.05	> 2.4	> 2.4	4,517	> 2.4	6,500
Sr.	<0.0001	0.0097	0.005367	0.011033	0.0104	0.016167

3.2 Microstructure of Al AC4B/Nano SiC Composites

Al_5FeSi phase is shown in Figure 1. which has a flat and sharp shape which is hard and brittle, where the phase can be formed due to the presence of a high Fe element in the composite. This phase is clearly seen in the composition of 0.3% of Nano SiC shown in figure 1 (c). The eutectic phase is also slightly seen where the phase will increase the hardness of the composite which is the residual of the Si element which does not form a phase with other elements[7]. Mg_2Si phase is also seen where Mg_2Si binary has a Chinese script form, which looks almost in all parts of the microstructure in each variation. The π-$Al_9FeMg_3Si_5$ phase also present in microstructure wherein the phase is brittle[8]. The phase formed can be seen more clearly in figure 2.

Al_5FeSi phase can be formed through a rapid cooling process with the reaction of $L + Al_8Fe_2Si \rightarrow Al + Al_5FeSi$ in which the phase has lower mechanical properties than Al_8Fe_2Si[2].The π-$Al_9FeMg_3Si_5$ phase can be formed due to the presence of Fe and also Mg elements with a content below 6%. The phase is formed by reaction $L \rightarrow Al+Si+Mg_2Si+ Al_9FeMg_3Si_5$[3,11].

The morphology of these phase which tend to have a sharp angles like the needle-like that can provide many points of high loading concentrations, so that final product of composite will be brittle and the mechanical strength will be decrease[10].

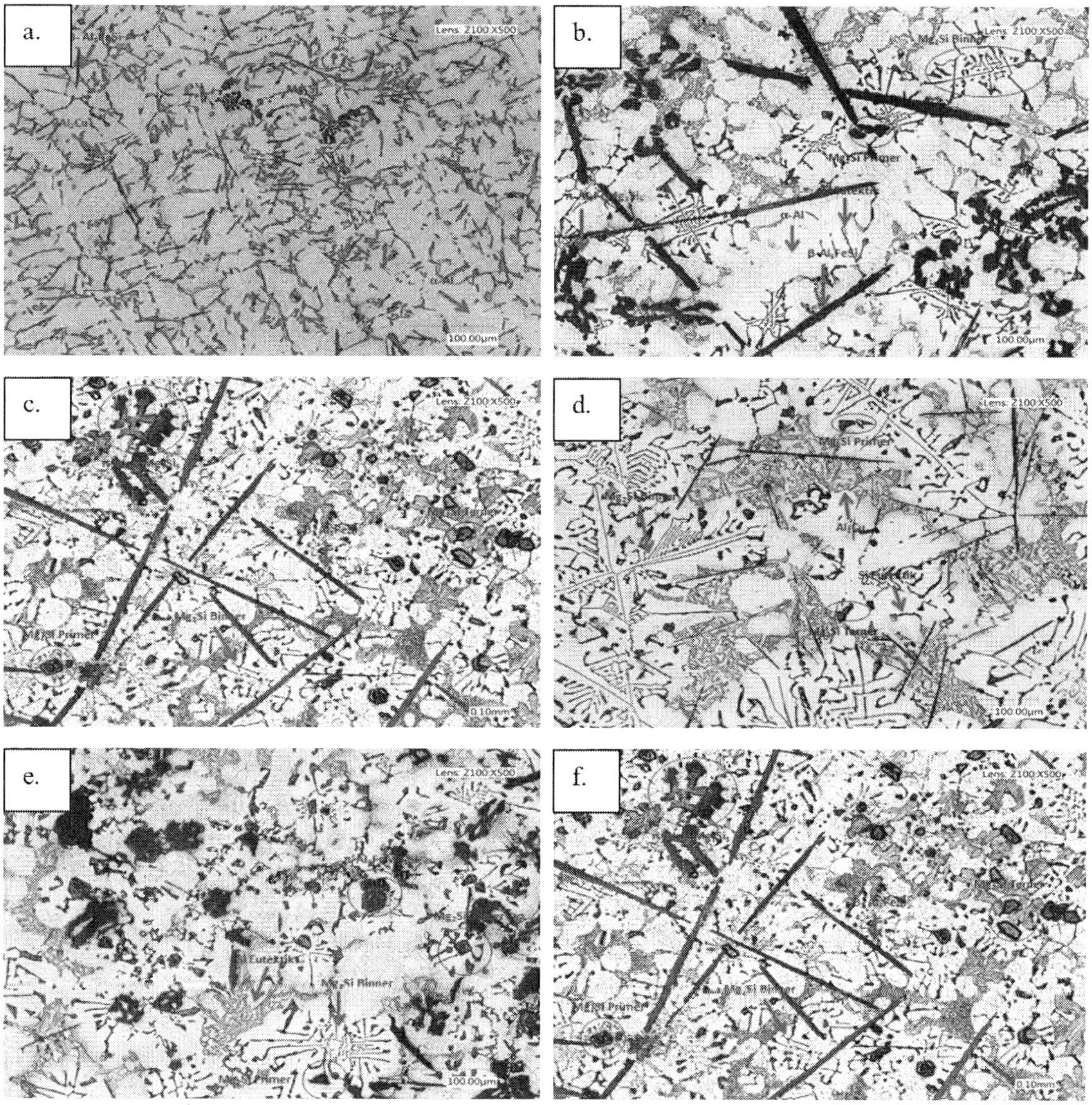

Figure 1. Microstructure of (a) Al AC4B as cast magnification of 500x and microstructure of composite Al AC4B / Si_3N_4 at volume fraction (b) 1%, (c) 3%, (d) 5%, (e) 7%, and (f) 10%.

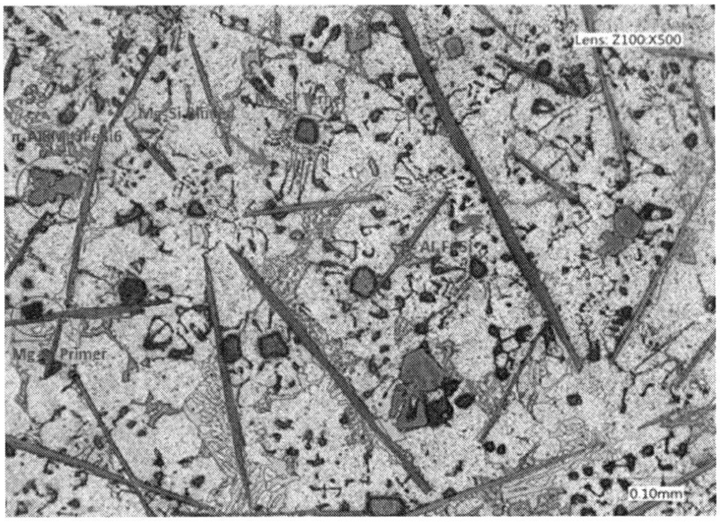

Figure 2. Microstructure of Al AC4B with addition of 0.3% volume fraction Nano SiC

3.3 SEM & EDS

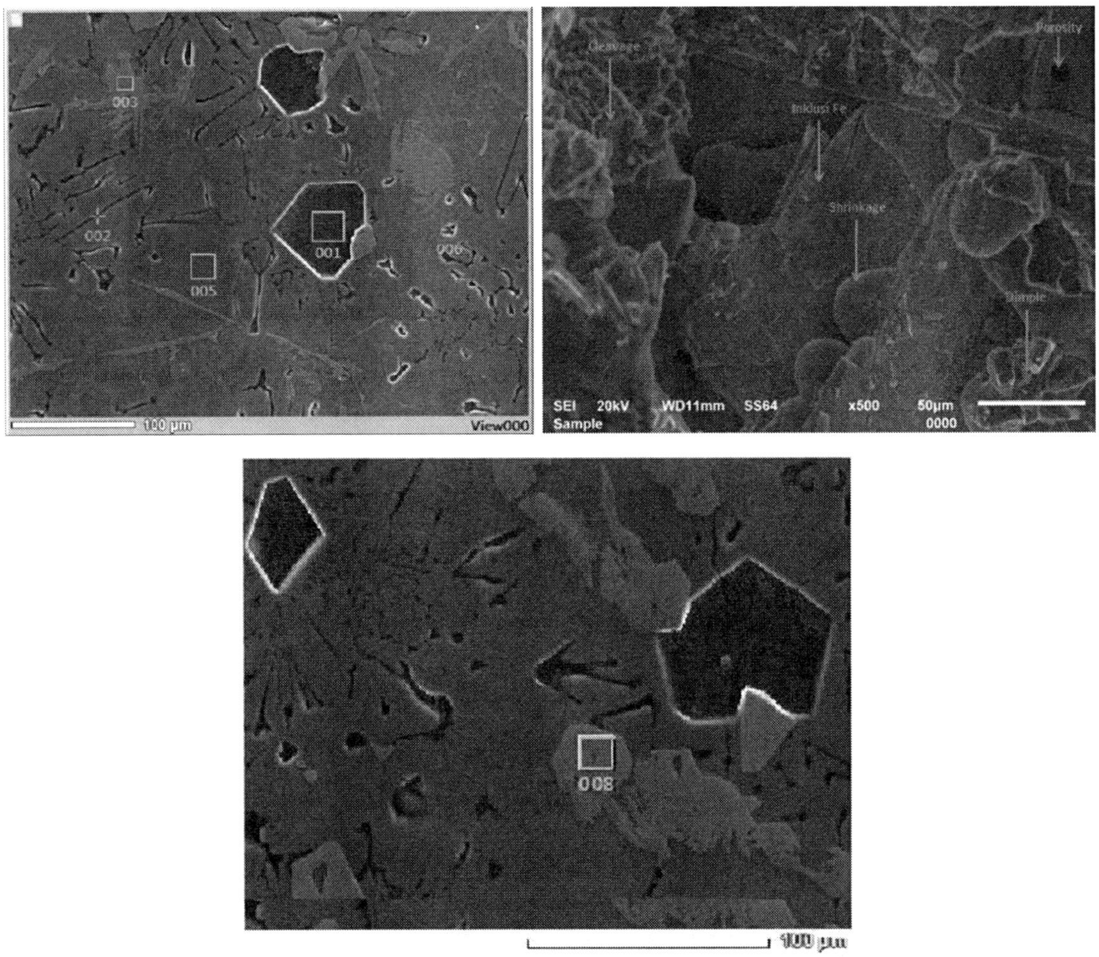

Figure 3. SEM and EDS observation

Table 3. EDS testing result

Points	Elemental Content (% wt)										Prediction Phases Formed
	B	C	O	Mg	Al	Si	Ti	Fe	Cu	Sr.	
1		1.31	45.95	28.5	7.5	23.22		0.04	0.22		Mg_2Si (Primary)
2			46.83	5.55	42.33	4.73		9.11	0.56		β-Al_5FeSi
3		1.41	46.52	1.29	44.72	4.93			1.13		Al_2Cu
5	10.88		54.58		34.08		0.06			0.4	α-Al
6		1.04	44.96	27.37	11.26	14,71		0.29	0.36		Al + Mg_2Si (Binary)
8		1.59	45.92	8.23	34.67	5.29		0.23	0.08		π- $Al_9FeMg_3Si_5$
11		0.78	46.29	4.95	46.54	1.44					Al + Mg_2Si (Binary)

SEM and EDS test were carried out at 7 different points. Besides, it was also observed on fracture formed on AC4B composites with the addition of Nano SiC reinforcement particles. In the results of SEM, it can be seen that the phases formed are the same as observing microstructural results. The dominant phase formed in the SEM results is Mg_2Si in the binary phase formed Chinese script which will increase material mechanical strength. Area number 8 shows the presence of elements Al, Fe, Mg, and Si which indicate the presence of π-$Al_9FeMg_3Si_5$ phase that has a grey colour. Al_2Cu phase is also detected in the EDS test results shown in area number 3, that contain the copper which will increase mechanical strength. β-Al_5FeSi that formed needle-like, which mean that phase have a high concentration and will decrease the mechanical strength[11].

The occurrence about fracture of this mixture as shown at figure 3, commonly found in composites, where the soft properties of the matrix and the brittle by reinforcement are combined into one material and this can increase the toughness of material. Order than that, there is porosity and shrinkage to the fracture surface, where the shaft and shrinkage occur due to poor casting process[9]. In the result of EDS test in the fractographic also found an inclusion in the form of Fe, because of the aluminium used for casting is in the form of scrap, so that there is a fairly high iron inclusion.

3.4 XRD

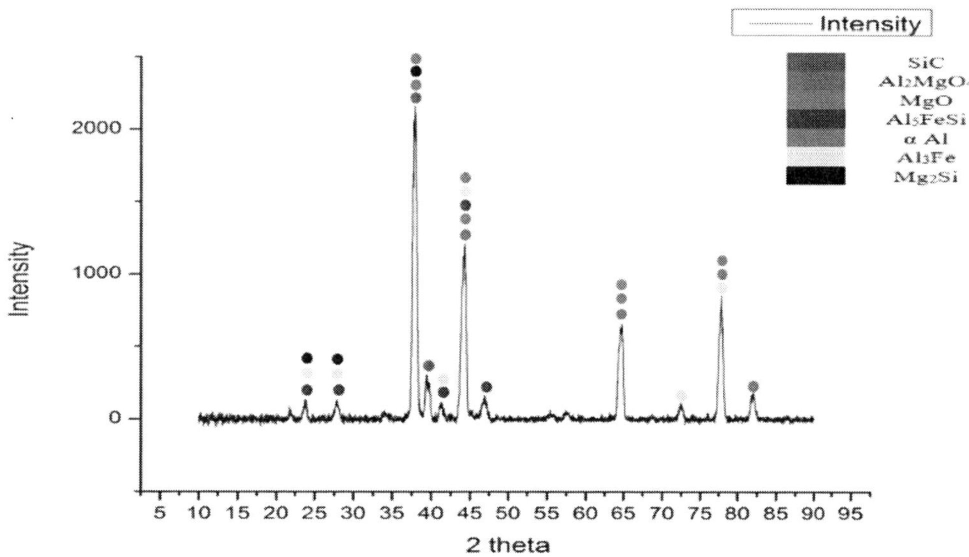

Figure 4. Results of XRD test on Al AC4B/Nano SiC composites of 0.3 % volume fraction

XRD testing on aluminium AC4B composite with Nano SiC particles was carried out on samples with composition of Nano SiC particle 0.3% Vf, where the samples has the highest impact resistance value of 0.0959 Joules/mm2. To process the data generated after testing, used the software X'Pert HighScore Plus oleh PANalytical B.V..

The results of XRD test in Figure 4 show the existence M_2Si phase which is quite high and hard is also present in the composite. In addition, there is a spinel $MgAl_2O_2$ phase which will increase contact between the AC4B matrix and Nano SiC, so that the gas layer on the surface of reinforced Nano SiC can be minimized where this can prevent the clumping of SiC[4]. The presence of Al_3Fe and Al_5FeSi phases is due to the presence of Fe content in the composite. π-$Al_9FeMg_3Si_5$ phase was formed due to the presence of Fe and Mg levels below 6%. MgO phase can be formed because of the magnesium reaction with oxygen in the environment[5].

3.5 Mechanical Properties of Al AC4B/Nano SiC Composites

The highest tensile strength was obtained by adding 0.3% volume of Nano SiC fraction with a tensile strength of 94.67 MPa as seen in figure 5 (b). Composite elongation is initially increase, and then it has decreased due to the strengthening of the grain refinement phenomenon, where Nano SiC can act as a grain refiner[1]. The more Nano SiC is added, the higher the elongation, but after 0.3%, it decreased due to the presence of Nano SiC agglomeration as shown in figure 5 (a). The increase of tensile strength in the addition of 0.3% of Nano SiC was caused by the orowan strengthening mechanism owing to the gathering of the reinforcement to form a loop between the Aluminium matrix AC4B and of Nano SiC which causes an increase of composite dislocation density[14]. Fluctuating tensile strength values are caused by the porosity of hydrogen gas formed on the composite, where hydrogen gas solubility in melting Aluminium will increase at high temperatures. While the decreasing elongation can be caused by the presence of the Mg_2Si phase which is brittle and also the porosity formed.

The addition of volume fraction of Nano SiC caused the hardness of the composite increase, with the highest hardness occurring in the addition of 0.3% of Nano SiC is 52.57 HRB while hardness at 0.25% of Nano SiC volume is 51.9 HRB as shown in figure 5 (c). the increase of hardness is caused by orowan strengthening mechanisms and also the formation of Mg_2Si and Al_2Cu phases which are hard[12]. Other that that, Nano SiC can act as a grain refiner, so the higher the addition of Nano SiC, the material will increase the mechanical strength[15], but as can see in the figure 5 (c) if the addition of Nano SiC exceed the optimum point the mechanical strength will decrease because of clustering by reinforced Nano SiC. Addition of Nano SiC particles caused composite toughness decrease which is indicated by the decreasing impact price according to figure 5 (d). This shows that the energy that can be absorbed by the composite decreases. The reduced impact price can be caused by the presence of brittle $Al_9FeMg_3Si_5$ phase.

Composite density should increase with the addition of Nano SiC owing to Nano SiC density higher than AC4B according to theoretical density calculations. However, the composite density decreases as shown in figure 5 (e). The decrease of the density value is related to the higher amount of porosity in the addition of Nano SiC in figure 5 (f) which will affect the composite density. This high porosity affects the fluctuations in the value of hardness, tensile strength, and also the toughness of the composite, which is caused by the trapping of enough hydrogen gas[13].

IOP Conf. Series: Materials Science and Engineering **909** (2020) 012037 doi:10.1088/1757-899X/909/1/012037

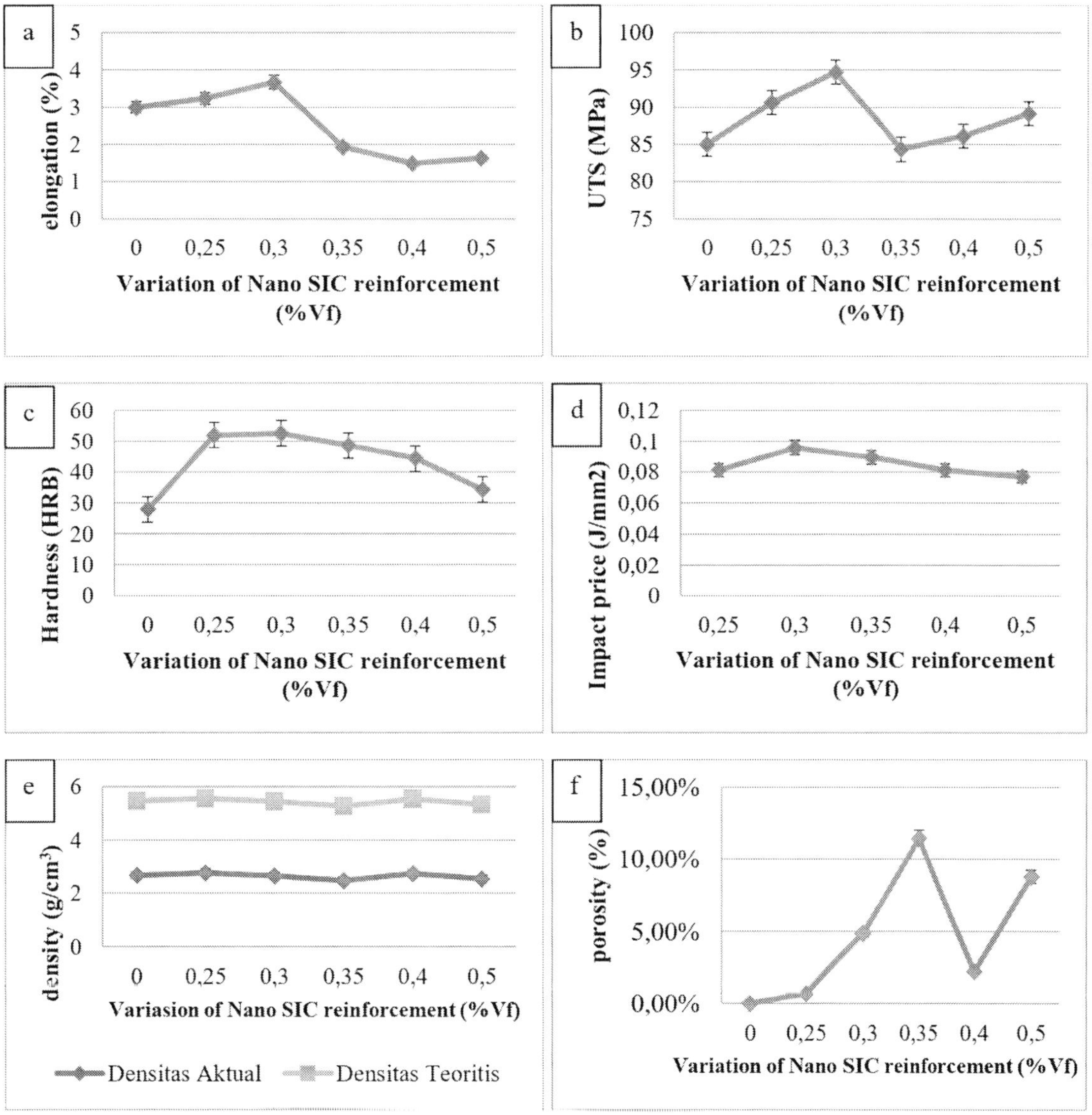

Figure 5. Effect of nano SiC volume fraction addition on tensile strength, elongation, hardness, impact resistant, density, and porosity

4. Conclusion

Nano SiC particles can increase the hardness and tensile strength of Aluminium through an orowan strengthening mechanism in which all the optimal points of mechanical strength are in the variation of 0.3%Vf reinforcement particles. Composite hardness is also influenced by the formation of Mg_2Si dan Al_2Cu phases. The presence of composite mechanical properties fluctuations because of the formation of porosity and also Nano SiC agglomeration. The agglomeration is caused by too much reinforcement of the incoming Nano SiC particles and also the uneven stirring. Besides that, there are things could be develop based on this research are, optimizing the parameter processing such as the temperature melting and pouring.

5. Acknowledgement
Authors would like to thank Directorate of Research and Community Services (DRPM), Universitas Indonesia for financial support through International Publications Index (PIT 9), 2019 with contract number: NKB-0042/UN2.R3.1/HKP.05.00/2019

6. Reference list
[1] A. Prasad Reddy, P. Vamsi Krishna, R. Narasimha Rao, and N. V. Murthy 2017 Silicon Carbide Reinforced Aluminium Metal Matrix Nano Composites-A Review *Mater. Today Proc.* **4** 3959–3971
[2] J. A. Taylor 2012 Iron-Containing Intermetallic Phases in Al-Si Based Casting Alloys *Procedia Materials Science* **1** 19-33
[3] Y. Liu, S. Kang and H. Kim 1999 The complex microstructures in an as-cast Al–Mg–Si alloy *Materials Letters* **41** 267-272
[4] W. Zhou and Z. M. Xu 1997 Casting of SiC reinforced metal matrix composites *Journal of Materials Processing Technology* **63** 358–363
[5] V. Aigbodion and S. Hassan 2007 Effects of silicon carbide reinforcement on microstructure and properties of cast Al–Si–Fe/SiC particulate composites *Materials Science and Engineering: A* **447** 355-360
[6] G.E. Totten, MacKenzie, D.Sott 2003 *Handbook of aluminium volume 1: Physical metallurgy and processes* New York: Marcel Dekker, Inc.
[7] D. S. Mackenzie 2003 *Handbook of aluminum volume 1: Physical metallurgy and processes* New York: Marcel Dekker, Inc.
[8] F. L. Mondolfo 1976 Aluminum alloys: structure and properties *Elsevier*
[9] C. Tekmen, I. Ozdemir, U. Cocen and K. Onel 2003 The mechanical response of Al–Si–Mg/SiCp composite: influence of porosity *Materials Science and Engineering* **360** 365-371
[10] F. A. Mirza and D. L. Chen 2015 A Unified model for the prediction of yield strength in particulate-reinforced metal matrix nanocomposites *Materials* **8** 5138-5153
[11] M. Alizadeh, M. H. Paydar, D. Terada and N. Tsuji 2012 Effect of SiC particles on the microstructure evolution and mechanical properties of aluminum during ARB process *Materials Science & Engineering* **540** 13-23
[12] M. Song 2009 Effects of volume fraction of SiC particles on mechanical properties of SiC/Al composites *Transactions of Nonferrous Metals Society of China* **19** 1400-1404
[13] S. B. Prabu, L. Karunamoorthy, S. Kathiresan and B. Mohan 2006 Influence of stirring speed and stirring time on distribution of particles in cast metal matrix composite *Journal of Materials Processing Technology* **171** 268–273
[14] D. Russo, E., M. Burrati, S. Veronelli 1981 Aluminium alloys composite plates US Patent no. 4426429
[15] T. Ma, Z. Chen, Z. Nie, and H. Huang 2013 Microstructure of Al-Ti-B-Er refiner and its grain refining performance *J. Rare Earths* **31** 622– 627

IOP Conf. Series: Materials Science and Engineering **909** (2020) 012038 doi:10.1088/1757-899X/909/1/012038

Characterization of aluminium AC4B/Nano TiC composite with the variation of volume fraction of nano TiC reinforced through stir casting process

M K Adam[1] **and A Z Syahrial**[1*]

[1] Department of Metallurgical and Materials Engineering, Faculty of Engineering, University of Indonesia, UI Campus, Depok, 16424, Indonesia

*Email: anne@metal.ui.ac.id;

Abstract. Characterization of aluminium ac4b/nano tic composite with variation of volume fraction of nano TiC reinforced through stir casting process has been studied. The aluminium matrix used was aluminium alloy AC4B, which contain silicon and copper as its main alloy. Furthermore, the addition of Nano TiC into AC4B composite can increase the tensile strength, ductility, and toughness of AC4B composite by refining the dendrite structure of the α-Al phase and forming super saturated solid phase, θ (Al_2Cu). In this study, AC4B/Nano TiC composite were made through stir casting with some variable parameters of Nano TiC reinforce composition of 0.25%, 0.3%, 0.35%, 0.4%, and 0.5% volume fraction to determine the optimum value of the mechanical properties of AC4B/Nano TiC composite. Stir casting process was chosen because it has several advantages, such as easy to use, flexible, and can be used to produce a large number of the products. It is known that AC4B/Nano TiC composite has optimum value of the mechanical properties when Nano TiC composition is 0,3% volume fraction with ultimate tensile strength of 132,31 MPa and the hardness of 55,18 HRB.

1. Introduction

Current technological developments are very fast because of seeing the need for new technologies that can meet human daily life needs. This rapid technological development is open the opportunities to the need for material that has mechanical properties that can compensate the sophistication of technology that is made. A material currently being developed is the composite materials.

Composite is a promising solution as a material that can compensate the sophistication of technology with its mechanical properties because composite is a material that made from two different material with different mechanical properties that can combine and improve the strength of the composite itself. Aluminium composite is one of the composite type that has bigger opportunities to be used as the material for the new technologies in the future because aluminium has the characteristics of low density, good strength, good wear and corrosion resistance, low melting point, and resilience.

In this study, the composite material used was a matrix composite material of aluminium AC4B with a reinforce in the form of nano TiC. The matrix composites of aluminium AC4B will be made through the stir casting process with variable parameter of nano TiC reinforce composition of 0.25%, 0.3%, 0.35%, 0.4% and 0.5% volume fraction so that it is expected to get the optimal value from the reinforce addition of Nano TiC to obtain optimal strength of aluminium matrix composites. The addition of Al-5Ti-1B aims to minimize the formation of dendritic grains by triggering *equiaxed* fine grain growth [1]. The addition of Al-15Sr aims to change the morphology of silicon eutectic which was originally in the form of a rough needle into a form of fibres that are spread evenly [2]. Beside, The addition of Al-15Sr can also increase the flowability of AC4B alloys [3]. Meanwhile, the addition of magnesium (Mg) aims to increase the elasticity of the aluminium matrix so that the distribution of particle reinforce is evenly distributed and prevents the occurrence of reinforcing particle agglomeration [4].

2. Methods

The study was begun with the fabrication process of composite samples of aluminium AC4B/Nano TiC. The AC4B alloy is melted at 850°C in the tilting furnace. The Nano TiC reinforce which has been pre-heating at 900°C in the muffle furnace for an hour, and the purification process to remove impurities using an ultrasonic vibrator are mixed into the melting AC4B alloy. Then, the melting AC4B alloy was stirred using a stirrer for 40 seconds and carried out the degassing process using argon gas for 2 minutes. Furthermore, 0.15%wt TiB, 0.04%wt Sr., and 5%wt Mg were added to the melting AC4B alloy. Then, it was stirred using a stirrer for 40 seconds and carried out the degassing process using argon gas for 2 minutes. The melting alloy was then poured into a mould which had been pre-heated at a temperature of 850°C in a muffle furnace for 7 minutes which was then cooled through an air-cooling process. Composites of aluminium AC4B/Nano TiC are made with variations of 0.25%, 0.3%, 0.35%, 0.4%, and 0.5%Vf. The composition of pure AC4B aluminium can be seen in Table 1.

Table 1. Al AC4B composition determined by optical electron spectroscopy

Element	Al	Si	Fe	Cu	Mg
%	86.5	7.65	1.01	2.51	1.05

Furthermore, destructive testing and identification of elements and compounds were carried out to determine the characteristics of the composite samples. Meanwhile, samples of aluminium AC4B/Nano TiC were tested through mechanical testing in the form of a tensile test using the Go Tech 27-7000 LA 10 machine with ASTM E8 standard, hardness test using the Rockwell B method with indenter in the form of a steel ball with a diameter of 1/16 inch according to the standard ASTM E18, and the Charpy impact test method with the ASTM E23 standard. Density testing was carried out to determine the effect of adding nano TiC on the composite density by using the Achimedes law and then the amount of porosity in the samples were calculated.

In addition, samples were characterized by metallographic testing, such as *Optical Microscope* (OM) to observe the shape of microstructure based on the ASTM E3-11 standard by using ETSA keller, SEM and EDS to identify phase and surface structure of AC4B/Nano TiC composites, OES to determine chemical composition, and XRD to determine the compounds formed.

3. Results and Discussion

3.1 Chemical Composition of AC4B/Nano TiC Composite

Table 2. OES Test Results of AC4B/Nano TiC Composite.

Chemical Elements	Composition of Nanto TiC (%Vf)					
	No reinforce	0.25% Nano TiC	0.3% Nano TiC	0.35% Nano TiC	0.4% Nano TiC	0.5% Nano TiC
Al	86.5	81.367	81.33	81.43	83.667	78.43
Si	7.65	6.76	7.397	5.967	7	7.227
Fe	1.01	5.373	4.53	4.353	3.513	4.65
Cu	2.51	2.057	1.95	2.02	1.97	1.99
Mn	0.233	0.214	0.278	0.195	0.178	0.183
Mg	1.05	>2.40	>2.40	4.89	>2.40	6.243
Ti	0.036	0.06	0.109	0.081	0.092	0.09
Sr	<0.0001	0.011	0.018	0.008	0.011	0.009

Based on the results of the OES test, there are discrepancies in some percentage of the chemical elements possessed by composite of aluminium AC4B/Nano TiC, such as the content of magnesium (Mg) and iron (Fe) which experience the increase of amount percentage. Magnesium (Mg) has increased due to the addition of magnesium levels which aims to increase the elasticity of aluminium to be able to wet the reinforce of Nano TiC. The Increase of the amount of magnesium content can affect the formation of several phases which can affect the mechanical properties of aluminium AC4B/Nano TiC composites, such as the formation of Mg_2Si phase owing to the change of main solidification reaction from Al-Si-Cu to Al-Si-Mg.

This change of main solidification reaction can occur due to the change of main chemical elements composition from the composition of copper (Cu) which initially has a content ranging from 2-4% to magnesium which has a much larger content, which ranges from 4-7% after the manufacturing process. Meanwhile, the increase in the amount of iron (Fe) is due to the presence of iron (Fe) elements which also dissolve into the melting AC4B alloys derived from stirrer made of steel during the fabrication process of aluminium AC4B/Nano TiC composite. The iron (Fe) content that is too high can reduce the ultimate tensile strength (UTS) and tenacity of the AC4B/Nano TiC composite by forming an iron intermetallic phase [5].

3.2 Microstructure of AC4B/Nano TiC composite

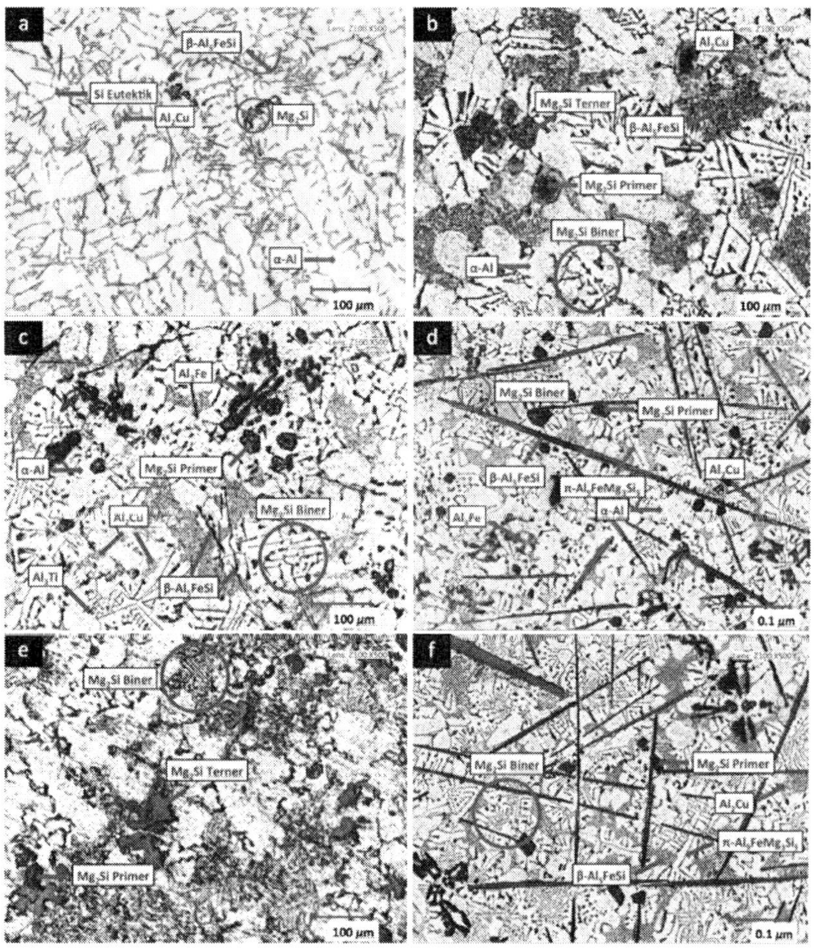

Figure 1. Observation Results of Microstructure of Aluminium AC4B/Nano TiC Composite: (a) AC4B with no reinforce; (b) 0.25%Vf; (c) 0.3%Vf; (d) 0.35%Vf; (e) 0.4%Vf; (f) 0.5%Vf.

Based on figure 1., it can be seen that there are several phases identified, namely α-Al phase, Eutectic Si phase shaped acicular. Mg_2Si phase, Al_2Cu phase, Al_3Ti phase and intermetallic phase of iron in the form of β-Al_5FeSi shaped sharp and flat, Al_3Fe phase, and π-$Al_9FeMg_3Si_5$ phase.

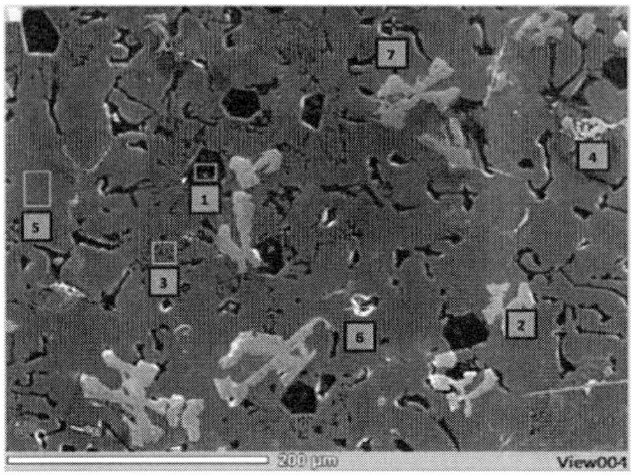

Figure 2. Results of SEM/EDS and fractography test for aluminium AC4B/Nano TiC composite of 0.3%

Table 3. Composition of Aluminium AC4B/Nano TiC Composite Elements.

	Element Content							Phase
Spot	C	Mg	Al	Si	Ti	Fe	Cu	
1	1.15	31.54	3.33	18.64	0.02	0.16	0.12	Mg_2Si Primary
2	0.89	0.89	38.88	2.81		12.65	1.51	Al_3Fe
3	0.39	4.56	46.04	2.24	0.06		0.13	Al_3Ti
4	1.86	0.95	23.86	0.34	0.03	0.59	39.14	Al_2Cu
5	0.41	0.56	50.89	0.45	0.09	0.01	0.06	α-Al
6	1.2	0.59	49.8	0.53	0.02	0.18	0.69	β-Al_5FeSi
7	1.85	21.36	16.99	14.06		0.06	0.38	Mg_2Si Primary

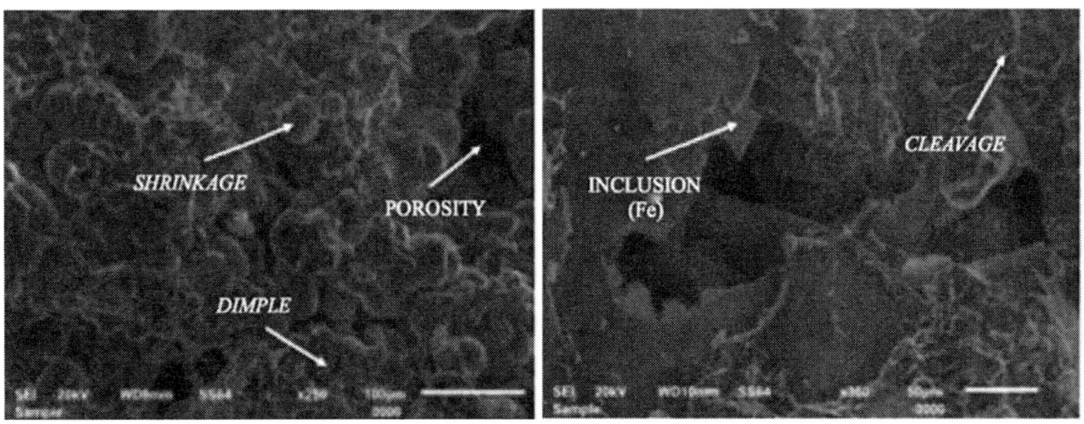

Figure 3. Results of fractography test for aluminium AC4B/Nano TiC composite of 0.3%

Besides, to ensure the phase correctness formed based on observations of microstructures as well as conformity between the phases formed and the elements contained within, several tests are carried out, such as SEM/EDS and XRD test which can be seen in figure 2. and figure 4.

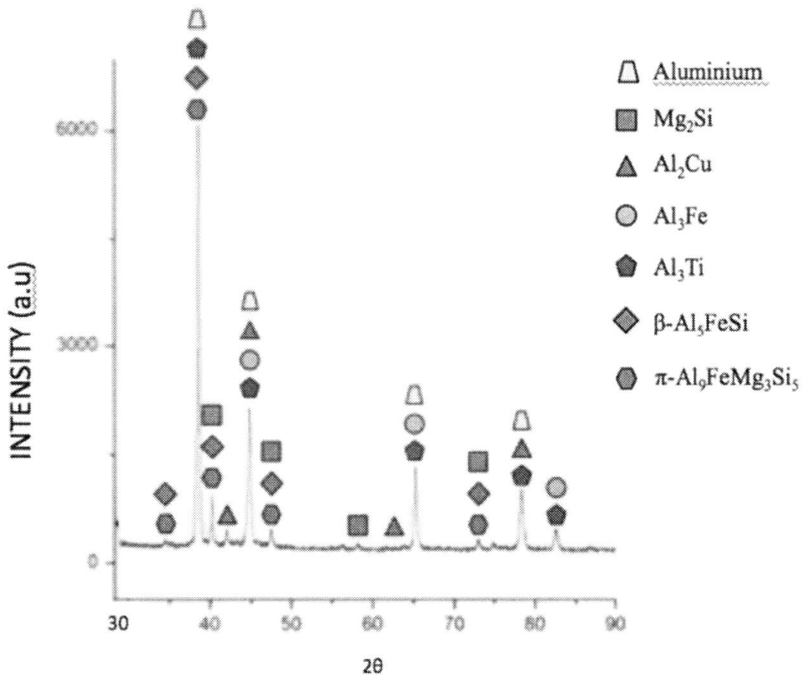

Figure 4. Results of XRD test of aluminium AC4B/Nano TiC composite of 0.3%

Based on the figure 4, it can be seen that there are several compounds that have been identified through XRD testing, including aluminium, Mg_2Si, Al_2Cu, Al_3Fe, Al_3Ti, β-Al_5FeSi, and π-$Al_9FeMg_3Si_5$. The identified Mg_2Si compound is a compound formed from the main solidification of Al-Si-Mg. Meanwhile, the Al_3Ti phase can be formed due to the addition of nano TiC amplifiers and TiB grain refining agents, where the Al_3Ti phase formed can be a nucleating agent for α-Al so that the α-Al phase which initially has a dendrite form will change to an equaixed form. Meanwhile, the Al_3Fe, β-Al_5FeSi, and π-$Al_9FeMg_3Si_5$ phases are formed due to the high content of magnesium and iron in the AC4B / Nano TiC aluminum composite.

Mg_2Si phase can be formed due to the high content of magnesium in which it is equal to 5.23%wt. This increase occurs due to the addition of magnesium levels of 5%wt. in which it aims to increase the elasticity of aluminium to be able to wet the reinforce particle. The increase of magnesium content which reaches 5.23% wt. cause a magnesium content being greater than the copper content in which it is only 1,997%wt. Thus, it causes a change of main solidification reaction from Al-Si-Cu to Al-Si-Mg. Therefore, Mg_2Si phase will be more dominant than Al_2Cu phase.

However, based on microstructural observations, it can be seen that in the AC4B/Nano TiC composite microstructure included an Al_2Cu phase although the main solidification reaction experiences a change from the Al-Si-Cu solidification reaction to the Al-Si-Mg solidification reaction. This Al_2Cu phase can be formed when done by a rapid cooling process was done in which it caused the solidification line would touch the eutectic line so that a decrease in temperature during solidification will allow the formation of Al_2Cu phase even though it is only in small amounts.

Iron intermetallic phase in the form of β-Al_5FeSi phase, Al_3Fe phase, and π-$Al_9FeMg_3Si_5$ phase can be formed because it was influenced by two factors, namely the high content of iron and magnesium which is equal to 4.48%wt. and 5.23%wt. with a low solidification rate. β-Al_5FeSi phase can be formed by reaction of L+Al_3Fe → Al+Al_8Fe_2Si+L → Al+Al_5FeSi. Based on these reactions, it can be seen that the reaction between Liquid and the intermetallic phase of Al_3Fe with a high solidification rate will form the Al_8Fe_2Si phase. However, when the solidification rate has not stopped, the Al_8Fe_2Si phase will react with Liquid to form a β-Al_5FeSi phase [6].

IOP Conf. Series: Materials Science and Engineering **909** (2020) 012038 doi:10.1088/1757-899X/909/1/012038

The formation of β-Al_5FeSi phase is influenced by the content of the elements iron and magnesium, where the iron content is above 2.4% wt. and magnesium content below 6% wt. with a low solidification rate can form β-Al_5FeSi phase which has a needle-like shape so that this phase has a relatively high strength.

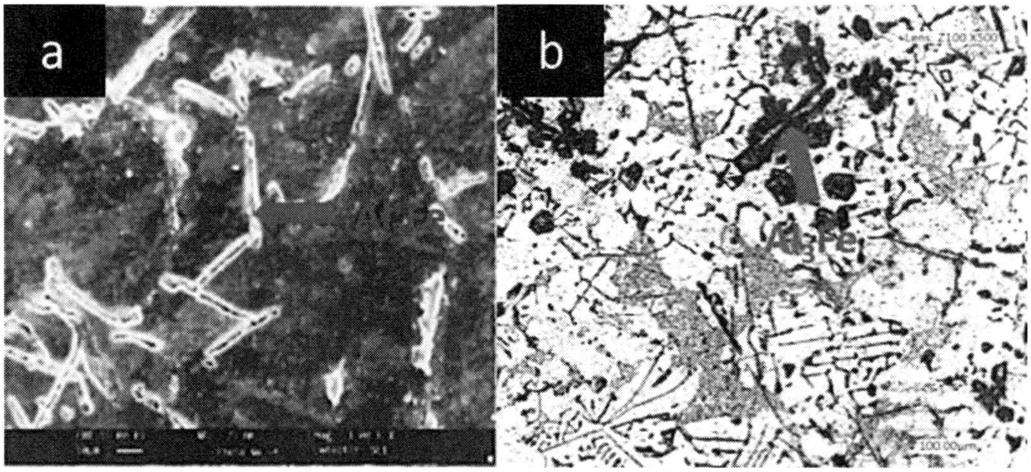

Figure 5. Comparasion of Al_3Fe microstructure forms: (a) Needle-shaped Al_3Fe forms; (b) Irregular-shaped Al_3Fe forms.

In addition, the intermetallic phase of Al_3Fe generally has a needle-shaped microstructure as shown in figure 5 (a). However, in this study the microstructure shape of the Al_3Fe intermetallic phase is irregular-shaped as shown in Figure 5 (b). The microstructural transformation of the Al_3Fe intermetallic phase is due to the high magnesium content in aluminium AC4B/Nano TiC composite, where the magnesium content is above 1.5% wt. able to change the microstructure of Al_3Fe from needle-shaped to irregular-shaped.

Meanwhile, Al_3Ti phase can be formed due to the addition of TiB grain refiner agent. The Al_3Ti phase formed will act as a nucleation agent for α-Al, so α-Al phase which initially has the form of dendrites will turn into an equiaxed form that can increase the mechanical strength of aluminium AC4B/Nano TiC composite.

3.3 Results of mechanical test of AC4B/Nano TiC composite

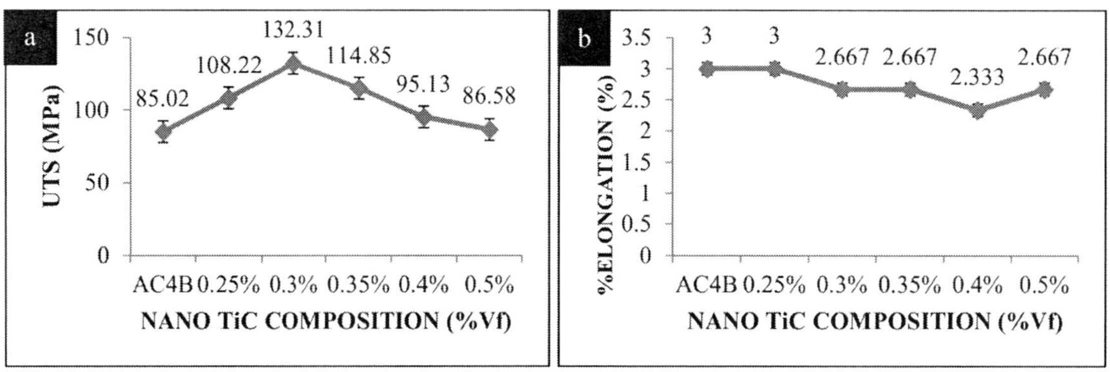

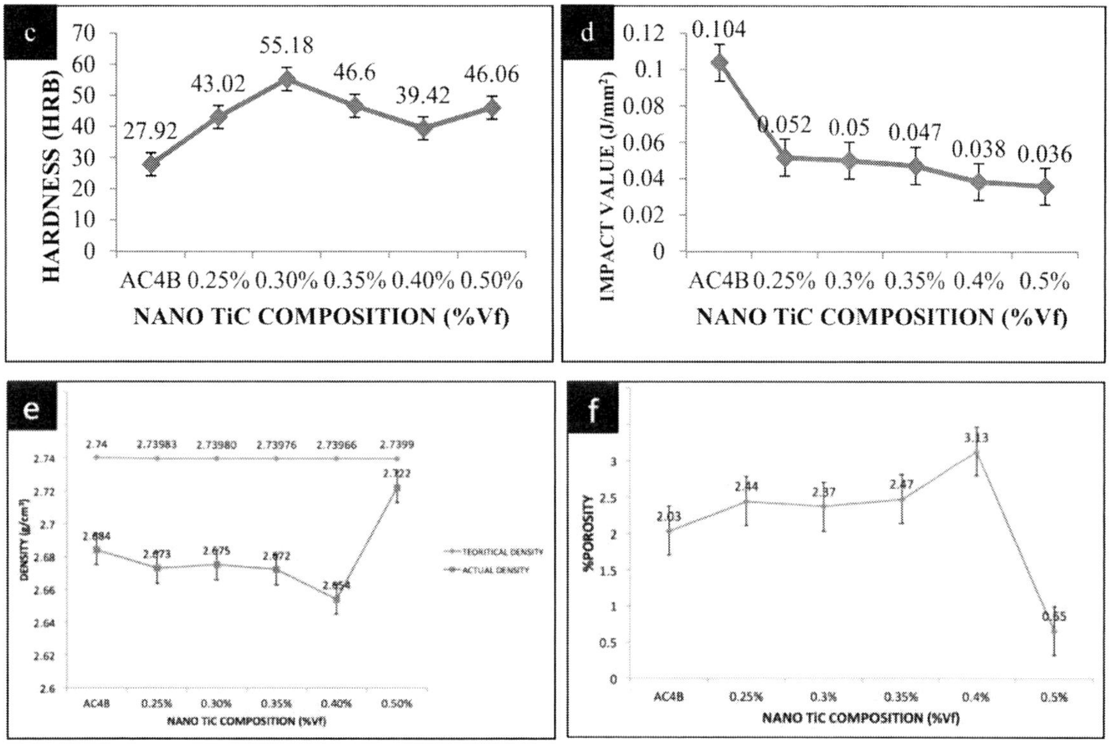

Figure 6. Results of mechanical test of aluminium AC4B/Nano TiC composite: (a) ultimate tensile strength, (b) %elongation, (c) hardness, (d) impact value, (e) density, and (f)%porosity.

Based on the results of the mechanical test, it can be seen that the addition of the Nano TiC reinforce can increase the mechanical strength of aluminium AC4B/Nano TiC composite by inhibiting dislocation movement by distributing the stress evenly in the composite matrix through three mechanisms, namely orowan strengthening mechanism, grain boundary strengthening mechanism through refining grain of α-Al phase and the precipitation hardening mechanism through the formation of saturated solids θ (Al_2Cu)[7]. This is shown in the results of the tensile test in figure 6 (a) and the hardness test results in figure 6 (c), In which the increase of number of nano TiC reinforce reaches 0.3%Vf. was able to produce the ultimate tensile strength (UTS) and the highest hardness value, which is 132.31 MPa and 55.18 HRB. The grain boundary strengthening mechanism occurred when the nano TiC reinforce acted as a nucleating agent that will change the structure of the dendrite of α-Al phase being more subtle (Equiaxed). Dendrite structure of α-Al phase which is refine can increase the resistance to crack propagation of aluminium AC4B/Nano TiC composite. Meanwhile, the precipitation hardening mechanism occurred when the nano TiC reinforce increases the formation through saturated solids θ (Al_2Cu) phase, in which saturated solids (Al_2Cu phase) formed will result in lattice and internal stress distortion which can inhibit dislocation movement [8].

In figure 6 (c), the hardness value of aluminium AC4B/Nano TiC composite from each variation of the volume fraction of nano TiC reinforce is higher than the hardness of AC4B without the reinforce. A higher hardness value is obtained because of the addition of reinforce particles, where the addition of reinforcing particles evenly into the aluminium matrix will distribute the stress evenly so that the dislocation movement is not focused in just one area.

In this study, an increase in the hardness value was obtained due to the formation of the θ (Al_2Cu) phase and the Mg_2Si phase. The addition of nano TiC reinforce is able to initiate the formation of phase θ (Al_2Cu) which is a saturated solid (Supersaturated Solid Solution) formed through the Precipitation Hardening mechanism. The saturated solid (Al_2Cu Phase) formed in lattice distortion so it can increase the hardness value of aluminium AC4B/Nano TiC composite. In addition, the formation of the Mg_2Si

phase can also increase the hardness value of the aluminium composite AC4B / Nano TiC, where the Mg_2Si phase is formed due to the addition of magnesium and a modifying agent in the form of strontium. Strontium will react with aluminium to Al_4Sr, then Al_4Sr will act as a nucleating agent for primary Mg_2Si so that the Mg_2Si phase formation will increase the hardness value of aluminium AC4B/Nano TiC composite in line with the addition of strontium. In other words, as the number of Mg_2Si phases is formed, the hardness value of the AC4B / Nano TiC aluminium composite will tend to increase.

However, in the variables 0.35%, 0.4%, and 0.5%Vf, there is a decrease in mechanical strength as the percentage of the increase of Nano TiC reinforce. This was caused by the uneven dispersion of nano TiC reinforce particles and the greater number of porosities in line with the increase of percentage of the volume fraction of nano TiC reinforce which can be seen in figure 6 (f).

In addition, in figure 6 (b), it can be seen that the greater the percentage of the volume fraction of the nano TiC reinforce can reduce the percentage of elongation of aluminium AC4B/Nano TiC composites, which is due to the increased agitation of aluminium AC4B/Nano TiC composites. The increase in aluminium AC4B composite agitation was caused by the formation of the Mg_2Si phase, the intermetallic phase of β-Al_5FeSi and intermetallic phase of π-$Al_9FeMg_3Si_5$, where these phases have brittle properties so these phases are able to reduce percentage of elongation of aluminium AC4B/Nano TiC composites [9].

In addition, the increase in the elasticity of aluminium AC4B/Nano TiC composite caused the aluminium AC4B/Nano TiC composite to be more difficult to absorb energy from impact loads, where it can be seen in figure 6 (d) which shows the increasing number of nano TiC reinforce, the impact prices from aluminium AC4B/Nano TiC composites tending to decrease.

In figure 6 (e), it can be seen that the actual density values tend to be lower than the theoretical density in each variation of the nano TiC reinforce. This is influenced by the presence of Mg_2Si and porosity found in aluminium AC4B/Nano TiC composites, where the Mg_2Si can be formed in aluminium AC4B/Nano TiC composites due to the addition of magnesium during the fabrication process of aluminium AC4B/Nano TiC composites by 5%wt. which is used to increase the wettability of aluminium to be able to form a good interface with reinforce nano TiC. Therefore, more Mg_2Si phases are formed which can affect the density of aluminium AC4B/Nano TiC composites.

For the presence of porosity in aluminium AC4B/Nano TiC composites, the percentage of the increase of porosity in aluminium AC4B/Nano TiC composites, the density value of aluminium AC4B/Nano TiC composites will tend to be more decrease. This is because porosity is gas/air trapped inside a solid and the material contained in the shaft has a lower density than material that is free from the shaft so that the material tends to be lighter than the material that is free from the shaft [10].

Based on the porosity test results in figure 6 (f), it can be seen that the decrease in the mechanical strength of aluminium AC4B/Nano TiC composite is due to the presence of increasingly large porosity [11]. In this study, porosity can be formed due to the non-optimal stirring process, the formation of *shrinkage* during the solidification process, and hydrogen evolution [12]. The evolution of hydrogen gas can occur due to the high pouring temperature, where the safe temperature limit that can be used to prevent the formation of hydrogen gas should be 675-700^{0}C. However, in this study the pouring temperature was around 800-850^{0}C. This is needed to prevent premature cooling during the stirring process of the reinforcement and pouring process into the mold. The effect of the presence of hydrogen gas in the melted AC4B alloy itself is able to reduce the tensile strength of the aluminium AC4B/Nano TiC composite because the material tends to be brittle [12].

4. Conclusion

1) Aluminium AC4B/Nano TiC composite with variations of the addition of volume fraction of Nano TiC reinforce has better mechanical properties than AC4B alloy without nano TiC reinforce particle. In addition, the addition of a nano TiC reinforce can increase the mechanical strength of aluminium composites through three mechanisms, namely the orowan strengthening mechanism, the Grain Boundary Strengthening mechanism, and the Precipitation Hardening mechanism.

2) Variable aluminium AC4B/Nano TiC composite of 0.3%Vf is an optimal variable because it has the highest mechanical strength among other variables, such as Ultimate Tensile Strength (UTS) amounted to 132.31 MPa, hardness of 55,18 HRB, and impact strength of 0.05 J/mm^2.
3) Microstructure observation of aluminium AC4B/Nano TiC composite shows that there are several phases formed, including phase of Mg_2Si Primary, Mg_2Si Biner, Mg_2Si Terner, Al_2Cu, Al_3Ti, Al_3Fe, β-Al_5FeSi, dan π-$Al_9FeMg_3Si_5$.
4) The addition of magnesium (Mg) reaches 5% wt. cause the magnesium content (Mg) is higher than the copper content (Cu) which is only 1.997% wt. thus causing a change in the main solidification reaction from Al-Si-Cu to Al-Si-Mg.
5) The change of the main solidification reaction from Al-Si-Cu to Al-Si-Mg causes the Mg_2Si phase to form easily. In addition, the addition of magnesium (Mg) which reaches 5% wt. also influences the formation of the intermetallic phase β-Al_5FeSi and the intermetallic phase π-$Al_9FeMg_3Si_5$ which is brittle and changes the microstructure of the Al_3Fe phase from needle-shaped to irregular-shape.
6) the addition of TiB grain refining agent can form the Al_3Ti phase which can be a nucleating agent for α-Al so that the α-Al phase which initially has a dendritic shape will change to an equaixed form.
7) The decrease of mechanical strength and percentage of elongation in aluminium AC4B/Nano TiC composite occurred due to a greater percentage of porosity, where the porosity can be formed due to the process of melting AC4B alloys that are too fast and the formation of *shrinkage* during the solidification process.
8) The decrease of mechanical strength and percentage of elongation in aluminium AC4B/Nano TiC composite occurred also due to the formation of Mg_2Si phase, β-Al_5FeSi, dan π-$Al_9FeMg_3Si_5$ which has brittle properties.

5. Acknowledgement
Authors would like to thank Directorate of Higher Education, Ministry of Higher Education and Research of Technology, Republic of Indonesia for financial support under PKRUPT fund with contract number: NKB-1073/UN2.R3.1/HKP.05.00/2019

6. References
[1] S. F. G. Timelli, G. Camicia 2014 Effect of grain refinement and cooling rate on the microstructure and mechanical properties of secondary Al-Si-Cu alloys *J. Mater. Eng. Perform* **23** 611–621
[2] M. A. Fallah 2009 Studi pengaruh kombinasi komposisi 0.02 wt.%Sr dan 0.063, 0.083 dan 0.108 wt.% Ti terhadap karakteristik paduan aluminium AC4B hasil proses low pressure die casting (LPDC)
[3] B. S.-P. J. Asensio-Lozano 2006 Effect of the addition of refiners and/or modifiers on the microstructure of die cast Al-12Si alloys *Scr. Mater.* **54** 943–947
[4] M. Al, Sahari, Zulfia, and Siradj 2009 Pengaruh Mg Terhadap Kekerasan Komposit Matriks *Semesta Tek* **4** 17–104
[5] S. Ji, W. Yang, F. Gao, D. Watson, and Z. Fan 2013 Materials science & engineering a effect of iron on the microstructure and mechanical property of Al – Mg – Si – Mn and Al – Mg – Si diecast alloys *Mater. Sci. Eng. A.* **564** 130–139
[6] A. D. Michael V Glazoff, Alexandra Khvan, Vadim S Zolotorevsky, Nikolai A Belov 2019 Casting aluminum alloys: Their physical and mechanical metallurgy
[7] S. V. Z. Ma, A. M. Samuel, F. H. Samuel, H. W. Doty 2008 A Study of Tensile Properties in Al-Si-Cu and Al-Si-Mg Aloys *Mater. Sci. Eng.* **490** 36–51
[8] D. Zhou, F. Qiu, and Q. Jiang 2015 The nano-sized TiC particle reinforced Al-Cu matrix composite with superior tensile ductility *Mater. Sci. Eng. A* **622** 189–193
[9] M. S. Salleh, M. Z. Omar, and J. Syarif 2014 The effects of mg addition on the microstructure and mechanical properties of thixoformed Al-5 % Si-Cu alloys *J. Alloys Compd.*
[10] S. N. Aqida, M. I. Ghazali, and J. Hashim 2004 Effects of porosity on mechanical properties of

metal matrix composite: an overview **40** 17–32

[11] K. A. Babu, P. Venkataramaiah, and K. D. Reddy 2018 Mechanical characterization of aluminium hybrid metal matrix composites synthesized by using stir casting process *Mater. Today Proc.* **5** 28155–28163

[12] B. Prabu, K. S.B., K. L., and M. S. 2006 Influence of stirring speed and stirring time on distribution of particles in cast metal matrix composite *J. Mater. Process. Technol.* **171** 268–273

IOP Conf. Series: Materials Science and Engineering **909** (2020) 012039 doi:10.1088/1757-899X/909/1/012039

2-D modelling of effect of free-stream turbulence on trailing edge vortex

Syaiful[1*], Anggie Restue Saputra[1], Nazaruddin Sinaga[1] and Bambang Yunianti[1]
[1]Departement of Mechanical Engineering, Diponegoro University, Semarang, Indonesia

[*]Email: syaiful.undip2011@gmail.com

Abstract Vorticity causes many adverse effects on the area near the surface, both on moving (turbine blade) and fixed objects. Therefore, this study attempts to investigate the effects of free-stream turbulence on trailing edge vortex on the NACA 0015 airfoil profile. This research was conducted using two-dimensional numerical simulations. To determine the interaction of vortex and free stream turbulence, two variations of turbulent intensity were given, namely 0.5% and 4.6% with Reynolds number of 1.6 x 10^5. The k-ε turbulence model was used in the current study. The angle of attack of the flow towards the airfoil was varied from 0° to 25° in conditions of increase and decrease in the horizontal position of the airfoil. The result was found that there is an agreement between the simulation and the experiment result. Increased turbulent intensity can delay the stall condition, the stall process begins to be observed after the 12° attack angle at 0.5% turbulent intensity while at turbulent intensity 4.6% the stall process was found after the 15° attack angle. From a velocity streamline, it was found that an increase in turbulent intensity can inhibit vortex growth.

1. Introduction

According to the projection of the International Energy Agency (IEA), up to 2030 world energy demand has increased by 45% or an average increase of 1.6% per year. Most or about 80% of the world's energy needs are supplied from fossil fuels [1]. Therefore, the use of new renewable energy needs to be improved. The wind is one of the potential energy sources that can be used as renewable energy. Wind turbines utilize mechanical energy from the wind to produce electrical power [2]. One crucial part of a wind turbine is the blade because this part is directly related to the flow of wind. The wind turbine blade has a particular blade profile called an airfoil that determines the primary performance of the wind turbine [3].

Various flow phenomena are found around the blade, including vortices. Vorticity can also be found in a variety of practical aerodynamic applications, such as wind turbine [4]. When the wind turbine rotates because of the fluid flow around it, vortices are generated. Vorticity causes many adverse effects on the near-field surface, both rotating and immovable objects [5]. The disadvantages caused by these vortices include high noise, vibration, mechanical fatigue, and erosion. These losses can occur in aerodynamically shaped objects, such as aircraft wings, helicopter propellers, ship propellers, and wind turbine blades.

Flow conditions and surface geometry influence the process of vortex formation. Reynolds number, flow attack angle, turbulent boundary layer, aspect ratio, and geometry of the surface are the parameters that affect the vortex formation process [6]. The method of formation, structure, and shape change of the vortex has been extensively studied. Meanwhile, vortex interaction with external disorders has not received much attention. Free-stream turbulence is an example of an external disorder that can affect vorticity [7].

Several studies have discussed the airfoil, vortex, and the effects of free-stream turbulence. Maldonado et al. (2015) conducted a review of the impact of free-stream turbulence on the aerodynamic performance of type S809 wind turbine blades with variations without and with turbulent grids (TU = 6.14%) [8]. From testing Maldando et al. (2015) note that an increase in turbulent intensity can increase the lift to drag (L/D) ratio and delay flow separation at high attack angles. The effect of turbulent intensity on the aerodynamic performance of horizontal wind turbines has been studied by Li et al.[9]. The results of their study indicated that an increase in turbulent intensity increases the lift coefficient at an angle of attack between 0° to 20°. Kim and Xi (2016) investigated the effect of turbulent intensity on the stall condition on the wind turbine blade [10]. They revealed that dynamic stall on the blade is generally found at an angle of attack greater than 18. Increased turbulent intensity from 0 to 10% results in an increase in the lift coefficient and drag coefficient at the angle of attack from 4° to 20°. Yao et al. (2012) studied numerically by comparing several turbulent models namely k-ε, RNG k-ε, SST4 model, Reynolds model [11]. Yao et al. observed that the lift coefficient from the numerical simulation results on all turbulent models had values that approached the experimental data. The effect of free-stream turbulence on the dynamic stall from wind turbine blades has been investigated by Kim and Xie (2016) [12]. Kim and Xie (2016) revealed that higher turbulent intensity reduces bubble separation, which causes an increase in the ratio of lift to drag (L/D) on static airfoils. Ahmadi-Baloutaki et al. (2015) investigated experimentally about the interaction of free-stream turbulence against near-field wing-tip vortex [7]. Ahmadi-Baloutaki et al. found that increasing turbulent intensity could increase the near-field wing-tip vortex decay rate and improve aerodynamic performance with static stall delay.

From previous studies, it can be seen that most of the results of the study reveal that free-stream turbulence can delay stall conditions. However, studies that focus on discussing the effects of free-stream turbulence on vortex that appear around the wings or blade on the turbine are very few. Therefore, the present study aims to discuss in detail the effect of free-stream turbulence on trailing edge vortex with a two-dimensional numerical simulation on the profile of the NACA 005 airfoil.

2. Description Of Physical Model

The profile of the NACA 0015 airfoil was used in this study. The NACA airfoil profile was developed by the U.S. National Aviation Advisory Board [11]. The airfoil profile used has a geometry with a chord length of 0.152 m, as can be seen in Figure 1. This airfoil has a symmetrical geometry with a maximum thickness of 15% of the length of the chord. The geometry used in this numerical simulation was the same as that used in Ahmad-Baloutaki et al. [7].

The numerical simulation parameter is determined by the Reynolds number: Re = ρuc/μ, where ρ is the air density, ρ = 1,225 kg/m^3; μ is dynamic viscosity, μ = 1.7894 × 10^{-5} kg/ms; c is the chord length of the airfoil, u is the velocity of fluid flow. The Reynolds number 1.6x10^5 was used in calculations with a Mach number of 0.046. The attack angle used was between 0° to 25° for cases of increase and decrease from the horizontal position of the airfoil.

3. Governing Equations and Boundary Conditions

The fluid was assumed to be steady state and incompressible with constant physical properties. Fluid flow was considered to be turbulent flow. The governing equations used to solve this case include:
The continuity equation:

$$\frac{\partial u}{\partial x}+\frac{\partial v}{\partial y}=0 \qquad (1)$$

The momentum equation:

$$\rho\vec{u}+\vec{\nabla}\cdot(\rho\vec{u}\vec{u})=-\vec{\nabla}p+\vec{\nabla}\cdot\left(\vec{\vec{\tau}}\right)+\rho\vec{g}+\vec{F} \tag{2}$$

where ρ is fluid density, P is static pressure, $\vec{\vec{\tau}}$ is stress tensor, and $\rho\vec{g}$ and $\vec{F}$are gravitational and external forces, respectively.
The boundary conditions for all surfaces are described as follows:

- In the inlet

$$u=u_\infty\cos\alpha,\ v=u_\infty\sin\alpha \tag{3}$$

- In the outlet

$$\frac{\partial u}{\partial x}=\frac{\partial v}{\partial x}=0 \tag{4}$$

- On the airfoil surface

$$u = v = 0 \tag{5}$$

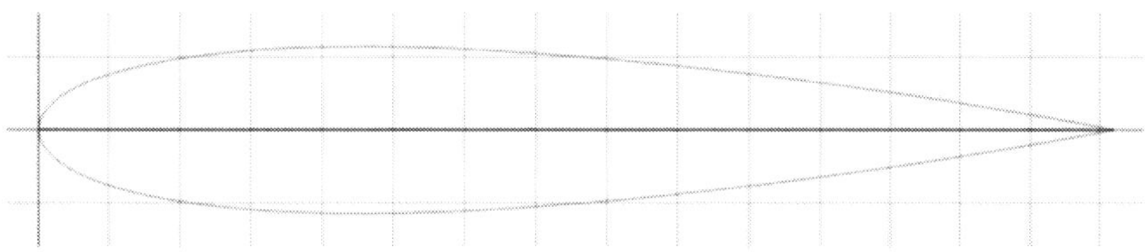

Figure 1. Geometry of NACA 0015 airfoil profile [12]

4. NUMERICAL METHOD

To solve the problem in a two-dimensional numerical simulation of this study, several assumptions were applied; namely, steady 2D flow, turbulent flow with Re = $1.6\text{x}10^5$, and fluid were considered as incompressible.

4.1. Grid Generation

Computational domain geometry was made in the form of a semicircular front and a rectangular back. The computational domain was made with the radius of a semicircular part of 50°C, the width and height of the rectangular sections of 50°C and 100°C, respectively, where C is the chord length of the airfoil as described in Figure 2.

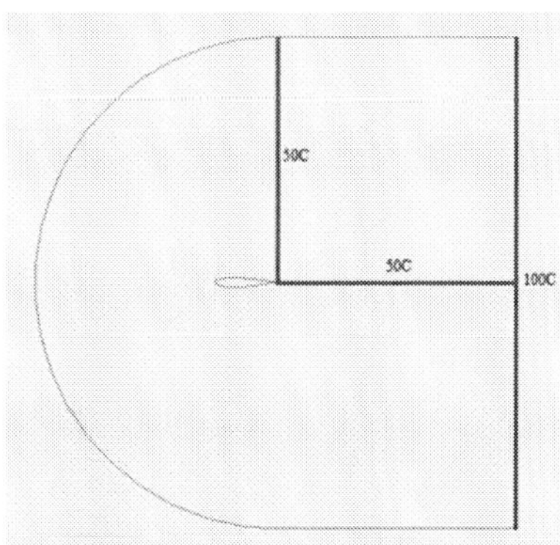

Figure 2. Geometry of computational domain

The type of grid used was C structured grid. The grid near the airfoil was closed to meet the turbulent model requirements [11]. The overall grid structure can be seen in Figure 3 and the grid structure close to the airfoil surface can be seen in Figure 4. The independent grid test was performed to obtain the minimum grid number, where the addition of the grid number will not affect the calculated parameters

[11]. From the independent grid test, it was found that the number of grids needed in this simulation was 116,000 as denoted in Table 1.

Figure 3. Grid structure of the computational domain

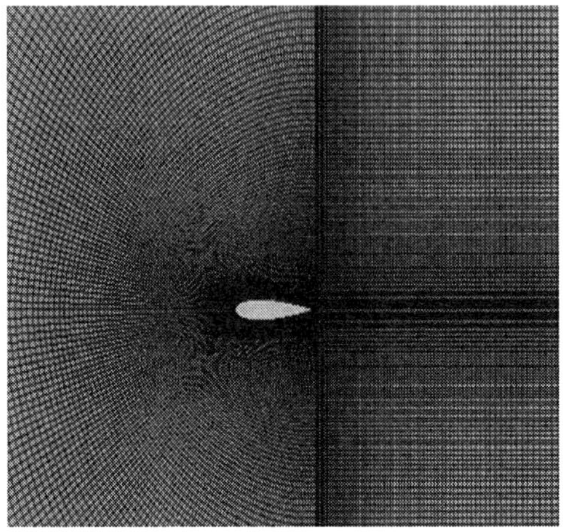
Figure 4. Grid structure near the airfoil

Table 1. Grid independent test

Number of grid	C_D (present simulation)	C_D (Ahmadi et al. [7])
47,950	0.0247	0.033
79,250	0.0268	
116,000	0.0282	
161,150	0.0283	

4.2. Turbulent Model
The turbulent model used in this numerical simulation was the *k-ε* standard with the standard wall function. Pressure-velocity coupling was solved by the SIMPLEC algorithm. The momentum equation was discretized by the upwind second order scheme, while the turbulent kinetic energy equation and dissipation ratio were discretized by the first order upwind scheme. The convergence criteria used were 10^{-5} for momentum equations, turbulent kinetic energy, and dissipation ratio.

5. Results and Discussion
5.1. Verification
To obtain accurate simulation results, the current numerical simulation results were compared with the experimental results of Ahmadi-Baloutaki et al. [7]. The equations used in calculating the aerodynamic coefficient of an airfoil are as follows:

Lift coefficient:
$$C_L=\frac{F_L}{0.5\rho U_\infty^2 C} \quad (6)$$

Drag coefficient:
$$C_D=\frac{F_D}{0.5\rho U_\infty^2 C} \quad (7)$$

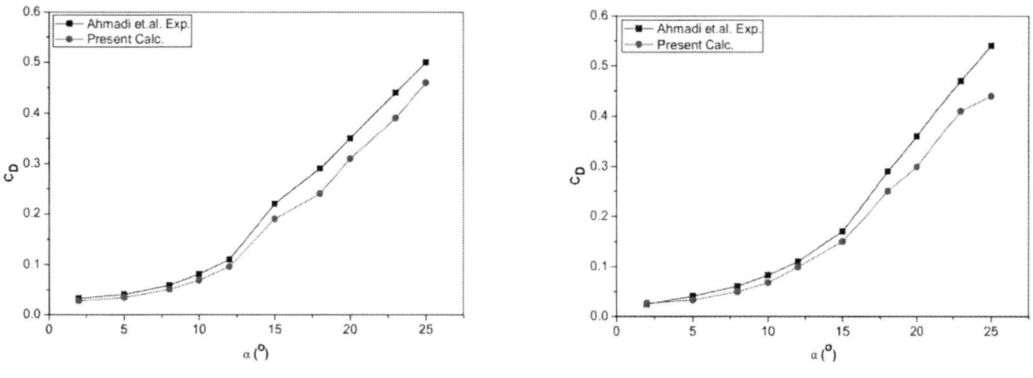

(a) (b)

Figure 5. Comparison of drag coefficient between simulation and experiment results in increasing attack angle for turbulent intensity (TI) of (a) 0.5% and (b) 4.6%

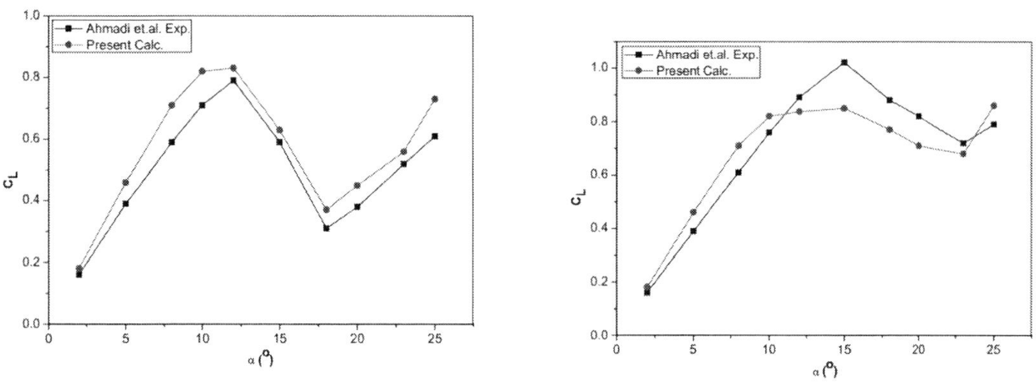

(a) (b)

Figure 6. Comparison of lift coefficient between simulation and experiment results in increasing attack angle for turbulent intensity (TI) of (a) 0.5% and (b) 4.6%

Figure 5 illustrates the comparison of the drag coefficients between the results of the present simulation and the experimental results of Ahmadi et al. (2015) [7]. The drag coefficient of the simulation results shows an agreement with that of the Ahmadi's et al. experiment. From Figure 5, it is found that the drag coefficient increases with increasing the attack angle at both turbulent intensities of 0.5% and 4.6%. A sharp increase in drag coefficient is observed with an increase in the attack angle for both turbulent intensities. As with the drag coefficient, the predicted coefficient of lift is observed as an agreement with the experimental results of Ahmadi et al. on increasing the attack angle as can be seen in Figure 6. The lift coefficient reaches the maximum value at the attack angle of 12° with the turbulent intensity of 0.5% and reaches the minimum value at the attack angle of 18°. At 4.6% turbulent intensity, the maximum value of the lift coefficient shifts at the attack angle 15° and the minimum value is found at the attack angle 23°.

5.2. Aerodynamic Performance

This aerodynamic performance is analyzed by considering the ratio of the lift force and the drag force to the increase in the attack angle as shown in Figure 7. As found in Figure 6, an increase in turbulent intensity results in a delay in stall condition as investigated by Ahmadi et al. [7]. From Figure 7 it is known that the higher turbulent intensity can increase the ratio of lift force and drag force (L/D) which indicates higher aerodynamic performance on the profile of the NACA 0015 airfoil. Increasing the ratio of lift force and drag force (L/D) due to the increase in turbulent intensity on static airfoils is in accordance with the statements from Maldonado et al (2015) [8] and Kim and Xie (2016) [12].

5.3. Velocity Streamline

The effect of free-stream turbulence on the trailing edge vortex can be observed through velocity streamline. Figure 8 shows the process of forming a trailing edge vortex seen from a velocity streamline. From Figure 8, it is known that an increase in turbulent intensity can affect the process of forming the trailing edge vortex.

Vortex begins to form at the 15° attack angle with a turbulent intensity of 0.5%. This can be observed with the secondary flow above the airfoil surface, however, the vortex still does not completely cover the airfoil surface. This vortex is called the leading edge vortex (LEV) because the direction of rotation is from the leading edge and clockwise (Prangemeier, 2010) [13]. This condition indicates the start of the stall condition indicated by the decrease in the coefficient of lift. The vortex continues to develop and the second vortex appears around the trailing edge of the airfoil from the 18° attack angle. This second vortex is called trailing edge vortex (TEV) because the direction of rotation is counterclockwise. Both of these vortices continue to grow until the 25° attack angle. At the 15° attack angle with a turbulent intensity of 4.6%, the vortex formed is relatively small compared to that of the turbulent intensity of 0.5%. The vortex begins to increase at the 18° attack angle but has not completely covered the airfoil surface. This causes the stall process to start which is indicated by the decrease in the coefficient of lift. The stall process can occur because of the flow separation which results in the formation of a vortex. The vortex continues to develop until a TEV is formed from the 23° to the 25° attack angles. However, LEV starts to disappear at the 25° attack angle.

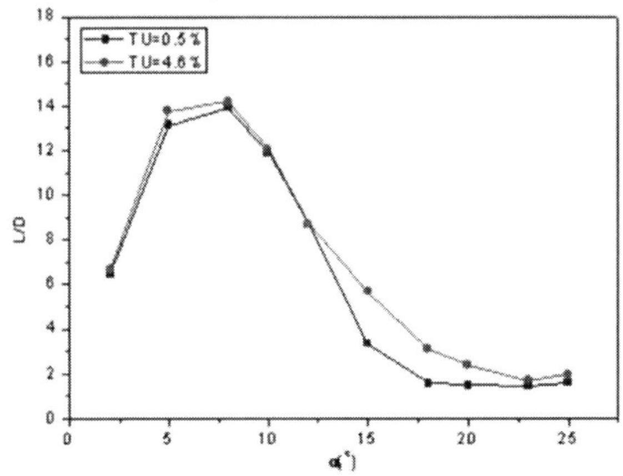

Figure 7. Effect of turbulent intensity on ratio of lift force and drag force in increase of attack angles

From the streamline, it can be seen that the process of forming a vortex is different in the increasing angle of attack. The higher the attack angle is, the greater the separation of flow. Flow separation can affect the process of vortex formation. From the streamline, it can be seen that the process of vortex formation is different from variations in turbulent intensity. The process of vortex formation is delayed at 4.6% turbulent intensity. At the 15° and -15° attack angles with a turbulent intensity of 0.5%, the vortex diameter formed is quite large, while the turbulent intensity of 4.6% vortex begins to form with a smaller diameter. The second vortex begins to form at the attack angles of 18° and -18° with the turbulent intensity of 0.5%, while the second vortex formed at the 23° attack angle at turbulent intensity of 4.6% as it is revealed by Leu et al. (20120 [14].

5.4. Vorticity

The increased turbulent intensity in the free stream region can affect vorticity. Comparison of growth and magnitude of vorticity with the turbulent intensity of 0.5% and 4.6% is shown in Figure 9. Vorticity at the trailing edge begins to form for the 18° attack angle with a turbulent intensity of 0.5%. This is indicated by the presence of vorticity that rolls in a circle on the trailing edge region. Vorticity on the trailing edge area has a negative value which indicates that the vorticity has a counterclockwise rotation (Prengemeier et al., 2010) [13] and (Panda and Zaman, 1994) [15]. The vorticity continues to develop until the 25° attack angle with different strengths. At the turbulent intensity of 4.6%, vorticity at the trailing edge region is observed at the attack angle 23°.

6. Conclusions

Investigation of interactions between free-stream turbulence against the trailing edge vortex through two-dimensional numerical simulations has been carried out. Comparisons have been made to the profile of the NACA 0015 airfoil for the Reynolds number 1.6 x 10^5 with 2° and 25° attack angles at0.5% and 4.6% turbulent intensity. From the results of numerical simulations, it was found that verification has shown an agreement between the results of the simulation and the experimental results. The increase of turbulent intensity can improve aerodynamic performance characterized by static stall delay. Increased turbulent intensity also delayed the process of forming the trailing edge vortex.

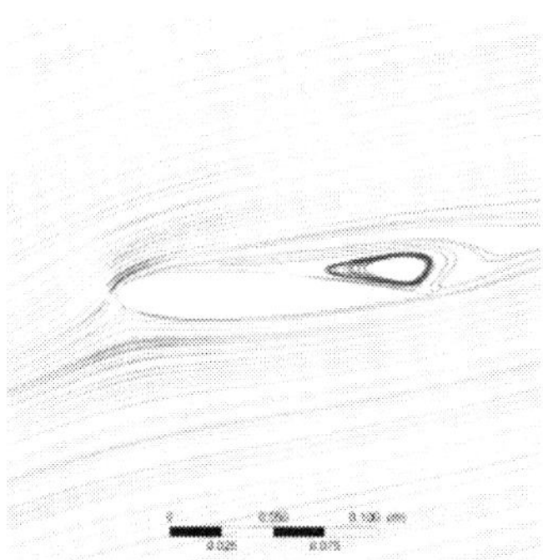

α = 15°, TI = 0.5%

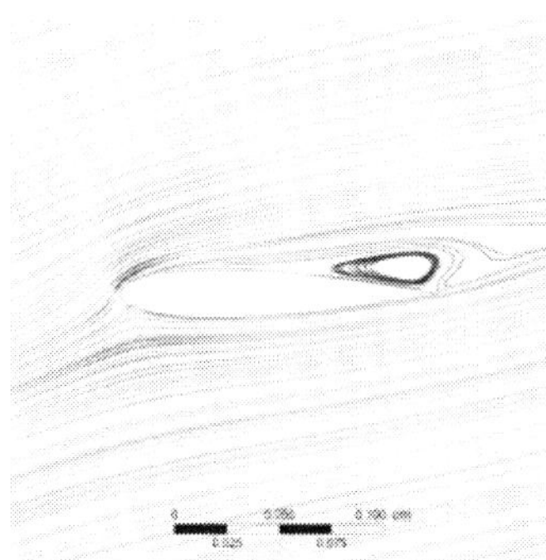

α = 15°, TI = 4.6%

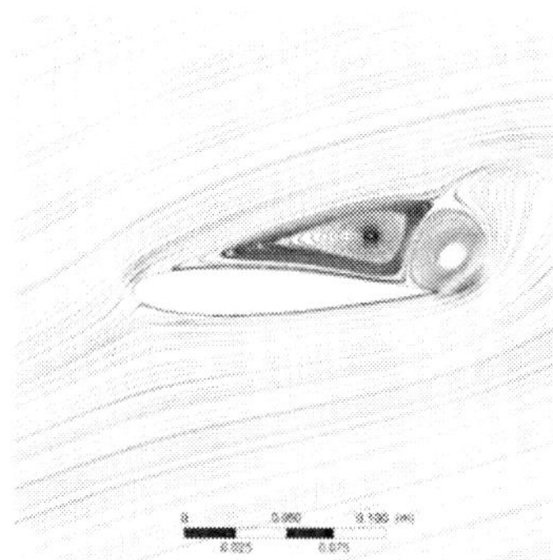

α = 18°, TI = 0.5%

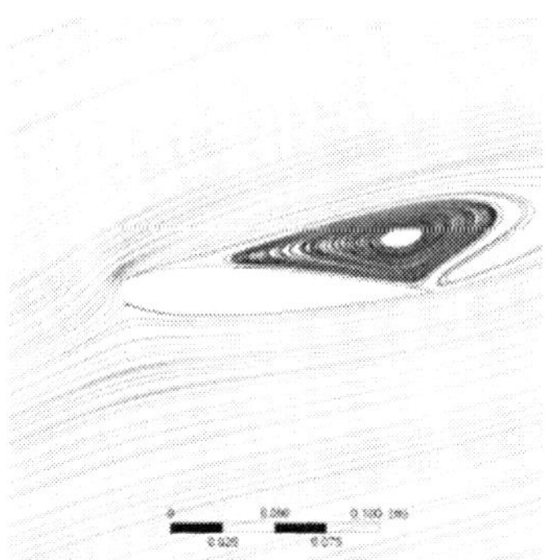

α = 18°, TI = 4.6%

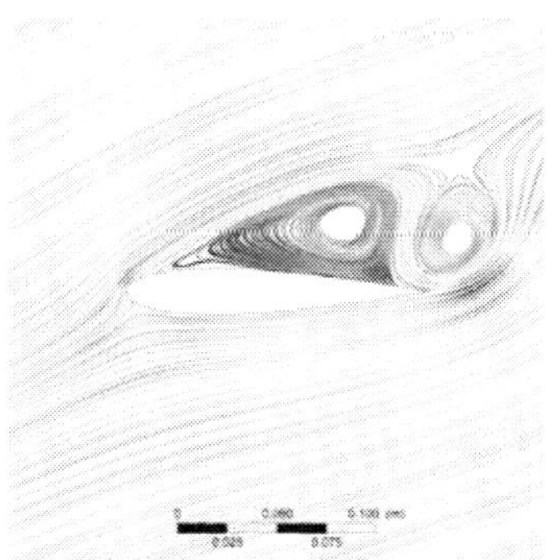

α = 20°, TI = 0.5%

α = 20°, TI = 4.6%

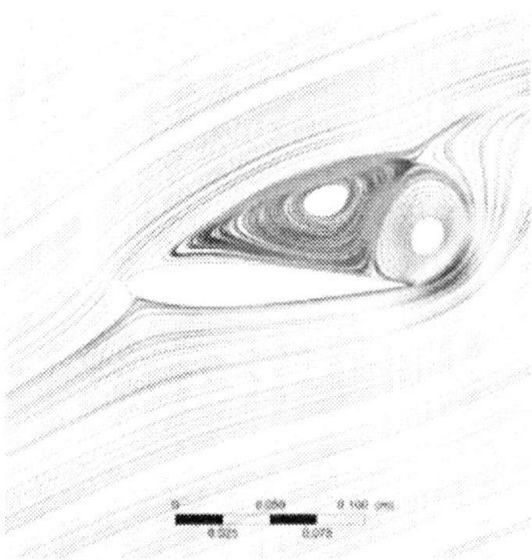

$\alpha = 23°$, TI = 0.5%

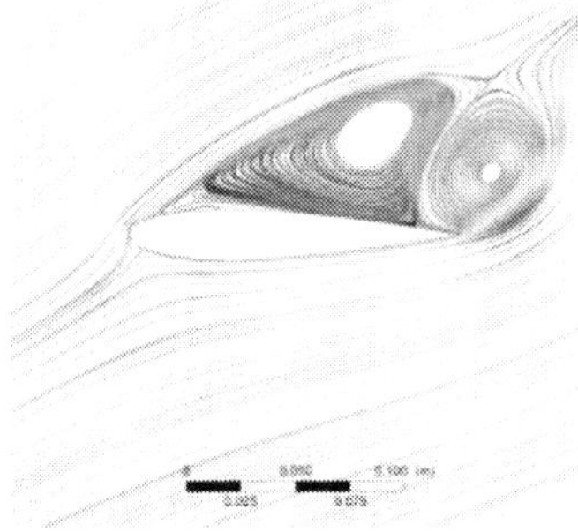

$\alpha = 23°$, TI = 4.6%

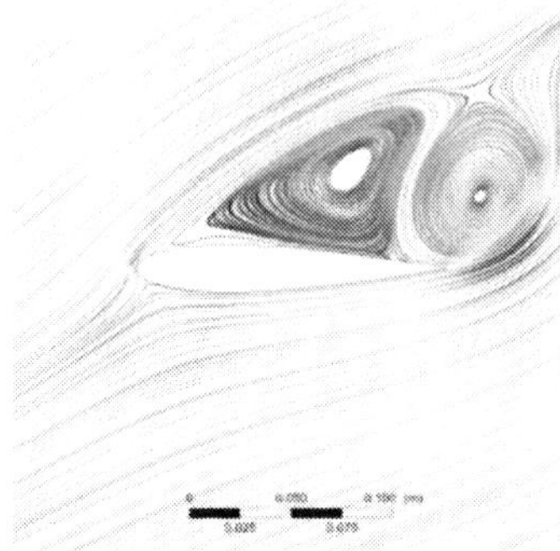

$\alpha = 25°$, TI = 0.5%

$\alpha = 25°$, TI = 4.6%

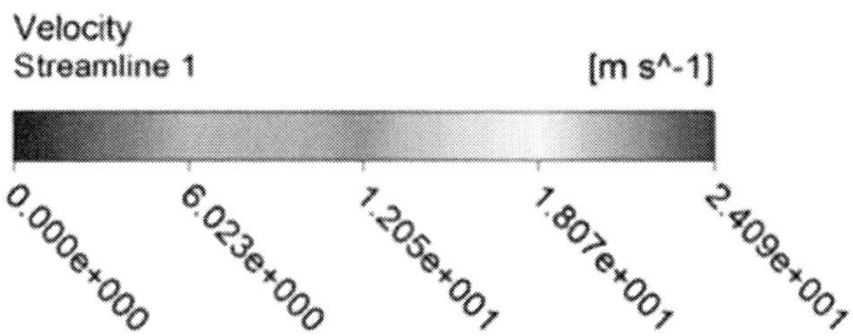

Figure 8. Velocity streamline for variation of attack angles at turbulent intensities of 0.5% and 4.6%

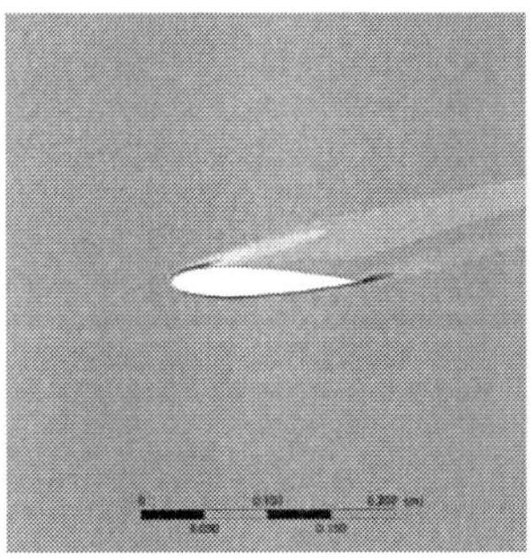

$\alpha = 15°$, TI = 0.5%

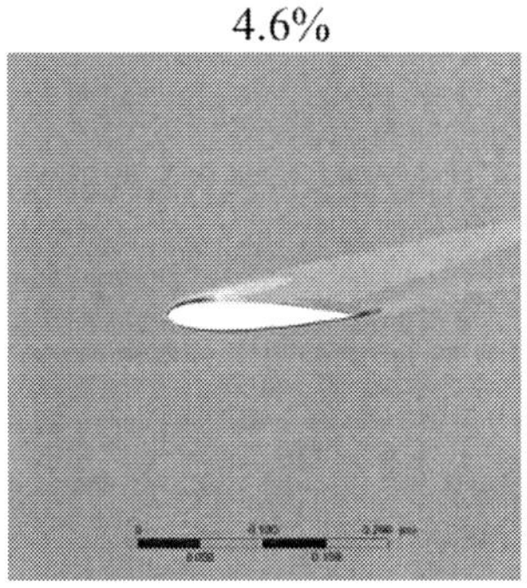

$\alpha = 15°$, TI = 4.6%

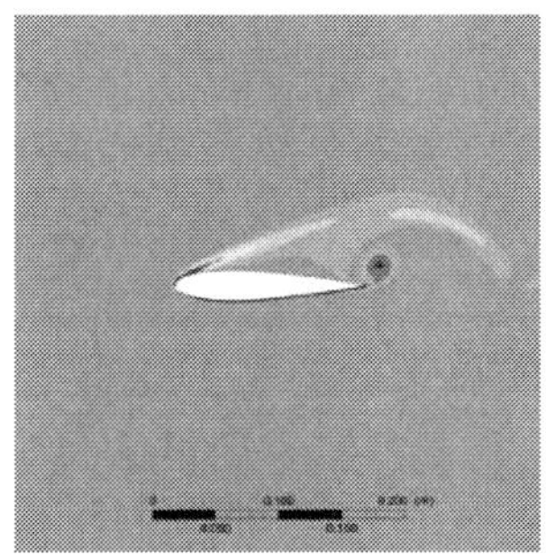

$\alpha = 18°$, TI = 0.5%

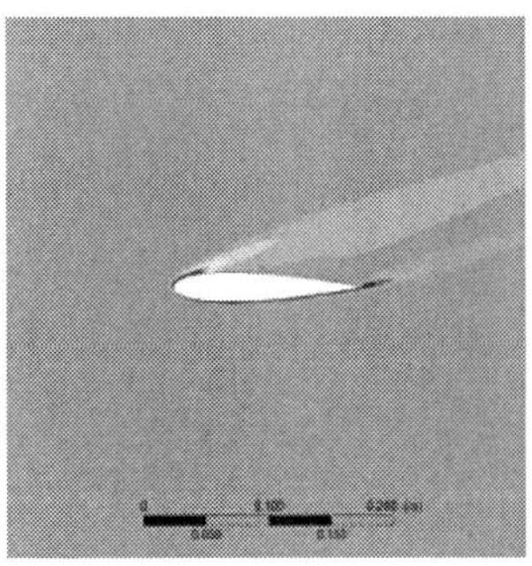

$\alpha = 18°$, TI = 4.6%

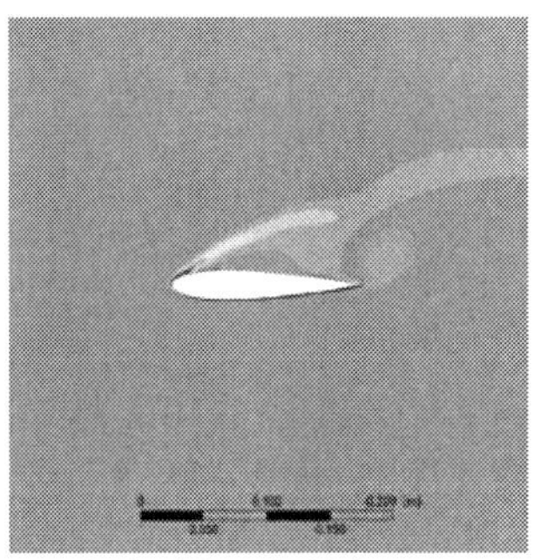

$\alpha = 20°$, TI = 0.5%

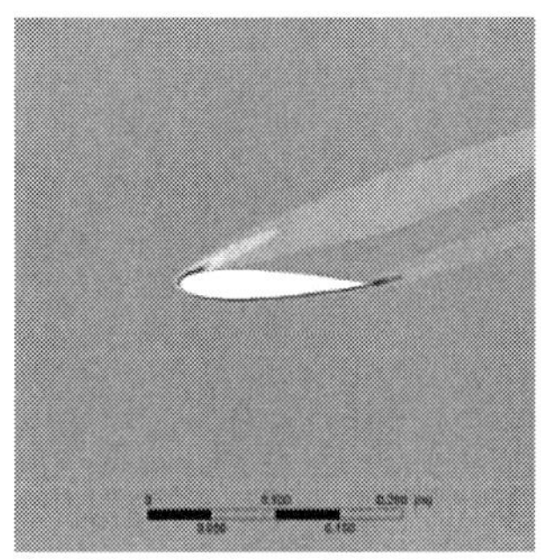

$\alpha = 20°$, TI = 4.6%

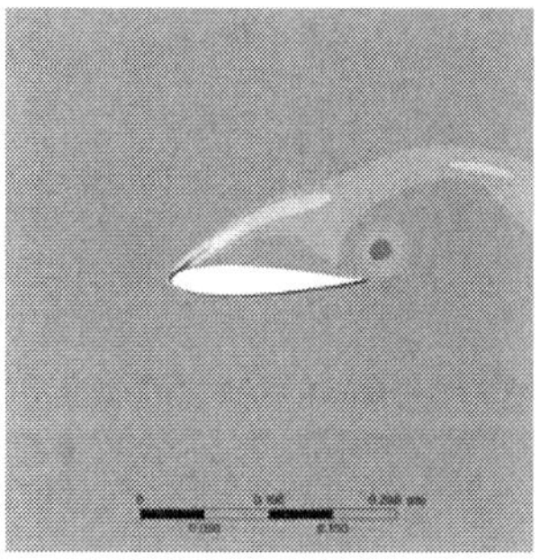

$\alpha = 23°$, TI = 0.5%

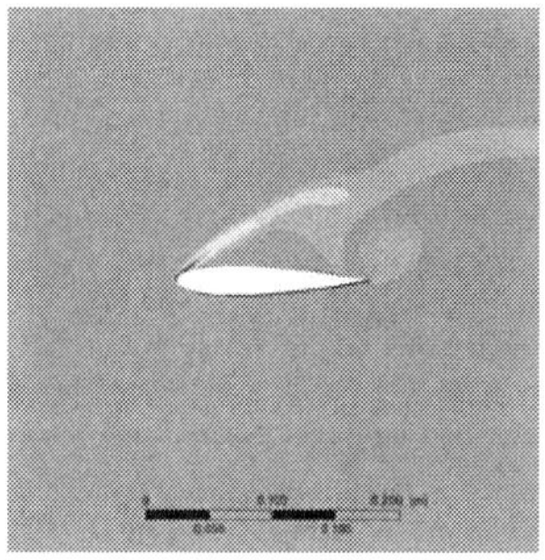

$\alpha = 23°$, TI = 4.6%

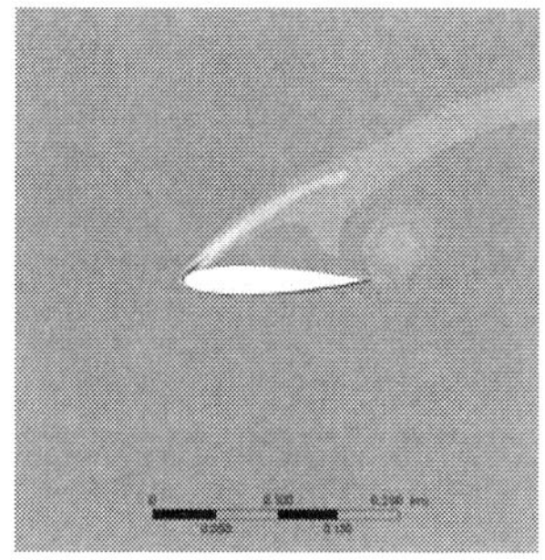

$\alpha = 25°$, TI = 0.5%

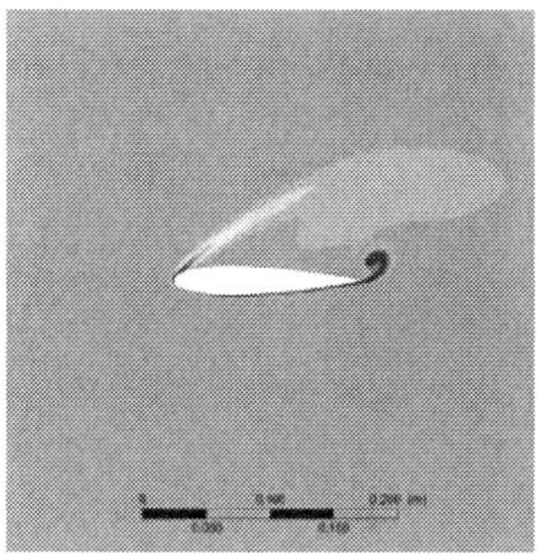

$\alpha = 25°$, TI = 4.6%

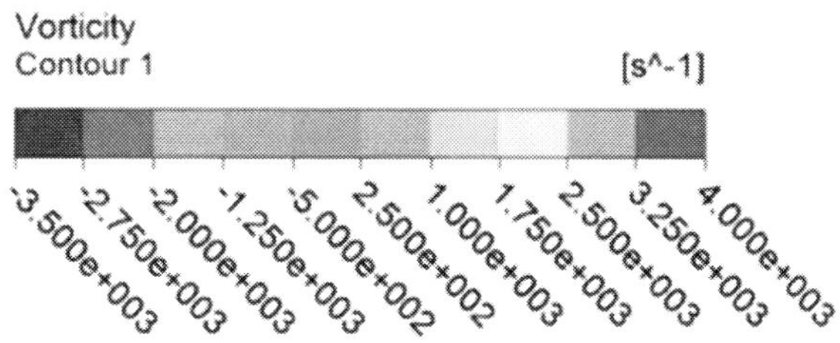

Figure 9. Vorticity for variation of attack angles at turbulent intensities of 0.5% and 4.6%

Acknowledgements
This work was supported by the Faculty of Engineering, Diponegoro University (DIPA FT 2019). The authors are grateful to all research members, especially Thermofluid Lab. and Energy Conversion Lab. of Mechanical Engineering of Diponegoro University.

References

[1] Ministry of Energy and Mineral Resources 2016 to 2030 World Energy Demand Increases 45% http://www.esdm.go.id/berita/37-umum

[2] Çengel YA, and Cimbala JM 2006 *Fluid Mechanics: Fundamentals and Applications* 1)Ed 2006 New York: McGraw-Hill

[3] Şahin İ., and Acir A 2015 Numerical and experimental investigations of lift and drag performances of NACA 0015 wind turbine airfoil *IJMMM* **3** 22-25.

[4] Qing'an Li, Jianzhong Xu, Takao Maeda, Yasunari Kamada, Shogo Nishimura, Guangxing Wu, and Chang Cai 2019 Laser doppler velocimetry (LDV) measurements of airfoil surface flow on a Horizontal Axis Wind Turbine in boundary layer *Energy* **183** 341-357.

[5] Kobra Gharali, Eshagh Gharaei, M. Soltani, and Kaamran Raahemifar 2018 Reduced frequency effects on combined oscillations, angle of attack and free stream oscillations, for a wind turbine blade element *Renewable Energy* **115** 252-259

[6] Giuni M., and Green R.B 2013 Vortex formation on squared and rounded tip *Aerospace Science and Technology* 29 191-199.

[7] Ahmadi-Baloutaki M., Carriveau R., and Ting D.S.K 2015 An experimental study on the interaction between free-stream turbulence and a wing-tip vortex in the near-field *Aerospace Science and Technology* **43** 395-405

[8] Maldonado V., Castillo L., Thormann A., and Meneveau C 2015 The role of free stream turbulence with large integral scale on the aerodynamic performance of an experimental low reynolds number S809 wind turbine blade *Journal of Wind Engineering and Industrial Aerodynamics* **142** 246-257

[9] Qing'an Li, Yasunari Kamada, Takao Maeda, Yusuke Nishida 2017 Experimental investigations of boundary layer impact on the airfoil aerodynamic forces of Horizontal Axis Wind Turbine in turbulent inflows *Energy* **135** 799-810.

[10] Yusik Kim and Zheng-Tong Xie 2016 Modelling the effect of freestream turbulence on dynamic stall of wind turbine blades *Computers and Fluids* **129** 53-66.

[11] Yao J., Yuan W., Wang J., Xie J., Zhou H., Peng M., and Sun Y 2012 Numerical simulation of aerodynamic performance for two dimensional wind turbine airfoils *Procedia Engineering* **31** 80-86

[12] Kim Ya, and Xie Z.T 2016 Modelling the effect of freestream turbulence on dynamic stall of wind turbine blades *Computers & Fluids* **129** 53-66.

[13] Prangemeier T., Rival D., and Tropea C 2010 The manipulation of trailing-edge vortices for an airfoil in plunging motion *Journal of Fluids and Structures* **26** 193-204

[14] Leu T.S., Yu J.M., Hu C.C., Miau J.J., Liang S.Y., Li J.Y., Cheng J.C., and Chen S.J 2012 Experimental study of free stream turbulent effects on dynamic stall of pitching airfoil by using particle image velocimetry *AMM* **225** 103-108

[15] Panda J., and Zaman K.B.M.Q 1994 Experimental investigation of the flow field of an oscillating airfoil and estimation of lift from wake surveys *Journal of Fluid Mechanics* **265** 65.

Minimization of transport costs in an industrial company through linear programming

V Prifti[1], I Dervishi[2], K Dhoska[3], I Markja[4] and A Pramono[5]

[1,2,3,4] Production and Management Department, Faculty of Mechanical Engineering, Polytechnic University of Tirana, Mother Teresa No.3, 1001 Tirana, Albania.
[5] University of Sultan Ageng Tirtayasa, Department of Metallurgy Engineering, Cilegon Banten 42435, Indonesia.

E-mail: vprifti@fim.edu.al

Abstract. Businesses and industries are very interested in optimizing processes and minimizing costs. One of the transportation models or problems to optimize relates to the case where the product output can be transported from the manufacturing plant to some warehouses or customers. The transport problem is a case of the linear programming problem. Linear programming algorithms have been used to solve the most difficult optimization problems. Linear programming has been used to manage the problems related to personnel assignment, engineering, distribution, banking, education, oil, transportation, etc. It has been widely used in various fields, like transport, telecommunications, health services, construction, public services, industry, etc.

The main purpose of the paper is to look at the problem of linear programming in detail by considering an example and try to solve the problem. The purpose of the transport problem in our case is to minimize the overall cost of transport from origin to destination by meeting supply and demand limits, in order to increase sales profit. This paper aims to determine the problem of minimizing the total cost of transport by determining the optimal distribution of the product, from the 2 sites of production to the 9 major distributors that are geographically dispersed. Management Science and Operational Research will be used to find the best solution for the transportation problem.

1. Introduction

Transportation problems are an integral part of linear programming, which accompanies us with daily activities in our life and mostly deals with the logistics sector. It helps to solve transportation problems from one origin to a relevant destination. The basic objective of the problem is to minimize transport costs from origin to destination.

In order to make strategic decisions the model is used for selecting the optimal transport routes in order to divide the production of different origins into many warehouses or centers of distribution. The transport model can be used in the other hand in making localization decisions. When two or more locations are being considered, the model helps in order to find a manufacturing plant, a new facility, or an office. Total costs of transport, delivery and production should be minimized by applying the model. Transport models play an important role in logistics and supply chain. The objective function is the minimized or maximized function. We have a linear programming problem in cases when the linear system model and objective functions are both linear equations. Moreover, linear programming algorithms are used to solve the most difficult optimization problems. In a transportation problem, are defined origins, the factories where the products are manufactured and which supply a required quantity of products to a certain number of destinations. The objective in a transportation problem is to fully meet the demand of the goods to the destination within the constraints of the production capacity at the least possible cost. Transportation problems deal with the problem of production planning and transportation given the quantity offered and the demand that different distribution locations and their product customers require.

2. Literature review

Nowadays mathematical methods solve many problems of operational planning for the transportation. Most of these transportation problems us linear programming method. Optimization means using existing technology and resources and at the best way possible. Better planning and execution results in optimization for a lot of problems. Quantitative models and mathematical tools such as linear programming allows a better result. Modern computing equipment can be used in this case.

Ahmed et al. (2017) proposed a new method to obtain an initial basic feasible solution for the transportation problems. The proposed method was illustrated with numerical examples and also a comparative study was done to justify the performance of the proposed method. It was noticed that the performance of proposed method was suitable for solving transportation problems.

Abdul et al. (2012) has shown a new method named ASM Method for finding an optimal solution for many transportation problems. By this method was established a numerical illustration and was yielded the best the result. This method required very simple arithmetical and logical calculation. This method was explored for those decision makers dealing with logistics and supply chain related issues.

Islam et al. (2012) has proposed an algorithm called Incessant Allocation Method to obtain a starting basic feasible solution for the transportation problems. Were solved a several numbers of numerical problems in order to justify the method and the results have shown that the proposed algorithm was effective in solving the problems of transportation.

Nigus and Tripti (2013) have analysed Vogel's Approximation Method and its modification due to Shimshak and Goyal in finding an initial solution to an unbalanced transportation problem. They have suggested a heuristic approach for balancing the unbalanced transportation problem and improving the Vogels Approximation Method.

Bazaraa et al. (2009) has presented concepts and techniques that are illustrated by numerical examples along with insights completed with detailed mathematical analysis and justification.

Muztoba (2014) has examined a real-world application of a transportation problem that involves transporting mosquito coil from company's warehouse to distributor's warehouse is modelled using linear programming in order to find the optimal transportation cost.

3. The variables, objective function, model and the problem with mathematical formulation

Assume a company has m origins and n destinations. A product will be moved from origin m to destination n. Every warehouse or manufacturing plant has a certain level of supply, and each retail market has a certain demand. The cost of transport between each pair of origin and destination is also given and these costs are assumed to be linear.

Total product supply is denoted $i = a$, where we have $i = 1,2,3,4, \ldots . m$
The total market demand is denoted $j = b$, where we have $j = 1,2,3,4, \ldots . n$
The cost of transport per unit of each product from origin m to destination n is expressed by cij, where we have $i = 1,2,3,4, \ldots m$ and $j = 1,2,3,4, \ldots n$.

3.1. The variables

The variables in this case of linear programming model of the transport problem will have the values for the number of units sent from an origin m to a destination n.

All the variables are presented by:

Xij- the size of the load from origin m to the destination n.
Where we have i = 1,2,3,4, … m and j = 1,2,3,4, … n. This is a series with m, n variables.

3.2. Objective function
The total cost of transportation is given by cij * xij. This is done when we say that the total functional cost is linear.

Collecting all i's and j's now gives the total shipping cost for all origin-destination pair combinations. The transport problem can be formulated as a linear programming problem model:
Minimization of: $\sum_{i=1}^{m}\sum_{j=1}^{n} X_{ij}C_{ij}$
Where:
x_{ij} - is the quantity of units that are moved from origin m to destination n.
c_{ij} - is the cost of moving a unit from origin m to destination n.

3.3. Limitations
Limitations are the conditions that compel the necessary capacity and demand to be met. In a problem of transport there is a drawback for each node.
We have:
a1 - Origin capacity indicator m
b1 - destination capacity indicator n

1. The capacity of each supplier can be:
$$\sum_{i=1}^{m} X_{ij} = a\,, \quad where\ i = 1,2,3 \dots m \tag{1}$$
2. The request for any destination can be:
$$\sum_{j=1}^{n} X_{ij} = b\,, \quad where\ j = 1,2,3 \dots n \tag{2}$$
3. Condition of non-negativity:
$$X_{ij} \geq 0\ for\ each\ i\ dhe \tag{3}$$

3.4. Transport model Formatting the title
Minimization of:
$$Z = \sum_{i=1}^{m}\sum_{j=1}^{n} X_{ij}C_{ij} \dots\dots \tag{4}$$
The supply capacity constraint
$$\sum_{i=1}^{m} X_{ij} \leq ai \qquad i = 1,2,3,4, \dots . m \tag{5}$$
The demand capacity constraint
$$\sum_{j=1}^{n} X_{ij} \geq bj \qquad j = 1,2,3,4 \dots . n \tag{6}$$
$$X_{ij} \geq 0\ for\ i = 1,2,3,4, \dots . m,\ j = 1,2,3,4, \dots . n \tag{7}$$
This is a linear programing with the variables m, n, and with m + n limiting variables and with m, n non-negative variables.
Where:
m = nr. of origins, n = nr. of destinations, ai = capacity of origins (in ton, liter, etc.), bj = capacity of destinations (in tonnes, liters, etc.), Cij = the cost of materials.

The existence of the equation below is a necessary and sufficient condition for the existence of a possible solution to the transport problem:
$$\sum_{i=1}^{m} a_i = \sum_{j=1}^{n} b_j \tag{8}$$
When this equation exists, we are dealing with a balanced transportation problem.

3.5. The problem of unbalanced transportation

If:

$$\sum_{i=1}^{m} a_i \neq \sum_{j=1}^{n} b_j \quad (9)$$

Transport problems are known as unbalanced transport problems.

Case 1:

The available capacity of the originals is greater than the capacity at the destination required.

$$\sum_{i=1}^{m} a_i > \sum_{j=1}^{n} b_j \quad (10)$$

Case 2:

The available capacity of the origins is less than the capacity at the destination required.

$$\sum_{i=1}^{m} a_i < \sum_{j=1}^{n} b_j \quad (11)$$

4. Representing the transport problem network

There is also a graphical representation of this problem.

Graphically, this problem is often seen as a network of m sources and n destinations, and a set of m, n "directed arcs". This network of transport problems is shown in the following figure.

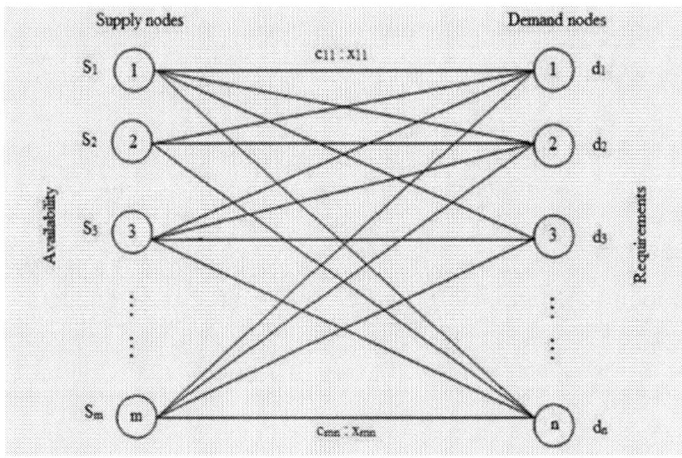

Figure 1. Source and destination nodes in the transport problem network

S1... ...Sn → origins

D1... ...Dn → destinations

C1... ...Cn → cost of transport

The arrows indicate the transport flows from source to destination.

4.1. Degradation of the transport problem

A transport problem is called degraded in the cases when the nr of filled cells is less than the nr of rows plus the nr of columns minus one (m + n - 1). We can be observed degradation during the initial allocation where the first entry in a row or column meets the row and column requirements, or when the added and subtracted values are equal, in cases when is applied the method of Stepping Stone. The basic solution is called degraded, when the transport problem with m origins and n destinations does not have m + n-1 positive base variables. Whenever the number of base cells is not more or equal than m + n - 1, the transport problem is degraded. To solve the degradation problem, are increased the positive variables with a lot of zero variables in such a way as to satisfy the basic variable m + n -1.

4.2. Solution algorithm for transport problem

The solution algorithm has the following steps:

Step 1. Formulation of the problem and place the data in matrix form. The objective function is the total cost of transport. The constraints are the demand and supply to each source and each destination.

Step 2. The basic solution can be reached using one of the methods:

North West Corner Rule, Least Cost Method, Vogel Approximation Method

The solution must meet all supply and demand constraints. The nr of positive partitions must be equal to m + n - 1, where m is the nr of rows and n is the nr of columns.

Step 3. Test the initial solution for optimality with the Stepping Stone Method or Modified Distribution Method

Step 4. Update the solution. Then repeat Step 3 until the optimal solution is reached.

4.3. Transportation Problem Methods

a. North West Corner

The North West Corner Rule is a method to compute the initial feasible solution of the transport problem. The name is given to this method because the basic variables are selected from the extreme left corner.

b. Least Cost Method

In the transport problem the Least Cost Method is used to obtain the initial feasible solution. The allocation begins with the cell which has the minimum cost. The lower cost cells are chosen over the higher-cost cell in order to have the least cost of transport.

c. Vogel's Approximation Method

VAM (Vogel's Approximation Method) is an iterative procedure calculated to find out the initial feasible solution of the transport problem. Here is taken into consideration the transport cost, but in a relative sense.

5. Case study of a transport problem in a company in Albania

This study aims to minimize the total cost of transportation of a company, from the two production sites, S1 and S2 for its many distributors throughout Albania, which are nine. The total factory capacity for S1 and S2 per day is 1736 and 2419 units. The company faces challenges in how to optimally distribute its products to 9 major distributors at a minimal trasportation cost. Each production site has its own supply limit and each customer has a specific demand at a time. The data is based on two manufacturing plants located in two different sites and distributes this manufactured product in 9 geographical regions. We have factories S1-S2 and destinations D1-D9

Table 1. Capacity and demand of sources and destinations

From/ To	D 1	D2	D3	D4	D5	D6	D7	D8	D9	Capacity
S1	90.79	88.21	82.08	68.99	30.59	424.91	30.6	13.87	70.85	1736
S2	288.74	37.6	176.41	72.95	114.32	173.09	73.37	38.61	239.2	2419
Demand	907	576	445	335	272	431	304	128	757	

5.1. Problem formulation:

Factory S1, Factory S2, Xij = Units moved from source *i* to destination *j*

Using the transport costs shown in the table above, we can write the following equations:

5.2. Cost minimization function:

We add all the values from the table, by multiplying the cost with the quantity:
$90.79X_{11} + 88.21X_{12} + 82.08X_{13} + 68.99X_{14} + 30.59X_{15} + 424.91X_{16} + 30.60X_{17} + 13.87 X_{18} + 70.85X_{19} + 228.74X_{21} + 37.60X_{22} + 176.41X_{23} + 72.955X_{24} + 114.32X_{25} + 173.09X_{26} + 73.37X_{27} + 38.61X_{28} + 239.2020X_{29}$

5.3. Capacity constraints:

First row: $X_{11} + X_{12} + X_{13} + X_{14} + X_{15} + X_{16} + X_{17} + X_{18} + X_{19} \leq 1736$
Second row: $X_{21} + X_{22} + X_{23} + X_{24} + X_{25} + X_{26} + X_{27} + X_{28} + X_{29} \leq 2419$

5.4. Demand constraints:

$X_{11} + X_{21} = 907$
$X_{12} + X_{22} = 576$
$X_{13} + X_{23} = 445$
$X_{14} + X_{24} = 335$
$X_{15} + X_{25} = 272$
$X_{16} + X_{26} = 431$
$X_{17} + X_{27} = 304$
$X_{18} + X_{28} = 128$
$X_{19} + X_{29} = 757$
$X_{ij} \geq 0$ for each i and j

5.5. The North West Method

Table 2. North West Method, First selection

From/ To	D 1	D2	D3	D4	D5	D6	D7	D8	D9	Capacity
	90.79	88.21	82.08	68.99	30.59	424.91	30.6	13.87	70.85	
S1	907									829
	288.74	37.6	176.41	72.95	114.32	173.09	73.37	38.61	239.2	
S2										2419
Demand	0	576	445	335	272	431	304	128	757	

First select the angle from the far left of the matrix, where are assigned as many units as possible within the bounds of demand and supply. We allocate 907 units in the first cell that has meet the demand of destination D1 but supply S1 is bigger. The supply capacity of the S1 is more with: 1736-907 = 829 units. This logic explained above is also followed for the other cells.

Table 3. North West Method, Final table of determinations:

From/ To	D 1	D2	D3	D4	D5	D6	D7	D8	D9	Capacity
	90.79	88.21	82.08	68.99	30.59	424.91	30.6	13.87	70.85	
S1	907	576	253							1736
	288.74	37.6	176.41	72.95	114.32	173.09	73.37	38.61	239.2	
S2			192	335	272	431	304	128	757	2419
Demand	907	576	445	335	272	431	304	128	757	

Summary of results of Variables and Values:
X_{11}: 907; X_{12}:576; X_{13}:253; X_{14},X_{15},X_{16},X_{17},X_{18},X_{19}:0
X_{21},X_{22}:0; X_{23}:192; X_{24}:335; X_{25}:272; X_{26}:431; X_{27}:304; X_{28}:128; X_{29}:0

Minimum cost total = 90.97 x 907 + 88.21 x 576 +82.08 x 253 +176.41 x 192 + 72.95 x 335 + 114.32 x 272 + 173.09 x 431 + 73.37 x 304 + 338.61 x 128 + 29.2 x 128 + 239.2 x 757 = ***564811.75***

The number of cells allocated is 10 equivalent to: m + n-1 = 2 + 9-1 = 10, indicating that this solution is not degraded.

Table 4. Least Cost Method, first table

From/ To	D 1	D2	D3	D4	D5	D6	D7	D8	D9	Capacity
	90.79	88.21	82.08	68.99	30.59	424.91	30.6	13.87	70.85	
S1								128		1608
	288.74	37.6	176.41	72.95	114.32	173.09	73.37	38.61	239.2	
S2										2419
Demand	907	576	445	335	272	431	304	0	757	

The minimum transport cost is in cell S1 D8. The maximum that can be determined in this cell is 128 units. Requirement for destination D8 is fulfilled, while factory S1 has a capacity of: 1736 -128 = 1608 units.

Table 5. Least Cost Method, final table of determinations

From/ To	D 1	D2	D3	D4	D5	D6	D7	D8	D9	Capacity
				68.99	30.59				70.85	
	90.79	88.21	82.08			424.91	30.6	13.87		
S1				335	272		304	128	697	1736
						173.09				
	288.74	37.6	176.41	72.95	114.32		73.37	38.61	239.2	
S2	907	576	445			431			60	2419
Demand	907	576	445	335	272	431	304	128	757	

Summary of results of Variables and Values:
X_{11},X_{12},X_{13}:0; X_{14}:335; X_{15}:272; X_{16}:0; X_{17}:304; X_{18}:128; X_{19}:697
X_{21}: 907; X_{22}:576; X_{23}:445; X_{24},X_{25}:0; X_{26}:431; X_{27},X_{28}:0; X_{29}:60

Minimum transport cost = 335 x 68.99 + 272 x 30.59 +304 x 30.6 +128 x 13.87 + 697 x 70.85 + 907 x 288.74 + 576 x 37.6 + 445 x 176.41 + 431 x 173.09 + 60 x 239.2 = 542893.36
The number of cells allocated is 10 equivalent to: m+n 1 = 2+9-1 =10, indicating that this solution is not degraded.

Table 6. Vogel's Method

From/ To	D 1	D2	D3	D4	D5	D6	D7	D8	D9	Capacity	Penal of the row
	90.79	88.21	82.08	68.99	30.59	424.91	30.6	13.87	70.85		
S1										1736	16.72
	288.74	37.6	176.41	72.95	114.32	173.09	73.37	38.61	239.2		
S2										2419	1.01
Demand	907	576	445	335	272	431	304	128	757		
Column Penalty	197.77	50.61	94.33	3.96	83.73	251.82	42.77	24.74	168.35		

The maximum penalty is in the sixth column at 251.82 and the minimum cost Cij is in column C26 = 173.09. The maximum determination we can make in this cell is 431. This satisfies the 6th column and the second row is completely saturated with a difference from 2419 to 1988: (S2 = 2419- 431 = 1988)

Table 7. Vogel's Method, first table

From/ To	D 1	D2	D3	D4	D5	D6	D7	D8	D9	Capacity	Penal of the row
S1	90.79	88.21	82.08	68.99	30.59	424.91	30.6	13.87	70.85	1736	16.72
S2	288.74	37.6	176.41	72.95	114.32	173.09 431	73.37	38.61	239.2	1988	1.01
Demand	907	576	445	335	272	0	304	128	757		
Column Penalty	197.77	50.61	94.33	3.96	83.73	--	42.77	24.74	168.35		

Table 8. Vogel's Method, final table

From/ To	D 1	D2	D3	D4	D5	D6	D7	D8	D9	Capacity
S1	90.79 907	88.21	82.08 72	68.99	30.59	424.91	30.6	13.87	70.85 757	1736
S2	288.74	37.6 576	176.41 373	72.95 335	114.32 272	173.09 431	73.37 304	38.61 128	239.2	2419
Demand	907	576	445	335	272	431	304	128	757	

Summary of results of Variables and Values:
X_{11}: 907; X_{12}:0; X_{13}:72; X_{14}, X_{15}, X_{16}, X_{17}, X_{18}:0; X_{19}:757
X_{21}: 0; X_{22}:576; X_{23}:373; X_{24}:335; X_{25}:272; X_{26}:431; X_{27}:304; X_{28}:128; X_{29}:0

Minimum cost total = 90.97 x 907 + 82.08 x 72 +176.4 x 373 + 72.95 x 335 + 114.32 x 272 + 173.09 x 431 + 73.37 x 304 + 38.61 x 128 + 70.85 x 757 + 37.6 x 576 = 386893.17
The number of cells allocated is 10 equivalent to: m+n-1 = 2+9-1 =10, indicating that this solution is not degraded.

Table 9. Summary table of the three methods: NWC, LCM, Vogel's

Variables	Value of NWC	Value of LCM	VOGEL'S Value
X_{11}	907	0	907
X_{12}	576	0	0
X_{13}	253	0	72
X_{14}	0	335	0
X_{15}	0	272	0
X_{16}	0	0	0
X_{17}	0	304	0
X_{18}	0	128	0
X_{19}	0	697	757
X_{21}	0	907	0
X_{22}	0	576	576
X_{23}	192	445	373
X_{24}	335	0	335
X_{25}	272	0	272
X_{26}	431	431	431
X_{27}	304	0	304
X_{28}	128	0	128
X_{29}	0	60	0
Minimum cost for each method	564811.75	542893.36	*386893.17*

All company costs are 10 ^ 3

Having solved the problem with all three methods then it turns out that the most efficient method is the Vogel's method. This is the method which is chosen as the most optimal as it results in a lower cost than the other two methods.

Table 10. Case Study transport table

From Source / To Destination	Dest 1	Dest 2	Dest 3	Dest 4	Dest 5	Dest 6	Dest 7	Dest 8	Dest 9	Capacity
Source 1	90.79 907	88.21	82.08 72	68.99	30.59	424.91	30.6	13.87	70.85 757	1736
Source 2	288.74	37.6 576	176.41 373	72.95 335	114.32 272	173.09 431	73.37 304	38.61 128	239.2	2419
Demand	907	576	445	335	272	431	304	128	757	

6. Conclusions and future work

The paper considered a real problem of linear programming in detail by taking an example in an Albania company. For the company under consideration the problem of minimizing transportation costs was solved by solving 3 methods: The North West Corner Method, the Least Cost Method, and the Vogel's Approximation Method. The calculations showed that based on the demand from the 9 geographical sites (detinations) and the capacity offered by the two manufacturing plants (sources), the most optimal solution turns out to be the one obtained by the Vogel's method at a minimum cost of 386893.17. Finally, the demand for the 3 geographical regions is covered by the S1 production plant, while the demand for the other 6 regions is covered by the S2 production plant.

In the future we can continue to apply the other steps of transportation methods in order to update and optimise the solution. Also, we can continue the research with experimenting different methods of operational research like simplex method, etc., and try to maximise the profit in an industrial enterprise.

References

[1] Ahmed M M, Sultana N, Khan A R and Uddin M S 2017 An innovative approach to obtain an initial basic feasible solution for the transportation problems *Journal of Physical Sciencies* 23-42

[2] Sudhakar V J, Arunsankar N and Karpagam T 2012 A new approach for finding an optimal solution for transportation problems *European Journal of Scientific Research* 254-257

[3] Islam M A, Haque M H and Uddin M S 2012 Extremum difference formula on total opportunity cost: a transportation cost minimization technique *Prime University Journal of Multidisciplinary Quest* 125-130

[4] Girmay N and Sharma T 2013 Balance an unbalanced transportation problem by a heuristic approach *International Journal of Mathematics and its applications* 13-19

[5] Ahmed M M, Khan A R, Uddin M S and Ahmed F 2016 A new approach to solve transportation problems *Open Journal of Optimization* 22-30

[6] Bazaraa M S and Jarvis J J 2009 *Linear Programming and network flows* (John Wiley and Sons 4rth edition)

[7] Winston W L 2004 *Operations Research: Applications and Algorithms* (Thomson Learning)

[8] Joshi V D and Gupta N 2011 Linear fractional transportation problem with varying demand and supply *Le Matematiche* 3-12

IOP Conf. Series: Materials Science and Engineering **909** (2020) 012041 doi:10.1088/1757-899X/909/1/012041

Improvement of insurance agents performance using data mining in OV agency

A N Habyba*, and R Fitriana
Department of Industrial Engineering, Trisakti University, Jakarta, Indonesia

*E-mail: anik@trisakti.ac.id

Abstract. One problem that occurs in the insurance industry is the high turnover rate of insurance agents. It makes many companies must pay a large fee for the recruitment and training of new agents. OV Agency had high insurance agent turnover rate. This company wants to improve the agents performance to avoid the turnover. This research aims to identify prospective insurance agents who have better performance at recruitment process by using classification task of data mining. There are three classification methods were used . They were decision tree (J48), Neural Network (MLP) and Naive Bayes. Based on the result, MLP has the best average accuracy value 84.2% and ROC area value 0.87 (good classification). MLP was chosen to predict the 25 insurance agent candidates and 3 agents were selected because they entered class A (better performance) in all performance criteria. The prediction results show that three agents are men, aged 25-32 years old, divorced, senior high school education and master degree, work status in other companies, and have work experience of 0-3 years. The OV Agency can consider these result in new agent recruitment process.

1. Introduction

The high turnover rate of insurance agents is a common problem for insurance companies like OV Agency. In the previous study the OV agency's agent performance was greatly influences their customer satisfaction. It was necessary to improve the agents service quality. Some agents were not considered to provide a detailed explanation of the product and causing customer misunderstanding [1]. The agents turnover became something that cannot be avoided by the OV Agency. The agent turnover make companies recruiting new agents frequently and also increasing the recruitment and training costs. If the companies could not predict the agents performance before recruit them, it could make other problems to the company. OV Agency must be able to select new agents with good performance in the future. Prevention of agent turnover can be done by predicting the prospective agents performance. The prediction can be done by using the old agents performance data. Based on this data, OV Agency can recruit only agents who are predicted to have good performance. The OV Agency assess agent performance based on length of service, total turnover, total policy, turnover persistency, and policy persistency. The agent performance can be divided into classes according to the company goals. Predictive data mining can be used to solve this problems.

Data mining has been widely used to assist insurance companies in processing their data to becomes useful information. Specifically for the use of classification and prediction, data mining has been used to predict consumer behavior, customers classification, the type of policy most desirable, segment performance and factors that will affect the insurance sector, also the classification of trends in organizational movements for successful/unsuccessful customer historical records [2]. In addition, data mining has also been used for the selection of insurance agents based on performance criteria for length of service, total policy, and policy persistency. Decision tree, discriminant analysis, and artificial neural networks were used in this previous study [3]. Furthermore naive bayesian classification is also used to predict the life insurance dataset where predictions can be made on customer perceptions of existing life insurance products [4]. Research to predict insurance fraud has also been carried out using data mining. [5] implemented naive bayesian, decision tree, and rule-based classification to predict and analyze fraud patterns in auto insurance data. And the results show that the reliable model to be used to detect fraud is another type of insurance. These various studies prove that data mining has many benefits in predictive research in the insurance industry and can also be applied by the OV Agency.

A study has been conducted to analyze the best classification algorithm in predicting slow learners in the education sector. Five classification algorithms such as multilayer perceptron, naive Bayes, SMO, J48, and REPtree are used to test student academic records performance. The results show that multilayer perceptron has the highest accuracy rate of 75%, then J48 69.73%, SMO 68.42%, REPTree 67.76%, and Naive Bayes 65.13% [6]. Further research needs to be done whether these results will be the same if applied to insurance agent data. The performance of OV Agency insurance new agents candidates can also be predicted using the classification algorithm. Furthermore, this study aims to identify prospective insurance agents who have better performance at recruitment process by using predictive data mining models using a classification based algorithm.

2. Research methodology

Data mining is also commonly known as knowledge discovery from data (KDD). Data mining is part of the knowledge discovery process which is an intelligent method that can be used to extract data patterns or information needed from large data sets [7]. The purpose of using data mining in this study was not only for classification but also prediction which is a high-level goal in the application of data mining. Prediction results can be used to see the possibilities that will occur in the future. In predictive modeling, systems will guess unknown values based on previously owned data patterns [8]. Predictive data mining in this study was used to predict the class value of the five insurance agent performance criteria, namely length of service, total turnover, total policy, turnover persistency and policy persistency. The value of each class from each insurance performance criteria can be seen in Table 1.

Table 1. Class of agents performance criteria

Agents performance criteria	Class
Length of service	Class A (length of service ≥7 years), Class B
Total turnover	Class A (turnover ≥ IDR 50000000), Class B
Total policy	Class A (policy ≥ 10), Class B
Persistency turnover	Class A (persistency ≥ 90%), Class B
Persistency policy	Class A (persistency ≥ 90%), Class B

Data collection was carried out using secondary data from the OV Agency. A total of 250 data from insurance agents owned by the company were used as training data and classified based on five performance criteria. The demographic data of insurance agents was used as an independent variable that will guess the performance of agents in the future. Six variables such as sex, age, marital status, academic level, employment status, and working experience are considered to affect the performance of insurance agents. This variable was chosen based on the results of previous studies [3] and adjusted for information that can be obtained from OV Agency. The classification was performed for each agent performance criteria and uses three classification methods in data mining, namely decision tree (JT8), ANN (multilayer perceptron), and naive Bayes in Weka (Waikato Environment for Knowledge Analysis) version 3.8.1. Weka is developed by the University of Waikato New Zealand as an open-source Java application that has many algorithms to analyze data sets. The C4.5 algorithm discovered by Ross Quinlan 1993 is called J48 in Weka [9]. This algorithm uses the concept of the gain ratio which is the development of information gain. So the decision tree becomes a method that can automatically determine the most important variables in classifying data into correct class categories. The resulting tree model is also easy to understand [10].

Multilayer perceptron (MLP) is one type of ANN that also includes supervised learning which is good for classification. Multilayer feedforward networks are the most important and most widely used type of ANN. MLP has the characteristics that the model includes non-linear, sigmoidal or hyperbolic activation functions, the network can have more than one hidden layer that can learn complex tasks, the network shows a high level of connectivity between layers [11]. MLP can be used to categorize more than two output classes, in contrast to decision trees and naive Bayes, naive Bayes are opportunity-based classification methods that add up the frequency and combination of values from a given dataset. There are two assumptions in naive Bayes that all variables are priority or equally important and are independent (Bayes theorem). Naive Bayesian Classifier is a classification method that can be compared with decision trees and neural networks. This method can have a minimum error rate if the assumptions used are accurately such as independence classes and enough available probability data [7]. This research compared three methods decision tree, MLP and Naive Bayes in OV Agency insurance agent classification.

Accuracy models were made that were formed from each method and observed that the demographic variables of the agent that most influenced the results of the data classification. The classification model accuracy is judged good also based on the values of correcting classified and ROC area. The ROC (Receiver Operating Characteristic) curve shows accuracy and visually compares classifications. ROC expresses the confusion matrix. ROC is a two-dimensional graph with false positives as horizontal lines and true positive as vertical lines [12]. Accuracy based on ROC as follows [13]:

a. The accuracy is 0.90 - 1.00 = excellent classification
b. The accuracy is 0.80 - 0.90 = good classification
c. The accuracy is 0.70 - 0.80 = fair classification
d. The accuracy is 0.60 - 0.70 = poor classification
e. The accuracy is 0.50 - 0.60 = failure

Predictive data mining was done to select the new OV Agency insurance agent based on information held in the recruitment process. The results of the prediction of the performance of new agents in the future were used to be considered as acceptance of new agents. A total of 25 new agent data were used as testing data in this study. The results of the three methods were compared and analyzed.

3. Results and discussion

3.1. Classification of insurance agents performance

The results of the data mining classification for the three methods based on five insurance agent performance criteria can be seen in Table 2. The goodness of the classification model can be seen from the ROC area and accuracy. The results show that the highest ROC area value is using multilayer perceptron in each performance criterion. The highest ROC area MLP value is 0.90 obtained from total turnover with an accuracy of 84.0%, these values are indicated by boldface. The ROC value of the area of 0.90 indicates good classification.

Naive Bayes classification results have the lowest ROC area and accuracy values for each performance criteria. The lowest yield is ROC area 0.59 and accuracy 57.6%. The value of ROC area 0.59 shows the accuracy of failure classification. These results also indicate if the Naive Bayes classification fails to classify the performance criteria, namely length of service correctly. The results of the classification of the three methods look good in the total policy performance criteria where the decision tree is 78.8%, MLP 88.4%, and Naive Bayes 72.4%. In this total policy performance criteria, Naive Bayes gets the highest accuracy value compared to the other four performance criteria, which is 72.4% and ROC area value 0.64. The value of ROC area 0.64 indicates the accuracy is poor classification. Overall results show that MLP can be used to predict the performance of new insurance agents well. The same results [14] also show that compared to J48, MLP is a better algorithm used for many data cases. MLP algorithm based on neural networks has good learning skills and is suitable for classification problems.

Table 2. Model performance of the classification (use training set)

Performance criteria	Classification methods	Class	TP rate	FP rate	Precision	Recall	F-measure	ROC area	Accuracy (correctly classified)
Length of service	J48	A	0.87	0.59	0.67	0.87	0.76	0.67	67.6%
		B	0.41	0.13	0.69	0.41	0.52	0.67	
	MLP	A	0.82	0.12	0.90	0.82	0.86	**0.83**	**84.4%**
		B	0.88	0.18	0.78	0.88	0.83	**0.83**	
	Naive Bayes	A	0.85	0.80	0.59	0.85	0.70	0.59	57.6%
		B	0.20	0.15	0.49	0.20	0.28	0.59	
Total turnover	J48	A	0.26	0.03	0.86	0.26	0.40	0.65	70.4%
		B	0.97	0.74	0.68	0.97	0.80	0.65	
	MLP	A	0.77	0.12	0.80	0.77	0.79	**0.90**	**84.0 %**
		B	0.88	0.23	0.86	0.88	0.87	**0.90**	
	Naive Bayes	A	0.27	0.12	0.59	0.27	0.37	0.61	65.2%
		B	0.88	0.73	0.67	0.88	0.76	0.61	
Total policy	J48	A	0.36	0.29	0.84	0.36	0.51	0.73	78.8%
		B	0.97	0.64	0.78	0.97	0.87	0.73	
	MLP	A	0.81	0.09	0.80	0.81	0.81	**0.88**	**88.4%**
		B	0.91	0.19	0.92	0.91	0.92	**0.88**	
	Naive Bayes	A	0.11	0.11	0.80	0.11	0.19	0.64	72.4%
		B	0.99	0.89	0.72	0.99	0.83	0.64	
Persistency turnover	J48	A	0.81	0.40	0.75	0.81	0.78	0.77	72.4%
		B	0.60	0.19	0.67	0.60	0.64	0.77	
	MLP	A	0.95	0.41	0.78	0.95	0.85	**0.85**	**80.4%**
		B	0.59	0.05	0.88	0.59	0.71	**0.85**	
	Naive Bayes	A	0.87	0.78	0.63	0.87	0.73	0.61	60.8%
		B	0.22	0.13	0.53	0.22	0.31	0.61	
Persistency policy	J48	A	0.82	0.28	0.79	0.82	0.80	0.85	77.6%
		B	0.72	0.18	0.76	0.72	0.74	0.85	
	MLP	A	0.94	0.29	0.81	0.94	0.87	**0.87**	**84.0%**
		B	0.71	0.06	0.91	0.71	0.80	**0.87**	
	Naive Bayes	A	0.74	0.63	0.60	0.74	0.66	0.63	57.6%
		B	0.37	0.26	0.53	0.37	0.44	0.63	

3.2. Prediction of new insurance agents performance

The purpose of OV Agency is to get a new insurance agent that is predicted to have good performance, namely class A. Prediction results of 25 new insurance agents used variables of sex, age, marital status, academic level, employment status, and working experience on each performance criterion showing results differently. The prediction result can be seen in Table 3. The highest accuracy multilayer perceptron method also shows the highest number of agents entering class A. At the length of service, as many as 11 agents are predicted to enter class A which has a work duration of ≥ 7 years. Based on the total turnover of 7 agents will reach a turnover of ≥ IDR 50,000,000 and 8 agents will reach ≥10 policies. Whereas based on the 15 agent turnover persistency and the 18 agent policy persistency will have a persistency of ≥ 90% or fall into class A.

Persistence is one of the determinants of agent performance that is very influential on the sustainability of insurance companies. The persistence value refers to the volume of turnover and policy that the agent can maintain. The persistency is obtained from the proportion of active policies remaining at the end of the period out of the total policies at the beginning of the period [15]. OV Agency has turnover data and non-active policies from each agent in order to monitor agent performance through persistency. High persistence value is a necessity that must be achieved by each agent for the continuity of the company, therefore OV Agency expects persistence ≥ 90% of each agent-owned and new agents to be recruited. Small persistence will have a negative impact on the company. If there are a lot of inactive policies or turnover, the company will have difficulty bearing the death rates of the existing policyholders. In addition, it also reduces the profits of the company, the increase in capital costs due to a decrease in revenue as well as the increase in operational costs and sales of each policy, the price of a new policy becomes expensive and more difficult to sell and policyholders will stop paying if it is too expensive [16].

Table 3. Prediction of each classification method

Performance criteria	Classification method	Predictive class	Total	Accuracy (correctly classified)
Length of service	J48	A	14	67.6%
		B	11	
	MLP	A	**11**	**84.4%**
		B	14	
	Naive Bayes	A	10	57.6%
		B	15	
Total turnover	J48	A	4	70.4%
		B	21	
	MLP	A	**7**	**84.0 %**
		B	18	
	Naive Bayes	A	0	65.2%
		B	25	
Total policy	J48	A	0	78.8%
		B	25	
	MLP	A	**8**	**88.4%**
		B	17	
	Naive Bayes	A	0	72.4%
		B	25	
Persistency turnover	J48	A	18	72.4%
		B	7	
	MLP	A	**15**	**80.4%**
		B	10	
	Naive Bayes	A	5	60.8%
		B	20	
Persistency policy	J48	A	8	77.6%
		B	17	
	MLP	A	**18**	**84.0%**
		B	7	
	Naive Bayes	A	14	57.6%
		B	11	

3.3. Multilayer perceptron prediction results

Multilayer perceptron was used to evaluate 25 new OV Agency insurance agent candidates. The results of the prediction of the class five performance criteria for each prospective insurance agent can be seen in Table 4. The results show that there are three agents who classified as class A in all performance criteria, namely Agent 1, Agent 15, and Agent 22. These three agents are men, aged 25-32 years old, divorced, senior high school education and master degree, work status in other companies, and have work experience of 0-3 years. These results are in accordance with MLP learning of training data, so the accuracy of the results is strongly influenced by the quality of the training data used.

Length of service is an important agent performance criteria because predicting agents who can have a long service life will reduce agent turnover. Agent turnover in an insurance company is caused by a number of things including high-pressure work that requires agents to understand customer needs and offer suitable products, poorly planned recruitment processes and insurance companies have agents from various sectors so that the likelihood that agents will switch jobs to other sectors is high [17]. MLP in this study was successfully used to predict the length of service as one of the important criteria for agent performance.

Table 4. Predicted class of insurance agent candidates

Insurance agent candidates	Length of service	Total turnover	Total policy	Persistency turnover	Persistency policy
1	**Class A**	**Class A**	**Class A**	**Class A**	**Class A**
2	Class B	Class B	Class B	Class A	Class A
3	Class B	Class B	Class B	Class A	Class A
4	Class A	Class B	Class B	Class B	Class A
:	:	:	:	:	:
:	:	:	:	:	:
15	**Class A**	**Class A**	**Class A**	**Class A**	**Class A**
16	Class B	Class A	Class B	Class B	Class A
17	Class A	Class B	Class A	Class B	Class B
:	:	:	:	:	:
:	:	:	:	:	:
22	**Class A**	**Class A**	**Class A**	**Class A**	**Class A**
23	Class B	Class B	Class A	Class B	Class A
24	Class B	Class B	Class B	Class B	Class B
25	Class B	Class B	Class B	Class B	Class B

4. Conclusion and recommendation

In this research, there are three classification methods used to train insurance agents data of OV Agency, they are decision tree (J48), Neural Network (MLP), and Naive Bayes. Based on the result, multilayer perceptron has the best average accuracy value 84.2% and ROC area value 0.87 (good classification). The accuracy values are high and have the highest number of agents entering class A for five agent performance criteria. MLP was chosen to predict the 25 insurance agent candidates and 3 agents were selected because they entered class A (better performance) in all performance criteria. The prediction results show that three agents are men, aged 25-32 years old, divorced, senior high school education and master degree, work status in other companies, and have work experience of 0-3 years. The OV Agency can consider these result in new agent recruitment process. The results were under MLP learning of training data, so the accuracy of the results is strongly influenced by the quality of the training data used. For future work more classification methods can be used and much more dataset from OV Agency should be taken.

Acknowledgments
The authors pay homage to the Faculty of Industrial Technology, Trisakti University for continuous encouragement, guidance, and support for their work. This paper is the research funded by Trisakti University in the academic year 2019/2020.

References

[1] Fitriana R, Kurniawan W and Anwar M R 2019 Measurement and proposal of improving marketing process to improve the quality of aftersales services with fuzzy quality function deployment and data mining methods in ov agency *IOP Conference Series: Materials Science and Engineering*
[2] Umamaheswari K and Janakiraman S 2014 Role of data mining in insurance industry *An Int. J. Adv. Comput. Technol.* **3** 961-6
[3] Cho V and Ngai E W T 2003 Data mining for selection of insurance sales agents *Expert Syst.* **20** 123–32
[4] Balaji S and K. Srivatsa S 2012 Naïve bayes classification approach for mining life insurance databases for effective prediction of customer preferences over life insurance products *Int. J. Comput. Appl.* **51** 22–6
[5] Bhowmik R 2011 Detecting auto insurance fraud by data mining techniques *J. Emerg. Trends Comput. Inf. Sci.* **2** 156–62
[6] Kaur P, Singh M and Josan G S 2015 Classification and prediction based data mining algorithms to predict slow learners in education sector *Procedia Computer Science* **57** 500–8
[7] Han J, Kamber M and Pei J 2012 *Data mining: concepts and techniques* (Waltham: Morgan Kaufmann)
[8] Sumathi S and Sivanandam S N 2006 *Introduction to data mining and its applications* (Heidelberg: Springer)
[9] Drazin S 2010 Decision tree analysis using weka *Mach. Learn. II, Univ. Miami* 1–3
[10] Kohavi R and Quinlan R 1999 Decision tree discovery *Handb. Data Min. Knowl. Discov.* **3** 267-276
[11] Kantardzic M 2019 *Data mining: concepts, methods, and algorithms* (Piscataway: IEEE Press)
[12] Vercellis C 2009 *Business intelligence: data mining and optimization for decision making* (West Sussex: John Wiley & Sons Ltd.)
[13] Gorunescu F 2011 *Data mining: concepts, models and techniques* (Heidelberg: Springer)
[14] Arora R and Suman S 2012 Comparative analysis of classification algorithms on different datasets using weka *Int. J. Comput. Appl.* *34* 21-25
[15] Life Insurance Council 2010 Persistency of life insurance policies – an exposure draft 1–5
[16] Capgemini 2013 Persistency management framework for life insurance 1–20
[17] Pathak S and Tripathi V 2010 Sales force turnover : an explatory study of the indian insurance sector *Manag. Univ. Primarosk. Fac. Manag. Koper.* **5** 3-19

IOP Conf. Series: Materials Science and Engineering **909** (2020) 012042 doi:10.1088/1757-899X/909/1/012042

Shipping cost optimization on the Indonesian sea tollway due to weather

M F Pradana[1]*, M I Hamdani[2] and B Noche[3]
[1] Department of Civil Engineering, University of Sultan Ageng Tirtayasa, Jl. Jend Sudirman Km 3 Cilegon 42435, Indonesia
[2] Mechanical Engineering Faculty, University of Duisburg-Essen, Forsthausweg 2, 47057 Duisburg, Germany
[3] Institute of Transport Systems and Logistics, University of Duisburg-Essen, Forsthausweg 2, 47057 Duisburg, Germany

* Email: mfakhruriza@untirta.ac.id

Abstract. Indonesian sea tollway main objective is to decrease price gap between the western and eastern part of Indonesia. Although, the price gap is still felt in eastern Indonesia, that means the objectives of this program has not yet reached its optimum mainly due to operational cost. This paper is trying to optimize shipping costs by reducing the speed with slow steaming due to weather. The method called Slow Steaming was added with the aim of optimizing the speed of the ship. In total, there are three scenarios created to achieve these research objectives. The first scenario used current operational data, while the second scenario use ship speed loss due to weather, and the last scenario used ship speed loss due to weather and slow steaming method. The results showed that reducing ship speed with scenarios two and three can optimize the total shipping cost. Decreasing the shipping speed due to weather and using 10% and 12% slow steaming method can reduce between 16,35% and 18,5% of total shipping costs depend on the route. It can be concluded that scenario III has the lowest shipping cost, but with a note that with different specific conditions of slow steaming for each route.

1. Introduction

Indonesia is one of the archipelagic nations with 17,504 islands. Sea transportation is significant for archipelagic nations since this is the best way to connect between locales from East to West Indonesia and it has a major job for economic activities [1]. Indonesia faces issues regarding sea transportation identified with port availability, delivering courses, costly coordination transportation costs, uneven distribution of logistics that does not reach the remote area. These factors affect the price gap between the western and eastern parts of Indonesia, which appeared by the number of Gross Domestic Product (GDP). The Eastern district delivered under 20 percent of GDP while the Western district contributes more than 80 percent [2].

To overcome this problem the Indonesian government through the Ministry of Transportation is implementing a sea tollway concept model. The Indonesian government began implementing this program in 2015, which began with 6 routes that developed in 2016 with the same number of routes but more ports were visited, in total 31 ports and in 2017 became 16 routes with 41 ports in implementing the sea tollway program, Indonesia faces numerous difficulties. In the first place, the backload brought from East to West was deficient. Second, the price gap is still felt in eastern Indonesia even though the objective of this program is to take care of this issue, but in reality, the effectiveness of this program has not yet reached its maximum. Third, the port performance, there are as yet numerous ports in Indonesia that must be upgraded again. Fourth is the utilization of data and technology in executing the sea tollway program [3]. Data sources for this program are as yet restricted and hard to access, in the future, it is hoped that these problems can be resolved and the effectiveness of this program can reach maximum potential.

Indonesia which is located geographically in the Pacific Circle has many natural events including hydrometeorological hazards including sea-level rise, flooding, and big waves. However, Indonesia consists of two-thirds of it from the sea and weather which factor plays a very important role in sea transportation. Weather at sea (wind and waves) affects the shipping process which directly impacts to ship performance. From Vessel Performance Optimisation home page stated that the shipping industry risks investing billions of dollars in ship optimization services that fail to deliver overall operational efficiency and performance due to a lack of understanding and appreciation of the weather, accounting for 80 percent of the impact on ship performance, warns shipping intelligence and Stratum Five insight specialists [4].

This ship performance optimization leads to shipping cost, which has several factors. The first one is wind; it has a big impact to ship speed. If the ship is in a headwind, it will lose speed and if it is in the following wind, it will gain speed. Wind also affects the height of the waves that can lead to speed reduction. Secondly, wave height is the second weather factor that influences ship performance. Waves can influence the boat's rolling and pitching movement. This movement will bring about diminished force from the boat's propeller and an expanded drag from the constant steering corrections. Wave height and direction impact to ship performance is similar to wind. The third factor is fuel quality. Fuel quality has an impact to ship machine, when the quality of fuel is poor it can damage the engine and cause speed reduction. The other factors are marine growth and current [5].

It can be concluded that ship speed has a big role to improve ship. Lowering speed means lowering fuel consumption, and this will lead to optimization of shipping costs. The changing prices of bunker fuel open the door for substantial cost savings by adjusting the sailing speed of ships. A large ship may be burning up to 100 000 USD of bunker fuel per day, which may constitute more than 75% of its operating costs [6]. One strategy to optimize shipping operation costs is to use slow steaming. Slow steaming corresponds to slower vessel speeds, has become popular to improve vessel fuel efficiency [7]. Shippers can benefit from moderate steaming through a reduced supply chain carbon footprint, but longer transit times will increase pipeline inventory costs. Although the operator has identified slow steaming as a solution for all stakeholders, the sender expressed concern.[8]. In addition, although operators believe that slower vessels can improve schedule reliability and further reduce the need for safety stock, speed is often more important than reliability for sea shipping [9]. By reducing the current speed to 10% and 12 % it will decrease the fuel consumption by 19% and 23% [10]. Since fuel consumption is one of the main parameters in shipping cost, that means that speed variable has a big influence on shipping cost, that is why speed optimization can be used to optimize shipping costs. The objective of this research is to optimize shipping costs by reducing the speed with slow steaming due to weather.

2. Methodology

Indonesian Sea Tollway program has 13 routes. This research is using 3 routes out of 13 routes from (the 2017th) sea tollway program, which consists of routes T1, T2, and T13. These routes can be analyzed to obtain the actual distance from one port to the other port by using Google Earth by referring to the existing routes from the Indonesian Government. The next step is to collect Indonesian weather data between 2014-2017, the map was developed by EMD International A/S, Denmark and financed by the Environmental Support Programme 3 (ESP3) / Danida.

After the actual distance and weather data are found, we use them to get the weather data (wind speed) with actual distance using Google Earth. The acquired data is comprehensive on the weather variation at sea because this research aims to face the real situation. Furthermore, the wave height and ship speed loss due to wind speed can be acquired by using the Beaufort Number Scale and referring to existing journals [11].

As stated in (1), maximum optimization can be obtained by adding, slow steaming. At a distance of 10% from the ship's departure from the port and a distance of 10% when they will reach the destination port, we will assume that the speed of the ship is 4 knots. So, in this study, the Slow Steaming process will be carried out at 90% of the total distance between the two ports [12] Once the vessel decreases its speed by 10 percent, the engine power will be reduced by 27 percent [13]. So, it can be assumed that at a 12 % reduction in speed, the engine power will be reduced by 32,4 percent because the relation from the speed, distance, and time is linear.

In this research, there are three scenarios. The first scenario uses real data on the field. The second scenario uses ship speed loss, while the third Scenario uses ship speed loss and slow steaming.

- Scenario I
  - Use the real data on the field.
- Scenario II
  - Using the ship speed loss with Beaufort Number
- Scenario III
  - IIIa. Adding 10% slow steaming from current ship speed after ship speed loss due to wind and wave height
  - IIIb. Adding 12% slow steaming from current ship speed after ship speed loss due to wind and wave height

There are several assumptions for this research. The first assumption is there is 2 area process of shipping areas, which are process at Sea and process at Port. Be that as it may, the limit for this research is only using the process at Sea, one of the reasons is data limitations for the process at Port. Then, for time charter vessels with the daily charter is rated in Dollar per day (US$/day) After that, the total shipping time per voyage is the sum of the total hours shipping process at sea and port After the data from three scenarios is obtained, the shipping total cost can be determined. The shipping costs are the sum of charter time rates, fuel costs, port charges, and container handling costs [14]. The variable that is used to reach the objective of this research is ship speed that influences on shipping cost, while the other is assumed to be constant. The formula for shipping cost is shown below

$$C_S^m = \sum_i \left[\alpha_{it} + O_t W_i + F_{it} + D_i^m \left(\frac{O_t}{V_t} + \frac{F_t}{V_t}\right)\right] + \sum_i \sum_j \left[\left(\beta_i + \frac{O_t}{R_i}\right) P_{ij}\right] \quad 1$$

Which:

$$\alpha_{it} = \sum_i Pl_{it} + To_{it} + L_{it} + B_{it} \quad 2$$

Where:

m : route
i : port of origin on route m
j : port of destination on route m
t : type of ship
α_{it} : fixed portion of port i charge for a ship of type t (US$)
O_t : average daily charter rate for a ship of type t (US$/day)
W_i : time a ship spends on the arrival and departure process in port i (day)
F_{it} : fuel cost in port i by a ship of type t (US$)
D_i^m : shipping distance between port i and port i + 1 on route m (nautical mile)
V_t : service speed for a ship of type t (knot)
F_t : fuel cost at sea for a ship of type t (US$)
β_i : average handling fee per TEU in port i (US$ per TEU)
R_i : average gross handling rate in port i (TEU per day)

P_{ij} : the number of containers shipped between port i and port j on route m (TEU)
Pl_{it} : pilotage for a ship of type t (US$)
To_{it} : towage for a ship of type t (US$)
L_{it} : anchoring fee for a ship of type t (US$)
B_{it} : berth occupancy charge for a ship of type t (US$)

The objective of the proposed model is to minimize shipping costs. The fixed variable (Λ_t^m) and the variable shipping cost (ϕ_t^m) for ship type t on the route m can additionally be indicated by simplifying the variables as:

$$\Lambda_t^m = \sum_i \left[\alpha_{it} + O_t W_i + F_{it} + D_i^m \left(\frac{O_t}{V_t} + \frac{F_t}{V_t}\right)\right] \quad 3$$

$$\phi_t^m = \sum_i \sum_j \left[\left(\beta_i + \frac{O_t}{R_i}\right) P_{ij}\right] \quad 4$$

Furthermore, the shipping cost equation can be simplified as

$$C_s^m = \Lambda_t^m + \phi_t^m \quad 5$$

The shipping cost function as denoted in equation (1) and simplified in equation (5) is a basic formula to calculate the total shipping costs in our excel sheet. Moreover, use unit cost (C_{ij}) as the variable cost of the one-unit container (TEU). Unit cost for every route within the arc (i, j) derived by dividing the total variable shipping cost (ϕ_t^m) by the number of containers carried from origin to destination (P_{ij}). While the fixed cost (Λ_t^m) reliant on the type of ship and route (X_{ij}). Thus, formulate the objective function (4) to determine the minimum shipping costs of distributing cargo on a certain route, as follows [10]:

$$Min \sum_{i,j \in N} \sum_{c \in D} \sum_{d \in D} [P_{ij} * C_{ij}] + \sum_{i,j \in N} \Lambda_t^m * X_{ij} \quad 6$$

The objective function refers to equation (4) and substitute the number of sailing frequency (f), subject to:

Connectivity constraint:

$$X_{ij} \in \{0,1\} \quad 7$$

$$\sum_{i \in N} X_{ij} \geq 1 \ \ \forall j \in N \quad 8$$

$$\sum_{j \in N} X_{ij} \geq 1 \ \ \forall i \in N \quad 9$$

Cargo allocation constraint:

$$\sum_{i \in N} P_{ij} = D_j \ \ \forall j \in N \quad 10$$

$$\sum_{j \in N} P_{ij} = S_i \ \ \forall i \in N \quad 11$$

$$P_{ij} \geq 0 \ and \ integer \ \forall \ i,j \in N \quad 12$$

Ship capacity constraint:

$$\sum_{i,j \in N} P_{ij} \leq \sum_{t \in T} U_t X_{ij} \qquad 13$$

Where:

P_{ij} : binary variable
: 1 = if ship sails from port i to port j using ship of type t
: 0 = otherwise

N : all nodes on route m

D_j : total demand of the container (TEU) at port j

S_i : total supply of the container (TEU) from port i

U_t : maximum TEU capacity of ship type t per voyage

Construct three constraints to gain the objective of this study. Firstly, develop connectivity constraint, equation (9) depicts that ship sails from port i to port j (1) or not (0). Equations (10) and (11) indicate that the sum of traveling routes should be equal or more than 1 since this multi-port-calling model is used to capture the shipping costs in the implementation of the Sea Tollway program of 2016 and 2017 where a vessel is allowed to visit port more than once in one route (pendulum service) [14].

3. Result and Discussion

In this research, there are 3 out of 13 routes that have been chosen. The data is provided by the Ministry of Transportation of the Republic of Indonesia. These 3 routes are chosen on the grounds that these routes have the most comprehensive data regarding ship and voyage information. These routes have the biggest hub (Tanjung Perak Port) in eastern Indonesia. The 3 routes have a different type of ship and destination port. The sea tollway routes can be seen in Figure 1.

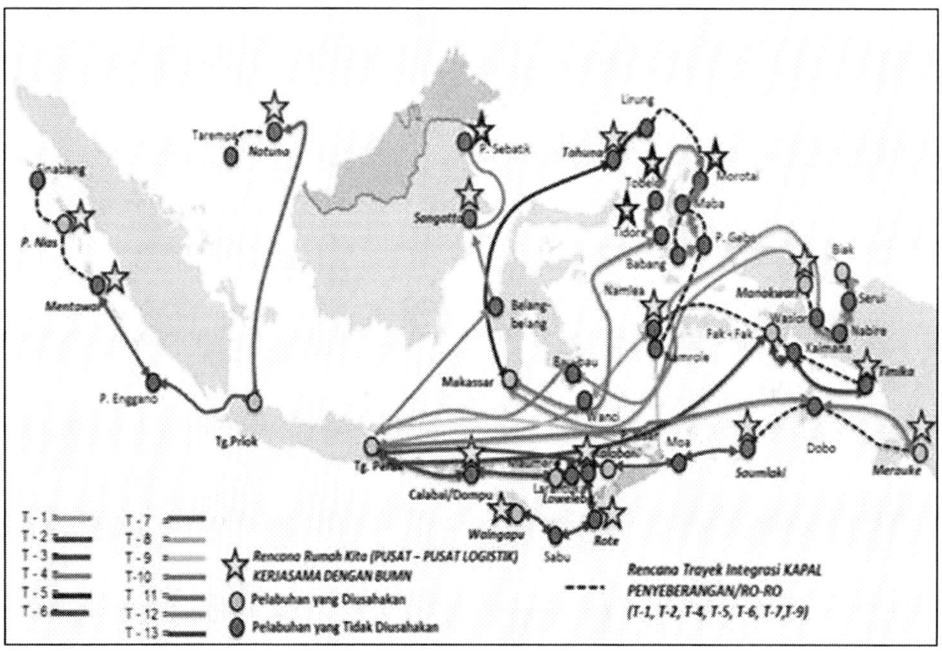

Figure 1. Route information for indonesian sea tollway program

3.1. Data

As mentioned above, 3 routes have been chosen. The routes are Tanjung Perak- Wanci-Namlea (T1), Tanjung Perak-Kalabahi-Saumlaki-Moa (T2), and Tanjung Perak- Fak-Fak-Kaimana-Timika (T13). For route T1, the ship utilized is KM. Nusantara Pelangi 101 that has 21 voyage targets and completed voyage only 9 in 2017.A Cargo ship, KM. Mentari Prakarsa is used for route T2 and has 17 voyage targets in 2017. The completed voyage by the route in 2017 is 7. The last route, T13 is using KM. Mentari Freedom that has completed 5 voyages in 2017 and 14 target voyages. Table 1 below summarizes these routes details used in this research.

Table 1. Route specification

Route	Port	Type of Ship	Target Voyage (2017)	Completed Voyage (2017)	Total Distance
T1	Tanjung Perak-Wanci-Namlea	KM Nusantara Pelangi 101	21	9	958
T2	Tanjung Perak-Kalabahi-Moa	KM Mentari Prakarsa	17	7	1158
T13	Tanjung Perak-Fak-Fak-Kaimana-Timika	KM Mentari Freedom	14	5	1604

Table 2. Ship specification

Type of Ship	DWT (ton)	GT (ton)	Capacity (TEU)	ME (HP)	AE (HP)	Vt (kts)
KM Nusantara Pelangi 101	3106	2997	115	2635	1137	11
KM Mentari Prakarsa	6207	4258	200	4788	1324	11
KM Mentari Freedom	5314	4303	192	4168	1331	11

3.2. Weather and ship speed loss

As stated in (2), the wave height and ship speed loss due to wind speed can be acquired by using the Beaufort Number Scale and referring to existing journals and research [11]. By using Google Earth, we can get the distance between each port. From this distance, weather data that was developed by EMD International A/S, Denmark is used and combined with real distance. From these data, the wind speed can be acquired and decide which Beaufort number is it. Then, take data from a journal that has ship speed loss due to Beaufort number. Since the wind speed data from contour is with range, not exact value and the Beaufort number is divided into the two limit, high breaking point, and low cutoff. From this limit, the average value of ship speed loss is taken and can be seen in Table 4. Refer to Table 4, the data from each port for Beaufort number, and the mean ship speed loss is shown.

Table 3. Beaufort number scale with wind speed and wave height

Beaufort Number	0	1	2	3	4	5	6	7
Wind speed (kts)	<1	1-3	4-6	7-10	11-16	17-21	22-27	28-33
Wave Height (m)	0.0	0.1	0.4	0.8	1.5	2	3	4
Ship Speed Loss	0,1922	0,2094	0,2610	0,3297	0,4501	0,5360	0,7079	0,9658

3.3. Shipping Cost

In this result, all scenarios are included in this calculation. From the results, can compare the value from each scenario and whether it answers the main objectives from this research or not. For route T1, all the data for each voyage is complete and can be seen from the graph that scenario 2 and scenario 3 have a lower amount of shipping costs. The voyage data for route T2 is not fully complete, the data for voyage 9 is not comprehensively. There are 2 sets of voyage data that are missing in route T13, they are voyage 5 and voyage 9. In route T2 and T13, the data for cargo volume (TEUs/Ton) for each port at a specified voyage is missing. But it doesn't have a big impact on total shipping cost since the only variable charter cost in port is missing. For route T2, the effect of missing voyage data to the total shipping cost only 0,01%-0,08%, and for the route T13, the effect is only 0,1%-0,6%. It can see from Figure 2, Figure 3, and Figure 4 that change one variable, which is speed can change the total shipping cost significantly.

Table 4. Mean ship speed loss due to wind and wave by using beaufort number

Route Port	T1-Wanci	T1-Namlea	T2-Kalabahi	T2-Moa	T2-Saumlaki	T13-Fak-Fak	T13-Kaimana	T13-Timika
Beaufort Number	3-4	2-4	2-4	2-4	4	3-4	3-4	2-4
Mean Ship Speed Loss (kts)	0,3484	0,3416	0,3407	0,4178	0,5730	0,3933	0,3709	0,3222

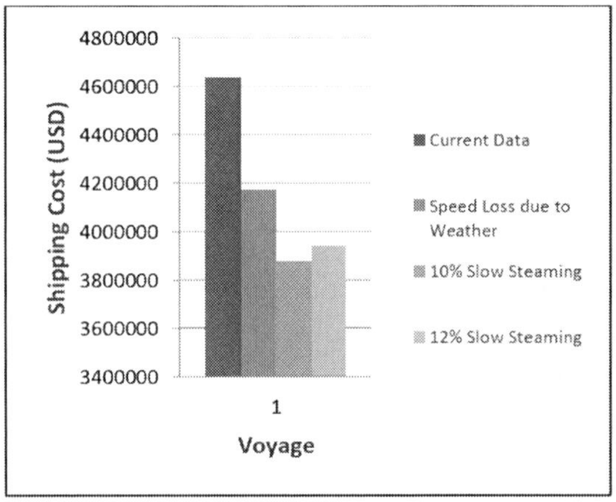

Figure 2. Mean shipping cost for route T1

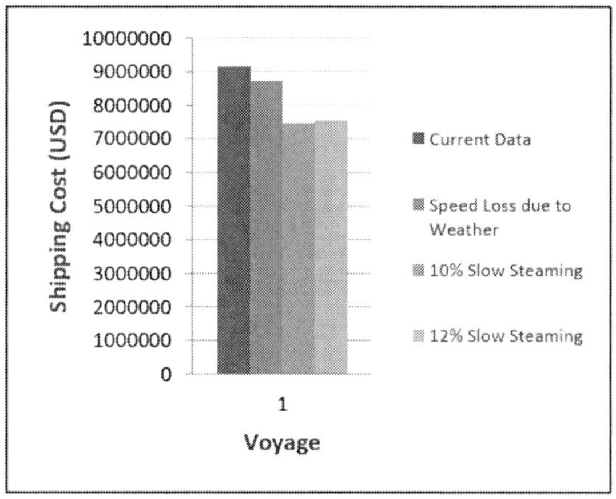

Figure 3. Mean shipping cost for route T2

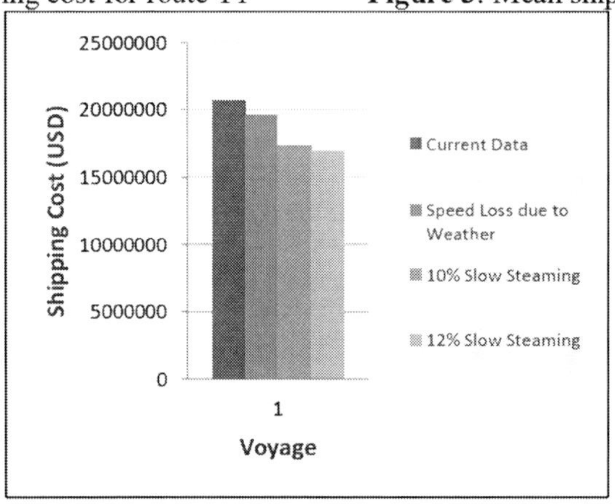

Figure 4. Mean shipping cost for route T13

4. Conclusion, Limitation and Future Research

In this research there are four results for each route based on three scenarios in which the one result was acquired by scenario I that use real data on the field, the other one by scenario II that use ship speed loss due to weather with Beaufort number, and the other two results by scenario III that added slow steaming after ship speed loss due to weather. Each different route has a different number of voyages, route T1 has 15 voyages, route T2 has 11 voyages and route T13 has 10 voyages.

From figure 2,3, and 4, it can be seen that the scenario I have the highest shipping cost for all routes. From closer observation from Figure 2, scenario III of the T1 route has the lowest shipping rate that reduces 16,35% of total shipping cost. For route T2, the result is similar to route T1, the lowest shipping cost is from scenario III that reduce 18,5% of total shipping cost. and it can be seen from Figure 3. The result from route T13 varies from Route T1 and Route T2, the lowest shipping cost is from Scenario III b, which added 12 % slow steaming from current speed to ship speed loss due to weather and it reduced 18,3% from the total shipping cost, it can be seen in Figure 4. It can be concluded that scenario III has the lowest shipping cost, but with a note that with different specific conditions of slow steaming for each route. These results answer the objectives of this research, which is to optimize shipping cost by reducing ship speed with slow steaming due to weather.

Although the objective of this research has already been accomplished, there is still a limitation of this research. The Beaufort Number with ship speed reduction from (3.2) is obtained by different ship models. The ship that is used as a reference is S175 and this data is compared to the existing method, Kwon's Method that use the same type of ship. Thus, the accuracy of the calculation in this research is limited. The next limitation is the data availability for each voyage is not completed. Lastly, the scope of this research is limited. This research focuses only on 3 out of 13 routes. These 3 routes may not represent all the databases for the sea tollway program. Therefore, future research is needed. The increase of the data availability in future research may be needed to increase the accuracy for the parameter that is used. Furthermore, the research for Beaufort Number of specific ships maybe uses for specific others research.

5. References

[1] Jusna and Nempung T 2016 Peranan transportasi laut dalam menunjang barang dan orang di kecamatan maligano kabupaten muna *Jurnal Ekonomi* **1**

[2] Central Statistical Bureau of Indonesia 2016

[3] Pradana MF and Noche B 2019 Prospect and challenges in developing indonesia sea tollway *The 2nd Int. Conf. on Food Security Innovation*

[4] Vessel Performance Optimisation (VPO) 2018 Available online: https://vpoglobal.com/2018/08/10/weather-accounts-for-80-per-cent-of-the-impact-on-vessel-performance/

[5] Stone C J 2010 Speed and consumption: factors affecting performance *Shipping & Transport International Magazine* **8**

[6] D Ronen 2011 The effect of oil price on containership speed and fleet size *Journal of the Operational Research Society* **62**

[7] L Cameron 2010 The big money in slow steaming *Canadian Business*

[8] Bonney J and Leach PT 2010 Slow boat from china Available online: https://www.joc.com/maritime-news/slow-boat-china_20100201.html

[9] Saldanha J P, Tyworth J E, Swan P F and Russell D M 2009 Cutting logistics costs with ocean carrier selection *Journal of Business Logistics* **30**

[10] Ichsan M, Pradana M F, and Noche B 2019 Estimation and optimization of the voyage energy efficiency operational indicator (EEOI) on Indonesian sea tollway corridors *IOP Conference Series: Materials Science and Engineering* **673**

[11] Kim M, Turan O, Hizir O, Incecik A and Day S 2016 A study on ship speed loss due to added resistance in a seaway *International Ocean and Polar Engineering Conference*

[12] Gusti A P and Semin 2016 The effect of vessel speed on fuel consumption and exhaust gas emissions *American Journal of Engineering and Applied Sciences* **9**

[13] Gusti A P and Semin 2018 Effect of ship speed on ship emissions *Asian Journal of Scientific Research* **11**

[14] Rumaji and Adiliya A 2019 Port maritime connectivity in south-east indonesia: a new strategic positioning for transhipment port for tenau kupang *The Asian Journal of Shipping and Logistic* **35**

IOP Conf. Series: Materials Science and Engineering **909** (2020) 012043 doi:10.1088/1757-899X/909/1/012043

Making of nusantara decorative motif creations necklace based on hdpe on type plastic waste

Desiana Nur Indra Kusumawati*, Widyo Wibisono and Wisnu Indra K
Al-Kamal Institute of Science and Technology Jl. Raya Al-Kamal No 2 Kedoya Selatan, Kebon Jeruk, Jakarta Barat, Indonesia 11520

*Email: desianakusumawati@ista.ac.id

Abstract. The problem of HDPE plastic waste from thick inorganic household waste that is difficult to destroy can damage the environment, the need for the utilization of HDPE plastic waste into products that are beneficial to humans. The purpose of this study was to determine the results of ornamental motif necklaces in the archipelago in accordance with the provisions of the aesthetic theory derived from HDPE plastic waste material that produces good quality so that it can attract consumers as users. The method used is a qualitative method based on the collection of literature data and information from the opinions of experts in their field of accessories to provide input or to the archipelago creation creations based on product theory and aesthetics. The results obtained are necklace products that have design elements derived from decorative motifs of the archipelago (Kawung and Tumpal) made from HDPE plastic waste that have benefits in fashion. The conclusion from the analysis based on theory and accessory expert opinion on the archipelago ornamental motif necklace especially kawung motif made from HDPE plastic waste can produce necklace creations that have good aesthetics and quality so that they can attract attention. consumers as users.

1. Introduction

Plastic waste has become a serious problem in the Jakarta area, but the amount of plastic waste continues to increase from used plastic at home, market activities, and other public facilities. Most plastic waste is produced from household and general waste generated from types of soap, fragrances, food, beverages such as mineral plastic bottles, shampoo, cooking oil, and others. Waste will continue to be produced and will never stop as long as humans remain. It can be imagined that the amount of waste generated by the inhabitants of this earth will increase. Waste itself is one of the consequences of human activity and the volume is directly proportional to the population. If not handled effectively and efficiently, the existence of waste in nature will certainly turn around destroying the surrounding life. Nature does have a big hand in automatic waste processing, especially in organic waste. However, nature's hard work in breaking down waste naturally is very unbalanced compared to the millions of tons of volume of waste produced each day [1].

The problem of waste is crucial. In fact, waste can be said to be a cultural problem because its impact is affected on various sides of life, especially such as big cities like Jakarta, Semarang, Surabaya, Bandung, Palembang, and Medan. The biggest source of waste is derived from settlements, the composition in the form of 75% consists of organic waste and the rest is inorganic waste. Organic waste has been widely used as material for composting, briquettes and biogas, but inorganic waste is still very minimal in its management. Inorganic waste is very difficult to be degraded or even completely degraded by nature, therefore a very large landfill is needed to balance this type of waste production. Inorganic waste that is most often found in the community is plastic waste. In 2008 the production of plastic waste for packaging reached 925,000 tons and around 80% of it has the potential to become hazardous waste for the environment (Kompas, 2009) [2].

Illegal recycling of plastic bottle packaging is so widespread that people destroy it first and then discard it so that it is not illegally recycled. Types of plastic bottles based on their use vary, if registered from the code contained in a plastic bottle. Likewise, when viewed from its shape, such as used plastic bottles, shampoo, deodorant clothing, liquid soap, and others. With a thick texture, glossiness, beautiful curves and colors are usually only reused for other things, for example, for detergent containers used for oil or used for detergent again. If properly processed, a plastic bottle with a thick texture or type of waste. HDPE plastic bottles can be used as attractive craft products. Much research has been produced from mineral plastic bottle waste, one of Indonesia's designers, Lenny Agustin, also processes mineral plastic bottle waste into necklace accessories made from pieces of mineral mineral waste bottles, then assembled, and then painted plastic bottles with permanent colors [3]. Based on recycled plastic bottle waste products made by several recycled craftsmen, craftsmen only process mineral plastic bottle waste. The craftsmen have not made a new breakthrough treating waste from plastic bottles, instead of mineral plastic bottles. Therefore, we researchers want to make a new breakthrough in processing plastic waste bottles with thick, sparkling and thick plastic bottles into necklace products for show or fashion where the necklace accessory is a visual supporter of costume in a performance or fashion art, and when viewed from In terms of research time, making necklace accessories is more efficient than making costumes, but its function as aesthetics in the performing arts or fashion is no less important than costumes and can arouse the interest of consumers or the public by wanting to use it and find out how to treat waste by recycling. So by recycling waste can run 4 principles, namely independent (managed by the community itself), productive (producing something of value), comprehensive (all waste can be overcome), and environmentally friendly (waste management does not pollute the environment) [4].

The design of accessories product design requires creativity and concept accuracy in designing necklace accessories for the performing arts or fashion, so it requires creativity that can arouse the desire of consumers or the public to use necklace products and find out waste treatment by recycling about how to make necklaces using the material of plastic bottle waste materials hard textured / HDPE, and the use of processing processes with simple techniques. Necklace accessories are based on inspiration, product theory, and easthetic. Based on the source of inspiration of the researchers in creative design in the manufacture of necklace accessories products made from hard textured plastic waste using traditional Nusantara motif design concepts in the form of Kawung and Tumpal motifs so that the necklace accessories have more appeal because it has special ethnic characteristics of the archipelago, then to Product theory is based on product quality, comfort, product durability, and based on aesthetic theory including design elements and design principles. The design elements used are based on shape, size, and color. Then with the principle of design based on proportion, balance, harmony, rhythm, transition, the center of attention.

The purpose of this study is intended for the community as consumers to find out the design process, the processing and the results of research objects in the creation of archipelago ornamental motifs that can be used as fashion with inorganic household HDPE waste material, then attract consumers as users.

1.1. Identification of problems.

How do you process hard plastic / HDPE bottles in making necklace accessories created by traditional Nusantara motifs as a support in performing arts and fashion so that they are attractive?

1.2. Restricting the problem.

1. How can the hard-pressed plastic bottle waste material be designed into an archipelago-patterned creative necklace accessory for performance and fashion art?

2. How to process the waste of hard-textured plastic bottles to be used as accessories for the creation of an interesting Nusantara motif based on product and aesthetic theories?

1.3. Research purposes

1. Explain that the waste of hard-textured plastic bottles can be reused as an accessory product in the archipelago's traditional patterned creations for performance art and fashion.
2. To find out how to process hard-textured plastic waste bottles to form a necklace product that matches the product and aesthetic requirements then can be used as an accessory for the performing arts and fashion

2. Plastic waste and theory

2.1. The following types of plastic (plastic bottles) based on the code:

a. PET (polyethylene thereptalate) is a polyester resin that is durable, strong, lightweight and easy to type when hot.
b. HDPE (High Density Polyethylene) is a tough, strong and stiff resin. Derived from petroleum.
c. PVC (Polyvinyl Chloride) is a tough and hard resin that is not affected by other chemicals.
d. LDPE (Low Density Polyethylene) is a plastic that is easily formed when hot, which is made from petroleum.
e. PP (Polypropylene) is a polymer plastic that is easily formed when it's hot.
f. (Polystyrene) is a polymer plastic that is easily formed when heated.
g. Others, most often products with code 7 are made from a mixture of 2 or more types of plastic (code 1 to code 6).

2.2. Product Theory and Product Aesthetics

2.2.1. Product Theory

According Pulat (1992) offers a product design concept to support efficiency and safety in the use of product design. The concept is design for reliability, comfort, length of time of use, ease of use, and efficiency in use [5].

According to Henrry Dreyfuss important goals in the development of new products are: Use of products that must be safe, pleasing product appearance, facilitate maintenance and repair, Product Durability, low costs in maintenance and production, research relationships in accordance with the philosophical through product quality.

2.2.2. Product Aesthetics

Aesthetic forms, consisting of intricate shape qualities, such as size, proportion, texture, color. The overall aesthetic quality or complete harmony (completeness) is called Unity. Unity has the following characteristics: Balance, contrast, harmony, accent, emphasis, proportion, simplicity, repetition, domination, rhythm, motion [6].

According to Katsoff, aesthetics is a branch of philosophy that deals with structural constraints and feelings of beauty, especially in art.

According to Hartoko in Kartika's work, aesthetics are the ability to see through sensing, perception, feeling, observation and scenery [7].

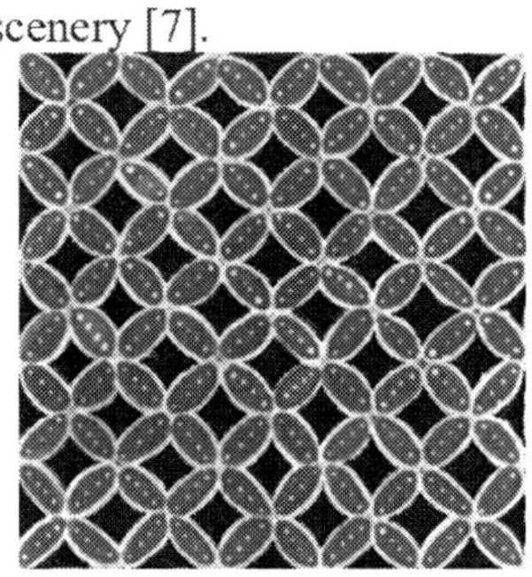

Figure 1. Kawung decorative motifs

Kawung in Sundanese means anen or fro. Because of that the kawung ornamental variety has a shape resembling a palm fruit that is cut transversely so that it looks four palm seeds. This ornamental variety has existed since the days of our ancestors, with evidence of this decoration used on Javanese Hindu temple statues. Decorative Variety kawung has the meaning of harmony in life in the world and the hereafter.

Figure 2. Tumpal decorative motif

Tumpal is an archipelago traditional ornamental variety which has the basic form of isosceles triangle. In further developments, this decorative variety has a variety of variations. Tumpal has been found in the decoration of temples in Indonesia. In addition to the temple decoration is also used to make wood carvings, and so on. The type of ornamental tumpal is also called untu walang which symbolizes fertility [8].

3. Research Methodology

3.1. Operational Objectives

Based on operational objectives, this study aims to obtain a general description of the design and manufacture of accessories for creative necklaces with decorative motifs of the archipelago using material from thick-textured plastic waste bottles.

3.2 Research Place and Time

3.2.1. Research location

This research was conducted in Jakarta.

3.2.2. Research Time

Study time is estimated to be 8 months.

3.3. Description of Research Settings

1. The initial step in the process of making an archipelago motif necklace from HDPE plastic bottle material taken by researchers is collected and chooses a plastic bottle that can be used as a supply of necklaces stacked with household waste0. Furthermore, the used plastic bottles are cleaned after being cleaned in scissors and separated from the plastic bottle caps and then taken in accordance with the source of inspiration, namely the archipelago decorative motifs of the Tumpal motif and the Kawung motif in combination into 3 types of models.
2. The next step is to conduct interviews with several experts in their fields then through data from the accessories experts before conducting an analysis based on product theory and aesthetics.
3. The author divides the necklace accessories products are divided into three types of design models by the authors based on the concept of Kawung and tumpal motifs.
4. Data from experts is then analyzed using products and aesthetics based on or input from some of the informants.
5. The author draws conclusions based on the best results after analyzing the product based on product theory and aesthetics and if needed deficiencies in each type of model it can be corrected for further improvement. Experts working in the field of accessories and design, among others: expert users of accessories and design, designers who understand about accessories, owners of the accessories industry and owners of making accessories courses.

IOP Conf. Series: Materials Science and Engineering **909** (2020) 012043 doi:10.1088/1757-899X/909/1/012043

3.4. Research Methods
This research uses a descriptive method using qualitative methods, according to Dantes. This research method uses research that discusses theories or phenomena according to what exists and is done to obtain information about the current situation [9].

3.5. Research Focus
In this research, the researcher took the focus of research on "Processing plastic bottles with HDDP in making Accessories for the creation of necklaces with Nusantara motifs taking the Kawung and Tumpal motifs"

3.6. Research Stage Procedure
In this study the field was approved and asked to collect data and documents needed in this study. Suryesna (2013) revealed the research in this study: discussing problems, research objectives, finding data, compiling designs, selecting materials through the product manufacturing process, interviewing interviewees, analyzing, concluding.

4. The Concept and Process of Making Necklace
4.1. The Concept of Making a Necklace
1. Making this design concept will look for alternatives to making works that will be created and look for the best based on theoretical analysis.
2. The alternative design of this form is certainly in accordance with the philosophical themes raised in the making of works based on aesthetic and philosophical values, namely the design of necklace creations with archipelago Kawung motifs and the motif of tumpal with the meanings that exist in archipelago decorative motifs, the archipelago motif is a distinct attraction.
3. The results of theoretical analysis based on the assessment of sources can determine the deficiencies that exist in the product so that it can be refined at a later stage so as to produce an ideal work of art.
4. Design development results sketches are divided into 3 model types and then get selected sketches based on input and assessment of sources analyzed based on aesthetic and product theories.
5. The best product conclusions, among other things in terms of artistic, product functions, ergonomics and manufacturing processes that are simple but can produce products that are of use value.

4.2. The process of making an archipelago motif necklace

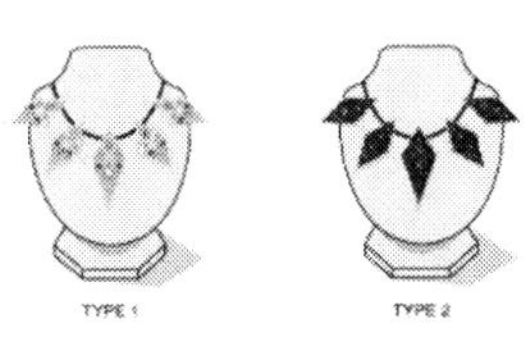

Figure 4. Tumpal motif (type 1 and 2)

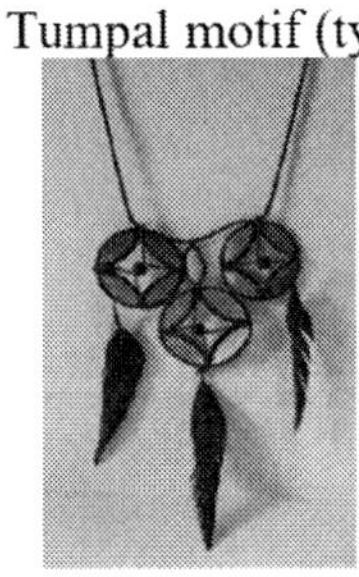

Figure 5. Kawung motif (type 3)

Table 1. The process of making an archipelago motif necklace

Design Concept	Material Selection	The process of making a necklace
Design Necklace tumpal motif and kawung motif	The selection of items that will be used is thick plastic waste (used shampoo bottles) that has been washed clean, other manufacturing tools such as cutters, scissors, glue, acrylic paint, necklace strings and hooks.	1. Cut the material manually with scissors into sheets by removing the top and bottom bottles. 2. Pattern making with a pencil and a ruler. 3. Cutting follows the existing pattern. 4. Sanding to be smooth so that it is comfortable when used. 5. Sticking stickers on type 1 and painting with acrylic paint on types 2 and 3 and cut other materials such as the necklace strap.

5. Inspirative sources/ Inspirative Concepts.

5.1. The design concept of Necklace Types 1 and 2 Icons Necklace

Tumpal motif (Theory of Semiotics) refers to the theory of pierce marks in pictures that can be classified as icons, indexes, symbols [10].

a. Icons

This necklace design icon consists of 5 pieces of the same pattern and is shaped like a triangle opposite the shape of a parallelogram. For the dominant color in the pattern is yellow and detailed For type 1 has more colors that are used more modern and contemporary by using stickers, whereas for type 2 only 2 dualistic colors (eastern aesthetics) are more philosophical.

b. Index

This tumpal motif necklace product contains several 5 pieces of geometric shaped parallelogram inside which are arranged, namely: Parallelogram as the main form of necklace decoration.

c. Symbols

The meaning contained in the decoration of a tumpal, also called untu walang, is a symbol that symbolizes fertility, which is characterized by a dualistic element symbolizing male and female elements in two sharp opposite angles.

5.2. Design concept necklace Type 3 Icon Kawung motif (Semiotics Theory)

a. Icon

The iconic arrangement of 3 circular shapes is a kawung motive using eastern philosophical aesthetics namely pattern 3, which is in accordance with the theories and concepts created from the start.

b. Index

Motif Kawung like palm fruit arranged in 3 forms.

c. Symbol

There is content taken from the shape of a necklace made from kawung motif in the form of a palm fruit. Motives that contain messages about humans have always been useful collections, like the palm tree in which all parts can be used. Is a symbol of dualism in the colors chosen for world life and the hereafter. The gold color was chosen because it was considered as the color of victory, prosperity and luxury / brilliance combined with black which is an elegant color and symbolizes elegance, this color can also make other colors look attractive. The interpretation of the 3 forms of kawung symbolizes eastern aesthetics in the form of tritangtu principles in the mindset of Indonesian people in general and Sundanese in particular.

6. Product analysis and discussion based on product theory and aesthetic theory

Table 2. Product Analysis and Discussion Based On Product Theory

No	Type Motif	Teory Product				Conclution Type	
		Product Quality	Comfort	Durability	Usability efficiency	Advantage	Defiency
1	Tumpal Motif necklace with sticker Adhesive motif	a. Having added value (+) b. The use of stickers doesnot naturally (-) c. Eliminate the philosophical of being modern(-) d. Satisfy consumers with color brilliance and safe material(+)	a. Safe in use because in the smoothing process. (+) b. Easy to use (+)	Sticker motifs are not durable (-)	easy in maintenance and repair, cheap repair costs (+)	5 point	3 point
2	Tumpal motif necklace with coloring on the motif detail	a. Having added value(+) b. The motif is more elegant with colors in classic nuance in accordace with the philosophical theme of the classic tumpal motif (+) c. Doesnot meet the overallhuman needs depending the customer of the wearer(-)	a. This necklace is safe to wear because the process is tidied up at the end of fine details even if the tip is pointed(+) b. The short neck strap will affect the user's skin and the coloring texture in the painting technique is not too smooth when in contact with the skin (-)	a. The motif is more durable because this motif is made with acrylic paints and brushes so that it has the characteristics of a classic artistic appearance incoloring (+) b. The use of rope from Woll material is not durable (-)	easy in maintenance and repair, cheap repair costs(+)	5 point	3 point

Table 2. Product Analysis and Discussion Based On Product Theory (Continue)

No	Type Motif	Teory Product				Conclution Type	
		Product Quality	Comfort	Durability	Usability efficiency	Advantage	Defiency
3	Kawung motif necklace with coloring on the motif cetail	a. The product or value has increased in the product when viewed from material of plastic waste into the form of necklace product(+) b. Motifs are more elegant with colors used in classics in accordance with the theme of classic kawung motif (-) Attracting consumers attention by natural colors, shapes and classic accents that are philosophical (+)	a. The use of a long rope is very beneficial because the paint is not smooth not on the user's skin but on clothes (+) The shape of round ornament details is more beneficial in terms of security because it is not sharp(+)	The motif is more durable because this motif is made with acrylic paints and brushes(+)	easy in maintenance and repair, cheap repair costs(+)	7 point	1 point

Then the conclusion can be drawn from the analysis based on this product theory that type 3 necklaces most meet the requirements of the product theory with the most number of conformities ((+)=YES) and the least number of discrepancies ((-) =NO)and in the sequence below are type 1 and type 2. Resource person:

1. Ratih K., S.Pd, (Costum Designer)
2. Kusnadi, S.Sn.,M.Ds., M.kom.(Product Designer)
3. Abidin, S.Sn., M.Ds.(Product Designer)
4. Dra Mariawati Nur (Culture and Course Owner)

Tabel 3. Product analysis and discussion based on aesthetic theories

No	Aesthetic	Model Type 1		Model Type 2		Model Type 3	
		Resources	Analysis	Resources	Analysis	Resources	Analysis
	Color	(-)More contrast to make it more interesting. (-)The color on the supporting accent is darker so that it looks contrasted.	In order to make the colors more attractive it has to be made, by way of supporting access it is made with a darker color. (-) 2 point.	(-) Too Monotonous (-), Less contrast.	The colors are too monotonous or lack contrast and tend to be dark, so there should be a contrasting color selection by combining bright and dark colors in a balanced portion. (-) 2 point.	(-)Must be more contrast	The color is better only has to be more contrast. (-) 1 point (+) 1 point Warna sudah lebih baik hanya harus lebih kontras. (-) 1 point (+) 1 point.
The Aesthetic Elements of Form consists of the qualities of instinctive forms, such as:	Texture	(+)(-)The use of stickers results in less visible texture.	Textures with the use of stickers make it less visible, so the texture looks smoother. (-) 1 point (+) 1 point	(+) (+)	Texture already visible Round and flat shape, round on the supporting accents and flat on the main shape (+) 2 point.	(+) (+)	Texture already visible The shape of threaded and flat, threaded on the neck strap supporting accents and flat on the main shape (+) 2 point.
	Size	(-), the form must be more varied in terms of size. (-), Less varied than the type of geometric size of the parallelogram.	The shape is less varied from the type of geometric size of the parallelogram used. (-) 2 point.	(+) (+)	Argonomic proportions are already visible in the center with an enlarged size. (+) 2 points.	(+) (+)	The shape is in accordance with the proportion (+) of 2 points

IOP Conf. Series: Materials Science and Engineering **909** (2020) 012043 doi:10.1088/1757-899X/909/1/012043

Tabel 3. Product analysis and discussion based on aesthetic theories (Continue)

No	Aesthetic	Model Type 1		Model Type 2		Model Type 3	
		Resources	Analysis	Resources	Analysis	Resources	Analysis
	Balance	(+) (+)	Simetris (+) 2 Point	(+) (+)	Simetris (+) 2 Point	(+) (+)	Simetris (+) 2 Point
	Contrast	(-) Must be more contrast (-) Too bright colors with less contrast are made on accent colors that are different from the main object	Color lacks contrast, because the accent and main object have colors that are too bright. (-) 2 point.	(-) The color of the motif must be made in (-) contrast. Tends to be dark, classic, natural.	motifs colour must be made in contrast, the colors tend to be dark, so less brilliant. (-) 2 point.	(+) (-) Color must be more contrasted.	Color must be more contrast. (+) 1 point (-) 1 point.
The aesthetic principle of a harmonious whole or a complete feeling is called Unity, which has the following characteristics:	Harmonious	(+) (+)	Harmonious between the size and placement of the main object and the neck strap accents. (+) 2 points	(+) (+)	Harmonious between the size and placement of the main object and the neck strap accents. (+) 2 points	(+) (+)	Harmonious between the size and placement of the main object and the neck strap accents. (+) 2 point
	Axis	(+) (+)	Already have a contrast accent on the rope and the main shape (+) 2 point	(+) (+)	Already have a content accent on the rope and the main shape (+) 2 point.	(-) For more feather accents, support bright colors and add a rather heavy accent. (-)More varied and mass or heavy.	For supporting accents are made in contrasting / bright colors so that the brilliance of the product looks attractive, and is more varied looking for shapes and accents that are mass (-) 2 points.

Tabel 3. Product analysis and discussion based on aesthetic theories (Continue)

No	Aesthetic	Model Type 1		Model Type 2		Model Type 3	
		Resources	Analysis	Resources	Analysis	Resources	Analysis
The aesthetic principle of a harmonious whole or a complete feeling is called Unity, which has the following characteristics	Emphasis	(-) Must be more specific on the theme. (-)The form has not accentuated the theme	Shapes should emphasize the theme of dualistic themes of men and women or fertility can be described with 2 colors and 2 opposite shapes (-) 2 point	(-) Size must be more varied (-) Almost the same size is less varied.	Shapes are so similar that they are less appealing. (-) 2 point.	(+) (+)	There is already an emphasis on the shape of the Oval theme. (+) 2 points
	Proportion	(+) (+)	There is already a proportion of ergonomic forms. (+) 2 points	(+) (+)	there is already a proportion of ergonomic forms. (+) 2 points	(+) (+)	There is already a proportion of ergonomic forms. (+) 2 points
	Simplicity	(+) (+)	already looks simple from the coloring and form of the material. (+) 2 points	(+) (+)	already looks simple from the coloring and form of the material. (+) 2 points	(+) (+)	already looks simple from the coloring and form of the material. (+) 2 point
	Repetition	(+) (+)	There is already a repetition of the object shape. (+) 2 points	(+) (+)	There is already a repetition of the object shape. (+) 2 points	(+) (+)	There is already a repetition of the object shape. (+) 2 points

7. Conclusion

Based on the results obtained in the analysis of the selection of archipelago ornamental theory with the motifs of tumpal and kawung, it can be concluded that the use of thick type plastic waste material / HDPE can produce a product with more value being a necklace product of archipelago motif creations. HDPE plastic waste material can produce a good necklace product by going through the steps of the design concept, the manufacturing process, to the theoretical analysis in order to produce a product that is in accordance with good product requirements. Step researcher made 3 alternative examples of design concepts presented by researchers in the form of archipelago motif necklace creations which were then analyzed based on the use of product theory and aesthetic theory, from the processed interviews with experts in the field of product design, costumes and course owners. Then more specific conclusions can be drawn, namely for type 3 with the concept of kawung motifs that can meet the requirements in the concept of design, product, and aesthetics with the need for a slight improvement from the deficiencies that exist in several points based on theoretical analysis. Archipelago necklace creations products Kawung motif on the results of research on type 3 models are superior in point through the stages of analysis in product theory and aesthetics so as to produce an archipelago creation necklace product in accordance with product and aesthetic requirements can be a special attraction for consumers.

References

[1] T. P. PS 2008 Penanganan dan pengolahan sampah Jakarta: Penebar Swadaya.

[2] H. P. Putra 2010 Pemanfaatan sampah plastik menjadi produk dan jasa kreatif *Jurnal Sains dan Teknologi Lingkungan, Fakultas Teknik Sipil & Perencanaan, Universitas Islam Indonesia Yogyakarta* **2** 21-31

[3] Republika.co.id. 4 September 2014 Lenny Agustina hadirkan kalung dari limbah botol plastik. *http://www.republika.co.id.*

[4] T. Pamungkas 2006 Bukan membuang tapi mengelola, dalam sampah dilema manusia modern dan krisis ekologi Yogyakarta: Balairung Jurnal Mahasiswa Universitas Gadjah Mada 39

[5] B. M. Pulat 1992 Fundamentals of Industrial Ergonomic, Oklahama: AT & T Network System

[6] Y. Safanayong 2006 Desain Komunikasi Visual terpadu, arte intermedia.

[7] Kartika 2007 Estetika, Bandung: Rekayasa Sains

[8] S. Yoyok RM 2015 Pendidikan seni budaya, Jakarta: Yudistira

[9] D. Nyoman 2012 Metode penelitian Yogyakarta: Andi Yogya

[10] S. Tinarbuko 2013 Semiotika komunikasi visual Yogyakarta: Jalasutra

IOP Conf. Series: Materials Science and Engineering **909** (2020) 012044 doi:10.1088/1757-899X/909/1/012044

Measurement of user satisfaction for web-base academic information system using end-user computing satisfaction method

Purwanto* and Deden Hedin P.B.
Al-Kamal Institute of Science and Technology
Jl. Raya Alkamal No. Kedoya Selatan, Kebon Jeruk
Jakarta Barat, Indonesia

*Email: pungpurwanto@gmail.com

Abstract. Academic Information System is a system that collects and processes data into useful information for users. This research aims to determine the level of user satisfaction of Web-Based Academic Information Systems at Al-Kamal Institute of Science and Technology. The research model uses quantitative by using a descriptive approach. The measuring of user satisfaction uses an End-User Computing Satisfaction (EUCS) method. The variables in this research are accuracy, content, format, timeliness, and ease of use. Data collection uses a questionnaire by 12 questions with the sum of data samples are 135 respondents. The results show that the overall level of user satisfaction averaged 2.99 at level 3 (neutral). The variables considered good are timeliness and ease of use, while for content, format, and accuracy still need improvement and redevelopment. The results of this research can be used as an evaluation material to develop the academic information systems in the future.

1. Introduction

Information Technology (IT) is one of the supporting tools for college activities. The use of IT is usually seen as a positive step to improve the business processes of a college. Therefore, the success of implementing IT is very dependent on human resources (HR). This factor is very important because the readiness of implementing IT has a big influence in determining the satisfaction of the users and the success of it [5].

Academic Information System at the college is a system that collects and processes data to be useful information for users [5]. The success of this system application depends on how much satisfaction the users have. So it needs to be analyzed.

The Al-Kamal Institute of Science and Technology (ISTA) located in West Jakarta is one of the tertiary institutions that has implemented one of the benefits of IT, i.e. the Web-based Academic Information System since 2010 [5]. This system is designed with several facilities to provide benefits and satisfaction to users. The End-User Computing Satisfaction (EUCS) model, develop by Doll et al, is used to measure user satisfaction with information systems. Thus, user satisfaction is a success indicator of information systems development. The EUCS has five component model. There are accuracy, content, format, timeliness, and ease of use for information systems applications. A questionnaire consisting of 12 questions was used as data collection [1].

In previous studies, there were several studies on the measurement of information system user satisfaction with the End-User Computing Satisfaction (EUCS) model, measuring the end-user computing satisfaction, including in End-User Computing: Concepts, Methodologies, Tools, and Applications [2]; Promoting radical innovation through the end-user computing satisfaction [3]; Determining User Satisfaction Factor in University Tuition Fee Systems Using End-User Computing Satisfaction (EUCS) [4]; Analysis of Acceptance and Satisfaction of Academic Information System Users Using Structural Equation Modeling [5]; User Satisfaction Factors in the Use of Learning Management Systems [6]; Testing User Satisfaction Using End-User Computing Satisfaction Method (EUCS) in Hospital Management Information System (SIMRS) (Case Study at dr. Dadi Tjokrodipo's Regional General Hospital) [7]. These studies have adopted the EUCS model from Doll et al.

Based on the aforementioned background of the study, this research is conducted with five variables measured content, format, ease of use, timeliness, and accuracy.

2. Research method

This research aims to determine the user satisfaction level of Web-Based Academic Information Systems at Al-Kamal Institute of Science and Technology. The research model uses quantitative by a descriptive approach. Data of the research is primary data, obtained by distributing questionnaires to users of Web-Based Academic Information Systems as many as 135 respondents consisting of students, lecturers, and academic staff. Data processing was performed using SPSS 23.

The variables in this research use instrument of the EUCS formulated by Doll et al [1] There are content, format, ease of use, timeliness, and accuracy.

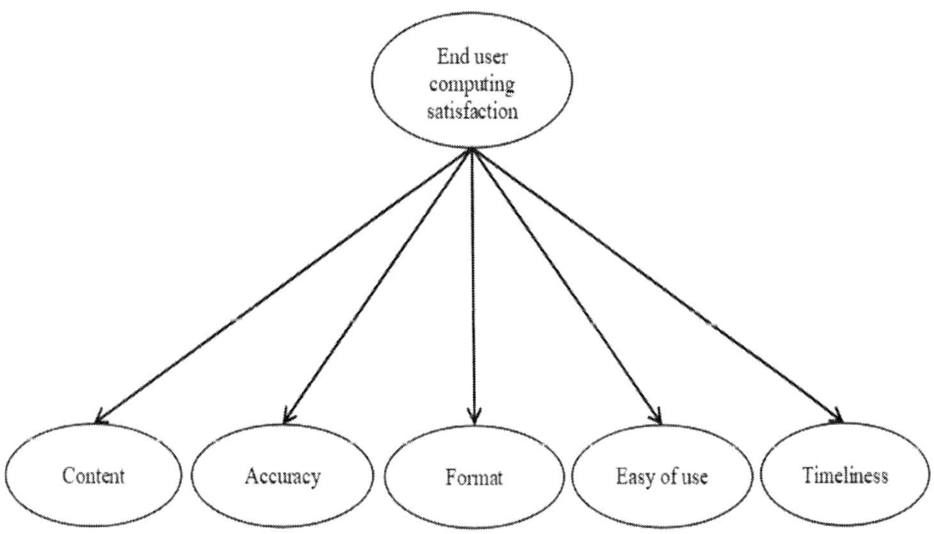

Figure 1. EUCSmodel by doll & torkzadeh

Doll et al [1] have compiled 12 items of the question for EUCS research and these questions were adopted and used as guidelines for making questionnaires for respondents [8,12].

Table 1. Questioner item

Variable	Question ID	Question Item
Accuracy	A1	Does thesystem be accurate?
	A2	Do you be satisfied withthe accuracy ofthe system?
Content	C1	Does the system produce information you need?

	C2	Does the content contain information you need?
	C3	Does the system produce reports about what you need?
	C4	Does the system produce enough information?
Easy of use	E1	Does the system be userfriendly?
	E2	Does the system be easy of use?
Format	F1	Are the output displayed in useful format?
	F2	Does the information be clear?
Timeliness	T1	Do you get the information ontime?
	T2	Does the system produce upto date information?

3. Results and discussion

The research data was obtained by distributing questionnaires to the use of web-basic academic information systems, which are lecturers, staff academic and students of ISTA with a research sample of 135 people. Then do the validity test and reliability test. Validity measurement is done by analyzing Product Moment correlation with decision-making criteria. If r-count>r-table, then the instrument is stated valid, and vice versa if r-count<r-table, then the instrument is stated invalid. If the probability (sig.2 tailed) < 0.05 then the instrument is stated valid, and vice versa if the probability (sig.2 tailed) > 0.05 then the instrument is stated invalid. While Reliability is a degree of consistency and stability of data or findings. To measure reliability using the CronbachAlpha statistical test, the variable can be said to be reliable if it gives a value $\alpha > 0.60$.

Table 2. Result of validity test

Variable	Question ID	r-count	r-table	Criteria
Accuracy	A1	0.904	0.142	Valid
	A2	0.918	0.142	Valid
Content	C 1	0.791	0.142	Valid
	C2	0.469	0.142	Valid
	C3	0.810	0.142	Valid
	C4	0.411	0.142	Valid
Ease of use	E1	0.752	0.142	Valid
	E2	0.750	0.142	Valid
Format	F1	0.921	0.142	Valid
	F2	0.925	0.142	Valid
Timeliness	T1	0.774	0.142	Valid
	T2	0.716	0.142	Valid

Based on table 2 shows that all instrument questions of this variables (format, ease of use, content, timeliness, and accuracy) have Pearson Product Moment correlation value greater than 0.142 (r table), so it can be concluded that the respondent's perception of the research instrument question is VALID.

Table 3. Result of reliability test

Variable	Cronbach'sAlpha (r count)	Alpha Cronbach	Result
Accuracy	0.83	0.60	Reliable
Content	0.87	0.60	Reliable

Ease of use	0.88	0. 60	Reliable
Format	0.84	0. 60	Reliable
Timeliness	0.89	0. 60	Reliable

Based on table 3 shows that all instrument questions of this variables (format, ease of use, content, timeliness, and accuracy) have Cronbach's Alpha value greater than 0.60 (Alpha Cronbach), so it can be concluded that the respondent's perception of the research instrument question is RELIABLE.

Determination of user satisfaction level for Web-Based Academic Information Systems, the researcher converts the statement of satisfaction and dissatisfaction on the user questionnaire into the following.

Table 4. Scale of user satisfaction level

Level	Score (0 - 5)	Satisfaction Level
1	0 - 1.5	Very Dissatisfied
2	1.6 - 2.5	Not Satisfied
3	2.6 - 3.0	Neutral
4	3.1 - 3.5	Satisfied
5	3.6 - 4.0	Very Strong

The score was obtained from the average value (mean) of the respondents' answers following the existing rating scale with a positive measurement scale.

The results of user satisfaction level with instrument questions of the variables (accuracy, format, ease of use, content, and timeliness) are presented in the descriptive statistics which can look in the following table.

Table 5. Measure result of user satisfaction level

Variable	N	Min	Max	Mean	Std Dev	Level	Predicate
Accuracy	135	1.00	5.00	2.81	0.91	3	Neutral
Content	135	1.25	4.00	2.90	0.58	3	Neutral
Ease of use	135	2.00	5.00	3.38	0.70	4	Satisfied
Format	135	1.00	5.00	2.81	0.92	3	Neutral
Timeliness	135	1.50	4.50	3.07	0.71	4	Satisfied
		Average		2.99			Neutral

Perception of respondents to the questions was raised for the accuracy variable to indicate the answer results:very dissatisfied (score 1) 10%, not satisfied (score 2) 26%, neutral (score 3) 41%,satisfied (score4) 19%and very strong(score5) 4%. The mean of the overall question is 2.81 which can be concluded that the satisfaction level of respondents to the Web-Based Academic Information System for the accuracy variable is at level 3 (Neutral).

Perception of respondents to the questions was raised for the content variable to indicate the answer results: very dissatisfied (score 1) 6%, not satisfied (score 2) 27%, neutral (score 3) 42%,satisfied (score 4) 21% and very strong (score 5) 4%. The mean of the overall question is 2.90 which can be concluded that the satisfaction level of respondents to the Web-Based Academic Information System for the content variable is at Level 3 (Neutral).

Perception of respondents to the questions was raised for the ease of use variable to indicate the answer results: very dissatisfied (score 1) 2%, not satisfied (score 2) 21%, neutral (score 3) 27%, satisfied (score 4) 37% and very strong (score 5) 13%. The mean of the overall question is 3.38 which can be concluded that the satisfaction level of respondents to the Web-Based Academic Information System for the ease of use variable is at Level 4(Satisfied).

Perception of respondents to the questions was raised for the format variable to indicate the answer results: very dissatisfied (score 1) 10%, not satisfied (score 2) 28%, neutral (score 3) 37%,satisfied (score4) 22%and very strong (score5) 3%. The mean of the overall question is 2.81 which can be concluded that the satisfaction level of respondents to the Web-Based Academic Information System for the format variable is at Level 3 (Neutral).

Perception of respondents to the questions was raised for the timeliness variable to indicate the answer results: very dissatisfied (score 1) 5%, not satisfied (score 2) 21%, neutral (score 3) 42%,satisfied (score4) 26%and very strong (score5) 6%. The mean of the overall question is 3.07 which can be concluded that the satisfaction level of respondents to the Web-Based Academic Information System for the timeliness variable is at level 4 (Satisfied).

4. Conclusions

Based on the result and discussion of the user satisfaction level of Web-based Academic Information System were obtained perception of respondents from the variables of EUCS methods can be concluded, that:

1) The user satisfaction level of Web-Based Academic Information Systems is an average value of 2.99 at level 3 (Neutral). That meaning, the perception of respondents assume to the Web-Based Academic Information Systems has not been satisfactory for users. The accuracy variable is an average value 2.81 at level 3 (Neutral), the content variable is an average value 2.90 at level 3 (Neutral), the ease of use variable is an average 3.38 at level 4 (Satisfied), the format variable is an average value 2.81 at level 3 (Neutral) and the timeliness variable is an average value 3.07 at level 4 (Satisfied).
2) The dominant factor of user satisfaction in the EUCS method on Web-Based Academic Information System is variable which has the highest average value as the ease of using a variable with a value of 3.38, while the variable which has lowest average value as the accuracy variable and the format variable which only reaches 2.81.
3) The variables that are considered good are timeliness and ease of use, while for accuracy, content, and format still need improvement and redevelopment.

References

[1] Doll, W.J & Torkzadeh, G 1988 The measurement of end-user computing satisfaction *MISQuarterly* **12** 259-274.

[2] Reynolds, R.A 2008 Measurement of end-user computing satisfaction in end-user computing concepts, methodologies, tools, and applications *IGI Global* 894-896

[3] Dominguez-Escrig, E. , Broch, F. F. M., Lapiedra, R., & Chiva, R 2018 Promoting radical innovation through end-user computing satisfaction *Industrial Management & Data Systems*

[4] Hidayah, N.A., & Rustamaji, E 2018 August Determining user satisfaction factors on university tuition fee systems using end-user computingSatisfaction (EUCS) *In 2018 6th International Conference on Cyber and IT Service Management (CITSM)*.1-5. IEEE.

[5] Purwanto, P 2016 Analisis penerimaan dan kepuasan pengguna sistem informasi akademik menggunakan structural equation modelling *journal.universitasbumigora.ac.id.*

[6] Tjong, Y., Sugandi, L., Nurshafita, A., Magdalena, Y., Evelyn, C., & Yosieto, N. S 2018 User satisfaction factors on learning management systems usage *In 2018 International Conference on Information Management and Technology (ICIMTech)*.11-14. IEEE.

[7] Cucus, A., & Halim, G 2019 Testing user satisfaction end-user computing satisfaction (EUCS) method in hospital management information system (SIMRS) *Cae Study at the Regional PublicHospital dr. A. Dadi Tjokrodipo assessment* **9**

[8] Pangri, M., & Umar, R 2019 Analisis sistem informasi perpustakaan menggunakan metode end-user computing satisfaction (EUCS) di Universitas Muhammadiyah Sorong. *Prosiding Semnastek* **1**

[9] Silistyani,N 2017 Analisis end-user computing satisfaction sistem pembuatan E-KTP di Kecamatan Kertasura Kabupaten Sukoharjo *Doctoral dissertation* STMIK Sinar Nusantara Surakarta

[10] Wana, F. & Abdillah, L. A 2019 Pengukuran kepuasan mahasiswa pengguna e-learning Universitas Bina Darma menggunakan metode end-user computing satisfaction (EUCS*) In Bina Darma Conference on Computer Science (BDCCS)* **5** 2222-2231

[11] bin Mohamed, A. Yasoa, M. R. B., bte Ilias, A., & Ghozali, M.F.B The study of end-user computing satisfaction (EUCS) on computerized accounting system (CAS) among paninsular Malaysia Public University: a survey in bursar's office

[12] Lim,K.S., Lim,J., &Heinrichs, J. H 2018 Validating an end-user computing satisfaction instrument: a confirmatory factor analysis approach using international data *Journal of International Technology and Information Management* **6**

[13] Rosalina,R,Utami,M.C.,Kumaladewi,N., &Putra, S. J 2017 The psychometric and interpretative analysis for assessing the end-user computingsatisfaction questionnaire *In 2017 5th International Conference on Cyber and IT Service Management (CITSM)* 1-6

[14] Doll,W.J., &Torkzadeh, G 1991 The measurement of end-user computingsatisfaction: theoretical and methodology issues MISquarterly 5-10

IOP Conf. Series: Materials Science and Engineering **909** (2020) 012045 doi:10.1088/1757-899X/909/1/012045

Optimization of bi-objective permutation flow shop scheduling with electricity cost consideration

B Kurniawan[1*] and S Fujimura[1]
[1] Graduate School of Information, Production, and Systems, Waseda University, Fukuoka, Japan

*Email: b.kurniawan@untirta.ac.id; b.kurniawan@fuji.waseda.jp;

Abstract. Increasing energy demand can create undesired problems for many governments worldwide. Several policies, such as time-of-use (TOU) tariffs, have been put in place to overcome such demand. The TOU policy's objective is to reduce electrical load during peak periods by shifting the use to off-peak periods. To that end, this paper addresses the bi-objective permutation flow-shop scheduling, minimizing total weighted tardiness and electricity costs. We propose a meta-heuristic algorithm based on SPEA2 to solve the problem. We conducted numerical experiments to evaluate the efficacy of the proposed algorithm by comparing it with NSGA-II. The results show that the proposed approach was more efficient compare with NSGA-II.

1. Introduction

Climate change and global warming are primarily due to the emissions of greenhouse gases. Emissions mainly resulted from the combustion of fossil fuels for economic and household activities. Because fossils fuels have been the primary source of energy generation, more and more greenhouse carbon dioxide will be emitted as the demand for electricity increases due to economic development and population growth.

The manufacturing sector consumes a massive amount of global energy consumption. In Japan, it consumes about 45% of the total energy [1]. Therefore, many countries have imposed several regulations to enforce the manufacturing sector to adopt energy-saving initiatives. One of the policies is time-of-use (TOU) tariffs. Energy suppliers charge different prices for different usage times. The TOU policy relocates the usage from the peak (higher prices) to off-peak (lower prices) periods [2].

Moon et al. [3] initiated the adoption of TOU policy in production scheduling. They proposed a hybrid genetic algorithm (HIGA) that outperformed standard genetic algorithm (GA). A multi-objective optimization (MOO) involving ecology, economic, and environmental of a flow-shop scheduling problem was solved by using simulated annealing [4]. Fang et al. [5] studied the effects of speed of machines a single machine scheduling. Ding et al. [6] proposed a mixed-integer programming (MIP) model for an unrelated parallel machine scheduling that minimizes the cost of electricity. They solved it using the column generation technique. Kurniawan et al. [7] investigated a bi-objective job-shop scheduling for minimizing the total weighted tardiness and electricity cost. They proposed a local search framework based on the distribution of elites to balance the exploration space towards both objectives.

This research studies bi-objective permutation flow-shop scheduling considering electricity cost that follows the TOU price scheme (FPSPTOU). The contributions from this research are as the followings. First, we presented the energy-conscious bi-objective permutation flow-shop scheduling under TOU (FPSPTOU). Second, we proposed a meta-heuristic algorithm based on the Strength Pareto Archive Evolutionary Algorithm 2 (SPEA2) to solve the bi-objective problem. Third, we compared the proposed algorithm with the Non-dominated Sorting Genetic Algorithm II (NSGA-II). Finally, this paper can inspire the manufacturing sector to embrace the TOU policy.

2. Related Works

The permutation flow-shop scheduling problem (PFSP) is a difficult combinatorial optimization problem. PFSP aims to seek the permutation of jobs such that a specific criterion is achieved. Garey has shown that PFSP is an NP-hard [8]. Several heuristic algorithms have been developed for solving FPSP, such as the CDS algorithm and NEH algorithm [9, 10].

For energy-aware FPSP, Schulz et al. [11] developed an iterated local search to minimize the makespan, energy consumption, and peak load. Chen et al. [12] have developed an innovative algorithm for distributed no-idle flow-shop scheduling. Fu et al. [13] proposed a chance-constraint approach to distributed PFSP with the total tardiness. A backtracking search algorithm was proposed to minimize the energy cost and makespan [14].

For scheduling under TOU, Moon et al. [3] proposed a hybrid GA to solve unrelated parallel machine scheduling with the sum of weighted makespan and energy cost. The problem is also solved by different approaches [15, 16]. Single machine scheduling under TOU also has been investigated extensively [2, 17-18]. Specifically, Wang et al. [19] studied a two-stage permutation flow-shop under TOU with machine states consideration. Zhang et al. [20] investigated a flow-shop scheduling problem and developed a MIP model. The objectives of their problem are to minimize the electricity cost and carbon emission for environment under TOU.

3. Formulation

The formulation of FPSPTOU is as follows. A feasible solution is represented as a permutation of n jobs, $\pi = \{\pi_1, \pi_2, \ldots, \pi_n\}$. Each job J_j consists of m stages of operation, and each stage is processed on machine M_i. Each job has a due date, d_j. Each job is processed on a machine with processing time p_{ij}. The sequence of jobs for each stage is the same. Once a machine processes a job, it cannot be interrupted until it finishes.

Time horizon of u periods is divided into several periods, $\mathcal{U} = \{U_1, U_2, \ldots, U_u\}$. Each period $U_v \in \mathcal{U}$ has an electricity price e_v. All jobs are available at time zero. Electricity cost incurs when a machine process a job. Machines can be idle to decrease electricity consumption. The total weighted tardiness and electricity cost are the objective that must be minimized.

It should be noted that the goals of MOO are to obtain a set of solutions rather than a single one. Therefore, the goal of FPSTOU is to find an approximate Pareto frontier, i.e., to find a set of non-dominated permutation of jobs.

Fig. 1 shows the TOU tariffs price scheme. The periods can be categorized based on electricity price. Periods 1–8 is off-peak periods. Periods 12–17 and 21–24 are mid-peak periods. The peak periods are 9–12 and 18–20.

For total weighted tardiness minimization, the jobs should be finished as soon as possible, whereas to minimize electricity cost, jobs should be processed in periods of low electricity prices, which may delay their start times. To balance these two conflicting objectives, we propose a meta-heuristic algorithm based on SPEA2.

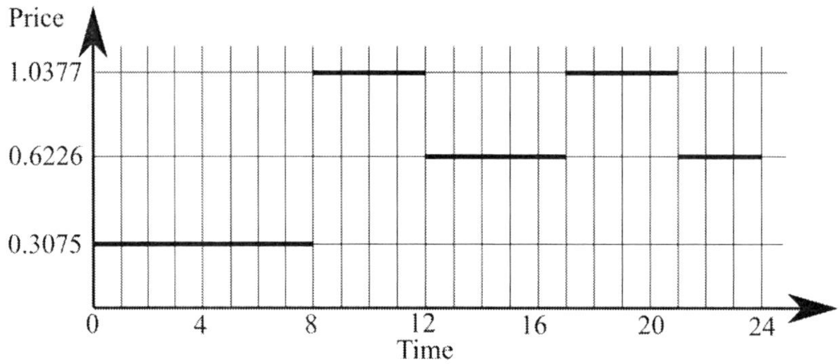

Figure 1. TOU price scheme

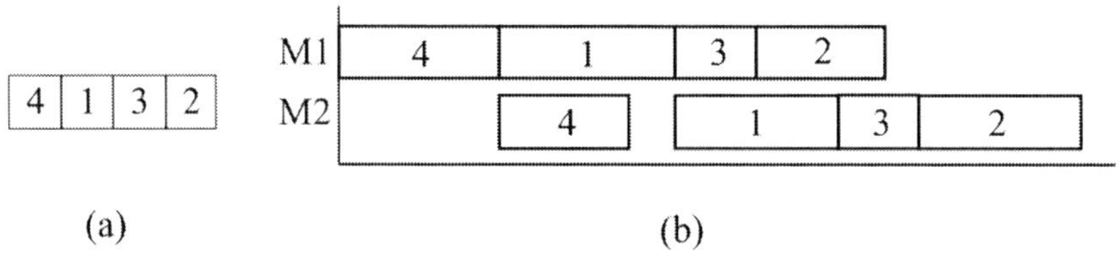

Figure 2. Encoding and decoding: (a) individual, (b) schedule

4. Proposed Method

The FPSTOU is an extension of FPSP. The inclusion of electricity cost objective into the FPSP makes the problem bigger and complex. Garey et al. [8] has proven FPSP is NP-hard, thus FPSPTOU is also NP-hard. Solutions from exact method such as branch and bound deteriorate as the problem bigger. Therefore, we propose a meta-heuristic based on SPEA2 [21]. In what follows, we explain the proposed method in detail.

4.1. Encoding

For FPSTOU, a job sequence is represented using an integer representation. The initial population is generated randomly. Fig. 2 shows an individual that consists of 4 jobs along with its corresponding schedule.

4.2. Genetic operators

A binary tournament is used to select parents. After that, a one-point crossover is conducted on a pair of parents. We perform job-swapping to mutate a randomly selected individual.

4.3. Overview of the proposed method

Let N and $\overline{N}$ be population size and archive size. Let T and t be the maximum number of iterations and its index. The procedure of the propose method are as follows.

At the initialization, $t = 0$, a population $\boldsymbol{P}_0$ is generated randomly. An empty archive $\overline{\boldsymbol{P}}_0$ is created to store the non-dominated solutions. Then, calculate the fitness of individuals in population $\boldsymbol{P}_t$ and archive $\overline{\boldsymbol{P}}_t$. In the next step, non-dominated solutions in population $\boldsymbol{P}_t$ and archive $\overline{\boldsymbol{P}}_t$ are copied into $\overline{\boldsymbol{P}}_{t+1}$. If the number of individuals in $\overline{\boldsymbol{P}}_{t+1}$ exceeds $\overline{N}$, remove individuals with lower fitness. Conversely, select dominated individuals from population $\boldsymbol{P}_t$ and archive $\overline{\boldsymbol{P}}_t$ into $\overline{\boldsymbol{P}}_{t+1}$ so that the number of individuals $\overline{N}$.

Perform tournament selection from individuals in $\overline{\boldsymbol{P}}_{t+1}$ to fill the pool. Then, perform crossover and mutation on individuals in the pool. Set the population created from crossover and mutation as $\boldsymbol{P}_{t+1}$. The procedure is repeated after stopping criterion is reached.

Table 1. Results for $f = 1.5$ using Taillard's instances [24]

$n \times m$	u	NSGA-II		Proposed method	
		$\overline{GD}$	$\overline{NDS}$	$\overline{GD}$	$\overline{NDS}$
20×5	1476	149.95	36.1	117.05	40.5
20×10	1666	418.39	17.5	119.13	23.3
20×20	2432	787.96	25.3	613.98	28.7
50×5	3065	49.26	24.8	25.81	30.6
50×10	3695	956.01	35.3	477.10	34.9
50×20	5077	185.01	21.4	216.66	18.2
Average		424.43	26.73	261.62	29.37

5. Numerical Experiments

To measure the efficacy of the proposed algorithm, we compare it with NSGA-II. The genetic operators of NSGA-II are the same as those of the proposed algorithm. All algorithms were implemented on a computer with a Dual core processor @2.4 GHz and 3 GB of RAM.

The number of non-dominated solutions, NDS, and the generational distance, GD [22], are used as the performance metrics. GD is calculated as follows

$$GD = \sqrt{\frac{\sum_{\theta=1}^{\gamma} \Delta_\theta}{\gamma}} \tag{1}$$

where γ denotes the number of non-dominated solutions, and Δ denotes the shortest Euclidean distance between the θ-th solution and reference set. For GD, lower values are preferred. On the contrary, the larger NDS the better.

The problem instances are taken from Taillard [23]. As for due date, it was generate by $d_j = r_j + \lfloor f * \sum_{i=1}^{m} p_{ij} \rfloor$ where f is constant, and r_j is the release time of job. In this study, $r_j = 0$ for all jobs. The weight is categorized as follows: 20% of jobs are very important ($w_j = 4$), 60% have average importance ($w_j = 2$), and 20% of jobs are of least importance ($w_j = 1$). The number of period is calculated as $u = \lfloor y * BM \rfloor$, where $y = [1.2,\ 1.5]$ and BM is the estimated makespan. The TOU price follows the pattern shown in Fig. 1.

The population size is set to 150, the crossover rate is set to 0.7, and the mutation rate is set to 0.1. The archive size is the same as population seize. Each instance is run for 10 replications. Each run is 50 s.

Table 1 shows the computational results for $f = 1.5$. The first column represent the problem size, consist of job and machine. The second column represents the number of periods. The third until sixth column show the metrics value of NSGA-II and the proposed method, respectively.

The proposed method outperformed NSGA-II in all performance metrics for all instances. In average, the proposed method also produced better results than those of NSGA-II. However, the differences were so small.

6. Conclusions

This study investigated a permutation flow-shop scheduling under TOU (PFSPTOU). The total weighted tardiness and electricity cost are the objectives that must be minimized. We proposed a meta-heuristic based on SPEA2 to solve the problem. Numerical results showed the proposed method outperformed the NSGA-II. For future research, a local search can be developed to improve the solutions quality. Moreover, genetic operators can be improved.

Acknowledgments
The authors wish to thank the anonymous referees for their constructive feedback. This research is supported by the Indonesia Endowment Fund for Education (Lembaga Pengelola Dana Pendidikan LPDP Indonesia).

References

[1] Okajima S and Okajima H 2013 *Energy Policy* **61** 574–586
[2] Kurniawan B, Gozali A A, Weng W and Fujimura S 2019 *2018 IEEE International Conference on Industrial Engineering & Engineering Management ed Varma A K (IEEM)* 137–141
[3] Moon J Y, Shin K and Park J 2013 *The International Journal of Advanced Manufacturing Technology* **68** 523–535
[4] Sharma A, Zhao F and Sutherland J W 2015 *Journal of Cleaner Production* **108** 256–270
[5] Fang K, Uhan N A, Zhao F and Sutherland J W 2016 *Annals of Operations Research* **238** 199–227
[6] Ding J Y, Song S, Zhang R and Chiong R 2016 *IEEE Transactions on Automation Science and Engineering* **13** 1138–1154
[7] Kurniawan B, Song W, Weng W and Fujimura S 2020 *Evolutionary Intelligence*
[8] Garey M, Johnson D and Sethi R 1976 *Math. Oper. Res.* **1** 117–129
[9] Campbell H G, Dudek R A and Smith M L 1970 *Management Science* **16** B630–B637
[10] Nawaz M, Enscore E E and Ham I 1983 *Omega* **11** 91–95
[11] Schulz S, Neufeld J S and Buscher U 2019 *Journal of Cleaner Production* **24** 421–434
[12] Chen J f, Wang L and Peng Z p 2019 *Swarm and Evolutionary Computation* **50** 100557
[13] Fu Y, Tian G, Fathollahi-Fardd A M, Ahmadi A and Chaoyong Z 2019 *Journal of Cleaner Production* **226** 515–525
[14] Lu C, Gao L, Li X, Pan Q and Wang Q 2017 *Journal of Cleaner Production* **144** 228–238
[15] Kurniawan B, Gozali A A, Weng W and Fujimura S 2018 *2017 IEEE International Conference on Industrial Engineering & Engineering Management ed Varma A K (IEEE)* 583–587
[16] Kurniawan B, Chandramitasari W, Gozali A A, Weng W and Fujimura S 2020 *IEEJ Transactions on Electrical and Electronics Engineering* **15** 208–217
[17] Shrouf F, Ordieres-Mere J, Garcia-Sanchez A and Ortega-Mier M 2014 *Journal of Cleaner Production* **67** 197–207
[18] Che A, Zeng Y and Lyu K 2016 *Journal of Cleaner Production* **129** 565–577
[19] Wang S, Zhu Z, Fang K, Chu F and Chu C 2017 *International Journal of Production Research*
[20] Zhang H, Zhao F, Fang K and Sutherland J W 2014 *CIRP Annals - Manufacturing Technology* **63** 37–40
[21] Zitzler E, Laumanns M and Thiele L 2001 Spea2: Improving the strength pareto evolutionary algorithm *Tech. rep. Swiss Federal Institute of Technology (ETH) Zurich*
[22] Veldhuizen D Multiobjective evolutionary algorithms: classifications, analyses, and new innovations
[23] Taillard E 2019 Scheduling instances http://mistic.heig-vd.ch/taillard accessed: 2019-06-25

IOP Conf. Series: Materials Science and Engineering **909** (2020) 012046 doi:10.1088/1757-899X/909/1/012046

Selection of supplier for the evaluation of procurement of special chemical using entropy method and topsis in xyz company

E R Wijaya[1*], Rudianto[1]

[1]Logistics Management of Agro Industry, ATIP Polytechnic, Padang, Indonesia

*Email: edorantou@kemenperin.go.id;

Abstract. In the process of procuring special materials, the company has more than one supplier for each item of goods, at XYZ companies selecting suppliers through process evaluation. In the evaluation process the supplier is chosen only based on the lowest price offered by the supplier, these criteria have not been able to ensure the selected supplier can meet other criteria such as quality, transportation system and so on. This study aims to determine whether the criteria that have the highest weight value in the selection of suppliers and to find out the best supplier in the purchase of special materials XYZ companies. In this study, the Entropy and TOPSIS methods are used to analyse the data.. Entropy method is used to calculate the weight of each criterion and alternatives, while the TOPSIS method is used to rank suppliers. The results showed that the criterion that had the highest weighting value in the selection of special material suppliers in the special material purchasing section of XYZ company was the criterion C3 and C12 with positive ideal solution value 0,0380 and 0,0389. Whereas the best supplier for PT XYZ's special material purchasing department is supplier E with a preference value of 0.5517.

1. Introduction

During this time the Special Material Purchase section of XYZ Company utilizes supplier companies to meet the needs of special goods / materials. Special materials that must be fulfilled are chemicals. This chemical is used in the laboratory for the production testing process. PT. XYZ selects suppliers through a tender process. This causes the supplier selection process through the tender process to not run as expected by the company. This discrepancy occurs because the selection of suppliers is only based on one type of criteria. This study wants to help XYZ companies determine the criteria for selecting the right supplier so that suppliers are selected through a tender evaluation process in accordance with company expectations._In the tender process the supplier is chosen only based on the lowest price offered by the supplier. The lowest price criteria itself is not necessarily able to ensure that the selected supplier can meet other criteria, such as quality, transportation systems, and so on. The company should also consider other criteria. Therefore, companies need to assess suppliers carefully and continuously because supplier assessments require a variety of criteria not just one criterion. Recently, the special materials are purchased from five suppliers. The supplier is A company (A), B company (B), C company (C), D company (D), and E company (E).

In general, companies have more than one supplier for each item of goods. The selection of the right supplier can guarantee the availability of goods needed by the company. The selection of suppliers that are based on low price quotes is no longer efficient to get maximum supply chain performance, companies must pay attention to other relevant criteria to achieve company goals. In terms of finding and choosing suppliers, it is necessary to determine the people who are interested in making the decision. Supplier selection is a multi-criteria problem which includes qualitative and quantitative factors. If several criteria are considered in a decision making process problem then the problem is known as Multi Criteria Decision Making (MCDM)[1,2,11–13,3–10]. Therefore we need a method that can include both of them in the measurement. In this case the method used for supplier selection is the TOPSIS (Technique For Order Preference By Similarity To Ideal Solution) Method[1,2,13,4–10,12].

TOPSIS was first introduced by Yoon and Hwang in 1981 to be used as a method for solving multi-criteria problems[1,2,4,5,10]. TOPSIS provides a solution of a number of possible alternatives by comparing each alternative with the best alternative and the worst alternative that exists among alternative problems[4,5]. Apply the TOPSIS method to select suppliers of cement, sand, iron, foundation stone and wood. And obtained 13 criteria in the selection of suppliers, namely the appropriateness of prices with the quality of goods produced (C1), the ability to provide discounts (discounts) on orders in a certain amount (C2), the suitability of goods with the specifications set (C3), the supply of goods without defects (C4), the ability to provide consistent quality (C5), the ability to deliver goods according to the agreed date (C6), the ability to handle the transportation system (C7), the accuracy and suitability of the quantities in the shipment (C8), the suitability of the contents of the package (C9), the ease of contact (C10), the ability to provide information clearly and easily to understand (11), the speed in responding to customer requests (C12), and responsive in resolving customer complaints (C13)[6,8,14]. Prior to ranking, all the criteria specified are weighted. Weighting is done using the entropy method[1,3,9,12]. And TOPSIS is used to rank the alternatives supplier.

2. Methodology

The stage of the research begins by making a research questionnaire based on literature studies and interviews with experts. Each question in the questionnaire was designed by setting one question for each criterion. Besides that, in the questionnaire also assessments were made for each supplier. The measurement scale used on the questionnaire is a Likert scale with a range from 1-10. The experts used as respondents in this study were the workers in the XYZ company specialized material purchasing division. The number of respondents consists of six people, each of whom has work experience in the field of purchasing and procurement.

After the questionnaire was successfully designed, it was then given to all respondents to be filled. The completed questionnaire was then processed using the entropy method to weight each criterion and the TOPSIS method to rank each alternative supplier. After that it is closed with conclusions and suggestions. The supplier selection model can be summarized in Figure 1.

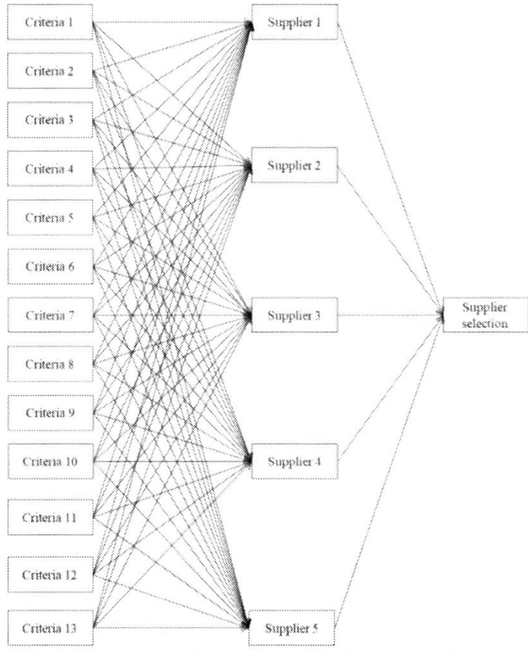

Figure 1. Supplier selection Model

2.1. Supplier selection

Supplier selection process is very important for the company because it will have an impact on the company's financial condition in the long run[12]. Supplier selection is the most frequent decision making problem. Quantitative and qualitative factors are needed to identify the right supplier according to the company's requirements and budget criteria. Raw materials provided by suppliers will have an impact on the products produced by the company. If the raw material supplied by the supplier is of high quality, the products produced by the company will be in accordance with customer requirements. Several approaches can be used to solve the problem of supplier selection decision making. However, in this study the entropy method is used for weighting criteria and TOPSIS to rank alternative suppliers to be chosen by the company. The supplier selection evaluation process is summarized in Figure 2.

2.2. Entropy

The entropy weighting method was first introduced by Shannon (2001)[12]. The term entropy is known in the concept of thermodynamics, which was later adopted for the world of information systems. Every sign of uncertainty in the communication process is called information entropy. If the level of information uncertainty is lower the higher the weighting value. Weight is the relative importance of the evaluation of a certain criterion[9]. The assessment of criteria weights is determined from the opinion of the decision maker.

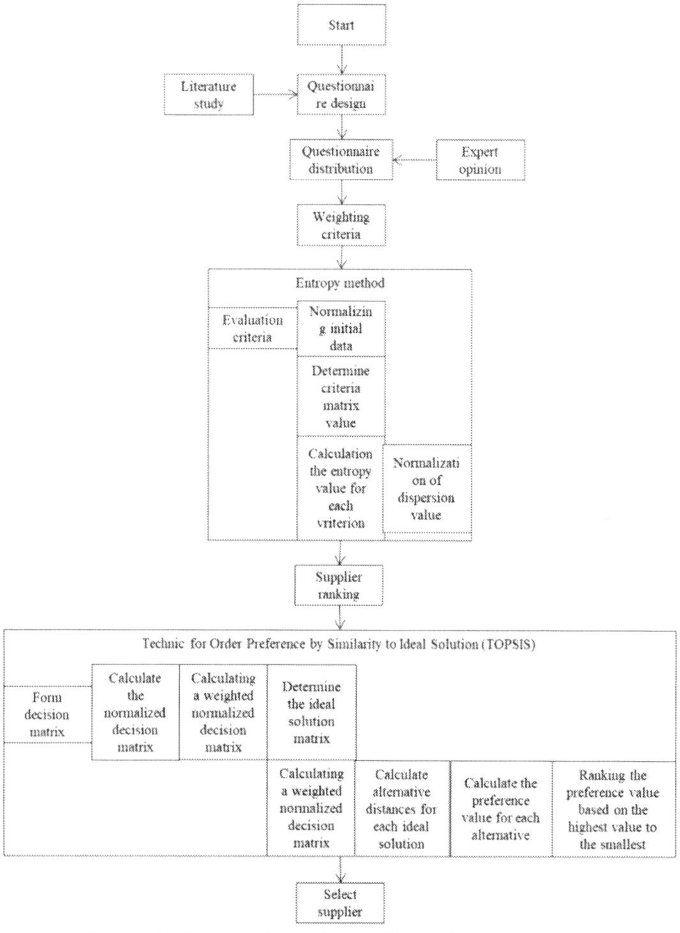

Figure 2. Evaluation of supplier selection process

If there are multiple decision makers, the assessment of criteria weights becomes more difficult. This is because each decision maker has a different view in assessing a criterion. Other than that, entropy weighting does not require criteria to have the same unit or range. Based on Maisari et al [9], the following are the stages of the entropy method in weighting a criterion.

- Evaluation criteria, each respondent provides an assessment for each criterion based on the level of importance in selecting suppliers.
- Normalizing the initial data, subtract the criterion value from the preferences of the respondents with the value of the maximum criteria importance level range. Give a notation k_{ij} for the subtraction result, i is the number of respondents and j is the number of criteria.
- Determine criteria matrix value. Denoted by a_{ij}.

$$a_{ij} = \frac{k_{ij}}{\sum_{i=1}^{m}\sum_{j=1}^{n} k_{ij}} \quad (1)$$

a_{ij} = criteria matrix value
k_{ij} = initial data (normalize criteria value)
i = number of respondent
j = number of criteria

- Calculation of the entropy value for each criterion. Denoted by E_i

$$E_j = \left[\frac{-1}{\ln m}\right] \sum_{j=1}^{n} \left[a_{ij} \ln a_{ij}\right] \quad (2)$$

- Dispersion calculation for each criterion. Denoted by D_j

$$D_j = 1 - E_j \quad (3)$$

- Normalization of dispersion value. Denoted by w_j

$$w_j = \frac{D_j}{\sum D_j} \tag{4}$$

2.3. TOPSIS

TOPSIS is a practical and simple decision-making method that uses performance measurements in a simple mathematical form[9]. The advantage of this method is that the optimal solution is obtained from the comparison between the best alternative and the worst alternative, thus narrowing the possibility of subjective judgments[9,12]. The following are the calculation steps using the TOPSIS method[9].

- Form a decision matrix. Denoted by x_{ij}
- Calculate the normalized decision matrix

$$r_{ij} = x_{ij} \Big/ \sqrt{\sum_{i=1}^{m} x_{ij}^2} \tag{5}$$

r_{ij} = normalized decision matrix
x_{ij} = decision matrix
i = number of criteria
j = number of alternative (supplier)

- Calculating a weighted normalized decision matrix, denoted by y_{ij}

$$y_{ij} = w_i r_{ij} \tag{6}$$

- Determine the positive ideal solution matrix and the negative ideal solution matrix
  - Positive ideal solution matrix is denoted by A^+

$$A^+ = (y_1^+, y_2^+, \dots, y_n^+) \tag{7}$$

$$y_1^+ = max(y_{11}, y_{12}, \cdots, y_{6j}) \tag{8}$$

  - Negative ideal solution matrix is denoted by A^-

$$A^- = (y_1^-, y_2^-, \dots, y_j^-) \tag{9}$$

$$y_1^- = min(y_{11}, y_{12}, \cdots, y_{6j}) \tag{10}$$

y_i^+ = the maximum value of the weighted normalized decision matrix
y_j^- = the minimum value of the weighted normalized decision matrix

- Calculate alternative distances for each ideal solution
  - Alternative distances for ideal positive solutions denoted by D_j^+

$$D_j^+ = \sqrt{\sum_{j=1}^{n} (y_i^+ - y_{ij})^2} \tag{11}$$

  - Alternative distances for ideal negative solutions denoted by D_j^-

$$D_j^- = \sqrt{\sum_{j=1}^{n} (y_{ij} - y_i^-)^2} \tag{12}$$

- Calculate the preference value of each alternative solution denoted by V_j

$$V_j = \frac{D_j^-}{D_j^+ + D_j^-} \tag{13}$$

- Ranking alternative solution preference value. Ranking is done by sorting the largest V_j value to the smallest value. V_j with the greatest value is the optimal solution or the best supplier chosen by the company.

3. Result

3.1. Numerical example of entropy method

3.1.1. Initial data

The questionnaire distributed to respondents was then summarized in a table or matrix. The table or matrix is the result of an assessment of six respondents for the available alternatives. The following matrix is preliminary data for alternative suppliers A. Initial data for other suppliers is shown in the appendix.

Table 1. Initial data questionnaire for alternative A

Respondent	Criteria												
	C1	C2	C3	C4	C5	C6	C7	C8	C9	C10	C11	C12	C13
1	7	8	7	7	8	8	8	8	8	7	7	7	8
2	8	5	7	6	8	7	7	7	8	9	6	6	6
3	7	7	7	7	9	7	7	7	7	8	7	7	7
4	8	8	8	8	7	8	8	9	8	8	8	8	7
5	7	7	8	7	8	8	7	6	6	7	7	8	7
6	7	7	7	7	9	7	7	7	7	7	7	7	7

3.1.2. Normalized initial data

The maximum value of the Likert scale used on the questionnaire is 10. In table 1, it can be seen that the respondent's value 1 for criterion 1 is 7 so the initial normalized value for C_{11} is -3.

- $C_{11} \sim 7 - 10 = -3$
- $C_{12} \sim 8 - 10 = -2$
- $\vdots$
- $C_{613} \sim 7 - 10 = -3$

Based on the initial data normalization calculation, the initial data is normalized, which then forms a normalized initial data matrix. Table 2 below provides an overview of the initial normalized data matrix for Alternative A.

Table 2. Initial normalized data matrix for alternative A

Respondent (R)	Criteria													Σ
	C1	C2	C3	C4	C5	C6	C7	C8	C9	C10	C11	C12	C13	
1	-3	-2	-3	-3	-2	-2	-2	-2	-2	-3	-3	-3	-2	-32
2	-2	-5	-3	-4	-2	-3	-3	-3	-2	-1	-4	-4	-4	-40
3	-3	-3	-3	-3	-1	-3	-3	-3	-3	-2	-3	-3	-3	-36
4	-2	-2	-2	-2	-3	-2	-2	-1	-2	-2	-2	-2	-3	-27
5	-3	-3	-2	-3	-2	-2	-3	-4	-4	-3	-3	-2	-3	-37
6	-3	-3	-3	-3	-1	-3	-3	-3	-3	-3	-3	-3	-3	-37
Σ	-16	-18	-16	-18	-11	-15	-16	-16	-16	-14	-18	-17	-18	-209

3.1.3. Criteria value

The formula for the criteria value is obtained from equation 1. An example of calculating the criteria value for criterion 1 by respondent 1 is as follows.

- $a_{11} \approx \frac{-3}{-209} = 0.0144$
- $a_{12} \approx \frac{-2}{-209} = 0.0096$
- $\cdots$
- $a_{613} \approx \frac{-3}{-209} = 0.0144$

Table 3 below provides an overview of the criteria value matrix for all respondents.

Table 3. The criteria value matrix for alternative A

R	Criteria value												
	C1	C2	C3	C4	C5	C6	C7	C8	C9	C10	C11	C12	C13
	a_{1n}	a_{2n}	a_{3n}	a_{4n}	a_{5n}	a_{6n}	a_{7n}	a_{8n}	a_{9n}	a_{10n}	a_{11n}	a_{12n}	a_{13n}
1	0.0144	0.0096	0.0144	0.0144	0.0096	0.0096	0.0096	0.0096	0.0096	0.0144	0.0144	0.0144	0.0096
2	0.0096	0.0239	0.0144	0.0191	0.0096	0.0144	0.0144	0.0144	0.0096	0.0048	0.0191	0.0191	0.0191
3	0.0144	0.0144	0.0144	0.0144	0.0048	0.0144	0.0144	0.0144	0.0144	0.0096	0.0144	0.0144	0.0144
4	0.0096	0.0096	0.0096	0.0096	0.0144	0.0096	0.0096	0.0048	0.0096	0.0096	0.0096	0.0096	0.0144
5	0.0144	0.0144	0.0144	0.0144	0.0096	0.0096	0.0144	0.0191	0.0191	0.0144	0.0144	00096	0.0144
6	0.0144	0.0144	0.0144	0.0144	0.0048	0.0144	0.0144	0.0144	0.0144	0.0144	0.0144	0.0144	0.0144

3.1.4. Entropy value

Equation 2 is used to calculate the entropy value for each criterion. The following is a calculation of the entropy value for alternative A. The number of respondents in this study is six people so that the value of m is 6. *Ln(6) = 1,7918* and *ln(0,0144) = -4,2404; ln(0,0096) = -4,6460*

- $E_1 = \left[\frac{-1}{ln(6)}\right] \sum_{i=1}^{6} [(0{,}0144\, x - 4{,}2404) + (0{,}0096\, x - 4{,}6460) + (0{,}0144\, x - 4{,}2405) + (0{,}0096\, x - 4{,}6460) + (0{,}0144\, x - 4{,}2405) + (0{,}0144\, x - 4{,}2404)]$

$$E_1 = \left[\frac{-1}{1{,}7918}\right] \sum_{i=1}^{6} [(-0{,}0611) + (-0{,}0446) + (-0{,}0611) + (-0{,}0446) + (-0.0611) + (-0{,}0611) \approx 0{,}1862]$$

- $E_2 = \left[\frac{-1}{ln(6)}\right] \sum_{i=1}^{6} [(0{,}0096\, x - 4{,}6460) + (0{,}0239\, x - 3{,}7339) + (0{,}0144\, x - 4{,}2405) + (0{,}0096\, x - 4{,}6460) + (0{,}0144\, x - 4{,}2405) + (0{,}0144\, x - 4{,}2404)]$

$$E_2 = \left[\frac{-1}{1{,}7918}\right] \sum_{i=1}^{6} [(-0{,}0446) + (-0{,}0892) + (-0{,}0611) + (-0{,}0446) + (-0.0611) + (-0{,}0611)] \approx 0{,}2019$$

- ⋮
- E_{13}

Table 4 below is a summary of the entropy value for all criteria of alternative A.

Table 4. Entropy value matrix for alternative A

Criteria	Entropy value (E_j)
1	0,1862
2	0,2019
3	0,1862
4	0,2035
5	0,1373
6	0,1770
7	0,1862
8	0,1837
9	0,1851
10	0,1664
11	0,2035
12	0,1943
13	0,2035

3.1.5. Dispersion value

Equation 3 is used to calculate the dispersion value for each criterion. The following is a calculation of the dispersion value for alternative A.

- $D_1 = 1 - E_1 \approx 1 - 0{,}1862 = 0{,}8138$
- $D_2 = 1 - E_2 \approx 1 - 0{,}2019 = 0{,}7981$
- $\vdots$
- $D_{13} = 1 - E_{13} \approx 1 - 0{,}2035 = 0{,}7965$

Table 5 below is a summary of the entropy value for all criteria of alternative A.

Table 5. Dispersion value matrix for alternative A

Criteria	Dispersion value (D_j)
1	0,8138
2	0,7981
3	0,8138
4	0,7965
5	0,8627
6	0,8230
7	0,8138
8	0,8163
9	0,8149
10	0,8336
11	0,7965
12	0,8057
13	0,7965
$\sum$	10,5852

3.1.6. Normalized dispersion value

Equation 4 is used to calculate the normalized dispersion value or criteria weights for each criterion. The following is a calculation of the normalized dispersion value for alternative A.

- $w_1 = \frac{D_1}{\sum D_j} \approx \frac{0{,}8138}{10{,}5852} \approx 0{,}0769$
- $w_2 = \frac{D_2}{\sum D_j} \approx \frac{0{,}7981}{10{,}5852} \approx 0{,}0754$
- $\vdots$
- $w_{13} = \frac{D_{13}}{\sum D_j} \approx \frac{0{,}7965}{10{,}5852} \approx 0{,}0752$

Table 6 below is a summary of the entropy value for all criteria of alternative A.

Table 6. Criteria weights value for alternative A

Criteria	Weights
1	0,0769
2	0,0754
3	0,0769
4	0,0752
5	0,0815
6	0,0778
7	0,0769
8	0,0771
9	0,0770
10	0,0788
11	0,0752
12	0,0761
13	0,0752

3.2. Numerical example of TOPSIS

3.2.1. Decision matrix

The decision matrix is a matrix with the order $m \times n$, where m is the number of alternatives (suppliers) and n is the number of criteria. The decision matrix value is taken from the normalized dispersion value or the result of weighting criteria for each alternative (supplier). Because suppliers number five, m equals five and criteria number 13, then n equals 13. Table 7 shows the decision matrix for all alternatives and criteria that have been determined using the entropy method.

Table 7. Decision matrix for all alternatives

	Criteria												
	C1	C2	C3	C4	C5	C6	C7	C8	C9	C10	C11	C12	C13
	w_1	w_2	w_3	w_4	w_5	w_6	w_7	w_8	w_9	w_{10}	w_{11}	w_{12}	w_{13}
A	0,0777	0,0756	0,077	0,0748	0,0762	0,0785	0,0777	0,0777	0,0756	0,0794	0,0763	0,07762	0,0771
B	0,0785	0,0778	0,0769	0,0754	0,0769	0,0786	0,0785	0,0769	0,0771	0,0786	0,0761	0,0747	0,0740
C	0,0785	0,0769	0,0794	0,0785	0,076	0,0751	0,0777	0,0759	0,0803	0,0763	0,0752	0,0736	0,0768
D	0,0767	0,0715	0,0771	0,0771	0,078	0,078	0,077	0,0771	0,0772	0,079	0,0771	0,0767	0,0777
E	0,0768	0,08	0,0816	0,0786	0,0724	0,0769	0,0768	0,0724	0,0731	0,0785	0,0761	0,0816	0,0761

3.2.2. Normalized decision matrix

Equation 5 is used to calculate the normalized decision matrix for all alternatives and criteria. The following is a normalized matrix calculation for alternative A and criterion 1, followed by the same method until alternative E and criterion 13. Table 8 summarizes the results of the normalized decision matrix for all alternatives and criteria.

- $r_{A1} = \frac{0{,}0777}{\sqrt{(0{,}0777)^2+(0{,}0785)^2+(0{,}0785)^2+(0{,}0767)^2+(0{,}0768)^2}} \approx r_{A1} = \frac{0{,}0777}{\sqrt{0{,}0302}} \approx 0{,}4475$
- $\vdots$
- $r_{E13} = \frac{0{,}0761}{\sqrt{(0{,}0771)^2+(0{,}074)^2+(0{,}0768)^2+(0{,}0777)^2+(0{,}0761)^2}} \approx r_{E13} = \frac{0{,}0761}{\sqrt{0{,}02915}} \approx 0{,}4457$

Table 8. Normalized decision matric for all alternatives

	Normalized decision matrix (r)												
	r_1	r_2	r_3	r_4	r_5	r_6	r_7	r_8	r_9	r_{10}	r_{11}	r_{12}	r_{13}
A	0,4475	0,4425	0,4391	0,4350	0,4488	0,4534	0,4481	0,4571	0,4408	0,4531	0,4480	0,4448	0,4516
B	0,4521	0,4553	0,4385	0,4385	0,4530	0,4540	0,4527	0,4524	0,4496	0,4485	0,4468	0,4361	0,4334
C	0,4521	0,4501	0,4528	0,4565	0,4477	0,4338	0,4481	0,4465	0,4682	0,4354	0,4416	0,4296	0,4498
D	0,4418	0,4185	0,4397	0,4484	0,4594	0,4505	0,4441	0,4535	0,4502	0,4508	0,4527	0,4477	0,4551
E	0,4424	0,4682	0,4653	0,4571	0,4265	0,4441	0,4429	0,4259	0,4262	0,4480	0,4468	0,4763	0,4457

3.2.3. Weighted normalized decision matrix

Equation 6 is used to determine the weighted normalized decision matrix for all alternatives and criteria. The following is a calculation for alternative weighted normalized decision A and criterion 1. Table 9 summarizes all the weighted normalized decision calculation values for all alternatives and criteria.

- $y_{A1} \approx 0{,}0777 \times 0{,}4475 = 0{,}0348$
- $y_{A2} \approx 0{,}0756 \times 0{,}4425 = 0{,}0334$
- $\vdots$
- $y_{E13} \approx 0{,}0761 \times 0{,}4457 = 0{,}0339$

Table 9. Weighted normalized decision matrix for all alternatives and criteria

	Weighted normalized decision												
	y_1	y_2	y_3	y_4	y_5	y_6	y_7	y_8	y_9	y_{10}	y_{11}	y_{12}	y_{13}
A	0,0348	0,0335	0,0338	0,0325	0,0342	0,0356	0,0348	0,0355	0,0333	0,0360	0,0342	0,0339	0,0348
B	0,0355	0,0354	0,0337	0,0331	0,0348	0,0357	0,0355	0,0348	0,0347	0,0353	0,0340	0,0326	0,0321
C	0,0355	0,0346	0,0360	0,0358	0,0340	0,0326	0,0348	0,0339	0,0376	0,0332	0,0332	0,0316	0,0345
D	0,0339	0,0299	0,0339	0,0346	0,0358	0,0351	0,0342	0,0350	0,0348	0,0356	0,0349	0,0343	0,0354
E	0,0340	0,0375	0,0380	0,0359	0,0309	0,0342	0,0340	0,0308	0,0312	0,0352	0,0340	0,0389	0,0339

3.2.4. Positive ideal solution matrix

Equation 8 is used to determine the value of a positive ideal solution for all criteria. The following is calculation of a positive ideal solution for alternative A with criterion 1 and criterion 2. The calculation is carried out to the end i.e. for alternative E criterion 13. The following table 10 summarizes the value of a positive ideal solution for all criteria.

- $y_1^+ = max\{0{,}0348; 0{,}0355; 0{,}0355; 0{,}0339; 0{,}0340\} \approx 0{,}0355$
- $y_2^+ = max\{0{,}0335; 0{,}0354; 0{,}0346; 0{,}0299; 0{,}0375\} \approx 0{,}0375$
- $\vdots$
- $y_{13}^+ = max\{0{,}0348; 0{,}0321; 0{,}0345; 0{,}0354; 0{,}0339\} \approx 0{,}0354$

Table 10. Positive ideal solution for all criteria

Positive ideal solution												
y_1^+	y_2^+	y_3^+	y_4^+	y_5^+	y_6^+	y_7^+	y_8^+	y_9^+	y_{10}^+	y_{11}^+	y_{12}^+	y_{13}^+
0,0355	0,0375	0,0380	0,0359	0,0358	0,0357	0,0355	0,0355	0,0376	0,0360	0,0349	0,0389	0,0354

3.2.5. Negative ideal solution matrix

Equation 10 is used to determine the value of a negative ideal solution for all criteria. The following is calculation of a negative ideal solution for criterion 1 and criterion 2. The calculation is carried out to the end i.e. for alternative E criterion 13. The following table 11 summarizes the value of a negative ideal solution for all criteria.

- $y_1^- \approx min\{0{,}0348; 0{,}0355; 0{,}0355; 0{,}0339; 0{,}0340\} = 0{,}0340$
- $y_2^- \approx min\{0{,}0335; 0{,}0354; 0{,}0346; 0{,}0299; 0{,}0375\} = 0{,}0299$
- $\vdots$
- $y_{13}^- \approx min\{0{,}0348; 0{,}0321; 0{,}0345; 0{,}0354; 0{,}0339\} = 0{,}0321$

Table 11. Negative ideal solution for all criteria

Negative ideal solution												
y_1^-	y_2^-	y_3^-	y_4^-	y_5^-	y_6^-	y_7^-	y_8^-	y_9^-	y_{10}^-	y_{11}^-	y_{12}^-	y_{13}^-
0,0339	0,0299	0,0337	0,0325	0,0309	0,0326	0,0340	0,0308	0,0312	0,0332	0,0332	0,0316	0,0321

3.2.6. Distances for positive ideal solution matrix

Equation 11 is used to determine the value of a distance for positive ideal solution for all alternatives. The following is a calculation for the distance for a positive ideal solution for alternative A. Table 12 below summarizes the values for the distance for a positive ideal solution for all alternatives.

- $$D_A^+ \approx \sqrt{\begin{matrix}(0{,}0355\text{-}0{,}0348)^2+(0{,}0375\text{-}0{,}0335)^2+(0{,}038\text{-}0{,}0338)^2+(0{,}0359\text{-}0{,}0325)^2+(0{,}0358\text{-}0{,}0342)^2+ \\ (0{,}0357\text{-}0{,}0356)^2+(0{,}0355\text{-}0{,}0348)^2+(0{,}0355\text{-}0{,}0355)^2+(0{,}0376\text{-}0{,}0333)^2+(0{,}036\text{-}0{,}036)^2 \\ +(0{,}0349\text{-}0{,}0342)^2+(0{,}0389\text{-}0{,}0339)^2+(0{,}0354\text{-}0{,}0348)^2\end{matrix}}$$

$= 0{,}00961$

IOP Conf. Series: Materials Science and Engineering **909** (2020) 012046 doi:10.1088/1757-899X/909/1/012046

- ⋮
- D_E^+

Table 12. Distances for positive ideal solution for all alternatives

D_A^+	D_B^+	D_C^+	D_D^+	D_E^+
0,00961	0,00961	0,00959	0,01044	0,00992

3.2.7. Distances for negative ideal solution

Equation 12 is used to determine the value of a distance for negative ideal solution for all alternatives. The following is a calculation for the distance for a negative ideal solution for alternative A. Table 13 below summarizes the values for the distance for a negative ideal solution for all alternatives.

- $$D_A^- \approx \sqrt{\begin{array}{c}(0{,}0339-0{,}0348)^2+(0{,}0299-0{,}0335)^2+(0{,}0337-0{,}0338)^2+(0{,}0325-0{,}0325)^2+(0{,}0309-0{,}0342)^2+\\(0{,}0326-0{,}0356)^2+(0{,}034-0{,}0348)^2+(0{,}0308-0{,}0355)^2+(0{,}0312-0{,}0333)^2+\\(0{,}0332-0{,}036)^2+(0{,}0332-0{,}0342)^2+(0{,}0316-0{,}0339)^2+(0{,}0321-0{,}0348)^2\end{array}}$$
  $=0{,}00905$
- ⋮
- D_E^-

Table 13. Distances for negative ideal solution for all alternatives

D_A^-	D_B^-	D_C^-	D_D^-	D_E^-
0,00905	0,00971	0,01039	0,0096	0,01221

3.2.8. Preference value matrix

Equation 13 is used to determine the value of preference matrix for all alternatives. The following is a calculation for the value of preference matrix for alternative A. Table 14 below summarizes the values for the value of preference for all alternatives.

- $V_A \approx \frac{D_A^-}{D_A^+ + D_A^-} \approx \frac{0{,}00905}{0{,}00961+0{,}00905} = 0{,}4849$
- ⋮
- V_E

Table 14. Preference value matrix

V_A	V_B	V_C	V_D	V_E
0,4849	0,5027	0,52	0,4791	0,5517

4. Conclusion

Supplier selection must be evaluated based on several criteria, not limited to just one criterion. Entropy and TOPSIS methods are able to provide optimal solutions to supplier selection problems in the procurement of specialty chemicals in XYZ companies. Based on the evaluation results of supplier selection criteria using the entropy method and TOPSIS it can be concluded that the C3 and C12 criteria are the most important criteria. These criteria are the conformity of the goods with the specified specifications and the speed criteria in responding to customer requests. While based on ranking using the TOPSIS method, five suppliers were selected as suppliers of specialty chemicals for XYZ companies. The suitability of goods with the specifications set (C3) and the speed in responding to customer requests (C12) criterion values are obtained from the value of a positive ideal solution matrix with values of 0.0380 and 0.0389. While the selection of suppliers E, is based on the final preference value of 0.5517. Therefore, so that the process of evaluating supplier selection through tenders is in line with company expectations, the company must consider criteria the suitability of goods with the specifications set (C3) and the speed in responding to customer requests (C12) for selecting suppliers.

In this study the entropy method is not able to provide an objective assessment to provide weight values for the overall criteria. However, these deficiencies can be covered by the TOPSIS method. Therefore, it is better that in the future the study includes fuzzy numbers to eliminate deficiencies that still exist in the entropy method.

References

[1] Dos Santos B M, Godoy L P and Campos L M S 2019 Performance evaluation of green suppliers using entropy-TOPSIS-F *J. Clean. Prod.* **207** 498–509

[2] Chauhan R, Singh T, Tiwari A, Patnaik A and Thakur N S 2017 Hybrid entropy – TOPSIS approach for energy performance prioritization in a rectangular channel employing impinging air jets *Energy* **134** 360–368

[3] Zyoud S H, Kaufmann L G, Shaheen H, Samhan S and Fuchs-Hanusch D 2016 A framework for water loss management in developing countries under fuzzy environment: Integration of Fuzzy AHP with Fuzzy TOPSIS *Expert Syst. Appl.* **61** 86–105

[4] Lima-Junior F R and Carpinetti L C R 2016 Combining SCOR® model and fuzzy TOPSIS for supplier evaluation and management *Int. J. Prod. Econ.* **174** 128–141

[5] Gupta H and Barua M K 2017 Supplier selection among SMEs on the basis of their green innovation ability using BWM and fuzzy TOPSIS *J. Clean. Prod.* **152** 242–258

[6] Arini D 2015 Analisis Pemilihan Vendor dengan menggunakan pendekatan metode fuzzy topsis di PT. Tripatra Engineers and Constructors *J. Ilm. Tek. Ind.* **3** 53–8

[7] Budianto A G 2017 Pemilihan green supplier berdasarkan fuzzy AHP dengan metode fuzzy topsis *J. Tek. Ind.* **17** 84

[8] Kurniawan D E 2016 Sistem pendukung keputusan pemilihan pemasok bahan baku menggunakan metode technique for order preference by similarity to ideal solution *J. Integr.* **8** 56–60

[9] Maisari K D, Andreswari D and Efendi R 2017 Implementasi metode TOPSIS dengan pembobotan entropy untuk penentuan calon penerima bantuan siswa miskin (BSM) apbd Kota Bengkulu (studi kasus: SMAN 8 Kota Bengkulu) *J. Rekursif* **5** 179–94

[10] Lima Junior F R, Osiro L and Carpinetti L C R 2014 A comparison between fuzzy AHP and fuzzy TOPSIS methods to supplier selection *Appl. Soft Comput. J.* **21** 194–209

[11] Wood D A 2016 Supplier selection for development of petroleum industry facilities, applying multi-criteria decision making techniques including fuzzy and intuitionistic fuzzy TOPSIS with flexible entropy weighting *J. Nat. Gas Sci. Eng.* **28** 594–612

[12] Mavi R K, Goh M and Mavi N K 2016 Supplier selection with shannon entropy and fuzzy topsis in the context of supply chain risk management *Procedia - Soc. Behav. Sci.* **235** 216–225

[13] Beikkhakhian Y, Javanmardi M, Karbasian M and Khayambashi B 2015 The application of ISM model in evaluating agile suppliers selection criteria and ranking suppliers using fuzzy TOPSIS-AHP methods *Expert Syst. Appl.* **42** 6224–6236

[14] Wardhani I K 2012 Seleksi supplier bahan baku dengan metode TOPSIS fuzzy MADM (Studi kasus PT. Giri Sekar Kedaton, Gresik) Surabaya

IOP Conf. Series: Materials Science and Engineering **909** (2020) 012047 doi:10.1088/1757-899X/909/1/012047

Management of information systems, implementation and their importance in Albanian enterprises

V Prifti[1*], I Markja[2], K Dhoska[3] and A Pramono[4]

[1,2,3] Production and Management Department, Faculty of Mechanical Engineering, Polytechnic University of Tirana, Mother Teresa No.3, 1001 Tirana, Albania.
[4] University of Sultan Ageng Tirtayasa, Department of Metallurgy Engineering, Cilegon Banten 42435, Indonesia.

*E-mail: vprifti@fim.edu.al

Abstract. Management information system collects and processes data (information) and provides it to managers at all levels who use it for decision making, planning, program implementation, and control. In addition to the global studies of management information system, this paper will analyze similar cases in Albanian enterprises. Albanian enterprises strongly need management information systems and the facilities that they bring to their management ensuring a better functioning and higher performance of different enterprise processes. In this paper was taken the case study of the Microsoft Dynamics NAV system, the implementation in "Infosoft Systems" enterprise in Albania and the impact of NAV system in some of the enterprise departments. Implementation phase of management information system is one of the most important stages which starts with the organization of the information systems management department and ends with the procurement of the relevant software. To get more information we used a questionnaire which was designed and sent to some enterprises in Albania. The purpose of this questionnaire was to learn how many of them use management information systems, what type of information system they use and the advantages or benefits that the implementation of information systems brings to enterprises in Albania. As a conclusion, management information systems have contributed to enterprises to put in place better functionality, which leads to increased enterprise performance, increased profits, simplified work processes in different departments.

1. Introduction

A management information system (MIS) is a computerized database organized and programmed in such a way that it produces regular reports on operations for each level of management in an enterprise. [1]. Also it can generate easily special reports from the system.

The main objective of MIS is to provide managers feedback on their performance so that managers can monitor the enterprise. The information displayed by MIS usually shows the "real" data versus the "expected" results and the results from previous year, it measures progress against goals. MIS receives data from enterprise units and functions. Some of the data are automatically collected from computer-related check-out counters and other data are collected at periodic intervals. Daily reports are pre-programmed and run at intervals or as per request, while others are executed using query language construction. Many advanced systems also monitor and display enterprise stock performance. The information system is an agreement between human, data, processes, and interfaces, interacting to support day-to-day operations in an enterprise, as well as support to solve problems and decision-making needs of managers and users. Also the information system can be described as a group of tools and methods that provide managers information for making decisions and for supporting different operations in an enterprise. Referred to literatures we have defined information systems as a set of different components that store, collect, disseminate and process information to support in an enterprise managers for decisions, job coordination, and also process control. Within the enterprise or in the environment in which it operates, information systems contain important information about things, people and locations. Information means data that is shaped in order to make sense and to be usable by humans.

In the enterprise, one of the key roles, from enterprise information systems is to provide applications for effective support to enterprise strategy. In the global market, this is done by developing services, capabilities or products that give the enterprise great importance over the competitors. In the enterprise, different information systems that are developed for supporting strategy decisions, can shape their position regarding the competitors and strategies. Any type of information system that help the enterprise with different decisions for the strategy, such as how to deal with competitive advantage or disadvantage and many other objectives from a stategic point of view, can be a strategic information system (example DSS, MIS, etc.).

2. Literature review

2.1 Classification of Information Systems (IS) and their fields

Karim (2011) stated that Management Information Systems (MIS) is the key factor to facilitate and to have in an organization an efficient decision making. His research explores the extent to which management information systems implemented to make successful decisions at two selected financial organizations.

Knight Moore (2005) and Fabunmi (2003) have shown that the limitations and deficiencies in the process of management information system performance are the main reason for diminishing the efficiency of decision-making process in the organization.

Whitten et al. (2004, p.12.), stated that "information is an arrangement of people, data, process, and information technology that interact to collect, process, store and provide as output the information needed to support an organization". The research indicates that information system is an arrangement of data, processes, groups and technology that act together to accumulate, process, store and provide information output needed to enhance and speed up the process of decision making. The case was in a Bank's information system.

Laudon and Laudon (2003) have defined MIS as "the study of information systems focusing on their use in business and management".

Baskerville and Myers (2002) have examined the MIS field and found a constant shift of MIS research from a technical focus to a technology-organizational and management-social focus.

Skyrius (2001) underlines the decision maker's attitudes towards different factors influencing the quality of business decisions; these factors include information sources, analytical tools, and the role of information technologies.

Liu and Young (2007) have defined key information models and the relationships in business decision support in three different scenarios. The authors proved that global businesses are in advance due to the Enterprise Applications System provided by modern IT tools such as Enterprise Resource Planning (ERP), Knowledge Management Systems (KMS) and Customer Relations Management (CRM) to enhance the efficiency and effectiveness of the process for decision making.

Becta (2005) describes an information system as "a system consisting of the network of all communication channels used within an organization".

Adebayo (2007) explained that the existence of MIS is needed to improve and enhance decision making on the issues affecting human and material resources.

Handzic (2001) has examined the impact of information availability on people's ability to process and use information for short/long term planning and in decision making. He stated that the better the availability of information, the better the impact on efficiency and accuracy of business decisions. [10]

The literatures also explore the importance of MIS in providing decision makers with facts, which consequently support and enhance the entire decision-making process. Furthermore, at the most senior level, MIS and DSS supply the data and required information to assist the board of directors and management levels to make an accurate and on time strategic decisions.

From the literatures presented, we can easily perceive that the importance of the role of both middle and top management to maintain a consistent approach to develop, use, and evaluate MIS systems within the institution. To financial institutions, MIS is used at various levels by top-management, middle and even by the operational staff as a support for decision making that aims to meet strategic goals and strategic objectives. [1]

2.2 Classification of IS based on Characteristics

In decision making in enterprises information used is categorized into three types:

• Strategic Information

Strategic information deals with the objectives of an enterprise with long-term policy decisions and controls whether these objectives are met to their level or not. Such as, buying a new plant or product, business diversification, etc.

• Tactical information

Tactical information deals with the information needed to control enterprise resources, such as budgeting, inventory level, productivity level, service level, quality control, etc.

• Operational information

Operational information can handle the proper performance of different operational tasks in the way it was planned or assigned to the various specific operators, special machines and specific job relocations for quality control operations.

2.3 Classification by Application

In terms of applications, the information can be categorized as:

• Planning information

This is the information used in enterprises for maintaining standard norms and specifications. Everything is planned here. This information is used in the strategic, tactical planning, and operation of any activity. Examples of such information are time standards, design standards.

• Control information

This information is needed for specific activities carried out by the system in order for enterprises to achieve their objectives. This information can be formal or informal and is used to control the achievement of objectives, the nature and use of important processes in a system. When this information has received any deviation from the established standards, the system must develop a decision that leads to checking the data regarding their objectives.

• Knowledge Information

Knowledge information is gained through experience and lessons learned from archival data and research studies. Knowledge is defined as "information about information".

• Organizational Information

Organizational information deals with the environment of an enterprise, where organizational objectives are met. An enterprise reduces its uncertainty by carefully collecting, managing and using this information. Everyone in the organization can use the information.

• Operational Information

This is specific information where the enterprise helps to perform its transaction functions. Mostly there are technical jobs for example, daily schedules at a manufacturing enterprise that refer to the detailed assignment of jobs to machines. This information is internal enterprise information mostly.

• Database of Information

The database of information is a collection of related data that is stored and managed to form databases. It stores large amounts of information that has multiple uses and applications. For example, for multiple users can be stored supplier information or the material specification.

3. Implementation and selection of Information Systems IS in an enterprise

The implementation plan is needed to help the user community understand the timeframe for installing the new system. Communicating the implementation plan to the user community helps users to prepare for change and makes them mentally prepared for it. The communication process can be complex in several stages. The senior can communicate the overall purpose of the new system and

then detailed information on staff can be left to the head of division. The communication process indicates the role that each employee must play in the implementation process. It includes the following steps:
1. Create a plan for different activities and their implementation
2. Setting deadlines for different activities with can be critical or not
3. Identification of major obstacles and their solutions
4. Communication of plan

3.1 Hardware Selection and Procurement
This is a very important step of the implementation process because it has to do with a lot of investments. The process of selecting and procuring equipment varies a lot from enterprise to enterprise depending on the size and the sector in which it operates.

After analyzing the experience of the enterprise sales managers with different vendors can be prepared the list of vendors or it may have been prepared on the basis of some vendor-accepted lists prepared by an organization or a regulatory body. Then is prepared the RFP by the implementation team based on the hardware of the new system. RFP must have complete technical details and information of hardware systems that are required, including here the format, specifications, performance expectations, also requirements for the warranty and service quality.

Once the RFP is prepared, it is sent through various communication modes to the simple vendor selection. The means of communication may be through an open advertisement in the electronic media, or it may be in the form of a letter to the vendors with a deadline for submission of the proposal.

Evaluation of the RFP is a very difficult process. Before the deadline, once the bids have been received, is done the preliminary check for fundamental errors. There are several evaluation methods. Evaluation can be cost-based or quality-only, or it can be a mixture of both cost and quality. Typically, a rating system which is results-based is used to rank vendor proposals. Then, based in this rating, a single vendor or a group of vendors are selected for shipping the hardware. With the selected group of vendors are held price and contract negotiations and then is signed the final contract after successful completion of negotiations. [5]

3.2 Software Procurement
The procurement of the system software is done in similar ways as procurement of hardware equipment. The only difference in the case of software procurement is that the choice of what software to support is already made in the design phase of system development and thus, the process of preparing the RFP is straightforward.

3.3 Database Creation
The new system to be implemented will have data records. Data storage is done with the database. These databases are relational database management systems which is a separate package of software applications.

3.4 User Training
The new system may not have a good use after is being installed, without proper user training. A bigger concept is implementation which is focused on installing and maintaining the transition process. A user training program is planned.

3.5 Creating physical infrastructure
Implementation of the new system may require physical infrastructure. The implementation team should ensure that system performance is not affected due to infrastructural difficulties.

Normally, since the new system is installed, the new system and the old system are both usable for a period to ensure that the company's performance will not have problems due to the transition. Later, when users have acquired more capabilities to handle the new system then the old system is discarded.

4. Case study-IS as strategic tool and implementation in an Albanian enterprise

NAV (Microsoft Dynamics) is an ERP (enterprise resource planning) application from Microsoft. The product is part of the Microsoft Dynamics family. The product aims to help finance, manufacture, manage customer relationships, supply chains, analyze and e-commerce for different size of enterprises and local affiliates from the large international groups. Microsoft announced in 2003 plans to develop Project Green which was a completely new ERP system. Later Microsoft decided to continue developing many ERP systems (Dynamics NAV, GP, SL AX). The four ERP systems that were lauched were with the same new role-based interface, SQL-based reporting and analysis, portals SharePoint-based, mobile clients Pocket PC-based, and with the integration with Microsoft Office. The product has changed name some times. "Navigator" was used originally in Denmark. Internationally it had the name "Navision". The name "Avista" was used in US. As of 2014 its name was "Microsoft Dynamics NAV". Prior to NAV 2013, Microsoft Dynamics NAV gave administrators the ability to use a database or Microsoft SQL server, such as DBMS. Now the exclusive database option for NAV is the SQL Server. Reporting of documents is based on the RDLC 2008 format. The reports are partially edited in the NAV Development Environment and in Visual Studio. NAV 2013 R2 includes a free report editor. Each report will be presented in preview screen formats, PDF, Word or Excel, depending on users' needs.NAV 2013 also supports OData format. With OData support in NAV 2013, Excel pivoting can now be done without knowledge of SQL specifics, restricted to those fields only available for RTC views.

Microsoft, with NAV 2009 introduced a whole new client interface which was called the Role Tailored Client. The RTC allows the customization of the NAV experience, based on the individuals job responsibilities and uses tool pages/menus (Profiles) and homepages (Roles). In a person's offices this can be a disadvantage for users have to re-login, change their role, or configure a different username to access the different RTCs will use. In multi-person offices when trying to help individuals who may have different screen layouts and also settings, lacking either of these options can lead to confusion. For users or groups are assigned roles. Individual users can customize their toolbar and navigation panel, or administrators can customize layouts for all users on a specific Profile; then they can disable individual customization for users in that profile. Some levels of customization are only available through the NAV Development Environment. The NAV client interface previously available in versions 5 and older was retained at NAV 2009, but renamed to Classic Client, making NAV 2009 the only "hybrid" version, offering both the Classic Interface and the RTC. Both Native and SQL databases are supported from the Classic Client. A SQL database is required form The RoleTailored Client. Additionally, SQL database identifications are not supported with the RoleTailored Client. In October 2012, was released NAV 2013, and was terminated support for Classic Client. Has been renamed RoleTailored Client with the name Windows Client. A built-in Web Client and SharePoint Client were added also. They are not required any special add-ons from the Web Client and it works on computers and mobile devices. and There are still available Repository building and database access that were previously available with Classic Client.

In Microsoft Dynamics NAV 2013 Web Client in relation to other Microsoft 3 ERP products, the Dynamics NAV sector is for small distribution and manufacturing companies. All product sales are via resellers approved from Microsoft who base their entire business on how many consulting hours can they do for any given installation, so only some installations have been "out of the box". The solution has a standard set of features. It can also be thought of as an "ERP System build set" if, at the end of the installation, you want to end up with any builder patch that was in the box attached to your system box. Development language like Pascal is easily accessible to the right developers and is designed for fast software customization.

In 2019, NAV reached 2.7 million users and 3500 certified partners worldwide. Dynamics 365 Business Central, which currently is a pure cloud application, will when it is combined with Dynamics NAV also be able to run on it's own server. With NAV 2018, we have probably seen the last version of Dynamics NAV and Dynamics NAV 2019 will instead go under the name Dynamics 365 Business Central on-premise.

Microsoft Dynamics NAV provides integrated functionality to provide support for: management of supply chain, manufacturing, distribution, managing customer relationships, sales and marketing, service management, management of human resources, project management, resource management, warehouse management, financial Management

4.1 Architecture of the software

NAV software consists of three main components. First a database server, a database that stores Microsoft Dynamics NAV data. Then the application server, a service that controls all aspects of the workflow of NAV. Also the client, user interface in Microsoft Dynamics NAV. NAV 2015 introduced a Client Chart and NAV 2016 introduced a Phone Client, Tablet Client.

Microsoft Dynamics NAV uses only one concurrent user licensing model. With NAV 2013, Microsoft introduced a new licensing model called "Permanent Licensing", which significantly simplifies the pricing structure. With Permanent Licensing, clients license the solution functionality and is provided access to that functionality by user licensing. User licenses are of two types: Full user or a heavily discounted limited user. The full user has access to the entire system, where as the restricted user only has read access to the system, in addition to the main writing tables, such as time sheets, storage and commentary, plus any three additional tables selected.

During a 3 month period we studied the Infosoft Systems enterprise. Infosoft Systems works with an ERP and specifically with Microsoft Dynamic NAV. Each department has access to the relevant NAV session, which contains updated daily data on the basis of work activity. This data needs to be converted to pdf in order for the application to accept the upload of the prepared files. After preparing the technical-economic offer, the document was converted to pdf. Having access to NAV to the Sales-Marketing & Sales session. We put in the Sales Manager's previously-designated identification code, code named ASL, and uploaded the bid to the NAV system. The document is left unapproved because the approval is given by the company manager or administrator. The technical-economic bid for a specific tender was easily identified in the NAV through the ASL code, without the need for additional additional data.

Through Microsoft Dynamics Certified Partners is available Microsoft Dynamics NAV. These value added resellers offer planning, implementation, customization and support services aimed at optimizing the solution to the specific needs of each customer., customers are entitled to purchase Microsoft Dynamics NAV licenses in advance regardless of how and where the software is located, by paying a monthly fee to the service provider. Since Infosoft Systems is a Microsoft Gold Partner, it has gained access to Microsoft Dynamic NAV.

There are two models of licensing:

1-Continuous licensing

2-Service Provider Subscription Licensing

Both models are designed to simplify the process of acquisition. Customers have the right to choose from two types of concurrent users, restricted and full users, and the option to give these users access to advanced functionality through the Extended Pack.

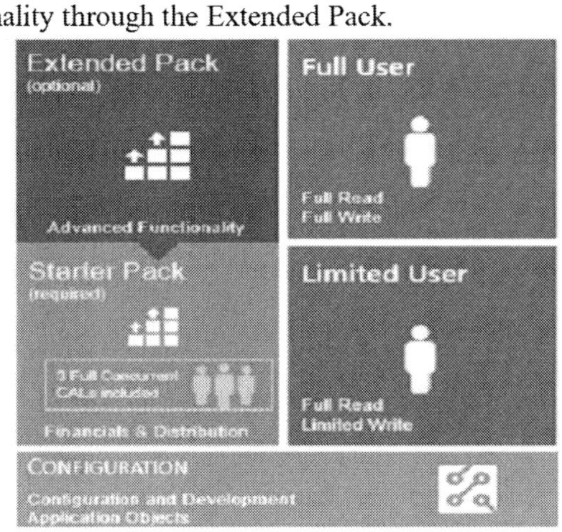

Figure 1. Continuous licensing, customization
(Source: Microsoft Dynamics NAV 2017 Licensing Guide | October 2016)

4.2 Initial and advanced NAV packages:

4.2.1 Initial Package.

Small and medium enterprises can quickly and cheaply use Microsoft Dynamics NAV. The basic startup package offers basic financial, distribution and professional services, plus three Full User licenses for a price. In the startup package, the functionality included is designed to help clients to gain control and insight into their finances and buy and sell items and materials. It also help them to pay and manage employees and also managing invoices generated by consultants. CRM deliver the highest quality of service and support and manage customer and supplier relationships.

4.2.2 Advanced Package.

Enterprises that require more advanced features can license the optional and add-on package. The advanced package enables clients to integrate financial management and distribution with additional features such as production, to support and control the production environment. Also storage, to manage the warehouse that supports operations.

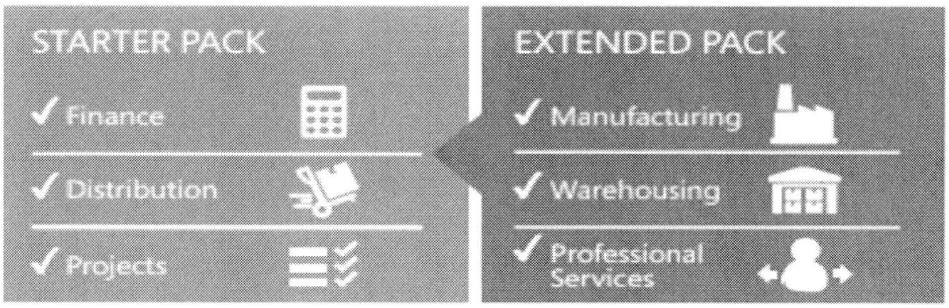

Figure 2. NAV packages, starter pack, extended pack
(Source: Microsoft Dynamics NAV 2017 Licensing Guide | October 2016)

Infosoft Systems enterprise in Albania, uses the Advanced Package and has chosen a subscription for a monthly fee. Implementation of the NAV took more than 6 months, while testing lasted up to one year. Once successful, the old system was put out of service. The NAV update is done every 3 months automatically. [9]

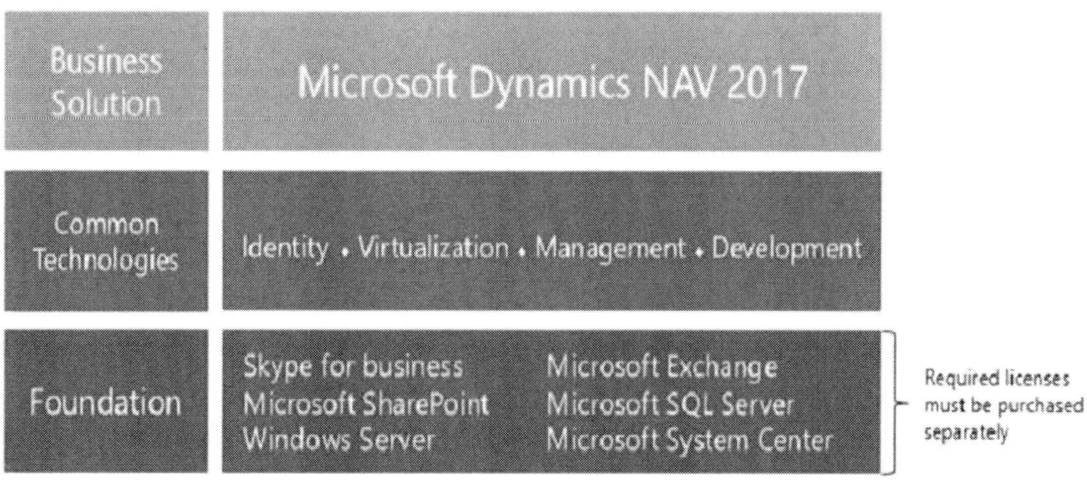

Figure 3. Licenses for additional software are not included with Microsoft Dynamics NAV licenses
(Source: Microsoft Dynamics NAV 2017 Licensing Guide | October 2016)

Licenses for additional software that may be required for the solution—such as Microsoft Windows Server, Microsoft SQL Server, and Microsoft SharePoint Server.

4.3 Impact of NAV in the Infosoft Systems enterprise

4.3.1 Account Planning

It is a powerful financial reporting tool that helps accountants and financial analysts extract the most essential business data from charts of accounts, budgets, cash flow accounts and cost types in financial reports, which are used to effectively monitor business situation and make a valuable contribution to business decision makers.

4.3.2 Selling & Marketing

Organize campaigns based on the contact segments the company defines. Define segments based on specific criteria, such as sales, contact profiles, and interactions and reuse existing segments or segmentation criteria. The Merge to Word function or other file formats is used to communicate with contacts in your segment.

4.3.3 Human resource. NAV efficiently manages the company's human resources

Relevant employee information is grouped and "tracked" and employee data is organized according to different types of information, such as experience, skills, education, training, etc. Personal data is stored, new job openings in the company are tracked and a list of potential candidates for these positions is drawn.

4.3.4 Workflow Continuity

In the event that for a significant sale payment to be received by two people then workflow continuity focuses on Microsoft Dynamics NAV on three key built blocks and short or long workflow processes. The steps associated with these three blocks are: Approval, Notification, and Process Automation.

4.3.5 Calendar

Calendars with working days and non-working days are set. Calendar entries will be used for date calculations for sales orders, purchase orders, transfer orders, service orders.

Benefits of NAV in the enterprise
Proven success in operating the enterprise
Allows greater control of the enterprise
Helps increase margin
Flexible and scalable
Easy to use
Reduce data entry time
Speeds up processes

Figure 4. Benefits of NAV in the enterprise

4.3.6 *Forecasting cash flow*

The cash flow forecast provides a forecast of how the company's liquidity and other treasury positions will evolve over time. It consists of bills and cash payments, money expected to be received and money expected to be paid, plus the liquid funds available. These elements together show a straightforward cash flow forecast.

5. Survay related with the implementation of Management Information Systems in enterprices in Albania

In this paper, we studed how enterprises in Albania use Management Information Systems in their enterprise operations in order to succeed in their markets.

A survay was developed with some enterprises in Albania. The questionnaire was developed with representatives of human resources and management level of these enterprises. The purpose of this

questionnaire was to learn how many of them use information systems, what kind of information systems they use, and the advantages they bring to enterprises which they have implemented.
The answers from the survey show that:
1. In total we received responses from 22 enterprises in Tirana, Albania.
2. 27% of enterprises are small enterprises, 53% are medium-sized enterprises and 20% are large enterprises (based on the number of employees: 1-9 employees: small-enterprise, 10-80 employees: medium-sized enterprise and with more than 80 employees: large enterprise.
3. 42% of enterprises belong to the telecommunications sector, 35% to the marketing sector, 16% to the tourism sector, 7% to other sectors.
4. 98% of enterprise have at least one computer in use
5. 82% of enterprises have internet access
6. 25% of enterprises use at least one ERP / CRM
7. Those who use ERP / CRM say that there are many advantages these systems bring such as: enterprise growth, facilitating enterprise processes, flexibility in data entry, better organization of jobs within a departament and fuller control over the enterprise.
8.Small enterprises do not see the implementation of ERP / CRM as necessary.
9. 25% of medium-sized enterprises have thought about implementing it but it a matter of time and finances until it is realized.

5.1 Results of the Graphic Questionnaire

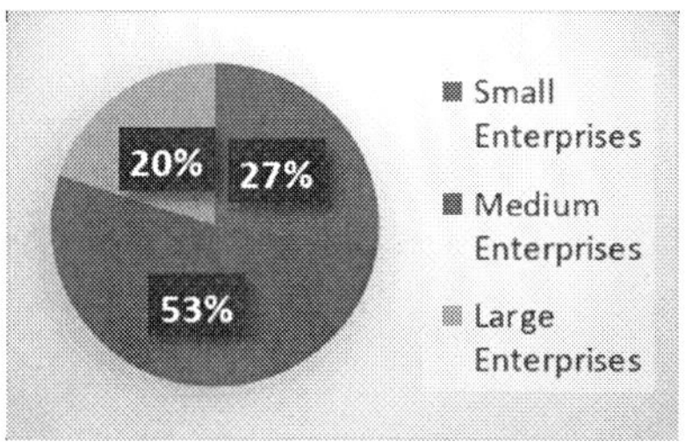

Figure 5. Percentage of enterprise types "

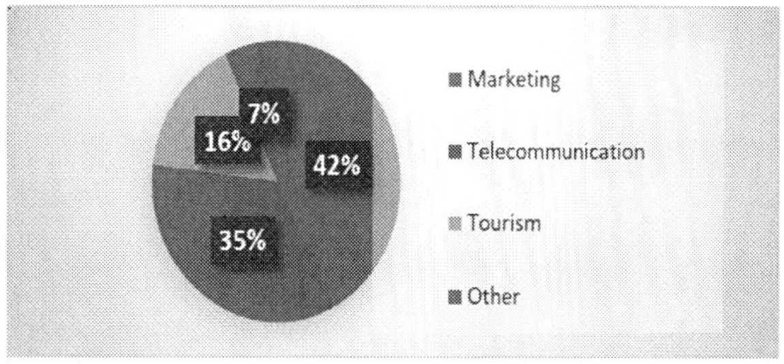

Figure 6. Percentage of enterprise sectors "

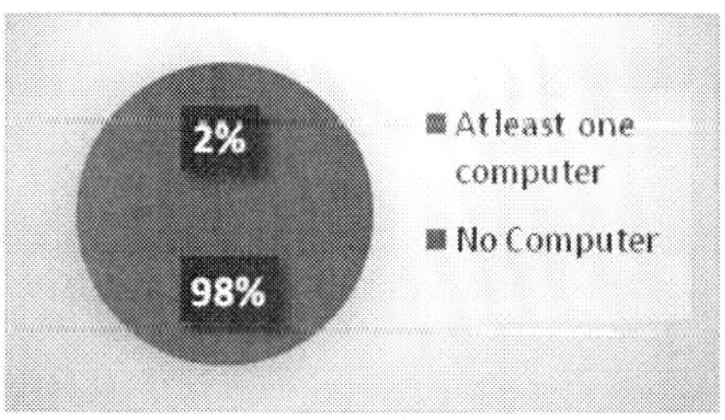

Figure 7. Use of computers '

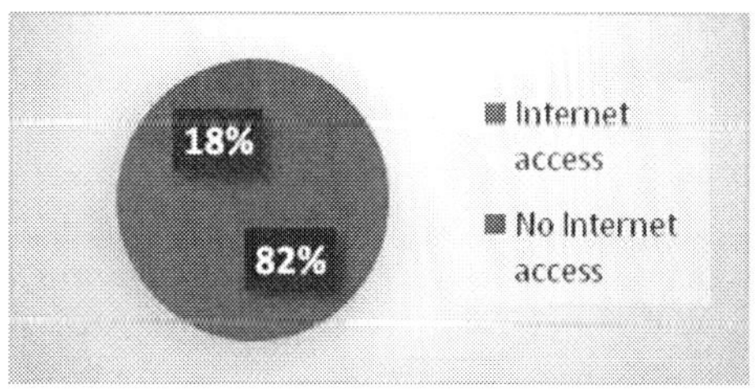

Figure 8. Internet access

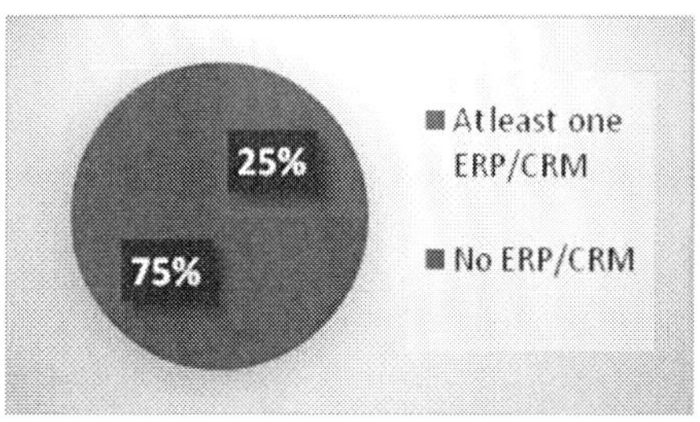

Figure 9. Use of ERP or CRM '

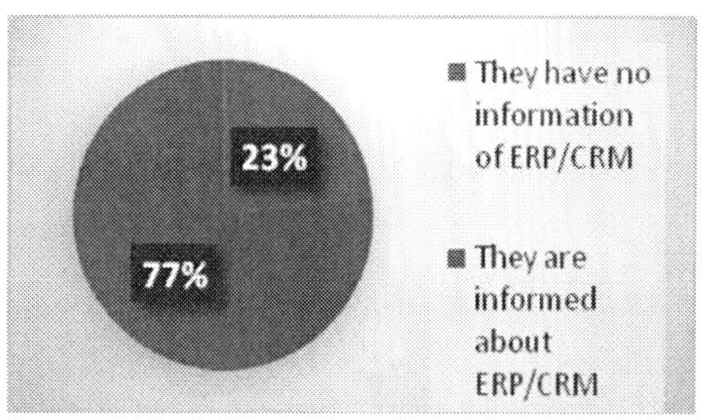

Figure 10. Information for ERP and CRM '

6. Conclusions and future work

Management of information systems brings benefits associated with achieving a higher level of efficiency and is essential in improving the performance of each enterprise. Managers have the information needed to identify enterprise strengths and weaknesses. It also improves the quality of decisions. Good information availability reduces uncertainty and allows managers to make more rational decisions based on reliable data. Communication between departments in a workplace is better. When managers, department heads, and employees share the same information, there is better communication between them to identify problem areas and find acceptable solutions.

Management is able to consider different alternatives to see possible outcomes before making decisions and commitments. Employees are more productive because they do not have to spend time collecting the data that the manager wants. The enterprise is more efficiently running by reducing and eliminating weaknesses and problem areas, thereby increasing the competitive advantage of the company over its rivals, and enhancing the competitive advantage of the company. With more data on customer needs, management is better able to improve customer service, marketing effectiveness, and promotional campaigns.

The need and importance of information systems in enterprises is increasing day by day in Albania. Using information systems in different businesses is turning into a strategy that is changing the way enterprises today think and operate. There are many advantages of implementing and using management information systems. They have helped enterprises for a better functionality, which leads to increased enterprise performance, increased profits, facilitated work in different departments (for mid-size and big enterprises).

From the survay with enterprises in Albania we saw that only a small percentage of enterprises use management information systems and this is due to two main reasons:

1. The high initial cost
2. Lack of information and uncertainty in the functioning of the management information systems

However, despite these two factors, Albanian enterprises are moving towards the trend of foreign enterprises when we take into account the use of information systems. Based on the main components of management information systems, starting with the basic ones, such as the computers, its use has become a huge necessity even for small enterprises that own at least one computer to perform different operations.

As in small, medium or large businesses, the advantages that management information systems brings are numerous, helping them grow day by day, it is certainly worthwhile to recommend new entrepreneurs to learn more about management information systems, understand the importance that they carry and to implement those systems in their enterprises.

For enterprises that have already implemented and operated a specific information sysem it is recommended that they develop more training on the latest management information systems updates, always with the primary objective of improving the current system they use.

In the future, we can continue the reseach with different surveys regarding the managent information systems, the different technologies that enterprises use and how they are implemented, data management, big data and their advantages and internet of things.

References

[1] Karoline B O, Julie R Olsen, Polyxeni V and Eli H 2018 ERP Systems in Multinational Enterprises: A literature Review of Post-implementation Challenges *Proc. Comp. Sci. Science Direct* 541 – 548

[2] Olga V V, Anna A K, Irina P F and Tatiana V B 2017 Implementation of an information management system for industrial enterprise resource planning *Rev Espacios* 23

[3] Fiona F N and Janet L L, 2015 Critical factors for successful implementation of enterprise systems *Business Process Management J.* 285

[4] Ermelinda K T IT 2011 Service Companies in Albania and Implementation of Information Technology Service Management *International J. of Computer Applications* 7-11

[5] Kozeta S and Ermelinda K T 2011 Information Systems and Information Technology as Strategic Tools - Their Use in Albanian Business *J. of Knowledge Management, Economics and Information Technology* 1-23
[6] Nebojsa D, Nebojsa Z and Boris S 2013 *Management of the Information Systems Implementation Project Annals* (Oradea University) 32-35
[7] Shari Ch and Tim S 2004 Managing User Resistance in Enterprise Systems Implementation Tenth Americas *Conf. on Information Systems* 149 – 153
[8] Katica S, Goran S, Sara H, Danijela P and Ilija S 2017 *The Role of ERP System In Business Process and Education* (Tehnički vjesnik) 711-719
[9] Microsoft Corporation 2016 *Microsoft Dynamics NAV 2017 Licensing Guide*
[10] Akram J K 2011 The significance of management information systems for enhancing strategic and tactical planning *J. of Information Systems and Technology Management* **8** 459-470

IOP Conf. Series: Materials Science and Engineering **909** (2020) 012048 doi:10.1088/1757-899X/909/1/012048

The implementation of lean construction and six sigma concepts in light brick installation: A case study in Cordova apartment project

Naniek Utami Handayani[1*], Mochamad Agung Wibowo[2], Anita Mustikasari[3], Ilham Wahyu Nurwidanto[1], Drajat Akbar Dilaga[1]

[1] Department of Industrial Engineering, Faculty of Engineering, Diponegoro University, Semarang, Indonesia 50275

[2] Department of Civil Engineering, Faculty of Engineering, Diponegoro University, Semarang, Indonesia 50275

[3] Department of Management, Faculty of Economic, Yogyakarta State University, Yogyakarta, Indonesia 55281

*Email: naniekh@live.undip.ac.id

Abstract. PT Adhi Perkasa Gedung is a construction company engaged in building construction that is a subsidiary of PT Adhi Karya (Persero). Sustainability has become a significant issue this decade, as well as construction projects. Construction companies are required to be able to manage their resources effectively and efficiently through the completion of environmentally friendly projects. PT Adhi Perkasa Gedung as the construction contractor for the Cordova Apartment, strives to minimize waste in light brick installation. The light brick is used as a bulkhead between rooms in the apartment. The problems that occur are the waiting time of light brick procurement, product defects, material handling, and installation process errors. The purpose of this study is to minimize waste in light brick installation using lean construction and six sigma methods. The results of this study indicate the existence of rework due to installation process errors, material quality.

1. Introduction

The Industrial Revolution 4.0 has required that many companies or industries perform their activities effectively and efficiently including companies engaged in construction. One of the industries that has a important impact toward the country is construction industry [1]. Based on Global Construction, the construction industry is intended to put up around 15 % of GDP by 2020 and is estimated in the next few years giving annual growth rate of 4 % [2]. Meantime, compared to other industries, many criticisms are directed toward the construction sector such as wasteful, being inefficient, and a high-risk industry [3]. Those problems on construction sector can be traced to the industry characteristics such as supply chain fragmentation, and also in project management issues like time overruns, cost overruns, safety issue and quality problems in construction project [3]. Whereas, construction companies are required to provide the best service to consumers. Construction services can be measured by delivering a product on time with agreed specifications and good quality. Also, a

IOP Conf. Series: Materials Science and Engineering **909** (2020) 012048 doi:10.1088/1757-899X/909/1/012048

successfull projects is when the project completion is efficient in terms of time and cost as well as achieving work efficiency.

PT. Adhi Persada Gedung is demanded in giving products based on customer expectations. To provide quality products at competitive prices and to manage their resources effectively and efficiently through the completion of environmentally friendly projects become one of their strategies. Cordova Apartment is one of PT. Adhi Persada Gedung Projects. In the development process, there are still various identified waste activities in terms of time and cost. The problems that occurred in the Cordova Apartment Projects is the light brick installation. Waiting time for light brick procurement, products defects, excessive inventory, inappropriate processing, and excessive transportation are several waste which often happens.

There are various methods that can be applied to reduce waste on every activity. One well-known method used is the Lean Construction and Six Sigma method. This method has a function to identify and minimize waste in an activity through a process of continuous improvement. Based on previous research, lean construction and six sigma have advantages in terms of financial improvement, quality improvement, and benefits for companies and consumers. Therefore, this study was conducted to minimize waste in light brick installation in Cordova Apartment using lean construction and six sigma methods.

2. Research Methods

2.1. Lean construction

The lean construction is first introduced by the International Group for Lean Construction in Finland in 1993 [4]. Koskela (2002) state that lean construction is approach to reduce materials waste, time, and effort to bring up the maximum value of material, time and effort. Other researchers defined that lean construction has objective to minimize waste, developing productivity, safety and health, and raising value in order to fulfill the costumer's needs [4]–[6]. Lean construction can be said as a production management strategy that has function to assure continuous improvement through dispose of waste of resources that are not meet the customer's need or do not add value [7], [8]. Previous research found that lean construction can set productive flows in motion to develop control systems with the aim of reducing cycle time, raising productivity and revenue and also maximize workforce encouragement during the project life cycle [9]. Nahmens et al. (2011) prove that lean construction has the positive impact on the sustainability such as well-performance of economic, social, and environment[10]. Also, another research show that the lean construction tools such as 5S methods and Last Planner System can significant improve the safety management [11]. In Egypt, lean construction can cut down 15.57 percent the completion time of a project [7]. Lean construction also has a significant impact on reduce construction period, develop project performance rising from lean construction principles, effective coordination and communication, increase the workforce productivity, and reduce reworks and zero value-adding activities [5], [12]. While, The big serious barriers toward lean construction are related with factors of leadership, the company culture, lack of knowledge in adapting the new system, and employee resistance [9].

2.2. Six sigma

Six sigma is a profit-maximizing technique achieved by meeting consumer satisfaction. It was first successfully implemented by the cellular company, 'Motorola' in 1986 and since then, other companies have also adopted it [13]. This technique follows a statistical approach and it is attained by minimizing variations and providing quality services or defect-free products. By reducing deficiencies, fewer materials are wasted and hence an optimal amount of raw materials is completely utilized. As six Sigma seeks to enhance the organization's capacities and satisfy the ever-changing consumer demands, it uses data in order to provide better solutions and statistically speaking, reaching the sixth sigma means that the achieved products or outputs have almost no defects. These are the kind of results that any organization would like to achieve [14].

Improving process performance and achieving high-levels of quality with identifying and disposing of the root causes of waste is becoming an objective of Six Sigma [15]. The Six Sigma has five phases that is Define, Measure, Analyze, Improve, and Control (DMAIC). DMAIC is particularly used to lead Six Sigma Implementation in order to reach company goal [16].

By imitating Six Sigma in construction, company can recognize and reduce waste that leads to customer satisfaction. Previous research also stated that by implementing six sigma in construction projects can increase customer satisfaction at the expected cost and can become business excellence [17], [18]. In addition, the term of continuous improvement and the DMAIC phase may enable company to develop customer values by delivering world-class projects [19], [20].

2.3. Lean six sigma

Manufacturing and other sector industry has known prevalent tool in order to develop operational excellence, namely Lean Six Sigma (LSS). The LSS method adopts the DMAIC phase of the Six Sigma and combined with Lean tools for improving the existing processes [21]. The LSS method is first introduced in the manufacturing sector [22]. After reaching a high achievement, many other sectors tried imitating this method to get better quality of their services and products [23].

There is several definition LSS based on previous researchers. According to Chen and Liu (2009), LSS method is a great strategy for employing a well-structured continuous improvement to effectively and efficient eliminate process variance and raise quality using statistical tools [24]. Laureani et all (2010) "LSS is a business improvement methodology that aims to maximize shareholders' value by improving quality, speed, customer satisfaction and costs: it achieves this through merging tools and principles from both Lean and Six Sigma"[25]. While, based on Burch et all (2016), LSS is a way that concern on improving and developing the finished products quality and continuous improvement by eliminating variation and reducing non-value-add activities [26][27].

The usage of the LSS method in the construction industry is started in 2000 [23]. Many researches has performed using the LSS method but the several did not perform the practical part in construction processes [28]. By eliminating waste, the LSS method can give process stabilization. In contruction project, the LSS could be used to eliminate variability between the customer expectations and his actual experience. Those variabilities are caused by many processes happened in construction such as project planning, material handling, materials supply chain, supplier, contractors and so on. Thus, the objective of Lean is to eliminate wastes and cycle times. In the other side, the aim of Six Sigma is to eliminate process variability. To sum up, based on those objective, the LSS method can provide a way to bring continuous improvement environment by eliminating process variability and cycle times and increasing quality [23].

2.4. Failure mode effect analysis (FMEA)

Failure mode and effects analysis (FMEA) is a technique used concerning on safety and reliability analysis. FMEA has an objective to define, recognize and reduce potential failures, error and problems in design, system, process, and service before sent to the customer [32,33]. Unlike other risk assessment standard, FMEA can evaluate the important potential risk. In addition, this method has significant effect in managing innovation on construction sectors.

Because the construction project is very expensive, the usage of FMEA will play a significant part in perform better. FMEA method recognize activities which can eliminate the chance of potential error and execute the implementation of activities in construction [31]. FMEA can be defined as a set of organized activities that are used to following goals [31]: recognizing and estimation of potential errors in a process or product or material or outcomes results from these errors; determination of activities which can eliminate probability of potential errors occurrences.

3. Results and Analysis

The research was conducted in collaboration with the Cordova Apartment. The research project has adopted the Lean Construction and Six Sigma method to identify or recognize the sources of waste to give the recommendations or suggestion for improvements. The Lean Six Sigma methodology is based on the five steps called the define, measure, analyze, improve and control (DMAIC) approach that are as follows:

3.1. The "define" phase

The "Define" phase concerns the identification of the problem. There are several waste that occurs in Cordova Apartment Develompment as shown on Table 1.

Table 1. The define phase

No	Waste	Description
1	Overproduction	PT. Adhi Persada Gedung does not produce light bricks but does build light bricks. In the construction process design has been set as a reference in the process of building light bricks, so there is no overproduction process.
2	Defect	The process of building the Cordova Apartment, damage that often occurs is a light brick broken or not according to size. The light brick broken is caused by the process of sending light bricks from suppliers to the project work site. *Critical to quality for defects* is the light brick has appropriate and do not break.
3	Unnecessary Inventory	The need for light bricks will be predicted by the procurement department every month using forecasting method. This method has a low accuracy so that the results is in excess inventory. Another result from excess inventory is the leadtime of light brick are longer so the impact is the light bricks damaged before installation, and increase in storage costs. *Critical to quality for unnecessary inventory* is product leadtime is not more than one month.
4	Inapropriate Processing	Workforces often engage in inefficient activities due to limited equipment and poorly trained workforces. Workforces often disobey work procedures (SOP) so the impact that result in the installation of light bricks is not passing the final quality test or must do rework. *Critical to quality for inappropirate processing* is not rework.
5	Excessive Transportation	The moving of materials and equipment from a storage area is still carried out quite a long distance. This results in an inefficient transportation system. *Critical to quality for excessive transportation* is a short distance in material moving.
6	Waiting	Waiting material often occurs in the work process of the Cordova Apartment project. *Critical to quality in waiting* is no more than 2500 minutes
7	Unnecesary Motion	Workforces at PT. Adhi Persada Gedung often rotates which causes understanding in light brick installation is not appropriate with SOP. In addition workforces often do work by smoking. *Critical to quality for unnecessary motion* is the workforce doing the installation appropriate with SOP.

3.2. The "measure" phase

The "measure" phase is about collecting the right data to visualize the current state of the process to help understand the causes behind the identified problem.

1. Overproduction

   In the installation of light brick is in accordance with SOP. Cordova Apartment has an area of 258.72 m^3 and requires 21473 units of light brick.

IOP Conf. Series: Materials Science and Engineering **909** (2020) 012048 doi:10.1088/1757-899X/909/1/012048

2. Defect
   Data on the number of defects that occur in light bricks at the Cordova Apartments in March to July 2019 is shown below:

Table 2. Defects number on March-July 2019

Defect Types	One lot	Defects	%Defects	DPMO	Level Sigma
Broken	3100	500	16,19	322580,65	2,00
Raw Material size is not appropriate	3100	300	9,67	193548,39	2,40

   From Table 2 can be seen that the defects with the lowest sigma level are in inappropriate size. It can be said that the defects are more frequent. Therefore, at the analysis phase will be identified why the light brick defects causes of the size is not appropriate.
3. Unnecessary Inventory
   There is an unnecessary inventory on the light bricks installation in the Cordova Apartment shown in the Table 3:

Table 3. Unnecessary inventory data March– July 2019

Month	Unnecessary Inventory	%Overproduction
March	150	4,84
April	100	3,23
May	170	5,48
June	130	4,19
Jule	140	4,52
Total	8400	13,5

4. Inappropriate Processing
   There is one type of inappropriate processing that is the rework process due to the light brick installation is not according to specifications. The rework that occurred due to defect products during March-July 2019 was 4 times dismantling with the number of light bricks which were dismantled 400 light bricks with one times the removal of light bricks equal to 75 minutes.
5. Excessive Transportation
   The excessive transportation in the project is as follows:

Table 4. Excessive transportation data

Start Position	Final Position	Distance (meter)
Material Storage	QC Department	15,09
QC Department	Light brick installation location	5,17

6. Waiting
   This measurement of this waste is when waiting the light bricks from the raw material warehouse to the Quality Control department to be inspected. One lot of the light brick which is 3000 light bricks and the waiting time for one light brick lot is one hour.
7. Unnecessary Motion
   There is several unnecessary movements of workforces such as smoking at work time, resting too long. It caused by workforces changes frequently.

3.3. *The "analyze" phase*

The analyze phase is carried out using the FMEA method. The use of the FMEA method is done to identify how the light brick, processes, or systems that exist in PT. Adhi Persada Gedung. The FMEA Table as shown below:

Table 5. The FMEA of light brick installation

Process Function Requirement	Potential Failure Mode	Potential Failure Effects	Sev	Potential Causes	Occ	Current Control	Det	RPN
Company policy	Excessive lead time	Holding cost is high	6	Forecasting error	5	Evaluation	6	180
Material procurement	Broken and an appropriate size material	Purchase of new material	6	Supplier selection error	2	Monitoring supplier	5	60
Installation process	Broke during installation	The product cannot be installed	9	Equipment and workforces	2	Light brick installation monitoring	5	90
Weathering	Uneven light brick wall	Rework	9	Oxygen levels are not right	3	Baking Monitoring	5	135
Material Handling	Material handling is too far	Increasing the production process time	6	Layout is not appropriate	10	Evaluation	6	360
Entire Production Process	Rework	Increasing failure cost	9	Defect	9	Increasing preventive cost	9	729
	Waiting is greater than the target	Increasing processing time and production costs	5	Machine breakdown	9	Uncontrollable	7	315

Based on FMEA identification, the biggest potential failure mode is rework, so analysis is needed using a fishbone diagram together with analysis to defect for broken bricks.

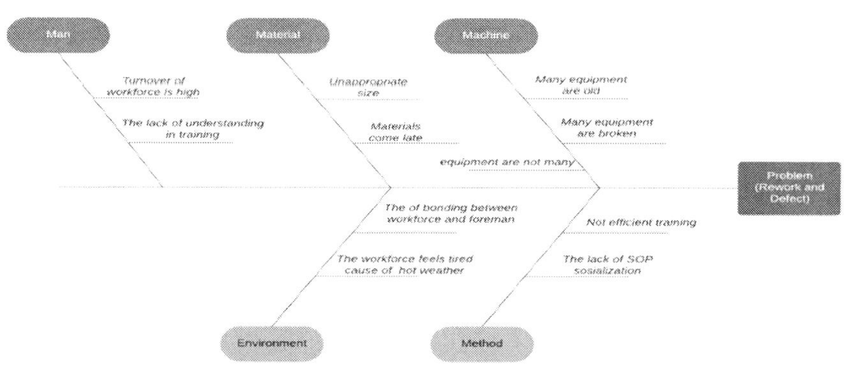

Figure 1. Fishbone diagram of rework and defect

Based on the analysis of fishbone diagrams, it is found that the rework caused by several factors. First, the equipment factor, the equipment is old, many are damaged and the numbers are small. Second, the human factor, many new workforces lack understanding in light brick installation training, the QC department should provide more treatment for workforces. Third is the material factor of the supplier, the material arrival from the supplier is sometimes late and the light brick size is not appropriate. Fourth, the method factor, lack of SOP socialization to workforces and less efficient training, the foremen and the QC department should be more agile to provide SOP light brick and improve training method so the rework is not happen. Last, the environmental factors, the lack of bonding the foreman and the workforce, foremen should create a comfortable situation toward the workforces.

3.4. *The "improve" phase*
The "improve" phase will try to find solutions and modifications in the process to eradicate the root. In this phase, we use PDCA (*Plan-Do-Check*-Action) to improve.

1. Plan
   PT Adhi Persada Gedung must arrange the installation of light brick planning carefully. The QC department have to inspect directly the light brick that comes from supplier. The inspection is based on the standard size of light brick which is 10x20x60 cm. Also, the QC department provides regular training for new workforces cause the turnover of the workforce is high.
2. Do
   PT Adhi Persada Gedung must prepare the light brick installation tools such as meter, rubber hammer, meter, thread, and water pass. The foreman conducts supervision for light brick installation workforces such as measuring using the meter and thread and stirring cement.
3. Check
   The foreman and the QC department do the controlling toward the light brick installation such as the straightness of the thread, the installation is according to the standard operation procedure (SOP) of the Cordova Apartment, and the density between glue in the marking line.
4. Action
   The foreman and the QC department write the report to the head of the Cordova Apartment concerning in occurrence happened in installing the light bricks including defects or reworking. Based on the report, it can be a future improvement. In addition, about the workforce, The Cordova Apartment must be strict on SOP, the workforce will be fired when do mistake 3 times in the light brick installation.

4. Conclusion
The conclusions of the research on the application of the lean six sigma method to the Cordova Apartment construction project are as follows:

1. DMAIC is conducted to identified the waste in light brick installation on Cordova Apartment. In the define phase, the VSM method is used to identify actions that are non-value added.
2. There is several waste in this case which is waiting, excess transportation, unnecessary inventory, unnecessary motion, and defects.
3. FMEA method is conducted to identify the biggest waste. Based on the FMEA, RPN value for a potential failure mode overproduction greater than safety stock is 168; leadtime exceeding the limit is 180; broken raw material or not in accordance with the size is 60; break during inspection is 90; uneven light brick is 135; the product moving disctance which far away is 360; rework is 729; waiting is greater than the target is 315. Based on the calculation of the RPN, the highest RPN is the rework. Fishbone diagram is used to identify the cause of the rework.
4. For the improvement, PDCA concept is suggested toward PT. Adhi Persada Gedung. In plan stage, PT Adhi Persada Gedung must arrange the installation of light brick planning carefully. In do stage, PT Adhi Persada Gedung must prepare the light brick installation tools. In check stage, the foreman and the QC department do the controlling toward the light brick installation. In action stage, the foreman and the QC department write the report to the head of the Cordova Apartment concerning in occurrence happened in installing the light bricks

Acknowledgement
This research was financially supported by PDUPT Grants from the Directorate of Research and Community Service (DRPM). Ministry of Research, Technology, and Higher Education, Republic of Indonesia.

References

[1] J. G. Sarhan, B. Xia, S. Fawzia, A. Karim, A. O. Olanipekun, and V. Coffey 2019 Framework for the implementation of lean construction strategies using the interpretive structural modelling (ISM) technique *Eng. Constr. Archit. Manag.*
[2] A. Francis and A. Thomas 2019 Exploring the relationship between lean construction and environmental sustainability: A review of existing literature to decipher broader dimensions *J. Clean. Prod.* 119913
[3] A. Tezel, L. Koskela, and Z. Aziz 2018 Current condition and future directions for lean construction in highways projects: A small and medium-sized enterprises (SMEs) perspective *Int. J. Proj. Manag.* **36** 267–286
[4] M. S. Bajjou and A. Chafi 2018 Lean construction implementation in the Moroccan construction industry *J. Eng. Des. Technol.*
[5] M. A. Marhani, A. Jaapar, and N. A. A. Bari 2012 Lean Construction: Towards enhancing sustainable construction in Malaysia *Procedia-social Behav. Sci.* **68** 87–98
[6] L. Koskela, G. Howell, G. Ballard, and I. Tommelein 2002 The foundations of lean construction *Des. Constr. Build. value* **291** 211–226
[7] U. H. Issa, 2013 Implementation of lean construction techniques for minimizing the risks effect on project construction time *Alexandria Eng. J.* **52** 697–704
[8] L. Song and D. Liang 2011 Lean construction implementation and its implication on sustainability: a contractor's case study *Can. J. Civ. Eng.* **38** 350–359
[9] A. P. Mano, S. E. G. da Costa, and E. P. de Lima 2020 Criticality assessment of the barriers to Lean Construction *Int. J. Product. Perform. Manag.*
[10] I. Nahmens and L. H. Ikuma 2012 Effects of lean construction on sustainability of modular homebuilding *J. Archit. Eng.* **18** 155–163
[11] A. M. Bashir, S. Suresh, D. Proverbs, and R. Gameson 2011 A critical, theoretical, review of the impacts of lean construction tools in reducing accidents on construction sites *Proceedings of 27th Annual ARCOM Conference, Bristol, UK, Association of Researchers in Construction Management* 5–7
[12] M. A. Marhani, A. Jaapar, N. A. A. Bari, and M. Zawawi 2013 Sustainability through lean construction approach: A literature review *Procedia Social Behav. Sci.* **101** 90–99
[13] L. B. M. Costa, M. Godinho Filho, L. D. Fredendall, and F. J. G. Paredes 2018 Lean, six sigma and lean six sigma in the food industry: A systematic literature review *Trends food Sci. Technol.* **82** 122–133
[14] B. R. Krishnan and K. A. Prasath 2013 Six Sigma concept and DMAIC implementation *Int. J. Business, Manag. Res.* **3** 111–114
[15] X. Zu, L. D. Fredendall, and T. J. Douglas 2008 The evolving theory of quality management: the role of Six Sigma *J. Oper. Manag.* **26** 630–650
[16] A. Cherrafi, S. Elfezazi, A. Chiarini, A. Mokhlis, and K. Benhida 2016 The integration of lean manufacturing, Six Sigma and sustainability: A literature review and future research directions for developing a specific model *J. Clean. Prod.* **139** 828–846
[17] K. Hussain, Z. He, N. Ahmad, and M. Iqbal 2019 Green, lean, six sigma barriers at a glance: a case from the construction sector of Pakistan *Build. Environ.* **161** 106225
[18] R. A. Stewart and C. A. Spencer 2006 Six-sigma as a strategy for process improvement on construction projects: a case study *Constr. Manag. Econ.* **24** 339–348
[19] M. Sinha, H. J. Harrington, F. Voehl, and H. Wiggin 2012 Applying TQM to the construction

industry *TQM J.*
[20] J. Ferng and A. D. F. Price 2005 An exploration of the synergies between Six Sigma, total quality management, lean construction and sustainable construction *Int. J. Six Sigma Compet. Advant.* **1** 167–187
[21] A. Sarkar, A. R. Mukhopadhyay, and S. K. Ghosh 2014 An outline of the 'control phase' for implementing lean six sigma *Int. J. Lean Six Sigma*
[22] B. Tjahjono *et al.* 2010 Six Sigma: a literature review *Int. J. Lean Six Sigma*
[23] Z. Dakhli, Z. Lafhaj, and A. Bos 2016 Experiencing lean six sigma in the French residential construction *Int. J. Lean Six Sigma*
[24] M. Chen and J. Lyu 2009 A Lean Six-Sigma approach to touch panel quality improvement *Prod. Plan. Control* **20** 445–454
[25] A. Laureani, J. Antony, and A. Douglas 2010 Lean six sigma in a call centre: a case study *Int. J. Product. Perform. Manag.*
[26] R. F. Burch V, L. Strawderman, and S. Bullington 2016 Global corporation rollout of ruggedised handheld devices: A Lean Six Sigma case study *Total Qual. Manag. Bus. Excell.* **27** 1–16
[27] V. R. Sreedharan and R. Raju 2016 A systematic literature review of Lean Six Sigma in different industries *Int. J. Lean Six Sigma*
[28] S. H. Han, M. J. Chae, K. S. Im, and H. D. Ryu 2008 Six sigma-based approach to improve performance in construction operations *J. Manag. Eng.* **24** 21–31
[29] Z. Wang, Y. Ran, Y. Chen, H. Yu, and G. Zhang 2020 Failure mode and effects analysis using extended matter-element model and AHP *Comput. Ind. Eng.* **140** 106233
[30] D. H. Stamatis 2003 Failure mode and effect analysis: FMEA from theory to execution *Quality Press*
[31] M. Bahrami, D. H. Bazzaz, and S. M. Sajjadi 2012 Innovation and improvements in project implementation and management; using FMEA technique *Procedia-Social Behav. Sci.* **41** 418–425

IOP Conf. Series: Materials Science and Engineering **909** (2020) 012049 doi:10.1088/1757-899X/909/1/012049

The manufacturing technology optimisation model: the crucial contribution of industrial mass-training in improving company performance

L.Sukarma[1], Franka Hendra[2,3,6] Riki Effendi[4] and E. Mohamad[5]

[1]Institut Sains dan Teknologi Nasional, Jakarta, Indonesia
.Jl. Kahfi II, Srengseng Sawah, Jagakarsa, Jakarta Selatan, Indonesia
[2]Razak Faculty of Technology and Informatics
Universiti Teknologi Malaysia, Jalan Sultan Yahya Petra, Kuala Lumpur
[3]Industrial Engineering,
Pamulang University, Jl. Surya Kencana no. 1, Pamulang,, Indonesia
[4]Department of Mechanical Engineering
Universtiy of Muhammadyah Jakarta.
[5]Universiti Teknikal Malaysia Melaka, Hang Tuah Jaya, 76100, Melaka, Malaysia.
Fakulti Kejuruteraan Pembuatan

Email: dosen01508@unpam.ac.id

Abstract : Fierce global competition, volatile changing market and the spread of advanced manufacturing technology have forced significant changes in the competitive arena of manufacturing industries. This manuscript presents the Manufacturing Technology Optimisation Model, which discusses the crucial role of industrial mass training in improving manufacturing company performance. This model provides insights toward the proper allocation of resources which will bring about optimum condition of improved company performance. Four lessons can be learnt from the model: (a) companies should focus first on improving their capability in effectiveness - doing the right things - before pursuing efficiency and adaptability; (b) the attainment of effectiveness can be accelerated by providing employees with on-the-job mass training on certain practices; (c) companies can continue to pursue the efficiency after they have excellent capability in effectiveness; (d) the pursuit of adaptability may simultaneously advance with that of efficiency, aimed at more responsive to customer requirements. A subsequent paper will demonstrate that selection of appropriate 'people-based' practices may accelerate improved company performance.

1. Introduction

Trends in competitive priorities, and hence strategies, change over time [1, 2]. As suggested by the principle of mass production, manufacturers in the 1960s favour cost reduction through the production of large quantities. Although pressures to reduce manufacturing cost are still huge up to this time, a low-cost production strategy is not itself capable of attracting and maintaining customers. In the post-industrial environment, the elements of competitive advantage have increased both in their variety and intensity [3].

This situation triggers the pursuit of excellence in manufacturing, which is characterized by an emergency paradigm of innovative production efficiency. In the course of time, methods to increase production efficiency without using buffer inventory, such as Total Quality Management (TQM), Just in Time Manufacturing (JIT), and Total Productive Maintenance (TPM) have been recognized as powerful tools for superior superior performance. These methods are best stated as person-based, steady, all-inclusive, continuous improvement [4], and are referred to as World-Class Manufacturing [5].

Based on a comprehensive survey of manufacturing management and improvement literature, Sukarma [6] developed a new perspective for achieving excellence in manufacturing or a new framework to the World Class Manufacturing (WCM). This framework combined principles of TQM, JIT and TPM and is complemented by a series of practices as well as methods for measuring and monitoring company performance aiming at attaining excellence in manufacturing [7]. The effectiveness of the WCM Framework had been confirmed empirically in [8]. It was revealed that the factory implementing TQM, JIT or TPM was superior to those who did not implement it, and the factory implementing all the paradigms would be completely superior to those who only used one or two approaches.es.

In order to assist manufacturers to obtain substantial benefits from the implementation of the WCM Framework, this manuscript presents a critical analysis of the Manufacturing Technology Optimisation Model. The model provides insights toward the crucial role of human resources in improving company performance, vis-a-vis effectiveness, efficiency and adaptability.

2. The WCM Framework and Practices

Sukarma [6] recommends a framework that integrates TQM, JIT and TPM, or basically the new WCM framework. This approach mainly stems from WCM's definition as "capturing the breadth and essence of fundamental changes taking place in industrial companies" [5]. The WCM logic is fully described in [7]. It brings together principles, practices and performance measurements related to TQM, JIT and TPM. In terms of achieving goals, this production paradigm shares complementary goals including improving quality (Q), delivering products on time (D), and reducing costs (C) through eliminating equipment-centered losses.

For these approaches to be effective, infrastructure or common practices are needed to support their execution [9, 10]. These common practices comprise new approaches to managing human resources, workplace, and suppliers, as well as practices for continuous improvement. These constitute the Infrastructure for implementing TQM, JIT and TPM.

The WCM framework embodies a set of 38 practices, four business performance indicators and 15 manufacturing performance indicators (Figure 1). Implemented in concert, these methods may also yield synergy [11, 12]. The use of TQM not only contributes directly to improved performance in quality, but also it indirectly leads to the improvement of others (just-in-time and cost performances). Likewise, JIT and TPM contribute to quality improvement by respectively reducing inventory, and hence exposing opportunities for process improvement, and to reducing defects due to equipment-related problems.

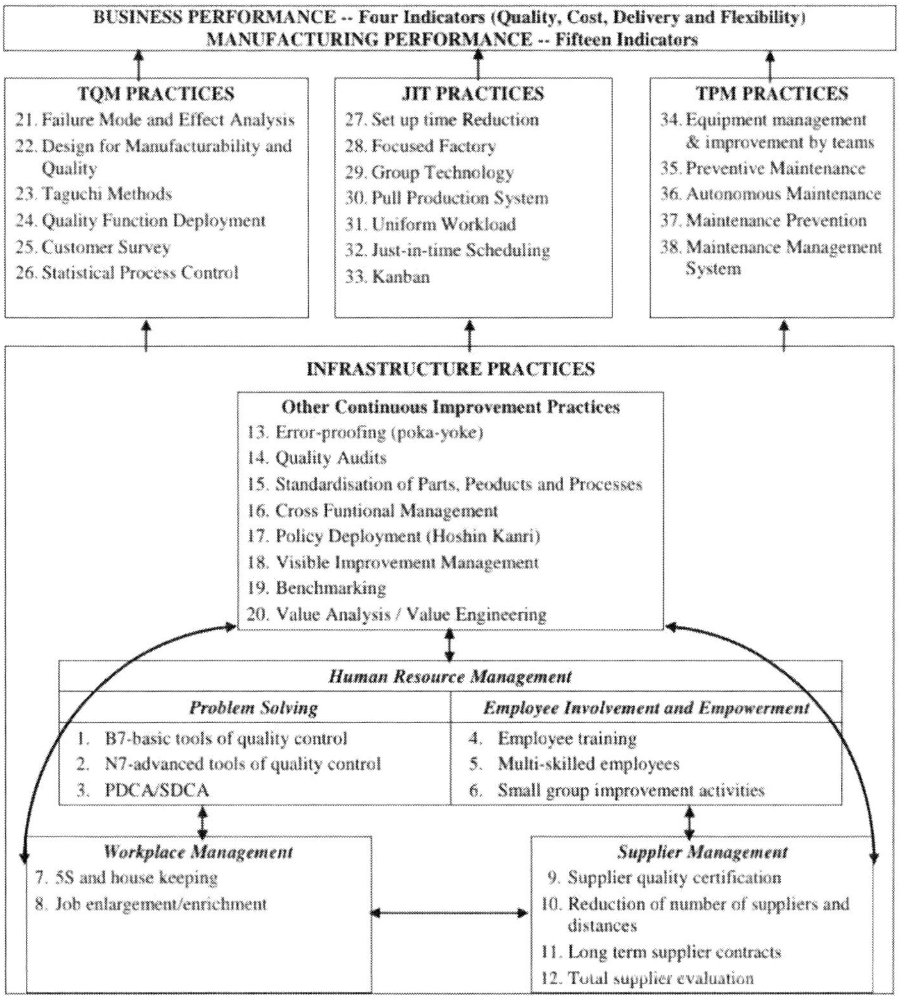

Figure 1. The WCM framework and practices [6]

To achieve standards of performance in manufacturing and enable them to support the core approach, the general practice of Infrastructure must be applied. Infrastructure practices are further classified into problem solving, employee engagement and empowerment, workplace management, supplier relations, and other continuous improvement practices. The implementation of the Quality Trilogy: quality planning, quality control, and quality improvement, to managing for quality is advised by [13]. They are interrelated processes aimed respectively at meeting quality goals during design, operations, and breaking through to remarkable levels of performance. In the manufacturing process, quality is built initially at the stage of product design, is realised by planning and administering activities necessary to make the product, and is achieved ultimately by listening continuously to the voice of the customer. Thus, quality management practices can be classified into Product Design, Process Management, and Customer Focus.

The JIT method is intended to make producers more responsive to customer demand by eliminating or minimizing all types of waste. If the manufacturer chooses to compete in shipping and costs, then JIT might be more appropriate. These practices include Set Up Reduction (SUR), focused factories, Group Technology (GT), pull production systems, uniform workloads, JIT scheduling, and Kanban.

TPM is implemented by extended responsibility for equipment maintenance to every operator, and is not just limited to the maintenance specialists. It can be seen as the logical extension of TQM [14]. In terms of installing the TPM program, [15] suggests that non-Japanese plants should take three distinct stages: planning and preparation, pilot project, and plant wide implementation.

This framework does not stand alone, companies implementing WCM techniques need to adopt performance measurement and monitoring methods which, in some cases are very different from those in traditional systems, as explained in the next section.

3. The Measures of Performance for The WCM Practices

In addition to using WCM practices, producers must modify their ways to measure performance in their efforts to achieve excellence in manufacturing [16]. Efforts towards the WCM environment are characterized by, among other things, highly competitive products, product mix and volume, short lead times and custom orders, and frequent introduction of new products [17]. The implications of applying the WCM method in measuring company performance comprehensively are described in [18].

As seen in Figure 1, there are two kinds of performance measurement proposed in the WCM framework: business and manufacturing performance. Business performance explains the company's ability to meet customer needs, so that most customers feel their measurement. Quality, Cost, Delivery, and Flexibility Performance are identified as business performance. This is a performance that explains customer satisfaction and, in many cases, has been proven to bring business financial success [19], and increase competitive advantage [20].

The definition of quality of Juran's as 'suitability for use' [21], for example, can be seen as business performance, because its valuation is largely based on the customer's point of view. For customers, quality includes among others all eight dimensions [22]. Therefore, companies must look for the needs of their 'right' customers, prioritize them, and try to satisfy them as much as possible.

Manufacturing performance designs company performance that can be measured in accordance with specified standards. It is a collection of benchmarks to assess its ability to satisfy customers by measuring the effectiveness of its resources. This is a performance that must be measured by the company and measured correctly. This paper proposes fifteen manufacturing performance as the main indicators of performance for companies that achieve manufacturing excellence, as described in [18]. Measuring of manufacturing performance must include three criteria: effectiveness, efficiency, and adaptability [23]. The three performance indicators are defined as follows:

Table 1. The meaning of performance indicators [23]

The Performance Indicators	The Meaning
Effectiveness (doing the right things)	The extent to which customer needs are satisfied
Efficiency (doing things right)	The extent to which the total resources in the company are used in an effective and economic way
Adaptability (ability to change)	The extent to which the company is prepared to handle changes in surrounding conditions (strategic awareness)

Furthermore, Table 2 presents a grouping of fifteen manufacturing performance according to these three indicators of performance.

Table 2. Manufacturing performances and indicators of performance

Num.	Manufacturing Performance	Indicators of Performance		
		Effectiveness	Efficiency	Adaptability
1	Process Defect	▲		
2	Product Return	▲		
3	Cost of Manufacturing		▲	

Num.	Manufacturing Performance	Indicators of Performance		
		Effectiveness	Efficiency	Adaptability
4	Cost of Maintenance		▲	
5	Turnover of Inventory		▲	
6	Delivery's On Time			▲
7	Production Lead Time			▲
8	Cycle Time			▲
9	Space of Effeciency		▲	
10	Equipment Availability	▲		
11	Equipment Performance Effeciency	▲		
12	Labour Productivity	▲		
13	Employee's Morale and Motivation	▲		
14	Accident Frequency	▲		
15	Efficiency of Capital Investment		▲	

According to [24], effectiveness is an indicator for assessing company activities related to doing the right thing at the right time, with the right quality. In Table 1, the effectiveness criteria are related to quality management measures (items 1 and 2), equipment maintenance (items 10 and 11), Human Resource Management (items 12 and 13), and frequency of accidents (items 14). Company performance measures are interrelated; in the sense that quality management performance is not only influenced by TQM practices but also by several other practices (specifically HRM and TPM practices).

Whereas traditional systems are primarily concerned with the efficiency of each individual work center, efficiency is defined as the ratio between the resources expected to be consumed and actually consumed [24]. The efficiency criteria in Table 2 are represented by items 3, 4, 5, 9 and 15. Items 3 and 4 are measures of efficiency in spending resources related to equipment manufacturing and maintenance. To get optimal quality costs, companies must allocate their resources properly. Items 5 and 15 are for measuring efficiency related to inventory and capital investment. Finally, item 9 is a measure of efficiency related to space use. Again, this performance measure reflects the results of applying several practices rather than a single technique.

Last but not least, producers must be sensitive and adapt (or nimble) to changing environments in order to survive and thrive in the market. Hamel and Prahalad [21] perspective of competition for the future as an arena to create and oppose opportunities that arise, not only to compare products and processes of competition and change their methods. Items 6 to 8 manufacturing performance (on-time delivery, production time, and cycle time).

The following section will discuss a theoretical perspective on how smart and appropriate allocation of resources will lead to the greatest of performance improvement vis-a-vis effectiveness, efficiency and adaptability.

4. The Manufacturing Technology Optimization Model

The basic premise of the Manufacturing Technology Optimization Model is that the application of WCM practices will lead to improved organisational performance ➔ $P = P_0 + \Delta P$. In this case, the resulting additional improvement (ΔP) relies heavily on the readiness of a company to allocate resources, which can usually be expressed in financial terms.

As explained previously, measures of performance comprise three indicators: effectiveness, efficiency, and adaptability. Assuming resources are allocated to improve company performance in these criteria, ΔP can be written as:

$$\Delta P = f(E_1 + E_2 + E_3) \quad (1)$$

Where E1, E2, and E3, represent company efforts, by way of allocating resources, aimed at improving performance respectively pertaining to effectiveness, efficiency, and adaptability.

Manufacturers should be concerned first with effectiveness (doing the right things), and then with efficiency (doing things right) [25]. Therefore, the priority is on E1. However, no previous reference is available about the order of importance of E2 and E3. It depends on the circumstances. Companies that choose to compete on price will allocate more resources via E2 rather than via E3, but those which wish to improve their speed of operation will select the other choice.

In fact, the contributions of E1, E2, and E3 are time-dependent, and (1) can further be expressed as

$$\Delta P = \Delta P\, C_1 + \Delta P\, C_2 + \Delta P\, C_3 \text{ and } C_1 + C_2 + C_3 = 1 \qquad (2)$$

where C_1, C_2, and C_3, represent the relative contributions to additional improvement on performance due to resource allocation aimed at improving effectiveness, efficiency, and adaptability respectively

Assuming that there is no negative impact on performance, the performance (P) will remain unchanged, or $\Delta P = 0$, unless there are additional improvements in one of the three, or at least C_1, C_2, or $C_3 > 0$. In addition to the amount of resources provided, the relative magnitudes of C_1, C_2, C_3 are determined by the initial performance (P_0) and driving forces from its customers and environment. High-performance corporations show improvement more slowly than low-performance companies. A company in a protective industry has less incentive to improve performance than that of a competitive industry.

Thus, (2) can be simplified into

$$\Delta P = \Delta P_1 + \Delta P_2 + \Delta P_3 \qquad (3)$$

where ΔP_1, ΔP_2, and ΔP_3, represent the additional improved performance with respect to effectiveness, efficiency, and adaptability respectively.

Equation (3) implies that overall improved firm performance (ΔP) is the sum of improvements in each component. Whatever the initial performance (P_0), the main objective of using WCM practices is to raise level of manufacturing performance on each dimension so that the overall system performance ($P_0+\Delta P$) will improve. This research seeks a better way to allocate resources in a business process in such a way that its additional improved performance (ΔP) can reach maximum.

In the case of allocating resources for improving management of people, improving management of equipment, and improving management of others, Equation (3) can be written as:

$$\Delta P = (r_1+r_2+r_3)\, \Delta P_1 + (r_1+r_2+r_3)\, \Delta P_2 + (r_1+r_2+r_3)\, \Delta P_3 \qquad (4)$$

Or

$$\Delta P = r_1\, (\Delta P_1+\Delta P_2+\Delta P_3) + r_2\, (\Delta P_1+\Delta P_2+\Delta P_3) + r_3\, (\Delta P_1+\Delta P_2+\Delta P_3) \qquad (5)$$

$r_1 + r_2 + r_3 = 1$

where

r_1: Relative of contribution to company performance due to improved management of *people*;

r_2: Relative of contribution to company performance due to improved management of *equipment*; and

r_3: Relative of contribution to company performance due to improved management of *others* (materials, energy, and information).

Equation (4) indicates that the additional improvement in each dimension results from a concerted effort in improving the management of people, equipment, and others simultaneously. Likewise, Equation (5) infers that improved management of people, equipment, and others may bring about improvement of company performance in all the three dimensions.

HRM plays an important role in managing infrastructure and supporting core approaches (TQM, JIT, and TPM). In fact, improved management of people contributes to improved management of equipment and others. In other words, r2 and r3 in Equation (5) contain people's contributions.

Separating people component from r2 and r3, then r1 + r2 + r3 = 1 can be written as

$$r_1 + (r_{21} + r_{22}) + (r_{31} + r_{32}) = 1$$

where
r_{21} and r_{31} are the contributions of improved management of "people" to improved management of equipment and others respectively;
r_{22}: the relative contribution to performance due to improved management of equipment after separating people component; and
r_{32}: the relative contribution to performance due to improved management of others after separating people component.

$$(r_1 + r_{21} + r_{31}) + r_{22} + r_{32} = 1$$

Or $$r_{people} + r_{equipment} + r_{others} = 1$$

where $r_{people} = r_1 + r_{21} + r_{31}$ is all "people" contributions

Really, we can significantly enhance people contribution when a company attempts to involve and empower employees in every aspect of management. In other words, r_{people} can be greater than both $r_{equipment}$ and r_{others}. In fact, a new approach to managing equipment suggests transfer of responsibility and authority from maintenance specialists to operators. With this new approach, the twelve types of resistance to managing workplace (5S), which originate from the shop-floor or clerical staff (dealing with the management of people), can be resolved [28].

Now, equation (5) becomes

$$\Delta P = r_{people}(\Delta P_1+\Delta P_2+\Delta P_3) + r_{equipment}(\Delta P_1+\Delta P_2+\Delta P_3) + r_{others}(\Delta P_1+\Delta P_2+\Delta P_3) \quad (6)$$

or

$$\Delta P = r_{people}(\Delta P_1 + \Delta P_2 + \Delta P_3) + (r_{equipment} + r_{others})\,\Delta P_1 + (r_{equipment} + r_{others})(\Delta P_2+\Delta P_3) \quad (7)$$

Thus, manufacturers should focus on the *first* and *second* parts of Equation (7). The *first* part implies that, if r_{people} can be made greater than both $r_{equipment}$ and r_{others}, improved management of people may lead to significant improvement in effectiveness, efficiency, and adaptability simultaneously. The *second* part infers that resource allocation aimed at improving the management of equipment and others should concentrate first on improving company capability in doing the right things – effectiveness (ΔP_1), before improving efficiency (ΔP_2) and adaptability (ΔP_3), the *third* part of the equation.

Is there any way to accelerate performance improvement in a reasonable time?

The answer is in the first part of Equation (7). People management is the most real and most easily influenced, but its success in improving company performance is determined by many factors. Education (studying in formal schools) and training (learning in the workplace) are the two most important factors in improving employee skills and in gaining sustainable competitive success [29].

If the contribution to additional performance due to improved management of people (people) is divided into education (p. 11), training (p. 12), etc. (p.13), then the first part of Equation (7) can be written as$\Delta P_{people} = (p_{11} + p_{12} + p_{13})(\Delta P_1 + \Delta P_2 + \Delta P_3)$ ....(8)

Since education and others (e.g. employee relations) can only lead to long-term improvement, the best way to improve company performance "quickly" is through industrial mass-training.

Then, Equation (8) can be written as:

$$\Delta P_{people} = p_{12}(\Delta P_1+\Delta P_2+\Delta P_3) + (p_{11} + p_{13})(\Delta P_1+\Delta P_2+\Delta P_3)$$

or

$$\Delta P_{people} = p_{12}\,\Delta P_1 + p_{12}(\Delta P_2+\Delta P_3) + (p_{11} + p_{13})(\Delta P_1+\Delta P_2+\Delta P_3) \quad (9)$$

Additional improvement of ΔP_{people} can be accelerated by multiplying the amount of industrial training, particularly concerning with improvement in effectiveness. Thus, Equation (9) becomes

ΔP_{people}	=	$k\, p_{12}\, \Delta P_1$	+	$p_{12}\,(\Delta P_2 + \Delta P_3)$	+	$(p_{11} + p_{13})$ $(\Delta P_1 + \Delta P_2 + \Delta P_3)$	
Improvement in performance due to improved management of people		Improvement in effectiveness due to training		Improvement in efficiency and adaptability due to training		Improvement in all indicators due to education and other people management	(10)

where k > 1

Figure 2 illustrates progress of company performance over time. In this case, for simplicity, only Equation (10), or the first part of (7), ΔP_{people}, is depicted. The whole of (7) can easily be visualised usi a similar logic.

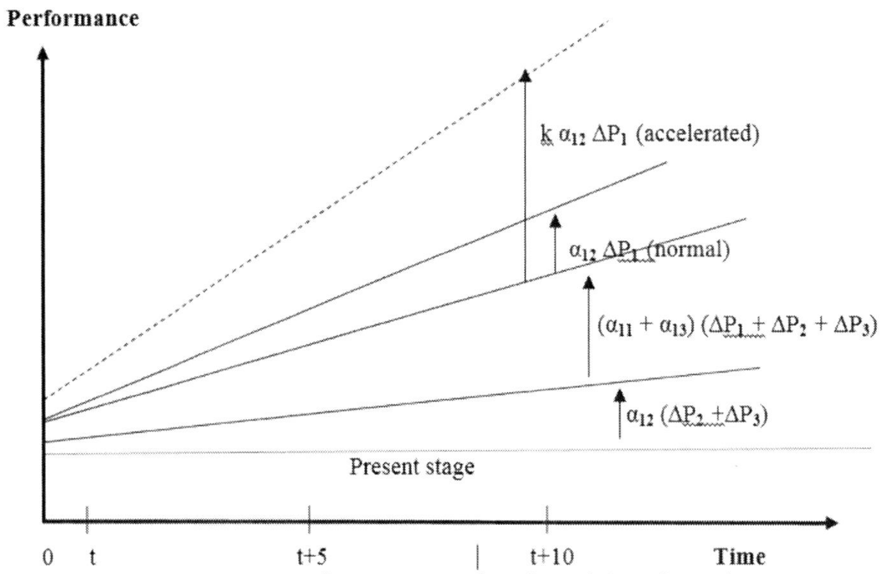

Figure 2 . Company performance as a function of time-dependent resource

5. Conclusions

From the Manufacturing Technology Optimization Model, manufacturers can learn several lessons: Firstly, a manufacturer should focus first on improving its capability on effectiveness before pursuing efficiency and adaptability. The effectiveness capability can be achieved when every employee has excellent skills: a) required by its products, manufacturing processes and customers; b) in improving and maintaining industrial house-keeping; c) in improving and maintaining the condition of his/her own equipment; and d) in problem solving.

Secondly, the achievement of the effectiveness can be accelerated by providing employees training on the job. Besides being less costly, on the job training is preferred because it provides 'hands-on' learning experience that facilitates learning transfer and can fit into the organisation's flow of activities [29].

Thirdly, the pursuit of improved company performance may then be continued by attempting to enhance the efficiency. This can be done, among others, using some practices of the core approaches. TQM expands its product life cycle by developing quality at source via product design; JIT extends its materials flow via supply chain management towards integration of suppliers; and last but not least, TPM continues its equipment life cycle via total maintenance system [30]. On the job training alone is not enough to achieve efficiency. This is especially true when a concurrent engineering methodology is applied, wherein, engineers should have expert skills and multiple expertises. Thus, formal education to some extent plays a central role in enhancing efficiency.

Fourthly, the pursuit of adaptability may proceed simultaneously with that of efficiency. The primary goal is to respond quickly to customer requirements in terms of products, lot sizes, and customised demands for individual customers, et cetera [13]. Adaptability indicates maintaining effectiveness and efficiency in a changing environment. Hence, in addition to the above efforts, companies have to use several other practices in order to improve its adaptive ability. JIT practices of the WCM Framework are among the core practices leading to its realisation. But their success has to be supported by the core practices under TQM and TPM as well as the Infrastructure practices (e.g. supplier partnerships).

Acknowledgments
The authors are grateful to the Institut Sains dan Teknologi Nasional Universiti Teknologi Malaysia and Pamulang University for the support given in completing this research paper.

References
[1] Skinner, W 1985 The Focused Factory in W. Skinner (1985) *Manufacturing the Formidable Competitive Weapon* (John Wiley & Sons, Inc.) 71-83.
[2] Gibson, P., G. Greenhalgh, and R. Kerr 1995 *Manufacturing Management: Principles and Concepts* (Chapman & Hall, London).
[3] Jaikumar, R. 1986 *Post-industrial Manufacturing* (Harvard Business Review) Nov-Dec.
[4] Arndt, G. 1990 An International Perspective on "Kaizen" in Manufacturing *Fifth International Conference on Manufacturing Engineering* (The Institute of Engineers, Australia).
[5] Schonberger, R. J. 1986 World Class Manufacturing: The Lessons on Simplicity Applied (The Free Press, New York).
[6] Sukarma, L. 2000 A Model for Guiding and Measuring Company Performance towards Achieving Manufacturing Excellence (PhD Thesis, The University of Wollongong).
[7] Sukarma, L., Azmi, H., and Abdullah, N.L. 2014 Applied Mechanics and Materials 564 *The Impact of World Class Manufacturing Practices on Company Performance: A Critical Review* 727-732.
[8] Sukarma, L. 2014 Applied Mechanics and Materials 660 *Enhancing Firm Competitiveness through Concurrent Implementation of TQM, JIT and TPM* 976-982.
[9] Sugimori, Y. K., Kusunoki, F. C., and Uchikawa S. 1977 Toyota Production System and Kankan System – Materialisation of Just-in-Time and Respect-for-Human System *International Journal of Production Research 15/6* 553-564.
[10] Harrison, A. 1992 Just-in-Time Manufacturing in Perspective (*Prentice Hall Int. Ltd.*, UK).
[11] Flynn, B. B., S. Sakakibara, and R. G. Schroeder 1995 Academy of Management Journal 38/5 *Relationship between JIT and TQM: Practices and Performance* 1325-1360.
[12] Nakamura, M., S. Sakakibara, and R. G. Schroeder 1999 Remade in America: Transplanting and Transforming Japanese Management Systems *Just-in-Time and Other Manufacturing Practices: Implications for U.S. Manufacturing Performance, in J. K. Liker, W. M. Fruin, and P. S. Adler (eds.), 1999* (Oxford University Press, New York).
[13] Juran, J.M. and H. Blackiston 1995 Universal Approach to Managing for Quality, in Shelton, Ken M. (ed) ***In Search of Quality*** (Executive Excellence Publishing, Provo, USA) 183-93.
[14] Arndt, G. 1995 From TQM to TPM: An Australian Perspective on Developments in Japan and Germany, Feature Paper at "TPM in Actions" Forum, Synergy Conventions, Sydney.
[15] Hartmann, E. H. 1992 Successfully Installing TPM in a Non-Japanese Plant (TPM Press, Inc., Pittsburgh, Pennsylvania).

[16] Eccles, R.G. 1995 The Performance Measurement Manifesto, in J. Holloway, J. Lewis, G. Mallory (Eds.) *Performance Measurement and Evaluation* (Sage Publications, London) 5-14.

[17] Maskell, B.H. 1996 Making the Numbers Count: The Accountant as Change Agent on the World Class Team (Productivity Press Inc., Portland, Oregon).

[18] Sukarma L. and Azmi H. 2015 Applied Mechanics and Materials 761 *The Measures of Performance for World Class Manufacturing Practices: A Critical Review* 545-549.

[19] Hausner A. and Arndt G. 1999 Linking Quality Management Practices and Business Performance, 3rd International Research Conference on Quality Management, Melbourne.

[20] Singh M. 2013 International Journal of Business and Management Invention *Product Quality for Competitive Advantage in Marketing* **2/6** 05-08.

[21] Juran, J. M. 1988 Quality Control Handbook (McGraw-Hill Publishing Co., New York).

[22] Garvin, D. A. 1987 Competing on the Eight Dimensions of Quality *Harvard Business Review* 101-109.

[23] Moseng, B. and Bredrup H. 1993 Production Planning and Control *A Methodology for Industrial Studies of Productivity Performance* **4/3**.

[24] Sink, S. and Tuttle T. 1989 Planning and Measurement in Your Organisation of the Future (Industrial Engineering and Management Press, Norcross, GA).

[25] Prahalad, C. K., & Hamel, G. 1994 Strategic management journal *Strategy as a field of study: why search for a new paradigm?* **15/S2** 5-16.

[26] Pall, G. A. 1987 Quality Process Management (Prentice-Hall, Inc., Englewood Cliffs, NJ).

[27] Riley, Jr., J. F. 1999 Process Management in Juran's Quality Handbook (The McGraw-Hill Companies, Inc., New York) p. 6.1 – 6.21.

[28] Hirano, H. 1990 5 Pillars of the Visual Workplace, the Sourcebook for 5S Implementation, (Productivity Press, Portland, Oregon).

[29] Schuler, R.S., Dowling, P.J., Smart, J.P., and Huber, V.L. 1993 Human Resource Management in Australia, Second Edition (Harper Educational Publishers, Pymble, NSW).

[30] Enkawa T. 1998 *Production Efficiency Paradigms: Interrelationship among 3T: TPM, TQC/ TQM and TPS (JIT)* 98 World-Class Manufacturing & JIPM-TPM Conference, Singapore

IOP Conf. Series: Materials Science and Engineering **909** (2020) 012050 doi:10.1088/1757-899X/909/1/012050

Knowledge base integration management system quality, safety and environmental to improve organizational performance in construction company

Retno Apriyati[1*], **Yusuf Latief**[1]

[1]Department of Civil Engineering, Faculty of Engineering, Universitas Indonesia, Kampus Baru UI Depok, Jawa Barat 16424, Indonesia

[*]Email: r.apriyati@gmail.com

Abstract. Development in Indonesia showed high progress in the economics but had low progress for social; institutional; and in environmental sectors. Furthermore, Indonesian Construction Company still inhibited by low competitiveness; along Safety system management application; safety record and bad occupational health. Based on Indication in above, it's necessary to apply an integrated management system between Quality, Safety and Environmental for manage the projects construction. The Implementation of an integrated management system, it's necessary to improve with knowledge base Integration Management System; due to accelerated targets achievement of integration and to learned cause of the failure of integration aspect. Therefore, this research aims to validated Knowledge Base Integration Management System of Quality, Safety and Environmental to improve organization's performance on Construction Service Company. This research was conducted with case studies and expert validation for the integration model, objectives and risk relationships.

1. Introduction

The construction is one kind of sector that helps increase Indonesia's economic growth. In 2018, the economic growth of 5.06%, the business of construction sector contributed 0.72% [1]. Because of this, the construction industry has provided a significant increase for national economic development. However, research for development in Indonesia between 2007 - 2012 showed high progress in the economic but low progress in the social and institutional and a decrease in the environmental. This shows that the sustainable development index in Indonesia only indicates sustainable development in the short term, not in the long term, development also raises problems in the form of a decrease in social resources, ie regions with more advanced development tend to have lower social capital. This also happened in the construction's sector in Indonesia [2]. Amount of burden on the economic sector from project construction in developing countries have a big impact on the community, environment and other social devices. So, it is necessary to continuous integration on project construction's management in thus countries [3]. Implement a parallel system that requires duplicates for documentation, running procedures, accessing, managing forms and other paperwork. It has proven difficult to install a separated management system that is equipped with quality, environment, and occupational health and safety, and to ensure their alignment with organizational strategies [4].

Based on the above phenomenon, it is necessary to apply an integrated management system between Quality, safety and Environmental in managing of construction projects. To optimize it, the system also supports the knowledge management system [5]. Therefore, there is a need to develop a Knowledge Base Integration Management System Quality, Safety and Environmental information system in Construction Services Companies to Improve Organizational Performance.

2. Literature Review

Process integration is a concept / methodology for combining several parts of a process or entire process to reduce the consumption of resources or harmful emissions to the environment [6]. The integration process consisting of process management, the process of creating value, and supporting processes have developments that are compatible with sustainable management processes that focus on risk management [7]. Process management integration was the integrating all activities in all aspects so that there is no duplication of activities in one single system [8]. The integration process has an effect on increasing business results, improving better decision making, as well as improving service quality based on the balance of service levels, customer and financial needs [9]. The management process can be integrated at different levels of application of management system standards [8].

The elements of process integration in accordance with the clauses contained in the management system standard published by ISO can be found in several studies. The elements of process integration are, the scope of the management system with organizational context input (1), leadership (2), management policy integration; including organizational reference norms, definitions and terminology (3), planning (4), supporting processes (5), Operations (6), Performance evaluation (7) and improvement (8) [5].

The management system integration process is inseparable from risk. These risks must be managed so as not to impede the objective achievements of the management system integration process. Therefore, risk management is needed to manage. Risk assessment is the main-key process of risk management [10. The risks that can occur the process of management system integration are, (1) Lack of financial resources; (2) Lack of employee knowledge of management system integration; (3) Lack of skilled employees; (4) Lack of employee motivation; (5) Lack of awareness of the benefits of integration; (6) Feel there are other priorities that are more important than management system integration; (7) demands from different stakeholders; (8) There are cultural differences between disciplines in each management system; (9) Uncertainty about the value of integrated management systems in the market; (10) Lack of examples of specific sector implementation (in this study the construction sector); (11) Short-term orientation; and (12) A huge effort is needed to implement an integrated management system [11].

3. Research Methodology

The first is literature review, In this part to identify indicators of models, targets and risk of integration management system Quality, Safety and Environmental. After review literature, all of sub indicators had collected should be verified by 5 (five) expert validation questionnaire. The expert is the directors, QMS Manager or Lecturer who have 10 years experiences minimum. The data that had been validated by experts is created for pattern recognition and then the knowledge base is developed.

4. Results

From previous studies and developments during expert validation, there are 46 dominant risks that must be managed with the right strategy. The proposed strategies in the form of causes, preventive actions, impacts, and corrective actions are summarized in the table below.

Table 1. Risk management strategy of integration system management [12]

Code	Risk of Integration System Management	Cause Variables	Preventive Action Variable	Impact Variable	Variable Corrective Action
X1.3.1. Component / Scope Clause					
X1.3.1.1	The complexity and differences between management systems	there are differences in understanding between management systems	improve related socialization and training	The identification of external and internal organizational issues is inaccurate	Conducting a review re- Related
X1.3.1.2	lack of inflationary considerations	lack of awareness to consider inflation in identifying issues	consider economic factors	The identification of external and internal organizational issues is inaccurate	Conducting a review re- Related
X1.3.1.3	Organizations in construction companies do not set job descriptions & standards / requirements that are integrated into 3 systems	lack of competence of the source of the power of human	wear source power human beings who are competent to define the job description & Standards / Requirements are integrated to the third system	Implementation of an optimal integrated management system	Assign Job Description appropriate scope of system management scope that is used
X1.3.1.4	Weak partner selection	limited partner availability	Pre- qualify partners	The implementation of an integrated management system is not optimal	Choosing a partner just who is considered more powerful
X1.3.1.5	Organizations in construction companies have not yet guaranteed that the Work Breakdown Structure (WBS) has taken into account quality, safety, health and environmental aspects	Lack of competence of the source of the power of human	Preparing WBS that have considered aspects of quality, K3, and environment	WBS is not composed with consideration of the three aspects of the	Conducting a review re- Related
X1.3.1.6	Companies are less able to produce	The process is not controlled ac	Compile SOP related	The organizational performance goal	Conducting a review re-

Code	Risk of Integration Syste	Cause Variables	Preventive Action Variable	Impact Variable	Variable Cor rective Actio
	a product that is quality and powerf ul competitiveness	cording to the mana gement system		was not achieved	Related
X1.3.1.7	Organizations in con struction companies have not been able to identify and analy ze important issues r elated to risk and opportunity related to the quality of safety and the environment	Lack of competence of the source of the power of human	Improve socializa tion and related tr aining	Risk organization is not in accordance with the process of the three systems of management of the	Doing identif ication reset
X1.3.1.8	Organizations in con struction companies have not been able to determine risk con trol related to the quality of safety, security, oc cupational health and the environment	Lack of competence of the source of the power of human	Improve socializa tion and related tr aining	Risks be not controll ed by either	Conducting a review re- Related
X1.3.1.9	Organization in the company of construction has not been able to identify and analy ze issues of importance related to Risk and Opportunity product friendly environment and save resource po wer energy	Lack of competence of the source of the power of human	Improve socializa tion and related tr aining	The company produces pr oducts that are not environmenta lly friendly	Doing identif ication reset
X1.3.1.10	Organizations in con struction companies have not been able to realize the confidence of workers in the company	lack of corporate ev ents that support co nsolidation between workers in the company	Doing events com panies are routinel y	The identification of exter nal and internal organization al issues is inaccurate	Involve work ers in the co mpany's HSE program
X1.3.1.11	The company's priori ty orientation is inacc urate	Understanding the scope of the different	Conduct related s ocialization	Identify the need and expectation of companies are not accurate	Conducting a review re- Related
X1.3.2. Component / Leadership Clause					
X1.3.2.1	The demands of stakehol der's interests are dif	Standard contracts are different for eac h stakeholder intere	Creating standard s and policies raw	Insurers are liable towards thi rd system was not ap	Conducting a review re- Related

Code	Risk of Integration Syste	Cause Variables	Preventive Action Variable	Impact Variable	Variable Cor rective Actio
	ferent	sts		propriate standard	
X1.3.2.2	An understanding of the duties and respon sibilities responsibilit ies are related to the person in charge SMMK3L stil l low	Lack of socialization unders tanding of the duties and respo nsibilities responsib ilities	Improve socializa tion and related tr aining	Insurers are liable not underst and related duties an d responsibilities res ponsibilities	Replacing th e person in charge who i s more aware and more co mpetent
X1.3.2.3	Lack of commitment to the level of top management t o contribute to the improvement of the comfort of the place of work and process innovation	The comfort of the workplace an d the innovation proc ess are considered not yet important	Conduct related s ocialization	The person in charge is less productive in c arrying out their duti es and responsibilitie s	Improving th e comfort of the place of work
X1.3.2.4	Lack of commitment to crafts manship quality high and stakeholder inter ests that support the delivery of the performance of the organization	Lack of awareness of the importance of h igh-quality workmanshi p	Making agreemen ts related to mockup standar ds	The quality of work that is produced does not conform with pla nning	Doing benc hmark, visual ization exam ples related workmanship quality high
X1.3.2.5	The lack of unity of view of the whole employees wh o already work in the bottom of the standard previously	Lack of socialization abo ut the importance of unity of views	do related socializ ation	the views of the workers are not equal or not equa l	do socializati on reset
X1.3.2.6	The integrated mana gement system has not been established as a goal of the company	lack of corporate m anagement commit ment	Conduct related s ocialization	The process of system manageme nt overlap overlap th at causes the process is too long	Incorporating an integrated management system as the goal of the company
X1.3.3. Component / Management Policy Integration Clause					
X1.3.3.1	The company does not have the QMS, QMS, and SMK3 certificates	Lack of corporate management comm itment	Conduct related s ocialization	Reference to the integration of ma nagement systems is not accurate	Conduct certi fication
X1.3.3.2	Development of ISO 2015 norms or standa rds, 8 Clauses in HLS and components are not	Lack of socialization is related to understanding the development of norms or standar	Conduct related s ocialization	Reference to the integration of ma nagement systems is not accurate	Carry out ISO certi fication for w orkers

Code	Risk of Integration Syste	Cause Variables	Preventive Action Variable	Impact Variable	Variable Cor rective Actio
	yet understood	ds			
X1.3.4. Component / Planning Clause					
X1.3.4.1	Lack of planning-oriented term long	Limited funds	Develop programs run leng th which efficientl y and in accordance with the funds that exist	The company's progr am is not directed	Arrange long - term progra ms according to stages and abilities
X1.3.4.2	The demands of stakehol der's interests are dif ferent	Levels of needs and interes ts and different	follow the direction of stake holder interests ar e primary (in case it is SOE)	difficult to get a varied project with the scope of the different	create a standard alter native in accordance with stakehol der interests are different
X1.3.4.3	Risk identification an d treatment have not been integrated	Risks each - each process is co mpartmentalized - box	Integrate risk iden tification	organizational perfor mance goals are not achieved	conduct a review the re lated
X1.3.4.4	The procedure of risk identification and ha ndling is not yet clear	Planning procedure s are not ready yet	Compile SOP related	identification and ha ndling of risks not working	agreed diagra m flow identi fication of risks that s imple before doing the manufacturin g procedure
X1.3.5. Components / Process Supporting Clauses					
X1.3.5.1	lack of source power that is competent	limited funds and have not been programmed with either	draw up a program run lengt h which efficientl y and in accordance with the funds that exist	organizational perfor mance goals are not achieved	make the program f ulfillment so urce of power that is competent in the term shor t
X1.3.5.2	lack of source power of financial	program preparatio n is not yet ripe	Setting up the program to the needs of the source of power financial ly with detailed	The organizational p erformance goal was not achieved	Make planni ng needs of source po wer financial ly term short and priorities

IOP Conf. Series: Materials Science and Engineering **909** (2020) 012050 doi:10.1088/1757-899X/909/1/012050

Code	Risk of Integration Syste	Cause Variables	Preventive Action Variable	Impact Variable	Variable Cor rective Actio
X1.3.5.3	lack of source power hum an beings who have co mpetence in the field of system manageme nt	the source of the power of man less to get the training and certific ation	Increased progra m of training and certif ication for the source of the power of man to determine the competence and t he criteria of competence	The criteria of competence that mad e not accurate	Creating a program ful fillment sour ce of power that is competent in the term shor t
X1.3.5.4	Organizations in com pany construction job description that acco mmodates the requirements of quality, health wor k, and the environment that can improve the performance of the organization	lack of awareness regard ing these requireme nts	Job description details are made more ful ly related to this matter	The criteria of competence that mad e not accurate	Socialization understandin g of the job descripti on to accom modate the three require ments of the
X1.3.6. Operational Components / Clauses					
X1.3.6.1	Organizations in ente rprise construction n ot clarify the requirements of quality, health and safety work , the environment for the design , purchase of products and servi ces	Not to assume that the K3 quality require ments, and the environment is i mportant to be prepared	Make the requirements are c lear about the quality, K3 and environment	Requirements not me t	Taking the benchmar k of the requirements of the good a nd true
X1.3.6.2	organizations in cons truction companies h ave not been able to monitor the agents that are in construction	The unavailability of methods for mon itoring agents	Compile SOP related	The action identified was not implemented	Conduct an audit for e valuation
X1.3.6.3	Lack of resources the power of man that competent	The source of the power of man less to get the training and certific ation	Improve training programs and rela ted certifications	The process of emergency is not handled with the appropriate	Improving co mpetencies s ource power human
X1.3.7. Components / Performance Evaluation Clause					
X1.3.7.1	Lack of auditors who have the competence	the source of the power of man less to get the	Enhancing trainin g programs and re lated certification	The audit process is not running with eith er	Conduct an audit by c ombining sen

Code	Risk of Integration Syste	Cause Variables	Preventive Action Variable	Impact Variable	Variable Cor rective Actio
		training and certific ation	s		ior auditors and junior auditors
X1.3.7.2	Evaluation of the results of follow- up audit results is still not quite right	Root cost analysis is not specific or not on ta rget	Ensuring the evaluation of the results of the follow -up results of the audit examined ba ck	System management is not effective	Conducting a review re- Related
X1.3.8.Component / Improvement Clause					
X1.3.8.1	Management review material is incomplet e	Lack of priority issues pe mbahansan on the management revie w	Make the agenda of major in manag ement reviews the appropriate stand ard system manag ement	Management review results are not on target	Doing repair s manageme nt review
X1.3.8.2	The results of the management revi ew are not relevant to the business processe s specified	Lack of priority issues pe mbahansan on the management revie w	Identify business processes that are defined and ad justed to review management mate rial	Management review results are not on target	Make improv ements to the results of manageme nt review
X1.3.8.3	The results of the management revi ew were not carried out	Lack of corporate management comm itment	Create an audit program for a sch eduled manageme nt review	Company performan ce has not improved	Carry out immediat e review resu lts

Perceive the relationship between information systems and the integration of the Quality, Safety and Environmental management system processes described above in previous studies, it is expected that with the development of a knowledge base Integration Management System Quality, Safety and Environmental organizational problem of it can be solved and performance can be improved and more efficient so that the objectives of the study where to improve the performance of the construction service company organization.

6. Conclusion

From this study it can be concluded (1) there are 8 clauses that can be integrated in the quality, safety and environmental system management process integrase, (2) there are 46 dominant risks that must be managed with the right strategy. The proposed strategy is in the form of causes, preventive actions, impacts, and corrective actions. By developing a process integration model, targets and risk relationships in the quality, safety and environmental system management process integration in a knowledge base website, it is hoped that the performance improvement of construction service company organizations can be achieved.

Acknowledgement

The Author would like to thank the all members which have fully supported the author in finishing this research. For companies in construction sector, especially contractors, to provide Knowledge Base Integration System Management information systems to improve organizational performance. And it

IOP Conf. Series: Materials Science and Engineering **909** (2020) 012050 doi:10.1088/1757-899X/909/1/012050

is hoped that in future research a more integrated Quality, Safety and Environmental process integrase system can be developed, as the online-based decision-making system is not only limited to the knowledge base.

References

[1] BPS-Statistics Indonesia 2018 Konstruksi Dalam Angka. *Konstruksi Dalam Angka*
[2] Alex O 2017 Pembangunan Berkelanjutan dan Hubungan dengan Modal Sosial di Indonesia *Sekolah Pasca Sarjana Institut Pertanian Bogor*
[3] Banihashemi S 2017 Critical Success factors (CSFs) for Integration of Sustainability into Construction Project Management Practices in Developing Countries *International Journal of Project Management.*
[4] Zeng S X Shi J J and Lou G X 2006 A synergetic model for implementing an integrated management system: an empirical study in China *Journal of Cleaner Production*
[5] Udin M R 2017 Pengembangan Model Integrasi Proses Sistem Manajemen Berkelanjutan pada Penerapan Sistem Manajemen Mutu, Keselamatan, Kesehatan Kerja dan Lingkungan dalam Pengelolaan Proyek Konstruksi *Universitas Indonesia*
[6] Klemes J J Varbanov P S and Kravanja Z 2013 Recent Developments in Process Integration. *Chemical Engineering Research and Design*
[7] Klute-Wenig S and Refflinghaus R 2015 Integrating Sustainability Aspect into An Integrated Management System *The Total Quality Management Journal*
[8] Forslund H 2015 Performance Management Process Integration in Retail Supply Chains *International Journal of Retail & Distribution Management*
[9] Pimenta M L Silva A L and Tate W L 2016 Characteristics of cross-functional Integration Processes: Evidence from Brazilian Organizations. *The International Journal of Logistics Management*
[10] Tian D Yang B Chen J and Zhao Y 2018 A multi-experts and multi-criteria risk assessment model for safety risks in oil and gas industry integrating risk attitudes *Knowledge-Based Systems*
[11] Rajkovic D Aleksic M Milicevic R and Cudic S 2008 IMS in SMES- Reasons, Advantages and Barriers on Implementation. *International Journal for Quality research*
[12] Siahaan N E 2019 Identifikasi Risiko dalam Integrasi Proses Manajemen yang Berdampak Terhadap Tujuan Komponen Sistem Manajemen *Universitas Indonesia*

Design and analysis of cell phone case from recycling plastic bottle

M T A Benhardy[1*], **A Solihin**[2], **E F Primaseta**[3] and **T R Sahroni**[4]

[1,2,3,4] Industrial Engineering Department, BINUS Graduate Program – Master of Industrial Engineering, Bina Nusantara University 11480, Jakarta, Indonesia

*Email: taufik@binus.edu

Abstract. Recognizing the large amount of plastic waste produced by humans today, recycling is one solution to reduce the accumulation of plastic waste. This study aims to analyze the recycling of HDPE (High Density Polyethylene) plastic waste into cell phone cases. This study compares the strength of recycled plastic cell case with other materials that are often used for cell phone cases in general. The methodology of this paper begins with the development of new 2D and 3D cellphone cases, and for processing with injection molding. The parameters are focused on the strength of the material. The simulation is presented using Autodesk Inventor with free fall analysis, and presented stress, displacement, main stress and strain. The results show that the results of stress analysis and material comparison between HDPE and other materials do not show significant differences in other words HDPE can be used as an alternative material for cell phone case materials.

1. Introduction

The problem of waste in Indonesia is a problem that has not been resolved to date [1]. Meanwhile, with the increase in population, it will also be followed by an increase in the volume of landfills generated from human activities.

The composition of waste generated from human activities is 60-70% organic waste and the rest is 30-40% non-organic waste, while the second highest composition of non-organic waste is 14% is plastic waste. Most plastic waste is a type of plastic bag or plastic bag in addition to plastic packaging. [2]

Figure 1 shows a significant increase in plastic production from 1950 to 2015.

Figure 1. Global plastic production per year (Geyer, 2017)

In 1950 the world's plastic production was only 2 million tons, but after that annual plastic production increased to 200 times reaching 381 million tons in 2015 [3]. Seeing this data is very dangerous if the plastic waste is not managed properly.

Plastic is an organic material that has the ability to be formed into various forms, when exposed to heat and pressure. Plastic can be in the form of bars, sheets or blocks, if in the form of products can be in the form of bottles, food wrappers, pipes, tableware, and others. Plastic compositions and materials are polymers and other additives. Polymers are composed of monomers that are bound by chemical bond chains [2].

The advantages of plastic compared to other materials include strong, lightweight, flexible, rust resistant, not easily broken, easily colored, easily shaped, and good heat and electricity insulators. Therefore, plastic waste has many advantages if it can be used as recycled products [4].

Plastic is still one of the main materials for making various equipment in addition to iron, wood and paper. Plastics are considered to have several advantages compared to other materials which are lighter, practical, durable and waterproof so that it is widely chosen by consumers [5]

Recognizing the need to reduce plastic waste that exists today, the industry is pushing towards a more sustainable and low-pollution industry [6] for this reason it is quite interesting to utilize HDPE type plastic packaging waste to be used as a base material for making cellphone cases. Utilizing plastic waste, in addition to material cost efficiency, can also help the industry become more sustainable, therefore this study aims to analyze the feasibility of cell phone case design using new materials from waste plastic packaging materials with parameters focused on the strength of the material presented by using Autodesk Inventor.

2. Materials and Methods

The material examined in this study is plastic packaging material made from HDPE (High density polyethylene). This type of material has strong, hard, and blurry characteristics, which are commonly used in beverage bottles, white milk bottles and jerry cans [7]. Before the manufacturing process, the material must be ensured prepared so that it is ready to be printed using molds, the steps that need to be taken are to clean the used bottles so that they are not contaminated with other materials and also all types of dirt, and separate bottles according to the type and color of each. Then puree the used plastic bottle that has been cleaned to a size of ± 2mm to facilitate the melting process using a plastic cutting machine, then grind the plastic that has been cut with clean water to ensure there is no contamination of other ingredients and then dry in the sun to dry and ready to be printed. The work steps follow the following steps [5]:

a) Prepare recycled plastic seeds ready for use and mix them with the color pigments
b) Put it into a mold with a heating system with a temperature of ± 230oC for approximately 10-15 minutes until the material melts and can be formed
c) Forward the recycled plastic ore melting material into the desired "case" cell phone
d) Cool the printed product so that it hardens for about 10 minutes according to the product size
e) Check the printed product to make sure the product prints properly

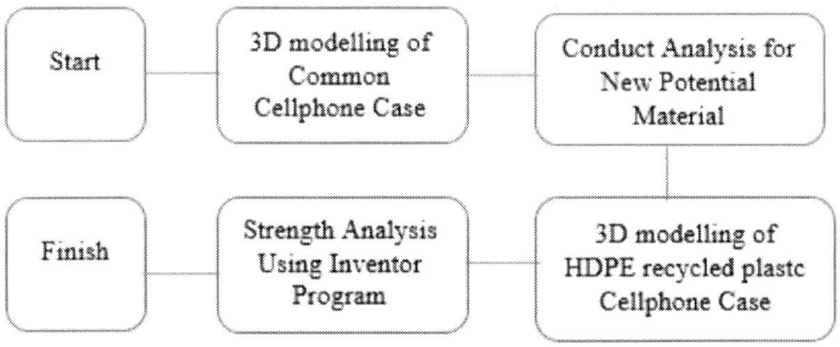

Figure 2. Development HDPE cell phone case plastic recycled stage flow

In HDPE Cellphone Case Plastic Recycled deveopment process starts from the initial stage of understanding 3D modeling of cell phone cases in general, then observes and analyzes the material potential by utilizing recycled plastic HDPE, then designs and analyzes the strength of the Palstic Recycled Cell Phone Case using an inventor program.

3. Results and Discussions

In previous research, HDPE waste used for Automotive Interior [8]. In this research, the design of the use of HDPE waste for other products was carried out. the result will generate the new product design that using HDPE plastic waste. The product was cell phone case that made of High-Density polyethylene or HDPE. Cell phone case materials generally use aluminum and polycarbonate materials but for this research using HDPE material. The comparison cellphone case material for plastic and aluminum as listed in Table 1.

Table 1. Comparison cellphone case material (Source: Hildenbrand, 2019) [9]

Plastic (Polycarbonate)	Aluminium
More affordable	Very strong
Easy to hold	Modern and sleek-looking
Easy to customize or decorate	Expensive
less expensive	

Product design used Xiaomi Redmi Note 5 cell phone as a model of case cell phone. The HDPE material was obtained from the recycle process of HDPE plastic waste. The cell phone case has a size of 16 x 7.5 x 9 cm with a standard design covering the back of the cell phone.

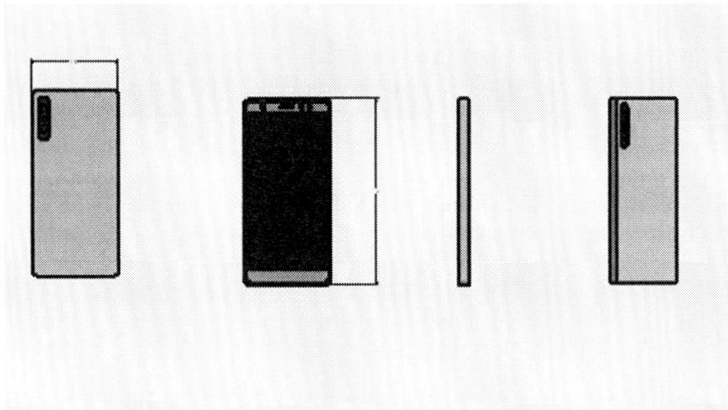

Figure 3. Cell phone case design

Figure 3 shows a model of a cell phone case as a result of the proposed by using the 3D Modeling Software Autodesk Inventor. In this study conducts free fall analysis test from the elevation of 3 meter. The test aims to determine the strength and stress level of the material. In this simulation shows the case fallen within a certain height and hit the floor in the bottom of the cell phone case. In addition, Figure 4 shows the force in bottom of the cellphone. The design is being further studied using CAD software simulation.

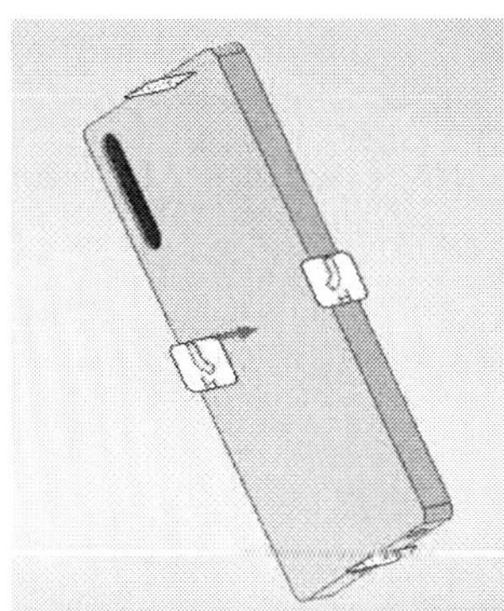

Figure 4. Force in bottom of cell phone case

The five force for this simulation such as 60 N, 70 N, 80, 90 N and 100 N. Yield Strength of HDPE material is achieved with 20.66 MPa and Ultimate Tensile Strength is reached at 13.77 MPa. The simulation using Autodesk Inventor software in the stress analysis section shows ability of the material in protecting the inside cellphone part. The following are the results of Free Fall Analysis test is listed in Table 2.

Table 2. Stress analysis result

Name	Force				
	60	70	80	90	100
Von Mises Stress (VMS)	0.291801 MPa	0.253575 MPa	0.2898 MPa	0.326025 MPa	0.36225 MPa
Displacement	0.000626361 mm	0.000730754 mm	0.000835148 mm	0.000939541 mm	0.00104393 mm
Contact Pressure	0.139717 MPa	0.162992 MPa	0.186279 MPa	0.209562 MPa	0.23286 MPa

The simulation result is safety factor for all force didn't have differences that is 15. The simulation focused on three parameters, Von Mises Stress (VMS), Displacement and Contact Pressure. In the Von Mises Stress section, the results obtained the force of 60 N is achieved at 0.21735 MPa, the 70 N force is noted with value of 0.253575 MPa, for 80 N force is achieved at 0.2898 MPa, the 90 N force is reached value of 0.326025 MPa, and for 100 N force is highlighted with 0.36225 MPa. Then the five forces it can be concluded that the increase of the force that applied, value of Von Mises Stress will increase too but the value of Von Mises Stress is not bigger than the yield strength, so the force applied to HDPE material is still on tolerance limits. In the Contact Pressure section, the results for a force of 60 N are 0.0599329 MPa, a 70 N force of 0.162992 MPa, an 80 N force of 0.186279 MPa, a 90 N force of 0.209562 MPa, a force of 100 N at 0.23286 MPa. Then the five forces it can be concluded that the increasing of the force applied, the value of Contact Pressure was increased. Afterwards in the Displacement section, the results for the force of 60 N are 0.000626361 mm, the 70 N force is 0.000730754 mm, the 80 N force is 0.000835148 mm, the 90 N force is 0.000939541 mm, the force is 100 N for 0.00104393 mm. Then from the five forces it can be concluded that the greater the force applied there will be an increased in displacement, but the change is not significant.

In the next stage, this study shows the simulation using a different cell phone case material (aluminum and polycarbonate). Aluminum and Polycarbonate are materials that are often used in cell phone cases. The Figure 5 to 7 show the results of the comparison between HDPE, Aluminum and Polycarbonate.

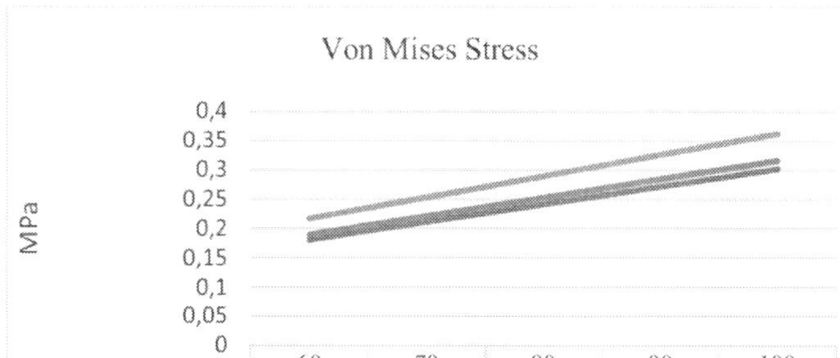

	60	70	80	90	100
Alumunium	0,181081	0,211261	0,241441	0,271621	0,301802
Polykarbonat	0,189743	0,221367	0,25299	0,284614	0,316238
HDPE	0,21735	0,253575	0,2898	0,326025	0,36225

Figure 5. Von mises stress comparison

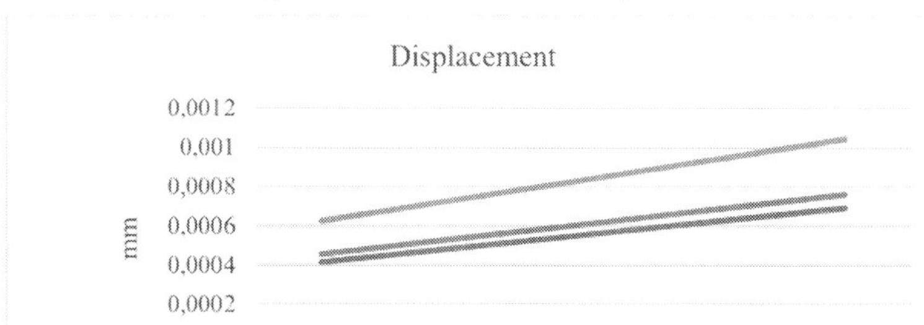

	60	70	80	90	100
Alumunium	0,000415755	0,000485047	0,000554339	0,000623632	0,000692924
Polykarbonat	0,000456786	0,000532917	0,000609048	0,000685179	0,00076131
HDPE	0,000626361	0,000730754	0,000835148	0,000939541	0,00104393

Figure 6. Displacement graph comparison

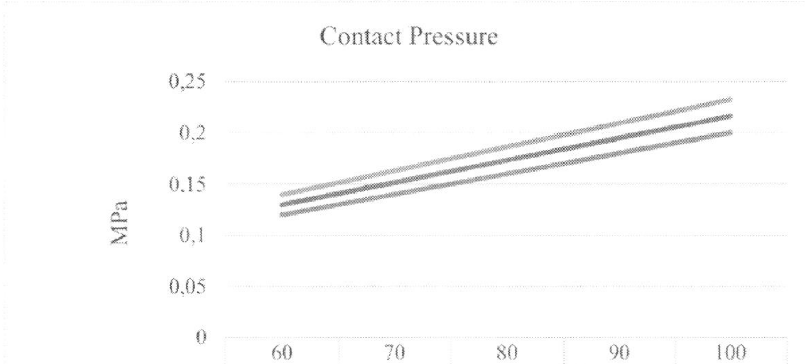

	60	70	80	90	100
Alumunium	0,12986	0,151504	0,173147	0,194791	0,216434
Polykarbonat	0,120045	0,140052	0,16006	0,180067	0,200075
HDPE	0,139712	0,162992	0,186279	0,209562	0,23286

Figure 7. Contact pressure graph comparison

Based on the results of the simulation, aluminum material has the lowest Von Mises Stress level compared to others while HDPE material has the largest Von Mises Stress. The difference between aluminum and HDPE of 0.060448 MPa for a force of 100 N. In aluminum materials have the smallest 1st Principal Stress while HDPE material has the largest 1st Principal than others. The difference between aluminum and HDPE of 0.08387 MPa for a force of 100 N. Then it was found that aluminum material has the shortest Displacement rate. While HDPE material has the longest displacement than others. The difference between aluminum and HDPE of 0.000351 mm for a force of 100 N. The polycarbonate material has the lowest Contact Pressure level while HDPE material has the highest Contact Pressure than others. The difference between aluminum and HDPE of 0.032785 MPa for a force of 100 N.

In addition, it is found that the value of von mises stress, displacement and contact pressure for HDPE is higher than other material (aluminum and Polycarbonate). However, the simulation also found that the differences among those materials are not significant. Therefore, the HDPE material could replace the existing cell phone case material.

4. Conclusion

The results of recycling HDPE waste can be used in making cell phone case. The result of recycle of HDPE plastic into a cellphone case proposed to reduce the spread of waste by managing it into a useful material for daily activities. The case has a size 16 x 7.5 x 9 cm. Based on the results of stress analysis in the Autodesk Inventor Program, if the force exerted is greater than the value of each measurement will increase. But it still visible because in VMS measurement, the result value smaller than yield. Based on comparison of stress analysis between the three materials, it can be concluded that aluminum is still the strongest material compared to polycarbonate and HDPE by having the smallest VMS, Displacement, and Contact Pressure. While HDPE material is the weakest material of the three materials, and other that that the difference is not significant. Therefore, the HDPE material is potential to be used as an alternative in a cellphone case.

References

[1] Mahyudin R 2017 Study of waste and environmental management problems at the final processing site (TPA) *Journal of Environmental Engineering* **3** 66-74
[2] Purwaningrum P 2016 Efforts to reduce the incidence of plastic waste in the environment *JTL* **8** 141-147
[3] Geyer R, Jambeck J R and Law K L 2017 Production, use, and fate of all plastics ever made *Science Advances* **3** 700-782

[4] Putra H P and Yebi Y 2010 Study of utilizing plastic waste into creative products and services *Journal of Environmental Science and Technology* **2** 1

[5] Abdelrasoul G N 2016 Nanocomposite scaffold fabrication by incorporating gold nanoparticles into biodegradable polymer matrix: synthesis, characterization, and photothermal effect *Materials Science and Engineering* **56** 305-310

[6] Desita Putri Pradani, Murtanti Jani Rahayu, Rufia Andisetyana Putri 2017 Classification of characteristics of industrial impacts in industrial affected settlement areas in cemani, sukoharjo regency *Architectural Journal* **15** 215-220

[7] Frizky Puja Rahmita 2015 Reports on drainage and environmental sanitation survey reports on the process of recycling plastic waste into a product *Sultan Ageng Tirtayasa University*

[8] Sanguanwong K, Nikzad M, Sbarski I and Masood S H 2018 Polymeric feedstock from post-consumer and post-industrial plastic wastes for automotive interior applications *IOP Conference Series: Materials Science and Engineering* **455** 012048

[9] Hildenbrand J 2019 Metal vs plastic vs glass which best material phones Retrieved from https://www.androidcentral.com/metal-vs-plastic-vs-glass-which-best-material-phones, accessed March 2019

Design and analysis of new box frame for online motorcycle transportation

Avid Christa Nugraha, Ferliadi R, Nurma Amelia A and T R Sahroni
Engineering Department, BINUS Graduate Program – Master of Industrial Engineering, Bina Nusantara University 11480, Jakarta, Indonesia.

*Email: taufik@binus.edu

Abstract. With the increasing needs and human's creativity, accompanied by the development of technological knowledge from various things, one example is the development of online-based transportation. Transportation is a tool in the form of a vehicle that purpose is to move an item from one place to another. In the current era of globalization, there are several companies that develop their business through online transportation, in addition to the shuttle service there is also a delivery or pick up of food/goods quickly that can provide convenience for the community. This paper describes the redesign of the box that motorcyclists could use in online transportation. The design concept is designed more innovative to design models and materials used are also more practical & efficient in terms of time in use both in terms of installation and dismantling of the box, the aim is to answer the problems of online transportation at this time, which are not only in terms of efficiency but also ergonomics. The research was conducted by observing the existing problems, then brainstorming in ergonomic box design and then making 3D modelling using the inventor's stress analysis method. to achieved a new innovative and ergonomic design for helping online transportation.

1. Introduction

In this globalization era, technology tends to develop more and more every time both from technological knowledge, tools used, communication networks and others, one of which can be seen from the ease of people doing their activities by using various technological tools such as smartphones, only by using smartphones, everyone can do various things such as shopping, looking for news, or traveling, with this ease the community can do all their activities in a short amount of time. Transportation is a tool used to move an item from one place to another, in this case using a vehicle. Basically transportation can determine the attractiveness of a city and can be used as a characteristic of the city, with the development of technology and human creativity, the use of transportation can be developed into a form of online applications, known as online transportation that can be applied and used on smartphones, but it can also be used as navigation to a place. So, this resulted in the emergence of several online transportation companies or better known as companies whose business activities are based online (e-commerce), which aims to be able to facilitate the public in traveling, shopping or shipping goods that must be done quickly [1].

The development of online transportation in Indonesia began in 2015, that year was a phenomenal time for the development of online transportation. Starting with the development of a mobile application called Go-Jek which was then followed by competitors such as Grab and especially Uber who had first entered the country in 2014. Entering 2016, the competition was getting tougher by launching several new features from each party is like a food delivery service [2]. If it is reviewed up to now the transportation company which initially only moves from online transportation has begun to spread its businesses with features such as non-cash payment methods that start using the application, which can further facilitate its users in the transaction. Not stopping there, they began to develop their business to the express / fast shipping feature which is very influential in today's e-commerce world. Due to the increasingly widespread distribution of e-commerce, thus causing a significant increase in demand for express delivery services to consumers [3]. To overcome the high demand, the solution obtained is to enlarge the scope of the shipment so that in one route there are several goods with different shipping destinations. In this case, it takes quite a long time compared to delivery by express or fast [4].

In the case of the development of the goods delivery feature contained in the online-based transportation application such as food delivery services or goods delivery services in a fast / lightning, there are several obstacles in the delivery process based on the results of a review that has been done previously to several online motorcycle taxi drivers who work in making deliveries food or goods, there are several complaints felt by the driver. Among them is there is no limit on taking the amount or weight limit of taking the order listed so that it can be difficult for the driver himself, then coupled with unpredictable weather conditions in the process of delivery or pickup of the goods so that it can be a problem for drivers who are taking and delivery of these goods both with a short distance or with the farther distance.

The purpose of making this box is to resolve the problem of the driver, which is not only in terms of efficiency but also in terms of ergonomics. Ergonomics becomes important in design not only because it meets design standards, but the aim is to prevent mistakes in the design process that will cause the design to be incompatible with the user's body. Therefore, ergonomic calculations must be considered and applied appropriately [5].

In the current situation, there are already several choices of boxes that can be used by drivers, but the existing box design feels less practical when mounted on a motorcycle. This is because the placement of the installed box does not have a suitable design, where the placement takes up quite a large amount of space in the seat of the motorcycle, therefore when the driver gets an order other than sending goods, the driver must remove the box and when the order process occurs the driver must re-install the box so that there is a less efficient process of time. These problems will be carried out by designing a multifunctional box that aims to facilitate the driver in sending goods ordered or also facilitate the use of the box in the use of other activities such as when the driver is delivering passengers without having to remove the box that is installed.

2. Method

Before doing this research, it has conducted several steps such as doing the observation on the current situation of box use, in this regard, it has formulated a problem to analyse box use in the current condition, then designing the right box model for online transportation users. Figure 1 shows a flowchart of the process a new concept of making box:

In general, the procedures of this paper are started in developing the design concept from sketching to detail design and follow the new product development concept. The detail procedures are as follow:

a) Identification of Problem: Firstly, doing the observation to online drivers for delivery order. What is the problem and inconvenience in delivery process?
b) Identification of User need: after doing the step 1, then conducted the brainstorm to start the ergonomics box design, and flexible. with purpose to facilitate online drivers in delivery order
c) Creation of Product Concept: doing the benchmarking with existing products
d) Product Specification: doing designing and determining of materials and size which will be used in the new concept of box design
e) Product Design: after doing step 4, then continues to 3D Modelling design using Inventor software

f) Product Concept Test: doing analyse stress by Inventor software to determine the maximum load which acceptable by new concept of box design
g) Product Analysis: doing the comparison between the new concept of box design and benchmarking result
h) Conclusions: the result of new concept box design which help to online drivers for delivery order

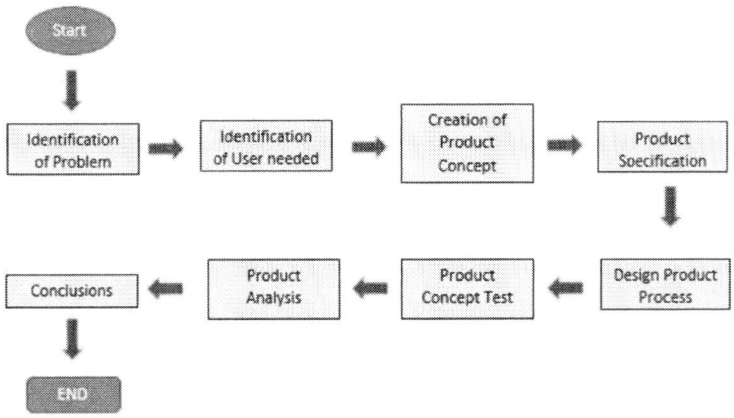

Figure 1. Flow chart of the process of making a new box concept

At the moment, most of delivery methods experienced by delivery company are used box as shown in Figure 2. However, this practice could lead unstable for the motor cycle driver when increase the speed and make manoeuvre during traffic jam or rush hour.

Figure 2. Fibber box

Figure 3 shows another box design that using fabric, it's also had weakness, due to the online drivers will difficult to ride with passenger, if they have a lot of goods to delivery in the same time

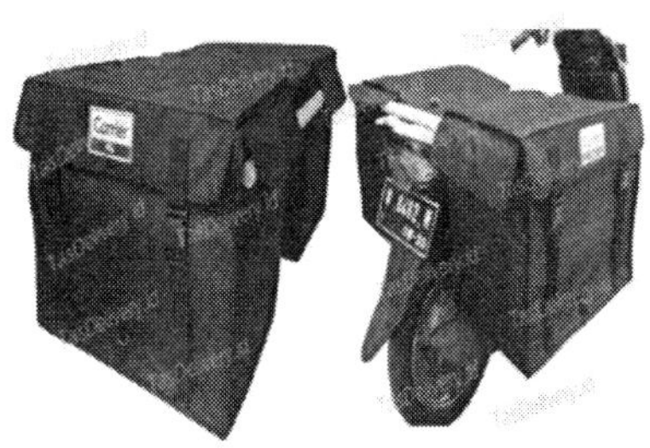

Figure 3. Side box

IOP Conf. Series: Materials Science and Engineering **909** (2020) 012052 doi:10.1088/1757-899X/909/1/012052

With the current condition of box design, it cannot give the easiness to online drivers due they have to unload the box when riding with passengers, and it's not flexible. Therefore, the purpose of research is to give the easiness to online drivers in riding with passengers and delivery order without take apart the box

3. Result and Discussion

This paper proposes a new product design that helps online motorcycle taxis or expedition services to send goods to customers easily and safely without having to take apart pairs in a long time and no need to feel worried during heavy rain. The box design that was designed also made it easy for online transportation to transport passengers without having to remove the box. This box design has a lock to make the frame strong. Here is a comparison analysis table of the new box concept with the previous design product.

Table 1. Comparison of current box vs new concept box.

Description	New Concept	Fibber Box	Side Box
Materials	Aluminium Frame with Parachute Fabric	Fiberglass Reinforced Plastic	Fabric
Size	505 X 320 X 325 mm	500 X 500 X 600 mm	500 X 300 X350 mm
Weight	3KG	8 – 10 KG	2,3 Kg
Capacity	52 liters	100 liters	50 liters/bag
Utility	- Bring a lot of goods - Flexible - No need to Disassembly assembly - High Visibility when driving - Waterproof materials - Stable when driving	- Bring a lot of goods - Permanent on Passengers seat - Waterproof Materials - Visibility when driving - Stable when driving	- Bring a lot of goods

Based on the comparison of Table 1. Prove that the latest box concept design has many functions and advantages compared to the previous box model. After doing a comparison of the box products, the stress analysis is carried out using the inventor software which can show an analysis of the results of the weight value of the goods carried and the speed of the motor that can affect the strength of the frame on the box.

The new box design concept is made more practical and makes it easier for drivers to fold the box when there are passengers so that passengers feel comfortable when riding a motorcycle. Figures 4, 5, 6 show the concept designed to be flexible and easy to process.

Figure 4. Folded view

Figure 5. Before folded view

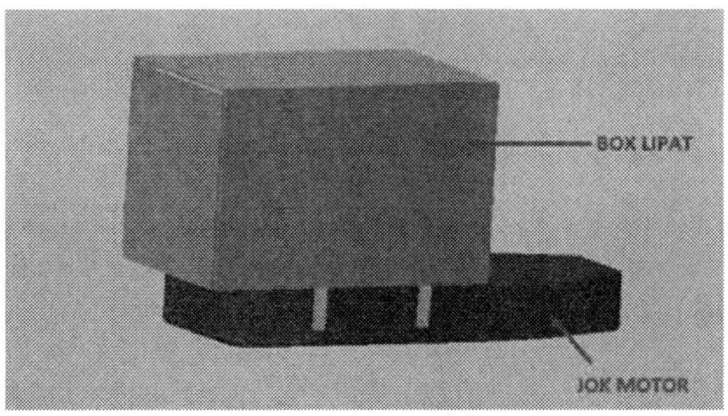

Figure 6. New concept box with parachute fabric

After designing the box, the maximum stress test was conducted on the Inventor yield strength software, from the data obtained as shown in Figure 7 which shows the value of 275 MPa which is the maximum stress that can be received by aluminum material used in the frame box and the ultimate tensile strength is achieved at 310 MPa. It is shown that the maximum force allowed to be applied in the box design is 275 MPa to avoid the aluminum material on the frame is damaged.

Table 2. Summary of product performance parameters

Name	Minimum	Maximum
Volume	21518700 mm^3	
Mass	3 kg	
Von Mises Stress	0.00000273086 MPa	128.943 MPa
1st Principal Stress	-40.4874 MPa	175.485 MPa
3rd Principal Stress	-138.124 MPa	39.9602 MPa
Displacement	0 mm	3.21379 mm
Safety Factor	2.13272 ul	15 ul
Stress XX	-50.0011 MPa	73.0341 MPa
Stress XY	-12.4238 MPa	35.8472 MPa
Stress XZ	-34.516 MPa	41.5779 MPa
Stress YY	-64.4678 MPa	81.6965 MPa
Stress YZ	-46.2858 MPa	58.6439 MPa
Stress ZZ	-124.101 MPa	135.958 MPa

Section stress of XX, XY, and XZ are shown on Figure 7, with a maximum stress value of 73.0341 MPa for section XX, 35.8472 MPa for part XY, and 41.5779 MPa for part XZ. It is found that all the figures are significantly safe to be applied in box frame design.

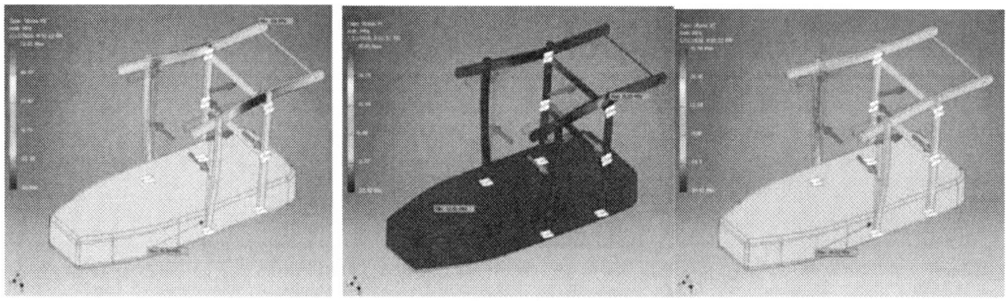

Figure 7. Section of XX, XY, XZ

The section of YY, YZ, and ZZ parts are shown in Figure 8, with a maximum stress value of 81.6965 MPa for YY, 58.6439 MPa for YZ, and 135.958 MPa for ZZ. The figure indicates the ability of the design to withstand several loads and to ensure to be safe when the design to be implemented in the motorcycle.

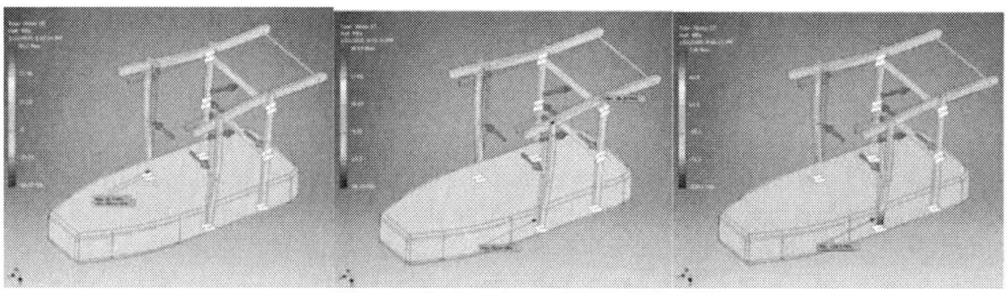

Figure 8. Section of YY, YZ, ZZ

Figure 7 shows the stress that given to part XX, XY, XZ, and figure 8 on the part YY, YZ, and ZZ still accepted by the product made by aluminum. Color that is shown in figure 7 and figure 8 shows that the product has not damaged when received stress on several sides. The blue color on the product is indicated the product in good condition, if the color turns into the red it means the stress that given to the product is over so that it can cause damage to the product. Figure 7 shows the stress that received on the side of the product, figure 8 shows the stress that received by-product on the backward

4. Conclusion

From the results of the analysis carried out for design, 3D modeling, and stress analysis, using stress analysis inventor yield strength software. it can be concluded that by making designs that can be folded could make it easier for motorists in the assembly and disassembly process of the box, and for the maximum stress analysis performed shows a value of 275 MPa which is the maximum stress that can be accepted by aluminum material used in the frame box and for ultimate tensile strength of 310 MPa to be able to show the maximum force before the aluminum material in the frame is damaged, in this case, the maximum load conditions that can be placed in the box are 50 kg for a speed of 72 km/h. As a result, a new design of the box frame is the potential to be applied for motorcycle online transportation.

References

[1] Nürnberg, Mariusz 2019 Analysis of using cargo bikes in urban logistics on the example of stargard. *Transportation Research Prodia,* **39** 360-369

[2] Aditya Hadi Pratama 2016 *Kilas balik perkembangan transportasi online sepanjang tahun 2016* https://id.techinasia.com (accessed on December 12, 2019)

[3] Morganti, Elenora, Saskia Seidel, Corinne Blanquart, Laetitia Dablanc, Barbara Lenz 2014 The impact of e-commerce on final deliverie alternative parcel delivery services in France and Germany. *Transportation Research Prodia.* **4** 178-190
[4] Punakivi, Mikko, Hannu Yrjola, Jan Holmstrom 2001 Solving the last mile issue reception box or delivery box. *International Journal of Physical Distribution & Logistics Management.* **31** 427-437
[5] Broberg, Ole 1997 Integrating ergonomics into the product development process. *International Journal of Industrial Ergonomics.* **19** 317-327

The measurement of customer satisfaction towards the service quality at xyz wholesale by using fuzzy service quality method

Chalis Fajri Hasibuan*

Industrial Engineering Departmenent, Faculty Of Engineering, Universitas Medan Area, Indonesia

*Email: chalisfajri@yahoo.co.id

Abstract. The service quality is the most important factor for the company, in order for the company can keep running and developing. The ways to improve service quality are to find out and understand the customers' desires. The quality service happens when the services or products meet the customers' expectations. This research was conducted at XYZ wholesaler to find out the quality service level and propose the attributes that need to be improved the quality of services. Furthermore, based on the gap value per attribute that needs to be prioritized by the grocery to being improved was the support facilities distance and the shopping area obtained the highest gap value of -4.08. Whereas the lowest gap value was in grocery location is 2.33. This matter showed that the grocery location was the attribute rated by the customers better than the other attributes. While based on the value per dimension, the dimension that needs to be prioritized by the grocery to being improved was the empathy dimension which obtained the highest gap value of -2.20. Whereas the lowest gap value was the assurance is 0.05. It showed that customer' assessments towards the assurance dimension better than the others dimension

1. Introduction

Satisfaction is a person's feelings of pleasure or dissapointment that result from comparing a product's perceived performance or outcome to expectations. If the performance falls short of expectations, the outcome is dissatisfied. If it matches expectations, the customer is satisfied or delighted [1].

At present every service provider is involved in increasingly fierce business competition, where each service provider must provide optimal services so as to increase customer satisfaction and be able to deal with competitors who are homogeneous service providers in their fields. Many service providers do not respond well to consumer complaints in the use of services, resulting in a decrease in the level of consumers who come to use the service. Consumers always want optimal service delivery [2][5].

Providing optimal services is one of the high and low factors of consumers who come in using these services. One of them is the wholesale service, the wholesale service is a large trading business. wholesale will have many visitors if it is able to offer more optimal services such as product completeness, price affordability, comfort of the place, cleanliness of the place, and friendliness of service. These factors can be taken into consideration for consumers to come and shop. Based on data from the number of customers from XYZ Wholesale in the last 6 months starting from September 2018 to February 2019, showed a decrease in consumers visiting. Based on the description above, the researcher wants to measure the level of satisfaction at the wholesaler.

IOP Conf. Series: Materials Science and Engineering **909** (2020) 012053 doi:10.1088/1757-899X/909/1/012053

2. Research Methods

The study was conducted by distributing Servqual questionnaires to 100 consumers. This Servqual questionnaire consists of 5 dimensions, tangibles, reliability, responsiveness, assurance and empathy. as for the stages in this study

1. Determination number of Samples, Sampling in this study using the Slovin formula
2. Validity Test, Validity testing can use the Product Moment equation
3. Reliability Test
4. Fuzzyfication and Defuzzyfication
5. Servqual Gap Calculation

3. Result and Discussion

In this study the data collected is a measurement of customer satisfaction

3.1. Determination number of sample

Sampling in this study using the Slovin formula. The sample size in this study was determined by the Slovin formula as follows:

$$n=\frac{5000}{(1+(5000)(0,1)^2)}=98,03\approx 100 \quad (1)$$

3.2. Validity test

Validity testing can use the Product Moment correlation equation, with the formula is as follows

$$r=\frac{n\sum xy-(\sum X)(\sum Y)}{\sqrt{[n\sum X^2-(X)^2][n\sum Y^2-(\sum Y)^2]}} \quad (2)$$

Table 1. Result validity test of satisfaction levels

No	Question	r_{Count}	Validity
1	P1	0,593	Valid
2	P2	0,719	Valid
3	P3	0,671	Valid
4	P4	0,593	Valid
5	P5	0,549	Valid
6	P6	0,559	Valid
7	P7	0,631	Valid
8	P8	0,553	Valid
9	P9	0,533	Valid
10	P10	0,593	Valid
11	P11	0,671	Valid
12	P12	0,673	Valid
13	P13	0,724	Valid
14	P14	0,515	Valid
15	P15	0,719	Valid
16	P16	0,593	Valid
17	P17	0,588	Valid
18	P18	0,618	Valid
19	P19	0,632	Valid
20	P20	0,616	Valid
21	P21	0,698	Valid
22	P22	0,725	Valid
23	P23	0,730	Valid
24	P24	0,691	Valid
25	P25	0,664	Valid
26	P26	0,645	Valid

Table 2. Validity test of importance level

No	Question	r_{Count}	Validity
1	P1	0,794	Valid
2	P2	0,514	Valid
3	P3	0,526	Valid
4	P4	0,781	Valid
5	P5	0,763	Valid
6	P6	0,748	Valid
7	P7	0,511	Valid
8	P8	0,569	Valid
9	P9	0,631	Valid
10	P10	0,758	Valid
11	P11	0,798	Valid
12	P12	0,535	Valid
13	P13	0,548	Valid
14	P14	0,735	Valid
15	P15	0,571	Valid
16	P16	0,794	Valid
17	P17	0,654	Valid
18	P18	0,763	Valid
19	P19	0,748	Valid
20	P20	0,739	Valid
21	P21	0,732	Valid
22	P22	0,710	Valid
23	P23	0,643	Valid
24	P24	0,739	Valid
25	P25	0,523	Valid
26	P26	0,594	Valid

3.3. Reliability test

Tabel 3. Reliabilitas testing

No	Variable	Score	Reliability
1	Satisfaction Levels	0,939	Reliable
2	Importance Level	0,950	Reliable

3.3. *Fuzzyfication and defuzzyfication*

The following is a formula for determining the Triangular fuzzy Number (TFN) [3][4].

a. Upper limit value

$$b_i = \frac{(b_{i2}n_1+b_{i3}n_2+b_{i4}n_3+b_{i5}n_4+b_{i5}n_5)}{n_1+n_2+n_3+n_4+n_5} \quad (3)$$

b. Middle Limit Value

$$a_i = \frac{(b_{i1}n_1+b_{i2}n_2+b_{i3}n_3+b_{i4}n_4+b_{i5}n_5)}{n_1+n_2+n_3+n_4+n_5} \quad (4)$$

c. Lower Limit Value

$$c_i = \frac{(b_{i1}n_1+b_{i1}n_2+b_{i2}n_3+b_{i3}n_4+b_{i4}n_5)}{n_1+n_2+n_3+n_4+n_5} \quad (5)$$

The average values of $a_i, b_{i,}c_i$ are defuzzyfication values which will be formulated as follows [6][7]:

$$d = \frac{a_i+b_i+c_i}{3} \quad (6)$$

Table 4. Recapitulation results level of satisfaction and defuzzyfication

Attribute	TFN			Defuzzyfication
	c_y	a_y	b_y	d_y
T 1	46,75	71,75	94,5	71,00
T 2	47,5	72,5	94	71,33
T 3	46,25	71,25	94	70,50
T 4	46,75	71,75	94,5	71,00
*R*1	47,75	72,75	94,5	71,67
R 2	51	76	95,5	74,17
R 3	44,75	69,75	92	68,83
R 4	49,25	74,25	93,7	72,42
R 5	46,5	71,5	93	70,33
R 6	46,75	71,75	94,5	71,00
Rp 1	46,25	71,25	94	70,50
Rp 2	46,75	71,75	95,2	71,25
Rp 3	47,25	72,25	94,5	71,33
Rp 4	45,75	70,25	92,7	69,58
Rp 5	47,5	72,5	94	71,33
A 1	46,75	71,75	94,5	71,00
*A*2	51,25	76,25	96,5	74,67
A 3	48,75	73,75	94,7	72,42
*A*4	49,75	74,75	95,5	73,33
A 5	49	74	95,2	72,75
E 1	46,25	71,25	93,5	70,33
E 2	45,75	70,75	94	70,17
E 3	48,25	73,25	94,7	72,08
E 4	45,25	70,25	93,2	69,58
E 5	45,75	70,75	93,5	70,00
E 6	47,25	72,25	94,2	71,25

Table 5. Recapitulation results level of importance and defuzzyfication

Attribute	TFN			Defuzzyfication
	c_x	a_x	b_x	d_x
T 1	47,2	72,2	91,5	70,33
T 2	46	71	90	69,00
T 3	52,7	77,7	93,2	74,58
T 4	46,7	71,7	91,2	69,92
*R*1	51,2	76,2	94	73,83
R 2	51,2	76,2	94,5	74,00
R 3	48	73	92	71,00
R 4	48,5	73,5	91,5	71,17
R 5	50	75	93,2	72,75

Attribute	TFN			Defuzzyfication
	c_x	a_x	b_x	d_x
R 6	48	73	92,2	71,08
Rp 1	48,2	73,2	92,5	71,33
Rp 2	48,2	73,2	92	71,17
Rp 3	50,7	75,7	92,5	73,00
Rp 4	47,0	71,7	91,3	70,05
Rp 5	51,7	76,7	94,5	74,33
A 1	47,2	72,2	91,5	70,33
*A*2	51,7	76,7	94	74,17
A 3	48	73	92,2	71,08
*A*4	51,2	76,2	94,5	74,00
A 5	51,7	76,7	94,5	74,33
E 1	50,7	75,7	93,7	73,42
E 2	51	76	94,2	73,75
E 3	50,2	75,2	92,5	72,67
E 4	50,2	75,2	93,5	73,00
E 5	49,7	74,7	91,2	71,92
E 6	48	73	91,2	70,75

3.4. *Servqual Gap Calculation*

Tabel 6. Gap calculation for each attribute

Attribute	Satisfaction Level d_{yi}	Importance level d_{xi}	Gap	Rank
T 1	71	70,33	0,67	5
T 2	71,33	69	2,33	1
T 3	70,5	74,58	-4,08	26
T 4	71	69,92	1,08	4
*R*1	71,67	73,83	-2,16	19
R 2	74,17	74	0,17	9
R 3	68,83	71	-2,17	20
R 4	72,42	71,17	1,25	3
R 5	70,33	72,75	-2,42	21
R 6	71	71,08	-0,08	11
Rp 1	70,5	71,33	-0,83	15
Rp 2	71,25	71,17	0,08	10
Rp 3	71,33	73	-1,67	17
Rp 4	69,58	70,05	-0,47	12
Rp 5	71,33	74,33	-3	22
A 1	71	70,33	0,67	5,5
*A*2	74,67	74,17	0,5	7
A 3	72,42	71,08	1,34	2
*A*4	73,33	74	-0,67	14
A 5	72,75	74,33	-1,58	16
E 1	70,33	73,42	-3,09	23
E 2	70,17	73,75	-3,58	25
E 3	72,08	72,67	-0,59	13
E 4	69,58	73	-3,42	24
E 5	70	71,92	-1,92	18

Attribute	Satisfaction Level d_{yi}	Importance level d_{xi}	Gap	Rank
E 6	71,25	70,75	0,5	7,5

4. Conclusion

The conclusions obtained from this study are as follows:

1. Based on the gap value of each attribute that causes a decrease in satisfaction by the party to do repairs is the distance of supporting facilities to the shopping place that gets the highest gap of -4.08. While the lowest gap is the location of the shop with a gap value of 2.33. This shows that the location of the store is an attribute that is valued by consumers better than other attributes.
2. Based on the gap value per dimension, the dimension that results in a decrease in satisfaction by the wholesaler for improvement is the empathy dimension that obtains the highest gap of -2.20. While the lowest gap is assurance, with a gap value of 0.05. This shows that consumers' assessment of the assurance dimension is better than other dimensions, besides that it also shows that the criteria in the empathy dimension need to be prioritized to be improved.

5. References

[1] Parasuraman, A., Zeithaml, V.A., & Berry, L.L. 1991 Refinement and reassessment of the SERVQUAL scale. *Journal of Retailing* **67** 420-450

[2] Parasuraman, A., Zeithaml, V.A., & Berry, L.L. 1994 Reassessment of expectations as a comparison standard in measuring service quality: implications for future research *Journal of Marketing* **58** 111-124

[3] Pepra, A.A., & Atara, B.A. (2014). Assessing patient's satisfaction using SERVQUAL model: Acase of Sunyani Regional Hospital Ghan International *Journal of Business and Social Research* **4(2),** 133-143.

[4] Martin. Lala Nilawati 2018 Model Fuzzy Mamdani Untuk Penilaian Tingkat Kepuasan Pelayanan Pengaduan Masyarakat *Jurnal Informatika* **5** 237-247

[5] Sharma and Mehta 2004 Expectations, performance evaluation, and consumers' perceptions of quality *Journal of Marketing* **57** 18-34

[6] H. Sholikah and S. W. Iriananda 2017 Analisis kepuasan pelanggan travel menggunakan metode fuzzy service quality *Journal of Information Technology and Computer Science (JOINTECS)* **2** 69-74

[7] N. F. Kartika and S. Suprayogi 2017 Implementasi fuzzy-service quality terhadap tingkat kepuasan layanan mahasiswa *Jurnal Ilmiah SISFOTENIKA* **7** 39-49

IOP Conf. Series: Materials Science and Engineering **909** (2020) 012054 doi:10.1088/1757-899X/909/1/012054

The application of LPT, CEGA, and PSO method on flow shop scheduling with parallel machine

Y Muharni[1*], A Irman[1], E Febianti[1], T E Rubyanti [2], N N Sofa [3]

[1] Department of Industrial Engineering, Engineering Faculty, Universitas Sultan Ageng Tirtayasa, Banten, Indonesia
[2] Corporate Management Office Dept., PT Chandra Asri Petrochemical
[3] Engineering Departement, PT Tridharma Kencana

*e-mail : yusraini@untirta.ac.id

Abstract. The production system in a fabrication company often face some problems, such as the inability to deliver orders to the customer at the due time. This situation could drive the company to bear the cost of penalties. This problem can be overcome if the production schedule arranged properly. One of the indicator optimal scheduling is minimize of makespan. This research investigates a company who dealing with fabrication structure and make to order production system with flow shop production lines and parallel machine arrangements. In order to solve the scheduling problem, several methods are applied consist of the heuristic approach of the Longest Processing Time (LPT) method and the metaheuristic approach using the Cross Entropy method combined with Genetic Algorithm to Cross-Entropy Genetic Algorithm (CEGA) and Particle Swarm Optimization (PSO). The algorithm was run in MATLAB software and the results show that proposed method using the Particle Swarm Optimization (PSO) and Cross Entropy Genetic Algorithm (CEGA) obtained makespan value of 8,394 minutes. The makespan for existing scheduling condition is 9,117 minutes. The proposed methods are shown a better results.

1. Introduction

The production system in a fabrication company often have some problems, one of the problem is when unable to deliver orders to the customer at the due time. This can cause the company to bear the cost of penalties. One alternative solution is improving the production schedule. The importance of production scheduling is to properly manage existing resources in order to create effectiveness and efficiency. One of the indicator optimal scheduling is minimize of makespan. This research investigates a fabrication company which the production system is make to order with flow shop production lines and parallel machine arrangements. One of the late deliveries occurred when making Steel Bridge B-60 which is one of the types of Truss Bridge. The Company has priority rules that Cross Grider (CG), Top Chord (TC), Bottom Chord (BC), Diagonal Chord (DG), and Stringer (ST) part produce at the beginning because it has the biggest dimension. This scheduling research is limited to schedule only five parts that have a permanent sequence of jobs in the company. This research will provide the smallest makespan of scheduling results with Permutation Scheduling. This research has some assumptions that the state of the engine when production is normal, no one is in the process of maintenance or shut down, all raw materials are available at the beginning of production, and transportation time includes set-up time and set-up time will be added with process time into total

IOP Conf. Series: Materials Science and Engineering **909** (2020) 012054 doi:10.1088/1757-899X/909/1/012054

time. This research was carried out with the heuristic approach of the LPT (Longest Processing Time) method and the metaheuristic approach using the CEGA (Cross Entropy Genetic Algorithm) and Particle Swam Optimization (PSO) using help of MATLAB software.

The purpose of this research is to propose a better sequence of job in the production process of Steel Bridge B-60, calculate makespan before and after the flow shop scheduling, and calculate the percent efficiency of makespan.

2. Literature review

Production scheduling is defined as the process of allocating resources or existing machines to run a set of tasks within a certain timeframe.[1]

Longest Processing Time_is one of the method for scheduling parallel machine to minimize makespan and mean flow time. The step is sort jobs from the largest processing time to the smallest ($t1 \geq t2 \geq \ldots \geq tn$) into the list and do scheduling based on such order. CEGA Cross Entropy - Genetic Algorithm Algorithm is the development of an algorithm which combine of Cross Entropy and Genetic Algorithm (CEGA) algorithms. The combination of this algorithm aims to expand the solution search on elite Cross Entropy (CE) samples when in optimal locality by adding a crossing mechanism and mutation in the Genetic Algorithm (GA) which serves to avoid the possibility of finding solutions stuck in the optimal local area, because the GA mechanism can produce new chromosomes that have the nature of maintaining the chromosome nature in the initial population.[2]

Cross Entropy (CE) method has advantage to solve combinatorial problem such scheduling problem[3]. As an adaptive importance sampling procedure, this method has a good reputation when applied in wide range complicated problem. The PSO algorithm was introduced by Dr. Eberhart and Dr. Kennedy in 1995, was an optimization algorithm that mimicked the processes that occur in the lives of populations of birds and fish in survival. PSO models search activities for the best solution in a solution space as the flight activity of groups of particles in a solution space.[4] The position of the particles in the solution space is a solution candidate that contains optimization variables. Each of these positions will be associated with a value called objective value or the value of fitness which is calculated based on the objective function of the optimization problem to be resolved [5] The application of PSO in solving scheduling flow shop problem has been applied broadly [6].

3. RESULTS AND DISSCUSSION

This is the total job time data that obtained by adding the setup and process time of each job on each machine, and multiply it by the number of parts.

Table 1. Total time of job

Machine / Job	Cutting (Minutes)	Fit-Up Machine (Minutes)	Wedding LMH (Minutes)	Strightening (Minutes)	Fit-Up Fab (Minutes)	Welding (Minutes)	Finishing (Minutes)
TB60-CG	89,06	131,84	419,34	155,12	284,79	288,60	76,38
TB60-TC	150,70	220,96	638,86	262,28	478,70	487,04	130,84
TB60-BC	160,56	244,12	707,76	279,16	532,80	526,60	135,08
TB60-DG	323,40	484,12	1545,00	562,24	1056,68	1058,92	274,64
TB60-ST	565,04	751,90	2431,32	975,14	1679,08	1678,08	313,40

Table 2. Available machine

No	*Machine Name*	*Quantity of Machine*
1	*Cutting Machine*	1
2	*Fit-Up Machine*	1
3	*Welding LMH*	2
4	*Straightening*	1
5	*Fit-Up Fab*	2
6	*Welding* (Manual)	3
7	*Finishing Station*	2

3.1 Existing Scheduling

This is the existing scheduling in the company with the sequence of job is 1-2-3-4-5. The Gantt Chart in Figure 1 shows the makespan for existing scheduling is 9117 minutes.

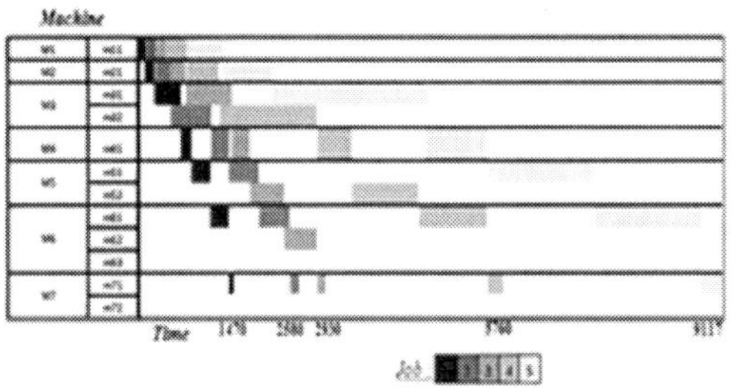

Figure 1. Gantt chart of existing scheduling

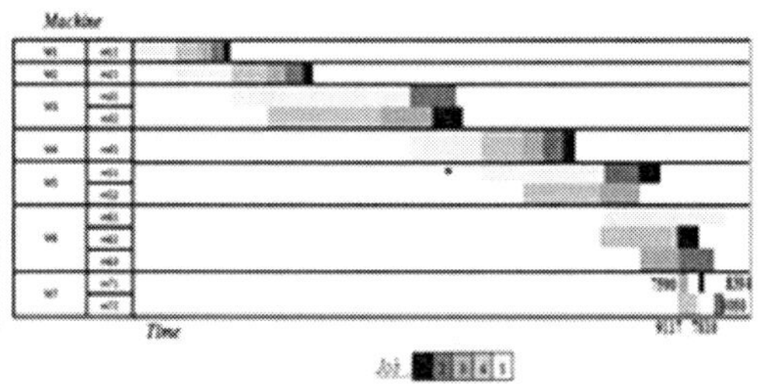

Figure 2. Gantt chart of LPT method scheduling

3.2 Scheduling with Longest Processing Time (LPT) method

This is the gantt chart of scheduling with LPT method with the sequence of job is 5-4-3-2-1. The makespan based on the gantt chart LPT scheduling above is 8394 minutes.

3.3 Scheduling Using Cross Entropy-Genetic Algorithm (CEGA) Method

1. Step 1: Select initial parameter
   Initial Sample (N) = 10, Elite Sample Ratio (ρ) for determining the elite sample = 3%, smoothness index.
   (α) = 0.5, cross over parameter (Pps) = 1, stopping criteria (β) = 0.0001
2. Step 2: Generate initial random population and calculate makespan as the objective function
   Generate initial random population with roulette wheel method and random value mechanism. The first step is count the total probability of each permutation (example = 5 job, 5! = 120 permutations with 120 probability) and generate 10 random number for select the population sample.

Table 3. Generate random sample

No	Random Value	Probability Value	Permutation	Makespan (Minutes)
1	0.026	0.025	12435	9120
2	0,178	0,179	23541	8705
3	0,238	0,238	24351	9031
4	0,290	0,290	31425	9117
5	0,313	0,313	31254	8794
6	0,411	0,411	35142	8554
7	0,456	0,457	42153	9429
8	0,531	0,533	45123	8717
9	0,570	0,567	51432	8394
10	0,676	0,677	54123	8394

3. Step 3: Select elite sample
   The formula of selecting elite sample is : ρ x N (1), Elite sample = 0.03 x 10 = 0.3 ≈ 1.So it will be 1 elite sample that will be still appear until the last step, it never be cross over or mutation. The elite sample is a job with the smallest makespan which is sample 9 with the sequence of job is 5-1-4-3-2.
4. Step 4: Calculate Linier Fitness Rangking (LFR)
   The purpose of calculate LFR is to select parents for crossover process. This is formula for calculate LFR:

$$\text{LFR}\left(\text{I(N-i+1)}\right)=\text{Fmax-(Fmax-Fmin)*}\frac{(i-1)}{(N-1)} \tag{2}$$

where $\text{Fmax}=\frac{1}{Z(1)}$; $\text{Fmin}=\frac{1}{Z(N)}$; $\text{Fmax}=\frac{1}{8394}=0.00119$, $\text{Fmin}=\frac{1}{9429}=0.00106$

5. Step 5: Select Cross Over Parents
   a. Select Parents 1
      Parents 1 in this cross over process is the elite sample which is the job with the smallest makespan. So the parents 1 is Sample 9 with the *job sequence is* 5-1-4-3-2.
   b. Select Parents 2
      This process using LFR value. If cumulative LFR bigger than random value, then the sample become parents 2.
6. Step 6: Cross Over Update Parameter
   Cross Over Update parameter is used to get the new parameter for stopping criteria evaluation.

$$\text{Pps(i)} = (1\text{-}\alpha)\text{*}(\frac{\overline{Z}e}{2 \text{ x Zbest}})\text{+(Pps*}\alpha) \tag{3}$$

$$\text{Pps(i)} =(1\text{-}0.5)\text{*}(\frac{8394}{2 \text{ x } 8394})\text{+(1*0.5), Pps(i)} = 0.75$$

   The new parameter update value will be used for a basic to determining whether a chromosome needs to be cross over or not.

7. Step 7: Cross Over. Cross Over methods that used in this research is 2-point order Cross Over

IOP Conf. Series: Materials Science and Engineering **909** (2020) 012054 doi:10.1088/1757-899X/909/1/012054

Table 4. Comparison of LFR and random value

Sample	LFR	Cumulative LFR	Notation	Random Value
S9	0.000119	0.10581	<	0.82211
S10	0.000118	0.21033	<	0.82211

Table 5. Generate random value for select cross over

Sample	Random	Pps	Conclusion
S9	0.95321	0.75	No *Cross Over*
S10	0.50262	0.75	*Cross Over*

To determine the part on parents to be crossing, two random numbers are generated for the crossing process in ceil or the parent gene. The random value is then converted to a round value, and it used as a delimiter of the sample portion that will be moved between parents.

Table 6. Cross over process

Cross Over Parents	Job Sequence	Random Value	Ceil (random *n)	Ceil Cross Over	Cross Over Result	Result
S9 VS S1	51432	R1= 0,22243	1,11215 = 2	51|4|32	51432	For sample S10
	12435	R2= 0,58040	2,90200 = 3	12|4|35	12435	For sample S6
S9 VS S1	51432	R1= 0,50240	2,51200 = 3	514|3|2	14352	For sample S2
	42153	R2= 0,62441	3,12205 = 4	421|5|3	42135	For sample S8

8. Step 8: Mutation
Mutations are intended to bring new individuals who are different from existing individuals. The mutation process will be chosen randomly and the genes on the site will be changed in value. If the random number (R) is smaller than the parameter mutation (Pm), the digit of the gene will be replaced, and vice versa.(2) The mutation parameter values are determined by the following formula:

$$Pm = \frac{P\_ps}{2} \tag{4}$$

After the value of the mutation (Pm) and N parameters set at is 10 samples, according to (2) the number of chromosome samples that will experience mutations can be calculated by the formula.

$$Na = Pm \times N \tag{5}$$

Then it can be seen that there are 4 chromosome samples that will undergo a mutation process. To determine the type of mutation on a chromosome, can be calculate as follows:

$$K = \text{ceil}\,(Rand\,K \times 3) \tag{6}$$

From these results, there are provisions as follows:
k=: flip mutation, k =2: swap mutation, k =3 slide mutation

Table 7. Mutation process result

Sample	Cross Over Result	Mutation Random	Mutation Probability (Pm)	Result	Random (ri,rj)	I,J Value with n = 5	Random K	K Value with k=3	Mutation process	Final Result
S9	51432	0,0316	0,375	Still S9						
S10	51432	0,00694	0,375	Mutation	0,14509 0,148895	0,72545=1 2,44475 = 3	0,03049	0,09147=1	Flip Mutation	54132
S6	12435	0,3455	0,375	Mutation	0,11152 0,64308	0,55760 = 1 3,21540 = 4	0,59564	1,786692=2	Swap Mutation	13425
S2	14352	0,94869	0,375	No Mutation						
S8	42135	0,55741	0,375	No Mutation						
S5	31254	0,79088	0,375	No Mutation						
S3	24351	0,1264	0,375	Mutation	0,12966 0,73908	0,64830 = 1 3,69540=4	0,13956	0,41868=1	Flip Mutation	25341
S4	31425	0,91209	0,375	No Mutation						
S1	12435	0,67069	0,375	No Mutation						
S7	42153	0,56825	0,375	No Mutation						

Table 8. Makespan From New Population

Sample	New Job Sequence	Makespan (Minutes)
S9	51432	8394
S10	54132	8394
S3	25341	8544
S5	31254	8794
S6	13425	9117
S4	31425	9117
S1	12435	9120
S2	14352	9329
S7	42153	9429
S8	42135	9680

9. Step 9: Calculate new objective function (makespan) from the new population
This is the result of calculating the value of new objective function (makespan) of the new sample.

Based on Table 9, it is known that the makespan value with the CEGA method shows that the minimum value is 8394 minutes with a job sequence of 5-1-4-3-2 and 5-4-1-3-2.

In scheduling the CEGA method using MATLAB software, the iterations are carried out 3 times by changing the parameters of the smooth coefficient value of low (0.1), medium (0.5) and high (0.9). The Gantt chart is built according to result from MATLAB. The Gantt chart of scheduling with the smallest makespan that is until S9 with a job sequence of 5-1-4-3-2 and sample S10 with a sequence of 5-4-1-3-2.

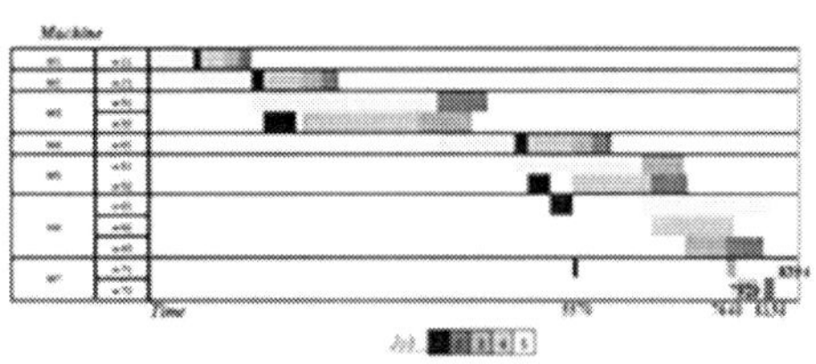

(a)

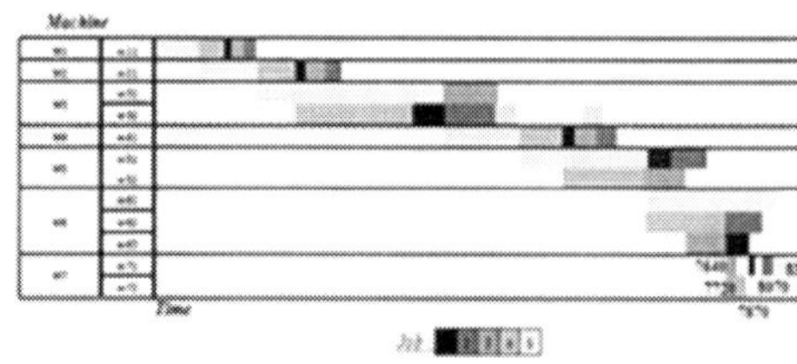

(b)

Figure 3. Gantt chart of the best CEGA scheduling, (a) sample 9th (b) sample 10th

4. Scheduling Using Particle Swam Optimization (PSO) Method

The result in this method obtained by running process in the MATLAB software, here are the stages of the PSO method based on the results of running iteration 1:

4.1. Determination of Initial Parameters

In this research the determination of the initial parameters used are: Iteration = 10, Particle Number (N) = 5, Maximum Velocity (vmax) = 0.8, Minimum Velocity (vmin) = 0.2, Inertia Weight (w) = 0.8, Cognitive Learning Factor (c1) = 2, Social Learning Factor (c2) = 2

4.2. Initialization

The initial position of the particle is taken from the existing condition or the existing job sequence of 1-2-3-4-5 then multiplied by 0.1 so that the position value for each particle are 0,1; 0,2; 0,3; 0,4 ; 0.5.

4.3. Fitness Value and Personal Best Determination

The fitness value in iteration 1 is obtained from the makespan value of the existing condition which is equal to 9.118 minutes. Personal best is a vector that describes the best position for a particle based on the best fitness value. Personal best value in this research is adjusted to the initial particles so that the personal best value for iteration 1 is the same as the initial particle sequence of 1-2-3-4-5 job sequences.

4.4. Global Best Determination

In the previous stage, it was explained that for the iteration 1 the best personal best is obtained from the initial sequence of particles so that the global best value is the same as the personal best value.

4.5. Velocity Update

Here is the velocity update in iteration 1: for particle 1 is 0.3914, for particle 2 is 0.2702, for particle 3 is 0.7200, for particle 4 is 0.2954, and for particle 5 is 0.0890. As according to the example calculation as follows:

Example Calculation (for particle 1 in iteration 1):

$$v11 = wvi + c1\ r1\ (pbest11 - x1) + c2\ r2(gbest11 - x1) \quad (7)$$

v_{11} = 0,8 (-0,908) + 2 . 0,201 (1 - 0,1) + 2 . 0,420 (1 - 0,1)

v_{11} = 0,3914

4.6. Position Update

Here is the update position in iteration 1 for particle 1 is 0.4914, for particle 2 is 0.4702, for particle 3 is 1.0200, for particle 4 is 0.6954, and for particle 5 is 0.5890. As according to the example calculation as follows:

Example Calculation (for particle 1 in iteration 1):

4.7. Criteria Checking

At this stage there is a graph that shows the results of the PSO algorithm, and for more details can be seen in the picture below:

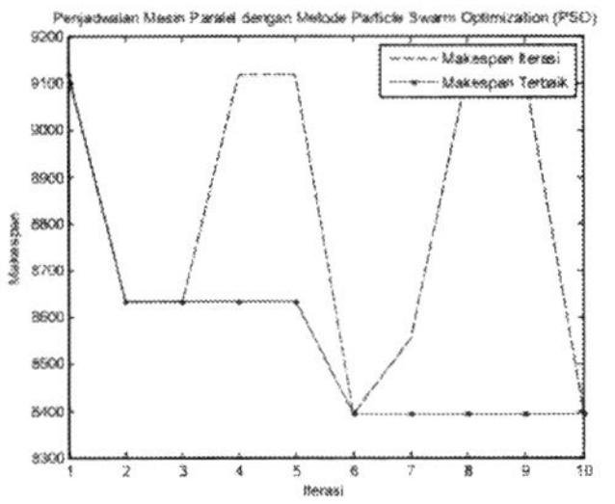

Figure 4. Makespan reduced during iteration by applying PSO Method

Based on the picture above, the best makespan value obtained from the fitness value in each iteration (10 iterations) is indicated by a blue line where it starts from the iteration 1 graph decreases to iteration 2 and then starts from iteration 2 to iteration 5 graph constant, then graph decrease at iteration 6 and constant until iteration 10, it indicates that from iteration 1 to iteration 6 still changes but after iteration 6 to iteration 10 the makespan value remains constant or unchanged. From the picture above it can also be seen that the determination of the criteria when to stop to get the best makespan value is at the iteration 6 with the makespan value of 8,394 minutes with the sequence ob 5-2-1-4-3. The results of the gantt chart obtained for the best iteration are 6 iterations this PSO method that have a sequence of job 5-2-1-4-3 is shown in Figure

Based on the gantt chart above, it can be seen that makespan scheduling using the PSO (Particle Swarm Optimization) method is 8,394 minutes.

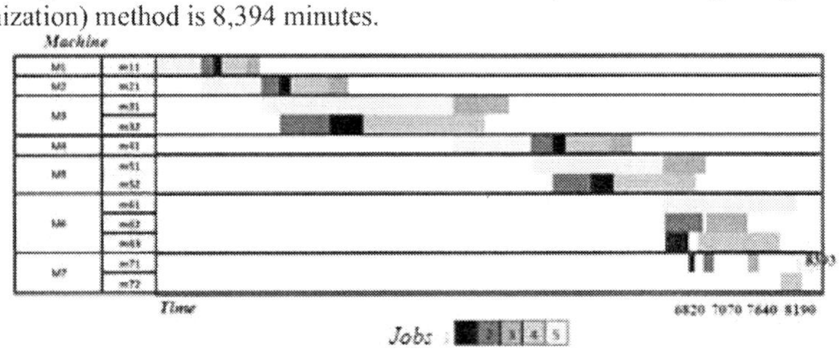

Figure 5. Gantt chart scheduling of PSO method (Particle Swarm Optimization) as the best iteration (iteration 6

4.8. Comparison Between Existing Scheduling with LPT, CEGA and PSO method

This is the table for Comparison Between Existing Scheduling with LPT, CEGA and PSO method

Table 9. Comparison makespan for eksisting, LPT, CEGA and PSO

No	Method	*Job Sequence*	Makespan (Minutes)	Efficiency
1	Existing	1-2-3-4-5	9117	[redacted]
2	LPT	5-4-3-2-1	8394	7,94%
3	CEGA	5-1-4-3-2	8394	7,94%
4	CEGA	5-4-1-3-2	8394	7,94%
5	PSO	5-2-1-4-3	8394	7,94%

$$\text{LPT Efficiency} = \frac{\text{Makespan Existing Method-Makespan LPT Method}}{\text{Makespan LPT Method}} \text{x} 100\%$$

$$= \frac{9117\text{-}8394}{9117} \text{ x } 100\% = 7{,}94\%$$

5. Conclusion

Based on the calculation through the heuristic method and the metaheuristic one, the minimum makespan has been achieved for the paralel machine scheduling in this research which is the 8394 minutes. The heuristic and the metaheuristic has shown their best performances. Since the metaheuristic algorithm such as CEGA and PSO have complicated procedure, for the managerial practising, the application of LPT method is recommended.

References

[1]. Baker KR, Trietsch D.Principles of sequencing and scheduling 2013

[2]. Lopez-Garcia P, Onieva E, Osaba E, Masegosa AD and Perallos A 2016 GACE: A meta-heuristic based in the hybridization of Genetic Algorithms and Cross Entropy methods for continuous optimization. Expert Syst Appl [Internet] 55 508–519. Available from: http://dx.doi.org/10.1016/j.eswa.2016.02.034

[3]. De Boer PT, Kroese DP, Mannor S, Rubinstein RY 2005 A tutorial on the cross-entropy method. Ann Oper Res. **134** 119–67.

[4]. Kuo IH, Horng SJ, Kao TW, Lin TL, Lee CL, Terano T, et al 2009 An efficient flow-shop scheduling algorithm based on a hybrid particle swarm optimization model. Expert Syst Appl [Internet] **36** (3 PART 2) 7027–32 Available from: http://dx.doi.org/10.1016/j.eswa.2008.08.054

[5]. Torabi SA, Sahebjamnia N, Mansouri SA, Bajestani MA 2013 A particle swarm optimization for a fuzzy multi-objective unrelated parallel machines scheduling problem *Appl Soft Comput J* [Internet] **13**12 4750–62. Available from: http://dx.doi.org/10.1016/j.asoc.2013.07.029

[6]. Muharni Y 2019 Production scheduling of bar mill using the combination of particle swarm optimization and Nawaz enscore ham for minimizing makespan in steel company Production Scheduling of Bar Mill Using The Combination of Particle Swarm Optimization and Nawaz Enscore Ham for Minimizing Makespan in Steel Company 030006

Fiscal incentive policy for coal downstreaming in Indonesia

Ragimun[1*], Imran Rosjadi[2]
[1]Researcher, Fiscal Policy Agency, Ministry of Finance of Indonesia
[2]Policy Analyst, Fiscal Policy Agency, Ministry of Finance of Indonesia

Email: ragimun@gmail.com

Abstract. This study aims to describe the coal downstreaming program which is expected to increase the added value of coal in Indonesia. Increasing the added value of coal such as the gasification project has a strategic role to increase state revenues and foreign exchange and is expected to have implications for increasing development in the region and nationally. The research method used is a quantitative and qualitative approach. The quantitative approach is used the accounting approach and benefit cost analysis (BCA), while the qualitative approach is used in the study of literature and focus group discussions. From the results of the discussions, the Government can provide coal gasification fiscal incentives to pioneering companies with the consideration that this downstreaming program will be able to spur state revenue as well as regional and national economic growth. Besides this downstreaming activities will be able to increase similar activities in Indonesia.

1. Introduction

Indonesia is a country endowed with abundant natural resources that store oil, gas, minerals, coal and others [1]. The potential yield of this mineral and coal is very large for state revenues both from taxes and non-taxes such as fixed fees, exploitation fees (royalties). Non-tax state revenue from natural resources currently has a tendency to increase. Contributions from oil and gas, minerals and coal reached more than IDR 90 trillion or accounted for 95 percent of natural resource revenues. The oil and gas and mining sector also has a 7.2 percent share as the main contributor to the Gross Domestic Product (GDP) after the trade, agriculture and construction services sector [2]. The realization of PNBP until 2018 for several natural resources such as copper, gold, tin nickel and silver has an increasing tendency. Realization of coal non-tax revenue was very dominant. In 2018, coal non-tax revenues amounted to IDR 21.85 trillion, an increase from 2017 which only amounted to IDR 18.60 trillion [3].

One of the downstream policies such as the export duty policy is intended as one of the restrictions on the export of raw minerals in the short term that will have a negative impact on the economy, in particular causing a decline in raw mineral exports and new investment in the mineral mining sector[4]. However, the export restriction policy will have a positive impact on the economy if supported by new investments in the mineral processing sector so that Indonesia can export processed mineral commodities [5].

In fact, the Government has provided many incentives for companies or Minerba businesses to produce or increase the added value of mining products by downstreaming. However, until now Minerba's derivative products, especially coal in the form of coal liquefaction process (gasification) have not been as expected.

This study aims to illustrate the opportunities for fiscal incentives for pioneer companies to increase the value added or downstream of coal. In addition, it can be seen the impact or benefits of providing fiscal incentives for the national economy, social benefits and labor absorption as well as tax and non-tax revenue. Furthermore, this research is expected to obtain recommendations to ensure the granting of coal downstream fiscal incentives.The goal application of fiscal incentives to all of coal product like cokies making, coal upgrading, coal briquetting, coal gasification, underground coal gasification, coal liquefaction, coal water mixtureare to increase value added coal, and altimately, coal mining companies hopefully could develop public settlement in underdevelop regions and public open space around the mine. Bycoal utilization for the Downstream Project, hopefully there are an increase in both the Central and Regional Tax Potentials caused by a new tax object for transactions from these activities.

2. Methods

The method used in this research is quantitative and qualitative methods. The quantitative method in this study used the accounting approach [6] and Benefit Cost Analysis (BCA) to create a benefit and cost scenario providing fiscal incentives for coal gasification, with the following formula[7] :

$$PV^i = \left(B_0^i - C_0^i\right) + \frac{\left(B_1^i - C_1^i\right)}{(1+r)} + \ldots + \frac{\left(B_T^i - C_T^i\right)}{(1+r)^T} \quad (1)$$

Where,

PV^i=Present Value project i

B_t^i = Benefit from project i at time t

C_t^i = Cost from project I at time t

The fiscal incentive scenario provided is tax allowance, tax exemption and royalties which can affect macroeconomic activities. The data used are data from the Central Bureau of Statistics and data from PT. Bukit Asam.To deepen the study also used explorative qualitative methods, by seeking information from policy makers about fiscal incentives to support coal gasification programs in Indonesia, through Focus Group Discussions and literature studies[8].

3. Result and Discussion

Natural Resources are all physical, chemical, biological and social things that shape the environment around us. According to Hunker[9], natural resources are all that come from the earth, biosphere, and atmosphere, whose existence depends on all human activities. All elements of the natural environment including soil, water, air, seeds, trees, sun, rivers are natural resources. One of Indonesia's natural resources which has not been optimally processed is coal.

Based on the BP Statistical Riview of World Energy report in 2019, Indonesia's coal reserves amounted to 3.1 percent of the total world coal reserves. While the countries that have the greatest natural wealth in the form of coal are the United States, Russia and China, as Figure 1.

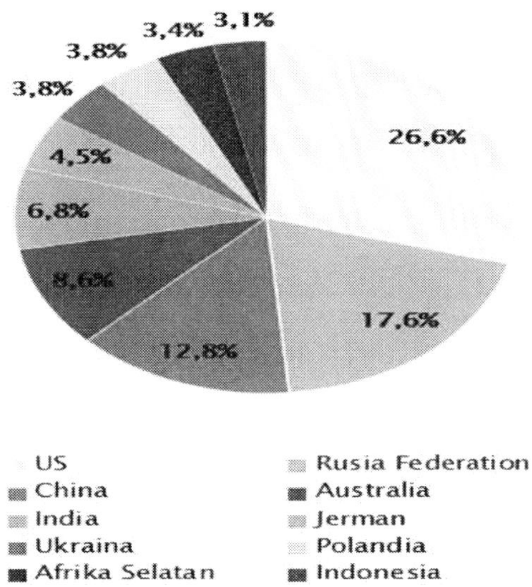

Figure 1. Coal reserves of several countries
Source: BP Statistical Reviews of World Energy, 2019 [10]

In Law number 30 of 2007 regarding energy, article 21 states that the use of energy is carried out among others by optimizing all potential energy resources, and prioritizing meeting the needs of the community. Furthermore, the Government Regulation (PP) governing coal is PP number 23 of 2010 concerning Implementation of Mineral and Coal Mining Business Activities. Whereas Article 84 Paragraph (1) states that the holder of production operation IUP and production operation IUPK must prioritize the needs of minerals and / or coal for domestic interests. Article 94 paragraphs (1) contains that for Production Operation IUP Holders and coal Production Operation IUPKs, they are obliged to carry out processing to increase the added value of coal produced either directly or in cooperation with companies, holders of IUPs and other IUPKs [11]. What is meant by processing in this provision includes coal crushing, coal washing, coal blending, coal upgrading, briquetting, coal liquefaction, coal gasification and Coal Water Mixing [12]. This can be described as follows:

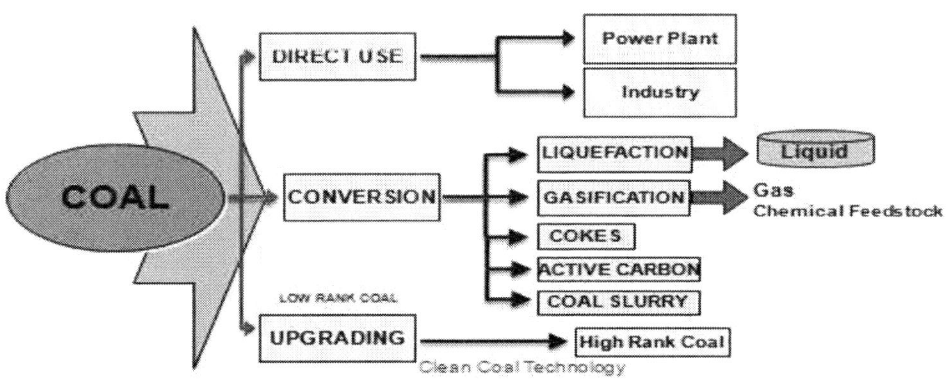

Figure 2. Derivation of coal downstream activities
Source: ESDM, 2016[13]

Downstreaming currently being developed includes coal gasification. At present there is only one State-Owned Enterprise company (BUMN) that has the opportunity to carry out gasification downstreaming, namely PT. Bukit Asam. The company will develop coal into gas (syngas) which is then converted into methanol and then made into Dimethyl Ether (DME) and Polyprpylene. There are two areas that will produce, namely in TanjungEnim, South Sumatra and Riau's role in the region. The two regions are expected to be able to produce DME of 400 KTA and 1,400 KTA. [14]

Currently coal production in Indonesia tends to be exported, whereas the target of national electricity power needs is always increasing. In the future, it is estimated that national coal consumption will increase along with the increasing need for fuel for power plants that still use coal. Indonesia has several countries to export to, namely India, China, South Korea, Japan, Taiwan, Europe and America. According to a report by The Oxford Institute for Energy Studies (OIES) in 2017 [15], it turned out that the last highest export of Indonesian coal occurred in 2013 and in the following years Indonesia's coal exports declined.

In accordance with the national coal production plan based on the National Medium Term Development Plan (RPJMN) for 2015-2019 and in accordance with Presidential Regulation number 2 in 2015, production in 2019 is 400 million tons. The percentage of domestic demand for national coal production is 60 percent in 2019 [16].

When viewed from the state revenue derived from non-tax Mineral and Coal state revenues for 2018 amounting to more than IDR 50 trillion or reaching 156 percent. While the revenue target of non-tax state revenue set in the 2018 State Budget for Development and Expenditure is only IDR 32.1 trillion. The composition of Minerba non-tax revenues in 2018 consists of Royalties of IDR 29.8 trillion or 59.6 percent, mining product sales of IDR 19.3 trillion or 38.6 percent, fixed contributions of IDR 0.5 trillion, and state revenues from services and information of IDR 0.4 trillion.

State revenues from non-tax state revenues from minerals and coal have been increasing since 2014. Successively in 2014 amounted to IDR 35.4 trillion, 2015 decreased by IDR 29.6 trillion, 2016 amounting to IDR 27.2 trillion. But in 2017 it has increased again to IDR 40.6 trillion and in 2018 it has risen to IDR 50 trillion. Furthermore, state revenue from non-tax state revenue from minerals and coal can be seen from several types of non-tax state revenue from minerals and coal such as fixed fees, exploitation fees in the form of royalties and royalties from the sale of mining products specifically for the Coal Mining Concession Work Agreement (PKP2B) [17].

In accordance with Regulation of the Minister of Energy and Mineral Resources number 17 of 2010, that certain types of coal used domestically can be sold at a price below the Coal Benchmark Price, after obtaining approval from the Director General. Coal for certain purposes can be sold at a price below the coal benchmark price, after obtaining approval from the Director General [18]. Certain types of coal are fine coal, reject coal and coal with certain impurities. Coal for certain purposes intended is coal used in the context of increasing the added value of coal carried out at the mouth of the mine, coal used by the Steam Power Plant whose electricity is only for the production process and for the development of disadvantaged areas. The price of certain types of coal and for certain purposes approved by the Minister of Energy and Mineral Resources is used in the calculation of Non-tax State Revenues. Furthermore, further provisions are regulated in Director General Regulation number 480 of 2014 [19].

The potential for increased value added to turn coal into gas is very high. Low-calorie coal in Indonesia has a great opportunity to develop or increase its added value. There are several companies that are interested in increasing value added, such as PT Bukit Asam, this company which will develop coal gasification in two projects. First, the coal gasification project located in TanjungEnim, South of Sumatra. This region has reserves of 363 million tons of GAR 4,200 Kcl / kg coal, annual coal consumption of 5.1 million tons plus a 2.9 million ton power plant. And the planned product is 400,000 tons of DME, 570,000 tons of Urea and 450,000 tons of Polypropyllene.

Second, the Coal gasification project in Peranap, Riau Sumatra. This region has coal reserves of 287 million tons of GAR 2715 Kcl / kg, annual coal consumption of 6,600,000 tons plus a power plant of 2,100,000 tons. The production plan is 1,400,000 DME.

3.1. Benefits of Coal Downstreaming for the Economy in Indonesia

The coal downstream program in Indonesia turned out to have an impact on increasing national economic activity in several sectors including the economic, industrial, taxation, and investment sectors.As an illustration, PT Bukit Asam, as one of the coal downstream pilot project companies in Indonesia, is expected to operate in 2022. The impact produced by increasing coal production and investment downstreaming to the national GDP is IDR 108.13 trillion from 2017 which is IDR 66.83 trillion. The impact on GDP in 2022 due to an increase in coal production of PT Bukit Asam is estimated at IDR 2.5 trillion and investment of IDR 42.59 trillion.

It is estimated that the output value of the benefits arising from coal downstreaming carried out by PT Bukit Asam amounted to USD 1,163,363,030, while the associated externality is estimated at USD 292,564,190 and the total value of social benefits obtained from downstreaming of PT BA is estimated at USD 478,592,253. It is estimated that the benefit coefficient from coal downstreaming (gasification) of PT Bukit Asam is equal to 0, 41.

For tax acquisition when downstreamed by PT Bukit Asam, it is estimated to be obtained in the amount of USD 564,954,468 which consists of externalities tax revenue estimated at USD 380,355,100, an excess payment is estimated at USD 64,019,216 and economic rent factors of USD 120,580.152.

Potential Non-Tax State Revenues (PNBP) derived from production fees (royalties), fixed fees (deadrent), forest taxes, and dividends after downstreaming are estimated at IDR 1.01 trillion, which consists of Non-Tax State Revenues for the central IDR 254, 01 billion, for the Regional Government of South Sumatra Province IDR 208.20 billion and the Regional Government of MuaraEnim for IDR 541.42 billion. It is estimated that several types of taxes will increase with coal gasification downstreaming. Some of the taxes that will increase include the Building Land Tax and the Surface Water Tax.

On the other hand, the downstreaming of coal into several derivative products has the consequence of an increase in investment in the coal gasification production process. The following is an illustration of the impact of downstreaming and investment for the MuaraEnim district GRDP. From this picture it can be seen that in 2017 (before there was downstreaming) the coal GRDP for MuaraEnim Regency amounted to IDR 47 trillion. Furthermore, after coal downstreaming was carried out in 2022, the estimated GRDP of coal production value is IDR 73 trillion and IDR 114 trillion from the coal downstream value or almost 3 times the increase in coal GRDP.

As an illustration also that with coal downstreaming there will be an increase in the rate of the economy, an increase in the value of production and an increase in investment and employment. Examples of cases in coal downstreaming at PT Bukit Asam located in MuaraEnim Regency. With this downstreaming, it is estimated that the GRDP of MuaraEnim Regency will increase which can absorb more than 5,000 workers. This increase in the number of workers comes from several sectors. The most dominant sector, among others, the coal sector, is expected to absorb a workforce of 2,403 people. This was followed by the chemical and plastic industry sector (306 people), the wholesale and retail trade sector (398 people), the company service sector and the building rental (459), the road transport sector (223) and several other sectors.

Coal downstreaming carried out by PT Bukit Asam is also expected to increase tax revenue due to the coal production process being the product of several other products such as products for urea (ammonia) fertilizer, DME as raw material for gas and polypropylene (PP) as plastic materials. From the simulation of Business Entity Tax and Individual Tax, Polypropylene downstream is targeted to contribute Corporate Income of USD 1,061,976 or IDR 15.5 billion. Whereas downstream DME increased Corporate Income Tax by USD 357,776 or by IDR 5.2 billion and for Urea increased Corporate Income by USD 253,673 or by IDR 3.7 billion. Likewise, the increase in Individual Income Tax will also increase.

Basically, the Government strongly supports business entities that will undertake downstreaming by providing various facilities and fiscal incentives. The provision of fiscal incentives can be given to companies that will undertake downstreaming by considering the positive impacts caused to the national and regional economy and to improve the standard of living and social life of the community.

As an illustration, the following shows the comparison of costs and benefits obtained by providing incentives. For example, if the Government reduces the royalty rate from 13.5% or IDR 208.94 billion is reduced to 0% for downstream business entities such as PT BA, it means that state revenue for the center is reduced by IDR 41.78 billion. While regional revenues (South Sumatra province and MuaraEnim regency) amounted to IDR 167.16 billion. However, from the incentives and then carried out downstreaming, it will increase benefits for the central and regional governments. The central government will be able to save foreign exchange by IDR 12.4 trillion because Indonesia is able to produce PP and DME and obtain additional foreign exchange of IDR 2.16 trillion because it can export fertilizer from ammonia from downstream coal.

For regions, the MuaraEnim regency's GRDP increased in the form of an additional coal production estimated at IDR 2.95 trillion and an additional investment of IDR 39.56 trillion. Likewise, there was an increase in job opportunities by 5,578 people and an estimated social benefit gained by MuaraEnim Regency was IDR 6.94 trillion.

When seen the effect on several aspects of the benefits obtained earlier the government should be able to provide fiscal incentives to pioneer companies such as PT Bukit Asam. Some of these aspects include an increase in the national economy and local governments such as increased state revenue, social benefits and increased employment and also a positive effect on the downstream exploitation areas.

4. Conclusions and Recommendation

The fiscal incentive scheme in the form of reducing royalties and taxes will in the short term reduce state revenues from the tax and non-tax revenue sector. But in the long run it will increase the national and regional economy such as an increase in GDP and GRDP by three times. Likewise, tax and PNBP receipts averaged 3 times. Besides that, it will provide social benefits and employment, estimated at more than 5000 workers.

To support the downstreaming program carried out by a pioneer company (PT Bukit Asam), the Government can provide fiscal incentives in the form of taxes (tax holidays, tax allowances), reduction of royalties, and other facilities. The success of pioneering companies in developing gasification downstreaming projects needs Government support in the form of fiscal incentive regulations. The gasification downstreaming that will be carried out by PT Bukit Asam is expected to encourage other companies to do the same.

References

[1] Alin H., Budy P. Resosudarmo 2004 Indonesia's optimal oil extraction rate: Application of dynamic optimization model *Indonesian Journal of Economics and DevelopmenT* **01** Juli 11-34

[2] Institute for Economic and Social Research, Faculty of Economics and Business, University of Indonesia 2006 Coal export levies policy *Final Report, LPEM FE UI.*

[3] Indonesian Fertilizers 2018 Internal rate of return study, Year of 2018

[4] Abidin M, Z 2014 Rationality and evaluation of establishing export levy in support of mineral management policies *Journal of Community Development* **6** 129 – 142

[5]Amir, H 2013 Economic impact analysis of the Indonesia mineral-export tax policy: CGE approach *The International Journal of Economic Policy Studies* **8** 1-22

[6] International Labour Office (ILO) and ILO Office for Indonesia 2013 The Impact of trade liberalization on bilateral relations between Indonesia and the three countries (China, India, and Australia) towards export-import performance, national output and job opportunities in Indonesia *Analysis of Social Accounting Matrix Simulation* (SAM) *and the SMART Model*

[7] Zulfiandri, et al 2017 Transaction cost and benefit cost on vertical integration of cacao small scale agroindustry value chain *Journal of Management and Agribusiness* **14** 187-197

IOP Conf. Series: Materials Science and Engineering **909** (2020) 012055 doi:10.1088/1757-899X/909/1/012055

[8] Antara, Made 2009 The Impact of government and tourist expenditures and private investment on Bali's economic performance: A Simulation of the accounting matrix model
[9] Cutter, Susan, Renwick & William H 2004 Exploitation, conservation, preservation, a geographic perspective on natural resource use (Fourth edition John W& Sons, Inc)
[10] BP Statistical Riview of World Energy 2019 Outlook 2019 London p. 36
[11]Industry Media, Ministry of Industry 2016 Realize downstream industry based on natural resources.01, year of 2016
[12]Hoiri, Syamsul 2010 Coal production funds and coal loyalty: arrangements, problems and alternative solutions Tesis Faculty of Law. University of Indonesia. p. 19-24
[13]Ministry Of Energy and Mineral Resources Republic of Indonesia.2017. Handbook of energy & economic statistics of Indonesia p. 38
[14]Ministry of Industry 2014 Revised customs duty rules. Sources: Seputar Indonesia Newspaper, Jakarta
[15]The Oxford Institute for Energy Studies (OIES) 2017 OIES's annual report 2017 University of Oxford United Kingdom
[16]Ministry of National Development and Planning 2016 Final report of the study on the achievement of the coal DMO target of 60% of National Production in 2019 (Directorate of Energy, Mineral and Mining Resources. Ministry of National Development and Planning, Jakarta)
[17] Sinaga, M.Bonar 2007 Economic sector linkages and income distribution in Java: Use social ccounting matrix approach
[18]Indonesian National Statistical Office 2016 Documentation for the Indonesian Social Accounting Matrix (SAM) from 2008
[19] Geological Agency Indonesian mineral and coal potential Bandung 2018

IOP Conf. Series: Materials Science and Engineering **909** (2020) 012056 doi:10.1088/1757-899X/909/1/012056

Effect of passenger presence towards driving performance level using kss and cnc indicators

M Mahachandra[1,2*], H Prastawa[1,2] and A H Mufid[1]

[1] Industrial Engineering Department, University of Diponegoro, Semarang, Indonesia
[2] Center for Biomechanics, Biomaterial, Biomechatronics, and Biosignal Processing, University of Diponegoro, Semarang, Indonesia

*Email : manik.mahachandra@ft.undip.ac.id

Abstract. Passenger presence does affect car driver performance while driving. However, a different type of conversation between driver and passenger might bring different effects on the performance. This car driver's performance can be measured by the level of sleepiness and the number of accidents experienced by the driver while driving. The level of sleepiness was obtained from the measurement of the driver's alert level using the Karolinska Sleepiness Scale. Furthermore, the number of accidents was obtained from the measurement of the performance level using the Crash-No Crash indicator. This study applied three experiments on a car simulation, i.e., drive alone, drive with a passive passenger, and drive with an active passenger. Results showed that driving alone experienced the highest level of sleepiness than others while driving with active passengers experienced the highest number of accidents than others.

1. Introduction

Distraction can be defined as anything that redirect the driver attention from the normative characteristics to navigate a vehicle. Distraction can also lower the performance and concentration level of the driver. According to [1], fifty-nine percent of accidents are caused by the number of distractions affecting the driver. The distraction usually come from outside the vehicle, such as billboards, collisions, or other incidents that attracts the driver attention [2]. However, distraction can also come from inside the vehicle. Some example of significant in-vehicle distractions are the use of smartphones and the presence of passengers [1]. The passenger presence can lower the performance and the concentration level of a driver, therefore increase the risk of traffic accidents. Driving with passenger increase the risk of vehicle collision by 96% [3].

Distraction caused by a passenger presence can be considered from the behavioural factor (the presence or the absence of passengers), gender, and age. In terms of behavioural factor, teen driver is likely to involve in traffic violation when driving alone, compared to when driving with passengers [3]. This is relevant with the studies conducted by [4], stated that the teen driver is likely to drive at a higher speed and take more risk when driving alone. While being with a male passenger, a male tends to drive at a higher speed with 21,7% higher risk-taking, compared to 5,5% chance while being with a female passenger. On the other hand, a female is likely to drive at a slower speed while being with the same gender with 15,5% driving risk, compared to 12% chance while being with a male passenger. It concludes that a greater risk involved when driving with the same gender of passenger, compared to driving with the opposite [4]. In terms of the age factor, a teen driver behaviour is being observed with the presence of adult passenger. Driving with adults has 67% less chance of being involved in an accident compared to when driving with peers, which has 18% of chance [3].

According to Oxford dictionary, a passenger is defined as a person who is travelling in a car, bus, train, plane or ship and who is not driving it or working on it. However, a passenger affects the behaviour of the driver indirectly. Driving with peers increases the risk-taking behaviour at a teenage driver, compared to when driving alone. Behaviour and mood of the passenger also affect the driver performance. As passenger actively interacting, while at the same time being in a good mood, the driver tends to speed up more [5].

Prior studies about the effects of passenger distraction toward driver performance have yet suggested the correlation between driver performance and the presence or absence of passenger. Observation about driver state without passenger includes the performance level in an undistracted environment. Meanwhile, the observation about passenger influence includes the implication on the driver performance with the presence of passenger, in which the experiment will be divided into two sections: passive passenger and actively engaged passenger. Passive passenger is limited to passenger who does not interact with the driver during the whole duration of experiment, while the active passenger is defined as someone who is actively engaged on conversation along with the driver during the observation.

This study will observe and analyse the implication of distraction caused by the presence of passenger, and therefore the driver-passenger optimum performance level can be concluded. The data will be collected from Karolinska Sleepiness Scale, which then converted into driver's alertness level. In addition, Crash-Near Crash data will be collected to determine the driver performance level.

2. Method

2.1. Study participants

This research was focused on the driving performance under influence of the presence of passenger. The participants consisted of 10 respondents age of 18-25 years-old. This age range is believed to contribute high number of vehicle collision [6]. The participants joined a simulation in which they can choose which route they will be driving. The condition in which passenger was involved would be adjusted as if it represented the real-life situation.

2.2. Alertness level

Alertness is defined as an ability to recognize unpredictable events. As cognitive neurosciences suggested, alertness is ability to stay intact on situation in a particular period of time [7]. KSS is a tool often used to measure the sleepiness level, in which sleepiness is a parameter to calculate person's alertness level [8]. This study uses 9 points KSS which have been translated to Indonesian, as applied by [9], in order to help the participant to understand the questionnaire better.

Table 1. Karolinska sleepiness scale used in this study

Scale	Description
1	Very alert and awake
2	
3	Alert and awake
4	
5	Not alert, nor sleepy
6	
7	Sleepy but alert
8	
9	Very sleepy and having difficulty to stay awake and alert

To evaluate the alertness level using KSS, the researcher considered the beginning and the end driving condition. Through the data, a gradient or the rate of change for the corresponding condition was determined.

2.3. Performance level

Performance is the result of the process of accomplishing a task or function for a certain goal [10]. According to [10], performance is the amount of effort generated by a person to perform a task. The driver performance was monitored using Crashes and Near Crashes (CNC) experienced by the driver [11]. Crashes is limited to the presence of physical contact with the object during the collision regardless of the vehicle speed, while near crash is similar to crash but with no physical contact. CNC can be used to obtain the performance level by calculating the number of crash and near crash. This performance level was calculated from the overall crash and near crash occur during the whole study duration.

2.4. Experimental design

This research is conducted using simulation and experimental study. According to [12], simulation is specified as the replication or visualization of how system will behave on a certain situation. Thus, it can be assumed that a simulation is a model containing some variable that can demonstrate a real-life situation. [13] stated that an experimental study is deliberately performed by researchers by applying a certain manipulation to the subject on which the implication will correspondingly be observed and measured. Simulation is carried on this research due to the risk when collecting data on the real traffic. The simulation utilizes a software called City Car Driving. This will be applied so that the participant will get the same feeling as driving in the real life.

The study was conducted using counter balance mechanism, whereas the first part of the study was conducted with the driver alone, and the second experiment was conducted using one driver and one passenger. On the second part of the study, passengers were divided into 2 categories: actively engaged passenger, and passive passenger. The experiment involving passive passenger was defined as Study 2A, while the active passenger was defined as 2B.

Timeline of the study was started since the arrival of the participants to the study location. First stage was the preparation, which took 10 minutes. During the preparation stage, the participants were given an inform consent and briefing about the data collection. The next stage was the data collection, which took 60 minutes to complete. In this stage, the participant drove through the driving simulator. CNC data was gathered to evaluate the performance level. CNC data was collected from the record generated by software city car driving, combined with manual data collection. The KSS questionnaire was only given by the researcher at the beginning and at the end of the study.

3. Results and Discussions

Laboratory experiments were run out for several days in the driving simulator. Some crash and near crash incidents were captured by video surveillance, as can be seen in Figure 1.

(a) (b)

Figure 1. Example of crash (a) and near crash (b)

3.1. Alertness level through KSS indicator

The rate of change in KSS values were calculated over the period of time, which then calculated into gradient. Figure 2 shows an example of the KSS alteration along the driving period. This alteration value for each individual was varied. The higher the gradient value, the higher the sleepiness experienced by the driver. The mean was then calculated from the variation of the gradient values gathered in each study. As shown in Figure 3, the mean gradient for Study 1 appeared to be the highest with the value of 1.027, and Study 2B was the lowest with the value of –0.828. It demonstrated that driver will experience the highest sleepiness when driving alone in a period of 1-hour drive, while the driver tended to be more awake when accompanied by a passenger.

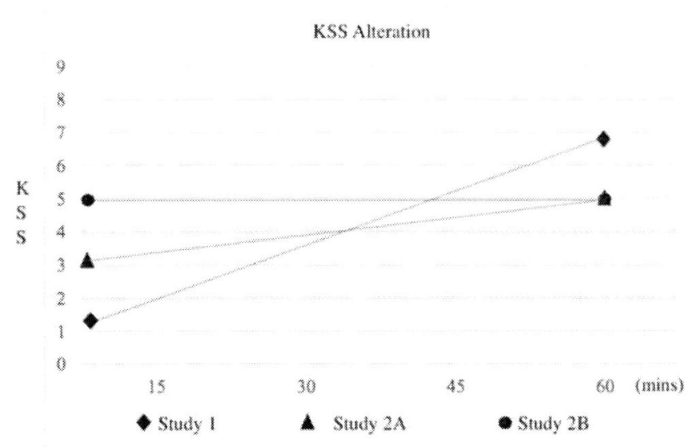

Figure 2. Example of KSS alteration and gradient of one participant

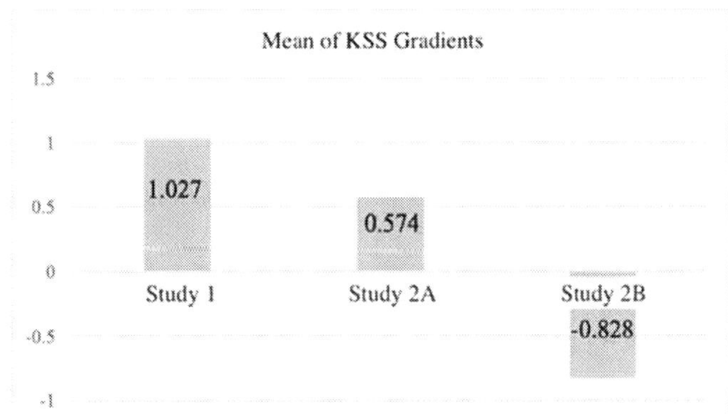

Figure 3. Comparison of mean of KSS gradients for each experiment

The mean gradients were relevant with the paired comparison result. Paired comparison result demonstrates that there was a significant difference between Study 1 and Study 2A, also Study 1 and Study 2B, which generate corresponding p-value of 0.000 ($t = 5.896$, $df = 9$) and 0.034 ($t = 2.494$, $df = 9$) respectively. On the other hand, Study 1 and Study 2A did not have significant difference, which represented by p-value of 0.176 ($t = 1.469$, $df = 9$). The result showed that the driver will experience higher sleepiness when driving alone compared to when accompanied by passenger and experienced far less sleepiness when accompanied by actively engaged passenger.

Paired comparison only showed the impact on the sleepiness level in each study, not the end state of each study. While the mean gradient and the paired comparison demonstrate that the driver on Study 2B was far less sleepy, therefore only Study 1 and Study 2A in which the end sleepiness state were considered. The mean end state of sleepiness level on Study 1 was higher compared Study 2A with the corresponding value of 3.6 and 3, respectively. The result shows that the driver's sleepiness level at the end of study was higher when driving alone, compared to when accompanied by passive passenger.

In this research, the alertness level was determined by considering the driver's sleepiness level. On another point of view, alertness level also can be indicated by the sleep quality, psychophysiology, work fatigue, and monotonous condition of someone [14]. From those few indicators, only the monotonous condition is relevant subject of this study. This was based on the fact that the driving simulation was completed in 60 minutes with free driving condition. That being said, it was possible for the participants to navigate on the same route condition repeatedly during a long period of time. This repeated situation was experienced when the participant is driving alone. Therefore, this further caused the driver to experience the highest sleepiness level compared to when driving with passenger. The implications could be due to the fact that during this study simulation, the driver only had an eye on the road instead of doing anything else for one hour.

This condition also stimulated the KSS value to increase, when comparing the value on the beginning and at the end of the simulation. The increased value of KSS was considered to be significant, as the initial mean of KSS was 4.5, compared to the final mean of 7.1. The overall mean of the specified condition was 5.8. Sleepiness level can be categorized as: alert (KSS 1-5), moderately sleepy ($6 \geq$ KSS ≤ 7), and very sleepy (KSS ≥ 8). When referred to the standard valuation of sleepiness level, it was categorized as moderately sleepy.

Meanwhile, the driver with active passenger was not exposed to monotonous condition that can lead to sleepiness during the study. The participant experienced different situation. Driver who was actively engaged in a conversation with the passenger experience the lowest amount of sleepiness [15]. The passenger was considered to cause a distraction towards the driver. During the simulation, the driver did not only pay attention, but also responded to the passenger conversation. Hence, the driver experienced only a small amount of monotonous condition. Furthermore, the presence of active passenger can lower the value of KSS during the simulation. The value decreased significantly, as the mean KSS on the beginning of the simulation was 5.6 and the mean at the end was 3.5. Therefore, the mean KSS of all time was 4.55. When referred to the classification mentioned previously, this can be categorized as alert.

3.2. Performance level through CNC indicator

In this study, the driving performance was estimated from the number of Crash and Near Crash occurred during the simulation [11]. CNC data was collected using dash camera installed in the back of the driver facing the display simulator.

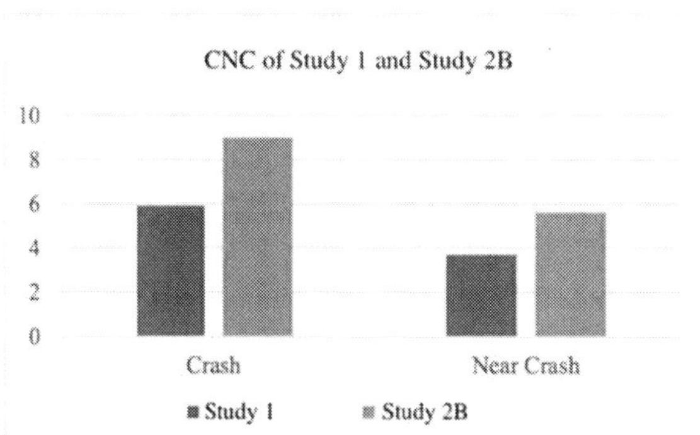

Figure 4. Comparison of crash and near crash incident frequencies between

CNC value of each participant differed in each study, so did the alteration. As an example, participant C experienced four crashes in Study 1, five crashes in Study 2A, and seven crashes in Study 2B. Further paired comparison was made to understand the implication of the presence of passenger in regard to driving performance in each study. Paired comparison showed that there was no significant difference between CNC on Study 1 and Study 2A, as well as Study 2A and Study 2B. The result showed the opposite when comparing Study 1 and Study 2B, where there was significant difference between both studies. Figure 4 graphs the comparison. The corresponding p-value of each Crash and Near Crash was 0.010 ($t = -0.328$, $df = 9$) and 0.006 ($t = -3.612$, $df = 9$), respectively. This infers that the presence of active passenger can increase the number of crash as well as near crash occurred compared to condition in which the driver is alone and with passive passenger. The result was relevant with [16], which stated that the passenger who is actively interacting with the driver can increase the risk of collision compared to the passive passenger.

When driving alone, the mean value of crash was 5.7, compared to 7.3 when driving with active passenger. Meanwhile the mean value of near crash was 3.7 when driving alone and 5.6 when driving with active passenger. There was 16% increase for crash and 19% increase of near crash when the driver is accompanied with active passenger. This correlated with [17] research which conclude that CNC value will be higher when accompanied with active passenger, compared to driving alone.

The case where the driver was solely present shows to have less crash and near crash. This was due to the fact that the driver will only focus on the road, and therefore minimizing the crash and near crash. In addition to that, the lone driver tends to have risk-averse towards the road situation [18], hence the low probability of crash and near crash. On the other hand, the active passenger showed to have higher crash and near crash number. Factors that contribute are the presence of distraction, lack of focus, and cognitive load. Distraction appeared when the passenger tried to engage in conversation. This distraction led to lack of focus, since the driver frequently diverted the attention away from the road to respond the passenger conversation. Hence, higher crash and near crash occurred. Furthermore, [18] stated that when accompanied with actively engaged passenger, the driver was likely to drive with higher risk-accepting. This implied that the presence of active passenger can arguably increase the probability of crash and near crash.

3.3. Problem solution

Participant was directed a simulation for one hour in free driving mode. To determine one hour of driving time seemed to imply monotonous activity for the driver. The road congestion was set to be 50% on the software to imitate the real traffic situation. Based on the respected condition, the problem solutions were designed to accommodate problem statement and repress the number of crash and near crash.

In order to prevent the sleepiness when driver is alone, some preventive measure can be implemented. First, when the driver is experiencing fatigue that leads to sleepiness, it is better to pull over and take a short break. The short break can be used to perform some light stretching and increase the water intake to lower the sleepiness and fatigue [19]. Second, driver can listen to music to lower the sleepiness using radio or cell phone while driving. Listening to music can induce the feeling of relaxation and arguably increase the attention level. Third, driver can prevent the fatigue by applying some aromatherapy essence in the car, such as peppermint as studied by [20], or ylang-ylang as studied by [21]. Aromatherapy could increase β wave on the brain to stimulate the alertness. Last, sleepy driver can also take simple, relatively cheap, easy to find countermeasure such as coffee, as concluded by [22].

According to Figure 4, to reduce the risk of vehicle collision, it is best to limit the driver-passenger conversation level. The study showed that the conversation with passenger appears to keep the driver awake but increase the risk of crash. That being said, it is not recommended to have continuous driver-passenger conversation when driving in order to minimize the danger. The conversation should be kept light and not involving any serious matter. The involved conversation can lead to lack of focus on the road. When passenger is able to obey this, the effect of distraction can be minimized. Furthermore, when distraction is kept low, the attention of the driver can be focused on the road solely. Hence, the lower risk of vehicle collision.

This study resulted on the sleepiness level was experienced higher when driving alone, compared to when accompanied by passenger. On the other hand, when passenger was actively engaged in conversation with driver, it reduced the sleepiness but at the same time increase the risk of vehicle collision. Based on the evaluation and problem solution mentioned previously, an optimum condition is examined. The optimum condition is therefore applied to driver with passenger but with minimum amount of interaction and light conversation. In this condition, the driver will be kept alert and promote the lower risk of vehicle collision.

4. Conclusion

The presence of active passenger (Study 2B) could affect the alertness level of the driver indicated by the KSS. The lone driver had higher significance compared to driver with active passenger (t = 5.896, df = 9, p-value = 0.000). Furthermore, driver with passive passenger had higher significance compared to driver with active passenger (t = 2.494, df = 9, p-value = 0.000). It could be concluded that active passenger can lower the driver's sleepiness level.

The presence of active passenger influenced the driving performance as examined from CNC indicator. Driver with active passenger had higher crash significance (t = -0.328, df = 9, p-value = 0.010) and near crash (t = -3.612, df = 9, p-value = 0.006), higher than the value when the driver was alone. There was 16% increase in crash and 19% increase in near crash when the driver was accompanied by active passenger. It could be concluded that the presence of active passenger could lead to higher risk of vehicle collision.

Acknowledgment

Authors thank for the research grant from the Ministry of Research, Technology, and Higher Education of Indonesia under the grant number 257-62/UN7.P4.3/PP/2019.

References

[1] Carney C, Harland KK, McGehee DV 2018 Examining teen driver crashes and the prevalence of distraction: Recent trends, 2007–2015 *Journal of Safety Research* **64** 21–7.
[2] Mahachandra M, Aji TW, Prastawa H, Suryoputro MR 2020 Assessing driver distraction on simulated driving *Int J Adv Sci Technol* **29(3)** 2714-20.
[3] Rosenbloom T, Perlman A 2016 Tendency to commit traffic violations and presence of passengers in the car *Transportation Research Part F: Traffic Psychology and Behaviour* **39** 10–8.
[4] Simons-Morton B, Lerner N, Singer J 2005 The observed effects of teenage passengers on the risky driving behavior of teenage drivers *Accident Analysis & Prevention* **37(6)** 973–82.
[5] Rhodes N, Pivik K, Sutton M Risky driving among young male drivers: The effects of mood and passengers 2015 *Transportation Research Part F: Traffic Psychology and Behaviour* **28** 65–76.
[6] NHTSA Traffic Safety Facts 2009 *National Highway Traffic Safety Administration.*
[7] Oken B, Zajdel D, Kishiyama S, Flegal K, Dehen C, Haas M et al 2006 Randomized, controlled, six-month trial of yoga in healthy seniors: effects on cognition and quality of life *Alternative Therapies in Health and Medicine* **12** 40-7
[8] Kaida, K., Takahashi, M., Akerstedt, T., Nakata, A., Otsuka, Y., Haratami, T., & Fukusawa, K 2006 Validation of the karolinska sleepiness scale against performance *Clinical Neurophysiology 117* 1574-1581
[9] Mahachandra, M., Sutalaksana, K., & Suryadi, K 2011 Sleepiness pattern of indonesian professional driver based on subjective scale and eye closure activity *International Journal of Basic & Applied Sciences* **11** 87-96.
[10] Rivai V *Manajemen Sumber Daya Manusia Untuk Perusahaan* 2004 PT. Raja Grafindo Persada.
[11] Dingus TA, Klauer SG, Neale VL, Petersen A, Lee SE, Sudweeks J, et al 2006 The 100-Car Naturalistic Driving Study: Phase II - Results of the 100-Car Field Experiment *PsycEXTRA Dataset 2 American Psychological Association (APA)*
[12] Syaefudin U, Syamsuddin A perencanaan pendidikan pendekatan komprehensif 2005 *PT Remaja Rosdakarya.*
[13] Jaedun A Metodologi Penelitian Eksperimen, Pelatihan, Penulisan Artikel Ilmiah 2011 *LPMP Propinsi Daerah Istimewa Yogyakarta.*

[14] Budiawan W, Prastawa H, Kusumaningsari A, Sari D 2016 Pengaruh monoton, kualitas tidur, psikofisiologi, distraksi, dan kelelahan kerja terhadap tingkat kewaspadaan *J@ti Undip : Jurnal Teknik Industri* **11** 37-44.

[15] Matthews RW, Kontou TG, Naweed A, Sargent C, Roach GD 2018 Driving when distracted and sleepy: The effect of phone and passenger conversations on driving performance *Chronobiology International* **35** 750–3

[16] McEVoy, S., Stevenson, M., & Woodward, M 2007 The contribution of passengers versus mobile phone use to motor vehicle crashes resulting in hospital attendance by the driver. *Accid Anal Prev* **11** 70-76.

[17] Simmons, B., & Ouimet, M 2017 Teen driving risk in the presence of passengers. *Research, practice, policy, and directions* 80-97.

[18] Bingham, R., Simmon - Morton, B., Pradhan, A., Li, K., Almani, F., Falk, E.,Albert, P. 2016 Peer passenger norms and pressure: Experimental effects. *Transportation Research* **41** 124-137

[19] Schmidt, E. A., Schrauf, M., Simon, M., Fritzsche, M., Buchner, A., & Kincses, W. E. 2009 Drivers' misjudgement of vigilance state during prolonged monotonous daytime driving. Accident Analysis & Prevention **41**1087- 1093.

[20] Mahachandra M, Yassierli, Garnaby ED 2015 The effectiveness of in-vehicle peppermint fragrance to maintain car driver's alertness. Procedia Manuf **4** 471-477.

[21] Moss, M., Hevitt, S., & Moss, L 2008 Modulation of cognitive performance and mood by aromas of pepermint and ylang-ylang *International Journal of Neurosciene* **118** 59-77.

[22] Mahachandra M, Munzayanah S, Yassierli 2017 The efficacy of one-time and intermittent intake of coffee as a countermeasure to sleepiness on partially sleep-deprived drivers. *Int J Technol* **8** 300-310

Identify product families using cluster analysis: case study in Passenger Car Radial (PCR) tire product

Rere Nugrahita[1*] **and Isti Surjandari**[1]
[1]Industrial Engineering Department, Faculty of Engineering, Universitas Indonesia, Kampus UI Depok, Indonesia

*E-mail: rere.nugrahita@ui.ac.id

Abstract. Manufacturing companies, such as tire manufactures are facing great challenges to cope with increased product variety which induced by customer demand. This variety lead to higher internal complexity in term of design and production. Thus, variety has to be well-managed in order to guarantee the positive outcome for company. One of the solution is to have a well-structured product family. In this research, products data are partitioned into clusters by applying cluster analysis for mixed-type data based on their general characteristic and component specification. Variants within cluster have similarities in term of characteristics and main product component used in production. By applying k-prototypes algorithm to handle these mixed type data, the data set is clustered and interpreted into eight different clusters using selected variables.

1. Introduction

Many manufacturing companies are increasing the number of product variety in order to stay competitive, tire manufacture is no exception. Increasing diversity in product is one of the effort to cover external factor such as, customer demand in term of product features, different needs in large number of market segments, certification and requirements due to regional issue [1]. High number of product variety induces opportunity and challenges at the same time. Product variety come with great potential for expanding market while increasing sales volume and revenues, but this best condition will only achieve if the product variety is well-managed by the company. In reality, offering more product variants to customer incurs more expenses to all stages of product including design, planning, production, inventory, selling and after sales services which lead to reducing acquired profit [1, 2]. Aside from that, product variety increase design and production complexity internally [1, 3, 4, 5], and also shorten product lifecycle which lead to the need of short period time in developing new product variants [3].

A product that seems simple such as tire, actually have broad product variety which is represented by thousands Stock Keeping Units (SKUs) number with hundreds of product lines, especially for Passenger Car Radial (PCR) tire. This variety needs to be well-managed to reduce complexity both internally and externally. One of essential prior knowledge which can support manufacturing companies in facing variety challenges is known as Product Variety Management (PVM). The main objective of PVM is to reduce complexity and its associated costs which induced by product variety [4]. Implementation of flexible manufacturing and advance information technologies has great impact in production system to deliver high level of variety in product at low cost as possible [1]. PVM strategies in acquire high flexibility classified into three main activities which are design, planning and manufacturing ranging from component, parts and product to entire enterprise [1]. Main enabler of PVM success are identifying commonalities of product and parts and grouping into families [6]. Product families represent characteristic which only belong to one group of similar product. Unclear defined product family at design stage, identified as the root cause for complexity in production system [3].

In this paper, product family is identified and interpreted using cluster analysis based on component specification that consists of mixed data between numeric and categorical data. Clustering method itself is proven to be well-performed and have been widely applied in many application of PVM related to design product architecture, product family and product platform [5]. Many previous research related to PVM using cluster analysis using clustering algorithm such as k-MM, hierarchical agglomerative, hierarchical complete linkage and k-means [5, 7, 8, 9], but mostly use numeric data as its input. Other research apply combination of two or more methods when different type of data is mixed, such as association rule, fuzzy c-means clustering and classification [10]. Clustering algorithm called k-prototypes algorithm can directly handle mixed type data of numeric and categorical [11], extension to this algorithm also has been done recently [12]. By applying k-prototypes clustering algorithm, this research has aim to grouping product variants based on product geometry and component specification which consists of total 31 variables and 1443 observed product. The paper sections are organized as follow. Section 2 presents the data collection, data pre-processing and methods in determining cluster variables. Section 3 describes how k-prototypes was applied using selected cluster variables. Finally, the paper is concluded in Section 4.

2. Materials and Methods

2.1. PCR tire product geometry and component specification

Product families, especially in tire product, can be identified through product geometry and component specification. In term of product geometry, dimension of section height, steel belts, plies, height of bead filler and bead construction are considered as parameters which differentiate one product with another. Similar in component specification, such parameters are code (for different material), height, angle or width of semi-finished component including, steel belts, plies, sidewall, bead, bead filler, inner liner, and other supporting material.

2.2. Research methods

Generally, the research methods is proceed as in Figure 1, each process will be described in next sub-section.

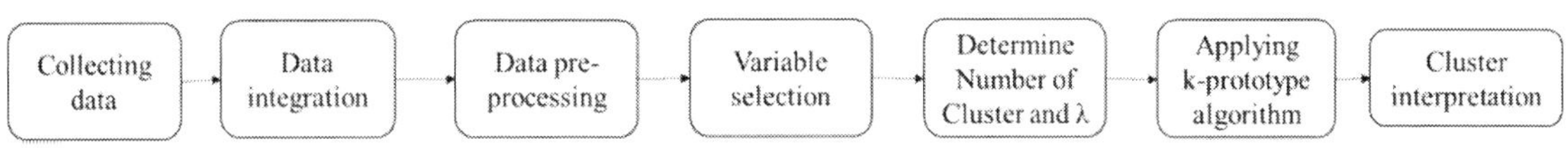

Figure 1. Research methods

2.2.1. Data integration and pre-processing

As collected data are separated in multiple data sets, data integration phase is needed to combine the data into a coherent data set before continue to data pre-processing. In data pre-processing, data cleaning is done which relates to several activities, such as eliminate irrelevant data item, omit the missing value, ensure the structure of a variable only consist of one type of data and standardize the numeric data using Z standardization. Output of the data pre-processing stage is a data set consists of 31 variables of 1443 observed product variants.

2.2.2. Variables Selection

Variable selection is the stage that must be done carefully before continue to process the data using clustering algorithm to avoid bias result and complex interpretation of cluster. This research use mixed type of numerical and categorical data. In term of numerical data, it is encouraged to examine the variables used for substantial multicollinearity. If multicollinearity is found, it is advised to include only cluster variables that are not highly correlated [13]. One of the way to detect multicollinearity is by examining correlation matrix of related variables and suppressed highly correlated variables. In term of categorical data, measuring the categorical level and approach by calculating permutation misclassification rate for each variable of the data using an existing tools in R knowingly as FeatureImpCluster, are done. Variable importance is interpreted by calculate the mean of misclassification rate over all iterations. The permutation misclassification rate of a variable is the number of wrong cluster assignments divided by the number of observations given a permutation of the variable.

2.2.3. K-Prototypes algorithm

An extension to k-means algorithm that can handle clustering process for both numeric and categorical data was developed by Huang [11] and widely known as k-prototype algorithm. The idea is integrate the k-means algorithm that works for purely numeric variables and k-modes algorithm for purely categorical variables. In k-prototypes algorithm, the cluster prototypes recalculate iteratively by the algorithm and reassign clusters. Clusters assignment aim to minimize cost function in Eq. 1 and use similarity measure $d(X, Q)$ in Eq. 5 between two mixed type objects [11].

$$P(W, \boldsymbol{Q}) = \sum_{l=1}^{k} (P_l^r + P_l^c) \tag{1}$$

Where P_l^r and P_l^c describe for numeric and categorical variables respectively.

$$P_l^r = \sum_{i=1}^{n} w_{i,l} \sum_{j=1}^{p} (x_{i,j}^r - q_{l,j}^r)^2 \tag{2}$$

$$P_l^c = \lambda \sum_{i=1}^{n} w_{i,l} \sum_{j=p+1}^{m} \delta(x_{i,j}^c, q_{l,j}^c) \tag{3}$$

Subject to,

$$\sum_{l=1}^{k} w_{i,l} = 1, \qquad 1 \le i \le n$$
$$w_{i,l} \in \{0,1\}, \quad 1 \le i \le n, 1 \le l \le k \tag{4}$$

Where W is an n x k partition matrix and $\boldsymbol{Q}$ is a set of objects in the same object domain, $\boldsymbol{Q}= \{Q_1, Q_2, \ldots, Q_k\}$. Similarity measure,

$$d(X_i, Q_l) = \sum_{j=1}^{p} \left(x_{i,j}^r - q_{l,j}^r\right)^2 + \lambda \sum_{j=p+1}^{m} \delta\left(x_{i,j}^c, q_{l,j}^c\right) \tag{5}$$

Where p is number of numeric and *m-p* is number of categorical variables which $V_1^r, V_2^r, \ldots, V_p^r, V_{p+1}^c, \ldots, V_m^c$. $x_{i,j}^r$ and $q_{l,j}^r$ are values of numeric variables and $x_{i,j}^c$ and $q_{l,j}^c$ are values of categorical variables for object i and the prototype of cluster l. The first term of Eq. 5 is the squared Euclidean distance measure on the numeric variables and the second term is the similarity matching measure on the categorical variables where,

$$\delta(a, b) = \begin{cases} 1, & if\ a \neq b \\ 0, & if\ a = b \end{cases} \tag{6}$$

The weight λ is used to avoid over-emphasizing either type of variables. The balance between both terms can be controlled by the parameter λ which has to be specified in advance, also for the number of clusters k. Higher values of lambda, $\lambda > 0$ will increase impact of the categorical variables, for $\lambda \sim 0$, the impact of the categorical variables decrease and vanishes at $\lambda = 0$ which only numeric variables are taken into account, like in k-means clustering. An extension to estimate the weight λ was introduced by dividing the average variance σ^2 of all numeric variables and the concentration h_{cat} of all categorical variables [12]. In term of number of clusters, researchers' perspective and qualitative judgement play significance role in determining number of cluster, but the number k can be estimated through scree plot, silhouette width, or index value [13, 14].

3. Result and discussion

3.1. Variables selection

In term of numeric variables, in order to detect multicollinearity, examining correlation matrix of related variables have a significance role. If any high correlated variables are found, either reduce the variables or use distance measure that takes into account the multicollinearity [13]. In this research, highly correlated variables are suppressed, hence numeric variables were reduced from 13 to 7 variables.

In term of categorical variables, level of factor is main concern for choosing variables to be included in clustering algorithm. It is recommended to choose categorical variables whose level number close to k cluster which these variables will have strong contribution and play significant role in determining clusters [15]. In opposite, categorical variables whose too high in number of factor level will spread out per clusters and this condition have to be avoided. In this research, 4 categorical whose level number less than twenty are chosen. So, the total number of variables that used in clustering algorithm is 11 variables as the code can be seen in Table 1.

Table 1. Selected variables to be included in clustering algorithm

Numeric Variables	Categorical Variables
NSH (Nominal Section Height)	BEAD (Bead Construction)
DW (Drum Width)	CAPC (Tread Compound)
BTB (Bead to Bead)	MC (Machine Type)
X1BW (First Belt Width)	SI (Speed Index)
BIC (Bead Inner Circle)	
HBF (Height of Bead Filler)	
SWW (Sidewall Width)	

3.2. *Cluster Result*

Prior to running the algorithm, value of λ and k need to be determined. As explained before, λ play significant role in balancing the effect of numeric and categorical to the cluster results. Value of λ that estimated in extended version of k-prototypes algorithm [12] can be used as starting point to examine the impact of each type of variables. The λ value is tuned to balance the effect of numeric and categorical variables. This research use λ value equal to twice of estimated λ, which is 0.233, since it will give balance impact of numeric and categorical variables as illustrated in Figure 2.

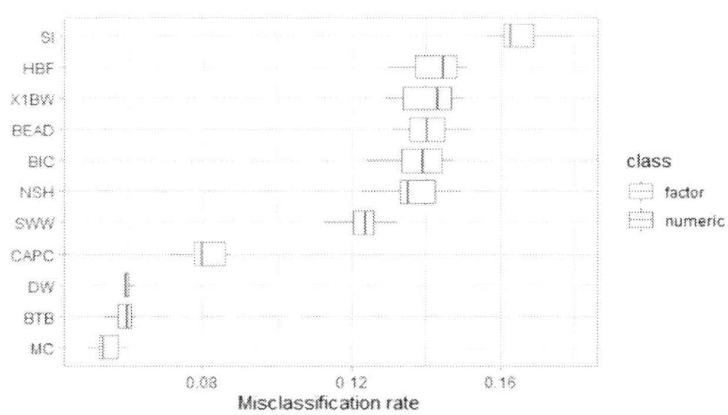

Figure 2. Visualization of balance impact of numeric and categorical variables using FeatureImpCluster in R

Number of cluster can be estimated through silhouette width calculation that run for certain range of cluster number. Mainly, there are two things need to be considered while estimating the number of cluster in clustering algorithm such k-prototypes which need value of k as an input . First is the silhouette width index which describe how close the distance between object inside the cluster and between clusters. Second, prior knowledge of researcher about the data also play significant role to ensure number of cluster will not too small or too high considering the difficulties in cluster interpretation stage. After calculating the silhouette index for k value ranging from 1 to 20 as illustrated in Figure 3, number of k=8 is considered as the most balance value in term of silhouette index and clusters representation of the data .

Figure 3. Silhouette index of clusters

3.3. *Clusters interpretation*

Examining the result of k-prototypes clustering algorithm, clusters can be interpreted as described in Table 2 below.

Table 2. Clusters interpretation and description

Cluster Number	Interpretation	Description
Cluster 1	Single middle series tire	Middle series and high speed tire, using single wire construction and expensive tread compound material built in one stage
Cluster 2	Small silica high series tire	High series and high speed tire with small rim, using box construction and silica tread compound
Cluster 3	LT high series tire	High series and low speed tire, close to light truck (LT) size, using single wire or big box construction and built in two stage
Cluster 4	Low series tire	Low series and ultra-high speed tire with large rim, using single wire construction and expensive silica tread compound built in two stage machine
Cluster 5	Box middle series tire	Middle series and high speed tire, using box construction and carbon black tread compound material built in two stage
Cluster 6	Small carbon high series tire	High series and high speed tire with small rim, using box construction and carbon black tread compound built in one stage
Cluster 7	SUV high series tire	High series and low speed tire, sport utility vehicle (SUV) size, using single wire or medium box construction and carbon black compound built in one stage
Cluster 8	Large single middle series tire	Middle series and high speed tire with large rim, using single wire construction and carbon black tread compound material built in two stage

4. Conclusion and future research

This research has been analyzed data of PCR tire product then clustering the data set into eight different clusters using k-prototypes clustering algorithm and interpret it in order to identify product families. It is well-proven that k-prototypes algorithm can handle clustering for mixed type data of numeric and categorical variables. In term of λ value, an existing developed package, FeatureImpCluster in R, is powerful in determining the suitable λ value through misclassification rate calculation and its boxplot visualization. For future research, this research can be extended as these clusters result can be used as an input for other research related to PVM, such as scheduling optimization, identify modular standardization and new variant design activities improvement in tire manufacture industry.

References

[1] ElMaraghy H, Schuh G, ElMaraghy W, Piller F, Schonsleben P, Tseng M and Bernard A 2013 *CIRP Annals Manuf. Tech.* **62** 629–52
[2] Jhonson P 2003 *Supply Chain Practice* **5** 54-67
[3] Schuh G, Rudolf S, Riesener M, Dölle C and Schloesser S 2017 *Procedia CIRP* **60** 344–9
[4] Schuh G, Reinhart G, Prote J P, Sauermann F, Horsthofer J, Oppolzer F and Knoll D 2019 *Procedia CIRP* **81** 874-9
[5] Hochdörffer J, Laule C and Lanza G 2017 *Proc. Int. Conf. on Industrial Engineering and Engineering Management* (Singapore: IEEE) pp 593-7
[6] Jiao J, Simpson T W and Siddique Z 2007 *J. Intell. Manuf.* **18** 5–29
[7] Zhihuang D and Scott M J 2007 *J. Intell. Manuf.* **18** 97-113
[8] Wallis R, Erohin O, Klinkenberg R, Deuse J and Stromberger F 2014 *Procedia CIRP* **23** 178-83
[9] Bohnen F, Maschek T and Deuse J 2011 *CIRP J. Manuf. Sci. Tech.* **4** 247-51
[10] Moon S and Simpson T W 2006 *Proc. Int. Des. Eng. Tech. Conf. & Comput. and Inf. Eng. Conf., (Philadelphia, USA)* pp 317-25
[11] Huang J Z 1998 *Data Mining and Knowledge Discovery* **2** 283-304.
[12] Szepannek G 2019 *The R journal* **10** 200-8
[13] Hair J F, Black W C, Babin B J and Anderson R E 2013 *Multivariate Data Analysis* (London: Pearson New International Edition) chapter 8 pp 415-474
[14] Wijayanto A, Suprapto Y K, and Wulandari D P 2019 *International Seminar on Intelligent Technology and Its Applications (Surabaya, Indonesia)* pp 210-5
[15] Khan S and Ahmad A 2013 *Expert system with Applications* **40** 7444-56

IOP Conf. Series: Materials Science and Engineering **909** (2020) 012058 doi:10.1088/1757-899X/909/1/012058

Minimize the potential failures in the wire rod production process using six sigma and multi attribute failure mode analysis method

Maria Ulfah[1*], **Ratna Ekawati**[2], **Nafila Amalia** [3]

[1,2,3] Department of Industrial Engineering, Engineering Faculty, Sultan Ageng Tirtayasa University, Jend. Sudirman 3 Street, Cilegon, Banten, Indonesia

*E-mail :_maria67_ulfah@yahoo.com

Abstract. Wire rod mill (WRM) plant is one of the factories in Company X that produces wire rod. In 2018 and 2019, WRM was recorded producing high quantity of rejected product exceeding the set standard. This study deploys systematic problem identification and improvement to the process. The methods used was Six Sigma method to evaluate current condition and process capability, followed by Multi Attribute Failure Mode Analysis (MAFMA) with Grey AHP approach to identify root cause and weighted failure process that will be put into priority of improvement. Based on the research results, the sigma value obtained is 4.297503 σ with DPMO value is 2,928.626. Improvements to the five highest of the causes of process failure is to improve the state of bearing roll, roll entry, hydraulic valve, an improvement on roll adjustment and improvement of inter-roll partings. It was concluded that the potential causes that the hot button bearing roll has the highest weight is 0.94569, so making it a top priority improvement.

1. Introduction

Wire Rod Mill (WRM) plant of Company X is known to have production process problems that resulting in nonconformance products. Repair time to overcome problematic process was causing delay in production and affecting the capability of achivening production target of the plant.

The WRM has determined that permissible product defect to be 0.2% and 0.25% in year 2018 and 2019 respectively. However, production data of steel wire WRM shows that in June 2018 defect product produced was 12,693 ton from target of 9,049 ton, while in January 2019 defect product produced was 158.12 ton from maximum target of 50.66 ton.

In this research, method to identify problem and proposal development to solve the production problems was deployed. The method was Six Sigma method, followed by Multi Attribute Failure Mode Analysis (MAFMA). Six Sigma method is known to be a method that focuses on the process and prevention of defects [1]. Follow up of Six Sigma method is Multi Attribute Failure Mode Analysis (MAFMA) which is an analytical technique that used to determine the cause of potential failure. MAFMA integrates the conventional aspects of FMEA with cost aspects, so that the causes of failure can be seen as influencing the costs. The AHP method is used for weighting against criteria or sub criteria and Greys are used as a comparison. Grey is developed as a quantitative approach to assisting decision making by examining the degree of similarity or difference between two sequences based on the relationship class [2].

Previous research by Elias [3] regarding quality control efforts for treable bridge type B1 parts in the upper end thicness in order to obtain the specifications desired by consumers with the DMAIC method approach at PT.Yamaha Indonesia. Another study was conducted by Kristyanto [4], regarding operational risk analysis in the sugar production process using MAFMA in the PG case study Kebon Agung Malang, producess nine critical risks which have the highest risk level out of 23 potential operational risks that disrupt the stability of the sugar production process. Anisa [5], that study identified the cause of stop line using the MAFMA method by determining the cause of stop line based on roll. The cause of potentia failure is bearing failure. Another study by Amalia [6] concerning the cause of potential failures in the wire rod production process using the six sigma method and identifying the cause of failure using MAFMA, and the latest research from M.Ulfah [7]. In this research, a proposed improvement is made to the problems that occur in the production process, which is to minimize the failure of the cold sheet steel production process at PT.X's Cold Rolling Mill by using lean six sigma and MAFMA with the fuzzy AHP approach.

2. Literature review

2.1. The six sigma concept

Six Sigma is a statistical concept that measures and reflects the capability of the actual process. Six Sigma is a method or technique of a process that is very orderly in terms of controlling and improving products where the system is very comprehensive and flexible [8]. At the state of Six Sigma value there are only 3.4 defects from a million occasions, meaning 99.99966% process perfection or zero defects.

Table 1. Benefit from achieving some level of sigma

Level of Sigma	DPMO	COPQ (Cost Of Poor Quality)
1-sigma	691,462 (very uncompetitive)	Can not be calculated
2-sigma	308,538 (industry average Indonesia)	Can not be calculated
3-sigma	66,807	25-40% of sales
4-sigma	6,210 (USA industry average)	15-25% of sales
5-sigma	233	5-15% of sales
6-sigma	3.4 (Industry world class)	< 1% of sales
Each increase or 1-sigma shift will provide a profit increase about 10% of sales		

Source: [9]

2.2. Analytic Hierarchy Process (AHP)

The Analytic Hierarchy Process (AHP) method was developed by Thomas Saaty in the 70's when at Warston School [9]. States that the AHP is a theory of the measurements used to the scale ratio either from the paired comparison of discrete or continuous [10]. AHP method is one of the methods used in the decision-making system by observing the factors of perception, preference, experience and intuation. AHP combines personal judgments and values in a logical way. Analytic Hierarchy Process (AHP) can solve complex multi criteria problems through hierarchy of decision. A complex problem can be interpreted as the criteria of a problems, the uncertainty of the problem structure, the uncertainty of opinion from the decision maker, more than one person of the decision maker, and the inaccuracy of available data.

2.3. Grey theory

Grey systems theory is a new methodology that focuses on research about problems involving small samples and incomplete information. Grey systems theory handles an uncertain system with partial information through generating, digging and retrieving useful information from what is available. In grey theory, according to the level of information, if the system information is fully known, the system is called a white system. If the information is unknown, the system is called a black system. A system with known information partly is called the grey system [11].

3. Methodology

Data collection was conducted in the production division and quality control division of Wire Rode Mill (WRM) plant of Company X which produce wire rod on December 19, 2019 until January 25, 2020.

1. Secondary data
   The data used are from the last two years starting from January 2018 until December 2019. Data collected to support the research includes production data which consisting of number of

reject products, production plans, production results, and percentage of production reject. In addition, production delay time due to defective product and delay cost data were also collected.

2. Primary data
   Primary data consists of interviews, brainstorming and questionnaire distribution. The questionnaire consists of two: Failure Mode Effects and Analysis (FMEA) and Multi Attribute Failure Mode Analysis (MAFMA). Troubleshooting path that is described in the form of a flow chart of MAFMA Calculation is shown in Figure 1.

The following were steps to implement the Six Sigma method:

1. Define:
   Identification and selection of a Six Sigma project, identification of physical flow of production process, and to identify the Critical-To-Quality (CTQ).
2. Measure:
   Calculation of DPMO number and sigma value.
3. Analyze:
   Identification of dominant defects using Pareto Diagram, analysis the cause of the failure process toward the dominant defects using Root Cause Analysis (5 Whys method), and analysis the causes of potential failure process using Multi Attribute Failure Mode Analysis (MAFMA).
4. Improve
   The proposed improvements using 5W + 1 H method.

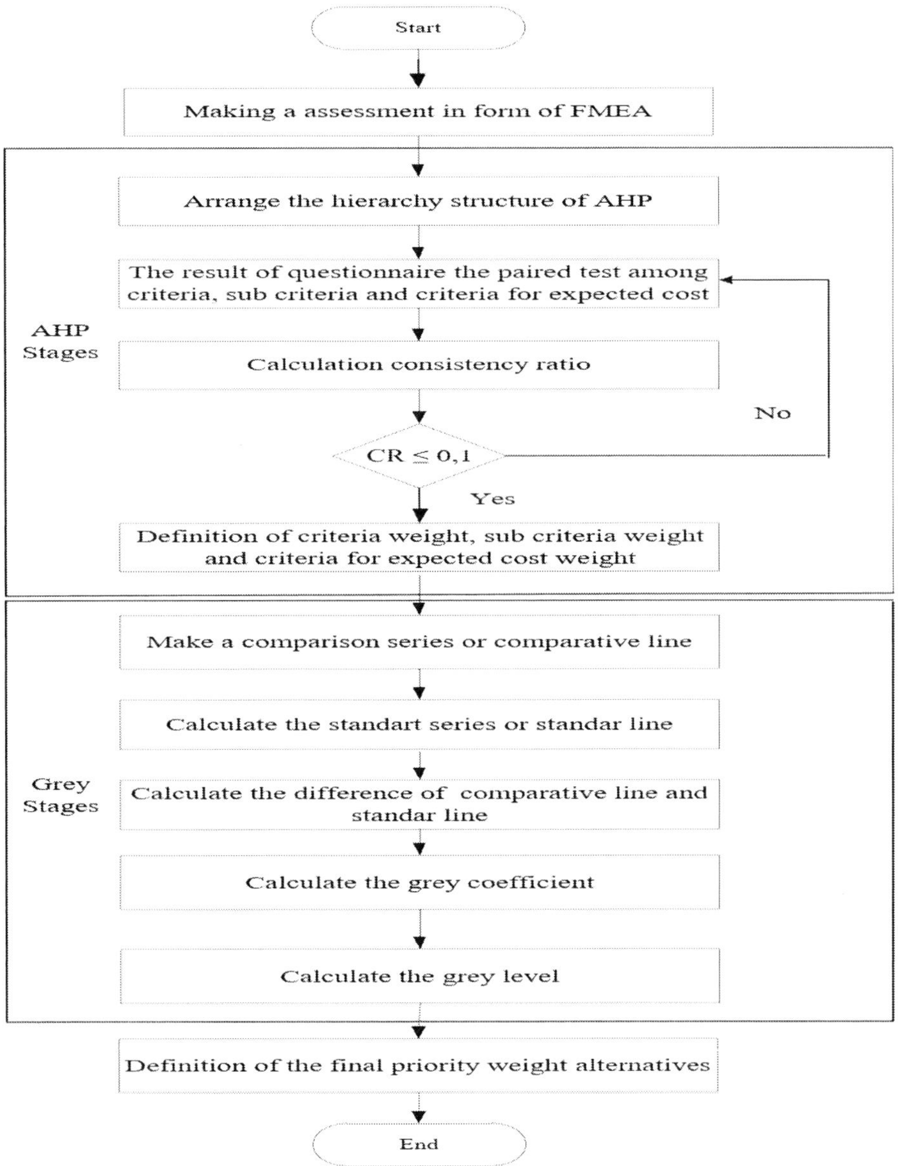

Figure 1. Flow chart of calculation MAFMA method

4. Result and discussion

In the implementation of Six Sigma concept consists of 5 stages namely DMAIC (Define, Measure, Analyze, Improve and Control). As the nature of the study is to find the cause of problems, the researchers limited the discussion up to the improve stage.

4.1. Define stage

Define is the stage for defining goals and backgrounds as well as identifying issues that need to be improved in order to achieve better quality performance.

1. Identify the selection of a six sigma project
   The Six sigma project is deployed based on the failure of the production process that causes rejected. The quantity of rejected products produced in the Wire Rod Mill during the year 2018 and 2019 was above the Company X set standard.
2. Identification of physical flow of production process

criteria, there are four criterias, namely severity, occurrence, detection and expected cost. This criterion is based on FMEA that has been developed into MAFMA, so there are additional expected cost criteria. In severity criteria, there are three sub criteria, namely potential damage, operation delay and time to repair. Classification of the sub criteria is based on an interview with the company parties. The structure of the MAFMA hierarchy is shown in Figure 2.

4.3.3. Determination of criteria weight on MAFMA hierarchy structure

The first step which taken in determining the criterion weight is conduct paired tests among criteria. Then, we will calculate the priority weight and value of Consistency Ratio or CR to know the consistency of the three respondents to the assessmen. Based on the calculation, it is known that the value of RI = 0.9 with n = 4 [10] will obtained the value of CR = 0.02244, there are indicating that the level of importance between the criteria of potential failure can be said respondents provide consistent assessment. This is because CR ≤ 0.1 [13].

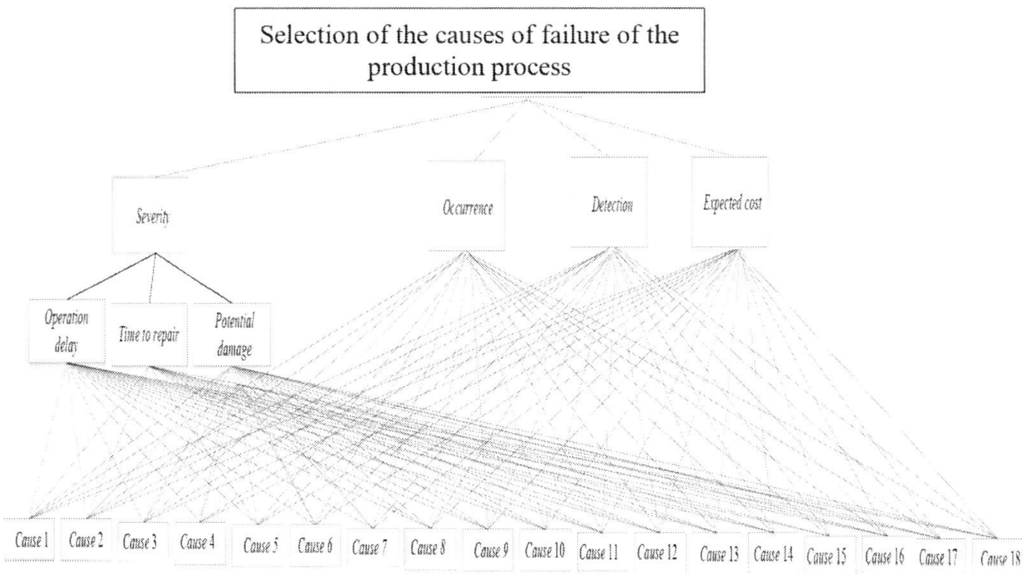

Figure 2. MAFMA hierarchy structure

Determination of alternative weight

a. Pairwise comparison test for criteria at expected cost
b. Build comparative series or comparative line
   To perform the processing by grey method, the data that were obtained on the FMEA results of the three respondents were calculated in comparative series or comparative series.
c. Define standard series or standard rows
   The standard that be appointed is the smallest value which found in severity, occurrence, and detection is 1.333. The following is the standard series or standard line of comparative or comparison line data Xo = [1.33333 1.3333 1.33333]
d. Determining the difference between comparative lines and standard rows
   The difference between a comparative line and a standard line was done by subtracting the value from the comparison series with the standard series.
e. Calculating the grey coefficient
   Grey coefficients were used to compare comparative sequences and reference sequences. In the grey coefficeint calculations, it is known that $\zeta \in [0, 1]$ is the coefficient used to control the resolution scale and ζ is normally a value of 0.5 [14].
f. Calculating grey level

The Grey level is used to show the degree of similarity between the comparative sequence and the reference sequence. Based on calculations using the AHP method, the weight of each criterion and sub criteria were obtained. The weight was used to calculate the grey level by multiplying it with the cause of failure data on the grey coefficient. From these results the total weight of potential causes of process failure based on criteria and sub criteria were obtained. For expected cost criteria, priority weight was processed by using AHP method while severity, occurrence, and detection criteria using grey method where priority weight is obtained from grey calculation.

g. Final determination of alternative priority weight

Table 2. Sequence of Failure Process Potency

Code	Cause	Total	Sequence
Cause 13	Hot roll button bearings	0.94569	1
Cause 3	Bearing broken	0.82369	2
Cause 9	The hydraulic valve is not working	0.81108	3
Cause 14	Roll entry cracked	0.78896	4
Cause 17	Parting is not perfect	0.78494	5

h. From the calculation of grey level, the final weight of improvement priorities was determined. From the eight sequence of causes of process failure, the most potential cause is cause number 13, which is hot button bearing roll. However, from the eighteen causes of the failure of wire rod production process, researcher took five of the highest potential failures that would be proposed for improvement.

4.4. Improve stage

After the root cause of the production process failure is identified based on the AHP grey approach, the next step is to find solution to the problem and reducing the cause of process failure. The step is improved using 5W + 1H method.

Improvements that can be made to minimize the failure of the process include:

1) Overcome the cause of hot roll bearing button by lubricating the roll on a regular basis, perform continuous inspection of roll usage and, make record of how often the hot button bearing roll,
2) Repair the damaged bearing by conducting continuous inspection and change the bearing regularly if it is not feasible to use and providing sufficient grease lubricant.
3) Repair non-working hydraulic valve by the careful installation of oil connection, replacing the damaged parts and the replacement must be in accordance with the specifications. Filtering of oil valve could be replaced.
4) Overcome the cracked roll entries by creating rolling tape monitoring record form, regular lubricating, and adjustment of roll appropriately,
5) Overcome imperfect parting between rolls by checking the bar size on a regular basis, revise the check sheet of the production division by adding description line 2. Measuring the size bar on line 2 should be careful and give a sanction for those who do not do their job well.

5. Conclusion

Based on this research, it can be concluded that quality characteristics or CTQ (Critical to Quality) on the wire rod production process are laps, overfill, underfill, scratch, scrappy, roll mark, coil multiple pieces, no center, cross roll / un round, crumpled and creep speed. The sigma value of the wire rod production process is 4.297503 which has reached the industry average of USA (± 4 sigma). Cause of failure of process including hot roll bearing button (0.94569), broken bearing (0.82369), hydraulic valve is not working (0.81108), roll entry cracked (0.78896), parting imperfect (0.78494), setting stopper control (0.75041), non-homogeneous spare part (0.69258), loose lock (0.66183), twist wear out (0.64855), poor adjustment (0.62455), guide less lubricant (0.60438), leakage cooler (0.59048), setting rotate bias (0.58745), base metal unclean (0.58355), lub lubricant roll (0.57084) and spare parts (0.57016). Several suggestion of improvement has been proposed.

References

[1] Ronald D Snee 2007 Discussion: development and use of statistical thinking: A New Era *Journal International Statistical Review* **67** 3
[2] Deng, Julong 2013 Control Problems of Grey Systems *International Journal System and Control Letter Elsevier* **1** 288-294
[3] Elias H P 2014 Upaya Pengendalian Kualitas Komponen Treable Bridge Tipe B1 Bagian Ketebalan Ujung Atas Guna Mendapatkan Spesifikasi Yang Di Kehendaki Konsumen Dengan Pendekatan Metode DMAIC di PT Yamaha Indonesia *Jurnal Teknologi dan Manajemen* **12** 2
[4] Kristyanto R, Sugiono and Rahmi Yuniarti 2015 Analisis risiko operasional pada proses produksi gula dengan menggunakan metode multi-attribute failure mode analysis (MAFMA) (studi kasus: pg. kebon agung malang) *Jurnal Rekayasa dan Manajemen Sistem Industri* **3** 3
[5] Annisa R 2016 Identifikasi penyebab stopline pada electrolytic tinning line menggunakan metode multi attribute failure mode analysis essay
[6] Amalia N 2017 Usulan perbaikan untuk mengurangi kegagalan proses potensial pada proses produksi batang kawat dengan pendekatan six sigma dan multi attribute failure mode analysis essay
[7] Ulfah M, D L Trenggonowati, R Ekawati and S Ramadhania 2019 The proposed improvements to minimize potential failures using lean six sigma and multi attribute failure mode analysis approaches *IOP Publishing Broad Exposure to Science and Technology 2019 (BEST 2019) IOP Conf. Series: Materials Science and Engineering* 673 (2019) 012082
[8] Yuvita, E 2017 Analisis pengendalian kualitas produk dengan metode six sigma pada pt mahakam media grafika di balikpapan *Ejournal Administrasi Bisnis* **5** 4 5 1241–1252
[9] Gaspersz Vincent 2002 P*edoman implementasi program six sigma terintegritas dengan ISO, 9001:2000, MBNQA dan HACCP* Jakarta: Penerbit PT. Gramedia Pusaka Utama
[10] Saaty TL 1999 Basic theory of the analytic hierarchy process: how to make a decision **93** 4 395-423
[11] Saaty TL 1994 *The analytical hierachy process* **6** (New York: McGraw-Hil)
[12] Novaline, Andalusiana Martha 2012 Pemilihan media periklanan dengan metode grey theory essay Jakarta : Fakultas Teknik Universitas Indonesia
[13] Braglia Marcello 2000 MAFMA: multi attribute failure mode analysis *International Journal Of Quality & Realibility Management* **17** 9 1017-1033
[14] Tseng and Tsaur 1994 The multiple criteria evaluation of grey relation *Journal of Grey System* **6** 2 87-108

IOP Conf. Series: Materials Science and Engineering **909** (2020) 012059 doi:10.1088/1757-899X/909/1/012059

Analysis of benzene exposure considering workers characteristic in the oil and gas industry

Yusita Attaqwa[1], Manik Mahachandra[1*], Heru Prastawa[1]

[1] Industrial Engineering Department, Faculty of Engineering, Diponegoro University, Semarang-Indonesia

*Email : manik.mahachandra@ft.undip.ac.id

Abstract. Operations and processes in the oil and gas industry have hazardous chemicals. Hence the possibility of having a work accident is high. Chemicals that exist are produced by activities related to the oil and gas industry processes, one of which is benzene. Because it has a severe impact on occupational health and safety, benzene exposure must be measured. Measurements were run through risk analysis to assess Hazard Index (HI) and make predictions of benzene exposure by focusing on the characteristics of workers, which can worsen the effect of the exposure. This study elaborated on several workers' characteristics from a literature study's human factors point of view. These characteristics were smoking, age, type of task, personal protective equipment use, exposure duration, regulations, hand washing habits, length of exposure, and nutritional status. Literature study results showed that regulatory characteristics, handwashing habits, and length of exposure were worsening the benzene exposure to workers. The other factors were in between agreed and disagreed in worsening the benzene exposure. These results perform as a base study in the further benzene analysis of oil and gas end distributor.

1. Introduction

Operations and processes in the oil and gas industry have hazardous chemicals; hence the possibility of having a work accident is high. Chemical work accidents in the course of carrying out work can threaten every worker to face occupational health and safety risks caused by his work. Case studies of workers in oil refineries investigating health hazards are critical to identifying health hazards for worker safety [1].

The oil and gas industry is the primary source of volatile aromatic hydrocarbons in the environment [2]. Benzene is the primary aromatic hydrocarbon compound whose presence is produced one of them by activities related to the oil and gas industry process. Benzene is a hazardous chemical classified as a class 1 carcinogen and mutagen that can be infected by humans and animals through dermal, oral, and inhalation exposure [3]. Benzene exposure to workplace workers mainly occurs through inhalation compared to dermal and oral [4]. The average benzene exposure in humans through inhalation has been reported to be around 50 to 80% [5]. Acute exposure to benzene concentrations can also affect fatigue, dizziness, headache, drowsiness, confusion, tremors, and loss of consciousness [7] [8]. Chronic exposure can cause more serious adverse health effects, such as myeloid leukemia, myeloma, reduced production of red and white blood cells from the bone marrow, decreased immune system, central nervous system damage, slow reflexes, liver and kidney failure, and cancer [8].

This explanation reveals the problems faced in the oil and gas industry regarding benzene exposure. This study, therefore, aims as literature review to understand benzene exposure to humans considering the characteristics of workers who can exacerbate benzene exposure to workers to determine risks endanger health and safety at work.

2. Literature Review

2.1. Industrial Hygiene

The Occupational Safety and Healthy (OSHA) defines Industrial Hygiene as a science aimed at anticipating, recognizing, evaluating, and controlling environmental factors from the workplace conditions that may cause illness, health problems, and well-being, or significant inconvenience and inefficiency among the worker. Industrial hygienists as occupational safety and health professionals controlling the work environment against occupational health hazards that arise as a result of or during work. Industrial hygiene monitoring and analysis to detect the extent of worker exposure and then base any hazard controls on actual exposure data [6].

The science ergonomics studies and evaluates focuses on the design of a working system where humans work a full range of tasks, including lifting, holding, pushing, walking, and reaching. All work systems consist of human components, machine components, and the environment that interact with each other. The relationship between humans and the environment is handled by industrial hygienists [9].

2.2. Benzene

Benzene (C_6H_6) is a cyclic organic chemical compound usually found in an environment that is flammable, colorless with a distinctive aromatic fragrance that is volatile into the air. This compound is a natural part of crude oil, gasoline, and cigarette smoke as a source of exposure [10].

Benzene exposure to humans is unavoidable because it requires measurement of benzene that occurs in the industrial oil and gas environment. The procedure for analyzing air samples refers to NIOSH 1501 (Hydrocarbons, Aromatic). Sampling and analysis can be done through the collection of benzene vapor in worker breathing zone by put down a personal sampling pump (Figure 1) that has been installed charcoal tube (Figure 2) near the respiratory tract of the worker. Subsequently, chemical analyzes were performed using gas chromatography in Figure 3 [10].

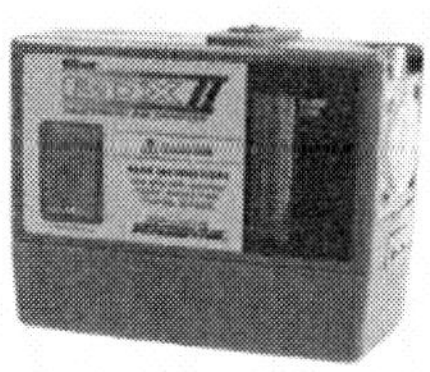

Figure 1. Personal sampling pump

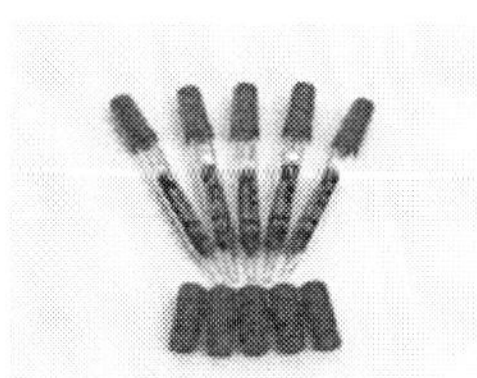

Figure 2. Charcoal tube

Figure 3. Gas chromatography

The Indonesian government's steps to minimize benzene exposure make regulations regarding limits on the value of chemicals that can expose workers while in the workplace. Based on the Minister of Manpower and Transmigration Regulation No. PER.13/MEN/X/2018 of 2011 concerning the Threshold Limit Value (TLV) of physical and chemical factors in the workplace, the recommended benzene TLV to prevent health impacts is 0.5 ppm. Measurements can be made with a risk analysis used to assess the Hazard Index (HI) [11].

IOP Conf. Series: Materials Science and Engineering **909** (2020) 012059 doi:10.1088/1757-899X/909/1/012059

2.3. *Worker Characteristics*

This study is based on several characteristics that are likely to affect benzene exposure. It is known from previous studies that smoking, age, type of task, PPE use, exposure duration, regulations, hand washing habits, length of exposure, and nutritional status are characteristics that can exacerbate benzene exposure in Table 1.

Table 1. Literature review result.

Location / **Characteristics**	+ / -	TBBM	SPBU	Oil Refinery
Smoking	+	[12]	[13] [14] [15] [16] [17] [18]	[17] [19] [20] [21] [22]
	-		[23] [24] [25] [26]	[27] [28]
Age	+	[12] [29]	[15] [18] [30]	[21] [22]
	-		[17] [23] [24] [25]	[20]
type of task	+	[12] [31] [32] [33]		[34]
	-	[35] [36]	[36]	[36]
PPE use	+		[37]	[22] [23]
	-		[26]	
Duration exposure	+		[15] [37]	[21] [39] [40]
	-		[15] [24] [25] [26]	[27] [28]
regulations	+	[42]		
hand washing habits	+		[37]	
length of exposure	+		[42] [43] [44]	[39] [40]
nutritional status	+			[21]
	-			[27]

(+) Effect; (-) No Effect

2.3.1. *Smoking habits*

Research agrees that the more significant number of cigarettes smoked affects the levels of benzene that enters the body. The information on petroleum distribution location when considered smokers, a positive relationship was found between personal air exposure levels of ethylbenzene levels in urine samples [12]. The information on service station BTX levels for exposure to smoking subjects is higher than nonsmokers [18] [13]. Smokers are found to be significantly higher than nonsmokers [15]. In smokers, the SPMA value and the value of benzene are significantly higher than nonsmokers [14] [16]. The information on oil refinery personal exposure that smoking has an impact on air exposure, exposure is higher in smokers than nonsmokers ($P < 0.001$) [17] [19]. Smokers and nonsmokers show statistically significant differences and confirm that smoking is a major confounding factor when assessing benzene exposure to work [20]. Smoking habits of oil refinery workers are mostly light smokers [21]. As many as 41 people or 83.7% of 49 respondents had smoking habits [22].

Research does not agree that the increasing number of cigarettes smoked affects the levels of benzene that enters the body. There were no significant differences in smoking characteristics in the oil service station [23] [24] [25] [26]. There were no significant differences between smokers and nonsmokers in the oil refinery between employees exposed to and not exposed to benzene [27]. Whereas the other study, there was no significant difference in ttMA levels in respondents with smoking habits ($p = 0.559$) [28].

2.3.2. Age

Research that agrees that the older the workforce is, the higher the risk of benzene poisoning. The information on petroleum distribution age grouping shows that health effects increase from the age of workers is 40 years. Research shows that age has a strong relationship with the hematological index on the health of oil workers [12] [29]. The information on the service station, the average age is 40.5 years [15]. The exposure group's average age was 38 years, ranging from 20 to 57 years [18]. The age of oil refinery workers is most in the age range of 49-52 years [21]. Whereas the other study, the average age of respondents in this study is the tankers is 36 years [22]. Data were obtained from a questionnaire about age in the group studied $p < 0.05$ [30].

Research that disagrees that the older the workforce is, the higher the risk of benzene poisoning. The information on the service station, there was no significant dependence of Benzene in the air on age characteristics [17] [23] [24] [25]. The information on oil refinery research shows no significant difference between subjects exposed to work and the age of the worker [20].

2.3.3. Types of Tasks

Research that agrees that the filling task, the higher the risk of benzene exposure. The information on petroleum distribution this type of task was found to be the most important characteristic associated with exposure [32]. Research that agrees to the older the workforce is, the higher the risk of benzene tanker loading is the highest risk occupation at this facility [12] [33]. The highest lighted occupations (8 hours) TWA is maintenance workers who can be directly affected by benzene [31]. Based on the information on oil refinery, the most predictive independent variable of benzene exposure is work assignment [34].

Research that does not agree that the type of filling task the higher the risk of benzene exposure. The information on petroleum distribution 8-hour TWA type of task for low benzene within the working limit of gasoline component exposure [35] [36].

2.3.4. Use of Personal Protective Equipment

Research that agrees that the use of PPE can reduce benzene exposure through inhalation. The information on the service station most workers still do not use personal protective equipment when doing work [22] [23] [37].

Research that does not agrees that the use of PPE can reduce benzene exposure through inhalation. The information on service station based on the results of the study, no significant relationship PPE usage of gas station operators to the exposure of benzene in SPBU, Tuak Daun Merah Subdistrict, Kupang City (p-value > 0.05) [26].

2.3.5. Exposure Duration

Research that agrees that the number of years worked will affect the intake of benzene exposure that enters the body of the worker. The information on the service station, the average exposure duration is 16 years and 23 years [15] [37]. The information on an oil refinery, there was a significant correlation of exposure duration (p-value < 0.05) [38]. The working period of oil refinery workers is the most in 22-25 years [21]. The safe working period for a lifetime (30 years) is 6.1 years [40].

Research does not agree that the number of years worked affects the intake of benzene exposure that enters the body of the worker. The information on the service station, there is no significant dependence of benzene in the air on the characteristics of the length of service [17] [24] [25] [26]. The information on an oil refinery, there was no significant difference between the exposure duration of work between employees who are exposed to and not exposed to benzene [27] [28].

2.3.6. Regulations

Research that agrees that rules which are not followed are higher at risk. Regulation is a factor influencing the concentration of benzene exposure in gasoline storage and distribution facilities in developing countries [45].

2.3.7. Handwashing Habits
Research that agrees that lousy handwashing habits have a higher risk. The habit of washing hands in the Ciputat Timur gas station operator in 2017 was weak in terms of only 28 people (38.4%) [37].

2.3.8. Length of Exposure
Research that agrees that the longer working hours that occur to workers, the higher the risk. There was a significant correlation of working hours to levels of t, t-muconic acid urine (all variables, p <0.05). The information on oil refinery this study concludes that the longer working hours that occur in workers, the higher levels of t, t-muconic acid urine [39]. The safe length of exposure for a lifetime of work (30 years) is 1.6 hours per day [40]. The information on service station Length of Exposure is 8 hours / day [42] [43] [44].

2.3.9. Nutrition Status
Research that agrees that the higher the weight will reduce the level of risk. The nutritional status of oil refinery workers is most in normal conditions [21]. Research that does not agree that the higher the weight will reduce the level of risk. There is no difference between abnormal nutritional status and significantly normal nutritional status between employees exposed to and not exposed to benzene [27].

3. Prospective Research
Perusahaan Pertambangan Minyak dan Gas Bumi (PERTAMINA) operates several TBBM (Terminal Bahan Bakar Minyak), one of which is the TBBM Semarang Group. According to The Agency for Toxic Substances and Disease Registry (ATSDR) [10], benzene is a hazardous chemical in oil that has a severe impact on health and safety. So the TBBM Semarang Group's environment will not be separated from the possibility of benzene exposure to workers, which can have a severe impact on work health and safety. Because it has a severe impact on occupational health and safety, benzene exposure must be measured. If more than the TLV has been determined, then corrective action is needed for health and safety. Measurements can be made for making predictions that will occur due to the presence of benzene exposure in the work environment by looking at the characteristics of workers who can exacerbate benzene exposure. When looking at literature studies for benzene exposure, some of the characteristics of workers must be considered. Some characteristics can worsen the occurrence of benzene exposure, affecting the health and safety of workers. So it is essential to conduct the assessment of benzene exposure in the TBBM Semarang Group.

4. Conclusion
Operations and processes in the oil and gas industry have hazardous chemicals, one of which is benzene that can be infected by humans. Hence, the possibility of a work accident is high for workers. After looking at the related review literature, it can be concluded that the study agreed and disagreed about the characteristics that could exacerbate benzene exposure to workers. Table 2 shows that (+) is a researcher who agrees and (-) is a researcher who disagrees. The characteristics of smoking, age, type of task, PPE use, exposure duration, and nutritional status have researchers who agree and disagree that these characteristics can exacerbate benzene exposure to workers. Regulatory characteristics, handwashing habits, and length of exposure were agreed to worsen benzene exposure to workers. So one of which PERTAMINA business units is the TBBM Semarang Group very important can be monitoring.

5. Reference
[1] Eyayo F 2014 Evaluation of occupational health hazards among oil industry workers: a case study of refinery workers *IOSR journal of environmental science, toxicology and food technology* **8** 22–53

[2] Rao P S, Ansari M F, Gavane A G, Pandit V I, Nema P, Devotta S 2006 Seasonal variation of tox ic benzene emissions in petroleum refinery *Environmental monitoring and assessment* **128** 32 3–8
[3] Li J, Lu S, Liu G, Zhou Y, Lv Y, She J, et al 2015 Co-exposure to polycyclic aromatic hydrocarb ons, benzene and toluene and their dose–effects on oxidative stress damage in kindergarten-ag ed children in Guangzhou, China *Science of the total environme*nt 524-525 74–80
[4] Tompa A, Jakab M G, Major J 2005 Risk management among benzene-exposed oil refinery work ers *International journal of hygiene and environmental health* **208** 509–16
[5] DECOS 2014 *Benzene, health-based recommended occupational exposure limit, No. 2014/03*. The hague: the health council of the netherlands.
[6] Industrial hygiene activities in construction 2006 Handbook of OSHA construction safety and heal th, second edition CRC Press 429–87
[7] Mitri S, Fonseca A, Otero U, Tabalipa M, Moreira J, Sarcinelli P 2015 Metabolic polymorphisms and clinical findings related to benzene poisoning detected in exposed brazilian gas-station w orkers *International journal of environmental research and public health* **12** 8434–47
[8] Li J, Lu S, Liu G, Zhou Y, Lv Y, She J, et al 2015 Co-exposure to polycyclic aromatic hydrocarb ons, benzene and toluene and their dose–effects on oxidative stress damage in kindergarten-ag ed children in Guangzhou, China *Science of the total environment* **5** 74–80
[9] Bridger R 2008 *Introduction to ergonomics* CRC Press
[10] Toxicological profile for benzene ATSDR's 2002 *Toxicological profiles* CRC Press
[11] Soemirat J 2000 *Lecture material health and environmental risk analysis.* KBK environmental health Department of environmental engineering Bandung institute of technology Bandung
[12] Heibati B, Godri Pollitt K J, Charati J Y, Ducatman A, Shokrzadeh M, Karimi A, et al 2018 Bio monitoring-based exposure assessment of benzene, toluene, ethylbenzene and xylene among w orkers at petroleum distribution facilities *Ecotoxicology and environmental safety* **149**19–25
[13] Scheepers P T J, de Werdt L, van Dael M, Anzion R, Vanoirbeek J, Duca R C, et al 2019 Assess ment of exposure of gas station attendants in sri lanka to benzene, toluene and xylenes *Environ mental research* 178:108670
[14] Carrieri M, Bonfiglio E, Scapellato M L, Maccà I, Tranfo G, Faranda P, et al 2006 Comparison o f exposure assessment methods in occupational exposure to benzene in gasoline filling-station attendants *Toxicology letters* **162**146–52
[15] Lovreglio P, Maffei F, Carrieri M, D'Errico M N, Drago I, Hrelia P, et al 2014 Evaluation of chro mosome aberration and micronucleus frequencies in blood lymphocytes of workers exposed to low concentrations of benzene *Mutation research/genetic toxicology and environmental muta genesis* **770** 55–60
[16] Lovreglio P, Doria D, Fracasso M E, Barbieri A, Sabatini L, Drago I, et al 2015 DNA damage an d repair capacity in workers exposed to low concentrations of benzene *Environmental and Mol ecular Mutagenesis* **57**151–158
[17] Fracasso M E, Doria D, Bartolucci G B, Carrieri M, Lovreglio P, Ballini A, et al 2010 Low air le vels of benzene: correlation between biomarkers of exposure and genotoxic effects *Toxicology letters* **192** 22–8
[18] *Genotoxicity in filling station attendants exposed to petroleum hydrocarbons* 2010 The annals of occupational hygiene Oxford University Press (OUP)
[19] Fustinoni S, Campo L, Mercadante R, Consonni D, Mielzynska D, Bertazzi P A 2011 A quantitati ve approach to evaluate urinary benzene and S-phenylmercapturic acid as biomarkers of low b enzene exposure *Informa UK Limited* **16** 334–45
[20] Carrieri M, Spatari G, Tranfo G, Sapienza D, Scapellato ML, Bartolucci GB, et al 2018 Biologic al monitoring of low level exposure to benzene in an oil refinery: Effect of modulating factors *Toxicology letters* **298** 70–5
[21] Indrawan D, Oginawati K 2014 Analisis paparan BTX terhadap pekerja di PT Pertamina RU IV cilacap *Jurnal tehnik lingkungan* 20(2):132–41

[22] Febrian N W, Rahardjo M, Nurjazull 2019 Environmental health risk analysis due to benzene exposure through inhalation of tank crew crews at PT Pertamina patra niaga *Journal Public Health* **7** 2356-3346
[23] Xiong F, Li Q, Zhou B, Huang J, Liang G, Zhang L, et al 2016 Oxidative stress and genotoxicity of long-term occupational exposure to low levels of BTEX in gas station workers *Internationa l journal of environmental research and public health* **13**1212
[24] Moro A M, Brucker N, Charão M F, Baierle M, Sauer E, Goethel G, et al 2017 Biomonitoring of gasoline station attendants exposed to benzene: Effect of gender *Mutation research/genetic to xicology and environmental mutagenesis* **8**131–9
[25] Krieg E F, Mathias P I, Toennis C A, Clark J C, Marlow K L, B'Hymer C, et al 2012 Detection o f DNA damage in workers exposed to JP-8 jet fuel *Mutation research/genetic toxicology and e nvironmental mutagenesis* **747** 218–27
[26] Nenotek J S 2019 *The relationship between work period, use of personal protective equipment, and smoking habits against benzene exposure to gas station operator officers in the kupang city tuak daun merah (TDM) village in 2019* Health polytechnic of the ministry of health of kupang Kupang
[27] Ramon, A 2007 *Analysis of Benzene Exposure to Blood Profile in Petroleum Processing Industry Workers* Thesis Postgraduate Program Environmental Health Masters Study Program Diponegoro University
[28] Sipayung L P, Suryanto D, Megawati E R 2016 Correlation between benzene exposure and the complete blood count overview of SPBU x and y employees *MKMI journal* **12**
[29] Ezejiofor T I 2016 Haematological indicators of exposure to petroleum products in petroleum refi ning and distribution industry workers in nigeria Journal of clinical toxicology **06**
[30] Goethel G, Brucker N, Moro A F, Charao M, Fracasso R, Barth A, et al 2014 Evaluation of genot oxicity in workers exposed to benzene and atmospheric pollutants *Mutation research/genetic t oxicology and environmental mutagenesis* **70** 61–5
[31] Chung E K, Jang J K, Koh D H 2017 A comparison of benzene exposures in maintenance and reg ular works at korean petrochemical plants *Journal of chemical health and safety* **24** 21–6
[32] Peretz C, Froom P, Pardo A, Goren A 2000 Exposure to benzene in fuel distribution installations: monitoring and prevention *Archives of environmental health an international journal* **55** 439–46
[33] Heibati B, Pollitt K J G, Karimi A, Yazdani C J, Ducatman A, Shokrzadeh M, et al 2017 BTEX e xposure assessment and quantitative risk assessment among petroleum product distributors *Ec otoxicology and environmental safety* **144** 445–9
[34] Scarselli A, Binazzi A, Di Marzio D 2011 Occupational exposure levels to benzene in italy: findi ngs from a national database *International archives of occupational and environmental health* **84** 617–25
[35] Pandya G H, Gavane A G, Bhanarkar A D, Kondawar V K 2006 Concentrations of volatile organ ic compounds (VOCs) at an oil refinery *International journal of environmental studies* **63** 337 –51
[36] Glass D 2000 Retrospective exposure assessment for benzene in the australian petroleum industry The annals of occupational hygiene **44** 301–20
[37] Khoir N F 2017 *Overview of safe work practices against benzene exposure to gas station operator workers in east ciputat region* UNDIP Semarang
[38] Gaffney S H, Burns A M, Kreider M L, Unice K M, Widner T E, Paustenbach D J, et al 2010 Occ upational exposure to benzene at the exxon mobil refinery in beaumont, txX (1976–2007) *inte rnational journal of hygiene and environmental health* **213**285–301
[39] Saadatuddaroini S, Keman S 2019 Korelasi masa kerja, jam kerja terhadap kadar t,t- mocanic ac id urin pekerja terpapar benzena di pertambangan minyak tradisional bojonegoro *The indonesi an journal of occupational safety and health* **8** 115

[40] Kartikasari D, Nurjazull, & Rahardjo M 2016 Health risk analysis of benzene exposure to workers of the petroleum processing industry laboratory section *Journal public health* **4** 2356-3346
[41] Ezejiofor T I N 2014 Risk assessment: re-appraisals for potential hazards in the operational envir onment and facilities of petroleum refining and distribution industry in nigeria - research and r eview Occupational medicine & health affairs *OMICS publishing group* **02**
[42] Salim R N 2012 *Health risk analysis of benzene exposure to employees at the 'x' pancoranmas depok gas station in 2011* Thesis Bachelor of public health study program Department of occupational safety and health Faculty of public health University of indonesia
[43] Hayat I 2013 *Analysis of health risk magnitude of benzene exposure to petrol station operator officers in the ciputat region in 2012* Repository of UIN Syarif Hidayatullah Jakarta
[44] Triyadi D, Nurjazull, & Dangiran H L 2016 Health risk analysis due to benzene exposure through inhalation at the public fuel station (SPBU) officers in the vicinity of diponegoro university semarang *Journal public health* **4** 2356-3346
[45] Nguema A F O, Tokai A, Thi Nguyen H, & Kojima N 2019 Analysing factors influencing occupa tional benzene exposure concentration in loading operations at gasoline storage and distributio n facility in developing countries *Asian Journal of Applied Sciences* **7**

Physical flow described as the flow of production in the manufacture of products from raw materials until becoming finished products. The physical flow identification was evaluated using SIPOC diagrams (Suppliers, Inputs, Process, Outputs, and Customers).

3. Identify Critical to Quality (CTQ)
   Critical to Quality (CTQ) is a characteristic of a product or service that fulfill the needs of the consumer namely: lacps, overfill, underfill, creep speed, crumpled, not center, cross roll/un round, other deft, scratch, scrappy, roll mark, and coil.
   Wire rod product consists of 12 types of quality characteristics that are divided into 3 types of criteria (surface defect, dimention defect, and mechanical defect). Characteristics of this quality was used as a benchmark to produce products with good quality.

4.2. Measure stage

Measure is the second step in the DMAIC method which is a follow up of the define step. This stage focuses on understanding the work of the selected process for improvement at this time. In Measure phase, calculation of the value of DPMO and sigma level were conducted. Based on the calculation, the average DPMO value is 2,928.626 and the average sigma value is 4.297503. It states that from a million opportunities there will be 2,928.626 the possibility that the process will cause defect and the level of achievement of Company X in producing wire rod is quite good, that is equal to 4.297503 which has reached the industry average of USA (approximately 4 sigma or nonconforming and level of achievement of Company X in producing wire rod).

4.3. Analyze stage

Analyze is the third step in the DMAIC method. Identification is done to determine the factors that influence the failures and type of factors are considered the most dominant in order to make improvements to the existing process.

1. Pareto diagram
   The Pareto diagram is a bar graph showing the problem based on the sequence of the occurrences number. Pareto diagram formed based on the principle that 80% of problems caused by 20% of the root of the problem. Therefore, the researcher takes 80% of the overall reject percentage on the wire rod product to be followed up by the cause of the reject or identify the failure of the process. The percentage of 80% are laps defects, creep speed defects, underfill defects, overfill defects and scrappy defects.
2. Root Cause Analysis (RCA)
   Root Cause Analysis is a problem-solving method that aims to identify the root cause of the problem from the occurrence of the problem. One of method in Root Cause Analysis (RCA) is 5 Whys. The 5 Whys were obtained from interviews with production division and synchronized with the quality control division because both of divisions are interconnected in handling the failure of production processes that cause product defects and production delay. Based on the results of 5 Whys, there are 18 causes of the failure process. Then, these indentified causes was used as input in subsequent data processing, namely Multi Attribute Failure Mode Analysis (MAFMA).
3. Multi Attribute Failure Mode Analysis (MAFMA)
   Multi Attribute Failure Mode Analysis (MAFMA) is a method that integrates the conventional Failure Mode and Effect Analysis (FMEA) by considering the economic aspects [12]. In this research, Multi Attribute Failure Mode Analysis (MAFMA) method uses Grey AHP approach.

4.3.1. Failure Mode and Effect Analysis (FMEA)

First FMEA team was established to fill out the FMEA questionnaire where the selected respondent is company parties that experienced in the scope of process failure ad its impact. Respondents consist of quality control technicians, employees of process control divisions and production supervisors.

4.3.2. Hierarchy structure preparation

The structure of MAFMA hierarchy consists of objectives, criteria, sub criteria and alternatives. The purpose of this hierarchical structure is the selection the causes of production process failure. On the

IOP Conf. Series: Materials Science and Engineering **909** (2020) 012060 doi:10.1088/1757-899X/909/1/012060

Detection system and security monitoring of the storage room of liquid oil gas with zoning method

Suzuki Syofian[1*], **Guswan**[2], **Timor Setiyaningsih**[3]
[1,2,3] Darma Persada University, Department Information of Technology

*Email: suzukiunsada@gmail.com

Abstract: Fires often occur due to leakage of the Liquid Petroleum Gas (LPG). Fuels in gaseous forms produce layers due to condensation with a negative effect when evaporated in free air. Fire usually starts with a small spark before inflaming largely. The frequent fires in homes and industrial sites that use LPG gas leads to huge losses of material and life. Therefore, cellular communication technology helps to facilitate the provision of information associated with the fire quickly. Communication is through voice and video calls, as well as via chats using the Telegram application, which provides bot system features from various cellular devices. With the zoning method, each signal sample of the communication is converted and characterized into a data vector with a fast calculation process. The result showed that the input data in the form of captured signals, extracted and processed, produces an output in the form of a warning or notification. Therefore, the application is expected to minimize the occurrence of fires.

1. Introduction

Cellular communication technology is a communication technology that has become part of human's communication today. Almost 90% of people have this cellular communication tool. The communication is done either by using voice, video call, or a chat through telegram application. The chat communication is reachable, and it has the service of system feature from all kinds of cellular communication device.

Gas as a supporting tool for household needs, restaurant, and hotel. If a leak occurs onto the gas, it will give a serious problem [1]. Therefore, it needs more extra handling, so that it will not cause a big loss to the users[2]. The gas sensor is a device that can detect a leak of the gas. Explosion caused by the leak of the gas can be anticipated by turning on the exhaust system to minimize the gas condensation. The existent of this incident can provide a warning alarm and a notification to the users. As we know that the fire comes from the triangle of flame (fuel, oxygen, and fire) [3]. from here, it can be detected from one of the triangles of flame, that is fuel or gas. Arduino Uno based gas leak detector with a GSM modem and buzzer as a notification[4], which can detect when a leak occurred, and can activate the handling system, and provide information to the users, so that fire can be anticipated as soon as possible [1][5]. This research is aimed to design and realize a device that can detect the leak in the LPG tube, provide early warning of the danger of the leak in the LPG tube, reduce the loss caused by the leak in the LPG tube.

2. Research Methods

System design and modeling is done to anticipate the occurrence of warehouse fires, in addition to providing information on gas leakage followed by a countermeasure / anticipation system, for example exsous system to reduce gas condensation or by activating a chimney pump to dump condensation of gas into the warehouse / free air at the top of the warehouse. The features of the LPG gas storage room detection and monitoring system are:

a. Detects fire in LPG storage room.

b. Detects the gas content in the LPG storage room.
c. Detects temperature and humidity in the LPG storage room.
d. Detects movement in the LPG storage room.
e. Detects vibrations in the LPG storage room.
f. Provides fire prevention by buzzers, exsous motors and water.
g. Sending monitoring data on the state of the gas storage room to the website and telegram.

Based on the predetermined features, the LPG gas storage and detection system is developed as follows:

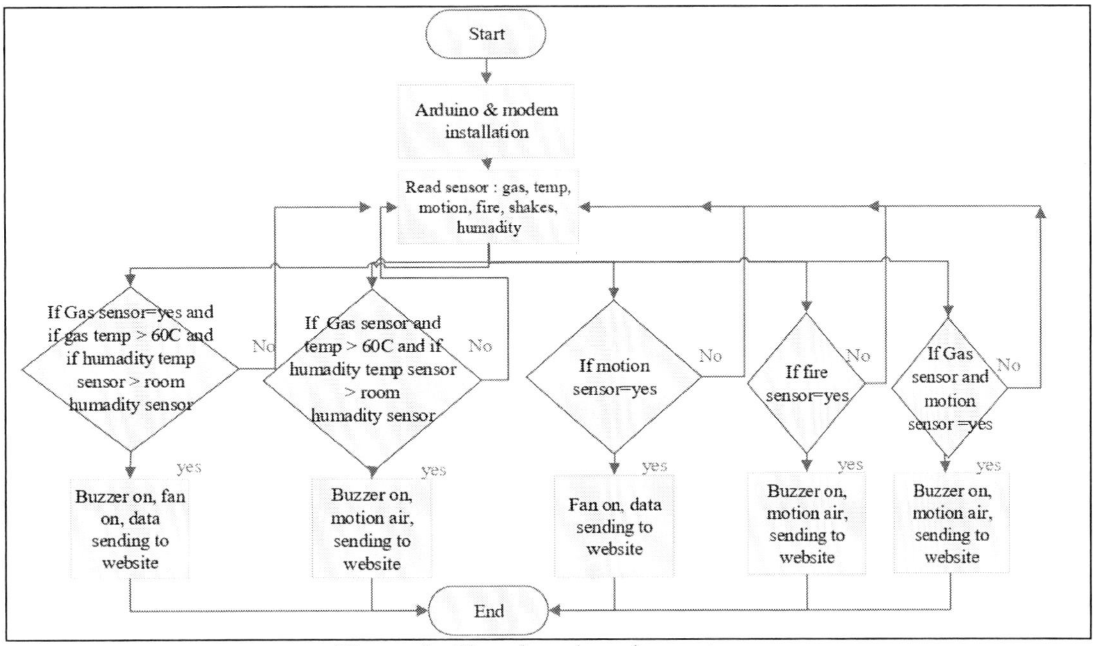

Figure 1. Flowchart detection system

2.1. Inputs needed

1. The MQ-2 sensor is a gas sensor
2. Sensor KY-015 DHT-1 is a temperature and humidity sensor

2.2. Output required

1. Buzzer is a warning sign
2. Relay activator exsous device and water pump

2.3. System integrator / minimum system

Arduino UNO

2.4. System programing

The command structure in Arduino outlines consists of 2 (two) parts, namely void setup and void loop. The void setup contains commands that will be executed only once since arduino is turned on while the void loop contains commands that will be executed repeatedly while arduino is turned on.

3. Results and Discussion

Figure 2. follows the use case diagram that shows the function of the application made, in the picture there are users who can open and view the fire sensor, gas sensor, temperature sensor, humidity sensor, motion sensor, and vibration sensor by logging in first. Figure 3. below is a monitoring display design, starting and main page when operating the program. In this monitoring page, there is a file menu that has the sub-menus of fire, gas, temperature, humidity, motion, and log out.

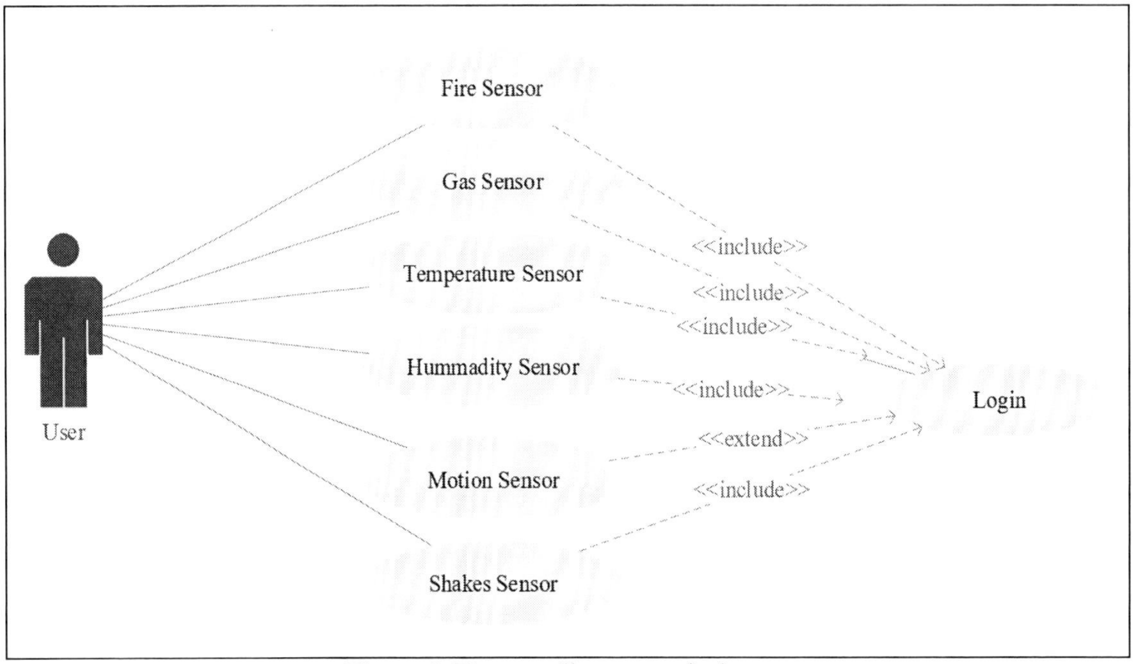

Figure 2. Use case diagram analysis

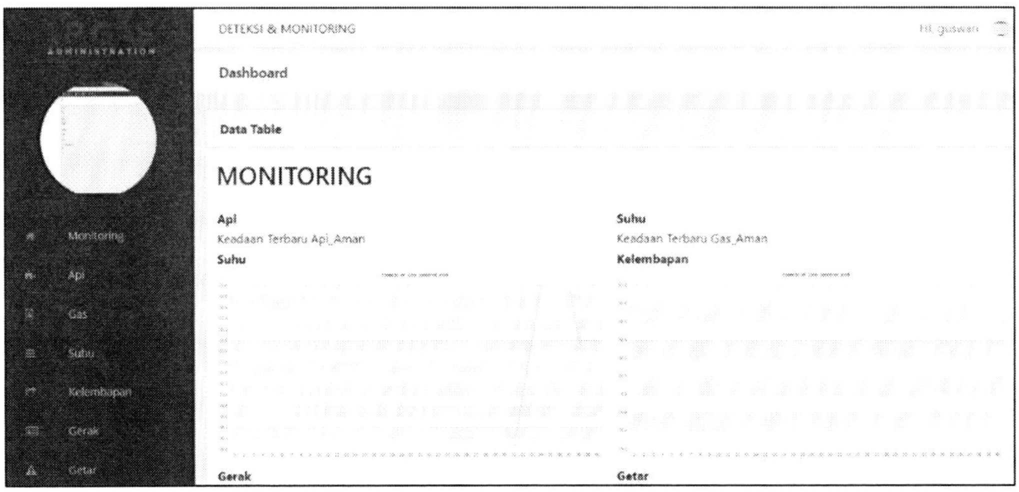

Figure 3. Detection and display monitoring

3.3. Database design

The database is very needed in making the detection and monitoring system security system for LPG gas storage room for storing the data. The following is a database design needed in the making of the LPG gas storage room security and detection system program.

3.4. Calculate sensor coverage

LPG warehouse is a place to store and unload LPG gas cylinders. Standard warehouse requirements are a minimum area of 400 m2 built from non-flammable material. The way to determine the distance covered by the sensor can be used APAR calculations, as in the equation below: [6]

- Sensor name = (Building Size (m ^ (2))) / (Minimum Protected Area (m ^ (2)))
- The hc-sr501 pir sensor has a limited range, which is a maximum of approximately 10 meters.
- Pir hc-sr501 motion sensor = (400)m ^ 2 / (10 m) = 40 m.

Based on the above calculation, the motion sensor can reach an area of 40 m. The Mq-2 gas sensor calculates the measurement distance by the amount of gas detected.

Measurement range:

- 200 - 5000ppm for LPG, propane
- 300 - 5000ppm for butane
- 5000 - 20000ppm for methane
- 300 - 5000ppm for Hydrogen
- 100 - 2000ppm for alcohol
- Ky-015 dht-11 sensor has a maximum range of approximately 20 meters.
- Pir hc-sr501 motion sensor = (400)m ^ 2 / (20 m) = 20 m
- Vibration sensor measurement with the amount of vibration.
- Robotdyn ir fire sensor detects the presence of fire or other infrared sources (fire or light source wavelengths in the range of 760 nm to 1100 nm can be detected)

3.5. Sensor Testing

3.5.1. Mq-2 sensor

In gas sensors, time is influential because the gas required is very large, therefore the distance needed is close so that the gas sensor can detect gas. except if the remote position of the gas produced is very small with a long time the gas sensor will not be active. Shown in Table 1. at a distance of 4 cm the sensor is inactive.

Table 1. Position data for gas

Distance (cm)	Time (Second)	Condition
1 cm	10 Second	Active
2 cm	10 Second	Active
3 cm	10 Second	Active
4 cm	10 Second	In-active

Table 2. Position obtaining gas

Distance (cm)	Time (Second)	Condition
1 cm	10 Second	Active
2 cm	10 Second	Active
3 cm	10 Second	Active
4 cm	10 Second	In-Active

3.5.2. Trials

To find out the work of the tools and websites that have been made and find out the response time of all sensors used in this tool.

3.5.3. Mq-2 sensor

In gas sensors, time is influential because the gas required is very large, therefore the distance needed is close so that the gas sensor can detect gas. Unless the remote position of the gas produced is very small with a long time the gas sensor will not be active, it is seen in Table 2. Obtaining gas position.

3.5.4. Robotdyn ir infrared fire sensor

In the fire sensor, the time is affected because the fire required is very large, therefore the distance needed is close so that the fire sensor can detect fire. except if the remote position of the fire produced is very small with a long time the fire sensor will not be active. As in the table Table 3. Data position obtain fire

3.5.5. Temperature sensor ky-015 dht-11

From the results of testing the DHT11 temperature sensor as an input module integrated into the Arduino Uno Microcontroller as a data processor when the temperature exceeds 35 degrees, it will output fan output and an active buzzer. As in table 4. Testing Temperature Sensor ky-015 dht-11.

Table 3. Position data to get fire

Distance (cm)	Time (Second)	Condition
8 cm	6 Second	Active
12,5 cm	6 Second	Active
22 cm	6 Second	Active
30 cm	6 Second	Active
40 cm	6 Second	In-active

Table 4. Trial of temperature sensor ky-015 dht-11

Condition	Output
Under 34 degress	Fan and *Buzzer* Inactive
Uper 35 degress	Fan and *Buzzer* Active

3.5.6. Pir hc-sr 501 motion sensor

From the results of testing the pir hc-sr 501 motion sensor as an input module that is integrated into the Arduino Uno Microcontroller as a data processor when the movement around the sensor, the output will be active. As in table 5. pir hc-sr 501 motion sensor trial test.

3.5.7. Vibration sensor

From the results of vibration sensor testing as an input module that is integrated into the Arduino Uno Microcontroller as data processing when shaking around the sensor, the buzzer output will be active. As in table 6. Trial of the vibration sensor trial.

Table 5. Testing of hc-sr 501 pir motion sensor

Condition	Output
No-Motion	Fan-Inactive
Motion	Fan-Active

Table 6. Trial of the vibration sensor

Condition	Output
No-Vibration	*Buzzer* In-active
Vibration	*Buzzer* Active

3.6. Output testing

3.6.1. Buzzer

The program on the microcontroller has been established that the microcontroller will issue a logic of 1 (5 Volts) only if the condition of the gas sensor, fire sensor, vibration sensor and temperature sensor detects gas leakage, fire, vibration and temperature increase and if the sensors are not detected then the output The microcontroller will be worth zero (0 Volt). When the program is run it is proven that the buzzer sounds when it is detected by the sensor. When the sensor does not detect, the buzzer will return to its original state (silent), as in table 7. Trial of the buzzer test.

IOP Conf. Series: Materials Science and Engineering **909** (2020) 012060 doi:10.1088/1757-899X/909/1/012060

3.6.2. Water motor

From the test results it can be seen that the fire sensor detects a fire, the water motor will be active. As in Table 8 Air Motor Trial.

Table 7. Trial of buzzer

Input Potential (Volt)	Information
0 v	*Buzzer* Inactive
4,8 v	*Buzzer* Active

Table 8. Water motor trial

Input Potential(Volt)	Information
0 v	Water Motor In-active
12 v	Water Motor Active

3.6.3. Fan

When a gas sensor, motion sensor, and temperature sensor are detected to detect a gas leak, there is movement and the temperature rises, the fan will turn on to reduce LPG gas levels in the test box. When LPG levels are not detected, the fan will stop spinning. As in table 9. Fan testing.

Table 9. Fan testing

Input Potential(Volt)	Information
0 v	Fan In-active
12 v	Fan active

3.7. Zoning method

Feature extraction is a process in which each signal sample will be converted into data vectors. The method to be used in the characterization process in identifying this signature is the zoning method. Like in Table 10. Zoning trial.

Table 10. Zoning trial

Sensor Input						Output				
Gas	Motion	Temp	Fire	Shakes	Humidity	Fan	Buzzer	Water Control	Website	Telegram
					Yes				✓	✓
				Yes			✓		✓	✓
	Yes					✓			✓	✓
		Yes				✓	✓		✓	✓
Yes						✓	✓		✓	✓
			Yes				✓	✓	✓	✓

4. Conclusion

Based on the results of research, design, and testing of systems and tools, the authors obtain conclusions as follows:

1. The application as a whole is running well and the test results are in accordance with the needs to detect, activate exsous and monitor.
2. The application aims to minimize the occurrence of LPG gas leakage in terms of report information

5. References

[1] Christian J and Komar N 2013 Prototipe sistem pendeteksi kebocoran gas LPG menggunakan sensor gas MQ2 board arduino duemilanove, buzzer, dan arduino gsm shield pada PT Alfa Retailindo *Ticom* **2** 58–64

[2] Putra M F, Kridalaksana A H and Arifin Z 2017 Rancang bangun alat pendeteksi kebocoran gas LPG dengan sensor MQ-6 berbasis mikrokontroler melalui smartphone android sebagai media informasi *Inform Mulawarman J Ilm Ilmu Kompu.* **12**

[3] Mandagi A and Immanuel S 2014 Penggunaan sensor gas MQ-2 sebagai pendeteksi asap rokok *J Tek dan Ilmu Komputer* **03** 1–7

[4] Santoso L H and Hasanah R S 2017 Rancang bangun sistem alarm kebakaran otomatis berbasis arduino uno menggunakan sensor cahaya dan sensor gas di teaching factory stt texmaco Subang *Jurnal Ilmiah Trend Tech* **2** 41

[5] Hasibuan M S, Syafriwel and Idris I 2019 Intelligent LPG gas leak detection tool with SMS notification *J Phys Conf Ser* 14–24

[6] Rizki, Lukman Handoko, Dany D 2018 Perancangan APAR dan ERP dengan simulasi pathfinder pada politeknik perkapalan negeri surabaya 337–340

IOP Conf. Series: Materials Science and Engineering **909** (2020) 012061 doi:10.1088/1757-899X/909/1/012061

Design of planning model for ERP system in warehouse management: an empirical study of public hospital in Indonesia

F D Utami[1*] W Puspitasari[1] M Saputra[1]

[1] Information system, Industrial Engginering Department, Telkom University, Bandung, Indonesia

*E-mail: fadhilahdianutami@student.telkomuniversity.ac.id

Abstract. Public Hospital is one of the complicated information in the healthcare industry. ERP provides a concept that integrates a whole series of business processes, so they are mutually sustainable as part of business management to achieve the goals of an organization. In line with the hospital's information system, ERP provides a solution for managing data information that runs in hospitals. As one of District Government Hospitals in Indonesia that provide health services, RSUD XYZ does not have an independent information system that manages the hospital's vital data specifically in warehouse management yet. Procurement and management of goods on a regular basis without a system managing in terms of inventory bring through redundancy, ineffective, and inefficient management of data in the warehouse. This research focuses on the planning model system for the public hospital in inventory management business processes, developed using OpenERP at Inventory Management module and QuickStart method. The results of this study is a design of a planning model for inventory management system in the public hospital. This research explains how the ERP system can help improve health services in the public hospital under business processes that should be in the healthcare industry.

1. Introduction

Health services need a complete system to manage business processes that are quite complex and vital, such as procedures and management data of patients, availability of doctors, management of employees, medicine stock information and medical tools until the management of invoices that take a long time to maintain. Like the other industries, the healthcare industry such as hospitals expects an accurate and quick management system that can make the organization's daily operations more effective and efficient.

Rumah Sakit Umum Daerah XYZ or RSUD XYZ is a public hospital that responsible as a type C hospital in Indonesia, which is responsible for one regional area, provide limited doctor specialist and referral from *Pusat Kesehatan Masyarat* (Puskesmas) (Putri et al., 2016). The absence of an information system to RSUD XYZ especially in the warehouse management division makes it difficult for RSUD XYZ to track every stock and high chance of double-entry. Moreover, RSUD XYZ use paper-based process and some in the excel to manage all of the information in the warehouse on RSUD XYZ, which makes it's difficult for warehouse staff to process warehouse data administration. The following table is types of items that are managed in the warehouse at RSUD XYZ.

Table 1. Item category of storable goods

Number	Item	Total Sub Item
1	Patient Food	67
2	Staff Food	6
3	Patient's food support	23
4	Office Stationery	144
5	Form	156
6	Cleaning Equipment	54
7	Electrical and Electronic Tools	66
8	Radiology Equipment	6
9	Laboratory Equipment	70
10	Pharmacy	2050

From the table above we can comprehend that the amount of goods in 10 items has exceeded 2500 items. Where each item has its procurement, management of when the good push and pull from warehouse and placement in the warehouse regularly. Without an information system that is able to manage terms in the warehouse in RSUD XYZ, warehouse staff will be overwhelmed by handling the data.

Information System is necessary because the amount of information that they rely on to carry out services is quite complex (Putri et al., 2016). Hence, the integration of different services and processes would benefit the healthcare organization and eradicate the traditional segregation of processes, which has resulted in the reduced sharing of information. Integration of data in hospitals between divisions will help the processing of existing data, because each data warehouse is created based on the needs of each other division in the hospital. Integrated systems would allow the healthcare organization to coordinate management and support the efficiency of patient care (Mucheleka & Halonen, 2015).

Larger organizations have been the main implementers of ERP systems. ERP systems provide solutions for many organizations that have successfully implemented them. The healthcare industries are also some sectors that could benefit from ERP systems implementation. ERP systems facilitate organizations using them with an opportunity to integrate individual information system while providing an organization with operational and technical benefits such as personnel reductions, service improvements, order improvements, cost reductions on logistics and procurements (Sanja, 2013)

2. Literature Review

2.1. ERP (Enterprise Resource Planning)

Enterprise Resource Planning is software intended to integrate business processes in manufacturing and production, finance and accounting, sales and marketing, and human resources into one software system (Hardjono et al., 2017). ERP collects information from all of the company's business functions that allow companies to have a much broader range of information. Enterprise Resource Planning is an application and database solution that automates and integrates information processing of business processes and business functions in an organization [5].

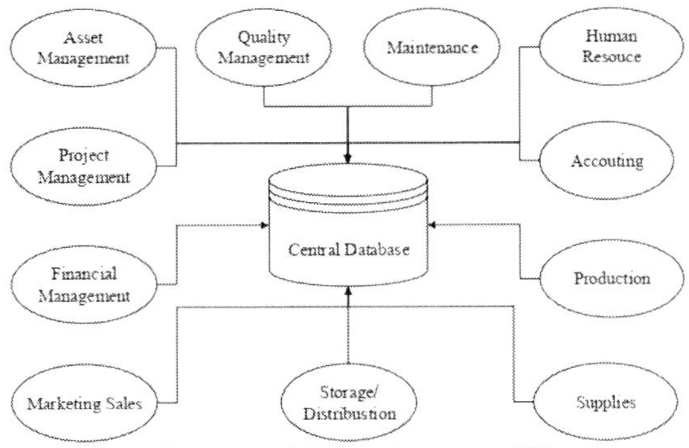

Figure 1. ERP subsystem [6]

Figure 1 describes general ERP subsystem model to combine and integrate business functions in one central database.

2.2. ERP system in healthcare industry
Manufacturing industries indeed use more ERP applications than the healthcare industry that deals with patients. ERP application usage in the healthcare industry is impeded due to the lack of standardization for ERP systems in the integration of the healthcare Industry. Apart from this challenge, ERP applications can still help health service organizations to integrate various hospital functions such as human resource management, patient scheduling, hospital decision making, and workflow management [7]. Besides, ERP systems have transcended the organization and tend to focus on resources, but ERP systems also help planning such as finance, controlling, operating management, analysing and reporting and decision making tasks(Sanja, 2013).

2.3. Warehouse management
Warehousing has an important role in the supply chain in every organization whatever materials it is operating with. Effectiveness of warehousing can affect the overall business of the company and cause reorganization in the company. Nowadays warehousing processes are rapidly changing due to information systems improvement. New technologies such as ERP systems, barcodes, computerized equipment, and many others have made warehousing efficient and accurate in its operations. Also, the educational level of human resources has developed, as personnel needs to be educated enough for computer-operated processes [8].

The basic function of the warehouse is to receive user orders, take the needed product and prepare to deliver the products to the customer. There are many ways to manage this process, but the whole process in a part of the warehouse follows the general phase of warehousing activities; Receiving, pre-packaging, put away, storage, order picking, and packaging, sortation, packing and Shipping, and shipment [9].

2.4. OpenERP
The open-source ERP software is ERP software that provides access and ways of working from its source code. Others can also add or subtract source code from the software. The main reasons for choosing an open-source ERP system are costs, flexibility, vendor independence and the generation of internal company knowledge [10]. Open source is open to the public, has the characteristics of flexibility, quality, adaptability, infrastructure, no hidden costs, scalable, the most important in the selection of open source ERP is suitable for all devices, such as PCs, mobiles, and others examples of such software are Odoo, Compiere, ERP5, Adempiere, OpenBravo and others.

OpenERP is an ERP software that is open source, python-based programming language. OpenERP is available in 18 languages and around 1500 developers have participated in the system development project. Until now there are more than 4000 applications that can be applied to OpenERP, based on OpenERP, the application can be classified into 6 major groups [8]:

1. Front-end apps: website builder, blog, e-commerce.
2. Sales management apps: CRM, point of sales, quotation builder.
3. Business operations apps: project management, inventory, manufacturing, accounting and purchase.
4. Marketing apps: mass mailing, lead automation, events, surveys, forums, live chat.
5. Human Resources apps: employee directory, enterprise social network, leaves management, timesheet, fleet management.
6. Productivity apps: business intelligence, instant messaging, notes.

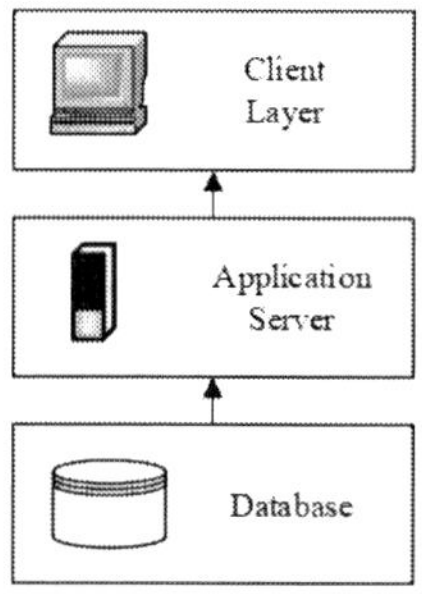

Figure 2. Architecture of OpenERP

Figure 2 shows architecture in OpenERP. Database in OpenERP uses PostgreSQL as the initial database for all of its functionality. The OpenERP server uses Python as a programming language. The OpenERP server application is released under the Aeffro GPL License. OpenERP uses web server technology to share client access. So, to access OpenERP, just use a standard browser. Besides, client access can be either as a server or as an end-user depending on the chosen point of view [10].

3. Methodology

The Method is a software implementation methodology that has 5 stages of implementation. QuickStart is recommended by Odoo for implementation in various project scopes.

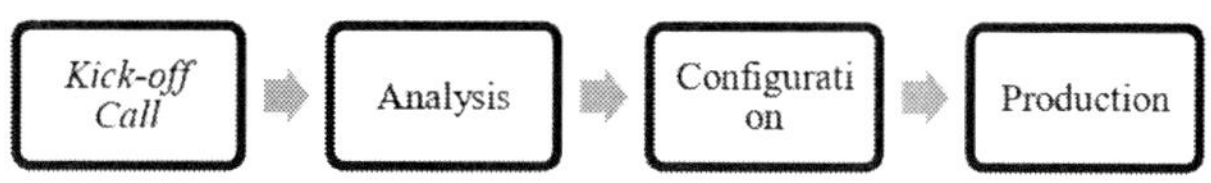

Figure 3. QuickStart methodology [10]

The following stages of the QuickStart implementation [10]:

3.1. Kick-off call

At this stage, the client is explained and planning the methodology steps along with the outputs that will be obtained in this project.

3.2. Analysis

At this stage, an analysis of the conditions of existing business processes is carried out and determines the needs and shortcomings of the current business processes so that they can be improved or developed into business processes.

3.3. Configuration

After the requirements are defined at the Analysis stage. At this stage, the application customization and configuration are performed according to the business processes and needs of the RSUD XYZ obtained from Gap Analysis.

IOP Conf. Series: Materials Science and Engineering **909** (2020) 012061 doi:10.1088/1757-899X/909/1/012061

3.4. Production
This section focuses on application implementation at RSUD XYZ, including the installation and configuration of applications that have been developed and tested.

4. Analysis and Design
In this study, the researcher focuses on implementing 2 phases in QuickStart methodology, start from Kick off Call until Analysis of the system OpenERP in RSUD XYZ. The Researcher using the methodology to see fit/gap analysis in the new design.

4.1. Kick off call
In the first phase, the researcher defines all the requirement that used to initiate a project, such as the design that involve stakeholders and determines the final objective of the project.

Table 2. Requirement to initiate the project

Environment	Development	Basic Concept
In the existing system at RSUD XYZ doesn't have an independent and integrated system that manage warehousing.	Develop an ERP system using OpenERP warehouse management module	-Enterprise Resource Planning -Warehouse Management -OpenERP functional & Technical Resource -OpenERP QuickStart

From table 2 the environment of business process in RSUD XYZ cannot yet fulfilled the business needs, by implementing ERP systems especially in warehouse division the purpose of RSUD XYZ can be fulfil. To achieve the purposed of the system the researcher uses 4 basic concepts to serve the suitable system.

4.2. Analysis
In this phase, the researcher analyzes the business needs, business processes that are currently running and the proposes business process with the OpenERP system.

4.2.1. Business needs of RSUD XYZ
The business need will explain requirements that are setting the direction of the design of the OpenERP system.

Table 3. Business needs of RSUD XYZ

System	Application	Data	Product & Service
Software that integrated to automates work	Software that people use as a tool to improve daily operational services	Accurate information and report to Government regulation	Deliver stockable product to user and proved a real-time service

Table 3 describe the business needs of RSUD XYZ in warehousing management can be fulfilled by implementing ERP System. Because the current system run offline in warehouse area causes process reporting take more time and staff overwhelm by repetitive work and high wrongdoing rate that lead the purposed business process cannot yet fulfill.

4.2.2. Business process with the current system
The business process existing will explain the problem that warehouse division found especially for the storable product. Warehouse management cannot be generated only by the warehouse division, but also integrated by several divisions. The procurement request process initiates by the user from each

division requests goods from the warehouse division by bringing request document. Then the warehouse division will deliver the documents to proceed by the document administration section, the person in charge, the administration, and lastly to the procurement department. Each process in purchasing will be attached one document from each section if the purchasing request approved. Warehouse staff ask for approval by coming directly to the concerned division

When goods arrived at the warehouse, there is paper-based documents that must be complete first before the goods can enter the warehouse as well as the goods for good receipt process. The storage rules do not yet have a business management process, so the warehouse is only a storage area and there is no management product and management warehouse or locations. The following are some pro that need to be considered in the existing inventory management process

Table 4. Business process existing of RSUD XYZ

Process	Existing Process
Purchase Requisition	- There are no terms for minimum stock or anticipation stock so the warehouse has difficulty in controlling the storable product. - Requests from users that become triggers for submission of procurement so that the concept of stock product is not actualized. - The process of purchase requests among divisions is still manual (paper-based) the document goes around manually so there is no guarantor that the document reaches the intended destination. - If there is one division that does not approve of procurement, there is no notification to the warehouse so the status of the item is unknown for warehouse staff.
Good Receipt and Good Issue	- No system regulates the number of products in real-time conditions, the recording is still done manually and repeatedly so that the amount of goods in and out is difficult to ascertain with the real condition. - repeated records document for Hospital formats and the government data formats - The Good Receipt and inspection of quality goods carried out by a separate team (Good Receiving and Quality Assurance team) so that there is repeated verification that causes the time for goods to enter the warehouse longer.
Storage rules	The warehouse only functions as a storage place, the warehouse does not have a layout rule and each item does not yet have an id number to identify.

From the table 4 There are 3 basic processes of the warehouse and the main problems that hamper the warehouse business process in providing services in the daily operation of the RSUD XYZ.

The following are the solutions that researcher offered against the existing business process that has been interfering at RSUD XYZ.

4.2.3. Business process target

The researcher will give the solution Business process proposes a solution based on problem described in table 4.

IOP Conf. Series: Materials Science and Engineering **909** (2020) 012061 doi:10.1088/1757-899X/909/1/012061

Table 5. Business process target of RSUD XYZ

Process	Solution
Purchase Requisition	- Configure product and term in the warehouse. So, the system will automatically give a notification to do the procurement process - The minimum stock standard is applied for each item. The procurement will be notified if the goods run out. So, when requesting goods, the user(divisions) will immediately get it without waiting for procurement - Using the OpenERP system based on information technology so that work can be executed to users who have authorization for a transaction. - a notification will update automatically in real-time when a transaction is executed
Good Receipt and Good Issue	- Product configuration where each item will have an item ID that will facilitate the recording of incoming and outgoing goods and the stock in the warehouse will be found in real-time - Make a report that can be printed in a variety of formats - The checking process is directly carried out by the purchasing section. the warehouse only accepts the handover of goods
Storage Rules	- Product Configuration Each Product will have an ID number to identify the terms of each item such as real-time quantities, stock sites, treatment of each item like expiration date and management of certain types of items. - Rules Configuration Each number of products will have a rule for safety stock and a notification will appear to enable the request for procurement

From table 5 the researcher proposes a solution for each problem described in table 2. Implementing an integrated information system is the key to the solution offered. Problems of procurement, management of goods in the warehouse, distribution of goods into the planning of the budget will be more easily managed with the help of information technology especially ERP.

4.3. Design

In this phase, the researcher will design the system with architecture system diagram, block system diagram and use case diagram.

4.3.1. Architecture system OpenERP in RSUD XYZ

Figure General Architecture system shows how OpenERP will implement in RSUD XYZ. This Architecture involves 4 actors from the warehouse management system.

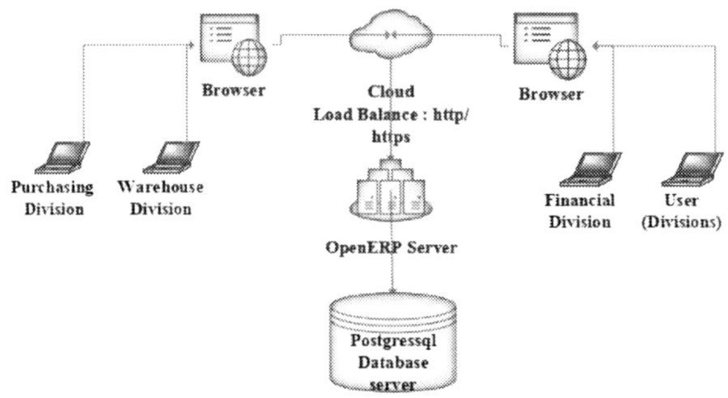

Figure 4. OpenERP Architectire for RSUD XYZ System

The OpenERP architecture compatible with modern browser such as Google Chrome, Mozilla Firefox, Microsoft Internet Explorer. It is platform-operating system- and browser-agnostic [11]. The open ERP web client sends focused procedure call instructions to the server when additional data is needed. To support this system, table VI describes the requirements that will be needed when implementing OpenERP in RSUD XYZ.

Table 6. Requirement of technological infrastructure [12]

Technology	Minimum Requirement
Hardware	- 2x8 CPU Server - 8 GB RAM - HDD 1 Terabyte - Network 1000Mbps
Software	- OS Ubuntu 18.04 LTS - PostgreSQL Server - Phyton version 3.5 - Apache web server

4.3.2. Block system

Design block system will show how this system integrate. The researcher designs a block system to explain the relation of integrated system among the division. Moreover, define the inputs and outputs such as the documents that are processed and the results of recording all transactions.

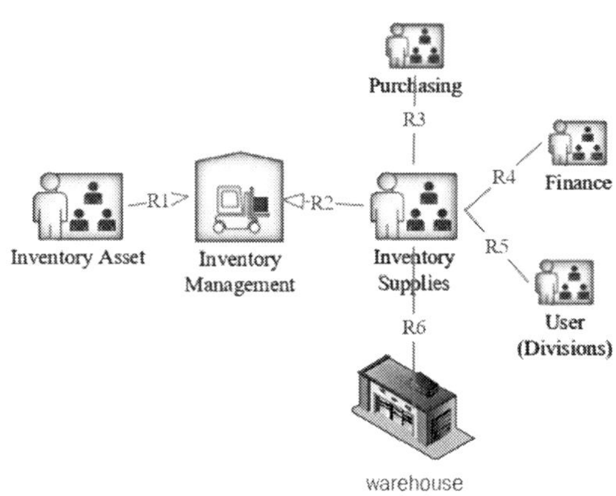

Figure 5. General block system for RSUD XYZ

Figure 5 Shows actors that integrated with the Warehouse Management module particularly for the storable product or Inventory Supplies. There are 5 Relationship, which we can itemize as follows.

- R1 and R2 (Inventory Supplies -Inventory Management – Inventory Asset)

The warehouse division manages 2 types of goods in inventory. The first is Inventory Assets such as assets in the field of nursing and medical. The second is Inventory supplies managing types of goods that are storable the category of goods can be seen in Table 1. This relationship will build Inventory Management in RSUD XYZ.

- R3 (Inventory Supplies - Purchasing)

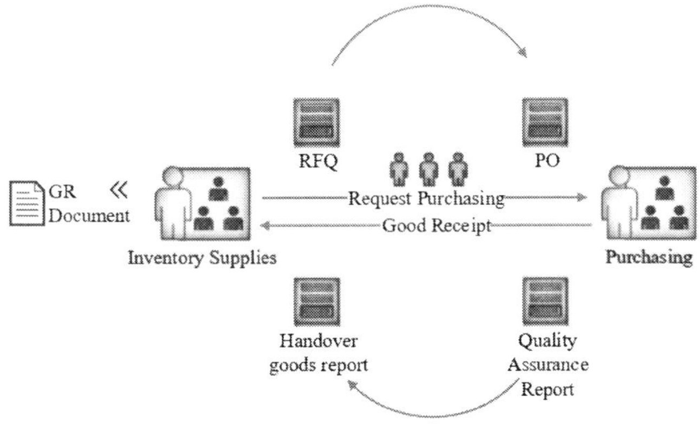

Figure 6. Inventory supplies - purchasing integrated system

Figure 6 shows how Inventory Supplies integrated with purchasing. When the OpenERP system of the warehouse has triggered because the goods have entered the minimum stock, the OpenERP system will issue a notification to Inventory Supplies. The Inventory Supplies will receive a purchasing request. The warehouse section will issue RFQ if it has been approved by:

a. Hospital Director who will provide a letter of disposition and direction to the Administration
b. Administration section confirms the approval of the purchase
c. In Charge of commitment will confirm the availability of the budget

If the three parties above agree to purchase, the completed RFQ has been completed and made, and ready to send to the Purchasing Section as a reference for creating Purchase Order to vendors. When the goods have been received by the purchasing department, the goods will be checked whether they are suitable with the Purchase order documents that they have sent. After that the goods will be forwarded to the warehouse for storage and a handover goods report will be made. Output from this process will be record in Good Receipt Document.

The Procurement request from warehouse division. Information gathered from this process including list of items that will needed by user automatically because in the system will set the safety stock for each item, and the status of the procurement will be updated to warehouse division what part of cycle it has been done. In the Good Receipt Process Each item handled with their need and recorded in OpenERP how the storable product pulled to the warehouse.

- R 4 (Inventory Supplies - Payable)

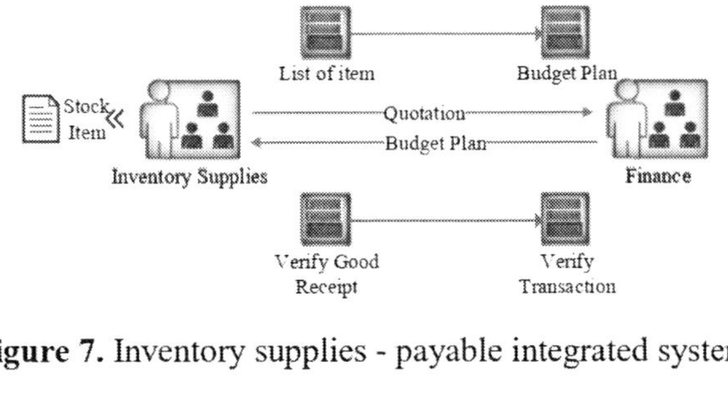

Figure 7. Inventory supplies - payable integrated system

Figure 7 shows how Inventory Supplies integrated with finance division. At the beginning of the fiscal year, there will be a list of needed items from each division so the Finance division can make a master budget plan for that year. In the same way for the Warehouse Division will request in the form of a list of goods to the finance department to make the budget for storable goods. The Finance Division will formulate the budget plan that will be used as a limit for the warehouse.

At the procurement cycle, the transaction will complete when finance does the payment to the vendor and before finance verifies the payment, warehouse division must have verified a good receipt first. The output from this process will be recorded in the stock item document.

- R 5 (Inventory Supplies – User (Divisions))

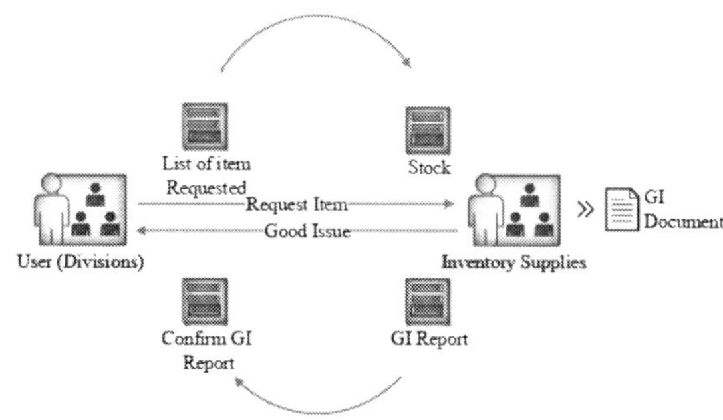

Figure 8. Inventory supplies – user (divisions) integrated system

From the figure 8 we can conclude, when the user loses a certain stock in his division, they will send requests of items to the warehouse to meet their needs. The warehouse division will check and update stock documents and give goods with the document of Good Issue for the goods coming out to the user. This transaction will be completed when the user confirms the Document of Goods Issue. Output from this process will be record in Good Issue Document.

When the user needs a storable product, warehouse division will provide a certain way to fulfill their needs without delay the confirmation because we already have a system that records the real amount in the warehouse

- R6 (Inventory Supplies - Warehouse)

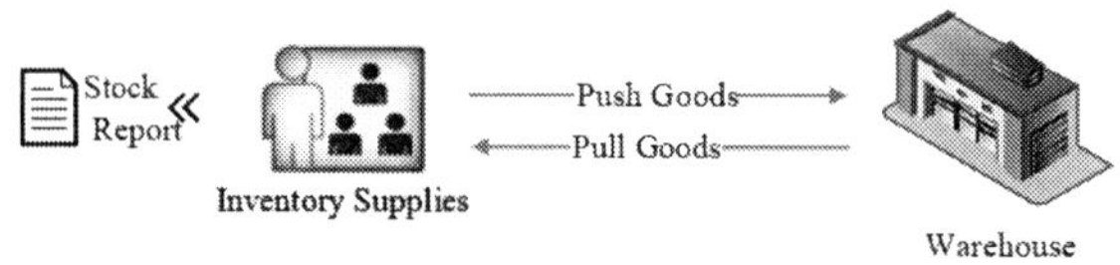

Figure 9. Inventory supplies - warehouse system integration

From figure 9 shows how storage allocation managed. Warehouse division will push and pull the good from the warehouse and the output will be record in Stock Report. Stock Report have information of a real-time stock in the warehouse. RSUD XYZ does not have yet a procedure of Business Process to managed the storable goods in the warehouse.

Existing Storage Allocation do not have yet business process so the researcher suggests as follow.

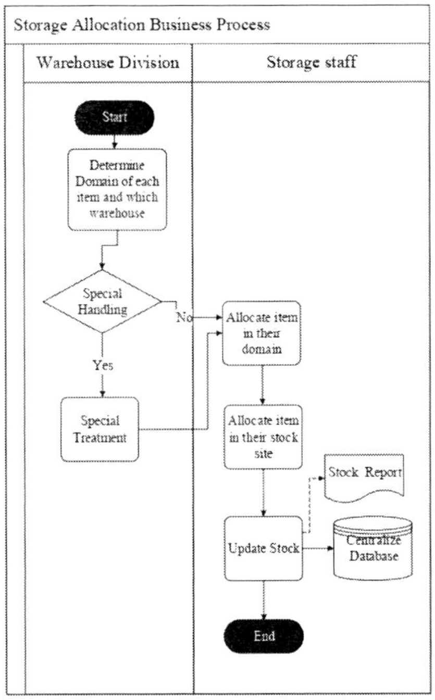

Figure 10. Storage allocation business process

Figure 10 shows the flow chart of storage allocation for each product. This system will determine how to manage an item the warehouse, when the item already have the ID number the storage staff will easily take and put items in their stock site. Advantages of using an Id number will make it easier to identify each item, such as managing location, the expiration date and make Stock Opname Report.

4.3.3. Use case

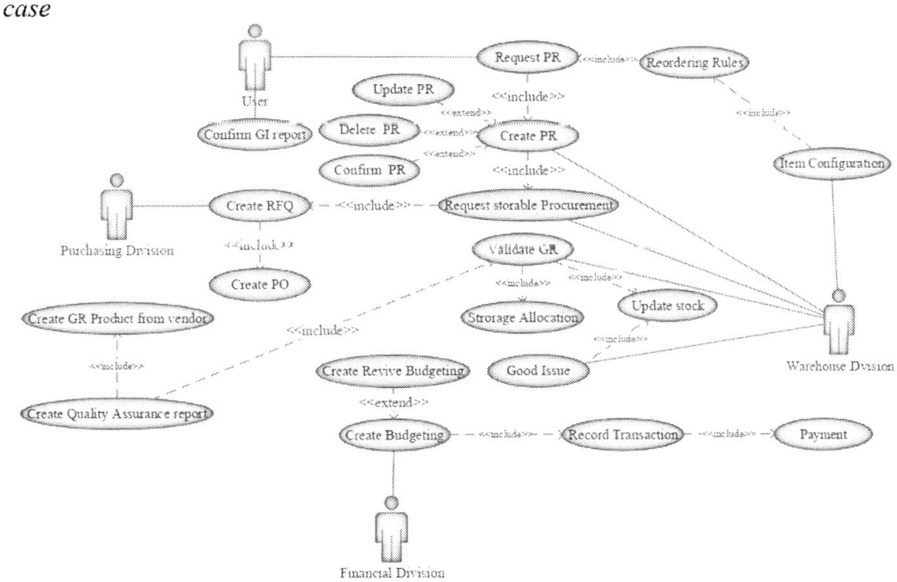

Figure 11. Use Case diagram

Figure 11 use case diagram designed to explain the workflow of the system. Also explained for each actor such as Warehouse Division, Purchasing Division, Financial Division and user of each division. Warehouse division can create, update, delete or confirm Purchase Requisition, do configuration of each item in the storage like manage ID number of an item, make reordering rules, request a procurement, manage good issue and good receipt. Accounting division do processes of payable like made budgeting plan and do the payment the payment when the goods have entered the warehouse. Purchasing division directly related to the vendor, Create RFQ and Purchase Order to vendor. User of each divisions can request list of storable products that their need and confirm Good Issue report.

5. Conclusion

The Conclusion of this research is designed of planning model for ERP system in warehouse management is the solution for RSUD XYZ. OpenERP as one of ERP software is compatible with the RSUD needs. The development of RSUD XYZ uses the QuickStart methodology, and this methodology can implement the completion process on time and the costs needed can be adjusted to the project. ERP also make their business process can be accessed effectively and efficiently with warehouse module in OpenERP with customization to improve service in daily operation's in RSUD XYZ. But still, the employee needs to learn using the new system. The researcher also give recommendation to carried out depth research regarding to another module which able to support the existing module that has been developed, it is better for quickstart methodology performed until the deployment to site and held training for OpenERP Warehousing Management user for effective and efficient planning and implementation

Acknowledgments

We thank you to the previous researcher of related researches and special thank you to the researcher team and ERP Laboratory.

References

[1] R. R. Putri, L. Andrawina, and R. W. Witjaksono 2016 Perancangan sistem asset management berbasis odoo dengan soft system methodology di rumah sakit muhammadiyah bandung *eProceedings Eng* **3** 3155–3162

[2] M. Mucheleka and R. Halonen 2015 ERP in healthcare international conference on enterprise information systems *Proceedings* **1** 162–171

[3] M. M. Sanja 2013 Impact of enterprise resource planning system in health care *International Journal Academic Research Business and Social Science* **3** 404–418

[4] C. Hardjono 2017 Perancangan dan implementasi ERP (enterprise resource planning) modul sales and warehouse management pada CV. Brada *eProceedings Eng* **4** 4983–4993

[5] C. E. &. A. D. Kristianti 2017 penerapan sistem enterprise resource planning: dampak terhadap kinerja keuangan perusahaan *Jurnal Akuntansi & Auditing Indonesia* **21**

[6] A. Garefalakis, G. Mantalis, E. Vourgourakis, K. Spinthiropoulos, and C. Lemonakis 2016 healthcare firms and the erp systems *Journal of Engineering science and Technology Review* **9** 139–144

[7] M. K. Mucheleka 2014 Enterprise resource planning systems in healthcare sector university of oulu

[8] V. Titova 2016 Implementation of enterprise resource planning service: 1c logistics system in the warehouse processes in construction companies

[9] A. W. D. &. W. R. W. Mustafri 2016 Pengembangan odoo modul warehouse pada gudang PT. Tarumatex menggunakan metode rapid application development *eProceedings of Engineering*

[10] C. Nafianto, W. Puspitasari, and M. Saputra 2019 Development of flexible production scheduling by applying gantt charts in manufacturing module open source ERP (case study cv. xyz) international conference on sustainable engineering and creative computing: new idea new innovation *Proceedings* 182–185

[11] Odoo ERP Modules Accessed 25 February 2020 *TenthPlanet Open Source Foundation* Available:https://tenthplanet.in/odoo/product/overview/

[12] D. Reis 2018 Development Essentials: Fast-track your Odoo development skills to build powerful business applications, 4th Edition, Packt Publishing Ltd

IOP Conf. Series: Materials Science and Engineering **909** (2020) 012062 doi:10.1088/1757-899X/909/1/012062

Evaluation criteria and ranking supplier raw materials waste paper with the methods ahp and topsis (case study paper company in west java)

Muhamad Sayuti[1*], Akda Zahrotul W[1], Robin Perdana[1]
[1]University of Buana Perjuangan Karawang

*Email: muhamad.sayuti@ubpkarawang.ac.id

Abstract. The study is done at a paper company in west java by doing evaluation of waste paper supplier of raw materials with the aim of gaining evaluation criteria supplier and ranking supplier .The process of making evaluation criteria supplier is done by using the method hierarchy (AHP) analytical process and ranking with the approach of the supplier technique for order preference by similarity to (TOPSIS) solution of a problem. according data processing by using both the main criterion used this method produced 6 , the quality criteria , delivery , the price of , resources , service , and company management and putting quality criteria as the criteria with the weighting for 30,7 % of as much as .While the supplier of ranking. the best supplier PT. CMN and PT. AMN supplier as the worst alternative supplier of 5. Made with a model evaluation criteria and ranking supplier is expected to supplier evaluation process useful for other types of businesses the same paper companies.

1. Introduction

PT .Pindo deli is a company working in the area of the production and sale of paper which is housed in west java and caused by its the basis of pulp .The current price of pulp and has increased from early january 2015 up to december 2018 .The pulp price hike , all pm (paper machine) will keep trying to improve quantity value its production so that the raw material supply pulp continue to decline .To anticipate the lack of raw materials pulp in frm (fiber recovery machine) payment then the company management decided to buy raw materials of lpg in 3-kg types of waste paper from the huge supplier.

The company formerly very difficult to determine the supplier in accordance with the criteria which had been in. determinedThe supplier is very useful for the company currently, because many suppliers having good products and prices competitive.Suppliers who sends goods into the, the following data suppliers:

Table 1. Suppliers in PT. Pindo Deli

No	Suppliers
1	PT. ANM
2	PT. AMI
3	PT. CMN
4	PT. KAR
5	PT. YAP

Source: corporate data

Supplier election is crucial to support company, about penyuplai because the improper can cause loss . It is better for him in huge supplier of less responsive to catch up with demand is the consequence which may arise is out of and excess supplies [1].In addition when lead time of long it can be huge supplier of disrupting the production process that result in delays in the process product and shipping of goods on customer .Hence by selecting huge supplier of proper is expected to minimize the risk to occur [2].

Table 2. Comparison of supplier selection methods.

Methods	Selection Criteria		Multiple Segment Aspiration Levels
	Qualitative	Quantitative	
Cost-point method	No	Yes	No
Weighted-point method	Yes	No	No
Supplier profile analysis	No	Yes	No
Dimensional analysis	Yes	No	No
Categorical models	Yes	No	No
Taguchi loss functiom	No	Yes	No
TOPSIS	Yes	No	No
LP	No	Yes	No
DEA	No	Yes	No
CPM	No	Yes	No
AHP	Yes	No	No
ANP	Yes	No	No
ANP+TOPSIS	No	Yes	No
AHP+MCGP	Yes	Yes	No
This proposed (Fuzzy TOPSIS+MSGP)	Yes	Yes	Yes

The company has the quality standards that requires the raw material waste paper does not contain levels of impurities up 10 %, and does not contain levels of moisture on 3 %. the solution for the problem the analysis needs to be done to evaluate which suppliers. company as expected

2. Literatur Review

In research evaluation suppliers give special attention to the academic and professional in its field.The suppliers can be used for some performance, suppliers of environment, of performance environment and waste management [3].

2.1. Supply Chain Management

Having logistics company in order to give efesiensi, cost of which engages the time shipping and compliance in the shipment of.The logistics to reduce production costs and capital costs to companies, especially the cost of cooperation with other. division. [4]. 'Supply chain management (scm) is a key factor strategies to improve the effectiveness of the company and the realization of the purpose of the company that better'.Companies are required to choose supply chain and logistics in its operation.Most companies trying to improve efficiency and effectiveness of the supply chain.An increase in business performance of the company can be done with the collaboration, suppliers the performance of shipping, customer service and reducing expenses logistics [5].

Supply chain or supply chain is the relationship between the companies or activities from the supply of goods or services from its place of origin came to the buyers and customers.Supply chain in a continuous, on cost and freight is a process in which a product created and delivered to consumers from the structural.A supply chain refer to an intricate network of relationships that maintains organization with a business to reach a source of production in conveying to consumers[6]

2.2. Decision Support System

Decisions are generally defined as a system that can provide the kind of problem solving skills and the capability to semi-structured issues. Specific communication, defined as a system that supports the work of a semi-structured manager and a group manager in solving problems by means of providing information or proposals lead to certain decisions [7].

2.3. Analytical Hierarchy Process (AHP)

AHP is a method to settle mcdm, an alternative to choose the best and to evaluate the performance criteria which has determined.Comprehensive ahp provides a frame to make the decision-making process and integrate different measures being a value judgment[8], [9]

The first step in a ahp is to determine the selection criteria have been set by the company.Ahp based on the assessment of the comparison with using the ratio which we have set.The results of the ratio can be summed up in a matrix use matrix normalization by calculating the Consistency Ratio (CR) [10], .

2.4.Technique for Order Preferences by Similarity to Ideal Solution (TOPSIS)

Topsis method developed by hwang ching-lai and yoon in 1981, and develop by yoon in 1987, and hwang liu in 1993. According to this technique, the best possible value will be closest to the positive-ideal solution and longest from negative-ideal solution.[11].

In a study conducted by [12], The supplier of engineering at levels after using ahp in continue using topsis, 63,14 %[12]. That must be considered on method topsis namely performance and the number of alternative [13] At an election suppliers use the most weight with topsis selection criteria and criteria with the criteria ahp [14].

A method of topsis very popular to assessment has raised the rating with the best choice and travel time that is brief from the choice of ideal and positive from a distance of the negative too far from the ideal at the same time [15].

3. Materials and methods

A method of ahp much in use by researchers who have data, mathematics for including the necessary data on method ahp very easily get in [16]. Ahp method used to measure from the qualitative and quantitative in decision-making [17]. This method can be affecting decision making in several levels of the hierarchy and used to measure the qualitative and quantitative some decision-making.

Table 3. The results of partial weight primary criteria and sub criteria

Primary Criteria	Weight	Sub Criteria	Partial Weight
		Raw Materials to Specifications	0,243
Quality	0,307	The Fulfillment of Standard	0,757
		Timeliness of Deliveries	0,250
Delivery	0,178	The distance location suppliers	0,750
		Price Fluctuation	0,198
Price	0,188	Ease payment methods	0,802
		Machine facilities	0,161
Resources	0,137	Experts Competence	0,839
		Ease supplier contact	0,492
Service	0,094	Response to the needs and complaint	0,508
		The reputation of suppliers	0,492
Company Management	0,039	Order against taxes	0,508

In assigning valueto global weight is to do multiplication of partial weight over it with weights

Table 4. The result of weight partial with weights a hirarchy

Primary Criteria	Weight	Sub Criteria	Partial Weight	Global Weight
Quality	0,307	Raw Materials to Specifications	0,243	0,075
		The Fulfillment of Standard	0,757	0,232
Delivery	0,178	Timeliness of Deliveries	0,250	0,045
		The distance location suppliers	0,750	0,134
Price	0,188	Price Fluctuation	0,198	0,037
		Ease payment methods	0,802	0,151
Resources	0,137	Machine facilities	0,161	0,022
		Experts Competence	0,839	0,115
Service	0,094	Ease supplier contact	0,492	0,046
		Response to the needs and complaint	0,508	0,048
Company Management	0,039	The reputation of suppliers	0,492	0,019
		Order against taxes	0,508	0,020

3.1. The consistency results in comparison

After the each weight and obtained, the main criterion used subkriteria then afterward calculated the ratio of konsistensinya to know whether the results obtained enough were consistent.As parameternya is the ratio of consistency to smaller or equal to 10 %.After the ratio of consistency or for all criteria and subkriteria, so next one is the consistency whole a hierarchy.

The calculation of the ratio of consistency (CR):

CR = CI/RCI

CI = (λ maks – n) / (n – 1)

CR = Consistency Ratio

CI = Consistency Index

RCI = Random Consistency Indec

N = number of orders

Value RCI depends of the order matrix (om) , the size of the cri can be seen in table 4 the following

Table 5. Index consistency random

OM	1	2	3	4	5	6	7	8	9	10	11	12	13	14	15
RI	0	0	0,58	0,9	1,12	1,32	1,41	1,45	1,49	1,49	1,51	1,48	1,56	1,57	1,59

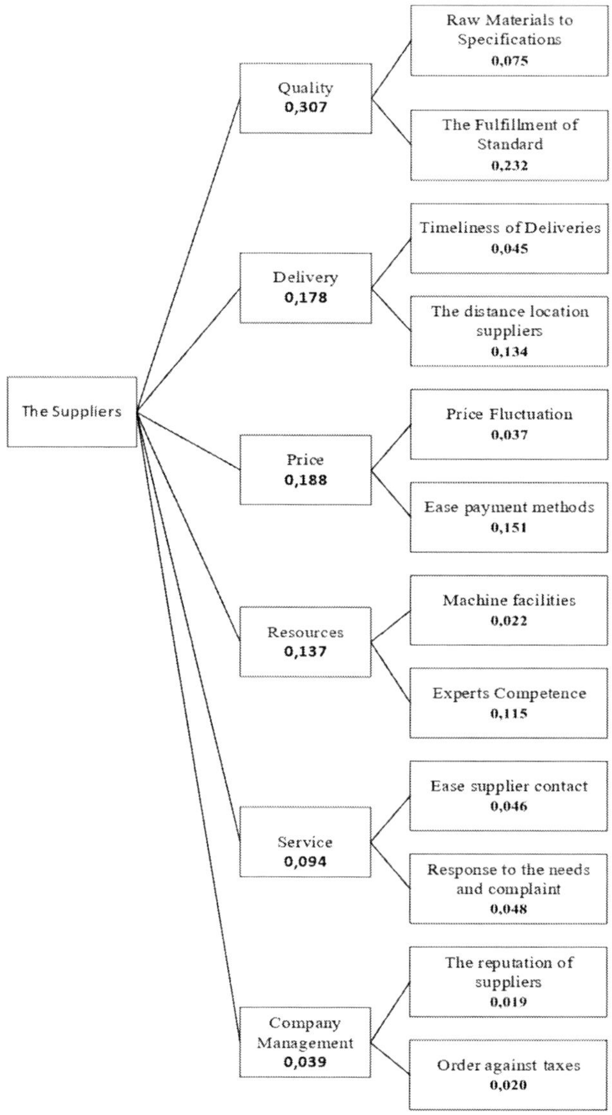

Figure 1. The evaluation of suppliers and weights

3.2. The Determination Of Rank Supplier To Calculation TOPSIS

After scoring supplier has been done , the next stage is performing calculations topsis to get rank supplier in order to know supplier assessment which that has the best performance and suppliers which have worst the assessment them to five supplier evaluated.

Tabel 6. Matrix criteria and sub criteria and weights and the supplier

Criteria and sub evaluation criteria												
	Quality		Delivery		Price		Resources			Service	Company Management	
	0,307		0,178		0,188		0,137			0,094	0,039	
Criteria and weight	Raw Materials to Specifica-tions	The Fulfillment of Standard	Time-liness of Delive-ries	The distance location suppliers	Price Fluctu-ation	Ease payment methods	Machine facilities	Experts Compe-tence	Ease supplier contact	Response to the needs and complaint	The reputation of suppliers	Order against taxes
	0,075	0,223	0,045	0,134	0.037	0,151	0,022	0,115	0,046	0,048	0,019	0,020
PT. ANM	2	3	2	5	4	3	5	5	3	3	1	1
PT. AMI	3	5	3	3	4	3	5	5	3	3	3	5
PT. CMN	5	5	4	5	4	5	5	5	5	5	5	3
PT. KAR	4	5	4	3	3	3	5	5	5	5	3	5
PT. YAP	4	5	4	3	1	5	5	5	5	5	5	5

3.3.Matrix For Decision Making

After decision made matrix, so the next step has the matrix to minimize the data and loosening in performing calculations topsis.

Table 7. Matrix for decision making

Criteria	Raw Materials to Specifica-tions	The Fulfillment of Standard	Timeliness of Deliveries	The distance location supp-liers	Price Fluctu-ation	Ease payment methods	Machine facilities	Experts Compe-tence	Ease supplier contact	Response to the needs and complaint	The reputation of suppliers	Order against taxes
Weight	0,075	0,223	0,045	0,134	0.037	0,151	0,022	0,115	0,046	0,048	0,019	0,020
PT. ANM	0,0179	0,0641	0,0115	0,0764	0,0194	0,0516	0,0098	0,0514	0,0143	0,0149	0,0023	0,0022
PT. AMI	0,0269	0,1068	0,0173	0,0458	0,0194	0,0516	0,0098	0,0514	0,0143	0,0149	0,0069	0,0108
PT. CMN	0,0448	0,1068	0,0230	0,0764	0,0194	0,0860	0,0098	0,0514	0,0238	0,0249	0,0114	0,0065
PT. KAR	0,0359	0,1068	0,0230	0,0458	0,0146	0,0516	0,0098	0,0514	0,0238	0,0249	0,0069	0,0108
PT. YAP	0,0359	0,1068	0,0230	0,0458	0,0049	0,0860	0,0098	0,0514	0,0238	0,0249	0,0114	0,0108

3.4.Determine The Matrix Is A Solution Ideal Positive And Negative Ideal Solution

Results of a solution ideal positive and negative ideal solution has been done by determining the greatest value of the matrix and determine the lowest score of the matrix so the result can be seen in table 6.

Identify the positive ideal solution (PIS) and negative ideal solution (NIS), respectively, as follow:

$$A+ = \{(\max v_{ij} \mid j \in J), (\min v_{ij} \mid j \in J'), i = 1, 2, 3, \ldots, m\} \quad (1)$$

$$= \{v1-, v2-, v3-, \ldots, vn-\}$$

$$A- = \{(\min v_{ij} \mid j \in J), (\max v_{ij} \mid j \in J'), i = 1, 2, 3, \ldots, m\} \quad (2)$$

= {v1-,v2-,v3-,...,vn-}

J = { j = 1, 2, 3,..., n

Table 8. Make the ideal solution of positive and negative ideal solution

Criteria	Raw Materials to Specifi-cations	The Fulfillment of Standard	Timelineless of Deliveries	The distance location suppliers	Price Fluctu-ation	Ease pay-ment methods	Machine facilities	Experts Competence	Ease supplier contact	Response to the needs and complaint	The reputation of suppliers	Order against taxes
S*I Positif	0,0448	0,1068	0,0230	0,0764	0,0194	0,0860	0,0098	0,0514	0,0238	0,0249	0,0114	0,0108
S'I Negatif	0,0179	0,0641	0,0115	0,0458	0,0049	0,0516	0,0098	0,0514	0,0143	0,0149	0,0023	0,0022

3.5. Calculate The Distance Positive Ideal Alternative Solutions

The next stage is calculate the distance an alternative to solution ideal positive the calculation on alternative about a point ideal positive presented in table 7.

$$D_i^- = \sqrt{\sum_{j-1}^{n} (y_{ij} - y_i^-)^2} \quad (3)$$

$$D_i^+ = \sqrt{\sum_{j=1}^{n} (y_i^+ - y_{ij})^2} \quad (4)$$

i= 1,2,...m

Table 9. Distance ideal alternative solutions Positive

Criteria	Raw Materials to Specifi-cations	The Fulfillment of Standard	Timeliness of Deliveries	The distance location suppliers	Price Fluctu-ation	Ease pay-ment methods	Mach-ine facili-ties	Experts Compet ence	Ease supp-lier con-tact	Response to the needs and complaint	The reputati on of supp-liers	Order against taxes	Total
PT. ANM	0,0007	0,0018	0,0001	0	0	0,0012	0	0	0,0001	0,0001	0,0001	0,0001	0,0649
PT. AMI	0,0003	0	0	0,0009	0	0,0012	0	0	0,0001	0,0001	0	0	0,0518
PT. CMN	0	0	0	0	0	0,0000	0	0	0	0	0	0	0,0043
PT. KAR	0,0001	0	0	0,0009	0	0,0012	0	0	0	0	0	0	0,0474
PT. YAP	0,0001	0	0	0,0009	0,0002	0,0000	0	0	0	0	0	0	0,0350

3.6.Calculate The Distance Negative Ideal Alternative Solutions

The next stage is an alternative to calculate the distance the ideal solution negative alternative the results of about a point ideal negative can be seen in table 9.

Table 10. Distance negative ideal alternative solutions

Criteria	Raw Materials to Specifica-tions	The Fulfillment of Standard	Time-liness of Delive-ries	The distance location suppliers	Price Fluctu-ation	Ease pay-ment methods	Machine facilities	Experts Compe-tence	Ease supplier contact	Response to the needs and complaint	The reputation of suppliers	Order against taxes	Total
PT. ANM	0	0	0	0,0009	0,0002	0	0	0	0	0	0	0	0,0338
PT. AMI	0,0001	0,0018	0	0,0000	0,0002	0	0	0	0	0	0	0,0001	0,0474
PT. CMN	0,0007	0,0018	0,0001	0,0009	0,0002	0,0012	0	0	0,0001	0,0001	0,0001	0	0,0728
PT. KAR	0,0003	0,0018	0,0001	0	0,0001	0	0	0	0,0001	0,0001	0	0,0001	0,0516
PT. YAP	0,0003	0,0018	0,0001	0	0	0,0012	0	0	0,0001	0,0001	0,0001	0,0001	0,0617

Table 11. The calculation on proximity relatively with the methods topsis

Supplier	S*I Positive	S'INegative	Relative	Ranking
PT. ANM	0,0649	0,0338	0,3427	5
PT. AMI	0,0518	0,0474	0,4779	4
PT. CMN	0,0043	0,0728	0,9438	1
PT. KAR	0,0474	0,0516	0,5213	3
PT. YAP	0,0350	0,0617	0,6382	2

From table 9 on can be seen that penyuplai with the largest of relatively PT. CMN, so that PT. CMN is rank first suppliers best.Meanwhile PT. ANM have value of least so relatively ranked last.

4. Conclusion

After all stages Analytical Hierarchy Process (AHP), so formed a suppliers evaluation criteria, with the main criterion used 6 the criteria: the quality (30,7 %), delivery (17,8 %), price (18,8%), resources (13,7 %), service (9,4 %) and company management (3,9 %). Then based on process with the ranking supplier *Technique For Order Preference By Similarity To Ideal Solution* (TOPSIS). Obtained suppliers and value proximity with the order ranging from the best namely PT. Citra Mega Nusantara (0,9438), PT. Yaputra Abadi Perkasa (0,6382), PT. Karya Agung Realty (0,5213), PT. Avfal Mega Indah (0,4779) dan PT. Arga Nusa Mandaya (0,3427)

References

[1] L. Merry, M. Ginting, and B. Marpaung 2013 Pemilihan supplier buah dengan pendekatan metode analytical hierarchy process (ahp) dan topsis : studi kasus pada perusahaan retail *J. Tek. dan Ilmu Komput.* **3** 45-58

[2] M. F. Limansantoso, J. Manajemen, F. Bisnis, and D. Ekonomika 2013 Pemilihan supplier produk calista dengan metode analytical hierarchy process (ahp) pada pt. Buana tirta utama -gresik *Calyptra J. Ilm. Mhs. Univ. Surabaya.* **3** 1-20

[3] H. C. Zhang and J. Li 2003 Using fuzzy multi-agent decision-making in environmentally conscious supplier management *CIRP Ann. - Manuf. Technol.* **52** 385-388

[4] H. Lödding 2013 Handbook of manufacturing control: Fundamentals, description, configuration.

[5] H. Purnomo and E. Santosa 2016 Kapabilitas pemasaran dan operasi serta pengaruhnya pada supply chain management dan kinerja perusahaan *J. Ekon. dan Bisnis* **17** 79-97

[6] J. Sumarauw, L. Kawet, and A. Subroto 2015 Evaluasi kinerja supply chain manajemen pada produksi beras di desa panasen kecamatan kakas *J. Ris. Ekon. Manajemen, Bisnis dan Akunt.* **3** 653-662

[7] E. Turban, J. Aronson, and T. Llang 2003 *Decision Support Systems and Intelligent Systems.*

[8] P. Ren, Z. Xu, and H. Liao, 2016 Intuitionistic multiplicative analytic hierarchy process in group decision making *Comput. Ind. Eng.* **101** 513-524

[9] Y. K. Fu 2019 An integrated approach to catering supplier selection using AHP-ARAS-MCGP methodology *J. Air Transp. Manag.* **75** 164-169

[10] Z. Xu and H. Liao 2014 Intuitionistic fuzzy analytic hierarchy process *IEEE Trans. Fuzzy Syst.* **22** 749-761

[11] R. Kamalakannan, C. Ramesh, M. Shunmugasundaram, P. Sivakumar, and A. Mohamed, 2020 Evaluvation and selection of suppliers using TOPSIS *Mater.*

[12] J. Chai, J. N. K. Liu, and E. W. T. Ngai 2013 Application of decision-making techniques in supplier selection: A systematic review of literature *Expert Syst. Appl.* **40** 3872-3885

[13] F. R. Lima Junior, L. Osiro, and L. C. R. Carpinetti 2014 A comparison between Fuzzy AHP and Fuzzy TOPSIS methods to supplier selection *Appl. Soft Comput. J.* **2** 1194-209

[14] A. Azimifard, S. H. Moosavirad, and S. Ariafar 2018 Selecting sustainable supplier countries for Iran's steel industry at three levels by using AHP and TOPSIS methods *Resour. Policy*

[15] K. Hwang, Ching-Lai; Yoon,1981 Multiple attribute decision making: methods and applications a state-of-the -art survey

[16] G. Rajesh and P. Malliga, 2013 Supplier selection based on AHP QFD methodology in *Procedia Engineering*

[17] T. Saaty 1997 *J. Math. Psychol.*

IOP Conf. Series: Materials Science and Engineering **909** (2020) 012063 doi:10.1088/1757-899X/909/1/012063

Optimization of supply chain operation cost and gas usage quantity using non-dominated sorting genetic algorithm II (NSGA-II) Method

A Ridwan[1*], **A Bahauddin**[1], **and M Naufal R P**[1]

[1] Department of Industrial Engineering, University of Sultan Ageng Tirtayasa, Banten, Indonesia

*Email: asep.ridwan@untirta.ac.id

Abstract. ABC plant is one of seven plants in XYZ industry which produce Cold Rolling Coil (CRC) and Cold Rolling Sheet (CRS). The problems faced by ABC plant in its supply chain network are the high overall costs of supply chain operations and the quantity of gas usage. The characteristics of this problem consist of more than one objective, the best method for solving this problem called multi-objective optimization. This research is aim to identify the decision variables that will be optimized and determine the optimal quantity of each decision variable using NSGA-II method. The parameters used in this research are the population size of 50, crossover probability with range 0.1-0.9, mutation probability 0.1 and 0.2 also the number of generation 50 and 100. The result shows that the selected parameters for this research are the crossover probability of 0.7, mutation probability of 0.1, number of generation 100. Four decision variables that are optimized, namely CRC produced (Lite, Medium, and Heavy) and CRC stored with the quantities simultaneously are 40,918, 112,479, 173,758, and 43,209 tons. This result will assist the ABC plant in minimizing the overall cost of supply chain (124,532,272 USD) and the quantity of gas usage (39,634,749,440 Kcal).

1. Introduction

The supply chain is a network consisting of companies that work together with the aim of making and distributing a product to end-user. In a supply chain network, it consists of three flow must be managed, namely information, finance, and material [1]. The method used in managing supply chain network is known as supply chain management [2]. ABC plant is one of seven plants in XYZ industry which produce CRC and CRS. This plant implements supply chain management under the supervision of the Supply Chain Improvement (SCI) division. As an actor in a supply chain network ABC plant must be able to see from a variety point of view, one of them is a business view (voice of business) [3]. In this view, the main priority that requires special attention is about the cost without compromising product quality and flexibility [4]. So, the ABC plant still able to fulfil the end user specification, it's called process capability [5]. Some researchers conduct the research about the cost reduction such as Wiliamson [6]; Ridwan and Noche [5]; and Roeck et al. [7]. Wiliamson [6] examining the transaction cost economics in outsourcing as an analysis of the basic unit and the procurement decision. Ridwan and Noche [5] decreasing the cost of poor quality of cargo handling in the port, consist of lost cargo, damaged cargo, equipment and transporter breakdown, and delay time. Roeck et al. [7] explaining there are six effects of distributed ledger technology (DLT) solutions to the supply chain transactions as a cost reduction. On the other side, an ABC plant must consider the impact resulting from the production process on the environment around the plant.

The problems faced by ABC plant in its supply chain network are the high overall costs of supply chain operations and the quantity of gas usage. Both problems are caused by the machine performance that is not optimal and sometimes ABC plant has to produce in quantities exceed the order. As a result of excessive production, the supply chain operation cost and the quantity of gas usage are high. The expenses include production, logistics, and savings. Production costs are divided into four variables, such as direct fixed cost, fixed allocations, raw materials, and conversions. Logistics cost is the cost of shipping CRC from ABC plant directly to end users while the storage cost is the cost for storing CRC in the ABC warehouse. Likewise, with the quantity of gas which consists of boiler natural gas, fuel, and process gas. Because of this problem, it is necessary to minimize both using the optimization model. The characteristics of this problem consist of more than one objective, so the best method for solving this problem called Multi Objective Optimization.

Multi-objective optimization is a completing method of optimization which consists of more than one objective function which is simultaneously optimized [8]. There are several methods in Multi-objective Optimization such as Genetic Algorithm (GA), Particle Swarm Optimization (PSO), Simulated Annealing, etc. Genetic Algorithm (GA) is a metaheuristics search algorithm which is based on the mechanism of biological evolution with the nature of searching for the possibilities of prospective solution to get the optimal solution for the problem [9]. GA can be implemented in various fields such as video game [8], maintenance scheduling [10], and also in supply chain [3]. In GA, there are several methods which often used, including Non-Dominated Sorting Genetic Algorithm (NSGA-II), Multi-objective Genetic Algorithm (MOGA), and Non-Dominated Ranking Genetic Algorithm (NRGA).

The focus of this research is to solve the optimization problem in the supply chain network of ABC plant not only from the voice of business but also by paying attention to the impact on the surrounding environment. In this research, we set two objective functions according to the problems that exist in the ABC plant, including minimizing the overall cost of supply chain operations and quantity of gas used as an innovation. We used the Non-Dominated Sorting Genetic Algorithm (NSGA-II) as completing method to solve this problem. NSGA-II is a genetic algorithm which uses fast non-dominated sorting procedure, an elitist-preserving approach, and a parameterless niching operator [11]. The consideration to use NSGA-II in this research is due to the advantages of this method compared to the other genetic algorithm and metaheuristics. Utama [3] compared NSGA-II with MOGA, and as a result, NSGA-II provides a better optimum value in achieving the objective function. Mousavi et al [12] also compared three metaheuristics methods including NSGA-II, NRGA, and PSO, with the results NSGA-II having an average the completion time is shorter than the other two metaheuristics methods.

Some example from previous researches using NSGA-II to solve supply chain optimization problem like in Utama [3] setting two objective functions namely minimizing the total cost of supply chain operations and inaccurate of shipping goods to costumers by calculating future and present value. Decision variables used in this research are the quantity of product delivery from producers to wholesalers and wholesalers to retailers. Bandyopadhyay and Bhattacharya [13] establishing three objective functions such as minimizing the total cost of the supply chain, the variance of the quantity of product order, and the total amount of products in inventory. In this research, the decision variables are the number of products stored, ordered, and sent. Another study, Chan et al. [14] setting objective function to minimize total cost which includes total production cost, transportation cost, storage cost, and penalty fee and the second objective is to reduce the percentage of the number of order by retailers within a particular time. The decision variable used in this study is the type of truck and the quantity of product demand from retailers.

Same with some previous researches mentioned above using the quantity of production as the decision variable. So, this research also uses the quantity of CRC produced consist of three types of CRC (Lite, Medium, and Heavy) and CRC stored as decision variables which will be optimized then. To obtain the optimal configuration for the quantity of CRC produced and stored, several trials were carried out on parameters with different values. The parameter which produces the best average fitness value will be selected and used as a reference to assist XYZ industry in achieving its goal to minimize the overall cost of supply chain operation and quantity of gas usage in ABC plant in certain time horizon.

2. Research Method

This research used quantitative approaches which in the process of solving problems using numerical data as a tool to analyze information under predetermined research objectives. The data used in this research include production cost, logistics cost, storage cost, the quantity of gas usage, production capacity, storage capacity, and also the proportion of production for each CRC product. After data has been collected, data processing start with determining decision variables, objective functions, and constraints. Then, this data presented in a mathematical model and testing the model using NSGA-II method. The NSGA-II process consists of several steps [15]. Before we start the NSGA-II procedures, the parameters are determined. The number of the population used is 50 with crossover probability in range 0.1-0.9, mutation probability 0.1 and 0.2; also the number of generations are 50 and 100. First, based on its range and constraints, a random population was initializing. Then, we evaluate chromosomes in population according to the objective functions. Once the population is evaluated, it is then sorted according to non-domination using fast sort algorithm into each front. The individuals in each front are assigned with rank. The rank one is assigned to individuals in the 1st front and so on. In each front, we measure the distance of individual with its neighbours called crowding distance.

After sorting the individuals, we selected parents using tournament selection and crowded comparison operator. Selected individuals are individuals with smaller rank and greater crowding distance than the other one. To produce offspring, selected parents will undergo crossover and mutation. The method used is a one-point crossover and swap mutation. Chromosome evaluation will also be experienced by the offspring population. The next step is to merge the initial population with the offspring population, so the population size becomes 2N. This merge population is again sorted based on non-domination, and crowding distance and only N individuals are selected. The new generation population will be filled by elements from each front until it exceeds the population size (N) on adding the final front. In this case, the elements in the final front will be sorted in descending order based on their crowding distance and then filled in new generation population until the population size reaches N. This procedure will be repeated until the number of generations is reached. After all NSGA-II procedures are completed, the final step is to measure the attainment of NSGA-II method based on execution time and objective functions attainment in each period. This research follows the steps, as shown in Figure 1.

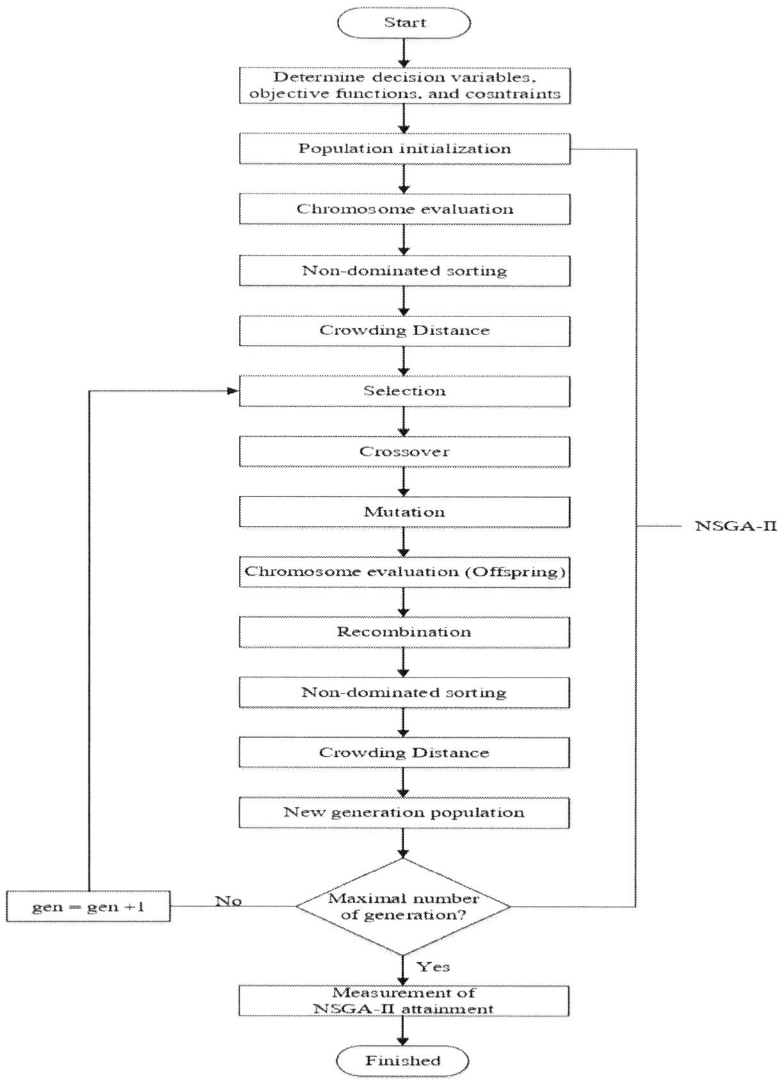

Figure 1. Steps of research flowchart

3. Input Data and The Mathematical Model

In this section, we will discuss the input data used in this research and mathematical model of the problems that exist in the ABC plant.

3.1. Input Data

The data used in this research include production cost consists of conversion cost, raw material cost, direct fixed cost, and fixed allocation cost. Then the logistics cost, storage cost, quantity of gas usage which includes boiler natural gas, fuel gas, and process gas, production and storage capacity as well as the proportion of production for each CRC product of the ABC plant in XYZ industry. Assume all of the data collected have been converted to the monthly average value.

3.2. The Mathematical Model

Mathematical model of the problem in ABC plant are presented in this section using the followingnotation include indexes, parameters, and decision variables and can be expressed in the numerical equations.

Sets

p	types of finish product $\{p = 1,2,..P\}$

Parameters

BK_p	conversion cost of product type p in a period
BBB_p	raw material cost of product type p in a period
BTL	direct fixed cost in a period
BTA	allocation fixed cost in a period
BP_p	production cost of product type p in a period
BL_p	logistics cost of product type p in a period directly to end-user
BS	storage cost in a period at ABC warehouse
GAB_p	quantity of boiler natural gas used for producing product type p in a period
GBB_p	quantity of fuel gas used for producing product type p in a period
GP_p	quantity of process gas used for producing product type p in a period
Qp	production capacity of ABC plant in a period
Qs	storage capacity of ABC warehouse in a period

Decision variables

CRC_L	quantity of CRC type *Lite* in a period
CRC_M	quantity of CRC type *Medium* in a period
CRC_H	quantity of CRC type *Heavy* in a period
CRC_f	quantity of CRC stored in ABC warehouse in a period

First of all, variables are converted into notation form. Next step is to make a mathematical model of the problem discussed in this research. The objective functions used in this research are minimized the overall cost of the supply chain and minimize the quantity of gas usage, which simultaneously expressed by Z_1 and Z_2. The first objective function (Z_1) can be seen, as shown below:

$$\boldsymbol{Min\ Z_1} = BP_p\,(CRC_L + CRC_M + CRC_H) + BL_p\,(CRC_{Lt} + CRC_{Mt} + CRC_{Ht}) + BS\,(CRC_f) \quad (1)$$

The objective function is shown in Eqs.(1) aim to minimize the overall cost of the supply chain in ABC plant that includes three components such as production cost, logistics cost, and storage cost. For production and logistics cost will be multiplied by the amount of production of three types of CRC while storage cost will be multiplied by the number of CRC stored in ABC warehouse. After knowing the value for each component, then the components are summed up to produce the overall cost of the supply chain in ABC plant. Since the production cost consists of four cost variables, it can be formulated, as shown in Eqs.(2):

$$BP_p = BK_p + BBB_p + BTA_p + BTL_p \quad (2)$$

The second objective function (Z_2) can be formulated as:

$$\boldsymbol{Min\ Z_2} = GAB_p\,(CRC_L + CRC_M + CRC_H) + GBB_p\,(CRC_L + CRC_M + CRC_H) + GP_p\,(CRC_L + CRC_M + CRC_H) \quad (3)$$

Eqs.(3) aims to minimize the quantity of gas usage, which include boiler natural gas, fuel gas, and process gas. The quantity of each type of gas will be multiplied by the amount of production of three types of CRC and summed up to produce a total quantity of gas usage. Also, there are constraints that indicate the capacity of ABC plant which formulated as:

Subject to:

$$CRC_L + CRC_M + CRC_H \leq Qp \quad (4)$$

$$CRC_f \leq Qs \quad (5)$$

$$CRC_L, CRC_M, CRC_H \geq 50 \quad (6)$$

$$CRC_f \geq 0 \quad (7)$$

Eqs.(4) is the constraint that shows the production capacity of ABC plant where the amount of CRC production must be smaller or equal to its production capacity. However, Qp is a composite constraint for three CRC products so to determine the production constraint of each type of CRC the production capacity will be multiplied by the proportion of production of each type of CRC. Eqs.(5) represented storage capacity of ABC warehouse while Eqs.(6) is the minimum quantity to produce each type of CRC (Lite, Medium, and Heavy) and Eqs.(7) and the lower bound for the number of CRC stored. In the next section, will be discussed regarding the results obtained using NSGA-II method.

4. Result and Discussion

In this research, the experimentation has been conducted in a PC with specification Pentium ®, Dual-Core CPU E5500 2.8 Ghz processor, 6 Gb memory, windows 10 Pro 64 bit, DirectX version 11.1. NSGA-II algorithm has been experimented using Matlab R2017a. The tuning parameters used in this research include population size of 50, crossover probability within range 0.1-0.9, mutation probability 0.1 and 0.2, and the number of generation 50 and 100. Metaheuristic algorithms are sensitive to their parameters, and a small change can affect the quality of the solution obtained [16]. After processing the data using NSGA-II method, the next step is to measure NSGA-II attainment based on objective function attainment and execution time. The execution time is the time required by the algorithm to obtain Pareto optimal solutions in one run [16]. In this research, we use several combination parameters which include nine values of crossover probability, two values of mutation probability, and two number of generations, so the total trials conducted in 36 trials. The result of trials, as shown in Table 1 below.

Table 1. NSGA-II trial results

Crossover Probability	Mutation Probability	Number of Generation	Cost of Supply Chain Operation (f1)	Quantity of Gas Usage (f2)	Execution Time (sec)
0.9			154,301,225	49,113,861,120	9.459
0.8			152,527,687	48,574,970,880	6.486
0.7			173,987,044	55,442,391,040	5.44
0.6			145,055,808	46,170,373,120	5.646
0.5	0.1	50	165,718,566	52,778,204,160	5.641
0.4			160,576,278	51,094,149,120	5.84
0.3			174,055,228	55,406,725,120	5.492
0.2			177,030,106	56,359,546,880	5.538
0.1			180,432,533	57,475,594,240	5.571
0.9			156,966,963	49,965,056,000	6.003
0.8			147,279,450	46,873,482,240	5.773
0.7			189,787,621	60,449,315,840	5.803
0.6			161,780,011	51,498,449,920	5.481
0.5	0.2	50	154,677,410	49,251,517,440	6.83
0.4			190,500,385	60,657,505,280	6.791
0.3			170,167,129	54,187,868,160	6.773
0.2			182,786,957	58,215,162,880	5.456
0.1			150,799,915	48,014,033,920	6.718

Crossover Probability	Mutation Probability	Number of Generation	Cost of Supply Chain Operation (f1)	Quantity of Gas Usage (f2)	Execution Time (sec)
0.9			151,247,600	48,160,942,080	12.358
0.8			154,136,884	49,073,879,040	13.954
0.7			123,469,572	39,293,660,160	15.727
0.6			161,651,694	51,476,633,600	14.69
0.5	0.1	100	163,561,864	52,077,685,760	11.471
0.4			153,820,427	48,962,380,800	12.917
0.3			133,471,941	42,480,952,320	14.058
0.2			153,303,861	48,834,718,720	10.916
0.1			150,752,295	47,966,458,880	12.402
0.9			158,821,945	50,557,767,680	11.102
0.8			136,703,873	43,501,276,160	10.936
0.7			161,694,269	51,468,661,760	10.842
0.6			159,148,211	50,657,597,440	11.21
0.5	0.2	100	158,838,047	50,586,833,920	10.988
0.4			159,988,147	50,907,837,440	12.04
0.3			151,384,471	48,207,032,320	13.408
0.2			171,773,478	54,671,800,320	12.168
0.1			167,296,018	53,258,557,440	12.961

From Table 1 above, we can find out the average fitness and execution time for each different parameter value. For the probability of mutation 0.1 with the number of generation 50, the optimal fitness is at the crossover probability 0.6 with an average fitness value of 145,055,808 USD (f1) and 46,170,373,120 Kcal (f2) with an execution time of 5.646 seconds. With the same number of generation but different mutation probability (0.2), the optimal fitness is at the crossover probability 0.8 with average fitness value of 147.279.450 USD (f1) and 46,873,482,240 Kcal (f2) with an execution time of 5.773 seconds. While in generation 100 with a mutation probability of 0.1, it can be seen that the optimal fitness is at the probability of crossing 0.7 with an average fitness value is 124,532,272 USD (f1) and 39,634,749,440 Kcal (f2) with an execution time 15.727 seconds and for the mutation probability 0.2 the optimal fitness is at the crossover probability 0.8 with an average fitness value of 136,703,873 USD (f1) and 43,501,276,160 Kcal (f2) with an execution time 10.936 seconds.

Also, from Table 1, when using different parameter values will produce different output because NSGA-II is very sensitive to its parameters [8][16]. First, by comparing two value mutation probability 0.1 and 0.2 for the same number of generation, the comparison will look like in the graph, as shown in Figure 2 and Figure 3 below.

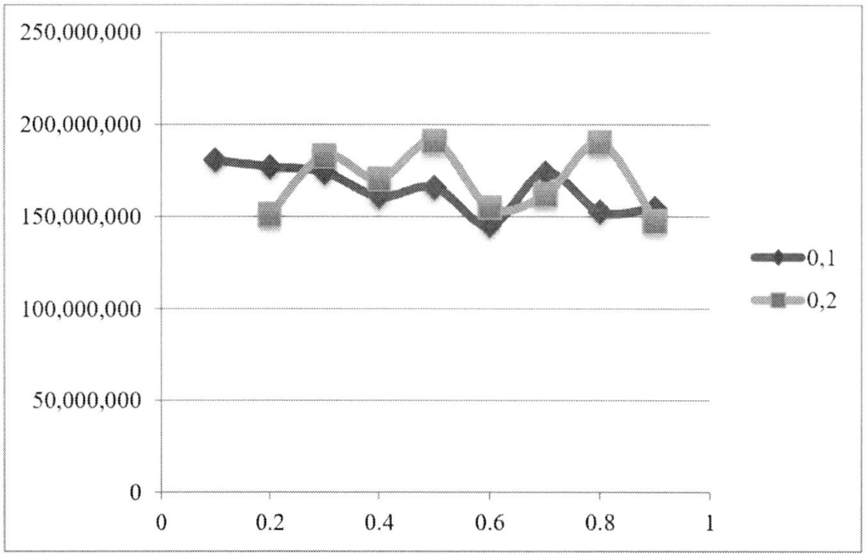

Figure 2. Fitness comparison graph for 50 generations

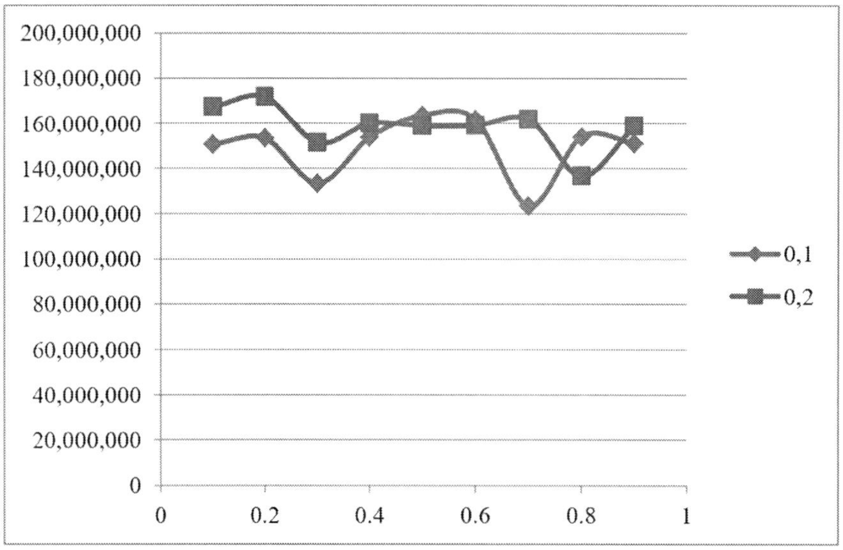

Figure 3. Fitness comparison graph for 100 generations

Based on the comparison graph for 50 generations, the mutation probability of 0.1 will produce more optimal fitness value than the mutation probability of 0.2. Likewise, for 100 generations, where the graph shows that the mutation probability 0.1 produces more optimal value than the mutation probability of 0.2. So the mutation probability of 0.1 will be chosen for this research. Another parameter that is compared is the number of generation. By using the chosen mutation probability for two different number of generation, namely in the 50th and 100th generations, the comparison is shown in Figure 4 below.

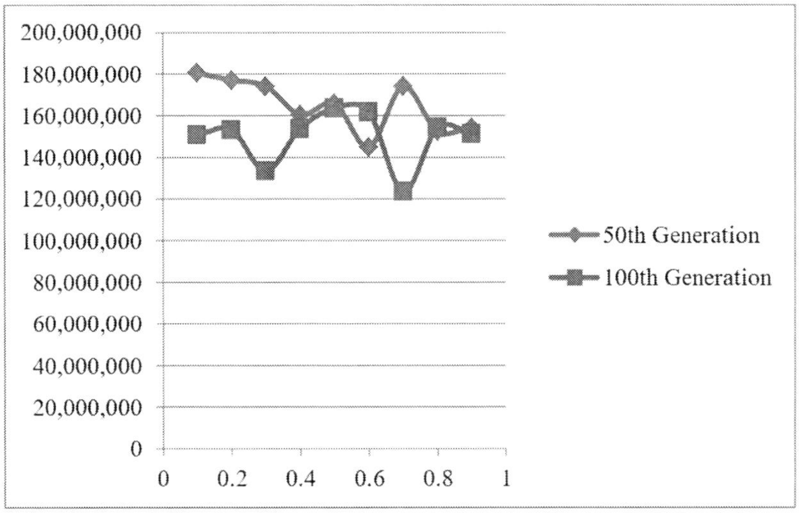

Figure 4. Fitness comparison graph for mutation probability 0.1

From Figure 4 above by using the chosen mutation probability, the average fitness value in 50th generation is greater than the average fitness value in the 100th generation. It means that the 100th generation has more optimal fitness value than the 50th generation according to the minimization function used in this research. After the probability of mutation and the number of generation is chosen, the next step is to find out the best crossover probability with mutation probability of 0.1 and the number of generation 100. From Table 1 above, shows that the crossover probability with the average fitness value is at the crossover probability of 0.7 with fitness value 124,532,272 USD (f1), and the quantity of gas usage is 39,634,749,440 Kcal (f2) with an execution time of 15.727 seconds. So, the chosen parameters for this research are the crossover probability of 0.7, mutation probability of 0.1 and the number of generation 100. And the output produced using these selected parameters, as shown in Table 2 below.

Table 2. Selected parameters output

Solution Population					Cost of Supply Chain Operation (f1)	Quantity of Gas Usage (f2)
Chromosome	Lite	Medium	Heavy	Stored		
1	4,067	73,038	954	39,423	62,997,548	19,983,104,000
2	426	57,745	39,923	37,578	70,994,158	22,441,984,000
3	12,092	48,855	66,175	51,896	70,994,158	22,441,984,000
4	44,119	62,074	67,753	26,290	84,212,560	26,776,320,000
5	32,316	91,512	87,467	19,763	90,075,984	28,568,832,000
6	69,913	54,376	14,789	21,446	70,994,158	22,441,984,000
7	2,887	70,659	75,160	51,159	75,899,722	24,060,416,000
8	72,656	54,948	52,582	32,886	75,899,722	24,060,416,000
9	46,112	30,872	130,138	66,307	76,990,186	24,514,048,000
10	71,172	151,859	10,406	31,588	84,212,560	26,776,320,000
11	57,710	21,201	155,364	71,283	84,212,560	26,776,320,000
12	60,546	105,317	134,441	61,023	85,173,770	27,016,960,000
13	77,927	71,861	192,123	25,038	85,173,770	27,016,960,000
14	40,844	165,612	227,858	52,613	86,209,298	27,401,984,000
15	61,514	150,088	276,799	20,629	89,828,164	28,568,832,000
16	69,736	51,298	65,502	53,230	90,075,984	28,568,832,000

Solution Population					Cost of Supply Chain Operation (f1)	Quantity of Gas Usage (f2)
Chromosome	Lite	Medium	Heavy	Stored		
17	10,215	89,326	98,920	16,367	96,253,546	30,608,128,000
18	77,291	49,633	102,378	27,291	98,199,442	31,161,856,000
19	37,761	83,907	110,707	31,350	100,374,378	31,938,304,000
20	20,059	118,277	107,043	51,027	107,100,158	34,156,544,000
21	20,814	72,126	171,767	53,261	107,100,158	34,156,544,000
22	74,623	126,029	85,426	38,969	113,613,516	36,204,288,000
23	24,777	147,011	116,038	44,805	113,613,516	36,204,288,000
24	51,927	195,659	55,785	40,522	113,943,126	36,204,288,000
25	49,634	155,005	114,352	28,082	122,639,702	38,981,376,000
26	22,246	33,859	274,343	22,212	123,920,428	39,440,384,000
27	6,800	87,083	236,248	61,434	123,920,428	39,440,384,000
28	24,323	103,469	213,339	64,298	133,670,138	42,662,144,000
29	46,889	119,520	175,199	46,121	144,634,832	45,971,456,000
30	13,909	111,150	220,699	31,294	145,251,176	46,148,608,000
31	55,208	33,124	323,516	57,669	149,517,930	47,560,960,000
32	47,478	84,221	295,564	48,947	149,517,930	47,560,960,000
33	3,174	138,366	290,748	75,173	149,517,930	47,560,960,000
34	16,813	199,210	261,633	7,742	153,663,812	48,884,736,000
35	48,972	150,050	282,736	60,283	155,381,572	49,510,656,000
36	7,368	149,267	102,721	13,878	155,381,572	49,510,656,000
37	42,123	149,864	138,213	6,772	155,638,212	49,592,576,000
38	68,905	146,186	114,890	78,410	161,797,352	51,480,576,000
39	63,936	190,684	85,452	65,726	161,893,872	51,480,576,000
40	14,723	72,521	306,631	62,681	161,893,872	51,480,576,000
41	69	207,312	190,270	8,585	166,231,338	53,055,744,000
42	68,823	84,994	279,123	13,948	168,463,280	53,699,840,000
43	15,604	107,551	331,036	26,577	168,759,806	53,854,208,000
44	44,540	187,570	222,489	79,935	170,484,106	54,404,608,000
45	42,371	81,766	354,559	64,912	174,749,276	55,778,048,000
46	38,353	127,335	315,297	40,986	174,798,676	55,778,048,000
47	68,268	178,326	267,037	32,442	174,798,676	55,778,048,000
48	61,419	187,400	280,240	55,475	178,261,100	56,887,040,000
49	77,347	198,362	277,835	61,363	178,738,310	56,887,040,000
50	55,120	196,459	288,248	79,747	188,946,146	60,297,728,000
Average	40,918	112,479	173,758	43,209	124,532,272	39,634,749,440

Based on Table 2, it can be seen the optimal quantity of each decision variable where the quantity used in this research is the average quantity. For CRC type Lite the optimal quantity is 40,918 tons, CRC type Medium 112,479 tons, CRC type Heavy 173,758 tons and quantity of CRC stored is 43,209 tons. Cost of supply chain operation is 124,532,272 USD, and the quantity of gas usage is 39,634,749,440 Kcal.

5. Conclusion

There are four decision variables used in this research including the quantity of CRC produced which consist of CRC type lite, medium, and heavy also the quantity of CRC stored in ABC warehouse with the quantity of each decision variables simultaneously is 40,918 tons, 112,479 tons, 173,758 tons, and 43,209 tons. Cost of supply chain operation is 124,532,272 USD, and the quantity of gas usage is 39,634,749,440 Kcal. By obtaining the quantity of each decision variable, it will assist the XYZ industry, especially ABC plant, in minimizing the overall cost of supply chain operation and the quantity of gas usage.

6. References

[1] Ridwan A Santoso M I Ferdinant P F and Ankarini R 2019 Design of strategic risk mitigation with supply chain risk management and cold chain system approach *IOP Conference Series: Materials Science and Engineering* **673** 1-6
[2] Azmiyati S and Hidayat S 2016 Supply chain performance measurement at pt louserindo magnificent permai using SCOR and FAHP models *Jurnal Al Azhar Indonesia Seri Sains dan Teknologi* **3** 163-170
[3] Utama P 2017 The Development of multiobjective supply chain optimization model by considering time value of money elements using genetic algorithm MOGA and NSGA-II. *Thesis* (Depok: Universitas Indonesia)
[4] Ridwan A Ekawati R and Novitasari A 2018 Quality control of the steel wire rod product by integration lean six sigma and taguchi method *MATEC Web of Conferences* **218** 1-7
[5] Ridwan A and Noche B 2017 Model of the port performance metrics in ports by integration six sigma and system dynamics *International Journal of Quality & Reliability Management* **35** 82-108
[6] Wiliamson O E 2008 Outsourcing: transaction cost economics and supply chain management *Journal of Supply Chain Management* **44** 5-16
[7] Roeck D Sternberg H and Hofmann E 2020 Distributed ledger technology in supply chains: a transaction cost perspective *International Journal of Production Research* **58** 2124-2141
[8] Aditama D 2017 Dynamic evolution behavior for non-player character on space shooter game using nsga-ii *Thesis* (Surabaya: Institut Teknologi Sepuluh November)
[9] Wardhani LK Safrizal M and Chairi A 2011 Optimizing the composition of freshwater fish feed ingredients using multi-objective genetic algorithm *Proceeding of National Seminar of Information Technology Application* 112-117
[10] Sutartono R T 2018 Evaluation of maintenance schedule of central industry with genetic algorithm *Thesis* (Yogyakarta: Universitas Islam Indonesia)
[11] Deb K Pratap A Agarwal S and Meyarivan T 2002 A Fast and elitist multi-objective genetic algorithm: NSGA-II *IEEE Transactions on Evolutionary Computation* **6** 182-197
[12] Mousavi SM Sadeghi J Niaki STA and Tavana M 2016 A Bi-objective inventory optimization model under inflation and discount using tuned pareto-based algorithms: NSGA-II, NRGA, and MOPSO.*Applied Soft Computing Journal* **43** 57-72
[13] Bandyopadhyay S and Bhattacharya R 2014 Solving a tri-objective supply chain problem with modified NSGA-II Algorithm *Journal of Manufacturing Systems* **33** 41-50
[14] Chan FTS Jha A and Tiwari M K 2016 Bi-objective optimization of three echelon supply chain involving truck selection and loading using NSGA-II with Heuristics Algorithm *Applied Soft Computing* **38** 978-987
[15] Jain V and Sachdeva G 2017 Energy, exergy, economic (3e) analyses and multi-objective optimization of vapor absorption heat transformator using NSGA-II technique *Energy Conversion and Management* **148** 1096-1113
[16] Pasandideh S H R Niaki S T A and Asadi K 2015 Bi-Objective optimization of a multi-product multi-period three-echelon supply chain problem under uncertain environments: NSGA-II and NRGA *Information Sciences* **292** 57-74

IOP Conf. Series: Materials Science and Engineering **909** (2020) 012064 doi:10.1088/1757-899X/909/1/012064

Measurement of psychological impact of industrial engineering students in fulfiil of online learning outcomes using NASA-TLX method

Ani Umyati*, Wahyu Susihono, Ade Sri Mariawati
Industrial Engineering, Universitas Sultan Ageng Tirtayasa, Indonesia

*email: ani.umyati@untirta.ac.id

Abstract. COVID-19 had a serious impact on the education sector in Indonesia, one of them is the application of online learning activities as a substitute for learning activities in the classroom. Viewed from the perspective of students, online learning is a new method and is not yet commonly used in universities including Sultan Ageng Tirtayasa University. So students still need time to adapt to the existing system. Each activity certainly causes consequences of workload, as well as its online learning activities conducted by students majoring in Industrial Engineering Untirta. This study will measure the psychological impact in the form of mental workload experienced by students majoring in Industrial Engineering Untirta in meeting online learning outcomes using the National Aeronautics and Space Administration Task Load Index (NASA-TLX) method. The online learning process carried out by Industrial Engineering students of Sultan Ageng Tirtayasa University has a psychological impact in the form of mental workload on these students. The results showed that 63.21% of students experienced mental workloads with moderate classification, 22.97% of students had workloads with heavy classifications while the rest experienced workloads with light classifications (13.51%). The most dominant indicators are Mental Demand, Temporal Demand, and Frustration Level.

1. Introduction

Corona Virus Disease - 2019 (COVID-19) which surrounded Indonesia seems unable to be muted. Its spread is massive and relatively fast makes people panic and fear. Everyone is forced to stay at home to break the chain of transmission of COVID-19. Mass agendas were omitted because COVID-19 also the term work from home (WFH) became popular. Not only that, but COVID-19 also had a serious impact on the education sector in Indonesia. One of them is the application of online learning activities as a substitute for learning activities in the classroom.

Since the beginning of March Industrial Engineering students at Untirta has been carrying out online learning activities, not only the process of delivering the material, even the Midterm Examination was also conducted online. This is due to the Government's policy on learning from home related to the existence of the COVID pandemic 19. However, to date, it has not been known how much psychological impact experienced by students in carrying out these online learning activities. Other paragraphs are indented.

Online learning is learning done face-to-face through the available platforms. The difference lies in the technical delivery of the material alone, while the assessment indicators and achievements to be achieved remain the same as ordinary learning activities. Taken from the perspective of students, online learning is a new method and is not yet commonly used in universities, including at the

University of Sultan Ageng Tirtayasa. There are several platforms used in the online learning process, including SPADA, ZOOM, SIAKAD, Whatsapp group, etc.

Each activity certainly causes consequences of workload, as well as its online learning activities conducted by students majoring in Industrial Engineering Untirta. One psychological impact caused is in the form of mental workload felt by students. The mental workload can be seen as an external independent variable in task demands, and mental workload is defined as an interaction between task demands and human capabilities or resources. Analysis of workloads can be useful in providing information about job demands by worker limitations, it can also be used for system optimization.

Online learning has several components involved, including learning content, hardware in the form of computer devices, interaction strategies, network infrastructure, and software used. Activities in this activity are more dominant involving mental processes because in this activity information processing takes place. This mental process includes the receipt of stimulus from the work system, the process of converting the stimulus into meaningful information, to making decisions that are in line with the information obtained (Iridiastadi & Yassierli, 2014).

The concept of mental workload leads to attentional demands experienced while performing cognitive tasks (O'Donnell in Matthews, et al., 2000). The concept of mental workload has become increasingly important since the development of semi-automatic and computerized technology. This condition makes humans must have the mental ability to process all information received, both information on manufacturing tasks and administrative tasks.

Although there is no specific definition of mental workload, the universally mental workload is based on the difference in the number of resources owned by the number of demands of work to be done (Sanders & McCormick, 1992). Also added by Haga, et al. (2002), Hacker (2005) and Attwood, et al. (2007) that mental workload is the degree of process capacity incurred during displaying tasks, and the concept of mental workload arises because of the information process. The information process involves the perception, interpretation, and process of information conveyed by sensory organs. From all these opinions it appears that mental workload exists in every type of work that results from the information process. There are various aspects that people understand when asked about mental workloads, including the amount of work that is a burden, the pressure of time, the level of effort, the success of meeting demands, the psychological and physiological consequences of the task. Besides mental workload is often associated with task difficulties. So the assessment of workload made by one person may reflect his assessment of the difficulty of the task, while others may reflect it as the level of effort that must be expended.

One method that can be used to measure mental workload is the National Aeronautics and Space Administration Task Load Index (NASA-TLX). This method is used to analyze the mental workload faced by someone who has to carry out various activities and work. This method was developed by Sandra G. Hart of NASA Ames Research Center and Lowell E. Staveland of San Jose State University in 1981. This method was developed based on the need for subjective measurements consisting of a scale of Nine factors (task difficulty, time pressure, type of activity, physical effort, mental effort, performance, frustration, stress, and fatigue). The nine factors are simplified into six, namely the needs of Mental Demand (MD), Physical Demand (PD), Temporal Demand (TD), Performance (P), and Frustration Level (FR). Therefore in this study, the measurement of mental workload will use the NASA-TLX method.

This study will measure the psychological impact experienced by industrial engineering students of the Sultan Ageng Tirtayasa University in online learning activities. This psychological impact can be seen from the mental workload score generated from the NASA-TLX questionnaire.

2. Methodology

In this study, a survey will be conducted on Industrial Engineering students of the Sultan Ageng Tirtayasa University about online learning activities that are being carried out. From the survey results, it can be seen the frequency of online learning and the platforms most widely used in online learning activities. While the psychological impact will be measured in the form of mental workload scores of

Industrial Engineering students in meeting online learning outcomes. The method used for measuring mental workload is NASA-TLX. Some steps in this research are explained below.

2.1. *Preliminary*
At this stage the preparation of the questionnaire will be used, there are some changes, especially in the questionnaire distribution technique. The questionnaire used was in the form of a NASA-TLX questionnaire that had been converted into a google form to facilitate the process of distributing questionnaires in the Covid-19 pandemic conditions. There are 2 groups of questions. The first group is the respondent's identity and the platform used in online learning activities, while the second group consisted of questions from the NASA-TLX Questionnaire. NASA-TLX questionnaire contains questions to weight 15 pairwise comparisons and gives a rating of 0 to 100 for each NASA-TLX indicator.

2.2. *Data Collection*
The distribution of questionnaires was carried out online through a network of social media groups found among Industrial Engineering students of Sultan Ageng Tirtayasa University. 74 respondents sent responses to the questionnaire that had been distributed and all of the data was then used for data processing in this study. Data collected in the form of respondent characteristic data include age, gender, and level of the student. Also, the main data in the form of weighting and rating of the indicators contained in the NASA-TLX questionnaire.

2.3. *Data Processing*
2.3.1. *Weighting*
In this section, respondents are asked to choose one of two indicators that are felt to be more dominant causing mental workload on these activities. The NASA-TLX questionnaire was given in the form of pair comparisons consisting of 15 pair comparisons. From this questionnaire, the number of tally from each indicator felt was the most influential. This tally amount will then be the weight for each mental load indicator.

2.3.2. *Rating*
In this section, respondents were asked to give a rating of the six indicators of mental workload. The rating given is subjective depending on the respondent's perception and the rating scale ranges from 0 to 100.

2.3.3. *Score Calculation*
To get a NASA-TLX mental workload score, the weights and ratings for each indicator are multiplied and then added up and divided by 15 (number of pairwise comparisons). as contained in the following formula

$$Skor = \frac{\Sigma(boot\ x\ rating)}{15} \quad (1)$$

2.3.4. *Mental Workload Classification.*
Based on the results obtained, an interpretation of the score results will be based on the following explanation *(Hart and Staveland, 1981)* :

a. > 80 indicates a heavy workload
b. 50 - 80 states the workload is moderate
c. < 50 stated the workload was rather light

IOP Conf. Series: Materials Science and Engineering **909** (2020) 012064 doi:10.1088/1757-899X/909/1/012064

3. Results and Discussion

3.1. Data characteristics respondents

In this study, all respondents were students of the Department of Industrial Engineering - Sultan Ageng Tirtayasa University who filled out the questionnaire online. The results of data collection 74 respondents are consisting of 35 men and 39 women who have an age range between 18 to 22 years. The percentage of respondents by gender can be seen in Figure 1, while the distribution of respondents by the level of study can be seen in Table 1.

The most widely used platform is the Zoom Application used by 69 respondents (93%), as many as 47 respondents stated using the Spada application and Google Classroom as many as 10 respondents. The Zoom application is an application used by almost all respondents, but respondents use more than one application in implementing online learning.

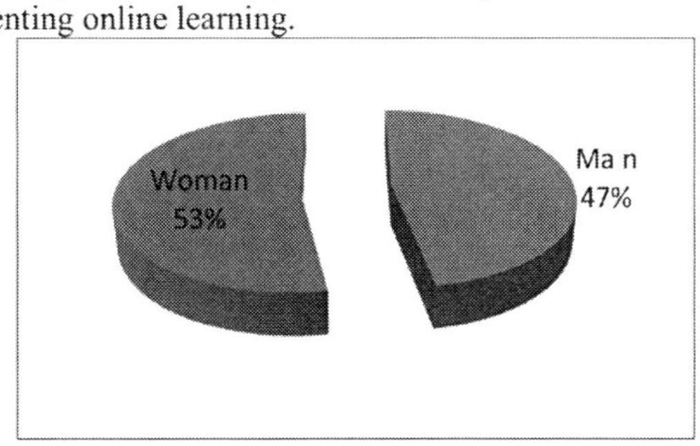

Figure 1. The percentage of respondents by gender

Table 1. Distribution of respondents by the level of study

No	Gender	Student Level			
		2016	2017	2018	2019
1	Man	7	10	13	5
2	Woman	3	18	12	6
	Total	10	28	25	11

From the results of data collection, it is known that respondents experienced several obstacles in implementing online learning. These complaints consisted of 72.8% of respondents experiencing complaints related to devices and signals, 2.8% of physical complaints, and 25.7% of psychological complaints in the form of difficulties in focusing and concentration during the online learning process.

3.2 *Mental Workload Score*

Based on the results of weighting and rating (Table 2) that has been done based on the results of the NASA TLX questionnaire, a mental workload score is felt by Industrial Engineering students during the online learning process. Based on the calculations performed, it is known that there are 10 students (13.51%) who have light mental workloads, 47 students (63.51%) who have moderate workloads, and 17 students (22.98%) students who have heavy workloads. While the average score of all respondents was 67.69 with a moderate mental workload classification.

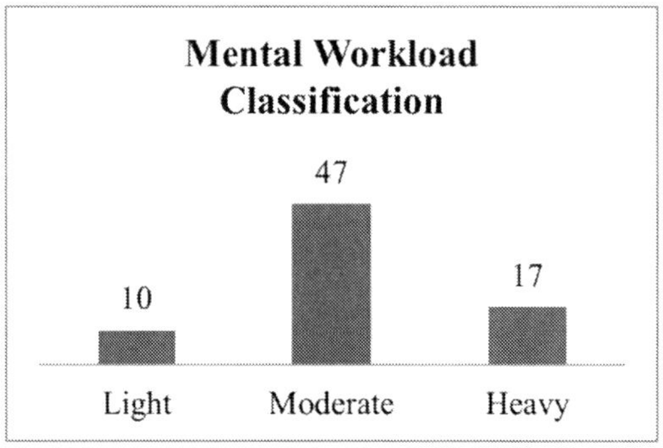

Figure 2. Mental Workload Classification

Figure 2 shows that 10 students feel a mental workload with a light classification, this is influenced by the frequency of students doing online learning in one week. Respondents who experienced a light mental workload classification had a frequency of online learning 5 to 8 times per week and most complained about complaints related to technical matters such as poor signal, quota usage costs, and matters related to online learning tools. The mental workload was the most experienced by respondents, as many as 47 students (63.51%), they have the frequency of doing online learning more often than those included in the classification of mild mental workloads of 8 to 25 times per week. Whereas for respondents who experienced mental workloads with heavy classifications totaling 12 people (22.97%) had the frequency of doing online learning more frequently than respondents who experienced moderate mental workloads.

Based on NASA TLX data processing, it can be seen as the most dominant indicator in the process of mental workload, as shown in Table 2 below.

Table 2. NASA-TLX scores

Respondent Number	Score	Respondent Number	Score	Respondent Number	Score	Respondent Number	Score	Respondent Number	Score
1	82,8	18	87,6	35	79	52	66,3	69	76
2	58	19	65,3	36	74,26	53	68	70	20,3
3	80,67	20	78	37	81,9	54	42	71	52,6
4	79,73	21	87,3	38	67,4	55	62,3	72	84
5	78,40	22	90,0	39	73,6	56	74	73	23,9
6	68,67	23	62,3	40	82	57	47,3	74	80,3
7	75,3	24	50	41	30,3	58	62,6		
8	40	25	64,3	42	62,6	59	80		
9	81,60	26	77	43	58	60	84,2		
10	84,00	27	74,86	44	69,0	61	89,9		
11	76,00	28	68	45	84	62	77,3		
12	78,00	29	60,3	46	45,3	63	60		
13	56,6	30	80,3	47	38,6	64	42,6		
14	62,3	31	59	48	18	65	77		
15	74	32	70	49	64	66	78,3		
16	73,3	33	74,6	50	79,7	67	74		
17	69,6	34	67,3	51	82,4	68	61		

Mental Demand (MD) and Temporal Demand (TD) are the biggest indicators, each at 20%. The mental demand indicator shows how much mental and perceptual activity (such as seeing, remembering, searching) needed in doing work. The MD aspect is dominant because the online learning process is related to the activity of seeing, and understanding the material delivered by the teaching lecturer. Whereas Temporal Demand is a very important dimension in completing a job. According to Sandra G Hart dan Lowell E Staveland (1988), Temporal Demand is a major factor in measuring workload and also has a high relationship with other dimensions. High temporal demand on students indicates that the pressure felt by students to complete their workload is related to time pressure. Due to the time online learning by using the application used there is a time limit so that sometimes the lecturer's explanation is only partially delivered.

The third dominant aspect is frustration level (FR) and effort (EF) of 19%, indicating that students in completing their work feel anxiety, stress, or severe stress. Because at the time of online learning applications that consume a lot of internet quota also become obstacles so that in the frustration aspect (FR) students do not feel satisfied. Because the learning process is done online, the concentration is not as good as learning in the classroom, also the home environment is sometimes not conducive, so students will feel effort (EF) or heavy mental work to complete the assignments.

The online learning process carried out is one of the forms of information processing in humans. in this activity, there is an interaction between humans and machines. One approach that can be used to understand human-machine interaction is to model how the human brain processes information. there are at least 3 major stages in processing information (Wickens et al. 2004). namely (1) understanding the information provided by the environment (2) processing the information at a higher level and (3) responding to that information. when online learning activities take place, students must be able to direct concentration on the lecture material that is being delivered by the lecturer. this activity requires high concentration and can be greatly disturbed by trivial matters such as the sound of surrounding conversation, signal interference, and so on. so it is not surprising if then the three aspects above become the most dominant factor in the occurrence of mental workload on students who are carrying out online learning activities.

Table 3. Comparison of NASA TLX score elements

Category	Score	Average	Percentage
MD	17032	210,2716	20%
PD	4314	53,25926	5%
TD	16661	205,6914	20%
OP	13090	161,6049	16%
EF	16014	197,7037	19%
FR	16067	198,358	19%

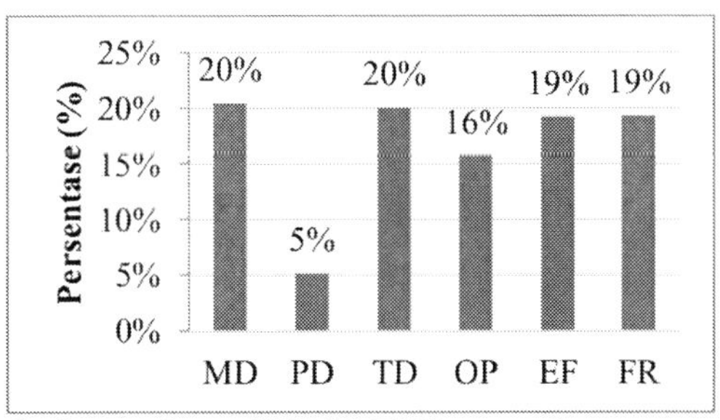

Figure 3. Comparison of NASA TLX score elements graphic

4. Conclusion

The online learning process carried out by industrial engineering students of the Sultan Ageng Tirtayasa University has a psychological impact in the form of mental workload on these students. The results showed that 63.21% of students experienced mental workloads with moderate classification, 22.97% of students had workloads with heavy classifications while the rest experienced light workloads (13.51%). The most dominant indicators are Mental Demand, Temporal Demand, and

Frustration Level. The three indicators are more influential on the mental activity experienced by students during online learning compared to physical activity. This is caused by constraints related to learning facilities and infrastructure, physical complaints, and psychological complaints.

References

[1] Iridiastadi, H 2014 Ergonomi suatu pengantar Bandung PT Remaja Rosdakarya.
[2] Wickens CD, and Hollands J 2004 An introduction to human factors engineering **2nd Ed** New Jersey USA Prentice-Hall
[3] Sanders M S, and McCormick E J 1992 Human factors in engineering and design New York McGraw-Hill Inc
[4] Attwood D, Baybutt P, Devlin C, Fluharty W, Hughes G, Isaacson D, Joyner P, Lee E, Lorenzo D, Morrison L, and Ormsby B 2007 Human factors methods for improving kinerjance in the process.
[5] Hacker W 2005 Handbook of work stress SAGE Publication Downloaded from http://www.Sageereference.com/hdbk_workstress.
[6] Haga S, Shinoda H, and Kokurun M 2002 Japanese psychology research *Japanese Psychological Association* Blackwell Publisher Ltd **44 (3)** 134-143
[7] Hancock PA, and Meshkati N 1988 Human mental workload North Holland *Elsevier Science Pubisher*

IOP Conf. Series: Materials Science and Engineering **909** (2020) 012065 doi:10.1088/1757-899X/909/1/012065

Study the use of shared power supply in mobile telecommunication station in Indonesia

B Budiman[1*], **F Farizal**[1] **and D S Gabriel**[1]

[1] Department of Industrial Engineering, Universitas Indonesia, Depok, Indonesia

*Email: budimanbagus93@yahoo.com

Abstract. Increasing need of telecommunications and fierce business competition makes telecommunication tower providers try to continue the improvement of their competitive advantage. Improving telecommunications services by providing telecommunications towers or base transceiver stations, is something that must be improved in order to meet the communication needs in Indonesia. Hence, improving network infrastructure is something that must be done to meet the changing times. Of course, in the context of improving network infrastructure, investment costs are one of few significant factors. Each telecommunications company must continue to innovate, so that network infrastructure can be improved with lower investment costs. In this study, we propose infrastructure sharing model of its power supply in order to reduce capital expenditure for Mobile Network initial investment. It is expected that the proposed model can be useful as a new business scheme for tower providers and operators for business development in the field of telecommunications infrastructure, particularly in accelerating the growth of BTS devices in Indonesia. In addition, the purpose of this research is to meet the increasing needs of telecommunications network users in Indonesia and to improve the business competitiveness of each network infrastructure provider and to improve the quality of Indonesian people in communication.

1. Introduction

Telecommunications is one of the most important aspects in civilizations. With the increasing number of internet users in Indonesia, the increasing need for telecommunications infrastructure in Indonesia is something that must be considered. To make sure that the service of traffic increase, and to work on the underlying business opportunity, this traffic growth has to be considered without increasing the network infrastructure costs in terms of both capital expenditure (CAPEX) and operational expenditure (OPEX) [1]. Amid this challenge, markets are in fierce competition and increasingly saturated, so that operators have to be conscientious of costs in order to keep their profitability margins [1]. Investing and improving telecommunications infrastructure must also pay attention to the costs that will be charged to consumers [2]. Costs for improving telecommunications services, especially for telecommunications infrastructure, will be charged to these consumers. This is a dilemma for telecommunications service providers, or Mobile Network Operators (MNO), to improve telecommunications infrastructure. Operators certainly do not want investment costs to improve telecommunications services which will then become a burden for consumers. In addition, Operators strive to increase competitive advantage with competitors, by providing the best service at the lowest price. This is certainly one of the factors that can hamper the growth rate of telecommunications infrastructure amid an urgent need for improvement in telecommunications infrastructure. Therefore, this study seeks to reduce capital expenditure incurred by operators so that the need for telecommunications infrastructure can develop without having to be borne by consumers.

In Indonesia, sharing infrastructure is used to be able to save costs related to the procurement of BTS equipment. On the other hand, tower providers provide towers and connections for electrical connections. Infrastructure sharing lies in the use of towers simultaneously.

In order to fulfil the demand of the telecommunication infrastructure, operators need to accelerate the growth of its infrastructure. For the operator side, one of the way that can be done to accelerate the growth of infrastructure is by sharing infrastructure. Infrastructure sharing is considered both in growth or new roll-out (e.g. new technology or additional coverage) and in consolidation (e.g. phase out of old technology, re-location) scenarios [3]. In previous studies, research on the utilization of power supply devices, a rectifier, has never been studied before. In several previous studies, the use of shared power supplies only reached electricity sources, the generators. In this study, the model of using a shared power supply by operators is expected to be a solution to reduce costs incurred by telecommunications operators to add services to telecommunications infrastructure. Hence, this model is expected to be used easily for all operators. The use of shared power supplies on BTS towers is designed by charging fees to tower providers. This research will then conduct an analysis using the cost benefit analysis method. This method will be used to determine the benefits for tower providers and operators.

2. Research description

2.1. Shared power supply concept

In a power supply scheme for BTS devices, several components are needed, including an electric power source (AC) as the current power supply system at the BTS backed by an online UPS which ensures uninterrupted operation, and rectifier as a converter AC current into DC [4]. This model of infrastructure sharing includes in passive infrastructure sharing. This sharing is limited to non-electronic infrastructure at the cell site. Passive sharing is also known as site sharing, and in this form of sharing, all tenants or operators agree to share available infrastructure such as buildings and easements, site space, towers and masts, power supply and transmission equipment [5]. Therefore, the scheme in this study can be included in passive sharing infrastructure.

Shared power supply scheme is an effort that can be done in order to do infrastructure sharing, or commonly called co-location. Different from the existing power supply scheme where the rectifiers and batteries are procured by each operator, rectifier and batteries in shared power supply scheme will be provided by Tower Provider, there will be additional rectifier modules to anticipate the amount of power needed for the operator. The scheme of shared power supply can be seen in Figure 1

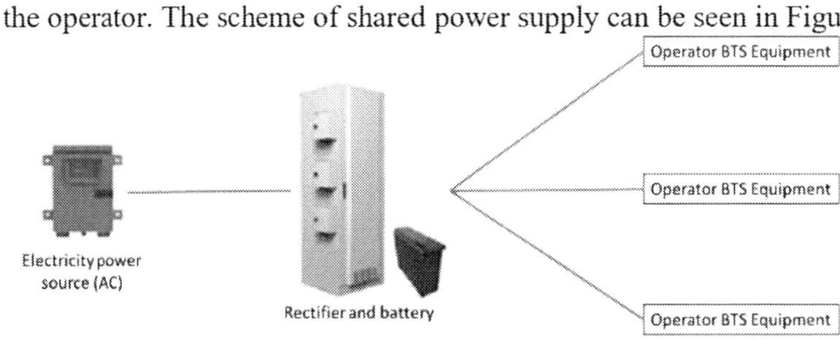

Figure 1. Scheme of shared power supply for BTS tower.

In this scheme, operator only issue capital expenditure on network devices (BTS and antennas). Thus, it is expected that operator's initial investment will be reduced by using shared power supply scheme. On the other hand, Tower Providers will increase the tower rental fees to compensate the purchase of rectifier. The important thing to be considered in preparing the rectifier is the capacity of the rectifier. Rectifier capacity must be able to accommodate the load of BTS devices from each operator.

2.2. Research methodology

Some methodologies used in this study are feasibility study, linear regression method, and budget estimation.

2.2.1. Feasibility Study

Feasibility Study of Shared Power Supply is conduct to assessing the investment that will be carried out by Tower Provider on a shared power supply scheme. Profitability assessment is done by calculating cost benefit analysis, NPV, PP and IRR. From these calculations it can be determined whether the shared power supply scheme is feasible to be carried out by Tower Provider [5]. Decision making is done by comparing the results of the profitability assessment between the existing scheme with the shared power supply scheme.

2.2.2. Linear regression method

Simple linear regression is a linear regression model with a single explanatory variable. Simple regression is based on a functional or causal relationship independent variable with one dependent variable. Simple linear regression analysis also can be used to test the nature of the relationship cause-and-effect between the independent variable (X) and the dependent variable (Y) [6]. In this study, regression method is used to predict the value of an unknown variable based on past observations of that variable and others variable. In this study, a simple linear regression equation is used to calculate the number of BTS devices for year-n.

2.2.3. Capital Budgeting

When the firms can continuously invests funds in assets and it will produce income and cash flow, these assets can represent the firm's capital [7]. These the capital budgeting can be used to analyse the budget allocation owned by Tower Provider to procure BTS devices each year. Capital budgeting is a normative approach to give rational selection of capital expenditure proposals. This process involves a determination of the size its budget and its financing, collect data and information on the viable alternatives investigating and identifying the prospective investment opportunities, classifying them according to some scheme, defining and estimating the cash flows, followed by economic analysis [9],[8]. The initial step taken is to calculate the budget for BTS equipment procurement using the equation:

$$\text{Procurement budget for BTS} = \text{total BTS equipment} \times (\text{CAPEX} + \text{OPEX}) \qquad (1)$$

After knowing the budget that is owned based on the existing scheme, from the budget owned by telecommunications operators each year, then the amount of BTS equipment procurement will be calculated if using the power supply scheme together with the equation:

$$\text{total BTS equipment} = \frac{\text{Operator's annual budget}}{\text{CAPEX+OPEX shared power supply scheme}} \qquad (2)$$

3. Finding and discussion

The research was conducted in 3 steps in order to find competitive advantages in using shared power supply which is the aim of this study, several findings were obtained:

3.1. CAPEX/OPEX analysis of shared power supply scheme

CAPEX used as an initial investment for both Tower Provider and Telecommunication Operator.

For Tower Provider's side, there is difference in value of CAPEX if using existing power supply scheme and shared power supply scheme, as seen below:

Table 1. Tower Provider CAPEX comparison between existing scheme and shared power supply scheme.

CAPEX existing scheme	IDR 1,158,763,900
CAPEX shared power supply scheme	IDR 1,418,163,900

The difference occurs because of shared power supply scheme carried out additional costs for shared power supply devices, a rectifier. And for Tower Provider annual OPEX:

Table 2. Tower Provider OPEX comparison between existing scheme and shared power supply scheme.

OPEX existing scheme	IDR 9,500,000
OPEX shared power supply scheme	IDR 105,300,968

The difference occurs because of Shared power supply scheme calculating electricity cost, which in the existing scheme is charged to telecommunication operator. Although the CAPEX and OPEX of shared power supply scheme is greater than existing scheme, Tower Provider can increase the rent fee to the Telecommunication Operator, which results a better investment feasibility assessment that obtained in this study. As for telecommunication operator side, existing scheme included rectifier in CAPEX. Whereas in the shared power supply scheme, rectifier as an infrastructure sharing devices is provided by Tower Provider.

Table 3. Operator CAPEX comparison between existing scheme and shared power supply scheme

CAPEX existing scheme	IDR 821,608,100
CAPEX shared power supply scheme	IDR 652,308,100

Table 4. Operator OPEX comparison between existing scheme and shared power supply scheme.

OPEX existing scheme	IDR 172,600,000
OPEX shared power supply scheme	IDR 196,600,000

Cost saving is the most important aspect when we need to take the decision for infrastructure sharing [3]. The difference, between the Operators OPEX compared to the existing scheme and the research scheme, is due to the rent cost. The electricity cost for existing scheme is charged to the operator, whereas in the research scheme, the electricity cost has already been borne by the tower provider as an integral part of the sharing power supply scheme. In this case, the operator does not need to think about electricity costs as an annual operational cost. However, the increase in OPEX occurred because the rental fees charged to operators increased, from IDR 10,000,000.00 per month to IDR 15,000,000.00

3.2. Feasibility Study of Shared Power Supply Investment for Tower Provider

Feasibility Study of Shared Power Supply conducted to provide an overview for Tower Provider about the proposed scheme investment. The calculation is done by using cash flow from existing scheme and shared power supply scheme. Cash flow components that calculated are CAPEX, OPEX, and revenue obtained from tower leases. From each of these cash flow, profitability is calculated using NPV, IRR, Payback Period, and B/C. The results will be used as a reference for the Tower Provider's stakeholder to carry out the investment or not.

Results for profitability assessment existing scheme and shared power supply scheme are:

Table 5. Profitability comparison between existing scheme and shared power supply scheme.

Scheme	NPV (IDR)	IRR	*Payback Period*	B/C
Existing	2,182,896,486.95	17.54%	7.95	2.6235
Shared Power Supply	3,502,082,504.71	20.62%	7.06	2.8899

From these results, can be seen that the profitability value for shared supply scheme is higher than existing scheme, as well as the payback period for this investment is shorter. So, the investment is feasible to be carried out by Tower Provider.

3.3. Utilization of Shared Power Supply for Telecommunication Operator

For operators, the utilization of shared power supply scheme can be seen from the perspective of their influence on the growth of the number of BTS equipment owned by telecommunications operators each year. Based on the growth data of BTS devices in 2011-2018, linear regression modelling is carried out for the next 20 years. The equation used is:

$$y=23,158.25+42,136.83x \tag{3}$$

By using the simple regression equation that has been obtained, forecasting the number of towers is done until 2030. Then the writer calculates the number of towers that can be obtained using a joint power supply scheme based on the annual budget allocated by operators in BTS procurement.

This calculation is based on the assumption that the Telecommunications Operator has an annual budget for BTS equipment procurement.

Table 6. Comparison total amount of BTS equipment.

Total amount of BTS equipment using existing scheme	Total amount of BTS equipment using shared power supply scheme
402,390	471,263.49
444,527	520,612.71
486,663	569,960.75
528,800	619,309.96
570,937	668,659.18
613,074	718,008.39
655,211	767,357.60
697,348	816,706.81
739,484	866,054.86
781,621	915,404.07
823,758	964,753.28
865,895	1,014,102.50

This, if the procurement of BTS equipment is carried out using a shared power supply scheme, there is an increasing number of annual BTS equipment by 17.12% compared to if the operator uses the existing scheme. Thus, the use of research schemes will affect the growth of the number of BTS operators.

4. Conclusion and Suggestions

4.1. Contribution

- For Tower Providers, from the results of a feasibility study on investment, the shared power supply scheme can be applied by Tower Provider with rent fees about IDR 15.000.000,00 because it produces greater profitability value than the existing scheme currently in use.
- For Telecommunication Operators, if the procurement of BTS equipment is carried out using a shared power supply scheme, there is an increasing number of annual BTS equipment by 17.12% compared to if the operator uses the existing scheme.;

4.2. Suggestions

- The need for a business strategy review for telecommunications tower providers to maximize the number of operators renting telecommunications towers using a sharing power supply scheme
- Need for sensitivity analysis on the implementation of sharing power supply schemes for future telecommunications technology
- The need for further study related to the impact of the implementation of the sharing power supply scheme at the level of business competition of telecommunications operators.

Acknowledgment

This work has been supported partially by PUTI Prosiding 2020 Grant funded by Directorate of Research and Development Universitas Indonesia under contract No. NKB-1146/UN2.RST/HKP.05.00/2020. Thank you.

References

[1] Khan A, Kellerer W, K Kozu and Yabusaki M 2011 Network sharing in the next mobile network: tco reduction, management flexibility, and operational independence *IEEE Communications Magazine* 134 - 142

[2] Meddour D E, Rasheed T and Gourhant Y 2011 On the role of infrastructure sharing mobile network operators in emerging markets *Computer networks* **55** 1576-1591

[3] Frisanco T and Ang R 2008 Infrastructure sharing and shared operations for mobile network operators: from a deployment and operations view *IEEE international conference on communications*

[4] Ajosh K, Sujit P, Rajan A, Aravind V and Raveendranathan K C 2010 A smart BTS power management system *In international conference on computational intelligence and communication networks*

[5] Ehiagwina F O, Fakolujo O A, Afolabi S A, Abdulraman Y A and Afolabi L O 2016 An overview of multi operator global systems for mobile communications base stations in the context of nigerian telecommunication sector *European journal of engineering research and science* **1:1** 50-55

[6] Indrajit R E 2016 *Analisa cost-benefit investasi teknologi informasi* Jakarta: The Preinexus

[7] Sugiyono 2017 *Metode penelitian kuantitatif, kualitatif, dan R&D* Bandung: Alfabeta CV

[8] Peterson P P and Fabozzi F J 2002 Capital budgeting: theory and practice Canada: John Wiley & Sons

[9] Gurnani C 1984 Capital budgeting: theory and practice, *The Engineering Economist* **30:1** 19-46

Implementation of lean manufacturing using waste assessment model model (WAM) in food industry (case study in usaha mikro kecil menengah (umkm) xyz)

E Febianti[1,2*], A Irman,[1] and M Juliana[1]
1Industrial Engineering Department, Sultan Ageng Tirtayasa University
[2]Indonesia center of excellence for food security (local food innovation), Universitas Sultan Ageng Tirtayasa

*E-mail: evi@untirta.ac.id

Abstract. The waste occurring on the production line leads to inefficient production and loss to the company. Based on the results of brainstorming with UMKM XYZ owners, it is known that there is a reduction in the production time of up to 20% of the production time is 4 hours per cycle. The waste that occurs will be identified and minimized using methods of Waste Assessment Model (WAM) and Fault Tree Analysis (FTA) for proposed repairs. The purpose of the research is to calculate the percentage of waste and know the dominant waste, knowing the repair to minimize the largest waste, and calculating the amount of production time. The results showed that the dominant waste was the transportation of 17,09% with improvements was to change the layout of the production, change the design of tools, expand the production floor, implement a more integrated fixed contract system and minimize the production process. The average time that can be minimized for donut products, bread, and molen is 490 seconds per cycle or 8.167 minutes per cycle.

1. Introduction

The production process is a stage of transforming inputs into outputs. The production process involves the entire resources of the company to be able to manage the production process becomes more efficient and effective. There are so many methods that can be used in managing effective and efficient production processes, such as scheduling [1], improving design facilities, or minimizing the waste that occurs on the production line led to the production process becomes inefficient and cause losses to the company.

Lean manufacturing is an approach that can identify and eliminate waste at the production process, so the output can be optimized and increased [2]. The concept of lean manufacturing requires the identification of waste classification to determine which activities are included in the value-added activity, non-value-added activity, and necessary but non-value-added activity. The next stage after the lean manufacturing concepts such as the identification of waste is planning improvements that can be made by the company such as eliminating some activities that are not needed at the production process.

UMKM XYZ is a small business operating in the food industry in Cilegon. UMKM XYZ products are donut, bread, brownies, sponge cake, and molen. The most produced product in UMKM XYZ that was the object of this research was a donut, bread, and molen. The production process in UMKM XYZ still used a simple and conventional production. Donut production takes 4 hours per cycle, while the bread requires 5 hours per cycle and molen takes 3 hours per cycle.

But the owners of UMKM XYZ, stated that the production time of three products decreased by 20% from the average time of the three products per cycle, or a decrease of 0.8 hours per cycle for each product.

Based on the problems, it is necessary to do an improvement in UMKM XYZ to reduce the decrease in production time. The focus of this research is the improvement in waste that often occurs in UMKM XYZ. The waste will be identified using Waste Assessment Model (WAM) methods and will be minimized using Fault Tree Analysis (FTA) [3] for improvements that can be done. WAM, the chosen method, can identify the dominant waste thoroughly, not only can identify the waste itself but also the identification based on the relationship between waste [4] [5].

This research begins by determining the standard time of each work element using a stopwatch. After that, the distribution of lean manufacturing questionnaires was carried out using the WAM method, and the most dominant waste was found in the UMKM XYZ. After learning the most dominant waste, the researcher will analyze the waste use VALSAT [6] and find the improvements using FTA (Fault Tree Analysis). The final result of this research is the proposed improvements that can be made to minimize waste and provides an illustration of the VALSAT future state of production in UMKM XYZ. The research objective to be achieved through this research are as follows:

1. Calculating the percentage of each waste that occurs in the production process in UMKM XYZ.
2. Knowing the most dominant waste from production processes in UMKM XYZ.
3. Knowing suggestions for improvements to minimize the largest existing waste in production processes in UMKM XYZ.
4. Calculating the amount of production time can be minimized when suggestions for improvements made.

2. Methodology

This research was used in qualitative and quantitative research. The qualitative research approach is an effective research approach to obtain specific information like cultural values, opinions, attitudes, and social context in a population [7]. While the approach of quantitative research is research that emphasizes the numerical data (numbers) processed with statistical methods.

The use of both of these approaches is caused by a data processing method used is the method Waste Assessment Model (WAM). WAM method using qualitative input (human perception) and changing the values obtained qualitative to quantitative for further processing using quantitative tools [5]. The object under investigation is the process of implementing lean manufacturing in production processes in UMKM XYZ.

From figure 1 we know that there are 5 data needed to be collected, they are sales data, production process data, flow diagrams, process maps, and data operations of observation time. The following is a flowchart of a lean manufacturing method WAM on research conducted in UMKM XYZ.

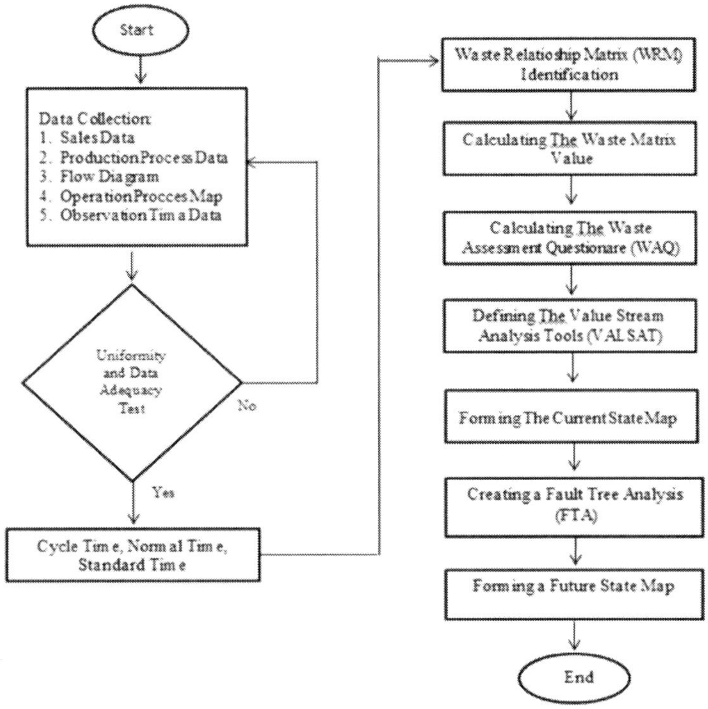

Figure 1. Research framework

3. Result and Analysis

3.1. Waste Assessment Model (WAM)

3.1.1. Waste Relationship Matrix (WRM)

Waste Relationship Matrix(WRM) is a matrix that is used to analyze the measurement criteria. The waste matrix describes the real relationship between the types of waste [8]. The recapitulation of the weighting of the Waste Relationship Matrix of this research can be seen in Table 1

Table 1. Weighted recaps of WRM

Notation	Respondents			Average	Conversion
	1	2	3		
O_I	11	4	11	9	O
O_D	7	12	5	8	O
O_M	14	14	9	12	I
O_T	11	18	9	13	E
O_W	5	8	6	6	O
I_O	11	4	10	8	O
I_D	8	4	3	5	O
I_M	6	6	8	7	O
I_T	9	10	7	9	I

Table 2. Weighted recaps of WRM (Continue)

Notation	Respondents			Average	Conversion
	1	2	3		
D_O	6	18	5	10	I
I_N	8	5	4	6	O
D_M	7	11	6	8	O
D_T	9	18	4	10	I
D_W	4	10	4	6	O
M_I	7	6	8	7	O
M_D	15	16	6	12	I
M_P	15	20	18	18	A
M_W	8	9	3	7	O
T_O	13	16	11	13	E
T_I	11	9	7	9	I
T_D	11	11	5	9	I
T_M	13	15	16	15	E
T_W	7	9	10	9	I
P_O	14	17	6	12	I
P_I	4	6	6	5	O
P_D	12	18	6	12	I
P_M	12	18	18	16	E
P_W	12	8	7	9	I
W_O	6	6	5	6	O
W_I	8	4	6	6	O
W_D	7	9	4	7	O

From the table above it can be seen the weighted average of Waste Relationship Matrix (WRM) of a third of respondents in every relationship of waste.

3.1.2 Waste Matrix Value

Waste Matrix Value is the matrix that contains letters from the conversion of weighted average WRM [9]. Waste Matrix Value of this research can be seen in Table 2.

Table 3. Waste conversion letter matrix value.

F / T	O	I	D	M	T	P	W
O	A	O	O	I	E	X	O
I	O	A	O	O	I	X	X
D	I	O	A	O	I	X	O
M	X	O	I	A	X	A	O
T	E	I	I	E	A	X	I
P	I	O	I	E	X	A	I
W	O	O	O	X	X	X	A

In Table 3 we know the relationship between waste with other waste. Letters in the second column of the third line are the relationship of waste from O (Overproduction) to I (Inventory) or could be interpreted waste affect waste Overproduction Inventory the relation is I (Important).

Each relationship between waste which is symbolized by the letter as shown in Table 5 is converted into a value based on Table 1 so that the results are as follows:

Table 4. Waste conversion value matrix value.

F / T	O	I	D	M	T	P	W	Total	%
O	A	O	O	I	E	X	O	36	15%
I	O	A	O	O	I	X	X	28	12%
D	I	O	A	O	I	X	O	34	14%
M	X	O	I	A	X	A	O	34	14%
T	E	I	I	E	A	X	I	44	18%
P	I	O	I	E	X	A	I	40	17%
W	O	O	O	X	X	X	A	22	9%
Total	38	36	40	40	30	20	34	238	100%
%	16%	15%	17%	17%	13%	8%	14%	100%	

From the table above is known from Overproduction waste percentage value of 15% means that the influence given Overproduction waste to other waste is 15%. While the percentage of waste to Overproduction of 16% means that the effect obtained Overproduction waste from other waste is 16%.

3.2. Waste Assessment Questionnaire (WAQ)

Waste Assessment Questionnaire (WAQ) was made to identify and allocate the waste that occurs in the production line [2] [5]. This questionnaire consists of 68 questions with the aim of determining the waste that occurs.

The final calculation of the Waste Assessment Questionnaire in this study is shown in Table 5. From Table 7 is known that the sequence of the most dominant waste in the production process of UMKM XYZ to the lowest is transportation (17.09%); motion (17.05%); defect (16.91%); overproduction (16.69%); inventory (13.60%); process (10.37%) and waiting (8.30%). Then, the most dominant waste that occurs in UMKM XYZ is waste transport as much as 17.09%.

Table 5. Calculation of final results WAQ

	O	I	D	M	T	P	W	Total
Score (YJ)	0.5709	0.6312	0.5818	0.5864	0.6056	0.6063	0.5188	-
Pj factor	0.0242	0.0178	0.0240	0.0240	0.0233	0.0141	0.0132	-
Results (YJ Final)	0.0138	0.0112	0.0140	0.0141	0.0141	0.0086	0.0069	0.0826
The final result (%)	16.69%	13.60%	16.91%	17:05%	17:09%	10:37%	8:30%	-
Ranking	4	5	3	2	1	6	7	

3.3. Value Stream Analysis Tools (VALSAT)

Selection tools on VALSAT performed by linkage of the seventh waste that was observed in UMKM XYZ used Waste Assessment Questionnaire (WAQ) method.

Table 6. Value stream analysis tools (VALSAT)

Weight of Waste	Waste	Process Activity Mapping	Supply Chain Matrix	Product Variety Funnel	Quality Filter Mapping	Demand Amplification Mapping	Decision Point Analysis	Physical Structure
0.1669	O	0.1669	0.5008		0.1669	0.5008	0.5008	
0.0830	W	1.5024	1.5024	0.1669		0.5008	0.5008	
0.1709	T	1.5024						0.1669
0.1037	P	1.5024		0.5008	0.1669		0.5008	
0.1360	I	0.5008	1.5024	0.5008		1.5024	0.5008	0.1669
0.1705	M	1.5024	0.1669					
0.1691	D	0.1669			1.5024			
Amount		6.8442	3.6725	1.1685	1.8363	2.5040	2.0032	0.3339
Ranked		1	2	6	5	3	4	7

The results of the calculation of the Value Stream Analysis Tool (VALSAT) above, the biggest weighting result was obtained in the Process Activity Mapping (PAM) tools section. Besides the use of tools Process Activity Mapping (PAM) is strongly correlated to waste Transportation of 1.5 and is very useful to eliminate waste Transportation.

3.4. *Fault Tree Analysis (FTA)*

Fault Tree Analysis (FTA) is an analysis method that can describe failures which is the cause of an undesired event and the probability of an undesired event [10].

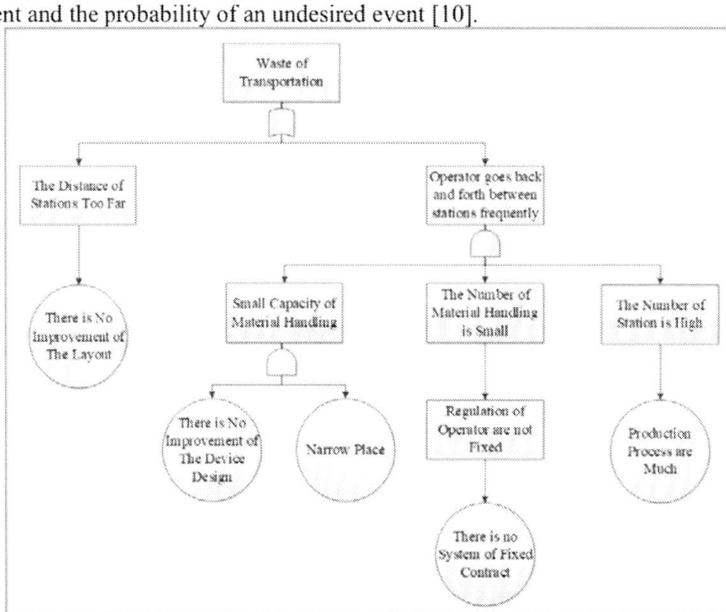

Figure 2. FTA of transportation waste in UMKM XYZ.

Based on the picture above it can be seen that the events that are not expected to cause waste transport of which is no improvement of the layout of the start of production, design tools have not changed from the start of production, a narrow place, there is no fixed contracts system and many production processes. The improvement can be given to minimize wastage of transport such as changing the layout of production, change the design of tools becomes larger capacity, expanding the production floor, applying a fixed contract system more clearly and integrated for employees and minimize the production process.

3.5. *Recaps before and after improvement*

Proposed improvements made to minimize waste by minimizing the production process can reduce the time of production of donuts, bread, and molen per cycle. Results improvement using Big Picture Mapping (BPM) tools and PAM of donuts, bread, and molen.

Table 7. Recaps before and after the improvement of total element work and lead

Product	Total Work Element (Event)			Lead Time (Second)			% Decrease
	Before	After	Difference	Before	After	Difference	
Donuts	42	37	5	19.865	19.327	538	2,7%
Bread	36	33	3	36.777	36.266	511	1,4%
Molen	34	31	3	121.970	121.549	421	0,3%

From Table 6 we know that recaps before and after improvements in UMKM XYZ for donut product have a difference of 5 elements of work (weigh the raw materials at the station weighing of raw materials, bring raw materials to the WIP kneading station 2, check the dough at the dough station, check the dough setting on the floured board at the rounding station and the perforation station) with a time of 538 seconds, bread product has a difference of 3 elements of work (weigh raw materials in the station weighing of raw materials, bring raw materials to WIP station weighing of raw materials 2 and check the dough at the dough station) with a time of 511 seconds, while the molen product has a difference of 3 elements of work (check dough at dough station, take the raw material in the winding station and set up raw material in the winding station) with a time of 421 seconds. The average time for all three products can be minimized until 490 seconds or 8.167 minutes per cycle.

4. Conclusion

The conclusion that can be taken from the research in UMKM XYZ are as follows:

1. The percentage of each waste that occurs in the production process in UMKM XYZ using methods Waste Assessment Model (WAM) is overproduction 16.69%, 13.60% inventory, defects 16.91%, 17.05% motion, transportation 17.09 %, process 10.37%, 8.30% waiting.
2. The most dominant waste among seven waste from production processes in UMKM XYZ is transportation, motion, defects, overproduction, inventory, process, and waiting.
3. Suggested improvements can be done to minimize waste transportation in UMKM XYZ is changing the layout of production, change the design of tools, expanding the production floor, applying a fixed contract system more clear and integrated for employees and minimize the production process.
4. The amount of production time can be minimized when the improvement suggestions made are donut product 538 seconds, bread product 511 seconds, molen product 421 seconds. The average time for all three products that can be minimized is 490 seconds or 8.167 minutes per cycle.

5. Acknowledgment

I want to thank my friends in Sultan Ageng Tirtayasa university who have taken the time to discuss. And thanks to the family (husband, children, and parents), who supported my research.

6. References

[1] A Irman, E Febianti, and U Khasanah 2019 Minimizing makespan on flow shop scheduling using Campbel Dudek and Smith, particle swarm optimization, and proposed heuristic algorithm doi: 10.1088/1757-899X/673/1/012099

[2] F Pasqualini and P A Zawislak 2005 Value stream mapping in construction: A case study in a Brazilian construction company

[3] B Vesely 2002 Fault tree analysis (fta): concepts and applications

[4] E Amrina, N T Putri, and D M Anjani 2019 Waste assessment using lean manufacturing in rubber production doi: 10.1088/1757-899X/528/1/012051

[5] I A Rawabdeh 2005 A model for the assessment of waste in job shop environments *Int. J. Oper. Prod. Manag.* doi: 10.1108/01443570510608619.

[6] N Zahrotun and I Taufiq 2018 Lean manufacturing: waste reduction using value stream mapping doi: 10.1051/e3sconf/20187307010.

[7] A P Pradana, M Chaeron, and M S A Khanan 2018 Implementasi konsep lean manufacturing guna mengurangi pemborosan di lantai produksi *OPSI* doi: 10.31315/opsi.v11i1.2196.

[8] H Henny and H R Budiman 2018 Implementation lean manufacturing using waste assessment model (wam) in shoes company doi: 10.1088/1757-899X/407/1/012077

[9] S B M Carneiro, I B Campos, D M De Oliveira, and J P B Neto 2012 Lean and green: a relationship matrix

[10] W E Vesely, F F Goldberg, N H Roberts, and D F Haasl 1981 Fault tree handbook doi: NUREG–0492

IOP Conf. Series: Materials Science and Engineering **909** (2020) 012067 doi:10.1088/1757-899X/909/1/012067

Identify factors that caused false and violation by motorcycle rider

Lovely Lady [1*], Teguh Try Mulyo[2], Kulsum Kumino[3]
[1,2,3]Industrial Engineering Department-Engineering Faculty, University of Sultan Ageng Tirtayasa, Indonesia

*e-mail: lady1971@gmail.com

Abstract. A motorcycle is a vehicle that is widely used by Indonesia community. However, increasing of motorcycle users were not followed by a good driving behaviour, the increasing was directly proportional to increase in the number of traffic accidents. The motorcycle is one of the dominant causes of the traffic accident in Indonesia. This research aims to identify the causes of the deviation and violation in driving that conducted by motorcyclists. The characteristics of driving and the type of deviation committed by motorcyclists collected using Driver Behavior Questionnaire (DBQ). The respondents were motor drivers who have a motorcycle driving license (SIM) in 8 districts in Cilegon. The data processing used the factor analysis method to identify the factors that caused traffic deviation by the motorcyclists. This research find seven factors that caused the deviation and violation by drivers, the dominant factor that explained 16.857 % deviation in riding is lack knowledge of traffic rules. Next factors were vehicle speed exceeds the threshold, the emotions, hasty, careless, not paying attention to the health, and not focus in driving.

1. Introduction

Motorcycles are private vehicles that are widely used by Indonesian people. Based on data from the central Bureau of Statistics (BPS) Indonesia in 2016, the development of the motorcycle rider in Indonesia was very high. The number of motorcycle was 105,150,082 units while the car was 14,580,666 units, truck was 7,063,433 units and bus was 2,486,898 units. Many people choose to ride a motorcycle because it is easy to drive, light and agile so it can easily overtake on the highway, and the price is affordable. However, the increasing of motorcycle rider is not followed by good driving behaviour. The increasing was directly proportional to the increasing of road accidents. Based on data from the Central Bureau of Statistics Indonesia in 2016, motorcycles were one of the factor that caused of high traffic accidents, it is demonstrated by 106,129 total accidents that involving motorcycles versus cars as number 98,021 from total highway accident in 2016.

The traffic accident was the culmination of a series of events, which in its last condition was shortly before the accident was preceded by the failure of the driver in anticipation of his surroundings including himself. Based on data from the World Health Organization (WHO) in 2004, there are three factors causing the traffic accident that endangers individuals. The first factor is human error, the second factor is the vehicle (machine error), and the third factor is the path (manufactured error). Among the three factors, human error is the greatest factor in the occurrence of traffic accidents.

Human error is an unnecessary decision by someone that can reduce or potentially reduce the effectiveness, security, or performance of a system. The possibility to make mistakes or human errors in a job will always exist and cannot be eliminated completely. This is due to human limitations. The occurrence of human error can be influenced by several factors including poor lighting, insufficient exercise or skills, poor machine design, high temperature in the work area, high noise level, no layout appropriate, too narrow distances, poor motivation, unsupportive equipment, poor machine operating and maintenance procedures, the degree of complexity of the work performed, and poor verbal communication. Pulat (1992) also presents several reasons for human errors such as bad behaviour, the provision of work over workers ability, the work stress, workplace and work methods that do not consider the factors of human beings, and the insufficient the training or skills. Human error can be derived from the behaviuor of riders or drivers on highways, perception, cross-patterns, driving skills, attention/concentration on highways, social problems, as well as the emotional problems of riders and drivers on the highway.

Unsafe actions are primarily caused of deviant mental processes such as underprivileged, lack of attention, poor motivation, carelessness, neglect and frivolity. System approach on the accidents is known error and human error can be predicted. Mistakes are seen as consequences not as causes [1]. Unsafe actions such as driving over speed limits (traffic speeds of more than 40 km/h) relate to high levels of accidents in Indonesia [2]. There are deviation performed by motorists Indonesia while riding using electronic map, it caused 9.6% human error [3]. Research on truck accidents in America showed interference during driving on the driver increased the chances of involved in a truck crash. The results of other the study also showed that when a truck increased it speed it was more likely to suffer severe injuries even death [4].

Cognitive failure is an error that occurs in a simple task which usually a person does it without mistakes. Statistical analysis show that the average cognitive level significantly contributes to occupational safety. Increased cognitive failures led to an increase in accidents. The score of cognitive failure is correlated with the level of error in driving, but not with the accident that someone suffered [5].

Warren Horison [6] researched drivers behaviour, he obtained the average score of false and violation that made by a person during the driving change between surveys. The research suggested that some aspects of the driver's behaviour change. There were no statistically significant changes of score on the Lapses and Error scales, but the Violation and Aggressive scores increased significantly. Based on Ali Tavakoli Kashani et al. [7] is known the frequent violations that drivers have done were ignored the restricted speed and overtook another ride unconsciously.

With regards to the high level of motorist accident and violation, this research aims to identify irregularities and violations carried out by motorists and identify the causal factors. The Manchester Driver Behaviour Questionnaire (DBQ) used to collect motorcycle rider. DBQ is a self-report questionnaire developed by Reason in 1990 as a measure of aberrant driving behaviours. This questionnair have used by Matus Sucha [8] in determining factor that affect driver behaviour. Martinussen [9] applied the DBQ in order to assess the problem areas within a driving population in Denmark.

2. Literature Review

Statistically data can be summarized into some significant factors that affect a system. Factor analyzing have used Principal Components Analysis (PCA), this is a tool to extract factors. Data from questionaires were condensed and summarized into more defined data sets using Principal Components Analysis (PCA). The general purpose of this step was to identify the common factor and explain their relationship to the observed data. Next steps had conduct to interpret Principal Components Analysis (PCA).

1. Conducting assumption tests to make sure that PCA can be used to analyze the data. The tests included: (a) Reliability test was performed and assessed by calculating Cronbach's alpha, which was a coefficient of reliability (internal consistency); (b) Sampling adequacy Kaiser-Meyer-Olkin (KMO) Measure of Sampling Adequacy was used to find out there is no uncertainty due to the prediction of communality problems; (c) Bartlett's Test of Sphericity,

with this test will get significant of confident base on α. Variables are ready to conduct PCA after pass three tests above.

2. Communalities. This is the proportion of each variable's variance that can be explained by the factors. Communalities explain how much factor which built can explain the variable. All of variables can be explained by the factors with value more than 0.5.
3. Extraction. Output of the extraction indicated the proportion of each variable's variance that can be explained by the retained factors. Variables with high values were well represented in the common factor space. However factors will be retained as we requested based on eigenvalue-greater-than-one.
4. Rotated Component Matrix was used to determine the member variable of any factors that are constructed. For ease of interpretation of the factor extracted, the principal components matrix often rotated with goal to achieve a simple structure. Varimax exploratory principal factor analysis method was conducted to assess the underlying structure for the original 29 variables of the questionnaire into small set of construct factors with minimum loss of information.

3. Methods

3.1 Questionnair

Driver Behaviour Questionnaire (DBQ) is a questionnaire used to determine driving behaviour. Questions in the DBQ use a four-level violation approach that covers the error, lapse, ordinary violation and aggressive aspects of violation. The questionnaire was developed by Reason, Manstead, Stradling, Baxter and Campbell (1990) as a tool for assessing deviant driving behaviour. The questionnaire has 28 questions of behavioural items. Respondents were asked to respond to each question by choice a point that appropriate with they behave. Respondents were asked to assess using a Likert scale of six points (1 = never; 2 = hardly ever; 3 = sometimes; 4 = quite often; 5 = often; 6 = most of the time).

Some aspects that affect the driver/rider behaviour are:

1. Lapse, it is a mistake that is not apparent when it behaves, associated with loss of concentration when setting the path to be taken to achieve a goal while driving.
2. Error, it is a distorted behaviuor or mistakes made accidentally. According to Reason et al (Stradling & Meadows, 2001), error is one example of the driver's fault in its execution even though it was previously planned by the driver.
3. Ordinary Violation, Driver made mistakes deliberately with the intention of law violating. According to Reason (2001) an ordinary violation is a form of behaviour that typically leads to aggressive driving.
4. Aggressive violation, driver make mistakes deliberately with the intent of breaking the law and can cause harm to other riders.

3.2 Respondents

The respondents of this research are motorcyclists in the Cilegon city area and they already have a driver's license. The study uses respondents with aged 17 – 55 years. When the questionnair be spread to the respondents, the researchers accompanying them to ensure that the respondents can fill the questionnaire without any mistakes and according to the rules given by the researcher. The number of respondents in this study was 100 people. A total of 74 were male respondents and 26 female respondents. Based on the age, a total of 22 respondents were at age of 17-21 years, 31 respondents were at age of 22-24 years, 35 respondents were at age 25-48, 4 respondents were at age of 49-51 years, and 8 respondents were at age of 52-55 years.

3.3 Location

This research was conducted in the city of Cilegon, Banten, and the research time was in March-April 2018. The spread of questionnaires was conducted in 8 sub-districts in Cilegon city, they are Cibeber, Jombang, Citangkil, Purwakarta, Ciwandan, Grogol, and Pulomerak

3.4 Data Processing

After the dissemination of questionnaires, the first phase of data processing is to identify the type of human error and violations that motorcyclists often do when driving. Then the data will be processed using factor analysis method to get the factors that affect the traffic violation on the motor driver. The following is steps of the data processing using factor analysis method.

1. Formulating correlation matrix, in using method of factor analysis, the first decision to be taken by researchers is analysing whether the existing data is sufficiently qualified in factor analysis. This first step is done by finding the correlation between the observed indicators.
2. Factor Extracting. The factor extraction is a method used to reduce data from some indicators to produce some factors that can explain the correlation between observed indicators.
3. Rotating. The factor rotation is done in order to obtain a simpler factor structure and to be easily interpreted.
4. Interpreting the factors. Interpreting the factors through knowing the variables that built it. Interpretation have used judgement, because of their subjective characteristic, the result can be different if done by others.

4. Result and Analysis

4.1 Research Variables

The variables in this study were formulated as factors or constructions. Construction is variables formed through the observed dimensions or observed indicators. Questions in the DBQ questionnaire use a four-level violation approach that covers the error, lapse, ordinary violation and aggressive aspects of violation. The four aspects of the offense is spelled out into indicators, and then referred as variables in the questionnaire, outlined in Table 1 as follows.

Table 1. List of variables in the driver behaviour questionnaire

No	Aspects	Indicators
1	*Error*	a) Ignore the speed when overtaking b) Not seeing mirrors before moving c) Forgot giving sign d) Overtaking another driver e) Almost bumped into another vehicle while turning f) Unaware of pedestrians g) Braking too fast h) Wrong turn
2	*Lapse*	a) Drive with wrong gear vehicle b) Enter the wrong path c) Forget where to park your vehicle d) Recognizing mistakes during driving e) Powering something else f) Wrong reading mark g) Can't remember the way h) Crashing during rewind
3	*Ordinarry Violations*	a) Ignoring speed on public roads b) Riding ahead when traffic lights turning red. c) Overtake rider from the left d) Drunk driving e) Ignoring the speed on highways f) Drive close to the car in front

Table 1. List of variables in the driver behaviour questionnaire (Continue)

No	Aspects	Indicators
4	*Aggresive Violations*	a) Showing pigue by honking b) Entering the closed track c) Exit the crossroads d) Angry against other riders/drivers e) Chasing other riders in an angry state f) Race away from red lights to beat other riders

4.2 Factor analysis

Data from the questionnaire of driving habits and deviations conducted by the respondent then be processed to obtain the factors causing deviations in riding. The stages of factor analysis are as follows:

1. Correlation test and feasibility of a variables

This stage have tested the correlation of variables that have been defined using the Barlett test (Bartleet's Test of Sphericity) and the Kaiser Meyer Olkin Measure of Sampling Adequacy (MSA). The Barlett test and MSA test were conducted to assess the feasibility of a variable that will be processed in factor analysis, the following criteria have used in analize stage:

a. Bartleet's Test of Sphericity

The Barlett test in factor analysis is to test the correlation between variables because the desired outcome in factor analysis is the presence of a high correlation between variables, has a high correlation if the value of Barlett calculated > Barlett table, or p-value (sig) < α = (0.05) indicates a high correlation between variables and the process can be resumed. The hypothesis for significance is:

Ho = no correlation

H1 = High correlation and sample is adequate for further analysis.

The criterion for significance is: If the value of sig > α = (0.05) then Ho accepted, sig < α (0.05) then Ho rejected.

In the first phase, have been collected 50 questionnaires from 50 respondents. This data was calculated to determine whether the data was decent or not. The results of Barlett test using this 50 samples got the value of KMO and Bartleet's Test of sphericity = 0.280 so that the factor analysis process can't be extended because it is not eligible because the value of KMO count < KMO table which 0.280 < 0.5. The data that have collected shows they have not correlation so cannot be processed further. Next step we collected 50 more samples, so we had 100 samples to determine the feasibility of data. By using this 100 data we got KMO value and Bartleet's Test of sphericity = 0.515, so that the factor analysis process can be extended because it had qualified because the value of the KMO calculated > KMO table where 0.515 > 0.5 or significance value (α) < = 0.05. So variables are proven to be correlated and can be further processed.

b. Measure of Sampling Adequacy (MSA) Test

An MSA test is a test used to measure the homogeneity of the variables and select variables so that only eligible variables can be processed further. The MSA test is performed several times to release variables that have MSA values below 0.5. Once an MSA-eligible variable is issued one by one then formed a variable that has a loading value of > 0.5. At this stage a seven variables are issued from a questionnaire with a loading value of < 0.5 indicating the homogeneity of the low variable.

2. Grouping Factor

Data from questionnair was condensed and summarized into more defined data sets using Principal Components Analysis (PCA). The general purpose of this step was to identify the common factor and explain their relationship to the observed data.

Grouping factor is an attempt to group variables to determine whether independent variables can be grouped into one or more factors. Grouping of factors is done by looking at whether to 28 independent variables can be simplified into one or several factors. After removing seven variables that have low homogeneity level in the MSA test, still remaining 21 variables that can be used to form the factor.

Communalities process got the proportion of each variable's variance that can be explained by the factors. Communalities output explain how much factor which built can explain the variable. All of variables can be explained by the factors with value more than 0.5.

Extraction process was the next step after Communalities stage, output of extraction process indicated the proportion of each variable variance that can be explained by the retained factors. Variables with high values were well represented in the common factor space. However in this research only the first seven factors will be retained as factors that caused riders making false and violence. All the seven factors had an Eigen value more than 1. Output of extraction process is shown in next Table.

Table 2. Factors output of extraction process

Factor	Initial Eigenvalues		
	Total	Variance (%)	Cumulative (%)
1	3.54	16.857	16.857
2	2.863	13.632	30.489
3	2.245	10.691	41.180
4	1.862	8.868	50.048
5	1.495	7.119	57.167
6	1.221	5.765	62.932
7	1.132	5.391	68.323

On this step we have got the variance of rider false and violation can be explained by seven factors. The factor 1 explained 16.857% of variance, the factor 2 explained 13.632 % of variance, the factor 3 explained 10.691% of variance, the factor 4 explained 8.868%, the factor 5 explained 7.119 %, the factor 6 explained 5.765%, and the factor 7 explained 5.391%. The total of seven factors explained 68.323% variance in rider false and violation. Then factor loading determined each independent variable that fit into one of the 7 construct factors. Rotated Component Matrix was used to determine the member variable of any factors that are constructed.

3. Factor Rotation

The variables that have been extracted than will pass through the rotation process because the placement of the variables are usually not yet precise or there were some variables that have not conform to the factors. The rotation process using Varimax methods have be done on the variables that have passed through the MSA test. The following table figured some factors formed after rotation process. Each factor had some variables as members of the factor. The number in the cross cell between column 'Factor' and row 'Variable' indicated contribution value of each variable into the factor. All factors formed have some forming variables, its explain in Table 3. The factors are :

Factor 1 has 5 forming variables
Factor 2 has 3 forming variables
Factor 3 has 3 forming variables
Factor 4 has 2 forming variables
Factor 5 has 3 forming variables
Factor 6 has 1 forming variables

Factor 7 has 4 forming variables
Table 3 figured all factors as result of rotation process.

Table 3. Factors as result of rotation process

Variable	Factor						
	1	2	3	4	5	6	7
X1			0.274				
X2							0.87
X5					0.845		
X6					0.678		
X7				0.699			
X8	0.493						
X9					0.338		
X12	0.769						
X13							0.259
X14	0.597						
X15							0.829
X17		0.562					
X19	0.694						
X20						0.849	
X21			0.762				
X23			0.719				
X24				0.631			
X25	0.326						
X26		0.591					
X27							0.558
X28		0.473					

The following table list the factors causing deviation and violation of the motor riders in Indonesia that has been processed through factor analysis. Each factor has been given the name according to its constructor variables.

Table 4. Factors caused deviation and violation of the motor riders

No	Factor	Variables
1	Lack of knowledge of traffic rules	Wrong turn
		Recognizing mistakes during driving
		Wrong riding mark
		Overtake rider from the left
		Exit the crossroads
2	Emotion	Ignoring speed on public roads
		Angry against other riders/drivers
		Race away from red lights to beat other riders
3	Vehicle speed exceeds safe limit	Ignore the speed when overtaking
		Ignoring the speed on highways
		Showing pigue by honking
4	Haste	Braking too fast
		Entering the closed track

Table 4. Factors caused deviation and violation of the motor riders

No	Factor	Variables
5	Careless	Almost bumped into another vehicle while turning
		Unaware of pedestrians
		Drive with wrong gear vehicle
6	No attention to health	Drunk driving
7	Not focus on driving	Not seeing mirrors before moving
		Powering something else
		Can't remember the way
		Chasing other riders in an angry state

Seven factors formed as caused of deviations and violations by motorists. The first factor 'Lack of knowledge of traffic rules' can explain 16.857% of rider false and violation. The cause of this factor is the number of driving licenses that were got in a fraudulent manner by motorist. The drivers got the SIM by paying some money to a person without passing the driving practice test and writing test. They didn't know about the traffic rules but already have got permit to drive. Many cases like cheating in getting the driving licenses resulted in the drivers did not read the theory about how to driving in highway, they did not understand the order in driving, so as not understand the rules in the cross and resulted in frequent irregularities in the way. This factor can be lowered by increasing the discipline in issue of the licence and removing the cheating in processing and issuing the licence.

Second, Third, and Fourth factor are influenced by the age of rider and culture. They are relatively young rider. There were 53% of respondents below 25 years old. Psychologically young age is still labile and emotional. This factor can be mitigated by enhancing the discipline on the highway and giving strict sanctions to disciplinary offenders. The sanctions rate also needs to be weighed down so that riders don't easily commit violations.

The fifth and seventh factors caused riders not master the true driving practice and they do not know the rules in driving. This condition is caused the riders did not pass the driving test and the writing test because the driving licence was process fraudulently, so they drove in highway as they were.

5. Conclusion

Processing data of rider behaviour with a factor analysis method acquired seven factors causing deviations and violations committed by the rider. Using factor analysing have got the dominant factor caused deviations and violations is lack of knowledge of traffic rules, that explained 16.857% of violation. Next factors were emotion of rider, speed of vehicle exceeds the threshold, hasty, careless, not paying attention to health, and does not focus in driving.

References

[1] Reason J 2000 Human error : models and management BMJ 320 4–6.

[2] Sulistio H 2018 Effect of traffic flow, proportion of motorcycle, speed, lane width, and the availabilities of median and shoulder on motorcycle accidents at urban roads in indonesia *The Open Transportation Journal* **12** 1–7.

[3] Lady L 2019 Effect of using electronic map while driving on human error probability *Proceeding of the International Conference on Industrial Engineering and Operation Management 2019 (IEOM 2019)*

[4] Mahdi M, Mashhadi R, Wulff SS, and Ksaibati KA 2018 Comprehensive study of single and multiple truck crashes using *The Open Transportation Journal* **701** 43–56.

[5] Allahyari T, et al 2008 Cognitive failures, driving errors and driving accidents *Int. J. Occup. Saf.*

Ergon **14** 2 149–158.

[6] Harrison, W 2009 Reliability of the driver behaviour questionnaire in a sample of novice drivers *Australian Road Safety Research, Policing and Education Conference Sydney*

[7] Kashani AT, Ravasani MS, Ayazi E 2016 Analysis of drivers' behaviour using manchester driver behaviour questionnaire based on roadside interview in Iran *International Journal of Transportation Engineering* **4** 1

[8] Matus Sucha M, Sramkova L, and Risser R 2014 The Manchester driver behaviour questionnaire: self-reports of aberrant behaviour among Czech drivers *Eur. Transp. Res. Rev* **6** 493–502 doi 10.1007/s12544-014-0147-z

[9] Martinussen, Marianne L., Blomqvist H, Liisa, Mette M., Özkan, Türker, Lajunen, Timo 2013 age, gender, mileage and the DBQ: the validity of the driver behaviour questionnaire in different driver groups *DTU Library* Link to article, DOI:10.1016/j.aap.2012.12.036.

IOP Conf. Series: Materials Science and Engineering **909** (2020) 012068 doi:10.1088/1757-899X/909/1/012068

Selection of IoT-based technology for electric smart meter on PLN Disjaya

Adi Mulyono[1*] and F. Farizal[1]**
[1]Department of Industrial Engineering, Universitas Indonesia, Depok, Indonesia
*Email: adi_st3@yahoo.com;

Abstract. With so many platforms or protocols developed by several organizations or institutions, such as LoRa, NB-IoT and SigFox that have their own advantages and limitations. While it takes a large and long-term investment to deliver IoT technology. Apart from that, losses caused by the use of conventional meters on electricity customers cause a decrease in income by electricity supply companies. In this study aims to choose the internet-based technology of things for electricity smart meters in the PLN Disjaya region between three LoRa, NB-IoT and Sigfox technologies. This research is doing network design, coverage analysis, capacity analysis and network utilization, cost benefit calculation and analysis, data rate speed and proprietary. From the analysis results obtained SigFox superior in terms of coverage, capacity and network utilization, cost benefits (NPV, IRR, Payback Period, BCR, ROI), LoRa superior in terms of proprietary and NB-IoT superior in terms of data rate speed.

1. Introduction

Internet of things concept first introduce on RFID development committee in 2000 [1]. Since then, internet of things related to object detection, finding, and control every day through internet, using RFID [2], wireless network [3], wide area network or other equipment [4]. Based on International Telcom Union (ITU), key from IoT are connection, human and information [5]. Generally, basic structure internet of things technology divided by 3-layer that are physical layer, network or cloud layer and intelligence layer. Many independent studies forecasting number device and income from internet of things technology and machine-to-machine industry for the next ten years. Number M2M device will be connected and electronic customer will be passed number people that used mobile phone, personal computer, laptops and tablet by year 2020. Its predicted number of devices connected with IoT technology will reach 20 billion devices by year 2025 [6].

Basically, wireless technology that supports IoT requires several parameters such as low power consumption, low cost, measurable, reliable, low latency, wide range and secure. According to [7] wireless technology divided by two, which is Long Range and Fixed & Short Range. Fixed & Short Range, wireless technology with low coverage area connectivity, limited bandwidth and long battery usage. For example, Fixed & Short-Range technologies are RFID, Bluetooth, Zigbee and WiFi. Long Range is a connectivity technology with a wide range of areas by using a variety of bandwidth and battery. Long range technology is divided into two namely 3GPP Standard and Non-3GPP Standard. 3GPP standard uses cellular technology and frequency, for example the IoT technology included in the 3GPP Standard is LTE-M, NB-IoT and E-GSM. Non-3GPP Standard uses wireless technology and un-license frequency. LoRa, SigFox and RPMA are examples of Non-3GPP Standard Long Range IoT technology. Figure 1 shows the distribution of IoT wireless technology.

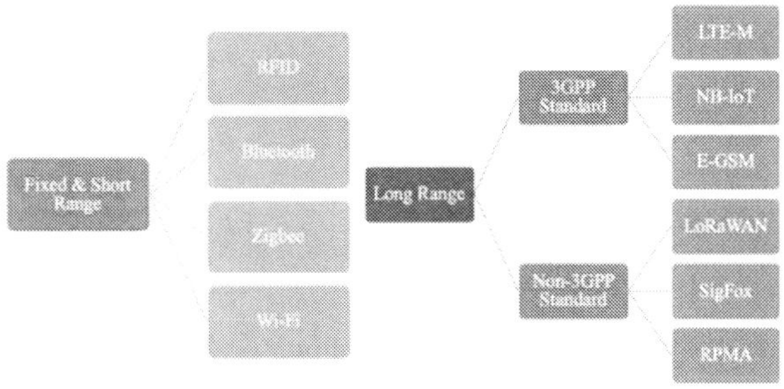

Figure 1. Shows the distribution of IoT wireless technology

One application of IoT technology is the application of the use of smart meters to electricity customers, which serves to record the consumption and use of electricity on a regular basis for monitoring and billing [8]. Figure 2 explains the comparison of traditional meter (electromechanical meter) systems with smart meters for electrical systems.

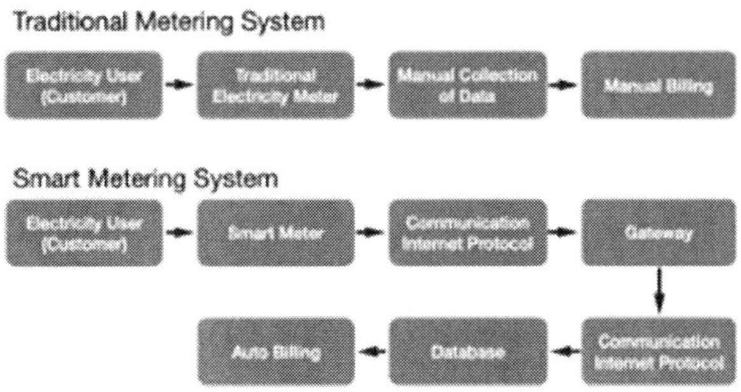

Figure 2. Traditional meter vs smart meter

According studies done by Analog Device Inc. USA find that accuracy from traditional electricity meter getting worst with the age of the metering. This is caused by several factor, such as environment, corrosion and mis-calibrated. By using smart meter, several advantages such as eliminate monthly manual reading, monitor electricity system real time, give responsive data for balancing electricity load, decreasing electricity stolen and eliminate loss happened from traditional electricity meter. Beside advantage, smart meter has disadvantage such as additional capex, personal training and implementation time smart meter.

From that disadvantage smart meter and many platform or protocol develop by organisation or institution such as SigFox, LoRa and NB-IoT, need to be choose one technology that best suited to be implement in smart metering in PLN Disjaya area. This research will do network planning for these three technologies that can covered PLN Disjaya area. From planning result, will conduct analysis coverage network, capacity and utilization network, cost benefit analysis, data rate speed and proprietary.

2. Research Methodology

In this research methodology using quantitative and study case selecting of IoT technology for smart meter in PLN Disjaya area, based on cost benefit analysis, starting from data collection, such as working area PLN Disjaya to get information about clutter classification and contour. Number PLN subscriber, income and exogen data, to do regression method and forecast PLN Disjaya subscriber and income. Continue with coverage and capacity prediction for every technology NB-IoT, SigFox and LoRa using Atoll software. PLN subscriber growth prediction using regression method. Next, calculation all implementation cost and all prediction benefit after using smart meter for every technology. Analysis compare IoT technology from coverage, capacity, cost benefit and economic evaluation.

3. Finding and Discussion

The research was conducted in 4 stepsin order to find IoT based technology for electric smart meter in PLN Disjaya area, which is the aim of this study, several findings were obtained:

3.1. PLN data collection and prediction

First, PLN Disjaya working area that cover all DKI Jakarta Province, Kota Bekasi, Kota Depok and some Kota Tangerang Selatan until Kota Tangerang. Plot this map into digital map from Badan Informasi Geospasial to get data clutter and contour. Figure 3 shown the map of PLN Disjaya working area and digital map clutter and contour. This data needed to calculated coverage prediction, since signal loss of electromagnetic wave depend contour and clutter classification.

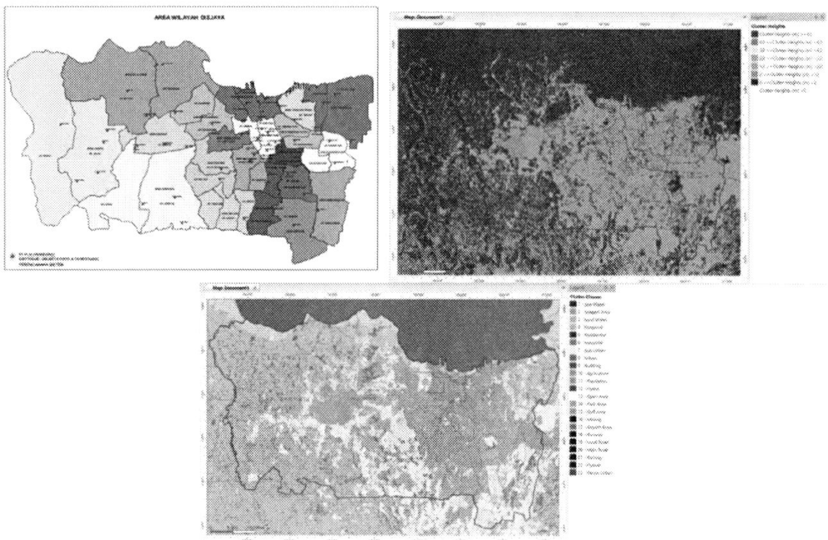

Figure 3. PLN Disjaya area map, contour and clutter classification

After that, we collect data for subscriber number and income from PLN Disjaya area starting from 2010 until 2018, there are 6 classification type of subscriber which are rumah tangga, industri, usaha and others. Figure 4 shown the subscriber PLN Disjaya based on type from 2010 to 2018.

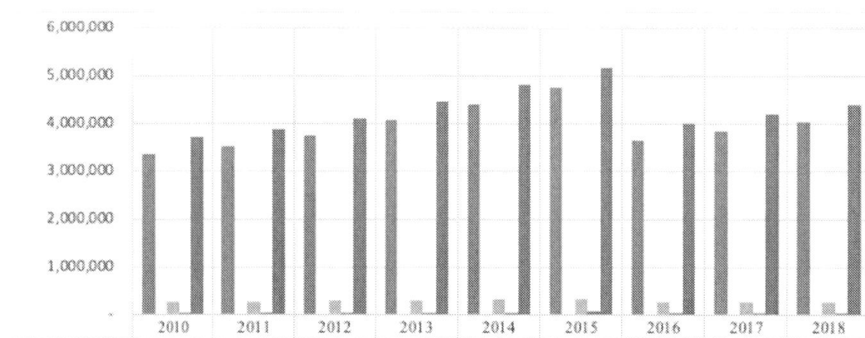

	2010	2011	2012	2013	2014	2015	2016	2017	2018
Rumah Tangga	3,382,334	3,516,485	3,742,965	4,083,307	4,394,705	4,736,763	3,668,360	3,856,239	4,040,310
Industri	10,770	11,001	11,129	11,374	11,509	11,644	6,382	6,266	6,134
Usaha	271,294	284,374	302,285	318,120	332,523	350,524	277,367	288,259	291,967
Others	56,266	57,068	59,615	62,863	64,853	68,313	52,873	54,601	56,655
Total	3,720,664	3,868,928	4,115,994	4,475,664	4,803,590	5,167,244	4,004,982	4,205,365	4,395,066

Figure 4. Data subscriber of PLN Disjaya area 2010 – 2018.

Figure 5 shown income PLN Disjaya based on type from 2010 until 2018. These data needed to forecast number subscriber and income PLN Disjaya area using multiple regression method.

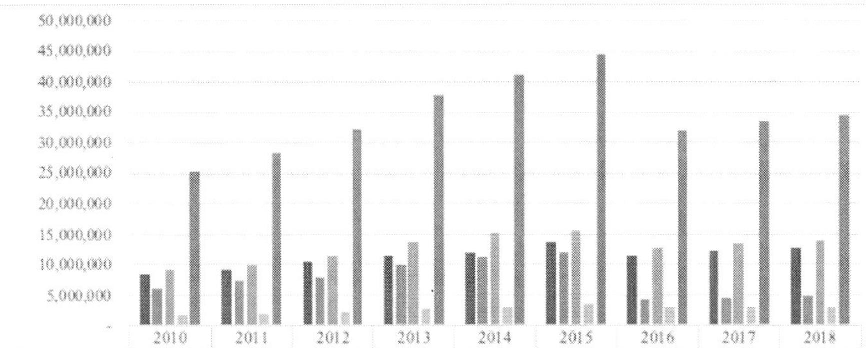

	2010	2011	2012	2013	2014	2015	2016	2017	2018
Rumah Tangga	8,428,878	9,284,464	10,546,39	11,501,57	11,955,38	13,731,38	11,598,48	12,350,97	12,685,95
Industri	6,167,578	7,262,265	8,015,787	9,907,075	11,121,24	11,886,02	4,396,444	4,668,625	4,878,220
Usaha	9,142,024	9,977,410	11,474,52	13,839,28	15,223,25	15,498,78	12,855,92	13,509,02	13,959,32
Others	1,671,636	1,953,235	2,181,313	2,682,772	2,979,990	3,428,653	2,993,201	3,016,333	3,124,278
Total	25,410,11	28,477,37	32,218,02	37,930,70	41,279,87	44,544,84	31,844,05	33,544,95	34,647,77

Figure 5. Data income of PLN disjaya area 2010 – 2018 (in million IDR)

Beside data above, data exogen also needed, such as total population in Jakarta province, Tangerang Selatan City, Tangerang Kabupaten, Tangerang City, Bekasi City and Depok City. Beside data total population, several data exogen such as growth production index, number of MSMEs, inflation and economic growth can be seen in table 1.

Table 1. Variable exogen data

Data Exogen	UoM	2010	2011	2012	2013	2014	2015	2016	2017	2018
Population	Million	241.8	245.1	248.5	251.8	255.1	258.4	261.6	264.6	267.7
Production Index Growth	%	4.82	4.1	4.12	6.01	4.76	4.76	4.01	4.74	4.05
MSMEs	Million	52.76	54.11	55.2	56.53	57.89	59.26	61.65	62.92	64.19
Inflation	%	6.96	3.79	4.3	8.38	8.36	3.35	3.02	3.61	3.13
Economic Growth	%	6.22	6.17	6.03	5.56	5.02	4.79	5.02	5.07	5.06

Simply linear regression model can be used to study relationship between two variables which are usually represented by y and x as representative of several population where y is the response variable or dependent variable and x is the control variable, the independent variable, or predictor variable. Simple equation in equation (1):

$$y = \beta_0 + \beta_1 x_1 + u \quad (1)$$

In addition to linear regression, there also multiple regression that used to see the relationship between the response variable and the predictor variable in which the predictor variable is owned by more than one. The multiple regression equation can be seen in the equation (2):

$$y = \beta_0 + \beta_1 X_1 + \beta_2 X_2 + \cdots + \beta_k X_k + u \quad (2)$$

By predicting the parameter coefficient values, OLS is obtained for multiple regression as equation (3):

$$y = \hat{\beta}_0 + \hat{\beta}_1 x_1 + \hat{\beta}_2 x_2 + \cdots + \hat{\beta}_k x_k \quad (3)$$

In this research, causal loop diagram of the relationship between each variable endogen and exogen can be shown in figure 6.

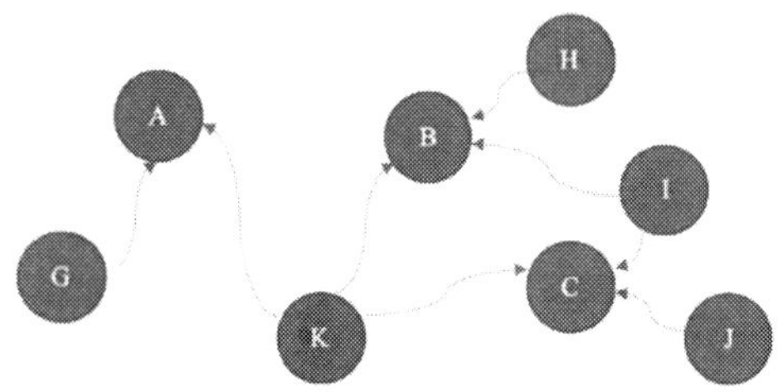

Variabel	Simbol	Keterangan	UoM
Endogen	*A*	Jumlah Pelanggan Rumah	Unit
	B	Jumlah Pelanggan Industri	Unit
	C	Jumlah Pelanggan Usaha	Unit
Eksogen	*G*	Jumlah Penduduk	Ribu
	H	Pertumbuhan Indeks Produksi	%
	I	Jumlah UMKM	Unit
	J	Inflasi	Unit
	K	Pertumbuhan ekonomi	%

Figure 6. Diagram causal loop connection between endogen and exogen data

After running regression with below condition in eViews software except for customer type others forecast with simple linear regression. Equation (4) shown for number subscribe rumah tangga, equation (5) shown for number subscribe industri and equation (6) shown for number subscribe usaha, where all equation validated with R-Square above 70%.

$$\ln A = 30.94804 - 1.337367 \ln G - 1.436916 \ln C \quad (4)$$
$$\ln B = 85.45697 + 0.20216 \ln H - 6.6603 \ln I - 2.093152 \ln K \quad (5)$$
$$\ln C = 29.67792 - 1.370846 \ln I + 0.015245 \ln J - 1.212393 \ln K \quad (6)$$

Table 2 shown the forecast data for PLN Disjaya subscriber in the year 2019 until 2028 for rumah tangga subscriber using equation 4, industry using equation 5 and usaha using equation 6, and for others using simple linear regression.

Table 2. Forecast PLN disjaya subscriber in 2019 – 2028.

Tahun	Rumah Tangga	Industri	Usaha	Lainnya	Total
2019	4,274,513	6,113	317,017	58,078	4,655,722
2020	4,361,031	5,686	324,549	57,846	4,749,112
2021	4,452,798	5,304	333,197	57,615	4,848,914
2022	4,549,885	4,961	342,974	57,384	4,955,204
2023	4,652,375	4,654	353,902	57,152	5,068,083
2024	4,760,366	4,376	366,010	56,921	5,187,673
2025	4,873,966	4,126	379,336	56,690	5,314,117
2026	4,993,300	3,899	393,925	56,458	5,447,583
2027	5,118,501	3,693	409,832	56,227	5,588,253
2028	5,249,718	3,506	427,115	55,996	5,736,334

Equation (7) shown for income rumah tangga, equation (8) shown for income industri and equation (9) shown for income usaha, where all equation validated with R-Square above 70 %.

$$\ln A1 = -15.834 + 0.220 \ln B1 + 2.867 \ln G - 0.002 \ln K \quad (4)$$
$$\ln B1 = 119.392 + 0.432 \ln H - 8.859 \ln I - 4.157 \ln K \quad (5)$$
$$\ln C1 = -17.880 + 0.323 \ln B1 + 2.674 \ln I + 0.075 \ln J - 0.184 \ln K \quad (6)$$

Table 3 shown forecast data for PLN Disjaya income in the year 2019 until 2028. All these data will be used to calculate number of smart meter than will be implemented (subscriber) and benefit of implemented smart mater (income).

Table 3. Forecast PLN disjaya income in 2019 – 2028 (in million idr).

Tahun	Rumah Tangga	Industri	Usaha	Others	Total
2019	13,882,564	5,675,076	16,088,009	3,617,617	39,263,265
2020	14,509,525	5,375,616	17,043,841	3,807,109	40,736,092
2021	15,172,921	5,114,701	18,094,079	3,996,601	42,378,302
2022	15,874,699	4,887,194	19,246,283	4,186,093	44,194,269
2023	16,616,930	4,688,820	20,508,828	4,375,585	46,190,162
2024	17,401,808	4,516,005	21,890,966	4,565,077	48,373,856
2025	18,231,664	4,365,754	23,402,894	4,754,569	50,754,881
2026	19,108,971	4,235,547	25,055,825	4,944,061	53,344,403
2027	20,036,346	4,123,259	26,862,079	5,133,552	56,155,237
2028	21,016,568	4,027,094	28,835,174	5,323,044	59,201,880

3.2. Coverage planning

In coverage planning, parameter determine from level of signal receiving power on the device or received signal strength indicator (RSSI). Equation to calculated the received power level or RSSI level is like in equation (7):

$$P_{rx} = P_{tx} + G_{system} - L_{system} - L_{channel} - M \quad (7)$$

Where, P_{rx} is receive signal level (RSSI) in dBm, P_{tx} is transmit power into the antenna, G_{system} is a combined signal amplifier in the system, L_{system} is loss caused by the system, $L_{channel}$ is signal propagation loss and finally M is Fading Margin.

To calculate the amount of loss caused by signal propagation the Cost-Hata equation is used. Cost-Hata loss propagation equation (8):

$$L = 46{,}3 + 33\log(f) - 13{,}82\log(hB) - a(hR,f) + [44{,}9 - 6{,}55\log(hB)]\log(d) + C \quad (8)$$

Where, for dense urban and urban:

$$a(hR,f) = \begin{cases} 8{,}29(\log(1{,}54hR))^2 - 1{,}1; 150MHz < f < 200MHz \\ 3{,}2(\log(11{,}75hR))^2 - 4{,}7; 200MHz < f < 1500MHz \end{cases} \quad (9)$$

While for sub-urban or medium population density:

$$a(hR,f) = (1{,}1\log f - 0{,}7)hR - (1{,}56\log f - 0.8) \quad (10)$$

And while for rural or less dense population:

$$L = L_{urban} - 2\left[\log\left(\frac{f}{28}\right)\right]^2 - 5{,}4 \quad (11)$$

Where L_{urban} is total loss propagation in urban area, hB is gateway antenna height, hR is receiver antenna height, d is propagation distance, f is frequency carrier and $a(hR,f)$ is correction factor antenna height. Especially for very densely population, C is 3 dB.

Using above equation, data contour and clutter classification, coverage planning for each IoT technology can be calculate using ATоll planning tolls with several adjustment parameter and specification from each technology as table 4.

Table 4. Parameter coverage planning IoT

Technology	Gateway Power Transmit (dBm)	Node Power Transmit (dBm)	Antenna Gain (dBi)	Node Gain (dB)	Antenna Height (m)
NB-IoT	26	23	11	2	30-40
LoRa	20	14	1-6	3	30-60
SigFox	20	14	6	2	30-60

The calculation result plan nominal coordinate gateway for each IoT technology can be shown in figure 7. From this result determine number of gateways needed to cover PLN Disjaya area, also average covered area for each technology gateway.

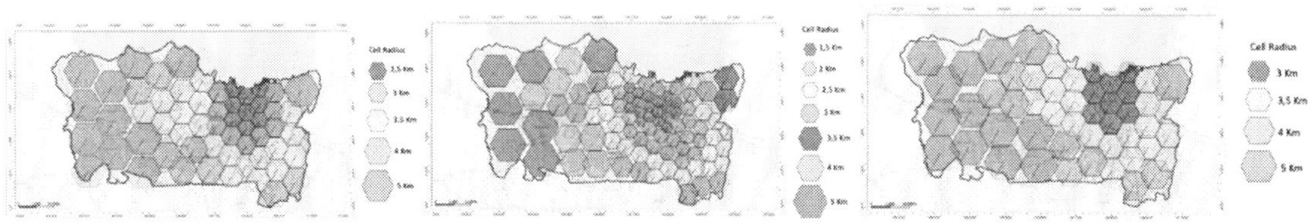

NB-IoT LoRa SigFox

Figure 7. Plan Gateway for PLN disjaya area for three technologies.

From table 5, shown that number of gateways needed SigFox less than NB-IoT and LoRa. Also, total area covered with RSSI uplink more than > -130dBm, SigFox higher than the others. This will be advantage, within the same covered area, the number gateway needed will be less.

Table 5. Coverage planning summary for smart meter iot in PLN disjaya

Technology	Covered Area (%)	Number of Gateway	Average Covered Area per Gateway (km)
NB-IoT	96,81	58	3.6
LoRa	93,90	92	2.5
SigFox	97.08	51	3.9

3.3. Capacity planning

In this research, each smart meter device sends data information with payload of 50 Bytes in an hour period. Network capacity should meet the requirement of spectrum for electricity smart meter in the area for the next ten years or 2028. Forecast number smart meter connected yearly can be used from table 5. Development full gateway assume need three years, where in first year 40% from total gateway and for the second and third year 30 % from total gateway needed.

For all three technology will depend on the number of devices connected in a network or gateway. From the number of devices serve in the network, knowing the maximum cell capacity and the number of cells, the utilization of the network can be known.

$$\text{Utilization } (\%) = \frac{\text{Total devices}}{\text{max cell capacity} \times \text{n cell}} \quad (12)$$

Table 6 shown the maximum capacity network, number device connected, number gateway, and utilization yearly for three technology. From this table, SigFox has the biggest maximum capacity compare to others.

Table 6. Network capacity and utilization NB-IoT, LoRA and SigFox

Year	Number PLN Customer	NB-IoT			LoRa			SigFox		
		Gate-way	Total Devices Covered	Network Utilization (Uplink)	Gate-way	Total Devices Covered	Network Utilization (Uplink)	Gate-way	Total Devices Covered	Network Utilization (Uplink)
2019	4,655,722	29	5,392,293	86.34%	46	5,439,639	85.59%	26	15,688,858	29.68%
2020	4,749,112	46	10,809,913	43.93%	74	9,863,217	48.15%	41	27,829,094	17.07%
2021	4,848,914	58	15,021,807	32.28%	92	13,112,638	36.98%	51	35,251,200	13.76%
2022	4,955,204	58	15,021,807	32.99%	92	13,112,638	37.79%	51	35,251,200	14.06%
2023	5,068,083	58	15,021,807	33.74%	92	13,112,638	38.65%	51	35,251,200	14.38%
2024	5,187,673	58	15,021,807	34.53%	92	13,112,638	39.56%	51	35,251,200	14.72%
2025	5,314,117	58	15,021,807	35.38%	92	13,112,638	40.53%	51	35,251,200	15.07%
2026	5,447,583	58	15,021,807	36.26%	92	13,112,638	41.54%	51	35,251,200	15.45%
2027	5,588,253	58	15,021,807	37.20%	92	13,112,638	42.62%	51	35,251,200	15.85%
2028	5,736,334	58	15,021,807	38.19%	92	13,112,638	43.75%	51	35,251,200	16.27%

With this result, SigFox has the lowest utilization compare to NB-IoT and LoRa, figure 8 shown the utilization graph yearly from 2019 – 2020 for three technology NB-IoT, LoRa and SigFox. This will be advantage to put more device beside electricity smart meter, such as health sensors, gas meter or water meter, since all three technologies capacity utilization still bellow 80% from maximum capacity to covered smart meter or sensor.

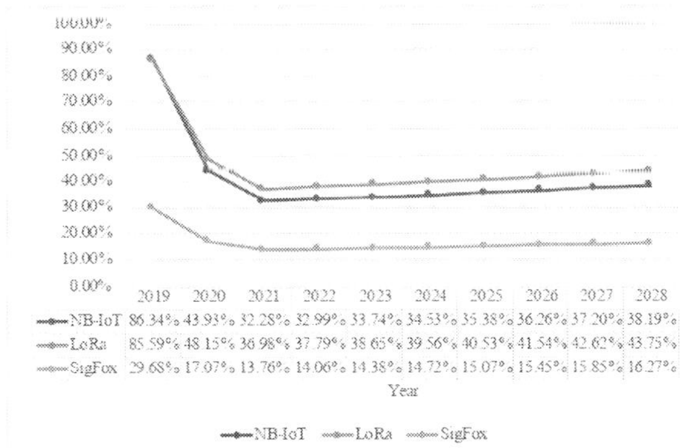

Figure 8. Utilization graph from 2019 – 2020 for NB-IoT, LoRa and SigFox.

3.4. Cost benefit analysis

In this research using cost benefit analysis, assuming 10 years (2019 until 2028) of implementation technology with forecasting using regression model for number of customers using data from 2010 until 2018. Several methods used in this research. Payback period (PP) method, this calculate how fast the return of the investment. Payback Period formula are shown as bellow

$$PP = \frac{CAPEX\ Investment}{Cash\ Flow} \tag{13}$$

Return on Investment (ROI) method, where to calculate percentage benefit from a project compare to cost that has been spend. ROI method using bellow formula

$$ROI = \frac{Total\ Benefit - Total\ Cost}{Total\ Cost} \quad (14)$$

Net Present Value (NPV) method, to calculate gap between investment value that has been spend with net profit upcoming years with set interest. NPV method using bellow formula

$$NPV = \sum_{i=1}^{t} \frac{Net\ Profit}{(1+r)^t} \quad (15)$$

Where t is year's period and r is interest.

Internal rate of return (IRR) method is to calculate the interest rate (r) which would equate the present value of the expected revenue received by the present value of the expenses for the investment project or when the NPV equal to zero. The value of IRR is calculated using a formula.

$$IRR = i_1 + \frac{NPV_1}{(NPV_1 + NPV_2)}(i_1 - i_2) \quad (16)$$

Depreciation method is to calculate the actual decrease in value of fair value of an asset, such as the decrease in value of factory equipment each year as it is used and wears, and second, the allocation in accounting statements of the original cost of the assets to periods in which the assets are used. In this research using straight line depreciation, where the formula as bellow.

$$SL = \frac{Total\ Cost\ Asset - Salvage\ Value}{Economic\ Life} \quad (17)$$

SL result can be shown as depreciation charge yearly.

In this research, benefit using smart meter compare to conventional meter, observe 396 existing customer using conventional metering, compare the monthly billing in February 2020 to March 2020 and discover that 33 customer has anomaly billing in March 2020 compare to February 2020. After that we check the existing conventional metering, found that 19 conventional metering not working properly. From these 19 customers, monthly billing in March 2020 less 1.23% form monthly billing in Feb 2020. This will be as based assuming benefit to be consider in the cash flow calculation 1.23% from income yearly PLN Disjaya. Where the forecast income in year 2019-2028 based on table 3.

Cost assumption will be divided by two, which is Capital Expenditure (CAPEX) and Operational Expenditure (OPEX). Table 7 shown breakdown for CAPEX and OPEX for these three-technology based on [10]. With the assumption implementation will be for ten years (2019 until 2028) with customer forecast based on the table 2, three-year implementation for network development, with configuration 40% from total gateway in first year, 30% in second year and 30% in third year. Another assumption is depreciation using straight line, lifetime 5 years and salvage value remaining 10%. All investment cost in first three years using loan from bank with the interest 6%.

Table 7. Cost assumption capex and opex technology iot

Cost	SigFox	LoRa	NB-IoT
CAPEX			
Equipment Cost	65,030,560	16,257,640	162,576,400
Installation Cost	20,641,667	20,641,667	20,641,667
Transmission Installation Cost	8,128,820	65,030,560	
Smart Meter Cost	153,773	170,031	218,803
Server & Application (Equipment & Installation) Cost	1,427,494,000	1,427,494,000	1,427,494,000
OPEX			
Site Leased Cost (Monthly)	10,000,000	10,000,000	10,000,000
Electricity Gateway Cost (Monthly)	1,250,000	1,250,000	1,250,000
Gateaway Connection Main & BackUp (Monthly)	6,120,000	6,120,000	6,120,000
Spectrum License Fee			2,354,400
O&M Cost (include spare part)	10%	10%	5%

With all those assumptions, then calculate for PP, NPV, BCR, ROI and IRR for these three technologies. Table 8 shown Comparison of Cost Benefit of IoT technology for Smart Meters at PLN Disjaya.

Table 8. Comparison of Cost Benefit of IoT technology for Smart Meters at PLN Disjaya.

Items	UoM	NB-IoT	LoRa	SigFox
NPV	IDR	891,110,453,323.56	1,091,142,456,200.17	1,440,159,425,136.23
BCR		2.65	2.59	2.81
IRR	%	16.30%	22.60%	29.83%
PayBack Periode	Years	7.23	6.53	5.88
ROI		2.24	2.02	2.29

From all those cost benefit parameters, found that SigFox technology has biggest NPV, BCR, IRR and ROI and the shortest Payback Period compare to LoRa and NB-IoT. One of the most influential factors in the cost benefit analysis is the price of the smart meter, this is because the number of smart meters used is very large, so a strategy is needed in implementing this smart meter.

4. Conclusion and Suggestions

4.1. Contribution

Based on all aspect data rate, coverage, capacity, network utilization and cost benefit analysis IOT technology from NB-IoT, LoRa and SigFox for electric smart meter IOT technology in PLN Disjaya area, SigFox technology has more advantage in coverage, capacity and cost benefit compare to NB-IoT and LoRa. NB-IoT only has advantage for the faster data rate and LoRa has advantage for proprietary (open source).

4.2. Suggestion

Additional smart meters that are used such as gas and water, or even sensors for health and other sensors. This is possible, because the predicted utilization of smart electricity meters in the 10th year of the three technologies is still less than 50% of the total network capacity if fully implemented in the third year. This can increase revenue for PLN Disjaya aside from the efficiency of electricity smart meters.

Acknowledgment

This work has been supported partially by PUTI Prosiding 2020 Grant funded by Directorate of Research and Development Universitas Indonesia under contract No. NKB-1146/UN2.RST/HKP.05.00/2020. Thank you.

Reference:

[1] Alansari Z, Soomro S, Belgaum M R, and Shamshirband S 2018 The rise of internet of things (IoT) in big healthcare data: review and open research issues *Advanced Computing and Intelligent Engineering* 675-685

[2] Al-Hunaity M F, Alshaer J, Dorgham O, and Farraj H 2015 Security model for communication and exchanging data in mobile cloud computing *International Journal of Computer Trends and Technology* **30** 138-146

[3] Alshaer J J 2016 Mobile object-tracking approach using a combination of fuzzy logic and neural networks *Global Journal of Computer Science and Technology*

[4] Belgaum M R, Soomro S, Alansari Z, Musa S, Alam M, and Su'ud M M 2017 Challenges: bridge between cloud and IoT *Engineering Technologies and Applied Sciences (ICETAS)* 1-5

[5] International Telecommunication Union 2012 Overview of the internet of things, series Y: global information infrastructure, internet protocol aspects and next-generation networks - frameworks and functional architecture models
[6] Gartner 2016 Gartner says 8.4 billion connected "Things" will be in use in 2017 up 31 percent from 2016
[7] Tabbane S 2017 IoT long range technologies: standards *International Telecommunication Union*
[8] Bimenyimana S, Asemota G N O 2018 Traditional vs smart electricity metering system: a brief overview *Marketing and Consumer Research* **46**
[9] Sforza F 2013 Communication system *US Patent US8406275 B2*
[10] Lin L 2019 Cost structure of iot connectivity services *KTH Royal Institute of Technology School of Electrical Engineering and Computer Science*
[11] Badan Pusat Statistik 2010 – 2019
[12] Direktorat Jenderal Ketenagalistrikan, Kementerian Energi dan Sumber Daya Mineral 2010 – 2018
[13] Sullivan W, Wicks E, Koelling C 2014 Engineering economy - sixteenth edition
[14] Giordano V, Onyeji I, Fulli G, Jimenez M S, Filiou C 2012 Guidelines for cost benefit analysis of smart metering deployment *Joint Research Center Institute for Energy and Transport*
[15] Department for Business, Energy & Industial Strategy 2019 Smart meter roll-out cost benefit analysis *UK Government*
[16] Pramudita D R, Nurcahyo R, Dachyar M 2019 Determinants of innovation strategy in indonesia telecommunication industry *Material Science and Engineering* **598**

IOP Conf. Series: Materials Science and Engineering **909** (2020) 012069 doi:10.1088/1757-899X/909/1/012069

Cost of quality system in passenger car plant: a methodology of implementation

W H Rahardjo[1*], **F Farizal**[1] **and D S Gabriel**[1]

[1] Department of Industrial Engineering, University of Indonesia, Indonesia

*Email: wuri.rahardjo@gmail.com

Abstract. The purpose of this research is to provide methodological adjustments or procedures for installation steps and the implementation of a quality cost system, including adjusting the quality cost calculation model for the automotive manufacturing industry, especially passenger car factories. A literature review approach is used and pilot studies are chosen for empirical aspects of research to enrich the mix of qualitative and quantitative data. Although the literature review shows that many experts write similar research, adjusting the quality cost model is still recommended. The results of the pilot study explain that the implementation of quality costs at an early stage is considered sufficient until reporting. Further research is advised to practice tracking quality costs. Research starts from preparation to reporting on quality costs. The scope of research is limited to quality costs in the operational area (production). The methodology for implementing a quality cost system can guide organizations to get an internal view of quality costs. This research confirms that organizations do need to modify the available quality cost models according to the organization's goals, needs, and environment. Thus, the simple models offered in this research are both valuable and contribute to quality economic literature.

1. Introduction

A highly competitive global business environment, forces producers to design and produce quality products and services and focus on meeting customer needs and expectations with costs that represent value for customers [1; 2]. Products that are well designed and produced correctly without errors may not be considered as quality products by customers if they do not meet their requirements [3]. This does not mean that there is no need to produce products or services that far exceed customer expectations, but that optimal production costs are always needed to meet these expectations [2]. Anyone who runs a company knows that these costs exist, but what they may not realize is how much of their expenses are directly related to quality costs [4]. Managers who are aware of expenses related to quality, make decisions that are different from managers who are not aware of expenses related to quality [5]. This has made quality cost reporting a useful tool in identifying potential areas for quality improvement, monitoring quality, and achieving optimal production costs so that businesses remain competitive [6].

However, previous research data shows that many industries around the world have not yet adopted quality cost reporting [7; 8; 9; 10]. For example, quality costs have not been widely adopted in Australia [11]. While 67% of companies in America and 66% of companies in Northeast England also have not yet implemented a quality cost system [12]. There are 85-90% of automotive manufacturing organizations in India not yet implementing quality cost systems [13]. 51 out of 84 or 60.71% of organizations in Malaysia have not yet implemented a quality cost system [14]. Data in the field shows that seven out of eight countries producing passenger car factories with the same production (brand X) have not yet installed and implemented a quality cost system. While the requirements of the ISO 9001 quality management system, ISO Technical Specification (ISO / TS) and International Automotive Task Force (IATF) 16949, clause 5.6 regarding management review that during management review an organization must monitor quality objectives, and report and evaluate the cost of poor quality regularly. It was found to be lacking in literature research that there is no quantitative evidence that shows an increase where ISO 9000 or IATF reduces quality costs. This makes there is a need to develop a quality cost system to complement the quality management system more significantly through quality cost training. In addition, to determine whether the quality management system is effective in improving the quality of work in the most profitable way, there is a need to test it through quality cost training.

Various reasons are cited for not adopting quality cost reporting and the most common reason is the lack of awareness and understanding of quality cost principles [10; 14; 11]. A number of reasons made the previous authors do a lot of related research regarding the steps of the installation procedure and the implementation of a quality cost system, including international standards. The process of installing and implementing a quality cost system is not always easy. Difficulties encountered during the process of installing and implementing quality costs one of which is the lack of literature sources regarding the methodology of installation and implementation of quality cost systems that approach the business practices of each organization. It is clear that each industry is required to modify the model or adjust the implementation model according to the needs and ease of application, which results in various quality cost structures [15; 16]. Likewise, to calculate the total quality cost. BS 6143-2 [17] has identified a list of quality cost elements under the categorization of Prevention-Appraisal-Failure. The list of elements only acts as a guideline for quality costing. Most quality experts say that a quality cost system must be tailored to each organization. Organizations that decide to manage their quality costs, must choose the appropriate model, which contains their quality cost elements and categories [18].

Therefore, this research proposes adjusting the model in the area of quality costs. One of them is the adjustment of steps or procedures in implementing a quality cost system and the second is the adjustment of categories and elements in estimating or measuring quality costs at passenger car factories.

2. Literature review

2.1. Literature related to research issues

The process of installing and implementing a quality cost system is not always easy. This process met with resistance from those involved because of the additional workload and from top management who might not be convinced of its usefulness. Many quality experts have realized this. Table 1 summarizes the research problems of some of the previous researchers. These problems require solutions so that the quality cost system can be widely applied to various industries throughout the world.

Table 1. Literature summary related to research issues

Researchers	Research Issues
Beamon [19] and Schmahl et al [20]	• Slow improvement. • Bureaucracy or complexity in a manner that maintains to deteriorate. • Changes in a single location generally tend to have large negative results in one or extra areas. • Management is in person concerned in quality problems solely throughout a significant crisis. • Management is coming up short on thoughts regarding how to reduce expenses further. *(continued)*

Researchers	Research Issues
Beamon [19] and Schmahl et al [20]	• Not all personnel are actively and in person concerned in pushing the organization's mission forward. • Many people and departments disagree concerning what ought to be the highest priority for the organization.
Pursglove and Dale [21]	• Lack of data that may be used for quality cost accounting functions. • Unrevised accounting system. • Lack of information concerning products that do not meet specifications for production.
Bottorff [22]	• Management does not assign adequate time and resources. • Because quality cost systems are typically not fully akin to existing product cost accounting systems, these systems usually require separate system administration. • Refinement of quality cost data does not directly make use of the data controversial. • By only using their own costs, or directly, quality can result in too little management investment in prevention activities. • Lack of coaching in quality concepts and group facilitation ways makes it troublesome to guide or participate in cross-functional groups. • The conception of quality is usually hampered by financial ideas. • When under pressure from the budgeting process, the exchange of dangerous quality costs is difficult to reject without strong commitment from top management.
Dale and Plunkett [23]	• Lack of resources and data; and • The present report is not in a structure that permits extracting quality cost data.
Roden and Dale [24]	• Company culture and employee attitudes towards quality cost systems do not seem to be contributory. • Lack of information and accountability makes it tough to gather quality cost data. • The complexness in the current method of accounting that cannot sort data by various divisions also makes it troublesome to measure quality cost data.
Dale and Wan [25]	• The victorious implementation of a cost of quality system is basically addicted to worker attitudes and enterprise culture. • There is no need to establish a new or sophisticated cost of quality system based on formal guidelines; simple systems can be very beneficial when integrated into existing enterprise practices. • It is also essential for personnel to aware that cost of quality is not an end in itself, but a means to an end. • The time to be had for personnel to participate in team conference related to cost of quality.
Rodchua [26]; Bamford and Land [27]; and Elridge et al., [28]	• Lack of data or trouble in gathering data. • Lack of commitment from the board of management. • Lack of understanding of quality cost principles.

(continued)

Researchers	Research Issues
Rasamanie and Kanapathy [29]	1. Lack of coordination with another departments. 2. Difficulties in gaining access to certain financial data. 3. Personalize the program (strong interest from only a few sponsors). 4. Difficulties with standardizing the company's quality cost system. 5. Lack of comparison opportunities, literature sources and consulting services. 6. Lack of top management support at the beginning stage. 7. Difficulties in identifying new quality improvement opportunities. 8. Identification of activities related to quality. 9. Difficulties with analyzing data collected. 10. Identification of quality cost items.
Chopra and Garg [13]	• Knowledge of how to measure this quality cost. • Understanding exactly what this information indicates after it is produced.
Trehan, Sachdeva, and Garg [15]	• Lack of management support. • The complexity of the manufacturing process. • Cannot be used for operational strategies. • Quality is part of organizational culture so measurement is not needed.

2.2. Literature related to methodology model of installation and implementation a quality cost system

Based on the problems above, it is not surprising to find so many experts in the field of writing quality costs about the installation process and implementation of quality cost systems. Table 2 summarizes the methodology models from some of the previous researchers. Thus, this research provides a solution through adjusting the methodology model of installation and implementing a quality cost system so that the quality cost system can be widely applied to various industries around the world.

Table 2. Literature summary related to research methodology model of installation and implementation a quality cost system

Researchers	Methodology Model of Installation and Implementation
Morse, et al. [30]	1. Obtain top management commitment and support 2. Form an installation team 3. Choosing an organizational segment as a prototype 4. Obtain cooperation and support of users and suppliers of information 5. Determine quality costs and quality cost categories 6. Identify quality costs in each category 7. Determine the source of quality cost information 8. Design cost reports and graphs 9. Establish procedures for collecting quality cost information 10. Collect data, prepare and distribute reports 11. Eliminating errors or bugs from the system 12. Expand the system
Low and Yeo [31]	1. Get support from top management 2. Form a quality cost committee 3. Determine quality costs 4. Determine the focus area 5. Get support from top management 6. Form a quality cost committee *(continued)*

Researchers	Methodology Model of Installation and Implementation
Low and Yeo [31]	7. Determine quality costs 8. Determine the focus area 9. Establish procedures for collecting quality costs 10. Get support from staff in the field 11. Training the personnel involved 12. Carry out, observe, and report regularly 13. Improve system costs 14. Expand system costs to include more operating areas
Mohanty and Tiwari [32]	1. Educate and get support from top management 2. Study of various quality cost approaches 3. Categorization of quality costs for the approach chosen 4. Identify quality cost elements according to IS / BIS 5. Determine and calculate cost elements 6. Establish methodology for data collection 7. Collecting quality cost data for each element 8. Prepare a summary of quality costs 9. Report and analyze for improvement 10. Presentation before top management 11. Establish a quality cost system and review and monitor
Tye, Halim and Ramayah [33]	1. Verify with factual costs 2. Get management commitment and support 3. Set up an installation team 4. Select organizational segments as prototypes 5. Conduct management presentations 6. Conduct a planned pilot program 7. Conduct education or training in all functions 8. Develop internal quality cost accounting procedures 9. Overall quality cost data collection and analysis 10. Reporting and use of quality costs
Chopra and Garg [13]	1. Form a quality cost team 2. Determine the quality cost base 3. Data collection related to quality cost elements 4. Placing quality cost elements in four types of categories; prevention costs, valuation costs, internal failure costs, external failure costs 5. Analysis of the difference in quality costs 6. Deciding how to reduce the level of quality costs 7. Presents the level of quality costs in terms of money 8. Changes approved by top management must be applied throughout the system
Chopra and Garg [1]	a) Model for calculating quality costs 1. Form a quality cost team 2. Determine the scope of work 3. Establish awareness among employees 4. Identification of quality related activities *(continued)*

Researchers	Methodology Model of Installation and Implementation
Chopra and Garg [1]	5. Formulate a methodology for determining the cost of money related to quality activities 6. Determine the cost of money for all these quality activities 7. Placing these activities under categories: prevention costs, valuation costs, internal failure costs, external failure costs b) Model for implementing quality costs 1. The current level of quality costs is taken as a reference and shown with a chart 2. Analyzing the current level of quality costs 3. The next plan of action is to reduce the current level of quality costs The quality cost team sends the report along with the proposed action plan to top management
Aniza [34]	1. Choose a process and determine its limits 2. Determine the resources and data needed 3. Collect the cost of each process using BS 6143 4. Analyze the results and implement improvements

3. Research Methodology and Implementation

3.1. Research Methodology

This research was conducted using qualitative and quantitative methods. Qualitative methods are used in literature studies and pilot studies to look at the problems and conditions of the implementation of the current quality cost system. In addition, literature studies are also used as material in making recommendations related to the theory of the implementation of quality cost systems and the theory of quality cost calculations. Quantitative methods are used in calculating quality costs. In this case, primary data were obtained from a passenger car manufacturing country, namely Indonesia. Data is collected and processed using a variety of methods, including discussion, communication, brainstorming, literature review references, Gantt Charts, spreadsheets, bar or pie charts. Stenecker [35] states that there are two approaches in implementing a cost system; that is, formal approaches and informal approaches. Stenecker [35] discusses more informal approaches because formal approaches have been well covered by other writers. The informal approach requires a minimum of management support. The aim is to conduct a pilot study to collect quality costs and then to present findings to top management. Ashford [36] believes that an informal approach is effective in proving the importance of quality costs.

3.2. Research Implementation

3.2.1. Case Study

The location of the research was conducted at a sole agent factory of a maker and assembler of premium scale passenger car products in Wanaherang Village, Gunung Putri District, Bogor 16965, West Java - Indonesia. The factory is headquartered in Stuttgart, Baden Württemberg, Germany and was established in Indonesia in 1970 which consists of two companies, namely a manufacturer and product assembler and main distributor responsible for marketing products. The company locally (until when this research was made) assembled six types of flagship vehicles in a 42-hectare factory along with the company that assembled commercial vehicles. The country of branch operations that has carried out the installation and implementation of quality costs is India. While the branch operational countries of Indonesia, Malaysia, Thailand, Vietnam, Brazil, Russia, and Egypt are branch operational countries that have not yet installed and implemented quality costs.

3.2.2. Adjustment Methodology Model of Installation and Implementation a Quality Cost System

The eight step model of adjusting the installation methodology and implementing the quality cost system offered in this research, has been adapted and implemented through a case study at the passenger car factory above. The activities for installing and implementing a quality cost system are shown in Table 3.

Table 3. Adjustment methodology model of installation and implementation a quality cost system in a passenger cars plant

Author	Methodology Model of Installation and Implementation
Rahardjo (2020)	1. Determine the area of installation and implementation of the quality cost system and determine its limits 2. Form a quality cost team 3. Arranging the activities of the installation and implementation of quality system costs 4. An initial introduction to the concept of quality costs 5. Understand together about the concept of quality costs with the quality cost team 6. Define and agree on quality cost elements 7. Collection of data sources and quality data costs 8. Finalization and reporting of quality cost systems

Step 1 - Based on references from ASQC Q9004-1 [37] regarding the main activities that have an impact on quality in the quality circle element, it is determined that the area to be installed and implemented in the quality cost system is the area of production or provision of services. Because this research in the manufacturing industry (factories that produce passenger cars), then the area is production. Step 2 - The quality cost team in the operational area is shown as in Figure 1. Step 3 - The composition of activities is made in the form of a Gantt chart as in Figure 2.

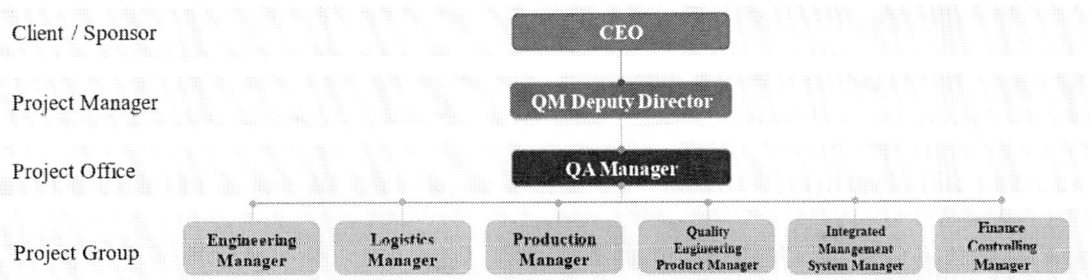

Figure 1. Quality cost team structure

Step 5 - A deeper understanding of the concept of quality costs is carried out by the quality cost team. Step 6 - Definition and collective agreement on quality cost elements results in an adjustment to the quality costing model as shown in Figure 3. Step 7 - The collection of quality cost data for 2019 is carried out through a spreadsheet instrument as in Table 4. Step 8 - Finalization and reporting of the 2019 quality cost system is displayed through pie and bar charts. A bar chart containing the trend of quality costs per month for 2019 is shown in Figure 4. Pie charts incorporated in the form of a dashboard containing the percentage of Preventive costs, Corrective costs, and Rejection costs are shown in Figure 5.

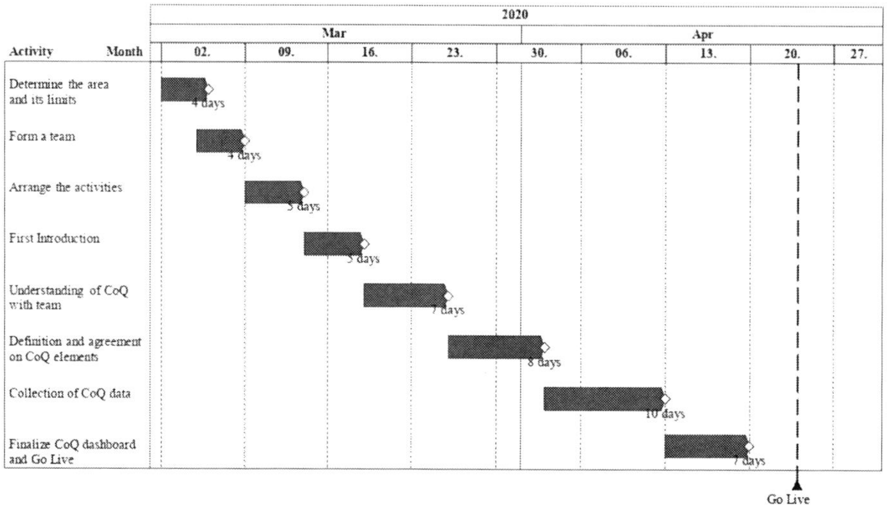

Figure 2. Installation activity and implementation of quality cost systems

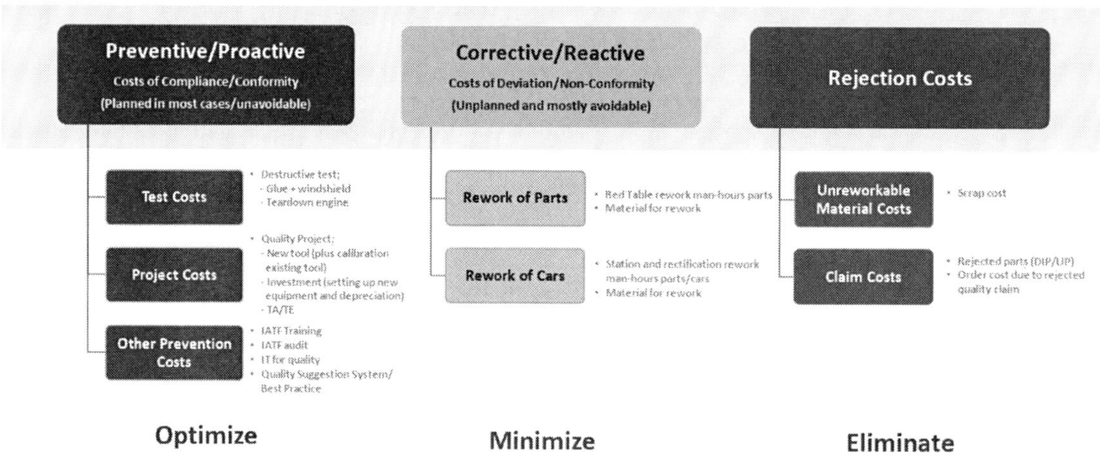

Figure 3. Adjustment of classification and quality cost elements

The cost of depreciation tools and equipment is the highest prevention cost, which is 54.78%. These costs include the cost of setting up new equipment and depreciation factor costs when the new equipment is used routinely. While one of the testing costs, Glue + Windshield, is the smallest prevention cost, which is 0.17%. While the cost of repairs, the cost of part rework is 57.84% and the cost of reworking the product in the repair area is 6.84%. Material ordering costs for damaged materials that are not accepted as quality claims to suppliers are the highest rejection costs, which is 84.73%. This cost does not include the cost of the damaged material itself. Because, the material cost is in the Rejected Parts element by 40.22%. Overall, the total quality cost of 2019 at the passenger car factory as a pilot study is shown that the Preventive fee is 58.90%, the Corrective fee is 1.25%, and the Rejection fee is 39.85%. The trend of quality costs per month in 2019 at passenger cars plant is shown that the average quality cost per product unit per month is 0.04% or 3,269,853 rupiah of the total quality cost in all operational areas.

Table 4. Quality cost data for 2019

No	Quality Cost Element	Total Quality Cost 2019
	Preventive/Proactive Costs (IDR)	**4.539.498.921**
	Preventive/Proactive Costs (IDR/Car)	**1.925.965**
	Destructive Test	
1	Glue + Windshield	7.823.757
2	Teardown Engine	110.666.727
	Project Cost	
3	New Tool and Equipment	10.370.000
4	Calibration Tool and Equipment	454.259.000
5	Depreciation Tool and Equipment	2.486.914.920
6	TA/TE	146.730.232
	Other Prevention Cost	
7	IATF Training	25.000.000
8	IATF Audit	69.500.000
9	IT for Quality	919.172.325
10	Quality Suggestion System/Best Practice	309.061.960
	Corrective/Reactive Costs (IDR)	**96.550.531**
	Corrective/Reactive Costs (IDR/Car)	**40.963**
	Rework of Parts	
11	Red Table Rework man-hours Parts	55.848.579
	Rework of Cars	
12	Station Rework man-hours Parts and or Cars	34.101.562
13	Rectification Rework man-hours Parts and or Cars	6.600.390
	Rejections (IDR)	**3.070.993.464**
	Rejections (IDR/Car)	**1.302.925**
	Un-rework-able Material Costs	
14	Scrap	32.525.311
	Claim Costs	
15	Rejected Parts (DIP/LIP)	1.235.013.818
16	Order Cost due to Rejected Quality Claims	1.803.454.335
	No. of Cars	**2357**
	Total Cost of Quality (IDR)	7.707.042.916
	Total Cost of Quality (IDR/Car)	3.269.853

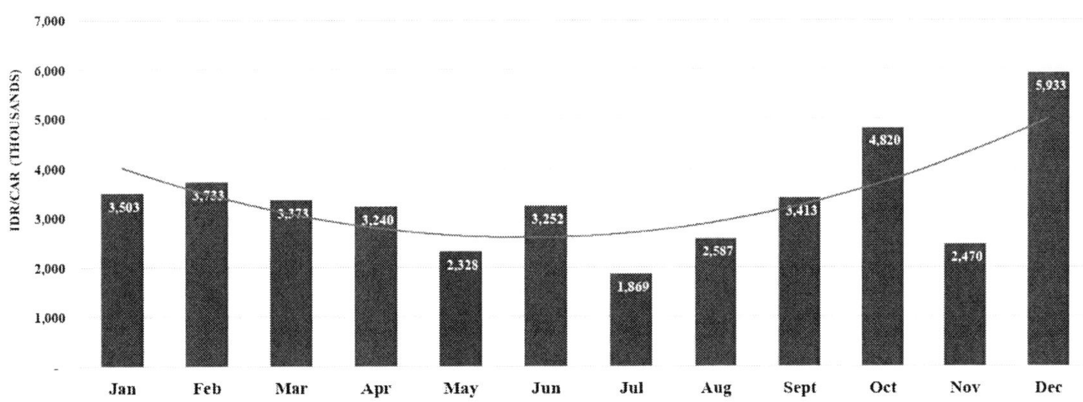

Figure 4. Display trends in quality cost per month in 2019

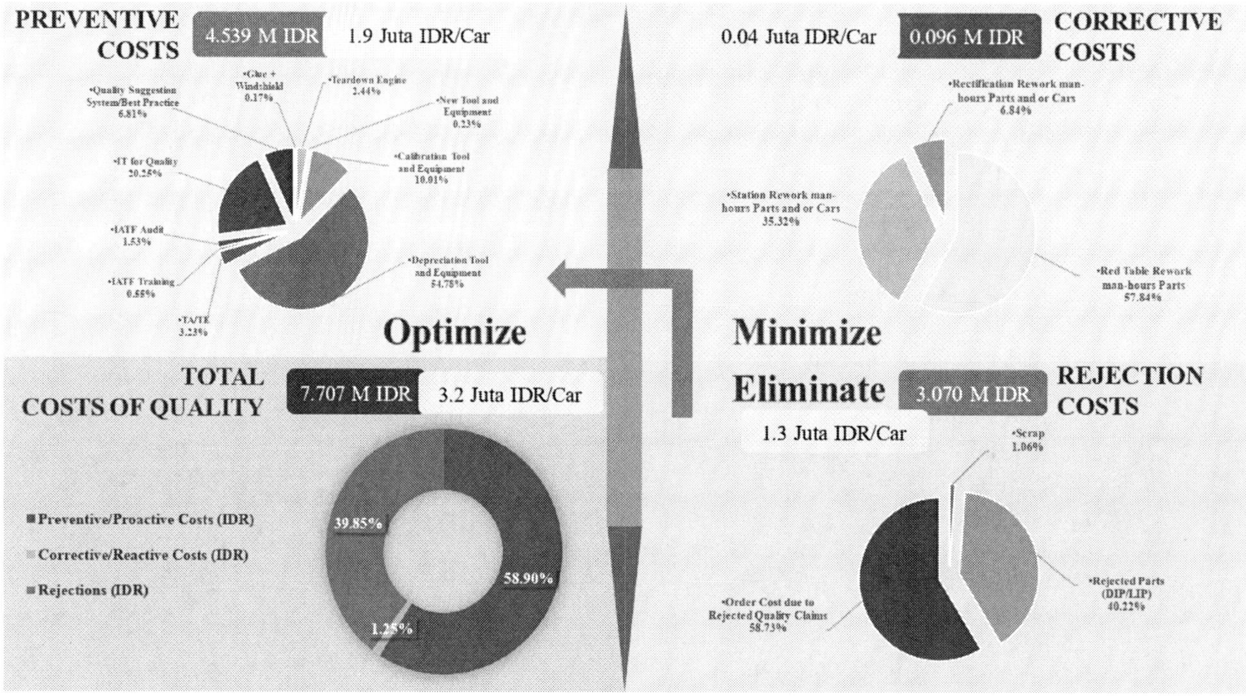

Figure 5. Dashboard cost of quality in passenger cars plant year 2019

4. Discussion and Finding

4.1. Discussion

This research established an adjustment of the installation methodology and implementation of a quality cost system at the passenger car factory. Although there is a lot of similar research as shown in the literature review, this methodology adjustment has its own novelty, differences and advantages for the automotive manufacturing industry. For example, the order of the process is different from the order of the results of previous studies. Although there are similarities between the sequence of processes from one author to another, differences still exist. Some authors begin the process of implementing quality costs by obtaining management commitment and support. There are also several writers starting with the establishment of a quality cost installation team. While the authors begin by determining the area of installation and implementation of the quality cost system and determine its limits. Then, a discussion of the methods used to carry out each process sequence. The majority of the literature discusses the definition of each process sequence or the purpose of each process sequence without further discussing how the sequence of processes is carried out. Although the method used is very simple, for example; discussion, or observation, or official meeting - discussion is needed in the literature, as a reference for organizations with an average expertise in this area of quality costs to still be able to implement quality costs.

Meanwhile, adjusting the quality cost category to Preventive-Corrective-Rejection. Preventive / Proactive is intended as a cost of compliance with quality requirements or costs of conformance to quality requirements. These costs are planned costs in most situations and conditions. This cost is also unavoidable. This cost should be optimized. The elements included are destructive tests, project costs, and other prevention costs. Corrective / Reactive is intended as a cost deviation of quality requirements or the cost of non-compliance with quality requirements. These costs are costs that are not planned in most situations and conditions. This cost is mostly avoided. This fee should be minimized. The element that includes it is rework. Rejection is nothing but a cost that should be eliminated. The elements included in this are scrap and part repurchase as well as part purchase costs are all irreparable losses.

The novelty and difference of the results of this research lies in its adjustment.

While the advantages of this research result lies in the discussion of how to make adjustments. In essence, each quality cost model has its own limitations due to the complex nature of the difficulties encountered. Thus, there is no one methodology model with the methods in it that are more important or more effective than the other methodologies - they all have their respective contributions to be played in each organization.

4.2. Finding

At the beginning of the construction of a new key performance indicator system, namely regarding quality costs, several things were done and some things were avoided, such as:

- The organization has clear objectives regarding the internal view of quality costs.
- The quality cost system starts from one area.
- The organization is not constrained by the traditional categorization of failure-assessment-failure-quality costs; the organization adjusts the categorization to be closer to standard business practices.
- Quality cost data collection starts with the easiest cost elements, for example internal failure costs.
- Quality data collection officers make friends with accountants from the finance department.
- Report only costs generated or supported by the finance department.
- As the first trial of a cost collection exercise, the quality cost team realized not to expect too much from the results of quality costing.

5. Conclusion

Installation and implementation of a quality cost system is not new for academic researchers. Many quality experts also realize that the purpose and use of quality costs are not always well understood by practitioners in the industry. Disclosing quality cost data is not to preach the sins of others who do not guarantee the victory of cooperation. Apart from that, although most industries realize the benefits of quality costs, only a few use those in organizations because of the difficulties encountered during the implementation process. Therefore, to help these industries, simple models are proposed to implement a quality cost system, which includes determining the quality cost elements and collecting quality cost data.

The methodological model with several methods in it is used by organizations to support the process of implementing quality costs. Each organization is recommended to make adjustments to their own quality cost systems through a pilot study approach - integrating into the organizational structure and accounting systems that exist in each organization. Thus, there is no one methodology model with the methods in it that are more important or more effective than the other methodologies - they all have their respective contributions to be played in each organization.

Products and services deemed quality by customers are products that meet customer expectations. However, that does not mean that manufacturers do not need to produce products and services that far exceed customer expectations. However, optimal production costs are always needed as costs that represent value for customers. Costs related to quality are costs that if managed properly provide a competitive and comparative advantage for the organization. A successful quality cost system, as a daily feature of the organization's management activities, takes a long time to build. It can take up to five years to achieve the credibility and expected benefits from the data displayed in the management information system. A measure of the status of quality cost organizations is if the feature is contained in a quality management system manual - which also complies with the requirements of the IATF management review clause for the automotive industry sector, both manufacturing and components.

6. Acknowledgement

This work has been supported partially by PUTI Prosiding 2020 Grant funded by Directorate of Research and Development Universitas Indonesia under contract No. NKB 1146/UN2.RST/HKP.05.00/2020. Thank you.

7. References

[1] Chopra A and Garg D 2012 Introducing models for implementing cost of quality system *The TQM J* **24** 498 – 504

[2] Teli S N, Majali V S, Bhusi U M, Surange V G 2014 Impact of poor quality cost in automobile industry *Int J Quality Eng and Technol* **4** 21 – 41

[3] Sailaja A, Basak P C, Viswanadhan K G 2018 Hybrid fuzzy MCDM model for effective utilisation of quality cost analysis in manufacturing firms *Int J Productivity and QM* **24** 219 – 241

[4] Teli S N, Majali V S, Bhusi U M, Surange V G 2012 Assessment of cost of poor quality for automobile industry *Int J of Eng Research and Applications* **2** 330-336

[5] Teli S N, Jagtap M M, Chanewar N 2017 Cost of quality applications and challenges: A review *Int J of Sci & Eng Research* **8**

[6] Setijono D and Dahlgaard J 2008 The value of quality improvements *Int J of Quality & Reliability Mgmt* **25** 292 – 312

[7] Yang C C 2008 Improving the definition and quantification of quality costs *TQM & Business Excellence* **19** 175 – 191

[8] Rapley C W, Prickett T W, Elliott M P 1999 Quality costing: a study of manufacturing organizations part 1: Case study and survey *TQM* **10** 85 – 93

[9] Oliver J, Qu W 1999 Cost of quality reporting some evidence of Australian *Int J of Applied QM* **2** 233 – 250

[10] Arvaiova M, Aspinwall E, Walker D 2009 An initial survey on the use of costs of quality programmes in telecommunications *The TQM J* **21** 59 – 71

[11] Wheldon B and Ross P 1998 Reporting quality costs: improvement needed Australian accountant **68** 54 – 6 in Rasamanie and Kanapathy The implementation of cost of quality reporting system in Malaysian manufacturing companies: difficulties encountered and benefits acquired *Int J of Business and Soc Sci* **2** 243-247

[12] Prickett T W and Rapley C W 2001 Quality costing: a study of manufacturing organizations part 2: main survey *TQM* **12** 211 – 222

[13] Chopra A and Garg D 2011 Integrated simple model for implementing CoQ program *Int J of Mgmt & Business Studies* **1** 57 – 60

[14] Sower E V, Quarles R, Broussard E 2007 Cost of quality usage and its relationship to quality system maturity *Int J of Quality and Reliability Mgmt* **24** 121 – 140

[15] Trehan R, Scahdeva A, Garg R K 2015 A comprehensive review of cost of quality *VIVECHAN Int J of Research* **6**

[16] Modhiya S and Desai D 2016 A review on cost of quality methodology and hidden costs in manufacturing industries *REST J on Emerging Trends in Modelling and Mfg* **2** 87 – 94

[17] BS 6143 1990 *Guide to the economics of quality: part 2, prevention, appraisal and failure model* (London: British Standards Institution)

[18] Alglawe A, Schiffauerova A, Kuzgunkaya O 2017 Analysing the cost of quality within a supply chain using system dinamics approach *TQM & Business Excellence*

[19] Beamon, B 1998 Supply chain design and analysis: models and methods *Int J of Prod Economics* **55** 281 – 294

[20] Schmahl K E, Dessouky Y, Rucker D 1997 Measuring the cost of quality: a case study in Teli, Majali, Bhusi, Surange Impact of poor quality cost in automobile industry *Int J Quality Eng and Technol* **4** 21 – 41

[21] Pursglove A B and Dale B G 1996 The influence of management information and quality

management systems on the development of quality costing *TQM* **7** 421 – 432

[22] Bottorff D 1997 COQ systems: the right stuff in Williams, van der Wiele, Dale Quality Costing:a management review *Int J of Mgmt Review* **1** 441 – 460

[23] Dale B G and Plunkett J J 1999 *Quality costing* 3rd ed (Hants: Gower Press)

[24] Roden S and Dale G B 2001 Quality costing in a small engineering company: issues and difficulties *The TQM Magazine* **13** 388 – 399

[25] Dale B G and Wan G M 2002 Setting up a quality costing system *Business Process Mgmt J* **8** 104 – 116

[26] Rodchua S 2009 Comparative analysis of quality costs and organizations sizes in the manufacturing environment *The QM J* **16** 34 – 42

[27] Bamford R D and Land N 2006 The application and use of the PAF quality costing model within a footwear company *Int J of Quality & Reliability Mgmt* **23** 265 – 278

[28] Eldridge S, Balubaid M, Barber D K 2006 Using a knowledge management approach to support quality costing *Int J of Quality & Reliability Mgmt* **23** 81 – 101

[29] Rasamanie M and Kanapathy K 2011 The implementation of cost of quality reporting system in Malaysian manufacturing companies: difficulties encountered and benefits acquired *Int J of Business and Soc Sci* **2** 243 – 247

[30] Morse W J, Roth H P, Poston K M 1987 *Measuring, planning and controlling quality costs* (Milwaukee: Institute of Management Accountants/ASQ Quality Press)

[31] Low P S and Yeo H K C 1998 A construction quality costs quantifying system for the building industry *Int J of Quality & Reliability Mgmt* **15** 329 – 349

[32] Mohanty R P and Tiwari M K 2007 Understanding costs of quality *The J of Eng Education*

[33] Tye L H, Halim H A, Ramayah T 2011 An exploratory study on cost of quality implementation in Malaysia: the case of Penang manufacturing firms *TQM & Business Excellence* **22** 1299 – 1315

[34] Aniza L A 2014 Development of quality cost models within a supply chain environment *Electronic Theses and Dissertations* 5267 https://scholar.uwindsor.ca/etd/5267

[35] Stenecker R G 1989 Attacking quality costs, in Campanella J *Quality costs: ideas and applications* a collection of papers (Milwaukee: Quality Press) **1** 117 – 126

[36] Ashford J L 1989 *The mgmt of uality in construction* (London: E & FN SPON an imprint of Chapman & Hall)

[37] ANSI/ASQC Q9004-1-1994 *Quality management and quality system elements-guidelines* (Milwaukee: ASQC Quality Press) 7 – 8

IOP Conf. Series: Materials Science and Engineering **909** (2020) 012070 doi:10.1088/1757-899X/909/1/012070

Evaluation of power plant reliability using index loss of load in the Suralaya power plant

Putro Ferro Ferdinant[1*] Laila Nurdiana[1] Ade Irman[1]
[1] Industrial Engineering Department, Faculty of Engineering, University of Sultan Ageng Tirtayasa

*Email: ferdinant@untirta.ac.id

Abstract. Suralaya Power Plant (SPP) is the largest capable electricity generation company in Indonesia, with a total installed capacity of 3400 MW. As the largest power plant in Indonesia, SPP holds a great responsibility as a supplier 17% of the electricity needs of the Java, Madura and Bali (JAMALI) systems. This makes the SPP also one of the national vital objects (*Obvitnas*). This paper will calculate the reliability index of the generator using the method of loss of load probability (LOLP) and loss of load expectation (LOLE) based on the value of the force outage rate (FOR). Based on data on failure and daily expenses in 2019, the reliability index of SPP for units 1-4 is 4,23 days per year or 1,23%. The higher values LOLP/LOLE are in 2019 are possibility of SPP has lost its burden for 2 days in a period of 2019 because of force majeure (blackout) which causes a reduced load. To improve power plant is that the electricity load in West Java can be fulfilled with the electricity flow.

1. Introduction

The power generation unit has a role as a generator of electricity for both household and industrial use. If a generating unit is damaged or in worst condition blackout happen, it can result in the unit not operating. If several large generating units experience this simultaneously, a load release or system loss can occur. If a generation unit often experiences load shedding, then the generation unit is not reliable in serving loads. Given the important role of generating units in the availability of electric power available, it is necessary to maintain the reliability of these generating units.

Reliability is the opportunity of an equipment to operate as planned well in a certain time interval and certain operating condition. The reliability of the electric power system is a measure of the level of service to the fulfillment of consumer electrical energy needs. There are four factors related to reliability, namely probability, working according to function, time period, and operating conditions [1,2].

The electricity generation system contained in the Suralaya power plant is a continuous process, which when a failure occurs in an engine component can result in the cessation of the machine so that the system's function is disrupted. Based on keruskaan data, it is known that during 2019 there were several engine failures which even caused the unit to stop operating. One of them is the leakage of boiler unit 3 in August. As the largest electric power generation service unit plus 50% of electricity demand produced at the Suralaya power plant and cover 17% of the electricity needs of the JAMALI system (Java Madura Bali). The Suralaya power plant should keep the generating unit in a reliable condition so that it can serve the needs of the power load required.

Based on the above problems, the company needs to know the probability that the operating units of the plant cannot serve the load. So that in the future the company can predict the needs of power reserves needed by consumers and can minimize the occurrence of things that are not desirable.

IOP Conf. Series: Materials Science and Engineering **909** (2020) 012070 doi:10.1088/1757-899X/909/1/012070

2. Methodology

In this paper to know the probability loss of load is using analytical technique from measures or indicators to find mathematical solutions. The system can be found resulted measures or indicators which appropriate to the evaluation technique based on assumptions and input data used in the models [6].

The probability of aggregate loss of capacity is calculated from basic probability principles and combining power of different generating unit from 1-4. The result can be presented in table of capacity outage probability. Another data needed is availability and unavailability of each generating unit.

The basic probability principles and combining the different generating units are used for calculating the probability of aggregate loss of capacity. The data and the results are usually represented in capacity outage probability tables. The success probability or availability and its complement, i.e., failure probability or unavailability of each generating unit are the input data.

In this paper is used the steps from define until suggestion. The flow chart research is presented in figure 1.

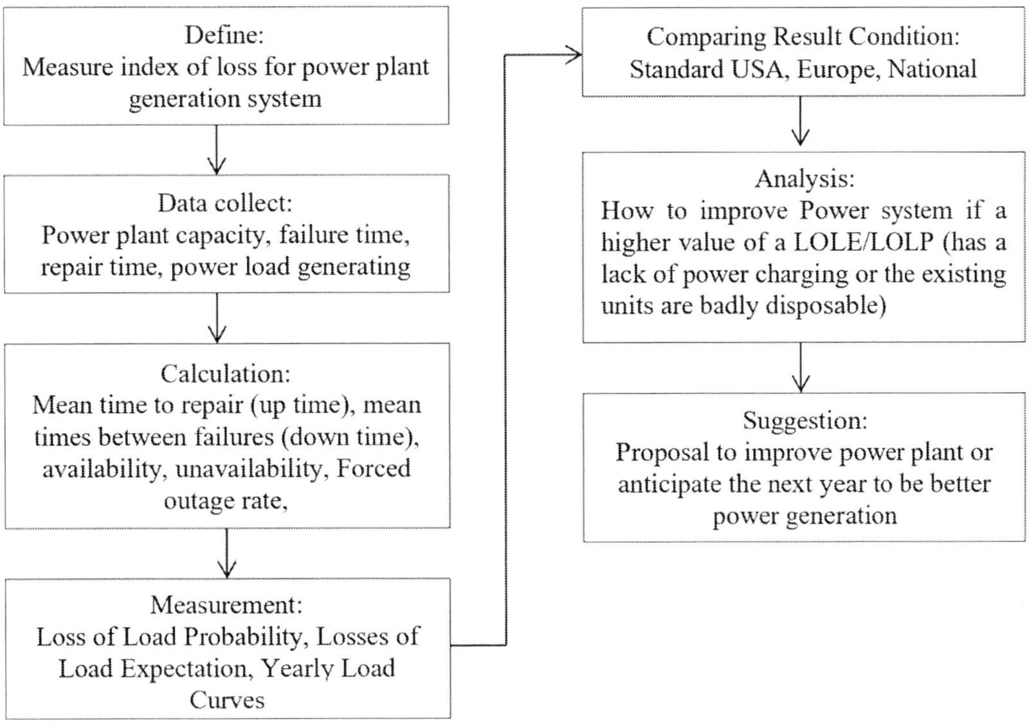

Figure 1. Flow chart

2.1. Availability and outage measures

In a Markov process, a component system especially in repairable electric power can be presented in two condition based on their life history there are functioning conditions (up) denoted by 1 and unavailable conditions (down) denoted by 0.[7].

For condition component fails, there is a transition from up state to down state. When a component repairs, there is a transition from down state to up state.

Thus when the component fails, it is said to undergo a transition from the as much as the down state, and conversely, whilst repairs are over, it is stated to return from the right down to the up state. This idea then allows to interpreting the idea of reliability in phrases of the fraction of overall time the aspect remains inside the up state. The duration of functioning length is also called the time-to-failure, and that of the duration under restore because the downtime.

The probabilistic technique to power system reliability analysis views the machine as a stochastic manner evolving over time. At any second the machine may alternate from one nation to another due to activities such as element outages or deliberate maintenance. Corresponding to a couple of states, say (i, j), there's a conditional opportunity of transition from the state i to the state j. All combinations of to be had and unavailable generating units are offered in tabular shape together with the calculated machine availability.

2.2. *Inherent availability*

The inherent availability, A_{inh}, is defined:

$$A_{inh} = \lim_{T\to\infty} A(T) = \frac{MTBF}{MTBF + MTTR} \quad (1)$$

The inherent availability is based totally at the failure distribution and repair-time distributions. It can therefore be viewed as an equipment design parameter, and reliability-maintainability trade-offs can be based totally on this interpretation [8].

2.3. *Loss of load probability*

A loss of load likelihood (LOLP) could be a probabilistic approach for determination of needed reserve. This approach examines the possibilities of synchronous outages of generating units that, along sides a model of daily peak-hour hundreds, confirm the quantity of days annually of expected capability shortages. [1]. Today, LOLP is that the most generally accepted approach within the utility business for evaluating generation capability needs. [2]

Loss of load happens whenever the system load exceeds the accessible generating capability. The LOLP is defined because the chances of the system load accessible generating capability underneath the belief that the height load is taken into account as constant through the day. The loss of load chance doesn't very signify a probability. It expresses statistically calculated price representing the share of hours or days in a very bound time-frame, once energy consumption cannot be coated considering the chance of losses of generating units. This point frame is typically one year, which might be painted as 100 percent of your time frame. In different words, the LOLP stands for associate expected proportion of hours or days p.a. of capability shortage. The LOLP truly doesn't signify a loss of load however rather for a deficiency of put in accessible capability.

The term LOLP is closely involving the term loss of load expectation (LOLE). If the interval used for the LOLP is expressed within the time units instead in proportion values, the LOLE is obtained rather than the LOLP. The generation system planners will appraise generation system dependableness and verify what proportion capability is needed to get a specified level of LOLP. As demand grows over time, extra generating units are enclosed during a means that the LOLP doesn't exceed the desired criterion. LOLP typically varies exponentially with load changes. Whereas the impact of random outages is evaluated probabilistically, regular outages are evaluated deterministically. Risk criteria comparable to proportion reserve and loss of target unit don't define systematically truth risk within the system.

2.4. *Loss of load probability definition*

Loss of one generating unit causes the expected risk of loss of power supply $E(t)$, with mathematical expectation defined as [2]:

$$E_i(t) = p_i . t_i \quad (2)$$

which p_i is the probability of loss of capacity, and t_i is the duration of loss of capacity in percent. Loss of load probability for the entire system is defined as a sum of all mathematical expectations for all units defined as [2]:

$$LOLP = \sum_{i=1}^{n} p_i . t_i \tag{3}$$

2.5. Loss of load probability during scheduled outages

The planner of power generation should to schedule planned outages throughout the year, as a results of the generating units ought to be typically maintained and inspected. Short maintenance technique is often updated. If a generating unit experiences a long-forced outage, the annual maintenance schedule for the power system could also be reshuffled to any improve system dependableness and to decrease the power system production costs [3].

Planned outage needs of power plants sometimes have a circular pattern [4]. The procedure schedules the maintenance of generating units, so accessible generation capability reserve is that the same for all weeks. This sort of procedure has the bottom LOLP.

2.6. Loss of load expectation

The loss of load expectation represents the chance that aggregates won't be ready the mandatory power consumption. The term LOLE is closely concerning the term LOLP. If the quantity used for the LOLP is expressed within the time units instead in proportion values, the LOLE is obtained rather than the LOLP. Figure one shows yearly load diagram [2].

The limit value of LOLE for a reliable provide is ten hour annually. In some European countries, the limit also can be settled between four and eight hour annually [2]. Most of the U.S. electrical power utilities are designed on the technical assumption that the overall accumulated time of provide interruptions (forced outages) ought to be no over one day in ten years for the developed countries [7]. In Indonesia with National power service commonplace could be a most of three annually for the Java-Bali electrical power system and five days annually for systems outside of Java [5].

Power system with a higher value of a LOLE contains a lack of power charging or the present units are badly disposable.

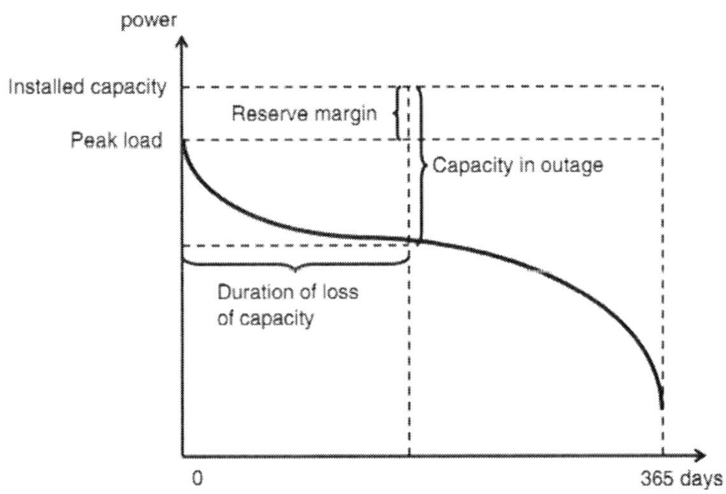

Figure 2. Yearly load diagram
[Source: Cepin (2011)]

2.7 Loss of load expectation definition

Loss of load expectation is obtained mistreatment the daily peak load variation curve. A selected capability outage contributes to the system by associate quantity up to the merchandise of the chance of existence of the actual outage and therefore the variety of time units (week, month or a year). The best application is that the use of the curve on yearly basis. Once employing a daily peak load variation curve on annual basic, the LOLE is in days each year [2].

$$LOLE = \sum_{i=1}^{n} p_i . t_i \quad (4)$$

where p_i is that the individual likelihood of capability in outage and t_i is the period of loss of power offer in days. Once the cumulative probability P_i is used, LOLE is defined as [2]:

$$LOLE = \sum_{i=1}^{n} P_i(t_i - t_{i-1}) \quad (5)$$

LOLE is additionally defined with a likelihood that consumption L would not throughout operating power capability C. [2]

$$LOLE = \sum_{i=1}^{n} P_i(C_i - L_{i-1}) \quad (6)$$

3. Result and Analysis

3.1. Power plant generation system and failure data

This research was conducted at 4 power plants in Suralaya Power Plant, namely Unit 1-4. Before calculating the generator reliability index, it must be known in ability the power capacity of each generator. The following is data on the capacity of generating units 1-4 in the Suralaya Power Plant.

Table 1. Power plant generation system

Unit	Capacity (MW)	Unit	Capacity (MW)
1	400	3	300
2	400	4	400
		Total	1600

In addition, it is also necessary to know the failure data from each unit in order to calculate the unavailability or force outage rate (FOR) value of each unit. Each unit has a primary machine there are coal feeder, damper, mill, turbine and boiler.

Table 2. Failure data

Unit	MTBF	MTTR	Availability	Unavailability
1	13564	7	0,999484194	0,000515806
2	2473	10	0,995972614	0,004027386
3	3136	4	0,998726115	0,001273885
4	2055	30	0,985611511	0,014388489

3.2. Loss of load probability calculation

The power system is composed of three generating units: 1, 2, 3 and 4, with their power capacity is same respectively 400 MW. The failure chance of every unit is given and also the success probability of each unit is given. The choice for unit failure chance is that the forced outage rate (FOR) or the inaccessibility of the unit. The choice for unit success chance is that the convenience of the unit.

Table 3 shows the outage state enumeration tables, which enumerates all of these states and the probability of each. For example, the probability that no generating unit is in outage, i.e., all units 1, 2, 3 and 4 are in service, is the product of the success probabilities of units 1, 2, 3 and 4. The probability of this state is the product of the unit failure probability for unit 1 multiplied with unit success probabilities of units 2,3,4. The result in monotonically ordered by increasing order of overall capacity of units in outage.

Evaluating the probability of not to supply a 400 MW load demand. The capacity of the four unit system is 1600 MW, the load could not be supplied if capacity of 400 MW or more is on outage 400 MW. The cumulative probability of 400 MW of more in outage is calculated as a data is presented from Table 4 is 0,000505692 + 0,003962337 + 0,001249854 + 0,014304903 = 0,020114119.

Table 3. Calculation outage state enumeration

Units in Outage	Unit in Service	Capacity in Outage (MW)	Capacity in service (MW)	Probability
-	1,2,3,4	0	1600	0,979885881
1	2,3,4	400	1200	0,000505692
2	1,3,4	400	1200	0,003962337
3	1,2,4	400	1200	0,001249854
4	1,2,3	400	1200	0,014304903
1,2	3,4	800	800	0,00000204
1,3	2,4	800	800	0,00000065
1,4	2,3	800	800	0,00000738
2,3	1,4	800	800	0,00000505
2,4	1,3	800	800	0,00005784
3,4	1,2	800	800	0,00001825
1,2,3	4	1200	400	0,000000002608
1,2,4	3	1200	400	0,000000029852
1,3,4	2	1200	400	0,000000009416
2,3,4	1	1200	400	0,000000073781
1,2,3,4	-	1600	0	0,000000000038
			Sum	1

Table 4. Contributions to loss of load probability

Capacity in Outage (MW)	Probability of determined capacity or more in outage	Time interval of Capacity in Outage (%)	Contributions of loss of load probability
0 or more	1	0	0
400 or more	0,020114119	61%	0,012178685
800 or more	0,00009133	98,63%	0,0000900812
1200 or more	0,000000115695	100%	0,0000001157
		Sum LOLP	0,012268882 or 1,23%

Table 3 is extended to Table 4 by adding the time intervals of capacities in outage and the contributions to find result of LOLP. To find the time interval, first, it is need the load data per day from one year. The load per day is presented in figure 2. Second, the load per day is sorting monotonically to get yearly load curve. In this curve we can find capacity outage and duration of loss capacity. The capacity outage is finding from difference between installed capacity and load capacity. The completed yearly load curve is presented in figure 3. The duration of loss capacity is finding from interval from first time until intersection of capacity outage at that time. For example for 400 MW outage or more we get time duration is 221 days or equivalent 61% of all of time interval (365 days).

Finally, The LOLP is the sum of contributions of loss of load probability of the Table 4. The LOLP for in this case is 0,012268882 or 1,23%.

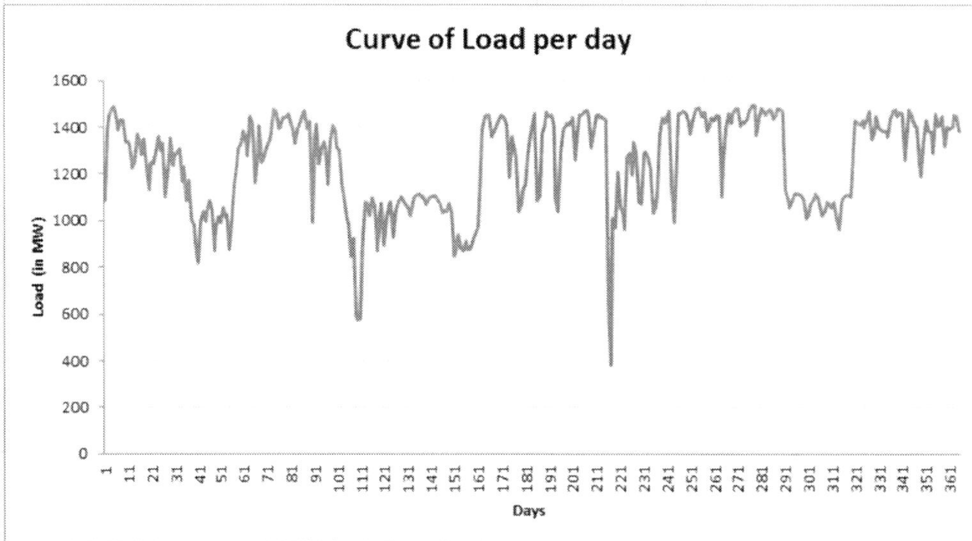

Figure 3. Curve of load per day

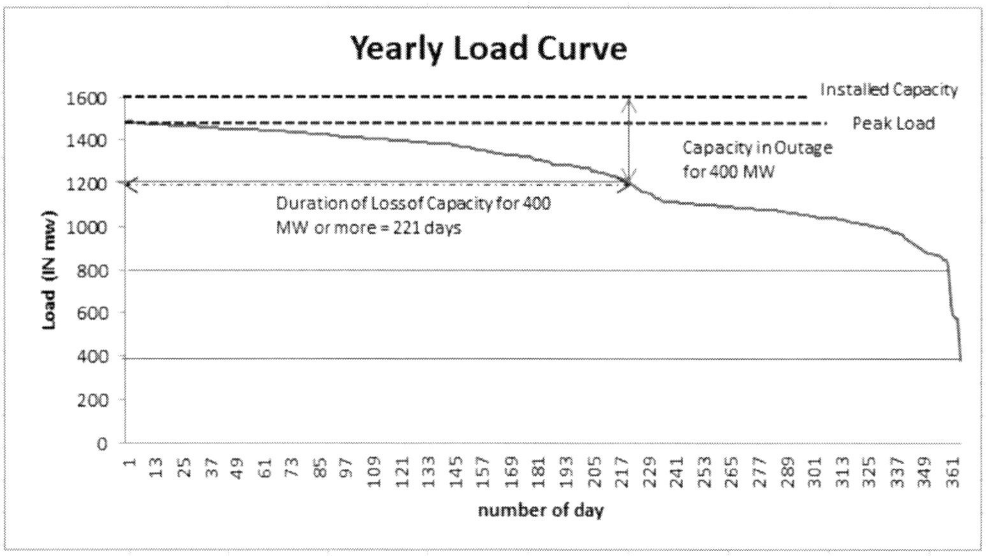

Figure 4. Yearly load curve

3.3. Loss of load probability calculation

The loss of load expectation represents the likelihood that aggregates won't be ready the required power consumption. The term LOLE is closely relating to the term LOLP. If the interval used for the LOLP is expressed within the time units instead in proportion values, the LOLE is obtained rather than the LOLP.

In LOLE calculation, the procedure is representing the power system in tables, including capacities, unavailabilities, and availabilities. A daily load diagram is given in figure 5.

The suralaya power plant generation system in this paper consists of four units from ten of the total units operated. Table 1 shows the corresponding generation system data. Figure 5 shows a daily load diagram. Sixteen combinations of generating units in outage power or in service. Notation 0 stands for outage of corresponding unit in the and notation 1 stands for operating unit in service. The completed outage stages are presented in table 5.

Table 3 is presented all unit outage combination probabilities are calculated and represented as cumulative probabilities. LOLE index is calculated from multiplication of expected load duration and cumulative outage probability for required state. The LOLE for in this case is 4,48 days.

Table 5. The outage stage

Unit 1	Unit 2	Unit 3	Unit 4	Capacity in Outage (MW)	Capacity in service (MW)	Probability of each capacity	Cumulative Probability
1	1	1	1	0	1600	0,979885881	1
1	1	1	0	400	1200	0,000505692	0,020114119
1	1	0	1	400	1200	0,003962337	0,019608427
1	0	1	1	400	1200	0,001249854	0,01564609
0	1	1	1	400	1200	0,014304903	0,0144
1	1	0	0	800	800	0,00000204	0,00009133
1	0	1	0	800	800	0,00000065	0,00008929
0	1	0	1	800	800	0,00000738	0,00008864
0	0	1	1	800	800	0,00000505	0,00008126
0	1	1	0	800	800	0,00005784	0,00007621
0	1	0	1	800	800	0,00001825	0,00001836
1	0	0	0	1200	400	0,000000002608	0,000000115695
0	1	0	0	1200	400	0,000000029852	0,000000113087
0	0	1	0	1200	400	0,000000009416	0,000000083235
0	0	0	1	1200	400	0,000000073781	0,000000073819
0	0	0	0	1600	0	0,000000000038	0,000000000038

Table 6. Contributions to loss of load expectation

Capacity in Outage (MW)	Probability of determined capacity or more in outage	Time interval of Capacity in Outage (days)	Contributions of loss of load expectation
0 or more	1	0	0
400 or more	0,020114119	221	4,445220191
800 or more	0,00009133	360	0,032879629
1200 or more	0,000000115695	365	0,00004223
		Sum LOLE	4,48 days

3.4. Comparison of index

From the result of LOLP and LOLE calculation, it can be compared with other standard. In this comparison, The Suralaya Power Plant index is below standard from European Countries and US Electric, but is above of developed country like India. If it is compared other power plant in Indonesia still relevant and argue with standard.

Table 7. Comparison of limit index LOLP/LOLE

Condition	Limit Index LOLP/LOLE
Suralaya Power Plant unit 1-4	LOLP=1,23% LOLE=4,23 days per years
Indonesia National Power Company (PLN)	3 days per year (Java-Bali) 5 days per year (Outside Java-Bali)
Cepin (2011)	10 h per year
Some European Countries	4-8 h per year
Most U.S Electric power utilities	No more than 1 day in 10 year
Developed country (Vijayamohanan, India)	20-25 days a year

4. Conclusions

The results obtained in this paper explain that the calculation of the generator reliability index at the Suralaya power plant using the loss of load probability (LOLP) method in 2019 obtained a value of 1,23% and loss of load expectation is 4,23 days. This condition is still accepted as a standard national power company. The reason why the higher values LOLP/LOLE are in 2019 are possibility of Suralaya PLTU has lost its burden, namely for 2 days in a period of 1 year. First in there was a decrease in load on April 10, 2019 to 576 MW. Second the cause of the blackout on August 4, 2019 was the blackout of Two SUTET circuits Ungaran-Pemalang 500 KVA spread to Depok-Tasikmalaya SUTET so that the areas of West Java, DKI Jakarta and Banten were extinguished simultaneously. At that time the load at the Suralaya power plant dropped to 384 MW. To avoid blackouts from happening again, PLN is currently regulating the flow of electricity from Java 7 and Java 8 to West Java. The goal is that the electricity load in West Java can be fulfilled with the electricity flow. In addition it is expected that the company can arrange maintenance schedules that are suitable to the usage load. In terms of generating capacity it is sufficient, only because of force majeure (blackout) which causes a reduced load which increases the LOLP / LOLE index. To Improve transmission lines is generally directed towards achieving a balance between generating capacity and power demand efficiently. In addition, it is also an effort to overcome channeling bottlenecks, improve service voltage and operating flexibility on subsystems used for balancing between subsystems.

5. References

[1] Cepin M. 2002 Optimization of safety equipment outages improves safety *Reliability Engineering System Safety* **77** 71–80
[2] Cepin, M. 2011 Assessment of power system reliability, methods and applications *Springer-Verlag London Limited*
[3] Syahrial, Sawitri.K., Gemahapsari, P. 2017 Studies on the reliability of the availability of power plants in the x regional network *ELKOMIKA Journal* **5** 93–105
[4] Apriyani, R., dan Thayib, R. 2015 Perhitungan loss of load probability (probabilitas kehilangan beban) sistem tenaga listrik di PT Pupuk Sriwidjaja *Jurnal Mikrotiga* **2** 22 – 27
[5] Gunadin, I.C., Muslimin, Z., dan Sudrajat, E. 2014 Studi keandalan ketersediaan daya perencanaan pembangkit listrik PT PLN sistem SULSELBAR tahun 2010-2020 *Semiar Nasional Riset Inovatif II*
[6] Billinton R, Allan R 1996 Reliability Evaluation of Power Systems *Plenum, New York*
[7] Vijayamohanan Pillai N. 2015 Loss of load probability of a power system *Journal of Fundamentals of Renewable Energy and Applications* **5**
[8] Ebeling, C,E. 2009 An introduction to reliability and maintainability engineering *3ed. McGraw Hill Companies.*

Sentiment analysis of tokopedia application review to service product recommender system using neural collaborative filtering for marketplace in Indonesia

Restu Meifitrah[1*], Irfan Darmawan[1], Oktariani Nurul Pratiwi[1]

[1] Information System, Faculty of Industrial Engineering Telkom University, Bandung

*Email: restumeifitrah@student.telkomunivesity.ac.id

Abstract. Tokopedia is one of the leading e-commerce companies in Indonesia and ranks second in the top 10 e-commerce in Indonesia in 2018 based on Statista data. Large companies like Tokopedia need to find out what users think of the products or services offered. User opinions on the Tokopedia application can be seen in the review column on the Google Play Store, but processing a review is not easy. To overcome this we need a sentiment analysis method or technique. Sentiment analysis is performed using the Naive Bayes algorithm. The results of positive sentiments obtained are used as a reference to maintain service quality and the results of negative sentiments can be used as an evaluation material in improving Tokopedia services and applications.

1. Introduction

1.1 Preliminery

Tokopedia is the most popular technology company in Indonesia with a mission to achieve economic equality digitally. Tokopedia allows individuals and business owners to open and manage online stores easily and for free. Tokopedia ranks second in the top 10 e-commerce sites in Indonesia still have the opportunity to occupy the first position and can continue to maintain its popularity by increasing the quality of service. To improve the quality of its services, Tokopedia can take advantage of user reviews on the Tokopedia application found on Google Playstore. Reviews contain lots of suggestions, praise and even user complaints. These reviews can be used as input for the Tokopedia service. To find out the tendency of comments and information contained in a review is not easy because it requires a method that can process the data quickly and automatically to categorize the reviews both positive and negative. The method that can be used to determine the tendency of comments in the review is to do sentiment analysis. To be able to do sentiment analysis of Tokopedia application user reviews, appropriate steps and algorithms are required.

To do sentiment analysis of Tokopedia application user reviews, appropriate methods and algorithms are needed. The algorithms used to do sentiment analysis are very much, which are often used include the Naive Bayes Classifier, Support Vector Machine, Decision Tree. Each algorithm has a different level of accuracy in each case. In a previous study by [1] about the Personality Classification of Twitter users found different levels of accuracy, are Naive Bayes 63%, KNN 60%, and SVM 61%. In research [2] about product analysis sentiments using machine learning by comparing several algorithms, namely Naive Bayes, SVM, and MaxEnt, the results were 81.33%, 78.67% and MaxEnt classifiers 72%. The advantage of Naive Bayes compared to other algorithms is that simple and intuitive Naive Bayes can produce good performance with little testing data [3]. For maximum results, a TF-IDF calculation is required. Based on considerations and observations from previous studies the author decides to do a sentiment analysis using the Naive Bayes algorithm.

1.2 Theoretical basis

1.2.1 Sentiment analysis

Sentiment Analysis or commonly called opinion mining is one branch of Text Mining research. Sentiment Analysis can be distinguished based on the data source, several levels that are often used in Sentiment Analysis research are Sentiment Analysis at the document level and Sentiment Analysis at the sentence level. Based on the level of data sources Sentiment Analysis is divided into 2 major groups namely Coarse grained Sentiment Analysis and Fine Grained Sentiment Analysis [4].

1.2.2 TF-IDF

The TF-IDF method is a method for calculating the weight of each word that is most commonly used in information retrieval. This method will calculate the value of Term Frequency (TF) and Inverse Document Frequency (IDF) on each word (token) in each document in the corpus [5].

$$\text{Tf-idf}(t,d) = \text{tf}(t,d) * \text{idf}(t)$$

1.2.3 Naive bayes algorithm

Naive Bayes Classifier is a simple probability classification method in which to apply the Bayes theorem with high independent assumptions. Referring to the basic concept of Naive Bayes classifier namely Bayes' Theorem which was first stated by Thomas Bayes [6]

$$P(C|D) = [\, P(D|C) \; x \; P(C) \,] \,/\, P(D)$$

Annotation :
D = Data training
C = Classification
$P(C|D)$ = Posterion Probability
$P(C)$ = Prior Probability
$P(D|C)$ = Probability of Data Occurrence
$P(D)$ = Opportunity classes in the data classification

1.2.4 Accuracy

Evaluation of accuracy needs to be done to ensure the accuracy of the results obtained. Evaluation can be measured using Confussion Matrix.

1.2.5 K-fold cross validation

Cross validation is one method that can be used to evaluate the performance of an algorithm. The dataset will be divided by the value of K, and will be tested alternately until all parts are met. In determining the K value, 10-fold validation is widely used because it is proven to produce a more good performance.

1.2.6 The state of the art

Here is a comparison table for algorithms in various studies that have been done before:

Table 1. The accuracy of various algorithms use in previous researches

No	Journal	Author	Accuracy			
			NB	SVM	DT	KNN
1.	Personality Classification Based on Twitter Text Using Naive Bayes, KNN and SVM	Bayu Yudha Pratama, Riyanarto Sarno	63 %	61 %	-	60 %
2.	Sentiment Analysis of Product using Machine Learning Technique: A Comparison among NB, SVM and MaxEnt	Monali Bordoloi , S.K. Biswas	81.33%	78,67%	-	
3.	Twitter Sentiment to Analyze Net Brand Reputation of Mobile Phone Providers	Nur Azizah Vidyaa , Mohamad Ivan Fanany , Indra Budi	78,90%	82,40%,	72.90%	-

In the first study by [1] Naive Bayes had the best 63% accuracy than SVM 61% and KNN 60%, this happened because SVM and KNN were considered to have difficulty separating word classes because the existing datasets were less accurate.

In the second study by [2] regarding mobile product review with several attempts using different algorithms also produced the highest level of accuracy by the Naive Bayes algorithm 81.33% compared to SVM and MaxEnt, but the three algorithms are still classified as algorithms that have the best level of accuracy.

For the tShird study [7] on the reputation of cellular phone providers giving different results from the two previous studies, here SVM 82.40% has a better accuracy rate compared to Naive Bayes 78.90% and Decision Tree 72.90%. Nevertheless, Naive Bayes still has a better level of accuracy compared to the Decision Tree.

2. Data Retieval

2.1 Dataset

Tokopedia application review data is obtained by doing web scraping on the google play store website. The reviews taken to serve as data in this study are the most recent reviews taken on February 10, 2020, totaling 6000 reviews.

2.2 Labeling

The data that has been obtained is then labeled manually. The given labels are 1, 2 and 3. Label 1 represents positive sentiment, label 2 represents negative sentiment, and label 3 represents neutral sentiment.

2.3 Text Preprocessing

Text preprocessing is the stage for preparing text data in a more structured form. The stages used in this research are punctuation removal, case folding, tokenizing, stopword removal, and stemming. Step of text preprocessing can be seen in this figure :

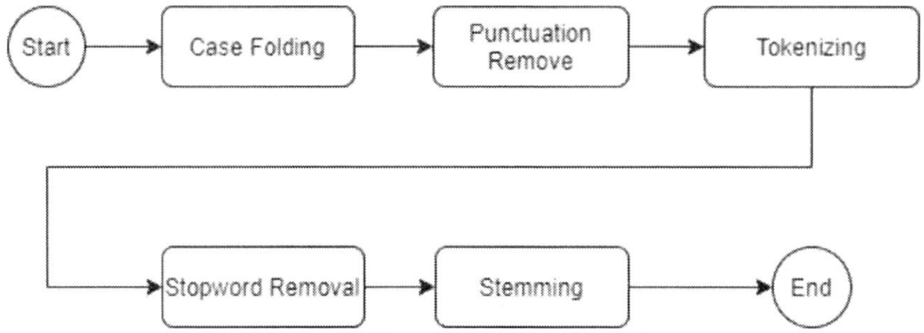

Figure 1. Step of text preprocessing

Based on the figure, it can be seen that the stages of the text preprocessing used in this study consisted of 5 stages, are case folding, punctuation removal, tokenize, stopword removal, and stemming.

2.3.1 Case Folding
Case folding is a process to change all letters to lower case .

Table 2. Example of case folding

Before	After
I LOVE shopping on this app, good items, and fast shipping	i love shopping on this app, good items, and fast shipping

2.3.2 Punctuation Removal
Punctuation removal is the process of removing symbols or punctuation contained in text.

Table 3. Example of punctuation removal

Before	After
i love shopping on this app, good items, and fast shipping	i love shopping on this app good items and fast shipping

2.3.3 Tokenizing
Tokenizing is the stage to separate input strings based on words that compose them using spaces.

Table 4. Example of tokenizing

Before	After	
i love shopping on this app good items and fast shipping	i love shopping on this app good	Items And Fast shipping

2.3.4 Stopword Removal
Stopword removal is the process of removing unnecessary words by checking the stoplist, the stoplist used is Indonesian which is taken from the NLTK library.

Table 5. Example of stopword removal

Before		After
i	good	love
love	items	shopping
shopping	and	app
on	fast	good
this	Shipping	items
app		fast
		shipping

2.3.5 Stemming
The stemming stage is done to return the words to their basic form words. This process is done by removing the prefix, suffix, or both. The stemming process is assisted by using literary python for Indonesian-language text.

Table 6. Example of stemming

Before	After
love	love
shopping	shop
app	app
good	good
items	item
fast	fast
shipping	ship

2.4 Text Processing
After all data review through the preprocessing stage will be classified using the Naive Bayes algorithm. The results obtained from this classification are calculated for accuracy, and their performance can be seen through the confusion matrix to ensure the validity of the results, the validation stage is performed using k-fold cross validation with a value of K= 10.

3. Result Analysis
3.1 Dataset
In this study the dataset containing user reviews will be classified into three labels that are done manually, 1 for positive labels, 2 for negative labels, and 3 for labels with neutral comments.

Table 7. Number of comment in dataset

Label	Number of Comment
1	2.670
2	2.123
3	1.207
Total	6.000

After text preprocessing, the amount of comment data is reduced, the change in the amount of comment data after the preprocessing stage can be seen in the following table:

Table 8. Number of comment after text preprocessing

Label	Number of Comment
1	2.552
2	2.121
3	1.187
Total	5.860

To process the data using an algorithm, the data is divided into training data and test data. The comparison ratio for training data and testing data is 80:20, which means 80% of the total data will be used as traning data and 20% of the total data will be used as testing data.

3.2 Analysis Sentiment

To do analysis sentiments, the data which has been processed through the preprocessing stage, is divided into training data and testing.

Table 9. Number of comment in data testing

No	Label	Number of Comment
1.	Positif	527
2.	Negatif	420
3.	Netral	225

After it did a selection of features using the tf-idf method then was processed using a naive bayes algorithm. Selection of features by using tf-idf can do a good job, the way it works is if the weight is gained, the higher the similarity rate of the document.

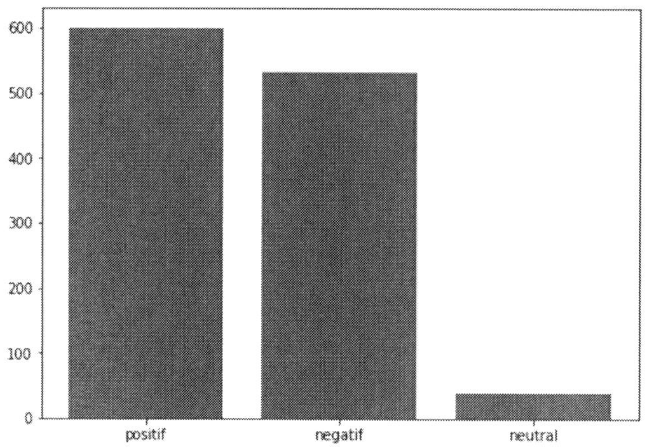

Figure 2. Amount of comment after analysis

Analysed comments amounted to 1172 comments made by 601 positive comments, 533 negative comments, and 38 neutral comments. From results obtained by user reviews tend to show positive comments, with an accuracy rate of 76%.

It can be seen that there is a considerable difference between the data of the results of the manual label and the data of the prediction results of the system. The biggest difference is in data with a neutral sentiment. This happens because of the misinterpretation of sentences, for example in neutral sentences there are positive words that make the system tend to read the data is positive

3.3 Accuracy

In this study the test calculates accuracy using a confusion matrix

Table 10. Confusion matrix

Actual	Predict		
	Positive	Negative	Netral
Positive	**489**	36	2
Negative	32	**384**	4
Netral	80	113	**32**

And below are the results of precision measurements, recall, f1 score using the Naive Bayes algorithm and TF-IDF feature extraction. We can see 79% precision value, 77% recall and 73% f1 score so we get an accurate value of 76%.

Table 11. Confusion matrix report

	Precission	Recall	F1-score
Positive	81%	93%	87%
Negative	72%	91%	81%
Neutral	84%	14%	24%
Avg	79%	77%	73%

3.4 Validation

To ensure that the accuracy value obtained is good and taken from the sample, validation is done by using the K-Fold method, in this method the data will be tested as many as the K value, the ideal K value used is 10. This means that 10 tests will be conducted. Following are the accuracy results obtained for the value of K = 10

Table 12. 10-fold scores

K	Score
1	0.815
2	0.796
3	**0.832**
4	0.815
5	0.747
6	0.691
7	**0.605**
8	0.715
9	0.773
10	0.774
Avg	0.76

The smallest accuracy value is 0.60 in the 7th experiment, and the greatest accuracy is in the 3rd experiment with a value of 0.83. With an average accuracy value obtained is 0.761

3.5 Visualization

The following is a picture that displays words that often appear in each class in this study.

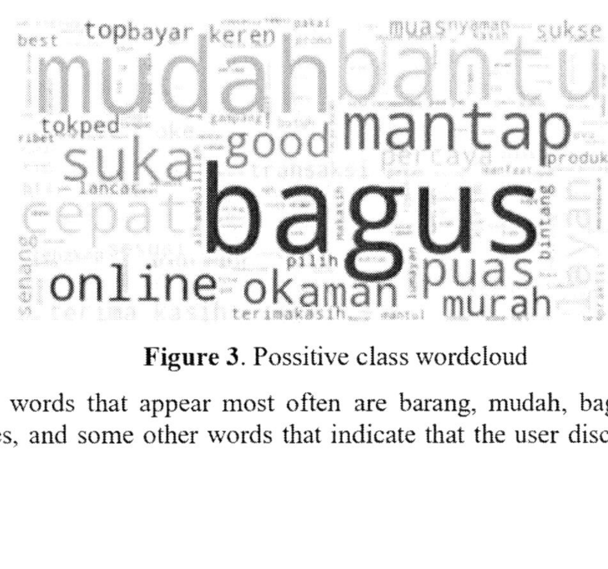

Figure 3. Possitive class wordcloud

In positive class, the words that appear most often are barang, mudah, bagus, cepat, aman, layan, toko, bantu, sukses, and some other words that indicate that the user discusses satisfaction when shopping because of the fast service process , the process of goods service is easy, safe, and good quality

Figure 4. Negative class wordcloud

The negative class shows the tendency of dissatisfied and disappointed user comments when buying and selling goods due to bad transactions, payment processing, and shipping with couriers. Users also complain about the process of downloading old applications and updating too often, and the absence of CS that serves complaints from users.

Figure 5. Netral class wordcloud

For neutral classes, users provide a lot of input to Tokopedia services, such as establishing a COD system for buying and selling transactions, choosing the right courier for fast delivery, shipping cost cuts. And many also provide support comments for Tokopedia for the better.

4. Conclussion

The conclusions that can be drawn from this research are as follows :

1. Sentiment analysis is done by the naive bayes algorithm and TF-IDF feature extraction
2. Based on the results of tests conducted, the accuracy of the Naive Bayes algorithm and TF-IDF is 76%. In validation with the 10-fold method, the highest accuracy is 83% and an average of 76%
3. User opinions about positive comments can be maintained and negative opinions can be an evaluation. In neutral comments there are many inputs for service improvement.

Suggestions for this research are expected in future studies to add Slankword to help the preprocessing process, and are expected to make improvements to writing errors.

IOP Conf. Series: Materials Science and Engineering **909** (2020) 012071 doi:10.1088/1757-899X/909/1/012071

Reference

[1] Pratama B Y and Sarno R 2015 Personality classification based on twitter text using naive bayes, KNN and SVM *International Conference on Data and Software Enginering* 170-174.

[2] S K B and Bordoloi M. 2018 Sentiment analysis of product using machine learning technique: a comparison among NB, SVM and MaxEnt International *Journal of Pure and Applied Mathematics* 71-83

[3] Pablo Gamallo and Marcos Garcia 2014 Citius: A naive-bayes strategy for sentiment analysis on english tweets *International Workshop on Semantic Evaluation* **8** 171-175

[4] Fink C R, Chou DS, Kopecky J J, and Llorens A 2011 Coarse and fine grained sentiment analysis of social media text *Johns Hopkins Apl Technical Digest* **30** 22-30

[5] Maarif A A 2015 Penerapan algoritma tf-idf untuk pencarian karya Ilmiah *Jurnal Teknik Informatika Fakultas Ilmu Komputer Universitas Dian Nuswantoro Semarang*.

[6] Ling J, Eka I P dan Oka T B 2014 Analisis sentimen menggunakan naive bayes classifier dengan seleksi fitur chi squere *E-Jurnal Matematika* **3** 92-99

[7] Vidya N A, Fanany M I, and Budi I 2015 Twitter sentiment to analyze net brand reputation of mobile phone providers *The Third Information Systems International Conference* **72** 519-526

Product recommender system using neural collaborative filtering for marketplace in indonesia

Arief Faizin[1] and Isti Surjandari[1]
[1]Industrial Engineering Department, Faculty of Engineering, Universitas Indonesia, Kampus UI Depok, Indonesia

E-mail: arief.faizin@ui.ac.id

Abstract. Marketplace has the potential growth in Indonesia indicated by the continued increase in the number of customers. However, the marketplace has some limitations to deliver personalized purchasing experience. Recommender system can support marketplace to overcome that limitations so that customer can find items or services based on their preferences. This study propose to develop product recommender system based on Neural Collaborative Filtering (NCF) algorithm. The product recommender system to be built is using implicit feedback data in the form of customer purchase data. Implicit feedback is reliable data for building recommendation system. The results have shown that NCF achieve the best performance and outperforms over the other collaborative filtering methods.

1. Introduction

Marketplace is the one of business sectors that has the great potential growth in Indonesia. According to definition, marketplace is an online site that contains many retailers to sell their products or services that can be similar to other retailers [1]. Tokopedia, one of the top marketplaces in Indonesia, has had a monthly user of 6.790.000 at Q4 2019 and increased by 45.91% from Q1 2017 [2]. However, that number is still around 2.5% of the total population of Indonesia that is 271.066.400 [3]. In fact, Google predicts that 52% of Southeast Asian e-commerce activities will be dominated by Indonesia in 2025 [4] and we also know that competition among the marketplace in Indonesia is getting tougher. There are Top 10 marketplaces in Indonesia including Shopee Indonesia, Tokopedia, Bukalapak, Lazada Indonesia, Orami, Blibli, JD.id, Sociolla, Zalora Indonesia, and Bhinneka [5].

However, some Indonesian society is still often shop offline. Based on survey data on Indonesian people who shop for groceries, 63% still choose to shop offline and 36% state that convenience is the most important factor regardless of online and offline [6]. Marketplace also has limitations which do not have staff who can help customers like in offline stores. Customers can get complete information and get convenience at purchasing experience through assistance from staff. The marketplace also has difficulty managing large number of products in a product catalogue. These limits can be overcome using features such as search and directory, but that not efficient because requires extra effort from the customer. Currently, marketplace have innovative solutions to overcome that problem using the recommendation system [7].

The recommendation system can recognize customer preferences and give suggestions of products that customers are interested in and tend to buy through the utilization of customer behavior data and product information data [8]. Research in the field of recommendation system is continuously improved. One of marketplace that has success in implementing a recommendation system is Amazon. Amazon utilize sales data that contains millions of products to identify potential products [9].

The recommender system algorithms are divide into collaborative filtering (CF), demographic filtering, content-based filtering, and hybrid filtering. In this study, we developed recommender system based on CF algorithm. CF is recommendation algorithm based on similarity of user past behaviour [10].

We used CF based on deep learning approach namely Neural Collaborative Filtering (NCF). NCF has been developed and showed the best performance compared to other collaborative filtering methods [11]. Some research has been done using NCF algorithm and shows the best performance such as commercial site recommendation [12] and e-learning recommendation [13]. Therefore, this paper proposed the implementation of NCF for product recommendation using implicit feedback data. We compare NCF with the conventional collaborative filtering method such as weighted matrix factorization (WMF) and bayesian personalized ranking (BPR) to measure the performance of the algorithm, and use some commonly adopted rank evaluation metrics consists of normalized discounted cumulative gain (NDCG), precision@k, and mean average precision (MAP).

2. Related Work

Recommender system generated by user feedback data. The user feedback can be collected either explicitly or implicitly. There are many research of product recommendation system. Research has been done in [14], [15], [16], [17], and [18]. However, That research mostly used explicit feedback based on users preference patterns from customer rating.

User interaction indirectly becomes input of implicit feedback data, so that the user is not disturbed or burdened, in contrast to explicit feedback data. Based on previous research, prediction algorithm that uses implicit feedback generate better prediction quality of user preferences than prediction algorithm that uses explicit ratings [19]. This paper uses implicit data in form of customer purchase data to generate recommender system model.

Next, we compare it with some recommender system algorithms such as WMF and BPR. In the study of Corso and Romani, there are some metrics that uses to evaluate the recommender system algorithm. In this study, the recommender system that will be built is using implicit feedback data, so the appropriate metrics are ranking metrics [20]. In this study we propose to use NDCG, precision@k, and MAP to compare the performance between algorithms.

3. Methods

3.1. Recommender System

The rapid growth of data causes users difficult to find products or services according to their needs and preferences. Recommender system can overcome that problems by suggesting products based on past behaviour of user [21]. There are 3 phases required to build a recommendation model [22] that depicted in Figure 1.

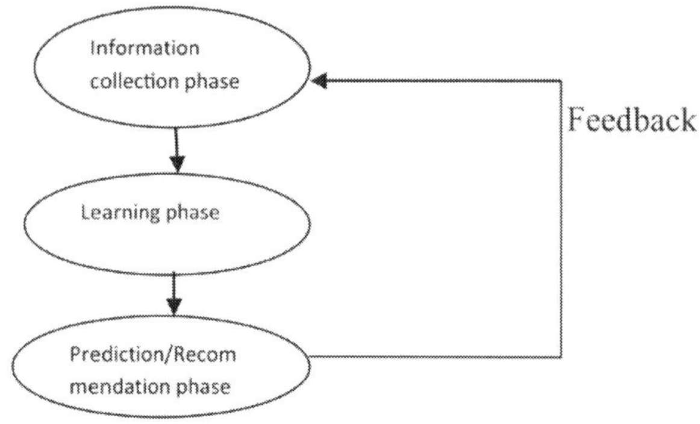

Figure 1. Recommendation phases.

1. Information collection phase is part where personal users information in form of user's attribute, behaviours or content is collected. There are two different input of recommender system, implicit feedback data and explicit feedback data. Explicit feedback data is user feedback that requires effort from user. Implicit feedback data is user feedback that automatically generated by system.
2. Learning phase is when the algorithm apply to learn the user feedback obtained in the previous phase.
3. Prediction/recommendation phase which is where the algorithm give recommendation or prediction result about user preferences.

3.2. Neural Collaborative Filtering
NCF is collaborative filtering algorithm that ensemble Matrix Factorization (MF) and Multi-Layer Perceptron (MLP) that collaborate non-linearity of MLP and linearity of MF and for modelling interaction of user and item. NCF can overcome the non-linearity problem by removing the inner product and replacing it with neural architecture. Figure 2 shows an overview of the NCF architecture [11].

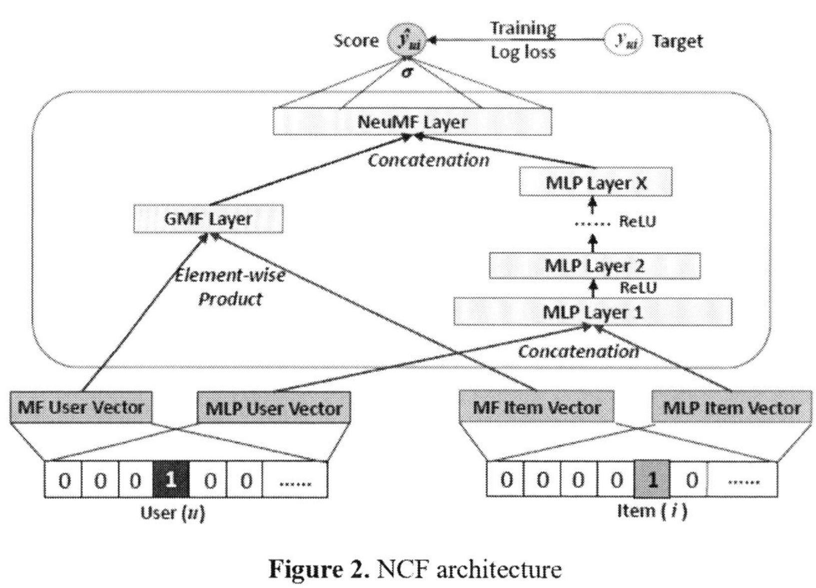

Figure 2. NCF architecture

The input layer consist of item feature vectors v_i^i and user feature vectors v_u^u which contains one hot encoded of items *i* and users *u* data. The next step is to get the latent vectors of item *i* and user *u* by transform the sparse vectors v_u^u and v_i^i into dense vector p_u and q_i by the embedding layer. User-item interactions are modeled separately between linear and nonlinear perspectives at the Neural CF layer, then the two previous models are combined and used as the final result. The calculations can be defined as Equation 1-3 [11,12]:

$$\emptyset^{GMF} = p_u^G \odot q_i^G \tag{1}$$

$$\emptyset^{MLP} = \alpha_X\left(w_X^T\left(\alpha_{X-1}\left(\dots\alpha_2\left(w_2^T\begin{bmatrix}p_u^M\\ p_u^M\end{bmatrix} + b_2\right)\dots\right)\right) + b_x\right) \tag{2}$$

$$\hat{y}_{ui} = \sigma\left(h^T\begin{bmatrix}\emptyset^{GMF}\\ \emptyset^{MLP}\end{bmatrix}\right) \tag{3}$$

Item embedding of the MLP and GMF parts are indicated in $q_i^G$and q_i^M. User embedding of the MLP and GMF are indicated in p_u^G and p_u^M. The output and hidden activation functions described as σ and α_X. The x-th layer's perceptron weight matrix and bias vector described as W_x and b_x. Cross entropy loss for implicit feedback is the NCF model optimization function, which is defined as Equation 4 [11,12].

$$L = -\sum_{(u,i)\in y\cup y^-} y_{ui}log\hat{y}_{ui} + (1-y_{ui})\log(1-\hat{y}_{ui}) \tag{4}$$

3.3. *Evaluation Metrics*

There are some popular ranking-based evaluation metrics to measure the performance for each recommender algorithm consists of precision@k, MAP, and NDCG [23], [24].

Precision@K is extension of a conventional precision metric. Precision@K is the top K list prediction items that actually consumed by the user divided by the K list generated by the model for each user in the data. The higher precision@k value indicates that the model generate more accurate predictions. The average of all user precision values indicates the final of precision value. The Precision@K calculation is calculated as shown as follows [23].

$$Precision@K = \frac{|C(K,u)|}{|K|} \tag{5}$$

MAP is very suitable for measuring performance for a ranking recommendation engine. The MAP value calculates how accurate the recommendations are in ranking items. The system will provide relevant advice among the first few recommended products if they have a high MAP score. The system will offer appropriate suggestions among the first few recommended products if they have a high MAP score. The first step in determining the MAP value is to calculate 'precision at the cutoff k' (P(k)) where precision is calculated by considering the list of recommendation subsets in order from first rank to k-th rank. The Average Precision (AP@k) can then be obtained using Equation 6. The MAP@k is calculated as follows [24].

$$AP@k = \frac{1}{\min(m,k)} * \sum_{i=1}^{k}(P(i) * rel(i)) \tag{6}$$

$$MAP@k = \frac{1}{|U|} * \sum_{u=1}^{|U|} (\frac{1}{m} * \sum_{i=1}^{k} (P_u(i) * rel_u(i)) \tag{7}$$

A larger MAP indicates better recommendation performance. The weighted rank of the recommended item is considered by NDCG metrics. Items that have a high ranking have a greater weight on the final result. NDCG is calculated by adding up the weighted rankings of items for users. The ranking list on the proposed model shows many items that are suitable if they have high NDCG values. NDCG can be obtained by calculating DCG(u) and Z(u) first. The value Z(u) indicates the user ideal value of DCG(u) which is calculated as the best item rating for the user. Furthermore NDCG calculations can be obtained using Equations 8. and 9.

$$DCG(u) = \sum_{j=1}^{n} \frac{2^{r(j)} - 1}{\log_2(1+j)} \tag{8}$$

$$NDCG(u) = \frac{DCG(u)}{Z(u)} \tag{9}$$

4. Experiment

4.1. Data preparation

The data was collected at one of the marketplaces in Indonesia in 2019. The data consists of 11.918 transaction data of implicit feedback in form of customer purchase data. The data has been used consists of customer data, product data, and customer purchase data.

4.2. Implementation

At the data processing stage, the data is divided into 2 parts, 0.70 is used as training data and 0.30 is used as testing data. Then the training data is trained using a neural collaborative filtering algorithm. First, we should set the parameter model on the NCF algorithm including learning rate, batch size, and the number of predictive factors with a value of k = 10 (recommended number of top items). The last hidden layer of NCF is as a predictive factors because it determines the model capabilities. We experimented the batch size of 256, learning rate of 0.001 and 4 trials of predictive factors of [8, 16, 32, and 64]. For WMF and BPR methods, the number of predictive factors is obtain from the number of latent factors. The model that is formed from the previous data processing is measured using the NDCG, Precision@10, and MAP metrics.

5. Result and Discussion

The result of evaluation shown in Table 1 below. The results show that the greater the predictive factor, the better the performance of the model.

Table 1. Evaluation Results

Metrics	NCF Predictive Factors				BPR Predictive Factors				WMF Predictive Factors			
	8	16	32	64	8	16	32	64	8	16	32	64
Prec@10	0.0176	0.0163	0.0195	0.0203	0.0201	0.0201	0.0201	0.0201	0.0041	0.0068	0.0115	0.0137
MAP	0.0779	0.0651	0.0898	0.0996	0.0397	0.0396	0.0397	0.0397	0.0095	0.0179	0.0413	0.0579
NDCG	0.0969	0.0840	0.1093	0.1194	0.0660	0.0660	0.0660	0.0660	0.0150	0.0266	0.0530	0.0697

Based on the evaluation results of the comparison in Table 1, NCF algorithm has shown the best performance on predictive factor of 64, this is indicated by the highest values of precision@10, MAP, and NDCG.

6. Conclusion

This paper studies the product recommender system for marketplace based on neural collaborative filtering. First, we generate the prediction model using NCF algorithm and then evaluate the performance of NCF method using ranking metrics consist of precision@k, MAP, and NDCG. Finally we compare NCF with BPR and WMF methods based on previous metrics. The results indicate that NCF shows best performance over the other methods.

References

[1] Nisafani A S, Wibisono A, and Revaldo M H T 2017 Analyzing the effectiveness of public e-marketplaces for selling apparel products in indonesia *Procedia Computer Science* **124** 274–279
[2] iPrice 2020 Online: https://iprice.co.id/insights/mapofecommerce/en/
[3] BPS 2020 Online: https://www.bps.go.id/statictable/2014/02/18/1274/proyeksi-penduduk-menurut-provinsi-2010---2035.html
[4] Freischlad N 2016 Online: https://www.techinasia.com/google-temasek-ecommerce-data-indonesia
[5] ECommerceIQ 2017 Online: https://ecommerceiq.asia/top-ecommerce-sites-indonesia/#1476946678150-d101b526-dcf7
[6] Snapcart 2020 Online: https://snapcart.global/indonesian-consumer-insights-customer-loyalty
[7] Hwangbo H, Kim Y S, and Cha K Y 2018 Recommendation system development for fashion retail e-commerce *Electronic Commerce Research and Applications* **28** 94–101
[8] Choi K, Yoo D, Kim G, and Suh Y 2012 A hybrid online-product recommendation system: combining implicit rating-based collaborative filtering and sequential pattern analysis *Electronic Commerce Research and Application* **11** 309–317
[9] Jiang B, Jerath K, and Srinivasan K 2011 Firm strategies in the "mid-tail" of platform-based retailing *Marketing Science* **30** 757-775
[10] Bobadilla J, Ortega F, Hernando A, and Gutiérrez A 2013 Recommender systems survey *Knowledge-Based Systems* **46** 109-132
[11] He X, Liao L, Zhang H, Nie L, Hu X, and Chua, T-S 2017 Neural collaborative filtering *In WWW* 173–182
[12] Li N, Guo B, Liu Y, Jing Y, Ouyang Y, Yu Z 2018 Commercial site recommendation based on neural collaborative filtering *UbiComp* 2018 138-141
[13] Deng X, Li H, and Huangfu F 2018 A trust-aware neural collaborative filtering for e-learning recommendation *Educational Sciences: Theory & Practice* **18**
[14] Wan M, Ni J, Misra R, and McAuley J, 2020 Addressing marketing bias in product recommendations *Proceedings of the 13th International Conference on Web Search and Data Mining* 618-626
[15] Dixit V S, and Gupta S 2020 Personalized recommender agent for e-commerce products based on data mining techniques *Advances in Intelligent Systems and Computing* 77-90
[16] Zhang X, Liu H, Chen X, Zhong J, and Wang D 2020 A novel hybrid deep recommendation system to differentiate user's preference and item's attractiveness *Information Sciences* 306-316

[17] Ding N, Lv J, and Hu L, 2019 Application of improved collaborative filtering algorithm in recommendation of batik products of miao nationality *IOP Conference Series: Materials Science and Engineering* **677**

[18] Bandyopadhyay S, and Thakur S S 2019 Product prediction and recommendation in e-commerce using collaborative filtering and artificial neural networks: a hybrid approach *Intelligent Computing Paradigm: Recent Trends* 59-67

[19] Palanivel K, and Sivakumar R 2010 A study on implicit feedback in multicriteria e-commerce recommender system *Journal of Electronic Commerce Research* **11** 140-156

[20] Del Corso, and G M Romani F 2019 Adaptive nonnegative matrix factorization and measure comparisons for recommender systems *Applied Mathematics Computation* **354** 164–179

[21] Acilar AM, and Arslan A 2009 A collaborative filtering method based on artificial immune network *Expert System Application 2009* **36** 8324–8332

[22] Isinkaye O, Folajimi Y O, and Ojokoh B A 2015 Recommendation systems: principles, methods, and evaluation *Egyptian Informatics Journal* **16** 261–273

[23] Zhao F, Shen Y, Gui X, and Jin H 2019 SDBPR: social distance-aware bayesian personalized ranking for recommendation *Future Generation Computer Systems* **95** 372–381

[24] Manning C D, Raghavan P, and Schütze H 2008 *An Introduction to Information Retrieval* (New York: Cambridge University Press)

IOP Conf. Series: Materials Science and Engineering **909** (2020) 012073 doi:10.1088/1757-899X/909/1/012073

Productivity measurement using Objective Matrix: case study in plate mill

N Wahyuni* **and R Alya**
Industrial Engineering Department, Faculty of Engineering, University of Sultan Ageng Tirtayasa, Jalan Jend. Sudirman Km 3, Cilegon, 42435, Indonesia

*E-mail: nrdwahyuni@gmail.com

Abstract. A decline of production level in December 2018 requires an evaluation. Productivity measurement is required to calculate performance. The purpose of this study is to measure the level of productivity using the objective matrix (omax). There were four steps to perform Omax. First step is select performance criteria, second is establish performance scales, third is rate the relative importance of the performance criteria, and the last is calculate the team productivity indicator. There were five ratios that become performance criteria, namely actual production compared to the target, material consumption, available time, makeup water consumption, and electricity consumption. The weight of each ratio is known in sequence from ratio 1 to ratio 5 were: 30%, 20%, 30%, 5%, and 15%. The performance index obtained the 3rd week has the highest productivity value, while the 4th week has the lowest productivity value, but the value is higher than standard value. From the value per ratio, we learned the ratio 1 has a fairly good performance, while the ratio 3 performance is considered less, because it is in the red zone.

1. Introduction

There was a decline in production levels in production line at XYZ, a company that manufactures plates, within a few weeks. Manager wanted to know the level of productivity to evaluate the resources they have. Resources as the input for production are generally materials, labour, and machines [1].

One of the most critical aspects of productivity measurement is the capability to acknowledge which factors are committed to changes in productivity [2]. Relationships such as enhancement, decline, or no changes in productivity can be caused by any figure of changes in input and output levels [2]. Partial productivity measures can uplift management at the source of productivity alteration [2].

In general, productivity is the ratio between output and input. Based on [3] the aim of productivity measurement is technology, efficiency, real cost savings, benchmarking production processes, and living standards. It is important to measure the productivity level to achieve those purpose. Depend on [4] there are four approach to do productivity measurement, that are: self-audit, extended audit, self-assessment, and benchmarking. Self-audit is based on questionnaire, extended audit is performed by expert, self-assessment focus on measuring of trends, and benchmarking is the process of comparing and measuring with another organisation [4].

This paper determination is to show the calculation of productivity level applying self-assessment approach. The intention is to practice continuous improvement. Objective matrix (Omax) is a tool to measure productivity which is widely used by some previous researchers. It helped to develop a composite index for the entire system or to individuals the entire system [5]. This evaluation model has an aspect which is the performance criteria of the work group integrated into the matrix [6].

Wahyuni [7] implementing Omax to measure the level of productivity in the injection area of a shoe factory. Faridz [8] calculated the optimal production of sugar using Omax. Sirajuddin [9] applying Omax to measure productivity in steel mills. In addition to being used in manufacturing, the Omax method can also be applied to the service sector. For example, done by [10] who measure productivity in emergency services in Turkey. Yosan [11] also used Omax to measure productivity in building maintenance services. Likewise [12] is using Omax to perform productivity in education. Omax also can be used to measure company performance, integrating with balance scorecard or human scorecard [13] [14].

The purpose of this study is to measure the level of productivity using the Omax method. Productivity will be translated into several criteria [5]. This article is expected to contribute to production managers in evaluating productivity levels. In addition, this article is expected to contribute as an enrichment for researchers and students.

2. Method

Depend on [15], there are four steps to perform Omax. First step is select performance criteria, second is establish performance scales, third is rate the relative importance of the performance criteria, and the last is calculate the team productivity indicator [15]. The selected productivity criterion is a calculation of productivity which then the level is determined to measure the productivity of the company [16]. The performance scales are attached at three levels [15]:

- Level 0: the lowest level of performance over a recent period under normal operating conditions.
- Level 3: the current baseline level of performance.
- Level 10: a stretch goal (a realistic estimate of the results) that the team can achieve with its expected capital allocation over the course of the project.

The project-management team determine weighting factors to each category to manifest their perceptions of its importance [15]. The amount of the weighting factors should equivalent one hundred percent [15]. The calculation of productivity indicators consists of three, namely [17]:

- Current, which is the result of measuring the productivity of the current period obtained from the sum of the values of each criterion
- Previous, i.e. the results of previous productivity measurement
- Index, which is an indication of changes in productivity that occur in the company. Value index productivity (IP) is obtained by the formula: IP = ((Current-Previous) / Previous) x 100%

3. Result and discussion

3.1. Performance criteria

Following are the results of data collection on consumption of raw materials and supporting materials at XYZ's production division. The following data are indicator data that influence productivity in the production division. The data collected was data in December 2018. This is a weakness of this article, the author did not get any other data besides the data for December 2018 due to limited research time.

Table 1. Data collection

Week	Production Target (metric ton) (A)	Production Actual (metric ton) (B)	Material Usage (kilogram) (C)	Plan Availability (%/metric ton) (D)	Air Make Up (x 10) (m^3/metric ton) (E)	Electricity (kilowatt) (F)
1	3662.11	3614.11	18000	721.29	11909	519667
2	3417.97	3348.54	18000	664.86	12806	513331
3	2441.41	2358.59	11000	621.03	8964	391935
4	4394.53	4357.74	18000	871.32	13692	607022

Table 2 presents the ratios that become criteria in measuring productivity. Ratio 1 is the ratio used as a determination of the productivity of the actual output of production against the production target which is made every week. This ratio compares actual production with production targets. Ratio 2 is the ratio used as a determination of the productivity of material consumption to the actual plate production made every week. This ratio compares material consumption with actual production. Ratio 3 is the ratio used as a determination of the productivity of the available time to the actual production of plates that are made every week. This ratio compares the time available with the actual production. Ratio 4 is the ratio used to determine the productivity of air make up consumption against the actual production of plates that are made every week. This ratio compares the consumption of air make up with actual production. Ratio 5 is the ratio used as a determination of the productivity of electrical energy consumption against the actual production of plates that are made every week. This ratio compares electrical energy consumption with actual production.

Table 2. Performance criteria

Ratio 1 (B/A)	Ratio 2 (C/B)	Ratio 3 (D/B)	Ratio 4 (E/B)	Ratio 5 (F/B)
0.9869	49.805	0.1996	32.951	143.7886
0.9797	53.755	0.1986	38.244	153.2998
09661	46.638	0.2633	38.005	166.1734
0.9916	41.306	0.1999	31.420	139.2974

3.2. Performance scale

After the ratio has been determined, the next step is determining the target matrix. In determining the target matrix there are three levels as a reference. Determination of the target matrix of the worst productivity value is expressed at level 0. The performance value obtained from the average value of each performance ratio that occurs during the observation is stated at level 3. Whereas for the determination of the target matrix the company expects in the future is stated at level 10. Table 3 below is a performance scale for minimum, standard, and maximum values. The minimum scale is obtained from the smallest ratio value, the standard value is the average value of the ratio, while the maximum value is the largest ratio value.

Minimum, standard, and maximum values that have been determined, then become the standard interval for each level of performance. In Omax there are 10 performance levels. The minimum value means level 0, the default value means level 3, and the maximum value means level 10. Here is a formula for creating intervals:

$$interval\ level\ 1\ and\ level\ 2 = \frac{Level\ 3 - level\ 0}{3-0} \quad (1)$$

$$interval\ level\ 4\ to\ level\ 9 = \frac{Level\ 10 - level\ 3}{10-3} \quad (2)$$

For example, in ratio 1:

- To determine interval level 1 and leve
- l 2: $\frac{0.9811-0.9661}{3} = 0.005$
- To determine interval level 4 to level 9: $\frac{0.9916-0.9811}{7} = 0.0015$

So, for level 1: 0.9661 + 0.005 = 0.9711 and for level 2: 0.9711 + 0.005 = 0.9761. And for level 4: 0.9811 + 0.0015 = 0.9826, for level 5: 0.9826 + 0.0015 = 0.9841, and so on.

IOP Conf. Series: Materials Science and Engineering **909** (2020) 012073 doi:10.1088/1757-899X/909/1/012073

Table 3. Performance scale

Performance scale	Ratio 1	Ratio 2	Ratio 3	Ratio 4	Ratio 5
Minimal (Level 0)	0.9661	4.1306	0.1986	3.1420	139.2974
Standard (Level 3)	0.9811	4.7876	0.2153	3.5155	150.6398
Maximal (Level 10)	0.9916	5.3755	0.2633	3.8244	166.1734

3.3. Performance rate

Weighting is the weight of each productivity criterion to total productivity. Each of the predefined criteria has a different effect on the level of the unit being measured. For this reason, it is necessary to include a weight which states the degree of importance (in percentage) which shows the relative influence of these criteria on the productivity of the work unit measured. Weighting is done subjectively by the production manager. Table below is the performance rate.

Table 4. Performance rate

Criteria	Weight (%)
Actual achievement production (Ratio 1)	30
Material consumption (Ratio 2)	20
Time availability (Ratio 3)	30
Air make up consumption (Ratio 4)	5
Electricity consumption (Ratio 5)	15

3.4. Productivity index

The measurement results will be used as an evaluation material for daily operational processes, so that company activities can be more efficient and effective. To find out how much the level of change in the performance of current productivity indicators against the standard value of productivity (first) and against the indicators of previous achievement. If the value of the change shows a positive value, then it can be said that there is a change that has a positive impact (increase), and vice versa if the value of the change shows a negative value, then it can be said that there is a change that has a negative impact (decrease).

Figure 1 is an objective matrix for measuring productivity in the first week. Seen in the picture, the values at each level (level 0 to level 10). The blue bar is the value in the ratio of 1 to 5 in the first week. For example, in the ratio 1, the value is 0.9869, the value approaches level 7, then, the score for the ratio 1 in the first week is 7. To calculate the performance value is a score of 7 times the ratio 1 weight of 30, obtained 210. The sum of the performance values the ratio of 1 to 5 for the first week is 300. Thus, the performance index is obtained at 18%.

Ratio 1	Ratio 2	Ratio 3	Ratio 4	Ratio 5	
0.9869	**49.805**	**0.1996**	**32.951**	**143.7886**	
0.9916	53.755	0.2633	38.244	166.1734	10
0.9901	52.915	0.2565	37.803	163.9543	9
0.9886	52.075	0.2496	37.362	161.7352	8
0.9871	51.235	0.2428	36.921	159.5161	7
0.9856	50.395	0.2359	36.479	157.2971	6
0.9841	49.556	0.2290	36.038	155.0780	5
0.9826	48.716	0.2222	35.597	152.8589	4
0.9811	47.876	0.2153	35.155	150.6398	3
0.9761	45.686	0.2097	33.910	146.8590	2
0.9711	43.496	0.2042	32.665	143.0782	1
[illegible]	[illegible]	[illegible]	[illegible]	[illegible]	0
7	6	0	2	1	Score
30	20	30	5	15	Weight
210	120	0	10	15	Value

Performance Indicator:	Current	Previous	Index
	355	300	18%

Figure 1. Objective matrix week 1

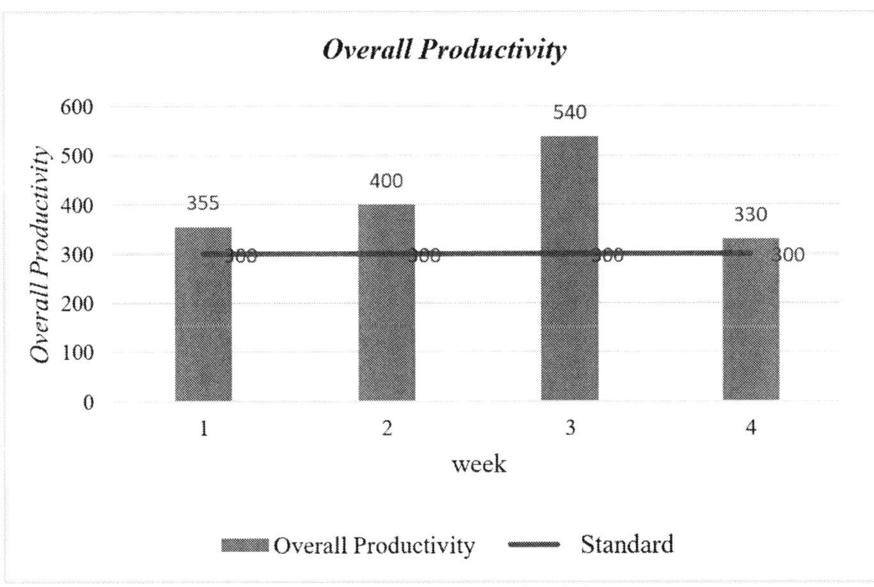

Figure 2. Overall productivity

If we pay attention to the achievement value of performance per week (figure 2), the achievement of production per week is above the standard productivity value of 300. As for the third week, the performance index value of 540 is the highest performance value in December 2018 with the index reaching 35%. This is influenced by the low material consumption, low air makeup consumption and using low electrical energy with the lowest plan availability for 4 weeks which shows that during the production process is going according to planning. And the lowest performance value was in the fourth week i.e. 330 with a performance index of -39%. In the fourth week experiencing some production problems.

Table 5 below shows the performance ratio 1 to ratio 5 based on Omax. The red color indicates below-standard performance (level 0 to level 2). Yellow is the standard performance ratio, meaning that the performance approaches the target to be achieved (level 3 to level 6). Green indicates the performance has reached the target (level 7 to level 10). In the third week there are 2 ratios in the red category which are the ratio 1 and ratio 2. This is because the supply of material runs out so that it affects the actual production. However, overall, this illustrates the value of performance in the third week is the best performance value. If we pay attention to performance per ratio, ratio 1 has a fairly good performance, while the ratio 3 performance is considered less, because it is in the red zone.

Table 5. Ratio's performance

Week	Ratio 1	Ratio 2	Ratio 3	Ratio 4	Ratio 5
1	0.9869	49.805	0.1996	32.951	143.7886
2	0.9797	53.755	0.1986	38.244	153.2998
3	0.9661	46.638	0.2633	38.005	166.1734
4	0.9916	41.306	0.1999	31.420	139.2974

4. Conclusion

Based on the calculation results and the description above, it can be concluded that the productivity of XYZ in December 2018 when viewed from the performance in a month, the 3rd week has the highest productivity value, while the 4th week has the lowest productivity value, but the value is higher than standard value. If seen from the value per ratio, the ratio 1 has a fairly good performance, while the ratio 3 performance is considered less, because it is in the red zone.

5. References

[1] Al-Darrab I A 2000 Relationships between productivity, efficiency, utilization, and quality *Work Study* **49** 97–104
[2] Misterek S D A, Dooley K J and Anderson J C 1992 Productivity as a performance measure *Int. J. Oper. Prod. Manag.* **12** 29–45
[3] OECD 2001 *Measuring Productivity - OECD Manual*
[4] Moseng R manager B 1995 Productivity measurement *Benchmarking-Theory and Practice* pp 248–60
[5] Marinkovic S Objectives matrix
[6] Yahya R, Mahachandra M and Handayani N U 2019 The mundel and objective matrix model of productivity measurement at PT Adi Perkapalan *IOP Conf. Ser. Mater. Sci. Eng.* **598**
[7] Wahyuni H C and Setiawan S 2017 Implementasi metode objective Mmatrix (OMAX) untuk pengukuran produktivitas pada PT.ABC *PROZIMA (Productivity, Optim. Manuf. Syst. Eng.* **1** 17
[8] Faridz E A 2011 Pengukuran dan analisis produktivitas produksi dengan metode objective matrix (OMAX) di PG. Krebet baru malang *Agrointek* **5** 80–7
[9] Sirajuddin, Katili P B and Jaya K C 2010 Pengukuran kinerja productivitas perusahaan dengan metode objective matrix (OMAX) 8
[10] Balkan D 2011 Enterprise Productivity measurement in services by OMAX (objective matrix) method and an application with Turkish emergency *Reser Conf. Product. Serv. Next Gen–Beyond Output/Input* 1–13
[11] Yosan I R B, Kholil I M and Soraya W Increasing productivity with objective matrix method case study on building maintenance management pio PT . XXX *Proceeding 9th Int. Semin. Ind. Eng. Manag.*
[12] Dervitsiotis K N 1995 The objectives matrix as a facilitating framework for quality assessment and improvement in education *Total Qual. Manag.* **6** 563–70
[13] Okfalisa, Anugrah S, Anggraini W, Absor M, Fauzi S S M and Saktioto 2018 Integrated analytical hierarchy process and objective matrix in balanced scorecard dashboard model for performance measurement *Telkomnika (Telecommunication Comput. Electron. Control.* **16** 2703–11

[14] Hardi J, Supardi D, Angelo C, Farhan N, CND F, Ahmad A, Jap L, Handoko I, Istijanto I and Rijanto A 2019 Human Resource Scorecard (HRSC) and objective matrix as performance measurement and performance booster method to optimize production performance *Int. J. Multicult. Multireligious Underst.* **6** 173

[15] C. E. Olstead, R. S. Kurland, J.T. Jackson J 1994 The objective matrix: a practitioner's perspective *Des. Manag. J.* 40–5

[16] Halpern L, Koren M and Szeidl A 2015 Imported inputs and productivity *Am. Econ. Rev.* **105** 3660–703

[17] Hamidah N H, Deoranto P and Astuti R 2013 Analisis produktivitas menggunakan metode objective matrix (OMAX): studi kasus pada bagian produksi Sari Roti PT Nippon Indosari Corpindo, Tbk Pasuruan productivity analysis using objective matrix (OMAX) method: case study on the production departement of *J. Teknol. Pertan. Anal. Produkt.* **14** 215–22

IOP Conf. Series: Materials Science and Engineering **909** (2020) 012074 doi:10.1088/1757-899X/909/1/012074

Food production performance measurement system using halal supply chain operation reference (SCOR) model and analytical hierarchy process (AHP)

I S Fauziyah*, A Y Ridwan and P S Muttaqin
School of Industrial and System Engineering, Telkom University, Bandung 40287, Indonesia

*Email: istisalmaf@gmail.com

Abstract. Food and beverage companies should be able to maintain and measure their performances to survive while the business competition condition is very dynamic. This study was conducted to design a performance measurement system for food production that combines the halal factors in its business process. Halal means authorized, legal, permitted, and allowed to any object or activity that is used or implemented in Islam. Therefore, designing and measure the performance in this study will use the Supply Chain Operation Reference (SCOR) model and Analytical Hierarchy Process (AHP). By using the SCOR model, we will get performance metrics that serve as indicators to observe and maintain the company's performance. Meanwhile, by using AHP, we will get the importance weight of each metric toward the company's performance score. The results of this study obtained 15 metrics with 3 halal metrics. The 15 metrics are divided into three performance attributes, which are 5 metrics in the reliability attribute, 9 metrics in the responsiveness attribute, and 1 metric in the cost attribute. The results of the overall performance score is 72.73. Those metrics are useful to maintain and improve the company's performance.

1. Introduction

Food is needed for the sustainability of life and as a value of worship from the perspective of Islam [1]. Around 85% of Muslims out of 255 million population in Indonesia makes Halal one of very important factors considered in running a food and beverage company [1][2]. Halal means authorized, legal, permitted, and allowed to any abject or activity that used or implemented in Islam [3]. Indonesia becomes a very attractive market for halal food because of its majority of the Muslim population [1].

The food and beverage industry in Indonesia continues to increase each year and its competitiveness is getting tight [4]. It was also mentioned as a business sector engaged in the food and beverage industry being one of the industries that contributed greatly to the Gross Domestic Product (GDP) of the non-oil and gas industry [5]. The contribution of the food and beverage industry has increased from 2017 to 2018. In 2017, the food and beverage industry has contributed 34.33%, and in 2018 the food and beverage industry has contributed 35.39% towards the GDP non-oil and gas industry.

To survive in conditions of dynamic business competition, companies are required to always excel in all dimensions of performance and the conditions of the company needed to be able to measure and observe the performance of the company itself [6]. Performance measurement becomes one of the biggest challenges for competitive companies these days as global competition and advanced technology make borders irrelevant and link business entities more closely, the supply chain is growing increasingly complex[7].

The research was conducted at a food and beverage company whose main activity was to produce ice cream. Business activities that have been running for two years have problems that have not been managed until now. The problem is that there is no standard to measure the performance for each production activity, so the company cannot identify the achievements that have been achieved when carrying out its business activities.

Based on the problem above, a performance measurement system for ice cream production will be developed using the SCOR model and the AHP method. Performance measurement for each production activity also considers halal factors and this is based on Government Regulations No.31 year 2019 concerning the implementing regulations of Law No. 33 year 2014 concerning Halal Product Guarantee [8]. The halal product guarantee in Indonesia has been regulated by The Assessment Institute for Foods, Drugs, and Cosmetics Indonesian Ulema Council (LPPOM MUI) using Halal Assurance System 23000 (HAS 23000) as an evaluation guidance for halal certification[1].

Previous research conducted by [9][10] has explained that the SCOR model can evaluate the company's operational efficiency from various aspects and performance measurement systems conducted using the SCOR model enables a company to objectively evaluate its competitiveness and weakness. Besides, the research conducted by [2][11] has explained that Indonesia is a country with a majority population that embraces Islam, so halal products are influential in daily life and are the main factors needed by most society, also a system that integrates its business processes and monitor processes must follow halal standardization. Furthermore, research conducted by [12] has explained that the AHP method is used to analyze the structure of the selection of strategies compiled based on the SCOR model and then determine the weight of each criterion. The AHP hierarchical structure tends to arrange elements in the system into different levels so that it is more easily understood by the decision-makers and the SCOR model is also a hierarchical model that contains the different levels of process [12]. Therefore, the SCOR model can be applied together with the AHP method to design a performance measurement system.

2. Methodology

To solve the problem in this study, it takes several phases to be done. These phases are data collection, mapping the metrics using the SCOR model, and determine the importance weight using AHP. The following figure shows the phases taken to solve the problem in this study.

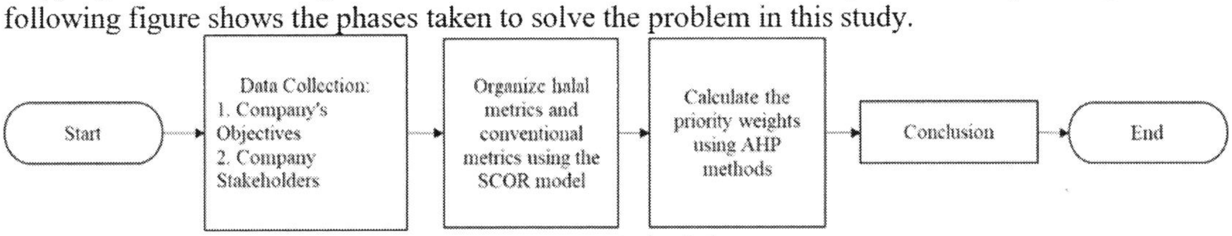

Figure 1. Methodology

2.1. Data Collection

The research began by collecting the data by conducting interviews with the company regarding the company objectives and stakeholders related to the ice cream production. Interviews were conducted with the CEO of the company who described that the company's objectives were divided into three points. Those company's objectives are meeting the production targets and quality standards, conducting all the business activities on time, and conducting all operational activities that fit with the budget cost.

2.2. Determine the metrics

After the data has been collected, the performance metrics are determined based on halal factors and the SCOR model. Furthermore, the performance metrics are submitted to the company for verification by the CEO. The following table shows halal metrics that will be used as part of the performance measurement system.

Table 1. Halal metrics

No.	Halal Metric
1.	Halal Raw Material [13]
2.	Product Warranty [2]
3.	Packaging [14]

Table 1 above shows that the first halal metric that will be used is halal raw material. The important production aspect of HAS is emphasizing there is no prohibited ingredients during process [1]. Raw material will go through a series of production processes and then the finished product will be delivered to customers, so the raw material that will be used in the production must be from halal sources [15]. The second halal metric is the product warranty. The company must ensure that no products are mixed with non-halal products [15].

The SCOR model has several levels of processes [16][17]. This study conducted until level 3. Level 1 defines the scope and content of the supply chain, and a supply chain performance target basis is determined in level 1 [18]. Level 2 defines the operating strategy, in which the process capability for the supply chain is determined. Level 3 defines the arrangement of individual processes, determines the ability to execute.

2.3. Measure the priority weight

The results of performance metrics that have been verified by the company will be calculated for each weight using the AHP method. Analytical Hierarchy Process (AHP) is a tool developed by Saaty in 1980 aimed at managing the qualitative and quantitative multi-criteria elements involved in decision-making behavior [19]. The steps in using AHP are as follows.

2.3.1. Identify the problem hierarchy

The problems that have been resolved are then broken down into several elements including criteria and sub-criteria, then arranged into a hierarchical form. Pairwise comparisons are made at the level of criteria and sub-criteria.

2.3.2. Determine the pairwise comparison of each criteria

All criteria and alternatives are assessed using pairwise comparisons. According to Saaty, a scale of 1 to 9 is used to express an opinion on each event [20].

2.3.3. Measure the consistency ratio

Calculate the Consistency Index (CI) with the following formula.

$$\mathrm{CI} = \frac{\lambda_{max} - n}{n-1} \quad (1)$$

Then the level of CI is calculated with the following formula.

$$\mathrm{CR} = \frac{CI}{RI} \quad (2)$$

CR is consistent when the value of CR $\leq$ 0, 1

Notations:

λ_{max} = maximum Eigen value

n = matrix size

RI = random consistency index [19]

2.3.4. Measure importance weights
After the pairwise comparison results are consistent, the next step is to measure the importance weights. Importance weights are calculated at the criteria and sub-criteria level.

2.4. Measure the final score
The measurement of performance final score is conducted by multiplying the importance weight of each criterion with the actual score of each company's performance. The following table shows the monitoring system of performance indicators [21].

Table 2. Monitoring system for performance indicators

Monitoring System	Work Indicators
<40	Poor
40 – 50	Marginal
50 – 70	Average
70 – 90	Good
>90	Excellent

3. Result and Discussion
To form performance metrics using the SCOR model, the first thing to do is to determine the company's objectives and interpret them into the SCOR performance attribute. The following table shows the company's objectives along with its performance attributes.

Table 3. SCOR performance attributes based on the company's objectives

Objectives	Performance Attribute
Meeting the production targets and quality standards.	Reliability
Conducting all the business process on time.	Responsiveness
Conducting all operational activities that fit with the budget cost	Cost

Based on the company objectives listed in Table 3 above, several performance attributes can be identified on the SCOR model that fit these objectives, there are reliability, responsiveness, and cost.

The reliability attribute indicates the ability to perform a determined tasks[17]. Therefore, the first objective corresponds to the reliability attribute.

Then, the responsiveness attribute indicates the speed at which the task is performed[17]. Therefore, the second objective corresponds to the responsiveness attribute.

Lastly, the cost attribute indicates the costs incurred to perform a task or process[17]. Therefore, the third objective corresponds to the cost attribute.

After pairing the company's objectives with the performance attributes in the SCOR model, the next step is to determine the performance metrics based on the performance attribute in Table 3 above. When the performance metrics have been determined, then measure the importance weights using the AHP method.

The following table shows halal and SCOR metrics.

Table 4. Metric's Code

Metric	Code
Forecast Accuracy	RL.1
Schedule Achievement	RL.2
Fill Rate	RL.3
Product Warranty (Halal Metric)	RL.H.1
Halal Raw Material (Halal Metric)	RL.H.2
Identify, Prioritize, and Aggregate Production Requirements	RS.1
Identify, Assess and Aggregate Production Resources	RS.2
Balance Production	RS.3
Establish Production Plans	RS.4
Issue Material	RS.5
Produce and Test	RS.6
Package	RS.H.1
Stage Finished Product	RS.7
Release Finished Product to Deliver	RS.8
Cost to plan make	CO.1

After determining the metrics based on the company's objectives, then measure the importance weight using the AHP method. As the importance weights are being determined, the next step is to measure the final score of performance metrics. The following table shows the results of performance final score.

Table 5. Performance Scores

Performance Attribute	Performance Attribute Weight	Metric	Actual Score	Metric Weight	Score	Final Score (Score x Performance Attribute Weight)
Reliability	0.236	RL.1	60	0.166	9.98	2.36
		RL.2	55	0.182	9.90	2.34
		RL.3	60	0.263	15.78	3.73
		RL.H.1	100	0.058	5.80	1.37
		RL.H.2	100	0.331	33.13	7.83
	Reliability Score				74.58	17.64
Responsiveness	0.514	RS.1	67	0.098	6.533	3.359
		RS.2	56	0.142	7.889	4.056
		RS.3	67	0.164	10.933	5.621
		RS.4	100	0.216	21.600	11.106
		RS.5	50	0.103	5.150	2.648
		RS.6	100	0.064	6.400	3.291
		RS.H.1	50	0.038	1.900	0.977
		RS.7	50	0.107	5.350	2.751
		RS.8	75	0.067	5.025	2.584
	Responsiveness Score				70.781	36.392
Cost	0.249	CO.1	75	1	75	18.701
	Cost Score				75	18.701
					Total Score	72.73

The performance attribute weights and metric weights were obtained from the measurement results based on the pairwise comparison of each criteria using AHP methods. A pairwise comparison of each criteria was assessed by the company. As for the actual score, it was obtained from the company's actual performance score. Then the score can be obtained by multiplying the actual score with the metric weight. Last, for the final score, it was obtained by multiplying the score with the performance attribute weights.

Based on the Table 5 above, the results show that the reliability attribute has obtained a score of 74. 58, the responsiveness attribute has obtained a score of 70.781, and the cost attribute has obtained a score of 75. The overall performance metrics have obtained a score of 72.73. Based on the monitoring system of performance indicators, the results of the overall performance score of 72.73 are qualified as Good[21].

4. Conclusion

Based on the results of the production performance measurement using the SCOR model, there are a total of 15 metrics that has been approved by the company, including 3 halal metrics. The 15 metrics are divided into three performance attributes, which are 5 metrics in the reliability attribute, 9 metrics in the responsiveness attribute, and 1 metric in the cost attribute.

The importance weight measurement and scoring using the AHP method revealed that the score for reliability attribute is 74.58, responsiveness attribute is 70.781, and the cost attribute is 75. As for the final score for the overall performance metrics is 72.73. Based on the monitoring system of performance indicators, the results of the overall performance score of 72.73 are qualified as Good, although the company still needs to improve and maintain the performance score so that the performance does not decrease.

References

[1] Adiarni N and Fortunella A 2018 The analysis of halal assurance system implementation (HAS 23000) in fried chicken flour product: a case study on XXX brand **149** 57–61

[2] Kusrini E, Qurtubi Q and Fathoni N H 2018 Design performance measurement model for retail services using halal supply chain operation reference (SCOR): a case study in a retail in indonesia *J. Adv. Manag. Sci.* **6** 218–21

[3] Chaudry M M, Hussaini M M, Jackson M A and Riaz M N 2000 Halal industrial production standards *Illinois: J&M Food Products Company*

[4] Kementerian Perindustrian 2019 Industri makanan dan minuman jadi sektor kampiun

[5] Kementerian Perindustrian 2017 Industri makanan dan minuman masih jadi andalan

[6] Cocca P and Alberti M 2010 A framework to assess performance measurement systems in SMEs *Int. J. Product. Perform. Manag.* **59** 186–200

[7] Sudaryanto and Bahri R 2007 Performance evaluation of supply chain using SCOR model: the case of PT. Yuasa, Indonesia *Proceeding, Int. Semin. Ind. Eng. Manag.* 49–55

[8] Republik Indonesia P 2019 Peraturan Pemerintah Republik Indonesia nomor 31 tahun 2019 tentang peraturan pelaksanaan Undang-Undang nomor 33 tahun 2014 tentang jaminan produk halal 1–32

[9] Jiang Y and Zhou L J 2011 The quantitative research on the index system of supply chain performance measurement based on SCOR *Adv. Intell. Soft Comput.* **112** 375–82

[10] Kuswandi R Y, Yanuar Ridwan A and El Hadi R M 2018 Development of monitoring reverse logistic system for leather tanning industry using scor model *Proceeding 2018 12th Int. Conf. Telecommun. Syst. Serv. Appl. TSSA 2018*

[11] Mefid K N, Ridwan A Y and Sari W P 2019 Global industry perspective of halal cosmetics applying sales and distribution process based on enterprise resources planning *Bull. Soc. Informatics Theory Appl.* **3** 61–8

[12] Kocaoğlu B, Gülsün B and Tanyaş M 2013 A SCOR based approach for measuring a benchmarkable supply chain performance *J. Intell. Manuf.* **24** 113–32

[13] Harwati and Yunita Pettalolo A N 2019 Halal criteria in supply chain operations reference (SCOR) for performance measurement: a case study *IOP Conf. Ser. Mater. Sci. Eng.* **505**

[14] Akbarizan, Lestari F, Ismail K, Hamid A B, Azwar B and Nazar H 2020 Managing halal foods through power collaboration between supply chain actors *Innov. Manag. Educ. Excell. Vis. 2020 Reg. Dev. to Glob. Econ. Growth*
[15] Nasaruddin R R, Fuad F, Mel M, Jaswir I and Hamid H A 2012 The importance of a standardized Islamic Manufacturing Practice (IMP) for food and pharmaceutical productions *Adv. Nat. Appl. Sci.* **6** 588–95
[16] Bolstroff P and Resenbaum R 2012 *Supply Chain Excellence*
[17] Gordon Stewart 2017 *Supply Chain Operations Reference Model*
[18] Waaly A N, Ridwan A Y and Akbar M D 2018 Development of sustainable procurement monitoring system performance based on supply chain reference operation (SCOR) and analytical hierarchy process (AHP) on leather tanning industry *MATEC Web Conf.* **204**
[19] Taherdoost H 2018 Decision making using the analytic hierarchy process (AHP); a step by step approach *Int. Journel Econ. Manag. Syst.* **2** 244–6
[20] Labib A 2014 *Introduction to the Analytic Hierarchy Process*
[21] Hasibuan A, Arfah M, Parinduri L, Hernawati T, Suliawati, Harahap B, Sibuea S R, Sulaiman O K and Purwadi A 2018 Performance analysis of supply chain management with supply chain operation reference model *J. Phys. Conf. Ser.* **1007**

IOP Conf. Series: Materials Science and Engineering **909** (2020) 012075 doi:10.1088/1757-899X/909/1/012075

Framework of risk priority and risk mitigation approach for palm sugar reverse supply chain

N Ummi[1], Marimin[2], E Noor[3], A Iskandar[4]
[1]Graduate Program of Agro-Industrial Engineering, Faculty of Agricultural Engineering and Technology, IPB University (Bogor Agricultural University), Bogor, Indonesia
[2]Department of Agro-Industrial Technology, Faculty of Agricultural Engineering and Technology, IPB University (Bogor Agricultural University), Bogor, Indonesia

E-mail: nurul.ummi@untirta.ac.id

Abstract. In the distribution of the palm sugar Supply Chain (SC) from producers to consumers, there are many palm sugar products quality damage, so a Reverse Supply Chain (RSC) activity is needed to minimize the loss and possible rocessed. There are some risks that occur in the RSC activities of palm sugar agricultural industry. The purpose of this study was to design a risk priority framework and risk mitigation as a reference for implementing RSC risk management in the palm sugar industry. The steps in the implementation of this research were the data and information collection, descriptive analysis, category selection, content evaluation and subsequently creating a framework. Approaches and methods for minimizing RSC risk based on the order of risk priorities and risk mitigation strategies were done by using Hybrid Interpretive Structural Modeling (ISM) - fuzzy Decision Making Trial and Evaluation Laboratory (DEMATEL) and House of Risk (HOR) 2. The results showed that the risk priority framework and the risk mitigation strategy are carried out through four stages starting from risk identification, risk classification, risk priority determination, and risk mitigation strategies.

1. Introduction

Palm sugar agroindustry is a business that processes the sugar palm juice from the Arenga tree (Arenga pinnata Merr) into shell palm sugar (SPS). The sugar palm juice is containing fructose, and sucrose, and other nutrients [1]. The sweet taste of palm sugar is caused by its carbohydrate content which reaches 11.8% [2]. This shell palm sugar contains a fairly high water content of 10-11% [3], when compared to the quality requirements of palm sugar in SNI 01-3743-1995 which a maximum value of 10%.

In the distribution of the palm sugar SC from producers to consumers, a lot of palm sugar products are damaged, both due to the buildup burden at the time of distribution or the length of product turnover. From the results of interviews with retailers also producers, it was found that the damage level of palm sugar is 10% -20%, and this damage will increase even to 30% if the palm sugar is not processed immediately. The types of damage that occur in palm sugar in the form of physical damage that is, cracked, destroyed, even melted. The durability of this palm sugar product ranges from 2-3 weeks and then this palm sugar will crack, melt, and be damaged.

For this reason, a method is needed in the management of palm sugar that is not absorbed by the market and is damaged, and how the sustainability of the palm sugar product becomes a very important issue. Returning food products into the SC for recovery or adding value, this approach is known as the RSC [4]; [5]. Figure 1 summarizes a description of the condition of the existing RSC in palm sugar.

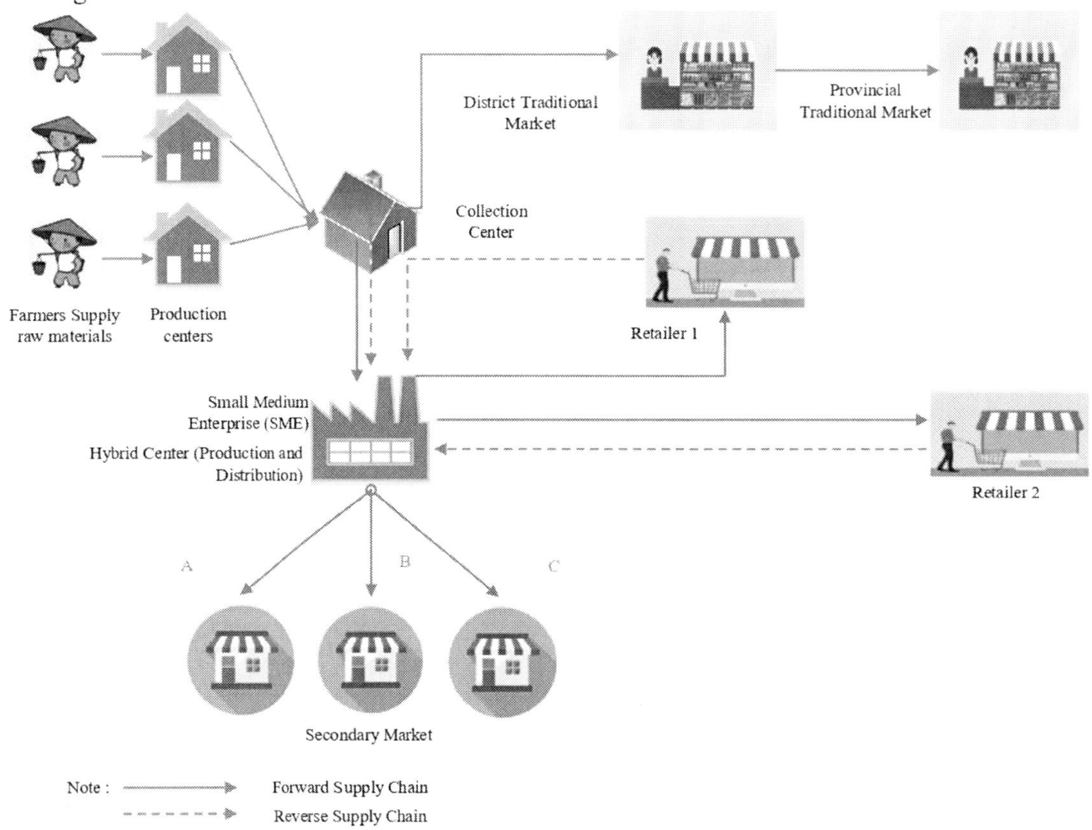

Figure 1. RSC on palm sugar

RSC is a series of activities needed to take products from consumers or other stages of the SC and reinsert them into the supply chain network to be reused, reprocessed, reproduced, recycled or even disposed of the right way efficiently and effectively [6], [7], [8]. In this way, the RSC contributes to the industry becoming more economically, socially, and environmentally sustainable. According to [9], Reverse Logistics (RL) is indispensable in the food industry due to the presence of waste products which in most cases can still be used.

Research on RSC has been discussed by several previous researchers [10], [11], [12], research on RSC Risks was carried out in the plastic industry, electronics, and glass manufacturing, but RSC risk Agri-food palm sugar has not been much studied.

While the main challenge in conducting RSC in the food industry is the risk of food safety. The smallest deviation from organoleptic characteristics can create food safety incidents [13]. Food safety assurance is a major factor as control of food products [14], for this, it requires steps in the design of methods that can be used to identify, minimize risks to the RSC in the food industry. The purpose of this study is to design a risk priority framework and risk mitigation as a reference in carrying out RSC risk management in the palm sugar industry.

The scope of this research is the identification of the types of risks from the literature study, both RSC risks and food supply chain risk, comparison of methods in determining the RSC risk, comparison of methods in determining risk mitigation, determining appropriate methods for priority risk and mitigation risks to the RSC of the palm sugar industry.

2. Method

The research method was consisted of four stages, i.e. data and information collection, descriptive analysis, category selection, and content evaluation.

2.1. Data and information collection

The secondary data was collected through literature search from Google Scholar search engine and other libraries such as in Emerald, Springer, Elsevier, Scopus, Wiley, Taylor & Francis, Inderscience, IEEE. Keywords used in article search are "Palm sugar agro-industry", "RSC", "Reverse Logistic (RL)", "RSC Risk", " The risk of Reverse Logistics in the agricultural industry ", Green Supply Chain (GSC) "," Agricultural industry SC "and" Agricultural industry SC risks", as summarized in Table 1 and Figure 2. Based on the search results using the 7 keywords above, 54 articles were obtained with the distribution of years from 2009 to 2019. From the results of 54 articles, only 53 articles that fit the topic to be studied, while one article is used as a reinforcement only.

Table 1. List of articles in the main research

Group	Reference
Palm sugar	[1], [2], [3]
RSC	[4], [5], [6], [7], [8], [9], [13], [15], [48], [49]
RSC Risk	[10], [11], [12], [16], [17], [18], [19], [20], [21], [22], [23], [24], [25], [26], [27], [28], [29], [30], [31], [32], [33]
SC	[36], [40], [41], [50], [52]
SC risk	[14], [37]
Agricultural industry SC risks	[34], [35], [38], [39], [53], [54]
GSC	[43], [44], [45], [46], [47]

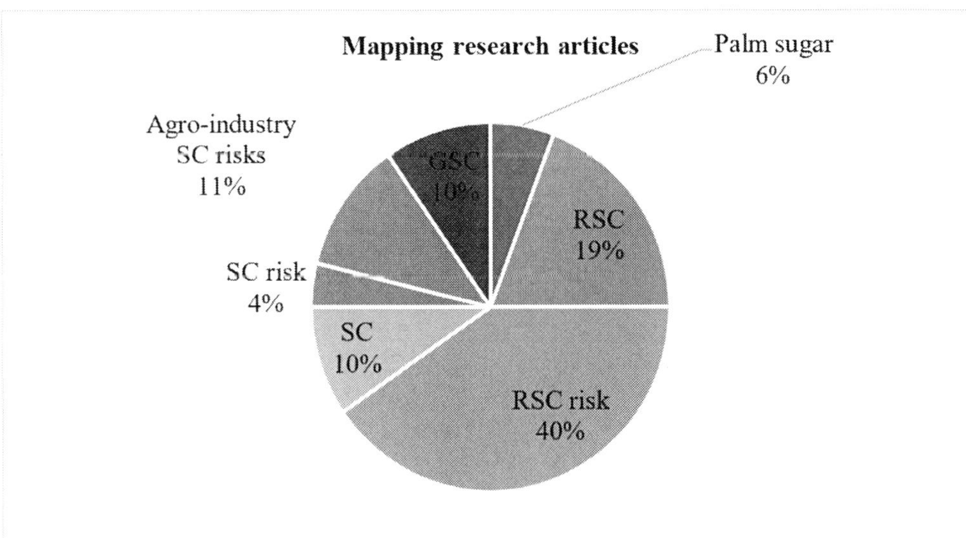

Figure 2. Mapping research articles

2.2. Descriptive analysis

Literature articles that have been obtained are then analyzed quantitatively, a description of the contents of the article, information about the distribution of the year the article was published, information about the distribution of the journal in which the article was published as described in Table 2,3,4 and Figure 3,4,5.

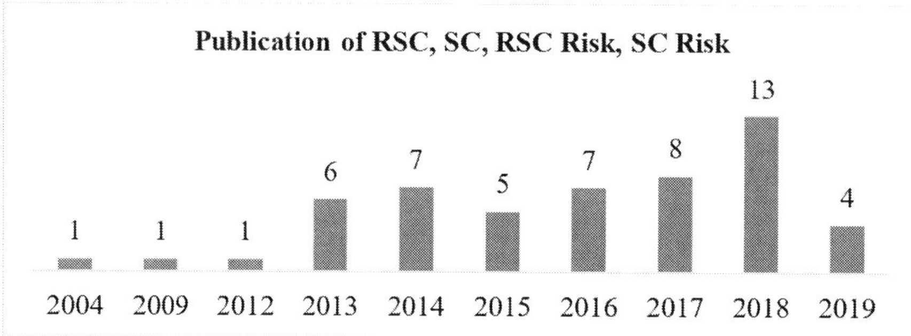

Figure 3. Distribution of research publication through the years

Table 2. The sum of articles published by major journals

Article Publish by Journals	Number of articles
American Journal of Engineering and Applied Sciences	1
Applied Mathematical Modelling	1
Applied Sciences	1
Applied Soft Computing	1
Benchmarking	2
Business Process Management Journal	1
Computer Aided Chemical Engineering	2
Computers an Industrial Engineering	2
Computers and Chemical Engineering	3
Decision Science Letters	1
Environmental Development	1
European Journal of Operational Research	2
International Journal of Supply Chain Management	2
International Journal of Applied Decision Sciences	1
International Journal of Integrated Supply Management	1
International Journal of Production Economics	3
International Journal of Production Research	2
International Journal of Research in Engineering and Technology	1
International Journal Supply Chain and Operations Resilience	1
International Strategic Management Review	1
Optimization	1
Proceedings of the palm sugar National Seminar	1
Renewable and Sustainable Energy Reviews	1
Sugar Tech	1
Supply Chain Forum	1
The Scientific World Journal	1
Transportation Research Part D: Transport and Environment	2
Journal Cleaner Production	8

Table 3. The number of articles published by books

Article Publish by Book	Number of articles
CRC Press Taylor & Francis Group	1

Table 4. The number of articles published by proceedings

Article Publish by Proceedings	Number of articles
MATEC Web of Conferences	2
IOP Conference Series : Materials Science and Engineering	2
IPTEK Journal of Proceedings Series	1
Transportation research procedia	1

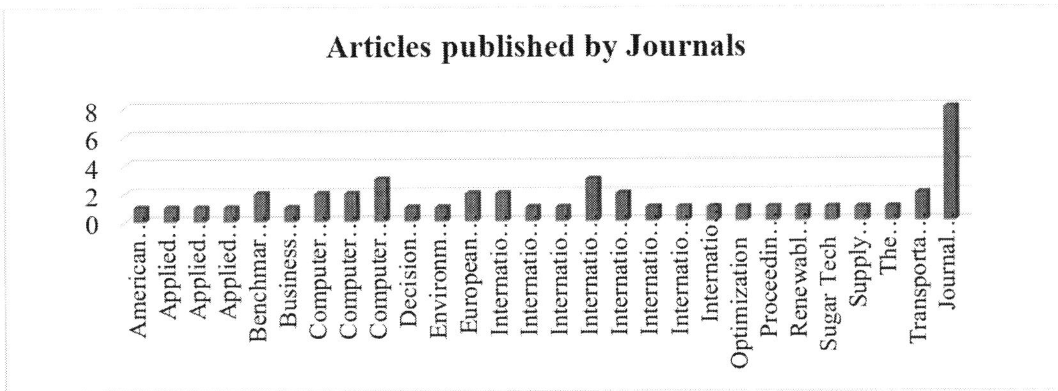

Figure 4. Articles published by journals

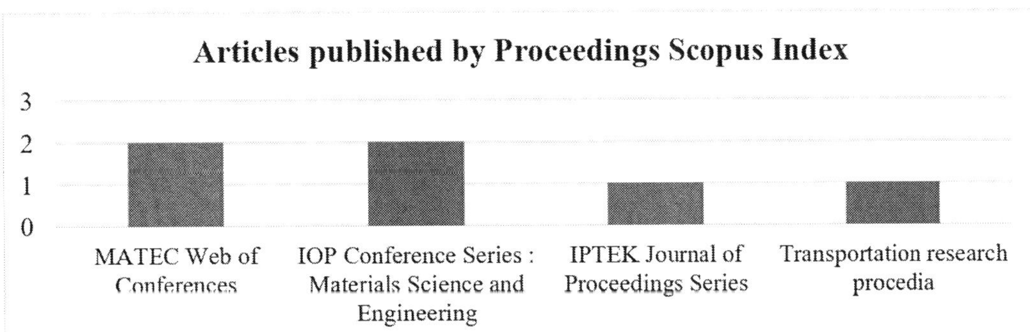

Figure 5. Articles published by proceedings

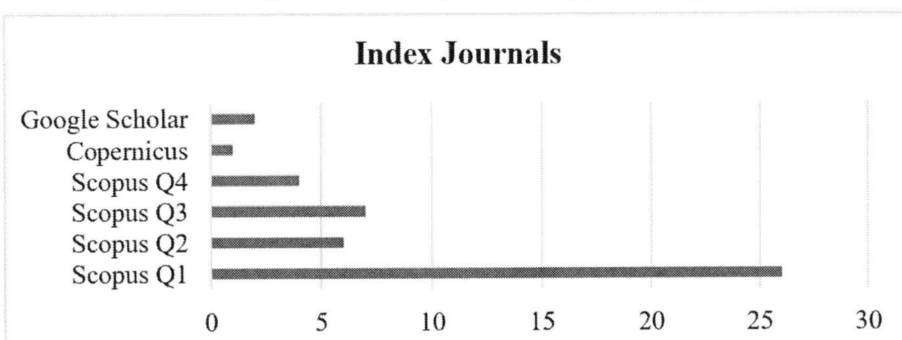

Figure 6. Index journal

2.3. Category selection

At this stage, articles are classified by topic. Referring to the main topic of this research on risk priority and risk mitigation, the articles obtained are classified on the topic, distribution of agro-industry supply chain articles, distribution of RSC articles, types of risks occurring in RSC, industry sectors, and the methods used. Based on this classification we can get the main flow of publications and research gaps.

2.4. Content evaluation

After the articles are collected, a descriptive analysis is performed and classified according to the research topic, then the material is evaluated for the articles by comparing the consensus between the articles by the researchers by consensus by checking the data bank spreadsheets of the articles. The content evaluation framework can be followed in Figure 7.

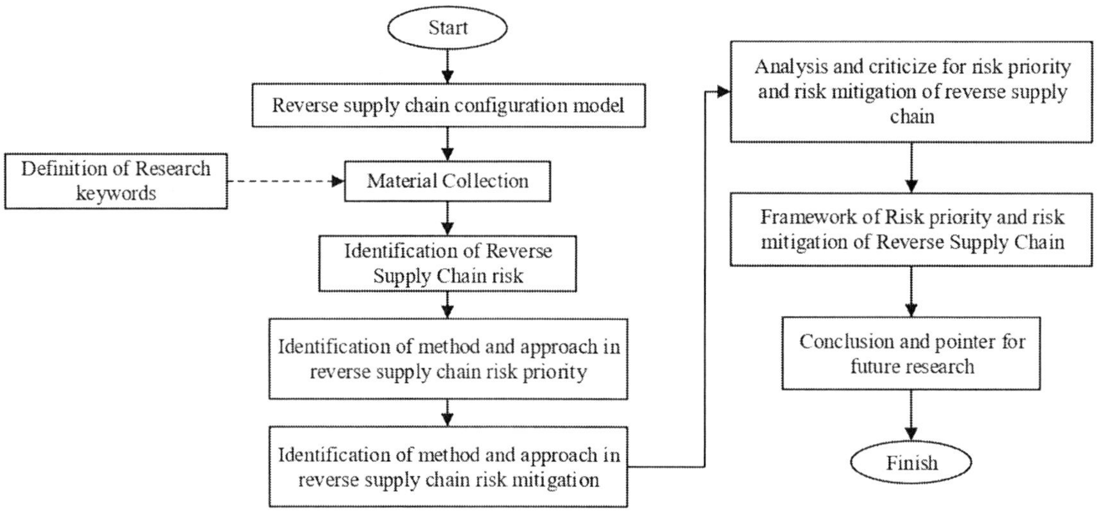

Figure 7. Content evaluation framework

3. Result and Discussions

3.1. Identification RSC risk

The risks of product demand uncertainty [24], production costs, volume, and evaluation of returned products need to be considered in planning multi-period and multi-product CLSC operations [20]. Some risks that occur in the plastic industry RSC activities are divided into two types, namely internal risk, and external risk. Internal risk consists of inventory risk, data management risk, time management risk, managerial risk, quantity risk, and cultural risk. While external risks are environmental and outsourcing risks, the risk of disruption [9]. According to [10], there are financial risks in the closed-loop supply chain (CLSC) planning.

According to [11], [21] In making recycled products in the plastics industry, there are risks in the form of uncertainties in raw material supply and uncertainty of consumer demand. From the results of the study [22], determining the optimal location of the RSC network also needs to consider transportation risk and the amount of product that must be transported. This study was conducted in the electronics trade. Research conducted in the automotive industry by [14] found that the purchasing patterns carried out by consumers for remanufacturing products in CLSC are influenced by the risks and benefits felt by consumers. From table 5, it can be seen the types of risk in RSC.

Table 5. Type of RSC risk

Author	Environment			Collection of return product			Supply Uncertainty	Demand Uncertainty	Process				Management							Financial
	A	B	C	D	E	F	G	H	I	J	K	L	M	N	O	P	Q	R	S	T
[11]	X	X	X		X										X	X	X	X	X	
[16]																				X
[17]	X													X						
[18]	X						X							X						
[19]		X		X																X
[20]				X		X				X										
[21]					X		X				X									X
[22]					X	X	X	X												X
[23]																				
[10]							X	X				X								
[12]							X	X							X					X
[24]							X	X												
[25]							X	X												
[26]							X	X												
[27]								X												
[28]									X						X					X
[29]													X							X
[30]								X												X
[31]								X												X
[32]																				X
[33]																				X

A :Disruption / Disaster ; B: Environment ; C: Culture; D: Collection of return product ; E: Quantity; F: Transportation; G: Supply Uncertainty; H:Demand Uncertainty; I: Marketing; J: Treatment; K: Quality; L: Capacity; M: Facility disputation; N:Operational; O: Inventory; P: Data processing; Q:Time management; R: Managerial; S: Outsourcing; T: Financial

3.2. Method and approach in RSC risk priority

Various risks that occur in Reverse Logistics need to be arranged and sorted by priority, research conducted in the plastic recycling industry is done by sorting the risk based on the level of loss and probability [11]. The Hybrid Multi-Criteria Decision Making Approach is Used to Prioritize Risks in Reverse Logistic Projects Based on Probability and Impact of Risk [11]. The Method of the Hybrid Multi-Criteria Decision Making (MCDM) Approach. Is an Integration Between Analytical Hierarchy Process (AHP) - Fuzzy Technique for order preference by similarity to an ideal solution (TOPSIS) and AHP - Preference Ranking Organization Method for Enrichment Evaluations (PROMETHEE), and AHP-Digraph Matrix, can be seen in Table 6.

Table 6. Method/approach reviews on reverse supply chain risk

Authors	Method and approach	Sector
[10]	Hybrid dan Chance constrained programming, GA, dan Monte Carlo Simulation	Manufacturing Industry
[11]	hybrid multicriteria decision making methods	Plastic Industry
[12]	MILP model, A two-stage stochastic, conditional value at risk (CVaR)	General Industry
[16]	Mathematical model, CVaR multi-period multi-objective two stage stochastic programming model with stochastic parameters	Undefined
[17]	Lagrangian relaxation algorithm	Glass industry
[18]	Game Theory	Manufacture
[19]	MILP	engine oil
[20]	MILP and solved by a mathematical programming Language, Sensitive analysis	Electronic
[21]	two stage stochastic, Fix-and-Relax Algorithm (FRA)	Undefined
[22]	Mathematical modeling, two-stage stochastic MILP model, CVaR	consumer goods company
[23]	MILP is utilized in the model and addressed by an optimization software	Electronic
[24]	Game-theoretical models	Undefined
[25]	Interval programming, Fuzzy programming	Computer
[26]	MILP, Multi-objective programming Stochastic programming	Manufacture
[27]	Mathematical modelling, Stochastic programming, Robust optimization	Automotive industry
[28]	Coordination activity	Truck engine
[29]	Hybrid meta-heuristic algorithm	Refrigerator
[31]	MILP formulation, expected net present value (ENPV), Pareto Optimal curve, augmented ε-constraint	Manufacturing industry
[31]	MILP	Undefined
[32]	Genetic Algorithm (GA)	electrical manufacturing
[33]	Mathematical programming, MAD, Var, CvaR	Undefined

The risk ordering model in the supply chain begins with the verification and risk mapping using the Supply Chain Operation Reference (SCOR) and House of Risk (HOR) methods [53], and then identifies the relationship between the risk of risk ranking by using the interpretative structural model (ISM) integration method [34]. This method can help decision-makers in choosing which risks should be handled first and which risks are handled then adjusted according to budget and resource constraints.

In determining risk priorities for perishable food SC, it is executed using the ISM method and risk priority number (RPN) [35]. In addition, the integration of the ISM-Dematel method can be used to determine the priority of the driving factors in implementing Agrifood Supply Chain sustainability [36]. Uncertainty, relevance, and subjectivity are inherent in the supply chain risk valuation process, therefore a comprehensive method to risk management is needed that takes into account the impact and the relationship between risk agents that are mutually integrated. This approach can be done using aggregated risk scores (ARS) by adding fuzzy logic. The use of fuzzy logic predicts uncertainty or ambiguity adequately [37] [38]. The failure mode effects critical analysis (FMECA) method is a risk valuation method for a product or activity using three valuation factors, namely severity, events, and detection. And also uses criticality analysis to rank the criticality of current risks [39].

3.3. Method and approach in RSC risk mitigation

Responsive risk mitigation strategies are carried out through a priority order of risk. According to [36], the formulation of risk mitigation strategies on perishable foods is carried out based on the risk priority number (RPN) value, then the priority of the mitigation strategy is determined by using a risk mitigation number (RMN). Risk mitigation analysis can be carried out at all levels, but by combining the RPN and RMN methods, the risk mitigation priority selection can be made, so that the risk mitigation strategy becomes more focused.

Determination of risk mitigation priorities can be done using the HOR method stage 2. HOR 2 is used to prioritize proactive actions that must be taken first, given the availability of resources, the level of difficulty in carrying out mitigation actions so that it can effectively reduce the risks that occur [52]. The combination of the balanced scorecard (BSC) tool can improve the analysis and determination of priorities of mitigation strategies in decision making, integration of risk mitigation strategies into the performance appraisal process with BSC can be used as an effective and proactive combination in strategy management [37]. Determination of risk mitigation priority strategies can also be done with the analytical network process (ANP) method [39]. The integration of HOR and ANP can be used to reduce SC risks and reduce the consequences of risks based on priorities [53], as described in Table 7.

Table 7. List method/approach risk priority and risk mitigation

Authors	Method/ Approach		Sector
	Risk priority	Risk Mitigation	Industry
[11]	Hybrid multicriteria decision making methods (AHP- Fuzzy Topsis, AHP-Promethe, AHP- Digraph)		Plastic
[34]	HOR - ISM		General Industry
[35]	ISM and Risk priority number (RPN)		Dairy Agro-industry
[36]	ISM- Fuzzy DEMATEL	RPN-RMN	Agro-I
[37]	Fuzzy Logic -Agregate risk score	BSC	Telecommunication
[38]	Fuzzy Logic		Dairy Agro-industry
[39]	FMECA		wooden toy industry
[52]	HOR 1	HOR 2	Fertilizer company
[53]	SCOR- HOR 1	HOR & ANP	Cassava agro-industry
[54]	SCOR- HOR 1	HOR 2	Agro-Industry

3.4. Analysis and criticize for RSC risk priority

Based on the results and discussion, it was found that the method used to determine risk priorities in the RSC palm sugar activities be able to do with Hybrid Multi-Criteria Decision Making, but the steps in this method are too long. The method of mapping risk with SCOR and determining risk priorities using HOR and ISM. This method is a combination of assessments in which risk identification uses HOR, then the risk is re-ordered using ISM. This method works twice in determining risk priorities. The RPN method, ARS measures risk in terms of the impact and occurrence of the risk, while FMECA measures risk by three factors, severity, incidence, and detection. These three methods (RPN, ARS, and FMEC) carry out quantitative risk measurement. Quantitative risk assessment is indeed good if it is supported by the availability of data and facts, not based on assumptions. Furthermore, to sort the results of the risk assessment the previous researchers used the ISM approach and critical analysis.

Risks that occur in the RSC activity of palm sugar are very complex and involve several stakeholders, for that we need a Intelligent Descision Support System (IDSS) that is used to make decisions appropriately [40] in minimizing the risk of RSC agroindustry of palm sugar. These risks need to be sorted by probability and risk impact by integrating ISM-Fuzzy DEMATEL. This method can analyze the elements of risk, identify relationships between elements of risk, and determine the ranking of risks in the form of a hierarchy and graph in order to facilitate the determination of risk mitigation. With the ISM-Fuzzy Dematel method, it enables decision-makers to be able to understand the relationship of related attributes in decision making, which can further increase the dependability of decisions [35].

The integrated ISM-fuzzy DEMATEL method [41] was created and used to identify and classify risks in the RSC activities. ISM is a systematic and interactive technique that depends on a group of independent experts and helps in knowing the interrelation between variables [42]. Fuzzy DEMATEL method be able to reveal the relationship between elements that influence other elements in uncertain conditions [43] [44]. ISM-Fuzzy-DEMATEL techniques are excellent to interpretive modeling techniques than other decision making, such as AHP, ANP, Graph theory, SEM modeling, Interpretive Ranking Process (IRP), etc.

The incorporation of the ISM method and Fuzzy-DEMATEL in this study is due to several things including [45], [46], [41]

1. ISM-Fuzzy-DEMATEL are able to describe the complex relationship between factors observed in decision making in an unpredictable environment.
2. The integration of the ISM-fuzzy-DEMATEL can show the cause and consequence relationship between the decision factors considered
3. Their integrated use can help in viewing the importance of the decision variables observed via well-explained diagrams (ISM-based hierarchical diagrams and cause-and-effect diagrams in fuzzy DEMATEL).
4. ISM can consider four possible relationships to analyze interactions between decision variables; However, it cannot differentiate the force of contextual relationships between decision variables using ISM. Meanwhile, Fuzzy-DEMATEL can explore more profound relationships and it moreover captures the power of contextual relationships by means of scale (from 0 to 4) under ambiguous environments.
5. ISM can be joint with DEMATEL to find out the direction and degree, in a multifaceted causal connection, thus, a complete feedback system thinks the importance and relationship between factors.

The DEMATEL method is appropriate for exploring multifaceted causal relations and sets of interrelated problems and is inclined to find ideal solutions in solving complex system problems. DEMATEL assists to get direct and indirect influence among criteria in a multi criteria decision problem. It also assists in calculating the relationships and strengths among the factors relevant [47]. The DEMATEL approach can help in classifying elements into groups of cause and effect [48]. However, DEMATEL utilization has limitations in dealing with issues of uncertainty, inaccuracy of data and prejudice related with human judgment [49]. Thus, we would be better off using fuzzy DEMATEL in this study [50], [51].

3.5. Analysis and criticize for RSC risk mitigation

Based on the findings of a strategy for risk mitigation that can be done using RMN, this strategy can direct decision-makers to choose the priority of mitigation to be carried out so that the mitigation strategy becomes more focused, but this strategy does not take into account performance appraisal. The weakness of this method can be corrected by integrating the BSC method for performance measurement. The RMN, BSC, and ANP methods, all three do not take into account the availability of resources and the level of difficulty in carrying out risk mitigation actions.

Based on the description above, the risk mitigation strategy in the RSC of palm sugar activities is carried out by using the HOR 2 method [52]. This method prioritizes proactive actions by considering the availability of resources and the level of difficulty so that risks can be reduced effectively [52]. HOR 2 is used to decide the initial activities that must be taken, given the different effectiveness and resources involved and the level of difficulty in carrying out a mitigation action. Companies must choose and management actions that are not so hard but can successfully decrease the possibility of risk factors happening [54].

3.6. Framework of RSC risk priority and risk mitigation of RSC

The framework of RSC risk priority and risk mitigation is depicted in Figure 8.

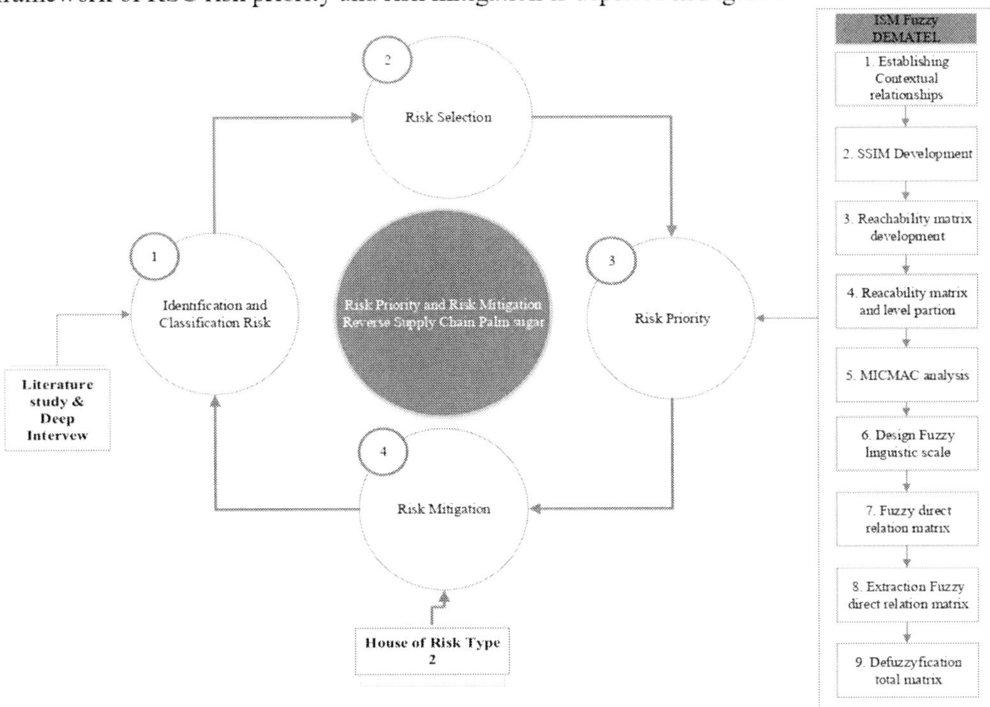

Figure 8. Framework of risk priority and risk mitigation RSC

The steps in determining risk priorities and risk mitigation in the RSC activities of the palm sugar agroindustry are as follows: First, identification and classification of risks that occur in the RSC activities of the palm sugar agroindustry (refer to literature studies and in-depth interviews with business actors and expert). Second, the selection of risks that occur in the RSC agroindustry of palm sugar, the risks were chosen are risks that have occurred or risks that are likely to occur. Third Determination of risk priorities using the ISM Fuzzy Dematel method with nine stages as shown in Figure 8. Finally, mitigate risk with the HOR type 2 approach.

The steps in HOR type 2 are as follows: first, determine the risk prevention measures based on the priority order of risk from highest to lowest. Note that one risk can be corrected by more than one action or one action can fix several risks, the second determines the assessment of risk prevention measures by considering the level of resource availability and the level of difficulty in carrying out preventive activities. the third calculates the total value of effectiveness against the level of difficulty ratio. Finally, determine the order of priority of actions based on the highest total effectiveness value.

4. Conclusions and Recommendations

The framework for determining risk priorities and risk mitigation strategies is carried out through four stages consisting of risk identification, risk classification, risk priority determination, and risk mitigation strategies. Risks that occur in the activity of RSC palm sugar is very complex and involves several stakeholders, these risks need to be sorted based on the probability and impact of risks,

Determination of the order of risk can be done by integrating the ISM-Fuzzy DEMATEL approach. This method can analyze the elements of risk and be able to describe the complex relationships between variables considered in decision making in an uncertain environment and can rank risks into hierarchical and graphical forms.

Mitigation strategies to minimize RSC risk based on the priority order of risk can be carried out using the HOR 2 method. HOR 2 method is expected to prioritize risk mitigation actions proactively by considering the availability of resources and the level of difficulty of risk management so that risks can be effectively reduced. The integration of the Hybrid ISM-Fuzzy DEMATEL-HOR 2 method, enables decision-makers to be able to understand the relationship of relevant attributes in decision making to determine risk priorities and risk mitigation strategies which can further improve the reliability of decisions.

For future research exploration that needs to be developed is RSC risk management in palm sugar agro-industry which is integrated with the optimization of RSC palm sugar products and sustainability of RSC palm sugar agro-industry.

Acknowledgments
This research is funded by Indonesia Endowment Fund for Education (LPDP) and Indonesia center of excellence for food security (local food innovation) University of Sultan Ageng Tirtayasa

Reference
[1] I. Victor and V. Orsat 2018 *Sugar Tech* **20** (1) 105–09
[2] M. Rumokoi 2004 *Proc. of the Aren National Seminar. Tondano. Research Institute for Coconut and other Palma Plants* **9**
[3] A. Iskandar, I. Yuliasih, and B. Haryanto 2014 *Int. J. Res. Eng. Technol* **2** (9) 33–44
[4] N. Kazemi, N. M. Modak, and K. Govindan 2018 *Int. J. Prod. Res* **7543** 1–24
[5] X. Bing, J. Bloemhof-Ruwaard, A. Chaabane, and J. van der Vorst 2015 *J. Clean. Prod* **103** 28–39
[6] S. M. Gupta 2013 *Reverse Supply Issues and Analyisis* (CRC Press Taylor & Francis Group)
[7] K. Govindan, H. Soleimani, and D. Kannan 2015 *Eur. J. Oper. Res* **240** (3) 603–26
[8] R. Frei, A. Bines, I. Lothian, and L. Jack, 2017 *Int. J. Supply Chain Oper. Resil* **2** (3) 246-66
[9] F. et al Gianfranco 2017 *World Conf. Transp. Res* **25** 695–15
[10] A. Nazari Gooran, H. Rafiei, and M. Rabani 2018 *Decis. Sci. Lett* **7** 13–24
[11] S. Senthil, K. Murugananthan, and A. Ramesh 2018 *J. Clean. Prod* **179** 716–30
[12] L. J. Zeballos and C. A. Méndez 2016 *Computer Aided Chemical Engineering* **39** 443–74
[13] I. P. Vlachos 2014 *Int. J. Integr. Supply Manag* **9**
[14] W. Septiani, Marimin, Y. Herdiyeni, and L. Haditjaroko 2016 *Int. J Sup. Chain. Mgt* **5** (2) 5164
[15] I. Russo, I. Confente, D. Scarpi, and B. T. Hazen *2019 J. Clean. Prod* **218** (1–2) 966-74
[16] M. Rahimi and V. Ghezavati 2018 *J. Clean. Prod* **172** 1567–81
[17] A. Jabbarzadeh, M. Haughton, and A. Khosrojerdi 2018 *Comput. Ind. Eng* **116** 178–91
[18] X. Han, H. Wu, Q. Yang, and J. Shang 2016 *Int. J. Prod. Econ* **182** 63–72
[19] M. M. Paydar, V. Babaveisi, and A. S. Safaei 2017 *Comput. Chem. Eng* **104** 38–55
[20] L. Thi, T. Doan, Y. Amer, S.-H. Lee, P. Nguyen, and K. Phuc 2017 *Am. J. Eng. Appl. Sci. Orig* **10** (4) 949–58
[21] S. Baptista, A. P. Barbosa-Póvoa, L. F. Escudero, M. I. Gomes, and C. Pizarro 2019 *Eur. J. Oper. Res* **274** (1) 91–107
[22] L. J. Zeballos, C. A. Méndez, and A. P. Barbosa-Povoa 2018 *Comput. Chem. Eng* **112** 211-38
[23] D. T. T. Linh, Y. Amer, S.-H. Lee, and P. N. K. Phuc 2018 *IPTEK J. Proc. Ser (*3) 30–36
[24] Y. He, 2017 *Int. J. Prod. Econ* **183** 39–52
[25] A. Mohajeri and M. Fallah 2016 *Transp. Res. Part D Transp. Environ* **48** 425–50
[26] G. Z. Saman Hassanzadeh Amin 2013 *Appl. Math. Model* **37** (6) 4165–76

[27] F. Mohammed, S. Z. Selim, A. Hassan, and M. Naqeebuddin 2017 *Transp. Res. Part D* **51** 146–172
[28] S. Zhao and Q. Zhu 2018 *J. Clean. Prod* **171** 1290–99
[29] J. Asl-Najafi, B. Zahiri, A. Bozorgi-Amiri, and A. Taheri-Moghaddam 2015 *Comput. Ind. Eng.*, **90** 414–28
[30] S. R. Cardoso, A. P. Barbosa-Póvoa, and S. Relvas 2016 *Comput. Chem. Eng* **85** 105–123
[31] S. R. Cardoso, A. P. F. D. Barbosa-Póvoa, and S. Relvas 2013 *Computer Aided Chemical Engineering* **32** 475–80
[32] R. Xiao, Z. Cai, and X. Zhang 2012 *Optimization* **61** (10) 1221–1251
[33] H. Soleimani, M. Seyyed-Esfahani, and G. Kannan, 2014 *Int. J. Prod. Res* **52** (6) 1843–67
[34] T. L. T. Nguyen, T. T. Tran, T. P. Huynh, T. K. D. Ho, A. T. Le, and T. K. H. Do 2018 *IOP Conference Series: Materials Science and Engineering* **337** (1) 12030
[35] S. Prakash, G. Soni, A. P. S. Rathore, and S. Singh 2017 *Benchmarking An Int. J* **24** (1) 2–23
[36] S. K. Mangla, S. Luthra, N. Rich, D. Kumar, N. P. Rana, and Y. K. Dwivedi 2018 *Intern. J. Prod. Econ* **203** 379–93
[37] A. Díaz-Curbelo, Á. M. Gento, A. Redondo, and F. Aqlan 2019 *Appl. Sci* **9** (24) 5329
[38] W. Septiani, Y. Herdiyeni, and L. Haditjaroko 2017 *Supply Chain Forum An Int. J* **17** (4) 218–30
[39] W. N. Tanjung, S. A. Atikah, S. Hidayat, E. Ripmiatin, S. S. Asti, and R. S. Khodijah 2019 *IOP Conference Series: Materials Science and Engineering* **528** (1) 12007
[40] M. Marimin, A. Wibisono, and M. A. Darmawan 2017 *Int. J. Supply Chain Manag* **6** (2) 60-74
[41] M. R. Mehregan, S. H. Hashemi, A. Karimi, and B. Merikhi 2014 *Int. J. Appl. Decis. Sci* **7** (3), 270–94
[42] J. N. Warfield, 1974 *IEEE Trans. Syst. Man. Cybern* (1) 81–7
[43] C.-W. Hsu, T.-C. Kuo, S.-H. Chen, and A. H. Hu 2013 *J. Clean. Prod* **56** 164–72
[44] R.-J. Lin 2013 *J. Clean. Prod* **40** 32–9
[45] H.-M. Chuang, C.-K. Lin, D.-R. Chen, and Y.-S. Chen 2013 *Sci. World J* **2013**
[46] L. S. dos Muchangos, A. Tokai, and A. Hanashima 2015 *Environ. Dev* **16** 76–89
[47] S. Gandhi, S. K. Mangla, P. Kumar, and D. Kumar 2015 *Int. Strateg. Manag. Rev* **3**(1–2) 96–109
[48] S. K. Mangla, K. Govindan, and S. Luthra *2016 J. Clean. Prod* **129** 608–21
[49] K. C. et al Chakraborty 2018 *Benchmarking An Int. J* **25**(7) 1992–17
[50] S. K. Patil and R. Kant 2014 *Appl. Soft Comput* **18** 126–35
[51] S. Luthra, K. Govindan, R. K. Kharb, and S. K. Mangla 2016 *Renew.Sustain. Energy Rev* **63**, 379–97
[52] I. N. Pujawan and L. H. Geraldin 2009 *Bus. Process Manag. J*
[53] N. Ummi, P. F. Ferdinant, A. Irman, and A. Gunawan 2018 *MATEC Web of Conferences* **218** 4022
[54] M. Ulfah, D. L. Trenggonowati, and F. Z. Yasmin 2018 *MATEC Web of Conferences* **218** 4023

IOP Conf. Series: Materials Science and Engineering **909** (2020) 012076 doi:10.1088/1757-899X/909/1/012076

Risk factors for rear-end collision: a systematic literature review

Fatin Saffanah Didin*, Hardianto Iridiastadi
Institut Teknologi Bandung

*E-mail: fatinsaffanahdidin@gmail.com

Abstract. Crash accidents on Indonesia toll road in 2019-2020 as much as 61% of accidents were a type of rear-end collision. That occurs at least two vehicles that act as crashing cars and cars in the hit, resulting in economic and social losses. There is not much research in Indonesia that discusses related to the disaster behind the run. The purpose of this review is to investigate human aspects while driving and gain potential development possible to reduce the rear-end collision. Search engine in this study is Science Direct, PubMed, Taylor & Francis Online, and IEEE. Once obtained, 826 articles carried out the process of checking, and the appropriate characteristics in this study, the result was 13 articles, further conducted a thorough review. Rear-end collision happened because of the contributions of one or all of the risk factor groups factors that have strong evidence ($p < 0.001$ or OR> 3), including drivers, vehicles, and the environment. This study will increase alertness to the risk factors of rear-end collision on toll roads. Future research intervention can use music, and the traffic density of the driver's domicile provides an effect on the behavior of drivers who drive with the type of minibus vehicles.

1. Introduction

The development of the infrastructure of toll roads makes increased use of vehicles in Indonesia. The number of accidents in Indonesia has increased gradually since 2017-2018 [1]. While in 2019-2020, 79 accidents occurred on toll roads or freeways with 61% overall the types of accidents are rear-end collision [2,3].

Crash collision on the freeway can occur when the driver interacts with the other vehicle continuously, then the driver accelerations or slows down the driving speed and performs track changes. A collision accident can occur on one vehicle or a single accident, and a continuous accident occurred at a minimum of two vehicles [4] [5]. According to the information from the World Health Organization (WHO), There are1.35 million people become victims of death caused by traffic accidents. Then the traffic accident caused injuries from low, moderate, and severe levels to death. From 20 to 50 million people suffered an unfatal injury by causing disability of its results [6].

The severity of an injury can be classified, include no injuries, possible injuries, suspected minor injuries, paralyzed injuries, suspected severe injuries, and fatal injuries [6]. Traffic accidents happened caused harm to the victims and roadways from both economic and social factors [7]. Based on the position of the vehicle, several types of collision accidents are the accident of the rear-end collision, front-shelf, side, and other crashes. However, the most common accident event and cause severe injuries and death [6]. Then based on vehicle operation type, there are two various vehicles, automatic vehicles as well as manual vehicles. The manual vehicle type has limitations on response times and driving errors that impact the rear-end collision [8]. Rear-end collision rises when the vehicle is too close to the other vehicle located in front of it [9] [10]. Based on the type of vehicle, a rear-end collision occurred between a small vehicle (passenger transport vehicle, van, pick up) crashing into other small vehicles, small vehicles crashing into large vehicles (trucks and buses), big vehicles crashing into small vehicles and big vehicles crashing into big vehicles [8,11].

Recently, not much research in Indonesia that discusses related to rear-end collision. Based on the review, this systematic literature review investigates the human aspects of driving relevant to the phenomenon of a rear-end collision in the last two years (2019-2020). Also, it sees potential development that has done to reduce the incidence of the rear-end collision.

2. Methods

2.1 *Literature search and data management*

The systematic literature review uses four electronic search databases from 2019 to 2020, using the keyword and advance search feature on-site. They are Science Direct, PubMed, Taylor & Francis Online, and IEEE (Institute of Electrical and Electronics Engineers). The keyword is the rear-ended collision, and the process review was using Microsoft Excel and Mendeley software. This systematic literature review search procedure was conducted based on a systematic rule of review of the University of Edinburgh and several other review literature research [12 – 14].

2.2 *Included and excluded Criteria*

In this systematic literature review, the rear-end collision focused on the vehicle with manual conditions. Here are the criteria of the selected article:

1. Connected to the rear-end collision
2. Year of article publication 2019-2020
3. Report on risk factors that cause rear-end collision
4. Using one of the investigation tools to examine the risk factors of the rear-end collision, such as experiments, simulations, and direct observation.

2.3 *Screening process*

Initially performed a paper search on the search site using the keyword in advance search feature. After obtaining the paper with the keyword, it checked, and the selection ensures that no paper duplicated. Screening process by checking the abstract of the paper, because the abstract can conduct the formulation of problems and conclusions on the research. Thus, by examining the abstract, a paper is potentially included in the criteria received in the systematic literature review of the rear-end collision.

2.4 *Evaluation process*

At this stage, evaluating the risk factors of rear-end collisions is seen from the study's significance. Criteria are said to have a strong relationship when having a minimum odds ratio value of 1.0. Besides, the significance value [P-value] also used to classify the risk factor [13]. The risk factors in the rear-hit accident are classified as follows:

1. Risk factors with strong evidence (odds ratio (OR) of > 3 or P-value < 0.001 ($P < 0.001$)
2. Risk factor with reasonable evidence (OR 1-3 or P-value < 0.05 ($P < 0.05$)
3. Risk factor with insufficient evidence (OR < 1 and P-value < 1 ($P > 0.05$)

3. Results

3.1 Literature search and evaluation

The search for the literature review of rear-end collision starting from the year 2019-2020 generates 830 articles from 4 types of search site publishers online research as in the Figure. 1. Each of the search results from databases online such as Science Direct 432 articles, IEEE 11 articles, Taylor & Francis Online 329 articles, and PubMed 58 articles. Then from a total of 830 articles, four articles duplicate the remaining 826 articles. Further checks related to the title and abstract 798 articles do not match the criteria. Six articles are still incomplete. There are 22 articles reviewed in total and resulted in 9 articles that did not conform to the criteria. For the result, the study of this literature uses13 articles that examine the risk factors of rear-end collision.

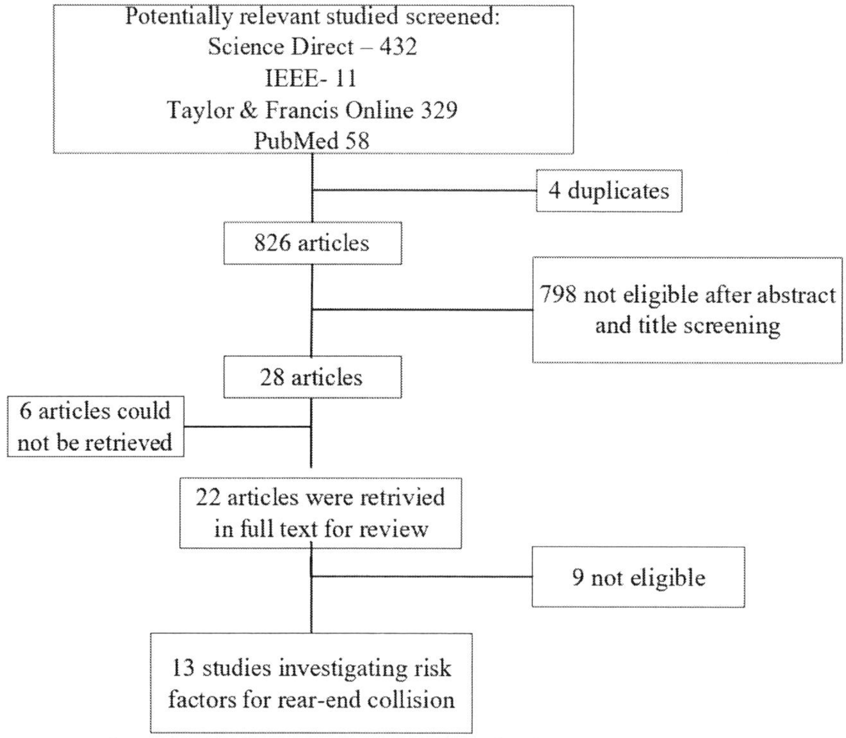

Figure 1. Stage of systematic literature review of studies investigating risk factors for rear-end collision

3.2 Investigation Tools for Risk Factors of Rear-end Collision

There are several practical investigative tools related to the risk factors of the rear-hit accident covering experiments [15 – 17], Simulation [18] [8], and direct observation [11]. Each method has advantages and disadvantages in its use in research [6].

a. Experiment

Experiments conducted for investigation of risk factors rear-end collision usually use vehicle simulation tools such as OKTAL CDS-650 [15]. A vehicle simulator equipped with the features of manual and automatic vehicle transmission, steering the power steering, there are gas pedals, brakes, TV screens, sound effects using a sound system, human models, and using video recorders [6,16]. Aside from the necessary experimental equipment, research has participant criteria that can participate in the experiment. Then the number of participants following the experiment is adjusted to the number of factors, levels, and conditions of the experiment. The number of participants following the experiment calculated using power statistics so that the data obtained sufficient for research. Participant's criteria in a rear-hit accident experiment seen from gender, age, and driving Experience [16] [15). Then the experimental conditions carried out the distraction using mobile phones (voice messaging and text messaging) by drivers while driving, differences in the combination of vehicles (cars crashing trucks and car crashing trucks), differences in traffic density situation, and the difference in weather conditions. Before conducting an experiment, participants filled in an approval statement about his participation in the experiment. The statement's content includes explanations of experiments from the objectives, experimental activity to be conducted, duration of the experiment, compensation of experimentation when there is, and possible risk of experimenting. The participants made in advance the adjustments to the simulators used before taking the actual experimental data. The resulting data of the experiment is processed and analyzed statistically by the researcher. Nevertheless, there are also types of experiments using used vehicles and human models to carry out rear-hit accidents, but these experiments are not repeatable and expensive [15][16][11][4].

b. Simulation
The simulation used in this study is generally a numerical simulation. This simulation can use to see the impact of injuries from rear-end collisions by using different parameters and combinations. The combination in the simulation can be the difference in speed, vehicle type, weather variation, traffic density, driving duration, and the order of the vehicle in the Journey [4][6][18]. Numerical simulation is one of the most powerful investigative tools. Nevertheless, numerical simulation requires sophisticated modeling, complicated analysis, and some natural conditions in the driver, vehicles, and environments are usually ignored [6].

c. Field Observation
On-the-field observation is done by observing ongoing vehicle journeys or using data recorded from the already occurred rear-end collision. The observation in this field provides complete information and details. Data obtained from field observations include genetic data such as human factors (gender, age, fatigue, sitting position or alcohol use), vehicle factors (vehicle type, vehicle speed, vehicle weight) and environmental factors (road conditions, weather conditions, lighting conditions) [6][19]. Direct observation uses in a particular area that records vehicles at specific points on a journey using sensors and video recorders (1.4). Observation can be done on two observation lines and observe days observation so that results can see what the highest probability of a rear-end collision can occur [4].

3.3 The parameters of rear-end collision

The parameters used to investigate the risk factor of rear-end collision are driving behavior, acceleration, retarding, response time, safety margin, Time Headway (TH), Time-to-Collision (TTC), Variable Speed Limits (VSL), Time Exposed Time-to-Collision (TET), Time Integrated Time-to-Collision (TITE), Total Travel Time (TTT) and Breaking Reaction Time (BRT). The definition of each parameter can explain as follows, driving behavior was observe during the drive in the road used camera [20]. The safety margin is a safety risk level that the driver feels by maintaining a secure distance while driving; this level of risk has an upper and lower limit for the level of risk received [21]. Acceleration and deceleration of vehicle speed observed during the vehicle running according to the research scenario conducted.

Response time is the time needed for the driver to respond to the stimulus of an incident when the driver is driving. TH is the time difference between the upcoming occurrences of two vehicles consecutively at a point of measurement on a track [20]. TTC is the required time for two vehicles to collide if both vehicles continue to use the same speed on the same track [5,20,22,23]. Variable speed limits are the vehicle speed limit set on a driving rule on the road [18]. TET is a summation of all the drivers used when approaching a vehicle in front of it or dangerous traffic conditions during the travel period studied [8,24]. TITE is the summation of entities TTC which is lower than the threshold that expects to express the average duration of a vehicle facing a critical situation or unsafe condition [22,25]. TTT is the primary indicator for assessing the operation and efficiency of the transportation system seen from the time spent by vehicles on the toll road and the number of vehicle-generated systems in the research period [26]. BRT is a vehicle braking reaction time performed by the driver when driving [15].

3.4 Risk Factors of Rear-End Collision

Based on a systematic literature review conducted, there are various risk factors of rear-end collision. In Table. 1 Risk factors are affecting the risk of rear-end collisions. The risk factor rear-end collision grouped into four groups: driver characteristics, vehicle characteristics, collision mechanisms, and environment. The strong evidence of risk factors over the driver includes driver emotions, male drivers, drivers under the age of 25, drivers over 64 years old, and driver response time. Risk factors that have sensible evidence are a distraction to the driver, driver behavior, driver's ability, gender, and the driver's age. Meanwhile, the risk factor with insufficient evidence is the driver of the female, who is over 64 years old.

In a risk factor group of vehicle characteristics that have strong evidence is a type of truck vehicle, the vehicle carries one or more connections, the vehicle does not have an airbag, year of vehicle production, and seat belts not used while driving. A risk factor with reasonable evidence is that the vehicle is not a passenger vehicle type, and its speed is more than 70 km/h. Then a risk factor collision mechanism that has strong evidence is a driver of text messaging while driving, a driver in the influence of drugs while driving, and a driver in the influence of alcohol while driving. A risk factor that has reasonable evidence is that the vehicle is hit by slowing or stopping, the vehicle continues ahead before the collision, the new vehicle starts to run or stops on the road before the collision, vehicles that hit the direction of the turn, the vehicle that hit the turn, the vehicles crashing to turn and the vehicle crashing changed lanes. The location of vehicle damage is in the direction of 7 hours, the location of the collision's impact on the vehicle body that hit in the direction of 7 hours, and the collision on the track with different the risk factors that have insufficient evidence is oscillation included. Then a group of environmental risk factors that have strong evidence is the road's characteristics and drive during the day at 07.00-18.00. Risk factors that have reasonable evidence are the weather, snowy or icy road surface, driving on weekdays, driving at night, daily traffic density, platoon ratio, and > 12 lux lighting level. A risk factor that has insufficient evidence is driving in spring and traffic conditions that are not too dense.

Table 1. Risk factors rear-end collision

Risk Factors Group	Risk Factor	Strong evidence	Reasonable evidence	Insufficient evidence	References
Driver Characteristic	Driver emotion	√			[9]
	Distraction		√		[9]
	Male Driver	√			[11],[6]
	Female Driver			√	[16]
	Age under 25	√			[11],[6]
	Age 64	√		√	[11],[15],[6]
	Driver Behavior		√		[20]
	Driver respons Time	√			[20]
	Driving Skill		√		[27]
	Gender		√		[27]
	Age		√		[27]
Vehicle Characteristic	Not a passanger vehicle		√		[28]
	Speed more than 70 km/h		√		[28],[27]
	Truck	√			[11],[27]
	Vehicle carrying one or more trail connections	√			[11]
	No airbag	√			[6]
	Not using seatbelt	√			[6]
	Year of vehicle production	√			[6]

Table 1. Risk factors rear-end collision

Risk Factors Group	Risk Factor	Strong evidence	Reasonable evidence	Insufficient evidence	References
Collision Mechanism	Vehicle action being hit slows or stops		√		[28]
	Vehicle continue to advance before collision		√		[11]
	Vehicle start or stopping on the road before collision		√		[11]
	Vehicle action that was hit by turning		√		[28]
	Vehicle crashing towards turn		√		[28]
	Vehicle changing track		√		[28]
	Damage impact in vehicle was hit body hit in the 1 clock direction		√		[28]
	Damage impact in vehicle crashing hit body hit in the 1 clock direction		√		[28]
	Collison on same line with different vehicle types		√		[4]
	Send Voice messaging while driving		√		[16]
	Texting while driving	√			[16]
	Driving under alcohol control	√			[6]
	Driving under drug control	√			[11],[6]
	Oscillation		√		[18]
	Driver Glace			√	[7]
Environment	Weather		√		[9],[27]
	Snowy or icy road surfaces		√		[28],[6]
	Weekdays		√		[11]
	Spring			√	[11]
	Night-time		√		[4]
	Uncongested traffic conditions			√	[15]
	Traffic density		√		[29],[27]
	Road Characteristic	√			[27]
	Daytime 7-18	√			[6]
	Platoon ratio		√		[5]
	Illumination level > 12 Lux		√		[5]

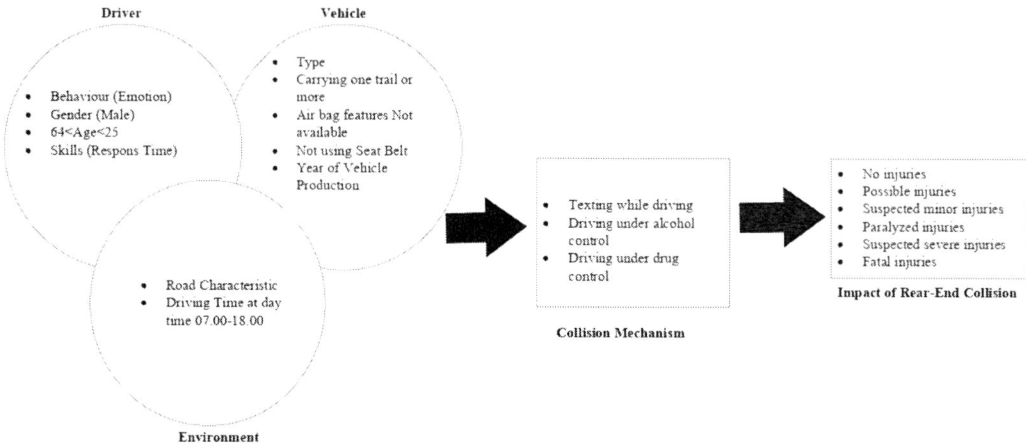

Figure 2. Mapping of risk factor relations, the rear-end collision mechanism and the impact of accidents contained

The mapping of risk factor relationships with strong evidence, a rear-hit accident mechanism, and the impact of accidents are present in the image. 2. The rear-end collision occurs influenced by three factors, namely drivers, vehicles, and the environment. Whereas when a rear-end collision occurs, the impact of the collision varies depending on the input causing the collision that occurred before. The impact of this collision began not to suffer injuries until fatal injuries even caused death. The three risk factors of a rear-end collision relate to each other because the driver drove a vehicle driving on the road with a particular environment—a critical driver of the driving system's controller.

4. Discussion

The rear-end collision occurs at least two vehicles having the vehicle crashing role, and the vehicle was hit [28]. Risk factors for rear-end collisions include driver characteristics, vehicle characteristics, accident mechanisms, and environmental mechanisms. In the driver's characteristic, there are different research results related to gender and driver's age risk factors. The male driver has strong evidence as a risk factor for a rear-hit accident [6,11], but there is insufficient evidence that the female driver as a risk factor for a rear-hit accident [16]. On the Age risk factor, which has strong evidence is a driver with age less than 25 years, this is because the driver's age under 25 years old has no control of emotions, the driving ability is quite good [6] [11] [9] [27].

In addition to the response time, the driver has strong evidence as a risk factor for rear-end collision; it relates to the driver's response when responding to the stimulus unexpected conditions of other vehicles and the environment while driving [20]. Response times, such as acceleration or retarding vehicles, relate to the driver's condition of driver characteristics. Each driving driver has a different perception of risk. Driver reaction and risk perception are easily affected by psychological conditions and environmental conditions [20]. Psychological conditions, such as stress, anxiety, and fatigue, are the impact of driving demands. Driver's demands are the driver's characteristic when facing a traffic density situation based on its ability [30.31]. Based on characteristics of the driver differ in this can be seen in the characteristics of drivers who come from 3 cities in Indonesia that have a level of traffic density and the type of car commonly faced by drivers in the city where the driver's residence. The driver's behavior on the toll road differs from that of drivers who dare to take the risk of following the road shoulder, moving on toll roads, and still driving safely. One of the psychological conditions is that the driver's emotion becomes one of the risk factors of an accidental rear crash that has strong evidence [9]. The driver's emotions affect driving conditions. The emotions of the driver can change in the form of positive and negative emotions. The driver's emotions depend on the task the driver belongs to. The driver who has terrible or unstable emotions will drive fewer maximums, such as not being able to control the speed [32]. Besides, the accident data that occurred on the Indonesian toll road in 2019-2020 showed that 24.05% of 79 accidents were caused by drivers who were unable to control their speed [2.3]. This shows that the findings of the review literature and the reality in Indonesia are quite relevant.

The cause of the second-highest accident on the Indonesia toll road in 2019-2020 is exhausted [2.3]. Fatigue that occurs in the driver can cause drowsiness, focus fatigue until half asleep at a particular driving time or long driving duration [33]. This fatigue condition affects the driver's response time to the traffic density environment. In the field condition of Toll road Indonesia, the driver likes to listen to music while driving to improve comfort, relieve stress, and increase focus. Besides music also, the driver usually listens to the radio. Music that the driver heard can raise the working mental business if the type of music does not match the driving preference [34]. Hence, determining the influence of music on the driver on the Indonesian toll road can be further investigated to reduce the risk of a rear-end collision.

On the risk factor of vehicle characteristics, the risk factor that has strong enough evidence is the type of truck vehicle, the vehicle carrying connections, vehicles not complete with the airbag, seatbelts, and vehicle production of a reasonably old year. Airbag and seat belts themselves have a role in minimizing the injuries experienced by the driver when the collision occurs [6] [11] [27]. The type of truck vehicles that usually do not carry the connection, carrying one connection, even more, has limited driving flexibility when compared to small vehicles. Some factors can be attributed to the risk factor of the rear-end collision mechanism when different types of vehicles are on one line [4]. Meanwhile, the accident on the Indonesian toll road 2019-2020 showed that 41.72% of the 79 accidents occurred in the type of minibus vehicle with the scenario of rear-end collision truck-minibusses and minibus-Minibus [2.3]. This type of minibus vehicle is also widely used in Indonesia, and this can be future researched on the accident behind the rear on the type of minibus vehicle.

Meanwhile, the rear-end collision mechanism, which is a text-messaging driver, has strong evidence as a risk factor for rear-end collision-related drivers that have insufficient evidence as a risk factor [16]. Nevertheless, the difference in research results is that there is a difference in the distraction of sending messages using mobile phones on Chinese drivers and western origin. The Chinese truck driver's occurrence of a change of sight on the mobile phone is only 4%-12.5%, while the driver in the West Country is 10%-20% [7]. The rear-crash accident has a risk factor with strong evidence associated with road characteristics [27]. A straight road characteristic is the type of road that often happens to the rear sprinter [28]. Then the risk factors of other rear-end collisions are driving during daylight hours [07.00 18.00], and the night is related to the existing lighting on the road and the traffic density, especially on weekdays [4] [5]. Data that occurred on the Indonesian toll road in the year 2019-2020 showed that 23% of the 79 accidents occurred in two times. Such as in the morning at 05.00-10.00, and 10% occurred during the day at 10.00-15.00. Data shows the findings of the review literature, and the fact in Indonesia is quite relevant for further investigation.

The systematic literature review conducted based on the value of the significant value of the P-value and odds ratio. Then this review shows that the risk factors have a link to the research, and there are differences in the results of previous research. Rear-end collisions occur due to contributions from one or all of the risk factors, including drivers, vehicles, and the environment. The collision mechanism occurs due to distraction in drivers ranging from internal distracting [drug and alcohol use] and external distracting such as mobile phone usage.

5. Conclusion

The systematic literature review shows that many factors are affected by the rear-end collision. Risk factors of rear-end collision divided into driver characteristics, vehicle characteristics, collision mechanisms, and environments. Accidents happened due to the contributions of one or all of the risk factor groups. Risk factors that have strong evidence and relevant to the accident condition in Indonesia toll road is the behavior [emotion], the response time of the driver so that the difficulty controlling the speed, type of vehicle in the form of the minibus, and time of driving at 05.00-18.00. While the intervention that can use in the future research of external intervention in the driver is music, and the traffic density at the driver's residence originated. This study expects to increase alertness to the risk factors of rear-end collisions on the freeway [toll]. The implications of this study know other risk factors as a preventive action strategy on a rear-end collision on the freeway [toll].

6. Acknowledgments

This study was funding by The Indonesia Endowment Fund for Education [LPDP]- The Ministry of Finance through doctoral program scholarship.

7. References

[1] Badan Pusat Statistik. Data Kecelakaan di Indonesia Tahun 2017-2018. 2020.
[2] Kompas.com. Data Kecelakaan di Jalan Tol Indonesia Tahun 2019-2020le [Internet]. 2020. Available from: https://www.kompas.com/
[3] Metro.tempo.co. Data Kecelakaan di Jalan Tol Indonesia Tahun 2019-2020 [Internet]. 2020. Available from: https://metro.tempo.co/
[4] Jo Y, Oh C and Kim S 2019 Estimation of heavy vehicle-involved rear-end crash potential using WIM data *Accid Anal Prev* **128** 103–113.
[5] Wang Y, Li Z, Ye Z and Zhang Y 2020 Nighttime safety evaluation for signalized intersections at the signal cycle level based on rear-end conflict models considering lighting and traffic conditions *Traffic Inj Prev* **21** 87–92.
[6] Chen F, Song M and Ma X 2019 Investigation on the injury severity of drivers in rear-end collisions between cars using a random parameters bivariate ordered probit model *Int J Environ Res Public Health* **16**
[7] Pipkorn L and Bianchi Piccinini G 2020 The role of off-path glances: A quantitative analysis of rear-end conflicts involving Chinese professional truck drivers as the striking partners *J Safety Res [Internet]* **72** 259–66
[8] Qin YY, He ZY and Ran B 2019 Rear-end crash risk of CACC-Manual driven mixed flow considering the degeneration of CACC systems *IEEE Access* **27** 140421–140429
[9] Khadim S, Riaz F, Jabbar S, Khalid S, Aloqaily M 2020 A non-cooperative rear-end collision avoidance scheme for non-connected and heterogeneous environment *Comput Commun* **150** 828–40
[10] Baldock MRJ, Long AD, Lindsay VL, Mclean AJ 2005 Rear end crashes Report documentation
[11] Shao X, Ma X, Chen F, Song M, Pan X, You K 2020 A random parameters ordered probit analysis of injury severity in truck involved rear-end collisions *Int J Environ Res Public Health*. **17**
[12] Piper ARJ 2013 How to write a systematic literature review : a guide for medical students How to write a systematic literature review : a guide for medical students
[13] Kusmasari W Sutalaksana IZ 2018 Risk Factors for Musculoskeletal Symptoms of Construction Workers : A Systematic Literature Review1–15.
[14] Talib M, Farhan A, Sadullah M, Anwar K, Kassim A 2019 Review article A review of behavioral issues contribution to motorcycle safety 1–13
[15] Chen T, Bai L, Sze NN 2019 Factors affecting the severity of rear-end conflicts: A driving simulator study. *ICTIS 2019 - 5th Int Conf Transp Inf Saf* **25203717** 1182–1187.
[16] Chen Y, Fu R, Xu Q, Yuan W 2020 Mobile phone use in a car-following situation: Impact on time headway and effectiveness of driver's rear-end risk compensation behavior via a driving simulator study *Int J Environ Res Public Health* **17** 1–17.
[17] Xu J, Li Y, Lu G, Zhou W 2009 International Journal of Impact Engineering Reconstruction model of vehicle impact speed in pedestrian – vehicle accident *Int J Impact Eng* **36** 783–738
[18] Zheng Y, Zhang G, Li Y, Li Z. Optimal jam-absorption driving strategy for mitigating rear-end collision risks with oscillations on freeway straight segments. Accid Anal Prev. 2020;135[May].
[19] Chen F, Chen S, Ma X 2018 Analysis of hourly crash likelihood using unbalanced panel data mixed logit model and real-time driving environmental big data. *J Safety Res*

[20] Zhang J, Wang Y, Lu G 2019 Impact of heterogeneity of car-following behavior on rear-end crash risk. Accid Anal Prev **125** 275–89
[21] Lu G, Cheng B, Wang Y, Lin Q 2013 A Car-Following Model Based on Quantified Homeostatic Risk Perception A Car-Following Model Based on Quantified Homeostatic Risk Perception
[22] Minderhoud MM, Bovy PHL 2001 Extended time-to-collision measures for road traffic safety assessment **33** 89–97.
[23] Hayward JC, Transportation P. Near-Miss Determination Through. :24–35.
[24] Li Y, Li Z, Wang H, Wang W, Xing L. 2017 on freeways **104** 137– 145.
[25] Bella F. Driver Behavior In Car-Following : A Driving Simulator Study. :545–55.
[26] Wang M, Daamen W, Hoogendoorn SP, Arem B Van, Wang M, Daamen W, et al 2016 Connected variable speed limits control and car-following control with vehicle-infrastructure communication to resolve stop-and-go waves vehicle-infrastructure communication to resolve stop-and-go waves. J Intell Transp Syst [Internet] **20** 559–572. Available from: http://dx.doi.org/10.1080/15472450.2016.1157022
[27] Zhang KI and Hassan M 2019 Investigating factors contributing to injury severity in work zone rear-end crashes *ICTIS 2019 - 5th Int Conf Transp Inf Saf* 1137–1145.
[28] Cicchino JB and Zuby DS 2019 Characteristics of rear-end crashes involving passenger vehicles with automatic emergency braking. Traffic Inj Prev [Internet]. **20** S112–S118. Available from: https://doi.org/10.1080/15389588.2019.1576172
[29] Gecchele G, Orsini F, Gastaldi M, Rossi R 2018 Freeway rear-end collision risk estimation with extreme value theory approach. A case study. Transp Res Procedia [Internet] **37** 195–202. Available from: https://doi.org/10.1016/j.trpro.2018.12.183
[30]Taylor P and Matthews G 2010 Theoretical Issues in Ergonomics Science Towards a transactional ergonomics for driver stress and fatigue 37–41.
[31]Tement S, Plohl N, Horvat M, Musil B, Jakus G, Sodnik J 2020 Driving demands , stress reactivity and driving behavior : An interactional approach. *Transp Res Part F Psychol Behav [Internet]* **69** 80–90. Available from: https://doi.org/10.1016/j.trf.2020.01.001
[32]Steinhauser K, Leist F, Maier K, Michel V, Pärsch N, Rigley P, et al 2018 Effects of emotions on driving behavior. *Transp Res Part F Psychol Behav [Internet]*. 2018;59:150–63. Available from: https://doi.org/10.1016/j.trf.2018.08.012
[33]Zhang G, Yau KKW, Zhang X, Li Y 2016 Traffic accidents involving fatigue driving and their extent of casualties *Accid Anal Prev [Internet]* **87** 34–42. Available from: http://dx.doi.org/10.1016/j.aap.2015.10.033
[34]Ünal AB, Waard D De, Epstude K, Steg L 2013 Driving with music : Effects on arousal and performance *Transp Res Part F Psychol Behav [Internet]* **21** 52–65. Available from: http://dx.doi.org/10.1016/j.trf.2013.09.004

IOP Conf. Series: Materials Science and Engineering **909** (2020) 012077 doi:10.1088/1757-899X/909/1/012077

Household solid waste composition and characterization in Indonesia Urban Kampong

Juntara Semilu Rosesar[1*] **and Gabriel Andari Kristanto**[1]
[1] Environmental Engineering, Civil Departement, University of Indonesia

*Email: juntara.semilu@ui.ac.id

Abstract. Household waste compositions and characteristics are the basic Information needed in planning a waste management system. The data are affected by many factors, such as the area characteristic and their income. As a part of city growth, urban Kampong is generally established due to urbanization and can be characterized by its density, low income residents, and poor sanitation. This research is conducted to analyze the composition and characteristics of solid waste produced in Urban Kampong. Cross sectional study approach was conducted to analyzed the household characteristics; the Indonesia standard of 19-3694-1994 is applied to examined wastes generated with 100 kg of waste samples was collected for 8 days in each urban kampong. The research result shows urban kampong waste composition is dominated by organic waste with 61,62%. While inorganic wastes are dominated by plastic (13,14%), multi-layered (5,52%) and paper (4,12%). The average of waste generation in urban kampong is 0,5 kg/person/day. This study is expected to be the baseline for improving waste management system in urban kampong in Indonesia.

1. Background

Solid waste has become the biggest issue in Indonesia that has not yet to be solved. According to the Ministry of Environment and Forestry and the Ministry of Industry in 2016, Indonesia's solid waste generation reached 65,2 million ton annually and most of them were not manageable [1]. These issues and challenges are also faced by other developing countries due to the incrase in population, living standards, and also changes in their lifestyles thus resulting in more solid waste generation [2]. Therefore the need for solid waste disposal and processing area has increased [3-7]. Other than that, unoptimized solid waste management systems can result in environmental pollution (air, water, and land) producing harm to-human health and the environment [8-11].

This waste related issue is also faced by many other urban kampong in big cities. The high rate of urbanization in urban areas is the main caused of daily waste volume increase [12-13]. At the same time, urban kampong is a highly populated area thus making waste management critical because of its direct effect on the population lives' quality [14]. The increasing urbanization phenomenon makes urban kampong prone to become slum area due to the imbalance between the increasing number of population versus good management on residential areas [15-16]. Aside from urbanization, slum areas also emerge as a consequence of less housing regulations creating illegal and unplanned areas in the city [17]. As the population increases, the number of households consequently increases thus resulting in a higher generation volume of household solid waste. Besides the high volume, the household solid waste composition has the potential of harming society's health [14, 18].

IOP Conf. Series: Materials Science and Engineering **909** (2020) 012077 doi:10.1088/1757-899X/909/1/012077

The composition and characteristics of waste are the basic information in planning a waste management system [19]. The appropriate management system according to UU 18/2008 is comprised of 2 main groups, the first one being waste reducing and the other is waste handling [14]. Many factors can effect the solid waste composition such as urbanization, GDP per capita, local culture like food of choice and other activities [13-14, 20-21]. Take for example the city of Taipei, where Taiwan, the high level of urbanization affects the society's behavior in using plastic increasing plastic waste annually [21]. In Indonesia, the household municipal waste consist of 60% organic and the rest are inorganic waste like plastics (±17%), papers (±7%), clothes (±0,4%) , rubbers (±0,1%) and metals (±1,5%) [22-23].

Previous studies have been conducted during the last decade regarding waste in slum area. For example, Sethi et al. (2013) in Jalandhar city, Punjab, India and Ramachandra et al (2018), Gonzalez et al. (2010) in Northern Mexico, Dorenfeld et al. (2012) in Namibia rural areas, Shah et al. (2012) in rural areas near Tekanpur in India; El-Messery et al. (2009) in Egypt rural areas; Mohammadi et al. (2012) in northern Iran, and De Medina-Salas (2013) in Cosautla'n_De Carvajal, Veracruz, Mexico. These research found organic waste was higher than inorganic waste like glasses, metals, plastics and others [24-31].

The lack of accurate specific information on waste compositions and characteristics resulting in the lack of development and planning in waste management in urban kampong. Furthermore, there is little to none attention to waste management in urban kampong from the local goverment. Urban kampong residents also face great health risks due to unmanage waste [14, 18]. The health risks can be increased due to the growth of many vectors that carry diseases from unmanaged waste piles like flies (fly-borne diseases), mosquitos (mosquito-borne diseases),and rodents (rodent-borne diseases) [18]. This study aims to analyze the quality and quantity of household solid waste in urban kampong in two big city in Indonesia. This research is not only important for developing waste management but also as a baseline for various strategies in improving sanitation in urban kampong areas.

2. Methods

2.1. Area of study

This research was conducted in an urban kampong in four areas, which were Kampong Markisa, Kampong Cimone, Kampong Gedong Pompa and Kampong Cikini Kramat (Gambar 1). Two of the kampongs are located in Jakarta City as the capital city of Indonesia while the other two located in the city of Tangerang. Kampong Gedong Pompa is located in North Jakarta, which lies close to the Java Sea. Having the lowest GDP in Jakarta, North Jakarta-has the greatest number of slum areas [32]. While Kampong Cikini Kramat is located in Central Jakarta's administration center with the highest population density of 5000 person / hectares.

Figure 1. Research location.

Kampong Markisa and Kampong Cimone are located in Karawaci sub-district, city of Tangerang. Both kampongs have the same characteristic that is both are slum settlements in Tangerang city. The kampong location is located near an industrial complex subsequently making them gain less attention from the local government. National Slum Upgrading Program (NSUP) recorded the total slum area in Tangerang city reaches 3,59 Ha where the Karawaci sub-district is included in the said area.

These four kampongs are categorized into a dense populated settlement which the road width are mostly less that 1 m. has This condition has become a challenge not only for waste collecting and transporting but also for fire protection and other disaster mitigation. Compare to the central city of Jakarta and Tangerang or surrounding residential areas, he four kampongs have inadequate sanitation infrastructure: limited access to clean water sources and availability of sewage. With no availability of clean water supplies and limited waste management from the local government only add insult to injury. The densely populated settlement is also one of the factors in the less number of adequate toilets for bathing, washing, and lavatory.

Figure 2. Slums kampong in Jakarta and Tangerang.

2.2. Urban kampong household characteristic
In this study, household characteristics were collected using cross sectional and systematic random sampling. Cross sectional study is conducted at one short period and usually used to predict a certain result prevalency for a specific population [33]. The sample size were calculated using the sample size formula for population proportion estimation using a complex sample design [34]. 150 respondents were interviewed from each kampong using 50% conservative coverage estimation, 8% precision and 95% confidence level.

2.3. Sampling waste generation and composition measurement method
Household solid waste generations and compositions were identified in temporary dumpsite (TPS) before finally transported into a landfill (TPA). Waste The Indonesian standard of waste sampling, referring to SNI 19-3694-1994 on collection method, waste generation sample measurement and urban waste composition were applied. The solid wastes collected from each household by the garbage collectors then transported TPS where 100 kg of random samples were taken to be analyzed for compositions and characteristics. The analysis of waste composition were conducted for 8 days. Data analysis method. Then, data were analyzed using the 2013 Microsoft excel program (Microsoft Corp) thereby wastecomposition, maximum and minimum and standard deviation standards could be acquired.

3. Result and discussion

3.1. Household and utilities

Urban kampong characteristics are identified as education level, occupation and household income (table 1). It is found that 72% of waste generated was transported by the garbage collectors to a TPS; while other 28% are dumped into the body of water, empty land, or burnedFrom the TPS the wastes were transported to the landfill wich is actually an open dumping (Figure 3).

Table 1. Household characteristics.

Components	Description	%
Education	Low Education Level	57.8%
	Higher Education Level	42.2%
Occupation	Farmer	1.7%
	Entrepreneur	27.8%
	Civil servant	2.5%
	Private Employee	32.0%
	Daily Labourer	29.8%
	Others	6.2%
Household Income	< Rp. 2.000.000	23.7%
	Rp. 2.000.000 - Rp. 4.000.000	35.8%
	Rp. 4.000.000 - Rp. 6.000.000	22.8%
	Rp. 6.000.000 - Rp. 8.000.000	8.2%
	Rp. 8.000.000 - Rp. 10.000.000	4.8%
	> Rp. 10.000.000	4.7%
Waste Processing	Collected by collector	79.2%
	Composting	0.3%
	Open Burning	0,9%
	Disposed to body of water	4.8%
	Random disposal	1.3%
	Others	13.5%

Based on table 1, the level of education from 4 kampongs is 57.8% , which is categorized as a low level of education with junior high school as its highest education level. As for occupation, private employees (32%) and daily labourers (29%) are the dominant ones. While household income is 35% dominated by income range of IDR 2 million - 4 million.

3.2. Household solid waste management

Field observation in this research was conducted for 8 days in each kampong, and the result shows that four kampongs hadn't been able to optimally manage the household solid waste generated. The waste separation and 3R (Reuse, Reduce, Recycle) process were rarely used. The lack of government attention in facilitating a good household waste management was the main contributor to this issue.

The flow of waste management in kampong urban was still using old paradigm: collect-transport-dispose.

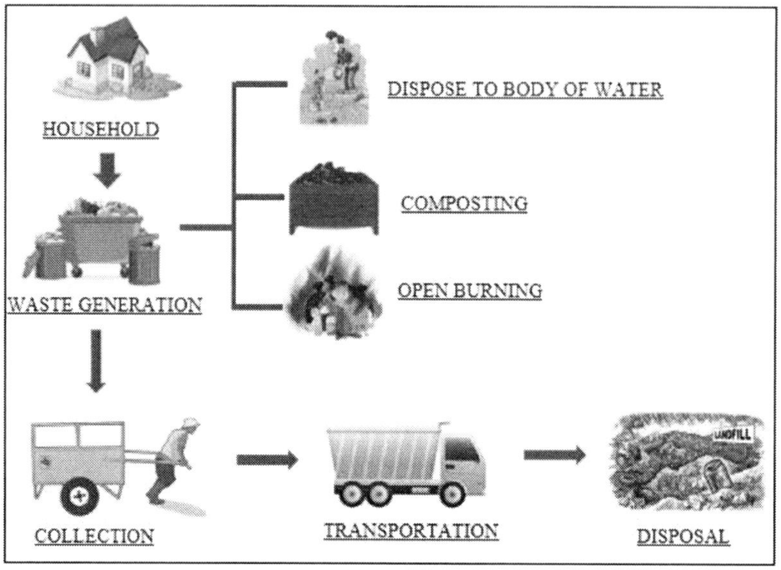

Figure 3. Household waste management system.

3.3. *Waste generation and composition*

In average, the urban kampong generates 0.5 Kg/person/day waste where kampong Cikini Kramat generates the highest number of waste which is 0.74 Kg/person/day (Table 2). As the previous research mentioned the higher the waste generation in urban areas, the higher the income and consumption [35]. However Kampong Cikini has the lowest household income compared to other kampongs. The high amount of waste generated is due to its location which lies in the town center because its inhabitants' low income and smaller living space make instant and take away foods as ideal consumption thus creating large amounts of packaging waste.

Table 2. Slum urban kampong waste generation.

Kampong	Waste Generation (Kg/person/day)	Monthly Income
Markisa	0.23	Rp. 2.000.000 – Rp. 4.000.000
Cimone	0.36	Rp. 2.000.000 – Rp. 4.000.000
Gedong Pompa	0.68	Rp. 2.000.000 – Rp. 4.000.000
Cikini Kramat	0.74	< Rp. 2.000.000
Average	0.50	

The waste generated in urban kampong was dominated by organic waste withan average of 61.23%. While plastic waste (non-multilayered) dominated inorganic waste with an average of 13.14%. The least number of wastes found was electronic waste with 0.11% and rubber with 0.25%. Organic Waste management using the composting system is difficult to implement due to the high price of the land and less space to utilize. The same goes for recycling processes, the low quality of waste makes it impossible to recycle. Thus making the transportation of waste from temporary dumpsite to the landfill the only reasonable method.

One interesting fact was the high number of small size multi-layered plastic waste generation with 5.25%. Multi-layered plastic is a plastic packaging waste created from Low Density Polyethylene (LDPE) which is normally used for snack packaging [36]. Multi-layered is categorized as a food-grade plastic consequently making this plastic used by many food products. Besides from snack packaging, multi-layered plastics found were also found in many consumer goods like small sized shampoo and conditioners, and personal hygine products. These small-size consumer goods are main items sold in the small shops or stalls owned by local people in urban kampong.

Table 3 shows comparison data between waste composition of Indonesia's urban kampong and other slum area cities. Slum area inhabitants are usually categorized as the low-income class [17] and it is apparent that the waste composition in an area with a low-income class is dominated by organic waste/food leftovers. In table 3 organic waste composition is at the range of 57% - 83,5%. This high level of organic waste is caused by low-income class mainly consuming traditional food like vegetables, meats, and rice. The different organic percentage in each country is affected by some factors like the society's activities, local food of choice and the income level [21]. Furthermore, the low income of the society is also an indicator of the low consumption of packaged products. This is evident from the percentage of plastic and paper in each country only showed less than 15 %. The low-income level in slum areas makes the society less consumptive and generates more organic waste.

In this study, the urban kampong inhabitants generate higher a plastic waste (13.14%) compared to other studies. The income of urban slum area inhabitants in Indonesia is slightly higher; IDR 2-4 million compared to other countries like China (1.234 CNY) and Vietnam (1.5 Juta VND). As a consequence, plastic waste generation in those two countries is less with 9.6% and 8.61% compared to Indonesia. Also, Indonesian society is still used to disposable plastic resulting in bigger plastic generation from total waste generated.

Table 3. Table comparison of waste composition to other countries.

Category	This research	Thu Dau Mot (Vietnam) [37]	Xinhaiyuan (China)[38]	Lahore (Pakistan) [39]	Slum Area (Nigeria) [40]	Ahvaz (Iran) [41]
	(%)	(%)	(%)	(%)	(%)	(%)
Organic	61.62	70.07	73.59	76.95	76	83.5
Paper/Card	4.12	5.61	6.52	1.41	6.6	5.2
Plastic	13.14	9.6	8.61	4.15	4	6,15
Metal	0.57	-	0.46	-	2.5	1.25
Rubber	0.25	-	-	0.02	-	-
Textile	2.90	-	1.9	0.43	1.4	1.4
Glass	0.82	3.63	2.87	0.73	3	1.5
Wood	1.51	-	-	-	-	-
Multi-layered	5.52	-	-	-	-	-
Hazardous waste	1.18	-	-	0.53	-	-
Other Waste	8,64	11.09	1.20	15.62	6.5	1
Total	100	100	100	100	100	100
Kg/person/day	0.5	-	-	0.39	0.44	1.49
Income/month	<4.000.000 IDR	<1.500.000 VND	1.234 CNY	<30.000 PKR	-	<300 USD
1 USD	13. 886 IDR	22.710 VND	6,8279 CNY	104,97 PKR	-	
Income (USD)	<288	<66.05	181.47	<285.79	-	

- = Not Reported; IDR (Indonesia Rupiah); VND (Vietnam Dong); CNY (China Yuan); PKR (Pakistan Rupee); USD (United States Dollar)

4. Conclusions

In this research, organic waste is still dominant in waste generated from slum kampong in Jakarta and Tangerang. Daily waste generation has an average of 0.5 Kg/person/day with two biggest compositions of Organic (61.62%) and Plastic (13.34). While the smallest waste composition categories are Rubber (0.25%), Metal (0.57) and Glass (0.82%). The low household income is one of the contributing factors in more organic waste composition. Moreover, the low level of waste management like recycling programs at household level creating unreduced waste before transported into landfills. Land space limitation and little to none attention from local government have been the crucial factor in less 3R facilities (reuse, reduce, recycle) like a waste bank or composting process

The availability of waste bank and composting can increase and push society's awareness in segregating waste from generated waste. The program can also add economic value from waste generated in slum areas. The main benefit from optimal waste management is making the residential area better and healthier; far from slum stigma although the area is densely populated.

5. References

[1] BPS-Statistic Indonesia 2018 Environment Statistics of Indonesia

[2] Rebehy, P. C. P. W. et al 2017 Innovative social business of selective waste collection in Brazil: Cleaner production and poverty reduction, *Journal of Cleaner Production* **154** 462–473

[3] Idris, A., Inanc, B. and Hassan, M. N 2004 Overview of waste disposal and landfills/dumps in Asian countries, *Journal of Material Cycles and Waste Management*, **6** , 104–110

[4] Hazra, T. and Goel, S 2009 Solid waste management in Kolkata, India: practices and challenges *Waste Management* **29** 470–478

[5] Al-Khatib, I. A. et al 2010 Solid waste characterization, quantification and management practices in developing countries. A case study: Nablus district - Palestine *Journal of Environmental Management* **91** 1131–1138

[6] Xue, B. et al 2011 An overview of municipal solid waste management in Inner Mongolia Autonomous Region, China *Journal of Material Cycles and Waste Management* **13** 283–292

[7] Menikpura, S. N. M., Gheewala, S. H. and Bonnet, S. 2012 Sustainability assessment of municipal solid waste management in Sri Lanka: problems and prospect *Journal of Material Cycles and Waste Management* **14** 181–192

[8] Sujauddin, M., Huda, S. M. S. and Hoque, A. T. M. R 2008 Household solid waste characteristics and management in Chittagong, Bangladesh *Waste Management* **28** 1688–1695

[9] Alavi Moghadam, M. R., Mokhtarani, N. and Mokhtarani 2009 B Municipal solid waste management in Rasht City, Iran *Waste Management* **29** 485–489

[10] Staley, B. F. and Barlaz, M. A 2009 Composition of municipal solid waste in the united states and implications for carbon sequestration and methane yield *Journal of Environmental Engineering* **135** 901–909

[11] Kurian Joseph et al. 2012 Integrated approach to solid waste management in Chennai: An Indian metro city *Journal of Material Cycles and Waste Management* **14** 75–84

[12] Wibowo, A. and Djajawinata, D 2005 *Integrated Waste Handling* 1–11

[13] Edjabou, M. E. et al 2015 Municipal solid waste composition: Sampling methodology, statistical analyses, and case study evaluation *Waste Management* **36** 12–23

[14] Damanhuri, E., Padmi, Tri 2018 *Integrated Waste Management*

[15] Yach, Derek. Mathews, Catherine. Buch, Eric 1990 Urbanisation and health: methodological difficulties in undertaking epidemiological research in developing countries *Soc Sci Med* **31** 507

[16] Montgomery, Mark R. 2009 Population Bulletin. Urban poverty and health in developing countries

[17] Uddin, N 2018 Assessing urban sustainability of slum settlements in Bangladesh: Evidence from Chittagong city *Journal of Urban Management* **7** 32–42

[18] Wilson, D.G 1997 *Handbook of Solid Waste Management*

[19] Villalba, L. et al 2020 Household solid waste characterization in Tandil (Argentina):

Socioeconomic, institutional, temporal and cultural aspects influencing waste quantity and composition *Resources, Conservation and Recycling* **152**

[20] Tchobanoglous, G., Theisen, H., Eliassen, R 1982 Desechos sólidos; principios de ingeniería y administración Serie Ambiente y Recursos Naturales Renovables.

[21] Chen, Y. C 2018 Effects of urbanization on municipal solid waste composition *Waste Management* **79** 828–836

[22] Azkha, N 2016 Analysis of waste generation, composition and characteristics in Padang city *Public Health Journal* **1** 14–18

[23] Ratya, H. and Herumurti, W 2017 Household Waste Generation and Composition in Rungkut Sub-district, Surabaya *ITS Engineering Journal* **6**

[24] Sethi et al 2016 Characterization of Municipal Solid Waste in Jalandhar City, Punjab, India *J Hazard Toxic Radioact Waste* **17** 97-106

[25] Ramachandra et al, 2018 Municipal solid waste: Generation, composition and GHG emissions in Bangalore, India. *Renewable and Sustainable Energy Reviews* **82**, 1122–1136

[26] Gonzalez, T.P et al 2010 Household solid waste characteristics and management in rural communities *Open Waste Management Journal* **3** 167–173

[27] Dorenfeld E et al 2012 *Improving solid and human waste management in rural Namibian communities*

[28] Shah R, Sharma US, Tiwari A 2012 Sustainable solid waste management in rural areas *Int J Theor Appl Sci* **4** 72–75

[29] El-Messery MA, Ismail GA, Arafa AK 2009 Evaluation of municipal solid waste management in Egyptian rural areas *J Egypt Public Health Assoc* **84** 52–71

[30] Mohammadi, A. et al 2012 Open access research article a survey on the rural solid wastes characteristics in North Iran (Babol) **2** 149–153

[31] De Medina-Salas L, Castillo-Gonzalez E, Romero-Lopez R 2013 Physical and chemical characteristics of municipal solid waste in a rural locality-study case: Cosautlan De Carvajal Veracruz, Mexico

[32] Putri, Siti Delima Amanda, R. A. and Chotib 2020 Detecting informal settlement development in Penjaringan Sub-District, North Jakarta, Indonesia *IOP Conference Series: Earth and Environmental Science* **43**, 0–9

[33] Levin, K. A 2006 Study design III: Cross-sectional studies, *Evidence-Based Dentistry* **7** 24–25

[34] Lemeshow, S. et al 1997 'Stanley Lemeshow, David W Hosmer Jr, Janelle Klar a'

[35] Taghipour, H. et al 2016 Characterizing and quantifying solid waste of rural communities *Journal of Material Cycles and Waste Management. Springer Japan* **18** 790–797

[36] Juliastuti, S. R. et al 2015 Multilayer ldpe (Low Density PolyEthilene) plastic packaging waste processing using pyrolisis microwave method *Prosiding Chemical Engineering National Seminar "Kejuangan" (2009)* 1–7

[37] Trang, P. T. T. et al 2017 The Effects of Socio-economic Factors on Household Solid Waste Generation and Composition: A Case Study in Thu Dau Mot, Vietnam *Energy Procedia. Elsevier B.V* **107** 253–258

[38] Qu, X. yan et al 2009 Survey of composition and generation rate of household wastes in Beijing, China *Waste Management* **29** 2618–2624

[39] Kamran, A., Chaudhry, M. N. and Batool, S. A. 2015 Effects of socio-economic status and seasonal variation on municipal solid waste composition: a baseline study for future planning and development *Environmental Sciences Europe. Springer Berlin Heidelberg* **27**

[40] Ogwueleka, T. C 2013 Survey of household waste composition and quantities in Abuja, Nigeria *Resources, Conservation and Recycling. Elsevier B.V* **77** 52–60

[41] Monavari, S. M. et al 2012 The effects of socioeconomic parameters on household solid-waste generation and composition in developing countries (a case study: Ahvaz, Iran) *Environmental Monitoring and Assessment* **184** 1841–1846

IOP Conf. Series: Materials Science and Engineering **909** (2020) 012078 doi:10.1088/1757-899X/909/1/012078

Development of Reverse Logistics Scenarios for Inorganic Waste Recovery in Grobogan Regency - Indonesia

M Hadiwidodo[1*], B P Samadikun[1], A I Putri[1], S Sumiyati[1], B S Ramadan[1]
[1]Department of Environmental Engineering, Universitas Diponegoro, Semarang City, Central Java, Indonesia 50275

*Email: bimastyaji@live.undip.ac.id

Abstract. Reverse logistics is a model of the flow of goods that goes inversely from consumers back to producers to increase the value of goods without wasting any part of the goods (recovery). In this study, the waste management system in Purwodadi District, Grobogan Regency, has been evaluated and propagated using a reverse logistic network system, which is considered more efficient. In-depth interviews and observations were made to understand the flow of waste in each level of recyclers. The flow of plastic, paper, and metal waste were evaluated in this study because it dominates the composition of inorganic waste in Grobogan Regency. The results of the scenario indicate the potential use of a reverse logistic system because it shortens the flow time of waste from consumers to factories. Besides, environmental losses caused by the long process of waste management can be reduced. It is necessary to carry out further evaluation related to economic and environmental benefits due to the development of a reverse logistic scenario for plastic, paper, and metal waste.

1. Introduction

The increase in population is always accompanied by an increase in the amount of inorganic waste. Furthermore, that increase is often not matched by a proper waste management system [1]. Lack of community understanding in managing waste tend to be unsafe and potentially pollute the environment. Therefore, the action is needed to reprocess plastic, paper, and metal waste to reduce adverse effects on the environment. One scenario that can be done is to find out the flow of the waste starts from the consumer and come back to the factory (reverse logistic scenario) [2]. Reverse logistics is a model that focuses on maintenance, recycling, product return, repair, and product disassembly [3]. A reverse logistics network for recycling waste with minimum costs and environmental impacts is needed in waste management [4]. Reverse logistics is an environmentally friendly waste recycling practice by reusing and reducing the amount of material used [5]. Plastic, paper, and metal waste will be reprocessed through reverse logistics network system to be raw material for factories.

The waste problem can occur in most cities, including in Grobogan Regency, Central Java, Indonesia. Purwodadi District, as a district in Grobogan Regency, was chosen as the study area since it has many waste recycling agents compared to other districts in Grobogan Regency. The amount of plastic, paper, and metal waste generated in Grobogan Regency is 14,887.26 kg/day, 13,894.76 kg/day, and 8,932.35 kg/day, respectively. Purwodadi District is also a district with the largest population in Grobogan Regency, which is 139,296 inhabitants [6]. The reverse logistic system in Indonesia has been carried out by the informal sector [7]. However, it has not yet been discovered how the system's impact on environment and waste management. This study attempts to reconstruct the reverse logistic network system in Purwodadi District, Grobogan Regency, Indonesia, through the evaluation of informal sector of waste management.

2. Research Methods

2.1 Sampling Methods

This research will be carried out in Purwodadi District, Grobogan Regency, Central Java, Indonesia. In this study, informal sector recycling actors such as scavengers, waste bank, collectors, dealers, and grinders/factories were interviewed to obtain relevant data. Figure 1 shows the administrative map of the Grobogan Regency and the study location. This research focuses on informal recycling agents to get information about the flow of inorganic waste into and out of business. That data is used to estimate suppliers and recipients of metal, paper, and plastic waste, each of which will look for the weight of managed, transported, and disposed the waste. The study began with a preliminary survey based on the reference to the rubble business data from the Grobogan District Regional Technical Implementation Unit, followed by selecting the sample location and collecting data using questionnaires and interviews. Since the detailed data of recycling agents are lack, then they are tracked using snowball sampling methods. As for the level of scavenger, there is no exact amount of data in Grobogan District. Then, scavenger surveys are conducted in a random sampling. The research method is illustrated in the Figure 2.

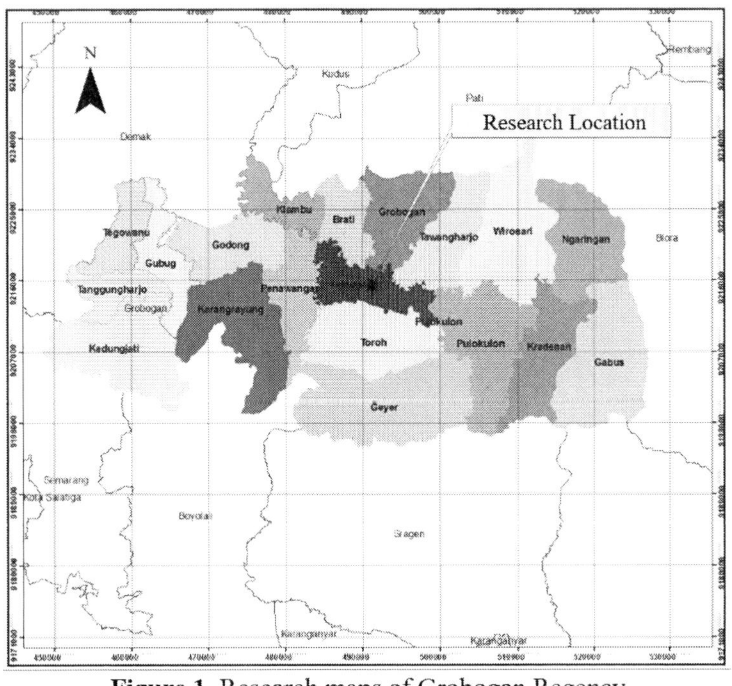

Figure 1. Research maps of Grobogan Regency

2.2. Data Analysis

Data on the amount of waste generation and the location of scavengers, waste banks, collectors I, collectors II, dealers, and grinders will be mapped into a waste management network in Purwodadi District. Identification of waste management flow in informal sector includes:

1. The route of the waste bank to collectors I, collectors II, dealers and grinders or factories
2. The collectors I route to collectors II, dealers, grinder and factory
3. The collectors II route sells their goods to the dealers, grinder, and factory
4. The dealers route sells goods directly to the factory
5. The grinder route sells its goods directly to the factory

The reverse logistics network scenario design is carried out to compare the existing conditions, with the optimization scenario with a centrally inorganic waste processing or known as material recovery facility (metal, paper, and plastic) in Purwodadi District. Scenarios to be made include scenario 1 based on existing conditions that are known from interviews with the perpetrators of metal, paper and plastic waste recycling in Purwodadi District and scenario two which are based on the optimization conditions of Purwodadi District waste collection by combining collectors I, collectors II and the dealers became material recovery facility (MRF).

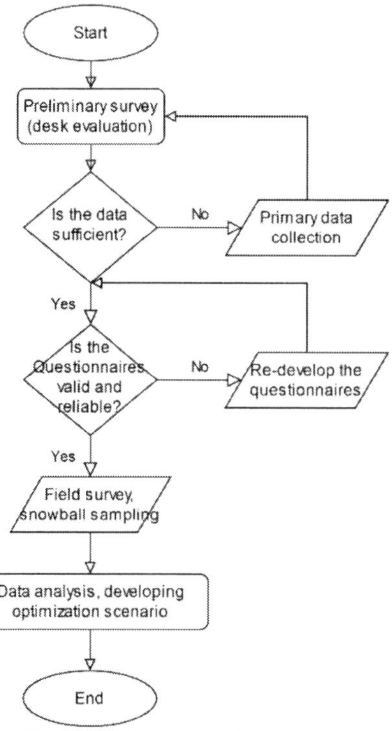

Figure 2. Research flow chart

3. Result and Discussion

3.1 Grobogan Inorganic Waste Management Practices

Waste in Purwodadi District comes from residential, shop, market, and street waste. Residents (waste bank) and informal sector sort and manage inorganic waste from those waste sources. Waste generation in Purwodadi District is 79,398.72 kg/day. This waste is dominated by organic waste (42.50%), and inorganic waste including plastic (18.75%), paper (17.50%), metal (11.25%), and others (10.00%). Waste sorting has not become a habit that can be implemented by the community. Scavengers or housing waste managers will later collect waste which still has economic value such as paper, plastic, and metal. At the same time, the residue will be taken by the waste transporters and waste managers who have collaborated with the community to the temporary waste storages. Waste collected at the temporary waste storages will be taken to the Ngembak landfill.

People who have sorted their waste, usually sell their waste to the waste bank located in their *Kelurahan*, or neighbourhood and untreated waste will be disposed of in the temporary waste storages. Waste can be disposed of directly by the residents to the nearest polling station or transported by officers with the obligation of citizens to pay the cost of waste transport. Besides residents, scavengers also have a role in waste reduction. Scavengers also take waste that has an economic value, which will be sold to waste collectors. Later the waste collectors sell their waste to the recycling industry, which processes the waste into raw material again.

The informal sector recycling actors, which include scavengers, waste banks, collectors I, collectors II, dealers, and grinders, play an essential role in reducing waste in Purwodadi District. In terms of management, the informal sector recycling actors are not related to the government. They set up this business at their own expense. Collectors do not always sell waste to the dealers but can also go to the grinder or factory. Therefore, the researchers classified recycling agents based on the weight of inorganic waste (paper, plastic, and metal) entering the warehouse per day, which can be seen in Table 1.

Table 1. Classification of recycling actors

Number	Recycling Actors	Weight of Waste Received per Day(kg/day)
1	Waste bank	≤ 30
2	Scavenger	≤ 50
3	Collectors I	50-500
4	Collectors II	500-1000
5	Dealers	>1000
6	Grinder	>1000

3.2 Existing Inorganic Waste Management Practices

After detailed observations and interviews in the field, there were found 16 inorganic waste recycling businesses and 14 waste banks in Purwodadi District that were used as research models. The 16 recycling businesses consist of 1 grinder, 5 dealers, 3 collectors II, and 7 collectors I. Meanwhile, among the 14 waste banks in Purwodadi District, there is 1 central waste bank in Purwodadi District. Therefore, scavengers are the critical factors in organizing inorganic waste recycling in the Purwodadi District. The scavenger sort waste which has economic value and does not. Later the disaggregated waste that has economic value will be sold to the closest collectors who have a purchase price they consider high [8].

A waste bank is a place of waste management that applies the 3R system. There are 14 waste banks in Purwodadi district, and one of them is the central waste bank. Waste generated from the waste bank is waste coming from the customer of the waste bank itself. Waste collected is waste that has an economic value in the form of inorganic waste (paper, plastic, and metal). The waste bank weighs once a month. The waste collection will be sorted according to types. The results of the collection of sorted waste will be sold to collectors of recycled waste from customers. Waste that is not sold will later be used as handicrafts in the form of decorations, which can later be marketed and sold to the public. At the waste bank, weighing, sorting, and shipping processes are sold to collectors.

Collectors I is the first level intermediary who will take inorganic waste (paper, plastic, and metal) in more quantities than scavengers and waste banks. The input of inorganic waste produced by collectors I per day is equal to 50-500 kg/day. Seven recycling businesses are collectors I in Purwodadi District. Waste generated from the collectors I is delivered by scavengers, waste banks, and the community. The activities that occur at the collectors I is weighing, buying waste from scavengers, waste banks and residents, sorting and shipping to larger businesses. Collectors I usually have a means of transportation in the form of a pickup car to transport its waste. Collectors I usually have a permanent scavenger to supply its waste.

Collectors II is a waste recycling business that has a purchasing capacity above Collectors I, in terms of the place and scope of waste supply services. Three recycling businesses become collectors II in Purwodadi District. Waste entering from collectors II comes from scavengers, waste banks, and collectors I. In collectors II, there is a weighing process, purchasing inorganic waste from scavengers, waste banks, or collectors I, sorting, and shipping to a more significant recycling business. Collector II usually has a small truck to transport waste from his business to other businesses for sale, or to partners who have collaborated to buy the waste.

Dealer is a waste recycling business with an inorganic waste purchasing capacity of above 1 ton per day. Five recycling businesses are used as dealers in Purwodadi District. The waste that comes from the city comes from scavengers, waste banks, collectors I, and collectors II. There is a weighing process, purchasing inorganic waste, storage, sorting, pressing, and shipping to the mill or grinder. For presses used are still simple box presses made of wood. The dealer uses a truck to transport waste that will be brought.

Grinder/factory is a final waste recycling business that processes inorganic waste. The selling price in the grinder is already high because it has been through a process of sorting, washing, and grinding. There are grinding machines that are driven by diesel engines. Inorganic waste milling business found in Purwodadi District is a plastic waste grinding business. Paper and metal waste grinding are not yet available in Purwodadi District or Grobogan District, so that the waste must be sent outside the city. The trade route for inorganic waste in Purwodadi District can be seen in Figure 3. From this figure, it can be seen that waste transported from scavengers and waste banks will be taken to collectors I, collectors II, and dealers. From the waste suppliers, the waste will be taken to the grinder and the waste processing factory according to their respective types.

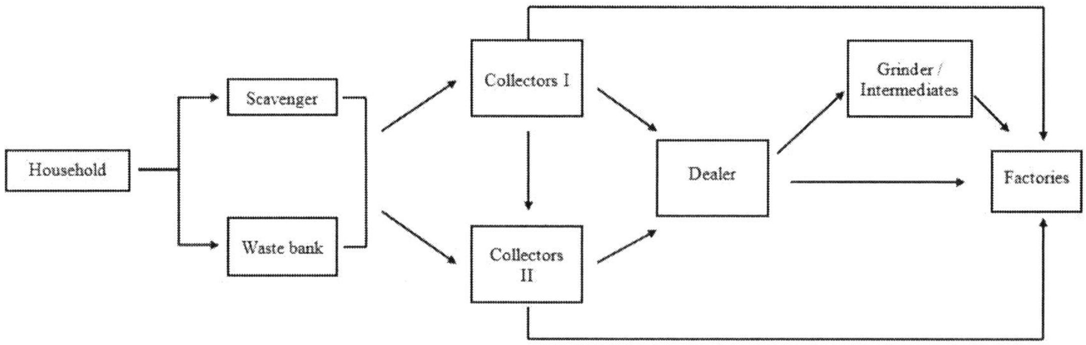

Figure 3. The role of recycling actors in Purwodadi District

3.3 Development of Reverse Logistic Scenario in Grobogan District

Recycled plastic waste by recyclers are in the form of beverage bottles, sauce bottles, ketchup bottles, cooking oil bottles, buckets (soap containers, basins), pipes, clear crackle, glass mineral water, bottle caps, children's toys, electronic devices, and household appliances. Whereas, for the type of paper waste, eight types of waste are traded by recyclers, including white paper, cardboard, duplex, books, newspapers, cement paper, opaque paper, and magazines. The types of metal waste sold by recyclers are iron B, iron A, aluminum cans, metal cans, zinc, copper, aluminum metal, brass, and stainless steel. The reverse logistics network in Purwodadi District has a diverse chain.

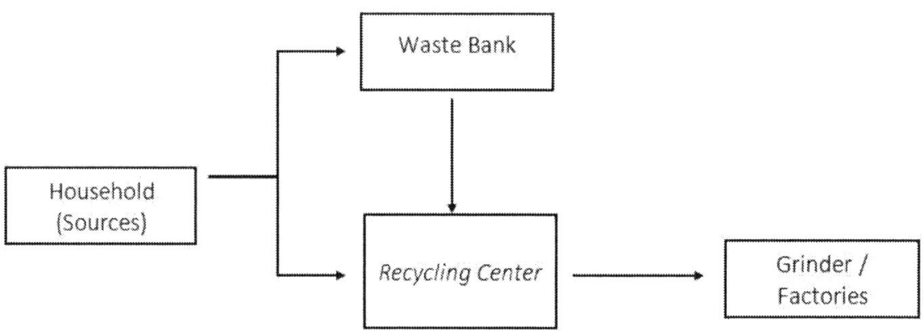

Figure 4. Reverse logistics network systems of inorganic waste in Purwodadi District

The flow of material can be from the waste bank to collectors, dealers, and factories, or from the dealers directly to the grinder. This condition happens due to differences in price quotations and agreements that have been set by sellers and buyers of inorganic waste (plastic, paper, and metal) or to partners who have collaborated to buy their waste. Recycling businesses are independently looking for buyers and suppliers of waste for them. However, the difference in price quotations can be reduced by creating a recycling unit (material recovery facility - commonly abbreviated as MRF) that can accommodate waste and maintain the stability of the selling and buying prices of each type of waste. Figure 4 shows the reverse logistics path for the three types of waste following the results of a survey that was conducted.

4. Conclusion

The design of reverse logistics scenarios in Purwodadi District is based on existing data from inorganic waste recycling (plastic, paper, and metal). In this study, there is a network map of inorganic waste management (plastic, paper, and metal) through the concept of reverse logistics. The scenario that has been designed has taken into account several factors, such as the sorting place, reprocessing unit, and the amount of waste managed. The scenario of combining the location of collectors I, collectors II, and dealers in a focus recycling centre whose main activities are weighing, sorting, pressing, and transporting is considered capable of optimizing the reverse logistic network system because centralizing waste management might be able to cut costs and greenhouse gas emissions generated. Further research is needed to understand the impact of the proposed optimization scenario. The use of this reverse logistic network system is a way to improve waste management systems in Purwodadi District, Grobogan Regency

Acknowledgment

Faculty of Engineering funded this research through RKAT Fakultas Teknik, Universitas Diponegoro, Fiscal Year 2020 scheme Number 330-04/UN7.P4.3/PM/2019.

References

[1] Damanhuri E and Padmi T 2010 *Diktat Pengelolaan Sampah*. Bandung: Institut Teknologi Bandung.
[2] Dyckhoff R L 2012 Networks in reverse logistics *Supply chain management and reverse logistics* Berlin: Springer.
[3] Srivastava S K 2008 Network Design for Reverse logistics *Omega* **36**(4): 535-548.
[4] Bing X 2014 Sustainable Reverse Logistics for Household Plastic Waste *Wagenigen: Wagenigen University*.
[5] Lambert S, Riopel D, and Abdul-Kadel W 2011 A reverse logistics decisions conceptual framework *Computers & Industrial Engineering* **61**(3):561-581.
[6] Badan Pusat Statistik Kabupaten Grobogan 2019 *Kabupaten Grobogan Dalam Angka 2018*.
[7] Steubing B, Boni H, Schluep M, Silva U, and Ludwig C 2010 Assessing computer waste generation in Chile using material flow analysis *Waste Management* **30**: 473–482.
[8] Samadikun B P, Rezagama A, Ramadan B S, Andarani P, Rumanti E D 2020 Understanding Informal Actors Of Plastic Waste Recycling In Semarang City *Jurnal Ilmu Lingkungan* **18**(1): 162-170.

IOP Conf. Series: Materials Science and Engineering **909** (2020) 012079 doi:10.1088/1757-899X/909/1/012079

Hybrid vector autoregression–recurrent neural networks to forecast multivariate time series jet fuel transaction price

Agung Bayu Aji* **and Isti Surjandari**
Department of Industrial Engineering, Faculty of Engineering, Universitas Indonesia, Kampus Baru UI Depok, Depok 16424, Indonesia

*E-mail : agung.bayu@ui.ac.id

Abstract. Fuel cost is the most contributed component in the operational cost of all transportation modes. In the aviation industry, jet fuel cost contributed to a percentage of 33.33% of the total airline operational costs. To increase efficiency in operational costs and the airline should have jet fuel price monitoring systems that can forecast the future price and give some strategy recommendations to airlines. In this research, we propose many multivariate time series-based predictive analytics as a tool for the airline to monitor and forecast the jet fuel price transaction based on jet fuel transaction price. We consider the global crude oil price and also global and local jet fuel prices in each airport. We also consider additional variables for the economical aspect that applied differently for each airport location. We examine two Recurrent Neural Network (RNN) algorithm, Long Short Term Memory (LSTM) and Gate Recurrent Units (GRU). For minimizing the weakness of LSTM and GRU, we combine each methods with Vector Autoregression (VAR). After forecasting results using VAR-LSTM and VAR-GRU, we get forecasting accuracy of 98.98% and 99.40% respectively.

1. Introduction

Nowadays, business competition in an information technology era drive all company to develop their business using the internet. The company can use the internet to get a blue ocean business, monitor the competitors, making a new strategy to compete in the market, etc. that will improve the usage of the internet. Increasing the usage of the internet can lead to an increase the number of data and bring us into a big data era [1]. To get the information from data and make a decision or new strategy, we can use data mining [2]. Data mining is a process to extract the information or knowledge from big data using many statistical techniques. Many techniques of data mining such as: (1) descriptive analysis, (2) estimation, (3) prediction, (4) classification, (5) clustering, and (6) association [3]. Data mining is already used in many industrial sectors like manufacturing, pharmacy, finance, transportation, etc.

In the transportation industry, one of the areas that data mining can applied to monitor and forecast the market price of each operational cost component that might help management to develop the strategy of business efficiency. According to a report submitted by Martino, et al. [4], Fuel cost is the most contributed in many transportation vehicles. Fig. 1 presents that for sea transportation, fuel cost is 50% of all operational costs. For air transportation, fuel cost is 33.33% of all operational costs. International Air Transport Association (IATA) as a world organization of airline reported in 2014 that fuel cost is contributed 33.3% from airline cost structure as shown in Fig.2 [5]. By using the Pareto theory, if we improve in the biggest component, we can get a significant impact on our goals. So, to give a significant impact to airline operational cost, we should make an efficiency strategy in jet fuel management system by using monitoring systems to estimate or forecast the jet fuel price, so the airline can make an efficiency strategy such as selecting the time and location to into-plane the jet fuel based on the forecasting results.

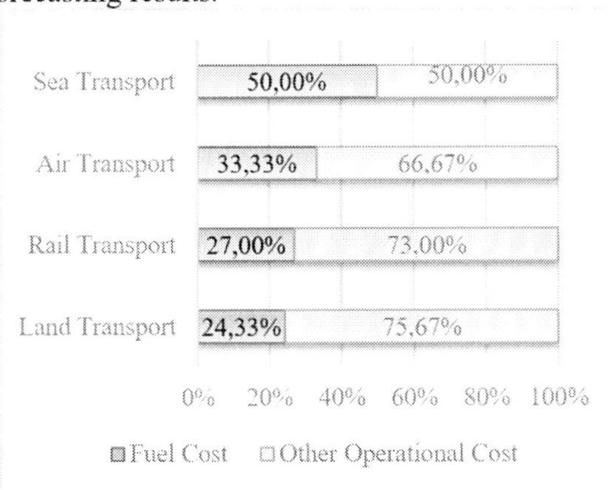

Figure 1. Fuel cost vs other operational cost in many types of vehicles [4]

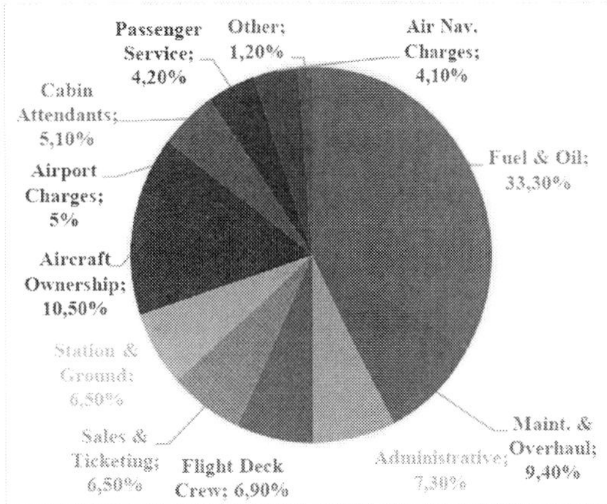

Figure 2. Airline cost structure [5]

According to many studies before, research about fuel price forecasting is common to forecast crude oil price or basic price of the derivative product of crude oil (such as Petrol, Diesel, LPG, Kerosene, Jet Fuel, etc.). The similarities of both prices is fluctuated in a time frame and standardized by an organization. The differences in crude oil prices can be applied worldwide, but derivative product price depends on locations refer to policy that applied by local authority [6].

For example, for jet fuel basic price, the several previous studies showed that jet fuel basic price is correlated with crude oil price and depends on its historical price [7] [8]. The Fig. 3 presents the comparisons of crude oil price (using Brent Spot Price) and jet fuel price (using MOPS Posting Price) prepared by Platts, Singapore in May 2013 until May 2020 that relatively have the same trends [9].

Figure 3. Jet fuel & crude oil price comparisons in usd/bbl [9]

Based on the real case in the airline industry, the airline should pay to a fuel supplier, not in jet fuel basic price, but the airline will negotiate for the scheme of jet fuel transaction price that different for each location [8]. For example, in Soekarno-Hatta International Airport, Indonesia (CGK), there are two price equations that can be applied by airline: (1) MOPS Posting Price + Differential Price, and (2) Pertamina Posting Price + Discount Price. The Pertamina Posting Price is the base price that already set by Pertamina as a fuel manufacturer in Indonesia for each airport in Indonesia [10]. So, the jet fuel transaction price will different, but depends on crude oil price and jet fuel basic price.

Considering that jet fuel transaction price is different in each location, it's possible to state that local economic conditions such as the exchange rate of local currency and the local inflation rate will affect jet fuel transaction price fluctuation. This is done considering the price of fuel transactions that are also influenced by currency hedging conducted by the airlines and fuel providers as well as the political-economic conditions that are applied in Indonesia. Where the political-economic decisions determined by the government will usually affect and be influenced by currency exchange rates and inflation rates. As well as stated in the IATA website presented in Fig. 4 that the trend of jet fuel price in USD and Euro that relatively have the same trends [9].

Figure 4. Jet fuel price comparison in usd and euro [9]

Considering the characteristics of crude oil price and jet fuel price is time series plotted, the previous studies using time-series forecasting-based predictive analytics to build monitoring and estimating tools. Khan et al. [6], that research price predictions for 4 petroleum products, namely Petrol, Diesel, LPG, and Kerosene in India, using ARIMA and Neural Network techniques using historical data of each petroleum products and crude oil prices as independent variables. For jet fuel price prediction, previous research has been conducted by Atems, et al. [7] by making a predictive analytics model of jet fuel price using the Vector Autoregressive (VAR) technique with the independent variable is crude oil prices and also using jet fuel historical data. For research on forecasting aircraft fuel transaction prices, previous research has been conducted by Firmansyah [11], where the independent variable used is Crude Oil Price (Brent Posting Price), Price from local suppliers (Pertamina Posting Price), and Basic Jet Fuel Price (MOPS Posting Price) and using multiple linear regression analysis with an accuracy of 93.68%.

One of the studies in forecasting oil prices that has a fairly high degree of accuracy is a study conducted by Guo [12], research conducted with deep learning methods, namely Long Term Short Memory (LSTM) and Gated Recurrent Unit (GRU) with a forecast accuracy level of 98.45% was obtained, but the research was only conducted using historical data (univariate). However, LSTM and GRU still have some weakness as the absence of a structured model and can be directly used for forecasting. In addition, in the LSTM there is an output gate that makes some stored data removed because it is considered too long or too far away. Whereas in the GRU there is no output gate, but calculations using all the data are very potential for bias [13].

Based on these explanations, this study will examine how predictions can be made using deep learning methods, LSTM and GRU, which according to prior research on time series forecasting can produce a sufficient level of accuracy. However, to minimize the risk of LSTM and GRU weakness, the optimum amount of historical data will be sought in advance to be entered into the model. One technique that can be used is to add a regression technique for multivariate time series data, the Vector Autoregressive (VAR) method. VAR used as a method in the feature selection process prior to the LSTM and GRU process because it can build a structured forecast model and can determine the optimal number of lags (historical data). In addition, this study will add several variables that have not been considered in previous studies but which affect jet fuel transaction price, namely inflation and local currency rates. The additional variable is expected to increasing forecast accuracy.

2. Research Methodology

In this research, we will discuss about jet fuel pricing, data collection and pre-processing, and forecast using many algorithm and combinations.

2.1. Jet Fuel Pricing

Fuel is the refinery product from crude oil and used by all transportation modes as the main material for the combustion and ignition systems in the engine and causes the transportation mode to move [8]. Fuel is an important component in the operation of a vehicle, not least in the aviation industry, which must use to good, quality, and standard because it involves aviation safety and security [4].

One of type of fuel product is Avtur (Aviation Turbine Fuel), or also known as Jet Fuel, which is a fuel with an external combustion system based on turbine gas or jet engines used by many commercial aircraft. One type of Avtur that is most often used by Indonesian airlines is the Avtur Jet A-1 series because it has an excess freezing point of up to -470C and this is needed when aircraft fly at cruising altitudes of 30,000 to 40,000 feet [11].

As stated in previous research that fuel costs are one of the biggest cost components of airlines [4] [5], one reason is the price of fuel which is very volatile. Compiled from several sources, except from political factors, fuel fluctuations can be caused by the following factors: (1) Crude oil price trends (2) Fueling location (crowds / frequency of flights, location of oil refineries, currency exchange rates, inflation), and (3) Fuel manufacturer monopoly.

Fuel manufacturer usually set the base price that applied in many locations in their country, then airline paying the fuel cost to supplier using manufacturer's base price that added by differential prices. For example, at Soekarno-Hatta International Airport, Indonesia (CGK), there are two price equations that can be applied by airline: (1) MOPS Posting Price + Differential Price, and (2) Pertamina Posting Price + Discount Price. The Pertamina Posting Price is the base price that already set by Pertamina as fuel manufacturer in Indonesia for each airport in Indonesia [10].

2.2. Data Collection

In this research, we will examine forecasting of jet fuel transaction price at CGK Airport in Tangerang, Indonesia. We will use data in the period January 2017 until April 2020 each dataset with details below:

Table 1. Dataset collection

Variable	Description	Variable Type
WTI	Crude Oil Price issued by West Texas Intermediate (US Approach)	Independent
Brent	Crude Oil Price issued by Brent (European Approach)	Independent
Gulf	Basic Jet Kerosene Fuel Price issued by Arabian Gulf Coast	Independent
MOPS	Basic Jet Kerosene Fuel Price issued by Platts Singapore	Independent
Pertamina	Local jet fuel supplier price in CGK Airport Indonesia	Independent
JISDOR	Indonesia Exchange Rate by Bank Indonesia	Independent
Inflation	Indonesia Inflation Rate by Bank Indonesia	Independent
Transaction	Indonesian Airline Jet Fuel Transaction Price	Dependent

The data used in this research we used 8 variables that are grabbed from 5 different sources as follows:

- Crude oil price dataset, we will use WTI Spot Price and Brent Spot Price from open data source that updated every business days [14].
- Basic jet fuel price dataset, that will use Gulf Coast Kero Jet Fuel Spot Price and MOPS Posting Price from Platts's website that updated every business days [15].
- Local jet fuel supplier posting price that in CGK airport will use Pertamina Posting Price from Pertamina's website that updated every 2 weeks in date 1st and 15th every month [10].

- Local economic aspect, using currency exchange rate, JISDOR rate from IDR to USD and local inflation rate from Bank Indonesia website as the Indonesian economic condition that updated every business days [16].
- Jet fuel transaction price of one of Indonesian airline dataset from their internal data source that updates every days.

2.3. Data Pre-processing

2.3.1. Data Integration
As data grabbed from 5 different sources with time series order, it is necessary to integrate all data into a coherent dataset. Data is integrated and adjusted in date order to become a database. So the database consists of 8 columns.

2.3.2. Data Cleansing and Selection
After integrating the data taken from several sources that has different dimensions, so it needs to be cleaning data to become a database that is ready for use. Stages of cleansing and selection of data carried out are as follows:

- Data selection with the shortest dimension, namely Pertamina Posting Price, which is valid from January 1, 2017. So, the data used is data from January 1, 2017 to April 30, 2020.
- Remove the data for international holidays, national holidays, and Saturdays and Sundays (non working days).
- Estimating incomplete data / has a value of 0, Refer to the characteristics of the data, if there is no price update, then the price in the previous period is still valid. So, for incomplete data will be filled with data in the previous period.

2.3.3. Data Transformation / Data Scaling
After cleansing and selecting the data, the transformation process is conducted by changing the scale of the data into the range of 0 to 1 per column. The 3 rows of already processed data shown in Table 2.

Table 2. First 3 rows of ready to processed data

Date	WTI	Brent	Gulf	MOPS	Pertamina	JISDOR	Inflation	Transaction
03-01-17	0.7810	0.5969	0.5773	0.6098	0.1350	0.0923	0.5	0.4193
04-01-17	0.7889	0.5906	0.5773	0.5968	0.1350	0.0903	0.5	0.4021
05-01-17	0.7933	0.5961	0.5825	0.5946	0.1350	0.0602	0.5	0.3953

After get the ready to processed data, data is divided into 2 groups. Data training is 80% of total amount data from first period dated from January 3rd, 2017 until August 29th, 2019 to build model. Then, next 20% data dated from August 30th, 2019 until April 30th, 2020 is grouped into data testing that used for validation and performance evaluation.

2.4. Data Processing
In this section, we proposed a novel model to forecast the jet fuel transaction price which is the hybrid of traditional multivariate time series forecasting method, Vector Autoregression (VAR) with two recurrent neural networks (RNN) algorithm, Long Short Term Memory (LSTM) and Gated Recurrent Units (GRU). The mapping of process flow will be shown in Figure 5.

In Figure 5, we can show that we will apply VAR as an optimization method for parameter and input before the LSTM and GRU process. VAR will optimize the parameter, build a model based on training data, and estimate the data as a feature selection process. So, it will be 2 steps of the training process. As a comparison, we will also forecast the jet fuel transaction price using LSTM (3 layers), GRU (3 layers), and VAR separately. In this research, the models of VAR, LSTM, GRU, VAR-LSTM, and VAR-GRU are operated in Python 3.5 via Google Collaboratory platform with the function tool of Keras and Tensorflow.

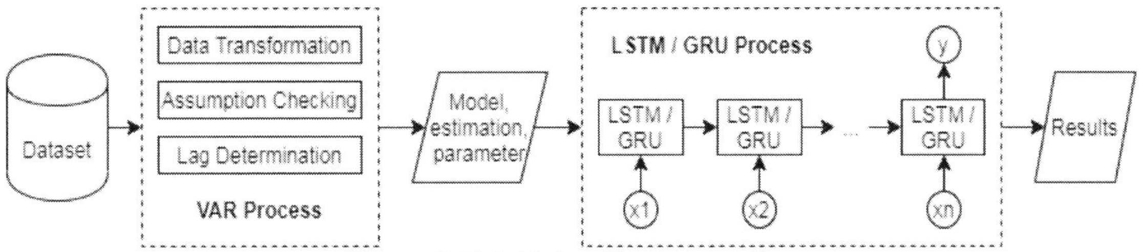

Figure 5. Hybrid algorithm process flow

2.5. Evaluation process

In assessing the performance of each algorithm, it is necessary to evaluate the level of accuracy of the predicted results. The following are some evaluation criteria that can be used to evaluate the performance of prediction accuracy [17]:

a. Mean Absolute Deviation (MAD):

$$MAD = \frac{1}{N}\sum_{i=1}^{N}|y_t - \hat{y}_t| \tag{1}$$

b. Root Mean Square Error (RMSE):

$$RMSE = \sqrt{\frac{1}{N}\sum_{i=1}^{N}(y_t - \hat{y}_t)^2} \tag{2}$$

c. Mean Absolute Percent Error (MAPE):

$$MAPE = \frac{1}{N}\sum_{i=1}^{N}\left|\frac{y_t - \hat{y}_t}{y_t}\right| \times 100\% \tag{3}$$

3. Results and Discussion

3.1. LSTM and GRU process

First, we will process using LSTM and tried for several epoch parameter configurations in the range of 500 to 1,000. For the input of neurons used is n = 70, and the batch size used is 128. The results of accuracy will be shown in Table 3.

Table 3. Performance evaluation of LSTM 3 layer

Epoch	MAD	RMSE	MAPE	Accuracy
500	202.38	187.99	0.1673	85.27%
600	593.72	989.95	0.1307	86.93%
700	411.16	298.41	0.1824	83.76%
800	498.95	371.40	0.1755	84.45%
900	446.57	217.56	0.1839	83.61%

1000	497.53	264.25	0.1859	83.41%

For next step, we will process using GRU and tried for several epoch parameter configurations in the range of 500 to 1,000. For the input of neurons used is n = 70, and the batch size used is 128. The results of accuracy will be shown in Table 4.

Table 4. Performance evaluation of GRU 3 layer

Epoch	MAD	RMSE	MAPE	Accuracy
500	573.44	833.29	0.1245	87.55%
600	541.86	806.93	0.1196	88.04%
700	513.29	794.27	0.1097	89.03%
800	374.27	842.51	0.1281	87.19%
900	385.91	795.10	0.1393	86.07%
1000	458.16	786.18	0.1498	85.02%

In processing data using LSTM and GRU, the forecasting results are obtained by plot and performance results as can be seen in sections Table 2 and Table 3. It can be seen that forecasting results with the best accuracy are obtained when the LSTM model is used with 3 layers and as many as 600 epochs which produce an accuracy of 86.93%. For forecasting with the GRU model, the best accuracy when the model is used with 3 layers and epoch is 700 which results in an accuracy of 89.03%. Figure 6 below shows a comparison graph plot between the forecasting done by the LSTM and GRU algorithm.

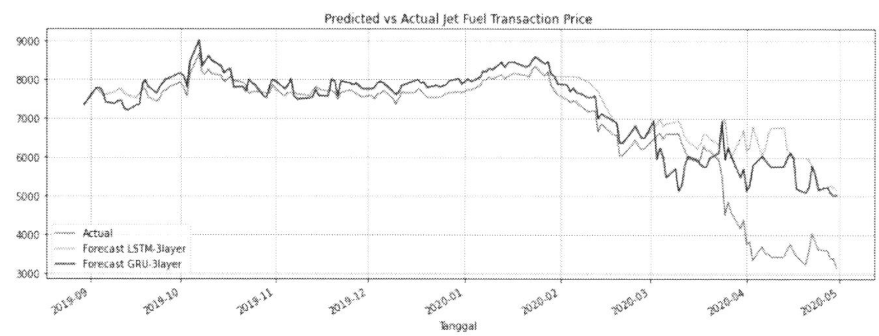

Figure 6. Graphical plot forecasting using LSTM 3 layer and GRU 3 layer

3.2. VAR, VAR-LSTM, and VAR-GRU process

In VAR process, we should determine the optimum lag number that will applied to LSTM and GRU in future steps. The determination will calculated by *Akaike information criterion* (AIC) and the results is in maximum lag order 2, with AIC number 2.255. Next, with maximum lag order 2, we can generate mathematics model like regression with VAR(2) as follows:

$$Trans_t = 1.04 - 1.36WTI_{t-1} - 1.75Brent_{t-1} + 41.29Gulf_{t-1} - 7.26MOPS_{t-1} + 0.20Pertamina_{t-1} - 0.04JISDOR_{t-1} + 2278.67Inf_{t-1} + 0.06Trans_{t-1} - 4.72WTI_{t-2} - 0.28Brent_{t-2} + 347.27Gulf_{t-2} - 0.62MOPS_{t-2} + 0.12Pertamina_{t-2} - 0.05JISDOR_{t-2} - 11347.37Inf_{t-2} - 0.05Trans_{t-2} + \varepsilon_t$$

Next, we will continue to LSTM and GRU process with input from VAR estimation. Figure 7 below shows a comparison graph plot between the forecasting done by the VAR, VAR- LSTM and VAR-GRU algorithm.

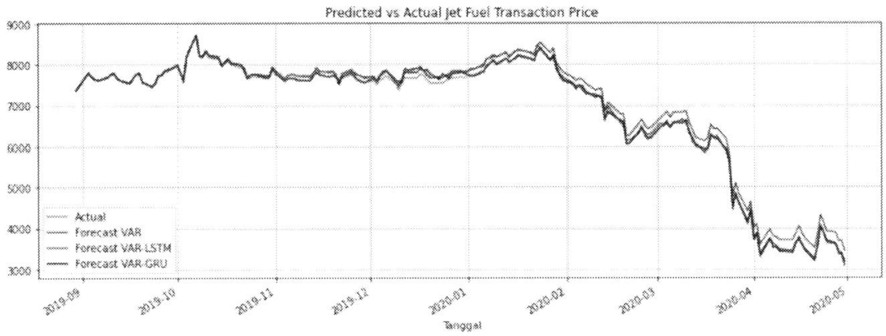

Figure 7. Graphical plot forecasting using VAR, VAR-LSTM, and VAR-GRU

The performance measurement for this 5 models as follows:

Table 5. Performance measurements for 5 models

Algorithm	MAD	RMSE	MAPE	Accuracy
LSTM-3layer	593.72	989.95	0.1307	86.93%
GRU-3layer	513.29	794.27	0.1097	89.03%
VAR	141.13	164.82	0.0252	97.48%
VAR-LSTM	68.21	81.52	0.0102	98.98%
VAR-GRU	40.50	62.53	0.0060	99.40%

In processing data using VAR-LSTM and VAR-GRU, the forecasting results obtained by plot and performance results can be seen in Table 4. It can be seen that forecasting results with the best accuracy are obtained when the VAR-LSTM model produces an accuracy of 98.98%. For forecasting with the VAR-GRU model, the best accuracy when the model produces an accuracy of 99.40%. Based on the graph plot shown in Figure 6 and 7, it can be seen that the experimental results for forecasting aircraft fuel transaction prices are better than the LSTM and GRU outputs. Even outputs are consistent from the beginning to the end of the testing data. The outputs of VAR-LSTM and VAR-GRU can also follow a downward trend in extreme conditions due to the Covid-19 pandemic that occurred twice, namely in February and mid-March 2020. By combining the VAR algorithm with the LSTM and / or GRU, causing instability experienced by forecasting LSTM and GRU can be minimized so that it can produce more accurate outputs with actual values.

Furthermore, if we look at the output of the VAR, the shortcomings can be seen that is along with the progress of the period, the wider bias is generated against the actual data. This turned out to also be minimized by combining VAR-LSTM and VAR-GRU so that the output produced would be more accurate. In addition, the level of performance of VAR-LSTM and VAR-GRU, the outputs become more convergent, which has a minimum difference between epoch accuracy. This proves that the instability of the LSTM and GRU which causes the output value to always change when running can be minimized by using a merge with the VAR method, although it has not been fully structured.

For comparisons of performance evaluation, based on Table 4, it can be seen that the best performance for all evaluation criteria is obtained by the VAR-GRU algorithm with total accuracy of 99.40%. This is because of the VAR-GRU algorithm has advantages over the other 3 algorithms as follows:

a. Has the advantage of LSTM, which has a memory cell in the forecasting process.
b. It has the advantage of GRU, that is, no information is released on the lag or previous period stored in the LSTM memory cell, so that the best value can be produced as an output.
c. Has the advantage of VAR, which is knowing the optimum amount of lag that is used, so that the data process will be more concise and optimum.

IOP Conf. Series: Materials Science and Engineering **909** (2020) 012079 doi:10.1088/1757-899X/909/1/012079

4. Conclusions

In this paper, we studied hybrid between Vector Autoregression (VAR) and Long Term Short Memory (LSTM) or Gated Recurrent Unit (GRU), to predict the multivariate time series problem, there is a jet fuel transaction price. We use jet fuel transaction price as a dependent variable that forecasted in time series using 7 independent variables such as 2 crude oil price, 2 basic jet fuel posting price, local supplier posting price, local currency exchange rate, and inflation rate. This is done considering the price of fuel transactions that are also influenced by currency hedging conducted by the airlines and fuel providers as well as the political-economic conditions that are applied in Indonesia. Where the political-economic decisions determined by the government will usually affect and be influenced by currency exchange rates and inflation rates. In addition to these two variables, the variables that are still being used are world crude oil prices and aircraft fuel prices that apply in the world and at every airport location.

The method that has been tried to improve the forecasting accuracy of 2 (two) RNN methods, namely Long Short Term Memory (LSTM) and Gated Recurrent Unit (GRU) in this study is to combine the Vector Autoregression (VAR) method as a feature selection process method. so that it can produce forecasting results with a higher degree of accuracy. Where for VAR-LSTM produces forecasting with an accuracy of 98.98%, while VAR-GRU produces forecasting with an accuracy of 99.40%. Especially with the Covid-19 pandemic which caused a significant decrease in the period from February to April 2020, VAR-LSTM and VAR-GRU remain sensitive to these changes and can produce forecasting with small bias.

For future research, we suggest developing a model with the addition of the variables which certainly can influence decisions when negotiations are carried out by airlines with fuel providers, such as planned a number of flights, a flight duration of each aircraft, and location of filling/intoplane. For example, with the implementation of Large-Scale Social Distancing during the Covid-19 pandemic in Indonesia, the demand for aircraft fuel will be low and automatically the price of aircraft fuel will go down as the number of flights decreases. For the methods on multivariate time series using neural network algorithms, it can also be combined with a metaheuristic algorithm to get a more optimal feature selection process.

References

[1] You Y 2019 Big data-driven innovation: an empirical research on sustainable development of business model *Advances in Intelligent Systems and Computing* **842** 907-914

[2] Fam S F, Ismail N and Shinyie W L 2019 The Magnitude of Big Data 5vs in Business Macroclimate *International Journal of Recent Technology and Engineering (IJRTE)* **8** 497-503

[3] Larose D T and Larose C D 2015 *Data Mining and Predictive Analytics* (New Jersey: John Wiley & Sons)

[4] Martino A, Casamassima G and Fiorello D 2019 *The impact of oil price fluctuations on transport and its related sectors* (Brussels: European Parliament's Committee on Transport & Tourism)

[5] International Air Transport Association (IATA) Annual Review 2014

[6] Khan M S, Saikia A and Tripathy S 2015 Prediction of Petroleum Price in India *International Journal of Research in Mechanical Engineering & Technology* **5** 11-16

[7] Atems B Bachmeier L and Williams C 2018 Do jet fuel price movements help forecast airline fares and the demand for air travel? *Applied Economics Letters* **26** 1-6

[8] Bazargan M 2010 *Airline Operations and Scheduling* (Surrey: Ashgate Publishing Limited)

[9] IATA 2020 *Jet Fuel Price Monitor* [Online] Available at: <https://www.iata.org/en/publications/economics/fuel-monitor/> [Accessed 22 February 2020]

[10] Pertamina Aviation 2020 *Pertamina Posting Price* [Online]. Available at: <https://www.pertamina.com/aviation/News.aspx?p=price> [Accessed 30 April 2020]

[11] Firmansyah D 2017 Perancangan dan Implementasi Business Intelligence dalam Memantau Pergerakan dan Prediksi Harga Bahan Bakar untuk Pemanfaatan Peluang dalam Pencapaian

Target Kinerja Harga Bahan Bakar Pesawat *MTI Thesis* Universitas Indonesia, Jakarta

[12] Guo J 2019 Oil price forecast using deep learning and ARIMA *International Conference on Machine Learning, Big Data and Business Intelligence (MLBDBI)* 241-247

[13] Santosa B and Umam A 2018 *Data Mining dan Big Data Analytics: Teori dan Implementasi Menggunakan Python & Apache Spark* (Yogyakarta: Penebar Media Pustaka)

[14] U.S. Energy Information Administration 2020 *Spot Prices For Crude Oil and Petroleum Products* [Online] Available at: https://www.eia.gov/dnav/pet/pet_pri_spt_s1_d.htm [Accessed 30 April 2020]

[15] S&P Global Platts 2020 *Jet Fuel – S&P Global Platts* [Online] Available at: https://www.spglobal.com/platts/en/oil/refined-products/jetfuel [Accessed 30 April 2020]

[16] Bank Indonesia 2020 *Bank Indonesia* [Online] Available at: https://bi.go.id/. [Accessed 30 April 2020]

[17] Huang L and Wang J 2018 Global crude oil price prediction and synchronization based accuracy evaluation using random wavelet neural network *Energy* **151** 875-888.

IOP Conf. Series: Materials Science and Engineering **909** (2020) 012080 doi:10.1088/1757-899X/909/1/012080

Exploiting online customer reviews for product design

Yanti Pasmawati[12*], Alva Edy Tontowi[1*], Budi Hartono[1], Titis Wijayanto[1]

[1]Departement of Mechanical and Industrial Engineering, Universitas Gadjah Mada, Yogyakarta, Indonesia
[2]Departement of Industrial Engineering, Universitas Bina Darma, Palembang, Indonesia

*E-mail: alvaedytontowi@ugm.ac.id

Abstract. Online customer reviews (OCRs) are the text features of a consumer feedback on the crowdfunding platform. Utilization of OCRs as online survey provide an opportunity to obtain extensive and fast data about product information, experience of use and services. OCRs can be important information to identify consumer desires for a product design. This research proposes to utilize OCRs as input in developing products for start-up product. The object of this research is technology products. 34,006 OCRs were collected by scraping from kickstarter website. Opinion pros and cons are classified so as to produce 26,993 positive reviews and 13,203 negative reviews. extracted data use text pre-processing and Term Frequency-Inverse Document Frequency (TF-IDF) is used to select words through the classification of product design attributes. Our results show that the functional attributes and followed by aesthetics to be an important role in product design because these attributes have high reviews that customers consider when choice and buying products. More than that, the pros and cons based on relative value of product design reviews contained in online customer reviews are consideration for maintain and evaluate product design attributes. So that technology product produced by start-up product can be successful in the online market.

1. Introduction

The development of the World Wide Web provides accessibility for internet users and the opportunity to have web pages. These opportunities are used by start-ups to reach a wide and fast market. At the moment, people's purchasing power is also increasing in the online market [1]. A crowdfunding based platform takes advantage of opportunities to help product start-ups compete in the online market. Start-up products use creative ideas to raise funds to be able to do mass production. Mass production can be done if crowdfunding investment reaches the goal. Failure is still unavoidable by start-ups. For example, Pasmawati et. al., [2] showed that only 7.35% of technology products were declared successful. Increased success occurred in 2020 by 20.66% [3]. The success of product design is very dependent on the initial concept at the conceptual design stage [4]. Product design is also a predictor of product preference [5].

Online surveys provide more information in a short amount of time, compared to conventional surveys that are limited. Online customer reviews (OCRs) are text features of feedback on crowdfunding platforms. OCRs in the form of information about user experience, the purchasing experience, and satisfaction with products and services [6]. Research on the use of OCRs for various applications has been carried out by several researchers, for example, sentiment analysis [7-10], product sales prediction [8,11], consumer purchase interest [12-15], consumer behavior [16-17], product development [18-20]. The object of research used is existing products that have reached the stage of mass production in e-commerce, marketplaces, and crowdfunding. However, this research has not yet discussed the use of OCRs for product start-up applications in product development efforts at the design stage. Design attributes for technology products that are needed and desired by consumers in the online market.

This study proposes how the use of qualitative OCRs can provide information to product start-ups as product design attributes in product development. Text pre-processing is used to extract unstructured data, then word selection uses the weight of the term frequency-inverses document frequency (TF-IDF). It is expected that by knowing the product design attributes, start-ups can compete with existing products to produce product designs according to consumer perceptions.

2. Method

OCRs are used as opportunities to identify product design attributes based on consumer perception. The Python 3.7 programming language is used as a tool for product design attribute analysis. This study went through 3 major stages outlined in Figure 1. The stages of the study began with data collection from the crowdfunding platform and the sentiment analysis process to find out the pros and cons of sentences. The next step is text pre-processing and text analysis.

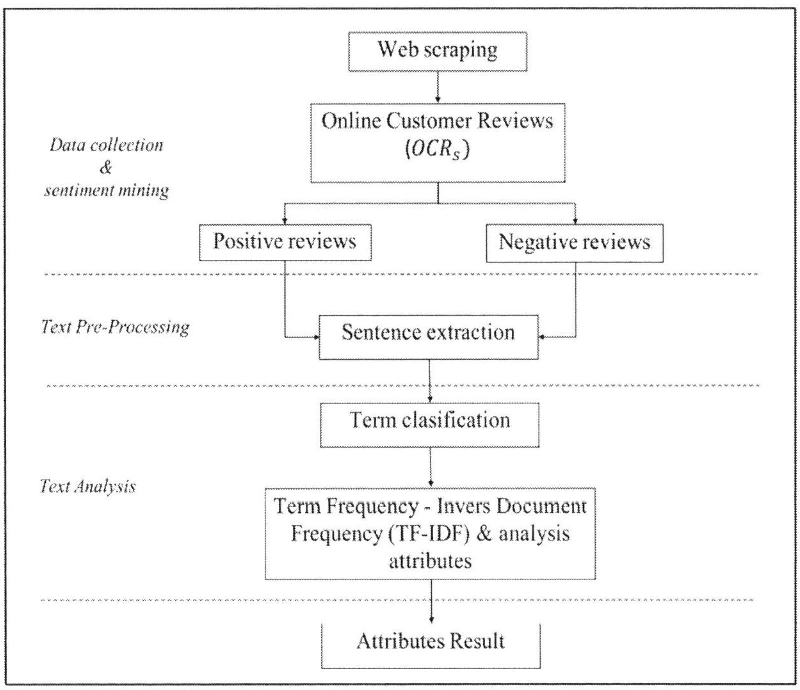

Figure 1. Research framework

2.1. Data Collection

Research data sourced from the kickstarter.com website. The object of research is technology products (drones, wearables, camera equipment, and 3d-printing). As many as 34,006 OCRs were taken by scraping from 100 technology products that had completed successful and failed campaigns. Natural language processing (NLP) is used for linguistic language analysis to find out sentiment reviews from OCRs. The OCRs data set was analyzed and yielded 26,993 positive reviews (U_S^+) and 13,203 negative reviews (U_S^-), as shown in Figure 2.

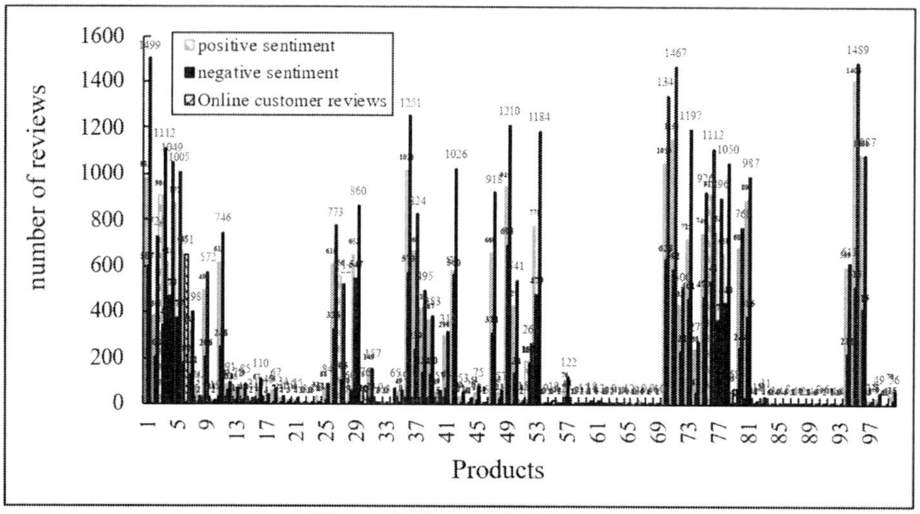

Figure 2. Research data set

2.2. Text Pre-Processing

Text pre-processing uses Python with the NLTK Library. Text pre-processing through several stages, namely tokenizing, lower capitalization, stopwords, stemming, and spelling normalization. Text pre-processing stage, the process of extracting unstructured words into structured words is carried out. Nouns in Parts of speech (POS) are used and are considered to be the most representative tokens of a document [21]. But this study also uses adjectives as semantic networks because not all nouns relate to attribute features.

2.3. Text Analysis

Text analysis conducted is extracting words and word selection. The word extraction is classified based on the product design attributes shown in Table 2. Whereas the data selection is based on the Term Frequency-Inverse Document Frequency (TF-IDF) weights. The data set used is the output of the spelling normalization process of U_S^+ and U_S^-. All extracted data is structured from U_S^+ and U_S^- as words and attributes are calculated TF-IDF automatically using Python software. TF-IDF is used to find the value representation of each document in the collection. The greater the weight calculation value obtained, the higher the level of document similarity to the query. TF-IDF as an interpretation of backer responses on product campaigns. TF-IDF determines the relative importance weights of words in U_S^+ and U_S^-. TF-IDF for each word added up based on the classification of attributes. Furthermore, TF-IDF was analyzed based on the relative values between TF-IDF on U_S^+ and U_S^-. The relative value of TF-IDF (N_rTF-IDF) < 0.5 means that the attribute has a negative opinion greater than a positive opinion. N_rTF-IDF $\geq$ 0.5 means that the attribute has a positive opinion that is greater than a negative opinion.

3. Result and Discussion

3.1. Text pre-processing

Data used for text pre-processing are U_S^+ and U_S^-. Before carrying out the text pre-processing stage, the data is classified based on the pro (U_S^+) review and the cons (U_S^-). The results of the text pre-processing steps are shown in Table 1. Import OCRs data from U_S^+ and U_S^- which are first saved in CSV files into python cells. Import and process data classified by type of sentiment review. The tokenizing stage can be done after or before the case folding. In this study, tokenizing was done before case folding because it was considered more effective. Sorting review sentences into separate words and followed by sorted frequency for each word. Case folding stages can also be done after tokenizing. After the review sentence has been divided into words along with the frequency of each word, a lower capitalization is performed. This process is considered more effective because lowercase conversions are structured words. At the stage of filtering/stopwords only removes Parts of speech (POS) such as the auxiliary verb, conjunction, pronoun, preposition, and interjection. This is because OCRs data sets use English. Words consisting of ≤ 1 or 2 or 3 are not automatically removed because there are attributes in the form of units. Numeric numbers are not omitted.

Table 1. Text pre-processing results

	Number of term	
	Positive	Negative
Tokenizing	834,104	413,563
Case folding (lower capitalization)	26,993	13,203
Filtering/stopwords	6,653	3,942
Stemming	6,653	3,942
Spelling normalization	6,653	3,942

In English text, the process of refining affixes such as prefix and suffix. The process of stemming by removing prefix can be done or also does not need to be done because it does not have the effect of being able to classify words. While suffix is not converted to basic words because there are some differences in meaning that can result in misclassification of attributes. For example, monitors are nouns, while monitored and monitoring are verbs. From the steaming process, there were 6,653 words from U_S^+ and 3,942 words from U_S^-. There are some texts from U_S^+ and U_S^- which have incorrect spelling. The spelling normalization process is carried out to improve spelling substitution. A total of 6,653 words U_S^+ and 3,942 words U_S^- at the end of the spelling, the normalization process is used to classify product design attributes at the feature extraction stage.

3.2. Feature Extraction

The extraction of words is classified using a manual check with Excel. Words are classified according to attribute groups. The results of the classification process of words on the product design attributes can be seen in Table 2 and Table 3.

Table 2. Term classification based on product design attributes

Positive reviews	Negative reviews	Design product attributes
black, blue, beautiful, beam, brand, color, logo, unique, display, digital, design, ...	appearance, blue, digital, metal, gold, display, unique, ...	aesthetics
ergonomics, compatibility, comfortable, lighting, tempature, vibrate, sounds, ...	comfort, glow, contrast, lighting, noice, noise, vibrate....	ergonomics
durability, adaptor, photochromic, smartscreen, stainless, video, ...	ability, control, access, detector, durability, setting, screen, ...	functional
allergenic, iritating, allergy, photokeratitis,	hazard, risk, ...	health
license, originally, authentic, copyright, certificate, ..	authenticate, copyright, creative, licence,	innovation
authority, authorized, law, legality, legally, regulation, certificate, ...	formal, legal, legality, ...	Regulation
safety, hazard, protection, dangerous, ...	danger, hazard, safety, ...	safety
ratification, accreditation, quality, standard, certification, method, calibration, ...	instruction, method, module, quality, ...	standard
trend, lifestyle, style, technology, high-tech,..	technology, millenial, ...	trend
reproduce, reproducible	environment, recycling, reproducing	environment
continue, sustainability	-	sustainable

Based on the results of the classification, it is known that there are no reviews that discuss the sustainable attribute of negative reviews. The classification of words/term is based on the definitions and synonyms of all the words contained in positive or negative reviews. Furthermore, the word is weighted based on TF-IDF.

3.3. *Product Design Attributes Based On TF-IDF*

TF-IDF attribute is the sum of TF-IDF synonym words that are included in the classification of product design attributes. TF-IDF on U_S^+ and U_S^-, as shown in Figure 3.

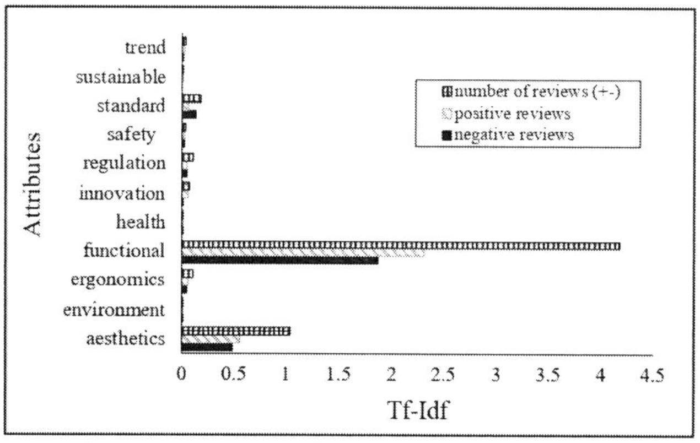

Figure 3. TF-IDF of product design attributes

The results showed that the TF-Idf values of positive reviews are in the range of 0.0002 to 2.3083. While the Tf-Idf values of negative reviews are in the range from 0.000 to 1.8667. Tf-Idf values that have $U_S^+ > U_S^-$ are aesthetics, ergonomics, functional, innovation, sustainable, and trend. Whereas $U_S^+ < U_S^-$ are environment, health, regulation, safety, and standard. Sustainable hasn't negative reviews from backers. If the Tf-Idf value of $U_S^+ > U_S^-$ means that the attribute has backers perception tend to be positive and vice versa. The higher TF-IDF, the greater the similarity in product design attributes. Tf-Idf also shows backer perceptions about product design in reviews. Figure 3 also shows the response rate of backers based on Tf-Idf. Tf-Idf value of backers response is 0.0025 - 4.1750. The highest response is functional and followed by aesthetics, standard, ergonomics, regulation, innovation, safety, trend, health, sustainable and environment. The next stage is to determine the product design attribute based on the number of the relative value (N_rTF-IDF). The results of the N_rTF-IDF calculation are shown in Figure 4.

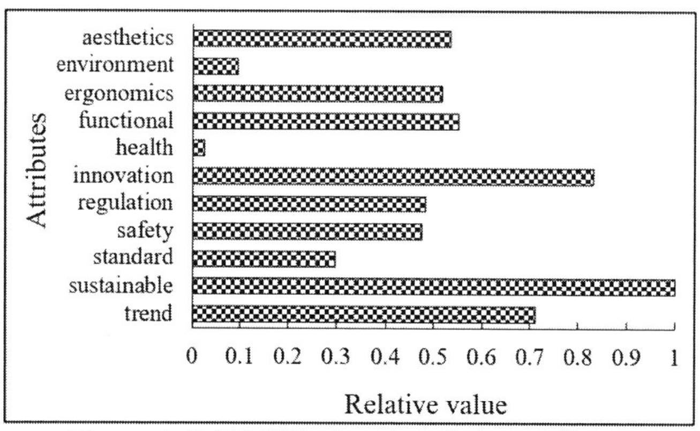

Figure 4. $N_r.TF\text{-}IDF$ of product design attributes

N_rTF-IDF of attributes ranges from 0.026 - 1.00. The highest Nridf of 1.00 was sustainable and followed by innovation, trend, functional, aesthetics, standard, ergonomics, regulation, safety, environment, and healthy. The high N_rTF-IDF, the high the level of positive responses. While the low N_rTF-IDF show that the low level of negative responses. Attributes that have N_rTF-IDF < 0.5, among other, regulation, safety, environment, and health. These attributes indicate that $U_S^+ < U_S^-$ or in other words have negative dominant backer perceptions. This attributes are attributes that are maintained in the development of product design. Attributes that have N_rTF-IDF > 0.5 are sustainable, innovation, trend, functional, aesthetics, standard, and ergonomic. Sustainable is an attribute that has the highest N_rTF-IDF, because sustainable only has pros perception without any cons. This Attributes indicated that have positive dominant perceptions. This attributes must be evaluated in developing product design.

4. Conclusion

Online customer reviews are reviews in the form of text, which has the potential to provide information and insight to the designer or start-up in product development. Our results show that functional attribute an very important role in product design because these attributes have the highest reviews and responses that customers consider when choosing and buying products. The attributes that must be evaluated in developing product design are regulation, safety, environment, and health because has negative dominant response. While, the attributes that must be maintained in the development of product design, among other, sustainable, innovation, trend, functional, aesthetics, standard, and ergonomic because has a positive dominant response. This evaluations used so that technology product produced by start-up product can be successful in the online market. This research only determines the attributes of product design based on OCRs. The future research, attribute validation needs to be done to predict success of technology products from start-up product in crowdfunding platform.customer reviews are reviews in the form of text, which has the potential to provide information and insight to the designer or start-up in product development. Our results show that functional and followed by aesthetics attributes an important role in product design because these attributes have high reviews and responses that customers consider when choosing and buying products. Moreover, aesthetics and functional are a priority for evaluated product design attributes so that technology product produced by start-up product can be successful in the online market. This research only determines the attributes of product design based on OCRs. The future research, attribute validation needs to be done to predict success of technology products from start-up product in crowdfunding platform.

Acknowledgments
This research has been supported by Research Grant RTA 2020 No. 2488/UN1.P.III/DIT-LIT/PT/2020 of Universitas Gadjah Mada.

References

[1] The Boston Consulting Group 2017 One - Fifth of the World's Internet Users Are in China and They Are Spending Close to $ 1 Trillion *https://www.bcg.com/d/press/28september2017-decoding-chinese-internet-172187* [Accessed:9 Jan2018]

[2] Pasmawati Y, Tontowi AE, Hartono B and Wijayanto T 2018 Determination of Favourite Market for 3D-Printer Product Based Online Platform *Int. Rev. Manag. Mark.* **8** 18–23

[3] Kickstarter 2020 Kickstarter Stats *https://www.kickstarter.com/help/stats* [Accessed:17April2020]

[4] Hambali A, Sapuan SM, Ismail N and Nukman Y 2009 Application of analytical hierarchy process in the design concept selection of automotive composite bumper beam during the conceptual design stage *Sci. Res. Essay* **4** 198–211

[5] Kumar M and Noble CH 2015 Beyond form and function : Why do consumers value product design? *J. Bus. Res.* **69** 613–20

[6] Jin J, Liu, Y, Ji P and Kwong CK 2018 Review on recent advances in information mining from big consumer opinion data for product design *Computing and Information Science in Engineering* **19**

[7] Bafna K and Toshniwal D 2013 Feature based summarization of customers reviews of online products *Procedia Comput Sci* **22** 142–51

[8] Fan ZP, Che YJ and Chen ZY 2017 Product sales forecasting using online reviews and historical

sales data: A method combining the bass model and sentiment analysis *J. Bus. Res.* **74** 90–100
[9] Ireland R and Liu A 2018 Application of data analytics for product design: Sentiment analysis of online product reviews *CIRP J. Manuf. Sci. Tech* **23** 128–44
[10] Li X, Wu C and Mai F 2019 The effect of online reviews on product sales: A joint sentiment-topic analysis *J. Inf. Manag.* **56** 172–84
[11] Decker R and Trusov M 2010 Estimating aggregate consumer preferences from online product reviews *Int. J. Res. Mark.* **27** 293–307
[12] Lee E and Shin SY 2014 When do consumers buy online product reviews? Effects of review quality, product type, and reviewer's photo *Comput. Human. Behav.* **31** 356–66
[13] Kostyra DS, Reiner J, Natter M and Klapper D 2016 Decomposing the effects of online customer reviews on brand, price, and product attributes *Int. J. Res. Mark.* **33** 11–26
[14] Zhang Y and Lin Z 2017 Predicting the helpfulness of online product reviews: A multilingual approach *Electron. Commer. Res. Appl.* **27** 1–10
[15] Zhuang M, Cui G and Peng L 2018 Manufactured opinions : The effect of manipulating online product reviews *J. Bus. Res.* **87** 24–35
[16] Bambauer-sachse S and Mangold S 2013 Do consumers still believe what is said in online product reviews? A persuasion knowledge approach *J. Retail. Consum. Serv.* **20** 373–81
[17] Luan J, Yao Z, Zhao F and Liu H 2016 Search product and experience product online reviews: An eye- tracking study on consumers' review search behavior *Comput. Human. Behav.* **65** 420–30
[18] Jin J, Ji P and Liu Y 2014 Prioritising engineering characteristics based on customer online reviews for quality function deployment *Journal of Engineering Design* **25** 303–324
[19] Jin J, Ji P and Gu R 2016 Identifying comparative customer requirements from product online reviews for competitor analysis *Engineering Applications of Artificial Intelligence* **49** 61–73
[20] Qi J, Zhang Z, Jeon S and Zhou Y 2016 Mining customer requirements from online reviews: A product improvement perspective *J. Inf. Manag.* **53** 951–63
[21] Feldman R and Sanger J 2007 The Text Mining Handbook: Advanced Approaches in Analyzing Unstructured Data *(New York: Cambridge University Press)* 1–16

IOP Conf. Series: Materials Science and Engineering **909** (2020) 012081 doi:10.1088/1757-899X/909/1/012081

Website library information system design using fast method

A Gunawan[1*], N Wahyuni[1], H Setiawan[1], PB Katili[1], R Azla[1]
[1]Department of Industrial Engineering, Universitas Sultan Ageng Tirtayasa, Jend. Sudirman Km. 3 Cilegon, Banten 42435, Indonesia

*E-mail: a68ar@untirta.ac.id

Abstract. This research was conducted at the Department of Industrial Engineering as part of the department at Sultan Ageng Tirtayasa University which has applied information technology to support every academic activity, information media, and services. Likewise the library which is part of the academic facilities in industrial engineering matters. However, in the current condition the library does not have an information system as a form of service improvement and information technology utilization. In service activities and the availability of information in the library it was found that industrial engineering students have not provided information system support, specifically regarding search reference books, and research. A general search for references for this book is handled by the industrial engineering library. Based on this problem, therefore an information system based website will be designed using a CMS (Content Management System) because it can meet the management system, management information system, and present information to users. The information system in this study was designed using the FAST (Framework for Application of System Technique) method in which there is an analysis using PIECES (Performance, Information, Economic, Control, Efficiency and Service) therein. This method provides a model for analyzing and defining user needs, choosing solutions and making an organized system for designing information systems.

1. Introduction

The need for information systems to support organizational processes and service systems is something that is needed to reach more users. The rapid development of technology, especially in the field of information technology and access to the Internet provides the right opportunity to develop better information services for each organization. The principles of information systems in an organization are the use of information systems that aim to add value to organizations that are strongly influenced by organizational structure, culture, needs, change and development.

The role of information systems in all aspects is very relevant to the era of the industrial revolution 4.0 which began to penetrate in Indonesia, especially with the presence of various platforms that provide convenience services for information, services will buy products or even in the academic field. In addition, the linkage of this system with the internet as a bridge of access to information also paves the way for any community that has a smartphone or laptop to find everything on the internet. The following are the objectives of the research, which are as follows: To find out the design of web-based information systems in the untirta industrial engineering library. Evaluation results from web-based information system design tests in the untirta industrial engineering library. Comparison of previous systems with information systems that have been designed.

2. Method

The design of library information systems in Untirta Industrial Engineering is based on the need for library-based service web sites to facilitate library data management, as well as access for users. In the current condition, library information systems do not yet exist in the industrial engineering department, so information system design is needed. The design of this information system uses open source media in the form of an open source Content Management System (CMS) wordpress with a local XAMPP server, using the PHP programming language, CSS and using a MySQL database. This design uses primary data in the form of interviews with Library Supervisors and Library Managers about what is needed to support a web-based library information system. There is also secondary data in the form of administrator account creation data, member data, book data, practical work report data, final project data in the Untirta Industrial Engineering Department. The following is the compilation process carried out in this study, there are 2 flow basins namely general research flow diagrams and data processing flow diagrams. (See Figure 1)

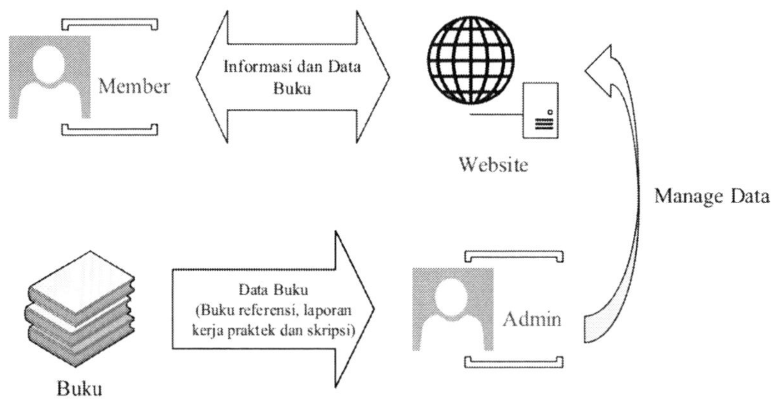

Figure 1. Process flow

Process Flow is a fundamental representation of a process that schematically describes the flow of data in the system. In this library information system, process flow provides a schematic picture between members, books, data flowing in the process and the admin. The following is a process flow from the library information system.

3. Result and Discussion

Data collected in this study are primary data and secondary data, where primary data is data that is taken directly and secondary data is data that is taken indirectly. Primary data collected were in the form of field observations, interviews with PIECES tools and brainstorming with the Untirta Industrial Engineering Library Board. In addition, the usability questionnaire was distributed to the Head of the Library, Library Management and Users of the Library System. Secondary data collected in this study is in the form of library information data, administrator accounts library members, books (reference books, practical work reports and thesis).

Processing library information systems requires administrator account data, library members, books, practical work reports and theses to support library information systems. Data is taken based on data that is already in the library of industrial engineering in the library management period 2019/2020.

In solving the problems of the library system berbasiswa web at the Department of Industrial Engineering Untirta used FAST method which consists of several phases including, definition of scope (Scope Definition), Problem Analysis (Problem Analysis), Needs Analysis (Requirement Analysis), Design Logical (Logical Design), Physical Design (Physical Design), Construction and Testing (Construction and Testing). Definition of scope is the initial identification phase in designing a system. In this phase the defined scope of s ystem library that will be in the making. The following is the scope contained in this design, as follows: The project name Website-Based Library Information System Design in Untirta Industrial Engineering Department. Software OpenSource CMS (Content Management System) Framework , Google Chrome Browser , PHP, HTML, CSS programming language, using a MySQL database, with local servers XAMPP (X, Apache, MySQL, PHP and Perl). Scope, There are several processes in the scope of the library including: Administrator and Member Login Processes, Process See Home, Profile, Book Data, Job Training Report Data and Thesis Data, Process of Input and Editing Homepage, Profile, Data Book, Data Job Training Report and Thesis Data, The Process of Adding Administrators and Members, The Process of Editing and Deleting Administrators and Members

This problem analysis phase is used to analyze problems that occur in the scope of research. In this research, PIECES tools are used. In analyzing the problem using the PIECES framework in this study five categories of PIECES are classified: performance, information, economics, control, efficiency, service. The following PIECES framework is used to define the problem:

Table 1. Defining the problem using the PIECES framework

PIECES	Defining the Problem
Performance	The data collection of books, practical work reports and thesis requires time, energy and data storage space
Information	Information on the results of data collection is stored quite a lot and is not absorbed well by the user
Economy	The use of paper is conventional and costs money
Control	The risk of loss of notebooks or data not stored on the system and lack of supervision from the responsible library
Efficiency	Data capture requires a long time and only one data collection path
Service	Data collection or recording of books, practical work reports, and theses are inflexible and unresponsive

The table defines the problems that occur in the existing system by classifying five categories of PIECES, namely performance, information, economics, control, efficiency, service. So that needs can be determined in accordance with the state of the system in order to achieve system design in the library information system problem.

After using the PIECES tools , a system translation is done by analyzing the old system (conventional) with the design of the information system under study, as follows:

Table 2. Analysis of the old system (conventional) with the design of information systems

Comparison	Conventional	Information Based System
Access	In the old system, members had to come directly to the library and search for the title of the book with the desired keywords manually	In this system, members only need to open the website and access information through the website on the condition of a login first

Efficiency	In the old system, the capacity of the room contained in the library did not allow all students to come and look for books whether available or not	In this system, users can search for relevant keywords related to the book sought and the system quickly provides book information
Promotion	In the old system, the library only had social media accounts for information facilities and relied on direct presence in the library to promote books and information	In this system, the library will be able to use social media accounts and for complete information facilities can rely on the website without the need for direct presence in the library for registration, filling out forms and promoting books and information

Based on the analysis of the problem that has been explained, then the system requirements are determined to form a library information system that can answer the problem. The following are some of the requirements for a library information system to be built including: Facilities for collecting data on books, practical work reports, and a thesis. Displaying book data, practical work reports, and theses to the admin and members in accordance with the access rights. There is a data search function in accordance with key data information both books, practical work reports, and thesis.

In needs analysis, there are 2 categories of needs, namely functional and non-functional needs. Functional requirements cover the function or service aspects that the system must provide, while non-functional features are additions (not including service functions) to support system utilities.

In the phase of functional and non-functional requirements, it aims to provide an overview of the system to be made. This stage is a step to model the library information system that needs to be designed. At this stage there are tools namely Process Flow, Context Diagrams, and Data Flow Diagrams (DFD) and Entity Relationship Diagrams (ERD). This tool is used to determine the data flow used in information systems starting from giving an overview of system processes, and overall system data flow. Entity Relationship Diagram (ERD) is used to determine the relationships between entities that exist in the system. There is also a Context Diagram which is also referred to as DFD Level 0 in this system. There are 2 entities involved, namely Admin as the head of the library or management and Members as library members. Below is the context diagram of the library information system. (Figure 2).

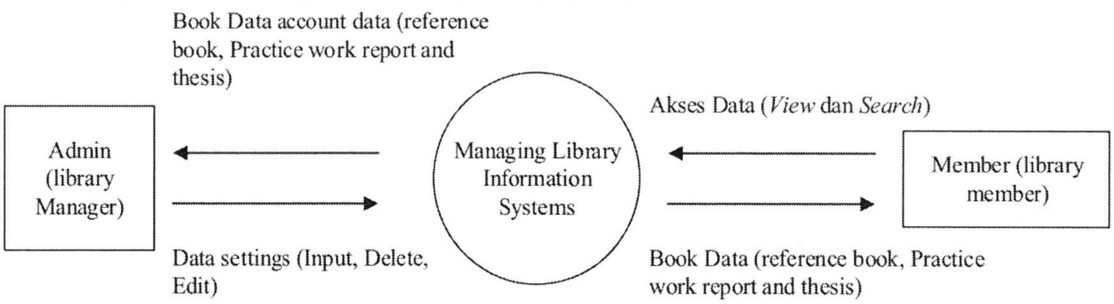

Figure 2. Context diagram (DFD Level 0)

Data Flow Diagrams (DFD) is a tool used to describe the flow of data that is in the system and a process carried out by the system. The following is DFD level 1 of the library information system:

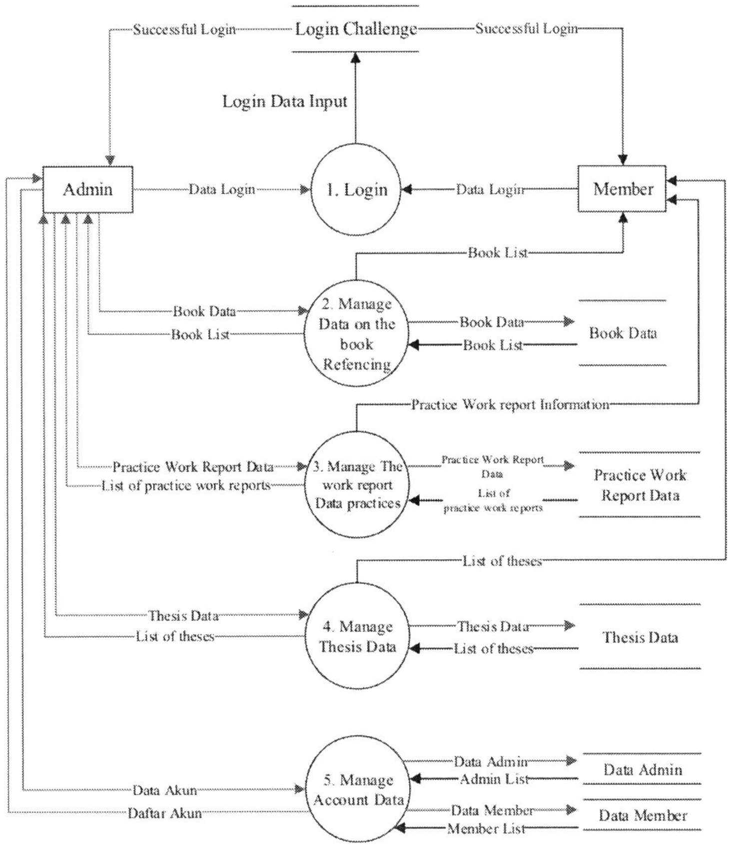

Information :
_______ : Admin
_______ : Member

Figure 3. DFD level 1

DFD level 1 above shows how the flow of admin and member libraries have a role in the system. The flow of data from each account can be seen according to information, where the admin can be seen through the color of blue arrows and members can be seen through the color of black arrows.

After DFD level 1, the next step is DFD level 2 of library information system design. DFD level 2 aims to provide a more in-depth picture of the system. The main process on the system will be broken down into subsystems. The following is DFD level 2 of library information system design.

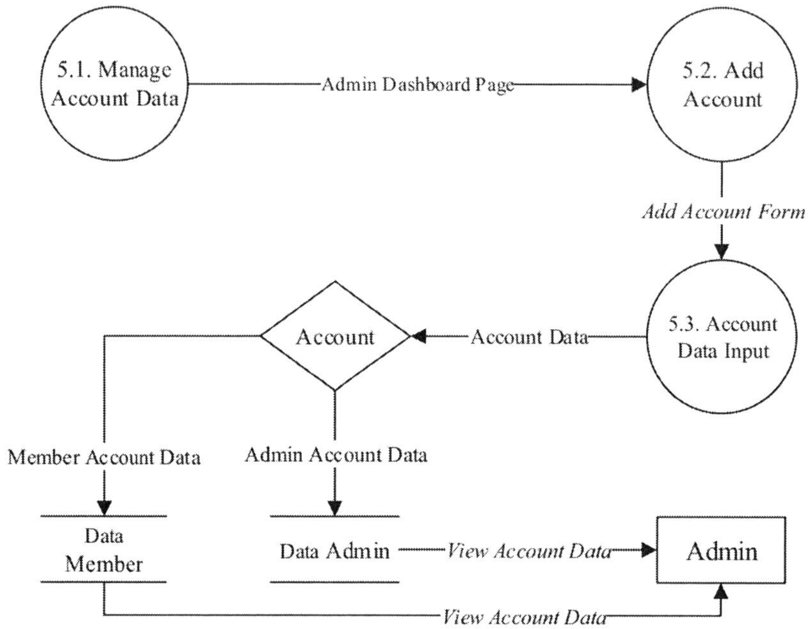

Figure 4. DFD level 2 thesis manage process

1. Process : 5 .1
   Process Name : Manage Account Data
   Enter : Account Menu
   Output : Open the Manage Data Account Page
   Process Summary:
   Admin chooses to manage account data then the manage account data page will open.
2. Process : 5 .2
   Process Name : Add Account
   Enter : Open the Add Account Form
   Output : Add Account Form
   Process Summary:
   After the Add Account Form opens, the admin can enter account data by filling out the available forms.
3. Process : 5 .3
   Process Name : Account Data Input
   Enter : Add Account Form
   Output : Account Data
   Process Summary:
   Account Add Form that has been filled according to the type of account that is admin or member. then click ok so that the account data is successfully input into the system.

Entity Relationship Diagram (ERD) is used to determine the relationships between entities that exist in the system. The following is the ERD of library information system design.

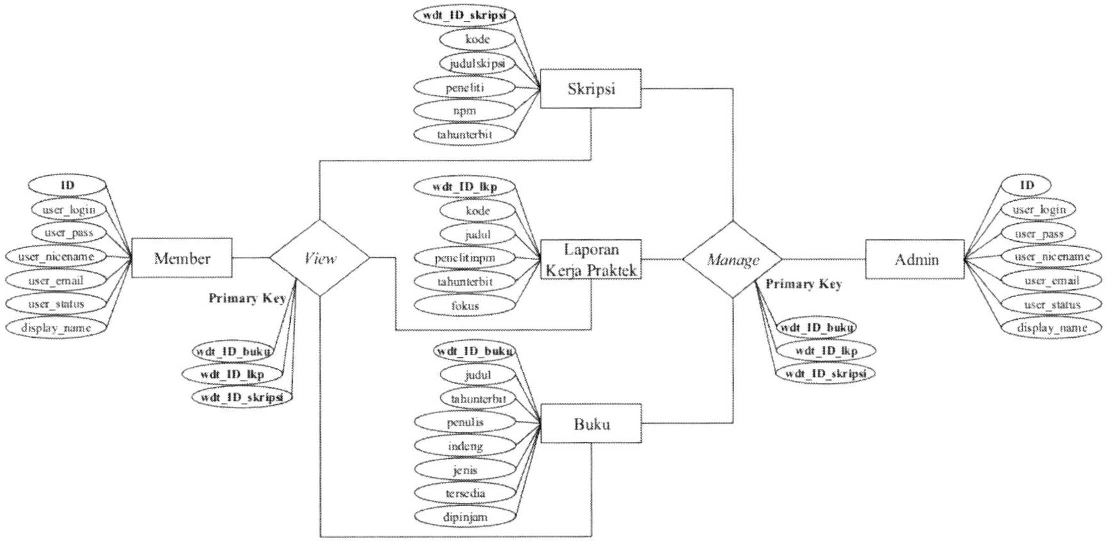

Figure 5. Entity relationship diagram

The following is the display of the home page in the library information system in the Untirta Industrial Engineering Department.

Figure 6. Home page display

On the home page there is information about the library information system, several menu items and there is a login column. The following is a profile page display on the library information system in the Untirta Industrial Engineering Department. On the profile page there is information about the untirta industrial engineering library, several menu items and a login column. The following is a display of book pages on the library information system in the Department of Industrial Engineering, Untirta. On the pages of the book there is information about the reference book data owned by the industrial engineering library. There are two roles namely admin and member. The role member only has access rights to information while the role admin can manage items, namely adding book items, editing book items, and deleting books. The following is a page display of practical work reports on the library information system in the Untirta Industrial Engineering Department. On the practical work report page, there is information about the practical work report data that is owned by the industrial engineering library. There are two roles namely admin and member. The role member only has access rights to information, while the role admin can do item management, namely the addition of practical work report books, edit work practice report items, and delete practical work report books.

The following is a dashboard page view on the library information system in the Untirta Industrial Engineering Department.

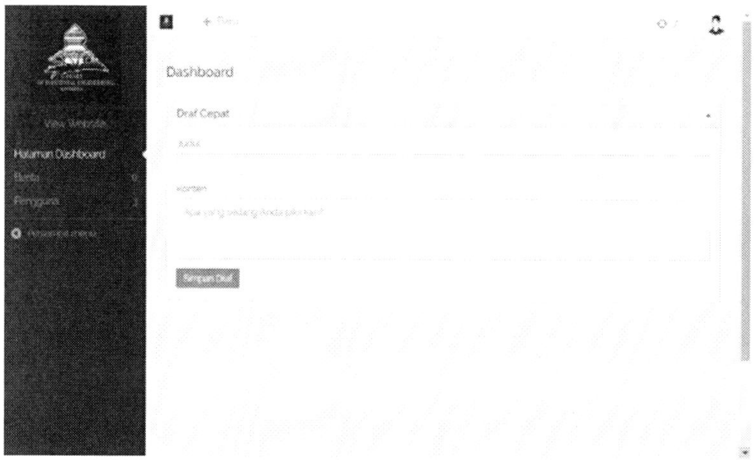

Figure 7. Dashboard page views

On the dashboard page there is information and a dashboard page menu. There are two roles namely admin and member. The role member only has access to the dashboard page and user profile while the role admin has access to the dashboard page, admin profile and account management in the form of adding accounts, editing accounts, and deleting accounts.

4. Conclution

Based on the research done, it can be concluded with the website of the library information system makes it easier for users of libraries in the process of obtaining book information, practice work reports, and thesis and members can find out what books are available in the library, facilitate the admin or the concierge Officer in managing the library data provide the ease of service process for its users.

References

[1] Elmasri, SBNR 2006 Fundamentals of database systems 4th ed San Francisco: Perason
[2] Foster, CE 2014 Software engineering a methodical approach New York USA: Apress
[3] Hermawan, E. & Supriyono, H 2016 Information system design PT. Amsa Based on web using wordpress
[4] Kadir, A 2013 Introduction to the revised edition of the information system Yogyakarta: Andi Offset
[5] Ministry of National Education 2010 School library management The Ministry of National Education
[6] Ladjamudin, BA-B 2005 Analysis and design of information systems Yogyakarta: Graha Science
[7] Ladjamudin, BA-B 2013 Analysis and design of information systems Yogyakarta: Graha Science
[8] Laudon, CK & Laudon, PJ 1968 Management information systems managing the digital firm
[9] Muhamad, DH, Wahyuni, N. & Febianti, E 2019 Design of information systems questionnaire evaluation of website-based teaching and learning process in the Department of Industrial Engineering Untirta *Journal of Industrial Services* **4 (2)** p.81-92

[10] Muhsin, A 2014 Application of technopreneurship to develop small industries through the use of e-commerce technology based on content management system case study on the self-sufficient gypsum *Telematics* **10 (2)** p.91-102

[11] Pramono, AABS & Utami, AW 2018 Design and development of industrial practice information systems in the web-based Department of Informatics Engineering UNESA *Journal of Information Management* **8 (1)** p.70-78

[12] Saputra, E 2014 Usability testing to measure the use of the Palembang City inspectorate website *Journal of Informatics Engineering* **5 (3)** p.60-68

[13] Suharyanto, EC, Chandra, EJ & Gunawan, EF 2017 Web-based integrated payroll information system design (case study at St. Elisabeth Hospital) *National Journal of Information Technology and Systems* **3** pp.225-32

[14] Tullah, R. & Hanafri, MI 2014 Evaluation of the application of information systems at the Jakarta LP3I Polytechnic using the pieces method *Journal of Sispotek Global* **4 (1)** pp.22-28

[15] Utama, WY, Alghofari, AK & Munawir, H 2014 Scientific publication manuscripts *Website-Based UMS Industrial Engineering Alumni Information System Design Using Wordpress Content Managing System (CMS)*

[16] Vixie, P 2014 Software engineering tutorial absolute beginners Tutorials Point (I) Pvt. Ltd

[17] Whitten, L. & Jeffert 2004 Systems design and analysis methods Yogyakarta: Andi

IOP Conf. Series: Materials Science and Engineering **909** (2020) 012082 doi:10.1088/1757-899X/909/1/012082

Development of set covering model for determining the open /closed facilities location and resizing capacity of fasilities

Utaminingsih Linarti*, Cahaya Annisa' Fatonah, Annie Purwani
Industrial Engineering, Faculty of Industrial Engineering, Universitas Ahmad Dahlan,Yogyakarta, Indonesia

*Email: utaminingsih.linarti@ie.uad.ac.id

Abstract. Set Covering model determines the location of facilities that can be used that minimize the cost of assigning facilities to the point of request with the limitation that each facility was used for a number points of request. If there is an imbalance between the volume of demand points and the capacity of the facilities, it is necessary to open/closed facilities. In fact, facility location problem can extended for the decision not only open/closed but rezising area the existing facilities that can be adjusted considering the avaibility area of facilities location and demand points. This case has not been accommodated on the basic model of set covering. Not all the existing facilities can be resized. Rezising the area of facility location was assumed discrete. The aims of this study was optimized the alghorithm of set covering model. The solution this study was devided into two stages. The first stage was screening facilities that can be resized with survey location. The second stage was set covering model which the facilities will be close/opened and resize the existing facilities.

1. Introduction

Theory and location modeling were first introduced by Weber [1] who revealed the problem of finding a single facility to minimize the total travel distance between the facility and a group of customers. Determination of the best location for one or more facilities so that customers or demand points can be served well [2]. According to Owen & Daskin [3] determining the location of facilities is very important for planning in various aspects for public and private facilities. This will affect the operational and logistical activities of a facility. Determination of the location of the facility can be determined from the size of the required costs [4] [5]. Besides based on the cost of determining the location of facilities can be determined through the distance from the demand point to the facility [6]. Hotelling [7] developed facility theory by studying the location of two facilities on one line. Customers choose the closest facility from two facilities and vendors to maximize their market share. Some real problems in determining the location of facilities both public and private include determining the location of the facility. In recent years, there have been many applications regarding the location of facilities such as public facilities, such as offices, post offices, fire engines, and hospitals) and private facilities (such as factories and other industrial facilities) [8] [9]. The mathematical model was developed by calculating the probability of a specific busy ambulance and validated by a simulation model carried out in determining the location of a facility [10].

In addition to the above facilities, one of the public facilities that requires optimal decision making in determining the location of facilities is a landfill. Where the model is, there is a temporary landfill located close to residents. Several studies regarding the choice of location of facilities related to the impossibility of these facilities close to residents or demand points because several factors related to health, odor, aesthetics, etc [11, 12, 13, 14]. This may be due to different models of waste management, especially household waste collection. In household waste collection, some countries use bin that was not permanent or easily moved to collect household waste and use compactors as a means of transported the garbage before being sent to landfills. Some countries use the permanent trash bin for collect household waste. Coffe & Coad [15] explain its in two categories, such as Haul Container System and Stationary Container System. In [11], the demand point was describe as people in campus building. Research on determining the location of waste disposal facilities that have been carried out using optimization techniques as a search for solutions.

The criteria used in determining the location of facilities using the set covering model include the location of the facility, the location of the demand point, the distance between the location of the facility and the location of the demand point, cost, capacity of the facility and others. This study assumes that the garbage disposal facility is a stationary container system in the form of a permanent building consisting of several types of facilities in accordance with the volume of capacity. Because they are stationary, each facility cannot be moved, and in one location the capacity of the facility can be adjusted with resized to determine the volume of new capacity for existing facilities. Adjustment of the area is certainly not possible for every location of the facility with the same size, because this depends on the availability of land that still exists in the facility area. Thus, determining the size of the resized area cannot be done simultaneously.

The demand point classification in this study assumes that waste can be collected through several types of sources not only from households but can also from other facilities such as offices, shopping centers, markets, education, recreation areas and others. Decision variables in this study not only determine which facilities will be opened or closed, but which old facilities can be expanded based on the type of facility (maximum capacity volume) and available area. In real conditions, it could be that with the request of the community or decision makers, some facilities do want to be closed, so that if an imbalance occurs between the availability of the capacity of the landfill and the amount of waste produced by the community.

So the purpose of this study is to show the concept of capacity expansion in the decision to open or close a facility based on the importance of decision making. Based on the conditions of developing criteria for the type of facility and facility decision variables that can be expanded, the solution to the location of the facility is discrete in two stages. Daskin [16] describes three conditions for the completion of a facility location namely planar, discrete and network. The first stage is direct screening to the location of the location of existing waste disposal facilities by determining the maximum area that can be expanded by measuring the length and width of the area, while the height is not expanded because it is related to the ability to move to transportation. This stage generates new volume capacity data for existing facilities. The second stage is developing a mathematical algorithm for the set covering model with the development of the criteria above. Completion of the set covering model is not simultaneous, using the LINGO 11.0 program.

2. Literatur Review

Facility location models are investigated in discrete space, planar continuous space, and particular in a network environment [17][6]. Its can be seen that the planar form will provide a large area without restrictions, while the discrete form of the facility has a location point. It is also different from the form of a network model that will provide a network from either the route or anything else in the facility.

According to Daskin [16] location theory can be used for discrete problems where there are only demand points and facility points without special requirements. One method is set covering. The set covering problem was firstly introduced by Toregas in 1970, but it was published by Toregas & ReVelle [18]. The problem of determining this location used to find the minimum number of facilities (from their location) so each request can be fulfilled optimally with distance and time. Set covering is a method that can be used to minimize the number of facilities that will cover demand. The covering set still only determines which facilities will be maintained. In addition, the facility point has been determined in advance based on the area covered [16]. Based on this, a model was developed to find out the allocation of demand to facilities and to limit the maximum distance between the demand points and facilities without firstly determining the facilities included in the coverage area.

The covering set in Daskin [16] also has the characteristic of using binary numbers (1,0) which can only determine whether demand can be allocated or not. In other function the binary will show which facilities will be maintained because they can cover demand and are closed when not covering demand. The impact of using binary numbers also makes each source point only able to allocate to one facility point. So, demand can be allocated when volume is less than capacity. In reality there are problems where the amount of capacity is greater indeed than demand but the size of the capacity of each facility is smaller than demand. Based on these problems, this journal develops a set covering model that requires additional research steps to be carried out before the set covering method is implemented when there is a problem that the volume to be allocated is greater than the capacity. If volume demand more than capacity of facility we can take a new facility or resize the facility[19]. Set covering model has not grouped the types of each demand and facility. So in this model development also develops the type of facilities and demand with the additional indexes. Based on figure 1 this model can be applied to discrete locations where facilities and demand points have points both latitude and longitude.

2. Numerical Model

2.1. Set Covering Model

Set covering model is one of the oldest from optimization problems in the literature of mathematical programming [20]. Set Covering is a model to determine the lowest cost of the placement of several facilities where each point of demand can be achieved by a minimum of one facility [16]. Set covering problem (SCP) is a mathematical model that is important for some problem solving applications [21]. Set covering problem classified as NP-Complete[22][23][24]. So, set covering is an optimization method that has a function to enable the number of needed facilities to uncover other facilities. Selected facilities will be able to cover requests for other facilities, so it will minimize the number of facilities that can reduce costs because the facility can cover all requests. According to Daskin [16].

2.2. Mathematical Model Development

The set covering model as one of the discrete location theory models has always been developed that determined facilities will be retained to be open/closed. The following mathematical model of set covering in this study from the set covering model of Daskin [16]. This study assumed that the facility has several types and the demand point has several categories. Its were not accomodated before in set covering model. There were three types of facilities and two categories of demand points. Its such as type K, type L and type M for types of facility and demand points I and demand points J for classifications of demand point.

The mathematical model in this study uses notations as follows:

I = Set of demand point type I with the index i
J = Set of demand point type J with the index j
K = Set of alternative point of facility type K with index k
L = Set of alternative point of facility type L with index l
M = Set of alternative point of facility type M with index m
C_k = capacity of facilities type K with index k (m^3/day)
C_l = capacity of facilities type L with index l (m^3/day)
C_m = capacity of facilities type M with index m (m^3/day)
V_i = volume of demand points type I with the index i (m^3/day)
V_j = volume of demand point type J with index j (m^3/day)
D_{max} = maximum distance between demand point with facility by seeing the ability of people walking by carrying load
D_{ik} = the distance between the demand points type I with index i and alternative facility locations type K with index k
D_{il} = the distance between the demand points type I with index i and alternative facility locations type L with index l
D_{im} = the distance between the demand points type I with index i and alternative facility locations type M with index m
D_{jk} = the distance between the demand points type J with index j and alternative facility locations type K with index k
D_{jl} = the distance between the demand points type J with index j and alternative facility locations type L with index l
D_{jm} = the distance between demand points type J with index j and alternative facility locations type M with index m
T_d = total mileage as distance of fulfillment
= $\{k,l,m|d_{ik}+ d_{il}+ d_{im} + d_{jk}+ d_{jl}+ d_{jm} \leq T_d \}$
all alternatives to facility locations includes the point of the demand points with index i (N_i) and j (N_j)

Variable decision:

X_k = (1, if the facilities type K with index k becomes the location of the can covered demand and 0, if the opposited)
X_l = (1, if the facilities type L with index l becomes the location of the can covered demand and 0, if the opposited)
X_m = (1, if the facilities type M with index m becomes the location of the can covered demand and 0, if the opposited)
Y_{ik} = (1, if the demand points type I with index i can be met by alternative location k and 0, if the opposited)
Y_{il} = (1, if the demand points type I with index i can be met by alternative location l and 0,if the opposited)
Y_{im} = (1, if the demand points type I with index i can be met by alternative location m and 0, if the opposited)
Y_{jk} = (1, if the demand points type J with index j can be met by alternative location k and 0, if the opposited)

Y_{jl} = (1, if the demand points type J with index j can be met by alternative location l and 0, if the opposited)

Y_{jm} = (1, if the demand points type J with index j can be met by alternative location m and 0, if the opposited)

The mathematical model set covering problem:

$$Minimize: \sum_{k \in K} C_k . X_k + \sum_{l \in L} C_l . X_l + \sum_{m \in M} C_m . X_m \quad (1)$$

$$Subject\ to \quad \sum_{k,l,m \in N_i} Y_{ik} + Y_{il} + Y_{im} \geq 1 \quad \forall i \in I \quad (2)$$

$$\sum_{k,l,m \in N_j} Y_{jk} + Y_{jl} + Y_{jm} \geq 1 \quad \forall j \in J \quad (3)$$

$$\sum_{i \in I} V_i . Y_{ik} + \sum_{j \in J} V_j . Y_{jk} \leq C_k . X_k \quad \forall k \in K \quad (4)$$

$$\sum_{i \in I} V_i . Y_{il} + \sum_{j \in J} V_j . Y_{jl} \leq C_l . X_l \quad \forall l \in L \quad (5)$$

$$\sum_{i \in I} V_i . Y_{im} + \sum_{j \in J} V_j . Y_{jm} \leq C_m . X_m \quad \forall m \in M \quad (6)$$

$$d_{\max} . X_k \geq \max \{d_k^i\} . Y_{ik} \quad \forall k \in K \quad (7)$$

$$d_{\max} . X_l \geq \max \{d_l^i\} . Y_{il} \quad \forall l \in L \quad (8)$$

$$d_{\max} . X_m \geq \max \{d_m^i\} . Y_{im} \quad \forall m \in M \quad (9)$$

$$d_{\max} . X_k \geq \max \{d_k^j\} . Y_{ik} \quad \forall k \in K \quad (10)$$

$$d_{\max} . X_l \geq \max \{d_k^j\} . Y_{jl} \quad \forall l \in L \quad (11)$$

$$d_{\max} . X_m \geq \max \{d_m^j\} . Y_{jm} \quad \forall m \in M \quad (12)$$

$$X_k \{0,1\} \quad \forall k \in K \quad (13)$$

$$X_l \{0,1\} \quad \forall l \in L \quad (14)$$

$$X_m \{0,1\} \quad \forall m \in M \quad (15)$$

$$Y_{ik} \{0,1\} \quad \forall k \in K \quad (16)$$

$$X_{il} \{0,1\} \quad \forall l \in L \quad (17)$$

$$X_{im} \{0,1\} \quad \forall m \in M \quad (18)$$

$$X_{jk} \{0,1\} \quad \forall k \in K \quad (19)$$

$$X_{jl} \{0,1\} \quad \forall l \in L \quad (20)$$

$$X_{jm} \{0,1\} \quad \forall m \in M \quad (21)$$

Equation (1) represents minimize the number of facilities consisting of three different types with the capacity of the facilities. Constraint (2) express that demand point type I and (3) express that demand point type J can at least be served by one facility such as sets of facilities type K, L and M. Constraint (4), (5) and (6) ensure that the sets of facilities type K, L and M have the capacity to accommodate both types of sets of demand points classification *I and J*. So, facility can't cover demand when the demand is more than its capacity. Constraint (7), (8), (9), (10), (11) and (12) restrict the maximum distance allocation from maximum mileage so the distance between the demand points and facilities is no more than the maximum distance for each sets of facilitis type K, L and M and demand points classification I and J. Constraint (13), (14) and (15) are binary constraint for decision variabel open/closed facility existing and facility that will be resized for facilities type K, L, and M. Then constraint (16), (17), (18) were binary constraint that defines determination sets of demand point type I that will be covered by certain facilities and constraint (19), (20) and (21) were binary constraint that defines determination sets of demand point type J that will be covered by certain facilities.

3. Application on Hypotetical Data and Disscussion

3.1. Hypotetical Data

The set covering model is tested through the application of hypothesis data. The process of changing the size of the facility area in this model that uses a discrete model which is done separately selected directly by researchers in the form of a point grid as in Figure 1. In real conditions, the grid points can be represented through the points of longitude and latitude. Figure 1 shows that there was three regions (A, B, and C) where each region has a location where demand points are from households (I) and one public facilities (J). There was I_1, I_2, I_3 for household and J_1, J_2, J_3 for other public facility. The demand points from household was assumed as a weight point as area, because of the difficulty of data retrieval if the location of the demand point was done for each house point for each household.

There are three types of facilities assumed in this study, namely facilities type K, L and M with differences in facility capacity. These three facilities are spread over three regions A, B and C. Region A has one type K (K_1) facility and one facility type L (L_1). Region B has one facility K (K_2) and one facility M (M_2). While region C has one facility K (K_3) and one facility L (L_2). In actual conditions, the number of facilities in one area can be more than one type of facility with more than one facility per facility. In real case the data application for this study can be describe fully for facility location decision for waste management in Yogyakarta [24]

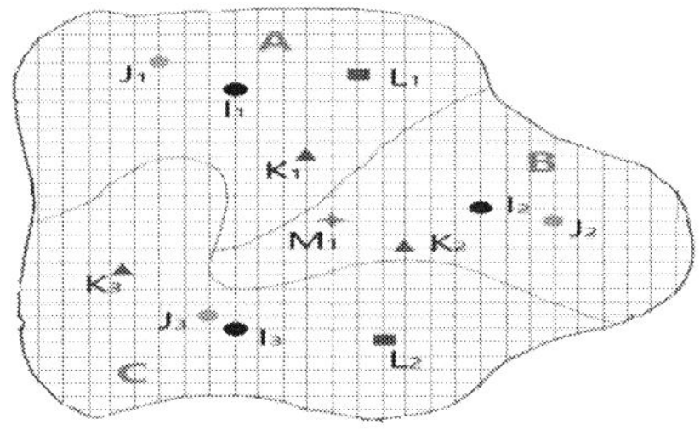

Figure 1. Discrete location in this model

Hypotetical data was carried out with distribution real case data with validation process firts, as follows:

3.1.1. Capacity of Facility

IOP Conf. Series: Materials Science and Engineering **909** (2020) 012082 doi:10.1088/1757-899X/909/1/012082

Facility capacity is the ability to accommodate a facility for the demand given to it. Hypothetical data on facility capacity in Table 1.

Table 1. Capacity of facilities

Facility	Volume (m^3)
K_1	5
K_2	7
K_3	1
L_1	1
L_2	2
M_1	2

3.1.2. Volume of Demand

Volume of demand points is the amount that will be allocated from the demand point to the facility. There are two classifications of demand points, namely households and other public facilities. Hypothetical data on demand volume for each demand points in Table 2 and Table 3.

Table 2. Volume of demand types I_i

Demand	Volume (m^3)
I_1	3
I_2	8
I_3	1

Table 3. Volume of demand ypes J_j

Demand	Volume (m^3)
J_1	1
J_2	3
J_3	1

3.1.3. Distance Facility Location and Demand Point

Distance is an important constraint in an allocation of demand to a facility in the set covering method [26]. So, distance is one of the needed data in solving problems with set covering. Distance measurement can use the distance and latitude distance approach in real conditions. Table 4 until Table 9 shown the distance data of facility locations each type and demand points each classification.

Table 4. Distance between K_k and I_i

	I_1	I_2	I_3
K_1	2	6	1
K_2	2	4	7
K_3	7	5	1

Table 5. Distance between L_l and I_i

	I_1	I_2	I_3
L_1	1	3	7
L_2	9	1	4

Table 6. Distance between M_m and I_i

	I_1	I_2	I_3
M_1	2	5	7

Table 7. Distance between K_k and J_j

	J_1	J_2	J_3
K_1	2	6	2
K_2	2	6	7
K_3	7	5	1

Table 8. Distance between L_l and J_j

	J_1	J_2	J_3
L_1	3	2	8
L_2	8	4	2

Table 9. Distance between M_m and J_j

	J_1	J_2	J_3
M_1	2	5	4

IOP Conf. Series: Materials Science and Engineering **909** (2020) 012082 doi:10.1088/1757-899X/909/1/012082

3.1.4. Maximum Distance

Set Covering Model solves the problem in minimizing the number of facilities needed to cover demand for all request node with the maximum service distance desired [17]. Maximum distance was the maximum distance allowed for demand to allocate to facilities. The maximum distance is assumed 5 m from the demand point to the facility location.

3.2. Disscussion

According to Hernandez [19], the resizing process in achieving goals in minimizing costs and the number of facilities to be allocated is important. The resizing process can be done based on consideration of extending costs and facility capabilities. The first stage for solution this problem was the screening location prosses. The screening process for selecting these facilities can be done using the researchers consideration factors such as the distance from the demand point to facility and availability of land. After Screening and resizing the selected facilities, it can proceed the second stage to optimization solution of the set covering model was developed. So the objectives of the method will be achieved in minimizing the number of facilities.

The set covering model that used to the basic model in the development of this model has the purpose of choosing the facilities to be opened or closed, as well as the model to be developed has the same purpose. This developed model can be used in non-commercial facilities that have rules for range such as the maximum distance that can be reached. Research on the determination of facilities that will remain open at an objective to minimize facilities has often been done as in the research of Hernandez [19] and Wang et al [5]. The development of the model was different from others, this model was developed discretely so that it requires the initial stages before processing was importance for the data using a mathematical model. Model suitability testing is done using hypothetical data.

The set covering model that has been developed in terms of demand points classification and facility type that was originally single into several types consisting of demand volume, facility capacity, distance between demand and facilities, and maximum distance will be processed using the LINGO 11.0 software. According to the results, infeasible results were obtained for determining the facilities to be opened to accommodate demand allocations from 6 facilities consisting of 3 type of facilities K, 2 type of facilities L and 1 type of facility M. If there was infeasible solution shows that there was no solution space for this problem or there was no area that meets all the constraints. This was because demand with a volume exceeding facility capacity will not be able to send or allocate to facilities, such as garbage from I_2 sources cannot allocate to any facility.

According to Table 10, Facility K_2 will be resized to 9 m^3 and L_2 to 3 m^3 and will be analyzed with the same software. The results of the analysis show that facilities K_1, K_2, L_1, and L_2 will still be opened. The K_1 facility will accommodate demand from J_1 and J_2. Demand from I_2 and J_3 will be allocated to K_2. Whereas for L_2 facilities will receive an allocation from I_1. The demand of I_3 will be allocated to L_1. The resizing process can be used for the set covering that has demand problems more than capacity. It also has the main purpose like the set covering purpose to minimize facilities that will still be placed. Allocation of demand from the point of demand to the facility can be seen in Table 10.

Table 10. Allocation of demand

No	Facility	Demand	Open/Closed Facility	Capacity (m^3)
1.	K_1	J_1, J_2	O	5
2.	K_2	I_2, J_3	O	7
3.	K_3	-	C	1

4.	L_1	I_1	O	1
5.	L_2	I_3	O	2
6.	M_1	-	C	2
				18

Based on the results of the analysis above, it can be seen that one of the-weaknesses of the set covering method and discrete location, when the total volume of demand and total capacity has the same value or below, but the capacity of each facility is smaller than the volume of demand will infeasible solution. This is because the use of binary numbers (1.0) in the set covering method which can only give a decision when the demand will allocate to the facility. There are only two choices, namely allocated or not seeing capacity. This problem can be dealt by adding new facility with capacities that are in accordance with the volume of demand that cannot be allocated yet or other than that, it can process with resizing possible facilities. The hypothetical data processing with this model were done by changing the size of the selected facility. This model can uses for the real situation problem if the volume of facility capacity not balanced with the volume of demand point in set covering model. The decision of facility location for the set covering model is not only open / closed facility with the addition of new facilities but still maintaining the existing facilities by conducting resized capacity.

4. Conclussion

Problems will arise when the demand point volume is greater than the capacity of existing facilities. The set covering model developed can represent a real problem where there are several classifications of demand points and several types of facilities that have different capacities and can adjust the size of existing facilities, not only the decision of open / closed facilities, but also can be changed the size of the area of the facility that has been exist so that a new volume is obtained. Model solutions will be obtained in two simultaneous steps. Stage one, namely resizing existing facilities can be requested to be done separately and selected directly by carrying out the screening process and stage 2 is optimizing the model of covering covering. Further research can be developed by searching for solutions simultaneously with the help of Geographic Information System (GIS) so that the location of demand points can be determined through each point of the household's household. In addition, GIS can also help determine the area of a facility that may be directly expanded in size and also measure distances with another approach, namely nekwork. So the set covering model is no longer discrete but network.

5. Reference

[1] Weber A, U 1929 Ber ben standort der industrien, tu Bingen Theory of the location of industries *University of Chicago Press*

[2] Gadegaard S.L 2016 Discrete location problems PhD disertation of Aarhus University Denmark https://pure.au.dk/portal/files/100556924/PhD_dissertation_Sune_Lauth_Gadegaard.pdf

[3] Owen, S.H, Daskin M.S 1998 Strategic facility location: a review *Journal of Operational Research* **111(3)** 423-447 https://doi.org/10.1016/S0377-2217(98)00186-6

[4] Farahani R.Z, Seifi M.S, Asgari N 2010 Multiple criteria facility location problems: a survey applied mathematical modelling **34(7)** 1689-1709 https://doi.org/10.1016/j.apm.2009.10.005

[5] Wang Q, Batta R, Badhury J, Rump S.M 2003 Budget constrained location problem with opening and closing of facilities *Computers & Operations Research* **30(13)** 2047-2069 https://doi.org/10.1016/S0305-0548(02)00123-5

[6] Drezner T 2014 A review of competitive facility location in the plane *Logistic Research* **7** 114 https://doi.org/10.1007/s12159-014-0114-z

[7] Hotelling H. Stability in Competition. *Economic Journal.* 1929 39(15) 41–57 DOI: 10.2307/2224214

[8] Başar A, Çatay B, Ünlüyurt T 2011 A taxonomy for emergency service station location problem *Optimization Letters* **6(6)** 1147-1160 https://doi.org/10.1007/s11590-011-0376-

[9] Abolian R, Berman O, Krass D 2007 Competitive facility location and design problem *European Journal of Operational Research* **182(1)** 40-62 https://doi.org/10.1016/j.ejor.2006.07.021

[10] Šarac D,Kopić M,Mostarac K, Kujačić M, Jovanović B 2016 Application of set covering location problem for organizing the public postal network *Traffic & Transportation* **28(4)** 403-413 https://hrcak.srce.hr/file/244272

[11] Mehr, Maryam Nikouei and Ronald G. McGarvey 2017 Planning solid waste collection with robust optimization location-allocation, receptacle type, and service frequency *Hindawi Advances in Operations Research* Article ID 2912483 14 pages https://doi.org/10.1155/2017/2912483

[12] Erkut, Erhan., Avraam Karagiannidis., George Perkoulidis., Stevanus A. Tjandra 2008 A multicriteria facility location model for municipal solid waste management in North Greece *European Journal of Operational Research* **187** 1402–1421

[13] Yadav, Vina., A.K. Bhurjee., Subhankar Karmakar., A.K. Dikshit 2017 A facility location model for municipal solid waste management system under uncertain environment *Science of the Total Environment* 603-604 760–771 Available online at www.elsevier.com/locate/scitotenv.

[14] Yu, Hao., Wei Deng Solvang 2017 A multi-objective location-allocation optimization for sustainable management of municipal solid waste *Environ Syst Decis* **37** 289

[15] Adeleke, Olawale Joshua 2017 Location-allocation-routing approach to solid waste collection and disposal *Thesis Submitted to The Department of Mathematics, College of Science and Technology, Covenant University, Ota*

[16] Daskin M.S 2008 Network and discrete location models, algorithms, and applications 2nd edition *New Jersey: John Wiley & Sons, Inc. Hoboken* 283-294

[17] Rajagopalan H.K, Saydam C, Xiao J 2008 A multiperiod set covering location model for dynamic redeployment of ambulances *Computers & Operations Research* **35(3)** 814-826

[18] Toregas C, ReVelle C 1972 Optimal location under time or distance constraints *Regional Science* **28** 133-143

[19] Hernandez D.R 2014 Capacitated delocation models for closing and resizing redundant branches *Madrid: Colegio Universitario de Estudios Financieros* http://biblioteca.cunef.edu/gestion/catalogo/doc_num.php?explnum_id=1398

[20] Yelbay B, Birbil S.I, Burbul K 2015 The set covering problem revisited: an empirical study of the value of dual information *Journal of Industrial & Management Optimization* **11(2)** 575-594

[21] Caprara A, Toth P, Fischetti M 2000 Algorithms for the set covering problem *Annals of Operations Research* **98(4)** 353–371

[22] Chen L, Crampton J 2009 Set covering problems in role-based access control *European Symposium on Research in Computer Security* Lecturer Notes in Computer Science **7889** 689-704 https://doi.org/10.1007/978-3-642-04444-1_42

[23] Garey M.R, Jhonsons D.S 2007 Computers and intractability: a guide to the theory of NP-completeness New York: W.H. Freeman and Company 1979 Mouthuy S, Deville Y, Dooms G. Global constraint for the set covering problem Actes troisième journées francophones de programmation par contraintes. France 1-10 http://becool.info.ucl.ac.be/pub/papers/jfpc2007_setcovering_en.pdf

[24] Fathonah, Cahaya Annisaa., Utaminingsih Linarti 2019 Penentuan open/closed fasilitas tempat pembuangan sampah dengan mempertimbangkan desain fasilitas dan sarana pendidikan formal Journal of Mathematics and Its Applications **16** 2019 27-38

[25] Serra D, Marianov V 2004 New trends in public facility location modeling *UPF Economics and Bussiness Working Paper*

IOP Conf. Series: Materials Science and Engineering **909** (2020) 012083 doi:10.1088/1757-899X/909/1/012083

Designing a computerization management production system in PT Akrilik Kurnia Kencana using barcode

Rudy Vernando Silalahi [1*], Ishak [2] and Marciello [3]
[1,2,3] Industrial Engineering Study Program, Pelita Harapan University, Tangerang, Indonesia

*E-mail : rudy.silalahi@uph.edu

Abstract. As technology develops, the production process of each industry also develops to become more sophisticated and modern. Barcode is a form of technological development in the production process and is useful for manufacturing companies to be able to store data so that it can be well integrated in a system. PT Akrilik Kurnia Kencana (AKK) is one of the companies engaged in manufacturing EPS (Expanded Polystyrene) and has a problem in finished goods warehouse where the company does not have a good system due to warehouse capacity inadequate. This study aims to identify current system implemented at PT AKK and develop a new system to solve the problems experienced by PT AKK. Data collection is done by the method of observation and interviews with the company to get the necessary data. After being collected, the data will be processed in such a way that it can be interpreted properly. Then, the data will be used as an input in designing a production management system-based application using the System Development Life Cycle (SDLC) waterfall model. By using this system, users can monitor the presence of each finished goods that has been produced and enter the warehouse through the information contained in the barcode.

1. Introduction

Manufacturing industry is a business field that produces a product through several production processes. As technology develops, the production process of each industry also develops to become more sophisticated and modern. This is very influential, because the industrial sector is one of the mainstay sectors in the Indonesian economy [1]. One form of application of technology in a production process is the use of barcodes in the production flow. Barcodes are useful for storing various information needed regarding related products such as manufacturing date, machine number, order quantity, and operator.

PT Akrilik Kurnia Kencana is a company engaged in manufacturing EPS (Expanded Polystyrene). This company has been established since 1990 and now has 80 employees working in 3 shifts for 24 hours. In a day, the company were able to produce around 2 tons of EPS which is divided into two types of products namely EPS Box and EPS Protector. The company has a smooth production process, but there are flaws in the production and warehouse division. According to the employees, the company's warehouse does not have enough area to accommodate a big number of daily productions, so the warehouse is not well-organized. The consequence of this is the number of finished goods that are damaged due to the existence of these products is not known by the company so that it becomes a potential loss for the company.

In addition, the cause of this problem is also because the company uses the Make to Order (MTO) and Make to Stock (MTS) methods simultaneously, so that the company stores too many items requested by customers as the company's safety stock. This is due to the absence of a warehouse system that is capable to organize and store the warehouse data, resulting in irregular placement of finished goods. Then, the company does not have a clear Standard Operating Procedure (SOP) for its application so a lot of operators and employees do not do their jobs effectively. To solve these problems, the company needs to develop a new kind of system in the company's warehouse that were able to record and organize finished goods warehouse properly in order to overcome the limited warehouse area.

2. Methodology

The following is the methodology used in this study:

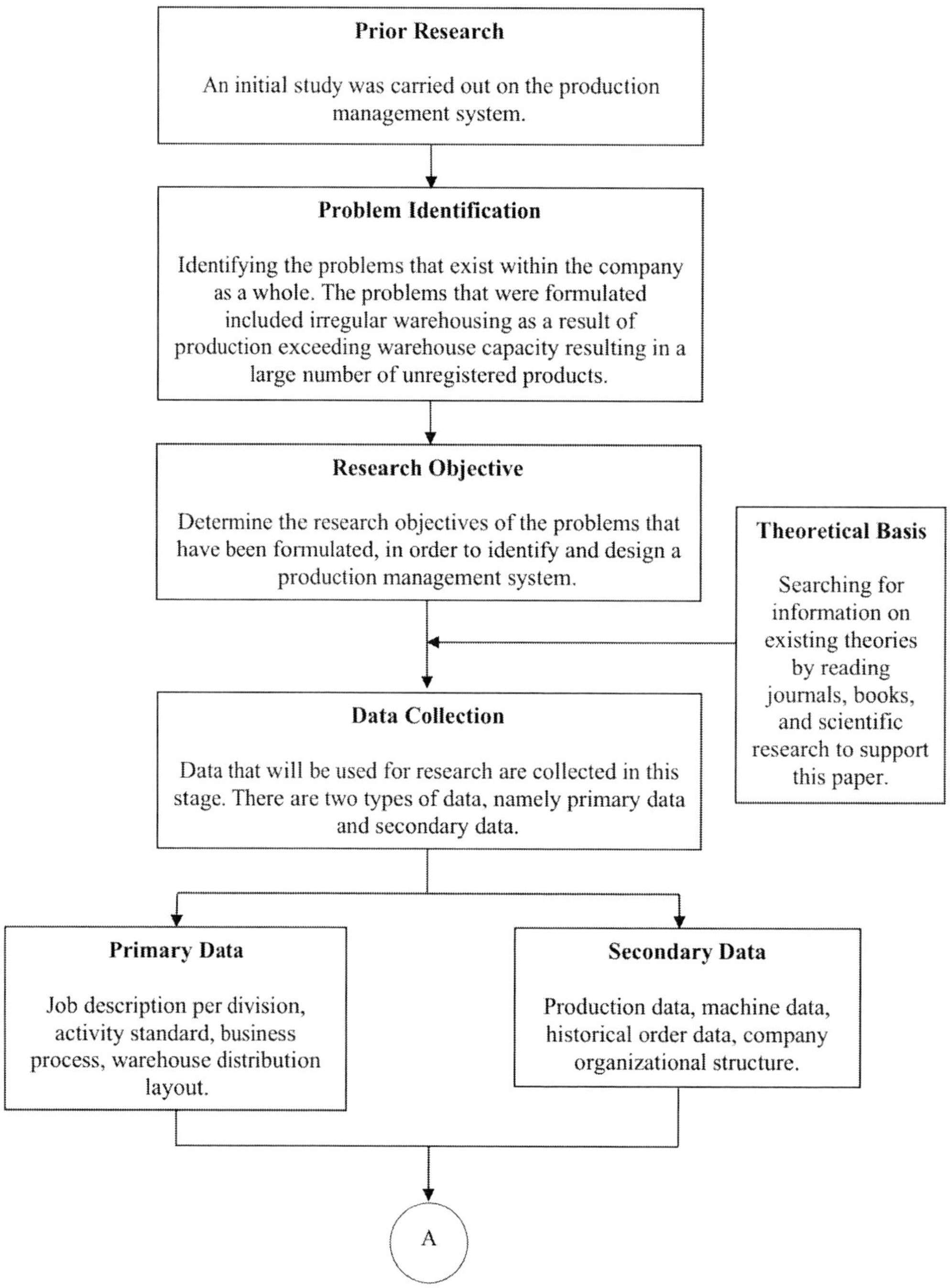

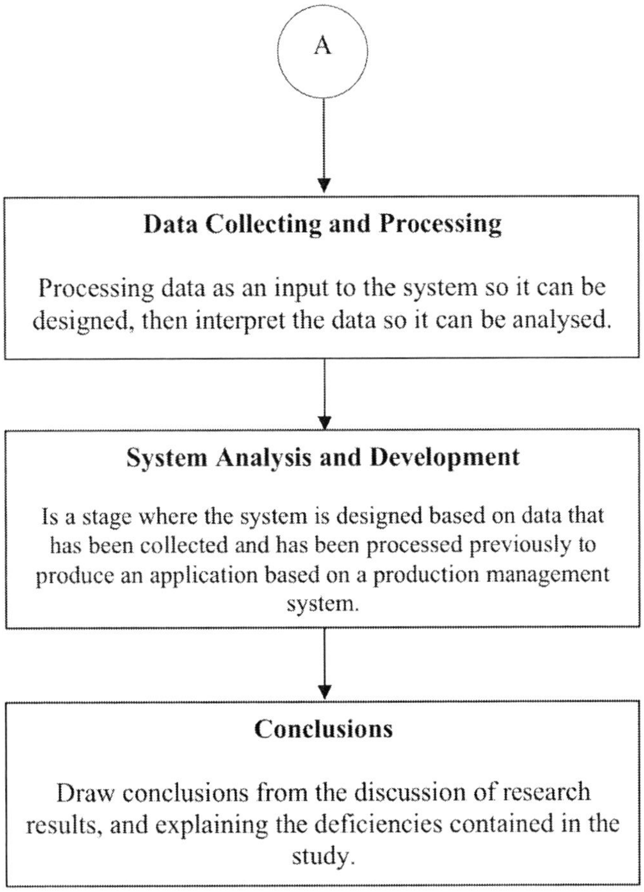

Figure 1. Methodology procedure scheme

2.1. Prior Research
In this section, the initial studies were conducted on problems experienced or the desire of company to improve performance in certain areas. This is done by making a direct visit to PT Akrilik Kurnia Kencana with the aim of seeing the situation and the actual conditions on the ground. Then, followed by a direct interview with the General Manager of the company, the head of production staff, the head of the warehouse staff, the staff of the production department, and the staff in the warehouse. After getting a picture of the problem, an initial literature study was conducted to ensure that this research could be carried out scientifically.

2.2. Problem Identification
The phase that will be carried out after getting the problem, is to formulate the problem clearly and concretely. This is necessary to find the root of the problem that is the cause of the problem while finding a way or solution to solve the problem. The problems experienced by the company are finished goods warehouses that do not have enough area to accommodate the production results, and warehouses that do not have information systems that result in many finished goods being damaged due to their presence unknown to the warehouse. This resulted in losses for the company and made it difficult for warehouse staff to find the needed stock. In addition, the company also complained about the productivity of the work of operators who did not meet company standards in general. This is caused by the fact that the company itself does not have clear standards or written rules regarding the activities that operators must carry out every day.

2.3. Research Objective

Determining the purpose of research is the stage carried out after successfully formulating the problem. The purpose of this research is to identify the current system and design an information system that can overcome the problems in PT Akrilik Kurnia Kencana's finished goods warehouse.

2.4. Theoretical Basis

In this stage, a theoretical search and scientific information on the elements of the topic being investigated are carried out. This is necessary so that this research has the support of statements or arguments from other studies that have been done before on similar topics in order to strengthen the basis of this research. The aim is that the results of this study can be scientifically proven by comparing the theories with the results from this study.

2.5. Data Collection

In collecting data needed in research, there are two types of data collection methods used, including:

2.5.1. Observation.

Data collection is carried out by looking directly at the condition of the company and recording the data needed. In this method, the data collected is primary data. Where some of the data needed to make observations are:

- Job Description for each division in the company.
- Activity standards that must be done by each division.
- Company's Business Processes.
- Warehouse Layout Allocation.

2.5.2. Interview.

Data collection is done by conducting interviews directly to the General Manager, production staff, production staff, and warehouse staff to get secondary data from the company about the company's condition and the problems experienced in the field. In addition, interviews were also conducted to obtain the necessary data, including:

- Data on all materials and finished goods used in production.
- Data on the number of production machines.
- Historical order data (Name, Address, Telephone Number, Order Date, Delivery Date, Quantity).
- Employee data and titles.
- Company Organizational Structure.

2.6. Data Collecting and Processing

In this chapter, data collection is needed based on the data collection methods described in the previous chapter. The data obtained will be processed in such a way so that it can be input into the system so that the data can be interpreted properly so that it can then be analyzed at the next stage.

2.7. System Analysis and Development

In this section, we will design an application based on a production management system using Microsoft Access. This system contains data that has been processed in the previous section so that it can describe the information needed by the system user easily. In addition, in this stage, the modelling and structural modelling behaviour and the interface layer of the system will be explained.

2.8. Conclusion

In this part, the overall conclusions will be drawn from the results of the study as well as the shortcomings that there are several suggestions included for newer studies that take similar topics.

IOP Conf. Series: Materials Science and Engineering **909** (2020) 012083 doi:10.1088/1757-899X/909/1/012083

3. Result and Discussion

The following is the result and discussion of the data used in this study:

3.1. Current System Analysis

In a company, of course there are several problems that result in a production process that is not running optimally because of the following problems:

Table 1. Current system problems

No.	Problems	Cause	Consequence
1.	No instructions or activity standards for each operator.	The company did not set any rules or standard operational procedure for specific activities.	Operator is not being productive because they are not working the way they should.
2.	Warehouse is not well organized and recorded.	Incapable warehouse area and the implemented system are not good.	A lot of finished goods that are not recorded went missing and caused a loss for the company.
3.	Inventory Staff having a hard time to search for certain finished goods.	The company only used several kind of coloured plastics to group and package the finished goods.	Some of the different customer orders are using the same coloured plastic as a packaging so it's difficult to differentiate each order.

3.2. Warehouse Layout Allocation

The area of the warehouse owned by the factory is around 100 m x 23.5 m. In the following figure, the warehouse layout will be divided into several blocks to illustrate the division of locations in the factory warehouse:

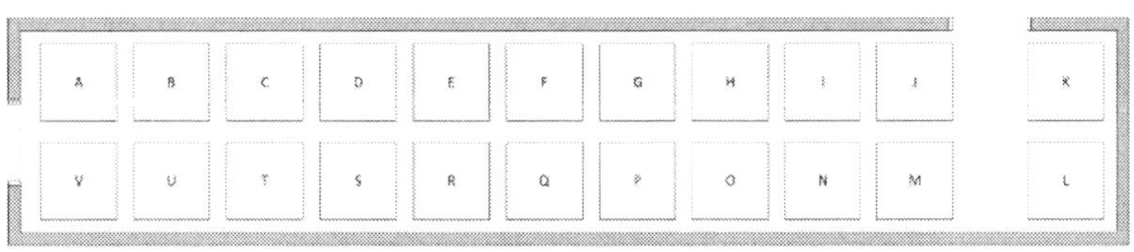

Figure 2. Factory warehouse layout block allocations

3.3. Planning

When developing, the first stage that must be passed is the planning stage. In this stage, a system feasibility study will be discussed as well as an analysis of the system requirements that need to be carried out to ensure this project is feasible. The following are some important points the objectives that must be achieved by this system, among others:

- Data storage of finished goods that are well recorded and ordered in the system in accordance with the input from the warehouse.
- The warehouse division can monitor stocks easily through display reports on the system and be able to distinguish between old and new stocks.
- The production division were able to record and track each finished goods in the warehouse.
- The sales division can view orders that have been input into the system and monitor the order's operation.

In this section, a feasibility analysis will also be conducted using two analytical methods, namely technical and operational feasibility analysis. Inside the technical feasibility analysis, it will show about how the condition of technical equipment owned by the company makes it possible to implement this system. The technical requirements in this case are divided into two types, which is hardware requirements and software requirements.

Table 2. Hardware requirements

Product Name	Specifications	Remarks
Laptop ASUS UX303UB	Processor: Intel® Core™ i7-6500 CPU @2.50 GHz (4 CPUs) Memory: 8192 MB RAM VGA: NVIDIA GeForce 940M	Laptop are used to run the system so that it can input the data and also required to see the output from system.
Mouse Wireless Logitech		Mouse is used for navigation on the laptop's screen
Router, Switch, Ethernet cable		Needed to establish an internet connection for the laptop.
Printer EPSON L360		Printer is needed to print the barcodes and reports.
Barcode Scanner EPPOS EP1020M	Lighting: 650 nm visible laser diode Decode ability: 1 Dimension & GS1 Databar Speed: 300 scan/s Laser Range (Length): 15 cm Laser Range (Width): 5 cm – 30 cm	Barcode scanner is needed so the system will be able to read the barcode label and display the stored information within it.

Table 3. Software requirements

Product Name	Specifications	Remarks
Operating System (OS)	Windows 7, 8, or 10	OS is needed for the laptop to work and capable to run the system
Microsoft Office (Microsoft Excel, Visio, dan Access)	Microsoft Office 2010 - 2016	Excel and Access are required to make data tables that is used to fill information on the system. Whereas Visio is needed to make various diagrams in designing the system
Adobe Acrobat Reader		Required so that the laptop is able to display reports from the system in PDF file types form.
DRPU Barcode Label Maker		Required to generate a barcode so that it can be printed and pasted on the product.

Furthermore, in determining the feasibility of the system operationally, there are several factors that can affect the feasibility of this system in company operations, including:

- The ability and knowledge of the user who will be the operator in carrying out the company's system activities as well as computerized systems.
- Availability of adequate human resources and have the desire to learn this new system.

When carrying out company activities, the information system is not a strange thing for the company so that enough employees and operators are familiar and able to run a system that has been computerized. The operators that have been chosen to run this system are the people who are doing the current activities usually using Microsoft Excel to do company data collection and are not accustomed to using Microsoft Access.

3.4. Requirements

There are two types of requirement that need to be identified include functional and non-functional requirements. The following are some points regarding functional requirements that must be met by the system:

- All users registered inside can log into the system using their respective usernames and passwords.
- All system users are able to open product data through the product menu on the system.
- Users can input product data in the form of item names, item codes, prices, and product information.
- Users can search products by product name.
- Users can change and delete product data.
- The system can calculate the amount of existing stock based on production data and sales data to customers.
- All system users can open and print sales reports through the report menu on the system.
- All system users can open and print product inventory reports through the report menu on the system.
- All system users can open and print production reports by date, weekly, monthly, and yearly through the report menu on the system.

The following are some points regarding non-functional requirements that must be met by the system:

- The system can be operated on a computer that uses OS Windows 7, 8, and 10 by using the Microsoft Access application
- The system can be connected to the printer so that reports can be printed.
- The system can be connected to a barcode scanner so that barcode scanning can be performed.
- The system must have a security system using a username and password to protect the data that is stored.
- The system must be able to accommodate data for at least one year during the system's operation so that an annual report can be printed.
- The system can be accessed at any time when needed.

3.5. Normalization

These are the steps to normalize a database, including:

3.5.1. Unnormalized Form (UNF).

Initial data set that still does not have a format, there is still repeated data and incomplete data. The data collected is purely based on the input that has been obtained.

3.5.2. 1st Normal Form (1NF).

Collection of data that are interconnected between one another, there is no repeating data, each data has only one understanding, and each table already has a specified Primary Key.

3.5.3. 2nd Normal Form (2NF).

A collection of data in which all attributes other than key attributes depend on the Primary Key but not more than one Primary Key.

3.5.4. 3rd Normal Form (3NF).

Data sets where the attribute is not a key that is in one field must not have transitive dependencies, that is, there is no data in the table that is associated with a field that is not a Primary Key. All data that is not a key in a relation depends only on the Primary Key in that relation.

The following is the system's data normalization table starting from UNF (unnormalized form) up to 3NF:

UNF	1NF	2NF	3NF
Kode Produk	TABEL PRODUK	TABEL PRODUK	TABEL PRODUK
Nama Produk	Kode Produk	Kode Produk	Kode Produk
No Mesin	Nama Produk	Nama Produk	Nama Produk
Remarks	Remarks	Remarks	Remarks
Harga Jual	Harga Jual	Harga Jual	Harga Jual
Stok	Stok	Stok	Stok
Nama Pelanggan	TABEL PRODUKSI	TABEL PRODUKSI	TABEL PRODUKSI
Alamat Pelanggan	Kode Produk	Kode Produksi	Kode Produksi
Email Pelanggan	Kode Karyawan	Kode Produk	Kode Produk
Telepon Pelanggan	Nama Karyawan	Kode Karyawan	Kode Karyawan
Tanggal Produksi	Jabatan	No Mesin	No Mesin
Jumlah Produksi	No Mesin	Tanggal Produksi	Tanggal Produksi
Lokasi	Tanggal Produksi	Jumlah Produksi	Jumlah Produksi
Tanggal Penerimaan	Jumlah Produksi	TABEL WAREHOUSE	TABEL WAREHOUSE
Tanggal Pengiriman	TABEL WAREHOUSE	Kode Produksi	Kode Warehouse
Tanggal Penjualan	Kode Produk	Kode Produk	Kode Produksi
Jumlah Penjualan	Kode Karyawan	Kode Karyawan	Kode Produk
Kode Karyawan	Tanggal Penerimaan	Lokasi	Kode Karyawan
Nama Karyawan	Tanggal Pengiriman	Tanggal Pengiriman	Lokasi
Jabatan	Lokasi	Tanggal Penerimaan	Tanggal Penerimaan
	Nama Karyawan	Jabatan	Tanggal Pengiriman
	Jabatan	Status	Status
	Status	TABEL PENJUALAN	TABEL PENJUALAN
	TABEL PENJUALAN	Kode Produk	Kode Penjualan
	Kode Pelanggan	Jumlah Penjualan	Kode Pelanggan
	Kode Produk	Tanggal Penjualan	Kode Produk
	Jumlah Penjualan	Nama Pelanggan	Jumlah Penjualan
	Tanggal Penjualan	Alamat Pelanggan	Tanggal Penjualan
	Nama Pelanggan	Email Pelanggan	TABEL KARYAWAN
	Alamat Pelanggan	Telepon Pelanggan	Kode Karyawan
	Email Pelanggan	TABEL KARYAWAN	Nama Karyawan
	Telepon Pelanggan	Kode Karyawan	Jabatan
		Nama Karyawan	TABEL PELANGGAN
		Jabatan	Kode Pelanggan
			Nama Pelanggan
			Alamat Pelanggan
			Email Pelanggan
			Telepon Pelanggan

Figure 3. Database normalization

3.6. Behavior Modelling

Behavior modelling is a model that describes and outlines aspects of internal logic that support a system to run. In this paper there are two diagram techniques that will be used to describe the system to be made, including use case diagrams and activity diagrams.

IOP Conf. Series: Materials Science and Engineering **909** (2020) 012083 doi:10.1088/1757-899X/909/1/012083

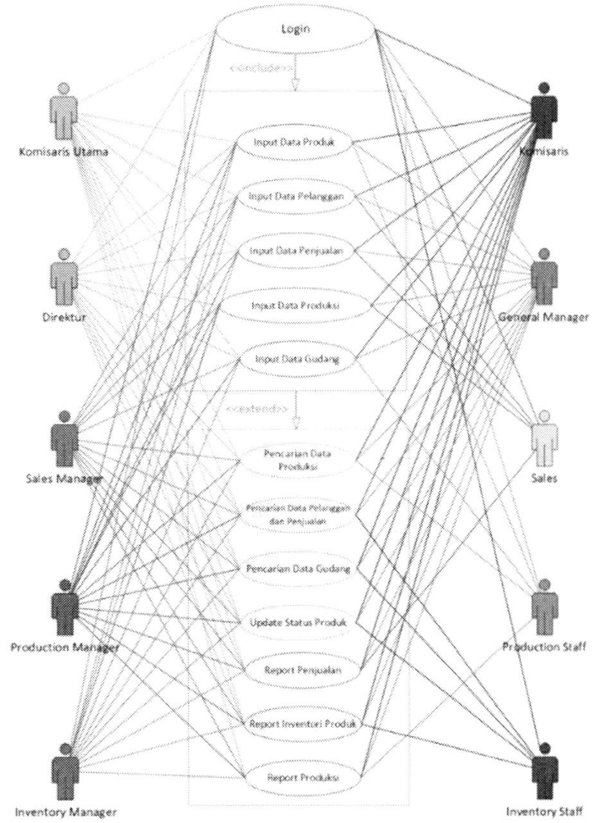

Figure 4. Use Case Diagram

Use case diagram is a diagram that aims to illustrate the functional needs of the system in the form of a series of actions (Use cases) that will be run by the system when interacting with users (actors). This diagram consists of use case tables that are used for each activity that can be carried out by the system, so that each activity has its own use case. The use case table is as follows:

Use case Name : Login		**ID :** 1	**Priority :** High
Actor : All user that have access to the system			
Description : This use case describes how the user can access the system through the security system using username and password to enter the system.			
Preconditions: 1. Users login were authenticated. 2. The user does not have username or password.			
Normal Course : 1. All users must fill in the name and title given to the manager. 2. The manager enters the user's profile into the system and creates a username and password. 3. The manager determines the Security Level for each user in order to limit certain user access to data in the system. 4. The user enters the username and password in accordance with what has been given by the manager into the system. 5. User logged in the system			
Postconditions: Employees successfully enter the system by using a username and password into the system			
Summary Inputs	**Source**	**Outputs**	**Destination**
Username	Manager	Login ke sistem	strEmpName
Password	Manager	Login ke sistem	strEmpPassword

Figure 5. Use Case Table

While on the other hand, Activity Diagram is a diagram that aims to describe the workflow (activity) or business process of a system.

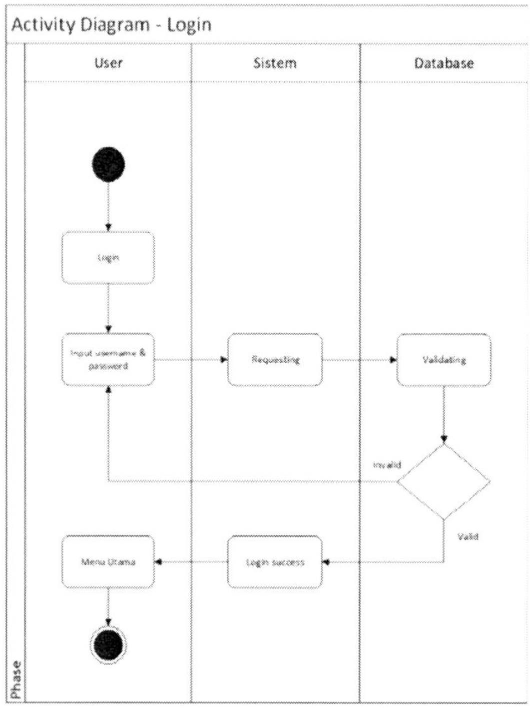

Figure 6. Activity Diagram

3.7. Structural Modelling

Structural Modelling is modelling that illustrates the contents of the data structure of the system in accordance with existing business processes. In this paper, the diagram technique that will be used is the class diagram. The following is a class diagram to illustrate the existing classes, including:

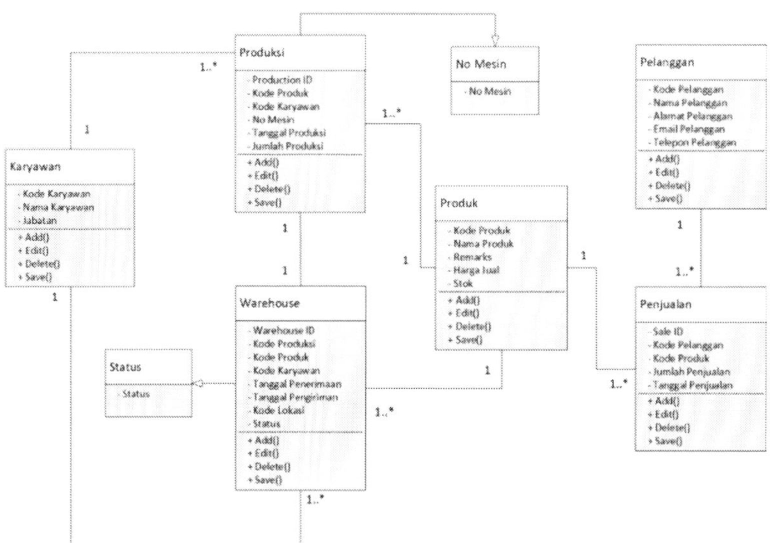

Figure 7. Class Diagram

IOP Conf. Series: Materials Science and Engineering **909** (2020) 012083 doi:10.1088/1757-899X/909/1/012083

3.8. Table Structure

Table structure is very necessary because it is useful in describing the contents of the table to be used such as the name of the data, the type of data, the size of the data, as well as a description of the data.

3.8.1. Product.

- Name: tblProducts
- Primary Key: Product_ID

Table 4. Product Table

No	Field Name	Data Type	Field Size	Remarks
1	Product_ID	Auto Number	Long Integer	Product Code
2	Product_Name	Short Text	50	Product Name
3	Remarks	Short Text	50	Remarks
4	Unit_Cost	Number	Double	Price
5	Adjusted_Qty	Number	Long Integer	Stock

3.8.2. Production.

- Name: tblProduction
- Primary Key: Production_ID_(Barcode)

Table 5. Production Table

No	Field Name	Data Type	Field Size	Remarks
1	Production_ID_(Barcode)	Number	Long Integer	Production Code
2	Machine_ID	Number	Long Integer	Machine Number
3	Qty_Prod	Number	Long Integer	Production Total
4	Date_Prod	Date/Time		Production Date
5	User_ID	Number	Double	Employee Code

3.8.3. Employee.

- Name: tblEmployees
- Primary Key: IngEmpID

Table 6. Employee Table

No	Field Name	Data Type	Field Size	Remarks
1	IngEmpID	Number	Long Integer	Employee ID
2	strEmpName	Short Text	10	Employee Name
3	strAccess	Short Text	20	Job
4	Username	Short Text	10	Username
5	strEmpPassword	Short Text	10	Password

IOP Conf. Series: Materials Science and Engineering **909** (2020) 012083 doi:10.1088/1757-899X/909/1/012083

3.8.4. Customer.

- Name: tblCustomers
- Primary Key: Customer_ID

Table 7. Customer table

No	Field Name	Data Type	Field Size	Remarks
1	Customer_ID	AutoNumber	Long Integer	Customer ID
2	Customer_First_Name	Short Text	25	Customer Name
3	Customer_Email	Short Text	30	Customer Email
4	Customer_Phone_Number	Short Text	15	Customer Phone
5	Customer_Address	Short Text	80	Customer Address

3.8.5. Warehouse.

- Name: tblWarehouse
- Primary Key: Warehouse_ID
- Foreign Key: Production_ID_(Barcode)

Table 8. Warehouse table

No	Field Name	Data Type	Field Size	Remarks
1	Warehouse_ID	AutoNumber	Long Integer	Warehouse ID
2	Production_ID_(Barcode)	Number	Long Integer	Production Code
3	Date_Received	Date/Time		Received Date
4	Location_ID	Short Text	3	Location ID
5	User_ID	Number		Employee Code

3.8.6. Sales.

- Name: tblSales
- Primary Key: Sale_ID
- Foreign Key: Customer_ID

Table 9. Sales table

No	Field Name	Data Type	Field Size	Remarks
1	Sale_ID	AutoNumber	Long Integer	Sale ID
2	Customer_ID	Number	Long Integer	Customer ID
3	Product_Name	Number	Long Integer	Product Code
4	Quantity_Sold	Number	Long Integer	Total Sales
5	Transaction_Date	Date/Time		Sales Date

3.9. Barcode

In this system, the use of barcodes is used as production codes. This is necessary so that the barcode can store data and information needed by warehouse staff who sometimes have difficulty finding products in the warehouse. The process is carried out by the production staff who will put a barcode sticker each time the product is finished packing, then then enter the relevant information into the production menu. Then the goods are handed over to the warehouse, then the warehouse storage data is re-entered into the system.

DEMO

0110190001

Figure 7. Barcode that are used in the system

The barcode used is Code 128 type and to determine the barcode value on the barcode the rule used is the first six digits is the date when the production is done with the DDMMYY format, then followed by the next four digits which means the product number produced at that day (1 - 9999). So overall the barcode uses 10 digits number. All of these numbers can be generated automatically using the DRPU Barcode Label Maker application or using data that was created in Microsoft Excel beforehand. System users must print this barcode every day before starting production activities.

4. Conclusion

The conclusion that can be drawn from this paper is the design of a system made for PT Akrilik Kurnia Kencana using Microsoft Access, after going through a system feasibility analysis it can be stated that the system designed is suitable for use.

Then operationally, the company has sufficient resources to operate the system within the company. This system as a whole is able to cover the deficiencies that exist in the system implemented by the company at this time, especially in the company's warehouse that has problems in storing finished goods, namely the product placement rules.

In addition, this system can also be useful for monitoring production results so that necessary information - information contained in a barcode can be input into this system.

References

[1] Ajie D M 2013 *Sistem Otomasi Perusahaan* (Bandung : Program Studi Perpustakaan dan Informasi Universitas Pendidikan Indonesia)

[2] Budihardjo M 2014 *Panduan Praktis Menyusun SOP* (Raih Asa Sukses)

[3] Dennis, Alan, Barbara H W, and Roberta M R 2012 *System Analysis and Design 5th Edition* (New Jersey : *John Wiley & Sons, Inc.*)

[4] Legowo, A I 2014 *Sistem* Informasi *Perpustakaan Berbasis Website dengan Menggunakan Barcode Scanner pada SMK Negeri 3 Kendal* (Semarang : Universitas Dian Nuswantoro)

[5] Putri, Nency E, Sovandi M, and Toni H 2015 *Aplikasi Berbasis Multimedia Untuk Pembelajaran Hardware Komputer* (Padang : Jurnal Edik Informatika STKIP PGRI Sumatera Barat)

[6] Qurniawan A 2009 *Prosedur Pelayanan Pembayaran Pensiun Pegawai Negeri Sipil (PNS) di PT.*

Taspen (Persero) Cabang Surakarta (Surakarta : Program Diploma III Manajemen Administrasi Fakultas Ilmu Sosial Dan Ilmu Politik Universitas Sebelas Maret)

[7] Ragunath P K, Velmourougan S, Davachelvan P, Kayalvizhi S, and Ravimohan R 2010 *Evolving A New Model (SDLC Model-2010) For Software Development Life Cycle (SDLC)* (Chennai : International Journal of Computer Science and Network Security)

[8] Rahman, Wilman, And Farhan A 2014 *Mengenal Berbagai Macam Software* (Tangerang : Surya University)

[9] Rainer, Kelly R, and Casey G C 2011 *Introduction to Information Systems* (New York : John Wiley & Sons, Inc)

[10] Tuteja, Maneela, and Gaurav D 2012 *A Research Study on importance of Testing and Quality Assurance in Software Development Life Cycle (SDLC) Models* (Kurukshtera : International Journal of Soft Computing & Engineering)

[11] Yuniartini, N P S 2013 *Pengaruh Modal,Tenaga Kerja Dan Teknologi Terhadap Produksi Industri Kerajinan Ukiran Kayu Di Kecamatan Ubud* (Bali : E - Jurnal Jurusan Ekonomi Pembangunan Fakultas Ekonomi Universitas Udayana)

[12] Widjaja J 2018 *Komputerisasi Sistem Koperasi Lembaga Daya Dharma Keuskupan Agung Jakarta* (Tangerang : Program Studi Teknik Industri Fakultas Sains dan Teknologi Universitas Pelita Harapan)

IOP Conf. Series: Materials Science and Engineering **909** (2020) 012084 doi:10.1088/1757-899X/909/1/012084

Performance evaluation of yarn raw materials supplier using fuzzy data envelopment analysis approach (case study Batik Fabric Company in Sleman)

Annisa Indah Pratiwi[1*], Muhamad Sayuti [2], Sukriyadi[3]
University of Buana Perjuangan Karawang[1]
University of Buana Perjuangan Karawang[2]
General Employee Automotive Parts Manufacturing[3]

* Email: annisa.indah@ubpkarawang.ac.id

Abstract. Evaluating supplier performance by measuring supplier performance is an important thing by the company to optimize costs and time. This study aims to evaluate supplier performance and determine the best supplier based on predetermined criteria. This research was conducted on Batic Fabric company by taking research object suppliers of raw material for 40 CD and 50 CM yarn. This study used 4 criteria and 3 month period data. Evaluation of supplier performance using the Fuzzy Data Envelopment Analysis (DEA) approach, the Fuzzy method is used to carry out linguistic assessments, while for the assessment efficiency is used by the DEA method. The result analysis was concluded that the efficient suppliers for 40 CD yarns were suppliers PTX1, PMT1 and AG1. PMT1 is the best supplier which has the highest super efficiency value of 10.01565 while the 2nd rank is occupied by the PTX1 supplier with a super efficiency value of 1.633666 and the third rank is AG1 supplier with a super efficiency value of 1,488659 and other suppliers are inefficient: PMT1, SC1, PDT1 and AR1. For 50 CM yarn, the best supplier is PTX2 which has an efficiency value of 1, while other suppliers are inefficient: PMT2, DLT2, and DLR2.

1. Introduction

Batik fabric company is a company that moves set a policy of quality in terms of textile firm as orientation in the business. The quality policy is "Customer Satisfaction Our Pride". From the quality policy it can be seen that customer satisfaction is one of the main objectives to be achieved by the batik factory. Customer satisfaction can be created after comparing the value performance obtained with the expected value performance. One of the performance parameters can be known from the timeliness and condition of ordered goods. The timeliness of completion, the quantity and quality of goods sent, can be determined and regulated through a good production management system. One of them is planning the procurement of raw materials. Batik Fabric company has many suppliers as suppliers of yarn raw materials. At the company there are still purchasing raw materials with low quality yarn. Many found defects in raw materials purchased such as too many neps. Defects in raw materials will result in poor quality of the finished product. In addition there is a delay in the arrival of yarn raw materials and the number of yarn quantities not in accordance with orders by the company. If this is ignored, it will hamper the production process and harm the company. Of the various types of yarn used, the 40 CD and 50 CM yarns are threads that are often used in the production process and experience changes in quality. According to Pujawan and Mahendrawathi (2010), supplier performance needs to be monitored continuously. This performance monitoring is important to be carried out as an evaluation material which can later be used to improve supplier performance or as a consideration for whether or not to look for alternative suppliers. Therefore, to maintain the stability of supplier performance, it is necessary to conduct an evaluation of the supplier's performance of yarn raw materials in batik factories by measuring supplier performance using the Fuzzy DEA approach.

Rouyende Gh & Erol in 2010 conducted a study entitled The DEA - FUZZY ANP Department Ranking Model Applied in Iran Amirkabir University. The study uses the DEA & Fuzzy ANP method as a tool to determine the ranking of each department that is efficient and inefficient so that it can be used as an evaluation material for improvement. In (2011) Dotoli & Falagario conducted research under the title A hierarchical model for optimal supplier selection in multiple sourcing contexts. The research uses DEA, TOPSIS & Linear Programming methods as a tool to show differences in the performance of each supplier based on multiple sourcing and order quantity calculations from each supplier.

2. Literature review

2.1. Performance evaluation supplier

According to Wibowo (2010), performance is doing work and the results achieved from the work, about what is done and how to do it. Performance measurement needs to be done to find out whether during the performance implementation there are deviations from the predetermined plan, whether the performance can be done according to the specified schedule, whether the performance results have been achieved as expected. Meanwhile the evaluation according to Wirawan (2009), is the process of gathering information concerning evaluation objects and assessing evaluation objects by comparing them with evaluation standards. The result is information about the evaluation object which is then used to make decisions about the evaluation object.

Performance evaluation is carried out to provide an assessment of the work results or work achievements obtained by the organization, team or individual. Performance evaluation will provide feedback on performance goals and objectives, planning and performance implementation processes. Performance evaluation can also be carried out on the process of assessment, review, and performance measurement. On the basis of performance evaluation steps can be taken to improve performance in the future (Wibowo, 2010). Supplier is also an organization that needs to be evaluated for its performance. This is very important to maintain business cooperation between company and suppliers and also to monitor and ensure products that are supplied in accordance with the specifications set by the company.

One of the fundamental aspects of supply chain management is performance management and continuous improvement. To create an effective performance management measurement system is needed that is able to evaluate the performance of the supply chain as a whole (Pujawan and Mahendrawathi, 2010). Vendor evaluation involves the process of finding potential vendors and determining the likelihood that they will become good suppliers (Heizer & Render, 2011). Supplier performance evaluation activities are part of supplier selection. Supplier selection is a strategic activity, especially if the supplier will supply critical items or will be used in the long term (Pujawan and Mahendrawathi, 2010). Many factors need to be considered in selecting suppliers. The selection process can be very complex because a company may have a number of capabilities in all fields or abilities that are so good in only a few fields [4].

2.2. Fuzzy Association Theory

The meaning of fuzzy (obscurity) is found in qualitative meanings which have relative values for different individuals. Fuzzy set theory in further development is an appropriate tool and technique for analyzing complex systems and is a decision making process of the uncertainty of a pattern caused by a blurring of randomness. This theory is built to solve problems where the description or description of activities, observations, and decision making are subjective, vague and inaccurate. The basic concept of fuzzy is the inaccuracy in defining a meaning expressed in linguistic form (Basuki et al, 2008). This form of linguistics is the naming of a group that represents a certain condition or condition by using natural language, such as young, middle age, and old [6].

According to Tettamanzi [6] suggested that fuzzy set theory is a mathematical framework used to represent uncertainty, ambiguity, inaccuracy, lack of information, and partial truth.

Kusumadewi et al (2006), states that the reasons for the use of fuzzy logic by most people are:
1. The concept of fuzzy logic is easy to understand because it is based on natural language.
2. Fuzzy logic is very flexible.
3. Fuzzy logic has a tolerance for incorrect data.
4. Fuzzy logic is able to model nonlinear functions that are very complex.
5. Fuzzy logic can build and apply the experiences of experts directly without having to go through the training process.
6. Fuzzy logic can work with conventional control techniques.

2.3. DEA (Data Envelopment Analysis)
The DEA approach was first developed theoretically by Charnes, Cooper and Rhodes in 1978. DEA is basically a linear programming based technique used to measure the relative performance of organizational units where the existence of multiple (multiple) inputs and outputs is difficult to compare. DEA identifies the relative units that use inputs in providing certain outputs in the most optimal way and DEA uses this information to form a frontier (efficiency) frontier of the available organizational unit data. DEA uses this efficient frontier to calculate the efficiency of other organizational units that are not on the efficient frontier so that it can provide information about units that do not use inputs efficiently [18].

In compiling develop the DEA model to evaluate supplier performance, input and output variables must be determined based on company considerations related to the objectives to be achieved (Rahman, 2012). DEA is a non-parametric approach which is basically a linear programming technique. DEA works with the identification step of the unit to be evaluated, the inputs needed and the output produced by the unit. Then form the border efficiency of the available data set and calculate the productivity value of the units that are not included in the efficiency border and identify which units do not use inputs efficiently, relative to the units with the best performance from the analyzed data set [8].

The relative efficiency values from the calculation of a DMU (Decision Making Unit) with the CRS model and output oriented are obtained from the following models:

$$\text{Max Em}=\frac{\sum_{j=1}^{J} v_{jm}\, y_{jm}}{\sum_{i=1}^{I} u_{im}\, x_{im}} \quad (1)$$

Subject to

$$0\leq \frac{\sum_{j=1}^{J} v_{jm}\, y_{jn}}{\sum_{i=1}^{I} u_{im}\, x_{in}} \leq 1; \quad n = 1, 2, K, N$$

v_{jm}, $u_{im} \geq 0$; i =1, 2, 3, K, I; j = 1, 2, K, J

symbol:

E_m = efficiency of m DMU
y_{jm} = number of output variables from m DMU
v_{jm} = the weight of the output
x_{im} = number of input variables from m DMU
u_{im} = the weight of the input
y_{jn} dan x_{in} = the number of *j output* and *i input* dari *n* DMU, *n* = 1, 2, 3, ...N
source: Ramanathan (2003, page 40).

The above formula cannot be solved using a linear program, the above formula can be changed to a linear program as follows:

$$\text{Max } z=\sum_{j=1}^{J} v_{jm}y_{jm} \quad (2)$$

Subject to

$$\sum_{i=1}^{I} u_{im}X_{im} = 1$$

$$\sum_{r=1}^{s} v_{jm}Y_{jn} - \sum_{i=1}^{m} u_{im}X_{in} \leq 0; \quad n=1,2, K, N$$

$v_{jm}, u_{im} \geq \varepsilon$; $i=1,2,K$, [13]

3. Materials and methods

The study focused on suppliers of 40 CD yarn for 6 suppliers and 50 CM yarn for 4 suppliers. Linguistic evaluation was carried out by the Head of the PPIC/GF (grain finishing) Section of 1 person, the Head of the Laboratory for testing thread 1 person and 1 person Head of Weaving Unit. The variables criteria used in this study are as follows:

1. Quality: Regarding the ability of suppliers to fulfil quality according to standards set by the company.
2. Price: Corresponds to the price level of yarn raw material offered by the supplier and has been approved by the company.
3. Delivery: Related to the ability to meet the delivery time of raw materials from suppliers to the company.
4. Order fulfilment: Related to the ability of suppliers to fulfil the quantity of yarn raw materials based on orders from company.

These criteria were obtained based on interviews with the company and also based on research by Li et al. (1997).

The research method that is used as follows:

1. Define the problem.
2. Define the criteria of variable by brainstorming and interview with Head of PPIC.
3. Collect the data by questionnaire.
4. The results of the questionnaire were processed with fuzzy, so we got the weight of each criterion each supplier.
5. Meanwhile the DEA data is prepared by determining the DMU input and output. The input and output variables used are as follows: price as input and Quality, Delivery, Order fulfilment as output.
6. Supplier evaluation also involves historical data over a specified period and uses price, quality, delivery and order fulfilment criteria. Furthermore, the data is aggregated with criteria weights. Supplier weight values are based on criteria that have been obtained, then determine the optimization model for supplier assessment.
7. Then processed by the DEA method using Lindo software which is a linear programming technique that calculates the output ratio of input from each production unit (DMU) whose results are called relative efficiency scores. The efficiency value indicates which supplier is the best.
8. The results obtained are the value of the efficiency of each supplier and then an analysis of these results.

3.1. Fuzzy processing

Linguistic variables are used to get judgment by decision makers. The implementation of triangular fuzzy number membership functions can classify the blurring that occurs in linguistic variables. The steps in the fuzzy method are as follows:

1. Determine the fuzzy number membership function of each decision maker by using a triangular fuzzy number.

Table 1. Membership function of triangular fuzzy number

No.	Variabel Lingustik	Triangular Fuzzy Number Membership (*l, m, u*)
1.	VL= Very Low	(0, 0.1, 0.3)
2.	L = Low	(0.1, 0.3, 0.5)
3.	M = Medium	(0.3, 0.5, 0.7)
4.	H = High	(0.5, 0.7, 0.9)
5.	VH= Very High	(0.7, 0.9, 1)

The membership function of the triangular fuzzy numb set on the weight variable above can be represented in the following figure:

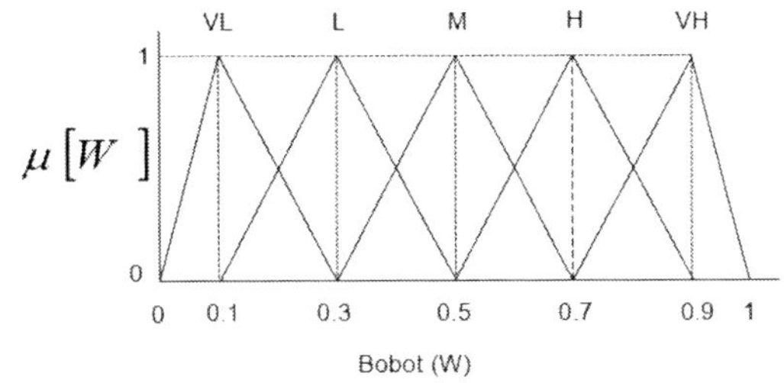

Figure 1. Fuzzy numbers for weights [6]

2. Tabulate the results of the assessment given in the first step to get an average value of each criterion.

Because there are more than one decision makers, it is necessary to do an aggregate calculation, an aggregate calculation is performed using the geometric average approach. This is used to find out one decision value from several questionnaires with the following formula (Kang, et al. 2010):

$$r_{ij}=(a_{ij1} \times a_{ij2} \times \ldots a_{ijk})^{1/k} \quad (3)$$

3. Calculate the single value of the results of the second step using the defuzzification of the center of gravity method with the following formula (Kang, et al. 2010):

$$F_{ij}=\frac{[(u_{ij}-l_{ij})+(m_{ij}-l_{ij})]}{3}+l_{ij}=\frac{l_{ij}+m_{ij}+u_{ij}}{3} \quad (4)$$

4. After calculating the defuzzification for each criterion, the next step is to normalize using the following formula:

$$\text{Weight}=\frac{\text{score each supplier}}{\text{total score all supplier}} \quad (5)$$

The following results are fuzzy processing:

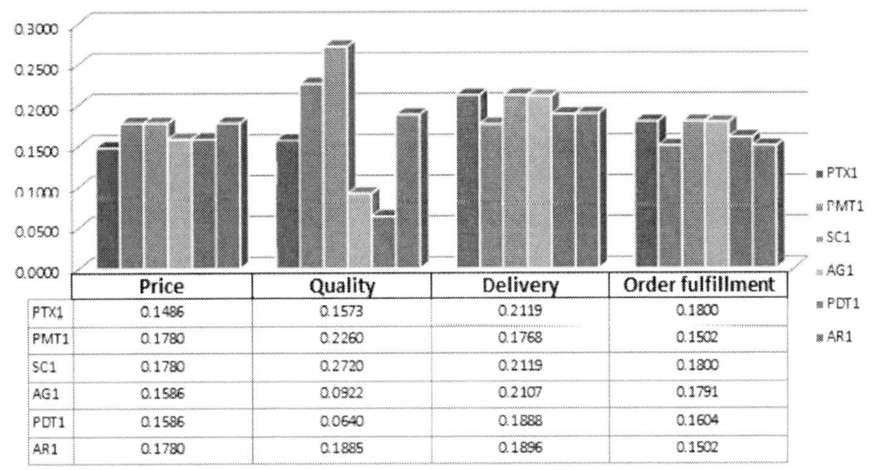

	Price	Quality	Delivery	Order fulfillment
PTX1	0.1486	0.1573	0.2119	0.1800
PMT1	0.1780	0.2260	0.1768	0.1502
SC1	0.1780	0.2720	0.2119	0.1800
AG1	0.1586	0.0922	0.2107	0.1791
PDT1	0.1586	0.0640	0.1888	0.1604
AR1	0.1780	0.1885	0.1896	0.1502

Figure 2. Weight criteria of Yarn Supplier 40 CD

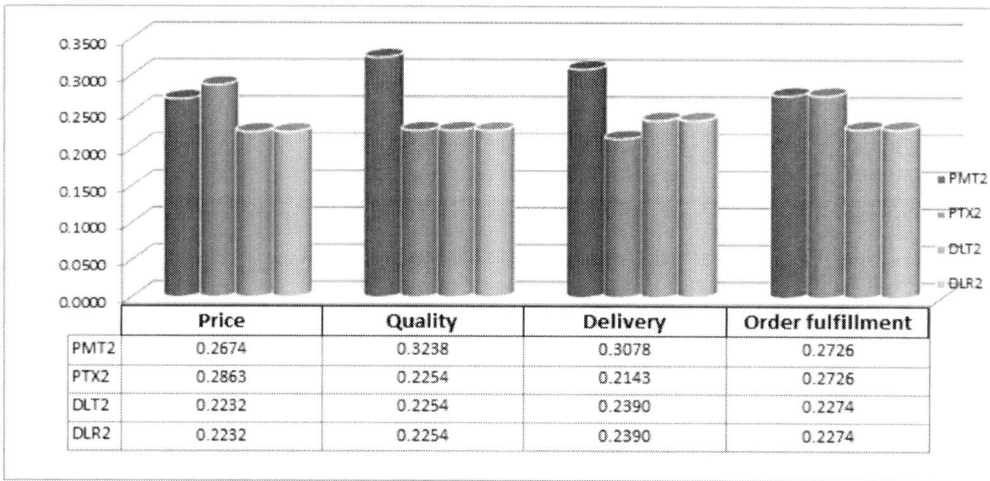

Figure 3. Weight criteria of Yarn Supplier 50 CM

3.2. Recapitulation of weight

Data obtained from the company for 3 months in the form of a percentage can be seen in tables 2 and 3 below:

Table 2. Company data recapitulation for 40 CD yarn

Supplier	Price	Quality	Delivery	Order fulfillment
PTX1	21,6716	13,9860	52	43
PMT1	5,1799	100	0	100
SC1	38,4292	0	16	51,1173
AG1	29,8194	24,4624	32	100
PDT1	4,8999	100	0	100
AR1	5,1099	0	0	50

Table 3. Company data recapitulation for 50 CM yarn

Supplier	Price	Quality	Delivery	Order fulfillment
PMT2	28,5550	0	8	100
PTX2	33,9450	0	28	61,6667
DLT2	28,2110	0	0	75
DLR2	9,2890	0	0	100

These four criteria are classified into input and output. The price is as input and the output are quality, delivery and order fulfillment.

3.3. DEA processing

Measurement of supplier efficiency is done by using the DEA method, which is a linear programming technique that calculates the ratio of output to input of each production unit (Decision Making Unit, DMU) whose results are called relative efficiency scores (Cooper, et al. 2002). In this study to measure supplier efficiency using the CRS (Constant Return to Scale) model and output oriented with the formula (2).

The DEA mathematical model of a unit can be formulated into a fractional linear program by making the input and output weights of the units concerned as decision variables (Palit et al, 2008).

The results of the fuzzy assessment are supplier weights based on criteria, then aggregated with historical data that has been calculated as a percentage and the following are the results:

Table 4. Data input and output for each supplier yarn 40 CD

Supplier	DMU	X1	Y1	Y2	Y3
PTX1	1	3,2199	2,2000	11,0163	7,7411
PMT1	2	0,9222	22,5973	0	15,0232
SC1	3	6,8420	0	3,3896	9,2024
AG1	4	4,7308	2,2562	6,7433	17,9069
PDT1	5	0,7774	6,3976	0	16,0416
AR1	6	0,9098	0	0	7,5116

Table 5. Data input and output for each supplier yarn 50 CM

Supplier	DMU	X1	Y1	Y2	Y3
PMT2	1	7,6359	0	2,4624	27,2553
PTX2	2	9,7178	0	5,9995	16,8074
DLT2	3	6,2954	0	0	17,0585
DLR2	4	2,0729	0	0	22,7447

Remarks:
X1 = value of input that is the purchase price of each supplier
Y1 = value of output which is the level of yarn quality of each supplier
Y2 = value of output which is the delivery time of each supplier
Y3 = value of output which is the level of order fulfillment from each supplier

The next stage after the data is aggregated, then relative efficiency is measured, but the optimization modeling is based on equation (2) first. DEA processing is done by inserting the optimization model that has been modeled before into the LINDO software and then completing it so that the efficiency values of each DMU will be obtained.

3.4. *Relative Efficiency Analysis*

Based on the results of DEA processing with LINDO software, the relative efficiency values of each DMU (supplier) are obtained. Here are the results of the calculation:

Table 6. Efficient and inefficient DMU for 40 CD yarns

Supplier	DMU	Efficient	inefficient
PTX1	1	1	
PMT1	2	1	
SC1	3		0.5139025
AG1	4	1	
PDT1	5		0.9298730
AR1	6		0.4194807

Table 7. Efficient and inefficient DMU for 50 CM yarns

Supplier	DMU	Efficient	inefficient
PMT2	1		0.5170738
PTX2	2	1	
DLT2	3		0.6258783
DLR2	4		0.8345056

From the table 6, it can be seen that DMU 1 has an efficiency value of 1, this means that the performance of DMU 1 (supplier PTX1) for 40 CD yarn is quite efficient while DMU-DMU which has an efficiency value of less than 1 is categorized as inefficient, as well as for the supplier yarn performance results 50 CM. Suppliers whose efficiency value is less than 1 are suppliers in their operational activities that have not been efficient in utilizing all of their potential capabilities to be able to produce maximum output. Suppliers who have low efficiency values, namely SC1 and AR1 for 40 CD yarn suppliers and DLT2 for 50 CM yarn suppliers. the cause of SC1 suppliers is inefficient, because in that period it is the rainy season and this causes flooding in various cities, one of which is in Jakarta and SC1 suppliers are located in Jakarta so that SC1 cannot operate optimally and as one of its impacts, it is unable to fulfill orders from Batik Fabric company. Besides the factors that influence is the occurrence of damage to transportation on the road so that slowing the delivery (delivery). Thus SC1 performance is less than the maximum. As for the AR1 supplier the inefficiency according to the planner is that there is too much wax in the thread so there is a need for prior communication before making a decision and as a result requires more time for negotiation and waits for a better yarn production than AR1. This will affect the performance of AR1 suppliers in partnering business with Batik Fabric factories. If this waxy substance is left alone it will disrupt the weaving production process in the batik fabric company, the disturbance is in the form of excessive thread adhesiveness. While the DLT2 supplier is inefficient according to the planner, which is the presence of too much honey cotton. Too much honey will cause disruption to the warping process, namely the thread will be hot and break easily so that it can hamper the production process in the Batik Fabric company. Therefore communication needs to be done first before making a decision and as a result requires more time for negotiation and waiting for a good yarn for yarn production results better than DLT2.

3.5. Super Efficiency Analysis

Based on the calculation of DEA with the CRS (Constant Return to Scale) model and output oriented, we get 3 suppliers of 40 CD threads with one efficiency value (table 4.5). To find out which of the three suppliers is the best, it is necessary to rank it using super efficiency calculations.

The basic concept of super efficiency is to allow the observed DMU efficiency to be greater than one or 100%. This is obtained by removing the associated constraints from the series of constraints or constraints p, so that there is no efficiency limit smaller than 1 for the p-DMU. Super efficiency only affects units (DMU) that are considered to be as efficient as the boundaries removed, which do not bind inefficient units because their efficiency is less than 1 or 100% [10].

The concept of super efficiency concept is almost the same as DEA CRS formulation, which differs only in output-input constraints, where in super efficiency removes the value of output-input constraints for the p-supplier or supplier being evaluated. The purpose of removing this obstacle is that the efficiency value of the supplier being evaluated can be greater than 1 or 100% with the aim of ranking suppliers that are already efficient [10].

Based on the results of calculations with LINDO software, we get the value of super efficiency and ranking for each supplier as shown in table 8 below:

Table 8. Super efficiency value and supplier ranking from 40 CD yarn

DMU	*Supplier*	Value of super efficient	Rank
DMU 1	PTX1	1,633666	2
DMU 2	PMT1	10,01565	1
DMU 4	AG1	1,488659	3

From table 8 it can be seen that for 40 CD threads, DMU 2 has the highest super efficiency value of 10,01565, this means that the performance of DMU 2 (supplier PMT1) is the best supplier by obtaining rank 1 while ranking 2nd is occupied by DMU 1 is the supplier of PTX1 with a super efficiency value of 1.633666 and the third rank is DMU 4 (AG1) with a super efficiency value of 1.488659.

Batik Fabric company should make the process of purchasing raw materials to efficient suppliers, in this case PMT1 for 40 CD yarn suppliers, and PTX2 for 50 CM yarn suppliers, because the supplier has an efficiency value of 1 based on DEA calculations with LINDO software.

Conclusion

1. The result is supplier performance level based on performance evaluation with Fuzzy DEA for 40 CD threads, namely suppliers PTX1, PMT1 and AG1, including efficient. Of the three suppliers, PMT1 is the best supplier because it has the highest super efficiency value of 10.01565 while the 2nd rank is occupied by the PTX1 supplier with a super efficiency value of 1.633666 and the third rank is AG1 supplier with a super efficiency value of 1, 488659. While other suppliers are inefficient namely SC1, PDT1 and AR1 with efficiency values respectively 0.5139025, 0.9298730 and 0.4194807. For 50 CM yarn, the best supplier is PTX2 because it has an efficiency value of 1, while other suppliers are inefficient, namely PMT2, DLT2, and DLR2 with efficiency values respectively 0.5170738, 0.6258783 and 0.8345056. Inefficient suppliers occur because they are unable to fulfill the orders of Batik Fabric company and delivery delays. For suppliers that fall into the efficient category it can be interpreted that is able to provide good performance for Batik Fabric company.
2. The factors that affect the performance of yarn raw material suppliers, namely quality, price, delivery, and order fulfillment. By evaluating the performance of supplier the company can obtain useful information for company management feedback in monitoring performance, diagnosing problems and identifying company successes.
3. The performance of suppliers of yarn raw materials can be known by measuring the valuation using the Fuzzy Data Envelopment Analysis approach where the Fuzzy method is used to carry out linguistic assessments so that the weighting value of each supplier is obtained, while for the supplier efficiency assessment the Data Envelopment Analysis (DEA) method is used.
4. As for discussion, the results of supplier performance evaluation based on research over a period of 3 months can be used as recommendations, namely maintaining existing suppliers, looking for alternative other suppliers or combining between suppliers. And suggestions for further research is to evaluate performance using Fuzzy DEA will be more perfect if it is equipped with other methods as a comparison and use of the Decision Support System (DSS) to design the application.

References

[1] Abidin, Z, & Endri 2009 Kinerja efisiensi teknis bank pembangunan daerah: pendekatan data envelopment analysis DEA *Jurnal Akuntansi dan Keuangan* **11**
[2] Cooper, W W, Seiford, L M, & Tone, K 2002 Data envelopment analysis *A Comprehensive Text with Models, Applications, References and DEA-Solver Software*
[3] Dotoli, M, & Falagario, M 2011 A hierarchical model for optimal supplier selection in multiple sourcing contexts *International Journal of Production Research*
[4] Heizer, J, & Render, B 2011 Manajemen operasi Jakarta: Salemba Empat
[5] Kang, H-Y, Lee, A H I, & Yang, C-Y 2010 A fuzzy ANP model for supplier selection as applied to IC packaging *Springer Science & Business Media, J Intell Manuf DOI 101007/s10845-010-0448-6*
[6] Kusumadewi, S, Hartati, S, Harjoko, A, & Wardoyo, R 2006 Fuzzy multi-attribute decision making fuzzy MADM Yogyakarta: Graha Ilmu
[7] Li, C C, Fun, Y P, & Hung, J S 1997 A new measure for supplier performance evaluation *IIE Transactions* **29** 753-758
[8] Nugroho, S W P, Sriyanto, & Chasanah, N 2011 Analisis efisiensi distribusi listrik unit pelayanan jaringan dengan metode data envelopment analysis *J@TI Undip* **VI** 47-56
[9] Palit, H C, Lienardo, H, & Widyadana, I G A 2008 Aplikasi kombinasi algoritma genetik dan data envelopment analysis pada penjadwalan flowshop multikriteria *Jurnal Teknik Industri Fakultas Teknologi Industri Universitas Kristen Petra* ***10***
[10] Permadhi, T R 2012 Analisis efisiensi distribusi pemasaran menggunakan metode data envelopment analysis studi kasus di UKM Kamties, Jetis Suruh, Donoharjo, Ngaglik, Sleman Universitas Islam Indonesia, Fakultas Teknologi Industri
[11] Pujawan, I N, & Mahendrawathi 2010 Supply chain management Surabaya: Guna Widya

IOP Conf. Series: Materials Science and Engineering **909** (2020) 012084 doi:10.1088/1757-899X/909/1/012084

[12] Rahman, F 2012 Pengukuran efisiensi supplier bambu dengan pendekatan data envelopment analysis studi kasus di sentra kerajinan bambu UD Dheling Asri Sendari Tirtoadi Mlati Sleman Universitas Islam Indonesia Fakultas Teknologi Industri

[13] Ramanathan, R 2003 An introduction to data envelopment analysis a tool for performance measurement

[14] Rickards, R C 2003 Setting benchmarks and evaluating balanced scorecards with data envelopment analysis *Benchmarking: An International Journal* **10** 226 - 245

[15] Rouyendegh, B D, & Erol, S 2010 The DEA – fuzzy ANP department ranking model applied in Iran Amirkabir University *Acta Polytechnica Hungarica* **74** 103-114

[16] Wibowo 2010 Manajemen kinerja Jakarta: Rajawali Pers

[17] Wirawan 2009 Evaluasi kinerja sumber daya manusia: teori aplikasi dan penelitian Jakarta: Salemba Empat

[18] Wulansari, R R 2010 Efisiensi relatif operasional puskesmas-puskesmas di kota semarang tahun 2009 Universitas Indonesia Depok

IOP Conf. Series: Materials Science and Engineering **909** (2020) 012085 doi:10.1088/1757-899X/909/1/012085

Analysis and strategy of supply chain risk mitigation using fuzzy failure mode and effect analysis (fuzzy fmea) and fuzzy analytical hierarchy process (fuzzy ahp)

Dyah Lintang Trenggonowati[1*] , Maria Ulfah[2] , Faula Arina[3], Chisti Lutfiah[4]
Industrial Engineering Department, Engineering Faculty, Sultan Ageng Tirtayasa University

*E-mail: dyahlintang@untirta.ac.id

Abstract. PT XYZ is a company engaged in steel manufacturing and technical assistance. An example of the risk that occurs in supply chain activities in this company is that 8% of the total items experience delays in the Request Order (RO) to Purchase Orders (PO) in 2019. Currently, PT XYZ does not have supply chain risk management to identify and mitigate the risks that occur within the company. Therefore, this study aims to identify risks, determine priority risks and provide priority risk mitigation strategies that occur in the supply chain at PT XYZ. The method used is Supply Chain Operation Reference (SCOR), Fuzzy Failure Mode and Effect Analysis (Fuzzy FMEA) and Fuzzy Analytical Hierarchy Process (FAHP). The results of the study found 29 risk events and 4 priority risk events. Based on 4 prioritized risk events, the criteria which have the highest weight are the lack of the process of checking the raw material to be produced have a weight value 0.650 with the mitigation strategy carried out on the criteria is the lack of the process of checking raw materials to be produced is to adjust the demand documents and raw materials available. For the supplier performance criteria is not good to have a weight value 0.265 with the mitigation strategy carried out is conducting direct inspection to supplier company. The criteria for lack of coordination between companies and suppliers have a weight value 0.085 with the mitigation strategy carried out is to coordinate with suppliers. The lowest weighting criteria is the absence of project demand / tonnage forecasting processes with weights 0,000 and mitigation strategies carried out is to carry out project demand / tonnage forecasting processes.

1. Introduction

The company in conducting the activities of production of goods or services is always faced with various problems or risks that exist and appears continuously. Problems that arise usually result from within the company and from outside the company. The more growing manufacturing companies in the world, it will result in very tight competition for each company. The company can face all existing competition if all activities in the company are running smoothly.

At supply chain activities in the company there are certain risks that affect each of its activities. Risk is a potential event or occurrence, both predictable and unpredictable, that could cause a negative impact on achieving the company's vision and mission [1]. Risk management in supply chain management activities within the company is usually referred to as Supply Chain Risk Management (SCRM). The benefits of SCRM are to be able to identify and assess supply chain disruptions and can reduce the negative impact of supply chain performance [2].

PT XYZ is a company engaged in steel manufacturing and technical assistance from Emoto, Japan. PT XYZ was established on March 17, 1984 which is located in Argawana Village, Pulo Ampel, Serang, Banten. Production activities at this company are steel manufacture and technical assistance.

IOP Conf. Series: Materials Science and Engineering **909** (2020) 012085 doi:10.1088/1757-899X/909/1/012085

In running its production process, there must be risks that preclude the activity, especially in supply chain activities. As an example of the risk occurring in supply chain activity in PT XYZ is a delay in Request Order (RO) process until Purchase Order (PO). Request order is list material or raw material created by user who need material or raw material, while the purchase order is the list of materials or raw materials that have been ordered to the supplier. The process of request order until the purchase order is for 3 days. In 2019, PT XYZ made a purchase of 8334 items. There are 630 items or an 8% delay.

Many previous studies [3], [4], [5] are discussed aims to identify and minimize possible process failures occurred in one of the case point, some example are warehouse, logistics, or manufacturing, used risk analysis one of the method FMEA or Fuzzy AHP to identify and minimize the possible process failures. But not many studies have used the Fuzzy FMEA combined with Fuzzy AHP methods to solve the problem about of the risk occurring in supply chain activity as happened in PT XYZ is a delay in Request Order (RO) process until Purchase Order (PO).

At present, PT XYZ does not yet have supply chain risk management to identify and mitigate risks that occur within the company. Based on examples of risks contained in the company, it is necessary to conduct research to identify risks using the method of Supply Chain Operations Reference (SCOR). SCOR divides supply chain processes into 5 core processes, which are plan, source, make, deliver, and return. The next step is to analyze and evaluate risks using the Fuzzy Failure Mode and Effect Analysis (Fuzzy FMEA) method.

Fuzzy FMEA is the development of a conventional Failure Mode and Effect Analysis (FMEA) method which displays flexibility for uncertainty due to vague information and subjective preference elements used in evaluating the failure modes that occur [6]. The next step is to mitigate supply chain risk that occurs within the company using the Fuzzy Analytical Hierarchy Process (FAHP) method. Because AHP data is verbal data, it is possible that the data is still blurry and qualitative. Therefore, the step to quantify the data is the Fuzzy Analytical Hierarchy Process (FAHP) method. Fuzzy Analytical Hierarchy Process (FAHP) is an analytical method developed from AHP. FAHP is able to overcome weaknesses in criteria that have more subjective traits in the AHP method. This method is a technique used to overcome uncertainty in problems that have many answers [7].

The purpose of this study is to identify the risks that occur in the supply chain at PT XYZ , determine priority risks in the supply chain at PT XYZ and provide mitigation strategies against priority risks occuring in the supply chain at PT XYZ.

2. Methodology

Risk is a danger, a consequence that can occur as a result of an ongoing process or an upcoming occurrence. Risk is the potential occurrence of an event, whether predictable and unpredictable, that could cause a negative impact on achieving the company's vision and mission [8].

Risk mitigation refers to minimizing risk after the risk arises. In other words, risk mitigation is a form of damage control. While the focus of risk mitigation is the action to be taken after the risk is realized, the company's risk mitigation strategies must be planned in advance, included or inserted in writing and are known by key people in an organization [9].

Supply Chain Management (SCM) is a series of activities including coordination, scheduling and control of procurement, production, inventory, and delivery of products or services to customers which includes daily administration, operations, logistics, and information processing ranging from customers to suppliers.[10].

The Supply Chain Operations Reference (SCOR) model was developed to provide a method of self-assessment and comparison of supply chain activities and performance as a cross-industry supply chain management standard. This model presents a business process framework, performance indicators, best practices and technology to support communication and collaboration between supply chain partners, so as to increase the effectiveness of supply chain management and the effectiveness of supply chain improvement. SCOR divides supply chain processes into 5 core processes, which are plan, source, make, deliver, and return [11].

Fuzzy FMEA is the development of a conventional FMEA method that displays flexibility for uncertainty due to vague information and subjective preference elements used in evaluating the failure modes that occur [12]. There are 3 main stages in fuzzy FMEA, namely [13]:

1. Fuzzification, the process of converting risk factors to severity, occurrence and detection into fuzzy.
2. Rule evaluation, contains knowledge from experts regarding the interaction of error modes and the effects caused in the form of fuzzy rules "if then".
3. Difuzzification, the process of creating ratings from fuzzy RPN to give the priority level of error mode

The following will explain the linguistic table and fuzzy number used to evaluate these factors and visualize the membership function of each factor. The table includes Table. 1 Fuzzy Severity, Table. 2 Fuzzy Occurrence, Table. 3 Fuzzy Detection and Table. 4 Fuzzy RPN [14].

Table 1. Fuzzy Severity Scale

Rating	Fuzzy Number
Very low	(3, 4, 5)
Minor	(2, 3, 4)
Very minor	(1, 2, 3)
None	(1, 1, 2)

(Source: Wang et al, 2009)

Table 2. Fuzzy Occurrence Scale

Rating	Fuzzy Number
Very high (VH)	(8, 9, 10, 10)
High (H)	(6, 7, 8, 9)
Moderate (M)	(3, 4, 6, 7)
Low (L)	(1, 2, 3, 4)
Remote (R)	(1, 1, 2)

(Source: Wang et al, 2009)

Table 3. Fuzzy Detection Scale

Rating	Fuzzy Number
Absolute uncertainty (AU)	(9, 10, 10)
Very remote (VR)	(8, 9, 10)
Remote (R)	(7, 8, 9)
Very low (VL)	(6, 7, 8)
Low (L)	(5, 6, 7)
Moderate (M)	(4, 5, 6)
Moderately high (MH)	(3, 4, 5)
High (H)	(2, 3, 4)
Very High (VH)	(1, 2, 3)
Almost certain (AC)	(1, 1, 2)

(Source: Wang et al, 2009)

Table 4. FRPN Value

Category	Parameter

VL	(0, 0, 25, 75)
VL-L	(25, 75, 125)
L	(75, 125, 200)
L-M	(125, 200, 300)
M	(200, 300, 400)
H	(400, 500, 700)
H-VH	(500, 700, 900)
VH	(700, 900, 1000, 1000)

(Source: Wang et al, 2009)

One of the disadvantages of AHP is that the judgment is too subjective and it is difficult to determine the weighting of each criterion. To deal with these weaknesses, a fuzzy AHP method was developed. Fuzzy AHP is a representation and existence of AHP by combining it with fuzzy set theory. The fuzzy concept used in the development of fuzzy AHP is Triangular Fuzzy Number (TFN). Triangular Fuzzy Number (TFN) is a fuzzy set theory that helps in measurements related to subjective judgment of humans using language or linguistics. The core of fuzzy AHP lies in paired comparisons which are described by the ratio scale related to the fuzzy scale. [15].

The following are the Fuzzy AHP stages [16]:

1. Arranging the Hierarchy of Problems. The hierarchy of problems is structured to assist the decision-making process by taking into account all the decision elements involved in the system.
2. Determination of Priorities. Compare the paired forms of all criteria for each hierarchy subsystem. Then calculate the eigenvalue and eigenvector.

3. Consistency

The formula of the consistency index is:

$$CI = (\lambda max - n) / (n - 1) \quad (1)$$

Where:
CI = consistency index
λmax = max eigenvalue
n = matrix order

If CI is zero (0), then the pairwise comparison matrix is consistent. The prescribed inconsistency limit is determined by using a Consistent Ratio (CR).

$$CR = \frac{CI}{RI} \quad (2)$$

Where:
CR = consistency ratio
RI = index random

If CR <0.1 then the value of pairwise comparisons on the criteria matrix is consistent.

4. AHP Fuzzification on Table 5, transform the AHP scale into Fuzzy numbers.

Table 5. Fuzzy Comparative Scale

AHP Scale	Fuzzy Scale Level	Fuzzy Scale Inverse
1	(1,1,1) If diagonal (1,1,3) If other	(1/3,1,1)
2	(1,2,4)	(1/4,1/2,1)
3	(1,3,5)	(1/5,1/3,1)
4	(2,4,6)	(1/6,1/4,1/2)
5	(3,5,7)	(1/7,1/5,1/3)
6	(4,6,8)	(1/8,1/6,1/4)
7	(5,7,9)	(1/9,1/7,1/5)
8	(6,8,9)	(1/9,1/8,1/6)
9	(7,9,9)	(1/9,1/9,1/7)

5. Calculates the value of fuzzy synthetic extent.

$$S_i = \sum_{j=i}^{m} M \otimes \left[\sum_{i=1}^{n}\sum_{j=1}^{m} M\right]^{-1} \tag{3}$$

To obtain M, the formula used is:

$$\sum_{j=i}^{m} M = \sum_{i=1}^{m} l, \sum_{i=1}^{m} m, \sum_{i=1}^{m} u \tag{4}$$

Where:
M = TFN number
m = number of criteria
j = column
i = row
l = lower value
m = medium value
u = upper value

6. Compare the possible levels between fuzzy numbers.

$$V(M_2 \geq M_1) = \begin{cases} 1, \text{if } m_2 \geq m_1 \\ 0, \text{if } l_1 \geq u_2 \\ \dfrac{(l_1 - u_2)}{(m_2 - u_2) - (m_1 - l_1)}, \text{other} \end{cases} \tag{5}$$

7. Vector weight

$$d'(A_i) = \min V(S_i \geq S_k) \tag{6}$$

$$W' = \left(d'(A_1), d'(A_2), \ldots, d'(A_n)\right) T$$

8. Normalization

(7)

$$W = \left(d(A_1), d(A_2), \dots, d(A_n)\right) T$$
$$d(A_n) = \frac{d'(A_n)}{\sum_{j=i}^{n} d'(A_n)}$$

3. Result And Discussion

Based on the the identification of risk events using the SCOR method there were 29 supply chain risk events that occurred at PT XYZ. The following Table. 6 and Table 7.are the result of risk measurement using the Fuzzy FMEA method with Matlab software:

Table 6. Result of Supply Chain Risk Defuzzification

Major Process	Risk Event	Code	Severity	Occurence	Detection	FRPN	Category
Plan	Determination of the number of project requests/tonnage is incorrect	E1	8	4	8	400	MH
	Forecasting projects are not as planned	E2	8	4	5	208,175	L-M
	Project completion is not on time	E3	9	2	8	208,175	L-M
	There is no project progress report	E4	8	2	9	208,175	L-M
	Overtime is not well planned	E5	8	3	8	208,175	L-M
Source	Raw materials that reject from suppliers	E6	10	3	9	300	M
	Raw materials do not match the order	E7	8	3	9	208,175	L-M
	Error choosing suppliers	E8	7	3	9	208,175	L-M
	Error in inspection process when item arrives	E9	9	2	9	208,175	L-M
	Supplier cannot fulfill order quantity	E10	7	2	8	208,175	L-M
	Raw materials available in the warehouse can no longer be used	E11	6	2	8	133,155	L
	Late issuance of purchase order letters	E12	7	2	8	208,175	L-M
Make	Delay in production schedule	E13	7	4	4	133,155	L
	Damage to the production machine	E14	5	3	5	24,5	VL
	Raw materials that are not available when they will be produced	E15	8	5	6	300	M
	Occupational accidents	E16	8	2	4	75	VL-L

Table 7. Result of Supply Chain Risk Defuzzification

Major Process	Risk Event	Code	Severity	Occurence	Detection	FRPN	Category
Make	The existence of reject products	E17	9	3	5	133,155	L
	Lack of raw materials	E18	7	5	5	208,175	L-M
	Lack of labor in the production section	E19	6	3	3	75	VL-L
	The production process is not going well because of additional projects	E20	6	5	5	133,155	L
Deliver	Delay in product delivery to customers	E21	3	4	8	75	VL-L
	Delay arrival of raw materials from suppliers	E22	3	3	7	24,5	VL
	The product is damaged while on the trip	E23	2	2	9	24,5	VL
	Limitations on transportation equipment for product delivery	E24	2	2	9	24,5	VL
Return	There are additional costs when the process of returning raw materials to suppliers	E25	3	4	7	75	VL-L
	Delay in the arrival of raw materials from suppliers	E26	7	4	7	300	M
	Suppliers who do not receive returned raw material	E27	3	2	9	75	VL-L
	Delay in the process of returning reject products from customers	E28	3	3	8	24,5	VL
	Delay in the process of sending replacement products to customer	E29	4	3	7	75	VL-L

The next stage is the risk evaluation stage using pareto diagrams and categorization. Here Figure 1. is a pareto diagram of supply chain risk:

IOP Conf. Series: Materials Science and Engineering **909** (2020) 012085 doi:10.1088/1757-899X/909/1/012085

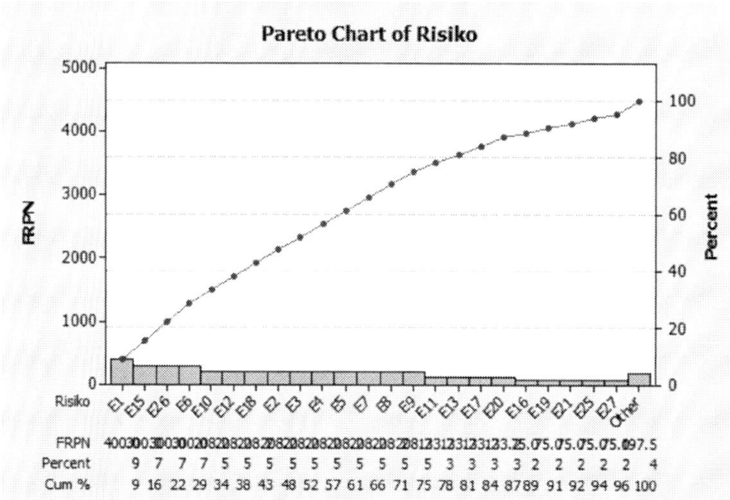

Figure 1. Risk Pareto Chart

Based on the Pareto risk of supply chain diagrams, categories can be divided based on the 80/20 priority value classification, where the priority of supply chain risk is around 80% of the problems caused by 20% of the causes. There are 16 priority risks included in the 80% are E1, E15, E26, E6, E10, E12, E18, E2, E3, E4, E5, E7, E8, E9, E11 and E13.

Then categorize the data into high categories based on 16 risk events. The following Table 8. is the results of determining supply chain risk categories:

Tabel 8. Penentuan kategorisasi risiko

No.	Code	FRPN	No.	Code	FRPN
1	E1	400	9	E9	208,175
2	E2	208,175	10	E10	208,175
3	E3	208,175	11	E11	133,155
4	E4	208,175	12	E12	208,175
5	E5	208,175	13	E13	133,155
6	E6	300	14	E15	300
7	E7	208,175	15	E18	208,175
8	E8	208,175	16	E26	300
Total			3648,06		
Average			228,004		
Standard Deviation			66,862		
Low Category			X<161,142		
Moderate Category			161,142 ≤ X < 294,866		
High Category			X ≥294,866		

Calculation example:

High category
$= X \geq (\mu + 1{,}0\sigma)$
$= X \geq (228{,}004 + (1 \times 66{,}862))$
$= X \geq 294{,}866$

The risks included in the high category are the determination of the number of project requests/inappropriate tonnage (E1), raw materials that reject from suppliers (E6), raw materials that are not available when produced (E15) and late arrival of raw materials replacing from suppliers (E26).

Then look for the causes of risk for 4 risk events to make it easier to mitigate risk. For the risk of determining the number of project requests / inappropriate tonnage (E1) the cause is the absence of a forecasting process carried out by related parties regarding project demand / tonnage for each month. In the event of risk of raw materials that reject from suppliers (E6) the cause is that supplier performance is not good. In the event of a risk of raw materials not available when it is being produced (E15) the cause is the lack of a process for checking raw materials to be produced. Then for the incident of the risk of late arrival of raw materials in lieu of supplier (E26) the cause is a lack of coordination between the company and the supplier.

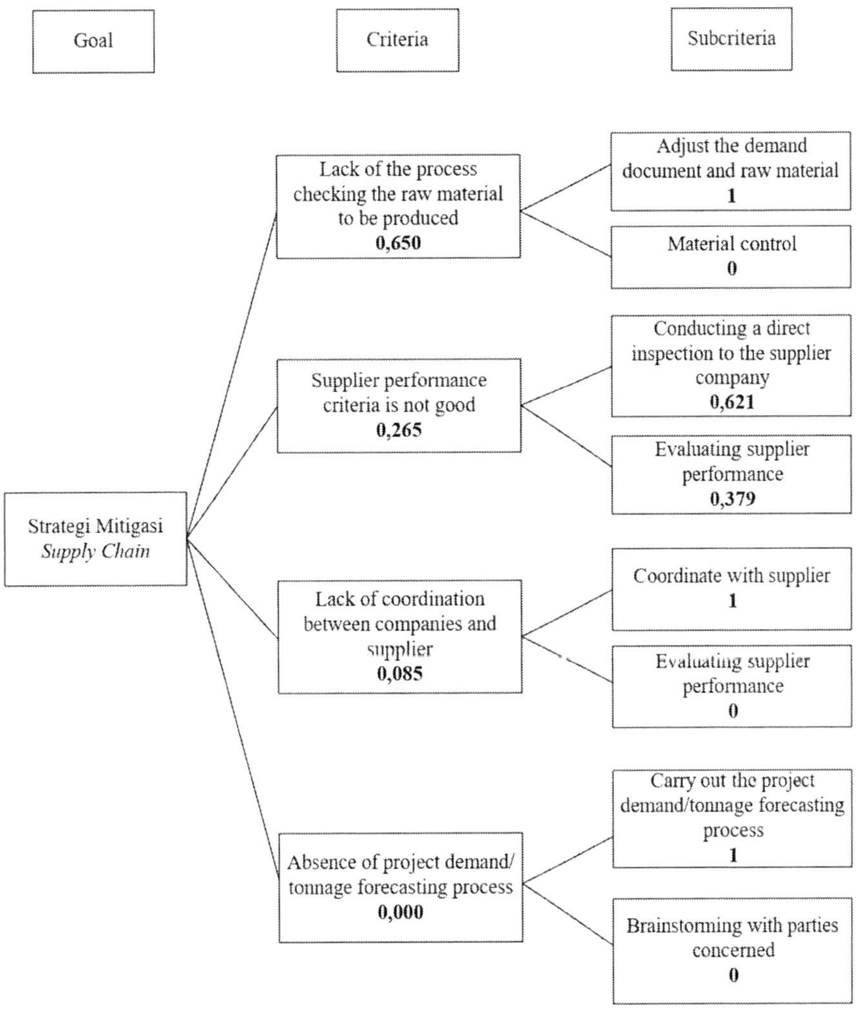

Figure 2. Criteria and subcriteria weighting

The next step is to carry out supply chain risk mitigation strategies using the FAHP method based on 4 risk causes. In Figure 2. is the result of weighting criteria and subcriteria that have been determined and assessed by the parties concerned.

The criteria are filled with the causes of risk and the subcriteria section is filled with risk mitigation steps. On the criteria for the absence of a project / tonnage demand forecasting process, the mitigation strategy undertaken is to carry out the project demand / tonnage forecasting process and brainstorm with the parties concerned. On the criteria of supplier performance is not good, the mitigation strategy that is carried out is evaluating supplier performance and conducting direct inspections to supplier companies. On the criteria for the lack of checking the raw material to be produced, the mitigation strategy undertaken is adjusting the demand documents and available raw materials and material controls. On the criteria for lack of coordination between companies and suppliers, the mitigation strategy undertaken is evaluating supplier performance and coordinating with suppliers.

On the criteria and subcriteria that have been determined, then the next step is to calculate consistency. The criteria for supply chain mitigation strategies have a CR value of 0.075. This CR value is less than 0.1, so the pairwise comparison matrix criteria are consistent. In the sub-criteria of the absence of project demand / tonnage demand forecasting processes, supplier performance sub-criteria are not good, sub-criteria lack of raw material checking process to be produced and sub-criteria lack of coordination between company and supplier have a CI value of 0, then pairwise comparison criteria are consistent.

Based on Figure 3, it can be seen that the criteria which have the highest weight are the lack of the process of checking the raw material to be produced with a weight = 0.650. The mitigation strategy carried out on the criteria is the lack of the process of checking raw materials to be produced is to adjust the demand documents and raw materials available with weights = 1. For the supplier performance criteria is not good to have a weight value = 0.265 with the mitigation strategy carried out is conducting direct inspection to supplier company with weight = 0.621. The criteria for lack of coordination between companies and suppliers have a weight value = 0.085 with the mitigation strategy carried out is to coordinate with suppliers with weights = 1. The lowest weighting criteria is the absence of project demand / tonnage forecasting processes with weights = 0,000 and mitigation strategies carried out is to carry out project demand / tonnage forecasting processes with weights = 1.

4. Conclusion

Based on the results of the research that has been done, the conclusions are as follows:

1. Risk identification occurring in the supply chain at PT XYZ are 29 risk events. In the plan process there are 5 risk events, the source process has 7 risk events, the make process has 8 risk events, in the deliver process there are 4 risk events and in the return process there are 5 risk events.
2. Priority risks in the supply chain at PT XYZ are 4 risk events, which are the risk of determining the number of project requests/inappropriate tonnage (E1), the risk of raw materials rejecting suppliers (E6), the risk of raw materials not available when they are to be produced (E15) and the risk of late arrival of raw materials in lieu of supplier (E26).
3. Mitigation strategies carried out on priority risks that occur in the supply chain at PT XYZ, namely for priority risk causes the lack of raw material checking process to be produced is to adjust the demand documents and available raw materials with a weight value of one (1). The mitigation strategy for the cause of risk is the supplier's poor performance priority is to conduct an inspection directly to the supplier company with a value of 0.621. The mitigation strategy that causes the risk of lack of coordination between companies and suppliers is coordinating with suppliers with a weight value of one (1). The priority risk mitigation strategy for the absence of a project demand / tonnage forecasting process is to carry out the project demand / tonnage forecasting process with a weighting value of one (1).

5. References

[1.] He, J, F Alavifard, D Ivanov, Jahani H. (2018). The International Journal of Management Science, DOI: 10.1016/j.omega.2018.08.008

[2.] Scheibe, K.P., Blackhurst, J. (2018). International Journal of Production Research, 56(1-2), 43-59.

[3.] Hassan, A., Purnomo, M.R.A., & Anugerah, A. R. (2019). Journal of Engineering, Design and Technology , vol. 18 no. 2. /doi.org/10.1108/JEDT-05-2019-0131

[4.] Roh, S. Y., Shin, Y. R., & Seo, Y. J. (2018). The Asian Journal of Shipping and Logistics, 34(4) (2018) 297-307

[5.] Butdee, S., Phuangsalee, P., (2019). Global Congress on Manufacturing and Management (GCMM – 2018). Procedia Manufacturing 30 (2019) 663 – 670
[6.] Braaksma, A. J. J., Klingenberg, W., & Veldman, J. (2013). International Journal of Production Research, 51(4), 1055–1071. https://doi.org/10. 1080/00207543.2012.674648.
[7.] Kou G., Chao X., Peng Y., Xu L., Chen Y. (2017). Studies in Informatics and Control, 26(2):131-142.
[8.] Nie, R.-x., Tian, Z.-p., Wang, X.-k., Wang, J.-q., & Wang, T.-l. (2018). Knowledge-Based Systems
[9.] Lücker, F., Seifert R.W. & Biçer I. (2019). International Journal of Production Research, 57(4), 1238-1249.
[10.]Ridwan, A., Santoso, M. I, Ferdinant, P.F., & Ankarni R. (2019) Materials Science and Engineering. Volume 673, Number 1, 673 012088, Desember 2019. DOI:10.1088/1757-899X/673/1/012088
[11.]Trenggonowati, D. L., Ridwan, A., & Anggraini, Z. D. (2018). Materials Science and Engineering. Volume 218, 04021, 2018. doi.org/10.1051/matecconf/201821804021
[12.]Ulfah, M., Trenggonowati, D.L., Ekawati, R., Ramadhania S. (2019) Materials Science and Engineering. Volume 673, Number 1, 673 012082, Desember 2019. DOI:10.1088/1757-899X/673/1/012082
[13.]Gargama, H., & Chaturvedi, S. K. (2011).. IEEE Transactions on Reliability, 60(1), 102–110.
[14.]Wang, Y.-M., Chin, K.-S., Poon, G. K. K., & Yang, J.-B. (2009).. Expert Systems with Applications, 36(2, Part 1), 1195-
[15.]Nazeri, A., & Naderikia, R. (2017). The International Journal of Advanced Manufacturing Technology, 92(9-12), 3749-3783.
[16.]Saaty, T.L. (1980) The Analytic Hierarchy Process, New York: McGraw Hill. International. (1996, 2000), Pittsburgh: RWS Publications.

IOP Conf. Series: Materials Science and Engineering **909** (2020) 012086 doi:10.1088/1757-899X/909/1/012086

Increased productivity using *lean service* (Case study: regional drinking water company x)

Kulsum [1,2*] **, Ade Irman**[1,2]**, Anwari. A**[1]

[1] Industrial Engineering Department, Engineering Faculty, Sultan Ageng Tirtayasa University

[2] Indonesia Center of Excellence for Food Security (Local Food Innovation), Sultan Ageng Tirtayasa University

*Email: kulsum@untirta.ac.id

Abstract. Constraints found at the stage of requesting new connection services in regional drinking water company X is the slow administration and process of connecting pipes. This is following the average length of the new connecting service time of 8 working days from the target of 6 working days. The existence of these obstacles affects the productivity of the company. One indicator of increased productivity can be seen from the value of the Process Cycle Efficiency (PCE). The results of the PCE evaluation before the proposed improvement obtained the results of the current state PCE value of 14.98% and the percentage of PCE after the proposed improvement of 25.46%. after the proposed improvement is carried out, it reduces the processing time of the New Connection which was originally 10242 minutes to 6013 minutes.

1. Introduction

Regional Drinking Water Company x is a regional company that has a concentration in the field of water supply services. The presence of a Regional Water Supply Company is regulated through the Law of the Republic of Indonesia No.5 of 1962 concerning Regional Enterprises. Regional Drinking Water Company is formed as a business unit owned by the regional government that provides services, organizes public benefits in the field of drinking water, and fosters revenue. So as a company, the Regional Drinking Water Company is required to have a performance-oriented to customer satisfaction, to have competence in the procurement and distribution of clean water, and to be efficient to obtain revenue.

Productivity is an important factor affecting the continuity and development company. Companies must be able to increase output by reducing or save input. Output produced by the company influenced by the waste in-process production[3]. Regional Water Supply Company X has 4 types of services, namely: new connection, renaming, re-opening of channels, and connection revocation. With most services, there are new connecting services from October 2018 to April 2019 with 832 SL (Channels). Regional Water Supply Company x must maintain good service consistency with customers, one of which is the speed of service. As the number of customers increases, several complaints are still often raised by customers at the request stage of new connection services, such as the slow process of administration and the pipe connection process. This is following the average length of the new connecting service time of 8 working days from the target of 6 working days. This has become a major challenge for the Regional Water Company in terms of administration and planning to improve service performance to overcome the increased water consumption. Service firms need to meet the specific requirements of each and every customer so that they can alwaystay ahead of their competitors [10]. But in practice, the service process offered is not always smooth or in accordance with the company's expectations. Not a few things that the company does not want and the consumer-related quality of service occurs.[6]

The service improvement step to improve performance is the application of the Lean concept to the Regional Water Company. Lean is lean because it provides a way to do better by using as little resources as possible - human effort, less equipment, less time and less space (less time) less space) by continuously approaching the desires that are desired by the customer [2]. Services are all actions or performance that can be offered by one party to another party which is essentially intangible and does not produce any ownership [7]. In this case, because the research was conducted in the service sector, the lean approach taken was lean service. The development of lean service concept is used by researchers to implement the concept of lean service in fast food restaurants, especially in the food company Jank Jank Wings to identify the waste that occurs in every process in this company and the process of improvement will be used the DMAIC method (define, measure, analyze, improve and control). In its development, the concept of lean service is still evolving and is examined in terms of the rationale and application of lean concepts into a service organization sector in terms of the foundation of existing lean services and analysis of withdrawals. the concept of the customer becomes a good service system [1]. Lean service is used as an approach to creating an effective internal service system so that important information can be ascertained to consumers or the public quickly and with effective service. Lean is an ongoing effort to eliminate waste and increase value-added products (goods and or services) to provide value to customers (customer value)[5]. By analyzing activities in the service process of the Regional Water Company, xi will find out which activities are included in the category of value-added activities and non-value-added activities. An analysis will be made of the service process at the Regional Water Company from the beginning to the end, to identify the waste contained in the Regional Water Company. Which is then analyzed and solved by applying lean concepts. Next, determine the proposed improvement using the 5W + 1H method. Therefore, in this study, the Lean service approach is used to make the Regional Water Company maintain its commitment as a regional company that has a concentration in the field of water supply and public service by the Law of the Republic of Indonesia No.5 of 1962 concerning Regional Enterprises.

2. Methodology

Service is the right of activity or series of invisible activities that occur due to interactions between consumers and employees or other rights provided by service providers that are intended to solve the problems of consumers or customers [9]. The service itself contains the importance of the meaning of communication. Lean is a continuous effort to eliminate waste or waste and increase the value-added of goods and services to provide value to customers or customer value. The purpose of lean is to increase customer value through a continuous increase in the ratio between value-added to waste (the value-to-waste ratio)[3]. Therefore, efforts to improve the production process always continue carried out with a systematic approach to identifying and eliminating waste or non-value added activities through radical continuous improvement to achieve six levels of performance to increase productivity [3].

Lean focuses on the goal of continuously turning waste into value from the prospective customer. This is a systematic approach precisely for the improvement process, error proofing, and reducing waste [4]. Lean service in other definitions is a set of methods and equipment designed to reduce costs, improve performance, reduce waiting times, and eliminate waste so that the quality of services produced can be maximized. Lean has also been applied to service companies to eliminate waste and to improve service to consumers by research [8]. The concept of lean, in standardization and continuous improvement procedures, is the basic thing in the continuity of the service process itself to improve performance in a company.

Waste can be defined as any work activity that does not add value to the process of transforming inputs into outputs along with the value stream mapping. From a lean perspective, all types of waste found throughout the value stream process, which transforms inputs into outputs must be eliminated to increase the value of products (goods or services) and subsequently increase customer value [5]. The type of waste in manufacturing is certainly different from the waste in the company or the service sector here is a general description of waste in services.

Table 1. Waste in service

No	Waste Service	Definition
1	Over production	The process of doing excessive service to consumers without considering the reciprocity
2	Waiting	The process is waiting for information to arrive or waiting for service order procedures
3	Defects	Error in the process of implementing a service
4	Excessive Transportation	Excessive displacement of the service process to consumers resulting in waste of time, effort and cost
5	Excess Inventory	Storage of service inventories or service procedures is too early than customer service needs
6	Inappropriate processing	Incompatibility of service methods to consumers
7	Unnecessary Motion	The treatment process is less effective and efficient for consumers

The 5W + 1H method (Kipling Method) has been used extensively and is considered an effective approach to gathering and presenting the information. 5W + 1H contains 6 basic questions in getting information: what, where, when, why, who, and how. A 5W + 1H analysis is carried out to explain the explanation in detail.

3. Result and Discussion

The data collected in this study consisted of primary and secondary data. Primary data is collected directly through observation activities, namely direct observation in the company that is the object of research and interviews with related parties the process of new connection services in the Regional Water Company x. Secondary data in the form of a company profile, organizational structure, service process flow, and other data from various references that support research.

Table 2. New connection service process

No	Process
1	New Connection Request
2	Form Processing
3	Cashier (Payment of Fees)
4	File Processing
5	Connection Installation
6	Retrieval File

From the data obtained, it can be seen that the recapitulation of new connection service waste is as follows:

Table 3. Recapitulation of new connection service waste

No	Waste	Weight			
		Head of Meter Installation	Head of Customer Development	Planning	Head of Meter Reading
1	Overproduction	4	4	5	7
2	Waiting	6	7	8	6
3	Defect	4	5	3	5
4	Excessive Transportation	8	6	6	5
5	Unnecessary Inventory	5	4	4	5
6	Inappropriate Processing	3	5	3	4
7	Unnecessary Motion	5	4	6	3
	Total	35	35	35	35

IOP Conf. Series: Materials Science and Engineering **909** (2020) 012086 doi:10.1088/1757-899X/909/1/012086

After getting the value of each waste that is there will be an appropriate tool selection using the Value Stream Analysis Tool. The value of each tool is obtained by multiplying the waste value in the results of the recapitulation of the questionnaire by the weight value in the Value Stream Analysis Tool table or VALSAT.

Table 4. Recapitulation of VALSAT calculation

Waste	Means Score	Process Activity Mapping	Suplay Chain Response Matrix	Production Variety Mapping	Quality Filter Mapping	Demand Aplication Mapping	Decision Point Analysis	Phsysical Structure A. Volume B. Value
Overproduction	5	5	15		5	15	15	
Waiting	6.75	60.75	60.75	6.75		20.25	20.25	
Defect	4.25	4.25			38.25			
Excessive Transportation	6.25	56.25						6.25
Unnescesarry Inventory	4.5	13.5	40.5	13.5		40.5	4.5	4.5
Innappropriate Processing	3.75	33.75		11.25	3.75		11.25	
Unnescesarry Motion	4.5	40.5	4.5					
Total	35	214	120.75	31.5	47	75.75	51	10.75
Rank		1	2	6	5	3	4	7

Based on the results of VALSAT calculations that have been done, it is found that the tool with the highest weight is Process Activity Mapping (PAM). Therefore the tool used to map activities in the new connecting service process is Process Activity Mapping.

The following is a recapitulation of the results of the current state and future state based on the value of the process cycle efficiency calculation:

Table 5. Summary of current state and future state comparison results

Category	Current State		Future State	
	Time	Persentace	Time	Persentace
VA	1534	14.98%	1531	25.46%
NVA	7252	70.81%	4347	72.29%
NNVA	1455	14.21%	135	2.25%
Total	10241	100%	6013	100%
PCE	14.98%		25.46%	

Then, PCE (*Value Added*) = PCE*future* – *PCEcurrent*
= 25.46 % - 14.98 %
= 10.48 %

It can be seen that there is an increase in efficiency in the process cycle. The amount of time VA (Value Added) in the current state condition is 534 minutes with a percentage of 14.98% and in the future state is 1531 minutes with a percentage of 25.46, the amount of time NVA (Non-Value Added) in the current state condition is 7252 minutes with a percentage of 70.81% and at a percentage of future state condition is 4347 minutes with a percentage of 72.29%, the amount of NNVA (Nacassery Non-Value Added) in the current state condition is 1455 minutes with a percentage of 14.21% and in the future state condition that is 135 minutes with a percentage of 2.25%. based on the Process Cycle

Efficiency value above, it is known that the efficiency of the new service process time has increased by 10.48%.

4. Conclusion

Based on the results of data processing and analysis that has been done, it can be concluded several conclusions include: Based on the weighting on the waste questionnaire obtained Value Added (VA) activity of 14.98% with a total time of 1534 minutes from 26 processes, the percentage of non-value-added NVA of 70.81% with a total time of 7252 minutes from 17 processes, and the percentage of necessary non-value-added (NNVA) of 14.21% with a total time of 1457 minutes from 9 processes. Critical waste results obtained in the New Connection Service Process are waste Waiting, waste Excessive Transportation, and Overproduction waste. Proposed improvement of 5W + 1H to reduce lead time by adding displays related to requirements, process flow, and all matters relating to new connection services and keep updating on new connection service process activities. Making HR planning for location survey officers will reduce some of the time of location survey delays, Updating pipeline data. With the installation of pipeline network data, it can be easier for officers, combining the activity of checking connection requirements, reducing the activity of printing documents as well as submitting the Official Report and RKB (Plans for Goods) and reducing the waiting time of the installation process queue. After the proposed improvement with the approach of 5W + 1H in the New Connection Service process, it was able to discuss the processing time of service which was originally 10241 minutes after the proposed improvement to 6013 minutes or reduced by 4228 minutes. Based on the proposed improvement of 5W + 1H, there is an increase in productivity from the value of PCE current state is 14.98% and the percentage of PCE after the proposed improvement is 25.46% or an increase of 10.48%.

5. References

[1] Arfmann, D. & Barbe, D. F. G. T. 2014 The value of lean in the service sector: a critique of theory & practice *Iternational Journal of Business and Social Science* **5** 18-24

[2] Dima, Lintya 2015 Use of the lean concept to improve the efficiency of outpatient pharmacy installation services at Bekasi's Anna Medika Hospital *Faculty of Public Health, University of Indonesia* **2**

[3] FaritsyA.Z.A, Suseno 2015 Increasing productivity company with using six sigma, lean and kaizen methods *Journal of Industrial Engineering* **10**

[4] Febianti E., Muharni Y, Mustikawati E.A. 2018 Lean healthcare approach and simulation forminimizing road service time at rsud x *Journal Industrial Servicess* **4**

[5] Gasperz, V., dan Fontana, A 2011 *Lean Six Sigma for Manufacturing and Service Industries* Bogor: Vichisto Publication

[6] J Hidayati, U P P Tarigan and U Tarigan 2020 Implementation of lean service to reduce lead time and non value added activity in a banking institution *IOP Conf. Series: Materials Science and Engineering* **505** 012076

[7] Kotler, K 2009 *Manajemen Pemasaran 1 Edisi ketiga belas* Jakarta: Erlangga

[8] Mollah Kalam, Munir Misbahul dan Sari Wulan 2011 improved service quality with the lean service approach method in transportation service companies (case study: PT. KAI DAOP 9 Surabaya) *Department of Industrial Engineering Surabaya Adhi Tama Institute of Technology National Seminar on Applied Science and Technology VI 2018*

[9] Ratna, Suminar. Et all 2017 Excellent services for parents of students in sempoa sip tc paramount summarecon *Pamulang University Secretariat Study Program* **4**

[10] Vignesh V, M.Suresh and S. Aramvalarthan 2016. Lean in service industries: A literature review *IOP Conf. Series: Materials Science and Engineering* **149** 012008

The ergonomic factor application for improvement of performance office staff

Hernadewita [1,*], Hendra [2], R. Kristianti[1], I. Asih[1], Dhimas S.[2], and E.N.S. Yuliani[1]

[1]Industrial Engineering Dept. University of Mercubuana, Indonesia

[2]Mechanical Engineering Dept. University of Sultan Ageng Tirtayasa Banten, Indonesia

Jl. Tedja Buana Building Menteng, Jakarta, Indonesia

*Email: hadeita@yahoo.com

Abstract. Comfortable workplace, healthy and safely is the goal of implementing occupational safety and health requirements. Ergonomics is one of the factors that can affect employee productivity. Achieving harmony between labor, equipment, environment, work methods and work processes is the most important application of ergonomics. The purpose of this study is to investigate the improvement of working conditions through the application of ergonomics in the workplace. Ergonomic testing is carried out using the Rapid Upper Limb Assessment (RULA) analysis method to evaluate worker posture on work machines or operating systems and to investigate reduction in upper extremities and Ergonomic Factor Policy Identification (SHORT), which is an ergonomic evaluation method that focuses on frequency, duration, weight and posture to identify ergonomic hazards faced by everyday workers using an assessment system. The results of ergonomic application analysis are on average at a low risk level. Improvements can be made to reduce the level of risk experienced by employees in accordance with ergonomic standards, including improving the design of tables and chairs, adjusting work positions, avoiding odd work positions, stretching muscles while doing work at the computer.

1. Introduction

Computer usage has increased dramatically in the last 20 years. Research shows that 56% of workers use computers while at work. Work-related musculoskeletal disorders (WRMDs) especially involving the neck area are common. Loss of productivity and lack of workers due to musculoskeletal pain complaints can be detrimental to employers. RULA is a commonly used tool for assessing the risk of posture for continuous work in a sitting position, in this way scores are calculated for each body part, among computer users RULA scores produce a pain level of 2, with 52% neck pain and lower back 52%, scores high mobility is associated with high neck pain complaints [1]. Computer users spend a lot of time sitting in static positions on computers, and many suffer from musculoskeletal problems. Office work using a computer with typing activities too long without rest can also cause pain when hands are used. This is known as carpal tunnel syndrome, other pain complaints in the fingers or wrists, the use of ergonomic computer tools and the mouse can also be used to reduce the risk of pain complaints while working with the computer.

The most common factor in computer work is typing frequency, repeated head movement from the keyboard to the monitor where more than 10 times in 1 (one) minute is included in repeated work.

Moreover, this is done over a long period of time, which can result in muscle and skeletal disorders (musculoskeletal disorders) due to static sitting postures in front of the computer. If this type of activity is carried out continuously it can lead to fatigue and injury.

Ergonomic science has been used to solve this problem to get minimum effect on this activity. Ergonomics is the study of human aspects in its work environment such as anatomy, physiology, psychology, engineering, management and design. Ergonomics requires a systematic study for people, workplaces and environments interact with the main purpose of adjusting the working environment with people. Ergonomic may affect in the workforce activity due to mismatch between workplaces (work methods, job placement, tools) and lifting of the workforce [2-4]. Potential ergonomic hazards based on Permanent Number 5 in 2018 include unsuitable work methods, work positions and posture, job design and workplace inconsistencies in employee anthropometry and lifting workloads. Complex interactions between aspects of work (work equipment, work procedures, processes or work systems and work environment with physical, and psychological/human/labor conditions) to tailor aspects of work to conditions in the field, so that they can work safely, comfortably, efficiently, and be more productive.

Office ergonomics is important to note because it is one of the potential dangers and risks that threaten employees in the office. Ergonomic hazards (biomechanical hazards) can come from work design, layout, or bad work. Ergonomic hazards are divided into three type namely work-related hazards such as consisting of duration (consist frequency, load, work order, work priority, and work posture), equipment-related hazards (consisting of dimensions, shape, design shape, and placement of facilities are used to support work) and hazards related to the environment or workplace (consisting of dimensions, areas and layout of the workplace). If the workplace, equipment, and work environment are not well planned, a variety of office effects can occur, such as eye irritation and fatigue (Astenophia) and neck muscle tension (headaches, frozen shoulders).

High frequency of using a computers that do not pay attention to the ergonomic side of the workplace results in a perceived risk to users. According to Watchman users experience excessive fatigue such as, headache, pressure, neck tension, back, arms, shoulders, muscle pain, and other parts directly related to computer work [5]. Static and continuous seating can increase the risk of pain in some parts of the body, as sitting can exert pressure and inhibit the flow of the body and thus reduce the nutrients that the joints absorb [6]. According to Watchman, complaints of high use of computers in the workplace occur due to problems with equipment or facilities, work layout, work environment conditions, or a combination of factors. Improper work layout influences may cause someone to experience poor work posture.

Office work has a very minimal variation of work movement, leading to static and prolonged work position. Working in the office includes work that requires a little muscle movement, but this type of muscle contraction can cause pain if sustained for a long time because the muscles will feel tense.

Government agencies that uses computers as one of the main tools in carrying out daily office activities. In this government almost everyone has a computer on their desk, with a daily use time of 8 hours. The duration of use of a computer can cause discomfort in the eyes, back, hands and feet. The state of the art work of computer workers requires ergonomic work chairs, comfortable chairs that fit the body posture, can be rotated in the right direction, a soft foam base, high and low chair rules. Proper ergonomics can increase productivity. Long sitting position in front of computer can cause pain in back and neck. It is also found that cumulative trauma disorder (CTD) is caused by excessive pressure on the top of the computer user [7], other complaints of discomfort in the neck, right shoulder, upper limbs or upper arms, wrists or fingers [8-9]. Not aware that their work position is awkward, they do not see the connection between the lower arm and upper arm position while operating the keyboard and the appearance of pain in the shoulder joint [10].

Some researcher show that 78.6% of computer workers experience MSD complaints influenced by Body Mass Index (BMI), duration of work, perception of work pressure and body posture, equipment risk factors contributing to MSD including chair height and chairs, working hours, rest periods, work pressure factors [11].

Office ergonomics is an application of ergonomics that covers the entire work environment and tools used such as computer equipment and chairs [12]. The application of ergonomics in the office focuses more on the dangers of using computers. Office hazards are generally caused by improper work posture, repetitive movement and long-term standing. The dangers of working in the office are also affected by the equipment used, including the mouse, keyboard, monitor, desk and computer chair. Each of these equipment has a prerequisite ergonomic condition, so that the user can use it comfortably. Using the mouse pad can reduce Dorsiflexion, the wrist belt on the mouse can provide comfort to the wrist and thus maintain a neutral hand posture [13]. Long-term use of computers can also cause eye irritation (computer vision syndrome/CVS). Works in front of computers for long periods of time, computer use without UV protection and lens use gaps make CVS worse [14]. CVS is characterized by dry and irritated eyes, eye fatigue, blurred vision, hot eyes, watery eyes, multiple eyesight, headache, sensitivity to light and changes in focus [15].

To determine and identify work postures of people using computers, a test is required, Rapid Upper Limb Assessment (RULA) and The Baseline Risk Identification of Ergonomic Factor (BRIEF). This method are used to measure risk factors for musculoskeletal disorders in the neck and upper body. The main focus of RULA assessment in detail is shoulder / upper arm, elbow / lower arm, wrist, neck and waist. RULA also considers the burdens and displacements made in its assessment. RULA also assesses whether the foot position is stable or not. RULA was developed by McAtamney and Corlett United Kingdom in 1993 [16]. The RULA Method demonstrates the effectiveness of ergonomic intervention on computer operators experiencing WRMSD complaints, Rapid Upper Limb Assessment (RULA) is a method of assessing posture, style, and movement of work activities related to use upper limbs (upper limbs) [17]. The Baseline Risk Identification of Ergonomic Factor (BRIEF) test can identify employee work postures that can be followed from the results of identifying work posture to minimize employee risk. BRIEF review is focuses on the frequency, duration, load and posture of the body to identify the ergonomic hazards faced by everyday workers using the rating system. In this study to test work position, components / office equipment, strength / pressure, repetition and office environment RULA and BRIEF are used in this research.

2. Method

Ergonomic method in this research using Rapid Upper Limb Assessment (RULA) and The Baseline Risk Identification of Ergonomic Factor (BRIEF) analyse to solve problem in the Office work. RULA was developed by Mc Atamney and Corlett United Kingdom in 1993 [16]. RULA calculates the ergonomic risk factors in a job where the job is to do a lot of work in a sitting or standing position without switching. The RULA method demonstrates the effectiveness of ergonomic intervention on computer operators experiencing WRMSD complaints. RULA is a method of assessing posture, style, and movement of work activities related to use upper limbs (upper limbs). This method was developed to investigate the risk of informalities experienced by an employee in carrying out work activities that employ top members. RULA calculates risk factors in the form of posture, energy / load, static work and repetition of work. The RULA method is a method of measuring upper body posture that is easy to understand and easy to implement because it has provided guidance on the evaluation procedure of each measured posture. This method also does not take a long time to evaluate. In addition, this method can also measure other ergonomic risk factors in the form of power / load, repetition and static time / work. The Baseline Risk Identification of Ergonomic Factor (BRIEF) test can identify employee work postures that can be followed from the results of identifying work posture to minimize employee risk. BRIEF Review is focuses on the frequency, duration, load and posture of the body to identify the ergonomic hazards faced by everyday workers using the rating system.

The BRIEF method uses three steps in its assessment, namely the assessment of ergonomic risk factors in the work environment, symptom review and medical examination results [18]. This method uses body posture charts and assessment tables to provide an assessment of the risk factors that employees will experience. The risk factors studied in this method are factors described by McPhee 'as external load factors which include amount of movement, static muscle work, strength, work posture determined by equipment and furniture, and restless work hours.

This study was carried out through ergonomic testing using the RULA and BRIEF methods, the investigation was carried out at the Office Staff with samples taken per level floor 2nd to 8th of Government building. The examination is carried out at the time of the official. Tests carried out with direct attention and image capture. Investigation data collection is done by filling out the RULA and

BRIEF assessment sheets. In filling out the RULA assessment sheet, he is assisted by looking at the position of the worker, the state of work ease, taking pictures with a camera and measuring the ease of work. Figure 1 show the workplace ergonomic and seating guide in the office.

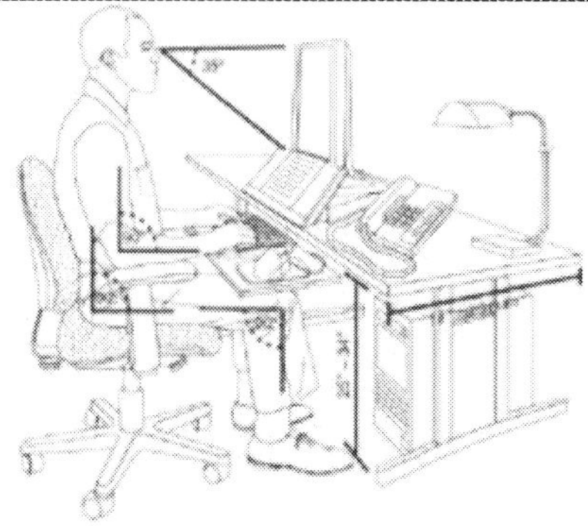

Good posture

Source: : Workplace Ergonomics Reference Guide

Good posture

Source: Seating Guide, Worksafe British Columbia 2009

Figure 1. Workplace Ergonomics and seating guide [10]

3. Result and Discussions

Analysis of the results of ergonomics testing using the RULA and BRIEF methods based on the results of the analysis of the Laboratory of the Directorate of Occupational Safety and Health as shown in Fig. 2 and 3. Figure 2 show the position working place of Ministry, Secretary and Employee in 2nd Floor. In Ministry Room no recommendation for improvement BRIEF due to support at the back and shoulder/arm is well and good. No risk in of wrist disturbance so this place it's good for activity. In the administration place there was no improvement but it ensure writing activity was separate from the operation area of computer. They have BRIEF for back is not support properly, have risk disturbance wrist, neck not likely to result in stiffness in the neck muscles.

Similarly with Administration place, General Secretary have BRIEF condition not supportedproperly such as rear posture is not supported properly, as personnel must reach the computer mouse, shoulder/arm position is not appropriate to the height of the table, the reach is too far, keyboard position is uneven and far from the wrist, risk of wrist disturbance. The legs are imperfect so the leg posture is bent inward. The recommendation is the writing work is separated from the computer desk, so there is enough space and reach for the hands, provides stitches/ keyboards that allow the wrist to be in neutral position as the keyboard is at the elbow level, re-position the monitor so that the horizontal operator can see straight, directs to rest periodically by moving the wrist to reduce static posture, slowly bend / bend your neck forward, as well as bend back and forth to relieve muscle cramps, put the foot down and provides training on ergonomic hazard awareness.

For other floor such as 3-6th floor they have similar case or observation with the Administration and Secretary Room. For 3th Floor Data and Information centre the recommendation is similar with the Secretary Room provides stitches/keyboards that allow the wrist to be in neutral position as the keyboard is at the elbow level, directs to rest periodically by moving the wrist to reduce static posture, move the cabinet under the table, so that there is enough leg space, slowly bend/bend your neck forward, as well as bend back and forth to relieve muscle cramps, put the foot down, provides training on ergonomic hazard awareness. Biro KLN Room have no recommendation for improvement.

IOP Conf. Series: Materials Science and Engineering **909** (2020) 012087 doi:10.1088/1757-899X/909/1/012087

Picture	Observation Results	Risk Value Legend	Recommendations for Improvement
	2nd Floor THE MINISTRY ROOM		
	BRIEF: 0 **The back** → Back-up posture is well supported at the back, so personnel feel comfortable when reaching for letters / books or phones at work desks. **Shoulder / arm** → Natural shoulder / arm position according to the height of the table, good reach. **Wrist** → Position of objects flat and adequate from wrist **Wrist posture for dynamic writing** → There is no risk of wrist disturbance. The foot is perfect so the posture does not bend inside. **Neck** → neck posture for the front (flexion) and the front of the spinal disc, not likely to result in stiffness in the neck muscles.		No Improvement
	2nd Floor ADMINISTRATION OF MINISTRY ROOM		
	**BRIEF: 1** **Back** → Rear posture is not supported properly, as personnel must reach the computer mouse. **Shoulder / arm** → Natural shoulder / arm position according to table height **Wrist** → Keyboard position is uneven and far from the wrist **Wrist posture to type firmly** → risk of wrist disturbance. The legs are imperfect so the leg posture is bent inward. **Neck** → neck posture for the front (flexion) and the front of the spinal disc, not likely to result in stiffness in the neck muscles.		There was no improvement, but it ensured writing activity was separate from the operating area of the computer.
	2nd Floor GENERAL SECRETARY ROOM		
	BRIEF: 2 **Back** → Rear posture is not supported properly, as personnel must reach the computer mouse. **Shoulder / arm** → Shoulder / arm position is not appropriate to the height of the table, the reach is too far. **Wrist** → Keyboard position is uneven and far from the wrist. **Wrist posture to type firmly** → risk of wrist disturbance. The legs are imperfect so the leg posture is bent inward.		The writing work is separated from the computer desk, so there is enough space and reach for the hands. Provides stitches / keyboards that allow the wrist to be in neutral position as the keyboard is at the elbow level. Re-position the monitor so that the horizontal operator can see straight. Directs to rest periodically by moving the wrist to reduce static posture. Slowly bend / bend your neck forward, as well as bend back and forth to relieve muscle cramps Put the foot down. Provides training on ergonomic hazard awareness.

Figure 2. Workplace and seating for 2nd Floor in Ministry Room.

In Figure 4 the observation result of 4-6th Floor show that Binapenta Room have similarly case with the General Secretary Room and Data and Information Centre Room. For administration of Sesitjen Room there is no improvement and Binalatas Room have recommendation such as to move the archives under the table, so that there is enough leg space, put the foot down, and provides training on ergonomic hazard awareness.

Picture	Observation Results	Risk Value Legend	Recommendations for Improvement
	3rd FLOOR DATA AND INFORMATION CENTER ROOM		
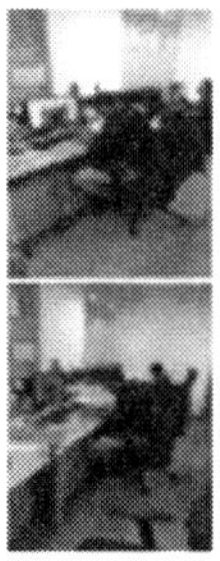	BRIEF: 2 **Back** → Rear posture is not supported properly, as personnel must reach the computer mouse. **Shoulder / arm** → Shoulder / arm position is not appropriate to the height of the table, the reach is too far. **Wrist** → Keyboard position is uneven and far from the wrist Wrist posture to type firmly → risk of wrist disturbance. The legs are imperfect so the leg posture is bent inward and space is limited		Provides stitches / keyboards that allow the wrist to be in neutral position as the keyboard is at the elbow level. Directs to rest periodically by moving the wrist to reduce static posture. Move the cabinet under the table, so that there is enough leg space. Slowly bend / bend your neck forward, as well as bend back and forth to relieve muscle cramps Put the foot down. Provides training on ergonomic hazard awareness.
	3rd FLOOR BIRO KLN ROOM		
	BRIEF: 1 **Shoulder / arm** → Shoulder / arm position is not appropriate to the height of the table, the reach is too far. **Wrist** → Keyboard position is uneven and far from the wrist Wrist posture to type firmly → risk of wrist disturbance.		There is no improvement, but keep writing activities separate from the computer operating area. By moving the archive stack, so that the writing work is done, so that there is sufficient space and reach Put the foot down.

Figure 3. Recommendation workplace and seating for 3th Floor in Information Centre and Biro KLN Room

Picture	Observation Results	Risk Value Legend	Recommendations for Improvement
	4th FLOOR BINAPENTA ROOM		
	BRIEF: 2 **Back** → Rear posture is not supported properly, as personnel must reach the computer mouse. Shoulder / arm → Shoulder / arm position is not appropriate to the height of the table, the reach is too far. **Wrist** → Keyboard position is uneven and far from the wrist **Wrist posture to type firmly** → risk of wrist disturbance. The legs are imperfect so the leg posture is bent inward and space is limited		Provides stitches / keyboards that allow the wrist to be in neutral position as the keyboard is at the elbow level. Directs to rest periodically by moving the wrist to reduce static posture. Move the cabinet under the table, so that there is enough leg space. Slowly bend / bend your neck forward, as well as bend back and forth to relieve muscle cramps Put the foot down. Provides training on ergonomic hazard awareness.
	5th FLOOR, ADMINISTRATION OF SESITJEN ROOM		
	BRIEF: 1 **Back** → The back posture is not well supported on the back. **Shoulder / arm** → Shoulder / arm position is not appropriate to the height of the table, the reach is too far. **Wrist** → Keyboard position is uneven and far from the wrist		No improvements, the room is ergonomic, but ensures adequate lighting for reading and writing. Provides training on ergonomic hazard awareness.
	6th FLOOR, BINALATAS ROOM		
	BRIEF: 1 The legs are imperfect so the leg posture is bent inward and space is limited		There is an improvement, which is to move the archives under the table, so that there is enough leg space. Put the foot down. Provides training on ergonomic hazard awareness.

Figure 4. Recommendation workplace and seating for 4-5th Floor

4. Conclusions

Based on the results of the ergonomic testing using the RULA and BRIEF methods are obtained that in this office have some improvement for writing work is separated from the computer desk, moving some cabinet to have enough space and reach for the hands, adequate lighting for reading and writing, provides stitches/keyboards to be in neutral position at the elbow level, re-position the monitor into the horizontal direction to easily operate and seen, directs to rest periodically by moving the wrist to reduce static posture, slowly bend/bend your neck forward, as well as bend back and forth to relieve muscle cramps and put the foot down.

References

[1] Sharan, D., Mohandoss, M., & Ranganathan, R. (2015). Co-relation of neck Rapid Upper Limb Assessment score with neck pain. August, 1–2.

[2] Yuliani, E. N. S. 2019 Development of Prototype Ergonomic Reflexology Vest (ERV) for Improving the Quality and Performance Features 1st International Conference of Health *Science & Technology* (ICOHETECH)

[3] Yuliani, E. N. S.,Tambunan. E.B.M., Reduce workload of security officers using NASA-LTX, SOFI Heart Rtae and Energy Expenditure Method, International Conference on Design, Engineering and Computer Sciences, IOP Conf. Series Material Science and Engineering 453, 2018.

[4] Yuliani, E. N. S. 2018 Measurement of Student's Mental Workload in Polytechnic using Nasa-TLX, Journal of Scientific and Engineering Research, 5 (12) 289-296.

[5] Watchman, G.R. 1997 Working Safely with Video Display Terminals (Department of Labor, New York, USA)

[6] Callaghan, J. P. and Nadine, M. D. 2002 Examination of The Flexion Relaxation Phenomenon in Erector Spine Muscles During Short Duration Slumped Sitting *Clinical Biomechanics* **17** 353-360.

[7] Goodman G, Kovach L, Fisher A, Elsesser E, Bobinski D, Hansen J. 2012 Effective interventions for cumulative trauma disorders of the upper extremity in computer users: practice models based on systematic review. Work. 2012; 42(1):153-172.

[8] Hoe, V., Urquhart, D., Kelsall, H., Zamri, E., & Sim, M. 2018 Ergonomic interventions for preventing work-related musculoskeletal disorders of the upper limb and neck among office workers (Review) Summary Of Findings For The Main Comparison. Cochrane Database of Systematic Reviews, 10, CD008570. https://doi.org/10.1002/14651858.CD008570.pub3.www.cochranelibrary.com

[9] Asih Media Yuniarti, Mukhammad Himawan Saputra, Dwi Helynarti Syurandhari, Fibriana, et al. 2017 Hubungan Lama Penggunaan Komputer Dengan Keluhan 310–314.

[10] Żabińska, I., Kuboszek, A., Sujova, E., & Zitnansky, J. 2018 Ergonomic Diagnosis of a Computer Workstation *Multidisciplinary Aspects of Production Engineering* **1(1)** 739–744. https://doi.org/10.2478/mape-2018-0093.

[11] Dinar, A., Susilowati, I. H., Azwar, A., Indriyani, K., & Wirawan, M. 2018 Analysis of Ergonomic Risk Factors in Relation to Musculoskeletal Disorder Symptoms in Office Workers *KnE Life Sciences* **4(5)** 16. https://doi.org/10.18502/kls.v4i5.2536.

[12] Kroemer, K.H.E., Kroemer, H.B., and Kroemer-Elbert, K.E. 2001 Ergonomics: How to Design for Ease & Efficiency (Prentice-Hall Inc., New Jersey).

[13] Liu, B. S., Huang, K. N., Chen, H. J., & Yang, K. C 2017 Ergonomic evaluation of new wrist rest on using computer mouse (Proceedings of the IEEE International Conference on Advanced Materials for Science and Engineering: Innovation, Science and Engineering, IEEE-ICAMSE 2016) 59–61. https://doi.org/10.1109/ICAMSE.2016.7840230.

[14] Ranasinghe, P., Wathurapatha, W. S., Perera, Y. S., Lamabadusuriya, D. A., Kulatunga, S., Jayawardana, N., & Katulanda, P. 2016 Computer vision syndrome among computer office workers in a developing country: An evaluation of prevalence and risk factors (BMC Research Notes) **9(1)** 1–9. https://doi.org/10.1186/s13104-016-1962-1.

[15] Gangamma, M., Poonam, & Rajagopala, M. 2010 A clinical study on "Computer vision syndrome" and its management with Triphala eye drops and Saptamrita Lauha. AYU (An International Quarterly Journal of Research in Ayurveda) **31 (2)** 236. https://doi.org/10.4103/0974-8520.72407

[16] N.A. Stanton, Stewart, R., Harris, D., Houghton, R.J., Baber, C., McMaster, R. 2006 Distributed situation awareness in dynamic systems: theoretical development and application of an ergonomics methodology *Journal Ergonomic* **49** Issue 12-13.

[17] Golebowicz, M., Levanon, Y., Palti, R., & Ratzon, N. Z 2015 Efficacy of a telerehabilitation intervention programme using biofeedback among computer operators *Ergonomics* **58 (5)** 791–802. https://doi.org/10.1080/00140139.2014.982210.

[18] Bramson JB, Smith S, Romagnoli G 1998 Evaluating Dental Office Ergonomic Risk Factors and Hazards *Journal of American Dental Association* **129(2)** 174-183.

IOP Conf. Series: Materials Science and Engineering **909** (2020) 012088 doi:10.1088/1757-899X/909/1/012088

Design of warehouse model with dedicated policy to minimize total travel costs: a case study in a construction workshop

Ade Irman[1]*, **Y Muharni**[1], **Andri Yusuf**[1]
[1]Industrial Engineering, University of Sultan Ageng Tirtayasa, Indonesia

***Email**: irman@untirta.ac.id

Abstract. Warehousing is one of the many important activities in a company. One of the important things in managing the warehouse is how to store finished or semi-finished goods, raw materials, spare parts, equipment, or other goods in the right location to minimize total travel costs inside the workshop. In a workshop construction company usually uses a raw material in a big size and the cost of moving a big size usually need a high cost also. Dedicated storage policy becomes an option to make a model where the material should be store inside the warehouse. A mathematical model as an Integer Linear Programming problem used to model the case problem. LINGO 17 used to solve the mathematical model in this research. The model can find the optimal solution in a fast time. The minimum total travel cost in this research resulted from LINGO is IDR 17.326.320.000. Minimizing warehouse operation cost will also minimize the total company operation cost, mean higher profit will be got by the company.

1. Introduction

There are so many functions of the warehouse; the main function is used as- temporary storage for goods or warehousing, it seems like a non-value-added activity, but the warehouse can also be used for some value-added activity, likes product packaging process inside a warehouse [1]. As warehousing, the warehouse can store finished or semi-finished goods, raw materials, spare parts, equipment, or other goods that support the activities of a company. The warehouse needs to be managed properly, so it can minimize the search time for goods to reduce overall warehouse operating costs.

There are several policies in creating a warehouse layout that is used to store products, here five common policies: There are random storage policy, dedicated storage policy, cube-per-order-index policy, class-based storage policy, and shared storage policy [2]. Each policy has advantages and disadvantages; a designer needs to maximize the advantages of the policy when he creates a layout.

In a workshop construction company usually uses a raw material in a big size and the cost of moving a big size usually need a high cost also. When some big size materials are placed in the same place, the level of difficulty in selecting and moving material will be high because it will stack each other, so in this case, we need a dedicated place for each material. Dedicated storage policy becomes an option in this case. Two important points in warehouse management, there are a time for searching and handling material; and the location where the materials should be stored [3].

The purpose of this research is to create a warehouse layout that minimizes total travel costs. There are so many methods to get a good layout that minimizes total travel costs, including optimization [4], heuristic, or metaheuristic algorithm [5], simulation, and others. Some more complex systems will be easier when describing it in the simulation model [6]. In this research, we used a linear programming model to get an optimal layout. Lingo 17 used to solve the mathematical model in this research.

IOP Conf. Series: Materials Science and Engineering **909** (2020) 012088 doi:10.1088/1757-899X/909/1/012088

2. Methodology

This section explains the data collection, which is needed in this research. The method, which is used in this research, also explains in this section.

2.1. Data Collection

There is some data, which is needed, including existing warehouse layout, types of materials, materials space requirement, frequency trip of the materials, the distance between storage spaces inside the warehouse, and the cost of moving materials. The warehouse has 24 storage spaces, which is used to store materials; it also has two doors as I/O points. There are 14 types of materials, which need to store inside the warehouse, the storage spaces requirement and the frequency trips of materials are shown in Table 1. The distance between storage spaces inside the warehouse and the cost of moving materials are shown in Tables 2 and 3. The cost of moving material is collected from an interview with the company, including operation cost, worker's salary, the overhead of material handling and building, assurance, and taxes.

Table 1. The storage spaces requirement and the frequency trips of materials

Type of Material	*n* Storage Spaces Requirement	Frequency Trips from I/O points	
		1	2
Material-1	4	13514	12205
Material-2	2	59	44
Material-3	1	5	3
Material-4	1	9647	3864
Material-5	1	43	10
Material-6	1	15421	1700
Material-7	4	27	3
Material-8	1	5	1
Material-9	4	11	8
Material-10	1	9	6
Material-11	1	32	21
Material-12	1	4	2
Material-13	1	3	1
Material-14	1	2	0

Table 2. The distance (in meter) between storage spaces with I/O points

I/O Point	Storage Space											
	1	2	3	4	5	6	7	8	9	10	11	12
1	6	6	6	6	12	12	12	12	18	18	18	18
2	36	36	36	36	30	30	30	30	24	24	24	24

Table 2. The distance (in meter) between storage spaces with I/O points ***(continue)***

I/O Point	Storage Space											
	13	**14**	**15**	**16**	**17**	**18**	**19**	**20**	**21**	**22**	**23**	**24**
1	24	24	24	24	30	30	30	30	36	36	36	36
2	18	18	18	18	12	12	12	12	6	6	6	6

Table 3. The cost of moving materials

Type of Material	Cost from I/O points (in IDR)	
	1	**2**
Material-1	23993.76	23993.76
Material-2	23993.76	23993.76
Material-3	4761	4761
Material-4	4761	4761
Material-5	4761	4761
Material-6	23993.76	23993.76
Material-7	23993.76	23993.76
Material-8	23993.76	23993.76
Material-9	23993.76	23993.76
Material-10	23993.76	23993.76
Material-11	23993.76	23993.76
Material-12	4761	4761
Material-13	4761	4761
Material-14	4761	4761

2.2. Method

We use a linear mathematical model to minimize the total cost of moving material, which satisfies a dedicated storage policy [2]. Then we use LINGO solver version 17 to solve the model. Below the formula:

Minimize

$$f(x) = \sum_{i=1}^{m}\sum_{j=1}^{n}\left[\frac{\sum_{k=1}^{p} C_{ik} f_{ik} d_{kj}}{S_i}\right] x_{ij} \quad (1)$$

Subject to

$$\sum_{j=1}^{n} x_{ij} = S_i \quad i=1,2 \ldots,m \quad (2)$$

$$\sum_{i=1}^{m} x_{ij} = 1 \quad j = 1,2 \ldots n \tag{3}$$

$$x_{ij} = (1,0) \quad i=1,2 \ldots,m \quad j = 1,2 \ldots n \tag{4}$$

Parameters,

C_{ik} Cost of moving material i trough I/O point k
f_{ik} Frequency trips of material i trough I/O point k
d_{kj} The distance of location j from I/O point k
S_i Storage spaces required by material- i

Decision variable

$$x_{ij} = \begin{cases} 1 & if\ material\ i\ store\ in\ location\ j \\ 0 & otherwise \end{cases}$$

Eq. (1) presents the objective function of the model is to minimize the total cost of moving all material trough I/O point k until p. The Eq. (2) guarantees the spaces required by materials. The Eq. (3) is an assignment problem. It has integer values for the result.

3. Result & discussion

The formulation in this research is the PILP (Pure Integer Linear Program) Problem. Another research used MILP (Mixed Integer Linear Program) [7]. We used a computer with intel i3 processors and 12 GB of RAM to run the LINGO in global solver mode. Then we have the result in 1 second of time. The minimum total travel cost resulted from LINGO is IDR 17.326.320.000. A partial result of the LINGO shown in figures one and Table 4.

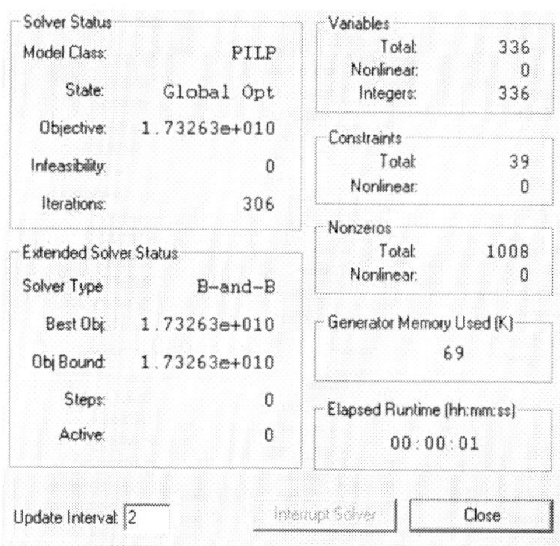

Figure 1. Summary of the LINGO

Table 4. The Lingo Result for Material's Storage Locations

Type of Material	Storage Location
Material-1	1; 3; 5; 8
Material-2	7; 10
Material-3	23
Material-4	2
Material-5	9
Material-6	4
Material-7	11; 12; 15; 16
Material-8	14
Material-9	17; 18; 19; 20
Material-10	13
Material-11	6
Material-12	24
Material-13	22
Material-14	21

4. Conclusion

The model can find the optimal solution in a fast time. This model is simple, but it can find the optimal solution; it also reliable for solving the real case with some assumptions. The minimum total travel cost in this research resulted from LINGO is IDR 17.326.320.000. Minimizing warehouse operation cost will also minimize the total company operation cost, mean higher profit will be got by the company.

References

[1] S. S. Heragu, L. Du, R. J. Mantel, and P. C. Schuur, "Mathematical model for warehouse design and product allocation," *International Journal of Production Research*. 2005, doi: 10.1080/00207540412331285841.

[2] S. S. Heragu 2008 *Facilities design, third edition*.

[3] C. Battista, a Fumi, F. Giordano, and M. M. Schiraldi 2011 Storage Location Assignment Problem: implementation in a warehouse design optimization tool *Proc. Conf. Breaking down barriers between Res. Ind.*

[4] M. K. Lee and E. A. Elsayed 2005 Optimization of warehouse storage capacity under a dedicated storage policy *Int. J. Prod. Res.* doi: 10.1080/13528160412331326496.

[5] Y. Muharni, Kulsum, and M. Khoirunnisa 2019 Warehouse Layout Designing of Slab Using Dedicated Storage and Particle Swarm Optimization in *IOP Conference Series: Materials Science and Engineering* doi: 10.1088/1757-899X/532/1/012003.

[6] A. Irman, D. Satria, N. M. Dewantari, Arliannur, and T. Ero 2019 Virtual traffic simulation for minimizing average of vehicle queue length in Serang city Indonesia in *IOP Conference Series: Materials Science and Engineering* doi: 10.1088/1757-899X/673/1/012095.

[7] G. Zhang, T. Nishi, S. D. O. Turner, K. Oga, and X. Li 2017 An integrated strategy for a production planning and warehouse layout problem: Modeling and solution approaches *Omega (United Kingdom)* doi: 10.1016/j.omega.2016.06.005.

IOP Conf. Series: Materials Science and Engineering **909** (2020) 012089 doi:10.1088/1757-899X/909/1/012089

Analysis of covariance for completely randomized design (case study: the life times of cutting tools used with lathes

Faula Arina[1*], Maria Ulfah [1], Dyah Lintang Trenggonowati [1]
[1]Department of Industrial Engineering, Universitas Sultan Ageng Tirtayasa, Cilegon , Indonesia

*Email: faula.arina71@gmail.com

Abstract. Analysis of covariance (ancova) is a blend of regression and analysis of variance (anova) used for upgrade precision of an experiment. Ancova can be used for all experimental design including completely randomized design. The completely randomized design is the response are randomly intrusted to treatments. Application ancova on completely randomized design in the manufacture is given to observe the effect of types of cutting tools of lifetimes of cutting tools . In this experiment covariate variable is lathe speed. Lathe speed affects tool life meaning that the involvement of the covariate variables in the covariance analysis model is appropriate. From the covariance analysis, it is also found that tool type influences tool life. The coefficient of variance in ancova is lower than anova. This proves that the ancova has a better level of precision than anova, so that ancova is more appropriate to use than anova.

1. Introduction

The analysis of covariance (ancova) is a blend of regression model and analysis of variance (anova). Ancova is a statistical analysis used in the experiments to determine variations resulting from one or more disturbance variables [1]. Engineers usually use blocking to handle disturbance variables. Blocking mainly reduce the noise that appear from disturbance variable by salve it as a special resource variability. This technique can only be used as the disturbance variables can be controlled.

The precision in an experiment can be improved through the use of assistive observation and a technique called covariance analysis. Covariance analysis is carried out based on the consideration that in reality there are certain variables which cannot be controlled, but which affect or correlate with the observed response variables.

Much of the theory and history of ancova is included in the Biometrics series [2,3]. See Huitema [4] for a detailed theories and practice of ancova for various designs. Several introductory linear models and experimental design provide this topic [5-8].

Covariance analysis can be applied in any experimental design including Completely Randomized Design (CRD). Based on this review, in this study used covariance analysis in the CRD. To clarify understanding, the CRD case is used in life times of cutting tools data [9].

We provide this paper as a tutorial to introduce ancova theory and application, complementing the types of experiments often used in quality improvement studies. The methodology is provided using an effect model building purpose that can be applied in CRD. The software uses the minitab language.

2. Research Methodology

2.1 Ancova for Completely Randomized Design

In many experiments, the response of an experiment are influenced not only by experimental factors that can be controlled, but also by uncontrolled variations. A model can be defined that links responses to controlled and uncontrolled variability, and the effect of every variants on the response of an experiment can be explored. To determine the controlled experimental factors of an uncontrolled experimental variable, the last is called covariate. Analysis of covariance is an analysis of experimental data as covariate variables and experimental factors are provide. Ancova compared the difference between the mean treatments, as in anova, while modeling between the covariate and the response of an experiment using a regression model.

The general ancova which has interesting a single experimental factor and a single covariate

$$b_{ij} = \mu + \alpha_i + \beta_i a_{ij} + \varepsilon_{ij} \tag{1}$$

where b_{ij} is the responses, a_{ij} is the covariate variables for the ijth observation, α_i is the experimental factors effect, β is the gradient of the regression between the covariate and the response of an experiment, and ε_{ij} is the random error of the ijth observation [4].

Table 1. Ancova for completely randomized design

Source	SS	DF	MS	F
Covariat	$(S_{ab})^2/S_{aa}$	1		
Faktor A	$SS_{Treat} = S_{bb} - (S_{ab})^2/S_{aa} - E_{bb} - (E_{ab})^2/E_{aa}$	$s-1$	$MS_{Treat} = \frac{SS_{Treat}}{s-1}$	$\frac{MS_{Treat}}{MSE}$
Error	$SS_E = E_{bb} - (E_{ab})^2/E_{aa}$	$n-s-1$	$MSE = \frac{SS_E}{n-s-1}$	
Total	S_{bb}	$n-1$		

The assumptions that must be resolved in using covariance analysis are as follows:

1. Covariate variables do not correlate with the treatment experiment
2. Linier relationship between response variables and covariate variables
3. Residual are normally distributed.

2.2 Coefficient of variation

The coefficient of difference is a coefficient that shows the accuracy of a conclusion or result obtained from an experiment [10].

$$CV = \frac{\sqrt{MSE_{adj}}}{\bar{b}} \tag{2}$$

where

CV = coefficient of variation

MSE_{adj} = mean square error adjusted

$\bar{b}$ = Average of all experimental data

The coefficient of variation shows the level of accuracy of an experiment. In general it can be said if the value of the coefficient of variation is smaller means that the level of accuracy is higher and the validity of the conclusions obtained from these experiments the better.

IOP Conf. Series: Materials Science and Engineering **909** (2020) 012089 doi:10.1088/1757-899X/909/1/012089

2.3 Data

Montgomery [9] represented an experiment where it is believed that the lifetimes of cutting tools used with lathe (b) depends on type of tool (the treatment) and possibly the speed of lathe (the covariate a). Table 2, data presented on the cutting tool life times. This study is focused on comparison type of cutting tools, type A and B.

Table 2. Cutting tool life times data [9]

Tool life times(hr)	Speed of lathe (rpm)	Type of tool	Tool life times(hr)	Speed of lathe (rpm)	Type of tool	Tool life times(hr)
18.73	610	A	30.16	670	B	18.73
14.52	950	A	27.09	770	B	14.52
17.43	720	A	25.40	880	B	17.43
14.54	840	A	26.05	1000	B	14.54
13.44	980	A	33.49	760	B	13.44
24.39	530	A	35.62	590	B	24.39
13.34	680	A	26.07	910	B	13.34
22.71	540	A	36.78	650	B	22.71

2.4 Analysis Method

The steps to be carried out in analysis of covariance in the completely randomized design are as follows:

1. Scatter plot of tool life and lathe speed
2. Conduct analysis of covariance.
3. Test the assumptions of covariance analysis
4. Coefficient of variation

3. Result and Discussion

Figure 1 shows the speed of lathe (a) and tool life times (b) linear relationship which has the same gradient for the two types of tools. The scenario would discuss modeling of covariate treatment without interactions in the model. Table 2 shows b cannot be compared directly using single-factor analysis of variance or the two-sample t -test. This is because b is a linear function from a . Although a can be set at the selected value, in this experiment they did not exist. So, the speed of lathe (a) must be treated as a covariate. The speed of lathe (a) and type of tool are not influenced by each other, so a can be considered a covariate and not another response.

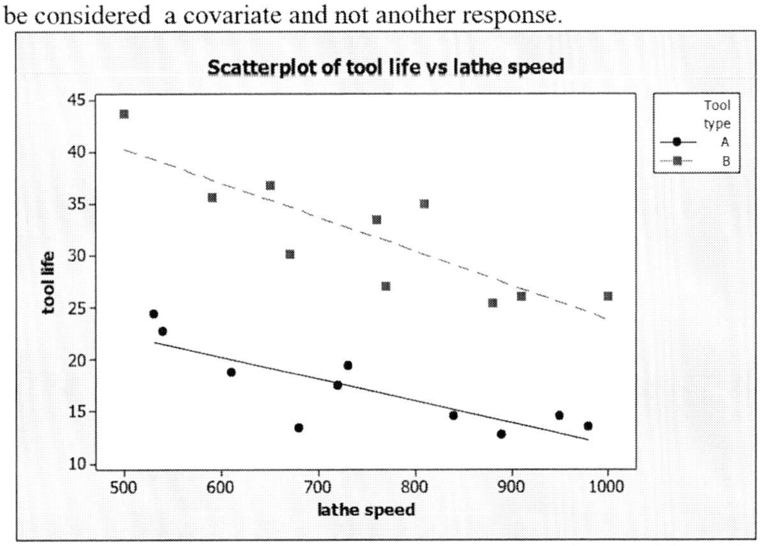

Figure 1. Scatter plot for cutting tool life times and lathe speed by tool type

Table 3 shows two anova tables for data of Table 2. The complete models including covariates in the first anova table. The reduced models does not include covariate in the second anova table. The anova table for the reduced models that SSE term and the degrees of freedom have been increased with the number of each for covariate in the first anova table. This shows the component error model can increase variability due to mess to enter important model terms.

Table 3. Anova for cutting tool life times data

Source	SS	DF	MS	F	P_value
Model A : Complete models, including covariates					
lathe speed	1418.04	2	709.02	76.73	0.00
type of tool					
Error	157.05	17	9.24		
Total	1575.09	19			
Model B : The Reduced models does not include covariate					
type of tool	1097.90	1	1097.90	41.41	0.00
Error	477.20	18	26.50		
Total	1575.10	19			

Table 4 is the anova table on the tool life times data. Consider that the covariate is very significant ($p < 0.00$). Figure 1 shows a statistically significant linear trend; that is, the lathe speed and the type of tool affects the lifetimes of the tools.

Model y = 44.488 -0.02661 lathe speed

Source	Coef	SE Coef	T	P
Constant	44.488	3.459	12.86	0.00
lathe speed	-0.02661	0.00452	-5.89	0.00
R squared	90.03%			

Figure 2. Effects model for cutting tool life data

Table 4 Anova for cutting tool life times data

Source	Seq SS	Df	Adj SS	Adj MS	F	P
lathe speed	293.01	1	320.17	320.17	34.66	0.00
Type of tool	1125.03	1	1125.03	1125.03	121.78	0.00
Error	157.05	17	157.05	9.24		
Total	1575.09	19				

Figure 2 suggest that relationship linier between the covariate variable and the response variable, The *P* value is 0.00, so we assesses the covariate effect as significant. Lathe speed effect, y = 44.488 -0.02661 lathe speed. The slope for this model is -0.02661, this too shows in Figure 1 the mean slope of the two regression lines. The proportion explained by the ancova model has increased to 90.03 %, mainly due to the contribution of lathe speed variable. Figure 3 suggest that the assumption using covariance analysis is residual are normally distributed.

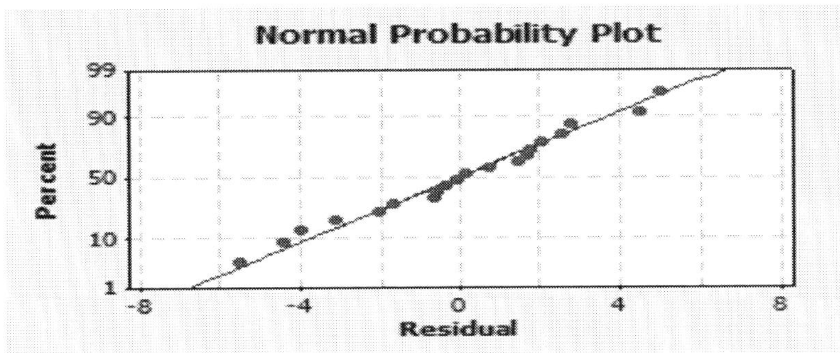

Figure 3. Residual plot for tool life times data

$\bar{y}_A = 17.11$ and $\bar{y}_B = 31.93$

The best tool type is tool type B because it produces a longer tool life.

$$CV_{anova} = \frac{\sqrt{MSE}}{\bar{y}_{..}} = \frac{\sqrt{26.5}}{24.519} \times 100\% = 21\%$$

$$CV_{ancova} = \frac{\sqrt{MSE_{adj}}}{\bar{y}_{..}} = \frac{\sqrt{9.24}}{24.519} \times 100\% = 12.4\%$$

From these calculations, the coefficient of variation after being corrected (anacova) is smaller than the coefficient of variation before being corrected (anova) which means the level of accuracy of the covariance analysis is better than the analysis of variance, so that the analysis of covariance is more appropriate to use than the analysis of variance.

4. Conclusion

We present an introductory analysis covariance for qualified engineers to help unlock the whole potency of this procedure which is often fit, but seldom used. Discussion of one factor design and examples with linear, not interacting covariates describe relationships regression analysis and anova to ancova. This initial study too begins an analysis of the apply and advantage of ancova.

Lathe speed affects tool life meaning that the involvement of the covariate variables in the covariance analysis model is appropriate. From the calculation of covariance analysis, it is also found that tool type influences tool life so that further tests are performed. From further tests that have been carried out, tool type B is the best tool type because it is able to produce a longer tool life.

The coefficient of variance in covariance analysis is lower than analysis of variance. This proves that the analysis of covariance has a better level of precision than analysis of variance, so that analysis of covariance is more appropriate to use than analysis of variance

References

[1] Silknitter KO Wisnowski JW and Montgomery DC 1999 The Analysis Of Covariance: A Useful Technique For Analysing Quality Improvement Experiment *Quality And Reliability Engineering International* **15(4)** 303316

[2] Cochran, WG 1957 Analysis of covariance: Its nature and uses Biometrics **13** 261281

[3] 'Special Issue on the Analysis of Covariance' 1982 *Biometrics* **38** 541753

[4] Huitema BE 1980 *The Analysis of Covariance and Alternatives: Statistical Methods for Experiments, Quasi-Experiments, and Single-Case Studies, 2nd Edition* (New York: Wiley)

[5] Myers RH and Milton JS 1999 *First Course* in The *Theory of Linear Statistical Models.*

(Boston : PWS-KENT Pub. Co)
[6] Mason RL Richard FG and James LH 2003 Statistical Design and Analysis of Experiments :With Applications to Engineering and Science (Canada : John Wiley & Sons Inc)
[7] Kutner MH Natchtsheim CJ Neter J Li W 2005 *Applied Linier Statistical Models Fifth Edition* (New York : McGraw-Hill Irwin)
[8] Montgomery DC 2013 *Design and Analysis of Experiments* (Singapore: John Wiley & Sons Inc)
[9] Montgomery D C Peck EA and Vining GG 2001 *Introduction to Linear Regression Analysis*, Third Edition (New York: John Wiley & Sons Inc)
[10] Farda NS Tatik W and Rita R 2017 Analisis Kovarian Pada Rancangan Bujursangkar Graeco Latin *Jurnal Gaussian* **6** Nomor 1 3140

IOP Conf. Series: Materials Science and Engineering **909** (2020) 012090 doi:10.1088/1757-899X/909/1/012090

Ergo-innovative design of cow slaughter tool

Hartomo Soewardi[1*], Eufrade Maulana Putra[2]

[1]Industrial Engineering Department, Faculty of Industrial Technology,
[2]Islamic University of Indonesia Yogyakarta Indonesia

*Email: eufradeputra@gmail.com

Abstract. Rough handling in the slaughtering operation can cause animal's stress resulting in the poor quality of meat because of the depleting muscle reserves to the lowest level. This condition will reduce the muscle lactic acid and increase the potential hydrogen on the meat above the normal level allowing a higher possibility for bacterial growth. Moreover, the rough slaughtering operation using the conventional technique may lead to slaughterer's accident because of the cow's mutiny. Thus, it is crucial to use the slaughtering tool that is safer and more comfortable to operate. This study aims to design the innovative slaughtering tool ergonomically and satisfying the user requirement. A survey was conducted to identify user criteria of the design. Theory of Inventive Problem Solving (TRIZ) was implemented as a fundamental basis to determine the inventive principle of the design, and the Quality Function Deployment was applied to develop the design specification. Non-parametric statistical analysis was done to test the hypothesis. This study reveals some identified inventive principles including dynamic quality, parameter changes, combination, and equipotentiality. Thus, the proposed design is valid to satisfy user need at 5% of significance level as it is proven to be safer, stronger, more practical, and more durable.

1. Introduction

The currently widespread animal's slaughtering technique, including the one for cows, is the manual slaughtering by tying the cow' body parts and legs before being knocked over by force. This commonplace method may harm the cows and result in many negative impacts. According to [1] rough handling in the slaughtering operation can cause animal's stress and result in the poor quality of meat. [2] and [3] confirm that the stressful animals before slaughtering will adversely affect the carcass quality i.e. slackening the supply of muscle glycogen, widely known as "Dark Firm Dry (DFD)". As a result, the DFD condition will reduce the content of lactic acid on the muscle and raise the potential hydrogen (pH) grade on the meat above the normal level resulting in higher possibility for bacterial growth [4]. Such condition is indicated by the darker color of the meat, stiffer and drier. This characteristic shows that the meat is not fresh and unhealthy.

On the other side, such conventional methods can also harm the slaughterers and may lead to an accident because the tie is usually not strong enough and may endanger the slaughterers during the tying process especially when the cow is frenzied. Such incident may also occur both in the slaughterhouse and during outdoor slaughtering, especially on the preparation process.

Another notable technique or method in the slaughtering process is known as stunning. This method is done with low shooting on the head of the cow to faint it [5]. However, this technique may still lead to harmful incident as the conventional method and may even kill the cow forbidding us from consuming its meat (haram) as indicated by [6] and [7]. Therefore, it is necessary to create a novel facility for safer slaughtering process to ensure that the quality of meat is fresh, healthy, and "halal".

On this basis, the objective of this study is to design the ergonomic and innovative slaughtering tools to satisfy the user requirement for safer and more comfortable cows' slaughtering process.

2. Research Method

2.1. Survey

The paper-based survey was conducted by distributing questionnaires to 30 respondents in an age range of 14 to 60 years old. The aim of this survey is to identify some attributes of the Assistance Appliance design for Slaughtering the Cow that the user is looking for.

2.2. Theory of Inventive Problem Solving (TRIZ)

TRIZ is a systematic method that provides a logical approach to develop creativity for innovation and inventive problem solving. This method produces an innovative solution of technical problems in the product development process [8]. This method will be used to find the inventive principle for design based on some contradictions identified between worsening features and improving features [9].

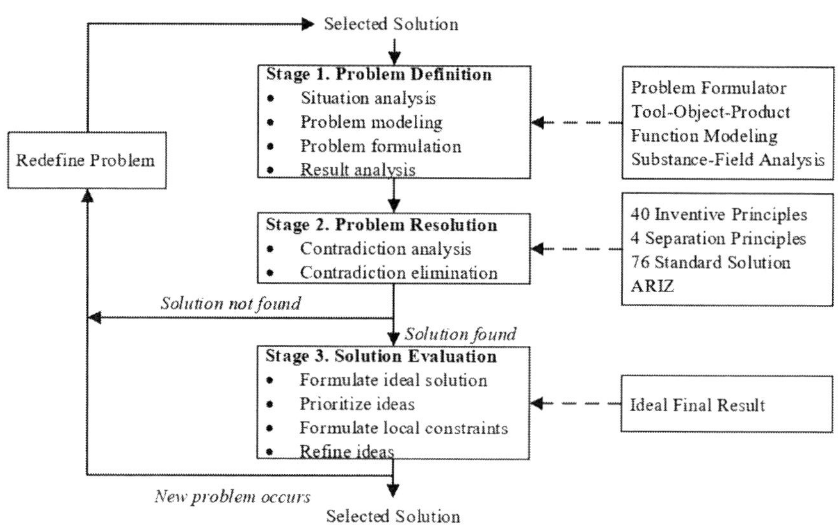

Figure 1. TRIZ problem solving steps [10]

Figure 1 explains the stages of TRIZ in finding the specific solution by referring to the inventive principle identified. These stages started from identification and analysis of the specific problems. Then it was done by determining contradiction between improving features and worsening features. The last stage was finding the solution according to the 40 principles [10].

2.3. Quality Function Deployment (QFD)

The method was used to plan and develop a structured product that allows the development team to clearly define the needs of consumers [11]. In the design process and/or product development, QFD was used for determining the specification of product concepts. House of Quality (HOQ) is a tool utilized to convert from what the customer wants into how to achieve it [12]. Subsequently [12] explains that this house encompasses six blocks. They are customer needs, competitive evaluation, technical requirement, relationship matrix between what and how, technical correlation matrix, and target specification. The relationship among them was determined by mapping process, which started from identifying customer requirements and transforming them to the technical requirements. This transformation describes the level of correlation; strong, moderate, and weak. The target specification was developed from the identified technical requirement as attribute of design. More details are presented in Figure 2.

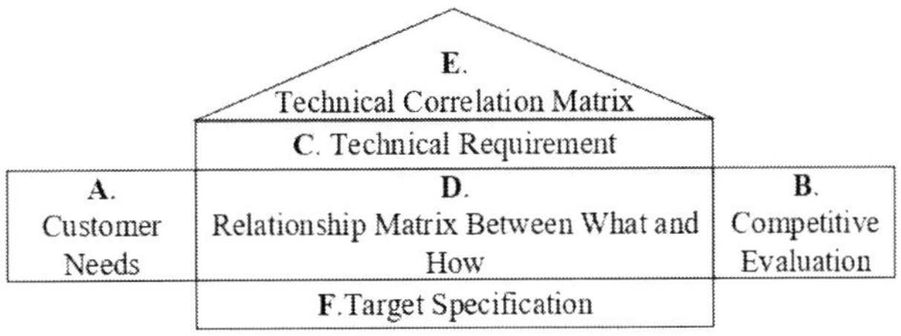

Figure 2. House of quality concept [12].

2.4. Apparatus

The following tools were used to support this study consist of Questionnaire. This questionnaire is to identify the attributes of the design, to determine the important rating, and to validate the proposed design. And IBM SPSS version 25 software is used to process some statistical data. As for Solidwork Software 2016 version to develop a virtual prototype of the proposed design.

2.5. Statistical Analysis

Non-parametric statistical analysis was conducted by using the Marginal Homogeneity method and the Wilcoxon method. The first method was applied to test a hypothesis with the specification of the developed design that satisfied the customer requirement. The second method was done to validate the hypothesis with the safer proposed design, more practical, more reliable and more durable than the existing design. 5% of significance level was used in these test [13].

3. Result and Discussion

3.1. Survey Result

Table 1 presents the four valid attributes of the design that consumers need. The attributes identified are strong, safe, practical, and durable. The strong attribute describes the ability of the slaughtering tool to withstand the motion of cow. It aims to calm the cow and avoid the state of mutiny for easy slaughter. The safe attribute of the design is required in order that the slaughtering operation does not endanger the slaughterers. In the common manual slaughtering process, the slaughterers may be harmed because the cow is not tied properly resulting in an uncontrollable motion. Meanwhile, the practical attribute of the tool is meant to ease the user in the slaughtering process. In addition, the durable attribute is required by the user because the tool can be used frequently and is not easy to damage.

Table 1. Consumer attribute.

Attribute
Safe
Strong
Practical
Durable

3.2. Functional Analysis

Functional analysis is described in Figure 3. This explains on the level of correlation between one variable to another in the slaughtering process. Such correlation covers normal (no risk), insufficient, and harmful attributes. Figure 3 shows that the roping operation on leg, head, and body by using rope is not rather sturdy since it risks the slaughterer because the binding is easy to loose and may expose the slaughterer to the reflex movements and kicks of the cow. Moreover, the process of cutting with a sharp knife does not have a significant risk on cow and slaughterer so that it is called the normal correlation.

Meanwhile, the means to restrain cow movement using the hollow box has a high risk because the leg of a cow can come out of the box and cause injury on the cow. The fall down process has a high risk because the head and body of the cow can hit the floor, causing injury and stress. Moreover, the water used to clean the floor also has a high risk, which results in the slippery floor so that both the cow and the slaughterer can slip and fall down.

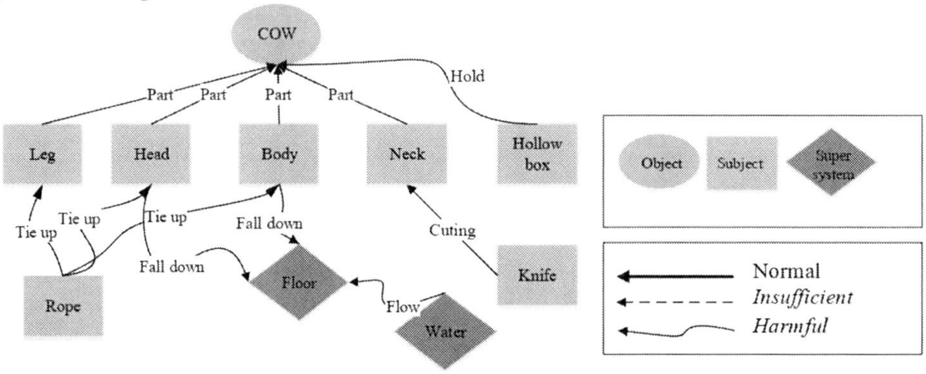

Figure 3. Function analaysis

3.3. Cause & Effect Chain Analysis

Figure 4 describes the Cause & Effect Chain Analysis. It is meant for identifying a root cause of the effect that is occurred in the pre-slaughtering process. It reveals five root causes namely are rough handling, hollow box, the roping process, the unmovable tool, and the easily damaged rope. Rough handling and the use of cow retaining hollow box can injure the cow and endanger the slaughterers. It is because the leg of the cow may come out of the box harming the slaughterers from being kicked. Meanwhile, the roping method used causes less practical operation of slaughtering because the slaughterers are required to tie many body parts of the cow causing complicated operation. Another drawback of the manual method is the tool used to fall the cow down that cannot be moved easily. Meanwhile, the rope cannot be used frequently because it is not forceful.

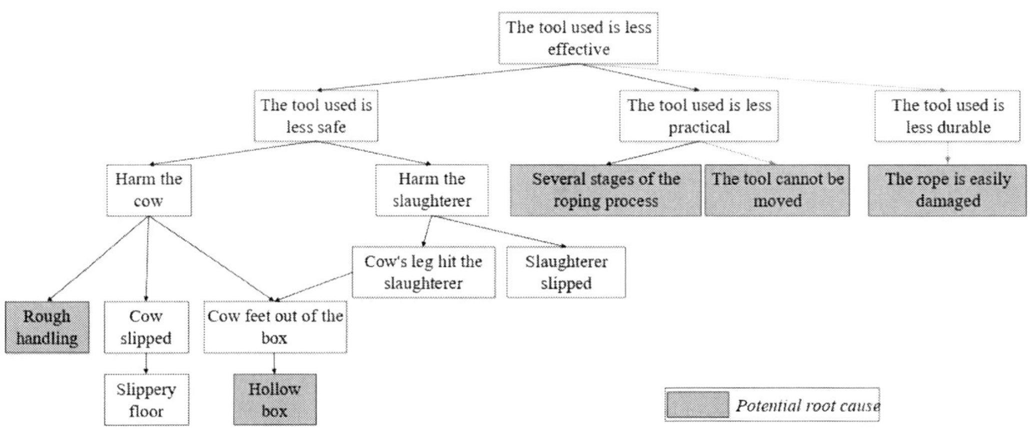

Figure 4. Cause and effect chain analysis

3.4. Develop the Inventive Principles and Specific Solution

Table 2 presents the inventive principle to develop the assistance appliance for slaughtering the cow that satisfies design attribute. It consists of engineering contradiction and physical contradiction. For safe attribute, it has an inventive dynamic principle and equipotentiality. The dynamic principle recommends adding the element of motion on the tool in order to drop the cow slowly and make the operation safer for the cow and the slaughterer. However, this method must require extra energy. An alternative to solve the problem is to install the double acting hydraulic system with single ended cylinder that has the maximum capacity of 600 kg with the system that can fall down the cow slowly and automatically (change dropping technique). While the equipotentiality principle recommends alignment in accordance with the use so it will be safe and practical. Thus, the improvement of the appliance from the hollow box is to provide two rooms; open space with the width of 63 cm for the clamp system and closed space with the width of 62 cm for the cow where the closed space is to avoid the cow's foot from slippage out of the box (modifying the box). In addition, the inventive principle for practical and strong attributes is parameters changes which recommends change on the objects to increase an effective use. Thus, to improve the appliance, it is necessary to change the roping technique with a clamp system with a length of 153cm, width of 30 cm, and height of 146 cm to withstand the movement of the cow firmly and practically (changing the roping technique). Meanwhile, the combination of inventive principle recommends combining two or more objects for two or more functions by installing 7-inch trolley wheel for mobility and a lock system for restraining the tool. Therefore, the slaughter process cannot endanger the cow and slaughterers (the movable tool design). In order to satisfy durable and strong attributes, it is necessary to have the parameter changes principle, which recommends replacing the rope with double acting hydraulic system with single ended cylinder that has a capacity of 600 kg for the process of dropping the cow and with a capacity of 200 kg for clamping, as well as a capacity of 50 kg for the cow head restraint system (change the roping tool). More details can be seen in Figure 5.

Table 2. TRIZ results

Attribute	Root of the Problem	Improving feature	Worsening feature	Inventive Principles
Safe	Rough handling	Force -10	Los of energy -22	Dynamic – 15
Safe and practical	Hollow box	Easy to roping	The cow's foot slips out of the box	Equipotentiality – 12
Practical and strong	Too many roping processes	Quantity of substance-26	Strength – 14	Parameter changes - 35
Practical and safe	The box cannot be moved	Reliability - 27	Object generated harmful factor – 31	Combination -5
Durable and Strong	The rope is easily damaged	To hold back the movement	Easy to damage	Parameter change – 35

Consumer Attribute / Technical Requirement	Change dropping technique	Modify box	Dimensions of the head place	The movable tool design	Change the roping technique	Strong use of materials	Change the roping tool
Safe	●	▲	▲	O		O	
Practical		▲	●	O	●		●
Strong					●	O	●
Durable						O	●
Target	Double Acting Hydraulic (Single Ended Cylinder) capacity	Length: 153cm, width: 62cm (close space) and 63cm (open space), height 146 cm.	Length: 55cm; Jaw Width: 31cm; Muzzle Width: 15 cm; 8 cm Jaw Height; Muzzle Height	Trolley Wheels with brakes 4 pics (7 inches)	[Clamp system + matte rubber 3 cm. Length: 153cm. Width: 61cm. Height: 146cm] [roll + matte rubber] [Full remote control]	Alloy Steel	Hidrotic Double Acting (Single Ended Cylinder) capacity 600kg, 200kg, and 50kg.

O Strong
● Moderate
▲ weak

Figure 5. House of quality

3.6. Conceptual Design

Figure 6 is a visual design of assistance appliance for slaughtering the cow based on the identification of consumer desires.

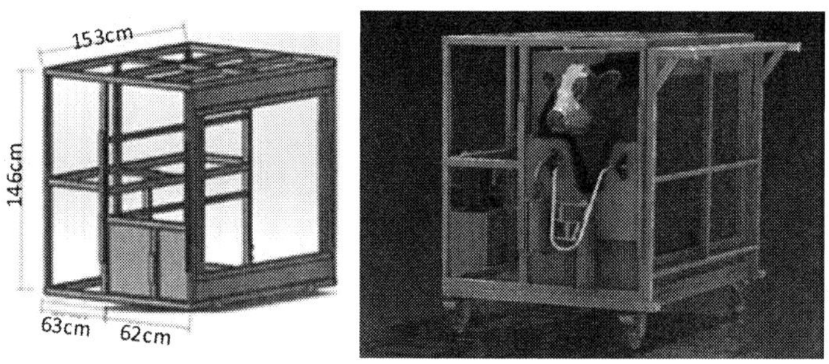

Figure 6. Design of assistance appliance for slaughtering

3.7. Statistical Analysis

3.7.1. Conformity Test

The conformity test is to identify the proposed design of the attributes of consumer desires. By using a significance level of 5%, the results of the conformity test are shown in Table 3 as follows:

Table 3. Result of marginal homogeneity

Attribute	Asymp Sig. (2-tailed)
Safe	0,371
Practical	0,109
Strong	0,139
Durable	0,637

The hypothesis at this conformity test is when the value of Asymp. Sig. (2-tailed)> 0.05, there is no significant differences between consumer need with the proposed design and vice versa. This can be seen in the results of the conformity test of the Asymp section. Sig. (2-tailed) the four attributes have Sig. > 0.05, which means that the proposed design is in accordance with the consumer need.

.3.7.2. Different Test
Table 4 presents the result of different-test used to identify differences between the proposed product with the existing products. This test uses a significance level of 5%, and has the hypothesis if Asymp. Sig. (2-tailed). <0.05, which means there are significant differences between these two tools.

Table 4. Result of the wilcoxon test

Attribute	Asymp Sig. (2-tailed)
Safe	0,000
Practical	0,000
Strong	0,000
Durable	0,000

The test result indicates of the four attributes having Asymp values. Sig. (2-tailed) <0.05. It can be said that the proposed tool has significant differences with the existing tools.

4. Conclusion
Based on the analysis of Ergo-Innovative Design of Cow Slaughter Tool, stated that consumer attribute identified of the design developed encompasses strong, safe, practical, and durable attributes. For the appropriate inventive principles to design the assistance appliance of slaughtering are dynamic, parameter changes, combination, and equipotentiality. Then the design parameters of the design are as follow: double acting hydraulic (single ended cylinder) capacity of 600 kg for slaughtering process assistance, box with the length of 153 cm, width of 125 cm, and height of 146 cm, dimensions of cow head restraint with the length of 55 cm, jaw width of 31 cm, muzzle width of 15 cm, jaw height of 8 cm, and muzzle height of 4 cm, wheel type: trolley wheel with the brake system, 4 pieces (7 inches), clamp system with the length of 153 cm, width of 61cm, height of 146 cm. Added with a matte rubber coated roll system, the wireless remote control system, alloy steel material, roping with a hydraulic system double acting (single ended cylinder) 100 kg capacity. Therefore, the developed assistance appliance design for slaughtering a cow is valid to conform the costumer requirement at 5% of significance level. This design is also valid to be better than the existing design.

References
[1] Warris PD, Brown SN, Adams SJM, et al. 1994 Relationship between subjective and objective assessments of stress at slaughter and meat quality in pigs pp 329-40
[2] Grandin T 1980 Mechanical, electrical and anesthetic stunning methods for livestock. International Journal for the Study of Animal Problems **1** pp 242-63
[3] Philip G Chambers and Temple Grandin 2001 Guidelines for humane handling, transport, and slaughter livestock Food and Agriculture Organization of the United Nations Regional Office for Asia and the Pacific
[4] Aletha Y M, Ida B N S, and I Ketut S 2016 Penilaian penerapan animal welfare pada proses pemotongan sapi di rumah pemotongan hewan mambal kabupaten Badung (Badung: Indonesia Medicus Veterinus)
[5] Temple Grandin 1994 Euthanasia and slaughter of livestock. Journal American Veterinary Medical Association. **204** pp 1354-60
[6] Temple Grandin 1994 Religious slaughter and animal welfare: A Discussion for Meat scientists CAB International
[7] Fuseini A, Toby GK, Phil JH, et al 2016 Halal stunning and slaughter : criteria for the assessment of dead animals. Meat Science **119** pp 132-7
[8] Imoh M. I, David P, and Robert P 2013 A review of TRIZ, and its benefits and challenges in practic. Technovation **33** pp 30-7
[9] San, Yeoh Teong 2009 TRIZ - Systematic innovation in manufacturing (Selangor: Firstfruit)

[10] Kah-Hin Chai, Jun Zhang and Kay-Chuan Tan 2005 A TRIZ-based method for new service design. Journal of Service Research. **8** pp 48-66
[11] Cohen L 1995 Quality function deployment: how to make QFD for you (New York: Addison-Wesley Publishing Campany)
[12] Chan LK and Wu M L 2002 Quality function deployment: a comprehensive review of its concepts and methods Jounal of Quality Engineering **1** pp 23-35
[13] Sheskin, D 2004 Handbook of parametric and non-parametric statistical procedures third edition (Washington: Cahpman & Hall /CRC)

IOP Conf. Series: Materials Science and Engineering **909** (2020) 012091 doi:10.1088/1757-899X/909/1/012091

Partial modeling of macroeconomic variables in industrial fields

N Valentika[1*], S Abdullah[2*], Ilmadi[1], S I U Chasanah[3], A R Nuha[4], M Huda[5] and V I Nursyirwan[6]

[1] Department of Mathematics, Universitas Pamulang, Jl. Puspitek Raya No.11, Serpong, Kota Tangerang Selatan, Banten 15417, Indonesia
[2] Department of Mechanical Engineering, Universitas Sultan Ageng Tirtayasa, Jl. Jenderal Sudirman Km 3 Cilegon, Banten 42435, Indonesia
[3] Department of Mathematics, Universitas Islam Negeri Sunan Kalijaga, Jl. Marsda Adisucipto, Yogyakarta 55281, Indonesia
[4] Department of Mathematics, Universitas Negeri Gorontalo, Jl. Jendral Sudirman No. 6 Kota Gorontalo 96128, Indonesia
[5] Department of Statistics, Universitas Bina Bangsa, Jl. Raya Serang – Jakarta Km 3 No. 1B, Penancangan, Cipocok Jaya, Kota Serang, Banten 42124, Indonesia
[6]Department of Accounting S1, Universitas Pamulang, Jl. Surya Kencana No.1, Pamulang Barat, Kecamatan Pamulang, Kota Tangerang Selatan, Banten 15417, Indonesia

*Email: abdullahsyarifayis@gmail.com

Abstract. This study modifies Catalbas's (2016) and Pratikto's (2012) research. The analysis used in this study is cross-correlation and analysis of the Vector Autoregression (VAR) method, Vector Autoregression in difference form (VARD) or Vector Error Correction Model (VECM) to test the relationship between variables with the ordinary least square (OLS) method and the causality test to test the relationship (causal relationship) between variables with the Granger Causality (GC) Test. Based on the correlation analysis, it is obtained that exports, imports, and interest rates have a negative and statistically significant relationship at the real level of 5% of the rupiah exchange rate. The model between exports and the rupiah exchange rate is the VAR (2) model. The model between imports and the rupiah exchange rate is a VECM with lag 3, there is 1 cointegration and the deterministic trend used is none intercept no trend. The model between the interest rate and the rupiah exchange rate is a VARD vector model with a lag 1. The model between inflation and the rupiah exchange rate is a VECM with lag 2 and the deterministic trend used is none intercept no trend and there is 1 significant cointegration at the 5% level.

1. Introduction

The analysis that needs to be done by investors is economic and capital market analysis, industry analysis, and company analysis. In this study, more focused on economic analysis and capital markets. The economic and capital market analysis aims to make decisions on the allocation of investment funds in several countries or domestically in the form of shares, bonds, or cash. Economic analysis needs to be done by investors reflecting the tendency for a strong relationship between what happens in the macroeconomic environment and the performance of a capital market. If estimating the cash flow, interest, and risk premium of security, it must consider macroeconomic analysis [1]. The environment that influences the company's daily operations is the macroeconomic environment. Investors' ability to understand and predict future macroeconomic conditions is very useful when making profitable investment decisions, so investors should be aware of various macroeconomic indicators that can help them understand and predict macroeconomic conditions [1]. One of the macroeconomic indicators in the preparation of the state budget (APBN) is the exchange rate of the rupiah against US $. Rupiah exchange rates are interrelated with other macro indicators. Factors in exports, imports, and inflation affect the exchange rate [2].

One of the studies related to macroeconomic variables can be read in [3]. The study uses multiple linear regression analysis. Based on [3], the consumer price index does not need to be considered in analyzing and estimating both imports and exports. Research on [4] analyzes exports, imports, and inflation of the rupiah exchange rate in Indonesia. The method used in that study is multiple linear regression. It found that export and import variables (net exports) had a negative and significant effect on the rupiah exchange rate variable against the US dollar while the inflation variable was not significant on the rupiah exchange rate variable against the US dollar. Selected macroeconomic variables and food security in Malaysia using VAR have been reaserched on [5].

Macroeconomic namely exchange rate change and inflation have been research on [6]. It research found an inertia effect of price increases in India. Research on macroeconomics is also discussed in [7]. It research found a significant relationship between exports and imports of crude steel and Iran's macroeconomic variables. Research on [8] found simultaneously, the economic growth of Indonesia's Riau Islands Province is affected by exports, imports, and investment.

The concentration of research on macroeconomic variables in Indonesia can be seen in [9]. This study uses the VAR method as its approach, which has impulse response and variance decomposition to obtain variable dynamics. The results of the Augmented Dickey-Fuller and Phillips-Perron root unit tests confirm that all stationary variables are in the level, therefore, the VAR level is the VAR method used. Empirical evidence shows that the effective exchange rate is significantly important in explaining export, import, and inflation movements. On [10], 1998: 1 - 2015: 3 for the period of nominal exchange rates in Turkey, has examined the relationship between imports and exports using quarterly data on the VAR model. Data were analyzed using variance decomposition, Granger Causality, and impulse response analysis. Causality According to the Granger test, there is a one-way relationship between imports and exports. There is no causal relationship between the nominal exchange rate and exports and imports. Details of the variance and impulse response analysis also confirm the results of the Granger causality analysis. These results indicate that the exchange rate does not have a significant effect on exports, imports, and the trade balance, and measures to limit imports will harm exports.

This research is a modification of [9] and [10] research, which is to combine, add variables and methods, partially model, and add correlation methods. The analysis used in this research is cross-correlation and Vector Autoregression (VAR), Vector Autoregression in difference form (VARD) or Vector Error Correction Model (VECM) to test the relationship between variables with the ordinary least square (OLS) method, and the causality test to test the relationship (causal relationship) between variables with the Granger Causality Test. This form of restricted access VAR is called VECM [11]. Macroeconomic variables used in this study are total export, import, inflation, interest rates, and rupiah exchange rate. Thus, this study model the Total Value of Exports, Imports, Inflation, and Interest Rates on the Rupiah Exchange Rate.

2. Research methods

In the research, there are five variables, namely: Amount of Export Value, Import, Level Inflation, Interest Rates, and Rupiah Exchange Rate. The data used are monthly data from January 2009 to December 2019. Types and sources of data are secondary data is data sourced from the publication of the official website of Bank Indonesia (BI) and the Central Statistics Agency (BPS).

The study approach in this study uses statistical testing using EViews software. The general steps taken are

1. Explore monthly data by looking at movements in the value of exports, imports, inflation rates, interest rates, and the rupiah exchange rate over a predetermined period.

2. Analyzing Cross-Correlation, by calculating the cross-correlation coefficient $\rho(X_t, Y_t)$ between two variables.
3. Check the stationary data in the mean of each variable via Augmented Dickey-Fuller (ADF) in the unit root test within the confirmation test framework.
4. Determination of the candidate lag can be done by looking at the lag that has the value of Akaike Information Criterion (AIC) or Schwarz Information Criterion (SC). Optimal lag in the VAR system is chosen with the Adj value. Highest R-Squared.
5. Conduct Granger causality (GC) tests on export, import, inflation, and interest rates variables against the rupiah exchange rate.
6. If the data is stationary, the model is VAR with the order p (VAR(p)). If the data is not stationary, Johansen's cointegration test. If cointegration rank $(r) = 0$, then use the VAR model indifference with the order p (VARD(p)). But if $r > 0$, then the model used is the VECM order p rank r.
7. Estimation of parameters in the VAR (p), VARD (p), or VECM models using EViews.
8. The model diagnostic test carried out to ensure the model obtained meets the assumptions and is appropriate to use.
9. At this stage, data predictions are made using monthly data in the past 2019.
10. Evaluation of forecasting by taking into account the relative errors produced with a tolerance limit of 5%.
11. Impulse Response Function (IRF) analysis and Forecast Error Decomposition Variance (FEDV) for models that meet the classical assumptions on both systems.

The flow of research can be described as follows.

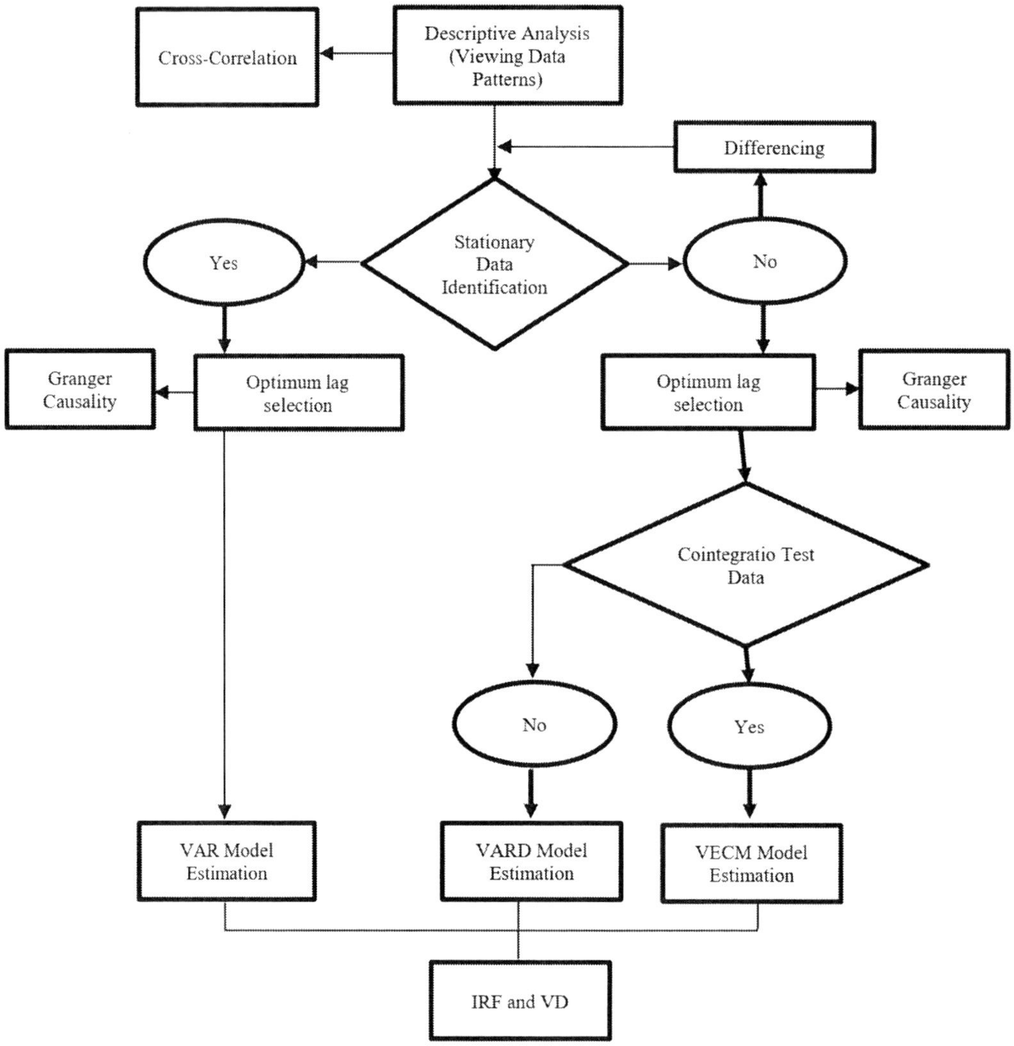

Figure 1. Research Flow.

3. Result and Discussion

3.1. Cross-Correlation

The correlation in this study is the Pearson Correlation. The correlations in this study are shown in Table 1.

Table 1. Cross-Correlation.

Export ↔ exchange rate	Import ↔ exchange rate	Inflation ↔ exchange rate	Interest rate ↔ exchange rate
-0.1881* (0.03081)	0.01102 (0.9002)	-0.2349* (0.006709)	-0.2115* (0.01492)

*significant at 5% level

Based on Table 1, it can be found that exports, imports, and interest rates are negatively related to the rupiah exchange rate at the 5% level.

3.2. Data stasionerity

Non-stationary data is included in the time series data problem. The data requires special treatment to be applied in time series analysis. This is due to the potential for false regression results [13]. ADF test is used to test stationary data or not. Stationary data test results at a level with a real level of 5% are shown in Table 2.

Table 2. Stationary data test results at a level with a real level of 5%.

Variable	t-Statistic	MacKinnon critical point 5% level	Prob.	Result
Import	-2.507625	-3.448021	0.3241	Non-stationary
Export	-3.948239	-3.444756	0.0128	Stationary
Rupiah Exchange Rate	-4.196178	-3.444756	0.0060	Stationary
Inflation	-2.853632	-3.445030	0.1813	Non-stationary
Interest Rate	-2.189573	-3.445030	0.4910	Non-stationary

Based on Table 2, it is found that the export variables and the rupiah exchange rate are stationary at 5% level. Stationary data test results at the first difference with a real level of 5% are shown in Table 3.

Table 3. Stationary data at the first difference with a real level of 5%.

Variable	t-Statistic	MacKinnon critical point 5% level	Prob.	Result
Import	-2.443620	-3.448021	0.3555	Non-stationary
Export	-11.86077	-3.445030	0.0000	Stationary
Rupiah Exchange Rate	-8.789074	-3.444756	0.0000	Stationary
Inflation	-7.901848	-3.445030	0.0000	Stationary
Interest Rate	-7.061221	-3.444756	0.0000	Stationary

Table 2 shows that the variables of exports, Rupiah exchange rate, inflation, and stationary interest rates at the first difference with a real level of 5%. Stationary data test results at the second difference with a real level of 5% are shown in Table 4.

Table 4. Stationary data test results at the second difference with a real level of 5%.

Variable	t-Statistic	MacKinnon critical point 5% level	Prob.	Result
Import	-6.864874	-3.448681	0.0000	Stationary
Export	-7.114187	-3.448021	0.0000	Stationary
Rupiah Exchange Rate	-14.27714	-3.445308	0.0000	Stationary
Inflation	-8.844436	-3.446168	0.0000	Stationary
Interest Rate	-17.44473	-3.445030	0.0000	Stationary

Based on Table 4 it is found that, all stationary variables at the second difference with a real level of 5%. After seeing the stationary data, the next step is determining the optimum lag.

3.3. Optimum lag

Determination of optimum lag through lag that has the value of AIC or SC. Optimal lag in the VAR system is also chosen with a model that meets all assumptions and lag that has Adj value. Highest R-Squared. The optimal lag for each model is shown in Table 5. After determining the optimal lag, cointegration testing is then determined.

Table 5. Optimal lag.

Model	Lag
Exports and Rupiah Exchange Rate	2
Import and Rupiah Exchange Rate	3
Interest Rates and Rupiah Exchange Rate	1
Inflation and Rupiah Exchange Rate	2

3.4. Cointegration test

Cointegration testing is carried out for all models except export models and the rupiah exchange rate. Cointegration test results are performed with the largest adj R-square model value and the VAR, VARD, or VECM models meet the assumption that errors are normal, errors do not have autocorrelation and errors are homogeneous. The cointegration test for the model is shown in Table 6.

Table 6. Cointegration test.

Model	Trend deterministic	Number of cointegration
Import and Rupiah Exchange Rate	None intercept no trend	1*
Interest rate and Rupiah Exchange Rate	-	No
Inflation and Rupiah Exchange Rate	None intercept no trend	1*

* significant at 5% level

Based on Table 6, it is obtained that the import model and the rupiah exchange rate and the inflation and rupiah exchange rate models use the VECM model, while the interest rate and rupiah exchange rate use the VARD model.

3.5. Export models and exchange exchange

The model between exports and the rupiah exchange rate is the VAR (2) model. The autoregressive vector model between exports and the rupiah exchange rate has 2 models: the first model is the exchange rate as the dependent variable and the second model is export as the dependent variable. Based on Breusch-Pagan-Godfrey, it was found that both models have homogeneous residuals at the 5% significance level. Based on the Breusch-Godfrey Serial Correlation LM Test, it was found that both models did not have autocorrelation in the error at the 5% significance level. As well as using the central limit proposition, it is found that the residuals approach the normal distribution.

The VAR model between exports and the rupiah exchange rate is as follows:

$$\text{exchange rate}_t = -0.841198 + 1.249865 \text{ exchange rate}_{t-1} - 0.247587 \text{ exchange rate}_{t-2} - 0.007795 \text{ export}_{t-1} + 0.042957 \text{ export}_{t-2} + v_{1t} \quad (1)$$

$$\text{export}_t = 5.491093 + 0.276099 \text{ export}_{t-1} + 0.521740 \text{ export}_{t-2} + 0.226299 \text{ exchange rate} -0.307998 \text{ exchange rate}_{t-2} + v_{2t} \quad (2)$$

The coefficient and significance of the autoregressive vector (VAR) model with the exchange rate as the dependent variable are shown in Table 7.

Table 7. Model (VAR) with the exchange rate as the dependent variable.

Variable	Coefficient
A constant	−0.841198*
exchange rate(−1)	1.249865*
exchange rate(−2)	−0.247587*
export(−1)	−0.007795
export(−2)	0.042957*

* significant at the 5% level

IOP Conf. Series: Materials Science and Engineering **909** (2020) 012091 doi:10.1088/1757-899X/909/1/012091

Table 7 reveals that the exchange rate variable in the first slowdown depends positively on the exchange rate and is statistically significant at the 5% level of 1.249865. The second level exchange rate variable has a negative relationship with the exchange rate and is statistically significant at the level of -0.247587. The export variable in the second lag has a positive relationship with the exchange rate at the 5% level of 0.042957. The coefficient and significance of the autoregressive vector (VAR) model with exports as the dependent variable are shown in Table 8.

Table 8. Autoregressive vector model (VAR) with export as the dependent variable.

Variable	Coefficient
A constant	5.491093*
export(−1)	0.276099*
export(−2)	0.521740
exchange rate(−1)	0.226299
exchange rate(−2)	-0.307998*

* statistically significant at the 5% level

Based on Table 8, it was found that the export variable in the first lag had a positive effect on exports at the 5% level of 0.276099. The exchange rate variable in the second lag has a negative relationship to exports and is statistically significant at the 5% level of -0.307998. The forecasting model for forecasting exports is shown in Table 9.

Table 9. Forecasting models for forecasting exports.

Month	Forecast value	Exact value	Relative Error
January 2020	23.32249	23.33569	0.06%
February 2020	23.32209	23.36666373	0.19%
March 2020	23.32169	23.36716144	0.19%
April 2020	23.32131	23.22167262	0.43%
May 2020	23.32093	23.07785497	1.05%

Based on Table 9, forecasting using a research model shows that each month the relative error is less than 5% as a tolerance limit. Thus, this research model is said to be good for forecasting exports if it uses past data from exports and the rupiah exchange rate. The forecasting model for forecasting exchange rates is shown in Table 10.

Table 10. Forecasting models to forecast exchange rates.

Month	Forecast value	Exact value	Relative Error
January 2020	9.511059	9.527501	0.17%
February 2020	9.511821	9.530694	0.20%
March 2020	9.512565	9.628693	1.21%
April 2020	9.513293	9.672024	1.64%
May 2020	9.514004	9.609532	0.99%

Based on Table 10, forecasting using the research model shows that each month of the error is relatively less than 5% as a tolerance limit. Thus, this research model is said to be good for forecasting the exchange rate if using past data from exports and the exchange rate.

The contribution of each variable to the shock it causes to the main endogenous variables observed is explained by forecasting error variance decomposition (FEVD). The proportion of other variables to explain the variability of the main endogenous variables of the study [12]. The Variance Decomposition of the VAR model (2) is shown in Table 11 and Table 12. The period in projecting this FEVD is 48 months (Four years).

Table 11. Variance Decomposition of export.

Period	Export	Rupiah Exchange Rate
1	100.0000	0.000000
2	99.77225	0.227748
3	99.82406	0.175943
4	99.83057	0.169430
5	99.83245	0.167548
6	99.80969	0.190309
⋮	⋮	⋮
48	85.41715	14.58285

Based on Table 11, an important source of export variation is shocking from exports themselves, while shocks from exchange rates are relatively small. In the first month of the shock, 100% of the export variation was contributed by the export itself. In the second month, 99.77% of export prediction value variations were contributed by exports and only 0.23% of export prediction value variations were explained by the exchange rate. And so on, in the 48th month of shocks, 85.41% of the variation in the predicted value of exports was contributed by exports and 14.58% of the variation in the prediction value of exports was explained by the exchange rate. The illustration for FEVD export from VAR (2) mode is shown in Figure 2.

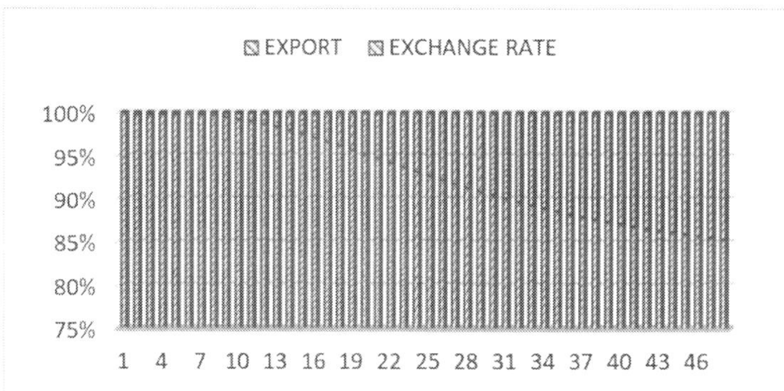

Figure 2. FEVD Export.

Table 12. Variance Decomposition of Rupiah Exchange Rate.

Period	Export	Rupiah Exchange Rate
1	1.190591	98.80941
2	0.866074	99.13393
3	2.369822	97.63018
4	3.763384	96.23662
5	5.721380	94.27862
6	7.647119	92.35288
⋮	⋮	⋮
48	39.53161	60.46839

Based on Table 12, an important source of exchange rate variations is shocking from the exchange rate itself, while shocks from exports are relatively small. In the first month of the shock, variations in the predicted exchange rate of 98.81% were contributed by the exchange rate and only 1.19% of the variation in the predicted exchange rate was explained by exports. In the second month of the shock, variations in the predicted exchange rate of 99.13% were contributed by the exchange rate and only 0.87% of the variation in the predicted exchange rate was explained by exports. And so on, in the 48th month of shocks, variations in the predicted exchange rate 60.46839% were contributed by the exchange rate, and 39.53161% variation in the predicted exchange rate was explained by exports. The illustration for the FEVD exchange rate from VAR mode (2) is shown in Figure 3.

Figure 3. FEVD Exchange Rate.

The IRF for this model is shown in Figure 4. The duration of the IRF used for this study is the next 48 months (four years).

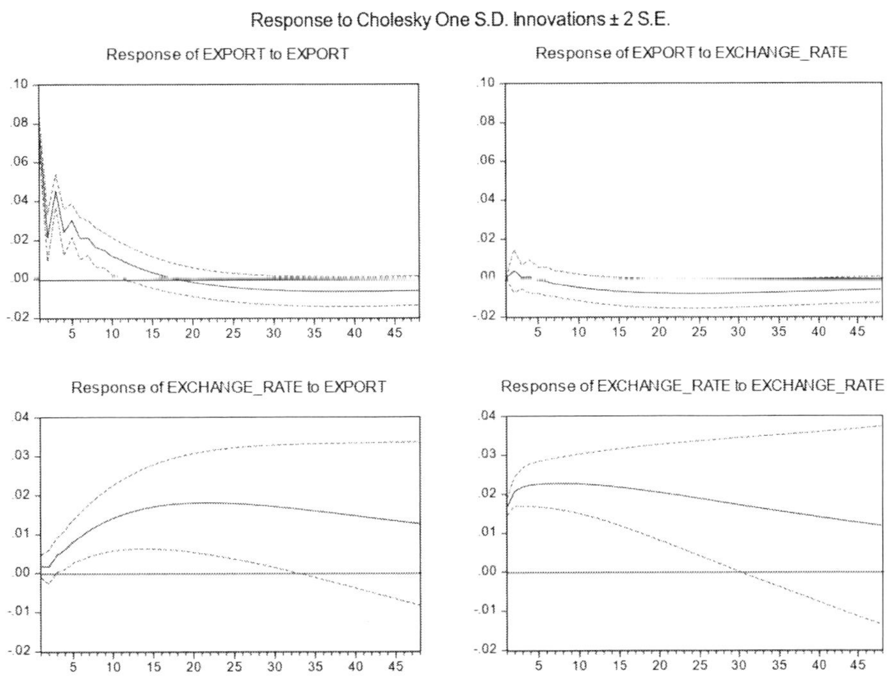

Figure 4. IRF for the model.

Figure 4 shows that an export shock of one standard deviation in the first month will cause an increase in exports of 7.5978 percent. Until the 18th month, export shocks were positively responded by exports themselves, even though the response was increasingly reduced. For example, in the 18th month, exports increased by only 0.0108 percent. But in the 19th month, shocks to exports caused exports to decrease by 0.0783 percent. From the 19th to the next 48 months, exports respond negatively to export shocks, and the longer these shocks cause exports to decrease in ever greater amounts. For example, in the 36th month, the shock caused a decline in exports of 0.6461 percent. In the 48th month, the shock caused a decline in exports of 0.6075 percent.

Based on Figure 4, it is obtained that the exchange rate shocks of one standard deviation have not been responded to by exports in the first month. In the second, third, and fourth month, the shocks were responded positively by exports respectively by 0.3771 percent, 0.0622 percent, and 0.0675 percent. However, in the fifth month, exchange rate shocks were responded negatively by exports, which was marked by a decline in exports of 0.1171 percent. The decline in exports increased in the long run until the 25th month. Starting from the 26th month to the 48th month, exports increased compared to the previous period. In the 48th month, the shock caused a decline in exports of 0.6073 percent.

Based on Figure 4, it was found that export shocks of one standard deviation in the first month responded positively by the exchange rate of 0.1829 percent. However, starting from the second month, export shocks responded positively by the exchange rate, which was marked by an increase in the exchange rate of 0.1694 percent. The increase in exchange rates continued to increase in the long run until the 22nd month. In the 23rd month, export shocks responded positively by the exchange rate marked by an increase in the exchange rate of 1.8047 percent. Starting in the 23rd month, the exchange rate decreased until the 48th period. Starting in the 47th month, the exchange rate responded positively to export shocks at around 1.2 percent.

Figure 4 shows that an exchange rate shock of one standard deviation in the first month will cause an exchange rate increase of 1.6664 percent. From the second month to the 21st month, the exchange rate responded positively to exchange rate shocks in the range of 1.2 percent. Starting in the 22nd month, the decline in the exchange rate will increase in the long run. Until its long-term period, the exchange rate responded positively to exchange rate shocks in the range of 1.1 percent. The GC test for exchange rates and exports is shown in Table 13.

Table 13. GC test for exchange rates and exports.

H_0	F-Statistic	Prob.
Rupiah Exchange Rate does not Granger cause export	2.47990	0.0879
Export does not Granger cause Rupiah Exchange Rate	9.17974	0.0002

Based on Table 13, the value of exports has a significant effect on the rupiah exchange rate at a significant level of 5%. Meanwhile, the rupiah exchange rate does not significantly influence the value of exports. Therefore, there is no two-way causality between the rupiah exchange rate and the export value.

3.6. *Model of import and Rupiah exchange rate*

The model between imports and the rupiah exchange rate is a VECM with lag 3, the deterministic trend used is none intercept no trend and there is 1 significant cointegration at the 5% level. The VECM model between imports and the rupiah exchange rate has 2 models: the first model is the exchange rate as the dependent variable and the second model is the import as the dependent variable. Based on Breusch-Pagan-Godfrey, it is found that only the model with imports as the dependent variable has a homogeneous residual at a real level of 5%. Based on the Breusch-Godfrey Serial Correlation LM Test, it was found that the model with import as the dependent variable contained no autocorrelation in the error at the 5% significance level. As well as using the central limit proposition, it is found that the residuals approach the normal distribution. Thus, the model with import as the dependent variable is said to be good.

Estimated VECM with import as target variable are as follows:

$$\begin{aligned}\Delta\text{import}_t = & -0.045767\text{ect}_{t-1} + 0.055499\Delta\text{exchange rate}_{t-1} - 0.497272\Delta\text{exchange rate}_{t-2} \\ & + 0.169064\Delta\text{exchange rate}_{t-3} - 0.668844\Delta\text{import}_{t-1} \\ & - 0.454356\Delta\text{import}_{t-2} - 0.157471\Delta\text{import}_{t-3}\end{aligned} \quad (3)$$

The import VECM estimation results that show the relationship of short-term variables are shown in Table 14.

Table 14. Import VECM estimation results showing short-term variable relationships.

Variable	Coefficient
CointEq1	−0.045767*
D(Exchange Rate(-1))	0.055499
D(Exchange Rate (-2))	−0.497272
D(Exchange Rate (-3))	0.169064
D(Import(-1))	−0.668844*
D(Import(-2))	−0.454356*
D(Import(-3))	−0.157471

* statistically significant at 5% level

Based on Table 14, it is found that in the short run, the statistically significant variable at the 5% level of import is the import in the first lag and imports in the second lag. Cointegration of errors that are statistically significant at 5% significance level and has a negative value indicates that there is an adjustment mechanism from the short term to the long term.

The cointegrating equation (long-run model) is as follows:

$$ect_{t-1} = 1.000000 exchange\ rate_{t-1} + 4.046333 import_{t-1} - 103.7762 \quad (4)$$

The import VECM estimation results that show the long-term relationship of variables are shown in Table 15.

Table 15. Import VECM estimation results showing long-term variable relationships.

Variable	Coefficient
C	−103.7762*
Exchange Rate(-1)	1.000000
Import(-1)	4.046333*

* statistically significant at 5% level

Based on Table 15, it was found that imports in the previous month had a positive and statistically significant effect at the real level of 5%, ie when there was an increase of one percent in the previous month's imports, it would increase imports this month by 4.046333 percent. The forecasting of this VECM model is shown in Table 16.

Table 16. Forecasting of the VECM model.

Month	Forecast value	Exact value	Relative Error
January 2020	23.41836	23.38133	0.16%
February 2020	23.41837	23.16978676	1.07%
March 2020	23.41837	23.314947	0.44%
April 2020	23.41837	23.25180652	0.72%
May 2020	23.41838	22.85649693	2.46%

Based on Table 16, forecasting using the VECM model shows that the relative error in each month is less than 5%. Thus, the VECM model is said to be good for forecasting imports. The GC test for exchange rates and imports is shown in Table 17.

Table 17. GC test for exchange rates and imports.

H_0	F-statistic	Prob.
D(Exchange Rate,2) does not Granger cause D(Import,2)	0.46580	0.7067

D(Import,2) does not Granger cause D(Exchange Rate,2)	1.59690	0.1937

Based on Table 17, the rupiah exchange rate does not significantly influence imports and imports do not significantly influence the exchange rate. Therefore, there is no two-way causality relationship between the rupiah exchange rate and imports.

3.7. Models for interest rate and exchange exchange

The model between the interest rate and the rupiah exchange rate is VARD with a lag 1. The VARD model between the interest rate and the rupiah exchange rate has 2 models: the first model is the exchange rate as the dependent variable and the second model is the interest rate as the dependent variable. Based on Breusch-Pagan-Godfrey, it is found that only the model with interest rates as the dependent variable has a homogeneous residual at a real level of 5%. Based on the Breusch-Godfrey Serial Correlation LM Test, it was found that the model with interest rates as the dependent variable had no autocorrelation in the error at the 5% significance level. And based on the central limit theorem it is found that the residuals approach the normal distribution. Thus, a model with an interest rate as the dependent variable is said to be good.

VARD between the interest rate and the rupiah exchange rate with the interest rate as the dependent variable is as follows:

$$D(\text{interest rate})_t = -0.000140 + 0.440618\ \text{D(interest rate)}_{t-1} + 0.009382\ D(\text{exchange rate})_{t-1} + v_{6t} \quad (5)$$

The coefficient and significance of the VARD with the interest rate as the dependent variable are shown in Table 18.

Table 18. Coefficients and significance of the VARD model with interest rates as the dependent variable.

Variable	Coefficient
A constant	-0.000140
D(interest rate(-1))	0.009382*
D(exchange rate(-1))	0.440618

*statistically significant at the 5% level

Based on Table 18, it was found that the previous interest rate variable had a positive relationship with current interest rates and was statistically significant at the 5% level of 0.009382. The forecasting model for interest rates is shown in Table 19.

Table 19. Forecasting model for interest rates.

Month	Forecast value	Exact value	Relative Error
January 2020	0.0498620	0.05	0.28%
February 2020	0.0496344	0.0475	4.49%
March 2020	0.0494068	0.045	9.79%
April 2020	0.0491791	0.045	9.29%
May 2020	0.0489515	0.045	8.78%

Based on Table 19, forecasting using the VARD (1) model shows that the first two months of the error are relatively less than 5% as the tolerance limit, while the following months the relative error is more than 5%. Thus, the VARD (1) model is said to be not good for predicting interest rates if it uses past data from interest rates and the rupiah exchange rate. The GC test for exchange rates and interest rates is shown in Table 20.

Table 20. GC test for exchange rates and interest rates.

H_0	F-Statistic	Prob.
D(Exchange Rate) does not Granger Cause D(Interest Rate)	1.49866	0.2231

D(Interest Rate) does not Granger Cause D(Exchange Rate)	16.5065	8.E-05

Based on Table 20, the interest rate has a significant effect on the rupiah exchange rate at a significant level of 5%. Meanwhile, the rupiah exchange rate has no significant effect on interest rates. Therefore, there is no two-way causality between the rupiah exchange rate and the interest rate.

3.8. Models of inflation and Rupiah exchange rate

The model between inflation and the rupiah exchange rate is a VECM with lag 2, the deterministic trend used is none intercept no trend and there is 1 significant cointegration at the 5% level. The VECM model between inflation and the rupiah exchange rate has 2 models: the first model is the exchange rate as the dependent variable and the second model is inflation as the dependent variable. Based on Breusch-Pagan-Godfrey, it is found that only the model with inflation as the dependent variable has a homogeneous residual at a real level of 5%. Based on the Breusch-Godfrey Serial Correlation LM Test, it was found that the model with inflation as the dependent variable did not have autocorrelation in the error at the 5% significance level. As well as using the central limit proposition, it is found that the residuals approach the normal distribution. Thus, the model with inflation as the dependent variable is said to be good.

Estimated VECM with inflation as the target variable is as follows:

$$\begin{aligned}\Delta \text{inflation}_t = &-0.006317 \text{ect}_{t-1} + 0.062503 \Delta \text{exchange rate}_{t-1} + 0.059444 \Delta \text{exchange rate}_{t-2} \\ &+ 0.381347 \Delta \text{inflation}_{t-1} - 0.186312 \Delta \text{inflation}_{t-2}\end{aligned} \quad (6)$$

The cointegrating equation (long-run model) is as follows:

$$ect_{t-1} = 1.000000 \text{exchange rate}_{t-1} + 19.62497 \text{inflation}_{t-1} - 10.18340 \quad (7)$$

The results of the inflation VECM estimation that show the short-term variable relationship are shown in Table 21.

Table 21. VECM estimation results for the short-run variable relationship.

Variable	Coefficient
CointEq1	−0.006317*
D(Exchange Rate(-1))	0.062503*
D(Exchange Rate (-2))	0.059444*
D(Inflation(-1))	0.381347*
D(Inflation(-2))	−0.186312*

* statistically significant at the 5% level

Based on Table 21, it is found that in the short run, all variables are statistically significant at the 5% level of inflation. Cointegration of errors that are statistically significant at 5% significance level and has a negative value indicates that there is an adjustment mechanism from the short term to the long term. The forecasting of the VECM model for predicting inflation is shown in Table 22.

Table 22. Forecasting the VECM model for predicting inflation.

Month	Forecast value	Exact value	Relative Error
January 2020	0.041223	0.0268	53.82%
February 2020	0.041223	0.0298	38.33%
March 2020	0.041223	0.0296	39.27%
April 2020	0.041223	0.0267	54.39%
May 2020	0.041223	0.0219	88.23%

Based on Table 22, forecasting using a research model shows that each month the error is relatively more than 5%. Thus, the VECM research model is said to be not good for predicting inflation. The GC test for exchange rates and inflation is shown in Table 23.

Table 23. GC test for exchange rates and inflation.

H_0	F-Statistic	Prob.
D(Exchange Rate) does not Granger Cause D(Inflation)	2.61816	0.0770
D(Inflation) does not Granger Cause D(Exchange Rate)	2.82092	0.0634

Based on Table 23, the rupiah exchange rate does not significantly influence inflation and inflation does not significantly influence the rupiah exchange rate. Therefore, there is no two-way causality between the rupiah exchange rate and inflation.

4. Conclusion

The conclusions of this study are:

1. Based on the correlation analysis, it is obtained that exports, imports, and interest rates have a negative and statistically significant relationship at the real level of 5% of the rupiah exchange rate.
2. Exports and the rupiah exchange rate using the VAR (2) model is said to be good for forecasting exports and the rupiah exchange rate because the relative error every month is less than 5%. Based on the GC, there is no two-way causality relationship between the rupiah exchange rate and the export value.
3. Imports and the rupiah exchange rate using the VECM with lag 3, there is 1 cointegration and the deterministic trend none intercept no trend is said to be good for predicting imports because the relative error at each month is less than 5%. Based on GC, there is no two-way causality relationship between the rupiah exchange rate and imports.
4. Interest rates and the rupiah exchange rate using the VARD (1) is said to be not good for predicting a relative error rate not less than 5% per month. Based on GC, there is no two-way causality relationship between the rupiah exchange rate and the interest rate.
5. Inflation and the rupiah exchange rate using the VECM with lag 2, the deterministic trend none intercept no trend and there is 1 cointegration said to be not good for predicting inflation because each month the relative error is more than 5%. Based on the GC, there is no two-way causality relationship between the rupiah exchange rate and inflation.

References

[1] Tandelilin E 2010 *Portofolio dan Investasi Teori dan Aplikasi* Yogyakarta: Kanisius
[2] Sukirno S 2013 *Makroekonomi Teori Pengantar* Edisi ketiga Jakarta: Rajawali Press
[3] Sumantria F and Latifah U 2019 The influence of interest rates, money circulation, inflation, and CPI against exports and imports in Indonesia 2012-2018 *Jurnal Ekonomi Pembangunan (JEP)* **17** 108-118.
[4] Silitonga RBR, Ishak Z, and Mukhlis 2017 Pengaruh Ekspor, impor, dan inflasi terhadap nilai tukar rupiah di Indonesia *Jurnal Ekonomi Pembangunan* **15** 53-39.
[5] Applanaidu SD, Bakar NAA, and Baharudin AH 2014 An econometric analysis of food security and related macroeconomic variables in Malaysia: A vector autoregressive approach (VAR) *UMK Procedia* **1** 93-102
[6] Yanamandra 2015 Exchange rate changes and inflation in India: What is the extent of exchange rate pass-through to imports? *Economic Analysis and Policy* **47** 57-68
[7] Todshki and Ranjbaraki 2016 The impact of major macroeconomic variables on Iran's steel import and export *Procedia Economics and Finance* **36** 390-398
[8] Kartikasari 2017 The effect of export, import, and investment to economic growth of Riau Islands Indonesia *International Journal of Economics and Financial Issues* **7** 663-667
[9] Pratikto R 2012 Dynamics of Indonesia's international trade a VAR approach *Procedia Economics and Finance* **4** 149-159
[10] Catalbas N 2016 The relationship among nominal exchange rate, import, and export in Turkey for the period 1988:1 to 2015:3. *International Research Journal of Applied Finance* **7** 11-25.
[11] Enders W 2004 *Applied Econometric Time Series* 2nd Edition John Wiley and Sons

[12] Firdaus M 2020 *Aplikasi Ekonometrika dengan E-Views, Stata, dan R*. Bogor: IPB Press
[13] Granger CWJ and Newbold P 1974 Spurious regressions in econometrics *Journal of Econometrics* **2** 111-120

Fabrication of fluorine-doped tin oxide by using Indonesian local stannic chloride precursors with spin coating method

E Yustanti[1,*], G Fahmi[1], L H Lalasari[2], T Arini[2], L Andriyah[2], A Subhan[3], F Firdiyono[2], and A Trenggono[1]
[1]Department of Metallurgical Engineering, Faculty of Engineering, Sultan Ageng Tirtayasa University, Jl. Jenderal Sudirman KM.03 Cilegon-Banten 42435, Indonesia
[2]Research Center for Metallurgy and Material, Indonesian Institute of Sciences (LIPI), Kawasan Puspiptek Serpong, Tangerang Selatan 15314, Indonesia
[3]Research Center for Physics–Indonesian Institute of Sciences (LIPI), Kawasan Puspiptek Serpong, Tangerang Selatan 15314, Indonesia

[*]Email: erlina.yustanti@untirta.ac.id

Abstract. Fluorine-doped tin oxide is a thin layer of transparent conductive oxide, which has a function as a semiconductor. Fabrication of fluorine tin oxide from this research was expected to replace indium-doped tin oxide. Indium-doped tin oxide function as a commercial transparent conductive oxide. The raw material of indium was limited, so the price of indium is higher than fluorine. The material used are Indonesia local tin (IV) Chloride, ammonium fluoride, and methanol. Conductive liquids has made by the sol - gel method. Sol gel liquids doped with ammonium fluoride to make a high conductivity. The transmittance value at the 1, 2, 3, 4, and 5 minutes deposition time were respectively 69.7; 43.6; 14.4; 14.1 and 34.7%. In this research, spin coating method under 3000 RPM on fix substrate temperature of 300°C. The results of the experiment shows, increased deposition time, make the thickness of the layer increased while resistivity and transmittance decreased. The optimum parameter for glass conductivity fabrication were obtained at 4 minutes time deposition, substrate temperature at 300°C has a resistivity of 125 k Ω, transmittance 14.1% and band gap energy 2.48 eV.

1. Introduction

Indonesia as a country that has the potential to use large solar energy is still not optimal in utilizing this type of energy. Based on data from the Ministry of Energy and Mineral Resources, solar radiation in Indonesia in the western region can reach 4.5 kWh/m^2/day, while in the eastern region it is even greater around 5.1 kWh /m^2/day [1]. A device that can convert solar heat radiation into energy is Dye Sensitize Solar Cell (DSSC). This research develops Transparent Conductive Oxide (TCO) with Fluorine Tin Oxide (FTO) type for fabrication DSSC. FTO was developed compared to Indium Tin Oxide (ITO) because indium as a coating material from thin films was relatively rare, the price of indium was also very expensive. This can be seen from the comparison of Indium and Fluorine reserves and production. Indium has a reserve of 6000 tons and production reaches only 475 tons/years while fluorine having reserves of 120 million tons and production reached 4 million tons/year [1].

Fabrication an FTO thin film in this research using the spin coating method. Spin coating method is one of the coating techniques for FTO using a spinning process [2]. Factors that influence the results of the spin coating method are rotation speed, temperature of substrate, and time of deposition.

2. Materials and Methods

The material used in this research was soda lime microscope glass substrate, stannic chloride ($SnCl_4$) from PT. Timah Industry Cilegon, NH_4F (98%, Merck), and methanol. Glass substrate was cleaned using commercial glass cleaning fluid. Cleans using an ultrasonic cleaner for 15 minutes. At first dissolve 8.1 ml of $SnCl_4$ with 91.82 ml of methanol. Furthermore, the addition of NH_4F doping was 0.2 grams. The solution was stirred for 30 minutes becomes homogeneous. Deposition using a simple spin coating machine as figure 1, with the distance of the solution from the glass substrate to the tip of the pipette were 2 cm. Conductive solutions are dropped vertically above the substrate surface, where the spin coating machine's rotating at 3000 RPM

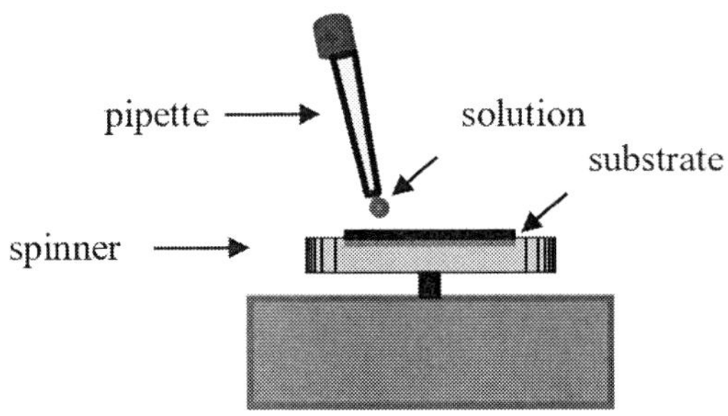

Figure 1. Scheme of the spin coating process

The repositioning is carried out in 2 repetitions. After the deposition process was completed. Spin spinning machine with a predetermined time variation of 1, 2, 3, 4, and 5 minutes. The coating process was carried out as many as 3 repetitions where each process pause was dried for 10 minutes in an oven with a temperature of 110°C. After the spin coating process was done, the coated glass substrate was fed into a muffle furnace at 300°C. Conductive Glass testing uses a multimeter (Sanwa), Scanning Electron Microscope Energy Dispersive Spectroscopy (JEOL-JSM 6390A), X-ray Diffraction/XRD (Shimadzu XRD-7000), UV-Vis spectroscopy (Thermo UV-Vis Genesys 10s), and Four Points Probe (FPP5000).

3. Results and Discussion

3.1 Identification of FTO Conductive Glass Resistivity

Testing to determine the resistivity value of the FTO conductive glass using a multimeter. Table 1 shows the resistivity of FTO at 300°C with deposition time variation. Based on Table 1, the largest resistivity was owned by 1 minute deposition time and the smallest was owned by 4 minutes. This is because drying done on spin coating does not coincide with the deposition process so it must be moved first to the top of the hotplate and requires a longer transition time. This caused premature clumping of the solution to the substrate.The factor caused high resistivity was increasing in layer thickness, so it can be increased of particle size and granular densification [3].

Table 1. The resistivity of FTO at 300°C with deposition time variation

Deposition time (min)	Resistivity (Ω)
1	12000 k
2	10071 k
3	8700 k
4	125 k
5	303 k

3.2 Identification of Morphological and Semi Quantitative Analysis of FTO

Figure 2 shows the results of SEM image of FTO with a magnification of 50,000 X in deposition time variation.

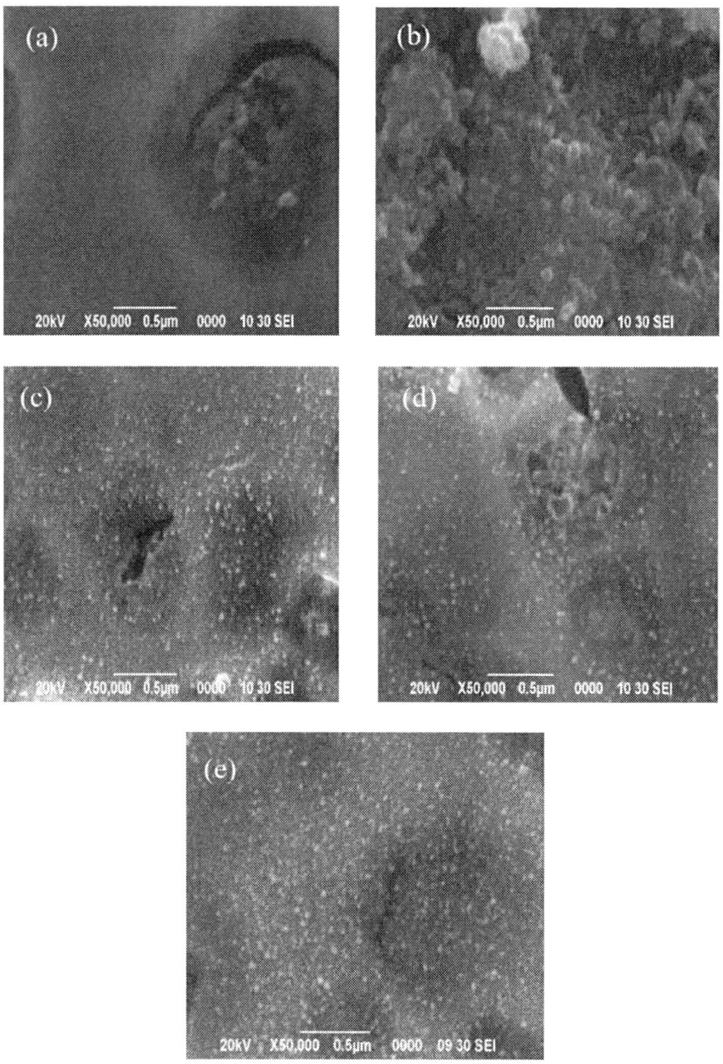

Figure 2. SEM image of FTO thin layers with variation deposition times (a) 1, (b) 2, (c) 3, (d) 4, (e) 5 minutes

Conductive glass results at 300°C in variation deposition times 1, 2, 3, 4, and 5 minutes have not uniform surface morphology. This can occur because the conductive solution that was deposited does not spread during the rotation. Conductive solutions stored in the same area were not evenly distributed. Factors that influence the results of spin coating are conductive solutions have high viscosity making it difficult to spread and tend to agglomerate in certain regions. The surface morphology of glass substrate on a spin coating method does have a rough texture because of the continuous agglomeration of the glass substrate [4]. Figure 2. Shows the formation of fine grains from deposition times 3 to 5 minutes. It is known that increased of deposition times, caused the size of fine grains becomes larger.

Table 2 is the result of conductive glass EDS with a variation of deposition time 1, 2, 3, 4, 5 minutes at 300°C.

Table 2. The result of semi quantitative EDS of FTO with deposition time variation.

Element	wt %				
	1 min	2 min	3 min	4 min	5 min
O	23.86	29.37	23.53	27.93	22.44
Na	3.98	2.19	2.16	1.29	-
Mg	1.38	0.83	-	-	-
Si	28.65	18.28	20.09	12.56	3.22
Ca	7.66	4.17	5.18	3.28	1.09
Sn	34.46	45.06	49.04	54.94	73.26

Table 2 shows, that the increasing wt % of Sn is related to the increasing of time deposition. As long as addition of wt% Sn element cause decrease of wt% of Na and Mg elements. At 5 minutes deposition time, Na was dissolved. For Mg, it starts to disappear in 3 minutes deposition times. The elements of Na and Mg were the composition of the glass substrate itself [5].

3.3 Identification of Diffraction Pattern on FTO Conductive Glass

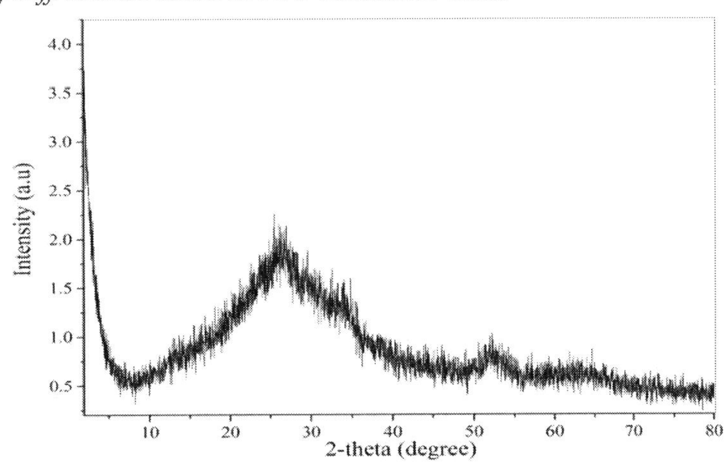

Figure 3. The XRD pattern of FTO at 300°C with 4 minutes deposition time

Based on Figure 3, the XRD characterization results showed that the FTO with the spin coating method did not detect the sharp peak, the peak intensity indicates that the structure of FTO still amorphous.

3.4 Identification of FTO Transmittance

Transmittance is the relative amount of light entering and exiting through the substrate. Figure 4 shows a graph of the comparison between deposition time and FTO transmittance.

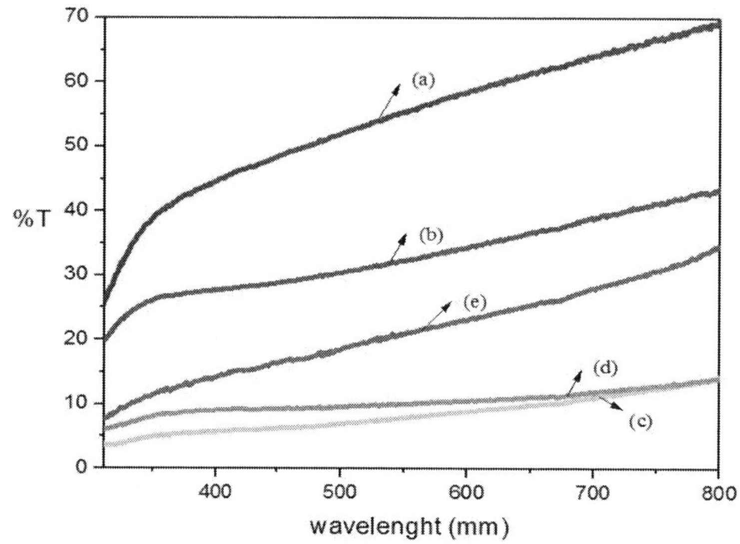

Figure 4. The graph of comparison between FTO transmittance and variation of deposition times (a) 1, (b) 2, (c) 3, (d) 4, (e) 5 minutes

Figure 4 shows that increasing variation of deposition times 1, 2, 3, 4, and 5 minutes, caused transmittance decreased respectively 69.7; 43.6; 14.4; 14.1 and 34.7%. This occurs because of the thickening of layer that attached to the substrate. Spin coating speed can also affect the thickness of the coating. Substrate spreading speed affects the centrifugal force in conductive solution and caused air turbulence during the rotation [6].

3.5 Identification of Band Gap Energy on FTO

Energy tape is a collection of lines at the same energy level, which will coincide and form a band [7]. Making this energy band gap graph uses the UV-VIS transmittance results which were included in the calculation using the Tauc Plot method. Figure 5 shows, the transmittance of FTO at 300°C during the spin coating process with deposition time variation.

The lowest energy band gap was achieved in 4 minutes deposition time. Energy band gap acts as a way to flow electricity more quickly. The longer the deposition time causes the band gap energy decreased. This indicates that the longer deposition time the energy band density decreases causing decreasing conductivity. The cause of the low band gap energy results can be caused by an increase in the thickness of the thin layer, the density of the layer and the homogeneity of the solution which is deposited on the substrate surface [8].

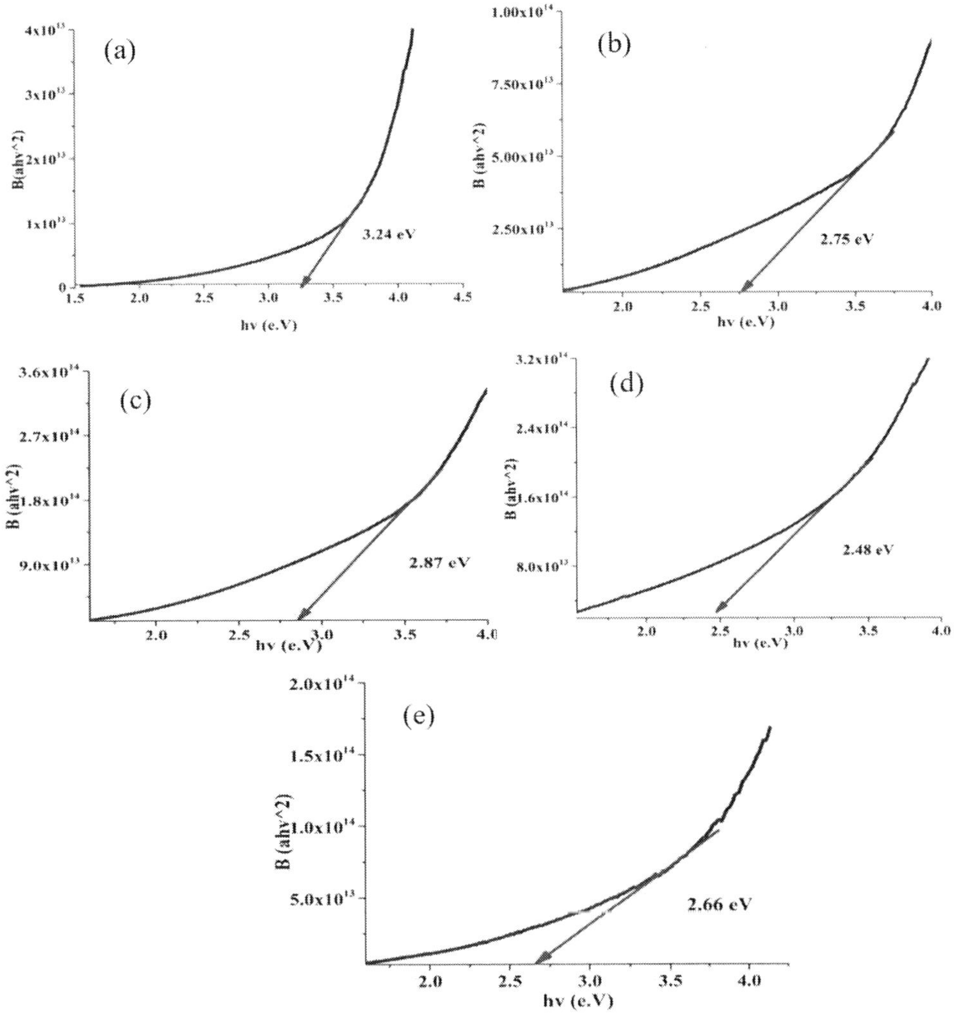

Figure 5. Band gap energy of spin coating method with variation of deposition times (a) 1, (b) 2, (c) 3, (d) 4, (e) 5 minutes

4. Conclusions

The results of the experiment shows, increased deposition time, make the thickness of the layer increased while resistivity and transmittance decreased. The optimum parameter for glass conductivity fabrication were obtained at 4 minutes time deposition, substrate temperature at 300°C has a resistivity of 125 k Ω, transmittance 14.1% and band gap energy 2.48 eV.

Acknowledgment

The authors gratefully acknowledge the support of the Research Center for the Metallurgy and Material, Indonesian Institute of Sciences (LIPI) Puspiptek Serpong for research facilities.

References

[1] Energy and Natural Resources Ministry of Indonesia 2015 Sunlight resources for PLTS in Indonesia
[2] Luurtsema G 1997 Spin Coating for Retangular Substrates EECS Department University of

California, Berkeley Technical Report No. UCB/ERL 97/49
[3] Akmal M H M, Warikh A R M, Azlan U A A, Azmi N A, Salleh M S and Kasim M S 2017 *Ceram Int.* 1- 9
[4] Chtouki T, Soumahoro L, Kulyk B, Bougharraf H, Erguig H, Ammous K and Sahraoui B 2017 *Mater Today Proc.* **4** 5146–53
[5] Karlsson S, Jonson B and Wondraczek L 2012 Glass Technology-European *J. Of Glass Sci. and Technol* Part A, **53(5)** 189-191(3)
[6] Anonymous 2013 Spin Coating Theory, Micro/Nano Technology Center, University of Louisville.
[7] Novianti R M and Utami S N 2016 Bandgap Semiconductor Pandjajaran University.
[8] Shinde N M, Deokate R J and Lokhande C D 2013 *J. Anal. App. Pyrolysis* **100** 12-16

AUTHOR INDEX